Springer Collected Works in Mathematics

For further volumes:
http://www.springer.com/series/11104

Ennio De Giorgi

(Courtesy of Foto Frassi, Pisa)

Ennio Giorgi

Selected Papers

Published with the support of Unione
Matematica Italiana and Scuola Normale
Superiore

Editors

Luigi Ambrosio · Gianni Dal Maso · Marco Forti · Mario Miranda
Sergio Spagnolo

Reprint of the 2006 edition

 Springer

Author
Ennio De Giorgi
   (1928 Lecce, Italy –
   1996 Pisa, Italy)
University of Pisa
Pisa, Italy

Editors
Luigi Ambrosio
SNS
Pisa, Italy

Gianni Dal Maso
SISSA
Trieste, Italy

Marco Forti
University of Pisa
Pisa, Italy

Mario Miranda
University of Trento
Trento, Italy

Sergio Spagnolo
University of Pisa
Pisa, Italy

ISSN 2194-9875
ISBN 978-3-642-40379-8   (Softcover)
       978-3-540-26169-8   (Hardcover)
DOI 10.1007/978-3-642-41496-1
Springer Heidelberg New York Dordrecht London

Library of Congress Control Number: 2012954381

The Mathematics Subject Classification (2000): 03-99, 35-99, 35B65, 49J45, 49Q05, 49Q15, 49Q20

Preface

The project of publishing some selected papers by Ennio De Giorgi was undertaken by the Scuola Normale Superiore and the Unione Matematica Italiana in 2000. The main motivations for this project have been the desire to make some of his classical papers, originally published in Italian, available to a large public and to display the broad range of his achievements and his entire intellectual path, as a problem solver and as a proponent of deep and ambitious mathematical theories.

We selected 43 papers, out of 152, and for 17 of them we decided to keep the original Italian version as well, in order to give a feeling of De Giorgi's original style.

In the development of this long project we have been helped by several collaborators. In particular we wish to thank Diego Pallara and Emanuele Paolini, for their many fundamental contributions, and Sisto Baldo, Giovanni Bellettini, Andrea Braides, Piero D'Ancona, Massimo Gobbino, Giacomo Lenzi, Roberto Monti, Maurizio Paolini, Paolo Tilli and Vincenzo Maria Tortorelli, who helped us in the typing and the translations of the papers.

We warmly thank Luis Caffarelli and Louis Nirenberg for their contributions to this volume.

Pisa, March 2005

Luigi Ambrosio
Gianni Dal Maso
Marco Forti
Mario Miranda
Sergio Spagnolo

Contents

Chapter 1

Biography

Ennio De Giorgi was born in Lecce on February 8, 1928. His mother, Stefania Scopinich, came from a family of navigators from Lussino, while his father, Nicola, professor of literature in the teachers' training school of Lecce, was an esteemed scholar in Arabic Language, History, and Geography. His father died an untimely death in 1930; his mother, to whom Ennio was particularly bound, lived until 1988.

In 1946, after his high school studies in Lecce, Ennio moved to Rome, where he started his university studies in Engineering. The following year he switched to Mathematics, and graduated in 1950 under the direction of Mauro Picone. Soon afterwards he obtained a fellowship at the Istituto per le Applicazioni del Calcolo, and in 1951 he became assistant to Picone at the Mathematical Institute "Guido Castelnuovo" of the University of Rome.

In 1958 he was awarded the Chair of Mathematical Analysis by the University of Messina, where he started in this post in December. In the autumn of 1959, following a proposal by Alessandro Faedo, he was hired by the Scuola Normale of Pisa, where he held the Chair of Algebraic and Infinitesimal Mathematical Analysis for almost forty years.

In September 1996 he was admitted to the hospital in Pisa. He underwent surgical treatment, and passed away on October 25.

1.1   Prizes and academic awards

In 1960 the Italian Mathematical Union awarded him the Caccioppoli Prize, founded in the same year. In 1973 the Accademia dei Lincei awarded him the Prize of the President of the Republic. In 1990 he received the prestigious Wolf Prize in Tel Aviv.

In 1983, during a solemn ceremony at Sorbonne, he received the degree *honoris causa* in Mathematics at the University of Paris. In 1992 the University of Lecce awarded him the degree *honoris causa* in Philosophy, of which he was particularly proud.

He was a member of the most important scientific institutions, in particular of the Accademia dei Lincei and of the Pontifical Academy of Sciences, where he had an active role until his last days. In 1995 he became a member of the Académie des Sciences de l'Institut de France and of the National Academy of Sciences of the United States.

1.2 Teaching and academic engagements

In all the activities he was engaged in, De Giorgi showed endeavour and availability well beyond the mere academic duties. He set only one limit on his collaboration: the firm refusal to take administrative or bureaucratic positions.

At the end of the 1950s, together with Enrico Magenes, Giovanni Prodi, Carlo Pucci and others, he founded an association of young researchers, named CONARM, that was the starting point for the future National Groups for Mathematics.

He was a member of the selection committee for the entrance examination of the Scuola Normale almost every year from 1960 and 1980, and many original problems proposed in that period show his imprint.

At the end of the 1960s he worked intensely on the Technical Committee of the Faculty of Sciences of the University of Lecce; ten years later he was a member of the Founding Committee of SISSA in Trieste.

In 1964 he entered the Scientific Council of the Unione Matematica Italiana and, later, the Director's Committee of the Istituto Nazionale di Alta Matematica. From 1979 on he offered his collaboration to CIMPA in Nice, an international centre that promotes teaching and research in Mathematics in developing countries.

In Pisa, De Giorgi used to give two courses each year, normally on Tuesday and Wednesday, from 11 a.m. to 13 p.m. The tone of these lectures was very relaxed, with frequent questions asked by participants. Sometimes the lecture was interrupted in the middle for about twenty minutes, and the whole class moved to a nearby café. Even if the details were not carefully prepared, his lectures were fascinating; those on Measure Theory given in the 1960s have become a classic. The notes of some of these courses have been written by his students, and carefully revised by him.

Beginning in the mid 1970s, De Giorgi reserved the Wednesday course to Foundations of Mathematics, while the other course continued to be devoted to the Calculus of Variations or to Geometric Measure Theory.

In 1967–68 he wanted to experience a "service course", namely Mathematics for Chemists. The notes of this course, written by Mario Miranda, are a model of essentiality and clarity; together with the (unfinished) textbook "Ghizzetti-De Giorgi" on Advanced Analysis, this is one of the very few books by Ennio.

1.3   The activities away from Pisa

At the end of the 1950s De Giorgi received pressing invitations from American universities; in particular, in 1960 Robert Oppenheimer repeatedly invited him to visit the Institute for Advanced Study in Princeton. But, maybe because of his scarce familiarity with the English language, he went to the United States only in 1964, when, accepting an invitation by Wendell Fleming, he spent four months between Brown and Stanford University.

His visits to Paris, however, were very frequent. Starting from the 1960s, he used to go to Paris almost every year for 3–4 weeks, invited by Jean Leray or Jacques-Louis Lions. In Paris Ennio felt at home. From these stays in Paris came his preference for *Le Monde*, which he used to read almost every day, even when he was in Italy.

In 1966 the International Congress of Mathematicians took place in Moscow, and Ivan G. Petrovskij invited De Giorgi to give one of the plenary lectures; but, after having prepared the text of his lecture, Ennio did not go. That text was read to the Congress by Edoardo Vesentini, and is an account of the most recent results on the theory of multi-dimensional minimal surfaces.

Several years later, in 1983, De Giorgi accepted an invitation to give a plenary lecture at the ICM in Warsaw. It was the time of Solidarność and Jaruzelski, and the Congress, already postponed for one year, took place in a very difficult political climate. Ennio began his lecture on Γ-convergence by manifesting a great admiration towards Poland. On the same occasion he publicly expressed one of his deepest beliefs, declaring that man's thirst for knowledge was, in his opinion, the "sign of a secret desire to see some ray of the glory of God".

De Giorgi was very interested, perhaps also for family tradition, in all that concerned far-away countries, like Brazil or Japan (where he never had the opportunity to go). In 1966 he accepted with enthusiasm Giovanni Prodi's proposal to spend some months as a teacher in the small University of Asmara, managed by Italian nuns. So, up to 1973, i.e., as long as the political situation in Eritrea made it possible, he spent one month in Asmara each year. When returning to Pisa, he used to speak at great length about his "African ventures".

In Italy Ennio had friends and former students almost everywhere. He was often away from Pisa for seminars or conferences, mainly in Pavia, Perugia, Naples, and Trento, not to mention, obviously, Rome and Lecce. He was a constant participant in the conferences on the Calculus of Variations held in Elba Island and in Villa Madruzzo, in Trento, where he felt particularly at ease. On these occasions he seemed to be indefatigable, promoting endless discussions and formulating new ideas and conjectures.

Beginning in 1988, when his first health problems appeared, Ennio began to spend long periods in Lecce, mainly in the summer, visiting his sister Rosa, his brother Mario, and their children and grandchildren. This was the occasion for this man, who was used to living alone in Pisa, lodging in a room at Collegio Timpano, to experience family life. These stays also gave him the opportunity for frequent meetings with his students and collaborators living in Lecce, either

on the beaches of the Adriatic or Ionian coasts of Salento, or in the department
which is now called after him.

1.4   Civil, political, and religious engagement

Ever since the first years he spent in Pisa, De Giorgi wanted to serve in several
charitable activities. For many years he took care of some poor families of
the town, which he used to visit together with his young students. He was
very generous, and his generosity was made more acceptable by the respect
he instinctively showed towards any person, regardless of his social or cultural
status.

In 1969 he undertook the task of teaching in an evening school for adults, at
the junior high school level. He sometimes used puzzle magazines as an auxiliary
teaching tool. The students appreciated his teaching, even if they found it a little
abstract.

Among the De Giorgi's commitment to social issues, the most important
was, without any doubt, the defense of human rights. This commitment, which
continued until the very last days of his life, started around 1973 with the cam-
paign to defend the Ukrainian dissident Leonid Plioutsch, who had been locked
up in a madhouse in Dniepropetrovsk. Thanks to the efforts of many scientists
all over the world, like Lipman Bers, Laurent Schwartz, and De Giorgi himself,
Plioutsch became a symbol of the struggle for freedom of opinion and he finally
was released in 1976. In Italy Ennio succeeded in involving in this battle hun-
dreds of persons with different political ideas. Later he continued his activity
to defend a great number of persons persecuted for their political and religious
ideas. He became an active member of Amnesty International, and was among
the founders of Amnesty's group in Pisa. He often took the opportunity to
illustrate and publicize the Universal Declaration of Human Rights.

Without taking active part in national politics, which he considered too far
from the great universal problems, De Giorgi always showed a lively interest
in the main themes that animated Italian life. In particular he took a public
position several times on the problem of abortion, on freedom of teaching, and
on the relationships between science and faith.

He was a deeply religious man. This is witnessed by the serenity he was able
to convey to those who were close to him in the hard trials of his last days. He
did not hide his religious beliefs, but his attitude towards a continuous search,
his natural curiosity, and his open-mindedness towards all ideas, made it easy
for him to have a constructive dialogue with others on these themes, too.

Besides St. John's *Apocalypse* and the *Book of Proverbs*, one of his favourite
books was *Pensées* by Pascal.

Chapter 2

The scientific work of Ennio De Giorgi

This list completes and updates the one contained in De Giorgi's obituary, appeared in 1999 on Boll. UMI, Sez. B, (8) **2**. We decided to include also some writings (as for instance [125]) that are not publications in a strict sense. De Giorgi, in particular in the last years of his life, used to circulated them among friends and colleagues, asking for opinions. We plan in the future to collect and to make available all these unpublished writings.

The publications appearing in this volume are marked with a star.

2.1 Complete list of De Giorgi's scientific publications

[1] E. DE GIORGI: Costruzione di un elemento di compattezza per una successione di un certo spazio metrico, *Atti Accad. Naz. Lincei Cl. Sci. Fis. Mat. Nat.* **(8) 8**, (1950), 302–304.12-195 (R. Arens).

[2] E. DE GIORGI: Un criterio generale di compattezza per lo spazio delle successioni, *Atti Accad. Naz. Lincei Cl. Sci. Fis. Mat. Nat.* **(8) 9**, (1950), 238–242.12-728, (V. L. Klee). See also Addenda and Errata, vol. 13, p.1139.

[3] E. DE GIORGI: Ricerca dell'estremo di un cosiddetto fuzionale quadratico, *Atti Accad. Naz. Lincei Cl. Sci. Fis. Mat. Nat.* **(8) 12**, (1952), 256–260.13-955 (L. M. Graves).

[4] E. DE GIORGI: Sulla sommabilità delle funzioni assolutamente integrabili, *Atti Accad. Naz. Lincei Cl. Sci. Fis. Mat. Nat.* **(8) 12** (1952), 507–510.14-257 (T. H. Hildebrandt)

[5] E. DE GIORGI: Compiuta ricerca dell'estremo inferiore di un particolare funzionale, *Rend. Accad. Sci. Fis. Mat. Napoli* **(4) 19** (1952), 29–41.14-291, 14-1278 (J. M. Danskin).

[6] E. DE GIORGI: Un nuovo teorema di esistenza relativo ad alcuni problemi variazionali, *CNR, pubbl. IAC no.371,* 1953.16-1127 (L. M. Graves).

[7]* E. DE GIORGI: Definizione ed espressione analitica del perimetro di un insieme, *Atti Accad. Naz. Lincei Cl. Sci. Fis. Mat. Nat.* (8) **14** (1953), 390–393.15-20 (L. C. Young).

[8] E. DE GIORGI: Un teorema sulle serie di polinomi omogenei, *Atti Accad. Sci. Torino Cl. Sci. Fis. Mat. Nat.* **87** (1953), 185–192.16-355 (J. Favard).

[9] E. DE GIORGI: Osservazioni relative ai teoremi di unicità per le equazioni differenziali a derivate parziali di tipo ellittico, con condizioni al contorno di tipo misto, *Ricerche Mat.* **2** (1953), 183–191.16-628 (L. Nirenberg).

[10]* E. DE GIORGI: Su una teoria generale della misura $(r-1)$-dimensionale in uno spazio ad r dimensioni, *Ann. Mat. Pura Appl.* (4) **36** (1954), 191–212. 15-945 (L. C. Young).

[11] M.PICONE-E.DE GIORGI: *Licões sobre una teoria das equações integrais lineares e suas applicaões segundo a orientação de Jordan-Hilbert.* Traduzio por Dante A. O. Martinelli, Escola politécnica da Universidade São Paulo (1954).

[12] E. DE GIORGI: Un teorema di unicità per il problema di Cauchy, relativo ad equazioni differenziali a derivate parziali di tipo parabolico, *Ann. Mat. Pura Appl.* (4) **40** (1955), 371–377.17-748 (F. G. Dressel).

[13] E. DE GIORGI: Un esempio di non unicità della soluzione di un problema di Cauchy, relativo ad un'equazione differenziale lineare di tipo parabolico, *Rend. Mat. e Appl.* (5) **14** (1955), 382–387.16-1119 (F. G. Dressel).

[14]* E. DE GIORGI: Nuovi teoremi relativi alle misure $(r - 1)$-dimensionali in uno spazio ad r dimensioni, *Ric. di Mat.* **4** (1955), 95–113.17-596 (L. C. Young).

[15]* E. DE GIORGI: Alcune applicazioni al Calcolo delle Variazioni di una teoria della misura k-dimensionale. *Atti V congresso UMI,* (Pavia-Torino, 1955), Cremonese, Roma, 1956.

[16]* E. DE GIORGI: Sull'analiticità delle estremali degli integrali multipli, *Atti Accad. Naz. Lincei Cl. Sci. Fis. Mat. Nat.* (8) **20** (1956), 438–441.18-489 (L. M. Graves).

[17]* E. DE GIORGI: Sulla differenziabilità e l'analiticità delle estremali degli integrali multipli regolari, *Mem Accad. Sci. Torino Cl. Sci. Fis. Mat. Nat.* (3) **3** (1957), 25–43.20-172 (C. B. Morrey).

[18]* E. DE GIORGI: Sulla proprietà isoperimetrica dell'ipersfera, nella classe degli insiemi aventi frontiera orientata di misura finita, *Atti Accad. Naz. Lincei Mem. Cl. Sci. Fis. Mat. Nat. Sez. I*, **(8) 5** (1958), 33–44.20-4792 (L. C. Young).

[19] E. DE GIORGI-A. GHIZZETTI: *Lezioni di Analisi Superiore: funzioni olomorfe, teoria dell'integrazione secondo Lebesgue, trasformata di Laplace.* Veschi, Roma, 1958.

[20] E. DE GIORGI: Considerazioni sul problema di Cauchy per equazioni differenziali di tipo parabolico di ordine qualunque, *Symposium on the numerical treatment of partial differential equations with real characteristic,* organized by the Provisional International Computation Centre (Rome, 1959), 136–139, Veschi, Roma, 1959.21-5083 (L. Nirenberg).

[21] E. DE GIORGI: Complementi alla misura $(n-1)$-dimensionale in uno spazio ad n dimensioni, *Seminario di Matematica della Scuola Normale Superiore di Pisa, 1960-61,* Editrice Tecnico Scientifica, 1961, 31 p.31-3896 (W. H. Fleming).

[22] E. DE GIORGI: Frontiere orientate di misura minima, *Seminario di Matematica della Scuola Normale Superiore di Pisa, 1960-61,* Editrice Tecnico Scientifica, 1961, 57 p.31-3897 (W. H. Fleming).

[23] E. DE GIORGI: Una estensione del teorema di Bernstein, *Ann. Sc. Norm. Sup. Pisa,* **(3) 19** (1965), 79–85.31-2643 (R. Osserman). E.C., *Ibidem,* 463.

[24] E. DE GIORGI-G. STAMPACCHIA: Sulle singolarità eliminabili delle ipersuperficie minimali, *Atti Accad. Naz. Lincei Rend. Cl. Sci. Fis. Mat. Nat.* **(8) 38** (1965), 352–357.32-4612 (L. Cesari).

[25] E. DE GIORGI-G. STAMPACCHIA: Sulle singolarità eliminabili delle ipersuperficie minimali, *Atti del convegno sulle equazioni alle derivate parziali* (Nervi, 1965), 55–58, Cremonese, Roma, 1966.34-3454 (U. D'Ambrosio).

[26]* E. DE GIORGI: Hypersurfaces of minimal measure in pluridimensional euclidean space, *Proc. Internat. Congr. Math.,* (Moscow, 1966), Izdat. "Mir", Moscow, 1968, 395–401.38-2646 (F. J. Almgren, Jr.).

[27] E. DE GIORGI: Maggiorazioni a priori relative ad ipersuperfici minimali, *Atti del convegno di Analisi Funzionale,* (Roma, 1968) Symposia Mathematica II, Academic Press 1969, 283–285.

[28]* E. DE GIORGI: Un esempio di estremali discontinue per un problema variazionale di tipo ellittico, *Boll. Un. Mat. Ital.* **(4) 1** (1968), 135–137.37-3411 (G. M. Ewing).

[29] E. DE GIORGI: *Corso di Analisi per Chimici.* Università di Pisa, 1967-68 (reprinted by De Salvio ed., Ferrara, 1969).

[30] E. BOMBIERI-E. DE GIORGI-M. MIRANDA: Una maggiorazione apriori per le ipersuperficie minimali non parametriche, *Arch. Rational Mech. Anal.* **32** (1969), 255–267.40-1898 (M. Pinl).

[31]* E. BOMBIERI-E. DE GIORGI-E. GIUSTI: Minimal cones and the Bernstein problem, *Invent. Math.* **7** (1969), 243–268.40-3445 (E. F. Beckenbach).

[32] E. DE GIORGI: *Teoremi di semicontinuità nel calcolo delle variazioni,* lectures given at the Istituto Nazionale di Alta Matematica, Roma, 1968-69; notes by U. Mosco, G. Troianiello, G. Vergara.

[33]* L. CATTABRIGA-E. DE GIORGI: Una formula di rappresentazione per funzioni analitiche in $\mathbf{R}^n$, *Boll. Un. Mat. Ital.* (4) **4** (1971), 1010–1014.45-7004 (M. Reade).

[34]* L. CATTABRIGA-E. DE GIORGI: Una dimostrazione diretta dell'esistenza di soluzioni analitiche nel piano reale di equazioni a derivate parziali a coefficienti costanti, *Boll. Un. Mat. Ital.* (4) **4** (1971), 1015–1027. 52-3702 (A. Haimovici).

[35] E. DE GIORGI: Sur l'existence de solutions analitiques d'équations à co-efficients constants, *Actes du colloque d'Analyse fonctionnelle* (Bordeaux, 1971), *Bull. Soc. Math. France,* **100** (1972), Suppl. Mém. 31-32, 133–135.50-7779 (Editors).

[36] E. DE GIORGI: Solutions analitiques d'équations à coefficients constants, *Sém. Goulaouic-Schwartz 1971-72: Équations aux dérivées partielles et analyse fonctionnelle, Exp. no.29,* École polytechnique, Centre de Math., Paris, 1972.53-997 (V. V. Olariu).

[37] L. CATTABRIGA-E. DE GIORGI: Sull'esistenza di soluzioni analitiche di equazioni a derivate parziali a coefficienti costanti in un qualunque numero di variabili, *Boll. Un. Mat. Ital.* (4) **6** (1972), 301–311.47-3799 (Z. C. Motteler).

[38] E. DE GIORGI-F. COLOMBINI-L. C. PICCININI: *Frontiere orientate di misura minima e questioni collegate,* Quad. Sc. Norm. Sup. Pisa, Editrice Tecnico Scientifica, Pisa, 1972.58-12647 (Editors).

[39] E. DE GIORGI: Problemi di superfici minime con ostacoli: forma non cartesiana, *Boll. Un. Mat. Ital.* (4) **8** (1973), suppl. no.2, 80–88.48-7093 (F. J. Almgren, Jr.).

[40]* E. DE GIORGI-S. SPAGNOLO: Sulla convergenza degli integrali dell'energia per operatori ellittici del secondo ordine, *Boll. Un. Mat. Ital.* (4) **8** (1973), 391–411.50-753 (M. Schechter).

[41] E. DE GIORGI: Introduction to minimal surface theory, *Global Analysis and its applications* (Lectures Internat. Sem. course, ICTP, Trieste, 1972) vol II, 43–45, Internat. Atomic Agency, Vienna, 1974.55-11138 (Editors).

[42] E. DE GIORGI: Sulle soluzioni globali di alcune equazioni differenziali, *Collection of articles dedicated to Giovanni Sansone on the occasion of his eighty-fifth birthday, Boll. Un Mat. Ital.* (4) 11 (1975), suppl. no.3, 77–79.53-3473 (E. Goldhagen).

[43]* E. DE GIORGI: Sulla convergenza di alcune successioni d'integrali del tipo dell'area, *Collection of articles dedicated to Mauro Picone on the occasion of his ninetieth birthday, Rend. Mat.* (6) 8 (1975), 277–294.11-233 (L. von Volfersdorf).

[44] L. CATTABRIGA-E. DE GIORGI: Soluzioni di equazioni differenziali a co-efficienti costanti appartenenti in un semispazio a certe classi di Gévrey, *Collection in memory of Enrico Bompiani, Boll. Un. Mat. Ital.* (4) 12 (1975), suppl. no.3, 324–348.54-10825 (L. Cerofolini).

[45] E. DE GIORGI-T. FRANZONI: Su un tipo di convergenza variazionale, *Atti Accd. Naz. Lincei Cl. Sci. Fis. Mat. Natur.* (8) 58 (1975), 842–850.56-6503 (S. Cinquini).

[46] E. DE GIORGI: Convergenza in energia di operatori ellittici, *Contributi del centro linceo interdisciplinare di scienze matematiche e loro applicazioni* 16 (1976)

[47]* E. DE GIORGI: Γ-convergenza e G-convergenza, *Boll. Un. Mat. Ital.* A (5) 14 (1977), 213–220.56-16551 (O. T. Alas).

[48] E. DE GIORGI-G. LETTA: Une notion générale de convergence faible pour des fonctions croissantes d'ensemble, *Ann. Scuola Norm. Sup. Pisa* (4) 4 (1977) 61–99.57-6357 (P. Gérard).

[49] F. COLOMBINI-E. DE GIORGI-S. SPAGNOLO: Existence et unicité des solutions des équations hyperboliques du second ordre à coefficients ne dépendant que du temps, *C. R. Acad. Sci. Paris Sér.* A-B 286 (1978), A1045–A1048.58-23093 (Summary).

[50] F. COLOMBINI-E. DE GIORGI-S. SPAGNOLO: Sur les équations hyper-boliques avec des coefficients qui ne dépendent que du temps, *Ann. Sc. Norm. Sup. Pisa Cl. Sci.* (4) 6 (1979), 511–559.81c:35077 (P. Ungar).

[51]* E. DE GIORGI: Convergence problems for functionals and operators, *Proc. Int. Meeting on Recent Methods in Nonlinear Analysis,* (Rome, 1978), E. DE GIORGI, E. MAGENES, U. MOSCO EDS, Pitagora, Bologna, 1979, 131–188.80k:49010 (C. Sbordone).

[52] E. DE GIORGI-T. FRANZONI: Su un tipo di convergenza variazionale, *Proceedings of the Brescia Mathematical Seminary* 3 (1979), 63–101.82b:49022 (M. Bernardi).

[53] E. DE GIORGI-L. MODICA: *Γ-convergenza e superfici minime,* Preprint Scuola Normale Superiore di Pisa, 1979.

[54] E. DE GIORGI: Quelques problèmes de Γ-convergence, *Computing Methods in applied sciences and engineering, Proc. Fourth Intern. Symp.* (Versailles, 1979), North-Holland, 1980, 637–643.83i:49029 (From the introduction).

[55] E. DE GIORGI: New problems in Γ-convergence and G-convergence, *Proc. Seminar "Free Boundary Problems"* (Pavia, 1979), Ist. Alta Mat., Roma, 1980, vol.II, 183–194.83e:49023 (G. Bottaro).

[56] E. DE GIORGI: Guido Stampacchia, *Atti Accad. Naz Lincei, Cl. Sci. Fis. Mat. Nat.* (8) **68** (1980), 619–625.83g:01051 (Editors.)

[57] E. DE GIORGI-G. DAL MASO-P. LONGO: Γ-limiti di ostacoli, *Atti Accad. Naz Lincei, Cl. Sci. Fis. Mat. Nat.* (8) **68** (1980), 481–487.82m:49013 (M. Biroli.)

[58] E. DE GIORGI-A. MARINO-M. TOSQUES: Problemi di evoluzione in spazi metrici e curve di massima pendenza, *Atti Accad. Naz. Lincei, Cl. Sci. Fis. Mat. Nat.* (8) **68** (1980), 180–187.83m:49052 (P. Marcellini.)

[59] E. DE GIORGI: Γ-limiti di ostacoli, *Recent Methods in Nonlinear Analysis and Applications,* Proc. SAFA IV (Napoli, 1980), A. CANFORA, S. RIONERO, C. SBORDONE, G. TROMBETTI EDS, Liguori, Napoli, 1981, 51–84.87f:49024 (N. Pacchiarotti).

[60]* E. DE GIORGI: Generalized Limits in Calculus of Variations, *Topics in Functional Analysis,* Quad. Sc. Norm. Sup. Pisa, Editrice Tecnico Scientifica, Pisa, 1981, 117–148.84b:49018 (E. Miersemann).

[61] E. DE GIORGI: Limits of Functionals and Differential Operators, *Analytic Solutions of Partial Differential Equations,* (Trento, 1981), L. CATTABRIGA ED., *Astérisque* **89-90** Soc. Math. France, 1981, 153–161.83j:49032 (From the introduction.)

[62] E. DE GIORGI-G. BUTTAZZO: Limiti generalizzati e loro applicazioni alle equazioni differenziali, *Matematiche (Catania)* **36** (1981), 53–64.85f:34027 (U. D'Ambrosio).

[63] E. DE GIORGI: Operatori elementari di limite ed applicazioni al Calcolo delle Variazioni. *Atti del convegno "Studio di problemi limite della Analisi Funzionale"* (Bressanone, 1981), Pitagora, Bologna, 1982, 101–116.

[64] E. DE GIORGI: Alcune osservazioni sui rapporti tra matematica pura e matematica applicata. *Atti del VI convegno AMASES,* Marina di Campo (Isola d'Elba, 1982), notes by A. Battinelli, 13–21.

[65] E. DE GIORGI: Nuovi risultati sulla convergenza dei funzionali, *Proc. Int. Conf. on Partial Differential Equations dedicated to Luigi Amerio for his* 70th *birthday* (Milano-Como, 1982), *Rend. Sem. Mat. Fis. Milano* **52** (1982), 167–173.86j:49034 (Summary).

[66] E. DE GIORGI-T. FRANZONI: Una presentazione sintetica dei limiti generalizzati, *Portugaliae matematica* **41** (1982), 405–436.86h:49016 (C. Sbordone).

[67] E. DE GIORGI-A. MARINO-M. TOSQUES: Funzioni (p, q) convesse, *Atti Accad. Naz Lincei, Cl. Sci. Fis. Mat. Nat.* **(8) 73** (1982), 6–14.85h:49029 (P. Marcellini).

[68] E. DE GIORGI-G. DAL MASO: Γ-convergence and the Calculus of Variations, *Mathematical Theory of Optimization* (Genova, 1981), T. ZOLEZZI ED., Lecture Notes in Math. 979, Springer 1983, 121–143.85d:49018 (L. Boccardo).

[69] E. DE GIORGI-M. DEGIOVANNI-M. TOSQUES: Recenti sviluppi della Γ-convergenza in problemi ellittici, parabolici ed iperbolici, *Methods of Functional Analysis and Theory of Elliptic Equations*, Proc. of the international meeting dedicated to the memory of Carlo Miranda (Napoli, 1982), D. GRECO ED., Liguori, Napoli, 1983, 103–114.86k:49013 (From the Introduction).

[70] E. DE GIORGI-G. BUTTAZZO-G. DAL MASO: On the Lower Semicontinuity of Certain Integral Functionals, *Atti Accad. Naz Lincei, Cl. Sci. Fis. Mat. Nat.* **(8) 74** (1983), 274–282.87a:49020 (B. Sh.Mordukhovich).

[71] E. DE GIORGI-M. DEGIOVANNI-A. MARINO-M. TOSQUES: Evolution equations for a class of nonlinear operators, *Atti Accad. Naz Lincei, Cl. Sci. Fis. Mat. Nat.* **(8) 75** (1983), 1–8.86c:34118 (Summary).

[72] E. DE GIORGI: On a definition of Γ-convergence of measures, *Multifunctions and Integrands, Proceedings of a conference on Stochastic Analysis, Approximation and Optimization*, (Catania, 1983), G. SALINETTI ED., Lecture Notes in Math. 1091, Springer 1984, 150–159.86h:49017 (L. Modica).

[73] E. DE GIORGI: G-operators and Γ-convergence, *Proc. Internat. Congr. Math.*, (Warsaw, 1983), PWN, Warsaw, and North-Holland, Amsterdam, 1984, 1175–1191.87e:49033 (G. Moscariello).

[74] E. DE GIORGI-M. FORTI: Premessa a nuove teorie assiomatiche dei Fondamenti della Matematica, *Quaderni del Dipartimento di Matematica* **45** Pisa, 1984.

[75] E. DE GIORGI: Some Semicontinuity and Relaxation Problems, *Ennio De Giorgi Colloquium*, (Paris 1983) P. KRÉE ED., Pitman Res. Notes in Math. 125, Pitman, 1985, 1–11. Remerciements, *Ibidem*, 188–190.88k:49019 (A. Salvadori).

[76] E. DE GIORGI: Su alcuni nuovi risultati di minimalità, *Atti del Convegno celebrativo del 1° centenario del Circolo Matematico di Palermo,* (Palermo, 1984), *Rend. Circ. Mat. Palermo* **(2) 8** Suppl., (1985), 71–106.88d:49016 (M. Biroli).

[77] E. DE GIORGI-G. DAL MASO-L. MODICA: Convergenza debole di misure su spazi di funzioni semicontinue, *Atti Accad. Naz Lincei, Cl. Sci. Fis. Mat. Nat.* **(8) 79** (1985), 98–106.89g:28008 (R. A. Alò).

[78] E. DE GIORGI: Nuovi sviluppi del calcolo delle variazioni, *Metodi di analisi reale nelle equazioni alle derivate parziali* (Cagliari, 1985), O. MONTALDO, A. PIRO, F. RAGNEDDA EDS, Dipartimento di Matematica, Cagliari, 1985, 83–90.

[79] E. DE GIORGI-M. FORTI: Una teoria quadro per i fondamenti della Matematica, *Atti Accad. Naz Lincei, Cl. Sci. Fis. Mat. Nat.* **(8) 79** (1985), 55–67.89g:03069 (N. C. A. Da Costa).

[80] E. DE GIORGI: Su alcuni indirizzi di ricerca nel calcolo delle variazioni, *Convegno celebrativo del centenario della nascita di M. Picone e di L. Tonelli* (Roma, 1985), Atti dei convegni lincei 77 (1986), 183–187.

[81] E. DE GIORGI-M. FORTI-V. M. TORTORELLI: Sul problema dell'autoriferimento, *Atti Accad. Naz Lincei, Cl. Sci. Fis. Mat. Nat.* **(8) 80** (1986), 363–372.89m:03046 (Summary).

[82] E. DE GIORGI: Some open problems in the theory of partial differential equations, *Hyperbolic Equations* (Padova, 1985), F. COLOMBINI, M. K. V. MURTHY EDS, Pitman Res. Notes in Math. 158, Longman, 1987, 67–73.88g:35004 (Contents.)

[83] E. DE GIORGI-G. DAL MASO-L. MODICA: Weak Convergence of Measures on Spaces of Semicontinuous Functions, *Proc. of the Int. Workshop on Integral Functionals in the Calculus of Variations* (Trieste, 1985), *Rend. Circ. Mat. Palermo,* **(2) 15** Suppl. (1987), 59–100.89i:46028 (C. Vinti).

[84] E. DE GIORGI-L. AMBROSIO-G. BUTTAZZO: Integral Representation and Relaxation for Functionals defined on Measures, *Atti Accad. Naz Lincei, Cl. Sci. Fis. Mat. Nat.* **(8) 81** (1987), 7–13.90e:49011 (T. Pruszko).

[85] E. DE GIORGI-L. AMBROSIO: Un nuovo tipo di funzionale del Calcolo delle Variazioni, *Atti Accad. Naz. Lincei, s.8* **82** (1988), 199–210.92j:49043 (F. Ferro).

[86] E. DE GIORGI-M. CLAVELLI-M. FORTI-V. M. TORTORELLI: A Self-reference Oriented Theory for the Foundations of Mathematics *Analyse Mathématique et Applications,* Gauthier Villars, Paris, 1988, 67–115.90f:03005 (C. F. Kielkopf).

[87] E. DE GIORGI-G. CONGEDO-I. TAMANINI: Problemi di regolarità per un nuovo tipo di funzionale del calcolo delle variazioni, *Atti Accad. Naz. Lincei, s.8* **82** (1988), 673–678.92k:49079 (T. Zolezzi).

[88]* E. DE GIORGI-L. AMBROSIO: New Functionals in Calculus of Variations, *Nonsmooth Optimization and Related Topics, Proc. of the Fourth Course of the International School of Mathematics held in Erice, June 20-July 1, 1988,* F. H. CLARKE, V. F. DEM'YANOV, F. GIANNESSI EDS, Ettore Majorana International Science Series: Physical Science, 43, Plenum Press, 1989, 49–59.90i:93003

[89] E. DE GIORGI: Su alcuni problemi comuni all'analisi e alla geometria, *Note Mat.* **9** (1989), Suppl., 59–71.93f:49036 (D. Scolozzi).

[90] E. DE GIORGI: Sviluppi dell'analisi funzionale nel novecento, *Il pensiero matematico del XX secolo e l'opera di Renato Caccioppoli,* Istituto italiano per gli studi filosofici, Seminari di Scienze, 5, Napoli, 1989, 41–59

[91] E. DE GIORGI: Introduzione ai problemi di discontinuità libera, *Symmetry in Nature. A volume in honour of Luigi A. Radicati di Brozolo,* vol. I, Scuola Norm. Sup., Pisa (1989), 265–285.

[92]* E. DE GIORGI-M. CARRIERO-A. LEACI: Existence theorem for a minimum problem with free discontinuity set, *Arch. Rational Mech. Anal.,* **108** (1989), 195–218. 90j:49002 (R. Klötzler).

[93] E. DE GIORGI: Conversazioni di Matematica – Anni accademici 1988/90, *Quaderni del Dipartimento di Matematica dell'Università di Lecce,* **2** 1990.

[94] E. DE GIORGI: Una conversazione su: Fondamenti della Matematica e "Teoria base". L'esempio della teoria 7×2, *Seminari di didattica 1988-89, Quaderni del Dipartimento di Matematica dell'Università di Lecce,* **3** (1990), 56–62.

[95] E. DE GIORGI: *Definizioni e congetture del giorno 2 maggio 1990,* Pisa.

[96] E. DE GIORGI: *Free discontinuity problems,* talk given at the Israel Math. Ass. Annual Congress 1990.

[97] E. DE GIORGI: Some conjectures on flow by mean curvature, *Methods of real analysis and partial differential equations,* (Capri, 1990) M. L. BENEVENTO, T. BRUNO, C. SBORDONE EDS, Liguori, Napoli, 1990. Revised edition in [114].

[98] E. DE GIORGI-M. FORTI: 5×7: A Basic Theory for the Foundations of Mathematics, *Preprint Sc. Norm. Sup. Pisa* **74** (1990).

[99] E. DE GIORGI: *Alcune congetture riguardanti l'evoluzione delle superfici secondo la curvatura media,* 1990.

[100] E. DE GIORGI: Qualche riflessione sui rapporti tra matematica ed altri rami del sapere. *Convegno Internazionale "Conoscenza e Matematica"* (Pavia, 1989) LORENZO MAGNANI ED., Marcos y Marcos, Milano, 241–249.

[101] E. DE GIORGI: Problemi con discontinuità libera, *Int. Symp. "Renato Caccioppoli",* (Naples, 1989), *Ricerche Mat.,* **40** (1991), Suppl., 203–214.96e:49039 (M. Carriero).

[102] E. DE GIORGI: Alcuni problemi variazionali della geometria, *Atti del convegno internazionale su " Recenti sviluppi in analisi matematica e sue applicazioni" dedicato a G. Aquaro in occasione del suo 70° compleanno* (Bari, 1990), *Confer. Sem Mat. Univ. Bari,* **237-244** (1991), 113–125.93k:49038 (C. Zalinescu).

[103] E. DE GIORGI: Some remarks on Γ-convergence and least squares methods, *Composite media and homogenization theory* (Trieste, 1990), G. DAL MASO, G. F. DELL'ANTONIO EDS, Birkhäuser, 1991, 135–142.92k:49022 (C. Zalinescu).

[104] E. DE GIORGI: *Riflessioni su alcuni problemi variazionali,* talk given at the Meeting "Equazioni differenziali e calcolo delle variazioni, Pisa, September 9-12, 1991.

[105] E. DE GIORGI: Free Discontinuity Problems in Calculus of Variations, *Frontiers in pure and applied Mathematics, a collection of papers dedicated to J. L. Lions on the occasion of his 60th birthday,* R. DAUTRAY ED., North Holland, 1991, 55–62.92i:49004 (D. Azé).

[106] E. DE GIORGI: *Varietà poliedrali tipo Sobolev,* talk given at the "Meeting on calculus of variations and nonlinear elasticity", Cortona, May 27-31, 1991.

[107] E. DE GIORGI: Congetture sui limiti delle soluzioni di alcune equazioni paraboliche quasi lineari, *Nonlinear Analysis: A tribute in honour of G. Prodi,* A. AMBROSETTI, A. MARINO EDS, Quaderni della Scuola Normale Superiore, Pisa, 1991, 173–187.94g:35115 (M. Ughi).

[108]* E. DE GIORGI: Conjectures on limits of some quasilinear parabolic equations and flow by mean curvature, *Partial differential equations and related subjects, Proceedings of a Conference dedicated to L. Nirenberg,* (Trento, 1992), M. MIRANDA ED., Pitman Res. Notes in Math. 269, Longman and Wiley, 1992, 85–95.93e:35004.

[109] E. DE GIORGI: Problemi variazionali con discontinuità libere, *Convegno internazionale in memoria di Vito Volterra* (Roma, 1990), Atti dei convegni lincei 92, 1992, 133–150.

[110] E. DE GIORGI: On the relaxation of functionals defined on cartesian manifolds, *Developments in Partial Differential Equations and Applications to Mathematical Physics*, (Ferrara, 1991), G. BUTTAZZO, G. P. GALDI, L. ZANGHIRATI EDS, Plenum Press, 1992, 33–38.94c:49041 (H. Parks).

[111] E. DE GIORGI: *Movimenti minimizzanti*, talk given at the meeting "Aspetti e problemi della matematica oggi, Lecce, October 20-22, 1992.

[112] E. DE GIORGI: *Un progetto di teoria unitaria delle correnti, forme differenziali, varietà ambientate in spazi metrici, funzioni a variazione limitata*. 1992-93.

[113] E. DE GIORGI: *Funzionali del tipo di integrali di Weierstrass - Osservazioni di Ennio De Giorgi su alcune idee di De Cecco e Palmieri, 1992-93*.

[114] E. DE GIORGI: Some conjectures on flow by mean curvature, *Methods of real analysis and partial differential equations*, (Capri, 1990) M. L. BENEVENTO, T. BRUNO, C. SBORDONE EDS, Quaderni dell'Accademia Pontaniana 14, Napoli, 1992.

[115] E. DE GIORGI: Movimenti secondo la variazione, *Problemi attuali dell'analisi e della fisica matematica, Simposio internazionale in onore di G. Fichera per il suo 70° compleanno* (Taormina, 1992), P. E. RICCI ED., Dipartimento di Matematica, Università "La Sapienza", Roma, 1993, 65–70.94k:49045 (T. Zolezzi).

[116]* E. DE GIORGI: New Problems on Minimizing Movements, *Boundary value problems for partial differential equations and applications, in honour of E. Magenes for his 70th birthday*, C. BAIOCCHI, J. L. LIONS EDS, Masson, 1993, 81–98.94i:35002 (Contents.)

[117] E. DE GIORGI: *Congetture riguardanti barriere, superfici minime, movimenti secondo la curvatura*, talk given in Lecce, November 4, 1993.

[118] E. DE GIORGI: *Congetture riguardanti barriere e sopramovimenti secondo la curvatura media*, talk given in Pisa, November 26, 1993.

[119] E. DE GIORGI: Recenti sviluppi nel calcolo delle variazioni, *Rend. Sem. Fac. Sci. Univ. Cagliari*, **63** (1993), 47–56. 95i:49002 (J. Saint Jean Paulin).

[120] E. DE GIORGI: *Complementi ad alcune congetture riguardanti le barriere*.

[121] E. DE GIORGI: *Introduzione alla teoria della misura e dell'integrazione*, talk given in Lecce, June 12, 1993.

[122]* E. DE GIORGI: Il valore sapienziale della Matematica. Lectio magistralis, *Ennio De Giorgi Doctor Honoris Causa in Filosofia,* Adriatica Editrice Salentina, Lecce, 1994, 19–28.

[123] E. DE GIORGI: New ideas in calculus of variations and geometric measure theory, *Motion by mean curvature and related topics,* (Trento, 1992), G. BUTTAZZO, A. VISINTIN EDS, de Gruyter, 1994, 633–69.95f:49063 (E. J. Balder).

[124] E. DE GIORGI: *Sugli assiomi fondamentali della matematica,* talk given in Napoli, 3 March, 1994.

[125] E. DE GIORGI: *Barriers, boundaries, motion of manifolds.* talk given in Pavia, March 18, 1994.

[126] E. DE GIORGI: *Barriere, frontiere, movimenti di varietà,* talks given in Pavia, March 19-26, 1994.

[127]* E. DE GIORGI: *Fundamental Principles of Mathematics,* talk given at the Plenary Session of the Pontificial Academy of Sciences, October 25-29, 1994. Proceedings Pontificiae Academiae Scientiarum, Scripta varia, n. 92, 269-277.

[128] E. DE GIORGI: *Complementi tecnici alla relazione di Ennio De Giorgi,* Pontificial Academy of Sciences, October 28, 1994.

[129] E. DE GIORGI: On the convergence of solutions of some evolution differential equations. *Set valued Analysis* **2** *(1994), 175-182.* 95m:35084 (V. A. Galaktnionov).

[130] E. DE GIORGI: Un progetto di teoria delle correnti, forme differenziali e varietà non orientate in spazi metrici, *Variational methods, nonlinear analysis and differential equations,* (Genova 1993), Proceedings of the international workshop on the occasion of the 75th birthday of J. P. Cecconi, M. CHICCO, P. OPPEZZI, T. ZOLEZZI EDS, ECIG, Genova, 1994.

[131] E. DE GIORGI-M. FORTI-G. LENZI: Una proposta di teorie base dei Fondamenti della Matematica, *Atti Accad. Naz Lincei, Cl. Sci. Fis. Mat. Nat.* **(9) 5** (1994), 11–22.95e:03037 (Summary).

[132]* E. DE GIORGI-M. FORTI-G. LENZI: Introduzione delle variabili nel quadro delle teorie base dei Fondamenti della Matematica, *Atti Accad. Naz Lincei, Cl. Sci. Fis. Mat. Nat.* **(9) 5** (1994), 117–128.95h:03021 (N. C. A. Da Costa).

[133] E. De Giorgi: Movimenti minimizzanti di partizioni e barriere associate, *talk given in Lecce, June 18, 1994.*

[134] E. De Giorgi: New conjectures on flow by mean curvature, *Nonlinear Variational Problems and Partial Differential Equations* (Isola d'Elba, 1990), A. Marino, M. K. Venkatesha Murthy eds, Pitman Res. Notes in Math. 320, Longman 1995, 120–128.96j:35112 (M. Ughi).

[135] E. De Giorgi: Su alcune generalizzazioni della nozione di perimetro, *Equazioni differenziali e calcolo delle variazioni* (Pisa, 1992), G. Buttazzo, A. Marino, M. V. K. Murthy eds, Quaderni U.M.I. 39, Pitagora, 1995, 237–250.

[136] E. De Giorgi: Problema di Plateau generale e funzionali geodetici, *Nonlinear Analysis – Calculus of variations* (Perugia 1993), *Atti Sem Mat. Fis. Univ. Modena,* **43** (1995), 285–292. 97a:49055 (S. Baldo).

[137] E. De Giorgi: *Congetture sulla continuità delle soluzioni di equazioni lineari ellittiche autoaggiunte a coefficienti illimitati,* talk given in Lecce, January 4, 1995.

[138] E. De Giorgi: *Congetture nel calcolo delle variazioni.* talk given in Pisa, 19 January, 1995.

[139] E. De Giorgi-M. Forti-G. Lenzi-V. M. Tortorelli: Calcolo dei predicati e concetti metateorici in una teoria base dei Fondamenti della Matematica, *Atti Accad. Naz Lincei, Cl. Sci. Fis. Mat. Nat.* **(9) 5** (1995), 79–92.

[140]* E. De Giorgi: Congetture riguardanti alcuni problemi di evoluzione, *A celebration of J. F. Nash Jr, Duke Math. J.* **81-(2)** (1995), 255–268.97f:35093 (M. Degiovanni).

[141] E. De Giorgi: *Fondamenti della Matematica. Teoria 1995.* talk given in Lecce, April 18, 1995.

[142] E. De Giorgi: *Congetture relative al problema della semicontinuità di certe forme quadratiche degeneri,* talk given in Pisa, July 17, 1995.

[143] E. De Giorgi: Movimenti di partizioni, *Variational Methods for discontinuous structures,* (Como 1994), R. Serapioni, F. Tomarelli eds, Birkhäuser, 1996, 1–4.97i:49001 (U. D'Ambrosio)

[144] E. De Giorgi: *Riflessioni su matematica e sapienza,* Quaderni dell'Accademia Pontaniana 18, Napoli 1996.98c:00002 (U. Bottazzini).

[145] E. DE GIORGI-G. LENZI: La teoria '95. Una proposta di teoria aperta e non riduzionistica dei fondamenti della matematica, *Rend. Accad. Naz. Sci. XL, Mem. Mat. Appl.* **20** (1996), 7–34.99c:03006 (Summary).

[146] E. DE GIORGI: *L'analisi matematica standard e non standard rivista in una nuova prospettiva scientifica e culturale.* talk given in Lecce, June 1996.

[147] E. DE GIORGI: Verità e giudizi in una nuova prospettiva assiomatica. *Con-tratto, rivista di filosofia tomista e filosofia contemporanea*, A CURA DI F. BARONE, G. BASTI, A. TESTI, volume *"Il fare della Scienza: i fondamenti e le palafitte"*, 1996, 233–252.

Posthumous Works

[148] E. DE GIORGI-M. FORTI: Dalla Matematica e dalla logica alla sapienza, *Pensiero Scientifico, fondamenti ed epistemologia* (Ancona, 1996), A. REPOLA BOATTO ED.), Quaderni di "Innovazione scuola", 29 (1996), 17–36.

[149] E. DE GIORGI: Su alcuni problemi instabili legati alla teoria della visione, *Atti del convegno in onore di Carlo Ciliberto* (Napoli, 1995), T. BRUNO, P. BUONOCORE, L. CARBONE, V. ESPOSITO EDS, La Città del Sole, Napoli, 1997, 91–98.

[150] E. DE GIORGI-M. FORTI-G. LENZI: Verso i sistemi assiomatici del 2000 in Matematica, Logica e Informatica, *Preprint Scuola Normale Superiore di Pisa* **26** 1996, and *Nuova civiltà delle macchine*, **57-60** (1997), 248–259.

[151]* E. DE GIORGI: Dal superamento del riduzionismo insiemistico alla ricerca di una più ampia e profonda comprensione tra matematici e studiosi di altre discipline scientifiche e umanistiche. *Atti Accad. Naz. Lincei Cl. Sci. Fis. Mat. Natur. Rend. Lincei (9) Mat. Appl.* **9** (1998), 71–80.

[152] G. BUTTAZZO-G. DAL MASO-E. DE GIORGI: Calcolo delle Variazioni, *Enciclopedia del Novecento, vol.XI,* suppl. II, Istituto della Enciclopedia Italiana, 1998, 832–848.

2.2 From the juvenile papers to Plateau's Problem and the solution of Hilbert's 19th Problem

2.2.1 The first articles

The first papers written by Ennio De Giorgi for scientific journals ([1]–[5] and [8]) were published in the years 1950–53. Each of them is the result of a strong student–teacher relationship with his advisor Mauro Picone, who was the director of the Istituto per le Applicazioni del Calcolo, where De Giorgi obtained his first job.

The papers [1] and [2] can be considered as the outcome of his active participation in a university course held by Picone in the academic year 1949–50. In both papers we read the remark "we follow the terminology adopted by Professor Picone in his course of Functional Analysis". As for the content of the papers, in the first one "there is shown how to obtain a limit point, in the space of closed subsets (under Hausdorff's metric) of a compactum, of any given sequence of closed subsets". This is the entire review by R. Arens in *Mathematical Reviews.*

The second paper considers the space Σ of sequences of real numbers and three elementary properties of distance functions on this space, which lead to an elementary characterization of compactness in Σ. The review for *Mathematical Reviews*, signed by V.L. Klee, Jr., as all those we shall refer to, is very detailed and contains some suggestions to simplify the original paper.

The paper [5] is a very careful study of the minimum problem for a one-dimensional quadratic functional. This was a problem presented by Picone in his book "Corso di Analisi Superiore–Calcolo delle Variazioni", published by the Circolo Matematico di Catania in 1922. The hasty review by J.M. Danskin seems to say that the paper contains a not short solution of a not difficult exercise.

The paper [3] considers a problem that was probably proposed by Picone in his course "Introduzione al Calcolo delle Variazioni", held in Rome in the academic year 1950–51. The problem is the minimization of a quadratic functional depending on vector–valued functions of one real variable. In the review L.M. Graves is not impressed by the results, as he writes "this paper contains a few remarks on quadratic variational problems".

The paper [4] presents a necessary and sufficient condition for the integrability of a quasi-continuous function, with respect to an elementary mass. For a general discussion on the problem the author refers to the monograph by M. Picone and T. Viola "Lezioni sulla teoria moderna dell'integrazione", Ediz. Scient. Einaudi, 1952. The review of the paper appeared just after that of the monograph; both were signed by T.H. Hildebrandt.

The paper [8] proves the holomorphy of the sum of a series of homogeneous polynomials, uniformly convergent in a real neighbourhood of the origin, under suitable boundedness conditions in the complex field. The origin of this paper is explained by the author, who writes "I prove a theorem that has been for a long time a conjecture of Professor Mauro Picone, who has often exposed it to his students; as far as I know this theorem has never had the simple proof I am going to present here". The review is by J. Favard.

2.2.2   The development of the theory of perimeters

In the second series of papers ([6], [7], [10], [14], and [18]) we see the exceptional abilities and the great self-confidence of the young De Giorgi. These gifts were acknowledged by Renato Caccioppoli, in the memorable meeting we will describe later.

The paper [6] is the text of a communication to the Congress of the Austrian Mathematical Society held in Salzburg from 9 to 15 September 1952. It contains a theorem à la Weierstrass, on the existence of maxima and minima of a real-valued function defined on the family of all measurable sets with equi-bounded perimeter in a given bounded set of the Euclidean n-dimensional space. The author remarks that the notion of perimeter is equivalent to the $(n-1)$-dimensional measure of the oriented boundary, introduced by Caccioppoli in two papers in 1952. The perimeter is computed analytically, in an original and extremely effective way, through the convolution product of the characteristic function of the set by a Gaussian function. In the same paper the author announces the validity of the isoperimetric inequality for every Lebesgue measurable set. The paper was reviewed by L.M. Graves, who did not notice the great interest of these results for the Calculus of Variations.

The connection of these new results with the Calculus of Variations was underlined by De Giorgi himself in the paper [7], submitted to the Accademia dei Lincei on 14 March 1953. In it he draws attention to the close connection between his results and the results announced by Caccioppoli. The paper was reviewed by L.C. Young, who did not see this connection, although he had expressed a very negative opinion on the papers by Caccioppoli, which he had reviewed before De Giorgi's paper.

This is not the only strange fact in the Caccioppoli–De Giorgi relationship, a recognized master and a young mathematician facing his first challenges. De Giorgi worked on his theory of perimeters, following with a remarkable self-confidence the path just seen by Caccioppoli, without having ever spoken with him. He exploited the papers by Caccioppoli, to which Picone, probably, had drawn his attention. In the academic year 1953–54, on the occasion of a lecture delivered in Rome, Caccioppoli met De Giorgi, who made him aware of his work on oriented boundaries. This meeting was recalled by Edoardo Vesentini during the funeral service for De Giorgi in Pisa, on October 27, 1996: "Before touching the mathematical aspects of De Giorgi's remark, Caccioppoli quoted a sentence by André Gide: nothing is more barbarous than a pure spirit; then, turning to Ennio, he added: it seems that you are an exception".

De Giorgi saw the confirmation of the acknowledgement of his great value in the review of the paper [10], which contains the detailed proof of all previously announced results. Young's review is not only positive, but it also contains a self-criticism of his previous negative opinion on Caccioppoli's work: "Although the author's definitions and theorems are given in precise terms, he is able to show that his definition of perimeter coincides with one proposed by Caccioppoli for $(r-1)$-dimensional boundary measure. This makes it possible to judge more clearly the precise scope of Caccioppoli's definitions". De Giorgi considered this review as the best one ever published in *Mathematical Reviews* on his work.

In the paper [14] De Giorgi analyses the geometrical properties of the boundaries of sets with finite perimeter. He singles out a part of the boundary, called the reduced boundary, which carries the measure associated with the gradient of the characteristic function of the set, and on which the total variation of this measure coincides with the $(n-1)$-dimensional Hausdorff measure. At each point of the reduced boundary De Giorgi proves also the existence of a tangent hyperplane to the boundary of the set.

These results, which are totally nontrivial and will be the starting point for the study of the regularity of the boundaries, when these are solutions of suitable variational problems, did not obtain due regard in the review by L.C. Young.

Instead Young underlined the value of the isoperimetric property of the hypersphere, which is proven in [18]. This paper concludes, in 1958, the cycle devoted to the theory of sets with finite perimeter, started at the Conference in Salzburg in September 1952.

2.2.3 The solution of Hilbert's 19th Problem

The four papers [15], [16], [17], and [26] are devoted to the solution of Hilbert's 19th Problem. The first one is the text of a communication to the fifth Congress of the Italian Mathematical Union, held in Pavia and Turin from 6 to 9 October 1955. This text is very important from the historical point of view. The title shows that, when De Giorgi decided to participate in the Congress, he intended to present his results on sets with finite perimeter, the isoperimetric inequality, and their applications to the Calculus of Variations. The first chapter of the paper is indeed devoted to these topics. The second chapter presents the regularity result for the minima of regular functionals of the Calculus of Variations. There is certainly a connection between the two chapters, since the proof of the regularity result uses the isoperimetric property of the hypersphere. But the fact that the solution of Hilbert's 19th Problem was not mentioned in the title confirms that De Giorgi had not yet obtained this result when he registered for the Congress.

The events which led to the proof of the regularity theorem, reported by Enrico Magenes in the Commemoration of De Giorgi at the Accademia dei Lincei, happened with breath-taking speed. In August 1955, during a hike near Pordoi Pass, in the Dolomites, De Giorgi was informed by Guido Stampacchia about the existence of the 19th Problem. He must have immediately seen the possibility of applying to the solution of this problem the results of his research on the geometry of subsets of multi-dimensional Euclidean spaces. Indeed he was able, in less than two months, to present his solution of Hilbert's Problem to the UMI Congress. This story points out an aspect of De Giorgi's scientific personality: a striking intuition, combined with the prodigious ability to obtain from it a complete proof, with all minor details. The other aspect of De Giorgi's personality is what we tried to outline in our description of his research on perimeters: long-term hard work accomplished in almost complete isolation.

The content of [15] was made more precise in the paper [16]. This was reviewed by L.M. Graves, who did not realize the outstanding value of the result.

C.B. Morrey, Jr. reviewed the paper [17], which contains all details of the proof of the regularity theorem. Morrey's words "the author proves the following important fundamental result" acknowledged that De Giorgi had entered the charmed circle of all-time great mathematicians.

The paper [26], which was written after more than a decade, contains the example of a discontinuous weak solution of a uniformly elliptic system. Therefore the question raised by Hilbert in his 19th Problem has a negative answer, if it is extended to vector-valued functions. This was clearly underlined in the review by G.M. Ewing.

2.2.4   The Plateau Problem and the Bernstein Theorem

The papers [21], [22], [23], [26], and [31] are devoted to the Plateau Problem.

The paper [21] is a collection of results in the theory of perimeters, which will be used in [22] to prove the almost everywhere analyticity of the minimal boundaries. This success is a striking example of what De Giorgi considered as the potential of the use of the theory of perimeters in the Calculus of Variations. He considered the regularity result for minimal surfaces as the victory in his most audacious scientific challenge. The technique used by De Giorgi to prove this result, based on the so-called excess (a way to measure locally the flatness of the hypersurface) and the so-called "excess decay lemma" was immediately adapted by W.K. Allard and F.J. Almgren to obtain the partial regularity of more general geometrical objects (varifolds, currents) and is by now widely known and used also in contexts quite far from the initial one: nonlinear equations and systems of parabolic and elliptic type, harmonic maps, geometric evolution problems, and so on. The papers [21] and [22] were reviewed by Wendell H. Fleming, who wrote for the second one: "A deep theorem about the regularity of solutions of the problem of minimum $(n-1)$-area in n-dimensional euclidean space $\mathbf{R}^n$ is proved".

In [23] the thesis of Bernstein's Theorem is proved for the solutions of the minimal surface equation over the whole three-dimensional Euclidean space. The classical Bernstein's Theorem was known in the two-dimensional case, with several different proofs; the simplest one was deduced from Liouville's Theorem for holomorphic functions of one complex variable. Fleming got a new proof of Bernstein's Theorem by showing that, for any number n of independent variables, the validity of Bernstein's Theorem is a consequence of the nonexistence of n-dimensional singular minimal boundaries in $\mathbf{R}^{n+1}$.

De Giorgi improves this result by showing that the validity of Bernstein's Theorem in dimension n is actually a consequence of the nonexistence of $(n-1)$-dimensional singular minimal boundaries in $\mathbf{R}^n$. This paper was reviewed by R. Osserman, who underlined the importance of the extension to the three-dimensional case, for which the classical two-dimensional methods cannot be used. Later on, F.J. Almbren and J. Simons ruled out the existence of singular n-dimensional minimal boundaries in $\mathbf{R}^{n+1}$ for $n \leq 6$, obtaining as a consequence of De Giorgi's result that Bernstein's Theorem holds for functions of seven independent variables.

The paper [26] is a text that De Giorgi prepared for the International Congress of Mathematicians held in Moscow in 1966. It contains an account of the most recent results on multi-dimensional minimal surfaces. The text was read to the Congress by Vesentini, since De Giorgi could not go to the Soviet Union. This paper was reviewed by F.J. Almgren, Jr., who underlined the fact that the majority of the results was obtained by De Giorgi himself.

The paper [31], in collaboration with Enrico Bombieri and Enrico Giusti, provides a dramatic improvement on the regularity theory for minimal boundaries and on Bernstein's Theorem: indeed, existence of singular seven-dimensional minimal boundaries is proven (among them, the so-called Simon's cone), and nontrivial entire solutions of the minimal surface equation in eight variables are built.

In the paper [24], written in collaboration with Guido Stampacchia, it is proven that every solution of the minimal surface equation, defined in $A \setminus K$, where A is an n-dimensional open set and K is a compact subset of A with Hausdorff $(n-1)$-dimensional measure zero, can be extended analytically to all of A. This result is a consequence of a refinement of the maximum principle for the solutions of the minimal surface equation. Weaker results had been obtained, by similar or totally different methods, by L. Bers, R. Finn, J.C.C. Nitsche, and R. Osserman, as observed by Lamberto Cesari in his review.

Later on, in the volume of the *Annali della Scuola Normale Superiore di Pisa* dedicated to Hans Lewy in 1977, an improvement of De Giorgi and Stampacchia's result was published by Mario Miranda: it is enough to assume that K is closed in A, instead of compact in A. This improvement is nontrivial; the proof is based on the Harnack inequality on minimal surfaces, proven in 1973 by Bombieri and Giusti, who had to use all results on minimal boundaries obtained by De Giorgi and his students.

In the short paper [27], not reviewed, De Giorgi conjectures the validity of the local estimate for the gradient of the solutions of the minimal surface equation in dimension n. This inequality was proven, soon afterwards, in the paper [30] by De Giorgi himself, in collaboration with Bombieri and Miranda. This estimate enabled him to prove the analyticity of the n-dimensional weak solutions of the minimal surface equation, thus extending De Giorgi's regularity theorem to a functional which is not regular in a strict sense.

2.3 Partial differential equations and the foundations of Γ-convergence

2.3.1 The examples of nonuniqueness

The paper [9] on nonuniqueness for the Laplace equation, published in 1953, can be considered as the first non-juvenile paper written by De Giorgi. Starting from a theorem proven in 1947 by Gaetano Fichera, he constructs a harmonic function in a half-disc D of the real plane, not identically zero, which vanishes on the curvilinear part of the boundary of D, and whose normal derivative vanishes almost everywhere on the rectilinear part. Even if it is not so widely known, this paper already contains some of the distinctive characters and recurrent themes of De Giorgi's mathematics. We see, for instance, his extraordinary ability to construct sophisticated functions using only elementary tools, like integrals, series, factorials, and convolutions. We see here for the first time the famous periodic functions with two values that will play an important role in many other examples, as well as the traces of harmonic functions that will be included later in the theory of analytic functionals. De Giorgi liked counterexamples and appreciated the fact that they can be displayed in few pages. He often repeated that, when one tries to prove some result, it is always convenient to devote part of the efforts to the search for the opposite result, i.e., for a possible counterexample.

In 1955 De Giorgi published another example [13] of *null solution* of a partial differential equation, but now in the context of Cauchy problems. He constructed an equation of the form

$$\partial_t^8 u + a_1(x,t)\partial_x^4 u + a_2(x,t)\partial_x^2 u + a_3(x,t)u = 0,$$

with regular coefficients, which admits a nontrivial regular solution, vanishing identically in the half-plane $\{t \le 0\}$.

This paper, which consists of few pages with no references, had a remarkable impact in the mathematical community, rousing in particular the interest of Carleman and the admiration of Leray. In 1966 the latter constructed other *contre-examples du type De Giorgi*. By a coincidence, which is not infrequent in the history of mathematics, a few months before De Giorgi, in Kraków, Andrzej Pliś constructed the first example of nonuniqueness for a *Kovalevskian system*. Unlike Pliś, De Giorgi never returned to this problem.

Curiously enough, among several papers on nonuniqueness which develop De Giorgi's techniques, there is one, in 1960, by a young American mathematician, Paul J. Cohen. He moved, quite soon, his interests to Logic, where he obtained fundamental results on the Continuum Hypothesis, for which he was awarded the Fields Medal. De Giorgi always had the greatest esteem of Cohen's work in Logic.

2.3.2   Gevrey classes and hyperbolic equations

De Giorgi calls *parabolic* the eighth-order equation in his nonuniqueness example, but, with respect to the evolution variable, it should be said (weakly) *hyperbolic*. The coefficients of this equation belong to the class of $\mathcal{C}^\infty$ functions, but do not belong to the class of analytic functions, otherwise Holmgren's uniqueness theorem would have been violated. In another paper ([12]), published in 1955, De Giorgi considers evolution equations of a more general type, and, using a family of intermediate function spaces between the two above-mentioned classes, he singles out the minimal regularity of the coefficients which ensures the uniqueness of the solution to the Cauchy problem. These intermediate spaces, now called *Gevrey classes*, are the spaces G^s of C^∞ functions whose jth derivatives grow not more than $(j!)^s$. What De Giorgi proves is the uniqueness result for all equations of the form $\partial_t^m u + \sum_{h,k} a_{hk}(t,x)\partial_t^k \partial_x^h u = 0$, with coefficients in G^s, under the condition that only terms with $sk + h \le m$ are present.

Several years later, in 1978, De Giorgi returned to this subject, writing, in collaboration with Ferruccio Colombini and Sergio Spagnolo, the paper [49] on second-order hyperbolic equations whose coefficients do not satisfy a Lipschitz condition with respect to the time variable. A new phenomenon is pointed out: for the solvability of the Cauchy problem, the small regularity in t of the coefficients can be compensated by a higher regularity in x of the initial data. In particular, if the coefficients of the equation are Hölder continuous, to have existence it is enough to choose the initial data in (suitable) Gevrey classes, while in the case of discontinuous coefficients one must take the initial data in the class of analytic functions.

Together with some conjectures contained in [82] and [140], which would deserve to be deeply investigated, these are the only papers that De Giorgi devoted to hyperbolic equations. However, he continued to show a special interest in this theory, mainly in a possible "variational approach" to the problem of global existence for nonlinear wave equations.

2.3.3 Analytic solutions of equations with constant coefficients

In the second half of the 1960s the community working in the general theory of partial differential equations was considering an apparently harmless problem, for which, however, the tools of Functional Analysis seemed to be inadequate: given an equation with constant coefficients $P(D)u(x) = f(x)$, with $f(x)$ *analytic* in the Euclidean n-dimensional space, is it possible to find some solution $u(x)$ which is *analytic* in the whole $\mathbf{R}^n$? The existence of some solution of class C^∞ is ensured by the Malgrange–Ehrenpreis Theorem, but the general nature of the operator does not allow one to assert that this solution is analytic, too.

De Giorgi had a particular sensitivity for real analytic functions, while he was less familiar with holomorphic functions. To these he preferred their real counterpart, harmonic functions, for their larger flexibility. In 1963, remembering the beautiful lectures given by Fantappié in Rome, he assigned a degree thesis on analytic functionals to a brilliant student of the Scuola Normale, Francesco Mantovani, who was to write a paper on this subject together with Sergio Spagnolo, before quitting a promising academic career to enter a Capuchin monastery and to leave finally for the Cape Verde Islands as a missionary. It is not surprising that, as soon as De Giorgi became aware of the problem of analytic surjectivity, he started thinking about it, although in a discontinuous way. One of his "theories" was that one should never completely abandon a difficult problem; it is convenient to keep it as a "guide problem", and to return to it from time to time.

Finally, in 1971, De Giorgi and Lamberto Cattabriga were able to solve the problem in [34], and the solution is rather surprising: in $\mathbf{R}^2$ every equation with constant coefficients with an analytic right-hand side has some analytic solutions, while in dimension $n \geq 3$ there are examples, as simple as the heat equation, for which there is no analytic solution. These examples were to completed later by Livio C. Piccinini. The existence proof is based on a representation formula for real analytic functions obtained in [33], which is the basis of the theory of harmonic and analytic functionals. The result was refined later in other papers with Cattabriga ([37] and [44]), while some conjectures on the subject are proposed in [36] and [42].

2.3.4   The foundations of Γ-convergence

The theory of Γ-convergence was developed starting with a simple, paradigmatic example that De Giorgi was discussing in the middle of the 1960s. Let us consider a family of ordinary differential equations depending on an integer k, of the form $D(a_k(x)Du(x)) = f(x)$, where the coefficients $a_k(x)$ are periodic functions taking two positive values, λ and Λ, alternatively in adjacent intervals of length 2^{-k}. Given an arbitrary interval $[a, b]$, we can associate with these equations the Dirichlet conditions $u(a) = u(b) = 0$, obtaining in this way, for every $f(x)$, a unique solution $u(x) = u_k(f, x)$. What happens when k tends to infinity? To answer the question, in this elementary case it is enough to write the solutions $u_k(f, x)$ explicitly: then one sees that they converge to the solution $u(f, x)$ of the equation (with constant coefficient) $a_0 D^2 u(x) = f(x)$, where $a_0 = 2\lambda\Lambda/(\lambda + \Lambda)$ is the *harmonic mean* of the values λ and Λ, or, equivalently, $1/a_0$ is the weak limit of the sequence $\{1/a_k(x)\}$. Since this conclusion is independent both of the interval $[a, b]$ and of the function $f(x)$, we can reasonably assert that $a_0 D^2$ represents the "limit" of the operators $\{D(a_k(x)D)\}$ as $k \to \infty$.

In 1967–68, developing De Giorgi's ideas, Spagnolo[1] introduced the notion of *G-convergence*, i.e., convergence of the Green's functions for elliptic operators of the form $A_k = \sum_{ij} D_j(a_{ij,k}(x)D_i)$; it is defined as the weak convergence, in suitable function spaces, of the sequence of the inverse operators $\{A_k^{-1}\}$. In 1973, with the paper [40] on the "convergence of energies", De Giorgi and Spagnolo show the variational character of G-convergence and its connection with the convergence of the energy functionals associated with the operators A_k. This allows one to compute, although not explicitly, the coefficients of the G-limit operator, and to study, in their natural framework, some homogenization problems in Physics proposed by Enrique Sanchez-Palencia.

Homogenization theory started with the study of electrostatic or thermal problems in composite materials. In the typical case only two phases are present, and in the main phase (for instance, iron) there are many small "impurities" of a different material (for instance, carbon), disposed in a regular fashion. From

---

[1] *Ann. Scuola Norm. Sup. Pisa (3)* **21** (1967), 657-699; *ibid. (3)* **22** (1968), 571-597.

the mathematical viewpoint, homogenization can be interpreted as a particular case of G-convergence, in which the operators A_k have *periodic* coefficients with periods tending to zero, i.e., $a_{ij,k}(x) = \alpha_{ij}(kx)$. The study of this subject had a remarkable development in the 1970s and 1980s, mainly in Italy, in France, in the Soviet Union, and in the United States, and is still object of active research. Besides the Pisa School, the first important contributions to the mathematical theory of homogenization are due to François Murat and Luc Tartar, in Paris.

In the paper [43] dedicated to Mauro Picone in 1975 on the occasion of his 90th birthday, De Giorgi abandons the "operational" notion of G-convergence, moving to the purely "variational" side. Instead of a sequence of differential equations, he considers now a sequence of *minimum problems* for functionals of the Calculus of Variations. Here he deals with integrals whose integrands are functions with linear growth in the gradient of the unknown function, like the area functional in Cartesian form. Without writing the corresponding Euler operators, De Giorgi establishes what is to be considered as the *variational limit* of this sequence of problems, and obtains also a compactness result. This is the starting point of Γ-convergence.

The formal definition of Γ-convergence of a sequence $\{f_k(x)\}$ of functions defined on a topological space X, and with real, or extended real, values, appears few months later in the paper [45] with Tullio Franzoni (see also [52]): $\{f_k\}$ Γ-converges to f if at every point x_0 of the space X these two conditions are satisfied: for every sequence of points $\{x_k\}$ converging to x_0 we have $\liminf_k f_k(x_k) \geq f(x_0)$, and moreover there exists at least one of these sequences $\{x_k\}$ for which $\{f_k(x_k)\}$ converges to $f(x_0)$.

This simple definition has a great importance, both for theory and applications. It includes the old notion of G-convergence as a particular case, and allows one to consider in a unified way many topological structures, and some important functorial notions. Moreover, Γ-convergence opens the way to a wealth of new problems, some of which were proposed by De Giorgi from the very beginning (see [51], [54], [55]), like establishing whether some families of functionals are Γ-compact, or least Γ-closed, or determining the asymptotic behaviour of particular sequences of variational problems (mainly of *homogenization* type). De Giorgi himself, who was usually very modest when speaking of his results, was very proud of this creation, and considered it as a conceptual tool of great moment.

2.4 Asymptotic problems in the Calculus of Variations

2.4.1 The developments of Γ-convergence

During the second half of the 1970s and the first half of the 1980s De Giorgi was engaged in the development of several technical aspects of Γ-convergence and in the promotion of the use of this theory in many asymptotic problems of the Calculus of Variations. In this period he stimulated the activity of a lively research group, by introducing fruitful ideas and original techniques, whose developments for specific problems are often left to students and collaborators.

In the paper [47] De Giorgi introduces the definition of the so-called multiple Γ-limits, i.e., Γ-limits for functions depending on more than one variable, which allow one to present the notion of G-convergence in a very general abstract framework. These notions are the starting point for the applications, developed by other mathematicians, of Γ-convergence techniques to the study of the asymptotic behaviour of saddle points in min-max problems and of solutions of optimal control problems.

In the paper [51] De Giorgi proposes some research lines where Γ-convergence can be applied and summarizes the main results proved by his school. He also formulates some interesting conjectures that have had a fruitful influence in this area for many years. In this paper he expresses, for the first time in an explicit way, some guidelines for the study of Γ-limits of integral functionals, that were already used in an implicit form in the paper [43].

We refer to what will be called the "localization method", which consists in the following strategy. If we want to study the Γ-limit of a sequence of integral functionals defined on an open set Ω of $\mathbf{R}^n$, it is not enough to examine the problem only on Ω, but it is more convenient to study it simultaneously on all open subsets of Ω. This leads us to consider the dependence of the functionals both on a function and on an open set. In the case of positive integrands, the Γ-limit turns out to be increasing with respect to the open set. It is precisely the study of the properties of the Γ-limit, considered as a set function, that enables us, in many cases, to represent it by an integral.

This is one of the motivations for the study of the general properties of increasing set functions and of their limits, that is developed in the paper [48], written in collaboration with Giorgio Letta. This paper also contains a careful study of the notion of integral with respect to an increasing (not necessarily additive) set function, and a general theorem which characterizes the functionals that can be represented by such an integral.

In the paper [53], with Luciano Modica, Γ-convergence is used to construct an example of nonuniqueness for the Dirichlet problem for the area functional in Cartesian form on the circle.

A different application of Γ-convergence is considered in the paper [57], with Gianni Dal Maso and Placido Longo, on the asymptotic behaviour of the solutions of minimum problems with unilateral obstacles for the Dirichlet integral $\int_\Omega |Du|^2 dx$. Given an arbitrary sequence of obstacles, satisfying a very weak bound from above, a subsequence is selected which admits a limit problem, in the sense that the solutions of the corresponding obstacle problems converge weakly to the solution of the limit problem. This asymptotic problem, which is obtained through the localization method for Γ-limits, is a minimum problem for a new integral functional, which may involve also some singular measures. The paper [59] is the text of De Giorgi's contribution to the conference SAFA IV (Naples, 1980), where these results had been presented.

In the papers [60] and [63] De Giorgi presents the theory of Γ-limits in a very general abstract setting, starting from the more elementary notion of operators of type G, based only on the order relation. He also explores the possibility of extending these notions to complete lattices. This project is made precise in the papers [62] and [66], written in collaboration with Giuseppe Buttazzo and Tullio Franzoni, respectively. The former also contains some general guidelines for the applications of Γ-convergence to the study of limits of solutions of ordinary and partial differential equations, including in this framework also optimal control problems.

In the papers [68], with Dal Maso, and [73], which is the text of his lecture at the International Congress of Mathematicians held in Warsaw in 1983, De Giorgi presents the main results of Γ-convergence theory and the most significant applications to the Calculus of Variations obtained by his school.

In the papers [73], [77], and [83] a general abstract framework is proposed for the study of Γ-limits of random functional, which permits one to attack stochastic homogenization problems with Γ-convergence techniques. The papers by Dal Maso and Modica had solved these problems in the case of equi-coercive functionals, using the fact that in this case the space of functionals under consideration is metrizable and compact with respect to Γ-convergence. These properties enabled the study of the convergence of the probability laws of sequences of integral functionals using the ordinary notion of weak convergence of measures. When the random functionals are not equi-coercive, the space of functionals can still be equipped with a topology related to Γ-convergence, but this topology is no longer metrizable, and, although it satisfies the Hausdorff separation axiom, it exhibits some pathological properties, which imply that the only continuous functions on this space are constant.

In the paper [73] De Giorgi proposes several notions of convergence for measures defined on the space of lower semicontinuous functions, and formulates some problems whose solution would enable one to determine the most suitable notion of convergence for the study of Γ-limits of random functionals. This notion of convergence is pointed out and studied in detail in the papers [77] and [83], with Dal Maso and Modica.

In the paper [103] De Giorgi proposes a method for the study of limits of solutions of linear partial differential equations of the first order of the form $A_j u = f$, based on the study of the Γ-limits of the functionals $\|A_j u - f\|_{L^2}^2$.

2.4.2 Semicontinuity and relaxation problems

The notes [32] of the course held by De Giorgi at the Istituto Nazionale di Alta Matematica in 1968–69, though never published, have had a remarkable impact in the field of semicontinuity problems for multiple integrals, and a wide circulation among the interested scholars. They contain the first proof of the (sequential) lower semicontinuity of the functional $F(u, v) = \int_\Omega f(x, u(x), v(x))\, dx$ with respect to the strong convergence of the function u and to the weak convergence of the function v, under the only hypotheses that the nonnegative integrand

$f(x, s, \xi)$ is jointly continuous with respect to the three variables (x, s, ξ) and convex in ξ. The result is obtained through an ingenious approximation from below of the integrand $f(x, s, \xi)$, which reduces the problem to the case of much simpler integrands.

The paper [70], with Buttazzo and Dal Maso, attacks the semicontinuity problem for the functional $\int_\Omega f(u(x), Du(x)) \, dx$ without assuming the semicontinuity of the integrand $f(s, \xi)$ with respect to s, except for $\xi = 0$. The only hypotheses are that $f(s, \xi)$ is nonnegative, measurable in s, convex in ξ, bounded in a neighbourhood of each point of the form $(s, 0)$, and such that $s \mapsto f(s, 0)$ is lower semicontinuous on $\mathbf{R}$.

In the paper [75] De Giorgi summarizes the main results of [70] and formulates some interesting conjectures on semicontinuity and relaxation problems for integral functionals depending on scalar- or vector-valued functions. Some of them have led to important results proved by other mathematicians.

The paper [84], written in collaboration with Luigi Ambrosio and Giuseppe Buttazzo, studies the relaxation problem for the functional $F(u) = \int_\Omega f(x, u(x)) \, d\lambda(x)$ defined for $u \in L^1(\Omega, \lambda; \mathbf{R}^n)$. It contains an integral representation result for the lower semicontinuous envelope of F in the space of bounded Radon measures on Ω, endowed with the topology of the weak* convergence.

2.4.3   Evolution problems for nondifferentiable functionals

At the beginning of the 1980s De Giorgi proposed a new method for the study of evolution problems in a series of papers written in collaboration with Antonio Marino and Mario Tosques, joined later by Marco Degiovanni. In [58] the notion of a solution to the evolution equation $u'(t) = -\nabla f(u(t))$ is extended to the case of functions f defined on metric spaces, through a suitable definition of slope of f and of the corresponding notion of steepest descent curve u. In this context several existence theorems are proved for steepest descent curves under very weak assumptions on the function f.

The paper [67] introduces the notion of (p, q)-convex function on a Hilbert space and proves an existence and uniqueness theorem for the corresponding evolution equations. These results are extended in [71] to nonconvex functions of more general type, and to a wide class of nonmonotone operators. Besides the classical case, where one considers the sum of a convex function and of a smooth function, these results can be applied to many other problems with a nonconvex and nondifferentiable constraint.

The theory of solutions of evolution equations developed in these papers enabled Degiovanni, Marino, Tosques, and their collaborators to extend the deformation techniques of nonlinear analysis to many different situations involving nondifferentiable functionals, and to obtain many multiplicity results for solutions of stationary problems with unilateral constraints.

2.5 The most recent developments in the Calculus of Variations

2.5.1 Free discontinuity problems

In 1987 De Giorgi proposed, in a paper written with Luigi Ambrosio [85] (translated into English in [88]), a very general theory for the study of a new class of variational problems characterized by the minimization of volume and surface energies. In a later paper [105] De Giorgi called this class "free discontinuity problems", referring to the fact that the set where the surface energies are concentrated is not a priori fixed, and can often be represented as the set of discontinuity points of a suitable auxiliary function. Problems of this kind are suggested by the static theory of liquid crystals and by some variational models in fracture mechanics.

Surprisingly, in the same years David Mumford and Jayant Shah proposed, in the framework of a variational approach to image analysis, a problem for which De Giorgi's theory is perfectly fit: the minimization of the functional

$$\int_{\Omega \setminus K} |\nabla u|^2 dx + \alpha \int_{\Omega \setminus K} |u - g|^2 dx + \beta \, \mathcal{H}^1(K \cap \Omega)$$

among all pairs (K, u) with K closed and contained in $\overline{\Omega}$ and $u \in C^1(\Omega \setminus K)$ (here Ω is a rectangle in the plane, α and β are positive constants, $g \in L^\infty(\Omega)$, while $\mathcal{H}^1$ is the one-dimensional Hausdorff measure).

The theory proposed by De Giorgi is based on the introduction of the new function space $SBV(\Omega)$ of special functions with bounded variation, whose study was developed later by Luigi Ambrosio in some subsequent articles. In the paper [92], written in collaboration with Michele Carriero and Antonio Leaci, De Giorgi proves the existence of solutions to the problem proposed by Mumford and Shah. Free discontinuity problems, which are still a research theme for many Italian and foreign mathematicians, are also the object of the papers [96], [101], and [110]. Many conjectures, still widely open, are presented in a paper dedicated to Luigi Radicati [91].

2.5.2 Mean curvature evolution

At the end of the 1980s De Giorgi was actively interested in a class of geometric evolution problems of parabolic type. The model problem is the evolution of k-dimensional surfaces Γ_t by mean curvature, where one requires that the normal velocity of Γ_t is equal to the mean curvature vector at each point of the surface; as this vector is related to the first variation of the k-dimensional measure of Γ_t, from a heuristic point of view this evolution can be regarded as the steepest descent curve for the area functional. In those years De Giorgi proposed several methods to define weak solutions of the problem and to compute approximate solutions; his ideas were developed later by several mathematicians.

The first method, proposed by De Giorgi in [97], [99], [107], [108], [133], and [134], is based on the approximation of the mean curvature motion in codimension one by the solutions of the parabolic equations

$$\frac{\partial u_\varepsilon}{\partial t} = \Delta_x u_\varepsilon + \frac{1}{\varepsilon^2} u_\varepsilon(1 - u_\varepsilon^2) \qquad u_\varepsilon(t,x) : [0, +\infty) \times \mathbf{R}^n \to (-1,1).$$

In this case the surfaces Γ_t are seen as the boundaries of suitable sets E_t, that can be approximated by the sets $\{u_\varepsilon(t,\cdot) > 0\}$ as ε tends to zero. A heuristic motivation for this convergence result is a theorem proven by Luciano Modica and Stefano Mortola in the first years of the developments of Γ-convergence. Indeed, the parabolic equations written above are precisely, up to a time rescaling, the steepest descent curves associated with functionals of Modica–Mortola type, converging to the area functional. It is therefore reasonable to expect the convergence for the steepest descent curves, too. In a paper dedicated to Giovanni Prodi [105], De Giorgi presents a series of conjectures on this subject; some of them have been proven independently by Piero De Mottoni, Michelle Schatzman, Xinfu Chen, Tom Ilmanen, Lawrence C. Evans, Halil Mete Soner, and Panagiotis Souganidis. Other important conjectures, which are still widely open, are presented few years later in a paper dedicated to John F. Nash [140] in a more general context which includes also hyperbolic equations and steepest descent curves associated with functionals depending on the curvature of the manifold.

Other methods proposed by De Giorgi are of very general nature, and can be applied, for instance, to the case of the evolution of surfaces of arbitrary dimension and codimension. We recall, for instance, the method of barriers, proposed in [117], [118], [120], [125], [126] and inspired by Perron's method, which enables us to define weak solutions starting from a suitable class of regular solutions; this method was developed later by two students of De Giorgi, Giovanni Bellettini and Matteo Novaga, who also compared it with the theory of viscosity solutions.

We recall also the idea of studying the geometric properties of a manifold Γ through the analytic property of the function $\eta(x,\Gamma) = \mathrm{dist}^2(x,\Gamma)$, presented for the first time in [91] and then in [126] in the context of evolution problems: using the function $\eta(t,x) = \eta(x,\Gamma_t)$ the mean curvature motion can be described by the system of equations

$$\frac{\partial}{\partial t}\frac{\partial \eta}{\partial x_i} = \Delta \frac{\partial \eta}{\partial x_i} \qquad \text{on } \{\eta = 0\} \qquad i = 1,\ldots,n.$$

This description is much simpler than the one obtained through the parametric representation of the surfaces, particularly in codimensions larger than one.

2.5.3   Minimizing movements

In 1992 De Giorgi proposed a general method for the study of steepest descent curves of a functional F in a metric space (X,d). The name used by De Giorgi, "minimizing movements", is due to the fact that the method consists in a recursive minimization scheme (corresponding in the classical cases to Euler's implicit method), inspired by a time discretization of the mean curvature motion introduced by Frederick J. Almgren, Jean E. Taylor, and Lihe Wang. De Giorgi's method is based on the recursive minimization of suitable perturbed functionals of the form $u \mapsto F(u) + G(u,v)$. In the classical case one takes $G(u,v) = d^2(u,v)/\tau$, where $\tau > 0$ is the time discretization parameter. Surprisingly, the method produces good results even if G is not the square of a distance function. Interesting results can be obtained also by choosing a nonsymmetric G. The first formulation of this method is contained in a paper dedicated to Enrico Magenes [116]. The theory is developed also in [111]. Applications to evolution problems for partitions are presented in [126] and [143].

2.5.4   Minimal surfaces in metric spaces

In 1993 De Giorgi returned to one of his favourite subjects, the theory of minimal surfaces, proposing in [136] a very general approach to the Plateau Problem. In this short paper, that was further developed in [130], De Giorgi is able to formulate the Plateau Problem in any metric space by using only the class of Lipschitz functions. This point of view is deeply innovative also for Euclidean spaces or Riemannian manifolds, where the classical theory of minimal surfaces developed by Herbert Federer and Wendell H. Fleming is based on the duality with differential forms. Luigi Ambrosio and Bernd Kirchheim have proved that the theory proposed by Giorgi extends the Federer–Fleming theory, and that the classical closure and rectifiability theorems for integral currents continue to hold in this much more general context. The papers [106], [110], and [135] are also devoted to problems of geometric measure theory.

2.6   The work on Foundations of Mathematics

2.6.1   The Seminar on Foundations

Stimulated by the problems and the difficulties that emerged from his teaching experience of basic courses at the University of Asmara, starting from the mid 1970s De Giorgi decided to transform one of his traditional courses at the Scuola Normale (the one held on Wednesday) into a seminar devoted to discussions on the Foundations of Mathematics with students and researchers interested in the subject. The idea was to involve not only specialists in Logic, but also, and mainly, researchers active in other areas, not only in Mathematics.

At the beginning the purpose was just to find a formulation of the usual set theoretic foundations, which could provide a clear and natural axiomatic basis on which one can graft the fundamental notions of Mathematical Analysis. According to this meaning, the Foundations, instead of providing a "sound basis" on which all mathematical constructions are supported, should provide an environment (an axiomatic framework) in which these constructions can be inserted; instead of being the root from which all trees of Science are born, they should prepare a network of paths to explore the forests of Science.

From the methodological point of view, De Giorgi followed the traditional axiomatic method used in classical Mathematics, trying to find the axioms among the most relevant properties of the objects under consideration, fully aware that the selected axioms, as well as other axioms possibly added to them, cannot exhaust all properties of the objects under consideration; his presentation was rigorous, but not bound to any formalism, even if he was willing to provide formalizations to be compared with the current foundational theories of formal type.

Gradually his reflections and the discussions in the Seminar and elsewhere led De Giorgi to elaborate and propose more and more general theories: in his approach to the Foundations it was essential to point out and analyse some concepts to be considered as fundamental, without forgetting that the infinite variety of reality can never be completely grasped, according to the warning

> *"There are more things in heaven and earth, Horatio,*
> *Than are dreamt of in your philosophy"*,

that Shakespeare's Hamlet gives to Horatio, and that De Giorgi had chosen to synthesize his philosophical position. Therefore the essential features of his theories can be summarized in four points:

- *nonreductionism*: every theory considers several kinds of objects, connected but not reducible to each other (even the most clever encoding of an object alters to some extent some of its fundamental features);

- *open-endedness*: every theory is open to the possibility of introducing naturally and freely new kinds of objects with their properties;

- *self-description*: the most important properties, relations, and operations involving the objects studied by the theory, as well as the assertions and predicates that can be formulated in the theory, must be objects of the theory;

- *semi-formal axiomatization*: the theory is exposed using the axiomatic method of traditional Mathematics; several formalizations of the theory of formal systems of Mathematical Logic are possible and useful to study different features of it, but none can be taken as definitive, since in each one part of the meaning of the original axioms is lost. Moreover, as one wants to reduce the number of meta-theoretic assumptions, finite axiomatizations are preferred.

Indeed, De Giorgi thought that a serious study of foundational problems would lead to discover the sapiential value of research, and underlined the importance of the nonformal mathematical rigour, well represented by the traditional axiomatic method, which allows one to present the theories in a clear way, and makes it possible to obtain the critical contribution of scholars of different disciplines.

In this activity he saw the height of his ethical idea of science, where the serene and open dialogue among scholars with different backgrounds, as well as the convivial nature of knowledge sharing, were regarded as factors of understanding, friendship, and respect for the fundamental freedoms, for whose defence the direct engagement (he never evaded) was not to be considered more important than the theoretical analysis. For this unitary humanistic vision of knowledge several essays contained in [144] are illuminating: it is not by chance that he used to quote the Universal Declaration of Human Rights of December 10, 1948 as a typical example of fundamental axiomatic system.

These themes became little by little one of his main research interests, developed for two decades, with proposals and elaborations with wider and wider range. He worked on these problems until the last days of his life, when he used to keep a notebook on the bedside table of his room in the hospital, to write the axioms he wanted to discuss with the friends who came to visit him.

2.6.2　The Free Construction Principle

At the beginning the discussions in the Seminar started from a traditional set theoretic basis, but the current set theories had soon to be substantially modified to match, at least partially, the above mentioned criteria: still set theories, but "in the ancients' style", hence:

- with *Urelemente*, objects that are not sets, in order to guarantee nonreductionism and open-endedness;

- with *large classes* that can also be *elements*, e.g., the universal class, the class of sets, the graphs of projections, and other classes that are useful for self-description;

- without *foundation*, to allow for self-membership or for more complex reflexivities.

In this period, however, the centrality of set theoretic notions was preserved; the requirement of a more natural and flexible theory, where the membership relation can model of any relation, led De Giorgi to formulate his "Free Construction Principle" in 1979:

It is always possible to construct a set of sets by freely assigning its elements through a suitable parametrization.

This is perhaps the most important technical contribution of De Giorgi to Mathematical Logic. Although the formal axioms which correspond to the main cases of the "Free Construction Principle" had been previously considered by several authors in the 1930s (Finsler) and '60s (Scott, Boffa, Hajek), the fact that its formulation was more general and more natural allowed for a deeper analysis and enabled to single out the *Antifoundation Axioms*, which are now considered as the most appropriate to the applications of set theory to Semantics and Computer Science. It is a paradox that the "Free Construction Principle" does not appear in any paper by De Giorgi. As he used to do for all technical developments of the Seminar, he preferred to leave the study of this subject to students and collaborators.[2]

---

[2]M. Forti, F. Honsell: Set theory with free construction principles, *Ann. Scuola Norm. Sup. Pisa Cl. Sci. (4)* **10** (1983), 493-522.

2.6.3   The Frame Theory and the Ample Theory

In the meantime De Giorgi was developing his ideas and came to a theory which decisively overcame the set theoretic reductionism, by requiring that Urelemente, instead of being simple *structureless atoms*, permit one to recover *naturally* many traditional notions of Mathematics (*natural numbers, operations, n-tuples*, etc.). While in the first published proposal [74] sets and classes would still be sufficient to encode all other objects, already in the Frame Theory [79] the use of many different kinds of objects, not reduced to their set theoretic encoding, is crucial to provide an internal presentation of the main relations, operations, and properties of the object considered in the theory.

De Giorgi provides here a first solution to the problem of *self-reference*: the classical antinomies can be overcome, even if not completely, by playing on the different keyboards provided by the different kinds of objects introduced in the theory. One has to consider, however, that even if this variety allows one, in practice, to insert each single object, it is not true, in general, that one can insert *simultaneously* all self-referential objects one would like to insert. De Giorgi formulates the problem in [81] and his efforts towards a *general* solution lead to the Ample Theory of 1987, elaborated in collaboration with Massimo Clavelli, Marco Forti, and Vincenzo M. Tortorelli [86], and published in a volume dedicated to Jacques-Louis Lions.

In the Ample Theory all the most important objects considered in Mathematics are present, like *natural and cardinal numbers, sets and classes, pairs and n-tuples, operations*), as well as some fundamental logical objects, like *relations, properties or qualities, propositions*; the axioms postulated permit a natural treatment of the usual mathematical theories and an almost complete inner description of the operations and relations considered. Of course this wealth raises the problem of the consistency of the theory. With an attitude that is typical of his research on Foundations, De Giorgi never attacks this problem directly, but restricts his attention to the analysis of possible antinomies, both classical and new, and develops the theory so that they can be avoided and transformed into limitation problems. Of course he stimulated students and collaborators to a deeper technical investigation, in order to obtain formal theories with richer models, whose consistency were comparable to the usual set theories.

In the case of the Ample Theory the consistency problem was studied deeply by a young student, Giacomo Lenzi. It turned out that De Giorgi's uncommon ability in the analysis of antinomies enabled him to avoid inconsistency. However, his desire to obtain the widest possible theory led him to walk near the abyss, so that many of the interesting extensions suggested in [86], as well as other more natural and apparently harmless extensions, turned out unexpectedly to be inconsistent: the open-endedness requirement was so violated.

2.6.4   The basic theories

This contrast between the width of a theory and its extensibility led De Giorgi to a decisive shift: no longer *general* theories of Mathematics as a whole, but *basic theories*, with few objects and less demanding axioms, sufficient to provide a light and flexible axiomatic framework, strongly self-descriptive and open to any kind of grafting, not only of mathematical notions, but also of notions from Logic and Computer Science (and, potentially, from any other *clear enough* discipline).

To accomplish this programme De Giorgi had to choose some fundamental objects and concepts, that, thanks to the previous analysis, were *qualities, relations, collections*, and *operations*, and to select few general axioms, sufficient to give their main properties. After the first proposals contained in [100], [94], [98], and [124], the Basic Theories were fully developed in [131].

The abandonment of all-embracing theories made it essential to plan specific works in order to graft the main notions of Mathematics and other disciplines on the basic theories: De Giorgi confined himself to reconsider the classical concept of *variable* of Mathematical Physics (in [132]) and to deal with the logical notions of *proposition* and *predicate* (in [139]), facing here for the first time the problematical notion of *truth*.

The programme of grafting on to the basic theories other interesting chapters of Mathematics, Logic, and Computer Science, which required more technical details, was assigned by De Giorgi to his collaborators, who could always rely on his contribution of discussions, proposals, and personal elaborations. Typical examples of "theories à la De Giorgi", that deserve to be mentioned and that he often quoted in his Seminar, are two general theories for the concepts of *collection, set*, and *function*, and for the concept of *operation*, elaborated by Marco Forti and Furio Honsell on the basis of his suggestions and of common discussions.

2.6.5   The theories of 2000

In the meantime De Giorgi's ideas evolved in the direction of releasing his proposals on Foundations from the original mathematical theories, to make them genuinely *interdisciplinary* (see [124], [127], and [128]). This is perhaps related also with the greater attention toward his theories shown by applied scientists (physicists, biologists, economists) rather than by mathematicians and logicians. A clear sign of this change is given by the fact that natural numbers lose the centrality they still had in all previous theories, while propositions and predicates play an essential role for self-description, and the notion of *truth* comes to the foreground. Only the concepts of *quality* and *relation* keep their status of undefined fundamental concepts: of course not in the reductionist sense that every other notion is eventually decomposed into quality and relations, but simply postulating that all other fundamental concepts taken into consideration (collections, operations, propositions, predicates, etc.) are introduced, qualified, and governed by suitable fundamental relations and qualities.

This new approach, anticipated in [145] and [151], was just starting in 1996, but De Giorgi launched out on this enterprise with enthusiasm and extremely interesting theories could have ripened (even before the new millennium, to which he used jokingly to ascribe them), if fate had not untimely cut short his life in October of that year. The papers [147], [148], and [150] are left as suggestions to be developed, even though they had already been superseded by De Giorgi's new line of research. Particularly rich in interesting problems and suggestions for further studies are the developments of logical calculus (in a semantic and not formal mode) and the recovery of the inner notion of (absolute) truth as a quality of some propositions, with the corresponding transformation of the Liar Antinomy, so congenial to De Giorgi's general ideas. An echo of the ideas discussed by De Giorgi in this last period can be found in a paper by Marco Forti and Giacomo Lenzi,[3] which develops these last proposals in a systematic way.

2.7 Remarks on some of the analytic works of Ennio De Giorgi, by Louis Nirenberg

This publication of selected works of De Giorgi by the Scuola Normale di Pisa and the Unione Matematica Italiana is a great service to the mathematical community—also because of the translations into English. This is a wonderful opportunity for young mathematicians to be exposed to some of the ideas of a great mathematician. De Giorgi was one of the most creative and original mathematical analysts of the second half of the last century.

My first acquaintance with his work came through John Nash. Nash had just completed his work [N] on Hölder regularity for solutions of second-order elliptic and parabolic partial differential equations (PDE) when he discovered the paper [17][4] by De Giorgi which had solved the same problem for elliptic equations. Nash asked me if I knew the paper of De Giorgi; I said I'd never heard of him. I finally met him in 1958 at a CIME conference in Pisa. I recall that he beat me at ping pong.

The problem settled the 19th problem posed by Hilbert in 1900—whether solutions of variational problems of the form

$$\min_u \int f\left(x, u, \nabla u\right) dx,$$

where u belongs to a certain class of functions, are smooth—and even real analytic in case f is analytic. Here the problem is assumed to be uniformly elliptic, i.e., the Euler equation

$$\sum \partial_{x_i} f_{u_i} - \partial_u f = 0$$

satisfies

$$c_0 |\xi|^2 \le \sum f_{u_i\, u_j} \xi_i\, \xi_j \le \frac{1}{c_0} |\xi|^2$$

for some $c_0 > 0$ and all $\xi \in \mathbf{R}^n$. The class of functions considered here have square integrable first derivatives in some domain Ω in $\mathbf{R}^n$.

---

[3] A General Axiomatic Framework for the Foundations of Mathematics, Logic and Computer Science, *Rend. Acc. Naz. Sci. XL, Mem. Mat. Appl. (5)* **21** (1997), 171–207.

[4] References [17] refer to the complete list of his publications; [N] refers to references at the end of this article.

Because of earlier well-known results for elliptic equations, the affirmative answer follows from the following basic result of De Giorgi, and Nash. It is for linear elliptic equations in divergence form

$$(2.1) \qquad \partial_i \left(\sum a_{ij}(x)\, \partial_j\, v \right) = 0, \quad \text{in } \Omega$$

where the a_{ij} are bounded measurable, and uniform ellipticity holds:

$$c_0 |\xi|^2 \le \sum a_{ij}(x)\, \xi_i\, \xi_j \le \frac{1}{c_0}\, |\xi|^2.$$

The result is that in any compact subdomain D of Ω, there is a Hölder continuity estimate for v: for some $0 < \alpha < 1$,

$$(2.2) \qquad |v|_{C^\alpha(D)} = \sup_{\substack{x,y \in D \\ x \ne y}} \frac{|v(x) - v(y)|}{|x - y|^\alpha} \le C \|v\|_{L^2(\Omega)}.$$

Here C depends only on n, c_0, D and Ω.

In the late 1930s C.B. Morrey had settled Hilbert's problem in the case $n = 2$. For higher n, the problem remained open until it was settled by De Giorgi. Apparently he heard about the problem in 1955 from G. Stampacchia and within a few months had solved it. (I think, though I am not sure, that I had suggested the problem to Nash.)

De Giorgi's proof is remarkable—see Caffarelli's comments here for some details. It makes use of the isoperimetric inequality for the level sets of the solution. Here, De Giorgi relied on his earlier striking work, in [6] and [7] on perimeters of general sets in $\mathbf{R}^n$, and his proof of the isoperimetric inequality for them. These papers clarified and generalized work by R. Caccioppoli.

Incidentally, it was M. Picone who persuaded both Cacciopoli and De Giorgi to switch to mathematics from engineering.

Establishing the estimate (2.2) may seem like a small step. But progress in PDE proceeds in this way. Someone finds a new inequality, or an improvement of a known one, and much development in the theory follows.

In 1960, J. Moser [Mos1] gave another proof of (2.2); in place of the isoperimetric inequality he used the Sobolev embedding theorem. In [Mos2] he obtained the (related) Harnack inequality for positive solutions.

In 1969, N.V. Krylov and M.V. Safonov [K-S1], [K-S2] established the estimate (2.2) for solutions of equations in non-divergence form, with bounded measurable coefficients

$$\sum a_{ij}(x)\, u_{x_{i+j}} + \sum f_i\, u + cu = 0,$$

and which are uniformly elliptic (as well as the Harnack inequality for positive solutions). Again, this result has had many consequences.

Some excellent books that go deeply into the subject are [G-T], [L-U], [L-S-U], [Mor] and [K]; they contain a wealth of material.

De Giorgi made many deep contributions in PDE and in the Calculus of Variations. Here I just mention a few. They are described in more detail in Chapter 2 of this volume.

[13] presents a counterexample to the uniqueness of solutions of the initial value problem for a linear differential equation with smooth coefficients. It has order eight. This question had been open for about half a century. (About the same time V. Plis also found a counterexample.) In [28], De Giorgi gave a remarkably simple example showing that no estimate like (2.2) was possible for vector-valued solutions of an elliptic system in divergence form.

Namely, in $\mathbf{R}^3$, $u = x/|x|$ has square integrable first-order derivatives and satisfies the Euler equation for the convex functional

$$\int \left[\sum \left(u^i_{x_i} + 3 \frac{x_i\, x_j}{|x|^2}\, u^i_{x_j} \right) \right]^2 + \sum_{i,j} \left[u^i_{x_j} \right]^2 .$$

For variational problems for vector-valued functions, Morrey, E. Giusti and M. Miranda and others, studied problems of partial regularity, to establish that the singular sets of solutions have at least some fixed codimension. They used ideas of De Giorgi in geometric measure theory.

Papers [30] and [31], concerned with minimal surfaces, are striking. In [30] with E. Bombieri and Miranda, interior bounds on first derivatives of weak solutions of the minimal surface equation

$$(2.3) \qquad \sum \partial_{x_i} \left[u_{x_i} \left(1 + |\nabla u|^2 \right)^{-1/2} \right] = 0$$

are obtained—from which analyticity then follows, in view of the estimate (2.2) etc.

In 1915, S. Bernstein had proved that the only solution of (2.3) on all of $\mathbf{R}^2$ is necessarily affine. This result was extended to three dimensions by De Giorgi, and then up to seven dimensions by J. Simons. Simons [S] also constructed a (locally) minimizing cone of eight dimensions—thus having a singularity. In [31], Bombieri, De Giorgi and Giusti showed that Simons' cone is also globally minimizing. They used the special symmetry of the cone to reduce the problem to one for ordinary differential equations. *Furthermore*, using the cone, i.e., its behaviour near infinity, and sub and super solutions of (2.3), they constructed a solution of (2.3) in $\mathbf{R}^8$ which is not affine.

With L. Cattabriga, in [34], De Giorgi proved that any constant coefficient PDE in $\mathbf{R}^2$, with real analytic right-hand side, possesses a real analytic solution. Beautiful.

De Giorgi studied weak limits of solutions of a sequence of variational problems: The weak limit is a solution of what problem? This leads to problems of homogenization and to the notions of G-convergence and Γ-convergence. These concepts are essential in many problems even today.

In connection with the problem of the flow of a hypersurface according to its mean curvature, De Giorgi suggested using as a suitable approximation, depending on some ϵ, solutions of a certain parabolic equation for a function u.

The hypersurface at time t is approximated by the boundary of the set where $u(t, \epsilon, \cdot) > 0$. In [107], and later in [140], De Giorgi presents a number of conjectures connected with gradient flow (steepest descent) for curves related to functionals depending on curvature of the manifold. (Also hyperbolic equations are considered.) Conjecture 1 in section 4 of [140] suggests a very interesting approach for solving the initial value problem for the wave equation with a nonlinear term involving a power of the function via a minimization problem for a function $u_\lambda(t, x)$. The integrand depends on a parameter λ. It is conjectured that as $\lambda \to \infty$, the minimizing function, which is easily seen to exist, tends to the desired solution of the initial value problem.

De Giorgi did much work on the theory of minimal surfaces. In [136] he presents a very general approach to the Plateau problem.

Over a period of years he worked on Foundations of Mathematics. But this goes far beyond my knowledge.

De Giorgi had remarkable intuition; He tended to see things geometrically, though his analytic technique was extremely powerful. He was a very modest, unassuming person. He was very active in problems of human rights, worldwide.

De Giorgi was truly revered by his colleagues, friends and students. People would consult him when they were stuck on some problem. In his quiet way he often suggested new approaches—which worked. We were friends for many years. Whenever I visited Pisa we would meet, dine together (not just at the mensa) etc. We never worked together. I'm sorry that I never came to him with a question. In my personal experience, the heart of Pisa was represented by Guido Stampacchia and Ennio De Giorgi.

Bibliography

[G-T] D. Gilbarg, W. S. Trudinger. Elliptic partial differential equations of second order, 2nd ed. *Grundlehren der Math. Wiss.* **224**, Springer-Verlag 1983.

[K-S1] N. V. Krylov, M. V. Safonov. An estimate of the probability that a diffusion process hits a set of positive measure. *Dokl. Aakad. Nauk SSSR* **245** (1979) 253–255; Eng. Translation: *Soviet Math. Dokl.* **20** (1979) 253–255.

[K-S2] ———— Certain properties of solutions of parabolic equations with measurable coefficients. *Izv. Akad. Nauk SSSR* **40** (1980) 149–175.

[K] N. V. Krylov. *Nonlinear elliptic and parabolic equations of the second order.* Math. and its applications, D. Reidel Pub. Co. 1987.

[L-U] O. A. Ladyzhenskaya, and N. N. Uraltseva, *Linear and quasi-linear elliptic equations.* Nauk. Moscow 1964; Engl. transl'n, Academic Press 1968.

[L-S-U] O. A. Ladyzhenskaya, V. Solonnikov, and N. N. Uraltseva, *Linear and quasi-linear equations of parabolic type.* Nauk. Moscow 1967; Engl. transl'n Am. Math. Soc. 1968.

[Mor] C. B. Morrey, *Multiple integrals in the calculus of variations.* Springer-Verlag 1966.

[Mos1] J. Moser, A new proof of De Giorgi's theorem concerning the regularity problem for elliptic differential equations. *Comm. Pure Appl. Math.* **13** (1960), 457–468.

[Mos2] ________On Harnack's theorem for elliptic differential equations. Ibid. **14** (1961) 577–591.

[N] J. Nash, Continuity of solutions of elliptic and parabolic equations. *Am. J. Math.* **80** (1958), 931–954.

[S] J. Simons, Minimal varieties in Riemannian manifolds. *Ann. of Math.* **88** (1968), 62–105.

2.8 De Giorgi's contribution to the regularity theory of elliptic equations, by Luis Caffarelli

In his celebrated paper [3], Ennio De Giorgi proved that if $w \in H^1_{\mathrm{loc}}$ is a weak solution to the elliptic equation

$$(2.4) \qquad \sum_{i,j} \frac{\partial}{\partial x_i}\left(a_{ij}(x)w_{x_j}\right) = 0,$$

where the coefficients $a_{ij}(x) = a_{ji}(x)$ satisfy the uniform ellipticity condition

$$(2.5) \qquad \nu|y|^2 \le \sum_{ij} a_{ij}(x)y_i y_j \le \nu^{-1}|y|^2, \quad y \in \mathbf{R}^n,$$

then w is locally Hölder continuous. This fundamental result represented the last—and perhaps the hardest—step in the solution of Hilbert's 19th Problem on the regularity of minimizers of multiple integrals in the calculus of variations. More precisely, suppose that u_0 is a local minimizer of the functional

$$J(u) = \int_\Omega F(\nabla u)\, dx,$$

where $F : \mathbf{R}^n \mapsto \mathbf{R}$ is a strictly convex function such that its second derivatives F_{ij} satisfy a uniform ellipticity condition of the kind (2.5). Hilbert's question was whether a minimizer u_0 is necessarily real analytic, under the assumption that F is real analytic.

The minimizer u_0 satisfies in a weak sense the Euler equation in divergence form

$$\sum_i \frac{\partial}{\partial x_i} F_i(\nabla u_0) = 0.$$

Moreover, it is possible to show that, if w is any first partial derivative of u_0, then w is a weak solution to the uniformly elliptic equation

$$(2.6) \qquad \sum_{i,j} \frac{\partial}{\partial x_i}\big(F_{ij}(\nabla u_0)w_{x_j}\big) = 0,$$

which is of the kind (2.4) with $a_{ij}(x) = F_{ij}(\nabla u_0(x))$. Then w is locally Hölder continuous by De Giorgi's theorem, and hence u_0 is locally $C^{1,\alpha}$ (note that, although F is real analytic by assumption, the composed function $F_{ij}(\nabla u_0(x))$ is *a priori* merely measurable, as long as no further regularity of u_0 is available except $u_0 \in H^1_{\mathrm{loc}}$). The information that ∇u_0 is C^α is fundamental, since it makes the coefficients $F_{ij}(\nabla u_0(x))$ in the equation (2.6) *more regular* (at least C^α) than known a priori: this gain of regularity of the coefficients, in turn, makes the solution w of class $C^{1,\alpha}$ (hence $u_0 \in C^{2,\alpha}$) by standard elliptic regularity theory, hence iterating this argument one obtains that u_0 is C^∞, and the analiticity of u_0 follows by classical results of Hopf [6] (see also [10]).

Besides providing the solution to Hilbert's 19th Problem, De Giorgi's paper [3] had an enormous impact in the theory on nonlinear elliptic equations, in homogenization theory, in the theory of free boundary problems and in the theory of degenerate diffusions, just to quote a few important research areas. This is due to the fact that the ideas underlying De Giorgi's regularity theorem have a universality that goes far beyond Hilbert's problem. These ideas can be appreciated at different levels, that I will shortly try to describe.

Dealing with the elliptic equation (2.6), it would seem natural to try to relate the coefficients $F_{ij}(\nabla u_0)$ with w (recall that w is any first partial derivative of u_0), and exploit the nonlinearity of the equation to obtain some smoothness of w. I believe that the first remarkable idea in De Giorgi's paper is the uncoupling of the coefficients $F_{ij}(\nabla u_0)$ from their dependence on u_0, thus treating (2.6) as a *linear* elliptic equation of the kind (2.4) without any smoothness assumption on the coefficients. More precisely, in [3] De Giorgi proves that a solution w to the elliptic equation (2.4) satisfies the *energy inequalities*

$$(2.7) \qquad \int_{B_r} \big|\nabla(w-k)^+\big|^2\, dx \le \frac{C}{(R-r)^2}\int_{B_R}\big|(w-k)^+\big|^2\, dx, \qquad k \in \mathbf{R},$$

where $(w-k)^+ = \max\{w-k, 0\}$ and $B_r \subset B_R$ are arbitrary concentric balls inside the domain where u solves (2.4), and C is some universal constant which depends only on the ellipticity constant ν in (2.5). De Giorgi realized that it is just the validity of the energy inequalities (both for w and $-w$, and at every scale r, R, k) that will provide the Hölder continuity of w. Indeed, in [3] the fact that w solves the elliptic equation (2.4) is used only to prove (2.7) for w and $-w$, then the Hölder continuity of w follows as a consequence of the energy inequalities.

The actual proof of the Hölder continuity of w is then divided into two parts: A "from measure to uniform" estimate and an oscillation lemma. The "from measure to uniform" estimate, after proper rescaling, can be stated as follows.

Let B_1 denote any ball of radius one inside the domain where w satisfies the energy inequalities (2.7), and let $B_{1/2}$ denote the concentric ball of radius one half. The estimate establishes that, if the L^2-norm $\|w^+\|_{L^2(B_1)}$ is small enough (say, smaller than a universal constant $C > 0$ depending only on the spatial dimension n and the ellipticity constant ν), then w^+ satisfies in the smaller ball $B_{1/2}$ the uniform estimate

$$\sup_{B_{1/2}} w^+ \leq 1.$$

The oscillation lemma then shows that, if the graph of $w_{|B_1}$ (the restriction of w to B_1) is trapped inside the cylinder $B_1 \times [-1, 1]$, then there exists a universal constant $\lambda = \lambda(n, \nu) > 0$ such that the graph of $w_{|B_{1/2}}$ is trapped in one of the two smaller cylinders $B_{1/2} \times [-1, 1 - \lambda]$ or $B_{1/2} \times [-1 + \lambda, 1]$. By properly scaling in the spatial variable, this means that letting

$$\operatorname{osc}(r) := \sup_{B_r} w - \inf_{B_r} w$$

(the oscillation of w in the ball B_r), one has

$$(2.8) \qquad \operatorname{osc}(r/2) \leq \theta \operatorname{osc}(r), \quad \theta := 1 - \frac{\lambda}{2} \in (0, 1),$$

i.e. the oscillation of w is reduced by a fixed amount $\theta < 1$ (a universal constant $\theta = \theta(n, \nu)$), on passing from a ball to the concentric ball of half the radius.

By induction, one obtains that $\operatorname{osc}(r/2^k) \leq \theta^k \operatorname{osc}(r)$, $k = 1, 2, \ldots$, hence the oscillation of w on B_ρ tends to zero as some power of the radius ρ, as $\rho \to 0$. Since this estimate is uniform as long as the centres of the balls stay away from the boundary of the domain, the local Hölder continuity of w easily follows.

Moreover, (2.8) yields also a Liouville theorem for global solutions: if w is a solution on $\mathbf{R}^n$ which grows slowly enough at infinity, then w is constant. In particular, suppose that w is a bounded global solution, and let

$$M := \sup_{\mathbf{R}^n} w, \quad m := \inf_{\mathbf{R}^n} w.$$

Then, passing to the limit as $r \to \infty$ in (2.8), one obtains $M - m \leq \theta(M - m)$, hence $M = m$ and w is constant.

Let me point out that the proof of the Harnack inequality for equations in non-divergence form of Krylov and Safonov [7], [8] follows this line and, in fact, a careful inspection of their proof suggests how to modify De Giorgi's argument to obtain Moser's Harnack inequality (see [5], [9]) for equations in divergence form.

Finally, let me comment that the "from measure to uniform" estimate is based on the interplay of two competing inequalities: one is a variant of the Sobolev inequality, which De Giorgi applies to suitable truncations of w (and which is proved in [3] by an original slicing argument, based on the isoperimetric inequality and a sort of ancestor of the coarea formula, proved therein as a lemma), and yields integral bounds for w in terms of ∇w; the other is the energy inequality (2.7), which provides some control of ∇w in terms of w, and can therefore be considered as a sort of reverse Sobolev inequality.

Since the energy inequality involves two different concentric balls, one might think that on trying an iterative argument starting, say, at the unit ball, the estimates are forced to diadically decay to the origin. But De Giorgi uses the fact that the two competing inequalities have different homogeneities to exploit a "nonlinear effect" that allows this iteration to converge within the ball $B_{1/2}$, instead of the origin.

This is a very powerful idea, extremely useful when two quantities compete with different homogeneities (for instance volume and area—one may think of $\|w\|_{L^2}$ as a volume and $\|\nabla u\|_{L^2}$ a surface area, the surface measure on the boundary of a domain and its harmonic measure, etc.).

The scope of this general idea can perhaps be illustrated by the following example. Let $E \subset \mathbf{R}^n$ be a set of locally finite perimeter, that is $P(E, B_R) < \infty$ for every ball B_R. The quantity $P(E, B_R)$, called the *perimeter of E inside B_R*, is a generalization of the notion of surface area, which coincides with the $(n-1)$-dimensional measure of $B_R \cap \partial E$ when E is a set with smooth boundary. The notion of *perimeter of a set* was introduced by De Giorgi in [1], [2], and turns out to be a basic tool in the weak formulation of several variational problems which involve a surface measure term. Indeed, De Giorgi himself relied on this notion of perimeter to give a weak formulation of the minimal surfaces problem (and to prove his famous regularity theorem for minimal surfaces in [4]). More precisely, we say that a set E of locally finite perimeter is area-minimizing (i.e. its boundary is a locally minimal surface) if E satisfies

$$(2.9) \qquad P(E, B_R) \le P(F, B_R)$$

for every measurable set F and every ball B_R such that the symmetric difference $E \triangle F$ is compactly contained inside B_R. Here the set F plays the role of a competitor, i.e. a local perturbation of E: then (2.9) says that the perimeter of E increases, after any local perturbation.

Under this assumption, one can prove the following fact: there exists a small constant $\varepsilon > 0$ (depending only on the dimension n) such that, if B_1 is any ball of radius one and

$$\mathrm{meas}(E \cap B_1) \le \varepsilon,$$

then

$$\mathrm{meas}(E \cap B_{1/2}) = 0,$$

where $B_{1/2}$ denotes the ball concentric with B_1 of radius one half (by scaling and taking the centre of the balls on the boundary of E, this lemma provides a uniform lower bound for the density of E at boundary points).

Trying to compare this lemma with the "from measure to uniform" estimate for the solutions of elliptic equations, one may regard the relative isoperimetric inequality

$$(2.10) \quad \mathrm{meas}(E \cap B_r) \le \frac{1}{2}\,\mathrm{meas}(B_r) \Rightarrow \mathrm{meas}(E \cap B_r)^{\frac{n-1}{n}} \le C(n)P(E, B_r)$$

(which is valid for any measurable set E and any ball B_r) as a substitute for the Sobolev inequality, whereas the analogue of the energy estimate (2.7) is given by the inequality

$$(2.11) \qquad P(E, B_r) \le \mathcal{H}^{n-1}(E \cap \partial B_r),$$

whose validity for almost every r follows from (2.9), choosing as a local perturbation of E the set $F = E \setminus B_r$ (heuristically, the perimeter decreases by $P(E, B_r)$ and increases by $\mathcal{H}^{n-1}(E \cap \partial B_r)$; here and above, $\mathcal{H}^{n-1}$ denotes $(n-1)$-dimensional Hausdorff measure). Letting

$$f(r) = \mathrm{meas}(E \cap B_r)$$

and observing that for almost every r

$$f'(r) = \mathcal{H}^{n-1}(E \cap \partial B_r),$$

if we further assume that

$$f(1) = \mathrm{meas}(E \cap B_1) \le \frac{1}{2}\,\mathrm{meas}(B_{1/2}),$$

then on combining (2.10) and (2.11) one obtains the differential inequality

$$f(r)^{\frac{n-1}{n}} \le C(n)f'(r), \quad \frac{1}{2} \le r \le 1.$$

Integrating this inequality one easily obtains that

$$f(1/2)^{\frac{1}{n}} \le f(1)^{\frac{1}{n}} - \frac{1}{2nC(n)},$$

hence $f(1/2) = 0$ as claimed, provided that $f(1)$ is small enough.

Bibliography

[1] E. DE GIORGI, *Su una teoria generale della misura $(r-1)$-dimensionale in uno spazio ad r dimensioni.*, Ann. Mat. Pura Appl. (4) **36** (1954), 191–213.

[2] E. DE GIORGI *Nuovi teoremi relativi alle misure $(r-1)$-dimensionali in uno spazio ad r dimensioni.* Ricerche Mat. **4** (1955), 95–113.

[3] E. DE GIORGI, *Sulla differenziabilità e l'analiticità delle estremali degli integrali multipli regolari.* Mem. Acc. Sc. Torino, **3** (1957), 25–43.

[4] E. DE GIORGI, *Frontiere orientate di misura minima*, Ed. Tecnico Scientifica, Pisa, 1961.

[5] E. DIBENEDETTO, N. TRUDINGER, *Harnack inequalities for quasiminima of variational integrals.* Ann. Inst. H. Poincaré Anal. Non Linéaire **1** (1984), no. 4, 295–308.

[6] E. HOPF, *Über den funktionalen, insbesondere den analytischen Charakter der Lösungen elliptischer Differentialgleichungen zweiter Ordnung.* Math. Zeitschrift, Band 34 (1932) 194-233.

[7] N. V. KRYLOV, M. V. SAFONOV, *A property of the solutions of parabolic equations with measurable coefficients.* (Russian), Izv. Akad. Nauk SSSR Ser. Mat. **44** (1980), no. 1, 161–175, 239.

[8] N. V. KRYLOV, M. V. SAFONOV, *An estimate for the probability of a diffusion process hitting a set of positive measure.* (Russian), Dokl. Akad. Nauk SSSR **245** (1979), no. 1, 18–20.

[9] J. MOSER, *On Harnack's theorem for elliptic differential equations.*, Comm. Pure Appl. Math. **14** (1961), 577–591.

[10] G. STAMPACCHIA, *Sistemi di equazioni di tipo ellittico a derivate parziali del primo ordine e proprietà dellle estremali degli integrali multipli.* Ricerche di Matematica, **1** (1952), 200-226.

Chapter 3

Selected papers

All papers have been typed and translated trying to reproduce as much as possible their original aspect. We only used common fonts and the same style for the titles, the abstracts and the bibliography.

In order to make some papers more immediately readable we replaced the old notation for the set-theoretic operations of union, intersection and symmetric difference, namely $A + B$, $A \cdot B$ and $A - B$, with the modern ones $A \cup B$, $A \cap B$, $A \setminus B$.

In a few cases, minor (and possibly misleading) misprints have been corrected without an explicit mention. However, all relevant changes concerning bibliography updates, errata corrige, etc. have been explicitly marked with an editorial footnote.

Definition and analytical expression of the perimeter of a set[‡][†]

Note by Ennio De Giorgi[*]

In this Note we will give a definition and an analytical expression of the perimeter of a set contained in a r-dimensional space. Such a definition is equivalent to the one given by CACCIOPPOLI[1] for the oriented boundaries (especially for domains and open sets) considered as sets of "dimensional elements"; it differs from other well known definitions of dimensional measure of a set of points (such as GROSS's and CARATHÉODORY's definitions, etc.). The analytical expression, however, allows to establish various rather hidden results concerning the notion of *perimeter*, that is the dimensional measure of a boundary. These results seem to me of basic importance in several problems of the theory of integration and in isoperimetric problems. In particular, I point out the result characterizing the perimeter as the lower limit of the perimeters of the approximating polyhedra, which yields a definition similar to LEBESGUE's definition of surface area. The proofs of the theorems that we present here will appear in a forthcoming more detailed paper.

1. Let S_r be a r-dimensional space; we denote by $x \equiv (x_1, \ldots, x_r)$ the generic point of S_r, and we indicate by $|x|$ the quantity $\sqrt{x_1^2 + \cdots + x_r^2}$; in the following, by a subset of S_r we always mean a Borel subset, and by a function we always mean a Baire function. Given any bounded function $f(x)$ defined on S_r and any positive number λ, we begin by setting

$$W_\lambda f(x) = \pi^{-r/2} \int_{S_r} e^{-|\xi|^2} f(x + \lambda\xi) \, d\xi_1 \cdots d\xi_r.$$

In general we denote by E a subset of S_r. It is possible to prove that, if E is a *polygonal domain*[2], the $(r-1)$-dimensional measure of the boundary of E, which will be briefly called *perimeter* of E, is given by the limit

$$(1) \qquad \lim_{\lambda \to 0} \int_{S_r} |\text{grad} \cdot W_\lambda \varphi(x|E)| \, dx_1 \cdots dx_r,$$

[‡]Editor's note: translation into English of the paper "Definizione ed espressione analitica del perimetro di un insieme" published in Atti Accad. Naz. Lincei Cl. Sci. Fis. Mat. Nat., **(8)** 14 (1953), 390–393.

[†]Research carried out at the Istituto Nazionale per le applicazioni del Calcolo.

[*]Communicated by the Correspondent R. Caccioppoli in the session of March 14, 1953.

[1]See R. Caccioppoli, *Misura e integrazione sugli insiemi dimensionalmente orientati*. The present "Rendiconti", ser. VIII, vol. XII, fasc. 1 and 2 (1952).

[2]By *polygonal domain* we mean a domain whose boundary is contained in a finite number of hyperplanes.

where $\varphi(x|E)$ stands for the characteristic function of the set E. We can now define the *perimeter* of any set E, which will be indicated by $P(E)$, as the limit in (1), which can be shown to exist (finite or infinite) in any case. Such a definition is illustrated and justified by the following theorems.

THEOREM I. – *Given a set $E \subset S_r$ such that the perimeter $P(E)$ is finite, there exists one and only one vector-valued set function $\mathbf{F}(B) \equiv (F_1(B), \cdots, F_r(B))$ satisfying the following conditions:*

a) *$\mathbf{F}(B)$ is a function defined on any set $B \subset S_r$, is countably additive and with finite total variation;*

b) *given any continuous function $g(x)$ in S_r having continuous first order derivatives, which is infinitesimal, together with its first order derivatives, of order not smaller than $|x|^{-(r+1)}$ as $|x| \to \infty$, it results*

$$\int_E \mathrm{grad} \cdot g(x) \, dx_1 \cdots dx_r = \int_{S_r} g(x) \, d\mathbf{F}.$$

This theorem can be inverted and completed with the interpretation of the expression in (1).

THEOREM II. – *Let a set $E \subset S_r$ be given; if there exists a function $\mathbf{F}(B)$ satisfying conditions a), b) of theorem I, then the perimeter of E is finite and coincides with the total variation in S_r of the function $\mathbf{F}(B)$.*

Theorem II allows to establish the equivalence between the definition of $(r-1)$-dimensional measure of the oriented boundary of a set given by CACCIOPPOLI and our definition of perimeter. Moreover, if the limit in (1) is finite, through a similar limit procedure we can directly define the function $\mathbf{F}(B)$ whose existence is ensured by theorem I.

More precisely, the following theorem holds.

THEOREM III. – *Let E be a set of finite perimeter, and let $\mathbf{F}(B)$ be the additive set function satisfying the conditions of theorem I. Then, for any set B such that the total variation of $\mathbf{F}(B)$ vanishes on the boundary of B, we have*

$$\lim_{\lambda \to 0} \int_B \mathrm{grad} \cdot W_\lambda \varphi(x|E) \, dx_1 \cdots dx_r = \mathbf{F}(B).$$

2. Let us denote now by Σ the space whose elements are the subsets of S_r endowed with the following distance: if $E_1, E_2 \subset S_r$, then the distance between E_1 and E_2 is $\mathrm{meas}\,(E_1 \cup E_2 \setminus E_1 \cap E_2)$, and let us define the notion of limit set using this notion of distance. Given an additive set function $F(\lambda|B)$ depending on a parameter λ, we say, according to the usual terminology, that $F(\lambda|B)$ *weakly converges* to a function $F(B)$ as $\lambda \to \lambda_0$ if, for any continuous function $g(x)$ in S_r which is infinitesimal as $|x| \to \infty$, we have

$$\lim_{\lambda \to \lambda_0} \int_{S_r} g(x) \, dF(\lambda|B) = \int_{S_r} g(x) \, dF.$$

THEOREM IV. – *Given a sequence of sets*

$$E_1, \ldots, E_n, \ldots$$

having equibounded perimeters, if

$$\lim_{n \to \infty} E_n = E$$

then also the perimeter $P(E)$ is finite; in addition we have

$$\min \lim_{n \to \infty} P(E_n) \geq P(E).$$

Moreover the functions $\mathbf{F}^{(1)}(B), \ldots, \mathbf{F}^{(n)}(B), \ldots$ satisfying the conditions of theorem I relatively to the sets $E_1, \ldots, E_n, \ldots$ weakly converge to the function $\mathbf{F}(B)$ relatively to the set E.

This theorem gives a criterion to recognize that a given set has finite perimeter; since the approximation of the sets is considered here with respect to the convergence in measure, the sets can be chosen with a high degree of freedom; in particular they can be chosen among polygonal domains. In this case a more expressive result is contained in theorem VI.

THEOREM V. – *Given a set $E \subset S_r$ (with $r \geq 2$), if E has finite perimeter, then one of the two following relations holds*

$$(2) \qquad \begin{cases} \operatorname{meas} E \leq [P(E)]^{r/(r-1)} \\[2ex] \operatorname{meas}(S_r \setminus E) \leq [P(E)]^{r/(r-1)}. \end{cases}$$

Relations (2) represent an isoperimetric-type inequality, since they establish a bound on the ratio between the $(r-1)$-th power of the measure of a set and the r-th power of its perimeter. Such a "large" inequality ensures the existence of a "sharp" inequality (which will be considered in a forthcoming paper). Clearly the class $\{\Pi\}$ of all polygonal domains is dense in the space Σ, i.e. any set E is a cluster point of polygonal domains. The following fundamental theorem holds.

THEOREM VI. – *The perimeter of a set is the lower limit of the perimeters of the polygonal domains which approximate E, that is we have*

$$\min \lim_{\Pi \to E} P(\Pi) = P(E).$$

3.   Beside the perimeter $P(E)$ of a set it is possible to define the projections of the perimeter on the coordinate planes, which will be denoted by $P_1(E), \ldots, P_r(E)$, and are defined by

$$P_h(E) = \lim_{\lambda \to 0} \int_{S_r} \left| \frac{\partial}{\partial x_h} W_\lambda \varphi(x|E) \right| \, dx_1 \cdots dx_r \qquad (h = 1, \ldots, r).$$

Concerning the projections of the perimeter, theorems similar to theorems I, II, III, IV hold, provided we replace the vector function $\mathbf{F}(B)$ with the scalar functions $F_1(B), \ldots, F_r(B)$ and the gradient with the partial derivatives.

It follows that, *given a set $E \subset S_r$, if, for any value of the index h, it is possible to find a sequence of polygonal domains approximating in measure the set E, whose perimeters have equibounded h-th projection, then the perimeter of E is finite.* We therefore establish a result stated by CACCIOPPOLI simply as a conjecture in the paper quoted above[3].

[3]Loc. cit. [1], Nota I, n. 8, nota 10.

Definizione ed espressione analitica del perimetro di un insieme[‡][†]

Nota di Ennio De Giorgi[*]

In questa Nota daremo una definizione ed espressione analitica del perimetro di un insieme contenuto in uno spazio ad r dimensioni. Tale definizione è equivalente a quella data da CACCIOPPOLI[1] per le frontiere orientate (specialmente per domini e insiemi aperti) riguardate come insiemi di "elementi dimensionali"; si discosta quindi dalle altre ben note (di GROSS, CARATHEODORY, ecc.) di misura dimensionale di un insieme di punti. La espressione analitica permette però di stabilire vari risultati alquanto riposti inerenti alla nozione di perimetro, cioè di misura dimensionale di un contorno. Tali risultati mi sembrano avere importanza essenziale in molte questioni relative alla teoria dell'integrazione ed ai problemi isoperimetrici. Segnalerò in particolare quello sul perimetro come minimo limite dei perimetri di poliedri approssimanti, che consente una definizione analoga a quella di LEBESGUE per l'area di una superficie. Le dimostrazioni dei teoremi che ora esponiamo seguiranno in un altro lavoro più esteso.

1. Sia S_r uno spazio ad r dimensioni il cui punto generico indicheremo con $x \equiv (x_1, \cdots, x_r)$, mentre indicheremo con $|x|$ la quantità $\sqrt{x_1^2 + \cdots + x_r^2}$; parlando nel seguito di insiemi contenuti in S_r e di funzioni ivi definite, intenderemo sempre riferirci ad insiemi di Borel e a funzioni di Baire. Cominciamo col porre, per ogni funzione $f(x)$ definita in S_r ed ivi limitata e per ogni numero positivo λ,

$$W_\lambda f(x) = \pi^{-r/2} \int_{S_r} e^{-|\xi|^2} f(x + \lambda\xi) \, d\xi_1 \ldots d\xi_r.$$

Indichiamo in generale con E un insieme contenuto in S_r. Si dimostra che, se E è un *dominio poligonale*[2], la misura $(r-1)$-dimensionale della frontiera di E, che chiameremo brevemente *perimetro* di E, è data dal limite

$$(1) \qquad \lim_{\lambda \to 0} \int_{S_r} |\text{grad} \cdot W_\lambda \varphi(x|E)| \, dx_1 \cdots dx_r$$

[‡]Editor's note: published in Atti Accad. Naz. Lincei Cl. Sci. Fis. Mat. Nat., (8) **14** (1953), 390–393.

[†]Lavoro eseguito presso l'Istituto Nazionale per le applicazioni del Calcolo

[*]Presentata nella seduta del 14 marzo 1953 dal Corrispondente R. Caccioppoli

[1]Vedi R. Caccioppoli, *Misura e integrazione sugli insiemi dimensionalmente orientati.* Questi "Rendiconti", ser. VIII, vol. XII, fasc. 1 e 2 (1952).

[2]Chiameremo *dominio poligonale* un dominio la cui frontiera sia contenuta in un numero finito di iperpiani.

ove con $\varphi(x|E)$ indichiamo la funzione caratteristica dell'insieme E. Dopo di ciò diremo *perimetro* di un insieme E qualunque e indicheremo con $P(E)$ il limite (1) che si prova esiste sempre (finito o infinito). Tale definizione è illustrata e giustificata dai teoremi che seguono.

TEOREMA I. – *Dato un insieme $E \subset S_r$ il cui perimetro $P(E)$ sia finito, esiste una ed una sola funzione vettoriale d'insieme $\mathbf{F}(B) \equiv (F_1(B), \ldots, F_r(B))$ soddisfacente le condizioni seguenti:*

a) $\mathbf{F}(B)$ *è una funzione definita in ogni insieme $B \subset S_r$, completamente additiva ed a variazione totale finita;*

b) *presa comunque una funzione $g(x)$ continua in S_r con le sue derivate parziali prime ed infinitesima, insieme a tali derivate, per $|x| \to +\infty$, d'ordine non inferiore a quello di $|x|^{-(r+1)}$, risulta*

$$\int_E \mathrm{grad} \cdot g(x)\, dx_1 \cdots dx_r = \int_{S_r} g(x)\, d\mathbf{F}.$$

Questo teorema si lascia invertire e completare con l'interpretazione della espressione (1).

TEOREMA II. – *Dato un insieme $E \subset S_r$, se esiste una funzione $\mathbf{F}(B)$ soddisfacente le condizioni a), b) del teorema I, il perimetro di E è finito e coincide con la variazione totale in S_r della funzione $\mathbf{F}(B)$.*

Il teorema II permette di stabilire l'equivalenza fra la definizione data da CACCIOPPOLI di misura $(r-1)$-dimensionale della frontiera orientata di un insieme e la nostra definizione di perimetro. Inoltre, se il limite (1) è finito, mediante un limite analogo si può definire direttamente la funzione $\mathbf{F}(B)$ la cui esistenza è assicurata dal teorema I.

Precisamente, sussiste il seguente

TEOREMA III. – *Dato un insieme E di perimetro finito e detta $\mathbf{F}(B)$ la funzione additiva di insieme soddisfacente le condizioni del teorema I, per ogni insieme B sulla cui frontiera sia nulla la variazione totale della funzione additiva $\mathbf{F}(B)$ risulta*

$$\lim_{\lambda \to 0} \int_B \mathrm{grad} \cdot W_\lambda \varphi(x|E)\, dx_1 \cdots dx_r = \mathbf{F}(B).$$

2. Indichiamo ora con Σ lo spazio avente come elementi gli insiemi di S_r e nel quale la distanza di due insiemi E_1 ed E_2 sia data dalla $\mathrm{mis}\,(E_1 \cup E_2 \setminus E_1 \cap E_2)$ e definiamo i limiti di insieme in base a questa definizione di distanza. Data poi una funzione additiva di insieme $F(\lambda|B)$, dipendente da un parametro λ diciamo, in accordo con la terminologia corrente, che $F(\lambda|B)$ converge *debolmente*, per $\lambda \to \lambda_0$, verso una funzione $F(B)$ se, per ogni funzione $g(x)$ continua in S_r ed infinitesima per $|x| \to \infty$, si ha

$$\lim_{\lambda \to \lambda_0} \int_{S_r} g(x)\, dF(\lambda|B) = \int_{S_r} g(x)\, dF.$$

Teorema IV. – *Data una successione di insiemi*

$$E_1, \ldots, E_n, \ldots$$

aventi perimetri equilimitati, se

$$\lim_{n \to \infty} E_n = E$$

anche il perimetro $P(E)$ è finito; si ha dippiù

$$\min \lim_{n \to \infty} P(E_n) \geq P(E);$$

inoltre le funzioni $\mathbf{F}^{(1)}(B), \ldots, \mathbf{F}^{(n)}(B), \ldots$ soddisfacenti alle condizioni del teorema I relativamente agli insiemi $E_1, \ldots, E_n, \ldots$ convergono debolmente verso la funzione $\mathbf{F}(B)$ relativa all'insieme E.

Questo teorema fornisce un criterio per riconoscere che un dato insieme ha perimetro finito; essendo gli insiemi approssimanti tenuti alla semplice convergenza in media, essi possono scegliersi con larga arbitrarietà e in particolare essere domini poligonali; in questo caso un risultato più espressivo è quello contenuto nel teorema VI.

Teorema V. – *Dato un insieme $E \subset S_r$ (con $r \geq 2$), se E ha perimetro finito risulta verificata una delle due relazioni*

$$(2) \qquad \begin{cases} \operatorname{mis} E \leq [P(E)]^{r/(r-1)} \\[2ex] \operatorname{mis}(S_r \setminus E) \leq [P(E)]^{r/(r-1)}. \end{cases}$$

Le (2) costituiscono una disuguaglianza di tipo isoperimetrico in quanto stabiliscono una limitazione del rapporto tra la potenza $(r-1)$-esima della misura di un insieme e la potenza r-esima del suo perimetro. Tale disuguaglianza "larga" ci assicura della esistenza di una disuguaglianza "stretta" (della quale ci occuperemo in un prossimo lavoro). Nello spazio Σ l'aggregato $\{\Pi\}$ dei domini poligonali è evidentemente denso, cioè ogni insieme E è elemento di accumulazione di domini poligonali. Sussiste il fondamentale

Teorema VI. – *Il perimetro di un insieme è il minimo limite di quelli dei domini poligonali approssimanti, si ha cioè*

$$\min \lim_{\Pi \to E} P(\Pi) = P(E).$$

3. Accanto al perimetro di un insieme si possono definire le sue proiezioni sui piani coordinati, proiezioni che indicheremo con $P_1(E), \ldots, P_r(E)$ e che sono definite dalle relazioni

$$P_h(E) = \lim_{\lambda \to \lambda_0} \int_{S_r} \left| \frac{\partial}{\partial x_h} W_\lambda \varphi(x|E) \right| dx_1 \cdots dx_r \qquad (h = 1, \ldots, r).$$

Per le proiezioni del perimetro valgono i teoremi analoghi ai teoremi I, II, III, IV, nei quali alla funzione vettoriale $\mathbf{F}(B)$ vanno sostituite le funzioni scalari $F_1(B), \ldots, F_r(B)$ e al gradiente vanno sostituite le derivate parziali.

Ne segue che, *dato un insieme $E \subset S_r$, se, per ogni valore dell'indice h, è possibile trovare una successione di domini poligonali approssimanti in media l'insieme E, i cui perimetri abbiano proiezioni h-esime equilimitate, il perimetro di E è finito.* Si viene così a stabilire un risultato dato come semplice presunzione da CACCIOPPOLI nel lavoro citato[3].

---

[3]Loc. cit. [1], Nota I, n. 8, nota 10.

On a general $(r-1)$-dimensional measure theory in a r-dimensional space[‡][†]

Memoir by Ennio De Giorgi (Rome)

Summary. In this work I introduce a definition of the $(r-1)$-dimensional measure of the oriented boundary of a Borel subset of an r-dimensional Euclidean space, and I prove some of the most remarkable properties of such a measure.

The measure theory developed in the present work[1] is concerned with boundaries of sets in an Euclidean space S_r; such boundaries are not regarded as mere sets of points, but as *oriented sets,* for which we define not only their absolute measure but also the relative measure of their projections on an arbitrary hyperplane. The foundations of this theory have been recently laid by R. CACCIOPPOLI[2] (in general for the $(r-h)$-dimensional measure); at the same time, and independently, I reached the same results, starting from a different point of view and with different aims. The aim of CACCIOPPOLI was to provide a general theory of integration of differential forms in several variables and a complete extension of the GREEN-STOKES formulas. On the other hand, my aim was initially to give a substantial generalization of some isoperimetric problems and I started from the GAUSS-GREEN formula as given.

In this way I establish a necessary and sufficient condition in order that such a formula holds relatively to a Borel set E: in this formula there is a vector-valued set function with bounded total variation whose differential plays the rôle of the "$(n-1)$-dimensional element" in the ordinary case. In the present paper the total variation of this additive function, which coincides with the $(r-1)$-dimensional measure in the sense of CACCIOPPOLI of the *oriented* boundary of E, will be called the *perimeter of the set E.* We provide an analytical expression of this perimeter through a limit of a volume integral containing a parameter. The procedure to get the above mentioned additive set function is extremely simple, and is based on a suitable approximation of the characteristic function of the set E by means of smooth functions, whose limit is the characteristic

---

[‡]Editor's note: translation into English of the paper "Su una teoria generale della misura $(r-1)$-dimensionale in uno spazio a r dimensioni", published in Ann. Mat. Pura Appl., (4) (**36**) (1954), 191–212.

[†]Research carried out at the Istituto Nazionale per le Applicazioni del Calcolo.

[1]The main results contained in this paper appeared in my note *Definizione ed espressione analitica del perimetro di un insieme,* "Rend. Accad. Naz. Lincei, Cl. Sc. fis. mat. nat.", Serie VIII, Vol. XIV, fasc. 3 (marzo 1953), pp. 390–393.

[2]See R. CACCIOPPOLI, *Misura e integrazione sugli insiemi dimensionalmente orientati,* "Rend. Accad. Naz. Lincei, Cl. Sc. fis. mat. nat.", Serie VIII, Vol. XII, fasc. 1, 2 (gennaio-febbraio 1952), pp. 3–11, 137–146.

function of the set E. The general GAUSS-GREEN formula is then obtained as the limit of the ordinary integration by parts formula.

The analytical expression of the perimeter allows to obtain, in a rather simple way, some fundamental results which seem not easy to get starting from the original CACCIOPPOLI's definitions. In my opinion, the most important of these results is the following: *the perimeter of a set E is the lower limit of the polyhedral domains approximating E in mean* (i.e., whose characteristic functions converge in $L^1(S_r)$ to the characteristic function of E). I also point out the isoperimetric property, stating that the measure of a set can be bounded by means of its perimeter. More precisely, the former turns out to be less than or equal to the $\left(\frac{r}{r-1}\right)$-th power of the perimeter.

The theory presented here is grounded on very different ideas from those at the basis of previous definitions of dimensional measure, such as the definitions of GROSS and of CARATHÉODORY, which however can not be applied to the above mentioned problems; for instance they can not be applied to the GAUSS-GREEN formula unless they are completed by providing some notion of *normal vector*, as we are going to do here. For what concerns the definition of CARATHÉODORY, one sees immediately that, in general, it yields a value for the measure of the boundary of a set which is larger than ours, and possibly infinite, when the perimeter is finite. Therefore all boundaries of finite measure in the sense of CARATHÉODORY have finite perimeter in our sense, and the converse is in general false.

Finally, I notice that the concepts exposed in the present paper lay the foundations of a substantially new treatment of certain isoperimetric problems, which will be studied in more detail in a forthcoming paper.

1. In order to simplify the exposition, by a subset of S_r we always mean in the following a BOREL subset, and by a function defined on (a subset of) S_r we always mean a BAIRE function.

Given two operators T_1, T_2 and a function $f(x)$, we will indicate by $T_1 T_2 f(x)$ the function obtained applying the operator T_1 to $T_2 f(x)$.

Let us consider now an Euclidean space S_r, whose generic point will be indicated by $x \equiv (x_1, \ldots, x_r)$ and let us define, for any value of the parameter λ, the operator W_λ by setting, for any bounded function $f(x)$ defined in the whole of S_r,

$$(1) \qquad W_\lambda f(x) = \pi^{-\frac{r}{2}} \int_{S_r} e^{-|\xi|^2} f(x + \sqrt{\lambda}\xi)\, d\xi_1 \ldots d\xi_r,$$

where $|\xi| = \sqrt{\xi_1^2 + \cdots + \xi_r^2}$ and where $(x + \sqrt{\lambda}\xi)$ denotes the point having coordinates $(x_1 + \sqrt{\lambda}\xi_1, \ldots, x_r + \sqrt{\lambda}\xi_r)$. From (1) it immediately follows

$$
\begin{aligned}
(2) \qquad W_\lambda f(x) &= (\pi\lambda)^{-\frac{r}{2}} \int_{S_r} e^{-\frac{|\xi|^2}{\lambda}} f(x + \xi)\, d\xi_1 \ldots d\xi_r \\
&= (\pi\lambda)^{-\frac{r}{2}} \int_{S_r} e^{-\frac{|x-\xi|^2}{\lambda}} f(\xi)\, d\xi_1 \ldots d\xi_r.
\end{aligned}
$$

From (2) one realizes that, for any bounded function $f(x)$ and any positive value of the parameter λ, the function $W_\lambda f(x)$ is continuous and bounded in the whole space S_r together with its partial derivatives of any order. If $f(x)$ is continuous and bounded on S_r and is differentiable with continuous and bounded first order partial derivatives, one immediately checks that

$$(3) \qquad \frac{\partial}{\partial x_h} W_\lambda f(x) = W_\lambda \frac{\partial f}{\partial x_h},$$

i.e., the operator W_λ commutes with the operator $\dfrac{\partial}{\partial x_h}$. It is also easily seen that, given two positive numbers λ, μ, we have

$$(4) \qquad W_\lambda W_\mu f(x) = W_{\lambda+\mu} f(x).$$

It is not difficult to show that, for any bounded function $f(x)$, we have

$$(5) \qquad \lim_{\lambda \to 0} W_\lambda f(x) = f(x)$$

at almost every point of S_r; if in addition $f(x)$ is continuous, then (5) holds uniformly in any closed and bounded subset of S_r. One can also check that the following relations hold:

$$(6) \qquad \int_{S_r} |W_\lambda f(x)|\ dx_1 \ldots dx_r \leq \int_{S_r} |f(x)|\ dx_1 \ldots dx_r,$$

$$(7) \qquad \lim_{\lambda \to 0} \int_{S_r} |W_\lambda f(x)|\ dx_1 \ldots dx_r = \int_{S_r} |f(x)|\ dx_1 \ldots dx_r.$$

Moreover, for any bounded set L, we have

$$(8) \qquad \lim_{\lambda \to 0} \int_L |W_\lambda f(x) - f(x)|\ dx_1 \ldots dx_r = 0.$$

It is clear that the above results still hold when $f(x)$, instead of a scalar function, is a vector-valued function with an arbitrary number of components.

Let us now fix a bounded scalar function $f(x)$ defined in the whole of S_r, and let us consider the integral

$$(9) \qquad \int_{S_r} |\operatorname{grad} W_\lambda f(x)|\ dx_1 \ldots dx_r;$$

since by (3), (4) we have, for any positive numbers λ, μ,

$$(10) \qquad \operatorname{grad} W_{\lambda+\mu} f(x) = \operatorname{grad} W_\mu W_\lambda f(x) = W_\mu \operatorname{grad} W_\lambda f(x),$$

recalling (6) we find that (for positive λ and μ)

$$(11) \qquad \int_{S_r} |\operatorname{grad} W_{\mu+\lambda} f(x)|\ dx_1 \ldots dx_r \leq \int_{S_r} |\operatorname{grad} W_\lambda f(x)|\, dx_1 \ldots dx_r.$$

Inequality (11) ensures that the integral in (9) is a nonincreasing function of λ and therefore there exists the limit

$$\text{(12)} \qquad \mathcal{I}[f(x)] = \lim_{\lambda \to 0} \int_{S_r} |\operatorname{grad} W_\lambda f(x)| \ dx_1 \ldots dx_r,$$

which can be finite or infinite.

Given a set E contained in S_r, we will define the *perimeter of the set E*, that will be denoted by $P(E)$, as

$$\text{(13)} \qquad P(E) = \mathcal{I}[\varphi(x|E)] = \lim_{\lambda \to 0} \int_{S_r} |\operatorname{grad} W_\lambda \varphi(x|E)| \ dx_1 \ldots dx_r,$$

where $\varphi(x|E)$ stands for the characteristic function of the set E.

2. Given a countably additive set function $F^{(\lambda)}(B)$ depending on a parameter λ, we will say that the function $F^{(\lambda)}(B)$ *weakly converges* to a function $F(B)$ as $\lambda \to \lambda_0$, if the equality

$$\text{(14)} \qquad \lim_{\lambda \to \lambda_0} \int_{S_r} g(x) \ dF^{(\lambda)} = \int_{S_r} g(x) \ dF$$

holds for any function $g(x)$ which is continuous in the whole space S_r and infinitesimal as $|x| \to \infty$.

We now recall a theorem of de LA VALLÉE POUSSIN[3] which can be stated as follows: *given a sequence of scalar set functions*

$$\text{(15)} \qquad \mu_1(B), \ldots, \mu_n(B), \ldots$$

defined on any set $B \subset S_r$, which are countably additive, nonnegative and equibounded, there exist an additive set function $\mu(B)$, which is nonnegative and bounded, and a subsequence of (15),

$$\text{(16)} \qquad \mu_{h_1}(B), \ldots, \mu_{h_n}(B), \ldots$$

such that

$$\text{(17)} \qquad \lim_{n \to \infty} \mu_{h_n}(B) = \mu(B)$$

for any set B such that $\mu(B)$ vanishes on the boundary of B.

From this theorem it folows easily that *any sequence of vector-valued set functions, defined on any set $B \subset S_r$, which are countably additive and have equibounded total variations, admits a weakly convergent subsequence.*

Using the above results we can now prove the following result.

THEOREM I. – *Given a bounded function $f(x)$ defined in the whole space S_r, if $\mathcal{I}[f(x)]$ is finite then there exists a unique vector-valued set function $\mathbf{F}(B) \equiv (F_1(B), \ldots, F_r(B))$ satisfying the following conditions:*

---

[3]See de LA VALLÉE POUSSIN, "Annales de l'Institut H. Poincaré", (1932), Vol. II, pp. 221–224.

a) $\mathbf{F}(B)$ *is defined on any set* $B \subset S_r$ *and is countably additive and with bounded total variation;*

b) *given any continuous function* $g(x)$ *in* S_r *having continuous first order derivatives, which is infinitesimal, together with its first order derivatives, of order not smaller than* $|x|^{-(r+1)}$ *as* $|x| \to \infty$, *it results*

$$(18) \qquad \int_{S_r} f(x)\operatorname{grad} g(x) \, dx_1 \ldots dx_r = \int_{S_r} g(x) \, d\mathbf{F}.$$

Proof. Let us pick a sequence of positive numbers converging to zero, which we denote by

$$(19) \qquad \lambda_1, \ldots, \lambda_n, \ldots$$

and consider the sequence of vector-valued set functions

$$(20) \qquad \mathbf{F}^1(B), \ldots, \mathbf{F}^n(B), \ldots$$

defined by

$$(21) \qquad \mathbf{F}^h(B) = -\int_B \operatorname{grad} W_{\lambda_h} f(x) \, dx_1 \ldots dx_r \qquad (\text{for } h = 1, 2, \ldots).$$

We have already seen that, from a sequence of vector-valued set functions, defined on any set $B \subset S_r$, which are countably additive and with equibounded total variations, it is possible to extract a weakly convergent subsequence; by the definition of $\mathcal{I}[f(x)]$, all the set functions in (20) have total variation not larger than $\mathcal{I}[f(x)]$ and therefore we can extract from the sequence in (20) a subsequence

$$(22) \qquad \mathbf{F}^{\nu_1}(B), \ldots, \mathbf{F}^{\nu_h}(B), \ldots$$

weakly converging toward a countably additive set function which will be denoted by $\mathbf{F}(B)$.

Hence, if we indicate by $\lim'$ the lower limit, we get

$$(23) \qquad \lim_{h \to \infty}{}' \int_{S_r} |d\mathbf{F}^{\nu_h}| \geq \int_{S_r} |d\mathbf{F}|.$$

Therefore, as all functions of the sequence in (22) have total variation not larger than $\mathcal{I}[f(x)]$, it follows

$$(24) \qquad \int_{S_r} |d\mathbf{F}| \leq \mathcal{I}[f(x)].$$

From (24) we see that $\mathbf{F}(B)$ satisfies condition a).

Let us pick now an arbitrary continuous function $g(x)$ in S_r having continuous first order derivatives, which is infinitesimal, together with its first order derivatives, of order not smaller than $|x|^{-(r+1)}$ as $|x| \to \infty$; we have

$$(25) \qquad \int_{S_r} g(x)\, d\mathbf{F} = \lim_{h \to \infty} \int_{S_r} g(x)\, d\mathbf{F}^{\nu_h}.$$

As $f(x)$ is a bounded function, by the definition of the operator W_λ it follows that the functions $W_{\lambda_{\nu_h}} f(x)$ are equibounded; recalling formulas (8) of n. 1 and the fact that $\lim_{h \to \infty} \lambda_h = 0$, we have

$$(26)$$
$$\lim_{h \to \infty} \int_{S_r} \left(W_{\lambda_{\nu_h}} f(x)\right) \operatorname{grad} g(x)\, dx_1 \ldots dx_r = \int_{S_r} f(x) \operatorname{grad} g(x)\, dx_1 \ldots dx_r.$$

On the other hand, from (21) it follows, performing an integration by parts,

$$(27) \qquad \int_{S_r} \left(W_{\lambda_{\nu_h}} f(x)\right) \operatorname{grad} g(x)\, dx_1 \ldots dx_r$$
$$= -\int_{S_r} g(x) \operatorname{grad} W_{\lambda_{\nu_h}} f(x)\, dx_1 \ldots dx_r = \int_{S_r} g(x)\, d\mathbf{F}^{\nu_h}.$$

Then equality (18) follows from (25), (26), (27), and the proof of the theorem is complete.

Recalling the definition of perimeter of a set, as a particular case of theorem I we get the following result.

THEOREM II. – *Given a set $E \subset S_r$, if the perimeter $P(E)$ of E is finite, there exists a vector-valued set function $\Phi(B) \equiv (\Phi_1(B), \ldots, \Phi_r(B))$ satisfying the following conditions:*

  a) $\Phi(B)$ *is defined on any set $B \subset S_r$ and is countably additive and with bounded total variation;*

  b) *if $g(x)$ is a continuous function in S_r having continuous first order derivatives, which is infinitesimal, together with its first order derivatives, of order not smaller than $|x|^{-(r+1)}$ as $|x| \to +\infty$, it results*

$$(28) \qquad \int_E \operatorname{grad} g(x)\, dx_1 \ldots dx_r = \int_{S_r} \varphi(x|E) \operatorname{grad} g(x)\, dx_1 \ldots dx_r$$
$$= \int_{S_r} g(x)\, d\Phi,$$

where $\varphi(x|E)$ denotes the characteristic function of the set E.

3. Theorem I can be inverted and completed by the following result.

THEOREM III. – *Given a bounded function $f(x)$, if there exists a vector-valued function $\mathbf{F}(B)$ satisfying conditions a), b) of theorem I, then $\mathcal{I}[f(x)]$ is finite and we have*

$$(29) \qquad \mathcal{I}[f(x)] = \int_{S_r} |d\mathbf{F}|.$$

Proof. Since clearly

$$\text{(30)} \qquad \operatorname{grad}_x e^{-\frac{|x-\xi|^2}{\lambda}} = -\operatorname{grad}_\xi e^{-\frac{|x-\xi|^2}{\lambda}},$$

where grad_x and grad_ξ denote the gradients with respect to $x \equiv (x_1, \ldots, x_r)$ and $\xi \equiv (\xi_1, \ldots, \xi_r)$ respectively, by condition b) of theorem I and recalling the definition of the operator W_λ, we have

$$
\begin{aligned}
|\operatorname{grad} W_\lambda f(\xi)| \;&=\; (\pi\lambda)^{-\frac{r}{2}} \left| \int_{S_r} \left(\operatorname{grad}_\xi e^{-\frac{|x-\xi|^2}{\lambda}} \right) f(x)\, dx_1 \ldots dx_r \right| \\
\text{(31)} \qquad &=\; (\pi\lambda)^{-\frac{r}{2}} \left| \int_{S_r} \left(\operatorname{grad}_x e^{-\frac{|x-\xi|^2}{\lambda}} \right) f(x)\, dx_1 \ldots dx_r \right| \\
&=\; (\pi\lambda)^{-\frac{r}{2}} \left| \int_{S_r} e^{-\frac{|x-\xi|^2}{\lambda}}\, d\mathbf{F} \right| \le (\pi\lambda)^{-\frac{r}{2}} \int_{S_r} e^{-\frac{|x-\xi|^2}{\lambda}} |d\mathbf{F}|.
\end{aligned}
$$

From (31) it follows

$$
\begin{aligned}
\int_{S_r} |\operatorname{grad} W_\lambda f(\xi)|\, d\xi_1 \ldots d\xi_r \;&\le\; (\pi\lambda)^{-\frac{r}{2}} \int_{S_r} d\xi_1 \ldots d\xi_r \int_{S_r} e^{-\frac{|x-\xi|^2}{\lambda}} |d\mathbf{F}| \\
\text{(32)} \qquad &=\; (\pi\lambda)^{-\frac{r}{2}} \int_{S_r} |d\mathbf{F}| \int_{S_r} e^{-\frac{|x-\xi|^2}{\lambda}}\, d\xi_1 \ldots d\xi_r \\
&=\; \int_{S_r} |d\mathbf{F}|
\end{aligned}
$$

and therefore, by the definition of $\mathcal{I}[f(x)]$, we have

$$\text{(33)} \qquad \mathcal{I}[f(x)] \le \int_{S_r} |d\mathbf{F}|.$$

Inequality (33) ensures that $\mathcal{I}[f(x)]$ is finite and therefore (24) holds; equality (29) follows from (33) and (24) and the proof of the theorem is complete.

As a particular case of the previous theorem we obtain the following result.

THEOREM IV. – *Given a set $E \subset S_r$, if there exists a function $\Phi(B)$ satisfying conditions a), b) of theorem II, then the perimeter of E is finite and we have*

$$\text{(34)} \qquad P(E) = \int_{S_r} |d\Phi|.$$

4. Theorems II, IV provide a necessary and sufficient condition for the existence of a function $\Phi(B)$ satisfying the hypotheses a), b) of theorem II; an expression of the function $\Phi(B)$ is given by the following theorem.

THEOREM V. – *Given a set $E \subset S_r$ of finite perimeter, let $\Phi(B)$ be the function satisfying hypotheses a), b) of theorem II. For any set L such that the total variation of the function $\Phi(B)$ vanishes on the boundary of L we have*

$$\text{(35)} \qquad \Phi(L) = -\lim_{\lambda \to 0} \int_L \operatorname{grad} W_\lambda \varphi(x|E)\, dx_1 \ldots dx_r.$$

Proof. If we denote by $\mathcal{F}L$ the boundary of L, the fact that the total variation of $\Phi(B)$ vanishes on the boundary of L can be expressed as

$$(36) \qquad \int_{\mathcal{F}L} |d\Phi| = 0.$$

Let ε be a positive number arbitrarily small; since $\Phi(B)$ has bounded total variation by condition a) of theorem II, and since (36) holds, it is surely possible to find two closed bounded sets C_1, C_2 which do not intersect $\mathcal{F}L$ and satisfy the conditions

$$(37) \qquad C_1 \subset L, \qquad \int_L |d\Phi| - \int_{C_1} |d\Phi| < \varepsilon,$$

$$(37') \qquad C_2 \subset (S_r \setminus L), \qquad \int_{(S_r \setminus L)} |d\Phi| - \int_{C_2} |d\Phi| < \varepsilon.$$

Let us denote by δ_1, δ_2 the distances of the sets C_1, C_2 from $\mathcal{F}L$ respectively, and let Γ be a circular domain centered at the origin and with radius smaller than the minimum between the two numbers δ_1, δ_2. We can certainly find a positive number λ_ε in such a way that, if $0 < \lambda < \lambda_\varepsilon$,

$$(38) \qquad (\pi\lambda)^{-\frac{r}{2}} \int_\Gamma e^{-\frac{|\xi|^2}{\lambda}} d\xi_1 \ldots d\xi_r > 1 - \varepsilon.$$

If we take a point $x^1 \in C_1$, the circular domain centered at x^1 and with radius equal to the radius of Γ is contained in L; on the other hand, if we take a point $x^2 \in C_2$, the circular domain centered at x^2 and with radius equal to the radius of Γ is contained in $(S_r \setminus L)$. Then from (38) we have

$$(39) \quad 1 > (\pi\lambda)^{-\frac{r}{2}} \int_L e^{-\frac{|x-\xi|^2}{\lambda}} d\xi_1 \ldots d\xi_r > 1 - \varepsilon, \qquad \text{for } x \in C_1,\ 0 < \lambda < \lambda_\varepsilon,$$

$$(39') \qquad (\pi\lambda)^{-\frac{r}{2}} \int_L e^{-\frac{|x-\xi|^2}{\lambda}} d\xi_1 \ldots d\xi_r < \varepsilon, \qquad \text{for } x \in C_2,\ 0 < \lambda < \lambda_\varepsilon.$$

From (39), (39') it follows

$$(40) \qquad \left| (\pi\lambda)^{-\frac{r}{2}} \int_L d\xi_1 \ldots d\xi_r \int_{C_1} e^{-\frac{|x-\xi|^2}{\lambda}} d\Phi - \int_{C_1} d\Phi \right|$$
$$< \varepsilon \int_{C_1} |d\Phi| \le \varepsilon \int_{S_r} |d\Phi| = \varepsilon P(E),$$

$$(40') \qquad \left| (\pi\lambda)^{-\frac{r}{2}} \int_L d\xi_1 \ldots d\xi_r \int_{C_2} e^{-\frac{|x-\xi|^2}{\lambda}} d\Phi \right| < \varepsilon P(E).$$

Now we observe that from (37), (37') it follows

$$(41) \qquad \left| \int_L d\Phi - \int_{C_1} d\Phi \right| \leq \int_{(L\setminus C_1)} |d\Phi| < \varepsilon,$$

$$(42) \qquad \int_{(S_r \setminus C_1 \setminus C_2)} |d\Phi| < 2\varepsilon,$$

whereas from (42) we deduce

$$(43) \qquad \left| (\pi\lambda)^{-\frac{r}{2}} \int_L d\xi_1 \dots d\xi_r \int_{(S_r \setminus C_1 \setminus C_2)} e^{-\frac{|x-\xi|^2}{\lambda}} d\Phi \right| < 2\varepsilon.$$

From (40), (40'), (41), (43) we infer that for any positive value of the parameter λ smaller than λ_ε we have

$$(44) \qquad \left| \int_L d\Phi - (\pi\lambda)^{-\frac{r}{2}} \int_L d\xi_1 \dots d\xi_r \int_{S_r} e^{-\frac{|x-\xi|^2}{\lambda}} d\Phi \right| < \varepsilon(2P(E) + 3).$$

As $\Phi(B)$ satisfies condition b) of theorem II, recalling (30) we have

$$
\begin{aligned}
(45) \qquad \operatorname{grad} W_\lambda \varphi(\xi|E) \ &= \ (\pi\lambda)^{-\frac{r}{2}} \int_E \operatorname{grad}_\xi e^{-\frac{|x-\xi|^2}{\lambda}} \, dx_1 \dots dx_r \\
&= \ -(\pi\lambda)^{-\frac{r}{2}} \int_E \operatorname{grad}_x e^{-\frac{|x-\xi|^2}{\lambda}} \, dx_1 \dots dx_r \\
&= \ -(\pi\lambda)^{-\frac{r}{2}} \int_{S_r} e^{-\frac{|x-\xi|^2}{\lambda}} \, d\Phi.
\end{aligned}
$$

From (44), (45) we can realize that, for any positive value of the parameter λ smaller than λ_ε, there holds

$$(46) \qquad \left| \Phi(L) + \int_L \operatorname{grad} W_\lambda \varphi(x|E) \, dx_1 \dots dx_r \right|$$
$$= \left| \int_L d\Phi + \int_L \operatorname{grad} W_\lambda \varphi(\xi|E) \, d\xi_1 \dots d\xi_r \right| < \varepsilon(2P(E) + 3).$$

As ε is an arbitrary positive number, (35) follows from (46), and the proof of the theorem is complete.

    5. The Euclidean space S_r, whose generic point will be denoted in the sequel by $x \equiv (x_1, \dots, x_r)$, can always be considered as the topological product of two spaces S_p ed S_{r-p}. We will denote by $y \equiv (y_1, \dots, y_p)$ the generic point of S_p, by $z \equiv (z_1, \dots, z_{r-p})$ the generic point of S_{r-p}, and by (y, z) the point of S_r having coordinates $(y_1, \dots, y_p, z_1, \dots, z_{r-p})$. Given a set $E^y \subset S_p$ and a set $E^z \subset S_{r-p}$, we denote by (E^y, E^z) the topological product of the sets E^y and E^z, that is the set of all points (y, z) when y varies in E^y and z in E^z. In a similar way as in the definition of the operator W_λ, we can define the operator W_λ^y by setting, for any bounded function $f(x) \equiv f(y,z)$ defined in S_r and any positive value of the parameter λ

$$(47) \qquad W_\lambda^y f(x) \equiv W_\lambda^y f(y,z) = \pi^{-\frac{p}{2}} \int_{S_p} e^{-|\eta|^2} f(y + \sqrt{\lambda}\eta, z)\, d\eta_1 \ldots d\eta_p.$$

For any function $g(x) \equiv g(y,z)$ which is differentiable with respect to the variables $y_1, \ldots, y_p$ we will denote by $\operatorname{grad}_y g(y,z)$ the vector of components $\left(\frac{\partial g}{\partial y_1}, \ldots, \frac{\partial g}{\partial y_p} \right)$.

Let us observe that all properties of the operator W_λ listed in n. 1 can be extended to the operator W_λ^y. Therefore, beside the functional $\mathcal{I}[f(x)]$, we can define the functional $\mathcal{I}_y[f(y,z)]$ by setting

$$(48) \qquad \mathcal{I}_y[f(y,z)] = \lim_{\lambda \to 0} \int_{S_p} |\operatorname{grad}_y W_\lambda^y f(y,z)|\, dy_1 \ldots dy_p.$$

Indeed, with an argument similar to the one used for the limit in (12), one can prove that the limit appearing in (48) is well defined. Let us now prove the inequality

$$(49) \qquad \mathcal{I}[f(x)] \equiv \mathcal{I}[f(y,z)] \geq \int_{S_{r-p}} \mathcal{I}_y[f(y,z)]\, dz_1 \ldots dz_{r-p}.$$

Recalling (6) we get that, for any pair of real positive numbers λ, μ there holds

$$(50) \qquad \begin{aligned}
& \int_{S_r} \left| W_\lambda \operatorname{grad}_y W_\mu^y f(y,z) \right|\, dy_1 \ldots dy_p\, dz_1 \ldots dz_{r-p} \\
={}& \int_{S_r} \left| W_\mu^y \operatorname{grad}_y W_\lambda f(y,z) \right|\, dy_1 \ldots dy_p\, dz_1 \ldots dz_{r-p} \\
\leq{}& \int_{S_r} \left| \operatorname{grad}_y W_\lambda f(y,z) \right|\, dy_1 \ldots dy_p\, dz_1 \ldots dz_{r-p} \\
\leq{}& \int_{S_r} \left| \operatorname{grad} W_\lambda f(x) \right|\, dx_1 \ldots dx_r \leq \mathcal{I}[f(x)].
\end{aligned}$$

From (50), passing to the limit as $\lambda \to 0$, we get, recalling (7),

$$(51) \qquad \int_{S_{r-p}} dz_1 \ldots dz_{r-p} \int_{S_p} \left| \operatorname{grad}_y W_\mu^y f(y,z) \right|\, dy_1 \ldots dy_p \leq \mathcal{I}[f(x)].$$

Therefore, from (51), taking into account (48), and using well known theorems on the exchange between the limit and the integral, we get (49).

6.

THEOREM VI. – *Given a set E contained in an euclidean space S_r (with $r \geq 2$), one of the two following inequalities is always satisfied:*

$$(52) \qquad \begin{cases} (\operatorname{meas} E)^{r-1} \leq (P(E))^r \\[2ex] (\operatorname{meas}(S_r \setminus E))^{r-1} \leq (P(E))^r. \end{cases}$$

Proof. Our theorem will be certainly proved if we show that, given an Euclidean space S_r (with $r \geq 1$) and a function $\varphi(x)$ which, at any point $x \in S_r$, takes either the value 0 or the value 1, one of the two following relations hold:

$$(53) \quad \begin{cases} \left(\int_{S_r} \varphi(x) \, dx_1 \ldots dx_r \right)^{r-1} \leq (\mathcal{I}[\varphi(x)])^r \\[2em] \left(\int_{S_r} (1 - \varphi(x)) \, dx_1 \ldots dx_r \right)^{r-1} \leq (\mathcal{I}[\varphi(x)])^r \end{cases}$$

provided, in case $r = 1$, we agree to assign the value 1 to the symbol ∞^0 and the value 0 to the symbol 0^0. We just begin by considering the case $r = 1$ and we notice that, if one of the two integrals $\int_{S_1} \varphi(x) \, dx$, $\int_{S_1} (1 - \varphi(x)) \, dx$ vanishes, one of the relations (53) is certainly valid. On the other hand, if

$$(54) \quad \int_{S_1} \varphi(x) \, dx = \int_{-\infty}^{+\infty} \varphi(x) \, dx \neq 0, \qquad \int_{-\infty}^{+\infty} (1 - \varphi(x)) \, dx \neq 0,$$

it is possible, by (5), to find two points $\overline{x}$, $\overline{\overline{x}}$, such that

$$(55) \quad \lim_{\lambda \to 0} W_\lambda \varphi(\overline{x}) = \varphi(\overline{x}) = 0, \qquad \lim_{\lambda \to 0} W_\lambda \varphi(\overline{\overline{x}}) = \varphi(\overline{\overline{x}}) = 1.$$

From (55) it follows

$$(56) \quad \mathcal{I}[\varphi(x)] \;=\; \lim_{\lambda \to 0} \int_{-\infty}^{+\infty} \left| \frac{d}{dx} W_\lambda \varphi(x) \right| \, dx$$

$$\geq \lim_{\lambda \to 0} \left| \int_{\overline{x}}^{\overline{\overline{x}}} \frac{d}{dx} W_\lambda \varphi(x) \right| \, dx = \lim_{\lambda \to 0} \left| W_\lambda \varphi(\overline{x}) - W_\lambda \varphi(\overline{\overline{x}}) \right| = 1.$$

Therefore, when (54) are verified, relations (53) are certainly satisfied; therefore, we have proved that, for $r = 1$, at least one of the relations (53) holds. Let us now prove that, given any positive integer m, if one of the relations (53) is valid for $r = m$, then the same is true for $r = m + 1$. If we show this, as (53) are clearly equivalent to (52), our theorem will be completely proved. Therefore, let us begin by observing that a space S_{m+1} can always be considered as the topological product of two spaces S_m and S_1; recalling the notation in n. 5, we will denote by $y \equiv (y_1, \ldots, y_m)$ the generic point of S_m and by z the generic point of S_1. Let now $\varphi(x) \equiv \varphi(y, z)$ be a function that, at any point $x \in S_{m+1}$, takes either the value 0 or the value 1 and let E_1, E_2, E_3 be the subsets of the space S_m characterized by the following properties:

$$(57) \quad \begin{cases} \int_{S_1} \varphi(y, z) \, dz \neq 0, \qquad \int_{S_1} (1 - \varphi(y, z)) \, dz \neq 0 \quad \text{for } y \in E_1 \\[2em] \int_{S_1} \varphi(y, z) \, dz = 0 \qquad\qquad\qquad\qquad\qquad\quad \text{for } y \in E_2 \\[2em] \int_{S_1} (1 - \varphi(y, z)) \, dz = 0 \qquad\qquad\qquad\qquad\quad\; \text{for } y \in E_3. \end{cases}$$

Clearly we have

$$(58) \qquad S_m = E_1 \cup E_2 \cup E_3.$$

As $\varphi(x) \equiv \varphi(y,z)$ can assume only the values 0, 1, using (57) we have $\varphi(x) \equiv \varphi(y,z) = 0$ at almost every point of $E_2 \times S_1$, while $\varphi(x) \equiv \varphi(y,z) = 1$ at almost every point of $E_3 \times S_1$. Denoting by μ the minimum between meas E_2 and meas E_3, for almost every value of z we have

$$(59) \qquad \int_{S_m} \varphi(y,z)\, dy_1 \ldots dy_m \geq \mu, \qquad \int_{S_m} (1-\varphi(y,z))\, dy_1 \ldots dy_m \geq \mu.$$

Since we have supposed that, for $r = m$, one of the relations (53) is always satisfied, we have, recalling the definition of $\mathcal{I}_y[\varphi(y,z)]$,

$$(60) \qquad \mathcal{I}_y[\varphi(y,z)] \geq \mu^{\frac{m-1}{m}}$$

for almost every value of z. On the other hand (49) ensures that

$$(61) \qquad \mathcal{I}[\varphi(x)] = \mathcal{I}[\varphi(y,z)] \geq \int_{-\infty}^{+\infty} \mathcal{I}_y[\varphi(y,z)]\, dz.$$

Therefore, if $\mu \neq 0$, that is, if both E_1 and E_2 have positive measure, it results by (60), (61)

$$(62) \qquad \mathcal{I}[\varphi(x)] = +\infty$$

and (53) are certainly satisfied.

Let us now consider the points $y \in E_1$; since for $r = 1$ one of the relations (53) is always satisfied, recalling the definition of $\mathcal{I}_z[f(y,z)]$ and taking into account (57), we get that, at any point $y \in E_1$, there holds

$$(63) \qquad \mathcal{I}_z[\varphi(y,z)] \geq 1.$$

From (63) it follows, recalling (49),

$$(64) \qquad \mathcal{I}[\varphi(x)] \geq \int_{S_m} \mathcal{I}_z[\varphi(y,z)]\, dy_1 \ldots dy_m \geq \int_{E_1} dy_1 \ldots dy_m = \text{meas } E_1$$

and therefore, if the set E_1 has infinite measure, (53) are certainly satisfied.

We now consider the case when E_1 has finite measure and E_2 has zero measure; then the set E_3 must have infinite measure and therefore, for almost every value of z, we have (recalling that at almost every point of the set $E_3 \times S_1$ we have $\varphi(x) = 1$)

$$(65) \qquad \int_{S_m} \varphi(y,z)\, dy_1 \ldots dy_m = +\infty.$$

Since we have supposed that at least one of the two relations (53) is always satisfied for $r = m$, we get, taking into account (65),

$$(66) \qquad \left(\int_{S_m} (1 - \varphi(y, z)) \, dy_1 \ldots dy_m \right)^{m-1} \leq (\mathcal{I}_y[\varphi(y, z)])^m$$

for almost every value of z. On the other hand, as E_2 has zero measure and $\varphi(y, z) = 1$ at almost every point of $E_3 \times S_1$, we have

$$(67) \qquad \int_{S_m} (1 - \varphi(y, z)) \, dy_1 \ldots dy_m \leq \operatorname{meas} E_1$$

for almost every value of z. From (64), (66), (67) it follows

$$
\begin{aligned}
(68) \qquad \mathcal{I}_y[\varphi(y, z)] \;&\geq\; \left(\int_{S_m} (1 - \varphi(y, z)) \, dy_1 \ldots dy_m \right)^{\left(1 - \frac{1}{m}\right)} \\
&\geq\; (\operatorname{meas} E_1)^{-\frac{1}{m}} \int_{S_m} (1 - \varphi(y, z)) \, dy_1 \ldots dy_m \\
&\geq\; (\mathcal{I}[\varphi(x)])^{-\frac{1}{m}} \int_{S_m} (1 - \varphi(y, z)) \, dy_1 \ldots dy_m.
\end{aligned}
$$

From (68), recalling (61) we have

$$
\begin{aligned}
(69) \qquad \mathcal{I}[\varphi(x)] \;&\geq\; \int_{S_1} \mathcal{I}_y[\varphi(y, z)] \, dz \\
&\geq\; (\mathcal{I}[\varphi(x)])^{-\frac{1}{m}} \int_{S_1} dz \int_{S_m} (1 - \varphi(y, z)) \, dy_1 \ldots dy_m \\
&=\; (\mathcal{I}[\varphi(x)])^{-\frac{1}{m}} \int_{S_{m+1}} (1 - \varphi(x)) \, dx_1 \ldots dx_{m+1}.
\end{aligned}
$$

From (69) it follows

$$(70) \qquad (\mathcal{I}[\varphi(x)])^{m+1} \geq \left(\int_{S_{m+1}} (1 - \varphi(x)) \, dx_1 \ldots dx_{m+1} \right)^m$$

and therefore we have proved that, if E_2 has zero measure and E_1 has finite measure, then the last relation in (53) holds for $r = m + 1$. With a similar argument one can prove that the first relation in (53) holds provided the set E_3 has zero measure and the set E_1 has finite measure. Since we have already proved that (53) holds when E_1 has infinite measure, or when the sets E_2 and E_3 have simultaneously nonzero measure, we can conclude that, for $r = m + 1$, one of relations (53) is always satisfied. The proof of our theorem is complete.

7. Given an Euclidean space S_r, whose generic point will be as usual denoted by $x \equiv (x_1, \ldots, x_r)$, we consider the class of all subsets of S_r; in this class we will introduce a metric by taking as distance between two sets E_1, E_2 the quantity

$$(71) \qquad \operatorname{meas} (E_1 \cup E_2 \setminus E_1 \cap E_2).$$

Let us denote by Σ the resulting metric space and let us agree that, when in this paper we will deal with limits of sets contained in an Euclidean space S_r, we will always consider such sets as elements of the space Σ.

Let us consider a sequence $E_1, E_2, \ldots$ converging to the set E; let us denote by $\Phi(B)$ the additive set function satisfying conditions a), b) of theorem II, whenever ths is defined, and let us denote by $\Phi^n(B)$ the analogous functions concerning the set E_n. Recalling the definition of weak convergence (n. 2) we can state the following theorem.

THEOREM VII. –   *Given a sequence of sets*

$$(72) \qquad E_1, \ldots, E_n, \ldots$$

converging to a set E, we have

$$(73) \qquad \lim_{n\to\infty}{}' P(E_n) \geq P(E).$$

If the sets of the perimeters $P(E_n)$ is bounded, the sequence

$$(74) \qquad \Phi^1(B), \ldots, \Phi^n(B), \ldots$$

weakly converges to $\Phi(B)$.

Proof. As usual, let us denote by $\varphi(x|L)$ the characteristic function of the generic set $L \subset S_r$. By the definition of distance between two elements of the metric space Σ, from the relation

$$(75) \qquad \lim_{n\to\infty} E_n = E$$

it follows

$$(76) \qquad \lim_{n\to\infty} \int_{S_r} |\varphi(x|E) - \varphi(x|E_n)|\, dx_1 \ldots dx_r = 0.$$

Therefore, recalling the definition of the operator W_λ, we have

$$(77) \qquad \lim_{n\to\infty} \operatorname{grad} W_\lambda \varphi(x|E_n) = \operatorname{grad} W_\lambda \varphi(x|E)$$

at any point $x \in S_r$ and for any value of the parameter λ. From (77) it follows, recalling the definition of perimeter of a set and well known theorems on the exchange between the limit and the integral,

$$
\begin{aligned}
(78) \qquad \lim_{n\to\infty}{}' P(E_n) \;&\geq\; \lim_{n\to\infty}{}' \int_{S_r} |\operatorname{grad} W_\lambda \varphi(x|E_n)|\, dx_1 \ldots dx_r \\[2mm]
&\geq\; \int_{S_r} \lim_{n\to\infty} |\operatorname{grad} W_\lambda \varphi(x|E_n)|\, dx_1 \ldots dx_r \\[2mm]
&=\; \int_{S_r} |\operatorname{grad} W_\lambda \varphi(x|E)|\, dx_1 \ldots dx_r.
\end{aligned}
$$

Note that (78) holds for any positive value of the parameter λ. From (78), passing to the limit as $\lambda \to 0$, we obtain (73). From (73) one can see that, if the set of all numbers $P(E_n)$ is bounded, then $P(E)$ is finite; in this case, if we denote by μ the supremum of all $P(E_n)$, we have, by theorem IV,

$$(79) \qquad \int_{S_r} |d\Phi| \le \mu, \qquad \int_{S_r} |d\Phi^n| \le \mu, \qquad (\text{for } n = 1, 2, \dots).$$

Given a continuous function $g(x)$ in S_r which is infinitesimal as $|x| \to \infty$, for each positive number ε we can certainly find a continuous function $g_\varepsilon(x)$ in S_r having continuous first derivatives, which is infinitesimal, together with its first order derivatives, of order not smaller than $|x|^{-(r+1)}$ as $|x| \to \infty$, such that

$$(80) \qquad |g(x) - g_\varepsilon(x)| \le \varepsilon.$$

By theorem II we have, recalling (76),

$$(81)\ \lim_{n \to \infty} \int_{S_r} g_\varepsilon(x) d\Phi^n = \lim_{n \to \infty} \int_{S_r} \varphi(x|E_n)\operatorname{grad} g_\varepsilon(x)\ dx_1 \dots dx_r$$

$$= \int_{S_r} \varphi(x|E)\operatorname{grad} g_\varepsilon(x)\ dx_1 \dots dx_r = \int_{S_r} g_\varepsilon(x)\ d\Phi.$$

Therefore, taking into account (79), (80), (81), we have

$$(82) \qquad \lim_{n \to \infty}{}'' \left| \int_{S_r} g(x)\ d\Phi^n - \int_{S_r} g(x)\ d\Phi \right| \le 2\varepsilon\mu;$$

as ε is arbitrary, we then have

$$(83) \qquad \lim_{n \to \infty} \left| \int_{S_r} g(x)\ d\Phi^n - \int_{S_r} g(x)\ d\Phi \right| = 0$$

and the weak convergence of the sequence (74) toward $\Phi(B)$ is proved.

     8. Given an Euclidean space S_r (with $r \ge 2$), let us consider the set $\{\Pi\}$ of all subdomains of S_r whose boundary is contained in the union of a finite number of hyperplanes. Such domains will be called *polygonal domains* (with a name similar to the one used for rectangular domains, circular domains, etc.). If in particular the space S_r reduces to the plane or to the ordinary space, the polygonal domains will be the polygons and the polyhedra, respectively. If we consider now the set $\{\Pi\}$ as a subset of the space Σ introduced in n. 7 it is clear that it is dense in Σ and, by theorem VII, we have, given an arbitrary set $E \subset S_r$,

$$(84) \qquad \lim_{\Pi \to E}{}' P(\Pi) \ge P(E).$$

The following result is more precise than the one given in (84).

THEOREM VIII. –   *Given a set $E \subset S_r$ (with $r \geq 2$) its perimeter equals the lower limit of the perimeters of the polygonal domains approximating E in $L^1(S_r)$; that is*

$$(85) \qquad \lim_{\Pi \to E}{}' P(\Pi) = P(E).$$

Proof. If $P(E)$ is infinite, from (84) it immediately follows (85) and our theorem is proved. On the other hand, if $P(E)$ is finite, by theorem VI it follows that one of the two sets E, $(S_r \setminus E)$ has finite measure. Let us suppose that E has finite measure; in this case its characteristic function $\varphi(x|E)$ is integrable and therefore, recalling (6) and (8) we have

$$(86) \qquad \lim_{\lambda \to 0} \int_{S_r} |W_\lambda \varphi(x|E) - \varphi(x|E)|\ dx_1 \ldots dx_r = 0.$$

Hence, given an arbitrarily small positive number ε we can certainly find a positive number λ such that

$$(87) \qquad \int_{S_r} |W_\lambda \varphi(x|E) - \varphi(x|E)|\ dx_1 \ldots dx_r \leq \varepsilon.$$

By the definition of the operator W_λ, the function $|\operatorname{grad} W_\lambda \varphi(x|E)|$ turns out to be bounded in S_r and therefore its supremum, which will be denoted by M, is finite. Given an arbitrary positive number $\eta < \frac{1}{4}$, let us consider the set L of all points of S_r for which

$$(88) \qquad W_\lambda \varphi(x|E) \geq \eta,$$

and let us prove that L is bounded. For any positive number ρ, let us indicate by $I_\rho(L)$ the neighborhood of L of radius ρ; let us also set $\bar{\rho} = \frac{\eta}{2M}$. Since (88) holds at points of L, and since

$$(89) \qquad |\operatorname{grad} W_\lambda \varphi(x|E)| \leq M,$$

at points of $I_{\bar{\rho}}(L)$, we have

$$(90) \qquad W_\lambda \varphi(x|E) \geq \eta - \frac{M\eta}{2M} = \frac{\eta}{2}.$$

Hence
$$(91)$$
$$\int_{S_r} |W_\lambda \varphi(x|E)|\ dx_1 \ldots dx_r \geq \int_{I_{\bar{\rho}}(L)} |W_\lambda \varphi(x|E)|\ dx_1 \ldots dx_r \geq \frac{\eta}{2} \operatorname{meas}\,(I_{\bar{\rho}}(L)).$$

From (91), taking into account (87) and recalling the fact that $\varphi(x|E)$ is integrable, one sees that the set $I_{\bar{\rho}}(L)$ must have finite measure, and therefore L must be bounded. Since $\varphi(x|E)$ is integrable and L is a bounded set, we can certainly find a positive number α sufficiently large such that, if T_α denotes the domain of all points satisfying

$$(92) \qquad |x_h| \leq \alpha \qquad (\text{for } h = 1, \ldots, r),$$

the following formulas are simultaneously satisfied:

$$(93) \qquad \int_{(S_r \setminus T_\alpha)} \varphi(x|E)\, dx_1 \ldots dx_r < \varepsilon,$$

$$(93') \qquad W_\lambda \varphi(x|E) < \eta \qquad (\text{for } x \in (S_r \setminus T_\alpha)).$$

In the space S_{r+1}, whose generic point is denoted by $(x_1, \ldots, x_r, y)$, let us consider the regular hypersurface Γ_1 defined by

$$(94) \qquad y = W_\lambda \varphi(x_1, \ldots, x_r|E) \equiv W_\lambda \varphi(x|E), \qquad x \in T_\alpha.$$

The function $W_\lambda \varphi(x|E)$ is continuous in S_r together with its first order derivatives, therefore we can certainly approximate the hypersurface Γ_1 with a hypersurface Γ_2 which is contained in the union of a finite number of hyperplanes and is represented by the equations

$$(95) \qquad y = g(x) \equiv g(x_1, \ldots, x_r), \qquad x \equiv (x_1, \ldots, x_r) \in T_\alpha,$$

where $g(x)$ is a continuous function satisfying the following conditions:

$$(96) \qquad 0 < g(x) - W_\lambda \varphi(x|E) < \eta,$$

$$(97) \qquad \int_{T_\alpha} |g(x) - W_\lambda \varphi(x|E)|\, dx_1 \ldots dx_r < \varepsilon,$$

$$
\begin{aligned}
(98) \quad \int_{T_\alpha} |\operatorname{grad} g(x)|\, dx_1 \ldots dx_r \;&<\; \int_{T_\alpha} |\operatorname{grad} W_\lambda \varphi(x|E)|\, dx_1 \ldots dx_r + \eta \\
&\leq\; \int_{S_r} |\operatorname{grad} W_\lambda \varphi(x|E)|\, dx_1 \ldots dx_r + \eta \\
&\leq\; P(E) + \eta.
\end{aligned}
$$

Since clearly $W_\lambda \varphi(x|E)$ is always nonnegative, by (96) the function $g(x)$ will be always positive, and therefore the set of all points $(x_1, \ldots, x_r, y)$ satisfying

$$(99) \qquad 0 \leq y \leq g(x), \qquad x \equiv (x_1, \ldots, x_r) \in T_\alpha$$

will be a polygonal domain $D \subset S_{r+1}$. Taking into account (93'), (96) we see that the function $g(x)$ is smaller than 2η at any point x belonging to the boundary of T_α. It follows that, for any real number $\theta \geq 2\eta$, the hyperplane $y = \theta$ intersects the boundary of the domain D only at points belonging to Γ_2. Let us denote by

$\rho(\theta)$ the $(r-1)$-dimensional measure (in an elementary sense) of the section of Γ_2 with the plane $y = \theta$ and let us indicate by Γ_2^* the portion of Γ_2 contained in the half space $y \geq 2\eta$.

Using elementary theorems on the measure of the sections of a set (theorems that we can certainly apply to the hypersurface Γ_2^* which, being contained in Γ_2, is in turn contained in the union of a finite number of hyperplanes) we will have

$$(100) \qquad \int_{\Gamma_2^*} \nu_y \, d\sigma = \int_{2\eta}^{+\infty} \rho(\theta) \, d\theta,$$

where $d\sigma$ stands for the r-dimensional measure on Γ_2^* and ν_y is the length of the orthogonal projection of the unit normal vector to the hypersurface Γ_2 on a hyperplane $y = 0$. Recalling that Γ_2 has equations (95) we have

$$(101) \qquad \int_{\Gamma_2^*} \nu_y \, d\sigma \leq \int_{\Gamma_2} \nu_y \, d\sigma = \int_{T_\alpha} |\operatorname{grad} g(x)| \, dx_1 \ldots dx_r.$$

From (98), (100), (101) it follows

$$(102) \qquad \int_{2\eta}^{+\infty} \rho(\theta) \, d\theta < P(E) + \eta$$

and therefore we have, a fortiori,

$$(103) \qquad \int_{2\eta}^{1-\eta} \rho(\theta) \, d\theta < P(E) + \eta.$$

Let us consider now, for any value of θ, the section of the domain D with the hyperplane $y = \theta$, which will be denoted by $\Pi(\theta)$. If we identify the hyperplane $y = \theta$ with the space S_r and therefore the generic point $(x_1, \ldots, x_r, \theta)$ of such a hyperplane with the point $(x_1, \ldots, x_r) \in S_r$, we find that, for almost every value of θ, the set $\Pi(\theta)$ is a polygonal domain, provided it is nonempty. Clearly, we have

$$(104) \qquad \begin{cases} g(x) \geq \theta & \text{for } x \in \Pi(\theta) \\[2ex] g(x) < \theta & \text{for } x \in (T_\alpha \setminus \Pi(\theta)). \end{cases}$$

Since, for $\theta \geq 2\eta$, the hyperplane $y = \theta$ intersects the boundary of D only at points belonging to Γ_2, for almost every value of θ between 2η and $(1-\eta)$ (values covering an interval, since by assumption $\eta < \frac{1}{4}$) the perimeter $P(\Pi(\theta))$ equals $\rho(\theta)$. Therefore there exists, by (103), a value $\bar{\theta}$ between 2η and $(1-\eta)$, such that

$$(105) \qquad P(\Pi(\bar{\theta})) = \rho(\bar{\theta}) < \frac{P(E) + \eta}{1 - 3\eta},$$

being also $\Pi(\bar{\theta})$ a polygonal domain of S_r.

On the other hand, by (104) we have

$$
(106) \quad
\begin{cases}
g(x) - \varphi(x|E) \geq \overline{\theta} > 2\eta & \text{for } x \in (\Pi(\overline{\theta}) \setminus E \cap \Pi(\theta)) \\[2mm]
\varphi(x|E) - g(x) \geq 1 - \overline{\theta} > \eta & \text{for } x \in (E \cap T_\alpha \setminus E \cap \Pi(\overline{\theta})),
\end{cases}
$$

while, from (87), (97) it follows

$$
(107) \qquad \int_{T_\alpha} |g(x) - \varphi(x|E)| \, dx_1 \ldots dx_r < 2\varepsilon.
$$

Hence, from (106), (107) we get

$$
(108) \qquad \operatorname{meas} (E \cap T_\alpha \cup \Pi(\overline{\theta}) \setminus E \cap \Pi(\overline{\theta})) < \frac{2\varepsilon}{\eta}.
$$

Since by (93)

$$
(109) \qquad \operatorname{meas} (E \setminus T_\alpha \cap E) < \varepsilon,
$$

we finally deduce

$$
(110) \qquad \operatorname{meas} (E \cup \Pi(\overline{\theta}) \setminus E \cap \Pi(\overline{\theta})) < \left(\varepsilon + \frac{2\varepsilon}{\eta} \right).
$$

Since the two numbers ε, η are arbitrary, formulas (105), (110) ensure that

$$
(111) \qquad \lim_{\Pi \to E}{}' P(\Pi) \leq P(E).
$$

Therefore, recalling (84), we obtain (85) and the proof of our theorem is complete.

From theorems VI, VIII one immediately sees that *our definition of perimeter of a set is equivalent to the definition given by Caccioppoli of $(r-1)$-dimensional measure of its oriented boundary.* Indeed, a set $E \subset S_r$ has finite perimeter if and only if it can be approximated in measure with polygonal domains whose boundaries have equibounded $(r-1)$-dimensional measure. In this case the perimeter $P(E)$ coincides with the total variation, in S_r, of the function $\Phi(B)$ satisfying conditions a), b) of theorem II, which, in turn, equals the $(r-1)$-dimensional measure of the oriented boundary of E in the sense of Caccioppoli (see loc. cit. [2]).

9. Beside the perimeter $P(E)$ of a set $E \subset S_r$ we can consider the *projections of the perimeter $P(E)$ on the coordinate hyperplanes,* which will be denoted by $P_1(E), \ldots, P_r(E)$ and are defined by

$$
(112) \quad P_h(E) = \lim_{\lambda \to 0} \int_{S_r} \left| \frac{\partial}{\partial x_h} W_\lambda \varphi(x|E) \right| \, dx_1 \ldots dx_r \qquad (\text{for } h = 1, \ldots, r).
$$

The existence of the limit in (112) can be proved in a similar way as in the proof of the existence of the limit in (13). Clearly, we have the relation

$$(113) \qquad \sum_{h=1}^{r} P_h(E) \geq P(E),$$

while, for any value of the index h, it results

$$(114) \qquad P(E) \geq P_h(E).$$

The following theorems hold.

THEOREM IX. –   *Given a set $E \subset S_r$, if $P_h(E)$ is finite for some value of the index h, then there exists a set function $\Phi_h(B)$ satisfying the following conditions:*

a) *$\Phi_h(B)$ is defined on any set $B \subset S_r$ and is countably additive and with bounded total variation.*

b) *given any continuous function $g(x)$ in S_r having continuous first derivatives, which is infinitesimal, together with its first order derivatives, of order not smaller than $|x|^{-(r+1)}$ as $|x| \to \infty$, it results*

$$(115) \qquad \int_E \frac{\partial g}{\partial x_h} \, dx_1 \ldots dx_r = \int_{S_r} g(x) \, d\Phi_h.$$

THEOREM X. –   *Given a set $E \subset S_r$, if there exists a function $\Phi_h(B)$ satisfying conditions a), b) of theorem IX, then $P_h(E)$ is finite and we have*

$$(116) \qquad P_h(E) = \int_{S_r} |d\Phi_h|.$$

THEOREM XI. –   *Let $E \subset S_r$ be a set such that $P_h(E)$ is finite for some value of the index h, and let $\Phi_h(B)$ be the function satisfying conditions a), b) of theorem IX. For any set L on the boundary of which the total variation of the function $\Phi_h(B)$ vanishes we have*

$$(117) \qquad \Phi_h(L) = -\lim_{\lambda \to 0} \int_L \frac{\partial}{\partial x_h} W_\lambda \varphi(x|E) \, dx_1 \ldots dx_r.$$

Given a sequence of sets $E_1, E_2, \ldots$ converging to the set E, for any value of the index h let us define the set function $\Phi_h(B)$ as in theorem IX, and let us denote by $\Phi_h^{(n)}(B)$ the analogous function for the set E_n. Then we have the following result.

THEOREM XII. –   *Given a sequence of sets*

$$(118) \qquad E_1, \ldots, E_n, \ldots$$

converging to a set E, we have

$$(119) \qquad \lim_{n\to\infty}{}' P_h(E_n) \geq P_h(E).$$

If for some value of the index h the sequence

$$(120) \qquad P_h(E_1), \ldots, P_h(E_n), \ldots$$

is bounded, then the sequence

$$(121) \qquad \Phi_h^{(1)}(B), \ldots, \Phi_h^{(n)}(B), \ldots$$

is weakly converging to $\Phi_h(B)$.

The proof of these theorems is entirely similar to the one of theorems II, IV, V, VII; it is enough to replace gradients with partial derivatives with respect to x_h in all arguments of the proofs.

Eventually, let us observe that from (113) and from theorems VIII, XII we immediately deduce a result (stated by Caccioppoli[4] simply as a conjecture) which, following the notation used in the present paper, can be stated as follows:

Given a set $E \subset S_r$, if for any value of the index h it is possible to construct a sequence of polygonal domains

$$(122) \qquad \Pi_1^h, \ldots, \Pi_n^h, \ldots$$

converging in $L^1(S_r)$ to the set E and such that the quantities

$$(123) \qquad P_h(\Pi_1^h), \ldots, P_h(\Pi_n^h), \ldots$$

are equibounded, then $P(E)$ is finite. Therefore the set E can be approximated in mean by a sequence of polygonal domains

$$(124) \qquad \Pi_1, \ldots, \Pi_n, \ldots$$

having equibounded perimeters

$$(125) \qquad P(\Pi_1), \ldots, P(\Pi_n), \ldots$$

---

[4]See loc. cit. [2], Nota I, n. 8, nota (10).

Su una teoria generale della misura $(r-1)$-dimensionale in uno spazio ad r dimensioni[‡][†]

Memoria di Ennio De Giorgi (a Roma)

Sunto. In questo lavoro do una definizione della misura $(r-1)$-dimensionale della frontiera orientata di un insieme di Borel contenuto in uno spazio euclideo ad r dimensioni e dimostro alcune fra le più notevoli proprietà di tale misura.

La teoria della misura sviluppata in questo lavoro[1] concerne le frontiere di insiemi di uno spazio euclideo S_r; tali frontiere non sono riguardate come semplici insiemi di punti ma come *insiemi orientati*, per i quali si definisce non solo la misura assoluta ma anche la misura relativa delle proiezioni su un qualunque iperpiano. I fondamenti di questa teoria sono stati posti recentemente da R. CACCIOPPOLI[2] (in generale per la misura $(r-h)$-dimensionale); contemporaneamente ed indipendentemente io ero pervenuto agli stessi risultati, partendo da un altro punto di vista e con intenti diversi. CACCIOPPOLI si è proposto di dare una teoria generale dell'integrazione delle forme differenziali in più variabili ed una estensione completa delle formule di GREEN-STOKES. Il mio scopo invece era inizialmente una generalizzazione sostanziale di certi problemi isoperimetrici e partivo dalla formula di GAUSS-GREEN come istanza a priori.

Stabilisco così una condizione necessaria e sufficiente perchè sussista una simile formula relativamente ad un insieme di Borel E: in tale formula compare una funzione vettoriale d'insieme a variazione limitata il cui differenziale adempie all'ufficio che ha l'"elemento $(n-1)$-dimensionale" nel caso ordinario. La variazione totale di tale funzione additiva, che si identifica con la misura $(r-1)$-dimensionale secondo CACCIOPPOLI della frontiera *orientata* di E, viene chiamata in questo lavoro *perimetro dell'insieme E*. Di questo perimetro si dà un'espressione analitica, mediante il limite di un integrale spaziale contenente un parametro. Il procedimento per giungere alla detta funzione additiva d'insieme è quanto mai semplice e si fonda su una conveniente approssimazione della funzione caratteristica dell'insieme E mediante funzioni regolari, come limite delle

---

[‡]Editor's note: published in Ann. Mat. Pura Appl., **(4)** **(36)** (1954), 191–212.

[†]Lavoro eseguito nell'Istituto Nazionale per le Applicazioni del Calcolo.

[1]I principali risultati contenuti in questo lavoro sono stati esposti nella mia nota *Definizione ed espressione analitica del perimetro di un insieme*, "Rend. Accad. Naz. Lincei, Cl. Sc. fis. mat. nat.", Serie VIII, Vol. XIV, fasc. 3 (marzo 1953), pp. 390–393.

[2]Vedi R. CACCIOPPOLI, *Misura e integrazione sugli insiemi dimensionalmente orientati*, "Rend. Accad. Naz. Lincei, Cl. Sc. fis. mat. nat.", Serie VIII, Vol. XII, fasc. 1, 2 (gennaio-febbraio 1952), pp. 3–11, 137–146.

quali si ritrova la funzione caratteristica dell'insieme E. La formula generale di GAUSS-GREEN si ottiene in tal modo come limite della formula ordinaria d'integrazione per parti.

L'espressione analitica del perimetro permette di ottenere con relativa facilità alcuni risultati fondamentali, cui non sembra agevole pervenire partendo dalle definizioni originali di CACCIOPPOLI. Il più importante a mio avviso è il seguente: *il perimetro di un insieme E è il minimo limite dei perimetri degli insiemi poliedrici approssimanti in media E* (cioè le cui funzioni caratteristiche tendano in media d'ordine 1 verso la funzione caratteristica di E). Segnalerò pure la proprietà isoperimetrica, per cui la misura di un insieme si limita mediante il suo perimetro, e precisamente risulta minore della potenza $\left(\frac{r}{r-1}\right)$-esima del perimetro stesso.

La presente teoria si ispira a concezioni affatto diverse da quelle che sono alla base di precedenti definizioni di misura dimensionale, p. es. quelle di GROSS e di CARATHÉODORY, le quali non sono peraltro applicabili alle questioni suaccennate, p. es. alla formula di GAUSS-GREEN, che dopo essere state completate da qualche definizione di *normale*, come appunto si è fatto finora. Per quanto riguarda la definizione di CARATHÉODORY, si vede subito che essa fornisce in generale, per la misura della frontiera di un insieme, un valore maggiore del nostro, eventualmente infinito, quando il perimetro è finito, sicchè tutte le frontiere di misura finita secondo CARATHÉODORY sono tali secondo la nostra definizione e non viceversa.

Avvertirò infine che i concetti esposti in questo lavoro sono alla base di una trattazione sostanzialmente nuova di certi problemi isoperimetrici, dei quali mi occuperò in un prossimo lavoro.

1. Conveniamo, per semplificare l'esposizione, che ogni volta che in questo lavoro si parla di insiemi contenuti in uno spazio euclideo S_r e di funzioni ivi definite, intenderemo sempre parlare di insiemi di BOREL e di funzioni di BAIRE. Stabiliamo inoltre che, dati comunque due operatori T_1, T_2 ed una funzione $f(x)$, indicheremo con $T_1 T_2 f(x)$ la funzione ottenuta applicando a $T_2 f(x)$ l'operatore T_1. Consideriamo ora uno spazio euclideo S_r, il cui punto generico indicheremo con $x \equiv (x_1, \dots, x_r)$ e definiamo, per ogni valore del parametro λ, l'operatore W_λ, ponendo, per ogni funzione $f(x)$ definita in tutto lo spazio S_r ed ivi limitata

$$(1) \qquad W_\lambda f(x) = \pi^{-\frac{r}{2}} \int_{S_r} e^{-|\xi|^2} f(x + \sqrt{\lambda}\xi)\, d\xi_1 \dots d\xi_r,$$

ove abbiamo posto $|\xi| = \sqrt{\xi_1^2 + \cdots + \xi_r^2}$ ed abbiamo indicato con $(x + \sqrt{\lambda}\xi)$ il punto di coordinate $(x_1 + \sqrt{\lambda}\xi_1, \dots, x_r + \sqrt{\lambda}\xi_r)$. Dalla (1) segue immediatamente

$$(2) \qquad \begin{aligned} W_\lambda f(x) &= (\pi\lambda)^{-\frac{r}{2}} \int_{S_r} e^{-\frac{|\xi|^2}{\lambda}} f(x + \xi)\, d\xi_1 \dots d\xi_r \\ &= (\pi\lambda)^{-\frac{r}{2}} \int_{S_r} e^{-\frac{|x-\xi|^2}{\lambda}} f(\xi)\, d\xi_1 \dots d\xi_r. \end{aligned}$$

Dalle (2) si vede che, per ogni funzione limitata $f(x)$ e per ogni valore positivo del parametro λ, la funzione $W_\lambda f(x)$ risulta continua e limitata in tutto lo spazio

S_r insieme a tutte le sue derivate parziali di qualsiasi ordine. Se $f(x)$ è continua e limitata in S_r ed ivi derivabile con derivate parziali prime continue e limitate, si verifica immediatamente che vale la

$$(3) \qquad \frac{\partial}{\partial x_h} W_\lambda f(x) = W_\lambda \frac{\partial f}{\partial x_h}(x),$$

cioè l'operatore W_λ è permutabile con l'operatore $\dfrac{\partial}{\partial x_h}$. Si vede pure facilmente che, dati due numeri positivi λ, μ, risulta

$$(4) \qquad W_\lambda W_\mu f(x) = W_{\lambda+\mu} f(x).$$

Non è difficile provare che, per ogni funzione $f(x)$ limitata, si ha, in quasi tutti i punti di S_r,

$$(5) \qquad \lim_{\lambda \to 0} W_\lambda f(x) = f(x);$$

se poi $f(x)$ è continua, allora la (5) è verificata uniformemente in ogni insieme chiuso e limitato contenuto in S_r. Si vede pure che valgono le relazioni

$$(6) \qquad \int_{S_r} |W_\lambda f(x)| \ dx_1 \ldots dx_r \le \int_{S_r} |f(x)| \ dx_1 \ldots dx_r,$$

$$(7) \qquad \lim_{\lambda \to 0} \int_{S_r} |W_\lambda f(x)| \ dx_1 \ldots dx_r = \int_{S_r} |f(x)| \ dx_1 \ldots dx_r;$$

si ha inoltre, per ogni insieme limitato L,

$$(8) \qquad \lim_{\lambda \to 0} \int_L |W_\lambda f(x) - f(x)| \ dx_1 \ldots dx_r = 0.$$

I risultati finora esposti valgono naturalmente anche se $f(x)$, invece di essere una funzione scalare, è una funzione vettoriale ad un numero qualunque di componenti.

Fissiamo ora una funzione scalare $f(x)$ definita in tutto S_r ed ivi limitata, e consideriamo l'integrale

$$(9) \qquad \int_{S_r} |\text{grad } W_\lambda f(x)| \ dx_1 \ldots dx_r;$$

poichè per le (3), (4) abbiamo, comunque si scelgano due numeri positivi λ, μ,

$$(10) \qquad \text{grad } W_{\lambda+\mu} f(x) = \text{grad } W_\mu W_\lambda f(x) = W_\mu \text{grad } W_\lambda f(x),$$

ricordando le (6) troviamo che, per λ e μ positivi, si ha

$$(11) \qquad \int_{S_r} |\text{grad } W_{\mu+\lambda} f(x)| \ dx_1 \ldots dx_r \le \int_{S_r} |\text{grad } W_\lambda f(x)| \, dx_1 \ldots dx_r.$$

La (11) ci assicura che l'integrale (9) è funzione non crescente di λ e quindi è certamente determinato il limite

$$(12) \qquad \mathcal{I}[f(x)] = \lim_{\lambda \to 0} \int_{S_r} |\operatorname{grad} W_\lambda f(x)| \; dx_1 \dots dx_r,$$

che potrà essere finito o infinito.

Dato un insieme E contenuto in S_r, definiremo il *perimetro dell'insieme E* che indicheremo con $P(E)$, ponendo

$$(13) \qquad P(E) = \mathcal{I}[\varphi(x|E)] = \lim_{\lambda \to 0} \int_{S_r} |\operatorname{grad} W_\lambda \varphi(x|E)| \; dx_1 \dots dx_r,$$

ove con $\varphi(x|E)$ abbiamo indicato la funzione caratteristica dell'insieme E.

2. Data una funzione d'insieme $F^{(\lambda)}(B)$, dipendente da un parametro λ, che sia completamente additiva, diremo che per $\lambda \to \lambda_0$ la funzione $F^{(\lambda)}(B)$ *converge debolmente* verso una funzione $F(B)$, se la relazione

$$(14) \qquad \lim_{\lambda \to \lambda_0} \int_{S_r} g(x) \, dF^{(\lambda)} = \int_{S_r} g(x) \, dF$$

è verificata per ogni funzione $g(x)$ continua in tutto lo spazio S_r ed infinitesima per $|x| \to \infty$.

Ricordiamo ora un teorema di de La Vallée Poussin[3] che può essere enunciato nella maniera seguente: *data una successione di funzioni scalari d'insieme*

$$(15) \qquad \mu_1(B), \dots, \mu_n(B), \dots$$

definite per ogni insieme $B \subset S_r$, completamente additive, mai negative ed equilimitate, è possibile trovare una funzione additiva d'insieme $\mu(B)$ mai negativa e limitata ed una successione subordinata alla (15),

$$(16) \qquad \mu_{h_1}(B), \dots, \mu_{h_n}(B), \dots$$

tali che si abbia

$$(17) \qquad \lim_{n \to \infty} \mu_{h_n}(B) = \mu(B)$$

per ogni insieme B sulla cui frontiera sia nulla $\mu(B)$.

Dal teorema ora enunciato si deduce facilmente che *da una successione di funzioni vettoriali d'insieme, definite per ogni insieme $B \subset S_r$, completamente additive ed aventi le variazioni totali equilimitate è sempre possibile estrarre una successione subordinata che converga debolmente.*

Servendoci dei risultati ora esposti possiamo dimostrare il

Teorema I. – *Data una funzione $f(x)$ definita in tutto lo spazio S_r ed ivi limitata, se è finito $\mathcal{I}[f(x)]$ esiste una ed una sola funzione vettoriale d'insieme* $\mathbf{F}(B) \equiv (F_1(B), \dots, F_r(B))$ *soddisfacente le condizioni seguenti:*

---

[3]Vedi de La Vallée Poussin, "Annales de l'Institut H. Poincaré", (1932), Vol. II, pp. 221–224.

a) $\mathbf{F}(B)$ *è definita per ogni insieme $B \subset S_r$ ed è completamente additiva ed a variazione totale limitata;*

b) *Per ogni funzione $g(x)$ continua in S_r insieme alle sue derivate parziali prime ed infinitesima insieme ad esse per $|x| \to \infty$, d'ordine non inferiore a quello di $|x|^{-(r+1)}$ risulta*

$$(18) \qquad \int_{S_r} f(x)\operatorname{grad} g(x)\, dx_1 \ldots dx_r = \int_{S_r} g(x)\, d\mathbf{F}.$$

Dim. Prendiamo una successione di numeri positivi tendente a zero, che indicheremo con

$$(19) \qquad \lambda_1, \ldots, \lambda_n, \ldots$$

e consideriamo la successione delle funzioni vettoriali d'insieme

$$(20) \qquad \mathbf{F}^1(B), \ldots, \mathbf{F}^n(B), \ldots$$

definite dalle relazioni

$$(21) \qquad \mathbf{F}^h(B) = - \int_B \operatorname{grad} W_{\lambda_h} f(x)\, dx_1 \ldots dx_r \qquad (\text{per } h = 1, 2, \ldots).$$

Abbiamo visto che da una successione di funzioni vettoriali d'insieme, definite per ogni insieme $B \subset S_r$, completamente additive ed aventi le variazioni totali equilimitate è sempre possibile estrarre una successione subordinata che converga debolmente; ora, per la definizione stessa di $\mathcal{I}[f(x)]$, le funzioni della successione (20) hanno tutte variazioni totali non superiori ad $\mathcal{I}[f(x)]$ e quindi è possibile estrarre dalla (20) una successione subordinata

$$(22) \qquad \mathbf{F}^{\nu_1}(B), \ldots, \mathbf{F}^{\nu_h}(B), \ldots$$

la quale converge debolmente verso una funzione additiva d'insieme che indicheremo con $\mathbf{F}(B)$.

Avremo allora, indicando con $\lim'$ il minimo limite,

$$(23) \qquad \lim_{h \to \infty}' \int_{S_r} |d\mathbf{F}^{\nu_h}| \geq \int_{S_r} |d\mathbf{F}|$$

e quindi, poichè le funzioni della successione (22) hanno variazioni totali non superiori ad $\mathcal{I}[f(x)]$, sarà

$$(24) \qquad \int_{S_r} |d\mathbf{F}| \leq \mathcal{I}[f(x)];$$

dalla (24) si vede che $\mathbf{F}(B)$ soddisfa la condizione a).

Presa ora arbitrariamente una funzione $g(x)$ continua in S_r con le sue derivate parziali prime ed infinitesima insieme ad esse per $|x| \to \infty$, d'ordine non inferiore a quello di $|x|^{-(r+1)}$, abbiamo

$$(25) \qquad \int_{S_r} g(x)\, d\mathbf{F} = \lim_{h \to \infty} \int_{S_r} g(x)\, d\mathbf{F}^{\nu_h}.$$

Poichè, per la definizione stessa dell'operatore W_λ, essendo limitata $f(x)$ risultano equilimitate le funzioni $W_{\lambda_{\nu_h}} f(x)$, abbiamo, ricordando le (8) n.1 e tenendo presente che $\lim_{h \to \infty} \lambda_h = 0$,

$$(26)$$
$$\lim_{h \to \infty} \int_{S_r} \left(W_{\lambda_{\nu_h}} f(x) \right) \operatorname{grad} g(x)\, dx_1 \ldots dx_r = \int_{S_r} f(x) \operatorname{grad} g(x)\, dx_1 \ldots dx_r.$$

Dalle (21) segue invece, mediante una integrazione per parti,

$$(27) \qquad \int_{S_r} \left(W_{\lambda_{\nu_h}} f(x) \right) \operatorname{grad} g(x)\, dx_1 \ldots dx_r$$
$$= -\int_{S_r} g(x) \operatorname{grad} W_{\lambda_{\nu_h}} f(x)\, dx_1 \ldots dx_r = \int_{S_r} g(x)\, d\mathbf{F}^{\nu_h}.$$

Dalle (25), (26), (27) segue la (18), e il nostro teorema è dimostrato.

Ricordando la definizione del perimetro di un insieme si trova, come caso particolare del teorema I, il seguente

Teorema II. – *Dato un insieme $E \subset S_r$, se il suo perimetro $P(E)$ è finito, esiste una funzione vettoriale d'insieme $\Phi(B) \equiv (\Phi_1(B), \ldots, \Phi_r(B))$ soddisfacente le condizioni seguenti:*

a) *$\Phi(B)$ è definita per ogni insieme $B \subset S_r$ ed è completamente additiva ed a variazione totale limitata;*

b) *Per ogni funzione $g(x)$ continua in S_r insieme alle sue derivate parziali prime ed infinitesima insieme ad esse per $|x| \to +\infty$, d'ordine non inferiore a quello di $|x|^{-(r+1)}$, risulta*

$$(28) \qquad \int_E \operatorname{grad} g(x)\, dx_1 \ldots dx_r = \int_{S_r} \varphi(x|E) \operatorname{grad} g(x)\, dx_1 \ldots dx_r$$
$$= \int_{S_r} g(x)\, d\Phi,$$

$\varphi(x|E)$ essendo la funzione caratteristica dell'insieme E.

3. Il teorema I può essere invertito e completato dal

Teorema III. – *Data una funzione limitata $f(x)$, se esiste una funzione vettoriale $\mathbf{F}(B)$ soddisfacente le condizioni a), b) del teorema I, allora $\mathcal{I}[f(x)]$ è finito e si ha*

$$(29) \qquad \mathcal{I}[f(x)] = \int_{S_r} |d\mathbf{F}|.$$

Dim. Dato che risulta evidentemente

$$\operatorname{grad}_x e^{-\frac{|x-\xi|^2}{\lambda}} = -\operatorname{grad}_\xi e^{-\frac{|x-\xi|^2}{\lambda}}, \tag{30}$$

ove con grad_x e grad_ξ indicheremo rispettivamente i gradienti fatti rispetto alle variabili puntuali $x \equiv (x_1, \dots, x_r)$, $\xi \equiv (\xi_1, \dots, \xi_r)$, per la condizione b) del teorema I risulta, ricordando la definizione dell'operatore W_λ,

$$
\begin{aligned}
|\operatorname{grad} W_\lambda f(\xi)| &= (\pi\lambda)^{-\frac{r}{2}} \left| \int_{S_r} \left(\operatorname{grad}_\xi e^{-\frac{|x-\xi|^2}{\lambda}} \right) f(x)\, dx_1 \dots dx_r \right| \\
&= (\pi\lambda)^{-\frac{r}{2}} \left| \int_{S_r} \left(\operatorname{grad}_x e^{-\frac{|x-\xi|^2}{\lambda}} \right) f(x)\, dx_1 \dots dx_r \right| \\
&= (\pi\lambda)^{-\frac{r}{2}} \left| \int_{S_r} e^{-\frac{|x-\xi|^2}{\lambda}}\, d\mathbf{F} \right| \le (\pi\lambda)^{-\frac{r}{2}} \int_{S_r} e^{-\frac{|x-\xi|^2}{\lambda}} |d\mathbf{F}|.
\end{aligned}
\tag{31}
$$

Dalle (31) segue

$$
\begin{aligned}
\int_{S_r} |\operatorname{grad} W_\lambda f(\xi)|\, d\xi_1 \dots d\xi_r &\le (\pi\lambda)^{-\frac{r}{2}} \int_{S_r} d\xi_1 \dots d\xi_r \int_{S_r} e^{-\frac{|x-\xi|^2}{\lambda}} |d\mathbf{F}| \\
&= (\pi\lambda)^{-\frac{r}{2}} \int_{S_r} |d\mathbf{F}| \int_{S_r} e^{-\frac{|x-\xi|^2}{\lambda}}\, d\xi_1 \dots d\xi_r \\
&= \int_{S_r} |d\mathbf{F}|
\end{aligned}
\tag{32}
$$

e quindi, per la definizione di $\mathcal{I}[f(x)]$, abbiamo

$$\mathcal{I}[f(x)] \le \int_{S_r} |d\mathbf{F}|. \tag{33}$$

Le (33) ci assicurano che $\mathcal{I}[f(x)]$ è finito e quindi vale la (24); dalla (33) e dalla (24) segue la (29) e il nostro teorema è dimostrato.

Come caso particolare del teorema ora esposto si trova il

Teorema IV. – *Dato un insieme $E \subset S_r$, se esiste una funzione $\Phi(B)$ soddisfacente le condizioni a), b) del teorema II, il perimetro di E è finito e si ha*

$$P(E) = \int_{S_r} |d\Phi|. \tag{34}$$

4. I teoremi II, IV ci danno una condizione necessaria e sufficiente per l'esistenza di una funzione $\Phi(B)$ soddisfacente le ipotesi a), b) del teorema II; una espressione della funzione $\Phi(B)$ è data dal

Teorema V. – *Dato un insieme $E \subset S_r$ avente perimetro finito, sia $\Phi(B)$ la funzione che soddisfa le ipotesi a), b) del teorema II; per ogni insieme L sulla cui frontiera sia nulla la variazione totale della funzione $\Phi(B)$ si ha allora*

$$\Phi(L) = -\lim_{\lambda \to 0} \int_L \operatorname{grad} W_\lambda \varphi(x|E)\, dx_1 \dots dx_r. \tag{35}$$

Dim. Indicando con $\mathcal{F}L$ la frontiera di L, potremo esprimere il fatto che la variazione totale di $\Phi(B)$ sulla frontiera di L sia nulla scrivendo

$$(36) \qquad \int_{\mathcal{F}L} |d\Phi| = 0.$$

Preso un numero positivo ε piccolo a piacere, poichè $\Phi(B)$ ha, per la condizione a) del teorema II, variazione totale limitata e poichè vale la (36), sarà certamente possibile trovare due insiemi chiusi e limitati C_1, C_2, non aventi punti comuni con $\mathcal{F}L$, i quali verifichino le condizioni

$$(37) \qquad C_1 \subset L, \qquad \int_L |d\Phi| - \int_{C_1} |d\Phi| < \varepsilon,$$

$$(37') \qquad C_2 \subset (S_r \setminus L), \qquad \int_{(S_r \setminus L)} |d\Phi| - \int_{C_2} |d\Phi| < \varepsilon.$$

Indicando rispettivamente con δ_1, δ_2 le distanze degli insiemi C_1, C_2 da $\mathcal{F}L$, sia Γ un dominio circolare avente il centro nell'origine e raggio minore del più piccolo fra i due numeri δ_1, δ_2; esisterà certamente un numero positivo λ_ε, tale che risulti, per $0 < \lambda < \lambda_\varepsilon$,

$$(38) \qquad (\pi\lambda)^{-\frac{r}{2}} \int_\Gamma e^{-\frac{|\xi|^2}{\lambda}} d\xi_1 \ldots d\xi_r > 1 - \varepsilon.$$

Preso comunque un punto $x^1 \in C_1$, il dominio circolare avente il centro in x^1 e raggio uguale al raggio di Γ è certamente contenuto in L; preso invece un punto $x^2 \in C_2$, il dominio circolare di centro in x^2 e raggio uguale a quello di Γ è contenuto in $(S_r \setminus L)$. Avremo allora per le (38)

$$(39) \quad 1 > (\pi\lambda)^{-\frac{r}{2}} \int_L e^{-\frac{|x-\xi|^2}{\lambda}} d\xi_1 \ldots d\xi_r > 1 - \varepsilon, \qquad \text{per } x \in C_1,\ 0 < \lambda < \lambda_\varepsilon,$$

$$(39') \qquad (\pi\lambda)^{-\frac{r}{2}} \int_L e^{-\frac{|x-\xi|^2}{\lambda}} d\xi_1 \ldots d\xi_r < \varepsilon, \qquad \text{per } x \in C_2,\ 0 < \lambda < \lambda_\varepsilon.$$

Dalle (39), (39') seguono le

$$(40) \qquad \left| (\pi\lambda)^{-\frac{r}{2}} \int_L d\xi_1 \ldots d\xi_r \int_{C_1} e^{-\frac{|x-\xi|^2}{\lambda}} d\Phi - \int_{C_1} d\Phi \right|$$
$$< \varepsilon \int_{C_1} |d\Phi| \le \varepsilon \int_{S_r} |d\Phi| = \varepsilon P(E),$$

$$(40') \qquad \left| (\pi\lambda)^{-\frac{r}{2}} \int_L d\xi_1 \ldots d\xi_r \int_{C_2} e^{-\frac{|x-\xi|^2}{\lambda}} d\Phi \right| < \varepsilon P(E).$$

Osserviamo ora che dalle (37), (37') seguono le

$$(41) \qquad \left| \int_L d\Phi - \int_{C_1} d\Phi \right| \leq \int_{(L\setminus C_1)} |d\Phi| < \varepsilon,$$

$$(42) \qquad \int_{(S_r\setminus C_1\setminus C_2)} |d\Phi| < 2\varepsilon,$$

mentre dalla (42) si ricava

$$(43) \qquad \left| (\pi\lambda)^{-\frac{r}{2}} \int_L d\xi_1 \ldots d\xi_r \int_{(S_r\setminus C_1\setminus C_2)} e^{-\frac{|x-\xi|^2}{\lambda}} d\Phi \right| < 2\varepsilon.$$

Dalle (40), (40'), (41), (43) si deduce che, per ogni valore del parametro λ positivo e minore di λ_ε, si ha

$$(44) \qquad \left| \int_L d\Phi - (\pi\lambda)^{-\frac{r}{2}} \int_L d\xi_1 \ldots d\xi_r \int_{S_r} e^{-\frac{|x-\xi|^2}{\lambda}} d\Phi \right| < \varepsilon(2P(E)+3).$$

Poichè $\Phi(B)$ soddisfa la condizione b) del teorema II, ricordando la (30) abbiamo

$$(45) \qquad \begin{aligned} \operatorname{grad} W_\lambda \varphi(\xi|E) &= (\pi\lambda)^{-\frac{r}{2}} \int_E \operatorname{grad}_\xi e^{-\frac{|x-\xi|^2}{\lambda}} \, dx_1 \ldots dx_r \\ &= -(\pi\lambda)^{-\frac{r}{2}} \int_E \operatorname{grad}_x e^{-\frac{|x-\xi|^2}{\lambda}} \, dx_1 \ldots dx_r \\ &= -(\pi\lambda)^{-\frac{r}{2}} \int_{S_r} e^{-\frac{|x-\xi|^2}{\lambda}} \, d\Phi. \end{aligned}$$

Dalle (44), (45) si vede che, per ogni valore positivo e minore di λ_ε del parametro λ, vale la

$$(46) \qquad \begin{aligned} & \left| \Phi(L) + \int_L \operatorname{grad} W_\lambda\varphi(x|E) \, dx_1 \ldots dx_r \right| \\ &= \left| \int_L d\Phi + \int_L \operatorname{grad} W_\lambda\varphi(\xi|E) \, d\xi_1 \ldots d\xi_r \right| < \varepsilon(2P(E)+3). \end{aligned}$$

Dalle (46), poichè ε è un numero positivo arbitrario, segue la (35) e il nostro teorema è dimostrato.

    5. Uno spazio euclideo S_r, il cui punto generico indicheremo sempre con $x \equiv (x_1, \ldots, x_r)$, può sempre essere considerato il prodotto topologico di due spazi S_p ed S_{r-p}. Indicheremo con $y \equiv (y_1, \ldots, y_p)$ il generico punto di S_p, con $z \equiv (z_1, \ldots, z_{r-p})$ il generico punto di S_{r-p}, con (y, z) il punto di S_r avente le coordinate $(y_1, \ldots, y_p, z_1, \ldots, z_{r-p})$. Dati un insieme $E^y \subset S_p$ ed un insieme $E^z \subset S_{r-p}$, indicheremo con (E^y, E^z) il prodotto topologico degli insiemi E^y ed E^z, cioè l'insieme descritto dal punto (y, z) al variare di y in E^y e di z in E^z. In maniera analoga a quella seguita per definire l'operatore W_λ, possiamo definire l'operatore W_λ^y ponendo, per ogni funzione $f(x) \equiv f(y, z)$ definita in S_r ed ivi limitata e per ogni valore positivo del parametro λ

$$(47) \qquad W_\lambda^y f(x) \equiv W_\lambda^y f(y, z) = \pi^{-\frac{p}{2}} \int_{S_p} e^{-|\eta|^2} f(y + \sqrt{\lambda}\eta, z)\, d\eta_1 \ldots d\eta_p.$$

Per ogni funzione $g(x) \equiv g(y, z)$ che sia derivabile rispetto alle variabili $y_1, \ldots, y_p$ indicheremo con $\operatorname{grad}_y g(y, z)$ il vettore di componenti $\left(\frac{\partial g}{\partial y_1}, \ldots, \frac{\partial g}{\partial y_p}\right)$.

È appena necessario osservare che tutte le proprietà dell'operatore W_λ che abbiamo visto nel n. 1 si estendono all'operatore W_λ^y. Possiamo perciò, accanto al funzionale $\mathcal{I}[f(x)]$, definire $\mathcal{I}_y[f(y, z)]$ ponendo

$$(48) \qquad \mathcal{I}_y[f(y, z)] = \lim_{\lambda \to 0} \int_{S_p} |\operatorname{grad}_y W_\lambda^y f(y, z)|\, dy_1 \ldots dy_p;$$

infatti si può provare che è determinato il limite che compare nella (48) con un ragionamento del tutto simile a quello usato per il limite (12). Dimostriamo ora la diseguaglianza

$$(49) \qquad \mathcal{I}[f(x)] \equiv \mathcal{I}[f(y, z)] \geq \int_{S_{r-p}} \mathcal{I}_y[f(y, z)]\, dz_1 \ldots dz_{r-p}.$$

Ricordando le (6) troviamo che, per ogni coppia di numeri reali positivi λ, μ vale la

$$(50) \qquad \int_{S_r} \left|W_\lambda \operatorname{grad}_y W_\mu^y f(y, z)\right|\, dy_1 \ldots dy_p\, dz_1 \ldots dz_{r-p}$$

$$= \int_{S_r} \left|W_\mu^y \operatorname{grad}_y W_\lambda f(y, z)\right|\, dy_1 \ldots dy_p\, dz_1 \ldots dz_{r-p}$$

$$\leq \int_{S_r} \left|\operatorname{grad}_y W_\lambda f(y, z)\right|\, dy_1 \ldots dy_p\, dz_1 \ldots dz_{r-p}$$

$$\leq \int_{S_r} |\operatorname{grad} W_\lambda f(x)|\, dx_1 \ldots dx_r \leq \mathcal{I}[f(x)].$$

Dalla (50), passando al limite per $\lambda \to 0$, si trova, ricordando la (7),

$$(51) \qquad \int_{S_{r-p}} dz_1 \ldots dz_{r-p} \int_{S_p} \left|\operatorname{grad}_y W_\mu^y f(y, z)\right|\, dy_1 \ldots dy_p \leq \mathcal{I}[f(x)]$$

e quindi dalle (51), tenendo presente la (48), per i noti teoremi sul passaggio al limite sotto il segno di integrale, si deduce la (49).

6.

Teorema VI. – *Dato un insieme E contenuto in uno spazio euclideo S_r (con $r \geq 2$), è sempre verificata una delle due diseguaglianze*

$$(52) \qquad \begin{cases} (\operatorname{mis} E)^{r-1} \leq (P(E))^r \\[2ex] (\operatorname{mis}(S_r \setminus E))^{r-1} \leq (P(E))^r. \end{cases}$$

Dim. Il nostro teorema sarà certamente dimostrato se faremo vedere che, dati uno spazio euclideo S_r (con $r \geq 1$) ed una funzione $\varphi(x)$ che in ogni punto $x \in S_r$ prenda il valore 0 oppure il valore 1, vale una delle due relazioni

$$(53) \quad \begin{cases} \left(\int_{S_r} \varphi(x)\, dx_1 \ldots dx_r \right)^{r-1} \leq (\mathcal{I}[\varphi(x)])^r \\[3ex] \left(\int_{S_r} (1 - \varphi(x))\, dx_1 \ldots dx_r \right)^{r-1} \leq (\mathcal{I}[\varphi(x)])^r \end{cases}$$

purchè, nel caso $r = 1$, si attribuisca il valore 1 al simbolo ∞^0 e il valore 0 al simbolo 0^0.

Cominciamo appunto col considerare il caso $r = 1$ e notiamo che, se è nullo uno dei due integrali $\int_{S_1} \varphi(x)\, dx$, $\int_{S_1} (1 - \varphi(x))\, dx$, certamente risulta verificata una delle relazioni (53). Se invece si ha

$$(54) \quad \int_{S_1} \varphi(x)\, dx = \int_{-\infty}^{+\infty} \varphi(x)\, dx \neq 0, \qquad \int_{-\infty}^{+\infty} (1 - \varphi(x))\, dx \neq 0,$$

sarà possibile, per la (5), trovare due punti $\overline{x}$, $\overline{\overline{x}}$, tali che si abbia

$$(55) \quad \lim_{\lambda \to 0} W_\lambda \varphi(\overline{x}) = \varphi(\overline{x}) = 0, \qquad \lim_{\lambda \to 0} W_\lambda \varphi(\overline{\overline{x}}) = \varphi(\overline{\overline{x}}) = 1.$$

Dalle (55) segue

$$(56) \quad \begin{aligned} \mathcal{I}[\varphi(x)] &= \lim_{\lambda \to 0} \int_{-\infty}^{+\infty} \left| \frac{d}{dx} W_\lambda \varphi(x) \right| dx \\[2ex] &\geq \lim_{\lambda \to 0} \left| \int_{\overline{x}}^{\overline{\overline{x}}} \frac{d}{dx} W_\lambda \varphi(x) \right| dx = \lim_{\lambda \to 0} \left| W_\lambda \varphi(\overline{x}) - W_\lambda \varphi(\overline{\overline{x}}) \right| = 1 \end{aligned}$$

e quindi, nel caso in cui si verifichino le (54), le relazioni (53) sono certamente soddisfatte; è così provato che, per $r = 1$, vale sempre almeno una delle relazioni (53). Mostriamo ora che, fissato comunque un intero positivo m, se per $r = m$ è sempre verificata una delle relazioni (53), ciò accade anche per $r = m + 1$; dopo di ciò, poichè le (53) equivalgono evidentemente alle (52), il nostro teorema sarà completamente dimostrato. Cominciamo perciò con l'osservare che uno spazio S_{m+1} può sempre essere considerato come il prodotto topologico di due spazi S_m e S_1; riprendendo le notazioni del n. 5, indicheremo con $y \equiv (y_1, \ldots, y_m)$ il generico punto di S_m e con z il generico punto di S_1. Sia ora $\varphi(x) \equiv \varphi(y, z)$ una funzione che, in ogni punto $x \in S_{m+1}$, prenda il valore 0 oppure il valore 1 e siano E_1, E_2, E_3 gli insiemi dello spazio S_m caratterizzati dalle proprietà seguenti:

$$(57) \quad \begin{cases} \displaystyle\int_{S_1} \varphi(y,z)\,dz \neq 0, \qquad \int_{S_1} (1 - \varphi(y,z))\,dz \neq 0 \quad \text{per } y \in E_1 \\[2ex] \displaystyle\int_{S_1} \varphi(y,z)\,dz = 0 \qquad\qquad\qquad\qquad\qquad\; \text{per } y \in E_2 \\[2ex] \displaystyle\int_{S_1} (1 - \varphi(y,z))\,dz = 0 \qquad\qquad\qquad\;\; \text{per } y \in E_3 \end{cases}$$

evidentemente si ha

$$(58) \qquad\qquad S_m = E_1 \cup E_2 \cup E_3.$$

Poichè $\varphi(x) \equiv \varphi(y,z)$ può prendere solo i valori 0, 1, in quasi tutti i punti di (E_2, S_1) si avrà, per le (57), $\varphi(x) \equiv \varphi(y,z) = 0$, mentre, in quasi tutti i punti di (E_3, S_1) abbiamo $\varphi(x) \equiv \varphi(y,z) = 1$. Indicando con μ la più piccola delle due quantità mis E_2 e mis E_3 per quasi tutti i valori di z si avrà

$$(59) \qquad \int_{S_m} \varphi(y,z)\,dy_1 \ldots dy_m \geq \mu, \qquad \int_{S_m} (1 - \varphi(y,z))\,dy_1 \ldots dy_m \geq \mu;$$

avendo supposto che, per $r = m$ fosse sempre verificata una delle relazioni (53) avremo, ricordando la definizione di $\mathcal{I}_y[\varphi(y,z)]$,

$$(60) \qquad\qquad \mathcal{I}_y[\varphi(y,z)] \geq \mu^{\frac{m-1}{m}}$$

per quasi tutti i valori di z. D'altra parte le (49) ci assicurano che

$$(61) \qquad \mathcal{I}[\varphi(x)] = \mathcal{I}[\varphi(y,z)] \geq \int_{-\infty}^{+\infty} \mathcal{I}_y[\varphi(y,z)]\,dz$$

e quindi se è $\mu \neq 0$, cioè se nessuno dei due insiemi E_2, E_3 ha misura nulla, risulta per le (60), (61)

$$(62) \qquad\qquad \mathcal{I}[\varphi(x)] = +\infty$$

e le (53) sono certamente verificate.

Consideriamo ora i punti $y \in E_1$; poichè per $r = 1$ è sempre verificata una delle relazioni (53), ricordando la definizione di $\mathcal{I}_z[f(y,z)]$ e tenendo presenti le (57), troviamo che, per ogni punto $y \in E_1$, vale la

$$(63) \qquad\qquad \mathcal{I}_z[\varphi(y,z)] \geq 1.$$

Dalla (63) segue, ricordando la (49),

$$(64) \qquad \mathcal{I}[\varphi(x)] \geq \int_{S_m} \mathcal{I}_z[\varphi(y,z)]\,dy_1 \ldots dy_m \geq \int_{E_1} dy_1 \ldots dy_m = \text{mis } E_1$$

e quindi, se l'insieme E_1 ha misura infinita, le (53) sono certamente verificate.

Esaminiamo ora il caso in cui E_1 ha misura finita ed E_2 ha misura nulla; avrà allora misura infinita l'insieme E_3 e quindi, per quasi tutti i valori di z, avremo (ricordando che in quasi tutti i punti dell'insieme (E_3, S_1) risulta $\varphi(x) = 1$)

$$(65) \qquad \int_{S_m} \varphi(y, z)\, dy_1 \ldots dy_m = +\infty.$$

Avendo supposto che almeno una delle due relazioni (53) fosse sempre verificata per $r = m$, troviamo, tenendo presenti le (65),

$$(66) \qquad \left(\int_{S_m} (1 - \varphi(y, z))\, dy_1 \ldots dy_m \right)^{m-1} \le (\mathcal{I}_y[\varphi(y, z)])^m$$

per quasi tutti i valori di z. D'altra parte, poichè E_2 ha misura nulla e in quasi tutti i punti di (E_3, S_1) si ha $\varphi(y, z) = 1$, avremo, per quasi tutti i valori di z,

$$(67) \qquad \int_{S_m} (1 - \varphi(y, z))\, dy_1 \ldots dy_m \le \operatorname{mis} E_1.$$

Dalle (64), (66), (67) segue

$$(68) \qquad
\begin{aligned}
\mathcal{I}_y[\varphi(y, z)] \; &\ge \; \left(\int_{S_m} (1 - \varphi(y, z))\, dy_1 \ldots dy_m \right)^{\left(1 - \frac{1}{m}\right)} \\[2mm]
&\ge \; (\operatorname{mis} E_)^{-\frac{1}{m}} \int_{S_m} (1 - \varphi(y, z))\, dy_1 \ldots dy_m \\[2mm]
&\ge \; (\mathcal{I}[\varphi(x)])^{-\frac{1}{m}} \int_{S_m} (1 - \varphi(y, z))\, dy_1 \ldots dy_m .
\end{aligned}$$

Dalle (68), ricordando la (61) si ha

$$(69) \qquad
\begin{aligned}
\mathcal{I}[\varphi(x)] \; &\ge \; \int_{S_1} \mathcal{I}_y[\varphi(y, z)]\, dz \\[2mm]
&\ge \; (\mathcal{I}[\varphi(x)])^{-\frac{1}{m}} \int_{S_1} dz \int_{S_m} (1 - \varphi(y, z))\, dy_1 \ldots dy_m \\[2mm]
&= \; (\mathcal{I}[\varphi(x)])^{-\frac{1}{m}} \int_{S_{m+1}} (1 - \varphi(x))\, dx_1 \ldots dx_{m+1};
\end{aligned}$$

dalle (69) segue

$$(70) \qquad (\mathcal{I}[\varphi(x)])^{m+1} \ge \left(\int_{S_{m+1}} (1 - \varphi(x))\, dx_1 \ldots dx_{m+1} \right)^m$$

e quindi è provato che, se ha misura nulla l'insieme E_2 ed ha misura finita E_1, la seconda delle relazioni (53) è verificata per $r = m + 1$. Se avesse avuto misura nulla l'insieme E_3 ed avesse avuto misura finita E_1 avremo potuto provare, con un ragionamento del tutto analogo a quello ora svolto la prima delle relazioni (53). Avendo già provato che le (53) sono verificate quando E_1 ha misura infinita, oppure hanno contemporaneamente misure diverse da zero gli insiemi E_2 ed E_3, possiamo concludere che, per $r = m + 1$, una delle relazioni (53) è sempre verificata; il nostro teorema è così completamente dimostrato.

7. Dato uno spazio euclideo S_r, il cui punto generico indichiamo al solito con $x \equiv (x_1, \ldots, x_r)$, consideriamo l'aggregato di tutti gli insiemi contenuti in S_r; in tale aggregato introdurremo una metrica prendendo come distanza di due insiemi E_1, E_2 la quantità

$$(71) \qquad \mathrm{mis}\,(E_1 \cup E_2 \setminus E_1 \cap E_2).$$

Chiamiamo Σ lo spazio metrico così ottenuto e stabiliamo che, quando in questo lavoro si parlerà di limiti di insiemi contenuti in uno spazio euclideo S_r, intenderemo sempre considerare tali insiemi come elementi dello spazio Σ.

Consideriamo una successione E_1, $E_2, \ldots$ convergente verso l'insieme E; indichiamo con $\Phi(B)$ la eventuale funzione additiva d'insieme verificante le condizioni a), b) del teorema II, e indichiamo con $\Phi^n(B)$ l'analoga funzione relativa all'insieme E_n. Ricordando la definizione di convergenza debole (n.2) possiamo enunciare il

TEOREMA VII. – *Data una successione di insiemi*

$$(72) \qquad E_1, \ldots, E_n, \ldots$$

convergente verso un insieme E, si ha

$$(73) \qquad \lim_{n \to \infty}{}' P(E_n) \geq P(E);$$

se è limitato l'insieme dei perimetri $P(E_n)$, la successione

$$(74) \qquad \Phi^1(B), \ldots, \Phi^n(B), \ldots$$

converge debolmente verso $\Phi(B)$.

Dim. Indicheremo al solito con $\varphi(x|L)$ la funzione caratteristica del generico insieme $L \subset S_r$. Per la definizione di distanze tra due elementi dello spazio metrico Σ, dalla relazione

$$(75) \qquad \lim_{n \to \infty} E_n = E$$

segue

$$(76) \qquad \lim_{n \to \infty} \int_{S_r} |\varphi(x|E) - \varphi(x|E_n)|\, dx_1 \ldots dx_r = 0.$$

Avremo allora, ricordando la definizione dell'operatore W_λ,

$$(77) \qquad \lim_{n \to \infty} \mathrm{grad}\, W_\lambda \varphi(x|E_n) = \mathrm{grad}\, W_\lambda \varphi(x|E)$$

per ogni punto $x \in S_r$ e per ogni valore del parametro λ. Dalle (77) segue, ricordando la definizione del perimetro di un insieme ed i noti teoremi sul passaggio al limite sotto il segno di integrale,

$$(78) \qquad \underset{n\to\infty}{\lim}{}' P(E_n) \;\geq\; \underset{n\to\infty}{\lim}{}' \int_{S_r} |\operatorname{grad} W_\lambda \varphi(x|E_n)|\, dx_1 \dots dx_r$$

$$\geq\; \int_{S_r} \lim_{n\to\infty} |\operatorname{grad} W_\lambda \varphi(x|E_n)|\, dx_1 \dots dx_r$$

$$=\; \int_{S_r} |\operatorname{grad} W_\lambda \varphi(x|E)|\, dx_1 \dots dx_r$$

e la (78) vale per ogni valore positivo del parametro λ. Dalla (78), passando al limite per $\lambda \to 0$, otteniamo la (73); dalla (73) si vede che, se l'insieme descritto da $P(E_n)$ al variare di n è limitato, $P(E)$ è finito; detto in questo caso μ l'estremo superiore dell'insieme descritto da $P(E_n)$ al variare dell'indice n, avremo, per il teorema IV,

$$(79) \qquad \int_{S_r} |d\Phi| \leq \mu, \qquad \int_{S_r} |d\Phi^n| \leq \mu, \qquad (\text{per } n = 1, 2, \dots).$$

Presa poi una funzione $g(x)$ continua in S_r ed infinitesima per $|x| \to \infty$, possiamo certamente, in corrispondenza ad ogni numero positivo ε, trovare una funzione $g_\varepsilon(x)$ continua in S_r insieme alle sue derivate parziali prime, infinitesima insieme ad esse, per $|x| \to \infty$, d'ordine non inferiore a quello di $|x|^{-(r+1)}$ tale che si abbia

$$(80) \qquad |g(x) - g_\varepsilon(x)| \leq \varepsilon.$$

Per il teorema II abbiamo, ricordando la (76),

$$(81)\; \lim_{n\to\infty} \int_{S_r} g_\varepsilon(x) d\Phi^n \;=\; \lim_{n\to\infty} \int_{S_r} \varphi(x|E_n)\operatorname{grad} g_\varepsilon(x)\, dx_1 \dots dx_r$$

$$=\; \int_{S_r} \varphi(x|E_n)\operatorname{grad} g_\varepsilon(x)\, dx_1 \dots dx_r = \int_{S_r} g_\varepsilon(x) d\Phi$$

e quindi, tenendo presenti le (79), (80), (81), si ha

$$(82) \qquad \underset{n\to\infty}{\lim}{}'' \left| \int_{S_r} g(x)\, d\Phi^n - \int_{S_r} g(x)\, d\Phi \right| \leq 2\varepsilon\mu;$$

data l'arbitrarietà di ε, abbiamo allora

$$(83) \qquad \lim_{n\to\infty} \left| \int_{S_r} g(x) d\Phi^n - \int_{S_r} g(x) d\Phi \right| = 0$$

e la convergenza debole della successione (74) verso $\Phi(B)$ risulta dimostrata.

    8. Dato uno spazio euclideo S_r (con $r \geq 2$), consideriamo l'aggregato $\{\Pi\}$ dei domini contenuti in S_r, la cui frontiera è contenuta nella somma di un numero finito di iperpiani; a tali domini daremo il nome di *domini poligonali* (con una denominazione affine a quelle spesso usate di domini rettangolari, circolari, ecc.). Se in particolare lo spazio S_r si riduce al piano o allo spazio ordinario, i domini poligonali saranno rispettivamente poligoni o poliedri. Se ora noi consideriamo l'aggregato $\{\Pi\}$ come insieme dello spazio Σ introdotto nel n. 7 è evidente che esso è denso nello spazio Σ e, per il teorema VII, si ha, comunque si prenda un insieme $E \subset S_r$,

$$(84) \qquad \lim_{\Pi \to E}{}' P(\Pi) \geq P(E).$$

Un risultato più preciso di quello fornito dalla (84) è dato dal

Teorema VIII. – *Dato un insieme $E \subset S_r$ (con $r \geq 2$) il suo perimetro è uguale al minimo limite dei perimetri dei domini poligonali che approssimano in media E; si ha cioè*

$$(85) \qquad \lim_{\Pi \to E}{}' P(\Pi) = P(E).$$

Dim. Se $P(E)$ è infinito dalla (84) segue immediatamente la (85) e il nostro teorema è subito provato; se invece $P(E)$ è finito, per il teorema VI avrà misura finita uno dei due insiemi E, $(S_r \setminus E)$. Supponiamo che abbia misura finita l'insieme E; in tal caso la sua funzione caratteristica $\varphi(x|E)$ è sommabile e quindi, ricordando le (6), (8) avremo

$$(86) \qquad \lim_{\lambda \to 0} \int_{S_r} |W_\lambda \varphi(x|E) - \varphi(x|E)|\, dx_1 \ldots dx_r = 0.$$

Preso un numero positivo ε piccolo a piacere sarà allora certamente possibile trovare un numero positivo λ tale che risulti

$$(87) \qquad \int_{S_r} |W_\lambda \varphi(x|E) - \varphi(x|E)|\, dx_1 \ldots dx_r \leq \varepsilon.$$

Per la definizione stessa dell'operatore W_λ, risulta limitata in S_r la funzione $|\operatorname{grad} W_\lambda \varphi(x|E)|$ e quindi sarà finito il suo estremo superiore che indicheremo con M. Preso arbitrariamente un numero positivo $\eta < \frac{1}{4}$, consideriamo l'insieme L formato dai punti di S_r nei quali si ha

$$(88) \qquad W_\lambda \varphi(x|E) \geq \eta,$$

e dimostriamo che esso è limitato. Per ogni numero positivo ρ, indicheremo con $I_\rho(L)$ l'intorno di raggio ρ di L; inoltre porremo $\overline{\rho} = \frac{\eta}{2M}$. Poichè nei punti di L vale la (88), essendo

$$(89) \qquad |\operatorname{grad} W_\lambda \varphi(x|E)| \leq M,$$

nei punti di $I_{\overline{\rho}}(L)$ avremo

$$(90) \qquad W_\lambda \varphi(x|E) \geq \eta - \frac{M\eta}{2M} = \frac{\eta}{2}$$

e quindi sarà

(91)
$$\int_{S_r} |W_\lambda\varphi(x|E)| \, dx_1 \ldots dx_r \geq \int_{I_{\overline{p}}(L)} |W_\lambda\varphi(x|E)| \, dx_1 \ldots dx_r \geq \frac{\eta}{2}\mathrm{mis}\,(I_{\overline{p}}(L));$$

dalla (91), tenendo presente la (87) e ricordando che $\varphi(x|E)$ è sommabile si vede che deve avere misura finita l'insieme $I_{\overline{p}}(L)$ e quindi L deve essere limitato. Essendo $\varphi(x|E)$ una funzione sommabile ed essendo L un insieme limitato, possiamo certamente trovare un numero positivo α abbastanza grande perchè, detto T_α il dominio formato dai punti soddisfacenti le relazioni

(92)
$$|x_h| \leq \alpha \qquad (\text{per } h = 1, \ldots, r),$$

siano contemporaneamente verificate le formule

(93)
$$\int_{(S_r \setminus T_\alpha)} \varphi(x|E) \, dx_1 \ldots dx_r < \varepsilon$$

(93')
$$W_\lambda\varphi(x|E) < \eta \qquad (\text{per } x \in (S_r \setminus T_\alpha)).$$

Consideriamo ora, in uno spazio S_{r+1} il cui punto generico sia $(x_1, \ldots, x_r, y)$, l'ipersuperficie regolare Γ_1 individuata dalle

(94)
$$y = W_\lambda\varphi(x_1, \ldots, x_r|E) \equiv W_\lambda\varphi(x|E), \qquad x \in T_\alpha.$$

La funzione $W_\lambda\varphi(x|E)$ è continua in S_r insieme alle sue derivate parziali prime, quindi è certamente possibile approssimare l'ipersuperficie Γ_1 con una ipersuperficie Γ_2, che sia contenuta nella somma di un numero finito di iperpiani e sia rappresentata dalle equazioni

(95)
$$y = g(x) \equiv g(x_1, \ldots, x_r), \qquad x \equiv (x_1, \ldots, x_r) \in T_\alpha,$$

essendo $g(x)$ una funzione continua soddisfacente le condizioni seguenti:

(96)
$$0 < g(x) - W_\lambda\varphi(x|E) < \eta,$$

(97)
$$\int_{T_\alpha} |g(x) - W_\lambda\varphi(x|E)| \, dx_1 \ldots dx_r < \varepsilon.$$

(98)
$$\begin{aligned}
\int_{T_\alpha} |\mathrm{grad}\, g(x)| \, dx_1 \ldots dx_r \;&<\; \int_{T_\alpha} |\mathrm{grad}\, W_\lambda\varphi(x|E)| \, dx_1 \ldots dx_r + \eta \\
&\leq\; \int_{S_r} |\mathrm{grad}\, W_\lambda\varphi(x|E)| \, dx_1 \ldots dx_r + \eta \\
&\leq\; P(E) + \eta.
\end{aligned}$$

Poichè evidentemente $W_\lambda \varphi(x|E)$ non è mai negativa, $g(x)$ sarà per le (96) sempre positiva e quindi l'insieme dei punti $(x_1, \ldots, x_r, y)$ che soddisfano le condizioni

$$(99) \qquad 0 \leq y \leq g(x), \qquad x \equiv (x_1, \ldots, x_r) \in T_\alpha$$

sarà un dominio poligonale $D \subset S_{r+1}$. Tenendo presenti le (93'), (96) vediamo che, in ogni punto x appartenente alla frontiera di T_α, la funzione $g(x)$ è minore di 2η; di conseguenza, per ogni numero reale $\theta \geq 2\eta$, l'iperpiano $y = \theta$ incontra la frontiera del dominio D soltanto in punti appartenenti a Γ_2. Indichiamo con $\rho(\theta)$ la misura $(r-1)$-dimensionale (intesa in senso elementare) della sezione di Γ_2 col piano $y = \theta$ e chiamiamo Γ_2^* la porzione di Γ_2 che è contenuta nel semispazio $y \geq 2\eta$.

Per i teoremi elementari sulla misura delle sezioni di un insieme (teoremi che possiamo certamente applicare all'ipersuperficie Γ_2^* la quale, essendo contenuta in Γ_2, è a sua volta contenuta nella somma di un numero finito di iperpiani) avremo

$$(100) \qquad \int_{\Gamma_2^*} \nu_y \, d\sigma = \int_{2\eta}^{+\infty} \rho(\theta) \, d\theta,$$

ove con $d\sigma$ indichiamo l'elemento di misura r-dimensionale su Γ_2^* e con ν_y la lunghezza della proiezione ortogonale su un iperpiano $y = \theta$ del vettore unitario normale alla ipersuperficie Γ_2. Ricordando che Γ_2 ha le equazioni (95) si ha

$$(101) \qquad \int_{\Gamma_2^*} \nu_y \, d\sigma \leq \int_{\Gamma_2} \nu_y \, d\sigma = \int_{T_\alpha} |\operatorname{grad} g(x)| \, dx_1 \ldots dx_r.$$

Dalle (98), (100), (101) segue

$$(102) \qquad \int_{2\eta}^{+\infty} \rho(\theta) \, d\theta < P(E) + \eta$$

e quindi si ha, a maggior ragione,

$$(103) \qquad \int_{2\eta}^{1-\eta} \rho(\theta) \, d\theta < P(E) + \eta.$$

Consideriamo ora, per ogni valore di θ, la sezione del dominio D con l'iperpiano $y = \theta$ e indichiamola con $\Pi(\theta)$; se identifichiamo l'iperpiano $y = \theta$ con lo spazio S_r e quindi il generico punto $(x_1, \ldots, x_r, \theta)$ di tale iperpiano col punto $(x_1, \ldots, x_r) \in S_r$, troviamo che, per quasi tutti i valori di θ, l'insieme $\Pi(\theta)$, se non è vuoto, è un dominio poligonale di S_r: si ha evidentemente

$$(104) \qquad \begin{cases} g(x) \geq \theta & \text{per } x \in \Pi(\theta) \\[2mm] g(x) < \theta & \text{per } x \in (T_\alpha \setminus \Pi(\theta)). \end{cases}$$

Poichè, per $\theta \geq 2\eta$, l'iperpiano $y = \theta$ incontra la frontiera di D soltanto in punti appartenenti a Γ_2, per quasi tutti i valori di θ compresi fra 2η ed $(1 - \eta)$ (valori che riempiranno un intervallo, essendo per ipotesi $\eta < \frac{1}{4}$) il perimetro $P(\Pi(\theta))$ risulterà uguale a $\rho(\theta)$. Esisterà allora per le (103) un valore $\overline{\theta}$, compreso fra 2η e $(1-\eta)$, tale che risulti

$$(105) \qquad P(\Pi(\overline{\theta})) = \rho(\overline{\theta}) < \frac{P(E) + \eta}{1 - 3\eta},$$

essendo inoltre $\Pi(\overline{\theta})$ un dominio poligonale di S_r.

D'altra parte per la (104) avremo

$$(106) \qquad \begin{cases} g(x) - \varphi(x|E) \geq \overline{\theta} > 2\eta & \text{per } x \in (\Pi(\overline{\theta}) \setminus E \cap \Pi(\theta)) \\[2ex] \varphi(x|E) - g(x) \geq 1 - \overline{\theta} > \eta & \text{per } x \in (E \cap T_\alpha \setminus E \cap \Pi(\overline{\theta})), \end{cases}$$

mentre dalle (87), (97) segue

$$(107) \qquad \int_{T_\alpha} |g(x) - \varphi(x|E)| \, dx_1 \ldots dx_r < 2\varepsilon;$$

quindi per le (106), (107) avremo

$$(108) \qquad \text{mis}\,(E \cap T_u \cup \Pi(\overline{\theta}) \setminus E \cap \Pi(\overline{\theta})) < \frac{2\varepsilon}{\eta}.$$

Essendo per la (93)

$$(109) \qquad \text{mis}\,(E \setminus T_\alpha \cap E) < \varepsilon,$$

avremo finalmente

$$(110) \qquad \text{mis}\,(E \cup \Pi(\overline{\theta}) \setminus E \cap \Pi(\overline{\theta})) < \left(\varepsilon + \frac{2\varepsilon}{\eta}\right).$$

Data l'arbitrarietà con cui possono essere presi i due numeri ε, η, le (105), (110) ci assicurano che vale la

$$(111) \qquad \lim_{\Pi \to E}{}' P(\Pi) \leq P(E),$$

quindi ricordando la (84), troviamo la (85) e il nostro teorema è dimostrato.

Dai teoremi VI, VIII si vede immediatamente che *la nostra definizione del perimetro di un insieme è equivalente alla definizione data da Caccioppoli della misura $(r-1)$-dimensionale della frontiera orientata dell'insieme stesso*. Infatti, perchè un insieme $E \subset S_r$ sia approssimabile in media mediante domini poligonali le cui frontiere abbiano misure $(r-1)$-dimensionali equilimitate, occorre e basta che il suo perimetro sia finito; in questo caso il suo perimetro $P(E)$ è uguale alla variazione totale in S_r della funzione $\Phi(B)$) che soddisfa le condizioni a), b) del teorema II, la quale variazione coincide a sua volta con la misura $(r-1)$-dimensionale secondo CACCIOPPOLI della frontiera orientata di E (vedi loc. cit.[2]).

9. Accanto al perimetro $P(E)$ di un insieme $E \subset S_r$ possiamo considerare le *proiezioni del perimetro* $P(E)$ sugli iperpiani coordinati, che indicheremo con $P_1(E), \ldots, P_r(E)$ e saranno definite dalle

$$(112) \qquad P_h(E) = \lim_{\lambda \to 0} \int_{S_r} \left| \frac{\partial}{\partial x_h} W_\lambda \varphi(x|E) \right| \, dx_1 \ldots dx_r \qquad (\text{per } h = 1, \ldots, r);$$

l'esistenza del limite (112) si prova in maniera del tutto analoga a quella seguita per provare l'esistenza del limite (13). Vale evidentemente la relazione

$$(113) \qquad \sum_{h=1}^{r} P_h(E) \geq P(E),$$

mentre, per ogni valore dell'indice h, risulta

$$(114) \qquad P(E) \geq P_h(E).$$

Sussistono i teoremi seguenti.

TEOREMA IX. – *Dato un insieme $E \subset S_r$, se per un certo valore dell'indice h risulta finito $P_h(E)$, esiste una funzione d'insieme $\Phi_h(B)$ soddisfacente le condizioni seguenti:*

    a) $\Phi_h(B)$ è definita per ogni insieme $B \subset S_r$ ed è completamente additiva ed a variazione totale limitata.

    b) Per ogni funzione $g(x)$ continua in S_r insieme alle sue derivate parziali prime ed infinitesima, insieme ad esse, per $|x| \to \infty$, d'ordine non inferiore a quello di $|x|^{-(r+1)}$, risulta

$$(115) \qquad \int_E \frac{\partial g}{\partial x_h} \, dx_1 \ldots dx_r = \int_{S_r} g(x) \, d\Phi_h.$$

TEOREMA X. – *Dato un insieme $E \subset S_r$, se esiste una funzione $\Phi_h(B)$ soddisfacente le condizioni a), b) del teorema IX, $P_h(E)$ è finito e risulta*

$$(116) \qquad P_h(E) = \int_{S_r} |d\Phi_h|.$$

TEOREMA XI. – *Sia dato un insime $E \subset S_r$ tale che, per un certo valore dell'indice h, risulta finito $P_h(E)$ e sia $\Phi_h(B)$ la funzione soddisfacente le condizioni a), b) del teorema IX; per ogni insieme L sulla cui frontiera sia nulla la variazione totale della funzione $\Phi_h(B)$ si ha allora*

$$(117) \qquad \Phi_h(L) = - \lim_{\lambda \to 0} \int_L \frac{\partial}{\partial x_h} W_\lambda \varphi(x|E) \, dx_1 \ldots dx_r.$$

Data la successione di insiemi $E_1, E_2, \ldots$ convergente verso l'insieme E, per ogni valore dell'indice h definiamo la funzione d'insieme $\Phi_h(B)$ come nel teorema IX, e indichiamo con $\Phi_h^{(n)}(B)$ l'analoga funzione per l'insieme E_n. Sussiste il

TEOREMA XII. – *Data una successione di insiemi*

$$(118) \qquad E_1, \ldots, E_n, \ldots$$

convergente verso un insieme E, si ha

$$(119) \qquad \lim_{n \to \infty}{}' P_h(E_n) \geq P_h(E);$$

se per un certo valore dell'indice h è limitata la successione

$$(120) \qquad P_h(E_1), \ldots, P_h(E_n), \ldots,$$

risulta debolmente convergente verso $\Phi_h(B)$ la successione

$$(121) \qquad \Phi_h^{(1)}(B), \ldots, \Phi_h^{(n)}(B), \ldots.$$

La dimostrazione di questi teoremi è perfettamente analoga a qella dei teoremi II, IV, V, VII; basta sostituire in tutte le considerazioni ai gradienti le derivate parziali rispetto a x_h.

Osserviamo infine che dalla (113) e dai teoremi VIII, XII, si deduce immediatamente un risultato (enunciato da CACCIOPPOLI[4] come semplice presunzione) che, seguendo le notazioni usate in questo lavoro può enunciarsi nel modo seguente:

Dato un insieme $E \subset S_r$, se, per ogni valore dell'indice h è possibile costruire una successione di domini poligonali

$$(122) \qquad \Pi_1^h, \ldots, \Pi_n^h, \ldots$$

convergente in media verso l'insieme E, tale che risultino equilimitate le quantità

$$(123) \qquad P_h(\Pi_1^h), \ldots, P_h(\Pi_n^h), \ldots$$

allora $P(E)$ è finito e quindi l'insieme E può essere approssimato in media mediante una successione di domini poligonali

$$(124) \qquad \Pi_1, \ldots, \Pi_n, \ldots$$

aventi i perimetri

$$(125) \qquad P(\Pi_1), \ldots, P(\Pi_n), \ldots$$

equilimitati.

---

[4]Vedi loc. cit. ^{2}, Nota I, n. 8, nota (10).

A uniqueness theorem for the Cauchy problem relative to linear partial differential equations of parabolic type[‡][†]

Ennio De Giorgi (Rome)

Dedicated to Mauro Picone, for his 70th birthday.

SUMMARY. *We prove the uniqueness theorem stated at the beginning of the paper.*

In this paper we prove a uniqueness theorem for the CAUCHY problem concerning linear differential equations whose coefficients are allowed to be non analytic. The study of such problems was initiated by M. PICONE [1], [2] and continued by several italian and foreign mathematicians of his school (see e.g. [3], [4], [5]). The phenomena encountered in this class of problems are essentially different from those pertaining to differential equations with analytic coefficients; for instance (see the remark at the end of the paper) there are cases when the uniqueness property for the Cauchy problem does not hold, differently from what happens in the case of analytic coefficients.

We can state our result as follows. For any real γ, denote with $[\gamma]$ the largest integer less or equal than γ. Then we have:

Let us consider the differential equation

$$(1) \qquad \frac{\partial^m u}{\partial t^m} = \sum_{h=0}^{m-1} \sum_{k=0}^{[\alpha(m-h)]} c_{hk}(x,t) \frac{\partial^{h+k} u}{\partial x^k \partial t^h}$$

and let $u(x,t)$ be a solution to (1) on the rectangle $R \equiv \{a_1 \leq x \leq a_2,\ t_1 \leq t \leq t_2\}$; assume that the derivatives of $u(x,t)$ appearing in (1) are continuous on R and that, on the line $\{a_1 \leq x \leq a_2;\ t = t_1\}$, $u(x,t)$ satisfies the initial conditions

$$(2) \qquad \frac{\partial^h u}{\partial t^h} \equiv 0 \qquad \textit{(for } h = 0, 1, \ldots, m-1\textit{).}$$

[‡]Editor's note: translation into English of the paper "Un teorema di unicità per il problema di Cauchy, relativo ad equazioni differenziali a derivate parziali di tipo parabolico", published in Ann. Mat. Pura Appl., **4 (40)** (1955), 371–377.

[†]Research carried out at the Istituto Nazionale per le Applicazioni del Calcolo.

Moreover, let α be a positive constant smaller than 1, let the functions $c_{hk}(x,t)$ be continuous on R and infinitely differentiable with respect to the variable x, and assume that there exists a positive number ρ such that

$$(3) \qquad \lim_{n \to \infty} \frac{\rho^n}{n!} \left| \frac{\partial^n c_{hk}}{\partial x^n} \right|^\alpha = 0 \qquad \textit{(for } 0 \leq h \leq m-1, \ 0 \leq k \leq \alpha(m-h))$$

uniformly in R; then $u(x,t)$ vanishes identically on R.

The proof of this result is based on a few lemmas which are proved in the following sections.

1. Given two arbitrary real numbers b_1, b_2 (with $b_1 \leq b_2$), we denote with R^* the rectangle defined by the conditions

$$(4) \qquad b_1 \leq x \leq b_2, \quad 0 \leq t \leq \frac{1}{5};$$

for any integer $h \geq 0$ and any real number $\beta \geq 1$, we shall denote by $\Phi\{\beta; h\}$ the class of all functions $\varphi(x,t)$ defined and continuous on R^*, infinitely differentiable with respect to the variable x and satisfying on R^* the conditions

$$(5) \qquad \left| \frac{\partial^n \varphi}{\partial x^n} \right| \leq \{(h+n)!\}^\beta \qquad \text{(for } n = 0, 1, 2, \ldots).$$

On the other hand, we shall denote by $\Psi\{\beta; h\}$ the class of functions $\psi(x,t)$ defined on R^*, which admit there a Taylor expansion of the form

$$(6) \qquad \begin{cases} \psi(x,t) = \displaystyle\sum_{n=h}^{\infty} t^n (n!)^{-(\beta+1)} \psi_n(x,t) \\[2ex] \psi_n(x,t) \in \Phi\left\{ \beta; n + \left[\dfrac{n}{\beta}\right] \right\}. \end{cases}$$

Recalling STIRLING's formula

$$(7) \qquad n! = n^n e^{-n} \sqrt{2\pi n}\,(1 + \varepsilon_n), \qquad \lim_{n \to \infty} \varepsilon_n = 0,$$

and the elementary inequality

$$(8) \qquad \left(1 + \frac{1}{\beta}\right)^{1+\beta} \leq 4 \qquad \text{(for } 1 \leq \beta < +\infty),$$

by (4), (5), (6) we easily deduce that, for any nonnegative integer σ, the series

$$(9) \qquad \sum_{n=h}^{\infty} t^n (n!)^{-(\beta+1)} \frac{\partial^\sigma \psi_n}{\partial x^\sigma}$$

converges in norm on R^*; hence $\psi(x,t)$ is continuous on R^* and differentiable of any order with respect to the variable x, with continuous derivatives.

2. **Lemma I.** – *Given two functions $\varphi_1(x,t)$, $\varphi_2(x,t)$ on R^*, if $\varphi_1 \in \Phi\{\beta; h\}$ and $\varphi_2 \in \Phi\{\beta; 0\}$, then the product $(h+1)^\beta \varphi_1(x,t)\varphi_2(x,t)$ belongs to $\Phi\{\beta; h+1\}$.*

Proof. From the well known formula

$$(10) \qquad \frac{d^n}{dx^n}(g(x)f(x)) = \sum_{l=0}^{n} \binom{n}{l} \frac{d^{n-l}g}{dx^{n-l}} \frac{d^l f}{dx^l},$$

choosing $g(x) = (1-x)^{-(h+1)}$, $f(x) = (1-x)^{-1}$, we easily deduce

$$(11) \qquad (n+h+1)! = (h+1) \sum_{l=0}^{n} \binom{n}{l} (h+n-l)!\, l!.$$

From (11), for $\beta \geq 1$, we obtain

$$(12) \qquad \{(n+h+1)!\}^\beta \geq (n+h+1)!\{(n+h)!(h+1)\}^{\beta-1} =$$

$$= (h+1)^\beta \sum_{l=0}^{n} \binom{n}{l} (n+h-l)!\, l!\{(n+h)!\}^{\beta-1} \geq (h+1)^\beta \sum_{l=0}^{n} \binom{n}{l} \{(n+h-l)!\, l!\}^\beta.$$

By (10), (12), recalling the definition of $\Phi\{\beta; h\}$, the proof of the lemma follows.

Lemma II. – *Let $\psi(x,t)$, $\varphi(x,t)$ be two functions defined on R^* and let h, s, k be three nonnegative integers; assume that $\psi(x,t) \in \Psi\{\beta; h\}$, $\varphi(x,t) \in \Phi\{\beta; 0\}$, $s\beta \leq k+1$. Then the function $\theta(x,t)$ defined as*

$$(13) \qquad \theta(x,t) = \int_0^t \frac{(t-\tau)^k}{k!} \frac{\partial^s}{\partial x^s}(\varphi(x,\tau)\psi(x,\tau))d\tau$$

belongs to $\Psi\{\beta; h+k+1\}$ (and consequently also to $\Psi\{\beta; h+1\}$).

Proof. By the definition of $\Psi\{\beta; h\}$, $\psi(x,t)$ admits a Taylor expansion of the form (6); by the convergence in norm of the series (9) we deduce

$$(14) \qquad \theta(x,t) = \sum_{n=h}^{\infty} \int_0^t \frac{(t-\tau)^k}{k!} \tau^n (n!)^{-(\beta+1)} \frac{\partial^s}{\partial x^s}(\varphi(x,\tau)\psi_n(x,\tau))d\tau.$$

Denoting by $\theta_{n+k+1}(x,t)$ the continuous function on R^* defined through the relation

$$(15) \qquad \begin{cases} \theta_{n+k+1}(x,t)t^{(n+k+1)}\{(n+k+1)!\}^{-(\beta+1)} \equiv \\ \qquad \equiv (n!)^{-\beta} \int_0^t \frac{(t-\tau)^k}{k!} \frac{\tau^n}{n!} \frac{\partial^s}{\partial x^s}(\varphi(x,\tau)\psi_n(x,\tau))d\tau, \end{cases}$$

we can rewrite (14) as follows

$$(14') \qquad \theta(x,t) = \sum_{\nu=h+k+1}^{\infty} t^\nu (\nu!)^{-(\beta+1)} \theta_\nu(x,\tau).$$

On the other hand, lemma I implies that

$$(16) \qquad (n+1)^\beta \varphi(x,t)\psi_n(x,t) \in \Phi\left\{\beta; n + \left[\frac{n}{\beta}\right] + 1\right\};$$

thus, recalling the definition of the class $\Phi\{\beta; h\}$ and the condition $s\beta \leq k+1$, we can write

$$(17) \qquad (n+1)^\beta \frac{\partial^s}{\partial x^s}(\varphi\psi_n) \in \Phi\left\{\beta; n + \left[\frac{n}{\beta}\right] + s + 1\right\} \subset$$

$$\subset \Phi\left\{\beta; \left[\frac{n+k+1}{\beta}\right] + n + 1\right\},$$

and finally

$$(18) \qquad \left\{\frac{(n+k+1)!}{n!}\right\}^\beta \frac{\partial^s}{\partial x^s}(\varphi\psi_n) \in \Phi\left\{\beta; \left[\frac{n+k+1}{\beta}\right] + n + k + 1\right\}.$$

By (15), (18) we obtain

$$(19) \qquad \theta_{n+k+1}(x,t) \in \Phi\left\{\beta; \left[\frac{n+k+1}{\beta}\right] + n + k + 1\right\}$$

and hence, by (14'), the proof of the lemma is achieved.

LEMMA III. – *Let us consider a sequence of functions*

$$(20) \qquad g_0(x,t),\ g_1(x,t),\ldots, g_n(x,t),\ldots$$

defined on R^*; if for each value of the index h, $g_h(x,t) \in \Psi\{\beta; h\}$, then, for any choice of the integer $s \geq 0$, the series

$$(21) \qquad \sum_{h=0}^{\infty} \frac{\partial^s g_h}{\partial x^s}$$

converges in norm on R^*.

PROOF. By the definition of the class $\Psi\{\beta; h\}$ it is possible to find a system $\{\varphi_{hk}(x,t)\}$ of functions satisfying

$$(22) \qquad \begin{cases} g_h(x,t) \equiv \sum_{k=h}^{\infty} t^k (k!)^{-(\beta+1)} \varphi_{hk}(x,t) \\[2em] \varphi_{hk}(x,t) \in \Phi\left\{\beta; \left[\frac{k}{\beta}\right] + \beta\right\}. \end{cases}$$

By (7), (8), (22) and by the definition of the class $\Phi\{\beta; h\}$ we obtain the convergence in norm of the double series

$$(23) \qquad \sum_{h=0}^{\infty} \sum_{k=h}^{\infty} t^k (k!)^{-(\beta+1)} \frac{\partial^s \varphi_{hk}(x,t)}{\partial x^s}$$

and hence, by (22), the proof of the lemma.

3. LEMMA IV. – *Let us consider the differential equation*

$$(24) \qquad \frac{\partial^m w}{\partial t^m} = \sum_{h=0}^{m-1} \sum_{k=0}^{[\alpha(m-h)]} c_{hk}^*(x,t) \frac{\partial^{h+k} w}{\partial t^h \partial x^k}$$

and let $w(x,t)$ be a solution of (24) on the rectangle R^*; assume that the derivatives of $w(x,t)$ appearing in (24) are continuous on R^* and that, on the line $\{t = \frac{1}{5};\ b_1 \leq x \leq b_2\}$, $w(x,t)$ satisfies

$$(25) \qquad \frac{\partial^h w}{\partial t^h} \equiv 0 \qquad \textit{(for } h = 0, 1, \ldots, m-1\textit{)}.$$

Moreover, let $0 < \alpha < 1$, assume that the functions $c_{hk}^*(x,t)$ and their derivatives of any order with respect to x are continuous on R^*, and that for any choice of the indices h, k, n the following inequalities hold on R^*:

$$(26) \qquad (m+1)^2 \left| \frac{\partial^n c_{hk}^*}{\partial x^n} \right|^\alpha \leq n!;$$

then $w(x,t)$ vanishes identically on R^*.

PROOF. We denote by H_{hk} the operators defined by

$$(27) \qquad H_{hk} f(x,t) = \int_0^t \frac{(t-\tau)^{m-h-1}}{(m-h-1)!} \frac{\partial^k}{\partial x^k} \{c_{hk}^*(x,\tau) f(x,\tau)\} d\tau,$$

while by H we denote the operator defined by

$$(28) \qquad (-1)^m H f(x,t) = \sum_{h=0}^{m-1} \sum_{k=0}^{[\alpha(m-h)]} (-1)^{h+k} H_{hk} f(x,t).$$

Setting $\beta = \frac{1}{\alpha}$, $\varphi_{hk}(x,t) = (m+1)^2 c_{hk}^*(x,t)$, by (26) the functions $\varphi_{hk}(x,t)$ belong to $\Phi\{\beta; 0\}$; hence, by lemma II, if $f(x,t) \in \Psi\{\beta; n\}$, we also have

$$(29) \qquad (m+1)^2 H_{hk} f(x,t) \in \Psi\{\beta; n+1\},$$
$$\text{for } 0 \leq h \leq m-1,\ 0 \leq k \leq \alpha(m-h),$$

and by (28), (29) we obtain

$$(30) \qquad H f(x,t) \in \Psi\{\beta; n+1\}.$$

Now, let $\psi(x,t)$ be a function on R^* such that

$$(31) \qquad \begin{cases} \psi(x,t) \in \Psi\{\beta; 0\} \\[2mm] \left(\dfrac{\partial^s \psi}{\partial x^s} \right)_{x=b_1} \equiv \left(\dfrac{\partial^s \psi}{\partial x^s} \right)_{x=b_2} \equiv 0 \qquad \textit{(for } s = 0,1,2,\ldots;\ 0 \leq t \leq \tfrac{1}{5}\textit{)}. \end{cases}$$

By (30) and lemma III, the series $\sum_{n=0}^{\infty} H^n \psi(x,t)$ converges on R^* and, denoting by $g(x,t)$ its sum, $g(x,t)$ is continuous on R^*, with continuous derivatives of any order with respect to x on R^*; moreover, again by lemma III, we have

$$(32) \qquad Hg(x,t) = g(x,t) - \psi(x,t),$$

while, by (31), we also have

$$(33) \qquad \left(\frac{\partial^s g}{\partial x^s}\right)_{x=b_1} \equiv \left(\frac{\partial^s g}{\partial x^s}\right)_{x=b_2} \equiv 0 \qquad (\text{for } s = 0,1,2,\ldots;\ 0 \leq t \leq \tfrac{1}{5}).$$

By (25), (27), (33) we easily have

$$(34) \quad (-1)^m \int_{R^*} \frac{\partial^m w}{\partial t^m} H_{hk} g(x,t)\, dx dt = (-1)^{h+k} \int_{R^*} g(x,t) c^*_{hk}(x,t) \frac{\partial^{h+k} w}{\partial t^h \partial x^k}\, dx dt$$

and hence, by (24),(28), (32),

$$(35) \qquad \int_{R^*} \frac{\partial^m w}{\partial t^m} \psi(x,t)\, dx dt = 0.$$

Since $\beta > 1$, the functions satisfying (31) form a complete system for the linear global approximation of continuous functions on R^*; hence by (35) the proof of the lemma is complete.

4. It is easy to deduce from lemma IV the uniqueness theorem stated at the beginning of this paper. Indeed, assume that all the assumptions of that theorem are satisfied; since the relations (3) are satisfied uniformly on R, it is certainly possible to find a positive number M such that, for any choice of the indices h, k, n, the following inequalities hold on R:

$$(36) \qquad \rho^n \left| \frac{\partial^n c_{hk}}{\partial x^n} \right|^\alpha \leq Mn!.$$

Denote by L the set of real numbers τ belonging to the interval (t, t_2) such that the relations (2) are not identically satisfied on the line $\{t = \tau;\ a_1 \leq x \leq a_2\}$; L is evidently an open set. If we assume by contradiction that $u(x,t)$ does not vanish identically on R, this implies that the set L is not empty and hence it possesses a greatest lower bound $\bar{\tau}$; since L is open and contained in the interval (t_1, t_2), $\bar{\tau}$ cannot belong to L and we must have $t_1 \leq \bar{\tau} < t_2$. Now, let $\beta \alpha = 1$ and define

$$(37) \qquad \begin{cases} w(x,t) = u\left(\rho^\beta x, \bar{\tau} + \sigma(\tfrac{1}{5} - t)\right), \\[2ex] c^*_{hk}(x,t) = (-\sigma)^{m-h} \rho^{-\beta k} c_{hk}\left(\rho^\beta x, \bar{\tau} + \sigma(\tfrac{1}{5} - t)\right), \\[2ex] b_1 = a_1 \rho^{-\beta}, \qquad b_2 = a_2 \rho^{-\beta}; \end{cases}$$

if we choose σ such that

$$0 < \frac{\sigma}{5} \le (t_2 - \overline{\tau}); \qquad (m+1)^2 \sigma^\alpha \left(1 + \frac{1}{\rho}\right)^m < \frac{1}{1+M}$$

the functions $w(x,t)$, $c^*_{hk}(x,t)$ fulfill the assumptions of lemma IV. Thus $w(x,t)$ vanishes identically on R^* and hence $u(x,t)$ vanishes on the rectangle $\{\overline{\tau} \le t \le \overline{\tau} + \frac{\sigma}{5};\ a_1 \le x \le a_2\}$; this contradicts the definition of $\overline{\tau}$ as the greatest lower bound of L and the contradiction is due to the assumption that $u(x,t)$ does not vanish identically on R.

REMARK. By comparing the uniqueness result just proved with a non-uniqueness example recently published (see [6]; in particular see the relations (12), (13), (14), (15) of that paper), we remark that, chosen an arbitrary positive number ε, the result fails to hold if, while keeping all the remaining assumptions unchanged, we replace (3) with

$$(3') \qquad \lim_{n\to\infty} \frac{\rho^n}{n!} \left| \frac{\partial^n c_{hk}}{\partial x^n} \right|^{\alpha(1-\varepsilon)} = 0.$$

Bibliography

[1] M. Picone, *Nuove determinazioni per gl'integrali delle equazioni lineari a derivate parziali*. Rend. Acc. Naz. dei Lincei, Vol. XXVIII, serie 6, fasc. 11, 1938.

[2] M. Picone, *Nuovi metodi d'indagine per la teoria delle equazioni lineari alle derivate parziali*. Rend. Seminario Matematico e Fisico di Milano, vol. XIII, 1939.

[3] D. Mangeron, *L'applicazione del metodo di Picone, della trasformata di Laplace ad intervallo d'integrazione finito, alla teoria delle equazioni a derivate parziali d'ordine qualunque*. Rend. Accad. d'Italia, Serie VII, vol. I, fasc. 1, 1939.

[4] C. Pucci, *Nuove ricerche sul problema di Cauchy*. Memorie Accad. delle Scienze di Torino, Serie 3, tomo I, parte I, 1953.

[5] L. Esteri, *Tesi di Laurea*. Roma, 1955.

[6] E. De Giorgi, *Un esempio di non unicità della soluzione del problema di Cauchy, relativo ad una equazione differenziale lineare a derivate parziali di tipo parabolico*. Rendiconti di Matematica e delle sue Applicazioni, Serie V, vol. XIV, fasc. 1-2, 1955.

An example of non-unique solution to the Cauchy problem, for a partial differential equation of parabolic type[‡][†]

Ennio De Giorgi (Rome)

In this paper I shall construct an example of a CAUCHY problem, relative to a partial differential equation of parabolic type, for which the uniqueness theorem does not hold.

More precisely, *I shall construct four functions $a(x,t)$, $b(x,t)$, $c(x,t)$, $w(x,t)$, continuous with their partial derivatives of any order on the strip T of the plane (x,t) defined by the relations*

$$0 \leq t \leq 1, \qquad -\infty < x < +\infty,$$

satisfying on T the following differential equation

$$(1) \qquad \frac{\partial^8 w}{\partial t^8} = a\frac{\partial^4 w}{\partial x^4} + b\frac{\partial^2 w}{\partial x^2} + cw$$

and such that, on the line $t = 0$, one has identically

$$(2) \qquad \frac{\partial^h w}{\partial t^h} = 0 \qquad (for \quad h = 0, 1, \ldots, 7),$$

w being not identically 0 in T.

The construction of the functions $a(x,t)$, $b(x,t)$, $c(x,t)$, $w(x,t)$ is divided in two steps.

In the first step one considers the strip T^* of the plane (x,t) defined by the conditions

$$0 \leq t \leq 7, \qquad -\infty < x < +\infty,$$

and one defines on that strip suitable functions $\alpha(x,t)$, $\beta(x,t)$, $\gamma(x,t)$, $u(x,t)$ continuous with continuous partial derivatives of any order on T^*, satisfying the differential equation

$$(3) \qquad \frac{\partial^8 u}{\partial t^8} = \alpha\frac{\partial^4 u}{\partial x^4} + \beta\frac{\partial^2 u}{\partial x^2} + \gamma u;$$

these functions are defined in terms of some parameters $\lambda, \mu, \omega, \tau$ which for the moment are not assigned.

[‡]Editor's note: translation into English of the paper "Un esempio di non unicità della soluzione di un problema di Cauchy, relativo ad un'quazione differenziale lineare di tipo parabolico", published in Rend. Mat. e Appl., **(5) 14** (1955), 382–387.

[†]Written during the Author's stay at the Istituto Nazionale per le Applicazioni del Calcolo.

In the second step one considers again the strip T, which is divided in an infinite number of partial strips; the functions $a(x,t)$, $b(x,t)$, $c(x,t)$, $w(x,t)$ appearing in (1), (2), are defined, in each partial strip, through the functions α, β, γ, u, computed for suitable choices of the parameters $\lambda, \mu, \omega, \tau$.

1. We begin by setting

$$f(t) = \int_0^t e^{\frac{1}{\xi^2 - \xi}}\, d\xi \left[\int_0^1 e^{\frac{1}{\xi^2 - \xi}}\, d\xi\right]^{-1};$$

it is clear that $f(t)$, defined and continuous with all its derivatives on the interval $(0,1)$, satisfies

$$(4) \qquad \begin{cases} f(0) = 0, \qquad f(1) = 1, \qquad 0 \le f(t) \le 1, \\[2ex] \left[\dfrac{d^h f}{dt^h}\right]_{t=0} = \left[\dfrac{d^h f}{dt^h}\right]_{t=1} = 0, \qquad\qquad (h = 1, 2, \ldots). \end{cases}$$

Given four arbitrary positive constants $\lambda, \mu, \omega, \tau$, the function $u(x,t)$, defined in T^* by

$$(5) \quad \begin{cases} \tau u = 1 + \lambda^{-1} f(t) \cos(\omega x) & \text{for } 0 \le t \le 1, \\ \tau u = 1 + \cos(\omega x)\lambda^{[2f(t-1)-1]} & \text{for } 1 \le t \le 2, \\ \tau u = 1 - 2f(t-2) + \lambda \cos(\omega x) - 2f(t-2)\cos(2\omega x) & \text{for } 2 \le t \le 3, \\ \tau u = -1 + \lambda \cos(\omega x) - 2\left[\frac{1+\mu}{2}\right]^{f(t-3)} \cos(2\omega x) & \text{for } 3 \le t \le 4, \\ \tau u = \mu - (1+\mu)f(5-t) + \lambda\cos(\omega x) - (1+\mu)f(5-t)\cos(2\omega x) \\ \qquad\qquad & \text{for } 4 \le t \le 5, \\ \tau u = \mu + \cos(\omega x)\mu^{2f(t-5)}\lambda^{f(6-t)} & \text{for } 5 \le t \le 6, \\ \tau u = \mu + \mu^2 f(7-t)\cos(\omega x) & \text{for } 6 \le t \le 7, \end{cases}$$

is, in view of (4), continuous with all its partial derivatives of any order on T^*; $u(x,t)$ clearly satisfies

$$(6) \quad \begin{cases} \dfrac{\partial^8 u}{\partial t^8} = \lambda^{-1}\cos(\omega x)\left[\dfrac{1}{\omega^2}\dfrac{\partial^2 u}{\partial x^2} + u\right]\dfrac{d^8 f}{dt^8} & \text{for } 0 \le t \le 1, \\[2ex] \dfrac{\partial^8 u}{\partial t^8} = \dfrac{1}{\omega^4}\dfrac{\partial^4 u}{\partial x^4}\lambda^{[1-2f(t-1)]}\dfrac{d^8}{dt^8}\lambda^{[2f(t-1)-1]} & \text{for } 1 \le t \le 2, \\[2ex] \dfrac{\partial^8 u}{\partial t^8} = \lambda^{-1}\cos(\omega x)\left[\dfrac{4}{3\omega^4}\dfrac{\partial^4 u}{\partial x^4} + \dfrac{16}{3\omega^2}\dfrac{\partial^2 u}{\partial x^2}\right]\dfrac{d^8}{dt^8}f(t-2) & \text{for } 2 \le t \le 3, \\[2ex] \dfrac{\partial^8 u}{\partial t^8} = \left[\dfrac{1}{12\omega^4}\dfrac{\partial^4 u}{\partial x^4} + \dfrac{1}{12\omega^2}\dfrac{\partial^2 u}{\partial x^2}\right]\left(\dfrac{1+\mu}{2}\right)^{-f(t-3)}\dfrac{d^8}{dt^8}\left(\dfrac{1+\mu}{2}\right)^{f(t-3)} \\ \qquad\qquad & \text{for } 3 \le t \le 4, \\[2ex] \dfrac{\partial^8 u}{\partial t^8} = \lambda^{-1}(1+\mu)\cos(\omega x)\left[\dfrac{2}{3\omega^4}\dfrac{\partial^4 u}{\partial x^4} + \dfrac{8}{3\omega^2}\dfrac{\partial^2 u}{\partial x^2}\right]\dfrac{d^8}{dt^8}f(5-t) \\ \qquad\qquad & \text{for } 4 \le t \le 5, \\[2ex] \dfrac{\partial^8 u}{\partial t^8} = \dfrac{1}{\omega^4}\dfrac{\partial^4 u}{\partial x^4}\mu^{-2f(t-5)}\lambda^{-f(6-t)}\dfrac{d^8}{dt^8}[\mu^{2f(t-5)}\lambda^{f(6-t)}] & \text{for } 5 \le t \le 6, \\[2ex] \dfrac{\partial^8 u}{\partial t^8} = \mu\cos(\omega x)\left[\dfrac{1}{\omega^2}\dfrac{\partial^2 u}{\partial x^2} + u\right]\dfrac{d^8}{dt^8}f(7-t) & \text{for } 6 \le t \le 7. \end{cases}$$

Now, if we define on the strip T^* the functions $\alpha(x,t)$, $\beta(x,t)$, $\gamma(x,t)$ by

$$
(7)\quad
\begin{cases}
\alpha = 0 & \text{for } 0 \leq t \leq 1, \\[2mm]
\alpha = \dfrac{1}{\omega^4}\lambda^{[1-2f(t-1)]}\dfrac{d^8}{dt^8}\lambda^{[2f(t-1)-1]} & \text{for } 1 \leq t \leq 2, \\[2mm]
\alpha = \dfrac{4}{3\omega^4}\lambda^{-1}\cos(\omega x)\dfrac{d^8}{dt^8}f(t-2) & \text{for } 2 \leq t \leq 3, \\[2mm]
\alpha = \dfrac{1}{12\omega^4}\left(\dfrac{1+\mu}{2}\right)^{-f(t-3)}\dfrac{d^8}{dt^8}\left(\dfrac{1+\mu}{2}\right)^{f(t-3)} & \text{for } 3 \leq t \leq 4, \\[2mm]
\alpha = \dfrac{2}{3\omega^4}\lambda^{-1}(1+\mu)\cos(\omega x)\dfrac{d^8}{dt^8}f(5-t) & \text{for } 4 \leq t \leq 5, \\[2mm]
\alpha = \dfrac{1}{\omega^4}\mu^{-2f(t-5)}\lambda^{-f(6-t)}\dfrac{d^8}{dt^8}[\mu^{2f(t-5)}\lambda^{f(6-t)}] & \text{for } 5 \leq t \leq 6, \\[2mm]
\alpha = 0 & \text{for } 6 \leq t \leq 7,
\end{cases}
$$

$$
(8)\quad
\begin{cases}
\beta = \dfrac{1}{\omega^2}\lambda^{-1}\cos(\omega x)\dfrac{d^8 f}{dt^8} & \text{for } 0 \leq t \leq 1, \\[2mm]
\beta = 0 & \text{for } 1 \leq t \leq 2, \\[2mm]
\beta = \dfrac{16}{3\omega^2}\lambda^{-1}\cos(\omega x)\dfrac{d^8}{dt^8}f(t-2) & \text{for } 2 \leq t \leq 3, \\[2mm]
\beta = \dfrac{1}{12\omega^2}\left(\dfrac{1+\mu}{2}\right)^{-f(t-3)}\dfrac{d^8}{dt^8}\left(\dfrac{1+\mu}{2}\right)^{f(t-3)} & \text{for } 3 \leq t \leq 4, \\[2mm]
\beta = \dfrac{8}{3\omega^2}\lambda^{-1}(1+\mu)\cos(\omega x)\dfrac{d^8}{dt^8}f(5-t) & \text{for } 4 \leq t \leq 5, \\[2mm]
\beta = 0 & \text{for } 5 \leq t \leq 6, \\[2mm]
\beta = \dfrac{1}{\omega^2}\mu\cos(\omega x)\dfrac{d^8}{dt^8}f(7-t) & \text{for } 6 \leq t \leq 7,
\end{cases}
$$

$$
(9)\quad
\begin{cases}
\gamma = \lambda^{-1}\cos(\omega x)\dfrac{d^8 f}{dt^8} & \text{for } 0 \leq t \leq 1, \\[2mm]
\gamma = 0 & \text{for } 1 \leq t \leq 6, \\[2mm]
\gamma = \mu\cos(\omega x)\dfrac{d^8}{dt^8}f(7-t) & \text{for } 6 \leq t \leq 7,
\end{cases}
$$

it is immediate to check that (3) is equivalent to the conditions (6); moreover, it is easy to verify that the functions $\alpha(x,t)$, $\beta(x,t)$, $\gamma(x,t)$ are, in view of (4), continuous with all their partial derivatives of any order on T.

2. Now, fixed two positive constants $\rho > \sigma > 1$, we consider, for any positive integer n, the functions obtained by $u(x,t)$, $\alpha(x,t)$, $\beta(x,t)$, $\gamma(x,t)$ when we assign the constants $\lambda, \mu, \omega, \tau$ appearing in (5), (6), (7), (8), (9) as follows:

$$
(10)\quad
\begin{cases}
\log \lambda = n!, \qquad \log \mu = -(n!)^\sigma, \qquad \omega = (n!)^{2\rho} \\[1mm]
\log \tau = 1 + (2!)^\sigma + \cdots + [(n-1)!]^\sigma, & \text{for } n = 2, 3, \ldots; \\[1mm]
\log \tau = 0, & \text{for } n = 1;
\end{cases}
$$

we shall denote these functions by $u_n(x,t)$, $\alpha_n(x,t)$, $\beta_n(x,t)$, $\gamma_n(x,t)$ respectively. It is easy to verify that the functions $u_n(x,t)$, $\alpha_n(x,t)$, $\beta_n(x,t)$, $\gamma_n(x,t)$ tend to 0 with all their derivatives as $n \to \infty$; indeed, by (4), (5), (7), (8), (9),

(10) we see that, for any nonnegative integer h, we can find a constant θ_h such that, for any integer $k \geq 0$ and any integer $n > 1$, one has

$$(11) \qquad \left| \frac{\partial^{h+k} u_n}{\partial t^h \partial x^k} \right| \leq \theta_h e^{-[(n-1)!]^\sigma} e^{n!} (2n!)^{2\rho k} (n!)^{\sigma h} \qquad \text{for } 0 \leq t \leq 7,$$

$$(12) \quad \begin{cases} \left| \dfrac{\partial^{h+k} \alpha_n}{\partial t^h \partial x^k} \right| \leq \theta_h (n!)^{-8\rho} (n!)^{8\sigma} & \text{for } t \in (0,2) \cup (3,4) \cup (5,7), \\[3mm] \left| \dfrac{\partial^{h+k} \alpha_n}{\partial t^h \partial x^k} \right| \leq \theta_h e^{-n!} (n!)^{2\rho k} & \text{for } t \in (2,3) \cup (4,5), \end{cases}$$

$$(13) \quad \begin{cases} \left| \dfrac{\partial^{h+k} \beta_n}{\partial t^h \partial x^k} \right| \leq \theta_h e^{-n!} (n!)^{2\rho k} & \text{for } t \in (0,3) \cup (4,7), \\[3mm] \left| \dfrac{\partial^{h+k} \beta_n}{\partial t^h \partial x^k} \right| \leq \theta_h (n!)^{-4\rho} & \text{for } 3 \leq t \leq 4, \end{cases}$$

$$(14) \qquad \left| \frac{\partial^{h+k} \gamma_n}{\partial t^h \partial x^k} \right| \leq \theta_h e^{-n!} (n!)^{2\rho k} \qquad \text{for } 0 \leq t \leq 7.$$

We consider now the functions $w(x,t)$, $a(x,t)$, $b(x,t)$, $c(x,t)$ defined on T through

$$(15) \quad \begin{cases} w(x,t) = u_n[x, 2^n 7(2^{1-n} - t)] & \text{for } 2^{-n} \leq t \leq 2^{1-n}, \\[3mm] a(x,t) = 2^{8n} 7^8 \alpha_n[x, 2^n 7(2^{1-n} - t)] & \text{for } 2^{-n} \leq t \leq 2^{1-n}, \\[3mm] b(x,t) = 2^{8n} 7^8 \beta_n[x, 2^n 7(2^{1-n} - t)] & \text{for } 2^{-n} \leq t \leq 2^{1-n}, \\[3mm] c(x,t) = 2^{8n} 7^8 \gamma_n[x, 2^n 7(2^{1-n} - t)] & \text{for } 2^{-n} \leq t \leq 2^{1-n}, \\[3mm] w(x,0) \equiv a(x,0) \equiv b(x,0) \equiv c(x,0) \equiv 0. \end{cases}$$

It is easy to check that, by (4), (5), (7), (8), (9), (10), the functions defined by (15) are continuous with all their derivatives of any order on the strip T, with the possible exception of the points on the line $t = 0$; moreover, keeping into account the relations (11), (12), (13), (14), one sees that the functions $a(x,t)$, $b(x,t)$, $c(x,t)$, $w(t,x)$ are continuous with all their partial derivatives of any order on the *whole* strip T, and vanish on the line $t = 0$ together with all their derivatives of any order; hence, by (3), these functions satisfy (1), (2), q.e.d.

[Communicated on July 26, 1954].

New theorems on $(r-1)$-dimensional measures in r-dimensional space[‡]

Memoir by Ennio De Giorgi (Rome)

This note is devoted to some new developments of the theory of $(r-1)$–dimensional measures in an r–dimensional space, introduced in the papers [1], [2], [3]; in particular, this note is strictly related to the paper [3], whose tools and results are used in a systematic way throughout.

The new results presented here are the foundations of a general theory of k–dimensional measure in an r–dimensional space, which I intend to develop in forthcoming papers.

1. Throughout, we shall always follow the definitions and the notations introduced in [3] and, in particular, we shall agree that, whenever we deal with sets in a Euclidean space and with functions there defined, we shall always mean *Borel* sets and *Baire* functions.

LEMMA I. – *Given a function $f(x)$, continuous with its first order partial derivatives of in the cubic domain T of the space S_r, assume that*

$$\int_T f(x)\, dx_1 \cdots dx_r = 0;$$

denoting by t the length of the sides of T [1] *we have*

$$(1) \qquad \int_T |f(x)|\, dx_1 \cdots dx_r \le t \sum_{h=1}^{r} \int_T \left| \frac{\partial f}{\partial x_h} \right| dx_1 \cdots dx_r.$$

Proof. We can suppose that T is the domain given by

$$(2) \qquad 0 \le x_h \le 1 \qquad \text{(with } h = 1,\ldots,r),$$

since we can always reduce to this case by a simple change of variables; moreover, the claim being trivial when $r = 1$, we can use induction and prove it for $r = s+1$, assuming it is true when $r = s$. Letting

$$g(x_1,\ldots,x_s) = \int_0^1 f(x_1,\ldots,x_s,x_{s+1})dx_{s+1}$$

[‡]Editor's note: translation into English of the paper "Nuovi teoremi relativi alle misure $(r-1)$-dimensionali in uno spazio a r dimensioni", published in Ric. di Mat., **4** (1955), 95–113.

[1]By a cubic domain of side t we mean a domain of the kind $[a_h \le x_k \le a_h + t, h = 1,\ldots,r]$.

and letting T^* denote the slice of T determined by the hyperplane $x_{s+1} = 0$, we have

$$(3) \qquad \int_T |f(x)|dx_1 \cdots dx_{s+1} \le \int_{T^*} |g(x_1, \ldots, x_s)|dx_1 \cdots dx_s +$$

$$+ \int_{T^*} dx_1 \cdots dx_s \int_0^1 |f(x_1, \ldots, x_s, x_{s+1}) - g(x_1, \ldots, x_s)|dx_{s+1} \le$$

$$\le \int_{T^*} \sum_{h=1}^s \left|\frac{\partial g}{\partial x_h}\right| dx_1 \cdots dx_s + \int_{T^*} dx_1 \cdots dx_s \int_0^1 \left|\frac{\partial f}{\partial x_{s+1}}\right| dx_{s+1} \le$$

$$\le \int_T \sum_{h=1}^{s+1} \left|\frac{\partial f}{\partial x_h}\right| dx_1 \cdots dx_{s+1}$$

and the lemma is proved.

LEMMA II. – *Let E be a set in S_r having both finite perimeter and finite measure, let ρ be a positive real number and let*

$$(4) \qquad\qquad T_1, \ldots, T_m, \ldots$$

be a sequence of cubic domains, all having sides of length ρ, whose interiors are pairwise disjoint, and whose union coincides with the whole space S_r. Then there exists a set R, consisting of a finite (possibly empty) union of domains of the sequence (4), satisfying

$$(5) \qquad\qquad \operatorname{meas}\left((E \cup R) \setminus (E \cap R)\right) \le \rho\sqrt{r}P(E),$$

where $P(E)$ denotes the perimeter of E.

Proof. Denoting by $\varphi(x; E)$ the characteristic function of E, let us consider the operator $W_\lambda f(x)$ given by (see [3], n. 1)

$$W_\lambda f(x) = (\pi\lambda)^{-r/2} \int_{S_r} e^{-\frac{|\xi|^2}{\lambda}} f(x+\xi)dx_1 \cdots dx_r \qquad \left(|\xi| = \sqrt{\xi_1^2 + \cdots + \xi_r^2}\right)$$

and let us define

$$(6) \qquad \begin{cases} \tau_h^* = \dfrac{\operatorname{meas}(E \cap T_h)}{\rho^r}, \quad \tau_h(\lambda) = \dfrac{1}{\rho^r} \displaystyle\int_{T_h} W_\lambda \varphi(x; E)dx_1 \cdots dx_r \\[4mm] \sigma_h^* = \displaystyle\int_{T_h} |\varphi(x; E) - \tau_h^*|dx_1 \cdots dx_r, \quad \sigma_h(\lambda) = \displaystyle\int_{T_h} |W_\lambda \varphi(x; E) \\ \qquad\qquad\qquad\qquad\qquad\qquad\qquad\qquad\qquad -\tau_h(\lambda)|dx_1 \cdots dx_r. \end{cases}$$

By Lemma I we have

$$(7) \qquad \sqrt{r}\rho \int_{S_r} |\operatorname{grad} W_\lambda \varphi(x; E)|dx_1 \cdots dx_r$$

$$\ge \rho \sum_{h=1}^r \int_{S_r} \left|\frac{\partial}{\partial x_h} W_\lambda \varphi(x; E)\right| dx_1 \cdots dx_r =$$

$$= \sum_{h=1}^\infty \sum_{k=1}^r \rho \int_{T_h} \left|\frac{\partial}{\partial x_k} W_\lambda \varphi(x; E)\right| dx_1 \cdots dx_r \ge \sum_{h=1}^\infty \sigma_h(\lambda).$$

Let us recall the properties of the operator W_λ and the definition of perimeter ([3], n. 1):

$$P(E) = \lim_{\lambda \to 0} \int_{S_r} |\operatorname{grad} W_\lambda \varphi(x; E)| \, dx_1 \cdots dx_r;$$

from (7), taking the limit as $\lambda \to 0$, we find

$$(8) \qquad \rho\sqrt{r}P(E) \geq \sum_{h=1}^{\infty} \sigma_h^* = 2\sum_{h=1}^{\infty} \rho^r \tau_h^*(1 - \tau_h^*).$$

Since E has finite measure, only finitely many domains T_h can satisfy

$$(9) \qquad \rho^{-r}\operatorname{meas}(E \cap T_h) = \tau_h^* > \frac{1}{2};$$

letting n denote their number, we can arrange the sequence (4) in such a way that the domains satisfying (9) be the first n. Then from (8) it follows that

$$(10) \qquad \begin{aligned} \rho\sqrt{r}P(E) \;&\geq\; \sum_{h=1}^{n} \rho^r(1 - \tau_h^*) + \sum_{h=n+1}^{\infty} \rho^r \tau_h^* = \\ &=\; \sum_{h=1}^{n} \operatorname{meas}(T_h \setminus E) + \sum_{h=n+1}^{\infty} \operatorname{meas}(T_h \cap E) \end{aligned}$$

hence, letting $R = \bigcup_{h=1}^{n} T_h$, (5) follows.

Now, let us recall the metric space Σ introduced in [3] n. 7; this space consists of the sets contained in S_r and the distance between two sets E_1, E_2, considered as elements of Σ, is equal to $\operatorname{meas}\big((E_1 \cup E_2) \setminus (E_1 \cap E_2)\big)$. Relying on lemma II and on well-known results on metric spaces, we shall prove the following

THEOREM I. – *Given a bounded set $L \subset S_r$ and a positive number p, the collection of all sets contained in L having perimeter less than p is compact in the space Σ.*

Proof. Let H denote the above mentioned collection. Since clearly the space Σ is complete, by a well-known compactness criterion (see [4]) it is enough to prove that H is *totally bounded* (i.e., for every positive number $\eta > 0$, H is contained in the union of finitely many sets whose diameter is less than η). To this end, we observe that, for any positive number ρ, we can always decompose S_r as the union of countably many cubic domains

$$(11) \qquad T_1, \ldots, T_n, \ldots,$$

having pairwise disjoint interiors and all of side ρ. Since by assumption L is bounded, the family of sets that can be decomposed as a finite (maybe empty) union of domains from the sequence (11) having nonempty intersection with L is finite; let

$$(12) \qquad R_1, \ldots, R_\nu,$$

denote such sets. By lemma II we can find, for every set $E \in H$, an index m satisfying

$$(13) \qquad \operatorname{meas}\big((R_m \cup E) \setminus (R_m \cap E)\big) \le \rho\sqrt{rp}, \quad 1 \le m \le \nu.$$

Hence, recalling that ρ is arbitrary, H is totally bounded.

2. Let us consider a sequence of vector-valued set functions

$$(1) \qquad \Theta^{(1)}(B), \ldots, \Theta^{(n)}(B), \ldots$$

defined for every set $B \subset S_r$ and countably additive; let

$$(2) \qquad \alpha_1(B), \ldots, \alpha_n(B), \ldots$$

denote their total variations on the generic set B. We shall say that the sequence (1) is *quasi convergent*, if there exist a vector-valued set function $\Theta(B)$ and a non-negative function $\alpha^*(B)$, both defined for every set $B \subset S_r$, countably additive and satisfying

$$(3) \qquad \lim_{n \to \infty} \Theta^{(n)}(L) = \Theta(L),$$

$$(3') \qquad \lim_{n \to \infty} \alpha_n(L) = \alpha^*(L),$$

for every bounded set $L \subset S_r$ such that

$$(4) \qquad \alpha^*(\mathcal{F}L) = 0.$$

It is easy to check that, letting $\alpha(B)$ denote the total variation of Θ on the generic set B, there holds

$$(5) \qquad \alpha(B) \le \alpha^*(B) \le \lim_{n \to \infty} \alpha_n(S_r).$$

From a well-known theorem of De La Vallée Poussin (see [3] n. 2) the following result follows.

THEOREM II. – *Given a sequence of vector-valued set functions, defined on every set $B \subset S_r$, countably additive and having equibounded total variations, it is possible to extract a subsequence which is quasi convergent.*

Proof. Denoting by $\Theta^{(n)}(B)$ the n-th vector-valued function of our sequence, every component of $\Theta^{(n)}(B)$ equals the difference of two scalar set functions defined for every set $B \subset S_r$, which are countably additive and non-negative, and which never exceed the total variation of $\Theta^{(n)}$ on the set B (recall that also $\Theta^{(n)}$ is a non-negative, countably additive set function). Applying the mentioned theorem of De La Vallèe Poussin to the sequences determined by these countably additive and non-negative set functions, we obtain our claim.

3. For every set $B \subset S_r$ and for every positive number ρ, we shall call (following [5]) *open neighborhood of radius ρ of the set B* (and we shall denote it by the symbol $I(B; \rho)$) the set of those points whose distance from B is less than ρ; if B is made of a single point x, its open neighborhood of radius ρ will also be denoted by the symbol $I(x; \rho)$.

For every set $E \subset S_r$ having finite perimeter there exists, as proved in [3] n. 2, a vector-valued set function $\Phi(B)$ satisfying the following conditions:

a) $\Phi(B)$, which is defined for every $B \subset S_r$, is countably additive and has finite total variation;

b) for every function $g(x)$, which is continuous in S_r together with its first order partial derivatives, and which is infinitesimal at infinity, together with its first order derivatives, of order not smaller than $|x|^{-(r+1)}$, there holds

$$\int_E \operatorname{grad} g(x)\, dx_1 \cdots dx_r = \int_{S_r} g(x)\, d\Phi.$$

The vector-valued function Φ will be called the *Gauss-Green function of the set E.*

Now we are in a position to prove the following

LEMMA III. – *Consider in the space S_r (with $r \geq 2$) a set E of finite perimeter and a point ξ; then, for every choice of the number ρ, the perimeter of $E \cap I(\xi; \rho)$ is finite. Moreover, denoting by $\varphi(x; B)$ the characteristic function of a set $B \subset S_r$, by $\Phi(B)$ the Gauss-Green function of E, and by $\Psi(B; \rho)$ the Gauss-Green function of $E \cap I(\xi, \rho)$, we have, for every set $B \subset S_r$ and for almost every ρ*

$$(1) \qquad \Psi(B; \rho) = \Phi[B \cap I(\xi; \rho)] + \int_{\mathcal{F}I(\xi;\rho)} \varphi(x; E \cap B) n(x)\, d\sigma,$$

where $n(x)$ is the outer unit vector normal to the spherical hypersurface $\mathcal{F}I(\xi; \rho)$ and $d\sigma$ is the element of $(r-1)$–dimensional measure on such hypersurface.

Proof. Since E has finite perimeter, according to theorems VII, VIII in [3] and theorem II we can find a sequence of polyhedral domains

$$(2) \qquad \Pi_1, \ldots, \Pi_m, \ldots$$

satisfying

$$(3) \qquad \lim_{m \to \infty} P(\Pi_m) = P(E),$$

$$(3') \qquad \lim_{m \to \infty} \operatorname{meas}\left((E \cup \Pi_m) \setminus (E \cap \Pi_m)\right) = 0$$

and, moreover, satisfying the following conditions: for almost every ρ, there holds

$$(4) \qquad \lim_{m \to \infty} \int_{\mathcal{F}I(\xi;\rho)} |\varphi(x; \Pi_m) - \varphi(x; E)|\, d\sigma = 0;$$

denoting by $\Phi^{(m)}(B)$ the Gauss-Green function of Π_m, the sequence

$$(5) \qquad \Phi^{(1)}(B), \ldots, \Phi^{(m)}(B), \ldots$$

is quasi convergent to $\Phi(B)$. One easily checks that, for every value of ρ, $\Pi_m \cap I(\xi; \rho)$ has finite perimeter, and that, letting $\Psi^{(m)}(B; \rho)$ denote the Gauss-Green function of $\Pi_m \cap I(\xi; \rho)$, there holds

$$(6) \qquad \Psi^{(m)}(B; \rho) = \Psi^{(m)}[B \cap I(\xi; \rho)] + \int_{\mathcal{F}I(\xi;\rho)} \varphi(x; B \cap \Pi_m) n(x) \, d\sigma.$$

From (6) it immediately follows that

$$(7) \qquad P[\Pi_m \cap I(\xi; \rho)] \le P(\Pi_m) + \int_{\mathcal{F}I(\xi;\rho)} d\sigma$$

and therefore, by (3), (3') and theorem VII in [3], the perimeter of $E \cap I(\xi; \rho)$ is finite.

Since the sequence (5) is quasi convergent to $\Phi(B)$, there exists a countably additive, non-negative set function $\alpha^*(B)$ such that, for every bounded set $L \subset S_r$ satisfying

$$(8) \qquad \alpha^*(\mathcal{F}L) = 0,$$

there holds

$$(9) \qquad \lim_{m \to \infty} \Phi^{(m)}(L) = \Phi(L).$$

Consider now the positive numbers ρ which satisfy (4) and

$$(10) \qquad \alpha^*[\mathcal{F}I(\xi; \rho)] = 0;$$

taken such a number and a bounded set L satisfying (8), from (8), (10) it follows that

$$(11) \qquad \alpha^*(\mathcal{F}[L \cap I(\xi; \rho)]) = 0$$

and therefore

$$(12) \qquad \lim_{m \to \infty} \Phi^{(m)}(L \cap I(\xi; \rho)) = \Phi(L \cap I(\xi; \rho)).$$

From (4), (6), (12) we then obtain

$$(13) \qquad \lim_{m \to \infty} \Psi^{(m)}(L; \rho) = \Phi[L \cap I(\xi; \rho)] + \int_{\mathcal{F}I(\xi;\rho)} \varphi(x; L \cap E) n(x) d\sigma;$$

hence, recalling (3') and theorem VII in [3], we find that (1) is satisfied for every set $B \subset S_r$ and for every positive number ρ satisfying (4), (10).

4. Consider in the space S_r (with $r \geq 2$) a set E of finite perimeter; as usual, let $\Phi(B)$ be the GAUSS-GREEN function of E, and let $\mu(B)$ be the total variation of Φ on a generic set $B \subset S_r$. We shall call *reduced boundary* of E (and we shall denote it by the symbol $\mathcal{F}^*E$) the set of all points $\xi \in S_r$ satisfying the following conditions:

for every positive number ρ there holds

$$(1) \qquad \mu[I(\xi;\rho)] > 0,$$

the limit

$$(2) \qquad n(\xi) = \lim_{\rho \to 0} \frac{\Phi[I(\xi;\rho)]}{\mu[I(\xi;\rho)]},$$

exists and is finite, and moreover

$$(3) \qquad |n(\xi)| = 1.$$

Following the terminology of [1] we can say that $\mathcal{F}^*E$ is made of the centers of the elements of the oriented boundary of E (see [1] n. 10).

Moreover, for every $\xi \in \mathcal{F}^*E$ we shall indicate by $A(\xi)$ the set of all points $x \equiv (x_1, \ldots, x_r)$ such that

$$(4) \qquad \sum_{h=1}^{r} n_h(\xi)(x_h - \xi_h) > 0,$$

and, similarly, we shall indicate by $B(\xi)$ the set of all points $x \equiv (x_1, \ldots, x_r)$ satisfying

$$(4') \qquad \sum_{h=1}^{r} n_h(\xi)(x_h - \xi_h) < 0;$$

finally, we shall consider the hyperplane made of all points $x \equiv (x_1, \ldots, x_r)$ such that

$$(5) \qquad \sum_{h=1}^{r} n_h(\xi)(x_h - \xi_h) = 0,$$

which shall be indicated by the symbol $S^*(\xi)$. Moreover, $L(\xi)$ will denote the set of all positive numbers τ satisfying the following property: for every integer $n \geq 1$, letting $\rho = \tau n$, for every set $B \subset S_r$ condition (1) of lemma III is satisfied and moreover

$$(6) \qquad \mu[\mathcal{F}I(\xi;\rho)] = 0.$$

Finally, for every positive integer m, ω_m will denote the measure of the unit ball in the space S_m. We point out that, from the definition of $\mathcal{F}^*E$ and by well-known theorems on set functions, it follows that

$$(7) \qquad \mu(S_r \setminus \mathcal{F}^*E) = 0.$$

Once this terminology has been introduced, we can prove the following

LEMMA IV. – *For every $\xi \in \mathcal{F}^*E$ there holds*

$$(8) \qquad \liminf_{\rho \to 0} \rho^{-r}\operatorname{meas}\left[E \cap I(\xi; \rho)\right] > 0,$$

$$(8') \qquad \liminf_{\rho \to 0} \rho^{-r}\operatorname{meas}\left[(S_r \setminus E) \cap I(\xi; \rho)\right] > 0,$$

$$(9) \qquad \limsup_{\rho \to 0} \rho^{1-r}P[E \cap I(\xi; \rho)] < +\infty.$$

Proof. According to the definition of $L(\xi)$, we have

$$(10) \qquad P[E \cap I(\xi; \rho)] = \mu[I(\xi; \rho)] + \int_{\mathcal{F}I(\xi;\rho)} \varphi(x; E)\, d\sigma$$

for every $\rho \in L(\xi)$; since, on the other hand, a set of null measure also has null perimeter, we obtain, using (1), (10), that

$$(11) \qquad \operatorname{meas}\left[E \cap I(\xi; \rho)\right] > 0$$

for every $\rho \in L(\xi)$ (hence also for every positive value of ρ).

For every $\rho \in L(\xi)$ and every function $f(x)$, which is continuous in S_r with its first order partial derivatives, which equals 1 at every point of $I(\xi; \rho)$ and which identically vanishes in $S_r \setminus I(\xi; 2\rho)$, we have, according to the definition of $L(\xi)$ and to property b) of the Gauss-Green functions,

$$(12) \qquad \int_{S_r} f(x)\, d\Psi(B; \rho) = \Phi[I(\xi; \rho)] + \int_{\mathcal{F}I(\xi;\rho)} n(x)\varphi(x; E)\, d\sigma = 0,$$

where, as usual, $\Psi(B; \rho)$ denotes the Gauss-Green function of $E \cap I(\xi; \rho)$.

On the other hand, by virtue of (1), (2), (3), there exists a positive number $\bar\rho$ such that, when $0 < \rho \leq \bar\rho$, there holds

$$(13) \qquad 2\left|\Phi[I(\xi; \rho)]\right| \geq \mu[I(\xi; \rho)];$$

hence, for every number ρ in the interval $(0, \bar\rho)$ which belongs to $L(\xi)$, we have, by virtue of (10), (12), (13),

(14)

$$P(E \cap I(\xi; \rho)) \leq \int_{\mathcal{F}I(\xi;\rho)} \varphi(x; E)\, d\sigma + 2\left|\int_{\mathcal{F}I(\xi;\rho)} \varphi(x; E)\, d\sigma\right| \leq 3\int_{\mathcal{F}I(\xi;\rho)} \varphi(x; E)\, d\sigma.$$

From (14), recalling theorem VI in [3], we obtain

$$(15) \qquad 3\int_{\mathcal{F}I(\xi;\rho)} \varphi(x; E)\, d\sigma \geq \left(\operatorname{meas}\left[E \cap I(\xi; \rho)\right]\right)^{1-\frac{1}{r}}.$$

The function $g(\rho) = \text{meas}\,[E \cap I(\xi;\rho)]$ is clearly an absolutely continuous function of ρ in the interval $(0,\overline{\rho})$, and (due to (11)) it is positive when $\rho \neq 0$; since, for almost every ρ, we have

$$(16) \qquad \frac{dg}{d\rho} = \int_{\mathcal{F}I(\xi;\rho)} \varphi(x;E)\,d\sigma,$$

keeping (15) into account, we have, almost everywhere in the interval $(0,\overline{\rho})$,

$$(17) \qquad \frac{1}{3} \leq [g(\rho)]^{\frac{1}{r}-1}\frac{dg}{d\rho} = r\frac{d}{d\rho}\left[g^{\frac{1}{r}}\right]$$

and hence, integrating,

$$(18) \qquad g(\rho) \geq \left(\frac{\rho}{3r}\right)^r.$$

From (18) we obtain (8); then (8') can be proved by a similar argument, since $\mathcal{F}^*E$ is the reduced boundary of $S_r \setminus E$ as well. Finally, from (14), recalling theorem VII of [3], we infer that, for *every* positive number $\rho \leq \overline{\rho}$, there holds

$$(19) \qquad P[E \cap I(\xi;\rho)] \leq 3\int_{\mathcal{F}I(\xi;\rho)} d\sigma = 3\rho^{r-1}\omega_{r-1}$$

and hence (9) holds true.

A consequence of lemma IV is the following

THEOREM III. – *For every point $\xi \in \mathcal{F}^*E$ and every positive number ε, there holds*

$$(20) \qquad \lim_{\rho \to 0} \rho^{1-r}\mu[I(\xi;\rho) \cap I(S^*(\xi);\varepsilon\rho)] = \omega_{r-1},$$

$$(21) \qquad \lim_{\rho \to 0} \rho^{-r}\text{meas}\,[I(\xi;\rho) \cap A(\xi) \cap E] = 0,$$

$$(21') \qquad \lim_{\rho \to 0} \rho^{-r}\text{meas}\,[I(\xi;\rho) \cap B(\xi) \cap (S_r \setminus E)] = 0.$$

Proof. We shall prove the theorem under the assumption that

$$(22) \qquad n_1(\xi) = 1; \quad n_h(\xi) = 0, \quad \text{per} \quad h = 2,\ldots,r;$$

hence the hyperplane $S^*(\xi)$ is parallel to the hyperplane $x_1 = 0$; this assumption is really not restrictive, since it is always possible to reduce to it by a simple rotation.

Since almost every positive number belongs to $L(\xi)$ and the functions of ρ which appear in (20), (21), (21') are continuous from the right, in order to prove such relations it suffices to prove that for every sequence

$$(23) \qquad \tau_1,\ldots,\tau_n,\ldots,$$

of positive numbers belonging to $L(\xi)$ and converging to 0, it is possible to extract from (23) a subsequence

$$(24) \qquad \rho_1, \ldots, \rho_n, \ldots$$

such that

$$(25) \qquad \lim_{n \to \infty} \rho_n^{1-r} \mu[I(\xi; \rho_n) \cap I(S^*(\xi); \varepsilon \rho_n)] = \omega_{r-1},$$

$$(26) \qquad \lim_{n \to \infty} \rho_n^{-r} \mathrm{meas}\,[I(\xi; \rho_n) \cap A(\xi) \cap E] = 0,$$

$$(26') \qquad \lim_{n \to \infty} \rho_n^{-r} \mathrm{meas}\,[I(\xi; \rho_n) \cap B(\xi) \cap (S_r \setminus E)] = 0.$$

If, for every positive number ρ, we denote by E_ρ^* the set obtained from E by the dilation centered at ξ of factor $1/\rho$ (that is, the correspondence which associates to a point of coordinates $(\xi_1 + y_1, \ldots, \xi_r + y_r)$ the point $(\xi_1 + y_1/\rho, \ldots, \xi_r + y_r/\rho))$, and we denote by $\Phi^*(B; \rho)$ the Gauss-Green function of E_ρ^* and by $\mu^*(B; \rho)$ its total variation on B, (25) reduces to

$$(27) \qquad \lim_{n \to \infty} \mu^* \left[I(\xi; 1) \cap I(S^*(\xi); \varepsilon); \rho_n \right] = \omega_{r-1}.$$

From (2), (22) we deduce that, for every positive number σ, we have

$$(28) \qquad \lim_{\rho \to 0} \frac{\Phi_1^* \left[I(\xi; \sigma); \rho \right]}{\mu^* \left[I(\xi; \sigma); \rho \right]} = 1.$$

Since by lemma IV the perimeters of the sets

$$(29) \qquad I(\xi; 2) \cap E_{\tau_1}^*, \ldots, I(\xi; 2) \cap E_{\tau_n}^*, \ldots$$

are bounded, being clearly

$$(30) \qquad P\left[I(\xi; 2) \cap E_{\tau_n}^* \right] = \tau_n^{r-1} P\left[E \cap I(\xi; 2\tau_n) \right],$$

from theorem I and theorem II we obtain that the sequence (24) can be chosen in such a way that, letting $D_n = I(\xi; 2) \cap E_{\rho_n}$, the sequence

$$(31) \qquad D_1, \ldots, D_n, \ldots$$

converges in mean to some set D and moreover, letting $\Psi^{(n)}(B)$ denote the Gauss-Green of D_n, the sequence

$$(32) \qquad \Psi^{(1)}(B), \ldots, \Psi^{(n)}(B), \ldots$$

is quasi convergent. By theorem VII in [3], the sequence (32) is quasi convergent to the Gauss-Green function of D, which we shall denote by $\Psi(B)$.

Since the numbers in the sequence (24) belong to $L(\xi)$, we have, for every positive integer n and for every set $B \subset S_r$, that

$$(33) \qquad \Psi^{(n)}\left[B \cap I(\xi;2)\right] = \Phi^*\left[B \cap I(\xi;2); \rho_n\right]$$

and hence, denoting $\alpha_n(B)$ the total variation of $\Psi^{(n)}$ on the generic set B,

$$(34) \qquad \alpha_n\left[B \cap I(\xi;2)\right] = \mu^*\left[B \cap I(\xi;2); \rho_n\right].$$

Since the perimeters of the sets (31) are bounded and hence the functions $\alpha_n(B)$ are equibounded, from (28), (33), (34) it follows that

$$(35) \qquad \lim_{n\to\infty}\left\{\alpha_n\left[I(\xi;2)\right] - \Psi_1^{(n)}\left[I(\xi;2)\right]\right\} = 0.$$

The quasi convergence of the sequence (32) guarantees the existence of a function $\alpha^*(B)$ such that, for every bounded set L satisfying

$$(36) \qquad \alpha^*(\mathcal{F}L) = 0,$$

there holds

$$(37) \qquad \lim_{n\to\infty}\alpha_n(L) = \alpha^*(L); \qquad \lim_{n\to\infty}\Psi^{(n)}(L) = \Psi(L);$$

from (35), (37) we then deduce that for every set $B \subset I(\xi;2)$ there holds

$$(38) \qquad \Psi_1(B) = \alpha^*(B) \geq 0,$$

$$(38') \qquad \sum_{h=2}^{r}\left|\Psi_h(B)\right| = 0.$$

From (38), (38'), recalling the definition of the operator W_λ and theorem II in [3], we obtain that

$$(39)\ 0 \leq \left|\frac{\partial}{\partial x_1}W_\lambda\varphi(x;D)\right| + \frac{\partial}{\partial x_1}W_\lambda\varphi(x;D) + \sum_{h=2}^{r}\left|\frac{\partial}{\partial x_h}W_\lambda\varphi(x;D)\right| \leq$$

$$\leq \left(1+\sqrt{r}\right)(\pi\lambda)^{-\frac{r}{2}}\int_{\mathcal{F}I(\xi;2)} e^{-\frac{|x-\xi|^2}{\lambda}}\left|d\Psi\right|,$$

and hence, in every closed set contained in $I(\xi;2)$, the condition

$$(40)\quad \lim_{\lambda\to0}\left(\left|\frac{\partial}{\partial x_1}W_\lambda\varphi(x;D)\right| + \frac{\partial}{\partial x_1}W_\lambda\varphi(x;D) + \sum_{h=2}^{r}\left|\frac{\partial}{\partial x_h}W_\lambda\varphi(x;D)\right|\right) = 0$$

is satisfied uniformly. Since (see [3] n. 1) $W_\lambda\varphi(x;D)$ converges in mean to $\varphi(x;D)$, from (40) we infer the existence of a non-increasing monotone function

$u(t)$, defined on the interval $(-2, 2)$, such that, at almost every point of $I(\xi; 2)$, there holds

$$(41) \qquad \varphi(x; D) \equiv \varphi(x_1, \ldots, x_r; D) = u(x_1).$$

The function $u(t)$ being monotone non-increasing, there exists a number γ in the interval $(-2, 2)$, such that

$$(42) \qquad u(t) = 1 \quad \text{if} \quad t < \gamma; \qquad u(t) = 0 \quad \text{if} \quad t > \gamma.$$

It is easy to prove that we must have $\gamma = 0$; indeed, from $\gamma < 0$ we would obtain that

$$(43) \qquad 0 = \text{meas} \left[D \cap I(\xi; |\gamma|) \right] = \lim_{n \to \infty} \text{meas} \left[I(\xi; |\gamma|) \cap E_{\rho_n}^* \right] =$$
$$= \lim_{n \to \infty} \rho_n^{-r} \text{meas} \left[E \cap I(\xi; |\gamma|\rho_n) \right],$$

thus violating (8) of lemma IV; similarly one can check that due to (8'), $\gamma > 0$ can be ruled out. Being $\gamma = 0$, by virtue of (41), (42), D can differ from the set $I(\xi; 2) \cap B(\xi)$ at most by a set of null measure; then we find that

$$(44) \qquad 0 = \text{meas} \left[D \cap A(\xi) \cap I(\xi; 1) \right] = \lim_{n \to \infty} \text{meas} \left[D_n \cap A(\xi) \cap I(\xi; 1) \right] =$$
$$= \lim_{n \to \infty} \text{meas} \left[E_{\rho_n}^* \cap A(\xi) \cap I(\xi; 1) \right] = \lim_{n \to \infty} \rho_n^{-r} \text{meas} \left[E \cap A(\xi) \cap I(\xi; \rho_n) \right],$$

and (26) follows; by a similar argument, one can prove (26').

To prove (27), which we have seen to be equivalent to (25), we start by observing that, since D differs from $I(\xi; 2) \cap B(\xi)$ at most by a set of null measure, $\Psi(B)$ coincides with the Gauss-Green function of $I(\xi; 2) \cap B(\xi)$. Therefore, for every positive number ε, we find

$$(45) \qquad \Psi_1 \left[I(\xi; 1) \cap I(S^*(\xi); \varepsilon) \right] = \omega_{r-1},$$

$$(45') \qquad \Psi_1 \left\{ \mathcal{F} \left[I(\xi; 1) \cap I(S^*(\xi); \varepsilon) \right] \right\} = 0$$

and hence, from (37), (38), (45), (45') it follows that

$$(46) \qquad \lim_{n \to \infty} \alpha_n \left[I(\xi; 1) \cap I(S^*(\xi); \varepsilon) \right] = \omega_{r-1},$$

from which, due to (34), (27) follows and hence (25) follows, as well. Since (25), (26), (26') imply the validity of (20), (21), (21'), our theorem is proved.

5. In order to complete the analysis of the properties of the reduced boundary of a set, we have to introduce some definitions of the $(r-1)$–dimensional measure theory for non-oriented sets in the r–dimensional Euclidean space. More precisely, given a set H in the space S_r (with $r \geq 2$), we shall say that H belong to the class Γ_{r-1} if there exists a function $f(x)$, defined in a domain A containing

$H \cup \mathcal{F}H$, which is continuous in A together with its first order partial derivatives and which satisfies

$$(1) \qquad f(x) = 0, \qquad |\operatorname{grad} f(x)| \neq 0$$

at every point of $H \cup \mathcal{F}H$.

Moreover, we shall say that a set belongs to Γ^*_{r-1} if it is possible to decompose it as the union of finitely or countably many sets, each belonging to Γ_{r-1}. It is easy to check that, given a closed and bounded set C belonging to Γ_{r-1}, its various $(r-1)$–dimensional measures according to MINKOWSKI, GROSS, CARATHÉODORY etc. coincide; their common value will be simply called $(r-1)$–*dimensional measure of C*. We shall also call $(r-1)$–*dimensional inner measure* of any set $M \subset S_r$ the supremum of the $(r-1)$–dimensional measures of all closed and bounded sets which belong to Γ_{r-1} and which are contained inside M. One can easily check that the $(r-1)$–dimensional inner measure is the *smallest* set function which is countably additive, non-negative (possibly infinite), which is defined on every set $M \subset S_r$ and which coincides with the $(r-1)$–dimensional measure elementarily defined, when the latter exists. It is also easy to see that, given a set $M \subset S_r$, there always exists a set G^* satisfying

$$(2) \qquad \operatorname{meas}_{r-1} G^* = \operatorname{meas}_{r-1} M; \qquad G^* \subset M, \qquad G^* \in \Gamma^*_{r-1}.$$

In the sequel, we shall need the following

THEOREM IV. – *Given a closed bounded set C, suppose that there exists a vector-valued function $\nu(x) \equiv (\nu_1(x), \dots, \nu_r(x))$ which is continuous in C, which never vanishes, and such that*

$$(3) \qquad \lim_{|x-y| \to 0} \sum_{h=1}^{r} \nu_h(x)(x_h - y_h)|x - y|^{-1} = 0$$

uniformly with respect to x, y in C. Then C belongs to the class Γ_{r-1}.

Proof. By a theorem of WHITNEY (see [6]), there exists a function $f(x)$, which is continuous in S_r together with its first order partial derivatives, satisfying at every point of C

$$(4) \qquad \frac{\partial f}{\partial x_h} = \nu_h(x); \qquad f(x) = 0.$$

Since by assumption the vector-valued function $\nu(x)$ does not vanish, (1) follows from (4) and the theorem is proved.

Now we go back to the set $\mathcal{F}^*E$ already considered in n. 4 and, keeping all the notations therein, we can prove the following

THEOREM V. – *For every set $B \subset \mathcal{F}^*E$ there holds*

$$(5) \qquad \operatorname{meas}_{r-1}(B) = \mu(B).$$

Proof. Since (21) and (21') of n. 4 are satisfied at every point of $\mathcal{F}^*E$, by well-known theorems on additive set functions we can decompose $\mathcal{F}^*E$ as the union of a set N satisfying

$$(6) \qquad\qquad \mu(N) = 0$$

and countably many closed bounded sets

$$(7) \qquad\qquad C_1, \ldots, C_n, \ldots$$

in each of which (21), (21') of n. 4 are satisfied uniformly. Let us focus on one of these sets, say C_1; taken a positive number $\varepsilon < 1$, we can find another number $\sigma < 1$ such that, when $\rho \leq 2\sigma$, for every point $\xi \in C_1$, we have

$$(8) \qquad \operatorname{meas}\left[I(\xi;\rho) \cap A(\xi) \cap E\right] < \frac{\varepsilon^r \omega_r \rho^r 2^{-r}}{4},$$

$$(8') \qquad \operatorname{meas}\left[I(\xi;\rho) \cap B(\xi) \cap E\right] > \frac{\omega_r \rho^r}{2} - \frac{\varepsilon^r \omega_r \rho^r 2^{-r}}{4}.$$

We now want to prove that, taken two points x ed y both contained in C_1 and having distance less than σ from each other, it necessarily holds

$$(9) \qquad |x-y|^{-1} \left| \sum_{h=1}^{r} n_h(x)(x_h - y_h) \right| < \varepsilon.$$

Indeed, let us suppose that (9) is violated and assume, for example, that

$$(10) \qquad \sum_{h=1}^{r} n_h(x)(y_h - x_h) > \varepsilon |x-y|;$$

since $\varepsilon < 1$ and since the sum in (10) represents the distance from y to the hyperplane $S^*(x)$, we have

$$(11) \qquad I(y; \varepsilon |x-y|) \subset A(x) \cap I(x; 2|x-y|).$$

Keeping (8), (8') into account and recalling that $|x-y| < \sigma$ and $\varepsilon < 1$, we find

$$(12) \qquad \operatorname{meas}\left[A(x) \cap E \cap I(x; 2|x-y|)\right] < \frac{\varepsilon^r \omega_r |x-y|^r}{4},$$

$$(12') \qquad \operatorname{meas}\left[E \cap I(y; \varepsilon |x-y|)\right] > \frac{\varepsilon^r \omega_r |x-y|^r}{4};$$

but (12), (12') would violate (11) and hence it is proved that (10) cannot hold true when $y \in C_1$, $x \in C_1$, $|x-y| < \sigma$.

A similar argument can be used to exclude that

$$(10') \qquad \sum_{h=1}^{r} n_h(x)(y_h - x_h) < -\varepsilon\,|x - y|$$

and hence (9) is established. Since ε is arbitrary, we can conclude that

$$(13) \qquad \lim_{|x-y|\to 0} |x - y|^{-1} \sum_{h=1}^{r} \nu_h(x)(x_h - y_h) = 0$$

is satisfied uniformly in C_1 and hence, by theorem IV, $C_1 \in \Gamma_{r-1}$; since the same argument can be repeated for every set in the sequence (7) we have proved that $(\mathcal{F}^*E \setminus N) \in \Gamma^*_{r-1}$. Keeping (2) and (6) into account and recalling that the union of two sets belonging to Γ^*_{r-1} still belongs to Γ^*_{r-1}, one can see that, for every set $B \subset \mathcal{F}^*E$, there exists a set B^* satisfying

$$(14) \qquad \mu(B^*) = \mu(B), \quad \mathrm{meas}_{\,r-1} B^* = \mathrm{meas}_{\,r-1} B, \quad B^* \subset B, \quad B^* \in \Gamma^*_{r-1}.$$

Recalling (14) and the definition of Γ^*_{r-1}, we can then conclude that, to prove (5), it suffices to prove that

$$(15) \qquad \mathrm{meas}_{\,r-1}(G) = \mu(G)$$

for every closed bounded set belonging to Γ_{r-1} and contained inside $\mathcal{F}^*E$.

By the definition of Γ_{r-1}, there exist a domain $A \supset G$ and a function $f(x)$, which is continuous in A together with its partial derivatives of the first order and which satisfies (1) at every point of G. By the continuity of the derivatives of $f(x)$, we can also find a domain D, contained in A and containing G, in such a way that at every point of D there holds

$$(16) \qquad |\mathrm{grad}\, f(x)| \neq 0.$$

Denoting by V the set of those points of D at which $f(x)$ vanishes, and letting, for every set $M \subset S_r$,

$$(17) \qquad \mathrm{meas}_{\,r-1}(M \cap V) = \gamma(M),$$

it is easy to check that at every point $x \in G$ there holds

$$(18) \qquad \lim_{\rho\to 0} \rho^{1-r} \gamma\left[I(x; \rho)\right] = \omega_{r-1}$$

and hence, by theorem III, since $G \subset \mathcal{F}^*E$, we find

$$(19) \qquad \lim_{\rho\to 0} \frac{\gamma\left[I(x; \rho)\right]}{\mu\left[I(x; \rho)\right]} = 1.$$

From (17), (19), observing that $G \subset V$, by well-known theorems on differentiation of additive set functions, (15) is established and our theorem is proved.

6. The arguments in nn. 3, 4, 5 were concerned with sets of finite perimeter in a space S_r, with $r \geq 2$; the case $r = 1$, which we have so far ignored, is indeed much simpler, since in order to characterize the sets of finite perimeter it suffices to establish the following

THEOREM VI. – *Given a set $E \subset S_1$ having finite perimeter, there exists a set E^* which is a finite union of intervals and which satisfies*

$$(1) \qquad \mathrm{meas}\left((E \cup E^*) \setminus (E \cap E^*)\right) = 0.$$

Proof. Denoting, as usual, by $\varphi(x; E)$ the characteristic function of E, by the properties of the operator W_λ (see [3] n. 1), $W_\lambda\varphi(x; E)$ converges in mean to $\varphi(x; E)$, in every bounded set, when $\lambda \to 0$. Therefore we can find a sequence of positive numbers

$$\lambda_1 > \lambda_2 > \cdots > \lambda_m > \cdots$$

and a set N of null measure such that

$$(2) \qquad \lim_{m\to\infty} W_{\lambda_m}\varphi(x; E) = \varphi(x; E) \qquad \text{for every} \quad x \in S_1 \setminus N.$$

Since, by the definition of $P(E)$, $W_\lambda\varphi(x; E)$ is a function whose total variation does not exceed $P(E)$ (for every choice of the positive number λ), also $\varphi(x; E)$, by virtue of (2), has a total variation on $S_1 \setminus N$ which does not exceed $P(E)$ (that is, for every choice of $(m + 1)$ numbers $x_0 < x_1 < \cdots < x_m$ belonging to $S_1 \setminus N$ there holds $\sum_{h=1}^m |\varphi(x_h; E) - \varphi(x_{h-1}; E)| \leq P(E)$).

Then there exists, for every point $x \in S_1$, the limit

$$(3) \qquad \varphi^*(x) = \lim_{\xi\to x^+} \varphi(\xi; E) \qquad (\text{in } S_1 \setminus N),$$

and the total variation of $\varphi^*(x)$ will not exceed $P(E)$; the set E^* of which $\varphi^*(x)$ is the characteristic function is then a finite union of intervals, open on the right. Since N has null measure, by (3) the functions $\varphi^*(x)$ and $\varphi(x; E)$ coincide everywhere, and the theorem is proved.

Bibliography

[1] R. Caccioppoli, *Misure e integrazione sugli insiemi dimensionalmente orientati.* Rend. Acc. Naz. dei Lincei, Serie VIII, Vol. XII, fasc. 1-2, gennaio-febbraio 1952.

[2] E. De Giorgi, *Definizione ed espressione analitica del perimetro di un insieme.* Rend. Acc. Naz. dei Lincei, Serie VIII, vol. XIV, fasc. 3, marzo 1953.

[3] E. De Giorgi, *Su una teoria generale della misura $(r - 1)$–dimensionale in uno spazio ad r dimensioni.* Annali di Matematica pura e applicata, Serie IV, Tomo XXXVI, 1954.

[4] M. PICONE, *Due conferenze sui fondamenti del calcolo della variazioni.* "Giornale della Matematiche" di Battaglini, S. IV, vol. 80, 1950–51.

[5] M. PICONE e G. FICHERA, *Trattato di Analisi Matematica.* Vol. I, Ed. Tuminelli, Roma, 1954.

[6] H. WHITNEY, *Analytic extensions of differentiable functions defined on closed sets.* Trans. of American Mathematical Society, Vol. 36, 1934.

Nuovi teoremi relativi alle misure $(r-1)$-dimensionali in uno spazio ad r dimensioni[‡]

Memoria di Ennio De Giorgi (a Roma)

Questa Memoria è dedicata ad alcuni nuovi sviluppi della teoria, esposta nei lavori [1], [2], [3] (v. Bibliografia) della misura $(r-1)$–dimensionale in uno spazio a r dimensioni; essa si collega in specie strettamente alla Memoria [3], di cui utilizza sistematicamente i procedimenti e risultati.

I nuovi risultati che qui faccio conoscere sono alla base di una teoria generale della misura k–dimensionale in uno spazio a r dimensioni, che svolgerò in successivi lavori.

1. Ci atterremo sempre alle definizioni ed alle notazioni usate in [3] e, in particolare, conserveremo la convenzione in base alla quale, ogni volta che parliamo di insiemi contenuti in uno spazio euclideo e di funzioni ivi definite, intendiamo sempre riferirci a insiemi di *Borel* e a funzioni di *Baire*.

LEMMA I. – *Data una funzione $f(x)$ continua con le derivate parziali prime nel dominio quadrato T dello spazio S_r, sia*

$$\int_T f(x)\, dx_1 \cdots dx_r = 0;$$

detta t la lunghezza dei lati di T [1] avremo allora:

$$(1) \qquad \int_T |f(x)|\, dx_1 \cdots dx_r \leq t \sum_{h=1}^{r} \int_T \left| \frac{\partial f}{\partial x_h} \right| dx_1 \cdots dx_r.$$

Dim. Possiamo supporre che T sia il dominio indicato dalle

$$(2) \qquad 0 \leq x_h \leq 1 \qquad (\text{per } h = 1, \ldots, r),$$

dato che a questo caso è sempre possibile ricondursi mediante un semplice cambiamento di variabili; inoltre, essendo il lemma banale per $r = 1$, potremo procedere per induzione e dimostrarlo per $r = s + 1$, supponendolo noto per $r = s$. Posto

$$g(x_1, \ldots, x_s) = \int_0^1 f(x_1, \ldots, x_s, x_{s+1})\, dx_{s+1}$$

[‡]Editor's note: published in Ric. di Mat., **4** (1955), 95–113.

[1]Per dominio quadrato avente lati di lunghezza t intendiamo un dominio del tipo $[a_h \leq x_k \leq a_h + t, h = 1, \ldots, r]$.

e detta T^* la sezione di T coll'iperpiano $x_{s+1} = 0$, si ha

$$(3) \qquad \int_T |f(x)|dx_1 \ldots dx_{s+1} \leq \int_{T^*} |g(x_1, \ldots, x_s)|dx_1 \cdots dx_s +$$

$$+ \int_{T^*} dx_1 \ldots dx_s \int_0^1 |f(x_1, \ldots, x_s, x_{s+1}) - g(x_1, \ldots, x_s)|dx_{s+1} \leq$$

$$\leq \int_{T^*} \sum_{h=1}^s \left|\frac{\partial g}{\partial x_h}\right| dx_1 \cdots dx_s + \int_{T^*} dx_1 \cdots dx_s \int_0^1 \left|\frac{\partial f}{\partial x_{s+1}} dx_{s+1}\right| \leq$$

$$\leq \int_T \sum_{h=1}^{s+1} \left|\frac{\partial f}{\partial x_h}\right| dx_1 \ldots dx_{s+1}$$

e il lemma è dimostrato.

LEMMA II. – *Sia E un insieme di S_r avente perimetro e misura entrambi finiti, sia ρ un numero reale positivo e sia*

$$(4) \qquad\qquad\qquad T_1, \ldots, T_m, \ldots$$

una successione di domini quadrati, tutti aventi i lati di lunghezza ρ, privi a due a due di punti interni comuni, la cui somma coincide con l'intero spazio S_r. Esiste allora un insieme R, formato dalla somma di un numero finito (eventualmente nullo) di domini della successione (4) e verificante la

$$(5) \qquad\qquad \text{mis}\left((E \cup R) \setminus (E \cap R)\right) \leq \rho\sqrt{r}P(E),$$

ove con $P(E)$ si indica il perimetro di E.

Dim. Detta $\varphi(x; E)$ la funzione caratteristica di E, consideriamo l'operatore $W_\lambda f(x)$ definito da (vedi [3], n. 1)

$$W_\lambda f(x) = (\pi\lambda)^{-r/2} \int_{S_r} e^{-\frac{|\xi|^2}{\lambda}} f(x + \xi)dx_1 \cdots dx_r \quad \left(|\xi| = \sqrt{\xi_1^2 + \cdots + \xi_r^2}\right)$$

e poniamo

$$(6) \quad \begin{cases} \tau_h^* = \dfrac{\text{mis}\,(E \cap T_h)}{\rho^r}, \quad \tau_h(\lambda) = \dfrac{1}{\rho^r} \displaystyle\int_{T_h} W_\lambda\varphi(x; E)dx_1 \cdots dx_r \\[3ex] \sigma_h^* = \displaystyle\int_{T_h} |\varphi(x; E) - \tau_h^*|dx_1 \cdots dx_r, \quad \sigma_h(\lambda) = \displaystyle\int_{T_h} |W_\lambda\varphi(x; E) \\ \hspace{6cm} -\tau_h(\lambda)|dx_1 \cdots dx_r. \end{cases}$$

Per il lemma I si ha

$$(7) \qquad \sqrt{r}\rho \int_{S_r} |\text{grad}\, W_\lambda\varphi(x; E)|dx_1 \cdots dx_r$$

$$\geq \rho\sum_{h=1}^r \int_{S_r} \left|\frac{\partial}{\partial x_h}W_\lambda\varphi(x; E)\right| dx_1 \cdots dx_r =$$

$$= \sum_{h=1}^\infty \sum_{k=1}^r \rho \int_{T_h} \left|\frac{\partial}{\partial x_k}W_\lambda\varphi(x; E)\right| dx_1 \cdots dx_r \geq \sum_{h=1}^\infty \sigma_h(\lambda).$$

Teniamo ora presente le proprietà dell'operatore W_λ e la definizione di perimetro ([3], n. 1):

$$P(E) = \lim_{\lambda \to 0} \int_{S_r} |\operatorname{grad} W_\lambda \varphi(x; E)| \, dx_1 \cdots dx_r;$$

dalla (7), passando al limite per $\lambda \to 0$, si ottiene

$$(8) \qquad \rho\sqrt{r}\, P(E) \geq \sum_{h=1}^{\infty} \sigma_h^* = 2 \sum_{h=1}^{\infty} \rho^r \tau_h^* (1 - \tau_h^*).$$

Poiché E ha misura finita, saranno in numero finito i domini T_h per cui risulta

$$(9) \qquad \rho^{-r} \operatorname{mis}(E \cap T_h) = \tau_h^* > \frac{1}{2};$$

detto n il loro numero, potremo sempre pensare di avere ordinato la successione (4) in modo che i domini verificanti la (9) siano i primi n. Dalla (8) segue allora la

$$(10) \qquad \begin{aligned} \rho\sqrt{r}\, P(E) \;&\geq\; \sum_{h=1}^{n} \rho^r (1 - \tau_h^*) + \sum_{h=n+1}^{\infty} \rho^r \tau_h^* = \\ &= \sum_{h=1}^{n} \operatorname{mis}(T_h \setminus E) + \sum_{h=n+1}^{\infty} \operatorname{mis}(E \cap T_h) \end{aligned}$$

e quindi, posto $\sum_{h=1}^{n} T_h = R$, la (5).

Riprendiamo ora in considerazione lo spazio metrico Σ introdotto in [3] n. 7; tale spazio aveva come elementi gli insiemi contenuti in S_r e la distanza di due insiemi E_1, E_2, considerati come elementi di Σ, era eguale alla mis $((E_1 \cup E_2) \setminus (E_1 \cap E_2))$. Facendo uso del lemma II e di noti teoremi sugli spazi metrici, proveremo il

TEOREMA I. – *Dati un insieme limitato $L \subset S_r$ ed un numero positivo p, l'aggregato degli insiemi contenuti in L ed aventi perimetro minore di p è compatto nello spazio Σ.*

Dim. Detto H l'aggregato considerato, poiché lo spazio Σ è evidentemente completo, per un noto criterio di compattezza (vedi [4]) basterà provare che H è *iperlimitato* (cioè, comunque si fissi un numero positivo $\eta > 0$, H è contenuto nella somma di un numero finito di insiemi aventi tutti diametro minore di η). A tale scopo osserviamo che, preso un numero positivo ρ arbitrario, è sempre possibile decomporre S_r nella somma di infiniti domini quadrati privi a 2 a 2 di punti interni comuni

$$(11) \qquad\qquad T_1, \ldots, T_n, \ldots$$

tutti aventi lati di lunghezza ρ. Poiché L è per ipotesi limitato, gli insiemi decomponibili nella somma di un numero finito (eventualmente nullo) di domini della successione (11) aventi punti comuni con L saranno essi stessi in numero finito; siano

$$(12) \qquad\qquad R_1,\dots,R_\nu,$$

tali insiemi. Per il lemma II è certo possibile, in corrispondenza ad ogni insieme $E \in H$, trovare un indice m verificante le

$$(13) \qquad \text{mis}\left((R_m \cup E) \setminus (R_m \cap E)\right) \leq \rho\sqrt{r}p, \quad 1 \leq m \leq \nu$$

quindi, data l'arbitrarietà di ρ, H è iperlimitato.

2. Consideriamo una successione di funzioni vettoriali di insieme

$$(1) \qquad\qquad \Theta^{(1)}(B),\dots,\Theta^{(n)}(B),\dots$$

definite per ogni insieme $B \subset S_r$ e completamente additive; siano

$$(2) \qquad\qquad \alpha_1(B),\dots,\alpha_n(B),\dots$$

le rispettive variazioni totali sul generico insieme B. Diremo che la successione (1) è *quasi convergente*, se esistono una funzione vettoriale d'insieme $\Theta(B)$ ed una funzione scalare mai negativa $\alpha^*(B)$, entrambe definite per ogni insieme $B \subset S_r$, completamente additive e verificanti le

$$(3) \qquad\qquad \lim_{n\to\infty} \Theta^{(n)}(L) = \Theta(L),$$

$$(3') \qquad\qquad \lim_{n\to\infty} \alpha_n(L) = \alpha^*(L),$$

per ogni insieme limitato $L \subset S_r$ e soddisfacente la

$$(4) \qquad\qquad \alpha^*(\mathcal{F}L) = 0.$$

È facile constatare che, detta $\alpha(B)$ la variazione totale di Θ sul generico insieme B, si ha sempre

$$(5) \qquad\qquad \alpha(B) \leq \alpha^*(B) \leq \lim_{n\to\infty} \alpha_n(S_r).$$

Da un noto teorema di DE LA VALLÉE POUSSIN (vedi [3] n. 2) segue il

TEOREMA II. – *Data una successione di funzioni vettoriali d'insieme, definite per ogni insieme $B \subset S_r$, completamente additive ed aventi variazioni totali equilimitate, è possibile estrarre da essa una successione subordinata quasi convergente.*

Dim. Detta $\Theta^{(n)}(B)$ la n-sima funzione vettoriale della nostra successione, ogni componente di $\Theta^{(n)}(B)$ è eguale alla differenza di due funzioni scalari d'insieme definite per ogni insieme $B \subset S_r$, completamente additive, mai negative e mai superiori alla variazione totale di $\Theta^{(n)}$ sull'insieme B (che è essa pure una funzione completamente additiva mai negativa). Applicando il citato teorema di De La Vallèe Poussin alle successioni descritte, al variare dell'indice n, da tutte questa funzioni d'insieme completamente additive e mai negative, si ottiene il nostro teorema.

3.   Per ogni insieme $B \subset S_r$ e per ogni numero positivo ρ, chiameremo (seguendo le denominazioni usate in [5]) *involucro aperto di raggio ρ dell'insieme B* e indicheremo col simbolo $I(B; \rho)$ l'insieme dei punti aventi da B distanza minore di ρ; se B è formato da un punto solo x, il suo involucro aperto di raggio ρ verrà indicato anche col simbolo $I(x; \rho)$.

Per ogni insieme $E \subset S_r$ avente perimetro finito, esiste, come dimostrato in [3], n. 2, una funzione vettoriale d'insieme $\Phi(B)$ soddisfacente le condizioni seguenti:

a) $\Phi(B)$, definita per ogni $B \subset S_r$, è completamente additiva ed a variazione totale finita;

b)   presa comunque una funzione $g(x)$ continua in S_r con le sue derivate prime, e con esse infinitesima all'infinito di ordine non inferiore a quello di $|x|^{-(r+1)}$, risulta

$$\int_E \operatorname{grad} g(x)\, dx_1 \cdots dx_r = \int_{S_r} g(x)\, d\Phi.$$

Chiameremo tale funzione vettoriale Φ la *funzione di Gauss-Green relativa all'insieme E*.

Ciò posto, dimostriamo il

LEMMA III. – *Siano dati nello spazio S_r (con $r \geq 2$) un insieme E di perimetro finito ed un punto ξ; allora, per ogni scelta del numero ρ, il perimetro di $E \cap I(\xi; \rho)$ risulta finito. Dette poi $\varphi(x; B)$ la funzione caratteristica del generico insieme $B \subset S_r$, $\Phi(B)$ la funzione di Gauss-Green relativa ad E, $\Psi(B; \rho)$ la funzione di Gauss-Green relativa ad $E \cap I(\xi, \rho)$, risulta, per ogni insieme $B \subset S_r$ e per quasi tutti i valori di ρ*

$$(1) \qquad \Psi(B; \rho) = \Phi[B \cap I(\xi; \rho)] + \int_{\mathcal{F}I(\xi;\rho)} \varphi(x; E \cap B) n(x)\, d\sigma$$

ove $n(x)$ è il versore della normale esterna alla ipersuperficie sferica $\mathcal{F}I(\xi; \rho)$ e $d\sigma$ è l'elemento di misura $(r-1)$–dimensionale su tale ipersuperficie.

Dim. Avendo E perimetro finito, per i teoremi VII, VIII di [3] e per il teor. II potremo trovare una successione di domini poligonali

$$(2) \qquad \Pi_1, \ldots, \Pi_m, \ldots$$

verificanti le

$$(3) \qquad \lim_{m \to \infty} P(\Pi_m) = P(E),$$

(3')
$$\lim_{m\to\infty} \operatorname{mis}\big((E \cup \Pi_m) \setminus (E \cap \Pi_m)\big) = 0$$

e soddisfacenti inoltre le condizioni seguenti: per quasi tutti i valori di ρ risulta

(4)
$$\lim_{m\to\infty} \int_{\mathcal{F}I(\xi;\rho)} |\varphi(x;\Pi_m) - \varphi(x;E)|\, d\sigma = 0;$$

detta $\Phi^{(m)}(B)$ la funzione di Gauss-Green relativa a Π_m, la successione

(5)
$$\Phi^{(1)}(B),\dots,\Phi^{(m)}(B),\dots$$

è quasi convergente verso $\Phi(B)$. Si constata immediatamente che, per ogni valore di ρ, $\Pi_m \cap I(\xi;\rho)$ ha perimetro finito e, detta $\Psi^{(m)}(B;\rho)$ la funzione di Gauss-Green relativa a $\Pi_m \cap I(\xi;\rho)$, risulta

(6)
$$\Psi^{(m)}(B;\rho) = \Psi^{(m)}[B \cap I(\xi;\rho)] + \int_{\mathcal{F}I(\xi;\rho)} \varphi(x;B \cap \Pi_m)n(x)\, d\sigma.$$

Dalla (6) segue immediatamente la

(7)
$$P[\Pi_m \cap I(\xi;\rho)] \leq P(\Pi_m) + \int_{\mathcal{F}I(\xi;\rho)} d\sigma$$

e quindi per le (3), (3') e per il teor. VII di [3], risulta finito il perimetro di $E \cap I(\xi;\rho)$.

Poichè la successione (5) è quasi convergente verso $\Phi(B)$, esisterà una funzione d'insieme completamente additiva e mai negativa $\alpha^*(B)$, tale che, per ogni insieme limitato $L \subset S_r$ soddisfacente la

(8)
$$\alpha^*(\mathcal{F}L) = 0,$$

risulti

(9)
$$\lim_{m\to\infty} \Phi^{(m)}(L) = \Phi(L).$$

Consideriamo ora i numeri positivi ρ che soddisfano la (4) e la

(10)
$$\alpha^*[\mathcal{F}I(\xi;\rho)] = 0;$$

presi uno di tali numeri ed un insieme limitato L verificante la (8), dalle (8), (10) segue

(11)
$$\alpha^*(\mathcal{F}[L \cap I(\xi;\rho)]) = 0$$

e quindi

(12)
$$\lim_{m\to\infty} \Phi^{(m)}(L \cap I(\xi;\rho)) = \Phi(L \cap I(\xi;\rho)).$$

Dalle (4), (6), (12) si deduce poi la

$$(13) \qquad \lim_{m \to \infty} \Psi^{(m)}(L; \rho) = \Phi[L \cap I(\xi; \rho)] + \int_{\mathcal{F}I(\xi;\rho)} \varphi(x; L \cap E) n(x) \, d\sigma$$

e quindi, tenendo conto della (3') e del teor. VII di [3], si trova che la (1) è certo verificata per ogni insieme $B \subset S_r$ e per ogni numero ρ positivo verificante le (4), (10).

4. Nello spazio S_r (con $r \geq 2$) fissiamo un insieme E di perimetro finito; sia al solito $\Phi(B)$ la funzione di Gauss-Green relativa ad E, $\mu(B)$ sia la variazione totale di Φ sul generico insieme $B \subset S_r$. Chiameremo *frontiera ridotta* di E ed indicheremo col simbolo $\mathcal{F}^*E$, l'insieme dei punti $\xi \subset S_r$ che verificano le condizioni seguenti:

per ogni numero positivo ρ si ha

$$(1) \qquad\qquad\qquad \mu[I(\xi; \rho)] > 0,$$

esiste determinato e finito il limite

$$(2) \qquad\qquad\qquad n(\xi) = \lim_{\rho \to 0} \frac{\Phi[I(\xi; \rho)]}{\mu[I(\xi; \rho)]},$$

risulta

$$(3) \qquad\qquad\qquad |n(\xi)| = 1.$$

Seguendo le denominazioni usate in [1] possiamo dire che $\mathcal{F}^*E$ è formata dai centri degli elementi della frontiera orientata di E (vedi [1] n. 10).

Per ogni $\xi \in \mathcal{F}^*E$ indicheremo poi con $A(\xi)$ l'insieme formato dai punti $x \equiv (x_1, \ldots, x_r)$ verificanti la

$$(4) \qquad\qquad\qquad \sum_{h=1}^{r} n_h(\xi)(x_h - \xi_h) > 0,$$

indicheremo invece con $B(\xi)$ l'insieme dei punti $x \equiv (x_1, \ldots, x_r)$ verificanti la

$$(4') \qquad\qquad\qquad \sum_{h=1}^{r} n_h(\xi)(x_h - \xi_h) < 0;$$

considereremo quindi l'iperpiano formato dai punti $x \equiv (x_1, \ldots, x_r)$ verificanti la

$$(5) \qquad\qquad\qquad \sum_{h=1}^{r} n_h(\xi)(x_h - \xi_h) = 0,$$

che verrà indicato col simbolo $S^*(\xi)$. Indicheremo poi con $L(\xi)$ l'insieme dei numeri positivi τ verificanti la condizione seguente: scelto comunque un intero

$n \geq 1$ e posto $\rho = \tau n$, è soddisfatta, per ogni insieme $B \subset S_r$, la relazione (1) del lemma III e si ha inoltre

$$(6) \qquad \mu[\mathcal{F}I(\xi; \rho)] = 0.$$

Infine, per ogni intero positivo m, indicheremo con ω_m la misura di un dominio circolare di raggio unitario dello spazio S_m. Osserviamo che, dalla definizione di $\mathcal{F}^*E$ e da noti teoremi sulle funzioni d'insieme segue la

$$(7) \qquad \mu(S_r \setminus \mathcal{F}^*E) = 0.$$

Stabilite queste notazioni passiamo alla dimostrazione del

LEMMA IV. – *Comunque si fissi un punto $\xi \in \mathcal{F}^*E$ si ha*

$$(8) \qquad \liminf_{\rho \to 0} \rho^{-r} \mathrm{mis}\,[E \cap I(\xi; \rho)] > 0,$$

$$(8') \qquad \liminf_{\rho \to 0} \rho^{-r} \mathrm{mis}\,[(S_r \setminus E) \cap I(\xi; \rho)] > 0,$$

$$(9) \qquad \limsup_{\rho \to 0} \rho^{1-r} P[E \cap I(\xi; \rho)] < +\infty.$$

Dim. Per la definizione di $L(\xi)$, risulta

$$(10) \qquad P[E \cap I(\xi; \rho)] = \mu[I(\xi; \rho)] + \int_{\mathcal{F}I(\xi;\rho)} \varphi(x; E)\, d\sigma$$

per ogni $\rho \in L(\xi)$; poichè d'altra parte un insieme di misura nulla ha anche perimetro nullo, deve essere, per le (1), (10),

$$(11) \qquad \mathrm{mis}\,[E \cap I(\xi; \rho)] > 0$$

per ogni $\rho \in L(\xi)$ (e quindi anche per ogni valore positivo di ρ).

Per ogni $\rho \in L(\xi)$ e per ogni funzione $f(x)$, continua in S_r assieme alle sue derivate parziali prime, che assuma il valore 1 in ogni punto di $I(\xi; \rho)$ e sia identicamente nulla in $S_r \setminus I(\xi; 2\rho)$ si avrà, per la definizione di $L(\xi)$ e per la proprietà b) delle funzioni di GAUSS-GREEN,

$$(12) \qquad \int_{S_r} f(x)\, d\Psi(B; \rho) = \Phi[I(\xi; \rho)] + \int_{\mathcal{F}I(\xi;\rho)} n(x)\varphi(x; E)\, d\sigma = 0,$$

ove al solito si indica con $\Psi(B; \rho)$ la funzione di GAUSS-GREEN relativa ad $E \cap I(\xi; \rho)$.

D'altra parte, per le (1), (2), (3) esiste un numero positivo $\bar{\rho}$ tale che, per $0 < \rho \leq \bar{\rho}$, risulta

$$(13) \qquad 2\,|\Phi[I(\xi; \rho)]| \geq \mu[I(\xi; \rho)];$$

quindi, per tutti i numeri ρ dell'intervallo $(0,\overline{\rho})$ appartenenti ad $L(\xi)$, avremo, in virtù delle (10), (12), (13),

(14)

$$P(E\cap I(\xi;\rho)) \leq \int_{\mathcal{F}I(\xi;\rho)} \varphi(x;E)\,d\sigma + 2\left|\int_{\mathcal{F}I(\xi;\rho)} \varphi(x;E)\,d\sigma\right| \leq 3\int_{\mathcal{F}I(\xi;\rho)} \varphi(x;E)\,d\sigma.$$

Dalle (14), ricordando il teor, VI di [3], si ottiene

$$(15) \qquad 3\int_{\mathcal{F}I(\xi;\rho)} \varphi(x;E)\,d\sigma \geq \left(\text{mis}\,[E\cap I(\xi;\rho)]\right)^{1-\frac{1}{r}}.$$

La funzione $g(\rho) = \text{mis}\,[E\cap I(\xi;\rho)]$ è evidentemente una funzione di ρ assolutamente continua nell'intervallo $(0,\overline{\rho})$, sempre positiva (in virtù della (11)) per $\rho \neq 0$; poiché, per quasi tutti i valori di ρ, si ha

$$(16) \qquad \frac{dg}{d\rho} = \int_{\mathcal{F}I(\xi;\rho)} \varphi(x;E)\,d\sigma,$$

tenendo conto della (15), avremo, in quasi tutto l'intervallo $(0,\overline{\rho})$,

$$(17) \qquad \frac{1}{3} \leq [g(\rho)]^{\frac{1}{r}-1}\frac{dg}{d\rho} = r\frac{d}{d\rho}\left[g^{\frac{1}{r}}\right]$$

e quindi, integrando,

$$(18) \qquad g(\rho) \geq \left(\frac{\rho}{3r}\right)^{r}.$$

Dalla (18) si deduce la (8); la (8') si prova con ragionamento del tutto analogo, dato che $\mathcal{F}^*E$ è anche frontiera ridotta di $S_r \setminus E$. Infine dalla (14), tenendo conto del teor. VII di [3], si deduce che, per *ogni* numero positivo $\rho \leq \overline{\rho}$, si ha

$$(19) \qquad P[E\cap I(\xi;\rho)] \leq 3\int_{\mathcal{F}I(\xi;\rho)} d\sigma = 3\rho^{r-1}\omega_{r-1}$$

e quindi vale la (9).

Dal lemma IV segue il

TEOREMA III. – *Comunque si fissino un punto $\xi \in \mathcal{F}^*E$ ed un numero positivo ε, si ha*

$$(20) \qquad \lim_{\rho\to 0} \rho^{1-r}\mu[I(\xi;\rho)\cap I(S^*(\xi);\varepsilon\rho)] = \omega_{r-1},$$

$$(21) \qquad \lim_{\rho\to 0} \rho^{-r}\text{mis}\,[I(\xi;\rho)\cap A(\xi)\cap E] = 0,$$

$$(21') \qquad \lim_{\rho\to 0} \rho^{-r}\text{mis}\,[I(\xi;\rho)\cap B(\xi)\cap (S_r\setminus E)] = 0.$$

Dim. Dimostreremo il teorema III nell'ipotesi che risulti

$$(22) \qquad n_1(\xi) = 1; \quad n_h(\xi) = 0, \quad \text{per} \quad h = 2, \dots, r;$$

e quindi l'iperpiano $S^*(\xi)$ sia parallelo all'iperpiano $x_1 = 0$; tale ipotesi non è sostanzialmente restrittiva, essendo sempre possibile ricondursi ad essa mediante una semplice rotazione.

Poiché quasi tutti i numeri positivi appartengono ad $L(\xi)$ e le funzioni di ρ che compaiono nelle (20), (21), (21') sono continue a destra, per provare tali relazioni basterà provare che, comunque si fissi una successione

$$(23) \qquad \tau_1, \dots, \tau_n, \dots,$$

di numeri positivi appartenenti ad $L(\xi)$ tendente a 0, si può estrarre dalla (23) una successione subordinata

$$(24) \qquad \rho_1, \dots, \rho_n, \dots$$

verificante le

$$(25) \qquad \lim_{n \to \infty} \rho_n^{1-r} \mu[I(\xi; \rho_n) \cap I(S^*(\xi); \varepsilon \rho_n)] = \omega_{r-1},$$

$$(26) \qquad \lim_{n \to \infty} \rho_n^{-r} \operatorname{mis}[I(\xi; \rho_n) \cap A(\xi) \cap E] = 0,$$

$$(26') \qquad \lim_{n \to \infty} \rho_n^{-r} \operatorname{mis}[I(\xi; \rho)_n \cap B(\xi) \cap (S_r \setminus E)] = 0.$$

Se, per ogni numero reale positivo ρ, indichiamo con E_ρ^* l'insieme ottenuto da E mediante l'omotetia di centro ξ e raggio $1/\rho$ (cioè la corrispondenza che ad un punto di coordinate $(\xi_1 + y_1, \dots, \xi_r + y_r)$ fa corrispondere il punto $(\xi_1 + y_1/\rho, \dots, \xi_r + y_r/\rho))$, mentre con $\Phi^*(B; \rho)$ indicheremo la funzione di GAUSS-GREEN relativa ad E_ρ^* e con $\mu^*(B; \rho)$ la sua variazione totale, su B, la (25) si traduce nella

$$(27) \qquad \lim_{n \to \infty} \mu^*[I(\xi; 1) \cap I(S^*(\xi); \varepsilon); \rho_n] = \omega_{r-1}.$$

Dalle (2), (22) si deduce poi che, per ogni numero positivo σ, si ha

$$(28) \qquad \lim_{\rho \to 0} \frac{\Phi_1^*[I(\xi; \sigma); \rho]}{\mu^*[I(\xi; \sigma); \rho]} = 1.$$

Poiché, per il lemma IV, i perimetri degli insiemi

$$(29) \qquad I(\xi; 2) \cap E_{\tau_1}^*, \dots, I(\xi; 2) \cap E_{\tau_n}^*, \dots$$

sono limitati, essendo evidentemente

$$(30) \qquad P\left[I(\xi; 2) \cap E_{\tau_n}^*\right] = \tau_n^{r-1} P\left[E \cap I(\xi; 2\tau_n)\right],$$

dal teor. I e dal teor. II si deduce che la successione (24) può essere scelta in modo che, posto $D_n = I(\xi; 2) \cap E_{\rho_n}$, la successione

$$(31) \qquad D_1, \ldots, D_n, \ldots$$

converga in media verso un insieme D e, detta $\Psi^{(n)}(B)$ la funzione di Gauss-Green relativa a D_n, risulti quasi convergente la successione

$$(32) \qquad \Psi^{(1)}(B), \ldots, \Psi^{(n)}(B), \ldots$$

Per il teor. VII di [3], la successione (32) sarà quasi convergente verso la funzione di Gauss-Green relativa a D che indicheremo con $\Psi(B)$.

Poiché i numeri della successione (24) appartengono ad $L(\xi)$, si avrà, per ogni intero positivo n e per ogni insieme $B \subset S_r$,

$$(33) \qquad \Psi^{(n)}[B \cap I(\xi; 2)] = \Phi^*[B \cap I(\xi; 2); \rho_n]$$

e quindi, detta $\alpha_n(B)$ la variazione totale di $\Psi^{(n)}$ sul generico insieme B,

$$(34) \qquad \alpha_n[B \cap I(\xi; 2)] = \mu^*[B \cap I(\xi; 2); \rho_n] \, .$$

Poiché i perimetri degli insiemi (31) sono limitati e quindi sono equilimitate le funzioni $\alpha_n(B)$, dalle (28), (33), (34) segue la

$$(35) \qquad \lim_{n \to \infty} \left\{ \alpha_n[I(\xi; 2)] - \Psi_1^{(n)}[I(\xi; 2)] \right\} = 0.$$

La quasi convergenza della successione (32) ci assicura che esiste una funzione $\alpha^*(B)$ tale che, per ogni insieme limitato L verificante la

$$(36) \qquad \alpha^*(\mathcal{F}L) = 0,$$

si abbia

$$(37) \qquad \lim_{n \to \infty} \alpha_n(L) = \alpha^*(L); \qquad \lim_{n \to \infty} \Psi^{(n)}(L) = \Psi(L);$$

dalle (35), (37) si deduce poi che, per ogni insieme $B \subset I(\xi; 2)$, risulta

$$(38) \qquad \Psi_1(B) = \alpha^*(B) \geq 0,$$

$$(38') \qquad \sum_{h=2}^{r} |\Psi_h(B)| = 0.$$

Dalle (38), (38'), tenendo presenti la definizione dell'operatore W_λ e il teorema II di [3], si deduce la

$$(39) \quad 0 \leq \left| \frac{\partial}{\partial x_1} W_\lambda \varphi(x; D) \right| + \frac{\partial}{\partial x_1} W_\lambda \varphi(x; D) + \sum_{h=2}^{r} \left| \frac{\partial}{\partial x_h} W_\lambda \varphi(x; D) \right| \leq$$

$$\leq (1 + \sqrt{r})(\pi\lambda)^{-\frac{r}{2}} \int_{\mathcal{F}I(\xi; 2)} e^{-\frac{|x-\xi|^2}{\lambda}} |d\Psi| \, ,$$

e quindi, in ogni insieme chiuso contenuto in $I(\xi;2)$, sarà verificata uniformemente la

$$(40) \quad \lim_{\lambda \to 0} \left(\left| \frac{\partial}{\partial x_1} W_\lambda \varphi(x;D) \right| + \frac{\partial}{\partial x_1} W_\lambda \varphi(x;D) + \sum_{h=2}^{r} \left| \frac{\partial}{\partial x_h} W_\lambda \varphi(x;D) \right| \right) = 0.$$

Poiché (vedi [3] n. 1) $W_\lambda \varphi(x;D)$ converge in media verso $\varphi(x;D)$, dalla (40) si deduce che esisterà una funzione monotona non crescente $u(t)$, definita nell'intervallo $(-2,2)$, tale che, in quasi tutti i punti di $I(\xi;2)$, risulti

$$(41) \qquad \varphi(x;D) \equiv \varphi(x_1,\ldots,x_r;D) = u(x_1).$$

Essendo la funzione $u(t)$ monotona non crescente, esisterà un numero γ contenuto nell'intervallo $(-2,2)$, tale che risulti

$$(42) \qquad u(t) = 1 \quad \text{per} \quad t < \gamma; \qquad u(t) = 0 \quad \text{per} \quad t > \gamma.$$

È facile provare che deve essere $\gamma = 0$; infatti se fosse $\gamma < 0$ avremmo

$$(43) \qquad 0 = \text{mis} \,[D \cap I(\xi;|\gamma|)] = \lim_{n \to \infty} \text{mis} \,[I(\xi;|\gamma|) \cap E^*_{\rho_n}] =$$
$$= \lim_{n \to \infty} \rho_n^{-r} \text{mis} \,[E \cap I(\xi;|\gamma|\rho_n)]$$

in contrasto con la (8) del lemma IV; analogamente si vede che, per la (8'), non può essere $\gamma > 0$. Essendo $\gamma = 0$, per le (41), (42), D potrà differire dall'insieme $I(\xi;2) \cap B(\xi)$ al più per un insieme di misura nulla; quindi avremo

$$(44) \qquad 0 = \text{mis} \,[D \cap A(\xi) \cap I(\xi;1)] = \lim_{n \to \infty} \text{mis} \,[D_n \cap A(\xi) \cap I(\xi;1)] =$$
$$= \lim_{n \to \infty} \text{mis} \,\left[E^*_{\rho_n} \cap A(\xi) \cap I(\xi;1)\right] = \lim_{n \to \infty} \rho_n^{-r} \text{mis} \,[E \cap A(\xi) \cap I(\xi;\rho_n)],$$

e la (26) è dimostrata; con procedimento analogo si prova la (26').

Per provare la (27), che abbiamo visto equivalere alla (25), cominciamo col notare che, poiché D differisce da $I(\xi;2) \cap B(\xi)$ al più per un insieme di punti di misura nulla, $\Psi(B)$ coinciderà con la funzione di GAUSS-GREEN relativa a $I(\xi;2) \cap B(\xi)$. Sarà pertanto, per ogni numero positivo ε,

$$(45) \qquad \Psi_1 \,[I(\xi;1) \cap I(S^*(\xi);\varepsilon)] = \omega_{r-1},$$

$$(45') \qquad \Psi_1 \,\{\mathcal{F}\,[I(\xi;1) \cap I(S^*(\xi);\varepsilon)]\} = 0$$

e quindi,dalle (37), (38), (45), (45') segue la

$$(46) \qquad \lim_{n \to \infty} \alpha_n \,[I(\xi;1) \cap I(S^*(\xi);\varepsilon)] = \omega_{r-1},$$

da cui per la (34) discendono la (27) e quindi la (25). Poiché le (25), (26), (26') avevano come conseguenza le (20), (21), (21') il nostro teorema è dimostrato.

5. Per completare l'esame delle proprietà della frontiera ridotta di un insieme, dobbiamo premettere alcune definizioni relative alla teoria della misura $(r-1)$–dimensionale degli insiemi non orientati dello spazio euclideo r–dimensionale. Precisamente, dato un insieme H dello spazio S_r (con $r \geq 2$), diremo che H appartiene alla classe Γ_{r-1} se esiste una funzione $f(x)$ definita in un campo A contenente $H \cup \mathcal{F}H$, continua in A con tutte le derivate parziali prime e verificanti le

$$(1) \qquad\qquad f(x) = 0, \qquad |\operatorname{grad} f(x)| \neq 0$$

in tutti i punti di $H \cup \mathcal{F}H$.

Diremo poi che un insieme appartiene a Γ^*_{r-1} se è possibile decomporlo nella somma di un numero finito o di una infinità numerabile di insiemi appartenenti a Γ_{r-1}. È facile constatare che, dato un insieme chiuso e limitato C appartenente a Γ_{r-1}, coincidono le sue misure $(r-1)$–dimensionali secondo Minkovski, Gross, Caratheodory ecc.; il loro comune valore verrà chiamato semplicemente *misura $(r-1)$–dimensionale di C*. Diremo poi *misura $(r-1)$–dimensionale interna* di un qualunque insieme $M \subset S_r$ l'estremo superiore delle misure $(r-1)$–dimensionali degli insiemi chiusi e limitati appartenenti a Γ_{r-1} e contenuti in M. Si vede facilmente che la misura $(r-1)$–dimensionale interna è *la più piccola* funzione d'insieme completamente additiva, mai negativa (eventualmente infinita), che sia definita per ogni insieme $M \subset S_r$ e coincida con la misura $(r-1)$–dimensionale elementarmente definita, quando quest'ultima esiste. È altresì facile vedere che, dato un insieme $M \subset S_r$, esiste sempre un insieme G^* verificante le

$$(2) \qquad\qquad \operatorname{mis}_{r-1} G^* = \operatorname{mis}_{r-1} M; \qquad G^* \subset M, \qquad G^* \in \Gamma^*_{r-1}.$$

Ha interesse per il seguito il

TEOREMA IV. – *Dato un insieme chiuso e limitato C, se esiste una funzione vettoriale $\nu(x) \equiv (\nu_1(x), \ldots, \nu_r(x))$ continua in C, mai nulla e tale che risulti*

$$(3) \qquad\qquad \lim_{|x-y| \to 0} \sum_{h=1}^{r} \nu_h(x)(x_h - y_h)\, |x-y|^{-1} = 0$$

uniformemente al variare di x, y in C, allora C appartiene alla classe Γ_{r-1}.

Dim. Per un teorema di Whitney (vedi [6]) esiste una funzione $f(x)$ continua in S_r con le derivate parziali prime e verificante nei punti di C le

$$(4) \qquad\qquad \frac{\partial f}{\partial x_h} = \nu_h(x); \qquad f(x) = 0.$$

Essendo per ipotesi mai nulla la funzione vettoriale $\nu(x)$, dalle (4) seguono le (1) e il teorema è dimostrato.

Possiamo ora riprendere in esame l'insieme $\mathcal{F}^*E$ considerato nel n. 4 e, conservando tutte le notazioni di tale numero, provare il

TEOREMA V. – *Per ogni insieme $B \subset \mathcal{F}^*E$ risulta*

(5)
$$\operatorname{mis}_{r-1}(B) = \mu(B).$$

Dim. Poiché le (21), (21') del n. 4 sono verificate in tutti i punti di $\mathcal{F}^*E$, per noti teoremi sulle funzioni additive d'insieme potremo decomporre $\mathcal{F}^*E$ nella somma di un insieme N verificante la

(6)
$$\mu(N) = 0$$

e di una infinità numerabile di insiemi chiusi e limitati

(7)
$$C_1, \ldots, C_n, \ldots$$

in ciascuno dei quali le (21), (21') del n. 4 sono verificate uniformemente. Fissiamo l'attenzione su uno di tali insiemi, per esempio C_1; preso un numero positivo $\varepsilon < 1$, sarà certo possibile determinare un numero $\sigma < 1$ tale che, per $\rho \leq 2\sigma$, comunque si prende il punto $\xi \in C_1$, risulti

(8)
$$\operatorname{mis}\,[I(\xi;\rho) \cap A(\xi) \cap E] < \frac{\varepsilon^r \omega_r \rho^r 2^{-r}}{4},$$

(8')
$$\operatorname{mis}\,[I(\xi;\rho) \cap B(\xi) \cap E] > \frac{\omega_r \rho^r}{2} - \frac{\varepsilon^r \omega_r \rho^r 2^{-r}}{4}.$$

Vogliamo ora dimostrare che, presi due punti x ed y entrambi contenuti in C_1 ed aventi distanza minore di σ, deve necessariamente valere

(9)
$$|x - y|^{-1} \left| \sum_{h=1}^{r} n_h(x)(x_h - y_h) \right| < \varepsilon.$$

Supponiamo infatti che la (9) non sia verificata e sia, per esempio,

(10)
$$\sum_{h=1}^{r} n_h(x)(y_h - x_h) > \varepsilon\,|x - y|;$$

essendo $\varepsilon < 1$ e poiché la somma che compare nella (10) rappresenta la distanza di y dall'iperpiano $S^*(x)$, sarà

(11)
$$I(y; \varepsilon\,|x - y|) \subset A(x) \cap I(x; 2\,|x - y|).$$

Tenendo conto delle (8), (8') e ricordando che $|x - y| < \sigma$ ed $\varepsilon < 1$, abbiamo le

(12)
$$\operatorname{mis}\,[A(x) \cap E \cap I(x; 2\,|x - y|)] < \frac{\varepsilon^r \omega_r |x - y|^r}{4},$$

(12')
$$\operatorname{mis}\,[E \cap I(y; \varepsilon\,|x - y|)] > \frac{\varepsilon^r \omega_r |x - y|^r}{4};$$

ma le (12), (12') sono in contrasto con le (11) e quindi è provato che la (10) non può essere verificata per $y \in C_1$, $x \in C_1$, $|x - y| < \sigma$.

Un analogo ragionamento può farsi per la

$$(10') \qquad \sum_{h=1}^{r} n_h(x)(y_h - x_h) < -\varepsilon |x - y|$$

e quindi la (9) è dimostrata. Data l'arbitrarietà di ε possiamo concludere che la

$$(13) \qquad \lim_{|x-y| \to 0} |x - y|^{-1} \sum_{h=1}^{r} \nu_h(x)(x_h - y_h) = 0$$

è verificata uniformemente in C_1 e quindi per il teor. IV, $C_1 \in \Gamma_{r-1}$; potendo lo stesso ragionamento ripetersi per ogni insieme della successione (7) è provato che $(\mathcal{F}^*E \setminus N) \in \Gamma_{r-1}^*$. Tenendo conto della (2) e della (6) e ricordando che la somma di due insiemi appartenenti a Γ_{r-1}^* appartiene ancora a Γ_{r-1}^*, si vede che, per ogni insieme $B \subset \mathcal{F}^*E$, c'è un insieme B^* verificante le

$$(14) \quad \mu(B^*) = \mu(B), \qquad \operatorname{mis}_{r-1} B^* = \operatorname{mis}_{r-1} B, \qquad B^* \subset B, \qquad B^* \in \Gamma_{r-1}^*.$$

Tenendo conto delle (14) e della definizione di Γ_{r-1}^*, possiamo allora concludere che, per provare la (5), basta far vedere che risulta

$$(15) \qquad \operatorname{mis}_{r-1}(G) = \mu(G)$$

per ogni insieme chiuso e limitato appartenente a Γ_{r-1} e contenuto in $\mathcal{F}^*E$.

Per la definizione di Γ_{r-1} esistono un campo $A \supset G$ ed una funzione $f(x)$ continua in A con le sue derivate parziali prime e verificante le (1) in tutti i punti di G. Per la continuità delle derivate di $f(x)$, potremo inoltre trovare un dominio D, contenuto in A ed al quale G sia interno, in modo che in ogni punto di D sia

$$(16) \qquad |\operatorname{grad} f(x)| \neq 0.$$

Detto V l'insieme dei punti di D nei quali $f(x)$ si annulla e posto, per ogni insieme $M \subset S_r$,

$$(17) \qquad \operatorname{mis}_{r-1}(M \cap V) = \gamma(M),$$

è facile constatare che in tutti i punti $x \in G$ si ha

$$(18) \qquad \lim_{\rho \to 0} \rho^{1-r} \gamma \left[I(x; \rho) \right] = \omega_{r-1}$$

e quindi, per il teor. III, essendo $G \subset \mathcal{F}^*E$, sarà

$$(19) \qquad \lim_{\rho \to 0} \frac{\gamma \left[I(x; \rho) \right]}{\mu \left[I(x; \rho) \right]} = 1.$$

Dalle (17), (19), tenendo conto del fatto che $G \subset V$, per noti teoremi sulla derivazione delle funzioni additive d'insieme segue la (15) ed il nostro teorema è dimostrato.

6. Le considerazioni svolte nei nn. 3, 4, 5, riguardavano insiemi di perimetro finito di uno spazio S_r, con $r \geq 2$; il caso $r = 1$, che finora abbiamo tralasciato, è molto più semplice, dato che a caratterizzare completamente gli insiemi aventi perimetro finito è sufficiente il

TEOREMA VI. – *Dato un insieme $E \subset S_1$, avente perimetro finito, esiste un insieme E^* formato dalla somma di un numero finito di intervalli e verificante la condizione*

$$(1) \qquad \mathrm{mis}\left((E \cup E^*) \setminus (E \cap E^*)\right) = 0.$$

Dim. Detta al solito $\varphi(x; E)$ la funzione caratteristica di E, per le proprietà dell'operatore W_λ (vedi [3] n. 1), $W_\lambda \varphi(x; E)$ converge in media verso $\varphi(x; E)$, in ogni insieme limitato, per $\lambda \to 0$. È quindi possibile trovare una successione di numeri positivi

$$\lambda_1 > \lambda_2 > \cdots > \lambda_m > \cdots$$

ed un insieme N di misura nulla tali che risulti

$$(2) \qquad \lim_{m \to \infty} W_{\lambda_m} \varphi(x; E) = \varphi(x; E) \qquad \text{per} \quad x \in S_1 \setminus N.$$

Poiché, per la definizione di $P(E)$, $W_\lambda \varphi(x; E)$ è una funzione a variazione totale non superiore a $P(E)$ (per ogni scelta del numero positivo λ), anche $\varphi(x; E)$ in virtù della (2) avrà variazione totale su $S_1 \setminus N$ non superiore a $P(E)$ (cioè, comunque si prendano $(m+1)$ numeri $x_0 < x_1 < \cdots < x_m$ appartenenti a $S_1 \setminus N$ si ha $\sum_{h=1}^{m} |\varphi(x_h; E) - \varphi(x_{h-1}; E)| \leq P(E)$).

Esiste allora determinato, per ogni punto $x \in S_1$, il limite

$$(3) \qquad \varphi^*(x) = \lim_{\xi \to x^+} \varphi(\xi; E) \qquad (\text{su } S_1 \setminus N),$$

e $\varphi^*(x)$ avrà variazione totale non superiore a $P(E)$; l'insieme E^* di cui $\varphi^*(x)$ è funzione caratteristica sarà pertanto la somma di un numero finito di intervalli aperti a destra. Avendo N misura nulla, per la (3) le funzioni $\varphi^*(x)$ e $\varphi(x; E)$ saranno eguali ovunque e quindi il teorema è dimostrato.

Bibliografia

[1] R. Caccioppoli, *Misure e integrazione sugli insiemi dimensionalmente orientati*. Rend. Acc. Naz. dei Lincei, Serie VIII, Vol. XII, fasc. 1-2, gennaio-febbraio 1952.

[2] E. De Giorgi, *Definizione ed espressione analitica del perimetro di un insieme*. Rend. Acc. Naz. dei Lincei, Serie VIII, vol. XIV, fasc. 3, marzo 1953.

[3] E. De Giorgi, *Su una teoria generale della misura $(r-1)$-dimensionale in uno spazio ad r dimensioni.* Annali di Matematica pura e applicata, Serie IV, Tomo XXXVI, 1954.

[4] M. Picone, *Due conferenze sui fondamenti del calcolo della variazioni.* "Giornale della Matematiche" di Battaglini, S. IV, vol. 80, 1950–51.

[5] M. Picone e G. Fichera, *Trattato di Analisi Matematica.* Vol. I, Ed. Tuminelli, Roma, 1954.

[6] H. Whitney, *Analytic Extensions of differentiable functions defined on closed sets.* Trans. of American Mathematical Society, Vol. 36, 1934.

Some applications of a K-dimensional measure theory to the calculus of variations[‡]

by Ennio De Giorgi (Roma)

The theory to which we refer has been presented for the first time by R. CACCIOPPOLI (see "Misura e integrazione sugli insiemi dimensionalmente orientati" Rend. Lincei Genn. Febbr. 1952) and studied by myself in various papers, some of them already published (see "Su una teoria generale della misura $(r-1)$-dimensionale in uno spazio ad r dimensioni", Ann. di Mat. Pura e Applicata 1954) and others forthcoming. On one hand, this theory allows to extend the notions of K-dimensional measure, K-dimensional tangent space, section, projection, boundary for oriented sets of a very general nature, and on the other hand, it allows one to give a notion of weak convergence of sequences of oriented sets that turns out to be quite useful in the calculus of variations. Indeed, many functionals of the calculus of variations are semicontinuous with respect to this convergence, and moreover, from any sequence of oriented sets having uniformly bounded measures and boundaries it is possible to extract a weakly convergent subsequence. As a consequence, it is possible to apply the direct methods of the calculus of variations to wide classes of problems (among them, the problem of Plateau, the study of the isoperimetric properties of the sphere, etc.) in order to prove the existence of the maximum or of the minimum.

The theory considered here is also useful in the study of the differentiability properties of the solutions of variational problems. In this connection, let me quote a result concerning a problem proposed to me by Prof. G. STAMPACCHIA. Let $u(x) \equiv u(x_1,\ldots,x_r)$ be a function defined in a subset A of the Euclidean space S_r, absolutely continuous on almost all the segments parallel to the coordinate axes, with first order partial derivatives that are square summable in A. Let $f(y_1,\ldots,y_r)$ be a real analytic function in S_r, satisfying, for every point $y \in S_r$ and for every vector $\lambda \equiv (\lambda_1,\ldots,\lambda_r)$, the inequalities

$$\mu|\lambda|^2 \leq \sum_{h,k}^{1,r} \frac{\partial^2 f}{\partial y_h \partial y_k}\lambda_h\lambda_k \leq M|\lambda|^2$$

with μ, M positive constants. If $u^*(x)$ is an extremal of the functional

$$\int_A f\Big(\frac{\partial u}{\partial x_1},\ldots,\frac{\partial u}{\partial x_r}\Big)\,dx_1\ldots dx_r$$

[‡]Editor's note: translation into English of "Alcune applicazioni al Calcolo delle Variazioni di una teoria della misura k-dimensionale", published in Atti V congresso UMI, (Pavia-Torino, 1955), Cremonese, Roma, 1956.

then $u^*(x)$ is real analytic in A (or, at least, there is a real analytic function in A which coincides with $u^*(x)$ almost everywhere).

The content of this communication has been partially published with details in the note "Sull'analiticità delle estremali degli integrali multipli", Rend. Acc. Naz. dei Lincei S. VIII, Vol. XX, fasc. 4 – Aprile 1956.

Alcune applicazioni al calcolo delle variazioni di una teoria della misura K-dimensionale[‡]

di Ennio De Giorgi (a Roma)

La teoria cui ci riferiamo è stata esposta per la prima volta da R. CAC-CIOPPOLI (v. "Misura e integrazione sugli insiemi dimensionalmente orientati" Rend. Lincei Genn. Febbr. 1952) e da me trattata in vari lavori già pubblicati (v. "Su una teoria generale della misura $(r-1)$-dimensionale in uno spazio ad r dimensioni", Ann. di Mat. Pura e Applicata 1954) o in via di pubblicazione. Tale teoria consente da una parte di estendere ad insiemi orientati molto generali le nozioni di misura K-dimensionale, spazio K-dimensionale tangente, sezione, proiezione, bordo, dall'altra di dare una nozione di convergenza debole di successioni di insiemi orientati assai utile nel calcolo delle variazioni. Infatti molti funzionali del calcolo delle variazioni sono semicontinui rispetto a tale convergenza, mentre da ogni successione di insiemi orientati aventi, coi loro bordi, misure equilimitate può estrarsi una successione subordinata debolmente convergente. Ne segue la possibilità di applicare a larghe classi di problemi (fra cui rientrano come casi particolari il problema di Plateau, lo studio delle proprietà isoperimetriche della sfera ecc ...) i metodi diretti del calcolo delle variazioni, per stabilire l'esistenza del massimo o del minimo.

La teoria considerata è utile anche nello studio delle proprietà differenziali delle soluzioni dei problemi variazionali; citerò in proposito un risultato relativo ad un problema propostomi dal Prof. G. Stampacchia. Sia $u(x) \equiv u(x_1, \ldots, x_r)$ una funzione definita in un campo A dello spazio euclideo S_r, assolutamente continua su quasi tutti i segmenti paralleli agli assi coordinati, avente derivate parziali prime di quadrato sommabile in A. Sia $f(y_1, \ldots, y_r)$ una funzione analitica in S_r, verificante per ogni punto $y \in S_r$ e per ogni vettore $\lambda \equiv (\lambda_1, \ldots, \lambda_r)$ la relazione

$$\mu|\lambda|^2 \le \sum_{h,k}^{1,r} \frac{\partial^2 f}{\partial y_h \partial y_k} \lambda_h \lambda_k \le M|\lambda|^2$$

con μ, M costanti positive. Se $u^*(x)$ è un'estremale del funzionale

$$\int_A f\left(\frac{\partial u}{\partial x_1}, \ldots, \frac{\partial u}{\partial x_r}\right) dx_1 \ldots dx_r,$$

allora $u^*(x)$ è analitica in A (o almeno esiste una funzione analitica in A e quasi ovunque uguale ad $u^*(x)$).

[‡]Editor's note: published in Atti V congresso UMI, (Pavia-Torino, 1955), Cremonese, Roma, 1956.

Il contenuto della comunicazione è stato in parte pubblicato per esteso nella nota "Sull'analiticità delle estremali degli integrali multipli", Rend. Acc. Naz. dei Lincei S. VIII, Vol. XX, fasc. 4 – Aprile 1956.

On the differentiability and the analyticity of extremals of regular multiple integrals[‡][†]

Memoir by Ennio De Giorgi[*]

Summary. We study the extremals of some regular multiple integrals: assuming that the first order derivatives exist and are square-summable, we show that they are Hölder continuous. It follows that the extremals are infinitely differentiable and real analytic.

In this paper I deal with the differentiability properties of the extremals of regular multiple integrals, and in particular with their analyticity. This topic has been the object of several investigations of both Italian and foreign mathematicians, and hence it appears to be quite difficult to give a complete bibliographical account. For this reason, we shall limit ourselves to quote a few papers where the reader can find further information. Let us only mention the results by Hopf [3][1], Stampacchia [9], Morrey [6], who give differentiability and analyticity results for less and less regular extremals. In particular, in [3] the existence and Hölder continuity of second order derivatives, in [9] of first order derivatives, in [6] the existence and continuity of first order derivatives is assumed. The results obtained by Stampacchia in [9] belong to another direction of research. He moves from existence theorems obtained by the direct methods of the calculus of variations, where solutions are found in very wide classes of functions, and studies the properties of these (a priori very little regular) solutions. Among other results, he proves the existence of square-summable second order derivatives, satisfying the Euler equation almost everywhere.

What was still missing, to my knowledge (with the exception of double integrals, see [2], [5], [7], [8], and some particular cases of multiple integral, as quadratic integrals, which give rise to linear Euler equations), were theorems which could bridge the gap between the results obtained in the first research line and those in the second, i.e., theorems ensuring that the solutions obtained by direct methods and studied in [9] satisfy the conditions required in [6]. The aim of this paper is to show a first theorem in this direction (see Theorem III[2]). Its proof is based on the study of some functions (characterized by certain integral inequalities) which are Hölder continuous (see Theorem I). Among the

---

[‡]Editor's note: translation into English of the paper "Sulla differenziabilità e l'analiticità delle estremali degli integrali multipli regolari", published in Mem. Accad. Sci. Torino Cl. Sci. Fis. Mat. Nat., **(3) 3** (1957), 25–43.

[†]Work performed at the Istituto Nazionale per le Applicazioni del Calcolo

[*]Presented by the Socio nazionale non residente MAURO PICONE in the meeting of April 25, 1957

[1]Numbers in brackets refer to the Bibliography at the end of this paper.

[2]This theorem has been presented in a talk at the U.M.I. meeting held in Pavia from 6 to 11 October, 1955 and also in the preliminary Note [1].

intermediate results, let us mention Theorem II, because it could be interesting also in other problems concerning elliptic partial differential equations.

This research has been suggested to me by some conversations with Prof. G. Stampacchia. I am grateful to him for the information and the advice he gave me, that have been very useful.

1. – Let us consider an open subset E of the Euclidean r-dimensional space S_r, and let us denote by $\mathcal{U}^{(2)}(E)$ the class of the functions $w(x)$ almost continuous in E which satisfy the following conditions:

1st)   *$w(x)$ is absolutely continuous on almost all segments contained in E parallel to the coordinate axes.*

2nd)   *$w(x)$ and its first partial derivatives are square-summable in every compact subset of E.*

Given a positive number γ, we denote by $\mathcal{B}(E;\gamma)$ the class of the functions $w(x)$ which, beside conditions 1st) and 2nd), satisfy also the following

3rd)   *Given $y \in E$ [3] (whose distance from $S_r \setminus E$ is denoted by $\delta(y)$) and given three numbers k, ϱ_1, ϱ_2 such that $0 < \varrho_1 < \varrho_2 < \delta(y)$ the inequalities*

$$(1) \qquad \frac{\gamma}{(\varrho_2 - \varrho_1)^2} \int_{A(k) \cap I(\varrho_2; y)} (w(x) - k)^2 \, dx_1 \ldots dx_r \geq$$

$$\geq \int_{A(k) \cap I(\varrho_1; y)} |\operatorname{grad} w|^2 \, dx_1 \ldots dx_r,$$

$$(1') \qquad \frac{\gamma}{(\varrho_2 - \varrho_1)^2} \int_{B(k) \cap I(\varrho_2; y)} (w(x) - k)^2 \, dx_1 \ldots dx_r \geq$$

$$\geq \int_{B(k) \cap I(\varrho_1; y)} |\operatorname{grad} w|^2 \, dx_1 \ldots dx_r,$$

hold, where $I(\varrho; y)$ denotes the ball with centre y and radius ϱ, $A(k)$ denotes the subset of E where $w(x) > k$, and $B(k)$ the subset of E where $w(x) < k$.

A first property of the class $\mathcal{B}(E;\gamma)$ just defined, which will be useful later, is given in the following

Lemma I. – *Let a sequence of functions*

$$(2) \qquad w_1(x), \ldots, w_n(x), \ldots$$

in $\mathcal{B}(E;\gamma)$ be given, with $|w_n(x)|^2$ summable in E for every n. If (2) converges in quadratic mean in E to a function $w(x)$, i.e.

$$(3) \qquad \lim_{n \to \infty} \int_E [w_n(x) - w(x)]^2 \, dx_1 \ldots dx_r = 0,$$

then $w(x)$ belongs to $\mathcal{B}(E;\gamma)$.

---

[3] The symbol $y \in E$ means: y belongs to E; the symbol $E \subset L$ means: E is contained in L.

Proof. Since the functions in (2) belong to $\mathcal{B}(E;\gamma)$, for every $y \in E$, $0 < \varrho_1 < \varrho_2 < \delta(y)$ we have

$$(4) \qquad \int_{I(\varrho_1;y)} |\mathrm{grad}\, w_n|^2 \, dx_1 \ldots dx_r \leq \frac{\gamma}{(\varrho_2 - \varrho_1)^2} \int_{I(\varrho_2;y)} |w_n(x)|^2 \, dx_1 \ldots dx_r$$

and then, by (3)

$$(5) \qquad \limsup_{n \to \infty} \int_{I(\varrho_1;y)} |\mathrm{grad}\, w_n|^2 \, dx_1 \ldots dx_r \leq$$
$$\leq \frac{\gamma}{(\varrho_2 - \varrho_1)^2} \int_{I(\varrho_2;y)} |w(x)|^2 \, dx_1 \ldots dx_r.$$

From (5) and the arbitrariness of y, ϱ_1, ϱ_2, we deduce that for every compact $C \subset E$, the sequence of the integrals of the norms of the gradients of the functions in (2) over C is bounded. From known properties of sequences of functions with square-summable first derivatives, it follows that $w(x)$ belongs to $\mathcal{U}^{(2)}(E)$.

In order to prove that $w(x)$ satisfies (1), it suffices to notice that, setting for any real number k and for any positive integer n

$$(6) \qquad w_n(x; k) = \begin{cases} w_n(x) - k & \text{if} \quad w_n(x) \geq k \\ 0 & \text{if} \quad w_n(x) \leq k, \end{cases}$$

the sequence

$$(7) \qquad w_1(x; k), \ldots, w_n(x; k), \ldots$$

converges in quadratic mean in E to the function $w(x; k)$ defined by

$$(8) \qquad w(x; k) = \begin{cases} w(x) - k & \text{if} \quad w(x) \geq k \\ 0 & \text{if} \quad w(x) \leq k. \end{cases}$$

For every $y \in E$, $0 < \varrho < \delta(y)$, we have thus

$$(9) \qquad \lim_{n \to \infty} \int_{I(\varrho;y)} [w_n(x; k)]^2 \, dx_1 \ldots dx_r = \int_{I(\varrho;y)} [w(x; k)]^2 \, dx_1 \ldots dx_r,$$

$$(10) \qquad \liminf_{n \to \infty} \int_{I(\varrho;y)} |\mathrm{grad}_x w_n(x; k)|^2 \, dx_1 \ldots dx_r \geq$$
$$\geq \int_{I(\varrho;y)} |\mathrm{grad}_x w(x; k)|^2 \, dx_1 \ldots dx_r;$$

from (9), (10), since by assumption the functions $w_n(x)$ belong to $\mathcal{B}(E;\gamma)$, we deduce that $w(x)$ satisfies (1). We could argue in the same way for (1') and then it is proved that $w(x) \in \mathcal{B}(E;\gamma)$.

2. – Assume that an open set $E \subset S_r$, a positive constant γ, a function $w(x) \in \mathcal{B}(E; \gamma)$ and a point $y \in E$ have been fixed; with the same notation as in Section 1, let us set

$$
\begin{aligned}
I(\varrho) &= I(\varrho; y) \\
A(k; \varrho) &= A(k) \cap I(\varrho; y) \\
B(k; \varrho) &= B(k) \cap I(\varrho; y).
\end{aligned}
$$

Under these assumptions, that will be kept throughout this section, let us prove the following

LEMMA II. – *There exists a constant β_1 such that for every ϱ, k, λ verifying the inequalities*

(1) $$0 < \varrho < \delta(y), \qquad k < \lambda,$$

we have

(2) $$\beta_1 \int_{[A(k;\varrho) \setminus A(\lambda;\varrho)]} \big|\operatorname{grad} w(x)\big| \, dx_1 \ldots dx_r \geq (\lambda - k)\big[\tau(k, \lambda; \varrho)\big]^{1 - \frac{1}{r}},$$

where $\tau(k, \lambda; \varrho) = \min\{\operatorname{meas} A(\lambda; \varrho), \ \operatorname{meas}[I(\varrho) \setminus A(k; \varrho)]\}$.

Proof. For every $L \subset S_r$, contained in the union of a finite number of hyperplanes and spherical surfaces, let us denote by $\mu_{r-1} L$ the $(r-1)$-dimensional measure of L, defined in an elementary way.

For every domain D in S_r we shall call *semilinear* in D the functions $g(x) \equiv g(x_1, ..., x_r)$ which are continuous in D and enjoy the following property: it is possible to divide D in the union of a finite number of domains, in each of which $g(x)$ is affine (i.e., constant or equal to a first order polynomial in the variables $x_1, \ldots, x_r$); geometrically, this property can be expressed by saying that the hypersurface in the $(r+1)$-dimensional space S_{r+1} defined by the equation

(3) $$x_{r+1} = g(x_1, ..., x_r); \qquad (x_1, ..., x_r) \in D$$

is contained in the union of a finite number of hyperplanes in S_{r+1}.

Under these conventions, let us prove the statement when $w(x)$ is semilinear in $[I(\varrho) \cup \mathcal{F}I(\varrho)]$; in this case, the boundaries $\mathcal{F}A(t; \varrho)$, $\mathcal{F}B(t; \varrho)$ of the domains $A(t; \varrho)$, $B(t; \varrho)$ are contained in the union of a finite number of hyperplanes and the spherical surface $\mathcal{F}I(\varrho)$. Moreover, for almost every t, we have

(4) $$\mu_{r-1}\mathcal{F}A(t; \varrho) = \mu_{r-1}[I(\varrho) \cap \mathcal{F}A(t)] + \mu_{r-1}[A(t) \cap \mathcal{F}I(\varrho)],$$

(4') $$\mu_{r-1}\mathcal{F}B(t; \varrho) = \mu_{r-1}[I(\varrho) \cap \mathcal{F}B(t)] + \mu_{r-1}[B(t) \cap \mathcal{F}I(\varrho)],$$

(5) $$\operatorname{meas} I(\varrho) = \operatorname{meas} A(t; \varrho) + \operatorname{meas} B(t; \varrho)$$

(6) $$\mu_{r-1}\mathcal{F}I(\varrho) = \mu_{r-1}[B(t) \cap \mathcal{F}I(\varrho)] + \mu_{r-1}[A(t) \cap \mathcal{F}I(\varrho)]$$

(7) $$I(\varrho) \cap \mathcal{F}B(t) = I(\varrho) \cap \mathcal{F}A(t).$$

Since $w(x)$ is semilinear, for every ξ we have

(8) $$\int_{A(\xi;\varrho)} \big|\operatorname{grad} w\big|\, dx_1 \ldots dx_r = \int_{\xi}^{+\infty} \mu_{r-1}\big[I(\varrho) \cap \mathcal{F}A(t)\big]\, dt.$$

On the other by the known[4] isoperimetric properties of hyperspheres, there is a constant $\alpha(r)$ such that

(9) $$\begin{cases} \big[\operatorname{meas} I(\varrho)\big]^{1-\frac{1}{r}} = \alpha(r)\mu_{r-1}\mathcal{F}I(\varrho) \\ \big[\operatorname{meas} A(t;\varrho)\big]^{1-\frac{1}{r}} \le \alpha(r)\mu_{r-1}\mathcal{F}A(t;\varrho) \\ \big[\operatorname{meas} B(t;\varrho)\big]^{1-\frac{1}{r}} \le \alpha(r)\mu_{r-1}\mathcal{F}B(t;\varrho) \end{cases}$$

and then, taking into account (4), (4') ,(5), (6), (7), (9), for almost every t we have

(10) $$2\alpha(r)\mu_{r-1}\big[I(\varrho) \cap \mathcal{F}A(t)\big] \ge \big[\operatorname{meas} A(t;\varrho)\big]^{1-\frac{1}{r}} +$$
$$+\big(\operatorname{meas}\big[I(\varrho) \setminus A(t;\varrho)\big]\big)^{1-\frac{1}{r}} - \big[\operatorname{meas} I(\varrho)\big]^{1-\frac{1}{r}}.$$

Setting $\tau(t;\varrho) = \min\big\{\operatorname{meas} A(t;\varrho),\ \operatorname{meas}\big[I(\varrho) \setminus A(t;\varrho)\big]\big\}$, we have $\operatorname{meas} I(\varrho) \ge 2\tau(t;\varrho)$, and (10) reads

(11) $$2\alpha(r)\mu_{r-1}\big[I(\varrho) \cap \mathcal{F}A(t)\big] \ge$$
$$\ge \big[\tau(t;\varrho)\big]^{1-\frac{1}{r}} + \big[\operatorname{meas} I(\varrho) - \tau(t;\varrho)\big]^{1-\frac{1}{r}} - \big[\operatorname{meas} I(\varrho)\big]^{1-\frac{1}{r}} \ge$$
$$\ge 2\big[\tau(t;\varrho)\big]^{1-\frac{1}{r}} - \big[2\tau(t;\varrho)\big]^{1-\frac{1}{r}}.$$

Setting

(12) $$\beta_1 = \frac{\alpha(r)}{1 - 2^{-1/r}}$$

and taking into account that for $\lambda \ge t \ge k$ the inequality $\tau(t,\varrho) \ge \tau(k,\lambda;\varrho)$ holds, from (8), (11), (12) we deduce (2).

Let us now prove the statement when $w(x)$ is not semilinear; then, by known results on linear approximation, we can find a sequence of semilinear functions in the domain $\big[I(\varrho) \cup \mathcal{F}I(\varrho)\big]$

$$w_1(x), w_2(x), \ldots, w_n(x), \ldots,$$

such that

(13) $$\lim_{n \to \infty} \int_{I(\varrho)} \big|w(x) - w_n(x)\big|\, dx_1 \ldots dx_r =$$
$$= \lim_{n \to \infty} \int_{I(\varrho)} \big|\operatorname{grad} w - \operatorname{grad} w_n\big|\, dx_1 \ldots dx_r = 0.$$

---

[4]Notice that here only domains whose boundaries are contained in the union of a finite number of hyperplanes and spherical hypersurfaces are involved.

Calling $A_n(t; \varrho)$ the set of points $x \in I(\varrho)$ where $w_n(x) > t$, for every pair of numbers k, λ (with $k < \lambda$), we set $\tau_n(k, \lambda; \varrho) = \min\{\text{meas } A_n(\lambda; \varrho),\ \text{meas}\,[I(\varrho) \setminus A_n(k; \varrho)]\}$.

An argument similar to the one used to prove (2) with $w(x)$ semilinear allows us to prove that

$$(14) \qquad \beta_1 \int_{[A_n(k;\varrho)\setminus A_n(\lambda;\varrho)]} |\text{grad } w_n|\, dx_1 \ldots dx_r \geq (\lambda - k)[\tau_n(k, \lambda; \varrho)]^{1-\frac{1}{r}}$$

for every positive integer n and for every pair of real numbers k, λ (with $k < \lambda$). Moreover, from (13) it follows, for almost every t, that

$$(15) \qquad \lim_{n\to\infty} \text{meas}\,\big[A_n(t; \varrho) \cup A(t; \varrho) \setminus A_n(t; \varrho) \cap A(t; \varrho)\big] = 0$$

and then, from (13), (14), (15) we deduce that (2) holds for almost every pair k, λ (with $k < \lambda$). It is also easily seen that if (2) is true for almost all pairs k, λ (with $k < \lambda$) then it is always true. For, it suffices to notice that the following inequalities hold

$$(16) \qquad \lim_{\eta\to 0^+} \int_{[A(k+\eta;\varrho)\setminus A(\lambda-\eta;\varrho)]} |\text{grad } w|\, dx_1 \ldots dx_r \leq$$

$$\leq \int_{[A(k;\varrho)\setminus A(\lambda;\varrho)]} |\text{grad } w|\, dx_1 \ldots dx_r,$$

$$(17) \qquad \lim_{\eta\to 0^+} \tau(k + \eta, \lambda - \eta; \varrho) \geq \tau(k, \lambda; \varrho).$$

Lemma III. – *There exists a constant β_2 such that, for every pair of real numbers ϱ, k such that*

$$(18) \qquad 0 < \varrho < \delta(y), \quad 0 < 2 \text{ meas } A(k; \varrho) \leq \text{meas } I(\varrho),$$

the inequality

$$(19) \qquad \beta_2 \int_{A(k;\varrho)} |\text{grad } w|^2\, dx_1 \ldots dx_r \geq$$

$$\geq \big[\text{meas } A(k; \varrho)\big]^{-\frac{2}{r}} \int_{A(k;\varrho)} (w(x) - k)^2\, dx_1 \ldots dx_r.$$

holds.

Proof. For every positive integer n, let λ_n be the minimum λ such that

$$(20) \qquad \text{meas } A(k; \varrho) \geq 2^{nr} \text{meas } A(\lambda; \varrho);$$

set also $\lambda_0 = k$. Then, obviously,

$$(21) \qquad k = \lambda_0 \leq \lambda_1 \leq \ldots \leq \lambda_n \leq \ldots$$

and, by the definition of the sets $A(\lambda; \varrho)$ and $B(\lambda; \varrho)$, we have

$$(22) \qquad \lim_{\varepsilon \to 0^+} \operatorname{meas} A(\lambda_n + \varepsilon; \varrho) = \operatorname{meas} A(\lambda_n; \varrho) \le 2^{-rn} \operatorname{meas} A(k; \varrho) \le$$

$$\le \operatorname{meas}\left[I(\varrho) \setminus B(\lambda_n; \varrho)\right] = \lim_{\varepsilon \to 0^+} \operatorname{meas} A(\lambda_n - \varepsilon; \varrho).$$

From (22) it follows that there is a sequence of sets

$$(23) \qquad A(k; \varrho) = D_0 \supset D_1 \supset \ldots \supset D_n \supset \ldots$$

such that for every n we have

$$(24) \qquad \operatorname{meas} D_n = 2^{-nr} \operatorname{meas} A(k; \varrho)$$

$$(24') \qquad \left[I(\varrho) \setminus B(\lambda_n; \varrho)\right] \supset D_n \supset A(\lambda_n; \varrho).$$

From Lemma II and (18), (22), (24') it follows

$$(25) \qquad \beta_1 \int_{(D_{n-1} \setminus D_n)} |\operatorname{grad} w| \, dx_1 \ldots dx_r \ge$$

$$\ge \lim_{\varepsilon \to 0_+} \beta_1 \int_{[A(\lambda_{n-1}+\varepsilon;\varrho) \setminus A(\lambda_n - \varepsilon;\varrho)]} |\operatorname{grad} w| \, dx_1 \ldots dx_r \ge$$

$$\ge 2^{-n(r-1)} (\lambda_n - \lambda_{n-1}) \left[\operatorname{meas} A(k; \varrho)\right]^{\frac{r-1}{r}}$$

and then, by Schwarz inequality and (24), (25), we have

$$(26) \qquad \beta_1^2 \int_{(D_{n-1} \setminus D_n)} |\operatorname{grad} w|^2 \, dx_1 \ldots dx_r \ge$$

$$\ge \left[\operatorname{meas}\left(D_{n-1} \setminus D_n\right)\right]^{-1} \left(\beta_1 \int_{(D_{n-1} \setminus D_n)} |\operatorname{grad} w| \, dx_1 \ldots dx_r\right)^2 \ge$$

$$\ge (\lambda_n - \lambda_{n-1})^2 \left[\operatorname{meas} A(k; \varrho)\right]^{\frac{r-2}{r}} 2^{-n(r-2)-r}.$$

From (23), (26) it follows

$$(27) \qquad \beta_1^2 \int_{A(k;\varrho)} |\operatorname{grad} w|^2 \, dx_1 \ldots dx_r \ge$$

$$\ge \sum_{n=1}^{\infty} (\lambda_n - \lambda_{n=1})^2 \left[\operatorname{meas} A(k; \varrho)\right]^{\frac{r-2}{r}} 2^{-n(r-2)-r},$$

and from (21), (23), (24), (24') it follows

$$\int_{A(k;\varrho)} (w(x) - k)^2 \, dx_1 \ldots dx_r = \sum_{n=1}^{\infty} \int_{(D_{n-1} \setminus D_n)} (w(x) - k)^2 \, dx_1 \ldots dx_r \le$$

$$(28) \quad \le \sum_{n=1}^{\infty} 2^{r-rn} (\lambda_n - \lambda_0)^2 \operatorname{meas} A(k; \varrho).$$

Let us denote by θ the maximum among the numbers

$$(29) \qquad (\lambda_n - \lambda_{n-1})^2 \, 2^{-n(r-2)-r} \qquad (n = 1, 2, \ldots).$$

Then, for every positive integer m, we have

$$(30) \qquad (\lambda_m - \lambda_0)^2 = \left[\sum_{n=1}^{m} (\lambda_n - \lambda_{n-1}) \right]^2 \leq \theta m^2 2^{m(r-2)+r},$$

and from (27) we deduce

$$(31) \qquad \left[\operatorname{meas} A(k; \varrho) \right]^{\frac{2-r}{r}} \beta_1^2 \int_{A(k;\varrho)} |\operatorname{grad} w|^2 \, dx_1 \ldots dx_r \geq \theta.$$

From (28), (30) it follows

$$(32) \qquad \int_{A(k;\varrho)} (w(x) - k)^2 \, dx_1 \ldots dx_r \leq \theta \operatorname{meas} A(k; \varrho) \sum_{n=1}^{\infty} n^2 2^{2(r-n)}$$

so that, setting

$$(33) \qquad \beta_1^2 \sum_{n=1}^{\infty} 2^{2(r-n)} n^2 = \beta_2,$$

from (31), (32), (33) inequality (19) follows.

LEMMA IV. – *For every positive $\sigma < 1$ there exists a positive $\theta(\sigma)$ such that, from*

$$(34) \qquad \operatorname{meas} A(k; \varrho) < \varrho^r \theta(\sigma), \quad 0 < \varrho < \delta(y), \quad -\infty < k < +\infty$$

it follows

$$(35) \qquad \operatorname{meas} A(k + \sigma c; \varrho - \sigma \varrho) = 0,$$

where

$$(36) \qquad c = \varrho^{-\frac{r}{2}} [\theta(\sigma)]^{-\frac{1}{2}} \left[\int_{A(k;\varrho)} (w(x) - k)^2 \, dx_1 \ldots dx_r \right]^{\frac{1}{2}}.$$

Proof. Let $\theta(\varrho)$ be the maximum θ such that the inequalities

$$(37) \qquad \operatorname{meas} I(\varrho - \sigma \varrho) \geq 2\theta \varrho^r$$

$$(38) \qquad \theta \leq \left[\frac{\sigma^2 2^{-(r+2)}}{1 + \gamma + \beta_2} \right]^r,$$

hold, where γ and β_2 are the constants in (1) of Section 1 and in (19); assuming (34), (36) we must prove that (35) hold. To this aim, we set

$$(39) \qquad \varrho_h = \varrho - \sigma\varrho + 2^{-h}\sigma\varrho, \qquad k_h = k + \sigma c - 2^{-h}\sigma c.$$

By property 3rd) of functions in $\mathcal{B}(E;\gamma)$ (see Section 1), we have

$$(40) \qquad \gamma 4^{h+1}\int_{A(k_h;\varrho_h)}(w(x)-k_h)^2\,dx_1\ldots dx_r \geq$$

$$\geq \varrho^2\sigma^2\int_{A(k_h;\varrho_{h+1})}|\operatorname{grad} w|^2\,dx_1\ldots dx_r,$$

and, by (34), (37), (39) and Lemma III we have

$$(41) \qquad \beta_2\int_{A(k_h;\varrho_{h+1})}|\operatorname{grad} w|^2\,dx_1\ldots dx_r\big[\operatorname{meas} A(k_h;\varrho_h)\big]^{\frac{2}{r}} \geq$$

$$\geq \beta_2\int_{A(k_h;\varrho_{h+1})}|\operatorname{grad} w|^2\,dx_1\ldots dx_r\big[\operatorname{meas} A(k_h;\varrho_{h+1})\big]^{\frac{2}{r}} \geq$$

$$\geq \int_{A(k_h;\varrho_{h+1})}(w(x)-k)^2\,dx_1\ldots dx_r.$$

From (40), (41) it follows

$$(42) \quad \beta_2\gamma\sigma^{-2}\varrho^{-2}4^{(h+1)}\big[\operatorname{meas} A(k_h;\varrho_h)\big]^{\frac{2}{r}}\int_{A(k_h;\varrho_h)}(w(x)-k_h)^2\,dx_1\ldots dx_r \geq$$

$$\geq \int_{A(k_h;\varrho_{h+1})}(w(x)-k_h)^2\,dx_1\ldots dx_r \geq$$

$$\geq \int_{A(k_{h+1};\varrho_{h+1})}(w(x)-k_{h+1})^2\,dx_1\ldots dx_r.$$

Moreover, since by (39) we know that

$$(43) \qquad \int_{A(k_h;\varrho_{h+1})}(w(x)-k_h)^2\,dx_1\ldots dx_r \geq \sigma^2 c^2 4^{-(h+1)}\operatorname{meas} A(k_{h+1};\varrho_{h+1}),$$

from (42), (43) it follows

$$(44) \qquad \sigma^4 c^2\varrho^2 2^{-4(h+1)}\operatorname{meas} A(k_{h+1};\varrho_{h+1}) \leq$$

$$\leq \beta_2\gamma\big[\operatorname{meas} A(k_h;\varrho_h)\big]^{\frac{2}{r}}\int_{A(k_h;\varrho_h)}(w(x)-k_h)^2\,dx_1\ldots dx_r.$$

Arguing by induction, let us now prove that for every nonnegative integer ℓ we have

$$(45) \qquad \operatorname{meas} A(k_\ell;\varrho_\ell) \leq \varrho^r\theta(\sigma)2^{-2r\ell}$$

$$(46) \qquad \int_{A(k_\ell;\varrho_\ell)} (w(x) - k_\ell)^2 \, dx_1 \ldots dx_r \leq \varrho^r \theta(\sigma) c^2 2^{-2r\ell}.$$

Inequalities (45), (46), for $\ell = 0$, follow immediately from (34), (36), (39); on the other hand, if (45), (46) hold true for $\ell = h$, they hold true for $\ell = h + 1$ as well, due to (38), (42), (44). From (45), letting $\ell \to \infty$ and using (39), we deduce (35) and the proof is complete.

Remark I. – If $w(x)$ belongs to $\mathcal{B}(E; \gamma)$, the function $-w(x)$ is in the same class. Hence, statements analogous to Lemmas II, III, IV, with $B(k; \varrho)$ replacing $A(k; \varrho)$, hold.

Remark II. – From Lemma IV it follows that, for every positive $p < \delta(y)$ the oscillation of $w(x)$ in $I(p)$ is finite[5]. Indeed, it is always possible to find two positive numbers ϱ, σ verifying

$$(47) \qquad \varrho - \sigma\varrho = p, \qquad\qquad 0 < \varrho < \delta(y)$$

and then to choose k and c in such a way that (34), (36) hold, so that, in almost all of $I(p)$, we have

$$(48) \qquad w(x) \leq k + \sigma c.$$

In the same way, by Remark I, we may state a lower bound for $w(x)$.

LEMMA V. – *There exists $\eta > 0$ such that for $0 < 4\varrho < \delta(y)$, the inequalities*

$$(49) \qquad (1 - \eta)\mathrm{osc}\,(w; 4\varrho) \geq \mathrm{osc}\,(w; \varrho),$$

hold, where osc *is the oscillation of $w(x)$ in $I(\varrho)$ (in the sense of footnote [5]).*

Proof. Calling μ_1, μ_2 the true least upper bound and greatest lower bound of $w(x)$ in $I(4\varrho)$, respectively, let us set

$$(50) \qquad \mathrm{osc}\,(w; 4\varrho) = \mu_1 - \mu_2 = \omega, \qquad \bar{\mu} = \frac{\mu_1 + \mu_2}{2}.$$

By the definition of the sets $A(k; \varrho)$, $B(k; \varrho)$ we know that one among the two inequalities

$$(51) \qquad 2 \,\mathrm{meas}\, A(\bar{\mu}; 2\varrho) \leq \mathrm{meas}\, I(2\varrho),$$

$$(51') \qquad 2 \,\mathrm{meas}\, B(\bar{\mu}; 2\varrho) \leq \mathrm{meas}\, I(2\varrho).$$

is true. Assume (51) and set, for every positive $\lambda \leq \dfrac{\omega}{4}$,

$$(52) \qquad D(\lambda) = A(\mu_1 - 2\lambda; 2\varrho) \setminus A(\mu_1 - \lambda; 2\varrho).$$

---

[5]The term oscillation must be understood in the sense of integration theory, as the difference between the *true least upper bound* of $w(x)$ in $I(p)$ (i.e., the smallest λ such that meas $A(\lambda; p) = 0$) and the *true greatest lower bound* of $w(x)$ in $I(p)$ (i.e., the greatest λ such that meas $B(\lambda; p) = 0$).

By Lemma II and (50), (51), (52) we infer

$$(53) \qquad \beta_1 \int_{D(\lambda)} |\operatorname{grad} w| \, dx_1 \ldots dx_r \geq \lambda \left[\operatorname{meas} A(\mu_1 - \lambda; 2\varrho)\right]^{\frac{r-1}{r}},$$

and by Schwarz inequality we infer

$$(54) \qquad \left(\int_{D(\lambda)} |\operatorname{grad} w| \, dx_1 \ldots dx_r\right)^2 \leq \operatorname{meas} D(\lambda) \int_{D(\lambda)} |\operatorname{grad} w|^2 \, dx_1 \ldots dx_r.$$

Moreover, by condition 3rd) in Section 1 we have

$$(55) \qquad \int_{D(\lambda)} |\operatorname{grad} w|^2 \, dx_1 \ldots dx_r \leq \int_{A(\mu_1 - 2\lambda; 2\varrho)} |\operatorname{grad} w|^2 \, dx_1 \ldots dx_r \leq$$

$$\leq \frac{\gamma}{4\varrho^2} \int_{A(\mu_1 - 2\lambda; 4\varrho)} (w(x) - \mu_1 + 2\lambda)^2 \, dx_1 \ldots dx_r \leq \gamma \varrho^{-2} \lambda^2 \operatorname{meas} I(4\varrho)$$

and then by (53), (54), (55) we obtain

$$(56) \qquad \operatorname{meas} D(\lambda) \geq \frac{\varrho^2}{\beta_1^2 \gamma} \left[\operatorname{meas} A(\mu_1 - \lambda; 2\varrho)\right]^{\frac{2r-2}{r}} \cdot \left[\operatorname{meas} I(4\varrho)\right]^{-1}.$$

Take an integer n such that

$$(57) \qquad \left[\theta\left(\frac{1}{2}\right)\right]^{\frac{2r-2}{r}} n\varrho^{2r} > \gamma \beta_1^2 \operatorname{meas} I(4\varrho) \operatorname{meas} I(2\varrho),$$

where $\theta\left(\frac{1}{2}\right)$ is the value of the function θ in (34) and (36) for $\sigma = \dfrac{1}{2}$, and set

$$(58) \qquad \eta = 2^{-(n+2)}.$$

Thus, we have

$$(59) \qquad \lambda \leq \frac{\omega}{4} \qquad \text{per} \qquad \lambda = 2^m \eta \omega, \qquad m = 1, \ldots, n$$

and then, by (56), we obtain

$$(60) \quad \operatorname{meas} D(2^m \eta \omega) \geq \frac{\varrho^2}{\beta_1^2 \gamma} \left[\operatorname{meas} A(\mu_1 - 2^m \eta \omega; 2\varrho)\right]^{\frac{2r-2}{r}} \cdot \left[\operatorname{meas} I(4\varrho)\right]^{-1} \geq$$

$$\geq \frac{\varrho^2}{\beta_1^2 \gamma} \left[\operatorname{meas} A(\mu_1 - 2\eta \omega; 2\varrho)\right]^{\frac{2r-2}{r}} \cdot \left[\operatorname{meas} I(4\varrho)\right]^{-1}$$

for $m = 1, \ldots, n$.

From (52), (60) it follows

$$(61) \qquad \operatorname{meas} I(2\varrho) \geq \sum_{m=1}^{n} \operatorname{meas} D(2^m \eta \omega) \geq$$

$$\geq n[\operatorname{meas} I(4\varrho)]^{-1} \frac{\varrho^2}{\gamma \beta_1^2} \left[\operatorname{meas} A(\mu_1 - 2\eta \omega; 2\varrho)\right]^{\frac{2r-2}{r}}$$

and then, by (57), (61) we deduce

$$(62) \qquad \operatorname{meas} A(\mu_1 - 2\eta\omega; 2\varrho) < \theta\left(\frac{1}{2}\right)\varrho^r < \theta\left(\frac{1}{2}\right)(2\varrho)^r.$$

Recalling that μ_1 is the true least upper bound of $w(x)$ in $I(4\varrho)$, by (62) we have

$$(63) \qquad \int_{A(\mu_1-2\eta\omega;2\varrho)} (w(x) - \mu_1 + 2\eta\omega)^2\, dx_1 \ldots dx_r < (2\varrho)^r (2\eta\omega)^2 \theta\left(\frac{1}{2}\right)$$

and then, by Lemma IV

$$(64) \qquad \operatorname{meas} A(\mu_1 - \eta\omega, \varrho) = 0,$$

i.e., the true least upper bound of $w(x)$ in $I(\varrho)$ does not exceed $(\mu_1 - \eta\omega)$ and (49) holds true. By Remark I, if we assume (51') instead of (51), we reach the same conclusion.

    3. – The lemmas proved in Section 2 allow us to prove the following result.

    THEOREM I. – *Each function $w(x) \in \mathcal{B}(E;\gamma)$ is uniformly Hölder continuous in every compact subset of E^6.*

    Proof. From Remark II and Lemma V in Section 2 we deduce that, for every $y \in E$, the oscillation of $w(x)$ in $I(\varrho;y)$ is infinitesimal as $\varrho \to 0$; as a consequence, for every $y \in E$ the limit

$$(1) \qquad \bar{w}(y) = \lim_{\varrho \to 0} \left[\operatorname{meas} I(\varrho; y)\right]^{-1} \int_{I(\varrho;y)} w(x)\, dx_1 \ldots dx_r$$

exists, and $\bar{w}(x)$ is continuous in E.

    Let us now set

$$(2) \qquad \alpha = -\log_4(1 - \eta).$$

Here η is the constant in Lemma V which, by (58), (57), (38), (37), (33), (12) in Section 2 is independent of y in E. Fix a compact set $C \subset E$ and a positive number p less than or equal to the distance of C from the boundary of E, and consider the set L whose elements are the numbers

$$(3) \qquad \frac{2|\bar{w}(x) - \bar{w}(y)|4^\alpha}{p^\alpha}$$

for

$$(4) \qquad 0 \le |x - y| \le p, \qquad y \in C,$$

where $|x - y|$ is the distance between x and y.

---

    [6] As usual, this must be understood in the sense of integration theory, i.e., either $w(x)$ itself is Hölder continuous, or there exists a Hölder continuous function coinciding almost everywhere with $w(x)$; analogous remarks are often understood in this paper (see [1] footnote [2]).

Since $\bar{w}(x)$ is continuous in E, the set L has an absolute maximum, which we shall denote by τ and then, fixed $y \in C$ we have, with the same notation as in Section 2 and in particular in Lemma V,

$$(5) \qquad \operatorname{osc}(w; \varrho) \leq \tau \varrho^{\alpha} \quad \text{for} \quad \frac{p}{4} \leq \varrho \leq p.$$

Notice that for every positive $\varrho < \dfrac{p}{4}$ there is an integer m such that

$$(6) \qquad \frac{p}{4} \leq 4^{m} \varrho < p$$

and then, by Lemma V and (2), (5), (6), we have

$$(7) \qquad \operatorname{osc}(w; \varrho) \leq (1 - \eta)^{m} \operatorname{osc}(w; 4^{m} \varrho) \leq \tau \varrho^{\alpha}.$$

By the arbitrariness of C and y the proof is complete.

THEOREM II. – *Let r^2 functions $a_{hl}(x)$, almost continuous in the open set $E \subset S_r$, be given. Assume that $a_{hl}(x) = a_{lh}(x)$ and that two positive numbers τ_1, τ_2 exist, such that the inequalities*

$$(8) \qquad \tau_1 |\lambda|^2 \leq \sum_{h,l}^{1,r} a_{hl}(x) \lambda_h \lambda_l \leq \tau_2 |\lambda|^2$$

hold for every $x \in E$ and for every vector $\lambda \equiv (\lambda_1, \ldots, \lambda_r)$. Let also $w(x)$ be a function in $\mathcal{U}^{(2)}(E)$ such that, for every compact set $C \subset E$ and for every function $g(x) \in \mathcal{U}^{(2)}(E)$, which vanishes in $(E \setminus C)$, we have

$$(9) \qquad \sum_{h,l}^{1,r} \int_E \frac{\partial g}{\partial x_h} a_{hl}(x) \frac{\partial w}{\partial x_l} \, dx_1 \ldots dx_r = 0.$$

Then, $w(x)$ belongs to $\mathcal{B}(E; \gamma)$, with $\gamma = \left(\frac{\tau_2}{\tau_1}\right)^2$, and then it is Hölder continuous in E by Theorem I.

Proof. Fix a point $y \in E$ and a real number k, and set, with the same notation as in Section 1,

$$(10) \qquad \begin{aligned} \varphi(x) &\equiv w(x) - k &&\text{if} \quad x \in A(k) \\ \varphi(x) &\equiv 0 &&\text{if} \quad x \in (E \setminus A(k)). \end{aligned}$$

It is easily checked that $\varphi(x)$ belongs to $\mathcal{U}^{(2)}(E)$ as well. Take a positive number $p < \delta(y)$ and a function $u(t)$ depending on t, continuous with its first derivative $u'(t)$ in the interval $[0, +\infty]$, and vanishing in $[p, +\infty]$, and set

$$(11) \qquad f(x_1, \ldots, x_r) = u(\sqrt{(x_1 - y_1)^2 + \ldots + (x_r - y_r)^2});$$

the function $\varphi(x) \cdot f(x)$ belongs to $\mathcal{U}^{(2)}(E)$ and vanishes in $[E \setminus I(p; y)]$; from (9) it follows

$$(12) \qquad \sum_{h,l}^{1,r} \int_E \frac{\partial(\varphi \cdot f)}{\partial x_h} a_{hl}(x) \frac{\partial w}{\partial x_l} \, dx_1 \ldots dx_r = 0.$$

162   E. De Giorgi

Since $f(x)$ vanishes identically in $[E \setminus I(p; y)]$, from (11), (12) we deduce

$$
(13) \quad \sum_{h,l}^{1,r} \int_0^p d\varrho \left[u'(\varrho) \int_{\mathcal{F}I(\varrho;y)} n_h a_{hl}(x)\varphi(x)\frac{\partial w}{\partial x_l}\, d\mu_{r-1} + \right.
$$
$$
\left. + u(\varrho) \int_{\mathcal{F}I(\varrho;y)} a_{hl}(x)\frac{\partial \varphi}{\partial x_h}\frac{\partial w}{\partial x_l}\, d\mu_{r-1} \right] = 0,
$$

where $n_1, \ldots, n_r$ are the components of the outward pointing unit normal to $\mathcal{F}I(\varrho; y)$ and $d\mu_{r-1}$ is the $(r-1)$-dimensional measure. From (13), integrating by parts, we deduce

$$
(14) \quad \sum_{h,l}^{1,r} \int_0^p d\varrho\, u'(\varrho) \left[\int_{\mathcal{F}I(\varrho;y)} n_h a_{hl}(x)\varphi(x)\frac{\partial w}{\partial x_l}\, d\mu_{r-1} - \right.
$$
$$
\left. - \int_0^\varrho dt \int_{\mathcal{F}I(t;y)} a_{hl}(x)\frac{\partial \varphi}{\partial x_h}\frac{\partial w}{\partial x_l}\, d\mu_{r-1} \right] = 0.
$$

By the arbitrariness of p and $u(t)$ we have, for almost every positive number $\varrho < \delta(y)$,

$$
(15) \quad \int_{\mathcal{F}I(\varrho;y)} \sum_{h,l}^{1,r} n_h a_{hl}\varphi(x)\frac{\partial w}{\partial x_l}\, d\mu_{r-1} = \int_{I(\varrho;y)} \sum_{h,l}^{1,r} a_{hl}\frac{\partial \varphi}{\partial x_h}\frac{\partial w}{\partial x_l}\, dx_1 \ldots dx_r.
$$

By (10), (15) we have then

$$
(16) \quad \int_{A(k)\cap\mathcal{F}I(\varrho;y)} \sum_{h,l}^{1,r} n_h a_{hl}(x)(w(x) - k)\frac{\partial w}{\partial x_l}\, d\mu_{r-1} =
$$
$$
\int_{A(k)\cap I(\varrho;y)} \sum_{h,l}^{1,r} a_{hl}(x)\frac{\partial w}{\partial x_h}\frac{\partial w}{\partial x_l}\, dx_1 \ldots dx_r,
$$

which by (8) implies

$$
(17) \quad \tau_2 \int_{A(k)\cap\mathcal{F}I(\varrho;y)} (w(x) - k)|\operatorname{grad} w|\, d\mu_{r-1} \geq
$$
$$
\geq \tau_1 \int_{A(k)\cap I(\varrho;y)} |\operatorname{grad} w|^2\, dx_1 \ldots dx_r.
$$

Setting now

$$
(18) \quad
\begin{aligned}
\psi_1(\varrho) &= \int_{A(k)\cap\mathcal{F}I(\varrho;y)} (w(x) - k)^2\, d\mu_{r-1}, \\[2mm]
\psi_2(\varrho) &= \int_{A(k)\cap\mathcal{F}I(\varrho;y)} |\operatorname{grad} w|^2 d\mu_{r-1},
\end{aligned}
$$

from (17), taking into account Schwarz inequality, we deduce

$$(19) \qquad \sqrt{\psi_1(\varrho)\psi_2(\varrho)} \geq \frac{\tau_1}{\tau_2} \int_0^\varrho \psi_2(t)dt.$$

Setting $\gamma = \left(\frac{\tau_2}{\tau_1}\right)^2$, from (19) and a Lemma due to Caccioppoli and Leray (see [4] on page 153), it follows

$$(20) \qquad \int_0^{\varrho_1} \psi_2(\varrho)\, d\varrho \leq \frac{\gamma}{(\varrho_2 - \varrho_1)^2} \int_0^{\varrho_2} \psi_1(\varrho)\, d\varrho,$$

for $0 < \varrho_1 < \varrho_2 < \delta(y)$. From (18) and (20) inequality (1) in Section 1 immediately follows. Since whenever $w(x)$ satisfies the hypotheses of the statement, the same holds for $-w(x)$, it is easily checked that (1') in Section 1 holds as well, so that

$$(21) \qquad w(x) \in \mathcal{B}(E; \gamma).$$

4. – We are now in a position to prove the announced analyticity theorem. To this aim, let us consider a function $f(p) \equiv f(p_1, \ldots, p_r)$, which we assume to be continuous in S_r together with its first and second order partial derivatives. Let us set

$$(1) \qquad f_{hk}(p) = \frac{\partial^2 f}{\partial p_h \partial p_k}, \quad f_h(p) = \frac{\partial f}{\partial p_h} \qquad (\text{for } h, k = 1, \ldots, r)$$

and assume that there are two positive numbers μ_1 and μ_2 such that for every $p \in S_r$ and for every vector

$$\lambda \equiv (\lambda_1, \ldots, \lambda_r),$$

we have

$$(2) \qquad \mu_1|\lambda|^2 \leq \sum_{h,k}^{1,r} f_{hk}(p)\lambda_h\lambda_k \leq \mu_2|\lambda|^2.$$

Given an open set $E \subset S_r$ and a function $u^*(x) \in \mathcal{U}^{(2)}(E)$, we say that $u^*(x)$ is *extremal in E for the integral functional*

$$(3) \qquad \mathcal{I}[u] = \int f\left(\frac{\partial u}{\partial x_1}, \ldots, \frac{\partial u}{\partial x_r}\right) dx_1 \ldots dx_r$$

if for every compact subset $C \subset E$ and for every function $g(x)$ which is continuous in E together with its first order derivatives and vanishes in $(E \setminus C)$ we have

$$(4) \qquad \sum_{h=1}^r \int_E \frac{\partial g}{\partial x_h} f_h\left(\frac{\partial u^*}{\partial x_1}, \ldots, \frac{\partial u^*}{\partial x_r}\right) dx_1 \ldots dx_r = 0.$$

The following result holds.

THEOREM III. – *Every extremal in E of the integral functional $\mathcal{I}[u]$ has first order derivatives uniformly Hölder continuous in every compact subset of E. Moreover, if $f(p)$ is real analytic in S_r, then every extremal is real analytic in E.*

Proof. By (1), (2) the first order derivatives of $f(p)$ satisfy an estimate of the form

$$(5) \qquad |f_h(p)| \leq \mu_2 |p| + c \qquad (h = 1, \ldots, r),$$

where $|p|$ is the distance between p and the origin of the coordinates and c is a positive constant. Since our extremal $u^*(x)$ belongs to $\mathcal{U}^{(2)}(E)$, both $u^*(x)$ and the functions

$$f_h \left(\frac{\partial u^*}{\partial x_1}, \ldots, \frac{\partial u^*}{\partial x_r} \right)$$

have first order derivatives which are square-summable in every compact subset of E.

From well-known theorems on the linear approximation it follows that (4) holds even if $g(x)$ belongs to $\mathcal{U}^{(2)}(E)$ and vanishes identically outside a compact subset of E.

Let us consider a bounded open set $H \subset E$ with positive distance from the boundary of E. Let σ be a positive number less than this distance and s a positive integer less than or equal to r. Setting, for every positive integer n,

$$(6) \qquad u_n(x) = u^* \left(x_1, \ldots, x_s + \frac{\sigma}{n}, \ldots, x_r \right)$$

clearly the functions $u_n(x)$ belong to $\mathcal{U}^{(2)}(H)$. Moreover, for every closed set $C \subset H$ and for every function $v(x) \in \mathcal{U}^{(2)}(H)$ vanishing in $(H \setminus C)$, we have

$$(7) \qquad \sum_{h=1}^{r} \int_H f_h \left(\frac{\partial u_n}{\partial x_1}, \ldots, \frac{\partial u_n}{\partial x_r} \right) \frac{\partial v}{\partial x_h} \, dx_1 \ldots dx_r = 0.$$

Furthermore, the functions

$$(8) \qquad w_n^*(x) = u_n(x) - u^*(x),$$

belong to $\mathcal{U}^{(2)}(H)$ as well, and the following relations hold:

$$(9) \qquad f_h \left(\frac{\partial u_*}{\partial x_1}, \ldots, \frac{\partial u_*}{\partial x_r} \right) - f_h \left(\frac{\partial u_n}{\partial x_1}, \ldots, \frac{\partial u_n}{\partial x_r} \right) + \sum_{k}^{1,r} a_{hk}^{(n)}(x) \frac{\partial w_n^*}{\partial x_k} = 0,$$

where we have set

$$(10) \qquad a_{hk}^{(n)}(x) = \int_0^1 f_{hk} \left(\frac{\partial u^*}{\partial x_1} + t \frac{\partial w_n^*}{\partial x_1}, \ldots, \frac{\partial u^*}{\partial x_1} + t \frac{\partial w_n^*}{\partial x_r} \right) dt.$$

From (2), (10) we deduce, for every $x \in H$ and for every vector λ,

$$(11) \qquad \mu_1 |\lambda|^2 \leq \sum_{h,k}^{1,r} a_{hk}^{(n)}(x) \lambda_h \lambda_k \leq \mu_2 |\lambda|^2.$$

From (4), (7), (9) it follows

$$(12) \qquad \sum_{h,k}^{1,r} \int_H a_{hk}^{(n)}(x) \frac{\partial w_n^*}{\partial x_k} \frac{\partial v}{\partial x_h} \, dx_1 \ldots dx_r = 0$$

for every function $v(x) \in \mathcal{U}^{(2)}(H)$ vanishing outside a closed subset of H. By Theorem II we may then conclude that the functions $w_n^*(x)$ belong to $\mathcal{B}(H;\gamma)$ with

$$(13) \qquad \gamma = \frac{\mu_2^2}{\mu_1^2}.$$

As a consequence, even the functions

$$w_n(x) = \frac{w_n^*(x)n}{\sigma}$$

belong to $\mathcal{B}(H;\gamma)$, and since the sequence

$$(14) \qquad w_1(x), \ldots, w_n(x), \ldots$$

converges in quadratic mean to $\dfrac{\partial u^*}{\partial x_s}$ in H (which by hypothesis is closed and has positive distance from the boundary of E) we can apply Lemma I to obtain that also $\dfrac{\partial u^*}{\partial x_s}$ belongs to $\mathcal{B}(H;\gamma)$. Since both H and s are arbitrary, we can conclude that all the first order derivatives of $u^*(x)$ belong to $\mathcal{B}(E;\gamma)$. At this point, from Theorem I it follows that they are Hölder continuous. Moreover, if $f(p)$ is real analytic, then $u^*(x)$ turns out to be real analytic as well, by [6, Teor. 9.2], which ensures the existence and the continuity of all the derivatives for C^1 extremals, and by the results of Stampacchia and Hopf (see [9, teor. VII]).

Bibliography

[1] E. De Giorgi, *Sull'analiticità delle estremali degli integrali multipli.* Rend. Lincei, Serie VIII, Vol. XX, fasc. 4, (1956).

[2] E. Hopf, *Zum analytischen Charakter der Lösungen regulärer zweidimensionalen Variations-probleme.* Math. Zeitschrift, Band 30 pp. 404-413 (1929).

[3] E. Hopf, *Über den funktionalen, insbesondere den analytischen Charakter der Lösungen elliptischer Differentialgleichungen zweiter Ordnung.* Math. Zeitschrift, Band 34 pp. 194-233 (1932).

[4] C. Miranda, *Equazioni alle derivate parziali di tipo ellittico*. Springer Verlag, Berlino (1955).

[5] C. B. Morrey, *Multiple integral problems in the calculus of variations and related topics*. Univ. California Publ. Math. (N.S.) Vol. I (1943).

[6] C. B. Morrey, *Second order elliptic systems of differential equations*. Annals of Math. Studies, N. 33, Princeton Un. Press (1954).

[7] M. Schiffman, *Differentiability and analyticity of solutions of double integral variational problems*. Annals of Math., Vol. 48 (1947).

[8] A. G. Sigalov, *On conditions of differentiability and analyticity of solutions of two dimensional problems of the calculus of variations*. Doklady Akad. Nauk. S.S.S.R., (N. S.) 85 (1952) (Math. Reviews, Vol. 14 n. 3).

[9] G. Stampacchia, *Sistemi di equazioni di tipo ellittico a derivate parziali del primo ordine e proprietà delle estremali degli integrali multipli*. Ricerche di Matematica, Vol. I, pp. 200-226 (1952).

Sulla differenziabilità e l'analiticità delle estremali degli integrali multipli regolari[‡][†]

Memoria di Ennio De Giorgi[*]

Sunto. Si studiano le estremali di alcuni integrali multipli regolari, supponendo nota a priori l'esistenza delle derivate parziali prime di quadrato sommabile; si dimostra il carattere hölderiano di tali derivate, da cui seguono l'indefinita differenziabilità e l'analiticità delle estremali.

In questo lavoro mi occupo delle proprietà differenziali e specialmente dell'analiticità delle estremali degl'integrali multipli regolari; tale argomento è stato oggetto di molte ricerche da parte di matematici italiani e stranieri, sicché appare assai difficile darne un quadro bibliografico completo; ci limiteremo quindi a citare qualche lavoro da cui il lettore potrà facilmente ricavare più ampie informazioni. Ricorderemo così i risultati di Hopf [3][1], Stampacchia [9], Morrey [6], che danno teoremi di differenziabilità ed analiticità per estremali sempre meno regolari: precisamente si richiede l'esistenza di derivate seconde hölderiane in [3], di derivate prime hölderiane in [9], di derivate prime continue in [6]. A un diverso indirizzo appartengono invece altri risultati ottenuti da Stampacchia in [9]: egli parte da teoremi di esistenza ottenuti coi metodi diretti del calcolo delle variazioni, nei quali le soluzioni vengono ricercate in classi assai ampie di funzioni, e studia le proprietà di queste soluzioni (a priori assai poco regolari) dimostrando, fra l'altro, l'esistenza di derivate parziali seconde di quadrato sommabile, soddisfacenti quasi ovunque l'equazione di Eulero.

Mancavano però, per quanto mi risulta, (qualora si escludano gli integrali doppi per i quali rinviamo il lettore a [2], [5], [7], [8] e qualche caso particolare di integrali multipli, come quello degl'integrali quadratici che danno luogo ad equazioni di Eulero lineari) teoremi che facessero per così dire da ponte fra i risultati ottenuti nel primo indirizzo e quelli ottenuti nel secondo, assicurando che le soluzioni trovate coi metodi diretti del calcolo delle variazioni considerate in [9] soddisfano le condizioni richieste in [6]; lo scopo di questo lavoro è appunto la dimostrazione di un primo teorema di questo tipo (precisamente il teor. III[2]).

[‡]Editor's note: published in Mem. Accad. Sci. Torino Cl. Sci. Fis. Mat. Nat., **(3) 3** (1957), 25–43.

[†]Lavoro eseguito nell'Istituto Nazionale per le Applicazioni del Calcolo

[*]Presentata dal Socio nazionale non residente MAURO PICONE nell'adunanza del 25 Aprile 1957

[1]Il numero fra parentesi è quello che compete al lavoro citato nella bibliografia riportata alla fine della Memoria.

[2]Questo teorema è stato comunicato oralmente al Congresso dell'U.M.I. che ha avuto luogo a Pavia fra il 6 e l'11 ottobre 1955 e poi esposto nella nota preventiva [1].

Tale dimostrazione è fondata sullo studio di alcune funzioni (caratterizzate da certe diseguaglianze integrali) delle quali col teor. I si prova il carattere hölderiano; fra i risultati intermedi noteremo il teor. II per l'interesse che può avere anche in altre questioni relative ad equazioni differenziali di tipo ellittico.

L'argomento di questa ricerca mi è stato suggerito da alcune conversazioni col prof. G. Stampacchia, che qui ringrazio per le informazioni ed i consigli che mi sono stati assai utili in questo lavoro.

1. – Nello spazio euclideo r-dimensionale S_r consideriamo un campo E ed indichiamo con $\mathcal{U}^{(2)}(E)$ la classe delle funzioni $w(x)$ quasi continue in E e soddisfacenti le condizioni seguenti:

1^a) *$w(x)$ è assolutamente continua su quasi tutti i segmenti paralleli agli assi coordinati contenuti in E.*

2^a) *$w(x)$ e le sue derivate parziali prime sono funzioni di quadrato sommabile in ogni insieme chiuso e limitato contenuto in E.*

Dato un numero positivo γ, chiameremo $\mathcal{B}(E; \gamma)$ la classe delle funzioni $w(x)$ che oltre alle condizioni 1^a) e 2^a) soddisfano la

3^a) *Comunque si fissino un punto $y \in E$ [3] (di cui $\delta(y)$ sia la distanza da $S_r \setminus E$) e tre numeri k, ϱ_1, ϱ_2 con $0 < \varrho_1 < \varrho_2 < \delta(y)$), si ha*

$$(1) \qquad \frac{\gamma}{(\varrho_2 - \varrho_1)^2} \int_{A(k) \cap I(\varrho_2; y)} (w(x) - k)^2 \, dx_1 \ldots dx_r \geq$$

$$\geq \int_{A(k) \cap I(\varrho_1; y)} |\operatorname{grad} w|^2 \, dx_1 \ldots dx_r,$$

$$(1') \qquad \frac{\gamma}{(\varrho_2 - \varrho_1)^2} \int_{B(k) \cap I(\varrho_2; y)} (w(x) - k)^2 \, dx_1 \ldots dx_r \geq$$

$$\geq \int_{B(k) \cap I(\varrho_1; y)} |\operatorname{grad} w|^2 \, dx_1 \ldots dx_r,$$

ove si indica con $I(\varrho; y)$ l'intorno di raggio ϱ di y, con $A(k)$ l'insieme dei punti di E ove è $w(x) > k$, con $B(k)$ l'insieme di quelli ove è $w(x) < k$.

Una prima proprietà della classe $\mathcal{B}(E; \gamma)$ ora definita che ci sarà utile in seguito è data dal

Lemma I. – *Assegnata una successione di funzioni*

$$(2) \qquad\qquad w_1(x), \ldots, w_n(x), \ldots$$

appartenenti a $\mathcal{B}(E; \gamma)$ ed aventi quadrato sommabile in E, se la (2) converge in media in E verso una funzione $w(x)$, cioè si ha

$$(3) \qquad \lim_{n \to \infty} \int_E [w_n(x) - w(x)]^2 \, dx_1 \ldots dx_r = 0,$$

---

[3] Il simbolo $y \in E$ significa: y appartiene ad E; quello $E \subset L$ significa: E è contenuto in L.

allora $w(x)$ appartiene a $\mathcal{B}(E; \gamma)$.

Dim. Infatti, poiché le funzioni della successione (2) appartengono a $\mathcal{B}(E; \gamma)$, sarà per $y \in E$, $0 < \varrho_1 < \varrho_2 < \delta(y)$

$$(4) \qquad \int_{I(\varrho_1;y)} |\operatorname{grad} w_n|^2 \, dx_1 \ldots dx_r \leq \frac{\gamma}{(\varrho_2 - \varrho_1)^2} \int_{I(\varrho_2;y)} |w_n(x)|^2 \, dx_1 \ldots dx_r$$

e quindi per la (3)

$$(5) \qquad \limsup_{n \to \infty} \int_{I(\varrho_1;y)} |\operatorname{grad} w_n|^2 \, dx_1 \ldots dx_r \leq$$

$$\leq \frac{\gamma}{(\varrho_2 - \varrho_1)^2} \int_{I(\varrho_2;y)} |w(x)|^2 \, dx_1 \ldots dx_r.$$

Dalla (5), per l'arbitrarietà di y, ϱ_1, ϱ_2, si deduce che, comunque si fissi un insieme chiuso e limitato $C \subset E$, gli integrali estesi a C delle norme dei gradienti delle funzioni (2) descrivono una successione limitata; ne segue, per note proprietà delle successioni di funzioni aventi derivate prime di quadrato sommabile, che $w(x) \in \mathcal{U}^{(2)}(E)$.

Per provare poi che $w(x)$ soddisfa la (1), basta osservare che, posto per ogni numero reale k e per ogni intero positivo n

$$(6) \qquad w_n(x; k) = \begin{cases} w_n(x) - k & \text{per } w_n(x) \geq k \\ 0 & \text{per } w_n(x) \leq k, \end{cases}$$

la successione

$$(7) \qquad w_1(x; k), \ldots, w_n(x; k), \ldots$$

converge in media in E verso la funzione $w(x; k)$ data dalle

$$(8) \qquad w(x; k) = \begin{cases} w(x) - k & \text{per } w(x) \geq k \\ 0 & \text{per } w(x) \leq k. \end{cases}$$

Si avrà pertanto, per $y \in E$, $0 < \varrho < \delta(y)$,

$$(9) \qquad \lim_{n \to \infty} \int_{I(\varrho;y)} [w_n(x; k)]^2 \, dx_1 \ldots dx_r = \int_{I(\varrho;y)} [w(x; k)]^2 \, dx_1 \ldots dx_r$$

$$(10) \qquad \liminf_{n \to \infty} \int_{I(\varrho;y)} |\operatorname{grad}_x w_n(x; k)|^2 \, dx_1 \ldots dx_r \geq$$

$$\geq \int_{I(\varrho;y)} |\operatorname{grad}_x w(x; k)|^2 \, dx_1 \ldots dx_r;$$

dalle (9), (10), poiché per ipotesi le funzioni $w_n(x)$ appartengono a $\mathcal{B}(E; \gamma)$, si deduce che $w(x)$ soddisfa la (1); analogamente si potrebbe ragionare per la (1') e quindi è dimostrato che $w(x) \in \mathcal{B}(E; \gamma)$.

2. – Pensiamo di aver fissato un campo $E \subset S_r$, una costante positiva γ, una funzione $w(x) \in \mathcal{B}(E; \gamma)$ ed un punto $y \in E$; riprendendo le notazioni del n. 1, poniamo

$$
\begin{aligned}
I(\varrho) &= I(\varrho; y) \\
A(k; \varrho) &= A(k) \cap I(\varrho; y) \\
B(k; \varrho) &= B(k) \cap I(\varrho; y).
\end{aligned}
$$

Stabilite queste ipotesi, che verranno conservate in tutto questo capitolo, passiamo alla dimostrazione del

LEMMA II. – *Esiste una costante β_1 tale che, per ogni terna di numeri ϱ, k, λ verificanti le*

$$
(1) \qquad\qquad 0 < \varrho < \delta(y), \qquad k < \lambda
$$

si abbia

$$
(2) \qquad \beta_1 \int_{[A(k;\varrho) \setminus A(\lambda;\varrho)]} \left| \operatorname{grad} w(x) \right| dx_1 \dots dx_r \geq (\lambda - k) \left[\tau(k, \lambda; \varrho) \right]^{1 - \frac{1}{r}},
$$

ove col simbolo $\tau(k,\lambda;\varrho)$ si indica il più piccolo dei due numeri

$$
\operatorname{mis} A(\lambda; \varrho), \qquad \operatorname{mis}[I(\varrho) \setminus A(k; \varrho)].
$$

Dim. Per ogni insieme $L \subset S_r$, che sia contenuto nella somma di un numero finito di iperpiani e di ipersuperfici sferiche, indicheremo con il simbolo $\mu_{r-1}L$ la misura $(r-1)$-dimensionale di L (elementarmente definita).

Per ogni dominio D di S_r chiameremo *semilineari* in D le funzioni $g(x) \equiv g(x_1, ..., x_r)$ continue in D che godono della proprietà seguente: è possibile decomporre D nella somma di un numero finito di domini, in ognuno dei quali $g(x)$ è lineare (cioè costante o uguale ad un polinomio di primo grado nelle variabili $x_1, ..., x_r$); geometricamente la proprietà ora enunciata può esprimersi dicendo che l'ipersuperficie dello spazio $(r + 1)$-dimensionale S_{r+1} di equazioni

$$
(3) \qquad\qquad x_{r+1} = g(x_1, ..., x_r); \qquad (x_1, ..., x_r) \in D
$$

è contenuta nella somma di un numero finito di iperpiani di S_{r+1}.

Stabilite queste convenzioni passiamo alla dimostrazione del nostro lemma nell'ipotesi che $w(x)$ sia semilineare nel dominio $[I(\varrho) \cup \mathcal{F}I(\varrho)]$; in tal caso le frontiere $\mathcal{F}A(t; \varrho)$, $\mathcal{F}B(t; \varrho)$ dei campi $A(t; \varrho)$, $B(t; \varrho)$ saranno sempre contenute nella somma di un numero finito di iperpiani e dell'ipersuperficie sferica $\mathcal{F}I(\varrho)$. Inoltre, per quasi tutti i valori di t, sarà

$$
(4) \qquad \mu_{r-1}\mathcal{F}A(t; \varrho) = \mu_{r-1}[I(\varrho) \cap \mathcal{F}A(t)] + \mu_{r-1}[A(t) \cap \mathcal{F}I(\varrho)],
$$

$$
(4') \qquad \mu_{r-1}\mathcal{F}B(t; \varrho) = \mu_{r-1}[I(\varrho) \cap \mathcal{F}B(t)] + \mu_{r-1}[B(t) \cap \mathcal{F}I(\varrho)],
$$

$$(5) \qquad \text{mis}\, I(\varrho) = \text{mis}\, A(t;\varrho) + \text{mis}\, B(t;\varrho)$$

$$(6) \qquad \mu_{r-1}\mathcal{F}I(\varrho) = \mu_{r-1}[B(t) \cap \mathcal{F}I(\varrho)] + \mu_{r-1}[A(t) \cap \mathcal{F}I(\varrho)]$$

$$(7) \qquad I(\varrho) \cap \mathcal{F}B(t) = I(\varrho) \cap \mathcal{F}A(t).$$

Per la semilinearità di $w(x)$ avremo poi, comunque si scelga il numero ξ,

$$(8) \qquad \int_{A(\xi;\varrho)} |\text{grad}\, w|\, dx_1 \ldots dx_r = \int_{\xi}^{+\infty} \mu_{r-1}\big[I(\varrho) \cap \mathcal{F}A(t)\big]\, dt.$$

D'altra parte, per le note[4] proprietà isoperimetriche delle ipersfere, esisterà una costante $\alpha(r)$ tale che risulti

$$(9) \qquad \begin{cases} \big[\text{mis}\, I(\varrho)\big]^{1-\frac{1}{r}} = \alpha(r)\mu_{r-1}\mathcal{F}I(\varrho) \\[2mm] \big[\text{mis}\, A(t;\varrho)\big]^{1-\frac{1}{r}} \le \alpha(r)\mu_{r-1}\mathcal{F}A(t;\varrho) \\[2mm] \big[\text{mis}\, B(t;\varrho)\big]^{1-\frac{1}{r}} \le \alpha(r)\mu_{r-1}\mathcal{F}B(t;\varrho) \end{cases}$$

e quindi, tenendo presenti le (4), (4') ,(5), (6), (7), (9), troviamo per quasi tutti i valori di t

$$(10) \qquad 2\alpha(r)\mu_{r-1}\big[I(\varrho) \cap \mathcal{F}A(t)\big] \ge \big[\text{mis}\, A(t;\varrho)\big]^{1-\frac{1}{r}} + $$
$$+\big(\text{mis}\, \big[I(\varrho) \setminus A(t;\varrho)\big]\big)^{1-\frac{1}{r}} - \big[\text{mis}\, I(\varrho)\big]^{1-\frac{1}{r}};$$

detto $\tau(t;\varrho)$ il più piccolo dei numeri $\text{mis}\, A(t;\varrho)$ e $\text{mis}\, \big[I(\varrho) \setminus A(t;\varrho)\big]$ sarà certo $\text{mis}\, I(\varrho) \ge 2\tau(t;\varrho)$ e la (10) diventa

$$(11) \qquad 2\alpha(r)\mu_{r-1}\big[I(\varrho) \cap \mathcal{F}A(t)\big] \ge$$
$$\ge \big[\tau(t;\varrho)\big]^{1-\frac{1}{r}} + \big[\text{mis}\, I(\varrho) - \tau(t;\varrho)\big]^{1-\frac{1}{r}} - \big[\text{mis}\, I(\varrho)\big]^{1-\frac{1}{r}} \ge$$
$$\ge 2\big[\tau(t;\varrho)\big]^{1-\frac{1}{r}} - \big[2\tau(t;\varrho)\big]^{1-\frac{1}{r}}.$$

Se ora poniamo

$$(12) \qquad \beta_1 = \frac{\alpha(r)}{1 - 2^{-1/r}}$$

e teniamo presente che, per $\lambda \ge t \ge k$, è certo $\tau(t,\varrho) \ge \tau(k,\lambda;\varrho)$ dalle (8), (11), (12) ricaviamo la (2).

Passiamo ora alla dimostrazione del nostro lemma nell'ipotesi che $w(x)$ non sia semilineare; allora, per noti teoremi sull'approssimazione lineare, potremo trovare una successione di funzioni semilineari nel dominio $\big[I(\varrho) \cup \mathcal{F}I(\varrho)\big]$

$$w_1(x), w_2(x), \ldots, w_n(x), \ldots,$$

---

[4] Si osservi che qui intervengono solo domini le cui frontiere sono contenute nella somma di un numero finito di iperpiani ed ipersuperfici sferiche.

verificanti le

$$(13) \qquad \lim_{n\to\infty} \int_{I(\varrho)} |w(x) - w_n(x)|\, dx_1 \ldots dx_r =$$

$$= \lim_{n\to\infty} \int_{I(\varrho)} |\operatorname{grad} w - \operatorname{grad} w_n|\, dx_1 \ldots dx_r = 0.$$

Detto $A_n(t; \varrho)$ l'insieme dei punti $x \in I(\varrho)$ nei quali è $w_n(x) > t$, per ogni coppia di numeri k, λ (con $k < \lambda$), indicheremo con $\tau_n(k, \lambda; \varrho)$ il più piccolo dei due numeri mis $A_n(\lambda; \varrho)$, mis $[I(\varrho) \setminus A_n(k; \varrho)]$.

Con ragionamento analogo a quello seguito per provare la (2) nel caso di $w(x)$ semilineare, si dimostra la

$$(14) \qquad \beta_1 \int_{[A_n(k;\varrho)\setminus A_n(\lambda;\varrho)]} |\operatorname{grad} w_n|\, dx_1 \ldots dx_r \ge (\lambda - k)[\tau_n(k, \lambda; \varrho)]^{1-\frac{1}{r}}$$

per ogni intero positivo n e per ogni coppia di numeri reali k, λ (con $k < \lambda$). D'altra parte dalle (13) si deduce, per quasi tutti i valori di t,

$$(15) \qquad \lim_{n\to\infty} \operatorname{mis}\left[A_n(t; \varrho) \cup A(t; \varrho) \setminus A_n(t; \varrho) \cap A(t; \varrho)\right] = 0$$

e quindi dalle (13), (14), (15) si deduce che vale la (2) per quasi tutte le coppie k, λ (con $k < \lambda$). Ma è facile vedere che se la (2) vale per quasi tutte le coppie k, λ (con $k < \lambda$) essa vale sempre; basta osservare che

$$(16) \qquad \lim_{\eta\to 0^+} \int_{[A(k+\eta;\varrho)\setminus A(\lambda-\eta;\varrho]} |\operatorname{grad} w|\, dx_1 \ldots dx_r \le$$

$$\le \int_{[A(k;\varrho)\setminus A(\lambda;\varrho)]} |\operatorname{grad} w|\, dx_1 \ldots dx_r,$$

$$(17) \qquad \lim_{\eta\to 0^+} \tau(k + \eta, \lambda - \eta; \varrho) \ge \tau(k, \lambda; \varrho).$$

Lemma III. – *Esiste una costante β_2 tale che, per ogni coppia di numeri ϱ, k soddisfacenti le condizioni*

$$(18) \qquad 0 < \varrho < \delta(y), \quad 0 < 2 \operatorname{mis} A(k; \varrho) \le \operatorname{mis} I(\varrho),$$

sia

$$(19) \qquad \beta_2 \int_{A(k;\varrho)} |\operatorname{grad} w|^2\, dx_1 \ldots dx_r \ge$$

$$\ge \left[\operatorname{mis} A(k; \varrho)\right]^{-\frac{2}{r}} \int_{A(k;\varrho)} (w(x) - k)^2\, dx_1 \ldots dx_r.$$

Dim. Per ogni intero positivo n, indichiamo con λ_n il più piccolo dei numeri λ che verificano la

$$(20) \qquad \operatorname{mis} A(k; \varrho) \ge 2^{nr} \operatorname{mis} A(\lambda; \varrho)$$

e poniamo $\lambda_0 = k$; sarà evidentemente

$$(21) \qquad k = \lambda_0 \leq \lambda_1 \leq \ldots \leq \lambda_n \leq \ldots$$

e, per la definizione degli insiemi $A(\lambda; \varrho)$, $B(\lambda; \varrho)$,

$$(22) \qquad \lim_{\varepsilon \to 0^+} \operatorname{mis} A(\lambda_n + \varepsilon; \varrho) = \operatorname{mis} A(\lambda_n; \varrho) \leq 2^{-rn} \operatorname{mis} A(k; \varrho) \leq$$
$$\leq \operatorname{mis}\big[I(\varrho) \setminus B(\lambda_n; \varrho)\big] = \lim_{\varepsilon \to 0^+} \operatorname{mis} A(\lambda_n - \varepsilon; \varrho).$$

Dalle (22) si deduce l'esistenza di una successione di insiemi

$$(23) \qquad A(k; \varrho) = D_0 \supset D_1 \supset \ldots \supset D_n \supset \ldots$$

tali che per ogni valore di n sia

$$(24) \qquad \operatorname{mis} D_n = 2^{-nr} \operatorname{mis} A(k; \varrho)$$

$$(24') \qquad \big[I(\varrho) \setminus B(\lambda_n; \varrho)\big] \supset D_n \supset A(\lambda_n; \varrho).$$

Dal lemma II e dalle (18), (22), (24') segue

$$(25) \qquad \beta_1 \int_{(D_{n-1} \setminus D_n)} |\operatorname{grad} w|\, dx_1 \ldots dx_r \geq$$
$$\geq \lim_{\varepsilon \to 0_+} \beta_1 \int_{[A(\lambda_{n-1}+\varepsilon;\varrho) \setminus A(\lambda_n - \varepsilon;\varrho)]} |\operatorname{grad} w|\, dx_1 \ldots dx_r \geq$$
$$\geq 2^{-n(r-1)} (\lambda_n - \lambda_{n-1}) \big[\operatorname{mis} A(k; \varrho)\big]^{\frac{r-1}{r}}$$

e quindi per la diseguaglianza di Schwarz e le (24), (25), si trova

$$(26) \qquad \beta_1^2 \int_{(D_{n-1} \setminus D_n)} |\operatorname{grad} w|^2\, dx_1 \ldots dx_r \geq$$
$$\geq \big[\operatorname{mis}(D_{n-1} \setminus D_n)\big]^{-1} \left(\beta_1 \int_{(D_{n-1} \setminus D_n)} |\operatorname{grad} w|\, dx_1 \ldots dx_r \right)^2 \geq$$
$$\geq (\lambda_n - \lambda_{n-1})^2 \big[\operatorname{mis} A(k; \varrho)\big]^{\frac{r-2}{r}} 2^{-n(r-2)-r}.$$

Dalle (23), (26) segue

$$(27) \qquad \beta_1^2 \int_{A(k;\varrho)} |\operatorname{grad} w|^2\, dx_1 \ldots dx_r \geq$$
$$\geq \sum_{n=1}^{\infty} (\lambda_n - \lambda_{n=1})^2 \big[\operatorname{mis} A(k; \varrho)\big]^{\frac{r-2}{r}} 2^{-n(r-2)-r},$$

mentre per le (21), (23), (24), (24') si ha

$$\int_{A(k;\varrho)} (w(x) - k)^2\, dx_1 \ldots dx_r = \sum_{n=1}^{\infty} \int_{(D_{n-1} \setminus D_n)} (w(x) - k)^2\, dx_1 \ldots dx_r \leq$$

$$(28) \quad \leq \sum_{n=1}^{\infty} 2^{r-rn} (\lambda_n - \lambda_0)^2 \operatorname{mis} A(k; \varrho).$$

Se ora indichiamo con θ il maggiore dei numeri

$$(29) \qquad (\lambda_n - \lambda_{n-1})^2\, 2^{-n(r-2)-r} \qquad\qquad (n = 1, 2, \ldots)$$

avremo, per ogni intero positivo m,

$$(30) \qquad (\lambda_m - \lambda_0)^2 = \left[\sum_{n=1}^{m} (\lambda_n - \lambda_{n-1}) \right]^2 \leq \theta m^2 2^{m(r-2)+r},$$

mentre per le (27) si ha

$$(31) \qquad \left[\operatorname{mis} A(k; \varrho) \right]^{\frac{2-r}{r}} \beta_1^2 \int_{A(k;\varrho)} |\operatorname{grad} w|^2\, dx_1 \ldots dx_r \geq \theta.$$

Dalle (28), (30) segue

$$(32) \qquad \int_{A(k;\varrho)} (w(x) - k)^2\, dx_1 \ldots dx_r \leq \theta \operatorname{mis} A(k; \varrho) \sum_{n=1}^{\infty} n^2 2^{2(r-n)}$$

e quindi, qualora si ponga

$$(33) \qquad \beta_1^2 \sum_{n=1}^{\infty} 2^{2(r-n)} n^2 = \beta_2,$$

dalle (31), (32), (33) segue la (19).

Lemma IV. – *Ad ogni numero positivo $\sigma < 1$ può associarsi un numero positivo $\theta(\sigma)$, tale che dalle*

$$(34) \qquad \operatorname{mis} A(k; \varrho) < \varrho^r \theta(\sigma), \quad 0 < \varrho < \delta(y), \quad -\infty < k < +\infty$$

segua

$$(35) \qquad \operatorname{mis} A(k + \sigma c; \varrho - \sigma \varrho) = 0,$$

ove si è posto

$$(36) \qquad c = \varrho^{-\frac{r}{2}} [\theta(\sigma)]^{-\frac{1}{2}} \left[\int_{A(k;\varrho)} (w(x) - k)^2\, dx_1 \ldots dx_r \right]^{\frac{1}{2}}.$$

Dim. Sia $\theta(\varrho)$ il più grande dei numeri θ che simultaneamente soddisfano le

$$(37) \qquad \operatorname{mis} I(\varrho - \sigma\varrho) \geq 2\theta\varrho^r$$

$$(38) \qquad \theta \leq \left[\frac{\sigma^2 2^{-(r+2)}}{1 + \gamma + \beta_2}\right]^r,$$

ove γ e β_2 sono le costanti che intervengono nella (1) del n. 1 e nella (19); supponendo verificate le (34), (36) dobbiamo provare che vale la (35) ed a tale scopo porremo

$$(39) \qquad \varrho_h = \varrho - \sigma\varrho + 2^{-h}\sigma\varrho, \quad k_h = k + \sigma c - 2^{-h}\sigma c.$$

Per la 3^a) proprietà delle funzioni della classe $\mathcal{B}(E; \gamma)$ (vedi n. 1), avremo

$$(40) \qquad \gamma 4^{h+1} \int_{A(k_h;\varrho_h)} (w(x) - k_h)^2 \, dx_1 \ldots dx_r \geq$$

$$\geq \varrho^2 \sigma^2 \int_{A(k_h;\varrho_{h+1})} |\operatorname{grad} w|^2 \, dx_1 \ldots dx_r,$$

mentre per le (34), (37), (39) e per il lemma III si ha

$$(41) \qquad \beta_2 \int_{A(k_h;\varrho_{h+1})} |\operatorname{grad} w|^2 \, dx_1 \ldots dx_r \left[\operatorname{mis} A(k_h; \varrho_h)\right]^{\frac{2}{r}} \geq$$

$$\geq \beta_2 \int_{A(k_h;\varrho_{h+1})} |\operatorname{grad} w|^2 \, dx_1 \ldots dx_r \left[\operatorname{mis} A(k_h; \varrho_{h+1})\right]^{\frac{2}{r}} \geq$$

$$\geq \int_{A(k_h;\varrho_{h+1})} (w(x) - k)^2 \, dx_1 \ldots dx_r.$$

Dalle (40), (41) segue

$$(42) \quad \beta_2 \gamma \sigma^{-2} \varrho^{-2} 4^{(h+1)} \left[\operatorname{mis} A(k_h; \varrho_h)\right]^{\frac{2}{r}} \int_{A(k_h;\varrho_h)} (w(x) - k_h)^2 \, dx_1 \ldots dx_r \geq$$

$$\geq \int_{A(k_h;\varrho_{h+1})} (w(x) - k_h)^2 \, dx_1 \ldots dx_r \geq$$

$$\geq \int_{A(k_{h+1};\varrho_{h+1})} (w(x) - k_{h+1})^2 \, dx_1 \ldots dx_r;$$

d'altra parte, essendo per le (39)

$$(43) \qquad \int_{A(k_h;\varrho_{h+1})} (w(x) - k_h)^2 \, dx_1 \ldots dx_r \geq \sigma^2 c^2 4^{-(h+1)} \operatorname{mis} A(k_{h+1}; \varrho_{h+1}),$$

dalle (42), (43) segue

$$(44) \qquad \sigma^4 c^2 \varrho^2 2^{-4(h+1)} \operatorname{mis} A(k_{h+1}; \varrho_{h+1}) \leq$$

$$\leq \beta_2 \gamma \left[\operatorname{mis} A(k_h; \varrho_h)\right]^{\frac{2}{r}} \int_{A(k_h;\varrho_h)} (w(x) - k_h)^2 \, dx_1 \ldots dx_r.$$

Proviamo ora per induzione che, per ogni intero non negativo ℓ, valgono le

$$(45) \qquad \operatorname{mis} A(k_\ell; \varrho_\ell) \leq \varrho^r \theta(\sigma) 2^{-2r\ell}$$

$$(46) \qquad \int_{A(k_\ell; \varrho_\ell)} (w(x) - k_\ell)^2 \, dx_1 \dots dx_r \leq \varrho^r \theta(\sigma) c^2 2^{-2r\ell};$$

le (45), (46) per $\ell = 0$ sono immediata conseguenza delle (34), (36), (39); d'altra parte se le (45), (46) sono verificate per $\ell = h$, lo sono pure per $\ell = h+1$ in virtù delle (38), (42), (44). Dalle (45) passando al limite per $\ell \to \infty$ e ricordando le (39) si ottiene la (35) e il lemma è dimostrato.

Osservazione I. – Poiché quando $w(x)$ appartiene alla classe $\mathcal{B}(E; \gamma)$ vi appartiene anche la funzione $-w(x)$, accanto ai lemmi II, III, IV sussistono i lemmi analoghi in cui intervengono gli insiemi $B(k; \varrho)$ in luogo degli insiemi $A(k; \varrho)$.

Osservazione II. – Dal lemma IV si deduce che, per ogni numero positivo $p < \delta(y)$ l'oscillazione di $w(x)$ in $I(p)$ è finita[5]. Infatti è sempre possibile trovare due numeri positivi ϱ, σ, che verifichino le

$$(47) \qquad \varrho - \sigma\varrho = p, \qquad\qquad 0 < \varrho < \delta(y)$$

e scegliere poi k e c in modo che siano verificate le (34), (36) e quindi sia, in quasi tutto $I(p)$,

$$(48) \qquad w(x) \leq k + \sigma c;$$

analogamente, per l'osservazione I, può stabilirsi una limitazione inferiore per $w(x)$.

Lemma V. – *Esiste un numero $\eta > 0$ tale che, per $0 < 4\varrho < \delta(y)$, si abbia sempre*

$$(49) \qquad (1 - \eta)\operatorname{osc}(w; 4\varrho) \geq \operatorname{osc}(w; \varrho),$$

ove osc è l'oscillazione di $w(x)$ in $I(\varrho)$ (intesa nel senso precisato dalla nota [5]).

Dim. Detti rispettivamente μ_1, μ_2 i veri estremi superiore ed inferiore di $w(x)$ in $I(4\varrho)$, poniamo

$$(50) \qquad \operatorname{osc}(w; 4\varrho) = \mu_1 - \mu_2 = \omega, \qquad \bar{\mu} = \frac{\mu_1 + \mu_2}{2};$$

per la definizione degli insiemi $A(k; \varrho)$, $B(k; \varrho)$ sarà certo verificata una delle due relazioni

$$(51) \qquad 2 \operatorname{mis} A(\bar{\mu}; 2\varrho) \leq \operatorname{mis} I(2\varrho),$$

---

[5] Il termine oscillazione va naturalmente inteso nel senso della teoria dell'integrazione, come differenza fra il *vero estremo superiore di $w(x)$ in $I(p)$* (cioè il più piccolo dei numeri λ per cui è $\operatorname{mis} A(\lambda; p) = 0$) e il *vero estremo inferiore di $w(x)$ in $I(p)$* (cioè il più grande dei numeri λ per cui è $\operatorname{mis} B(\lambda; p) = 0$).

$$(51') \qquad 2\,\mathrm{mis}\,B(\bar{\mu};2\varrho) \leq \mathrm{mis}\,I(2\varrho).$$

Supponiamo verificata la (51) e poniamo, per ogni numero positivo $\lambda \leq \dfrac{\omega}{4}$,

$$(52) \qquad D(\lambda) = A(\mu_1 - 2\lambda;2\varrho) \setminus A(\mu_1 - \lambda;2\varrho);$$

per il lemma II e le (50), (51), (52) sarà

$$(53) \qquad \beta_1 \int_{D(\lambda)} |\mathrm{grad}\,w|\,dx_1 \ldots dx_r \geq \lambda \big[\mathrm{mis}\,A(\mu_1 - \lambda;2\varrho)\big]^{\frac{r-1}{r}},$$

mentre per la diseguaglianza di Schwarz sarà

$$(54) \qquad \left(\int_{D(\lambda)} |\mathrm{grad}\,w|\,dx_1 \ldots dx_r\right)^2 \leq \mathrm{mis}\,D(\lambda) \int_{D(\lambda)} |\mathrm{grad}\,w|^2\,dx_1 \ldots dx_r.$$

Per la condizione 3^a) del n. 1 si ha poi

$$(55) \qquad \int_{D(\lambda)} |\mathrm{grad}\,w|^2\,dx_1 \ldots dx_r \leq \int_{A(\mu_1 - 2\lambda;2\varrho)} |\mathrm{grad}\,w|^2\,dx_1 \ldots dx_r \leq$$

$$\leq \frac{\gamma}{4\varrho^2} \int_{A(\mu_1 - 2\lambda;4\varrho)} (w(x) - \mu_1 + 2\lambda)^2\,dx_1 \ldots dx_r \leq \gamma\varrho^{-2}\lambda^2 \mathrm{mis}\,I(4\varrho)$$

e quindi per le (53), (54), (55) sarà

$$(56) \qquad \mathrm{mis}\,D(\lambda) \geq \frac{\varrho^2}{\beta_1^2 \gamma}\big[\mathrm{mis}\,A(\mu_1 - \lambda;2\varrho)\big]^{\frac{2r-2}{r}} \cdot \big[\mathrm{mis}\,I(4\varrho)\big]^{-1}.$$

Prendiamo ora un intero n verificante la

$$(57) \qquad \left[\theta\left(\frac{1}{2}\right)\right]^{\frac{2r-2}{r}} n\varrho^{2r} > \gamma\beta_1^2 \mathrm{mis}\,I(4\varrho)\mathrm{mis}\,I(2\varrho),$$

ove $\theta\left(\frac{1}{2}\right)$ è la funzione che compare nelle (34), (36) calcolata per $\sigma = \dfrac{1}{2}$, e poniamo

$$(58) \qquad \eta = 2^{-(n+2)};$$

avremo allora

$$(59) \qquad \lambda \leq \frac{\omega}{4} \qquad \text{per} \qquad \lambda = 2^m\eta\omega, \qquad m = 1,\ldots,n$$

e quindi per le (56) sarà

$$(60) \qquad \mathrm{mis}\,D(2^m\eta\omega) \geq \frac{\varrho^2}{\beta_1^2\gamma}\big[\mathrm{mis}\,A(\mu_1 - 2^m\eta\omega;2\varrho)\big]^{\frac{2r-2}{r}} \cdot \big[\mathrm{mis}\,I(4\varrho)\big]^{-1} \geq$$

$$\geq \frac{\varrho^2}{\beta_1^2\gamma}\big[\mathrm{mis}\,A(\mu_1 - 2\eta\omega;2\varrho)\big]^{\frac{2r-2}{r}} \cdot \big[\mathrm{mis}\,I(4\varrho)\big]^{-1}$$

per $m = 1,\ldots,n$.

Dalle (52), (60) segue

$$(61) \qquad \operatorname{mis} I(2\varrho) \geq \sum_{m=1}^{n} \operatorname{mis} D(2^m \eta \omega) \geq$$

$$\geq n[\operatorname{mis} I(4\varrho)]^{-1} \frac{\varrho^2}{\gamma \beta_1^2} \left[\operatorname{mis} A(\mu_1 - 2\eta\omega; 2\varrho) \right]^{\frac{2r-2}{r}}$$

e quindi dalle (57), (61)

$$(62) \qquad \operatorname{mis} A(\mu_1 - 2\eta\omega; 2\varrho) < \theta\left(\frac{1}{2}\right) \varrho^r < \theta\left(\frac{1}{2}\right)(2\varrho)^r.$$

Ricordando che μ_1 è il vero estremo superiore di $w(x)$ in $I(4\varrho)$, si ha per la (62)

$$(63) \qquad \int_{A(\mu_1 - 2\eta\omega; 2\varrho)} (w(x) - \mu_1 + 2\eta\omega)^2 \, dx_1 \ldots dx_r < (2\varrho)^r (2\eta\omega)^2 \theta\left(\frac{1}{2}\right)$$

e quindi per il lemma IV

$$(64) \qquad \operatorname{mis} A(\mu_1 - \eta\omega, \varrho) = 0,$$

cioè il vero estremo superiore di $w(x)$ in $I(\varrho)$ non supera $(\mu_1 - \eta\omega)$ e quindi vale la (49); alla stessa conclusione saremmo arrivati, per l'osservazione I, supponendo che invece della (51) fosse verificata la (51').

3. – I lemmi stabiliti nel n. 2 consentono la dimostrazione del seguente:

TEOR. I. – *Ogni funzione $w(x) \in \mathcal{B}(E; \gamma)$ è uniformemente hölderiana in ogni insieme chiuso e limitato contenuto in E*[6].

Dim. Dall'osservazione II e dal lemma V del n. 2 si deduce che, per ogni punto $y \in E$, l'oscillazione di $w(x)$ nell'intorno $I(\varrho; y)$ è infinitesima per $\varrho \to 0$; ne seguono l'esistenza in ogni punto $y \in E$ del limite

$$(1) \qquad \bar{w}(y) = \lim_{\varrho \to 0} \left(\int_{I(\varrho;y)} w(x) \, dx_1 \ldots dx_r [\operatorname{mis} I(\varrho; y)]^{-1} \right)$$

e la continuità di $\bar{w}(x)$ in E.

Poniamo ora

$$(2) \qquad \alpha = -\log_4(1 - \eta),$$

ove η è la costante che compare nel lemma V e che, per le (58), (57), (38), (37), (33), (12) del n. 2 è indipendente dalla scelta del punto y in E. Fissato un

---

[6]Al solito la frase va intesa nel senso della teoria dell'integrazione, cioè $w(x)$ è hölderiana essa stessa oppure esiste una funzione hölderiana quasi ovunque eguale a $w(x)$; analoghe considerazioni sono spesso sottintese nel corso di questa memoria (cfr. [1] nota [2]).

insieme chiuso e limitato $C \subset E$ ed un numero positivo p minore della distanza di C dalla frontiera di E, consideriamo l'insieme L descritto dalla quantità

$$\text{(3)} \qquad \frac{2|\bar{w}(x) - \bar{w}(y)|4^\alpha}{p^\alpha}$$

per

$$\text{(4)} \qquad 0 \le |x - y| \le p, \qquad y \in C,$$

ove $|x - y|$ è la distanza dei punti x, y.

Per la continuità di $\bar{w}(x)$ in E, l'insieme L avrà un massimo che indicheremo con τ e quindi, fissato un punto $y \in C$ avremo, riprendendo le notazioni del n. 2 e in particolare del lemma V,

$$\text{(5)} \qquad \operatorname{osc}(w; \varrho) \le \tau \varrho^\alpha \quad \text{per} \quad \frac{p}{4} \le \varrho \le p.$$

Ma, comunque si fissi un numero positivo $\varrho < \dfrac{p}{4}$, esisterà un intero m tale che si abbia

$$\text{(6)} \qquad \frac{p}{4} \le 4^m \varrho < p$$

e quindi, per il lemma V e le (2), (5), (6),

$$\text{(7)} \qquad \operatorname{osc}(w; \varrho) \le (1 - \eta)^m \operatorname{osc}(w; 4^m \varrho) \le \tau \varrho^\alpha;$$

data l'arbitrarietà di C ed y il teorema è dimostrato.

TEOR. II. – *Siano date r^2 funzioni $a_{hl}(x)$ quasi continue nel campo $E \subset S_r$; sia sempre $a_{hl}(x) = a_{lh}(x)$ ed esistano due numeri positivi τ_1, τ_2 tali che, per ogni punto $x \in E$ e per ogni vettore $\lambda \equiv (\lambda_1, \ldots, \lambda_r)$ si abbia*

$$\text{(8)} \qquad \tau_1 |\lambda|^2 \le \sum_{h,l}^{1,r} a_{hl}(x) \lambda_h \lambda_l \le \tau_2 |\lambda|^2;$$

sia poi $w(x)$ una funzione della classe $\mathcal{U}^{(2)}(E)$ tale che, per ogni insieme chiuso e limitato $C \subset E$ e per ogni funzione $g(x) \in \mathcal{U}^{(2)}(E)$ ed identicamente nulla in $(E \setminus C)$, si abbia

$$\text{(9)} \qquad \sum_{h,l}^{1,r} \int_E \frac{\partial g}{\partial x_h} a_{hl}(x) \frac{\partial w}{\partial x_l}\, dx_1 \ldots dx_r = 0.$$

Allora $w(x) \in \mathcal{B}(E; \gamma)$, con $\gamma = \left(\frac{\tau_2}{\tau_1}\right)^2$, e quindi è hölderiana in E per il teor. I.

Dim. Fissati arbitrariamente un punto $y \in E$ ed un numero reale k, poniamo, riprendendo le notazioni usate nel n. 1,

$$\text{(10)} \qquad \begin{aligned} \varphi(x) &\equiv w(x) - k & &\text{per} \quad x \in A(k) \\ \varphi(x) &\equiv 0 & &\text{per} \quad x \in (E \setminus A(k)); \end{aligned}$$

è facile constatare che $\varphi(x)$ appartiene ancora ad $\mathcal{U}^{(2)}(E)$. Presi poi un numero positivo $p < \delta(y)$ ed una funzione $u(t)$ della variabile t continua con la derivata prima $u'(t)$ nell'intervallo $[0, +\infty]$, ed identicamente nulla in $[p, +\infty]$, poniamo

$$(11) \qquad f(x_1, \ldots, x_r) = u(\sqrt{(x_1 - y_1)^2 + \ldots + (x_r - y_r)^2});$$

la funzione $\varphi(x) \cdot f(x)$ appartiene evidentemente alla classe $\mathcal{U}^{(2)}(E)$ ed è identicamente nulla in $[E \setminus I(p; y)]$; ne segue per la (9)

$$(12) \qquad \sum_{h,l}^{1,r} \int_E \frac{\partial(\varphi \cdot f)}{\partial x_h} a_{hl}(x) \frac{\partial w}{\partial x_l}\, dx_1 \ldots dx_r = 0.$$

Essendo $f(x)$ identicamente nulla in $[E \setminus I(p; y)]$ dalle (11), (12) segue

$$(13) \qquad \sum_{h,l}^{1,r} \int_0^p d\varrho \left[u'(\varrho) \int_{\mathcal{F}I(\varrho;y)} n_h a_{hl}(x)\varphi(x) \frac{\partial w}{\partial x_l}\, d\mu_{r-1} + \right.$$
$$\left. + u(\varrho) \int_{\mathcal{F}I(\varrho;y)} a_{hl}(x) \frac{\partial \varphi}{\partial x_h} \frac{\partial w}{\partial x_l}\, d\mu_{r-1} \right] = 0,$$

ove $n_1, \ldots, n_r$ sono i coseni direttori della normale esterna ad $\mathcal{F}I(\varrho; y)$ e $d\mu_{r-1}$ è l'elemento di misura $(r - 1)$ dimensionale. Dalla (13) integrando per parti si ha

$$(14) \qquad \sum_{h,l}^{1,r} \int_0^p d\varrho\, u'(\varrho) \left[\int_{\mathcal{F}I(\varrho;y)} n_h a_{hl}(x)\varphi(x) \frac{\partial w}{\partial x_l}\, d\mu_{r-1} - \right.$$
$$\left. - \int_0^\varrho dt \int_{\mathcal{F}I(t;y)} a_{hl}(x) \frac{\partial \varphi}{\partial x_h} \frac{\partial w}{\partial x_l}\, d\mu_{r-1} \right] = 0,$$

quindi, per l'arbitrarietà di p ed $u(t)$ avremo, per quasi tutti i numeri positivi $\varrho < \delta(y)$,

$$(15) \qquad \int_{\mathcal{F}I(\varrho;y)} \sum_{h,l}^{1,r} n_h a_{hl}\varphi(x) \frac{\partial w}{\partial x_l}\, d\mu_{r-1} = \int_{I(\varrho;y)} \sum_{h,l}^{1,r} a_{hl} \frac{\partial \varphi}{\partial x_h} \frac{\partial w}{\partial x_l}\, dx_1 \ldots dx_r.$$

Dalle (10), (15) si ha poi

$$(16) \qquad \int_{A(k) \cap \mathcal{F}I(\varrho;y)} \sum_{h,l}^{1,r} n_h a_{hl}(x)(w(x) - k) \frac{\partial w}{\partial x_l}\, d\mu_{r-1} =$$
$$\int_{A(k) \cap I(\varrho;y)} \sum_{h,l}^{1,r} a_{hl}(x) \frac{\partial w}{\partial x_h} \frac{\partial w}{\partial x_l}\, dx_1 \ldots dx_r,$$

da cui per la (8)

$$(17) \qquad \tau_2 \int_{A(k)\cap\mathcal{F}I(\varrho;y)} (w(x) - k)|\mathrm{grad}\, w|\, d\mu_{r-1} \ge$$

$$\ge \tau_1 \int_{A(k)\cap I(\varrho;y)} |\mathrm{grad}\, w|^2\, dx_1 \ldots dx_r.$$

Se ora poniamo

$$\psi_1(\varrho) = \int_{A(k)\cap\mathcal{F}I(\varrho;y)} (w(x) - k)^2\, d\mu_{r-1},$$

(18)

$$\psi_2(\varrho) = \int_{A(k)\cap\mathcal{F}I(\varrho;y)} |\mathrm{grad}\, w|^2 d\mu_{r-1},$$

dalle (17), tenendo presente la diseguaglianza di Schwarz, si ricava

$$(19) \qquad \sqrt{\psi_1(\varrho)\psi_2(\varrho)} \ge \frac{\tau_1}{\tau_2} \int_0^\varrho \psi_2(t)dt;$$

posto $\gamma = \left(\frac{\tau_2}{\tau_1}\right)^2$, dalla (19), per un lemma di Caccioppoli-Leray (vedi [4] pag. 153), si ottiene

$$(20) \qquad \int_0^{\varrho_1} \psi_2(\varrho)\, d\varrho \le \frac{\gamma}{(\varrho_2 - \varrho_1)^2} \int_0^{\varrho_2} \psi_1(\varrho)\, d\varrho,$$

per $0 < \varrho_1 < \varrho_2 < \delta(y)$. Dalle (18), (20) segue subito la (1) del n. 1 e poiché quando $w(x)$ soddisfa le ipotesi del teorema esse sono anche soddisfatte da $-w(x)$, è facile convincersi che vale anche la (1') del n. 1 e quindi

$$(21) \qquad w(x) \in \mathcal{B}(E;\gamma).$$

4. – Passiamo finalmente alla dimostrazione dell'annunciato del teorema di analiticità e a tale scopo cominciamo col considerare una funzione $f(p) \equiv f(p_1,\ldots,p_r)$ continua in S_r insieme alle sue derivate parziali prime e seconde; poniamo

$$(1) \qquad f_{hk}(p) = \frac{\partial^2 f}{\partial p_h \partial p_k}, \quad f_h(p) = \frac{\partial f}{\partial p_h} \qquad (\text{per } h, k = 1,\ldots,r)$$

e supponiamo che esistano due numeri positivi μ_1 e μ_2 tali che, per ogni punto $p \in S_r$ e per ogni vettore

$$\lambda \equiv (\lambda_1,\ldots,\lambda_r),$$

si abbia

$$(2) \qquad \mu_1|\lambda|^2 \le \sum_{h,k}^{1,r} f_{hk}(p)\lambda_h \lambda_k \le \mu_2|\lambda|^2.$$

Dati un campo $E \subset S_r$ ed una funzione $u^*(x) \in \mathcal{U}^{(2)}(E)$, diremo che $u^*(x)$ è *estremale in E dell'integrale*

$$(3) \qquad \mathcal{I}[u] = \int f\left(\frac{\partial u}{\partial x_1}, \dots, \frac{\partial u}{\partial x_r}\right) dx_1 \dots dx_r$$

se, per ogni insieme chiuso e limitato $C \subset E$ e per ogni funzione $g(x)$ continua in E con le derivate prime ed identicamente nulla in $(E \setminus C)$ si ha

$$(4) \qquad \sum_{h=1}^{r} \int_E \frac{\partial g}{\partial x_h} f_h\left(\frac{\partial u^*}{\partial x_1}, \dots, \frac{\partial u^*}{\partial x_r}\right) dx_1 \dots dx_r = 0.$$

Vale allora il

Teor. III. – *Ogni estremale in E dell'integrale $\mathcal{I}[u]$ ha le derivate parziali prime uniformemente hölderiane in ogni insieme chiuso e limitato contenuto in E: se poi $f(p)$ è analitica in S_r, ogni estremale è analitica in E.*

Dim. Per le (1), (2) le derivate prime di $f(p)$ soddisferanno una limitazione del tipo

$$(5) \qquad |f_h(p)| \leq \mu_2 |p| + c \qquad (h = 1, \dots, r)$$

ove $|p|$ è la distanza di p dall'origine delle coordinate e c è una costante positiva; poiché la nostra estremale $u^*(x)$ appartiene alla classe $\mathcal{U}^{(2)}(E)$, anche le funzioni

$$f_h\left(\frac{\partial u^*}{\partial x_1}, \dots, \frac{\partial u^*}{\partial x_r}\right)$$

avranno allora, al pari delle derivate di $u^*(x)$, quadrato sommabile in ogni insieme chiuso e limitato contenuto in E.

Ne segue, per noti teoremi sull'approssimazione lineare, che la (4) sarà verificata anche nel caso in cui $g(x)$ appartiene ad $\mathcal{U}^{(2)}(E)$ ed è identicamente nulla al di fuori di un insieme chiuso e limitato contenuto in E.

Consideriamo ora un campo limitato $H \subset E$ avente distanza positiva dalla frontiera di E e siano σ un numero positivo minore di tale distanza, s un intero positivo non superiore ad r. Posto, per ogni intero positivo n,

$$(6) \qquad u_n(x) = u^*\left(x_1, \dots, x_s + \frac{\sigma}{n}, \dots, x_r\right)$$

$u_n(x)$ appartiene evidentemente ad $\mathcal{U}^{(2)}(H)$ e, per ogni insieme chiuso $C \subset H$ e per ogni funzione $v(x) \in \mathcal{U}^{(2)}(H)$ ed identicamente nulla in $(H \setminus C)$, si ha

$$(7) \qquad \sum_{h=1}^{r} \int_H f_h\left(\frac{\partial u_n}{\partial x_1}, \dots, \frac{\partial u_n}{\partial x_r}\right) \frac{\partial v}{\partial x_h} dx_1 \dots dx_r = 0.$$

D'altra parte, posto

$$(8) \qquad w_n^*(x) = u_n(x) - u^*(x),$$

anche $w_n^*(x)$ apparterrà ad $\mathcal{U}^{(2)}(H)$ e sarà

$$(9) \qquad f_h\left(\frac{\partial u_*}{\partial x_1}, \ldots, \frac{\partial u_*}{\partial x_r}\right) - f_h\left(\frac{\partial u_n}{\partial x_1}, \ldots, \frac{\partial u_n}{\partial x_r}\right) + \sum_k^{1,r} a_{hk}^{(n)}(x)\frac{\partial w_n^*}{\partial x_k} = 0,$$

ove si è posto

$$(10) \qquad a_{hk}^{(n)}(x) = \int_0^1 f_{hk}\left(\frac{\partial u^*}{\partial x_1} + t\frac{\partial w_n^*}{\partial x_1}, \ldots, \frac{\partial u^*}{\partial x_1} + t\frac{\partial w_n^*}{\partial x_r}\right) dt.$$

Dalle (2), (10) si deduce, per ogni $x \in H$ e per ogni vettore λ,

$$(11) \qquad \mu_1|\lambda|^2 \leq \sum_{h,k}^{1,r} a_{hk}^{(n)}(x)\lambda_h\lambda_k \leq \mu_2|\lambda|^2,$$

mentre dalle (4), (7), (9) segue la

$$(12) \qquad \sum_{h,k}^{1,r} \int_H a_{hk}^{(n)}(x)\frac{\partial w_n^*}{\partial x_k}\frac{\partial v}{\partial x_h}\, dx_1 \ldots dx_r = 0$$

per ogni funzione $v(x) \in \mathcal{U}^{(2)}(H)$ ed identicamente nulla fuori di un insieme chiuso contenuto in H; per il teor. II possiamo allora asserire che $w_n^*(x)$ appartiene a $\mathcal{B}(H;\gamma)$ con

$$(13) \qquad \gamma = \frac{\mu_2^2}{\mu_1^2}.$$

Apparterranno allora a $\mathcal{B}(H;\gamma)$ anche le funzioni

$$w_n(x) = \frac{w_n^*(x)n}{\sigma}$$

e, poiché la successione

$$(14) \qquad w_1(x), \ldots, w_n(x), \ldots$$

converge in media verso $\dfrac{\partial u^*}{\partial x_s}$ nell'insieme H (che abbiamo supposto limitato ed avente distanza positiva dalla frontiera di E) per il lemma I anche tale derivata appartiene a $\mathcal{B}(H;\gamma)$. Data l'arbitrarietà con cui sono stati scelti il campo H e il numero s possiamo allora concludere che tutte le derivate prime di $u^*(x)$ appartengono a $\mathcal{B}(E;\gamma)$: il loro carattere hölderiano segue allora dal teor. I e, nel caso in cui $f(p)$ sia analitica, $u^*(x)$ risulta tale per il teor. 9.2 di [6], che assicura l'esistenza delle derivate continue di qualsiasi ordine per le estremali di classe C^1, e per i risultati di Stampacchia e Hopf (vedi [9] teor. VII).

Bibliografia

[1] E. De Giorgi, *Sull'analiticità delle estremali degli integrali multipli.* Rend. Lincei, Serie VIII, Vol. XX, fasc. 4, (1956).

[2] E. Hopf, *Zum analytischen Charakter der Lösungen regulärer zweidimensionalen Variations-probleme.* Math. Zeitschrift, Band 30 pp. 404-413 (1929).

[3] E. Hopf, *Über den funktionalen, insbesondere den analytischen Charakter der Lösungen elliptischer Differentialgleichungen zweiter Ordnung.* Math. Zeitschrift, Band 34 pp. 194-233 (1932).

[4] C. Miranda, *Equazioni alle derivate parziali di tipo ellittico.* Ed. Springer-Verlag, Berlino (1955).

[5] C. B. Morrey, *Multiple integral problems in the calculus of variations and related topics.* Univ. California Publ. Math. (N.S.) Vol. I (1943).

[6] C. B. Morrey, *Second order elliptic systems of differential equations.* Annals of Math. Studies, N. 33, Princeton Un. Press (1954).

[7] M. Schiffman, *Differentiability and analyticity of solutions of double integral variational problems.* Annals of Math., Vol. 48 (1947).

[8] A. G. Sigalov, *On conditions of differentiability and analyticity of solutions of two dimensional problems of the calculus of variations.* Doklady Akad. Nauk. S.S.S.R., (N. S.) 85 (1952) (Math. Reviews, Vol. 14 n. 3).

[9] G. Stampacchia, *Sistemi di equazioni di tipo ellittico a derivate parziali del primo ordine e proprietà delle estremali degli integrali multipli.* Ricerche di Matematica, Vol. I, pp. 200-226 (1952).

On the isoperimetric property of the hypersphere in the class of sets whose oriented boundary has finite measure[‡][†]

Memoir by Ennio De Giorgi[*]

Summary. In a note (see [2] in the references) on some problems in Calculus of Variations, I claimed, among other results, that the isoperimetric property of the circle, of the sphere (and more generally of the hypersphere in a space of arbitrary dimension) is still valid provided we assume the notion of measure of its oriented boundary in the sense of Caccioppoli (see [1], [4]). This result (as well as the other results listed in [2]) is a consequence of the theorems proved in [4], [5], but its proof requires some further considerations, which are the subject of the present paper.

1. We begin by recalling some definitions and results contained in [4], [5], which will be used in the sequel. We denote by S_r an r-dimensional Euclidean space, whose generic point will be indicated by $x \equiv (x_1, \dots, x_r)$, and by $|x|$ the distance of x from the origin of the coordinates (that is $\sqrt{x_1^2 + \cdots + x_r^2}$); given two points $x \equiv (x_1, \dots, x_r)$, $y \equiv (y_1, \dots, y_r)$, we denote by $(x - y)$ the point $(x_1 - y_1, \dots, x_r - y_r)$; finally, whenever we mention a set contained in S_r and a function defined on it, we always mean a Borel set and a Baire function, respectively. Given a set $E \subset S_r$ and setting, for $\lambda > 0$,

$$(1) \qquad \psi(x; \lambda) = \frac{1}{\sqrt{\pi^r \lambda^r}} \int_E e^{-\frac{|x - \xi|^2}{\lambda}} \, d\xi_1 \dots d\xi_r,$$

we define *perimeter* of E the limit

$$(2) \qquad \lim_{\lambda \to 0} \int_{S_r} |\operatorname{grad}_x \psi(x; \lambda)| \, dx_1 \dots dx_r,$$

which will be denoted by the symbol $P(E)$. Then the following theorem holds (see [4], theorems II, IV).

I. *Given a set $E \subset S_r$, the perimeter $P(E)$ is finite if and only if there exists a vector-valued set function $\Phi(B) \equiv [\Phi_1(B), \dots, \Phi_r(B)]$ satisfying the following conditions:*

[‡]Editor's note: translation into English of the paper "Sulla proprietà isoperimetrica dell'ipersfera, nella classe degli insiemi aventi frontiera orientata di misura finita", published in Atti Accad. Naz. Lincei Mem. Cl. Sci. Fis. Mat. Nat. Sez. I, (8) 5 (1958), 33–44.

[†]Research carried out at the Istituto Nazionale per le Applicazioni del Calcolo.

[*]Communicated in the session of january 11, 1958, by the Member M. PICONE

a) *$\Phi(B)$ is defined on any set $B \subset S_r$, is countably additive and with finite total variation.*

b) *Given any continuous function $g(x)$ in S_r having continuous first derivatives, which is infinitesimal, together with its first derivatives, of order not smaller than $|x|^{-(r+1)}$ as $|x| \to \infty$, it results*

$$(3) \qquad \int_E \frac{\partial g}{\partial x_h} \, dx_1 \ldots dx_r = \int_{S_r} g(x) \, d\Phi_h \qquad \text{(for } h = 1, \ldots, r\text{)}.$$

In this case the perimeter $P(E)$ coincides with the total variation in S_r of the function $\Phi(B)$, that is we have[1]

$$(4) \qquad P(E) = \int_{S_r} |d\Phi|.$$

The function $\Phi(B)$ in the above theorem will be called the *Gauss-Green function corresponding to the set E* (see [5] n. 3).

Notice that, when E is a *polygonal domain* (that is a domain whose boundary is contained in the union of a finite number of hyperplanes) and its boundary $\mathcal{F}E$ is bounded, than also the perimeter of E is finite, and equals the $(r-1)$-dimensional measure of $\mathcal{F}E$ (defined in an elementary way). In this case the Gauss-Green function $\Phi(B) \equiv [\Phi_1(B), \ldots, \Phi_r(B)]$ corresponding to the set E is given by the formulas

$$(5) \qquad \Phi_h(B) = \int_{B \cap \mathcal{F}E} n_h \, d\mu_{r-1}, \qquad \text{for } h = 1, \ldots, r,$$

where n_h is the cosine formed by the outer normal to $\mathcal{F}E$ with the x_h-axis, while $d\mu_{r-1}$ is the $(r-1)$-dimensional measure on $\mathcal{F}E$. In order to be convinced of this fact, it is enough to compare (3), (5) with the usual Gauss-Green formulas relatively to E.

Given a sequence of (scalar or vector-valued) functions

$$(6) \qquad F_1(B), F_2(B), \ldots, F_n(B), \ldots$$

defined on any set $B \subset S_r$ and countably additive, we say that the sequence (6) *weakly converges* to a function $F(B)$ if, for any continuous function $g(x)$ in S_r which is infinitesimal as $|x| \to \infty$, we have

$$(7) \qquad \lim_{n \to \infty} \int_{S_r} g(x) \, dF_n = \int_{S_r} g(x) \, dF.$$

We will denote by Σ the metric space whose elements are the subsets of S_r and such that the distance between two sets E_1, E_2 is given by meas $(E_1 \cup E_2 \setminus E_1 \cap$

---

[1] We recall that, given a (scalar or vector-valued) set function $\psi(B)$, defined on any set $B \subset S_r$ and countably additive, we denote by the symbol $\int_B |d\psi|$ the total variation of ψ in the set B.

E_2). Then the following theorems hold (see [4] theorems VII, VIII, [5] theorem I).

II. *Given a sequence of subsets of S_r*

$$\text{(8)} \qquad E_1, \ldots, E_n, \ldots$$

converging (in the metric space Σ) to a set E, we have

$$\text{(9)} \qquad \min \lim_{n \to \infty} P(E_n) \geq P(E).$$

If the set of the perimeters $P(E_n)$ is bounded, the sequence of the Gauss-Green functions corresponding to the sets in (8) weakly converges to the Gauss-Green function corresponding to E.

III. *Given a set $E \subset S_r$ (with $r \geq 2$) its perimeter $P(E)$ equals the lower limit of the perimeters of the polygonal domains Π approximating E (in the space Σ), that is we have*

$$\min \lim_{\Pi \to E} P(\Pi) = P(E).$$

IV. *Given a bounded set $L \subset S_r$ and a positive number p, the class of all subsets of L having perimeter $\leq p$ is compact in the space Σ.*

2. Let us now turn to the proof of three simple theorems which, in a certain sense, make more precise or complete some of the results of n. I.

I. *Let*

$$\text{(1)} \qquad E_1, \ldots, E_n, \ldots$$

be a sequence of subsets of S_r having bounded perimeters and converging (in the space Σ) to a set E and satisfying the limit relation

$$\text{(2)} \qquad \lim_{n \to \infty} P(E_n) = P(E).$$

Let $\Phi(B)$ be the Gauss-Green function corresponding to E, $\Phi^{(n)}(B)$ the Gauss-Green function corresponding to E_n, $\mu(B)$ the total variation of Φ in B, and $\mu_n(B)$ the total variation of $\Phi^{(n)}$. We then have

$$\text{(3)} \qquad \lim_{n \to \infty} \mu_n(B) = \mu(B)$$

for any set B whose boundary satisfies the condition

$$\text{(4)} \qquad \mu(\mathcal{F}B) = 0.$$

Proof. Given a set B satisfying (4) and a number $\varepsilon > 0$, by known theorems in integration theory we can find a continuous vector-valued function $g(x) \equiv [g_1(x), \ldots, g_r(x)]$ in S_r, which is infinitesimal as $|x| \to \infty$ and satisfies the following conditions:

$$\text{(5)} \qquad |g(x)| \leq 1 \quad \text{for } x \in B; \qquad |g(x)| = 0 \quad \text{for } x \in (S_r \setminus B)$$

$$(6) \qquad \sum_{h=1}^{r} \int_{S_r} g_h(x)\, d\Phi_h \geq \int_B |d\Phi| - \varepsilon = \mu(B) - \varepsilon.$$

On the other hand, by theorem II n. I we have

$$(7) \qquad \lim_{n \to \infty} \sum_{h=1}^{r} \int_{S_r} g_h(x)\, d\Phi_h^{(n)} = \sum_{h=1}^{r} \int_{S_r} g_h(x)\, d\Phi_h.$$

Hence, since by (5) it follows

$$(8) \qquad \sum_{h=1}^{r} \int_{S_r} g_h(x)\, d\Phi_h^{(n)} \leq \int_B |d\Phi^{(n)}| = \mu_n(B),$$

we have, comparing (6), (7), (8),

$$(9) \qquad \min \lim_{n \to \infty} \mu_n(B) \geq \mu(B) - \varepsilon.$$

As ε is arbitrary, (9) can be replaced by the inequality

$$(9') \qquad \min \lim_{n \to \infty} \mu_n(B) \geq \mu(B).$$

Since the boundary of $(S_r \setminus B)$ coincides with $\mathcal{F}B$, beside (9') we have the similar relation

$$(10) \qquad \min \lim_{n \to \infty} \mu_n(S_r \setminus B) \geq \mu(S_r \setminus B),$$

while from (2) it follows

$$(11) \quad \lim_{n \to \infty} [\mu_n(B) + \mu_n(S_r \setminus B)] = \lim_{n \to \infty} \mu_n(S_r) = \lim_{n \to \infty} P(E_n)$$
$$= P(E) = \mu(S_r) = \mu(B) + \mu(S_r \setminus B).$$

Finally, from (9'), (10), (11), relation (3) follows.

II. *If k is a real number, $f(y) = f(y_1, \ldots, y_{r-1})$ is a function defined in an interval T of the space S_{r-1} (with $r \geq 2$), which is uniformly Lipschitz continuous and everywhere larger than k, then the domain D of the space S_r consisting of all points $(x_1, \ldots, x_r)$ which satisfy the conditions*

$$(12) \qquad (x_1, \ldots, x_{r-1}) \in T, \qquad k \leq x_r \leq f(x_1, \ldots, x_{r-1})$$

has finite perimeter. Moreover, if H is a set contained in $(T \setminus \mathcal{F}T)$, H^* is the set of all points $(x_1, \ldots, x_r)$ satisfying the conditions

$$(13) \qquad (x_1, \ldots, x_{r-1}) \in H, \qquad x_r = f(x_1, \ldots, x_{r-1}),$$

and $\mu(B)$ is the total variation of the Gauss-Green function corresponding to D in the set B, we have

$$(14) \qquad \mu(H^*) = \int_H \sqrt{1 + \left(\frac{\partial f}{\partial y_1}\right)^2 + \cdots + \left(\frac{\partial f}{\partial y_{r-1}}\right)^2}\, dy_1 \ldots dy_{r-1}.$$

Proof. Clearly, the domain D can be approximated by a sequence of polygonal domains

$$(15) \qquad\qquad D_1, \ldots, D_n, \ldots,$$

such that D_n consists of all points $(x_1, \ldots, x_r)$ satisfying the conditions

$$(16) \qquad k \le x_r \le f_n(x_1, \ldots, x_{r-1}), \qquad (x_1, \ldots, x_{r-1}) \in T$$

and where the functions $f_1(y), \ldots, f_n(y), \ldots$ are continuous in T, are $> k$ and satisfy the relations

$$(17) \qquad \lim_{n \to \infty} f_n(y) = f(y) \quad \text{uniformly in } T,$$

$$(18) \qquad \lim_{n \to \infty} \int_T \left| \frac{\partial f}{\partial y_h} - \frac{\partial f_n}{\partial y_h} \right| dy_1 \ldots dy_{r-1} = 0 \qquad \text{for } h = 1, \ldots, r-1.$$

Recalling that the perimeter of a polygonal domain coincides with the $(r-1)$-dimensional measure of its boundary and taking into account (17), (18), it is easy to see that the perimeters of the domains in (15) are bounded.[2] Therefore, by theorem II n. I, also D has finite perimeter. If we take now a function $g(x) \equiv g(x_1, \ldots, x_r)$ which is continuous in S_r together with its first partial derivatives and vanishes for $(x_1, \ldots, x_{r-1}) \in \mathcal{F}T$, it is easy to show that

$$(19) \qquad \int_D \frac{\partial g}{\partial x_h} \, dx_1 \ldots dx_r$$
$$= - \int_T g[y_1, \ldots, y_{r-1}, f(y)] \frac{\partial f}{\partial y_h} \, dy_1 \ldots dy_{r-1} \quad (\text{for } h = 1, \ldots, r-1),$$

$$(19') \qquad \int_D \frac{\partial g}{\partial x_r} \, dx_1 \ldots dx_r$$
$$= - \int_T \{ g[y_1, \ldots, y_{r-1}, f(y)] - g(y_1, \ldots, y_{r-1}, k) \} \, dy_1 \ldots dy_{r-1}.$$

Indeed, to this purpose it is enough to write the Gauss-Green formulas relative to the polygonal domain D_n and to the function $g(x)$ and to pass to the limit as $n \to \infty$ (taking into account (17), (18)).

If we now denote by $\Phi(B) \equiv (\Phi_1(B), \ldots, \Phi_r(B))$ the Gauss-Green function corresponding to D, by V a subset of H, by V^* the set of all points satisfying the conditions

$$(20) \qquad (x_1, \ldots, x_{r-1}) \in V, \qquad x_r = f(x_1, \ldots, x_{r-1}),$$

---

[2] Indeed we find, by means of elementary considerations,

$$P(D_n) = \operatorname{meas} T + \int_{\mathcal{F}T} [f_n(y) - k] d\mu_{r-2} + \int_T \sqrt{1 + \left(\frac{\partial f_n}{\partial y_1} \right)^2 + \cdots + \left(\frac{\partial f_n}{\partial y_{r-1}} \right)^2} \, dy_1 \ldots dy_{r-1}.$$

from (19), (19') we deduce, taking into account the arbitrariness of $g(x)$ and the definition of Gauss-Green function,

$$(21) \qquad \Phi_h(V^*) = -\int_V \frac{\partial f}{\partial y_h}\, dy_1 \ldots dy_{r-1} \qquad \text{(for } h = 1, 2, \ldots, r - 1\text{)},$$

$$(21') \qquad\qquad\qquad \Phi_r(V^*) = \operatorname{meas} V.$$

Therefore, since (21), (21') are valid for any set $V \subset H$, formula (14) is proved.

III. *Given two finite perimeter subsets E_1, E_2 of the space S_r, let $\Phi(B)$, $\Psi(B)$ be their Gauss-Green functions, respectively. Let A be an open subset of S_r satisfying the condition*

$$(22) \qquad\qquad\qquad E_1 \cap A = E_2 \cap A.$$

Then for any $B \subset A$ we have

$$(23) \qquad\qquad\qquad \Phi(B) = \Psi(B).$$

Proof. Let $g(x)$ be any continuous function in S_r having continuous first derivatives, which is infinitesimal, together with its first derivatives, of order higher than $|x|^{-(r+1)}$ as $|x| \to \infty$, and is identically zero in $(S_r \setminus A)$. By (22) and by the definition of Gauss-Green function we have

$$
\begin{aligned}
(24) \qquad \int_{S_r} g(x)\, d\Phi \;&=\; \int_{E_1} \operatorname{grad} g(x)\, dx_1 \ldots dx_r \\
&=\; \int_{E_2} \operatorname{grad} g(x)\, dx_1 \ldots dx_r = \int_{S_r} g(x)\, d\Psi.
\end{aligned}
$$

Equalities (24) hold for any continuous function $g(x)$ in S_r having continuous first derivatives, which is infinitesimal, together with its first derivatives, of order higher than $|x|^{-(r+1)}$ as $|x| \to \infty$, and is identically zero in $(S_r \setminus A)$. By known theorems on linear approximation of functions, it follows that the first term equals the fourth term in (24), also under the assumption that $g(x)$ is a bounded function in S_r identically zero in $(S_r \setminus A)$. In particular, $g(x)$ can be the characteristic function of a set $B \subset A$ and therefore the theorem is proved.

3. Given a set $E \subset S_r$ (with $r \geq 2$) and a hyperplane I of S_r, we will say that E is *pointwise normal with respect to I* if, given any line orthogonal to I, the intersection of such a line with E is either a segment, or a point, or the empty set. On the other hand, we will say that E is *normal in mean* (or simply *normal*) with respect to I if it is equivalent[3] to a pointwise normal set. A set which is pointwise normal with respect to every hyperplane is obviously convex, hence a set which is simply normal in mean with respect to every hyperplane is equivalent to a convex set.

In this section we shall prove a theorem on symmetric normal sets with respect to a hyperplane which, together with theorem IV of n. I, has a crucial role

---

[3]Two sets E_1, E_2 are said to be equivalent if $\operatorname{meas}(E_1 \cup E_2 \setminus E_1 \cap E_2) = 0$.

in the proof of the isoperimetric property of the hypersphere. To this purpose, it is convenient to introduce the following lemma.

I. *Let $\alpha(B)$, $\gamma(B)$ be two vector-valued set functions defined on any $B \subset S_r$, countably additive and with finite total variation. If*

$$(1) \qquad [\alpha(B); \gamma(B) - \alpha(B)] \geq 0 \ ^4$$

for any set $B \subset S_r$, then there also holds

$$(2) \qquad 2 \int_B |d\gamma| \left(\int_B |d\gamma| - \int_B |d\alpha| \right) \geq \left(\int_B |d(\gamma - \alpha)| \right)^2$$

for any set $B \subset S_r$.

Proof. From (1) it immediately follows

$$(3) \qquad |\gamma(B)|^2 \equiv [\gamma(B); \gamma(B)] \geq |\alpha(B)|^2 + |\gamma(B) - \alpha(B)|^2.$$

Hence, if we let $\mu(B)$ be the total variation of γ in the set B and

$$(4) \qquad \varphi(x) = \frac{d\gamma}{d\mu}, \qquad \psi(x) = \frac{d\alpha}{d\mu},$$

we will have, from known theorems on differentiation of set functions,

$$(5) \qquad |\varphi| = 1, \qquad 1 - |\psi|^2 = |\varphi|^2 - |\psi|^2 \geq |\varphi - \psi|^2, \qquad |\psi| \leq 1,$$

$$(6)$$
$$\int_B |d\gamma| = \int_B d\mu, \qquad \int_B |d\alpha| = \int_B |\psi| \, d\mu, \qquad \int_B |d(\gamma - \alpha)| = \int_B |\varphi - \psi| \, d\mu.$$

From (5) it follows

$$(7) \qquad 2 \int_B (1 - |\psi|) \, d\mu \geq \int_B (1 - |\psi|^2) \, d\mu \geq \int_B |\varphi - \psi|^2 \, d\mu,$$

while, by Schwarz inequality, we have

$$(8) \qquad \int_B |\varphi - \psi|^2 \, d\mu \int_B d\mu \geq \left(\int_B |\varphi - \psi| \, d\mu \right)^2.$$

Therefore (2) follows from (6), (7), (8).

Let us now pass to the proof of the stated theorem concerning symmetric normal sets with respect to a hyperplane.

II. *Let E be a subset of the space S_r (with $r \geq 2$) having finite perimeter and finite measure. For any point $y \equiv (y_1, \ldots, y_{r-1})$ of the space S_{r-1}, we denote by $f(y) \equiv f(y_1, \ldots, y_{r-1})$ the linear measure of the intersection of E with the line $R(y)$ of S_r (whose generic point will be always indicated by $(x_1, \ldots, x_r)$) having*

---

^{4}Given two vectors $u \equiv (u_1, \ldots, u_n)$, $v \equiv (v_1, \ldots, v_n)$ we denote by $[u; v] = \sum_{h=1}^n u_h v_h$ their scalar product.

equations $x_1 = y_1, \ldots, x_{r-1} = y_{r-1}$. Let L be the set[5] of all points $(x_1, \ldots, x_r)$ for which

$$(9) \qquad f(x_1, \ldots, x_{r-1}) > 2x_r > -f(x_1, \ldots, x_{r-1}).$$

Then

$$(10) \qquad P(L) \le P(E).$$

If equality holds in (10), then E is normal with respect to the hyperplane $x_r = 0$ and, denoting by $\mu(B)$ the total variation of the Gauss-Green function corresponding to E in the set B, and by $\nu(B)$ the total variation of the Gauss-Green function corresponding to L, we have, for any $M \subset S_{r-1}$,

$$(11) \qquad \mu(M \times S_1) = \nu(M \times S_1),$$

where $(M \times S_1)$ is the set of all points $(x_1, \ldots, x_r)$ satisfying the conditions

$$(12) \qquad (x_1, \ldots, x_{r-1}) \in M, \quad -\infty < x_r < +\infty.$$

(i.e., the topological product of M and S_1).

Proof. By theorem III n. I there exists a sequence of polygonal domains

$$(13) \qquad E_1, \ldots, E_n, \ldots,$$

satisfying the conditions

$$(14) \qquad \lim_{n \to \infty} \operatorname{meas}(E_n \cup E \setminus E_n \cap E) = 0, \qquad \lim_{n \to \infty} P(E_n) = P(E).$$

Moreover, we can suppose that, for any integer $n > 0$, the normal to the boundary of E_n is never parallel to the hyperplane $x_r = 0$. This is surely possible since, if for some value of n this assumption is not satisfied, it can always be achieved by performing an arbitrary small rotation of the domain E_n.

Moreover, let us introduce the following notation: D_n is the polygonal domain of the space S_{r-1} consisting of all points y such that the line $R(y)$ has nonempty intersection with E_n, $f_n(y)$ is the linear measure of such an intersection, and L_n is the polygonal domain consisting of all points $(x_1, \ldots, x_r)$ satisfying the conditions

$$(15) \qquad f_n(x_1, \ldots, x_{r-1}) \ge 2x_r \ge -f_n(x_1, \ldots, x_{r-1}); \qquad (x_1, \ldots, x_{r-1}) \in D_n.$$

Taking into account the first of relations (14) and the definition of L and L_n, one can immediately verify that

$$(16) \qquad \lim_{n \to \infty} \operatorname{meas}(L \cup L_n \setminus L \cap L_n) = 0.$$

---

[5]The geometric meaning of the construction of L is clear; L is obtained by E through a *normalization* and a *symmetrization* with respect to the hyperplane $x_r = 0$. Similar observations hold for the sets L_n defined by (15).

Let us now fix a value of the index n and let us denote by $\Phi(B) \equiv [\Phi_1(B), \ldots, \Phi_r(B)]$ the Gauss-Green function corresponding to E_n, by $\Psi(B) \equiv [\Psi_1(B), \ldots, \Psi_r(B)]$ the Gauss-Green function corresponding to L_n, by $\mu_n(B)$ the total variation of Φ in B, and by $\nu_n(B)$ the total variation of Ψ. In addition, for any set $M \subset S_{r-1}$, let $\alpha_h(M)$ be the total variation of $\Psi_h(B)$ in $(M \times S_1)$, and $\gamma_h(M)$ be the total variation of $\Phi_h(B)$.

In order to obtain some properties of the functions α_h, γ_h which will be useful in the sequel, let us begin by observing that, since E_n is a polygonal domain of S_r and since the normal to $\mathcal{F}E_n$ is never parallel to the hyperplane $x_r = 0$, $f_n(y)$ is continuous in D_n, vanishes on $\mathcal{F}D_n$ and the domain D_n can be decomposed in a finite number of polygonal domains $G_1, \ldots, G_m$ having the following properties:

a) For any positive integer $k \leq m$, the number of points where the line $R(y)$ intersects $\mathcal{F}E_n$ is constant with respect to y in the interior of G_k.

b) Denoting by $p(k)$ such a number (which is even and ≥ 2) and letting $g_{k,1}(y)$,
$g_{k,2}(y), \ldots, g_{k,p(k)}(y)$ be the r-th coordinates of the $p(k)$ points belonging to both $R(y)$ and $\mathcal{F}E_n$ (considered in increasing order), the functions $g_{k,1}(y), \ldots, g_{k,p(k)}(y)$ are Lipschitz continuous in the interior of G_k.

Using the proprties a), b) and recalling the definition of $f_n(y)$, one can immediately see that the following inequalities

$$(17) \qquad \left| \frac{\partial f_n}{\partial y_h} \right| \leq \sum_{l=1}^{p(k)} \left| \frac{\partial g_{k,l}}{\partial y_h} \right|, \qquad \text{for } y \in (G_k \setminus \mathcal{F}G_k), \qquad h = 1, \ldots, r-1$$

hold. Recalling (5), n. I and taking into account the fact that the normal to $\mathcal{F}E_n$ is never parallel to the hyperplane $x_r = 0$, we find the following expressions for the functions α_h, γ_h, μ_n, ν_n :

$$(18) \qquad \gamma_h(M) = \sum_{k=1}^{m} \int_{M \cap G_k} \sum_{l=1}^{p(k)} \left| \frac{\partial g_{k,l}}{\partial y_h} \right| \, dy_1 \ldots dy_{r-1} \qquad \text{for } h = 1, \ldots, r-1,$$

$$(19) \qquad \alpha_h(M) = \sum_{k=1}^{m} \int_{M \cap G_k} \left| \frac{\partial f_n}{\partial y_h} \right| \, dy_1 \ldots dy_{r-1} \qquad \text{for } h = 1, \ldots, r-1,$$

$$(20) \qquad \gamma_r(M) = \sum_{k=1}^{m} p(k) \operatorname{meas}(M \cap G_k), \qquad \alpha_r(M) = 2 \sum_{k=1}^{m} \operatorname{meas}(M \cap G_k),$$

$$(21)$$

$$\mu_n(M \times S_1) = \sum_{k=1}^{m} \int_{M \cap G_k} \sum_{l=1}^{p(k)} \sqrt{1 + \left(\frac{\partial g_{k,l}}{\partial y_1} \right)^2 + \cdots + \left(\frac{\partial g_{k,l}}{\partial y_{r-1}} \right)^2} \, dy_1 \ldots dy_{r-1},$$

$$(22) \quad \nu_n(M \times S_1) = \sum_{k=1}^{m} \int_{M \cap G_k} \sqrt{4 + \left(\frac{\partial f_n}{\partial y_1}\right)^2 + \cdots + \left(\frac{\partial f_n}{\partial y_{r-1}}\right)^2} \, dy_1 \ldots dy_{r-1}.$$

If we introduce the vector-valued set functions $\alpha \equiv [\alpha_1(M), \ldots, \alpha_r(M)]$, $\gamma \equiv [\gamma_1(M), \ldots, \gamma_r(M)]$, from (18), (20), (21) it follows

$$(23) \quad \mu_n(M \times S_1) \geq \int_M |d\gamma|,$$

and from (19), (20), (22) it follows

$$(24) \quad \nu_n(M \times S_1) = \int_M |d\alpha|.$$

From (17), (18), (19), (20) we deduce the relations

$$(25) \quad 0 \leq \alpha_h(M) \leq \gamma_h(M) \qquad \text{for } h = 1, \ldots, r, \qquad M \subset S_{r-1},$$

and we see that, denoting by H_n the set of all points $y \in S_{r-1}$ such that the line $R(y)$ meets $\mathcal{F}E_n$ at more than two points, we always have

$$(26) \quad 0 \leq \alpha_r(M) \leq \gamma_r(M) - 2\text{meas}\,(H_n \cap M).$$

From (25) we deduce the relation

$$(27) \quad [\alpha(M); \gamma(M) - \alpha(M)] = \sum_{h=1}^{r} \alpha_h(M)(\gamma_h(M) - \alpha_h(M)) \geq 0,$$

while from (26) it follows

$$(28) \quad \int_M |d(\gamma - \alpha)| \geq |\gamma(M) - \alpha(M)| \geq 2\text{meas}\,(H_n \cap M).$$

From theorem I and (23), (24), (27) we have

$$(29) \quad \mu_n(M \times S_1) \geq \nu_n(M \times S_1)$$

and, recalling (28),

$$(30) \quad \mu_n(M \times S_1)[\mu_n(M \times S_1) - \nu_n(M \times S_1)] \geq 2[\text{meas}\,(H_n \cap M)]^2.$$

In particular, for $M = S_{r-1}$, inequalities (29) and (30) become

$$(31) \quad \mu_n(S_r) = P(E_n) \geq \nu_n(S_r) = P(L_n)$$

$$(32) \quad P(E_n)[P(E_n) - P(L_n)] \geq 2(\text{meas}\, H_n)^2.$$

From (31), passing to the limit as $n \to \infty$ and taking (14), (16) and theorem II, n. I, into account, inequality (10) follows. Moreover, from (32) we see that, in order that the equality holds in (10), the following relations must be simultaneously satisfied:

$$(33) \qquad \lim_{n\to\infty} \operatorname{meas} H_n = 0, \qquad \lim_{n\to\infty} P(L_n) = P(L).$$

Recalling the definition of H_n, from (14) and the first of (33) we see that E must be normal with respect to the hyperplane $x_r = 0$. On the other hand, from (14), (16) and the last of (33) we deduce, recalling also theorem I, n. 2,

$$(34) \qquad \lim_{n\to\infty} \mu_n(M \times S_1) = \mu(M \times S_1), \qquad \lim_{n\to\infty} \nu_n(M \times S_1) = \nu(M \times S_1),$$

for any set M which satisfies the conditions

$$(35) \qquad \mu[\mathcal{F}(M \times S_1)] = 0, \qquad \nu[\mathcal{F}(M \times S_1)] = 0.$$

Therefore, recalling (29),

$$(36) \qquad \mu(M \times S_1) \geq \nu(M \times S_1).$$

Since (36) is valid for any set $M \subset S_{r-1}$ satisfying (35), it is also valid for any $M \subset S_{r-1}$.

On the other hand, if the equality holds in (10), by theorem I, n. I we get

$$(37) \qquad \mu(S_r) = \nu(S_r),$$

and therefore, by the additivity of the functions μ, ν, inequality (36) can be identically satisfied only if (11) is satisfied.

4. Let us now prove the above quoted isoperimetric property, which is expressed by the following theorem:

I. *Given a hypersphere C of the space S_r (with $r \geq 2$), for any set $B \subset S_r$ satisfying the condition*

$$(1) \qquad \operatorname{meas} B = \operatorname{meas} C,$$

the following isoperimetric relation holds:

$$(2) \qquad P(B) \geq P(C).$$

Moreover, in (2) we have equality only if B itself is a hypersphere (or is equivalent to a hypersphere).

Proof. In view of theorem III, n. I, to prove that (2) holds for any set B satisfying (1) it is enough to show that, given an arbitrary polygonal domain Π of finite measure and a hypersphere C satisfying the condition

$$(3) \qquad \operatorname{meas} \Pi = \operatorname{meas} C,$$

it results

$$(4) \qquad P(\Pi) \geq P(C). \;^{6}$$

---

[6] Since Π is a *polygonal domain*, (4) is well known; we give here a new proof of this fact, showing how this assertion fits in the general framework of our research.

To this aim let us observe that, as the polygonal domain Π has finite measure, it is necessarily bounded and therefore there exists a hypersphere C^* centered at the origin of the coordinates and of radius large enough which contains Π. If we denote by Γ the class of all sets B satisfying

$$(5) \qquad \operatorname{meas} B = \operatorname{meas} \Pi, \qquad P(B) \le P(\Pi), \qquad B \subset C^*,$$

by theorems II, IV of n. I, it follows that the functional $P(B)$ has a minimizer in the class Γ. Let us indicate by E one of these sets of minimal perimeter, and let us compare it with the set L whose construction is described in theorem II, n. 3. It is easy to realize that L still belongs to the class Γ and therefore, taking into account the minimality of E and theorem II, n. 3, we get

$$(6) \qquad P(L) = P(E).$$

From (6), using again theorem II, n. 3, we deduce that E is normal with respect to the hyperplane $x_r = 0$. On the other hand, the minimality property of E is still valid for any set obtained by rotating E around the origin, hence E is normal with respect to any hyperplane of the space S_r, that is (see n. 3) E is a convex set (or is equivalent to a convex set).

Since the perimeters of two equivalent sets coincide, we can also suppose that E is a convex set which can be represented through the relations

$$(7) \qquad f_1(x_1, \ldots, x_{r-1}) \le x_r \le f_2(x_1, \ldots, x_{r-1}), \qquad (x_1, \ldots, x_{r-1}) \in D,$$

where, by well known properties of convex sets, D is a convex domain of S_{r-1}, and $f_1(y)$, $f_2(y)$ are two functions defined for $y \in D$, which are uniformly Lipschitz continuous in any interval T contained in $D \setminus \mathcal{F}D$.

If E is represented by (7), the set L constructed with the procedure described in theorem II, n. 3 will be represented through the relations

$$(8) \qquad f_2(x_1, \ldots, x_{r-1}) - f_1(x_1, \ldots, x_{r-1}) < 2|x_r|, \qquad (x_1, \ldots, x_{r-1}) \in D.$$

Let us denote by $\mu(B)$, $\nu(B)$ the total variations in B of the Gauss-Green functions corresponding to E and L respectively. Let T be an interval contained in $D \setminus \mathcal{F}D$, and let M be a set contained in $(T \setminus \mathcal{F}T)$; by theorems II, III, n. 2, we have

$$(9) \qquad \mu(M \times S_1) = \int_M \left(\sqrt{1 + \left(\frac{\partial f_1}{\partial y_1}\right)^2 + \cdots + \left(\frac{\partial f_1}{\partial y_{r-1}}\right)^2} \right.$$
$$\left. + \sqrt{1 + \left(\frac{\partial f_2}{\partial y_1}\right)^2 + \cdots + \left(\frac{\partial f_2}{\partial y_{r-1}}\right)^2} \right) dy_1 \ldots dy_{r-1},$$

$$(10) \qquad \nu(M \times S_1)$$
$$= \int_M \sqrt{4 + \left(\frac{\partial f_2}{\partial y_1} - \frac{\partial f_1}{\partial y_1}\right)^2 + \cdots + \left(\frac{\partial f_2}{\partial y_{r-1}} - \frac{\partial f_1}{\partial y_{r-1}}\right)^2} \, dy_1 \ldots dy_{r-1}.$$

From theorem II, n. 3 and using (6) it follows

$$(11) \qquad \mu(M \times S_1) = \nu(M \times S_1)$$

Equalities (9), (10), (11) can be simultaneously satisfied only if at almost every point of M we have

$$(12) \qquad \frac{\partial f_1}{\partial y_1} + \frac{\partial f_2}{\partial y_1} = \cdots = \frac{\partial f_1}{\partial y_{r-1}} + \frac{\partial f_2}{\partial y_{r-1}} = 0.$$

From the arbitrariness of T and M, formulas (12) are then satisfied at almost every point of the convex domain D and therefore the sum $f_1(y) + f_2(y)$ is constant in D, that is the domain E is symmetric with respect to the hyperplane passing through its barycenter and parallel to the hyperplane $x_r = 0$.

Taking into account once again that the minimality property enjoyed by E holds for any set obtained from E through a rotation around the origin of the coordinates, we conclude that E is normal and symmetric with respect to all hyperplanes passing through its barycenter, and therefore E is a hypersphere. As E has minimal perimeter in the class Γ and since Π belongs to Γ, (4) is proved.

From (4), as we have already observed, (2) follows. To prove that, if the equality holds in (2) then B is a hypersphere (or is equivalent to a hypersphere), it is enough to observe that, in this case, B has minimal perimeter in the class of all sets having the same measure of C and hence the arguments considered above for E can be repeated for B.

Bibliography

[1] R. CACCIOPPOLI, *Misura e integrazione sugli insiemi dimensionalmente orientati*. Rend. Acc. Naz. dei Lincei, Serie VIII, Vol. XII, fasc. 1-2, gennaio-febbraio 1952.

[2] E. DE GIORGI, *Un nuovo teorema di esistenza relativo ad alcuni problemi variazionali*. Pubbl. n. 371 (serie II) dell'Istituto Nazionale per le Applicazioni del Calcolo.

[3] E. DE GIORGI, *Definizione ed espressione analitica del perimetro di un insieme*. Rend. Acc. Naz. dei Lincei, Serie VIII, Vol. XIV, fasc. 3, marzo 1953.

[4] E. DE GIORGI, *Su una teoria generale della misura $(r-1)$-dimensionale in uno spazio ad r dimensioni*. Annali di Matematica pura e applicata, serie IV, Tomo XXXVI, 1954.

[5] E. DE GIORGI, *Nuovi teoremi relativi alle misure $(r-1)$-dimensionali in uno spazio ad r dimensioni*. Ricerche di Matematica, vol. IV, 1955.

Sulla proprietà isoperimetrica dell'ipersfera, nella classe degli insiemi aventi frontiera orientata di misura finita[‡][†]

Memoria di Ennio De Giorgi[*]

Sunto. In una comunicazione (vedi [2] della bibliografia) su alcuni problemi di calcolo delle variazioni, ho affermato fra l'altro che la proprietà isoperimetrica del cerchio, della sfera (e più in generale dell'ipersfera di uno spazio ad un numero qualunque di dimensioni) sussiste ancora quando come "perimetro" di un insieme si assuma la misura della sua frontiera orientata secondo Caccioppoli (vedi [1], [4]). Questo risultato (al pari degli altri esposti in [2]) discende dai teoremi provati in [4], [5], ma la sua dimostrazione richiede delle ulteriori considerazioni, che formano l'argomento del presente lavoro.

1. Cominciamo col richiamare definizioni e risultati contenuti in [4], [5], di cui faremo uso in seguito. Sia S_r uno spazio euclideo a r dimensioni, il cui punto generico indicheremo con $x \equiv (x_1, \ldots, x_r)$, mentre con $|x|$ indicheremo la distanza di x dall'origine delle coordinate (cioè la $\sqrt{x_1^2 + \cdots + x_r^2}$); dati due punti $x \equiv (x_1, \ldots, x_r)$, $y \equiv (y_1, \ldots, y_r)$, con $(x - y)$ indicheremo il punto $(x_1 - y_1, \ldots, x_r - y_r)$; infine, parlando d'insiemi contenuti in S_r e di funzioni ivi definite, intenderemo sempre riferirci ad insiemi di Borel e funzioni di Baire. Dato un insieme $E \subset S_r$ e posto, per $\lambda > 0$,

$$(1) \qquad \psi(x; \lambda) = \frac{1}{\sqrt{\pi^r \lambda^r}} \int_E e^{-\frac{|x - \xi|^2}{\lambda}} \, d\xi_1 \ldots d\xi_r,$$

chiameremo *perimetro* di E il limite

$$(2) \qquad \lim_{\lambda \to 0} \int_{S_r} |\mathrm{grad}\,_x \psi(x; \lambda)| \, dx_1 \ldots dx_r,$$

che verrà indicato col simbolo $P(E)$; vale allora il seguente teorema (vedi [4], teor. II, teor. IV).

I. *Dato un insieme $E \subset S_r$, perchè sia finito il suo perimetro $P(E)$ occorre e basta che esista una funzione vettoriale di insieme $\Phi(B) \equiv [\Phi_1(B), \ldots, \Phi_r(B)]$ soddisfacente le condizioni seguenti:*

[‡]Editor's note: published in Atti Accad. Naz. Lincei Mem. Cl. Sci. Fis. Mat. Nat. Sez. I, **(8) 5** (1958), 33–44.

[†]Lavoro eseguito nell'Istituto Nazionale per le Applicazioni del Calcolo.

[*]Presentata nella riunione dell'11 Gennaio 1958 dal socio M. PICONE

a) $\Phi(B)$ è definita per ogni insieme $B \subset S_r$, completamente additiva ed a variazione totale finita.

b) Per ogni funzione $g(x)$ continua in S_r insieme alle sue derivate parziali prime ed infinitesima insieme ad esse per $|x| \to \infty$ d'ordine non inferiore a quello di $|x|^{-(r+1)}$, si ha

$$(3) \qquad \int_E \frac{\partial g}{\partial x_h} \, dx_1 \ldots dx_r = \int_{S_r} g(x) \, d\Phi_h \qquad (\text{per } h = 1, \ldots, r).$$

In questo caso il perimetro $P(E)$ risulta eguale alla variazione totale in S_r della funzione $\Phi(B)$, cioè risulta

$$(4) \qquad P(E) = \int_{S_r} |d\Phi|^1.$$

Alla funzione $\Phi(B)$ considerata nel teorema ora enunciato daremo il nome di *funzione di Gauss-Green relativa all'insieme E* (vedi [5] n.3).

Notiamo che, quando E è un *dominio poligonale* (cioè un dominio la cui frontiera è contenuta nella somma di un numero finito d'iperpiani) e la sua frontiera $\mathcal{F}E$ è limitata, allora anche il perimetro di E è finito, risultando eguale alla misura $(r-1)$-dimensionale di $\mathcal{F}E$ (elementarmente definita). In tal caso la funzione di Gauss-Green relativa ad E, $\Phi(B) \equiv [\Phi_1(B), \ldots, \Phi_r(B)]$ è data dalle formule

$$(5) \qquad \Phi_h(B) = \int_{B \cap \mathcal{F}E} n_h \, d\mu_{r-1}, \qquad \text{per } h = 1, \ldots, r,$$

ove n_h è il coseno formato con l'asse x_h dalla normale esterna ad $\mathcal{F}E$, mentre $d\mu_{r-1}$ è l'elemento di misura $(r-1)$-dimensionale su $\mathcal{F}E$; per convincersene basta confrontare le (3), (5) con le ordinarie formule di Gauss-Green relative ad E.

Data una successione di funzioni (scalari o vettoriali)

$$(6) \qquad F_1(B), F_2(B), \ldots, F_n(B), \ldots$$

definite per ogni insieme $B \subset S_r$ e completamente additive, diremo che la successione (6) *converge debolmente* verso una funzione $F(B)$ se, per ogni funzione di punto $g(x)$ continua in S_r ed infinitesima per $|x| \to \infty$, si ha

$$(7) \qquad \lim_{n \to \infty} \int_{S_r} g(x) \, dF_n = \int_{S_r} g(x) \, dF.$$

Indicheremo poi con Σ lo spazio metrico avente come elementi gl'insiemi di S_r e nel quale la distanza di due insiemi E_1, E_2 sia data dalla mis $(E_1 \cup E_2 \setminus E_1 \cap E_2)$. Valgono allora i seguenti teoremi (vedi [4] teor. VII, VIII, [5] teor. I).

---

[1] Ricordando che, data una funzione d'insieme $\psi(B)$ (vettoriale o scalare), definita per ogni insieme $B \subset S_r$ e completamente additiva, col simbolo $\int_B |d\psi|$ indichiamo la variazione totale di ψ sull'insieme B.

II. *Data una successione d'insiemi contenuti in S_r*

$$(8) \qquad\qquad E_1, \ldots, E_n, \ldots$$

convergente (nello spazio metrico Σ) verso un insieme E, si ha

$$(9) \qquad\qquad \lim_{n\to\infty}{}' P(E_n) \geq P(E);$$

se è limitato l'insieme dei perimetri $P(E_n)$, la successione delle funzioni di Gauss-Green relative agl'insiemi (8) converge debomente verso la funzione di Gauss-Green relativa ad E.

III. *Dato un insieme $E \subset S_r$ (con $r \geq 2$) il suo perimetro $P(E)$ è uguale al minimo limite dei perimetri dei domini poligonali Π che approssimano E (nello spazio Σ), cioè si ha*

$$\lim_{\Pi\to E}{}' P(\Pi) = P(E).$$

IV. *Dati un insieme limitato $L \subset S_r$ ed un numero positivo p, l'aggregato degl'insiemi contenuti in L ed aventi perimetro $\leq p$ è compatto nello spazio Σ.*

2. Passiamo ora alla dimostrazione di tre semplici teoremi che, in un certo senso, precisano o completano alcuni risultati esposti nel n. I.

I. *Sia data una successione d'insiemi di S_r aventi perimetri limitati*

$$(1) \qquad\qquad E_1, \ldots, E_n, \ldots$$

convergente (nello spazio Σ) verso un insieme E e verificante la relazione di limite

$$(2) \qquad\qquad \lim_{n\to\infty} P(E_n) = P(E);$$

siano $\Phi(B)$ la funzione di Gauss-Green relativa ad E, $\Phi^{(n)}(B)$ quella relativa ad E_n, $\mu(B)$ la variazione totale di Φ su B, $\mu_n(B)$ quella di $\Phi^{(n)}$. Si ha allora

$$(3) \qquad\qquad \lim_{n\to\infty} \mu_n(B) = \mu(B),$$

per ogni insieme B la cui frontiera $\mathcal{F}B$ soddisfa la condizione

$$(4) \qquad\qquad \mu(\mathcal{F}B) = 0.$$

Dim. Dati un insieme B soddisfacente la (4) ed un numero $\varepsilon > 0$, per noti teoremi della teoria dell'integrazione si può trovare una funzione vettoriale di punto $g(x) \equiv [g_1(x), \ldots, g_r(x)]$, continua in S_r, infinitesima per $|x| \to \infty$ e soddisfacente le seguenti condizioni:

$$(5) \qquad |g(x)| \leq 1 \qquad \text{per } x \in B; \qquad |g(x)| = 0 \qquad \text{per } x \in (S_r \setminus B)$$

$$(6) \qquad \sum_{h=1}^{r} \int_{S_r} g_h(x)\, d\Phi_h \geq \int_B |d\Phi| - \varepsilon = \mu(B) - \varepsilon.$$

D'altra parte, per il teor. II n. I abbiamo

$$(7) \qquad \lim_{n \to \infty} \sum_{h=1}^{r} \int_{S_r} g_h(x)\, d\Phi_h^{(n)} = \sum_{h=1}^{r} \int_{S_r} g_h(x)\, d\Phi_h$$

e quindi, essendo per le (5)

$$(8) \qquad \sum_{h=1}^{r} \int_{S_r} g_h(x)\, d\Phi_h^{(n)} \leq \int_B |d\Phi^{(n)}| = \mu_n(B),$$

abbiamo, confrontando le (6), (7), (8),

$$(9) \qquad \lim_{n \to \infty}{}' \mu_n(B) \geq \mu(B) - \varepsilon;$$

data l'arbitrarietà di ε alla (9) può sostituirsi la disuguaglianza

$$(9') \qquad \lim_{n \to \infty}{}' \mu_n(B) \geq \mu(B).$$

Poichè la frontiera di $(S_r \setminus B)$ coincide con $\mathcal{F}B$, accanto alla (9') sussiste la relazione analoga

$$(10) \qquad \lim_{n \to \infty}{}' \mu_n(S_r \setminus B) \geq \mu(S_r \setminus B),$$

mentre dalle (2) segue

$$(11) \quad \lim_{n \to \infty} [\mu_n(B) + \mu_n(S_r \setminus B)] \;=\; \lim_{n \to \infty} \mu_n(S_r) = \lim_{n \to \infty} P(E_n)$$
$$=\; P(E) = \mu(S_r) = \mu(B) + \mu(S_r \setminus B);$$

dalle (9'), (10), (11) segue infine la (3).

II. *Se k è un numero reale, $f(y) = f(y_1, \ldots, y_{r-1})$ una funzione definita in un intervallo T dello spazio S_{r-1} (con $r \geq 2$), ivi uniformemente lipschitziana e sempre maggiore di k, allora il dominio D dello spazio S_r formato dai punti $(x_1, \ldots, x_r)$ che soddisfano le condizioni*

$$(12) \qquad (x_1, \ldots, x_{r-1}) \in T, \qquad k \leq x_r \leq f(x_1, \ldots, x_{r-1})$$

ha perimetro finito. Se poi H è un insieme contenuto in $(T \setminus \mathcal{F}T)$, H^* è l'insieme dei punti $(x_1, \ldots, x_r)$ che soddisfano le condizioni

$$(13) \qquad (x_1, \ldots, x_{r-1}) \in H, \qquad x_r = f(x_1, \ldots, x_{r-1}),$$

$\mu(B)$ è la variazione totale nell'insieme B della funzione di Gauss-Green relativa a D, si ha

$$(14) \qquad \mu(H^*) = \int_H \sqrt{1 + \left(\frac{\partial f}{\partial y_1}\right)^2 + \cdots + \left(\frac{\partial f}{\partial y_{r-1}}\right)^2}\; dy_1 \ldots dy_{r-1}.$$

Dim. Il dominio D può evidentemente essere approssimato mediante una successione di domini poligonali

$$(15) \qquad\qquad D_1, \dots, D_n, \dots,$$

tale che D_n sia formato dai punti $(x_1, \dots, x_r)$ che verificano le condizioni

$$(16) \qquad k \le x_r \le f_n(x_1, \dots, x_{r-1}), \qquad (x_1, \dots, x_{r-1}) \in T$$

e le funzioni $f_1(y), \dots, f_n(y), \dots$, siano continue in T, sempre $> k$ e soddisfino le relazioni

$$(17) \qquad \lim_{n \to \infty} f_n(y) = f(y) \quad \text{uniformemente in } T,$$

$$(18) \qquad \lim_{n \to \infty} \int_T \left| \frac{\partial f}{\partial y_h} - \frac{\partial f_n}{\partial y_h} \right| dy_1 \dots dy_{r-1} = 0 \qquad \text{per } h = 1, \dots, r-1.$$

Ricordando che il perimetro di un dominio poligonale coincide con la misura $(r-1)$-dimensionale della sua frontiera e tenendo conto delle (17), (18), si constata facilmente che i perimetri dei domini (15) sono limitati[2] e quindi, per il teor. II n. I, anche D ha perimetro finito. Presa poi una funzione $g(x) \equiv g(x_1, \dots, x_r)$ continua in S_r con le sue derivate parziali prime ed identicamente nulla per $(x_1, \dots, x_{r-1}) \in \mathcal{F}T$, si provano facilmente le relazioni

$$(19) \qquad \int_D \frac{\partial g}{\partial x_h} \, dx_1 \dots dx_r$$

$$= -\int_T g[y_1, \dots, y_{r-1}, f(y)] \frac{\partial f}{\partial y_h} \, dy_1 \dots dy_{r-1} \quad (\text{per } h = 1, \dots, r-1),$$

$$(19') \qquad \int_D \frac{\partial g}{\partial x_r} \, dx_1 \dots dx_r$$

$$= -\int_T \{g[y_1, \dots, y_{r-1}, f(y)] - g(y_1, \dots, y_{r-1}, k)\} \, dy_1 \dots dy_{r-1};$$

basta a tale scopo scrivere le formule di Gauss-Green relative al dominio poligonale D_n ed alla funzione $g(x)$ e passare al limite per $n \to \infty$ (tenendo presenti le (17), (18)).

Se ora indichiamo con $\Phi(B) \equiv (\Phi_1(B), \dots, \Phi_r(B))$ la funzione di Gauss-Green relativa a D, con V un insieme contenuto in H, con V^* l'insieme dei punti $(x_1, \dots, x_r)$ che verificano le condizioni

$$(20) \qquad (x_1, \dots, x_{r-1}) \in V, \qquad x_r = f(x_1, \dots, x_{r-1}),$$

---

[2]Si trova infatti, con considerazioni elementari,

$$P(D_n) = mis\,T + \int_{\mathcal{F}T} [f_n(y) - k] d\mu_{r-2} + \int_T \sqrt{1 + \left(\frac{\partial f_n}{\partial y_1}\right)^2 + \cdots + \left(\frac{\partial f_n}{\partial y_{r-1}}\right)^2} \, dy_1 \dots dy_{r-1}.$$

dalle (19), (19') si ricava, tenendo conto dell'arbitrarietà di $g(x)$ e ricordando la definizione di funzione di Gauss-Green,

$$(21) \qquad \Phi_h(V^*) = -\int_V \frac{\partial f}{\partial y_h}\, dy_1 \dots dy_{r-1} \qquad (\text{per } h = 1, 2, \dots, r-1),$$

$$(21') \qquad\qquad\qquad \Phi_r(V^*) = \operatorname{mis} V$$

e quindi, valendo le (21), (21') per ogni insieme $V \subset H$, la (14) è dimostrata.

III. *Dati due insiemi E_1, E_2 dello spazio S_r, aventi entrambi perimetro finito, siano $\Phi(B)$, $\Psi(B)$ le rispettive funzioni di Gauss-Green; sia A un campo di S_r verificante la condizione*

$$(22) \qquad\qquad\qquad E_1 \cap A = E_2 \cap A.$$

Per ogni $B \subset A$ si ha allora

$$(23) \qquad\qquad\qquad \Phi(B) = \Psi(B).$$

Dim. Sia $g(x)$ una funzione continua in S_r con le sue derivate parziali prime, infinitesima insieme ad esse per $|x| \to \infty$, d'ordine superiore a quello di $|x|^{-(r+1)}$, identicamente nulla in $(S_r \setminus A)$; per la (22) e per la definizione stessa di funzione di Gauss-Green avremo

$$(24) \qquad \int_{S_r} g(x)\, d\Phi \;=\; \int_{E_1} \operatorname{grad} \cdot g(x)\, dx_1 \dots dx_r$$
$$= \int_{E_2} \operatorname{grad} \cdot g(x)\, dx_1 \dots dx_r = \int_{S_r} g(x)\, d\Psi.$$

Poichè le (24) valgono per ogni funzione $g(x)$ continua in S_r con le derivate parziali prime, infinitesima con esse, per $|x| \to \infty$, d'ordine superiore a quello di $|x|^{-(r+1)}$, identicamente nulla in $(S_r \setminus A)$, per noti teoremi sull'approssimazione lineare delle funzioni esse saranno verificate (limitatamente ai due estremi dell'uguaglianza) anche nell'ipotesi che $g(x)$ sia una funzione limitata in S_r ed identicamente nulla in $(S_r \setminus A)$; in particolare $g(x)$ potrà essere la funzione caratteristica di un insieme $B \subset A$ e quindi il teorema è dimostrato.

3. Dati un insieme $E \subset S_r$ (con $r \geq 2$) ed un iperpiano I di S_r, diremo che E è *puntualmente normale rispetto ad I* se, comunque si fissi una retta ortogonale ad I, l'intersezione di tale retta con E è un segmento, o un punto, o l'insieme vuoto; diremo invece che E è *normale in media* (o *semplicemente normale*) rispetto a I se è equivalente[3] ad un insieme puntualmente normale. Un insieme che sia puntualmente normale rispetto a ogni iperpiano è evidentemente convesso, quindi un insieme che sia semplicemente normale in media rispetto a ogni iperpiano sarà equivalente ad un insieme convesso.

---

[3] Due insiemi E_1, E_2 sono detti equivalenti se $\operatorname{mis}(E_1 \cup E_2 \setminus E_1 \cap E_2) = 0$.

In questo numero dimostreremo un teorema sugli insiemi normali e simmetrici rispetto ad un iperpiano che, insieme al teorema IV del n. I, ha un ruolo essenziale nella dimostrazione della proprietà isoperimetrica dell'ipersfera; a tale scopo conviene premettere il seguente lemma.

I. *Siano date due funzioni vettoriali d'insieme $\alpha(B)$, $\gamma(B)$, definite per ogni $B \subset S_r$, completamente additive e a variazione totale finita; se per ogni insieme $B \subset S_r$ è verificata la relazione*

$$(1) \qquad [\alpha(B); \gamma(B) - \alpha(B)] \geq 0 \ {}^{4}$$

allora vale anche la relazione

$$(2) \qquad 2 \int_B |d\gamma| \left(\int_B |d\gamma| - \int_B |d\alpha| \right) \geq \left(\int_B |d(\gamma - \alpha)| \right)^2$$

per ogni insieme $B \subset S_r$.

Dim. Dalla (1) segue immediatamente

$$(3) \qquad |\gamma(B)|^2 \equiv [\gamma(B); \gamma(B)] \geq |\alpha(B)|^2 + |\gamma(B) - \alpha(B)|^2$$

e quindi, detta $\mu(B)$ la variazione totale di γ sull'insieme B e posto

$$(4) \qquad \varphi(x) = \frac{d\gamma}{d\mu}, \qquad \psi(x) = \frac{d\alpha}{d\mu},$$

avremo, per noti teoremi sulla derivazione delle funzioni d'insieme,

$$(5) \qquad |\varphi| = 1, \qquad 1 - |\psi|^2 = |\varphi|^2 - |\psi|^2 \geq |\varphi - \psi|^2, \qquad |\psi| \leq 1,$$

$$(6)$$
$$\int_B |d\gamma| = \int_B d\mu, \qquad \int_B |d\alpha| = \int_B |\psi| \, d\mu, \qquad \int_B |d(\gamma - \alpha)| = \int_B |\varphi - \psi| \, d\mu.$$

Dalle (5) segue poi

$$(7) \qquad 2 \int_B (1 - |\psi|) \, d\mu \geq \int_B (1 - |\psi|^2) \, d\mu \geq \int_B |\varphi - \psi|^2 \, d\mu,$$

mentre, per la disuguaglianza di Schwarz, si ha

$$(8) \qquad \int_B |\varphi - \psi|^2 \, d\mu \int_B d\mu \geq \left(\int_B |\varphi - \psi| \, d\mu \right)^2,$$

e quindi dalle (6), (7), (8) segue la (2).

Passiamo ora alla dimostrazione dell'annunciato teorema relativo agl'insiemi normali e simmetrici rispetto ad un iperpiano.

---

[4]Dati due vettori $u \equiv (u_1, \dots, u_n)$, $v \equiv (v_1, \dots, v_n)$ indichiamo il loro prodotto scalare col simbolo $[u; v] = \sum_{h=1}^{n} u_h v_h$.

II. *Sia dato un insieme* E *dello spazio* S_r *(con* $r \geq 2$*), avente perimetro e misura finiti; per ogni punto* $y \equiv (y_1, \ldots, y_{r-1})$ *dello spazio* S_{r-1}*, indichiamo con* $f(y) \equiv f(y_1, \ldots, y_{r-1})$ *la misura lineare dell'intersezione di* E *con la retta* $R(y)$ *dello spazio* S_r *(di cui* $(x_1, \ldots, x_r)$ *sia sempre il punto generico) avente le equazioni* $x_1 = y_1, \ldots, x_{r-1} = y_{r-1}$. *Detto* L *l'insieme[5] formato dai punti* $(x_1, \ldots, x_r)$ *per i quali si abbia*

$$(9) \qquad f(x_1, \ldots, x_{r-1}) > 2x_r > -f(x_1, \ldots, x_{r-1}),$$

risulta sempre

$$(10) \qquad P(L) \leq P(E).$$

Se nella (10) vale il segno di eguaglianza, allora E *è normale rispetto all'iperpiano* $x_r = 0$ *e, dette* $\mu(B)$ *la variazione totale della funzione di Gauss-Green relativa ad* E *sull'insieme* B*,* $\nu(B)$ *la variazione di quella relativa ad* L*, si ha, per ogni insieme* $M \subset S_{r-1}$*,*

$$(11) \qquad \mu(M \times S_1) = \nu(M \times S_1),$$

ove $(M \times S_1)$ *è l'insieme (prodotto topologico di* M *ed* S_1*) formato dai punti* $(x_1, \ldots, x_r)$ *che verificano le condizioni*

$$(12) \qquad (x_1, \ldots, x_{r-1}) \in M, \quad -\infty < x_r < +\infty.$$

Dim. Per il teorema III n. I esiste una successione di domini poligonali

$$(13) \qquad E_1, \ldots, E_n, \ldots,$$

verificante le condizioni

$$(14) \qquad \lim_{n \to \infty} \operatorname{mis} (E_n \cup E \setminus E_n \cap E) = 0, \qquad \lim_{n \to \infty} P(E_n) = P(E).$$

Inoltre supporremo che, per ogni intero $n > 0$, la normale alla frontiera di E_n non sia mai parallela all'iperpiano $x_r = 0$; ciò è sicuramente lecito, poichè se per qualche valore di n tale ipotesi non fosse soddisfatta, ad essa potremmo sempre ricondurci sottoponendo il dominio E_n ad una rotazione di ampiezza tanto piccola quanto si vuole.

Indicheremo poi: con D_n il dominio poligonale dello spazio S_{r-1} formato dai punti y tali che la retta $R(y)$ abbia intersezione non vuota con E_n, con $f_n(y)$ la misura lineare di tale intersezione, con L_n il dominio poligonale formato dai punti $(x_1, \ldots, x_r)$ che soddisfano le condizioni

$$(15) \qquad f_n(x_1, \ldots, x_{r-1}) \geq 2x_r \geq -f_n(x_1, \ldots, x_{r-1}); \qquad (x_1, \ldots, x_{r-1}) \in D_n.$$

---

[5] È chiaro il significato geometrico della costruzione di L; L è ottenuto mediante una operazione di *normalizzazione* e *simmetrizzazione* rispetto all'iperpiano $x_r = 0$. Analoghe osservazioni valgono per gl'insiemi L_n definiti dalle (15).

Tenendo conto della prima delle (14) e della definizione di L ed L_n, si verifica subito la relazione

$$(16) \qquad \lim_{n \to \infty} \mathrm{mis}\,(L \cup L_n \setminus L \cap L_n) = 0.$$

Fissiamo ora un valore dell'indice n ed indichiamo con $\Phi(B) \equiv [\Phi_1(B), \ldots, \Phi_r(B)]$ la funzione di Gauss-Green relativa ad E_n, con $\Psi(B) \equiv [\Psi_1(B), \ldots, \Psi_r(B)]$ la funzione di Gauss-Green relativa ad L_n, con $\mu_n(B)$ la variazione totale di Φ su B, con $\nu_n(B)$ quella di Ψ; inoltre, per ogni insieme $M \subset S_{r-1}$, siano $\alpha_h(M)$ la variazione totale di $\Psi_h(B)$ in $(M \times S_1)$, $\gamma_h(M)$ la variazione totale di $\Phi_h(B)$.

Per stabilire alcune proprietà delle funzioni α_h, γ_h che ci saranno utili nel seguito, cominciamo con l'osservare che, essendo E_n un dominio poligonale di S_r e poichè la normale a $\mathcal{F}E_n$ non è mai parallela all'iperpiano $x_r = 0$, $f_n(y)$ è continua in D_n e nulla su $\mathcal{F}D_n$ ed il dominio D_n può decomporsi in un numero finito di domini poligonali $G_1, \ldots, G_m$ godenti delle seguenti proprietà:

a) Per ogni intero positivo $k \le m$, il numero dei punti in cui la retta $R(y)$ incontra $\mathcal{F}E_n$ è costante al variare di y nell'interno di G_k.

b) Detto $p(k)$ tale numero (che è pari e ≥ 2) e dette $g_{k,1}(y), g_{k,2}(y), \ldots, g_{k,p(k)}(y)$ le r-sime coordinate dei $p(k)$ punti comuni ad $R(y)$ ed $\mathcal{F}E_n$ (prese in ordine crescente), le funzioni $g_{k,1}(y), \ldots, g_{k,p(k)}(y)$ sono lipschitziane nell'interno di G_k.

Dalle ipotesi a), b), ricordando la definizione di $f(y)$, si vede subito che valgono le disuguaglianze

$$(17) \qquad \left| \frac{\partial f_n}{\partial y_h} \right| \le \sum_{l=1}^{p(k)} \left| \frac{\partial g_{k,l}}{\partial y_h} \right|, \qquad \text{per } y \in (G_k \setminus \mathcal{F}G_k), \qquad h = 1, \ldots, r-1.$$

Ricordando le (5), n. I e tenendo presente che la normale ad $\mathcal{F}E_n$ non è mai parallela all'iperpiano $x_r = 0$, si trovano per le funzioni α_h, γ_h, μ_n, ν_n le espressioni

$$(18) \quad \gamma_h(M) = \sum_{k=1}^{m} \int_{M \cap G_k} \sum_{l=1}^{p(k)} \left| \frac{\partial g_{k,l}}{\partial y_h} \right| \, dy_1 \ldots dy_{r-1} \qquad \text{per } h = 1, \ldots, r-1,$$

$$(19) \qquad \alpha_h(M) = \sum_{k=1}^{m} \int_{M \cap G_k} \left| \frac{\partial f_n}{\partial y_h} \right| \, dy_1 \ldots dy_{r-1} \qquad \text{per } h = 1, \ldots, r-1,$$

$$(20) \qquad \gamma_r(M) = \sum_{k=1}^{m} p(k)\mathrm{mis}\,(M \cap G_k), \qquad \alpha_r(M) = 2 \sum_{k=1}^{m} \mathrm{mis}\,(M \cap G_k),$$

(21)

$$\mu_n(M \times S_1) = \sum_{k=1}^{m} \int_{M \cap G_k} \sum_{l=1}^{p(k)} \sqrt{1 + \left(\frac{\partial g_{k,l}}{\partial y_1} \right)^2 + \cdots + \left(\frac{\partial g_{k,l}}{\partial y_{r-1}} \right)^2} \, dy_1 \ldots dy_{r-1},$$

$$(22) \quad \nu_n(M \times S_1) = \sum_{k=1}^{m} \int_{M \cap G_k} \sqrt{4 + \left(\frac{\partial f_n}{\partial y_1}\right)^2 + \cdots + \left(\frac{\partial f_n}{\partial y_{r-1}}\right)^2} \; dy_1 \ldots dy_{r-1}.$$

Introdotte le funzioni vettoriali d'insieme $\alpha \equiv [\alpha_1(M), \ldots, \alpha_r(M)]$, $\gamma \equiv [\gamma_1(M), \ldots, \gamma_r(M)]$ dalle (18), (20), (21) segue

$$(23) \qquad \mu_n(M \times S_1) \geq \int_M |d\gamma|,$$

dalle (19), (20), (22) segue

$$(24) \qquad \nu_n(M \times S_1) = \int_M |d\alpha|,$$

dalle (17), (18), (19), (20) si deducono le relazioni

$$(25) \qquad 0 \leq \alpha_h(M) \leq \gamma_h(M) \qquad \text{per } h = 1, \ldots, r, \qquad M \subset S_{r-1},$$

e si vede che, detto H_n l'insieme dei punti $y \in S_{r-1}$ tali che la retta $R(y)$ incontri $\mathcal{F}E_n$ in più di due punti, è sempre

$$(26) \qquad 0 \leq \alpha_r(M) \leq \gamma_r(M) - 2\mathrm{mis}\,(H_n \cap M).$$

Dalle (25) si deduce la relazione

$$(27) \qquad [\alpha(M); \gamma(M) - \alpha(M)] = \sum_{h=1}^{r} \alpha_h(M)(\gamma_h(M) - \alpha_h(M)) \geq 0,$$

mentre dalle (26) segue

$$(28) \qquad \int_M |d(\gamma - \alpha)| \geq |\gamma(M) - \alpha(M)| \geq 2\mathrm{mis}\,(H_n \cap M).$$

Per il teor. I e le (23), (24), (27) si ha

$$(29) \qquad \mu_n(M \times S_1) \geq \nu_n(M \times S_1)$$

e, ricordando le (28),

$$(30) \qquad \mu_n(M \times S_1)[\mu_n(M \times S_1) - \nu_n(M \times S_1)] \geq 2[\mathrm{mis}\,(H_n \cap M)]^2;$$

in particolare, per $M = S_{r-1}$, le (29), (30) diventano

$$(31) \qquad \mu_n(S_r) = P(E_n) \geq \nu_n(S_r) = P(L_n)$$

$$(32) \qquad P(E_n)[P(E_n) - P(L_n)] \geq 2(\mathrm{mis}\,H_n)^2.$$

Dalle (31), passando al limite per $n \to \infty$ e tenendo presenti le (14), (16) ed il teor. II, n. I, segue la (10); inoltre dalle (32) si vede che, perchè nella (10) valga il segno di eguaglianza, debbono essere simultaneamente verificate le relazioni

$$(33) \qquad \lim_{n \to \infty} \operatorname{mis} H_n = 0, \qquad \lim_{n \to \infty} P(L_n) = P(L).$$

Ricordando la definizione di H_n, dalla (14) e dalla prima delle (33) si vede che E deve essere normale rispetto all'iperpiano $x_r = 0$. Dalle (14), (16) e dalla seconda delle (33) si deduce invece, ricordando il teor. I, n. 2,

$$(34) \qquad \lim_{n \to \infty} \mu_n(M \times S_1) = \mu(M \times S_1), \qquad \lim_{n \to \infty} \nu_n(M \times S_1) = \nu(M \times S_1),$$

per ogni insieme M che verifichi le condizioni

$$(35) \qquad \mu[\mathcal{F}(M \times S_1)] = 0, \qquad \nu[\mathcal{F}(M \times S_1)] = 0,$$

e quindi, ricordando le (29),

$$(36) \qquad \mu(M \times S_1) \geq \nu(M \times S_1);$$

la (36), valendo per ogni insieme $M \subset S_{r-1}$ che soddisfi le (35), sarà addirittura verificata per ogni $M \subset S_{r-1}$.

D'altra parte, se nella (10) vale il segno di eguaglianza, per il teor. I, n. I sarà

$$(37) \qquad \mu(S_r) = \nu(S_r),$$

e quindi, per l'additività delle funzioni μ, ν, la (36) può essere identicamente soddisfatta solo se lo è la (11).

4. Passiamo ora a provare l'annunciata proprietà isoperimetrica espressa dal seguente teorema:

I. *Data un'ipersfera C dello spazio S_r (con $r \geq 2$), per ogni insieme $B \subset S_r$ che soddisfa la condizione*

$$(1) \qquad \operatorname{mis} B = \operatorname{mis} C,$$

vale la relazione isoperimetrica

$$(2) \qquad P(B) \geq P(C)$$

e nella (2) si ha il segno di eguaglianza solo se B è anch'esso un'ipersfera (o è equivalente ad un'ipersfera).

Dim. Per provare che la (2) vale per ogni insieme B verificante la (1) basta, in virtù del teor. III, n. I, provare che, fissati arbitrariamente un dominio poligonale Π di misura finita ed una ipersfera C che verifica la condizione

$$(3) \qquad \operatorname{mis} \Pi = \operatorname{mis} C,$$

risulta

$$(4) \qquad P(\Pi) \geq P(C)^6$$

Osserviamo perciò che il dominio poligonale Π, avendo misura finita, è necessariamente limitato e quindi esiste un'ipersfera C^* di centro nell'origine delle coordinate e raggio abbastanza grande per contenere Π; detto Γ l'aggregato degli insiemi B che verificano le condizioni

$$(5) \qquad \operatorname{mis} B = \operatorname{mis} \Pi, \qquad P(B) \leq P(\Pi), \qquad B \subset C^*,$$

per i teoremi II, IV del n. I, il funzionale $P(B)$ ammette minimo nella classe Γ. Indichiamo con E uno degli insieme aventi perimetro minimo e confrontiamolo con l'insieme L costruito nel modo indicato nel teor. II, n. 3; è facile constatare che L appartiene ancora alla classe Γ e quindi, tenendo conto della proprietà di minimo di E e del teor. II, n. 3, si trova

$$(6) \qquad P(L) = P(E).$$

Dalle (6), sempre per il teor. II, n. 3, si deduce che E è normale rispetto all'iperpiano $x_r = 0$; d'altra parte la proprietà di minimo goduta da E vale altresì per ogni insieme ottenuto da E mediante una rotazione intorno all'origine, quindi E risulta normale rispetto a tutti gli iperpiani dello spazio S_r, cioè (vedi n. 3) è un insieme convesso (o equivalente ad un insieme convesso).

Poichè i perimetri di insiemi equivalenti sono uguali, possiamo addirittura supporre che E sia un dominio convesso rappresentato dalle relazioni

$$(7) \qquad f_1(x_1, \ldots, x_{r-1}) \leq x_r \leq f_2(x_1, \ldots, x_{r-1}); \qquad (x_1, \ldots, x_{r-1}) \in D,$$

ove, per note proprietà degli insiemi convessi, D è un dominio convesso di S_{r-1}, $f_1(y)$, $f_2(y)$ sono due funzioni definite per $y \in D$, uniformemente lipschitziane in ogni intervallo T contenuto in $D \setminus \mathcal{F}D$.

Se E è rappresentato dalle (7), l'insieme L costruito col procedimento indicato nel teor. II, n. 3 sarà rappresentato dalle relazioni

$$(8) \qquad f_2(x_1, \ldots, x_{r-1}) - f_1(x_1, \ldots, x_{r-1}) < 2|x_r|, \qquad (x_1, \ldots, x_{r-1}) \in D.$$

Dette rispettivamente $\mu(B)$, $\nu(B)$ le variazioni totali in B delle funzioni di Gauss-Green relative ad E e ad L, siano T un intervallo contenuto in $D \setminus \mathcal{F}D$, M un insieme contenuto in $(T \setminus \mathcal{F}T)$; per i teoremi II, III, n. 2, sarà

$$(9) \qquad \mu(M \times S_1) = \int_M \left(\sqrt{1 + \left(\frac{\partial f_1}{\partial y_1}\right)^2 + \cdots + \left(\frac{\partial f_1}{\partial y_{r-1}}\right)^2} \right.$$
$$\left. + \sqrt{1 + \left(\frac{\partial f_2}{\partial y_1}\right)^2 + \cdots + \left(\frac{\partial f_2}{\partial y_{r-1}}\right)^2} \right) dy_1 \ldots dy_{r-1},$$

---

[6]La disuguaglianza (4), *essendo Π un dominio poligonale*, è ben nota; ne diamo qui una nuova dimostrazione che vale ad inquadrarla, subordinandola al punto di vista generale della ricerca.

$$(10) \qquad \nu(M \times S_1)$$

$$= \int_M \sqrt{4 + \left(\frac{\partial f_2}{\partial y_1} - \frac{\partial f_1}{\partial y_1}\right)^2 + \cdots + \left(\frac{\partial f_2}{\partial y_{r-1}} - \frac{\partial f_1}{\partial y_{r-1}}\right)^2} \, dy_1 \ldots dy_{r-1}.$$

Dal teor. II, n. 3 e dalle (6) segue poi

$$(11) \qquad\qquad\qquad \mu(M \times S_1) = \nu(M \times S_1)$$

e le (9), (10), (11) possono essere simultaneamente verificate solo se in quasi tutti i punti di M si ha

$$(12) \qquad\qquad \frac{\partial f_1}{\partial y_1} + \frac{\partial f_2}{\partial y_1} = \cdots = \frac{\partial f_1}{\partial y_{r-1}} + \frac{\partial f_2}{\partial y_{r-1}} = 0.$$

Data l'arbitrarietà di T, M, le (12) risultano allora verificate in quasi tutti i punti del dominio convesso D e quindi la somma $f_1(y) + f_2(y)$ è costante in D, cioè il dominio E è simmetrico rispetto all'iperpiano, parallelo a quello $x_r = 0$, passante per il baricentro di E.

Tenendo sempre presente che la proprietà di minimo goduta da E vale per ogni insieme ottenuto da E mediante una rotazione intorno all'origine delle coordinate, si conclude che E è normale e simmetrico rispetto a tutti gl'iperpiani passanti per il suo baricentro e quindi è un'ipersfera; avendo E perimetro minimo nella classe Γ cui appartiene anche Π, la (4) è dimostrata.

Dalla (4), come abbiamo osservato, segue la (2); per provare poi che, se nella (2) vale il segno di eguaglianza, allora B è un'ipersfera (o è equivalente ad un'ipersfera), basta osservare che in tal caso B ha perimetro minimo nella classe degl'insiemi aventi misura eguale a quella di C e quindi i ragionamenti fatti sopra per E possono ripetersi per B.

Bibliografia

[1] R. CACCIOPPOLI, *Misura e integrazione sugli insiemi dimensionalmente orientati*. Rend. Acc. Naz. dei Lincei, Serie VIII, Vol. XII, fasc. 1-2, gennaio-febbraio 1952.

[2] E. DE GIORGI, *Un nuovo teorema di esistenza relativo ad alcuni problemi variazionali*. Pubbl. n. 371 (serie II) dell'Istituto Nazionale per le Applicazioni del Calcolo.

[3] E. DE GIORGI, *Definizione ed espressione analitica del perimetro di un insieme*. Rend. Acc. Naz. dei Lincei, Serie VIII, Vol. XIV, fasc. 3, marzo 1953.

[4] E. DE GIORGI, *Su una teoria generale della misura $(r-1)$-dimensionale in uno spazio ad r dimensioni*. Annali di Matematica pura e applicata, serie IV, Tomo XXXVI, 1954.

[5] E. De Giorgi, *Nuovi teoremi relativi alle misure $(r-1)$-dimensionali in uno spazio ad r dimensioni.* Ricerche di Matematica, vol. IV, 1955.

Complements to the $(n-1)$-dimensional measure theory in a n-dimensional space[‡]

Ennio De Giorgi

In this seminar we shall expose some complements to the theory of Caccioppoli sets, which serve as tools in the investigation of area-minimizing oriented boundaries; these complements mainly deal with the relationships between the classical theory of the Gauss-Green formulas and the theory of Caccioppoli (see, in particular, theorem IV n.3, III n.4).

1.

Let us introduce some notation which will be used throughout. For every positive number n we shall denote by $\mathbf{R}^n$ the n-dimensional Euclidean space: in particular, $\mathbf{R} = \mathbf{R}^1$ is the real line. We shall always endow $\mathbf{R}^n$ with the linear structure: more precisely, given two points $x = (x_1, \ldots, x_n)$ and $y = (y_1, \ldots, y_n)$ we let

$$(1) \qquad x + y = (x_1 + y_1, \ldots, x_n + y_n), \qquad x - y = (x_1 - y_1, \ldots, x_n - y_n).$$

Given a point $x = (x_1, \ldots, x_n)$ and a real number α we let

$$(2) \qquad \alpha x = (\alpha x_1, \ldots, \alpha x_n);$$

given two points $x = (x_1, \ldots, x_n)$ and $y = (y_1, \ldots, y_n)$, we denote their scalar product with the symbol

$$(3) \qquad \langle x, y \rangle = \sum_{h=1}^{n} x_h y_h,$$

and we define

$$(4) \qquad |x| = (\langle x, x \rangle)^{\frac{1}{2}} = \sqrt{\sum_{h=1}^{n} x_h^2}.$$

Given a function f, differentiable in an open set $A \subset \mathbf{R}^n$, we let

$$(5) \qquad D_h f(x) = \frac{\partial f}{\partial x_h}(x), \qquad Df(x) = \operatorname{grad} f(x) = (D_1 f(x), \ldots, D_n f(x));$$

[‡]Editor's note: translation into English of the paper "Complementi alla misura $(n-1)$-dimensionale in uno spazio ad n dimensioni", Seminario di Matematica della Scuola Normale Superiore di Pisa 1960-61, Editrice Tecnico Scientifica, 1961.

given two functions $f(x)$, $g(x)$ defined in $\mathbf{R}^n$ we denote their convolution with the symbol

$$(6) \qquad (f * g)(x) = \int_{\mathbf{R}^n} f(x - \xi) g(\xi) \, d\xi.$$

Accordingly to [2], [3] we agree that, whenever we introduce sets in $\mathbf{R}^n$ and functions there defined, we shall always mean Borel sets and Baire functions.

For every set $E \subset \mathbf{R}^n$ we denote by $\mathcal{F}E$ its boundary and by $\overline{E} = E \cup \mathcal{F}E$ its closure; for every point $x \in \mathbf{R}^n$ and for every positive number ρ, we denote by the symbols

$$(7) \qquad A_\rho(x), \qquad C_\rho(x) = \overline{A_\rho(x)}$$

respectively the open and the closed ball of radius ρ centered at x. If x coincides with the origin of the coordinates, in place of $A_\rho(x)$, $C_\rho(x)$ we shall simply write A_ρ, C_ρ.

Denoting by $\operatorname{meas} E$ the Lebesgue measure of the generic set $E \subset \mathbf{R}^n$, we shall denote by the symbol

$$(8) \qquad \omega_n = \operatorname{meas} A_1 = \operatorname{meas} C_1$$

the measure of the unit ball in $\mathbf{R}^n$.

Given some relations $\alpha, \beta, \gamma, \ldots$ we denote by the symbol

$$(9) \qquad \{x \, : \, \alpha, \beta, \gamma, \ldots\}$$

the set of those x satisfying these relations. Given a set E and a function $f(x)$ defined in E, we denote by the symbol

$$(10) \qquad f(E) = \{f(x) \, : \, x \in E\}$$

the image of E under f, i.e., the set described by $f(x)$ as x ranges through E.

For every set $E \subset \mathbf{R}^n$ and every integer $k > 0$, we denote by the symbol $\mathcal{H}^k(E)$ the k-dimensional Hausdorff measure of E, which, as well known (see e.g. [4] n.2,3), is defined by the formula

$$(11) \qquad \mathcal{H}^k(E) = 2^{-k} \omega_k \lim_{\rho \to 0} \inf \left\{ \sum_{h=1}^{\infty} \operatorname{diam}(E_h)^k \, : \, E \subset \bigcup_{h=1}^{\infty} E_h, \right.$$

$$\left. \operatorname{diam}(E_h) < \rho \quad \text{for} \quad h = 1, 2, \ldots \right\}.$$

The Hausdorff measure is a completely additive set function, never negative (maybe infinite) and, when $k = n$, it coincides with the usual Lebesgue measure.

From (11) it can be seen that if, for some k, $\mathcal{H}^k(E)$ is finite, then $\mathcal{H}^{k+1}(E) = 0$.

It can also be checked that, for sufficiently smooth manifolds, the Hausdorff measure coincides with the elementarily defined measure; more precisely, if T

is a closed rectangle in $\mathbf{R}^k$, if $f_1(y),\ldots,f_n(y)$ are n functions there defined and Lipschitz continuous (which is trivially true when the f_i are continuous with their first order derivatives) and if the map f associating the generic point $y \in T$ with the point $f(y) = (f_1(y),\ldots,f_n(y)) \in \mathbf{R}^n$ is injective, then the set $E = f(T)$ has a Hausdorff measure given by

$$(12) \qquad \mathcal{H}^k(E) = \mathcal{H}^k(f(T)) = \int_T \sqrt{\det \|g_{hi}\|}\, dy,$$

where the entries of the square matrix $\|g_{hi}\|$ are given by

$$(13) \qquad g_{hi} = \sum_{l=1}^{n} \frac{\partial f_l}{\partial y_h} \frac{\partial f_l}{\partial y_i}, \qquad h,i = 1,\ldots,k.$$

Then we define the measure $\mathcal{H}^0$ in the following way: if E is made by finitely many points, then $\mathcal{H}^0(E)$ is the number of these points; otherwise, $\mathcal{H}^0(E) = +\infty$.

Given a vector valued set function $a(E) = (a_1(E),\ldots,a_n(E))$ defined for every Borel set of $\mathbf{R}^n$, completely additive and bounded, its total variation

$$(14) \quad \sigma(E) = \sup\left\{ \sum_{h=1}^{\infty} |a(E_h)| \; : \; E = \bigcup_{h=1}^{\infty} E_h, \quad E_h \cap E_k = \emptyset \quad \text{for} \quad h \neq k \right\}$$

is a completely additive and bounded set function. We will let, for $E \subset \mathbf{R}^n$ and f defined in E,

$$(15) \qquad \int_E |da| = \sigma(E), \qquad \int_E f\, d\sigma = \int_E f\, |da|.$$

Finally, for every set $E \subset \mathbf{R}^n$, we denote by the symbol $\varphi(x, E)$ the characteristic function of E, i.e., the function equal to 1 in E and equal to 0 in $\mathbf{R}^n \setminus E$.

2.

Let us recall, with slight modifications, some notions from [2], [3], [5].

Assume $E \subset \mathbf{R}^n$ and $n \geq 2$; for every positive number λ, let

$$(1) \qquad \varphi_\lambda(x) = (\pi\lambda)^{-\frac{n}{2}} \exp\left(\frac{-|x|^2}{\lambda} \right) * \varphi(x, E);$$

then there exists, finite or infinite, the limit

$$(2) \qquad P(E) = \lim_{\lambda \to 0} \int_{\mathbf{R}^n} |D\varphi_\lambda(x)|\, dx,$$

which is called the perimeter of E.

Then we have the following theorem (see [2], theorem 2.4):

I – *A necessary and sufficient condition for $P(E)$ to be finite, is the existence of a vector valued, completely additive and bounded function $a(B)$, defined for every set $B \subset \mathbf{R}^n$, satisfying the generalized Gauss–Green formulae*

$$(3) \qquad \int_E Dg\, dx = - \int_{\mathbf{R}^n} g(x)\, da$$

for every function $g(x)$ defined in $\mathbf{R}^n$, continuous together with its first order partial derivatives, and having compact support. In this case, there holds

$$(4) \qquad P(E) = \int_{\mathbf{R}^n} |da|.$$

The equation (3) can be rephrased saying that the function a is the gradient, in the sense of distributions, of $\varphi(x, E)$; we shall stress this letting

$$(5) \qquad \int_B D\varphi(x, E) = a(B), \qquad \int_B f(x)\, D\varphi(x, E) = \int_B f\, da,$$

$$(5') \qquad \int_B |D\varphi(x, E)| = \int_B |da|, \qquad \int_B f(x)\, |D\varphi(x, E)| = \int_B f\, |da|,$$

$$(5'') \qquad \int_B D_h\varphi(x, E) = a_h(B), \qquad \int_B f(x)\, D_h\varphi(x, E) = \int_B f\, da_h, \quad h = 1, \ldots, n.$$

The perimeter $P(E)$ is a lower semicontinuous functional with respect to integral convergence; more precisely, we have the following theorem (see [2], theorem VII):

II – *Assume $\{E_h\}$ is a sequence of sets in $\mathbf{R}^n$ and E a set in $\mathbf{R}^n$ satisfying*

$$(6) \qquad \lim_{h\to\infty} \int_{\mathbf{R}^n} |\varphi(x, E) - \varphi(x, E_h)|\, dx = \lim_{h\to\infty} (\operatorname{meas}(E \setminus E_h) + \operatorname{meas}(E_h \setminus E)) = 0.$$

Then we have

$$(7) \qquad \liminf_{h\to\infty} P(E_h) \geq P(E).$$

Among the sequences converging to a set E, particularly relevant are those made of polyhedral domains.

More precisely, calling *polyhedral domain* any set $E \subset \mathbf{R}^n$ which is the closure of an open set and whose boundary $\mathcal{F}E$ is contained in a finite union of hyperplanes of $\mathbf{R}^n$, we have the following theorem:[‡]

III – *If E has finite perimeter, then there exists a sequence of polyhedral domains $\{E_h\}$ satisfying*

$$(8) \qquad P(E) = \lim_{h\to\infty} P(E_h), \qquad \lim_{h\to\infty} (\operatorname{meas}(E \setminus E_h) + \operatorname{meas}(E_h \setminus E)) = 0.$$

The sets which can approximated in mean by polyhedral domains having bounded perimeter were introduced by Caccioppoli in [1]; such sets, according

---

[‡] Editor's note: see Theorem VIII in "Su una teoria generale della misura $(r-1)$-dimensionale in uno spazio ad r dimensioni", Ann. Mat. Pura Appl., (4) **36** (1954) 191–212, in this volume

to theorems II, III, coincide with the sets of finite perimeter, which are therefore called *Caccioppoli sets.*

In view of some applications in the calculus of variations, beside the semicontinuity theorem II, the following compactness theorem is of interest (see [3], theorem I):

IV – *Given a bounded set $L \subset \mathbf{R}^n$, a positive number p and a sequence of sets $\{E_h\}$ satisfying*

$$(9) \qquad E_h \subset L, \qquad P(E_h) \leq p, \qquad h = 1, 2, \ldots$$

there exist a set $E \subset \mathbf{R}^n$ and a subsequence $\{E_{h_i}\}$ such that

$$(10) \qquad \lim_{i \to \infty} \left(\operatorname{meas}(E \setminus E_{h_i}) + \operatorname{meas}(E_{h_i} \setminus E) \right) = 0.$$

In order to investigate the local properties of the boundaries of Caccioppoli sets, we shall need the following theorem (see [3], lemma III):

V – *Assume $E \subset \mathbf{R}^n$ is a Caccioppoli set and let ξ be a point in $\mathbf{R}^n$. Then, for every positive number ρ, the set $E \cap A_\rho(\xi)$ is a Caccioppoli set and, moreover, for every $B \subset \mathbf{R}^n$ and for almost every positive number ρ*

$$(11) \quad \int_B D\varphi(x, E \cap A_\rho(\xi)) = \int_{E \cap A_\rho(\xi)} D\varphi(x, E) + \int_{\mathcal{F}A_\rho(\xi)} \varphi(x, E \cap B)\, \nu(x)\, d\mathcal{H}^{n-1},$$

where $\nu(x)$ is the inner normal vector to $\mathcal{F}A_\rho(\xi)$.

We observe that from (11) it follows that

$$(12) \quad \int_B |D\varphi(x, E \cap A_\rho(\xi))| = \int_{B \cap A_\rho(\xi)} |D\varphi(x, E)| + \int_{\mathcal{F}A_\rho(\xi)} \varphi(x, E \cap B)\, d\mathcal{H}^{n-1}.$$

Then (11) can be rewritten as

$$\int_B D\big(\varphi(x, E)\varphi(x, A_\rho(\xi))\big) =$$

$$(11') \qquad = \int_B \varphi(x, A_\rho(\xi)) D\varphi(x, E) + \int_B \varphi(x, E)\, D\varphi(x, A_\rho(\xi)).$$

With an argument similar to that used to prove (11), (12), one proves that

$$(13) \quad \int_B D\varphi(x, E \setminus A_\rho(\xi)) = \int_{E \setminus A_\rho(\xi)} D\varphi(x, E) - \int_{\mathcal{F}A_\rho(\xi)} \varphi(x, E \cap B)\, \nu(x)\, d\mathcal{H}^{n-1},$$

$$(14) \quad \int_B |D\varphi(x, E \setminus A_\rho(\xi))| = \int_{B \setminus A_\rho(\xi)} |D\varphi(x, E)| + \int_{\mathcal{F}A_\rho(\xi)} \varphi(x, E \cap B)\, d\mathcal{H}^{n-1}.$$

Given a Caccioppoli set $E \subset \mathbf{R}^n$, we call *reduced boundary* of E, and we denote it by $\mathcal{F}^*E$, the set of those points ξ satisfying the following three conditions: for every positive number ρ there holds

$$(15) \qquad \int_{A_\rho(\xi)} |D\varphi(x, E)| > 0,$$

there exists the limit

$$(16) \qquad \lim_{\rho \to 0} \frac{\int_{A_\rho(\xi)} D\varphi(x, E)}{\int_{A_\rho(\xi)} |D\varphi(x, E)|} = \nu(\xi),$$

and there holds

$$(17) \qquad |\nu(\xi)| = 1.$$

Amongst the properties of the reduced boundary, we recall those given by the following theorems (see [3] theorem III, [5]):

VI – *For every Caccioppoli set $E \subset \mathbf{R}^n$ and for every point $\xi \in \mathcal{F}^* E$ there holds*

$$(18) \qquad \lim_{\rho \to 0} \rho^{1-n} \int_{A_\rho(\xi)} |D\varphi(x, E)| = \omega_{n-1},$$

$$(19) \qquad \lim_{\rho \to 0} \rho^{-n} \mathrm{meas}\left(A_\rho(\xi) \cap E\right) = \lim_{\rho \to 0} \rho^{-n} \mathrm{meas}\left(A_\rho(\xi) \setminus E\right) = \frac{\omega_n}{2}.$$

VII – *For every Caccioppoli set E and for every set B in $\mathbf{R}^n$ there hold*

$$(20) \qquad \begin{aligned} \int_B D\varphi(x, E) &= \int_{B \cap \mathcal{F}^* E} \nu(x)\, d\mathcal{H}^{n-1}, \\ \int_B |D\varphi(x, E)| &= \mathcal{H}^{n-1}(B \cap \mathcal{F}^* E), \\ P(E) &= \mathcal{H}^{n-1}(\mathcal{F}^* E), \end{aligned}$$

where $\nu(x)$ is given by (16).

In order to avoid writing the limit (16) and the similar ones giving the components $\nu_1(x), \ldots, \nu_n(x)$ of the vector $\nu(x)$, we shall let

$$(21) \qquad \nu(x) = \frac{D\varphi(x, E)}{|D\varphi(x, E)|}, \qquad \nu_h(x) = \frac{D_h\varphi(x, E)}{|D\varphi(x, E)|} \qquad \text{for} \quad h = 1, \ldots, n.$$

This notation is motivated by the observation that from (20) it follows that, for every B and every function $f(x)$ such that the following integrals are meaningful,

$$(22) \qquad \begin{aligned} \int_B f(x)\, d\varphi(x, E) &= \int_{B \cap \mathcal{F}^* E} f(x)\, \nu(x)\, d\mathcal{H}^{n-1} = \\ &= \int_{B \cap \mathcal{F}^* E} f(x)\, \frac{D_h\varphi(x, E)}{|D\varphi(x, E)|} \cdot |D\varphi(x, E)| = \\ &= \int_B f(x)\, \frac{D_h\varphi(x, E)}{|D\varphi(x, E)|} \cdot |D\varphi(x, E)|. \end{aligned}$$

Moreover, note that from (19) it follows that

$$(23) \qquad \mathcal{F} E \supseteq \overline{\mathcal{F}^* E} \supseteq \mathcal{F}^* E$$

and hence, if E is a Caccioppoli set, there holds by (20)

$$(24) \qquad \mathcal{H}^{n-1}(\mathcal{F}E) \geq P(E).$$

The unconditioned validity of (24) is then guaranteed by a remark of Federer (see [5]), who observes that, if $\mathcal{H}^{n-1}(\mathcal{F}E)$ is finite, then E is a Caccioppoli set.

Another interesting consequence of (20) is obtained observing that, given a Caccioppoli set E and a point ξ in $\mathbf{R}^n$, if for some positive ρ we have

$$(25) \qquad \mathcal{F}^*E \cap A_\rho(\xi) = \emptyset,$$

then we have, by virtue of (4), (5):

$$(26) \qquad \int_E Dg(x)\,dx = \int_{\mathbf{R}^n} \varphi(x,E)\,Dg(x)\,dx = 0$$

for every function $g(x)$ continuous together with its first order partial derivatives and having compact support in $A_\rho(\xi)$: therefore the characteristic function $\varphi(x,E)$ is constant on almost every all of $A_\rho(\xi)$. It follows that, letting

$$(27) \qquad \varphi^*(x) = \varphi(x,E), \quad \text{for} \quad x \in \overline{\mathcal{F}^*E},$$

$$(27') \qquad \varphi^*(x) = \lim_{\rho \to 0} \frac{\text{meas}\left(E \cap A_\rho(\xi)\right)}{\omega_n \rho^n}, \quad \text{for} \quad x \in \mathbf{R}^n \setminus \overline{\mathcal{F}^*E},$$

$\varphi^*(x)$ turns out to be the characteristic function of a set E^* satisfying

$$(28) \qquad \mathcal{F}E^* \subset \overline{\mathcal{F}^*E}, \quad \text{meas}\left((E^* \setminus E) + \text{meas}\left(E \setminus E^*\right)\right) = 0.$$

But, from the previous definitions, sets differing by negligible sets have the same perimeter and the same reduced boundary; therefore, from (23), (28) we have

$$(29) \qquad \mathcal{F}E^* = \overline{\mathcal{F}^*E^*} = \overline{\mathcal{F}^*E}.$$

3.

We now investigate the relationships between the classical and the generalized Gauss formulae. To this aim, it will be convenient to introduce some definitions. Supposing $n \geq 2$, we shall say that a set $S \subset \mathbf{R}^n$ is a *locally regular hypersurface* if, for every point $\xi \in S$, there exist an open set A containing ξ and a function $f(x)$ defined in A and there continuous together with its first order partial derivatives, such that

$$(1) \qquad S \cap A = \{x : x \in A, \quad f(x) = 0\}$$

and that, for every $x \in A$,

$$(2) \qquad |Df(x)| \neq 0.$$

If, for every $\xi \in S$, the open set A and the function $f(x)$ can be chosen in such a way that $f(x)$, beside the previous conditions, has continuous partial derivatives of every order in A, we shall say that S is a *locally regular hypersurface of class* C^∞. If, moreover, for every $\xi \in S$, the set A and the function $f(x)$ can be chosen in such a way that $f(x)$ is analytic in A, we shall say that S is a *locally analytic hypersurface*. We shall say that a set $E \subset \mathbf{R}^n$ is a *regular domain* if it is the closure of an open set and its boundary $\mathcal{F}E$ is a bounded locally regular hypersurface.

We have the following theorem:

I – *If E is a regular domain in $\mathbf{R}^n$, its perimeter $P(E)$ is finite and*

$$(3) \qquad \mathcal{F}^*E = \mathcal{F}E, \qquad P(E) = \mathcal{H}^{n-1}(\mathcal{F}E);$$

moreover, the vector

$$(4) \qquad \frac{D\varphi(x, E)}{|D\varphi(x, E)|}$$

coincides with the inner normal vector $\nu(x)$ elementarily defined.

Proof. From (12) n.1 it follows that the Hausdorff measure coincides on $\mathcal{F}E$ with the $(n-1)$-dimensional measure elementarily defined. Therefore, the classical Green formulae can be written as

$$(5) \qquad \int_E Dg\,dx = -\int_{\mathcal{F}E} g(x)\,\nu(x)\,d\mathcal{H}^{n-1}$$

for every function $g(x)$ continuous together with its first order partial derivatives and having compact support.

Comparing (5) with the generalized formulae appearing in theorem II n.1, we have, recalling (5), (5') n.2

$$(6) \qquad \int_B D\varphi(x, E) = \int_{\mathcal{F}E} \varphi(x, B)\,\nu(x)\,d\mathcal{H}^{n-1}$$

for every $B \subset \mathbf{R}^n$, whence our theorem follows.

An easy extension of theorem I is obtained recalling that, given two vector valued set functions $a(B) = (a_1(B), \dots, a_n(B))$ and $\beta(B) = (\beta_1(B), \dots, \beta_n(B))$, defined for every $B \subset \mathbf{R}^n$, completely additive and bounded in an open set $A \subset \mathbf{R}^n$, if for every function $g(x)$ of compact support in A which is continuous together with its first order partial derivatives, there holds

$$(7) \qquad \int_A g(x)\,da = \int_A g(x)\,d\beta$$

then, for every set B contained in A, there holds

$$(8) \qquad a(B) = \beta(B).$$

As a consequence, we have

II –   *Given two Caccioppoli sets E, L and an open set A in $\mathbf{R}^n$, if*

$$(9) \qquad\qquad A \cap E = A \cap L,$$

then for every $B \subset A$ there holds

$$(10) \qquad \int_B D\varphi(x, E) = \int_B D\varphi(x, L), \qquad \int_B |D\varphi(x, E)| = \int_B |D\varphi(x, L)|$$

and hence

$$(10') \qquad\qquad A \cap \mathcal{F}^* E = A \cap \mathcal{F}^* L.$$

Proof. It suffices to consider the generalized Gauss–Green formulae of theorem II n.1, concerning functions $g(x)$ with compact support in A.

III –   *Assume E is a Caccioppoli set and A is an open set in $\mathbf{R}^n$; suppose that $A \cap \mathcal{F}E$ is a locally regular hypersurface and that each of its points is a limit both of points interior to A and of points interior to $A \setminus E$. Then*

$$(11) \qquad\qquad \mathcal{F}^* E \cap A = \mathcal{F}E \cap A$$

and for every point $\xi \in \mathcal{F}E \cap A$, the inner normal vector $\nu(x)$ elementarily defined coincides with the vector

$$(12) \qquad\qquad \frac{D\varphi(\xi, E)}{|D\varphi(\xi, E)|}.$$

Proof. It suffices to repeat the argument used to prove theorem III n.1, restricting ourselves to consider functions $g(x)$ supported in A.

Theorem III allows us to clarify the relationships between classical and generalized Green formulae for domains whose boundary may have singularities (like corners, cusps etc.). More precisely, we shall say that a set $E \subset \mathbf{R}^n$ is a *quasi regular domain* if it is the closure of an open set and there exists a closed set N, having Hausdorff measure $\mathcal{H}^{n-1}(N) = 0$, such that the set $\mathcal{F}E \setminus N$ is a locally regular hypersurface.

From theorem III and (24) n.2 we obtain the following corollary:

IV –   *If E is a quasi regular domain in $\mathbf{R}^n$, there holds*

$$(13) \qquad\qquad P(E) = \mathcal{H}^{n-1}(\mathcal{F}E).$$

If moreover $P(E)$ is finite, then

$$(14) \qquad\qquad \mathcal{H}^{n-1}(\mathcal{F}E \setminus \mathcal{F}^* E) = 0$$

and for almost every $\xi \in \mathcal{F}E$ the inner normal vector elementarily defined coincides with the vector

$$(15) \qquad\qquad \frac{D\varphi(\xi, E)}{|D\varphi(\xi, E)|}.$$

Among quasi regular domains are, of course, polyhedral domains, for which (13), (14) can indeed be obtained comparing the classical Green formulae, certainly valid for such domains, with the generalized Green formulae of theorem II n.1.

As a consequence we have the following theorem:

V – *Given two Caccioppoli sets E, L, there holds*

$$(16) \qquad P(E) + P(L) \geq P(E \cup L) + P(E \cap L).$$

Proof. By theorem II we can find two sequences $\{E_h\}$, $\{L_h\}$ of polyhedral domains satisfying

$$(17) \qquad \lim_{h \to \infty} \big(\operatorname{meas}(E_h \setminus E) + \operatorname{meas}(E \setminus E_h)\big) =$$
$$\lim_{h \to \infty} \big(\operatorname{meas}(L_h \setminus L) + \operatorname{meas}(L \setminus L_h)\big) = 0,$$

$$(18) \qquad \lim_{h \to \infty} P(E_h) = P(E), \qquad \lim_{h \to \infty} P(L_h) = P(L).$$

By theorem IV we have

$$(19) \qquad \mathcal{H}^{n-1}(\mathcal{F}E_h) = P(E_h), \qquad \mathcal{H}^{n-1}(\mathcal{F}L_h) = P(L_h).$$

On the other hand, we have

$$(20) \qquad \mathcal{F}(E_h \cup L_h) \subset \mathcal{F}E_h \cup \mathcal{F}L_h, \qquad \mathcal{F}(E_h \cap L_h) \subset \mathcal{F}E_h \cup \mathcal{F}L_h,$$

$$(21) \qquad \mathcal{F}(E_h \cup L_h) \cap \mathcal{F}(E_h \cap L_h) \subset \mathcal{F}E_h \cap \mathcal{F}L_h,$$

and hence from (20), (21), by the additivity of the Hausdorff measure, we have

$$(22) \quad \mathcal{H}^{n-1}\big(\mathcal{F}(E_h \cup L_h)\big) + \mathcal{H}^{n-1}\big(\mathcal{F}(E_h \cap L_h)\big) \leq \mathcal{H}^{n-1}\big(\mathcal{F}E_h\big) + \mathcal{H}^{n-1}\big(\mathcal{F}L_h\big).$$

From (19), (22), recalling (24) n.2, it follows that

$$(23) \qquad P(E_h \cup L_l) + P(E_h \cap L_h) \leq P(E_h) + P(L_h)$$

and hence, by theorem II n.2, (16) is established.

From theorem V and the trivial equality

$$(24) \qquad P(E) = P(\mathbf{R}^n \setminus E)$$

we infer that the family of all Caccioppoli sets is closed under the operations of union, intersection and difference.

4.

Since in our study we shall simultaneously consider a space $\mathbf{R}^n$ and a space $\mathbf{R}^m$ (with $n = m + 1$), beside the notation introduced in n.1 we shall adopt the following one.

If $y = (y_1, \ldots, y_m) \in \mathbf{R}^m$, $t \in \mathbf{R}$ we let

$$(1) \qquad (y, t) = (y_1, \ldots, y_m, t) \in \mathbf{R}^n.$$

We shall always denote by $E \times L$ the Cartesian product of two sets E, L. Still denoting by $C_\rho(\xi)$, $A_\rho(\xi)$ the closed and open balls in $\mathbf{R}^n$, centered at x and with radius ρ, we will use the symbol $G_\rho(y)$ to denote the closed ball in $\mathbf{R}^m$, centered at y and with radius ρ; as usual, if y coincides with the origin, we will write G_ρ in place of $G_\rho(y)$.

Now we prove two lemmata.

I – *Let ρ, ε, σ be three numbers satisfying*

$$(2) \qquad 0 < \rho, \qquad 0 < 2\varepsilon < 1, \qquad 0 < 2\sigma < 1, \qquad 0 < \frac{2\sigma}{(1-\sigma)^2} < \frac{\varepsilon^2}{4}.$$

Let $n \geq 2$ be an integer, set $m = n - 1$, and let E be a Caccioppoli set in $\mathbf{R}^n$ satisfying the condition

$$(3) \qquad \mathcal{F}^* E \cap A_\rho = \mathcal{F} E \cap A_\rho,$$

such that the set $\mathcal{F} E \cap A_\rho$ is a locally regular hypersurface of class C^∞. Suppose, moreover, that at every point $x \in \mathcal{F} E \cap A_\rho$

$$(4) \qquad \frac{D_n \varphi(x, E)}{|D\varphi(x, E)|} > 1 - \sigma,$$

and that there exists a point ξ satisfying

$$(5) \qquad \xi \in \mathcal{F} E, \qquad |\xi| < \sigma \rho.$$

Then there exist an open set $B \subset \mathbf{R}^m$ and a function $f(y)$ defined in B, continuous together with its first order partial derivatives, such that

$$(6) \qquad \mathcal{F} E \cap A_\rho = \{(y, t) : y \in B, \quad t = f(y)\}, \qquad B \supset G_{\rho - \varepsilon\rho}$$

and, moreover, such that

$$(7) \qquad |Df(y)| < \varepsilon, \qquad |f(y)| < \varepsilon\rho.$$

Proof. If, for every point $x \in \mathcal{F} E$, we denote by $\nu(x) = (\nu_1(x), \ldots, \nu_n(x))$ the inner normal vector elementarily defined, whose existence follows from (3) and theorem VI n.2, by theorem III n.3 we have

$$(8) \qquad \nu_n(x) = \frac{D_n \varphi(x, E)}{|D\varphi(x, E)|} > 1 - \sigma > 0,$$

and hence every straight line parallel to the x_n axis will meet $\mathcal{F} E \cap A_\rho$ at most at one point.

Therefore, there exist a set $B \subset \mathbf{R}^m$ and a function $f(y)$ defined in B, satisfying

$$(9) \qquad \mathcal{F}E \cap A_\rho = \{(y,t) : y \in B, \quad t = f(y)\}.$$

By (8) and by well known theorems on implicit functions, the set B is open and the function $f(y)$ is continuous in B with its partial derivatives of any order. Its derivatives are given by

$$(10) \qquad D_h f(y) = \frac{\nu_h(y, f(y))}{\nu_n(y, f(y))} \quad \text{for} \quad h = 1, \ldots, m = n - 1.$$

Comparing (2), (4), (8), (10) we find

$$(11) \qquad |Df(y)| = \frac{\sqrt{1 - \nu_n^2}}{\nu_n} < \frac{\sqrt{2\sigma}}{1 - \sigma} = \sqrt{\frac{2\sigma}{(1-\sigma)^2}} < \frac{\varepsilon}{2}.$$

Now consider the point ξ satisfying (5); being $\sigma < 1$, we have $\xi \in \mathcal{F}E \cap A_\rho$ and hence we can find a point η such that

$$(12) \qquad \eta \in B, \qquad (\eta, f(\eta)) = \xi.$$

From (2), (5), (12) it follows that

$$(13) \qquad |\eta| \leq |\xi| < \sigma\rho < \frac{\varepsilon}{4}\rho, \qquad |f(\eta)| \leq |\xi| < \frac{\varepsilon}{4}\rho$$

and hence, being $2\varepsilon < 1$ by assumption, the intersection of B with $G_{\rho-\varepsilon\rho}$ is not empty. Having already proved (9), in order to complete the proof of (6) it will be sufficient to show that

$$(14) \qquad \mathcal{F}B \cap G_{\rho-\varepsilon\rho} = \emptyset.$$

Arguing by contradiction, let us suppose that (14) is violated; denoting by λ the distance from η to the closed set $G_{\rho-\varepsilon\rho} \setminus B$, let η^* denote a point of this set such that $|\eta - \eta^*| = \lambda$. Let L denote the line segment with endpoints η and η^*; by (11), there exists the limit

$$(15) \qquad \tau^* = \lim_{y \to \eta^*} f(y) \quad \text{along} \quad L.$$

Recalling (13), we have

$$(16) \qquad |\tau^* - f(\eta)| < \frac{\varepsilon\lambda}{2} \leq \frac{\varepsilon}{2}(|\eta| + |\eta^*|) < \frac{\varepsilon}{2}\rho,$$

which implies, again by (13), that

$$(17) \qquad |\tau^*| \leq |f(\eta)| + \frac{\varepsilon\rho}{2} < \varepsilon\rho.$$

Now the point $\xi^* = (\eta^*, \tau^*)$ belongs to $\mathcal{F}E$, being the limit of points $(y, f(y))$ belonging to such set; on the other hand, being $\eta^* \in G_{\rho-\varepsilon\rho}$ by construction, ξ^* belongs, by virtue of (17), to the ball A_ρ. It follows, by (9), that η^* belongs to the open set B, against the assumption that $\eta^* \in \mathcal{F}B$. This contradiction arises from assuming (14) to be false, hence (14) is proved.

Finally, the first part of (7) follows from (11), whereas the second part is an easy consequence of (6), (11), (13).

II – *Given a positive number ρ and a sequence $\{f_h(y)\}$ of functions defined on the ball $G_\rho \in \mathbf{R}^m$, assume that the functions $\{f_h(y)\}$ are equi-uniformly Lipschitzian in G_ρ, i.e., there exists a constant λ such that*

$$(18) \qquad |f_h(y) - f_h(y')| \leq \lambda |y - y'|$$

for every h and every pair of points y, y' in G_ρ. Moreover, suppose that

$$(19) \qquad \lim_{h,k\to\infty} \int_{G_\rho} |f_h(y) - f_k(y')|\, dy = 0.$$

Then the sequence $\{f_h\}$ converges uniformly in G_ρ to a Lipschitzian function $f(y)$.

If, for almost every point $y \in G_\rho$ there exists the limit

$$(20) \qquad \lim_{h\to\infty} Df_h(y),$$

then we have

$$(21) \qquad Df(y) = \lim_{h\to\infty} Df_h(y)$$

almost everywhere in G_ρ.

Proof. The assumptions (18), (19) guarantee the equi-uniform continuity and boundedness of the functions $f_h(y)$; hence, recalling (19), the sequence $\{f_h(y)\}$ is uniformly convergent in G_ρ to a continuous function $f(y)$. Then from (18), (19) it follows that

$$(22) \qquad |f(y) - f(y')| \leq \lambda |y - y'| \quad \text{for} \quad y, y' \in G_\rho.$$

On the other hand, if $g(y)$ is a function continuous together with its partial derivatives of the first order supported in G_ρ, we clearly have (being, by (18), $|Df_h(y)| \leq \lambda$)

$$(23) \quad \int_{G_\rho} g(y)\, Df(y)\, dy = -\int_{G_\rho} f(y)\, Dg(y)\, dy = \lim_{h\to\infty} \int_{G_\rho} f_h(y)\, Dg(y)\, dy =$$

$$= \lim_{h\to\infty} \int_{G_\rho} g(y)\, Df_h(y)\, dy = \int_{G_\rho} g(y) \left(\lim_{h\to\infty} Df_h(y) \right) dy.$$

From (23), by the arbitrariness of $g(y)$, we infer that (21) holds true almost everywhere in G_ρ.

THEOREM III – *Given a Caccioppoli set and an open set A in $\mathbf{R}^n$ (with $n > 1$), assume that*

$$(24) \qquad \mathcal{F}E \cap A = \mathcal{F}^*E \cap A$$

and that, moreover, the vector

$$(25) \qquad \nu(x) = \frac{D\varphi(x, E)}{|D\varphi(x, E)|}$$

is a continuous function in $\mathcal{F}E \cap A$. Then $\mathcal{F}E \cap A$ is a locally regular hypersurface.

Proof. Choose a point $\xi^* \in \mathcal{F}E \cap A$; we can assume that

$$(26) \qquad \nu_n(\xi^*) = \frac{D_n\varphi(\xi^*, E)}{|D\varphi(\xi^*, E)|} = 1$$

and that ξ^* coincides with the origin of the coordinates, since we can always reduce to this case by a rotation and a translation.

Chosen two positive numbers ε, σ satisfying (2) of theorem I, by the continuity of $\nu(\xi)$ we can find a positive number p such that

$$(27) \qquad C_{4p} \subset A, \qquad \nu_n(x) = \frac{D_n\varphi(x, E)}{D\varphi(x, E)} > 1 - \sigma \quad \text{for} \quad x \in C_{4p} \cap \mathcal{F}E.$$

Then consider a function $g(x)$, continuous in $\mathbf{R}^n$ with all of its derivatives of any order, satisfying

$$(28) \qquad g(x) = 0 \quad \text{for} \quad |x| \geq p, \qquad g(x) > 0 \quad \text{for} \quad |x| < p,$$

$$(29) \qquad \int_{\mathbf{R}^n} g(x)\, dx = \int_{C_p} g(x)\, dx = 1.$$

For every integer $h > 0$, let

$$(30) \qquad \psi_h(x) = h^n\, g(h\, x) * \varphi(x, E).$$

A necessary and sufficient condition for a point $x \in C_{3p}$ to satisfy

$$(31) \qquad 0 < \psi_h(x) < 1$$

is that the point x have distance to $\mathcal{F}E$ less than p/h; indeed, the necessity is trivial by (28), (29), whereas the sufficiency follows from (27), (28), (29), keeping into account (24) and (18) n.2 of theorem VI n.2.

In particular, being $\xi^* \in \mathcal{F}E \cap A$, $|\xi^*| = 0$, we have

$$(32) \qquad 0 < \psi_h(\xi^*) < 1.$$

Now let

$$(33) \qquad b_h = \psi_h(\xi^*),$$

$$(34) \qquad F_h = \{x \ : \ \psi_h(x) = b_h\},$$

$$(35) \qquad E_h = \{x \ : \ \psi_h(x) > b_h\}.$$

By (30), $\psi_h(x)$ is continuous in $\mathbf{R}^n$, hence by (34), (35) we have

$$(36) \qquad \mathcal{F}E_h = F_h.$$

From (28), (29), (30), (32), (33), (35) it follows that, when the distance from x to $\mathcal{F}E$ is no less than p/h, there holds

$$(37) \qquad \varphi(x, E_h) = \varphi(x, E)$$

and hence

$$(38) \qquad \lim_{h \to \infty} \varphi(x, E_h) = \varphi(x, E) \quad \text{for} \quad x \in \mathbf{R}^n \setminus \mathcal{F}E.$$

By (24), (27) and theorem VII n.2, the set $\mathcal{F}E \cap C_{4p}$ (clearly contained in $\mathcal{F}E \cap A$) has finite Hausdorff measure $\mathcal{H}^{n-1}(\mathcal{F}E \cap C_{4p})$ and hence (see n.1)

$$(39) \qquad \operatorname{meas}\left(\mathcal{F}E \cap C_{4p}\right) = 0.$$

From (38), (39) we obtain

$$(40) \qquad \lim_{h \to \infty} \int_{C_{4p}} |\varphi(x, E) - \varphi(x, E_h)| \, dx = 0.$$

Now consider a point $x \in F_h \cap C_{3p}$; by (28), (29), (30), (32), (33), (34), the distance from x to $\mathcal{F}E$ is less than p/h and hence recalling (2), (24), (27), (28), (30) and theorem VI n.2 we have

$$(41) \qquad D_n \psi_h(x) = \int_{\mathbf{R}^n} h^n g\big((h(x - \xi)\big) \, D_n \varphi(\xi, E) =$$

$$= \int_{C_{4p}} h^n g\big((h(x - \xi)\big) \, D_n \varphi(\xi, E) \geq (1 - \sigma) \int_{C_{4p}} h^n g\big((h(x - \xi)\big) \, |D\varphi(\xi, E)| > 0.$$

On the other hand, since for every $x \in C_{3p}$

$$(42) \qquad D_n \psi_h(x) \leq |D\psi_h(x)| = \left| \int_{\mathbf{R}^n} h^n g\big((h(x - \xi)\big) \, D\varphi(\xi, E) \right| =$$

$$\left| \int_{C_{4p}} h^n g\big((h(x - \xi)\big) \, D\varphi(\xi, E) \right| \leq \int_{C_{4p}} h^n g\big((h(x - \xi)\big) \, |D\varphi(\xi, E)|$$

we obtain, combining (41), (42), that

$$(43) \qquad |D\psi_h(x)| > 0, \qquad \frac{D_n\psi_h(x)}{|D\psi_h(x)|} > 1 - \sigma > 0 \quad \text{for} \quad x \in F_h \cap C_{3p}.$$

From (43) we see that $F_h \cap A_{3p}$ is a locally regular hypersurface; furthermore, since $g(x)$ (and hence also $\psi_h(x)$) are continuous in $\mathbf{R}^n$ with all of their partial derivatives of any order, $F_h \cap A_{3p}$ is a locally regular hypersurface of class C^∞.

Recalling the properties of the Hausdorff measure discussed in n.1, it can be seen that the measure $\mathcal{H}^{n-1}$ coincides with the elementary $(n-1)$-dimensional measure on every locally regular hypersurface. It follows that every closed set $K \subset F_h \cap A_{3p}$ has a finite $(n-1)$-dimensional Hausdorff measure and, in particular, there holds

$$(44) \qquad \mathcal{H}^{n-1}(F_h \cap C_{2p}) < +\infty.$$

On the other hand, by (36),

$$(45) \qquad \mathcal{F}(E_h \cap C_{2p}) \subset (F_h \cap C_{2p}) \cup \mathcal{F}C_{2p}$$

and hence, by (24) and by the additivity of the Hausdorff measure,

$$(46) \qquad P(E_h \cap C_{2p}) \leq \mathcal{H}^{n-1}(E_h \cap C_{2p}) < +\infty,$$

i.e. $E_h \cap C_{2p}$ is a Caccioppoli set. By (36) we also have

$$(47) \qquad A_{2p} \cap F_h = A_{2p} \cap \mathcal{F}(E_h \cap C_{2p}),$$

whereas from (34), (35), (43) we see that every point of $A_{2p} \cap F_h$ is a limit point of points interior to E_h and of points interior to $\mathbf{R}^n \setminus E_h$. By theorem III n.3 it follows that

$$(48) \qquad A_{3p} \cap F_h = A_{2p} \cap \mathcal{F}^*(E_h \cap C_{2p}),$$

$$(49) \qquad \frac{D\varphi(x, E_h \cap C_{2p})}{|D\varphi(x, E_h \cap C_{2p})|} = \frac{D\psi_h(x)}{|D\psi_h(x)|} \quad \text{for} \quad x \in A_{2p} \cap F_h.$$

On the other hand, recalling that ξ^* coincides with the origin of $\mathbf{R}^n$, by (2), (32), (33), (34), (43), (47), (48), (49) and theorem IV n.1 we can find an open set B_h in $\mathbf{R}^m$ (with $m = n - 1$) and a function $f_h(y)$ continuous in B_h with all of its partial derivatives of any order, such that

$$(50) \qquad A_{2p} \cap F_h = \{(y,t) : y \in B_h, \quad t = f_h(y)\}, \qquad B_h \supset G_{2p-2\varepsilon p} \supset G_p,$$

$$(51) \qquad |Df_h(y)| < \varepsilon, \qquad |f_h(y)| < \varepsilon p \quad \text{for} \quad y \in G_p.$$

On the other hand, from (34), (35), (43), (50) it follows that

$$(52) \qquad A_{2p} \cap E_h \cap (G_p \times \mathbf{R}) = A_{2p} \cap \{(y,t) : y \in G_p, \quad t > f_h(y)\}$$

and hence, for every pair of positive integers h, s

$$(53) \qquad \int_{A_{2p}} |\varphi(x, E_h) - \varphi(x, E_s)| \, dx > \int_{G_p} |f_h(y) - f_s(y)| \, dy,$$

from which it follows, recalling (40), that

$$(54) \qquad \lim_{h,s \to \infty} \int_{G_p} |f_h(y) - f_s(y)| \, dy = 0.$$

From (51), (54), by theorem IV n.2, we deduce that the sequence $\{f_h(y)\}$ converges uniformly in G_p to a function which satisfies

$$(55) \qquad \lim_{h \to \infty} f_h(y) = f(y), \qquad |f(y) - f(y')| \le \varepsilon |y - y'| \quad \text{for} \quad y, y' \in G_p.$$

Letting

$$L = \{(y,t) \;:\; y \in G_p, \quad t > f(y)\}, \quad F = \{(y,t) \;:\; y \in G_p, \quad t = f(y)\},$$

from (40), (52), (55) we obtain

$$(57) \qquad \int_{A_p} |\varphi(x, E) - \varphi(x, L)| \, dx = 0.$$

Recalling (24), (27) and theorem VI n.2, from (56), (57) we find

$$(58) \qquad \mathcal{F}E \cap A_p = \mathcal{F}^* E \cap A_p = F \cap A_p.$$

Letting

$$(59) \qquad Y = \{y \;:\; (y, f(y)) \in A_p\},$$

the set Y, by the continuity of $f(y)$, is open and (58) becomes

$$(60) \qquad \mathcal{F}E \cap A_p = \mathcal{F}^* E \cap A_p = \{(y,t) \;:\; y \in Y, \quad t = f(y)\}.$$

Moreover, by (51) we have

$$(61) \qquad Y \supset G_{p-\varepsilon p}.$$

Now choose a point $y^* \in Y$ and let

$$(62) \qquad x_h = (y^*, f_h(y^*)), \quad x^* = (y^*, f(y^*)).$$

Then we have by (50), (55), (58)

$$(63) \qquad \lim_{h \to \infty} |x_h - x^*| = 0, \qquad x_h \in F_h, \qquad x^* \in \mathcal{F}E \cap A_p = \mathcal{F}^* E \cap A_p.$$

Let us compute $Df_h(y^*)$; by (34), (43), (50) we have

$$(64) \qquad D_i f_h(y^*) = \frac{D_i \psi_h(x_h)}{D_n \psi_h(x_h)} \quad \text{for} \quad i = 1, \ldots, m = n - 1.$$

On the other hand, letting

$$(65) \qquad \gamma_h(x) = \int_{\mathbf{R}^n} h^n \, g\big(h(x-\xi)\big) \, |D\varphi(x,E)|$$

by (42), (43), (63) we have

$$(66) \qquad \gamma_h(x_h) > 0,$$

whereas from (30) it follows that

$$(67) \qquad D\psi_h(x_h) = \int_{\mathbf{R}^n} h^n \, g\big(h(x-\xi)\big) \, D\varphi(x,E).$$

Comparing (25), (65), (66), (67) and recalling (28) we find by theorem VII n.2

$$(68) \qquad \left| \frac{D\psi_h(x_h)}{\gamma_h(x_h)} - \gamma(x^*) \right| < \sup\left\{ |\nu(x) - \nu(x^*)| \; : \; x \in \mathcal{F}^*E, \quad |x - x_h| < \frac{p}{h} \right\}.$$

From (68), by the continuity of $\nu(x)$ and by (27), (63), (64), we have that

$$(69) \qquad \lim_{h \to \infty} D_i f_h(y^*) = \frac{\nu_i(x^*)}{\nu_n(x^*)} \quad \text{for} \quad i = 1, \ldots, m = n - 1.$$

By the arbitrariness of y^*, we can then conclude that

$$(70) \qquad \lim_{h \to \infty} D_i f_h(y) = \frac{\nu_i(y, f(y))}{\nu_n(y, f(y))} \quad \text{for} \quad y \in Y, \quad i = 1, \ldots, m.$$

From (55), (61), (70) we deduce, by virtue of theorem II, that, almost everywhere in $G_{p-\varepsilon p}$, we have

$$(71) \qquad D_i f(y) = \frac{\nu_i(y, f(y))}{\nu_n(y, f(y))} \quad \text{for} \quad i = 1, \ldots, m = n - 1.$$

But then, since $f(y)$ satisfies the Lipschitz condition (55) and the right hand side of (71) is continuous, this equality will be satisfied in all of $G_{p-\varepsilon p}$ and hence $f(y)$ turns out to be continuous in $G_{p-\varepsilon p}$ together with its first order partial derivatives. Recalling (60), (61), we deduce that $\mathcal{F}^*E \cap A_{p-\varepsilon p} = \mathcal{F}E \cap A_{p-\varepsilon p}$ is a locally regular hypersurface. Finally, thanks to the arbitrariness of ξ^*, the proof is completed.

Bibliography

[1] R. Caccioppoli, *Misura ed integrazione sugli insiemi dimensionalmente orientati.* Rend. Lincei VIII, vol. 12, (1952).

[2] E. De Giorgi, *Su una teoria generale della misura $(r-1)$–dimensionale in uno spazio ad r dimensioni.* Ann. di Matematica, S. IV, vol. 36 (1954).

[3] E. De Giorgi, *Nuovi teoremi relativi alle misure $(r-1)$-dimensionali in uno spazio ad r dimensioni.* Ricerche di Matematica, vol. IV (1955).

[4] H. Federer, *Measure and area.* Bull. of the Am. Math. Soc., vol. 58 (1952), pp. 306–378.

[5] H. Federer, *A note on the Gauss-Green theorem.* Proc. of the Am. Math. Soc., vol. 9 (1958).

Area-minimizing oriented boundaries[‡]

Ennio De Giorgi

The investigation of area-minimizing oriented boundaries is the natural extension, to the framework of the theory of Caccioppoli dimensionally oriented sets, of the classical theory of the Plateau problem.

In this work we mainly rely on the results collected in [1]; besides the papers cited in [1], we mention the memoirs [2], [3], [4], containing a wide bibliography and interesting analogies of techniques and results.

An investigation of these analogies will be carried out elsewhere, in the framework of a general comparison between the several theories of measure and integration on dimensionally oriented sets, which seem to be connected with the theory of Caccioppoli.

1.

Throughout the paper, we adopt the notation introduced in [1]. Consider two sets E, L in $\mathbf{R}^n$ (with $n \geq 2$). We suppose that E is a Caccioppoli set, i.e. its perimeter $P(E)$ is finite, and we introduce three functions of the pair E, L.

Let

$$\text{(1)} \qquad Q(E, L) = \inf\Big\{ P(B) \ : \ B \subset \mathbf{R}^n, \quad B \setminus L = E \setminus L \Big\},$$

$$\text{(2)} \qquad \Theta(E, L) = Q(E, L) - \int_{\mathbf{R}^n \setminus L} |D\varphi(x, E)|,$$

$$\text{(3)} \qquad \Psi(E, L) = P(E) - Q(E, L).$$

We shall say that the set E is area-minimizing in L if there holds

$$\text{(4)} \qquad \Psi(E, L) = 0,$$

that is, if

$$\text{(4')} \qquad P(E) = \min\Big\{ P(B) \ : \ B \subset \mathbf{R}^n, \quad B \setminus L = E \setminus L \Big\}.$$

[‡]Editor's note: translation into English of the paper "Frontiere orientate di misura minima", Seminario di Matematica della Scuola Normale Superiore di Pisa 1960-61, Editrice Tecnico Scientifica, 1961. This translation takes also into account some misprints corrected in a later version of (essentially) the same paper, presented in "Frontiere orientate di misura minima e questioni collegate", Quad. Sc. Norm. Sup. Pisa, Editrice Tecnico Scientifica, Pisa, 1972.

From (1), (3), it immediately follows that

$$(5) \qquad Q(E, L) \geq Q(E, L'), \quad \Psi(E, L) \leq \Psi(E, L') \quad \text{whenever} \quad L \subset L'$$

and hence, since clearly

$$(6) \qquad \Psi(E, L) \geq 0,$$

if a set E has minimal boundary on L' then the same is true for every $L \subset L'$.

From Theorem II n.3 in [1] it follows that, if the set L is closed, then

$$(7) \quad \Theta(E, L) = \inf \left\{ \int_L |D\varphi(x, B)| \ : \ B \subset \mathbf{R}^n, \ P(B) < +\infty, \ B \setminus L = E \setminus L \right\},$$

$$(8) \qquad \Psi(E, L) = \int_L |D\varphi(x, E)| - \Theta(E, L).$$

An existence theorem for sets with minimal boundary is the following.

I – *Given a Caccioppoli set E and a bounded set L in $\mathbf{R}^n$, there exists a Caccioppoli set M satisfying*

$$(9) \qquad M \setminus L = E \setminus L, \qquad P(M) = Q(E, L);$$

the set M has minimal boundary in L.

Proof. According to the definition of $Q(E, L)$, we can find a sequence of sets $\{M_h\}$ such that

$$(10) \qquad M_h \setminus L = E \setminus L, \quad P(E) \geq P(M_h), \quad \lim_{h \to \infty} P(M_h) = Q(E, L).$$

Letting

$$(11) \qquad Z_h = (M_h \setminus E) \cup (E \setminus M_h),$$

Theorem V n.3 and (24) n.3 in [1] imply that

$$(12) \qquad P(Z_h) \leq P(M_h \setminus E) + P(E \setminus M_h) \leq 2P(E) + 2P(M_h) \leq 4P(E),$$

whereas, according to (10), (11), there holds

$$(13) \qquad Z_h \subset L.$$

Applying Theorem IV n.2 in [1] to the sequence $\{Z_h\}$, we can prove the existence of a subsequence $\{Z_{h_i}\}$ and a set Z satisfying

$$(14) \qquad Z \subset L, \quad \lim_{i \to \infty} \big(\text{meas}\,(Z \setminus Z_{h_i}) + \text{meas}\,(Z_{h_i} \setminus Z)\big) = 0.$$

Moreover, from (11) we find that

$$(15) \qquad M_h = (E \cup Z_h) \setminus (E \cap Z_h)$$

and hence, letting

$$(16) \qquad M = (E \cup Z) \setminus (E \cap Z)$$

by (14) we have

$$(17) \qquad M \setminus L = E \setminus L, \quad \lim_{i \to \infty} \left(\text{meas}\,(M \setminus M_{h_i}) + \text{meas}\,(M_{h_i} \setminus M) \right) = 0.$$

Using Theorem II n.2 in [1] and (10), from (17) it easily follows that

$$(18) \qquad P(M) \le Q(E, L);$$

combining (18) with the trivial bound

$$(19) \qquad P(M) \ge Q(M, L) = Q(E, L),$$

the proof is completed.

2.

Here we prove some theorems which will allow us to characterize the behaviour of a set with minimal boundary in a neighborhood of a point.

I – *Given two Caccioppoli sets E, F in $\mathbf{R}^n$, for almost every positive number ρ there holds*

$$(1) \qquad \int_{\mathcal{F}C_\rho} |\varphi(x, E) - \varphi(x, F)| \, d\mathcal{H}^{n-1} \ge |\Theta(E, C_\rho) - \Theta(F, C_\rho)|.$$

Proof. It suffices to prove the inequality

$$(2) \qquad \int_{\mathcal{F}C_\rho} |\varphi(x, E) - \varphi(x, F)| \, d\mathcal{H}^{n-1} \ge \Theta(F, C_\rho) - \Theta(E, C_\rho),$$

since the sets E, F play a symmetric role.

From (13), (14) n.2 in [1] we have*

$$(3) \quad \int_{\mathcal{F}C_\rho} |D\varphi(x, E \setminus C_\rho) - D\varphi(x, F \setminus C_\rho)| = \int_{\mathcal{F}C_\rho} |\varphi(x, E) - \varphi(x, F)| \, d\mathcal{H}^{n-1},$$

for almost every positive number ρ; chosen a number ρ such that (3) holds true, according to Theorem I n.1 we can find a set M satisfying

$$(4) \qquad M \setminus C_\rho = E \setminus C_\rho, \qquad Q(E, C_\rho) = P(M).$$

---

*In analogy with the notation (5), (5'), (5") n.2 in [1], if

$$\alpha(B) = \int_B D\varphi(x, E), \quad \beta(B) = \int_B D\varphi(x, F)$$

we will let

$$\int_B |D\varphi(x, E) \pm D\varphi(x, F)| = \int_B |d(\alpha \pm \beta)|.$$

234   E. De Giorgi

From Theorem I n.2 and Theorem II n.3 in [1] we find that

$$P(M) = \int_{\mathbf{R}^n \setminus C_\rho} |D\varphi(x, E)| + \int_{\mathcal{F}C_\rho} |D\varphi(x, E \setminus C_\rho) + D\varphi(x, M \cap C_\rho)| +$$

$$(5)$$

$$+ \int_{C_\rho \setminus \mathcal{F}C_\rho} |D\varphi(x, M)|,$$

and, similarly, letting

$$(6) \qquad Z = (F \setminus C_\rho) \cup (M \cap C_\rho),$$

$$P(Z) = \int_{\mathbf{R}^n \setminus C_\rho} |D\varphi(x, F)| + \int_{\mathcal{F}C_\rho} |D\varphi(x, F \setminus C_\rho) + D\varphi(x, M \cap C_\rho)| +$$

$$(7)$$

$$+ \int_{C_\rho \setminus \mathcal{F}C_\rho} |D\varphi(x, M)|.$$

From (3), (5), (7) it follows that

$$P(Z) - P(M) \le \int_{\mathbf{R}^n \setminus C_\rho} |D\varphi(x, F)| - \int_{\mathbf{R}^n \setminus C_\rho} |D\varphi(x, E)| +$$

$$(8)$$

$$+ \int_{\mathcal{F}C_\rho} |\varphi(x, E) - \varphi(x, F)| \, d\mathcal{H}^{n-1}$$

and hence, since clearly

$$(9) \qquad Q(F, C_\rho) \le P(Z),$$

from (4) and from the definition of the function Θ, (2) is established.

II – *Assume that E is a polyhedral domain in $\mathbf{R}^n$ having finite perimeter and let ρ, r, p be three numbers such that*

$$(10) \qquad 0 < \rho < r, \qquad p > \int_{C_r \setminus C_\rho} |D\varphi(x, E)|.$$

Then we have

$$\int_\rho^r \mathcal{H}^{n-2}(\mathcal{F}E \cap \mathcal{F}C_t) \, dt \le \int_{C_r \setminus C_\rho} |D\varphi(x, E)| -$$

$$(11)$$

$$- \frac{1}{2p} \left(\int_{\mathcal{F}C_\rho} \left| \varphi(\frac{r}{\rho} x, E) - \varphi(x, E) \right| d\mathcal{H}^{n-1} \right)^2.$$

Proof. We let, for every $x \in \mathcal{F}^* E$,

$$(12) \qquad \nu(x) = \frac{D\varphi(x, E)}{|D\varphi(x, E)|}.$$

When $|x| \neq 0$ the vector $\nu(x)$ can be uniquely decomposed as the sum of a vector $\beta(x)$, lying on the straight line joining x to the origin, and a vector $\gamma(x)$ orthogonal to this line, i.e.,

$$(13) \qquad \nu(x) = \beta(x) + \gamma(x), \quad |\langle \beta(x), x \rangle| = |\beta(x)| \, |x|, \quad \langle \gamma(x), x \rangle = 0,$$

and hence

$$(14) \qquad 1 = |\nu(x)|^2 = |\beta(x)|^2 + |\gamma(x)|^2.$$

By Theorem III n.3 in [1], $\nu(x)$ coincides with the inner normal vector elementarily defined and hence, recalling that on $\mathcal{F}E$ the Hausdorff measure reduces to the elementarily defined $(n-1)$-dimensional measure, by elementary arguments one finds (recalling that E is a polyhedral domain) that

$$(15) \qquad \int_{\mathcal{F}E \cap (C_r \setminus C_\rho)} |\gamma(x)| \, d\mathcal{H}^{n-1} = \int_\rho^r \mathcal{H}^{n-2}(\mathcal{F}E \cap \mathcal{F}C_t) \, dt.$$

One also finds, by elementary computations, that

$$(16) \qquad \int_{\mathcal{F}E \cap (C_r \setminus C_\rho)} |\beta(x)| \, d\mathcal{H}^{n-1} \geq \int_{\mathcal{F}C_\rho} \left| \varphi(\tfrac{r}{\rho} x, E) - \varphi(x, E) \right| \, d\mathcal{H}^{n-1};$$

indeed, it suffices to observe that, if at some point x belonging to the boundary of C_ρ we have $\varphi(\tfrac{r}{\rho} x, E) \neq \varphi(x, E)$, then necessarily there exists a number t satisfying

$$(17) \qquad 1 \leq t \leq \frac{r}{\rho}, \quad tx \in \mathcal{F}E$$

and recall that, since $\mathcal{F}E$ is contained in a finite number of hyperplanes, $\mathcal{H}^{n-1}(\mathcal{F}E \cap \mathcal{F}C_\rho) = 0$. On the other hand, by (14) we have

$$\int_{\mathcal{F}E \cap (C_r \setminus C_\rho)} |\gamma(x)| \, d\mathcal{H}^{n-1} = \int_{\mathcal{F}E \cap (C_r \setminus C_\rho)} \sqrt{1 - |\beta(x)|^2} \, d\mathcal{H}^{n-1} \leq$$
$$(18)$$
$$\leq \mathcal{H}^{n-1}\big(\mathcal{F}E \cap (C_r \setminus C_\rho)\big) - \frac{1}{2} \int_{\mathcal{F}E \cap (C_r \setminus C_\rho)} |\beta(x)|^2 \, d\mathcal{H}^{n-1},$$

whereas from (10) it follows that

$$p \int_{\mathcal{F}E \cap (C_r \setminus C_\rho)} |\beta(x)|^2 \, d\mathcal{H}^{n-1} \geq \mathcal{H}^{n-1}\big(\mathcal{F}E \cap (C_r \setminus C_\rho)\big) \times$$
$$(19)$$
$$\times \int_{\mathcal{F}E \cap (C_r \setminus C_\rho)} |\beta(x)|^2 \, d\mathcal{H}^{n-1} \geq \left(\int_{\mathcal{F}E \cap (C_r \setminus C_\rho)} |\beta(x)| \, d\mathcal{H}^{n-1} \right)^2.$$

From (15), (16), (18), (19), taking into account Theorem VII n.2 and Corollary IV n.3 in [1], one obtains (11).

III – *Assume E is a polyhedral domain in $\mathbf{R}^n$ having finite perimeter and let ρ, r, p be three numbers such that*

$$(20) \qquad 0 < \rho < r, \qquad p > \int_{C_r \setminus C_\rho} |D\varphi(x, E)|.$$

Then we have

$$(21) \quad r^{1-n}\,\Theta(E, C_r) - \rho^{1-n}\left(\Theta(E, C_r) - \int_{C_r \setminus C_\rho} |D\varphi(x, E)|\right) \geq$$
$$\geq \frac{r^{1-n}}{2p}\left(\int_{\mathcal{F}C_\rho} \left|\varphi\left(\frac{r}{\rho}x, E\right) - \varphi(x, E)\right| d\mathcal{H}^{n-1}\right)^2.$$

Proof. Since E is a polyhedral domain, the function

$$(22) \qquad f(t) = t^{2-n}\mathcal{H}^{n-2}\left(\mathcal{F}E \cap \mathcal{F}C_t\right)$$

achieves a minimum in the interval $\{t \ : \ \rho \leq t \leq r\}$; indeed, it is continuous when $n > 2$, whereas when $n = 2$ it only takes integral values.

Let z be a point where such a minimum is achieved and let

$$(23) \qquad b = f(z).$$

Then by Theorem II we have

$$(24) \qquad \int_{C_z \setminus C_\rho} |D\varphi(x, E)| \geq \int_\rho^z f(t) t^{n-2}\, dt \geq \frac{b(z^{n-1} - \rho^{n-1})}{n-1},$$

$$(25) \qquad \int_{C_r \setminus C_z} |D\varphi(x, E)| \geq \frac{b(r^{n-1} - z^{n-1})}{n-1}.$$

Now consider the two cones

$$(26) \qquad M = \{t \cdot x \ : \ x \in \mathcal{F}C_z \cap E, \quad 0 \leq t \leq 1\},$$

$$(27) \qquad F = \{t \cdot x \ : \ x \in \mathcal{F}C_z \cap \mathcal{F}E, \quad 0 \leq t \leq 1\},$$

and let

$$(28) \qquad B = M \cup (E \setminus C_z).$$

One can easily check that

$$(29) \qquad B \setminus C_z = E \setminus C_z, \qquad \mathcal{F}B \cap C_z \subset F.$$

On the other hand, from (22), (23), (27), recalling once again that E is a polyhedral domain, by elementary computations one finds that

$$(30) \qquad \mathcal{H}^{n-1}(F) = \frac{z}{n-1}\mathcal{H}^{n-2}\left(\mathcal{F}C_z \cap \mathcal{F}E\right) = \frac{z^{n-1}b}{n-1},$$

and hence, since by (29)

$$(31) \qquad \mathcal{F}B \subset \mathcal{F}E \cup F,$$

by (24) and the additivity of the Hausdorff measure, B is a Caccioppoli set.

By (29), (30) and Theorems VII n.2 and II n.3 in [1], we have

$$\int_{C_r \setminus C_z} |D\varphi(x, B)| = \int_{C_r \setminus C_z} |D\varphi(x, E)|,$$

$$(32)$$

$$\int_{C_z} |D\varphi(x, B)| \le \mathcal{H}^{n-1}(F) = \frac{z^{n-1}b}{n-1}.$$

Since $z \le r$, (29) implies that

$$(33) \qquad B \setminus C_r = E \setminus C_r,$$

hence from (7) n.1, taking (24) and (32) into account, we find that

$$(34) \quad \Theta(E, C_r) \le \int_{C_r \setminus C_z} |D\varphi(x, E)| + \frac{z^{n-1}b}{n-1} \le \int_{C_r \setminus C_\rho} |D\varphi(x, E)| + \frac{\rho^{n-1}b}{n-1}.$$

On the other hand, from the definition of z and b, we have

$$(35) \qquad \frac{b(r^{n-1} - \rho^{n-1})}{n-1} \le \int_\rho^r \mathcal{H}^{n-2}\left(\mathcal{F}C_t \cap \mathcal{F}E\right) dt,$$

and hence, by Theorem II,

$$(36) \qquad \frac{b(r^{n-1} - \rho^{n-1})}{n-1} \le \int_{C_r \setminus C_\rho} |D\varphi(x, E)|$$

$$- \frac{1}{2p} \left(\int_{\mathcal{F}C_\rho} \left| \varphi\left(\frac{r}{\rho}x, E\right) - \varphi(x, E) \right| d\mathcal{H}^{n-1} \right)^2.$$

On combining (34) and (36), one obtains (21).

IV – *Given a Caccioppoli set $E \subset \mathbf{R}^n$ and a sequence of Caccioppoli sets $\{E_h\}$ satisfying*

$$(37) \qquad P(E) = \lim_{h \to \infty} P(E_h), \quad \sum_{h=1}^{\infty} \int_{\mathbf{R}^n} |\varphi(x, E) - \varphi(x, E_h)| \, dx < +\infty,$$

for almost every positive number t there hold

$$(38) \qquad \int_{C_t} |D\varphi(x, E)| = \lim_{h \to \infty} \int_{C_t} |D\varphi(x, E_h)|, \quad \Theta(E, C_t) = \lim_{h \to \infty} \Theta(E_h, C_t).$$

Proof. From the second assumption in (37) it follows that, for almost every positive number t,

$$(39) \qquad \sum_{h=1}^{\infty} \int_{\mathcal{F}C_t} |\varphi(x, E) - \varphi(x, E_h)| \, d\mathcal{H}^{n-1} < +\infty,$$

and hence

$$(40) \qquad \lim_{h \to \infty} \int_{\mathcal{F}C_t} |\varphi(x, E) - \varphi(x, E_h)| \, d\mathcal{H}^{n-1} = 0.$$

From (40) and Theorem I, the second claim in (38) follows immediately.

On the other hand, by Theorem I and (11), (12), (13), (14) n.2 in [1], we have

$$(41) \qquad \int_{C_t} |D\varphi(x, E)| = P(E \cap C_t) - \int_{\mathcal{F}C_t} \varphi(x, E) \, d\mathcal{H}^{n-1},$$

$$(42) \qquad \int_{\mathbf{R}^n \setminus C_t} |D\varphi(x, E)| = P(E \setminus C_t) - \int_{\mathcal{F}C_t} \varphi(x, E) \, d\mathcal{H}^{n-1},$$

$$(41') \qquad \int_{C_t} |D\varphi(x, E_h)| = P(E_h \cap C_t) - \int_{\mathcal{F}C_t} \varphi(x, E_h) \, d\mathcal{H}^{n-1},$$

$$(42') \qquad \int_{\mathbf{R}^n \setminus C_t} |D\varphi(x, E_h)| = P(E_h \setminus C_t) - \int_{\mathcal{F}C_t} \varphi(x, E_h) \, d\mathcal{H}^{n-1},$$

for every h and almost every positive t.

On the other hand, by Theorem II n.2 in [1] and the second assumption in (37), we have

$$(43) \qquad P(E \cap C_t) \leq \liminf_{h \to \infty} P(E_h \cap C_t), \qquad P(E \setminus C_t) \leq \liminf_{h \to \infty} P(E_h \setminus C_t),$$

whereas, by (41), (42), (41'), (42'), and by Theorem I n.2 in [1], there holds

$$(44) \qquad \begin{aligned} P(E) &= \int_{\mathbf{R}^n} |D\varphi(x, E)| = \\ &= P(E \cap C_t) + P(E \setminus C_t) - 2 \int_{\mathcal{F}C_t} \varphi(x, E) \, d\mathcal{H}^{n-1}, \end{aligned}$$

$$(44') \qquad \begin{aligned} P(E_h) &= \int_{\mathbf{R}^n} |D\varphi(x, E_h)| = \\ &= P(E_h \cap C_t) + P(E_h \setminus C_t) - 2 \int_{\mathcal{F}C_t} \varphi(x, E_h) \, d\mathcal{H}^{n-1}. \end{aligned}$$

Since from (40)

$$(45) \qquad \lim_{h\to\infty} \int_{\mathcal{F}C_t} \varphi(x, E_h)\, d\mathcal{H}^{n-1} = \int_{\mathcal{F}C_t} \varphi(x, E)\, d\mathcal{H}^{n-1},$$

combining (37), (43), (44), (44'), (45), one finds that

$$(46) \qquad \lim_{h\to\infty} P(E_h \cap C_t) = P(E \cap C_t), \qquad \lim_{h\to\infty} P(E_h \setminus C_t) = P(E \setminus C_t);$$

hence, from (41), (42), (41'), (42'), (45), (46), the first claim in (38) does follow.

From Theorem III and Theorem IV we easily obtain the following

V – *Given a Caccioppoli set* $E \subset \mathbf{R}^n$, *there exists a set* $N \subset \mathbf{R}$ *of null Lebesgue measure such that, if* p, ρ, r *are three numbers satisfying*

$$(47) \qquad 0 < \rho < r, \quad \rho \in \mathbf{R} \setminus N, \quad r \in \mathbf{R} \setminus N, \quad p > \int_{C_r \setminus C_\rho} |D\varphi(x, E)|,$$

then

$$(48) \qquad r^{1-n}\,\Theta(E, C_r) - \rho^{1-n}\left(\Theta(E, C_r) - \int_{C_r \setminus C_\rho} |D\varphi(x, E)| \right) \ge$$

$$\ge \frac{r^{1-n}}{2p} \left(\int_{\mathcal{F}C_\rho} \left| \varphi\left(\frac{r}{\rho}x, E \right) - \varphi(x, E) \right| d\mathcal{H}^{n-1} \right)^2.$$

Proof. By Theorem III n.2 in [1], we can find a sequence of polyhedral domains $\{E_h\}$ satisfying

$$(49) \qquad \lim_{h\to\infty} P(E_h) = P(E), \qquad \sum_{h=1}^{\infty} \int_{\mathbf{R}^n} |\varphi(x, E) - \varphi(x, E_h)|\, dx < +\infty.$$

By (49) and Theorem IV, there exists a negligible set $N \subset \mathbf{R}$ such that, for every positive number $t \in \mathbf{R} \setminus N$, there hold

$$(50) \qquad \lim_{h\to\infty} \int_{\mathcal{F}C_t} |\varphi(x, E) - \varphi(x, E_h)|\, d\mathcal{H}^{n-1} = 0,$$

$$(51) \qquad \lim_{h\to\infty} \Theta(E_h, C_t) = \Theta(E, C_t),$$

$$(52) \qquad \lim_{h\to\infty} \int_{C_t} |D\varphi(x, E_h)| = \int_{C_t} |D\varphi(x, E)|.$$

Applying Theorem III to the polyhedral domains E_h and taking (50), (51), (52) into account, one immediately obtains (48).

VI – *Assume E is a Caccioppoli set in $\mathbf{R}^n$, and let p, q, r, ρ be four numbers satisfying*

$$(53) \qquad 0 < \rho < r < q, \quad p > \int_{C_r \setminus C_\rho} |D\varphi(x, E)|, \quad \Psi(E, C_q) = 0.$$

Then

$$(54) \qquad r^{1-n} \int_{C_r} |D\varphi(x, E)| - \rho^{1-n} \int_{C_\rho} |D\varphi(x, E)| \geq$$

$$\geq \frac{\rho^{2(n-1)}}{2pr^{n-1}} \left| r^{1-n} \int_{C_r} D\varphi(x, E) - \rho^{1-n} \int_{C_\rho} D\varphi(x, E) \right|^2.$$

Proof. By Theorem V n.2 in [1] there exists a negligible set $N \subset \mathbf{R}$ such that, for every positive number $t \in \mathbf{R} \setminus N$, there holds

$$(55) \qquad \int_{\mathbf{R}^n} D\varphi(x, E \cap C_t) = \int_{C_t} D\varphi(x, E) + \int_{\mathcal{F}C_t} \varphi(x, E) \cdot \nu(x) \, d\mathcal{H}^{n-1},$$

where $\nu(x)$ is the inner normal vector to $\mathcal{F}C_t$; if $g(x)$ is any function, continuous in $\mathbf{R}^n$ together with its first partial derivatives, having compact support and being equal to 1 inside C_t, by Theorems I n.2 and II n.3 in [1] we have

$$(56) \qquad \int_{\mathbf{R}^n} D\varphi(x, E \cap C_t) = \int_{C_t} D\varphi(x, E \cap C_t) = \int_{E \cap C_t} Dg(x) \, dx = 0,$$

and hence (55) reduces to

$$(57) \qquad \int_{C_t} D\varphi(x, E) = - \int_{\mathcal{F}C_t} \varphi(x, E) \, \nu(x) \, d\mathcal{H}^{n-1}.$$

Taken two positive numbers $\rho, r \in \mathbf{R} \setminus N$, from (57) we obtain, after elementary computations,

$$(58) \qquad \int_{\mathcal{F}C_\rho} \left| \varphi(x, E) - \varphi\left(\frac{r}{\rho}x, E\right) \right| d\mathcal{H}^{n-1} \geq$$

$$\geq \left| \int_{C_\rho} D\varphi(x, E) - \frac{\rho^{n-1}}{r^{n-1}} \int_{C_r} D\varphi(x, E) \right|.$$

Then from (2), (3), (5), (6) n.1, it can be seen that, if (53) is satisfied, then

$$(59) \qquad \Theta(E, C_r) = \int_{C_r} |D\varphi(x, E)|.$$

Combining (58), (59) with (48) of Theorem V, we conclude that (54) holds true for almost every pair of numbers ρ, r satisfying (53); but, due to well known theorems on the limits of integrals, we have for every t

$$
\begin{aligned}
\lim_{\varepsilon \to 0^+} \int_{C_{t+\varepsilon}} |D\varphi(x,E)| &= \int_{C_t} |D\varphi(x,E)|, \\
\lim_{\varepsilon \to 0^+} \int_{C_{t+\varepsilon}} D\varphi(x,E) &= \int_{C_t} D\varphi(x,E),
\end{aligned}
$$

(60)

and hence (54), being true for almost every pair ρ, r satisfying (53), is in fact true for every such pair.

VII – *Assume E is a Caccioppoli set and L is an open set in $\mathbf{R}^n$, such that*

(61)
$$
\Psi(E,L) = 0,
$$

and let ξ be a point interior to L. Then, for every positive number ρ smaller than the distance from ξ to $\mathcal{F}L$, there holds

(62)
$$
\Theta(E,C_\rho(\xi)) = \int_{C_\rho(\xi)} |D\varphi(x,E)| \leq \frac{n\,\omega_n}{2}\rho^{n-1}.
$$

If, moreover, $\xi \in \overline{\mathcal{F}^*E}$, then, besides (62), we also have

(63)
$$
\Theta(E,C_\rho(\xi)) = \int_{C_\rho(\xi)} |D\varphi(x,E)| \geq \omega_{n-1}\rho^{n-1}.
$$

Proof. By (2), (3), (5), (6) of n.1 and by (61) we have

(64)
$$
\Psi(E,C_\rho(\xi)) = 0, \qquad \int_{C_\rho(\xi)} |D\varphi(x,E)| = \Theta(E,C_\rho(\xi)).
$$

Without loss of generality we can assume that ξ is the origin of the coordinates, since we can always reduce to this case by a translation; applying Theorem I in the two cases $F = \mathbf{R}^n$ and $F = \emptyset$, for almost every positive number ρ we find that

$$
\begin{aligned}
\Theta(E,C_\rho) &\leq \int_{\mathcal{F}C_\rho} \varphi(x,E)\,d\mathcal{H}^{n-1}, \\
\Theta(E,C_\rho) &\leq \int_{\mathcal{F}C_\rho} (1-\varphi(x,E))\,d\mathcal{H}^{n-1},
\end{aligned}
$$

(65)

and hence

(66)
$$
\Theta(E,C_\rho) \leq \frac{1}{2}\mathcal{H}^{n-1}(\mathcal{F}C_\rho) = \frac{n\,\omega_n}{2}\rho^{n-1}.
$$

Recalling that for every ρ we have

(67)
$$
\lim_{\varepsilon \to 0^+} \int_{C_{\rho+\varepsilon}} |D\varphi(x,E)| = \int_{C_\rho} |D\varphi(x,E)|
$$

and that for ρ smaller than the distance from ξ to L (64) is satisfied, we conclude that, for every ρ which is smaller than the distance from ξ to $\mathcal{F}L$, (62) holds true.

Now we prove (63) under the assumption that $\xi \in \mathcal{F}^*E$; supposing, as usual, that ξ coincides with the origin of the space $\mathbf{R}^n$, besides a number ρ smaller than the distance from ξ to $\mathcal{F}L$, consider two other numbers t, b, both smaller than such distance, satisfying

$$(68) \qquad 0 < t < \rho < b.$$

From Theorem VI we have

$$(69) \qquad \rho^{1-n} \int_{C_\rho} |D\varphi(x, E)| \geq t^{1-n} \int_{C_t} |D\varphi(x, E)|$$

and hence, taking the limit as $t \to 0$ and recalling Theorem VI n.2 in [1], one obtains (63).

Finally, if $\xi \in \overline{\mathcal{F}^*E} \setminus \mathcal{F}^*E$, for every ρ smaller than the distance from ξ to $\mathcal{F}L$ and every positive number $\lambda < \rho$ there exists a point ξ^* satisfying

$$(70) \qquad \xi^* \in \mathcal{F}^*E, \qquad |\xi - \xi^*| < \lambda.$$

According to what we have already proved, we have

$$(71) \qquad \int_{C_\rho(\xi)} |D\varphi(x, E)| \geq \int_{C_{\rho-\lambda}(\xi^*)} |D\varphi(x, E)| \geq (\rho - \lambda)^{n-1}\omega_{n-1},$$

and hence (63) follows from the arbitrariness of λ.

VIII – *Assume E is a Caccioppoli set and L is an open set in $\mathbf{R}^n$, such that*

$$(72) \qquad \Psi(E, L) = 0.$$

Let ξ be a point in L, and let ρ, r be two positive numbers, both smaller than the distance from ξ to $\mathcal{F}L$, and such that $0 < \rho < r$. Then
(73)

$$\left| \rho^{1-n} \int_{C_\rho(\xi)} D\varphi(x, E) - r^{1-n} \int_{C_r(\xi)} D\varphi(x, E) \right| \leq \sqrt{n\omega_n}\, e^n \times$$

$$\times (\log r - \log \rho + 1) \left(r^{1-n} \int_{C_r(\xi)} |D\varphi(x, E)| - \rho^{1-n} \int_{C_\rho(\xi)} |D\varphi(x, E)| \right)^{1/2}.$$

Proof. Suppose, as usual, that ξ coincides with the origin of the coordinates and, denoting h the smallest integer which is not smaller than $\log r - \log \rho$, let

$$(74) \qquad \rho_k = re^{-k}, \quad \text{for} \quad k = 0, 1, \ldots, h-1; \qquad \rho_h = \rho.$$

By Theorem VI we have

$$(75) \qquad \begin{aligned} \rho^{1-n} \int_{C_\rho} |D\varphi(x, E)| &\leq \rho_k^{1-n} \int_{C_{\rho_k}} |D\varphi(x, E)| \leq \\ &\leq \rho_{k-1}^{1-n} \int_{C_{\rho_{k-1}}} |D\varphi(x, E)| \leq r^{1-n} \int_{C_r} |D\varphi(x, E)|, \quad k = 1, 2, \ldots, h, \end{aligned}$$

whereas, from the definition of h, we have

$$(76) \qquad h < (\log r - \log \rho + 1), \qquad \rho_k \leq \rho_{k-1} \leq e\rho_k, \quad \text{for} \quad k = 1, 2, \ldots, h.$$

By Theorem VII it clearly holds

$$(77) \qquad \rho_{k-1}^{n-1} \frac{\omega_n n e^2}{2} > \int_{C_{\rho_{k-1}}} |D\varphi(x, E)| - \int_{C_{\rho_k}} |D\varphi(x, E)|, \quad k = 1, 2, \ldots, h.$$

From (75), (76), (77), applying Theorem VI, we find that

$$(78) \qquad \left| \rho_k^{1-n} \int_{C_{\rho_k}} D\varphi(x, E) - \rho_{k-1}^{1-n} \int_{C_{\rho_{k-1}}} D\varphi(x, E) \right| \leq$$

$$\sqrt{n\omega_n} e^n \left(r^{1-n} \int_{C_r} |D\varphi(x, E)| - \rho^{1-n} \int_{C_\rho} |D\varphi(x, E)| \right)^{1/2}$$

and hence, from (74) and (76), we obtain (73).

IX – *Suppose E is a Caccioppoli set in $\mathbf{R}^n$. Then for every $\xi \in \mathbf{R}^n$ and every positive number p we have*

$$(79) \qquad \left| \int_{C_p(\xi)} D\varphi(x, E) \right| \leq \omega_{n-1} p^{n-1}.$$

If, moreover,

$$(80) \qquad \xi \in \overline{\mathcal{F}^* E}, \qquad \Psi(E, C_p(\xi)) = 0,$$

then

$$(81) \qquad \int_{C_p(\xi)} \varphi(x, E) \, dx \geq \frac{\omega_{n-1} p^n}{n}, \qquad \int_{C_p(\xi)} (1 - \varphi(x, E)) \, dx \geq \frac{\omega_{n-1} p^n}{n}.$$

Proof. Choose ξ. For almost every positive p, we have (see (55), (56), (57))

$$(82) \qquad \left| \int_{C_p(\xi)} D\varphi(x, E) \right| = \left| \int_{\mathcal{F} C_p} \varphi(x, E) \nu(x) \, d\mathcal{H}^{n-1} \right|$$

where $\nu(x)$ is the inner normal vector. On the other hand, we clearly have

$$(83) \qquad \left| \int_{\mathcal{F} C_p} \varphi(x, E) \nu(x) \, d\mathcal{H}^{n-1} \right| \leq \omega_{n-1} p^{n-1},$$

and hence (79) holds for almost every positive number p.

Observing that for every positive p there holds

$$(84) \qquad \lim_{\varepsilon \to 0^+} \int_{C_{p+\varepsilon}(\xi)} D\varphi(x, E) = \int_{C_p(\xi)} D\varphi(x, E),$$

we conclude that (79) is satisfied for every positive number p.

Now suppose that (80) is satisfied; by Theorem VII, we have

$$(85) \qquad \Theta\big(E, C_\rho(\xi)\big) \geq \omega_{n-1}\rho^{n-1} \quad \text{for} \quad \rho < p.$$

Recalling Theorem I, and comparing E with the empty set $\emptyset$ and with the whole space $\mathbf{R}^n$, for almost every positive number ρ we find that

$$\Theta\big(E, C_\rho(\xi)\big) \leq \int_{\mathcal{F}C_\rho} \varphi(x, E)\, d\mathcal{H}^{n-1},$$

(86)

$$\Theta\big(E, C_\rho(\xi)\big) \leq \int_{\mathcal{F}C_\rho} (1 - \varphi(x, E))\, d\mathcal{H}^{n-1}.$$

Combining (85) and (86) with the identities

$$\int_{C_p} \varphi(x, E)\, dx = \int_0^p d\rho \int_{\mathcal{F}C_\rho} \varphi(x, E)\, d\mathcal{H}^{n-1},$$

(87)

$$\int_{C_p} (1 - \varphi(x, E))\, dx = \int_0^p d\rho \int_{\mathcal{F}C_\rho} (1 - \varphi(x, E))\, d\mathcal{H}^{n-1},$$

one obtains (81).

3.

We shall now establish some properties of regular surfaces approximating minimal surfaces; these properties will be obtained by a comparison argument with the graph of suitable harmonic functions. To this end, we introduce the following lemma:

I – *Let $m \geq 1$ be an integer, let $\rho > 0$ and, as usual (see [1] n.3), let G_ρ denote the closed ball in $\mathbf{R}^m$ centered at the origin, of radius ρ. Let $u(y)$ be a function which is continuous in G_ρ together with its partial derivatives of every order, and which is harmonic in $G_\rho \setminus \mathcal{F}G_\rho$.*

Denoting by

$$(1) \qquad q = \omega_m^{-1}\rho^{-m} \int_{G_\rho} Du(y)\, dy$$

the mean value of the gradient of u on G_ρ, we have for every positive number $\alpha < 1$

$$(2) \qquad \int_{G_{\alpha\rho}} \big(|Du|^2 - |q|^2\big)\, dy \leq \alpha^{m+2} \int_{G_\rho} \big(|Du|^2 - |q|^2\big)\, dy.$$

Proof. By well known theorems on harmonic functions, $u(y)$ can be expanded as an infinite sum of homogeneous harmonic polynomials

$$(3) \qquad u(y) = \sum_{h=0}^{\infty} V_h(y),$$

such that the series in (3) and the series

$$(3') \qquad Du(y) = \sum_{h=1}^{\infty} DV_h(y)$$

are uniformly convergent in G_ρ. For every h, the polynomial V_h is homogeneous of degree h and hence, by well known properties of homogeneous polynomials, we have

$$(4) \qquad \int_{G_\rho} \langle DV_h, DV_k \rangle \, dy = \int_{G_{\alpha\rho}} \langle DV_h, DV_k \rangle \, dy = 0 \quad \text{if} \quad h \neq k,$$

$$(5) \qquad \int_{G_\rho} DV_h \, dy = \int_{G_{\alpha\rho}} DV_h \, dy = 0 \quad \text{if} \quad h \geq 2.$$

From (3'), (4), (5) it follows, recalling (1), that

$$(6) \qquad q = DV_1(y), \qquad \int_{G_\rho} \left(|Du|^2 - |q|^2 \right) dy = \sum_{h=2}^{\infty} \int_{G_\rho} |DV_h|^2 \, dy,$$

$$(7) \qquad \int_{G_{\alpha\rho}} \left(|Du|^2 - |q|^2 \right) dy = \sum_{h=2}^{\infty} \int_{G_{\alpha\rho}} |DV_h|^2 \, dy,$$

and hence, by the homogeneity of the polynomials V_h, we obtain (2).

II – *Suppose $\{\beta_h\}$, $\{\gamma_h\}$ are two sequences of positive numbers satisfying*

$$(8) \qquad \lim_{h \to \infty} \beta_h = 0, \qquad \lim_{h \to \infty} \gamma_h = 0, \qquad \gamma_h < 1.$$

Suppose, as usual, $m \geq 1$, $\rho > 0$, and let $\{w_h(y)\}$ be a sequence of functions defined in $G_\rho \subset \mathbf{R}^m$, continuous with their partial derivatives of any order.

For every h, let $u_h(y)$ be the function, continuous in G_ρ with its partial derivatives of any order, which solves the Dirichlet problem

$$(9) \quad u_h(y) = w_h(y) \quad \text{for} \quad y \in \mathcal{F}G_\rho, \qquad \triangle u_h(y) = 0 \quad \text{for} \quad y \in G_\rho \setminus \mathcal{F}G_\rho.$$

Moreover, define for every h

$$(10) \qquad q_h = \rho^{-m} \omega_m^{-1} \int_{G_\rho} Dw_h \, dy = \rho^{-m} \omega_m^{-1} \int_{G_\rho} Du_h \, dy.$$

Suppose that the inequalities

$$(11) \qquad |Dw_h(y)| \leq \gamma_h, \quad \text{for} \quad y \in G_\rho, \quad h = 1, 2, \ldots$$

$$(12) \qquad \int_{G_\rho} \left(\sqrt{1 + |Dw_h|^2} - \sqrt{1 + |q_h|^2} \right) dy \leq \beta_h, \quad \text{for} \quad h = 1, 2, \ldots$$

$$(13) \qquad \int_{G_\rho} \left(\sqrt{1 + |Dw_h|^2} - \sqrt{1 + |Du_h|^2} \right) dy \leq \gamma_h \beta_h, \quad \text{for} \quad h = 1, 2, \ldots$$

are satisfied. Then, for every positive number $\alpha < 1$, there holds

$$(14) \qquad \limsup_{h \to \infty} \beta_h^{-1} \int_{G_{\alpha\rho}} \frac{|Dw_h - q_h|^2}{2} \, dy \leq \alpha^{m+2}.$$

Proof. Given two numbers b, c satisfying the conditions

$$(15) \qquad c > 0, \qquad c + b > 0,$$

from the Taylor's formula it follows that

$$(16) \qquad \sqrt{c + b} - \sqrt{c} - \frac{b}{2\sqrt{c}} \leq 0.$$

If, moreover, in place of (15), we assume the stronger conditions

$$(17) \qquad 2 \geq c \geq 1, \qquad c + b \geq 1,$$

besides (16) we also have

$$(18) \qquad \sqrt{c + b} - \sqrt{c} - \frac{b}{2\sqrt{c}} \geq -\frac{b^2}{8} \geq -\frac{b^2}{2\sqrt{c}}.$$

Using (16), (18) and recalling (8), (10), (11) we find

$$(19)$$

$$0 \geq \int_{G_\rho} \left(\sqrt{1 + |Dw_h|^2} - \sqrt{1 + |q_h|^2} - \frac{|Dw_h|^2 - |q_h|^2}{2\sqrt{1 + |q_h|^2}} \right) dy \geq$$

$$\geq -\int_{G_\rho} \left(|Dw_h|^2 - |q_h|^2 \right)^2 \cdot \left(2\sqrt{1 + |q_h|^2} \right)^{-1} dy \geq -\gamma_h^2 \int_{G_\rho} \frac{|Dw_h|^2 - |q_h|^2}{2\sqrt{1 + |q_h|^2}} \, dy$$

for every value of the index h. Since by assumption $0 < \gamma_h < 1$, (19) can be rewritten as

$$\int_{G_\rho} \left(\sqrt{1 + |Dw_h|^2} - \sqrt{1 + |q_h|^2} \right) dy \leq \int_{G_\rho} \frac{|Dw_h|^2 - |q_h|^2}{2\sqrt{1 + |q_h|^2}} \, dy \leq$$

$$(20)$$

$$\leq \frac{1}{1 - \gamma_h^2} \int_{G_\rho} \left(\sqrt{1 + |Dw_h|^2} - \sqrt{1 + |q_h|^2} \right) dy.$$

In a similar way, one finds that

$$(21) \qquad \int_{G_\rho} \left(\sqrt{1 + |Du_h|^2} - \sqrt{1 + |q_h|^2} \right) dy \leq \int_{G_\rho} \frac{|Du_h|^2 - |q_h|^2}{2\sqrt{1 + |q_h|^2}} \, dy.$$

On the other hand, by (19) and by well known properties of harmonic functions, we have

$$(22) \qquad 0 \le \int_{G_\rho} |Du_h - Dw_h|^2 \, dy = \int_{G_\rho} \left(|Dw_h|^2 - |Du_h|^2 \right) \, dy,$$

and hence combining (20), (21), (22) with (8), (12), (13) one finds, recalling that $\lim_{h\to\infty} |q_h| = \lim_{h\to\infty} \gamma_h = 0$

$$(23) \qquad \lim_{h\to\infty} \beta_h^{-1} \int_{G_\rho} |Du_h - Dw_h|^2 \, dy = 0.$$

Therefore, for every positive number $\alpha \le 1$,

$$(24) \qquad \lim_{h\to\infty} \beta_h^{-1} \int_{G_{\alpha\rho}} |Du_h - Dw_h|^2 \, dy = 0.$$

Moreover, from (8), (12), (20), we deduce that

$$(25) \qquad \limsup_{h\to\infty} \beta_h^{-1} \int_{G_\rho} \left(|Dw_h|^2 - |q_h|^2 \right) \, dy \le 2$$

and hence, by (22), we also deduce that

$$(26) \qquad \limsup_{h\to\infty} \beta_h^{-1} \int_{G_\rho} \left(|Du_h|^2 - |q_h|^2 \right) \, dy \le 2.$$

Recalling (10) and Theorem I, from (26) we obtain

$$(27) \qquad \limsup_{h\to\infty} \beta_h^{-1} \int_{G_{\alpha\rho}} \left(|Du_h|^2 - |q_h|^2 \right) \, dy \le 2\alpha^{m+2}.$$

On the other hand, as u_h is harmonic, from (10) it follows that, for every positive number $\alpha < 1$,

$$(28) \qquad q_h = (\rho\alpha)^{-m} \omega_m^{-1} \int_{G_{\alpha\rho}} Du_h \, dy,$$

and hence

$$(29) \qquad \int_{G_{\alpha\rho}} \left(|Du_h|^2 - |q_h|^2 \right) \, dy = \int_{G_{\alpha\rho}} |Du_h - q_h|^2 \, dy.$$

Combining (24), (27) and (29), one finally obtains (14).

III – *Assume $\{E_h\}$ is a sequence of Caccioppoli sets in $\mathbf{R}^n$ (with $n \ge 2$), let t be a positive number and assume that, for every h, there hold*

$$(30) \qquad \mathcal{F}^* E_h \cap A_t = \mathcal{F} E_h \cap A_t, \qquad E_h \cap \mathcal{F} E_h \cap A_t = \emptyset.$$

Moreover, assume that $\mathcal{F}E_h \cap A_t$ is a locally regular hypersurface of class C^∞, and that

$$(31) \qquad \lim_{h \to \infty} \left(\inf \left\{ \frac{D_n\varphi(x, E_h)}{|D\varphi(x, E_h)|} \; : \; x \in \mathcal{F}E_h \cap A_t \right\} \right) = 1.$$

Assume, further, that we are given a sequence of positive numbers $\{\beta_h\}$ and a sequence $\{\xi_h\}$ of points of $\mathbf{R}^n$ satisfying

$$(32) \qquad \int_{C_t} |D\varphi(x, E_h)| - \left| \int_{C_t} D\varphi(x, E_h) \right| \le \beta_h, \qquad \xi_h \in \mathcal{F}E_h,$$

$$(33) \qquad \lim_{h \to \infty} \beta_h = 0, \qquad \lim_{h \to \infty} |\xi_h| = 0, \qquad \lim_{h \to \infty} \beta_h^{-1}\Psi(E_h, C_t) = 0.$$

Then we have, for every positive number $\alpha < 1$,

$$(34) \qquad \limsup_{h \to \infty} \beta_h^{-1} \left(\int_{C_{\alpha t}} |D\varphi(x, E_h)| - \left| \int_{C_{\alpha t}} D\varphi(x, E_h) \right| \right) \le \alpha^{n+1}.$$

Proof. By Theorem I n.4 in [1] we can find a sequence of open sets $\{B_h\}$ in $\mathbf{R}^m$ (with $m = n - 1$), a sequence of functions $\{w_h\}$ (with w_h continuous with all its partial derivatives of any order) and a sequence of positive numbers $\{\gamma_h\}$, satisfying, for every integer h large enough,

$$(35) \quad \gamma_h < 1, \quad \mathcal{F}E_h \cap A_t = \{(y, z) : y \in B_h, \quad z = w_h(y)\}, \quad B_h \supset G_{t-\gamma_h t},$$

$$(36) \quad |Dw_h(y)| < \gamma_h, \quad |w_h(y)| < \gamma_h t \quad \text{for} \quad y \in G_{t-\gamma_h t}, \quad \Psi(E_h, C_t) \le \beta_h \gamma_h$$

and such that

$$(37) \qquad \lim_{h \to \infty} \gamma_h = 0.$$

Now consider two numbers α, ρ, satisfying

$$(38) \qquad 0 < \alpha t < \rho < t.$$

When h is large enough, from (37) we obtain that

$$(39) \qquad \rho < t - \gamma_h t,$$

and hence we have, by Theorems VII n.2 and III n.3 in [1] and by (30), (35), that

$$(40) \qquad \int_{A_t \cap (G_\rho \times \mathbf{R})} |D\varphi(x, E_h)| = \int_{G_\rho} \sqrt{1 + |Dw_h|^2} \, dy,$$

$$(41) \qquad \int_{A_t \cap (G_{\alpha t} \times \mathbf{R})} |D\varphi(x, E_h)| = \int_{G_{\alpha t}} \sqrt{1 + |Dw_h|^2} \, dy.$$

Resuming the definitions (9), (10) of Theorem II, we obtain that

$$(42) \qquad \left| \int_{A_t \cap (G_\rho \times \mathbf{R})} D\varphi(x, E_h) \right| = \int_{G_\rho} \sqrt{1 + |q_h|^2} \, dy$$

and hence, since clearly

$$(43) \qquad \begin{aligned} \int_L |D\varphi(x, E_h)| - \left| \int_L D\varphi(x, E_h) \right| &\leq \\ \leq \int_M |D\varphi(x, E_h)| - \left| \int_M D\varphi(x, E_h) \right| & \qquad \text{if} \quad L \subset M, \end{aligned}$$

from (32), (40), (42), we see that (12) is satisfied.

Now consider the set T_h defined by

$$(44) \qquad T_h \setminus \big(A_t \cap (G_\rho \times \mathbf{R})\big) = E_h \setminus \big(A_t \cap (G_\rho \times \mathbf{R})\big),$$

$$(45) \qquad T_h \cap A_t \cap (G_\rho \times \mathbf{R}) = A_t \cap \big\{ (y, z) \,:\, y \in G_\rho, \quad z > u_h(y) \big\}.$$

From (30), (31), (35), (39), we deduce that, if h is large enough, then

$$(46) \qquad E_h \cap A_t \cap (G_\rho \times \mathbf{R}) = A_t \cap \big\{ (y, z) \,:\, y \in G_\rho, \quad z > w_h(y) \big\}$$

and hence, combining (44), (45), (46), we find that

$$(47) \quad (E_h \setminus T_h) \cup (T_h \setminus E_h) \subset \big\{ (y, z) \,:\, y \in G_\rho, \, (z - w_h(y)) \cdot (z - u_h(y)) \leq 0 \big\}.$$

Since $u_h(y)$ solves (9) and hence it achieves its maximum and its minimum on $\mathcal{F} G_\rho$, where it coincides with $w_h(y)$, by (36), (39), the set in the right-hand side of (47) is closed and is contained inside A_t; recalling the Theorems VII n.2, II, III n.3 in [1], we can then conclude, by (9), (44), (45), (46), (47), that for h large enough

$$(48) \qquad P(E_h) - P(T_h) = \int_{G_\rho} \sqrt{1 + |Dw_h|^2} \, dy - \int_{G_\rho} \sqrt{1 + |Du_h|^2} \, dy.$$

On the other hand, by (44), (45), (36), and by the definition of the function Ψ, we see that

$$(49) \qquad \Psi(E_h, C_t) \geq P(E_h) - P(T_h),$$

and hence (13) is satisfied provided h is large enough.

On the other hand, (33), (35), (36), (37) guarantee that, when h is large enough, (8) and (11) are necessarily fulfilled, and hence, applying Theorem II, we can conclude that

$$(50) \qquad \limsup_{h \to \infty} \beta_h^{-1} \int_{G_{\alpha t}} \frac{|Dw_h - q_h|^2}{2} \, dy \leq \left(\frac{\alpha t}{\rho} \right)^{n+1}.$$

Letting

$$(51) \qquad q_h^* = (\alpha t)^{-m} \omega_m^{-1} \int_{G_{\alpha t}} Dw_h \, dy,$$

one finds, after an elementary computation, that

$$(52) \qquad \int_{G_{\alpha t}} \frac{|Dw_h|^2 - |q_h^*|^2}{2} \, dy = \int_{G_{\alpha t}} \frac{|Dw_h - q_h^*|^2}{2} \, dy \leq \int_{G_{\alpha t}} \frac{|Dw_h - q_h|^2}{2} \, dy.$$

On the other hand, by an argument similar to that used to prove (42), one can prove that

$$(53) \qquad \left| \int_{A_t \cap (G_{\alpha t} \times \mathbf{R})} D\varphi(x, E_h) \right| = \int_{G_{\alpha t}} \sqrt{1 + |q_h^*|^2} \, dy,$$

and hence, combining (41), (50), (51), (52), and recalling (16) of Theorem II, one finds that
$$(54)$$

$$\limsup_{h \to \infty} \beta_h^{-1} \left(\int_{A_t \cap (G_{\alpha t} \times \mathbf{R})} |D\varphi(x, E_h)| - \left| \int_{A_t \cap (G_{\alpha t} \times \mathbf{R})} D\varphi(x, E_h) \right| \right) \leq \left(\frac{\alpha t}{\rho} \right)^{n+1}.$$

From (54) and (43), it follows that

$$(55) \qquad \limsup_{h \to \infty} \beta_h^{-1} \left(\int_{C_{\alpha t}} |D\varphi(x, E_h)| - \left| \int_{C_{\alpha t}} D\varphi(x, E_h) \right| \right) \leq \left(\frac{\alpha t}{\rho} \right)^{n+1},$$

and hence, by (38) and by the arbitrariness of ρ, (34) is established.

4.

The results so far established for regular hypersurfaces which are close to minimal surfaces allow us to attack the problem of the regularity of such surfaces.

To this aim, let us establish some lemmata.

I – *Given an open set $A \subset \mathbf{R}^n$ (with $n \geq 2$), a function f continuous in A together with its first order partial derivatives, satisfying the condition*

$$(1) \qquad D_n f(x) > 0 \quad for \quad x \in A,$$

for every set $E \subset A$ there holds

$$(2) \qquad \int_E |Df(x)| \, dx = \int_{-\infty}^{+\infty} d\lambda \, \mathcal{H}^{n-1}\left(\{x : x \in E, \quad f(x) = \lambda\}\right).$$

Proof. The theorem is trivial when $f(x)$ is linear, i.e. of the kind

$$(3) \qquad f(x) = \sum_{h=1}^{n} a_h x_h + b$$

for suitable constants a_h, b. One can easily reduce to this case under the assumption that there exists a point ξ and a positive number ρ satisfying

$$\text{(4)} \qquad A \supset A_{2\rho}(\xi) \supset C_\rho(\xi) \supset E;$$

indeed, it suffices to consider the map τ which associates the generic point $x = (x_1, \dots, x_n) \in A$ with the point $\tau(x) = (x_1, \dots, x_{n-1}, f(x))$ and recall the classical formulas concerning the change of variable in the integrals and the differentiation of implicit functions and (12) n.1 in [1].

In the general case, one can always find a sequence of sets $\{E_h\}$ and a sequence of positive numbers $\{\rho_h\}$ satisfying

$$\text{(5)} \qquad \begin{aligned} &C_{\rho_h}(\xi_h) \subset A_{2\rho_h}(\xi_h) \subset A, \quad E_h \subset C_{\rho_h}(\xi_h), \\ &E = \bigcup_{h=1}^{\infty} E_h, \quad E_h \cap E_k = \emptyset \quad \text{for} \quad h \neq k. \end{aligned}$$

Then there holds, for what we have just proved,

$$\text{(6)} \qquad \int_{E_h} |Df(x)|\, dx = \int_{-\infty}^{+\infty} d\lambda\, \mathcal{H}^{n-1}\left(\{x \ : \ x \in E_h, \quad f(x) = \lambda\}\right)$$

and hence, summing over h, (2) is proved.

II – *Assume $\{L_h\}$ is a sequence of Caccioppoli sets in $\mathbf{R}^n$ $(n \geq 2)$, let ε, ρ be two positive numbers and let $\{\eta_h\}$ be a sequence of positive numbers such that*

$$\text{(7)} \quad \operatorname{meas}(L_h \cap C_{\alpha\rho}) \cdot \operatorname{meas}(C_{\alpha\rho} \setminus L_h) \geq \varepsilon(\alpha\rho)^{2n}, \quad 0 < \alpha \leq 1, \quad h = 1, 2, \dots$$

$$\text{(8)} \quad \int_{C_\rho} |D\varphi(x, L_h)| - \left| \int_{C_\rho} D\varphi(x, L_h) \right| \leq \eta_h, \qquad \Psi(L_h, C_\rho) = 0, \quad h = 1, 2, \dots$$

$$\text{(9)} \qquad \sum_{h=1}^{\infty} \eta_h < +\infty.$$

Then, for every positive number $\alpha < 1$, there holds

$$\text{(10)} \qquad \limsup_{h \to \infty} \eta_h^{-1} \left(\int_{C_{\alpha\rho}} |D\varphi(x, L_h)| - \left| \int_{C_{\alpha\rho}} D\varphi(x, L_h) \right| \right) \leq \alpha^{n+1}.$$

Proof. We can suppose that

$$\text{(11)} \qquad \int_{C_\rho} |D\varphi(x, L_h)| - \int_{C_\rho} D_n\varphi(x, L_h) \leq \eta_h,$$

since we can reduce to this case by suitable rotations.

Now consider the functions

$$(12) \qquad g_h(x) = \sqrt{\pi^{-n}} \, \eta_h^{-3n} \exp(-|x \eta_h^{-3}|^2),$$

$$(13) \qquad f_h(x) = g_h(x) * \varphi(x, L_h \cap C_\rho),$$

and choose two numbers b, q satisfying

$$(14) \qquad 0 < b < q < \rho.$$

Denoting B_h the set of those points in C_b whose distance to $\overline{\mathcal{F}^* L_h}$ is smaller than η_h^2, from (28), (29) n.2 in [1] and from (9) we obtain for large enough[†] h

$$(15) \qquad \{x : \ x \in C_b, \ \ |f_h(x) - \varphi(x, L_h)| \geq \eta_h^2\} \subset B_h \subset C_b,$$

whereas from (7), (9), for large enough h, it follows that

$$(16) \qquad B_h \neq \emptyset.$$

Chosen a number h such that (15), (16) hold true together with

$$(17) \qquad \rho - 2\eta_h^2 > q > b + 2\eta_h^2, \qquad \eta_h < 1,$$

since $\overline{\mathcal{F}^* L_h} \cap C_b$ is bounded, we can choose finitely many points

$$(18) \qquad y_1, \ldots, y_k$$

in such a way that

$$(19) \qquad y_i \in \overline{\mathcal{F}^* L_h} \cap C_b \quad \text{for} \quad i = 1, \ldots, k, \qquad |y_i - y_r| > 2\eta_h^2 \quad \text{for} \quad i \neq r,$$

$$(20) \qquad \bigcup_{i=1}^{k} C_{2\eta_h^2}(y_i) \supset \overline{\mathcal{F}^* L_h} \cap C_b.$$

From (17), (20), it follows that

$$(21) \qquad B_h \subset \bigcup_{i=1}^{k} C_{3\eta_h^2}(y_i),$$

whereas from (19) we have

$$(22) \qquad C_{\eta_h^2}(y_i) \cap C_{\eta_h^2}(y_r) = \emptyset \quad \text{for} \quad i \neq r$$

and hence, by (17) and (19),

$$(23) \qquad \int_{C_\rho} |D\varphi(x, L_h)| \geq \sum_{i=1}^{k} \int_{C_{\eta_h^2}(y_i)} |D\varphi(x, L_h)|.$$

[†]Stating that some property is satisfied for large enough h means that the set of those positive integers h such that this property is not satisfied is finite.

From (8) and Theorem IX n.2 we obtain that

$$(24) \qquad \int_{C_\rho} |D\varphi(x, L_h)| \leq \omega_{n-1}\rho^{n-1} + \eta_h \quad h = 1, 2, \dots$$

whereas, by Theorem VII n.2, we have that

$$(25) \qquad \int_{C_{\eta_h^2}(y_i)} |D\varphi(x, L_h)| \geq \omega_{n-1}\eta_h^{2(n-1)} \quad i = 1, \dots, k.$$

Combining (23), (24), (25), we find

$$(26) \qquad k \leq \left(\frac{\eta_h}{\omega_{n-1}} + \rho^{n-1}\right) \eta_h^{2(1-n)}$$

and hence from (21)

$$(27) \qquad \text{meas}\,(B_h) \leq \omega_n 3^n \eta_h^2 \left(\rho^{n-1} + \frac{\eta_h}{\omega_{n-1}}\right).$$

Now consider the functions

$$(28) \qquad u_h(x) = \int_{C_q} g_h(x - \xi)|D\varphi(\xi, L_h)|,$$

$$(29) \qquad w_h(x) = \int_{\mathbf{R}^n \setminus C_q} g_h(x - \xi)|D\varphi(\xi, L_h \cap C_\rho)|,$$

$$(30) \qquad z_h(x) = \int_{C_q} g_h(x - \xi)\big(|D\varphi(\xi, L_h)| - D_n\varphi(\xi, L_h)\big).$$

By (13) and Theorem II n.3 in [1] we have

$$(31) \qquad |D_n f_h(x) + z_h(x) - u_h(x)| \leq w_h(x),$$

$$(32) \qquad |Df_h(x)| \leq u_h(x) + w_h(x).$$

From (12), (13), (17), (19), (21), (25) it follows that

$$(33) \qquad u_h(x) \geq \sqrt{\pi^{-n}} \exp\left(-\left(4\eta_h^{-1}\right)^2\right) \omega_{n-1}\eta_h^{-n-2} \quad \text{for} \quad x \in B_h.$$

On the other hand, since $\overline{\mathcal{F}^*L_h} \cap C_q$ is bounded, we can choose finitely many points

$$(34) \qquad y_1, \dots, y_s$$

in such a way that conditions similar to (19), (20), (22) are satisfied, namely

$$(35) \qquad C_q \cap \mathcal{F}^* L_h \subset \bigcup_{i=1}^{s} C_{2\eta_h^8}(y_i),$$

$$(36) \qquad C_{\eta_h^8}(y_i) \cap C_{\eta_h^8}(y_r) = \emptyset \quad \text{for} \quad i \neq r, \qquad y_i \in \mathcal{F}^* L_h \cap C_q.$$

Then, analogously to (25), we have

$$(37) \qquad \int_{C_{2\eta_h^8}(y_i)} |D\varphi(x, L_h)| \geq \omega_{n-1}(2\eta_h^8)^{n-1},$$

$$\int_{C_{\eta_h^8}(y_i)} |D\varphi(x, L_h)| \geq \omega_{n-1}(\eta_h^8)^{n-1}.$$

Moreover, we have by (11) and (36)

$$(38) \qquad \int_{C_{\rho-q}(y_i)} |D\varphi(x, L_h)| - \int_{C_{\rho-q}(y_i)} D_n\varphi(x, L_h) \leq \eta_h \quad i = 1, \ldots, s$$

and hence, by Theorem IX n.2 we find, as in (24), that

$$(39) \qquad (\rho - q)^{1-n} \int_{C_{\rho-q}(y_i)} |D\varphi(x, L_h)| \leq \omega_{n-1} + \eta_h(\rho - q)^{1-n} \quad i = 1, \ldots, s.$$

From (37), (38), (39), relying on Theorem VIII n.2, it follows that

$$(40) \qquad \int_{C_{2\eta_h^8}(y_i)} \big(|D\varphi(x, L_h)| - D_n\varphi(x, L_h)\big) \leq c_h^*(2\eta_h^8)^{n-1}$$

where we have set

$$(41) \qquad c_h^* = \eta_h(\rho - q)^{1-n} + e^n\big(\log(\rho - q) - 8\log\eta_h + 1\big)\sqrt{n\omega_n}|\rho - q|^{1-n}\eta_h.$$

On the other hand, by (17), (28), (29), (30), (35), (36) and by Theorem II n.3 in [1], there holds

$$(42) \qquad u_h(x) + w_h(x) \geq \sum_{i=1}^{s} \int_{C_{\eta_h^8}(y_i)} g_h(x - \xi)|D\varphi(\xi, L_h)|,$$

$$(43) \qquad z_h(x) \leq \sum_{i=1}^{s} \int_{C_{2\eta_h^8}(y_i)} g_h(x - \xi)\big(|D\varphi(\xi, L_h)| - D_n\varphi(\xi, L_h)\big)$$

and hence, letting

$$(44) \qquad c_h = \sup\left\{ \frac{g_h(x - \gamma)}{g_h(x)} : x \in \mathbf{R}^n, \quad |x| < \rho + b, \quad |\gamma| < 2\eta_h^8 \right\},$$

combining (37), (40), (42), (43), we obtain that

$$(45) \qquad \left(u_h(x) + w_h(x)\right) c_h^* c_h^2 \frac{2^{n-1}}{\omega_{n-1}} \geq z_h(x) \quad \text{for} \quad x \in B_h.$$

Then we observe that by Theorem II n.3 and VII n.2 in [1], recalling (24), there holds

$$(46)$$
$$P(L_h \cap C_\rho) = \int_{\mathbf{R}^n} |D\varphi(x, L_h \cap C_\rho)| \leq \int_{A_\rho} |D\varphi(x, L_h)| + \mathcal{H}^{n-1}(\mathcal{F}C_\rho) \leq$$
$$\leq \rho^{n-1}(\omega_n n + \omega_{n-1}) + \eta_h \quad \text{for} \quad h = 1, 2, \ldots$$

and hence from (9), (12), (13), (14), (15), (29) it follows that

$$(47) \qquad \lim_{h \to \infty} \left(\sup\{|w_h(x)| \cdot \exp(\eta_h^{-5}) : x \in B_h\}\right) = 0.$$

From (9), (33), (41), (44), (45), (47), after straightforward computations, it follows that

$$(48) \qquad \lim_{h \to \infty} \left(\sup\left\{\frac{w_h(x)}{u_h(x)} : x \in B_h\right\}\right) = 0,$$

$$(49) \qquad \lim_{h \to \infty} \left(\sup\left\{\frac{z_h(x)}{u_h(x)} : x \in B_h\right\}\right) = 0,$$

and hence by (31), (3), for large enough h we obtain

$$(50) \qquad D_n f_h(x) > 0 \quad \text{for} \quad x \in B_h.$$

On the other hand, by (13), (14), (28), (29) and by Theorem II n.3 in [1] we have

$$(51) \qquad |Df_h(x)| = \left|\int_{\mathbf{R}^n} g_h(x - \xi) D\varphi(\xi, L_h \cap C_\rho)\right| \leq u_h(x) + w_h(x)$$

and hence from (31), (48), (49), it follows that

$$(52) \qquad \lim_{h \to \infty} \left(\inf\left\{\frac{D_n f_h(x)}{|Df_h(x)|} : x \in B_h\right\}\right) = 1.$$

Now consider the function $\tau(x)$ defined by

$$(53) \qquad \tau(x) = b - |x| \quad \text{if} \quad |x| < b, \qquad \tau(x) = 0 \quad \text{if} \quad |x| \geq b.$$

From (12), recalling that $\tau(x)$ is 1-Lipschitzian, we obtain by simple computations

$$(54) \qquad |g_h * \tau(x) - \tau(x)| \leq \eta_h^3 \int_{\mathbf{R}^n} |\xi| \exp(-|\xi|^2) \, d\xi, \quad x \in \mathbf{R}^n.$$

Moreover, from (12), (13) it follows that

$$
(55) \quad
\begin{aligned}
\int_{\mathbf{R}^n} \tau(x)\,|Df_h(x)|\,dx &\leq \int_{\mathbf{R}^n} g_h * \tau(x)\,|D\varphi(x, L_h \cap C_\rho)| \leq \\
&\leq \int_{\mathbf{R}^n} \tau(x)\,|D\varphi(x, L_h \cap C_\rho)| + \int_{\mathbf{R}^n} |\tau(x) - g_h * \tau(x)|\,|D\varphi(x, L_h \cap C_\rho)|
\end{aligned}
$$

and hence by (46), (54)

$$
(56) \quad \limsup_{h \to \infty} \eta_h^{-2} \left(\int_{\mathbf{R}^n} \tau(x)\,|Df_h(x)|\,dx - \int_{\mathbf{R}^n} \tau(x)\,|D\varphi(x, L_h \cap C_\rho)| \right) \leq 0.
$$

On the other hand, by Theorem II n.3 in [1] we have, recalling (53),

$$
(57) \quad
\begin{aligned}
\int_{\mathbf{R}^n} \tau(x)\,|D\varphi(x, L_h \cap C_\rho)| &= \int_0^b dt \int_{C_t} |D\varphi(x, L_h \cap C_\rho)| = \\
&= \int_0^b dt \int_{C_t} |D\varphi(x, L_h)|,
\end{aligned}
$$

$$
(58) \quad \int_{\mathbf{R}^n} \tau(x)\,|Df_h(x)|\,dx = \int_0^b dt \int_{C_t} |Df_h(x)|\,dx,
$$

and hence from (56), (57), (58)

$$
(59) \quad \limsup_{h \to \infty} \eta_h^{-2} \left(\int_0^b dt \int_{C_t} |Df_h(x)|\,dx - \int_0^b dt \int_{C_t} |D\varphi(x, L_h)| \right) \leq 0.
$$

From (15), recalling the definition of B_h, it can be seen that for large enough h

$$
(60) \quad \{x \,:\, \eta_h^2 \leq f_h(x) \leq 1 - \eta_h^2, \quad x \in C_t\} \subset B_h \setminus \mathcal{F}B_h \quad \text{for} \quad t < b
$$

and hence, recalling (50) and Theorem I, we find that

$$
(61) \quad \int_{C_t} |Df_h(x)|\,dx \geq \int_{\eta_h^2}^{1-\eta_h^2} d\lambda \, \mathcal{H}^{n-1} \left(\{x \,:\, x \in C_t, \quad f_h(x) = \lambda\} \right),
$$

and thus

$$
(62) \quad
\begin{aligned}
\int_0^b dt \int_{C_t} |Df_h(x)|\,dx &\geq \\
&\geq \int_{\eta_h^2}^{1-\eta_h^2} d\lambda \int_0^b dt \, \mathcal{H}^{n-1} \left(\{x \,:\, x \in C_t, \quad f_h(x) = \lambda\} \right).
\end{aligned}
$$

By (62) we can find a number λ_h satisfying

$$
(63) \quad
\begin{aligned}
\int_0^b dt \, \mathcal{H}^{n-1} \left(\{x \,:\, x \in C_t, \quad f_h(x) = \lambda_h\} \right) &\leq \\
&\leq \frac{1}{1 - 2\eta_h^2} \int_0^b dt \int_{C_t} |Df_h(x)|\,dx,
\end{aligned}
$$

$$(64) \qquad \eta_h^2 < \lambda_h < 1 - \eta_h^2.$$

Since by (50), (60), (64) the set $\{x \,:\, x \in C_t, \quad f_h(x) = \lambda_h\}$ is closed, bounded and contained in a locally regular hypersurface of class C^∞, having finite $n-1$-dimensional Hausdorff measure, it follows by (24) n.2 in [1] that the set

$$(65) \qquad E_h = \{x \,:\, x \in C_b, \quad f_h(x) > \lambda_h\}$$

is a Caccioppoli set, provided h is large enough. On the other hand, by Theorem III n.3 in [1] and by (50), (65) we have

$$(66) \qquad \mathcal{F}E_h \cap A_b = \mathcal{F}^* E_h \cap A_b = \{x \,:\, x \in C_b, \quad f_h(x) = \lambda_h\}.$$

Hence we have found that, for large enough h and for every positive number $t < b$, the assumptions (30) of Theorem III n.3 are fulfilled; from (52), (60) it then can be seen that also the assumption (31) of that theorem is fulfilled.

Now let us denote by μ an integer such that all the claims so far proved for large enough h, are in fact true for $h \geq \mu$. Since by (9), (14), (24), (59), the sequence of integrals

$$(67) \qquad \int_0^b dt \int_{C_t} |Df_h(x)| \, dx$$

is bounded uniformly with respect to h, we have, by (9), (63), (66) and by Theorem VII n.2 in [1]

$$(68) \qquad \sum_{h=\mu}^{\infty} \eta_h^{-1} \left(\int_0^b dt \int_{C_t} |D\varphi(x, E_h)| - \int_0^b dt \int_{C_t} |Df_h(x)| \, dx \right) < +\infty,$$

and hence by (9), (59), the infinite sum

$$(69) \qquad \sum_{h=\mu}^{\infty} \eta_h^{-1} \left(\int_0^b dt \int_{C_t} |D\varphi(x, E_h)| - \int_0^b dt \int_{C_t} |D\varphi(x, L_h)| \right) < +\infty$$

is well defined (either finite or equal to $-\infty$). Recalling (9), (15), (27), (64), (65), we find that

$$(70) \qquad \begin{aligned} &\sum_{h=\mu}^{\infty} \eta_h^{-1} \int_0^b dt \int_{\mathcal{F}C_t} |\varphi(x, E_h) - \varphi(x, L_h)| \, d\mathcal{H}^{n-1} \\ &= \sum_{h=\mu}^{\infty} \eta_h^{-1} \int_{C_b} |\varphi(x, E_h) - \varphi(x, L_h)| \, dx < +\infty. \end{aligned}$$

From the definition and the main properties of the functions Θ, Ψ, discussed in n.1, from Theorem I n.2 and from (9) it follows that, for almost every positive number $t < b$,

$$(71)$$
$$\int_{C_t} |D\varphi(x, L_h)| = \Theta(L_h, C_t) \leq \Theta(E_h, C_t) + \int_{\mathcal{F}C_t} |\varphi(x, E_h) - \varphi(x, L_h)| \, d\mathcal{H}^{n-1}$$
$$= \int_{C_t} |D\varphi(x, E_h)| + \int_{\mathcal{F}C_t} |\varphi(x, E_h) - \varphi(x, L_h)| \, d\mathcal{H}^{n-1} - \Psi(E_h, C_t).$$

Since by definition we have $\Psi(E_h, C_t) \geq 0$, from (69), (70), (71), we obtain that

$$(72) \qquad \sum_{h=\mu}^{\infty} \eta_h^{-1} \int_0^b \Psi(E_h, C_t)\, dt < +\infty,$$

$$(73) \qquad \sum_{h=\mu}^{\infty} \eta_h^{-1} \int_0^b dt \left| \int_{C_t} |D\varphi(x, E_h)| - \int_{C_t} |D\varphi(x, L_h)| \right| < +\infty.$$

From (70), (72), (73), it follows that for almost every positive number $t < b$

$$(74) \qquad \lim_{h \to \infty} \eta_h^{-1} \int_{\mathcal{F}C_t} |\varphi(x, E_h) - \varphi(x, L_h)|\, d\mathcal{H}^{n-1} = 0,$$

$$(75) \qquad \lim_{h \to \infty} \eta_h^{-1} \left| \int_{C_t} |D\varphi(x, E_h)| - \int_{C_t} |D\varphi(x, L_h)| \right| = 0,$$

$$(76) \qquad \lim_{h \to \infty} \eta_h^{-1} \Psi(E_h, C_t) = 0.$$

From (74), recalling (57) n.2, we find that

$$(77) \qquad \lim_{h \to \infty} \eta_h^{-1} \left| \int_{C_t} D\varphi(x, E_h) - \int_{C_t} D\varphi(x, L_h) \right| = 0,$$

and, since when $t < \rho$ there clearly holds

$$(78) \qquad \int_{C_\rho} |D\varphi(x, E)| - \left| \int_{C_\rho} D\varphi(x, E) \right| \geq \int_{C_t} |D\varphi(x, E)| - \left| \int_{C_t} D\varphi(x, E) \right|,$$

by (8) we can find a sequence of positive numbers $\{\beta_h\}$ such that (32), (33) n.3 are satisfied together with

$$(79) \qquad \lim_{h \to \infty} \eta_h^{-1} \beta_h = 1.$$

The possibility of finding a sequence of points $\{\xi_h\}$ satisfying (32), (33) n.3 follows immediately from (7), (9), (74), hence for almost every positive number $t < b$ the assumptions of Theorem III n.3 are fulfilled. Therefore, (34) n.3 holds true, and hence from (75), (77), (78) it follows that, for almost every pair of numbers t, α satisfying

$$(80) \qquad 0 < t < b, \qquad 0 < \alpha < 1,$$

we have

$$(81) \qquad \limsup_{h \to \infty} \eta_h^{-1} \left(\int_{C_{\alpha t}} |D\varphi(x, L_h)| - \left| \int_{C_{\alpha t}} D\varphi(x, L_h) \right| \right) \leq \alpha^{n+1}.$$

From (81), recalling the arbitrariness of α, b and observing that the difference

$$(82) \qquad \int_{C_{\alpha t}} |D\varphi(x, L_h)| - \left| \int_{C_{\alpha t}} D\varphi(x, L_h) \right|$$

is a non decreasing function of t, we obtain that (10) is established.

From Lemma II one easily obtains the following corollary:

III – *For every integer $n \geq 2$, there exists a number $\sigma(n)$ such that, if E is a Caccioppoli set in $\mathbf{R}^n$, if ξ is a point in $\mathbf{R}^n$ and t is a positive number, and if, moreover,*

$$(83) \qquad \xi \in \overline{\mathcal{F}^*E}, \qquad \Psi(E, C_{2t}(\xi)) = 0,$$

$$(2t)^{1-n} \left(\int_{C_{2t}} |D\varphi(x, E)| - \left| \int_{C_{2t}} D\varphi(x, E) \right| \right) \leq \sigma(n),$$

then[‡]

$$t^{1-n} \left(\int_{C_t} |D\varphi(x, E)| - \left| \int_{C_t} D\varphi(x, E) \right| \right)$$

$$(84) \qquad \leq \frac{1}{2}(2t)^{1-n} \left(\int_{C_{2t}} |D\varphi(x, E)| - \left| \int_{C_{2t}} D\varphi(x, E) \right| \right).$$

Proof. Clearly, we can prove the theorem under the assumption that

$$(85) \qquad |\xi| = 0, \qquad t = 1.$$

From this assumption and from (83) it follows, by Theorem IX n.2, that

$$(86) \qquad \operatorname{meas}(E \cap C_{2\alpha t}) \cdot \operatorname{meas}(C_{2\alpha t} \setminus E) \geq (2\alpha t)^{2n} \left(\frac{\omega_{n-1}}{n} \right)^2.$$

Comparing (83), (86) with the assumptions of Theorem II, it can be seen that, if our claim were false then, letting $\rho = 2t$, it would be possible to find a sequence of sets $\{E_h\}$ and a sequence of positive numbers $\{\beta_h\}$ satisfying the assumptions of Lemma II but not its thesis, and this is a contradiction.

IV – *Assume E is a Caccioppoli set in $\mathbf{R}^n$ ($n \geq 2$), ρ is a positive number, and ξ is a point in $\mathbf{R}^n$, such that*

$$\xi \in \overline{\mathcal{F}^*E}, \qquad \Psi(E, C_\rho(\xi)) = 0,$$

$$(87) \qquad \int_{C_\rho(\xi)} |D\varphi(x, E)| - \left| \int_{C_\rho(\xi)} D\varphi(x, E) \right| \leq \rho^{n-1}\sigma(n),$$

where $\sigma(n)$ is the constant mentioned in Corollary III. Then

$$(88) \qquad \xi \in \mathcal{F}^*E$$

[‡]Editor's note: in the original version of the paper the inequality (84) contains $\sigma(n)/2$ in the right hand side. However the stronger statement given here is the one that is actually proved, and that leads by iteration to the regularity of the reduced boundary

and, for every positive number $t < \rho$, there holds

$$(89) \qquad \left| \frac{D\varphi(\xi, E)}{|D\varphi(\xi, E)|} - \frac{\int_{C_t(\xi)} D\varphi(x, E)}{\int_{C_t(\xi)} |D\varphi(x, E)|} \right| \leq \eta(n)\sqrt{\frac{t}{\rho}},$$

where $\eta(n)$ is a constant depending only on the dimension n.

Proof. Clearly, it suffices to prove the claim in the case where

$$(90) \qquad \rho = 1, \qquad |\xi| = 0.$$

Then we have from Corollary III

$$(91) \qquad \int_{C_{2^{-h}}} |D\varphi(x, E)| - \left| \int_{C_{2^{-h}}} D\varphi(x, E) \right| \leq \sigma(n) \cdot 2^{-hn}.$$

Now define, for every number α satisfying

$$(92) \qquad 1 \leq 2\alpha \leq 2,$$

and for every positive integer h,

$$(93) \qquad u = \frac{\int_{C_{2^{-h}}} D\varphi(x, E)}{\int_{C_{2^{-h}}} |D\varphi(x, E)|}, \qquad v = \frac{\int_{C_{\alpha 2^{-h}}} D\varphi(x, E)}{\int_{C_{\alpha 2^{-h}}} |D\varphi(x, E)|}.$$

From (91), (93) it follows by Theorem IX n.2 that

$$(94) \qquad \begin{aligned} \langle u, v \rangle \int_{C_{\alpha 2^{-h}}} |D\varphi(x, E)| &- \int_{C_{\alpha 2^{-h}}} |D\varphi(x, E)| = \\ = \int_{C_{\alpha 2^{-h}}} &(\langle u, D\varphi(x, E)\rangle - |D\varphi(x, E)|) \geq \\ \geq \int_{C_{2^{-h}}} &(\langle u, D\varphi(x, E)\rangle - |D\varphi(x, E)|) = \\ = |u| \cdot \left| \int_{C_{2^{-h}}} \right. &\left. D\varphi(x, E) \right| - \int_{C_{2^{-h}}} |D\varphi(x, E)| \geq \\ \geq -\sigma(n)\, 2^{-hn} &- (2^h - 2^h |u|)\omega_{n-1} 2^{-hn}. \end{aligned}$$

Let us distinguish the two possibilities

$$(95) \qquad \frac{\sigma(n)}{\omega_{n-1}} 2^{-h} \leq \frac{1}{2^{n+1}},$$

$$(96) \qquad \frac{\sigma(n)}{\omega_{n-1}} 2^{-h} \geq \frac{1}{2^{n+1}}.$$

In the former case, we have by Theorem VII n.2 and (92)

$$(97) \qquad \int_{C_{\alpha 2^{-h}}} |D\varphi(x, E)| \geq \int_{C_{2^{-(h+1)}}} |D\varphi(x, E)| \geq \omega_{n-1} 2^{-(h+1)(n-1)},$$

and hence by (94), (95), (97) the scalar product $\langle u, v \rangle$ is positive, since by Theorem VII n.2 and (91), (93)

$$(98) \qquad 2^h - 2^h |u| \leq \frac{\sigma(n)}{\omega_{n-1}}.$$

From (94), (97), (98) it follows that

$$(99) \qquad 1 - \langle u, v \rangle \leq 2^{-h} \frac{\sigma(n)}{\omega_{n-1}} 2^n$$

and hence, since by (93) we have $|u| \leq 1$, $|v| \leq 1$, we see that

$$(100) \qquad |u - v| \leq 2\sqrt{\frac{\sigma(n)2^n}{\omega_{n-1}}} \cdot \sqrt{2^{-h}} < 4\sqrt{2^{-h}}\sqrt{\frac{\sigma(n)2^n}{\omega_{n-1}}},$$

whereas, if (96) occurs, then (100) is trivial.

Now consider two numbers t, s satisfying

$$(101) \qquad 0 < 8t < s < 1,$$

and choose two integers k, r such that

$$(102) \qquad 2^{-k} \geq s > 2^{-(k+1)} > \cdots > 2^{-k-r} > t \geq 2^{-k-r-1}.$$

Repeatedly choosing in (93), (100)

$$(103) \qquad h = k, \qquad \alpha = s\, 2^k,$$

$$(104) \qquad h = k, \ldots, k + r - 1, \qquad 2\alpha = 1,$$

$$(105) \qquad h = k + r, \qquad \alpha = t\, 2^{k+r},$$

one finds

$$(106) \qquad \left| \frac{\int_{C_t} D\varphi(x, E)}{\int_{C_t} |D\varphi(x, E)|} - \frac{\int_{C_s} D\varphi(x, E)}{\int_{C_s} |D\varphi(x, E)|} \right| \leq \eta(n)\sqrt{s},$$

where we have set

$$(107) \qquad \eta(n) = 8\sqrt{2}\sqrt{\frac{\sigma(n)\, 2^n}{\omega_{n-1}}} \cdot \sum_{i=0}^{\infty} \sqrt{2^{-i}}.$$

From (106) we infer the existence of the limit

$$(108) \qquad \lim_{t \to 0^+} \frac{\int_{C_t} D\varphi(\xi, E)}{\int_{C_t} |D\varphi(\xi, E)|};$$

by (90), (91) and by Theorem VII n.2 this limit vector has unitary length and hence (88) is established; finally, (89) is an easy consequence of (90), (106).

V – *Assume E is a Caccioppoli set in $\mathbf{R}^n$ $(n \geq 2)$, ρ is a positive number, and ξ is a point in $\mathbf{R}^n$, such that*

$$(109) \qquad \Psi(E, C_\rho(\xi)) = 0, \qquad \xi \in \overline{\mathcal{F}^*E},$$

$$\int_{C_\rho(\xi)} |D\varphi(x, E)| - \left| \int_{C_\rho(\xi)} D\varphi(x, E) \right| \leq \sigma(n) \rho^{n-1}.$$

Then we have

$$(110) \qquad \overline{\mathcal{F}^*E} \cap C_{\rho 2^{-n}}(\xi) = \mathcal{F}^*E \cap C_{\rho 2^{-n}}(\xi)$$

and the vector

$$(111) \qquad \frac{D\varphi(x, E)}{|D\varphi(x, E)|}$$

is a continuous function in $\mathcal{F}^*E \cap C_{\rho 2^{-n}}(\xi)$.

Proof. For every $y \in \overline{\mathcal{F}^*E} \cap C_{\rho 2^{-n}}(\xi)$ there holds

$$(112) \qquad C_{\rho 2^{-n}}(y) \subset C_{\rho 2^{1-n}}(\xi);$$

by Theorem III we have

$$(113) \qquad \int_{C_{2^{-n}\rho}(y)} |D\varphi(x, E)| - \left| \int_{C_{2^{-n}\rho}(y)} D\varphi(x, E) \right| \leq$$

$$\leq \int_{C_{2^{1-n}\rho}(\xi)} |D\varphi(x, E)| - \left| \int_{C_{2^{1-n}\rho}(\xi)} D\varphi(x, E) \right| \leq \sigma(n)\, 2^{n(1-n)} \rho^{n-1}$$

and hence (110) follows from Theorem IV.

We now prove the continuity of the function $\frac{D\varphi(x,E)}{|D\varphi(x,E)|}$. By (113) and Theorem IV we have

$$(114) \qquad \frac{D\varphi(y, E)}{|D\varphi(y, E)|} = \lim_{s \to 0^+} \frac{\int_0^s dt \int_{C_t(y)} D\varphi(x, E)}{\int_0^s dt \int_{C_t(y)} |D\varphi(x, E)|}$$

uniformly in the closed set $C_{2^{-n}\rho}(y) \cap \mathcal{F}^*E$; since the function of y

$$(115) \qquad \frac{\int_0^s dt \int_{C_t(y)} D\varphi(x, E)}{\int_0^s dt \int_{C_t(y)} |D\varphi(x, E)|}$$

is continuous in $\mathcal{F}^*E$, the theorem is proved.

THEOREM VI – *Given an open set A and a Caccioppoli set E in $\mathbf{R}^n$ $(n \geq 2)$, if E has minimal boundary on A then $\mathcal{F}^*E \cap A$ is a locally regular hypersurface.*

Proof. Chosen a point $\xi \in \mathcal{F}^*E \cap A$, from the definition of reduced boundary and by (5), (6) n.1 we can find a positive number ρ satisfying the assumptions of Theorem IV. It then follows from Theorem IV and Theorem III n.4 in [1] that the set

$$(116) \qquad\qquad \mathcal{F}^*E \cap A_{2^{-n}\rho}(\xi)$$

is a locally regular hypersurface. Since ξ is arbitrary, the proof is complete.

Bibliography

[1] E. De Giorgi, *Complementi alla teoria della misura $(n-1)$-dimensionale in uno spazio n-dimensionale.* Seminario di Matematica della Scuola Normale Superiore di Pisa (1961).

[2] H. Federer, W.H. Fleming, *Normal and Integral Currents.* Annals of Mathematics, Vol. 72 N.3 November 1960.

[3] E.R. Reifenberg, *The Plateau Problem.* Acta Mathematica 104:1,2 (1960).

[4] W.II. Fleming, *On the oriented Plateau Problem.* Amer. Math. Soc. Notices Vol. 8 n.3 June 1961 pag. 273.

An extension of Bernstein theorem[‡]

Ennio De Giorgi

In a recent paper FLEMING conjectured that the BERNSTEIN theorem concerning area-minimizing hypersurfaces in the 3-dimensional Euclidean space should also be true in the case of hypersurfaces of any dimension (see [1] n. 5).

In this paper we prove this conjecture in the case of hypersurfaces in the 4-dimensional Euclidean space when they are representable as $x_4 = f(x_1, x_2, x_3)$. The main results are contained in theorems I, II.

In the sequel I will always make use of the definitions and of the notation introduced in [2], [3], [4].

1. LEMMA I. – *Let $m \geq 2$ be a given integer and let $f(x)$ be a function defined on $\mathbf{R}^m \setminus \{0\}$, which is assumed to be continuous with continuous partial derivatives and positively homogeneous of degree 1; this means that for any $x \in \mathbf{R}^m \setminus \{0\}$ and for any real number $t > 0$, the following equality holds:*

$$(1) \qquad f(tx) = tf(x).$$

Let the set

$$(2) \qquad E = \{(x, z) : \ x \in \mathbf{R}^m \setminus \{0\}, z \in \mathbf{R}, z \leq f(x)\}$$

be a subset of $\mathbf{R}^{m+1}$ having locally bounded perimeter and such that

$$(3) \qquad \psi(E, K) = 0$$

for any compact set $K \subset \mathbf{R}^{m+1}$.

Then, in $\mathbf{R}^m \setminus \{0\}$, $f(x)$ is equal to a polynomial of degree 1.

Proof. From the definition of $\psi(E, K)$ and from well known theorems of the calculus of variations it immediately follows the Euler equation (see e.g. [4], pages 158–160)

$$(4) \qquad (|Df|^2 + 1)\Delta_2 f - \sum_{h,k}^{1,m} D_h f \cdot D_k f \cdot D_{hk} f = 0.$$

Fix a positive integer $s \leq m$ and set

$$(5) \qquad w = D_s f$$

[‡]Editor's note: translation into English of the paper "Una estensione del teorema di Bernstein", published in Ann. Sc. Norm. Sup. Pisa, **(3) 19** (1965), 79–85.

then from (4) we have

$$(6) \qquad (|Df|^2 + 1)\Delta_2 w - \sum_{h,k}^{1,m} D_h f \cdot D_k f \cdot D_{hk} w + \sum_{i=1}^{m} b_i \cdot D_i w = 0$$

where b_i are continuous functions with continuous partial derivatives whose explicit computation is not necessary. On the other hand, from the homogeneity of f and from (5) we have

$$(7) \qquad \max\{w(x);\ |x| = 1\} = \max\{w(x);\ |x| > 0\}$$

whence, from (6), (7), from the maximum principle (which is valid for elliptic linear differential equations of the same kind of (6)), we can conclude that w is constant, and hence that the derivatives of f are constant.   q.e.d.

In the following lemmas II, III, IV, and in theorem I, we shall always assume that the following assumptions are satisfied:
n is an integer greater than 2; the set $E \subset \mathbf{R}^n$ has locally bounded perimeter; α is a point in $\mathbf{R}^n$ satisfying the condition

$$(8) \qquad |\alpha| \equiv \left(\sum_{h=1}^{n} \alpha_h^2 \right)^{\frac{1}{2}} = 1;$$

for any point $x \in \mathcal{F}^*E$ we have

$$(9) \qquad \sum_{h=1}^{n} \alpha_h \frac{D_h \varphi(x, E)}{|D\varphi(x, E)|} \geq 0;$$

for any compact set $K \subset \mathbf{R}^n$ we have

$$(10) \qquad \psi(E, K) = 0.$$

LEMMA II. – *If we further require that ξ is a point in $\mathcal{F}^*E$, that t is a nonzero real number and that*

$$(11) \qquad \sum_{k=1}^{n} \alpha_k \frac{D_k \varphi(\xi, E)}{|D\varphi(\xi, E)|} > 0,$$

then it follows

$$(12) \qquad (\xi + \alpha t) \in \mathbf{R}^n \setminus \mathcal{F}_e E.$$

Proof. Using well known results about boundaries having minimal measure (see e.g. [2], thm. XIII, XIV), from (10) and (11) we deduce that there exists a neighborhood A of ξ such that

$$(13) \qquad A \cap \mathcal{F}^*E = A \cap \mathcal{F}_e E$$

and, moreover, (see [4] p. 157), $\mathcal{F}^*E \cap A$ is a locally smooth hypersurface whose tangent plane in ξ does not contain the straight line in the direction of the vector $(\alpha_1, ..., \alpha_n)$. Then, for any given $\rho > 0$, there exist two real numbers η, σ such that

$$(14) \qquad \operatorname{meas}\left[C_\sigma(\xi - \eta\alpha) \cap E\right] = 0$$

$$(15) \qquad \operatorname{meas}\left[C_\sigma(\xi + \eta\alpha) \setminus E\right] = 0$$

$$(16) \qquad 0 < \eta < \rho, \qquad 0 < \sigma < \rho.$$

On the other hand, using (8), (9) we have, for any given $\tau > 0$ and for almost every $x \in \mathbf{R}^n$,

$$(17) \qquad \varphi((x + \alpha\tau), E) \geq \varphi(x, E)$$

whence from (14), (15), (16), (17) and the arbitrariness of ρ and τ we get (12). q.e.d.

Lemma III. – *Let ξ be a point of $\mathcal{F}^*E$ and let $\{\xi_h\}$ be a sequence of points satisfying the following conditions:*

$$(18) \qquad \lim_{h \to \infty} \xi_h = \xi, \quad \xi_h \in \mathcal{F}^*E, \quad \sum_{i=1}^{n} \alpha_i \frac{D_i\varphi(\xi_h, E)}{|D\varphi(\xi_h, E)|} > 0 \quad \text{for } h = 1, 2, ...$$

Then we have

$$(19) \qquad \sum_{i=1}^{n} \alpha_i \frac{D_i\varphi(\xi, E)}{|D\varphi(\xi, E)|} > 0.$$

Proof. Equation (19) is clearly satisfied when

$$(20) \qquad \sum_{i=1}^{n} \alpha_i \frac{D_i\varphi(\xi, E)}{|D\varphi(\xi, E)|} = 1;$$

it is then enough to prove the theorem when equation (20) is not satisfied. If (20) is not true, after a suitable rotation, we can suppose that

$$(21) \qquad \alpha \equiv (\alpha_1, ..., \alpha_n) = (1, 0, ..., 0), \qquad \frac{D_n\varphi(\xi, E)}{|D\varphi(\xi, E)|} < 0.$$

Set $m = n-1$, then from (10) and using well known results concerning boundaries of minimal measure (see [4] pp. 157–160), it is possible to find a neighborhood A of ξ, an open connected set $B \subset \mathbf{R}^m$ and an infinitely differentiable function $f(y)$ defined on B, such that

$$(22) \qquad A \cap \mathcal{F}^*E = \{(y, z); \; y \in B, z = f(y)\}$$

and the Euler equation (4) will be satisfied. On the other hand, from (9), (21), (22), we always have

$$(23) \qquad\qquad D_1 f \geq 0,$$

while, by setting

$$(24) \qquad \xi = (\eta, \zeta), \quad \xi_h = (\eta_h, \zeta_h), \quad \eta \in \mathbf{R}^m, \quad \eta_h \in \mathbf{R}^m, \quad \zeta \in \mathbf{R}, \quad \zeta_h \in \mathbf{R}$$

with a sufficiently large h we have, recalling (18), (21), (22),

$$(25) \qquad\qquad \eta_h \in B, \quad D_1 f(\eta_h) > 0, \quad \eta \in B.$$

Recalling (4), (5) and the maximum principle, from (23) and (25) we obtain

$$(26) \qquad\qquad D_1 f(\eta) > 0$$

whence, using (21), (22), (24), we get (19). \qquad q.e.d.

LEMMA IV. – *Let E be a cone having its vertex at the origin of $\mathbf{R}^n$, in other words for any $t > 0$ we have*

$$(27) \qquad\qquad \varphi(tx, E) = \varphi(x, E).$$

Furthermore, let us assume that

$$(28) \qquad\qquad \mathcal{F}_e E \setminus \mathcal{F}^* E \neq \emptyset.$$

Then $(\mathcal{F}_e E \setminus \mathcal{F}^* E)$ contains at least one straight half-line of $\mathbf{R}^n$.

Proof. If for every $x \in \mathcal{F}^* E$ we had

$$(29) \qquad\qquad \sum_{h=1}^{n} \alpha_h \frac{D_h \varphi(x, E)}{|D\varphi(x, E)|} = 0,$$

then for all $t \in \mathbf{R}$ and for almost every $x \in \mathbf{R}^n$ we would have

$$(30) \qquad\qquad \varphi(x + t\alpha, E) = \varphi(x, E).$$

From (30) we conclude that if a point $\bar{x}$ were contained in $\mathcal{F}_e E \setminus \mathcal{F}^* E$, then the whole line $\{\bar{x} + t\alpha;\ t \in \mathbf{R}\}$ would be contained in $\mathcal{F}_e E \setminus \mathcal{F}^* E$.

Hence we can suppose that (29) is not satisfied at any point $\mathcal{F}^* E$; after a suitable rotation we can suppose that

$$(31) \qquad\qquad \alpha \equiv (\alpha_1, ..., \alpha_n) = (0, ..., 0, 1).$$

From (31), using Lemma II, hypothesis (10) and well-known results concerning oriented boundaries of minimal measure (see [4] pp. 157–160), we obtain

$$(32) \qquad \left\{ x;\ x \in \mathcal{F}^* E, \frac{D_n \varphi(x, E)}{|D\varphi(x, E)|} > 0 \right\} = \{(y, z);\ y \in B, z = f(y)\},$$

where B is an open subset of $\mathbf{R}^{n-1}$ and $f(y)$ is a continuous function with continuous partial derivatives defined on B.

Moreover, since E is a cone of $\mathbf{R}^n$ with its vertex at the origin, then B is itself a cone of $\mathbf{R}^{n-1}$ and f is a homogeneous function of degree 1. If $B = \mathbf{R}^m$, or if $\mathcal{F}B$ contains only the origin, then by lemma I the function f would be a polynomial of degree 1 and hence E would be a half-space, in contradiction with hypothesis (28). Consequently $\mathcal{F}B$ is forced to contain infinitely many points besides the origin; since B is a cone and since f is a homogeneous function, we can find a sequence of points $\{\eta_h\}$ in the space $\mathbf{R}^{n-1}$ and a point $(\eta, \tau) \in \mathbf{R}^n$ such that we have

$$(33) \qquad \eta_h \in B, \quad \lim_{h \to \infty} (\eta_h, f(\eta_h)) = (\eta, \tau), \quad |(\eta, \tau)| = 1, \quad \eta \in \mathcal{F}B.$$

It is easy to prove that

$$(34) \qquad (\eta, \tau) \in \mathcal{F}_e E \setminus \mathcal{F}^* E.$$

Indeed from (32), (33) it follows (see [2], theorem XIV)

$$(35) \qquad (\eta, \tau) \in \overline{\mathcal{F}^* E} = \mathcal{F}_e E;$$

if we had

$$(36) \qquad (\eta, \tau) \in \mathcal{F}^* E$$

then by lemma III we would have

$$(37) \qquad \frac{D_n \varphi((\eta, \tau), E)}{|D\varphi((\eta, \tau), E)|} > 0$$

so that η would be a point of B, which is in contrast with hypothesis (33). From (33), (34), recalling that E is a cone with the vertex at the origin of $\mathbf{R}^n$, we deduce that the whole half-line from the origin through the point (η, τ) is contained in $(\mathcal{F}_e E \setminus \mathcal{F}^* E)$.   q.e.d.

2. From lemma IV we deduce an extension of the Bernstein theorem in the 4-dimensional Euclidean space which can be stated either under hypotheses of global type (like those used in Caccioppoli set theory), or under hypotheses of classical type. The first formulation is contained in the following

Theorem I. – *Let E be a set of $\mathbf{R}^4$ with locally bounded perimeter satisfying conditions (8), (9), (10); moreover suppose $\mathcal{F}^* E \neq \emptyset$. Then E is equivalent to a half-space.*

Proof. We can assume that the origin of $\mathbf{R}^4$ is contained in $\mathcal{F}^* E$. From theorems X, XII of [2] there exists (and is finite) the limit

$$(1) \qquad b = \lim_{\rho \to \infty} \rho^{1-n} \int_{C_\rho} |D\varphi(x, E)|$$

and moreover

$$(2) \qquad\qquad b \geq w_{n-1}$$

where $w_{n-1} = \text{meas}\,\{x;\ x \in \mathbf{R}^{n-1}, |x| \leq 1\}$.

If $b = w_{n-1}$, from theorem X of [2] it follows that E is equivalent to a cone with vertex at the origin and then, since the origin of $\mathbf{R}^4$ is contained in $\mathcal{F}^*E$, using well-known results about oriented boundaries of minimal measure (see [4] p. 157) the set is equivalent to a half-space.

If $b > w_{n-1}$, using theorem VI of [3] we can find a sequence $\{\rho_h\}$ of positive numbers and a set L such that, setting

$$(3) \qquad\qquad E_h = \{\rho_h x;\ x \in E\},$$

we have[‡]

$$(4) \qquad \lim_{h\to\infty} \rho_h = 0, \qquad \lim_{h\to\infty} \left[\text{meas}\,((L \setminus E_h) \cap K) + \text{meas}\,((E_h \setminus L) \cap K)\right] = 0$$

for any compact set $K \subset \mathbf{R}^n$. From theorem VII of [3] and from theorems VI, X, XIV of [2], we conclude that L is a cone with vertex at the origin and satisfying the requirements of lemma IV; hence the set $\mathcal{F}_e L \setminus \mathcal{F}^*L$ contains at least one straight half-line, in contradiction with a theorem of TRISCARI (see theorem XI of [4]). We must then conclude that $b = w_{n-1}$.   q.e.d.

From theorem I, following the same reasoning of [1] n. 5, we immediately deduce the next theorem.

THEOREM II. – *If $f(x_1, x_2, x_3)$ is a continuous function defined on $\mathbf{R}^3$, with continuous partial derivatives of any order and satisfying the Euler equation*

$$(5) \qquad \sum_{h=1}^{3} \frac{\partial}{\partial x_h} \left(\frac{\dfrac{\partial f}{\partial x_h}}{\sqrt{1 + \sum_{k=1}^{3} \left(\dfrac{\partial f}{\partial x_k}\right)^2}} \right) = 0,$$

then f is a polynomial of degree 1.

Bibliography

[1] W. H. FLEMING, *On the oriented Plateau problem*. Rend. Circolo Mat. Palermo, **9** (1962), 69–89.

[2] D. TRISCARI, *Sulle singolarità delle frontiere orientate di misura minima,* Ann. Scuola Norm. Sup. Pisa (3), **17** (1963), 349–371.

[‡]Editor's note: formula (4) below has been modified according to an errata corrige included in the same volume

[3] D. Triscari, *Sull'esistenza di cilindri con frontiera di misura minima,* Ann. Scuola Norm. Sup. Pisa (3), **17** (1963), 387–399.

[4] D. Triscari, *Sulle singolarità delle frontiere orientate di misura minima nello spazio euclideo a 4 dimensioni,* Le Matematiche, **18** (1963), 139–163.

Hypersurfaces of minimal measure in pluridimensional Euclidean spaces[‡]

Ennio De Giorgi

In this talk I shall deal with some recent developments of the theory of minimal hypersurfaces which are embedded in the n-dimensional Euclidean space (with $n > 3$). I shall not deal with the results for surfaces in $\mathbf{R}^3$ (or, more generally, for 2-dimensional varieties in $\mathbf{R}^n$) which are dealt with, for example, in the recent papers [4], [22] and in [21], and in the numerous references there given.

However, I think it is worthwhile to observe that many typical methods in the theory of minimal surfaces (conformal representation, ...;) do not seem to work well when we pass to the many-dimensional case. This, of course, has led many researchers in this field to different approaches and methods.

The following exposition, even with the restrictions mentioned at the beginning, will definitely lack completeness and will be rather vague: I apologize for that. However, I hope that what I am saying will prove to be useful as a brief indication for non specialists of some lines along which the theory is developing.

I shall deal mainly with the Plateau problem in Cartesian form, with the extension of Bernstein's theorem to the n-dimensional case, and with removable singularities of minimal hypersurfaces.

I will add some final remarks on some recent formulations of the general Plateau problem which are interesting in themselves and have shown their relevance also for the problem in Cartesian form.

The Plateau problem in Cartesian form may be expressed as follows: let Ω be an open bounded set in $\mathbf{R}^n$ and let $g(x)$ be a real continuous function on the boundary $\partial\Omega$ of Ω: does there exist then a continuous function $u(x)$ on the closure $\overline{\Omega}$ of Ω with continuous and summable first derivatives on Ω and such that

$$g(x) = u(x), \quad \text{for every } x \in \partial\Omega, \tag{1}$$

which minimizes the integral

$$\int_\Omega \sqrt{1 + \sum_{h=1}^n \left(\frac{\partial u}{\partial x_h}\right)^2} \quad ? \tag{2}$$

[‡]Editor's note: published in Proc. Internat. Congr. Math. (Moscow, 1966), Izdat "Mir", Moscow, 1968, 395–401.

It is easy to show that if such a function $u(x)$ does exist, then it is unique and, due to a result of Morrey (see [20]), it is continuous in Ω with its derivatives of every order and satisfies the Euler equation

$$(3) \qquad \sum_{h=1}^{n} \frac{\partial}{\partial x_h} \left(\frac{\dfrac{\partial u}{\partial x_h}}{\sqrt{1 + \sum_{k=1}^{n} \left(\dfrac{\partial u}{\partial x_k}\right)^2}} \right) = 0.$$

Conversely, any solution of equation (3) with the boundary condition (1), having summable first derivatives in Ω is also a solution of the variational problem. Moreover, as Hopf has shown (see [15]), the solution is analytic in Ω.

Many theorems have been given ensuring the existence of the minimum for the functional (2) under fairly general hypotheses.

If Ω is convex and $g(x)$ is regular enough, Gilbarg, Hartman, Miranda, Stampacchia (see, respectively, [12], [14], [17], [26]) have given a solution which may be expressed as follows. We say that $g(x)$ satisfies the B.S.C. (bounded slope condition) condition if two vector functions $a(\eta)$ and $b(\eta)$ exist on $\partial\Omega$ in such a way that there is a constant p such that

$$(4) \qquad |a(\eta)| \leq p, \qquad |b(\eta)| \leq p,$$

and

$$(5) \qquad \sum_{i=1}^{n} a_i(\eta)(x_i - \eta_i) \leq g(x) - g(\eta) \leq \sum_{i=1}^{n} b_i(\eta)(x_i - \eta_i)$$

for every x, $\eta \in \partial\Omega$; then, if $g(x)$ satisfies the B.S.C. condition the functional (2) has a minimum which satisfies condition (1) (i.e. there exists a solution of equation (3) with condition (1)); moreover, the solution is uniformly lipschitzian on $\overline{\Omega}$. Hartman (see [13]) has proved that the B.S.C. condition is equivalent to a $(n+1)$-points condition, natural extension of the 3-points condition. Miranda (see [17]) has dealt with the case of a continuous function $g(x)$ on a uniformly convex domain Ω: his formulation of the problem, in addition, is weaker. He has shown that there exists a continuous function on $\overline{\Omega}$, satisfying the boundary condition (1), which minimizes the Lebesgue area of the hypersurface

$$(6) \qquad x_{n+1} = u(x_1, ..., x_n); \qquad (x_1, ..., x_n) \in \overline{\Omega}.$$

His proof makes use of a uniform approximation of the boundary data with functions satisfying the B.S.C. condition. The last result has been sharpened by the author himself (see [18]), who has shown that, in the case $n = 3$, the solution $u(x)$ is analytic on Ω (so it turns out to be a solution of (3) with condition (1)).

It is worth noticing that while the usual methods for uniformly elliptic problems achieve the regularization of solutions which are lipschitzian on Ω, the same methods do not seem to be adequate to deal with continuous functions in the problem of minimizing the Lebesque area: so, some results from the theory of the general Plateau problem had to be used (see [27]).

An extension of Miranda's results to the case $n = 4$ has been obtained by Almgren using some results of [2]. As far as I know, no result is available when $n > 4$.

Quite recently Jenkins and Serrin (see [16]) have considered the case of non convex domains; they have shown that the essential hypothesis on Ω is that it should have the boundary with mean curvature of constant sign.

As to the extension of the Bernstein's theorem, De Giorgi (see [6]) has shown that, if $n = 3$, every function which is solution of (3) on the whole of $\mathbf{R}^n$ is actually a first degree polynomial. The general setting of the problem follows the line of Fleming [10] and uses some results related to the general Plateau problem (see [27]). De Giorgi's result has been extended to the case $n = 4$ by Almgren (see [2]). To my knowledge, the possible extensions of the Bernstein's theorem are an open question.

As to the problem of removable singularities, Finn has proved (see [9]) that if a function solves the equation (3), then it cannot have isolated singularities. Using the same methods, De Giorgi and Stampacchia (see [7]) have extended this result showing that if A is an open set in $\mathbf{R}^n$ and if K is a compact subset of A whose $(n-1)$-dimensional Hausdorff measure is zero, then every function which solves equation (3) on $A \setminus K$ can be analytically extended to the whole of A.

I believe that the problem of removing singularities in the solutions of equation (3) when K is closed in A (not necessarily compact) and of zero $(n-1)$-dimensional measure is still open, e.g. if K is the intersection of a straight line with A and $n > 2$.

I turn now from the Cartesian form of the Plateau problem to the same problem in general form. I should say that the parametric approach to the problem for varieties of dimension higher than 2 seems to be rather difficult. Thus many authors have been led to the formulation of the general Plateau problem in various classes of generalized varieties; among the early papers along these lines I recall those of Caccioppoli (see [3]) and Young (see [28]).

The ideas of Caccioppoli have influenced in turn the work of De Giorgi, Miranda, Triscari. In Miranda's most recent formulation of the problem (see [19]), the study of oriented minimal hypersurfaces has led to the search for the minimum of the integral of $|\operatorname{grad} g(x)|$ where $g(x)$ is a function whose derivatives are measures (in the sense of distribution theory). More precisely, denoting by $\int_K |Dg|$ the total variation on K of the vectorial measure $Dg = (D_1g, \ldots, D_ng)$, gradient of the function $g(x)$, we consider an open set $A \subset \mathbf{R}^n$ and a set $E \subset \mathbf{R}^n$ whose characteristic function $\varphi(x, E)$ has derivatives which are measures. The set E is said to have oriented boundary of minimum measure on A if and only if, for every compact set $K \subset A$, $\varphi(x, E)$ is the minimum of the integral $\int_K |Dg|$ in the class of all functions $g(x)$ whose derivatives are measures and such that, for every $x \in \mathbf{R}^n \setminus K$, $g(x) = \varphi(x, E)$. Then the following regularization theorem holds: the part of the boundary of E contained in A consists of analytic hypersurfaces and of a singular set (which may be empty) whose $(n-1)$-dimensional Hausdorff measure is zero.

More generally, Federer and Fleming (see [8], [10]) have also considered the k-dimensional varieties in $\mathbf{R}^n$ with $0 \le k \le n$. They have considered the k-dimensional currents in $\mathbf{R}^n$ (that is, linear functionals on the space of k-th order differential forms) and they have defined the usual boundary operator as dual of the exterior differential. Moreover, they have defined a norm $M(T)$, called the mass of the current T, which, in the case of regular oriented varieties, coincides with their elementary measure. Rectifiable currents are then defined as the elements of the closure with respect to the norm M of the set of all regular oriented varieties. In this setting, the general Plateau problem is reduced to the search for rectifiable currents with prescribed boundary and minimum mass, and the authors have given existence theorems. In the case $k = n-1$, except on the boundary, a minimal current is decomposable into analytic hypersurfaces and a singular set (which may be vacuous) whose k-dimensional measure is zero.

Finally, Young (see [28]) has considered functions of a more general type than currents: precisely, his generalized varieties V are linear functionals defined on real functions $f(x, y)$, where $x \in \mathbf{R}^n$, y is a k-vector, f is continuous and homogeneous in the sense that $f(x, ty) = tf(x, y)$ for $t \ge 0$, such that, if $f(x, y) \ge 0$ for every (x, y), then $V(f) \ge 0$; then he considers minimizing and extremal varieties for many variational problems and gives a general cone-inequality.

All what has been said so far is related to the concept of oriented variety.

Non-oriented varieties have been considered by Reifenberg (see [23]). He calls k-dimensional variety a non-oriented set, takes as area the k-dimensional Hausdorff measure and defines the notion of "variety with given boundary" through Čech homology. Following a theorem on the existence of the minimum area and some partial results on regularization (see [23]), Reifenberg proves (see [24], [25]) that a minimal variety, with the exception of the boundary, is decomposable into analytic varieties and a singular set (possibly empty) with zero k-dimensional measure.

A somewhat different approach to non-oriented varieties may be found in the papers by Ziemer (see [30]) and Fleming (see [11]). The latter author defines flat chains as the elements of the completion of polyhedral chains with coefficients in a metric finite abelian group, with respect to the Whitney norm

$$W(P) = \inf\{M(Q) + M(R) : \ P = Q + \partial R\}$$

where P, Q are polyhedral chains of dimension k, R is of dimension $k+1$, ∂R is the boundary of R, M is the elementary measure.

Finally, I wish to present briefly Almgren's theory of *varifolds*.

A k-dimensional varifold V is a non-negative functional defined over the space of k-th order differential forms which are continuous and have compact support, such that

(i) $V(r\varphi) = |r|V(\varphi)$, for every real r;

(ii) $V(\varphi + \psi) \le V(\varphi) + V(\psi)$;

(iii) $V(f \wedge \varphi + g \wedge \varphi) = V(f \wedge \varphi) + V(g \wedge \varphi)$, when f and g are non-negative functions.

Almgren gives a solution to the general Plateau problem in the context of varifolds and relates it to solutions obtained in other settings.

The same author also deals with the regularization problem for minimal varieties in the theory of the general Plateau problem, considered from various points of view. In [2] he proves that 3-dimensional minimal surfaces in $\mathbf{R}^4$ are 3-dimensional real analytic submanifolds of $\mathbf{R}^4$, except perhaps at their boundaries, where a minimal surface is meant to be:

(i) an oriented frontier of least 3-dimensional measure, see [5], p. 3;

(ii) a minimal 3-dimensional integral current, see [8], 9.1;

(iii) a minimal flat 3-chain over the group of integers mod 2, see [11];

(iv) a proper minimal surface of Reifenberg (see [23]) with boundary containing a cyclic subgroup of the 2-dimensional Čech homology group with coefficients in the group of integers mod 2 of the boundary set.

From these results the aforementioned regularization theorem for minimal varieties in Cartesian form and the extension of Bernstein's theorem follow.

Scuola Normale Superiore,
Pisa, Italy

Bibliography

[1] ALMGREN, F.J., *The theory of varifolds.* Inst. Adv. Study, Princeton; Brown Univ. (1965).

[2] ALMGREN, F.J., *Some interior regularity theorems for minimal surfaces and an extension of Bernstein's theorem*, Ann. of Math. **84** (1966), 277–292.

[3] R. CACCIOPPOLI, *Misura ed integrazione sugli insiemi dimensionalmente orientati.* Atti Accad. Naz. Lincei Rend. Cl. Sci. Fis. Mat. Natur. (8), **12** (1952), 137–146.

[4] S.S. CHERN, *Minimal surfaces in an Euclidean space of N dimensions*, Symp. on diff. and combin. topology; ed. Cairns, Princeton, 1965.

[5] E. DE GIORGI, *Frontiere orientate di misura minima*, Scuola Normale Superiore, Pisa, 1961 (the results of this paper may be found, with complete proofs, also in paper [19]).

[6] E. DE GIORGI, *Una estensione del teorema di Bernstein*, Ann. Scuola Norm. Sup. Pisa, **19** (1965), 79–85; 463.

[7] E. DE GIORGI, G. STAMPACCHIA, *Sulle singolarità eliminabili delle iper-superficie minimali*, Atti Accad. Naz. Lincei Rend. Cl. Sci. Fis. Mat. Natur. (8), **38** (1965), 352–357.

[8] H. FEDERER, W.H. FLEMING, *Normal and integral currents*, Ann. of Math., **72** (1960), 458–520.

[9] R. FINN, *On partial differential equations (whose solutions admin no isolated singularities)*, Scripta Mathematica, **26** (1963), 107–115.

[10] W.H. FLEMING, *On the oriented Plateau problem*, Rend. Circ. Matem. Palermo, **9** (1962), 69–90.

[11] W.H. FLEMING, *Flat chains over finite coefficient groups*, Trans. Amer. Math. Soc., **121** (1966), 60–86.

[12] D. GILBARG, *Boundary value problems for non-linear elliptic equations in n variables*, Proc. Symp. on nonlinear problems, Univ. Wisconsin, Madison, 1962.

[13] P. HARTMAN, *On the bounded slope condition*, Pacific J. Math., **18** (1966), 495–511.

[14] P. HARTMAN, G. STAMPACCHIA, *On some nonlinear elliptic differential functional equations*, Acta Math., **115** (1966), 271–310.

[15] E. HOPF, *Über den Funktionalen, insbesondere den analytischen Charakter der Lösungen elliptischer Differentialgleichungen zweiter Ordnung*, Math. Zeit., **34** (1932), 194–233.

[16] H. JENKINS, J. SERRIN, *The Dirichlet problem for the minimal surface equation in higher dimensions*, J. Reine Angew. Math., **229** (1968), 170–187.

[17] M. MIRANDA, *Un teorema di esistenza e unicità per il problema dell'area minima in n variabili*, Ann. Scuola Norm. Sup. Pisa, **19** (1965), 233–249.

[18] M. MIRANDA, *Analiticità delle superfici di area minima in* $\mathbf{R}^4$, Atti Accad. Naz. Lincei Rend. Cl. Sci. Fis. Mat. Natur. (8), **38** (1965), 632–638.

[19] M. MIRANDA, *Sul minimo dell'integrale del gradiente di una funzione*, Ann. Scuola Norm. Sup. Pisa, **19** (1965), 627–665.

[20] C.B. MORREY, *Second order elliptic systems of differential equations*, Ann. of Math. Studies, no. 33, Princeton (1954), 101–159.

[21] J.C.C. NITSCHE, *On new results in the theory of minimal surfaces*, Bull. Amer. Math. Soc., (1965), 195–270.

[22] R. OSSERMAN, *Le théorème de Bernstein pour des systèmes*, C. R. Acad. Sci. Paris, **262** (1966), 571–574.

[23] E.R. REIFENBERG, *Solution of the Plateau problem for m-dimensional surfaces of varying topological type*, Acta Mathem., **104** (1960), 1–92.

[24] E.R. REIFENBERG, *An epiperimetric inequality related to the analitycity of minimal surfaces*, Ann. of Math., **80** (1964), 1–14.

[25] E.R. REIFENBERG, *On the analitycity of minimal surfaces*, Ann. of Math., **80** (1964), 15–21.

[26] G. STAMPACCHIA, *On some regular multiple integral problems in the calculus of variations*, Comm. Pure Appl. Math, **16** (1963), 383–421.

[27] D. TRISCARI, *Sulle singolarità delle frontiere orientate di misura minima*, Ann. Scuola Norm. Sup. Pisa 3, **17**, 349–371 (1963).

[28] L.C. YOUNG, *Surfaces paramètriques gènèralisèes*, Bull. Soc. Math. France, **79** (1951), 59–84.

[29] L.C. YOUNG, *Contours in generalized and extremal varieties*, J. Math. and Mech., **11** (1962), 615–646.

[30] W.P. ZIEMER, *Integral currents mod 2*, Trans. Amer. Math. Soc., **105** (1962), 496–524.

Removable singularities of minimal hypersurfaces[‡]

Ennio De Giorgi, Guido Stampacchia[*]

In 1950 L. Bers [1] established that a minimal surface representable in the form $u = u(x, y)$ cannot have isolated singularities. Various proofs of this result and extensions to solutions of more general equations than the minimal area equation where given by R. Finn, J. Nitsche, R. Osserman (see [2] for related references).

In this note we shall prove that *any minimal hypersurface representable in the form $u = u(x_1, x_2, ..., x_n)$ cannot have singularities in a compact set having zero capacity of order 1 or, which is the same, in a compact set having zero $(n-1)$-dimensional Hausdorff measure*[*].

This property of minimal surfaces does not have an analog for solutions of general linear or nonlinear elliptic equations, because no a priori bound on the solution is required here.

The problem of removable singularities for elliptic equations was recently considered by J. Serrin [7].

Let A be an open subset of $\mathbf{R}^n$; a real analytic function $u(x)$ defined in A is a solution of the minimal surfaces equation in A if, setting

$$W[u] = \left(1 + \sum u_{x_i}^2\right)^{1/2}, \qquad W_i[u] = \frac{u_{x_i}}{W[u]}$$

we have[†]:

$$(1) \qquad \int_A W_i[u]\eta_{x_i}\ dx = 0$$

for any infinitely differentiable function η with compact support in A (and hence for any Lipschitz function η with compact support in A).

We remark, in this respect, that $u(x)$ is analytic in A if it is locally Lipschitz in A (see Theorem 3.2).

Let us consider now an open set Ω in $\mathbf{R}^n$ and a compact set N in Ω having zero capacity of order 1 (§1).

The Theorem we are considering says that a solution of (1) in $\Omega \setminus N$ can be extended to a solution of (1) in Ω. The proof consists in showing that the

---

[‡]Editor's note: translation into English of the paper "Sulle singolarità eliminabili delle ipersuperficie minimali", published in Atti Accad. Naz. Lincei Rend. Cl. Sci. Fis. Mat. Nat., **(8) 38** (1965), 352–357.

[*]Note presented by the Socio M. Picone in the meeting of March 13, 1965.

[*]It is a known result [7] that a compact set has zero capacity of order 1 if and only if its $(n-1)$-dimensional Hausdorff measure is zero.

[†]Here and in the sequel we understand implicit summation over repeated indices.

properties of *maximum, minimum* and *comparison* of solutions of (1) in Ω are still valid for solutions of (1) in $\Omega \setminus N$.

1. Compact sets with zero capacity of order 1

Let K be a compact subset of $\mathbf{R}^n$; we shall call capacity of order 1 of K the infimum of

$$\int |v_x|\, dx \qquad \left[|v_x| = \left(\sum_i |v_{x_i}|^2\right)^{\frac{1}{2}}\right]$$

in the class of (infinitely differentiable) Lipschitz functions with compact support and such that $v \geq 1$ on K ([‡]).

LEMMA 1.1 (see [7] p. 285). – *If N is a compact set having zero capacity of order 1, there exists a sequence $\{v^{(s)}\}$ of Lipschitz functions with compact support satisfying the following conditions:*

$$
\begin{array}{lll}
(i) & 0 \leq v^{(s)} \leq 1 & \textit{everywhere} \\
(ii) & v^{(s)} = 1 & \textit{in an open set containing } N \\
(iii) & \lim_{s \to \infty} \|v_x^{(s)}\|_{L^1} = 0 & \\
(iv) & \lim_{s \to \infty} v^{(s)}(x) = 0 & \textit{almost everywhere in } \mathbf{R}^n.
\end{array}
$$

Proof. Let $w^{(s)}$ be a sequence of Lipschitz functions with compact support such that $w^{(s)} \geq 1$ on N and $\|w^{(s)}\|_{L^1} \to 0$.

Hence, the function $2w^{(s)}$ is > 1 in an open set containing N and the function $v^{(s)}$, obtained by truncating $2w^{(s)}$ from below to zero and from above to one, satisfies conditions (i), (ii), (iii). Since $v^{(s)}$ has compact support, we have (cf. for example [6])

$$\|v^{(s)}\|_{L^{n/(n-1)}} \leq C\|v_x^{(s)}\|_{L^1},$$

where C is a constant depending only on n. We deduce that $v^{(s)}$ goes to zero in measure. Therefore, taking a subsequence of $\{v^{(s)}\}$, which we shall still denote by $\{v^{(s)}\}$, we have condition (iv).

COROLLARY (see [7]). – *A compact set N having zero capacity of order 1 has also zero n-dimensional Lebesgue measure.*

Indeed N is contained in the set where (iv) is not satisfied.

LEMMA 1.2. – *If N_1 and N_2 are two compact sets having zero capacity of order 1, then $N_1 \cup N_2$ has zero capacity of order 1.*

Proof. Let $\{v_i^{(s)}\}$, $i = 1,2$, be two sequences of Lipschitz functions with compact support such that

$$v_i^{(s)} \geq 1 \quad \text{in } N_i \qquad (s = 1,2,...)$$

$$\|(v_i^{(s)})_x\|_{L^1} \to 0.$$

[‡]Here the notion of capacity has, like in [7], a different meaning with respect to what is usually assumed in potential theory.

The sequence $v^{(s)} = \max\{v_1^{(s)}, v_2^{(s)}\}$ of Lipschitz functions with compact support is such that

$$v^{(s)} \geq 1 \quad \text{on } N_1 \cup N_2$$

and

$$\|v_x^{(s)}\| \leq \|(v_1^{(s)})_x\|_{L^1} + \|(v_2^{(s)})_x\|_{L^1} \to 0.$$

LEMMA 1.3 – *If N is a compact set with zero capacity of order 1 contained in a connected open set Ω, then $\Omega \setminus N$ is connected.*

Proof. If this were not the case we could find two disjoint open sets A_1 and A_2 such that $A_1 \cup A_2 = \Omega \setminus N$ and $\partial A_1 \cap \partial A_2 \subset N$ (§).

We shall then consider a sequence $\{v^{(s)}\}$ like in Lemma 1.1 and construct a new sequence $\{w^{(s)}\}$ obtained by setting, for example:

$$\begin{aligned} w^{(s)} &= v^{(s)} && \text{in } A_2 \\ w^{(s)} &= 1 && \text{in } A_1 \cup N. \end{aligned}$$

Since $\|w^{(s)}\|_{L^1(\Omega)} \leq \|v^{(s)}\|_{L^1}$, it would follow that either A_1 or A_2 have zero measure, which is a contradiction since they are open sets.

2. Maximum property

We shall prove the following lemmas which generalize some results of [8, §4] and [5].

LEMMA 2.1. – *Let $u(x)$ be a solution of (1) in $\Omega \setminus N$ where N is a compact set contained in Ω, with zero capacity of order 1. Then, if Ω' is a bounded open set such that $N \subset \Omega'$ and $\overline{\Omega}' \subset \Omega$, we have:*

$$\sup_{\Omega' \setminus N} |u| \leq \max_{\partial \Omega'} |u|..$$

Proof. We set $\Phi = \max_{\partial \Omega'} |u|$ and let $\theta(t)$ be an infinitely differentiable function defined on $(-\infty, +\infty)$ such that

$$\begin{aligned} \theta(t) &= 0 && \text{if } |t| \leq \Phi, \\ \theta(t) &\neq 0 && \text{if } |t| > \Phi, \\ |\theta(t)| &\leq 1, && 0 \leq \theta'(t) \leq 1. \end{aligned}$$

We then set in (1)

$$\eta = \begin{cases} (1 - v^{(s)})\theta(u) & x \in \Omega' \\ \\ 0 & x \in \Omega \setminus \Omega', \end{cases}$$

---

§With $\partial \Omega$ we shall denote the boundary of Ω and with $\overline{\Omega}$ the closure of Ω.

where $v^{(s)}$ is the sequence of Lemma 1.1. This is possible because η is a Lipschitz function in Ω whose support is contained in $\Omega \setminus N$. We then have

$$\int_{\Omega'} [1 - v^{(s)}] \theta'(u) W_i[u] u_{x_i} \, dx - \int_{\Omega'} v^{(s)}_{x_i} W_i[u] \theta(u) \, dx = 0$$

whence

$$\int_{\Omega'} [1 - v^{(s)}] \theta'(u) W_i[u] u_{x_i} \, dx \leq \int_{\mathbf{R}^n} |v^{(s)}_x| \, dx.$$

Since the integrand of the left-hand side is nonnegative we have

$$0 \leq \int_{\Omega'} [1 - v^{(s)}] \theta'(u) W_i[u] u_{x_i} \, dx \leq \int_{\mathbf{R}^n} |v^{(s)}_x| \, dx.$$

Taking the limit as $s \to +\infty$, using Fatou lemma and (iii), (iv) of Lemma 1.1, we have

$$\int_{\Omega'} \theta'(u) W_i[u] u_{x_i} \, dx = 0..$$

Taking into account Lemma 1.3 applied to each connected component of Ω', it follows that $|u| \leq \Phi$ almost everywhere in Ω'.

LEMMA 2.2. – *Let $u(x)$ and $z(x)$ be two solutions of (1) in $\Omega \setminus N$, where N is a compact set contained in Ω with zero capacity of order 1. Let Ω' be a bounded open set such that $N \subset \Omega'$ and $\overline{\Omega}' \subset \Omega$; then if $u(x) \leq z(x)$ on $\partial \Omega'$, it follows that*

$$u(x) \leq z(x)$$

almost everywhere in Ω'.

Proof. Let $v^{(s)}$ be the sequence of Lemma 1.1 and set

$$\eta = \begin{cases} (1 - v^{(s)})[\min(u, z) - u] & \text{if } x \in \Omega' \\ 0 & \text{if } x \in \Omega \setminus \Omega'. \end{cases}$$

The function η is Lipschitz with support contained in $\Omega \setminus N$ and hence it can be used in equation (1) and in the corresponding one for the solution $z(x)$.

Denoting by B the subset of Ω' where $u(x) \geq z(x)$ we then have

$$\int_B (1 - v^{(s)}) W_i[u](z_{x_i} - u_{x_i}) \, dx + \int_B W_i[u](z - u) v^{(s)}_{x_i} \, dx = 0$$

$$\int_B (1 - v^{(s)}) W_i[z](z_{x_i} - u_{x_i}) \, dx + \int_B W_i[z](z - u) v^{(s)}_{x_i} \, dx = 0.$$

Subtracting the first equation from the second one and using Lemma 2.1, we have

$$\int_B (1 - v^{(s)})(W_i[z] - W_i[u])(z_{x_i} - u_{x_i}) \, dx \leq \left(\max_{\partial \Omega'} |z| + \max_{\partial \Omega'} |u| \right) \|v^{(s)}_x\|_{L^1}.$$

The function $(W_i[z] - W_i[u])(z_{x_i} - u_{x_i})$ is nonnegative and it is zero only if $z_{x_i} = u_{x_i}$ $(i = 1, ..., n)$.

Reasoning as in the proof of the previous Lemma we find that

$$\int_B (W_i[z] - W_i[u])(z_{x_i} - u_{x_i})\, dx = 0,$$

whence it follows that the derivatives of the function $\min(u, z) - u$ vanish a.e. in Ω' and hence, by Lemma 1.3, we have $u(x) \leq z(x)$ a.e. in Ω'. The Lemma is thus proved.

COROLLARY. – *With the same assumptions of Lemma 2.2 we have*

(3)
$$\sup_{\Omega' \setminus N} |u - z| \leq \max_{\partial \Omega'} |u - z|.$$

Proof. Indeed, by setting $\Phi = \max_{\partial \Omega'} |u - z|$, the functions $z + \Phi$ and $z - \Phi$ are solutions of (1) in $\Omega \setminus N$. Then, since

$$z - \Phi \leq u \leq z + \Phi \qquad \text{on } \partial \Omega',$$

by the previous Lemma it follows that

$$z - \Phi \leq u \leq z + \Phi$$

almost everywhere in Ω', which is (3).

3. Removable singularities

Our aim is to prove the result announced in the introduction, and precisely we shall prove the following

THEOREM 3.1. – *Let $u(x)$ be a solution of (1) in $\Omega \setminus N$ where N is a compact set contained in Ω, with zero capacity of order 1. Then $u(x)$ is analytic in Ω.*

We shall first prove the following

LEMMA 3.1. – *Under the same hypotheses of Theorem 3.1, the function $u(x)$ can be extended to a locally Lipschitz function in Ω.*

Proof. Let Ω' be a bounded open set such that $N \subset \Omega'$ and $\overline{\Omega}' \subset \Omega$; fix d such that the set
$$E = \{x \in \Omega \mid \operatorname{dist}(x, \partial \Omega') \leq d\}$$
does not intersect $N \cup \partial \Omega$.

Now set
$$M = \max_E |u_x|.$$

Let τ be an arbitrary vector such that $|\tau| < d$. The function $g(x) = u(x + \tau)$ is a solution of (1) in $\Omega' \setminus N_\tau$ where N_τ denotes the set N translated by τ.

The set $N \cup N_\tau$ has zero capacity of order 1 (Lemma 1.2). By the Corollary of Lemma 2.2, we have

(4)
$$\sup_{x,\, x+\tau \in \Omega' \setminus N} |u(x + \tau) - u(x)| \leq \max_{y \in \partial \Omega'} |u(y + \tau) - u(y)| \leq M|\tau|.$$

The function $u(x)$ is then Lipschitz in $\Omega' \setminus N$ and hence it can be extended to a Lipschitz function in Ω' with the same Lipschitz constant.

Proof of Theorem 3.1. The Theorem follows from Lemma 3.1 and from the following Theorem.

THEOREM 3.2. – *A locally Lipschitz solution of (1) in an open set Ω is analytic in the interior of Ω.*

Proof. Let Ω' be a bounded open set such that $\overline{\Omega}' \subset \Omega$ and let K be the maximum of the first derivatives of the solution u in Ω'.

Let

$$g(t) = \sqrt{1 + t^2} + c\alpha(t)t^2$$

where $\alpha(t)$ is an infinitely differentiable function for $0 \leq t < +\infty$ such that $\alpha(t) \equiv 0$ for $t \leq 2K$, $\alpha(t) \equiv 1$ for $t \geq 3K$. It is easy to show that one can choose a positive number c in such a way that

$$f(p) = g(|p|) = \sqrt{1 + p^2} \qquad \text{for } |p| \leq K$$

where $p = (p_1, ..., p_n)$ and

$$(5) \qquad m|\xi|^2 \leq f_{p_i p_j}(p)\xi_i\xi_j \leq M|\xi|^2 \qquad 0 < m < M < +\infty.$$

The solution u of (1) is then also a solution of the equation

$$\int_{\Omega'} f_{p_i}(u_x)\eta_{x_i} \, dx = 0$$

for any infinitely differentiable function η with support in Ω', where f satisfies (5). Whence it follows that u is analytic [3].

Bibliography

[1] L. BERS, *Isolated singularities of minimal surfaces.* Ann. of Math. **53**, 364–386 (1951).

[2] R. COURANT, D. HILBERT, *Methods of Mathematical Physics, II.* Interscience Publ., New York, 1962.

[3] E. DE GIORGI, *Sulla differenziabilità e l'analiticità delle estremali degli integrali multipli regolari.* Mem. Accad. Ser. Torino **143**, 25–43 (1957).

[4] W. H. FLEMING, *Functions whose partial derivatives are measures.* Illinois Journal of Mathematics **4**, 452–478 (1960) (theorem 4.3, page 457).

[5] M. MIRANDA, *Un teorema di esistenza e unicità per il problema dell'area minima in n variabili.* (to appear) ‡

---

‡Editor's note: Ann. Scuola Norm. Sup. Pisa **19**, 233–249 (1965).

[6] L. Niremberg, *On elliptic partial differential equations.* Ann. Scuola Norm. Sup. Pisa **13**, 115–162 (1959) (lecture II).

[7] J. Serrin, *Local behavior of solutions of quasi-linear equations.* Acta Mathematica **111**, 247–302 (1964).

[8] G. Stampacchia, *On some regular multiple integral problems in the Calculus of variations.* Comm. pure and applied Math. **XVI**, 383–421 (1963).

Added in proof. - Prof. Serrin kindly told us that Theorem 3.1 in the case of two variables has been proved by J. Nitsche in a paper to be published soon in the *Bulletin of the American Mathematical Society* ‡

---

‡Editor's note: J. C. C. Nitsche *On new results in the theory of minimal surfaces.* Bull. Amer. Math. Soc. 71 (1965).

An example of discontinuous extremals for a variational problem of elliptic type[‡]

Ennio De Giorgi (Pisa)[*]

Summary. We provide an example of discontinuous extremals for an integral whose integrand function is a positive definite quadratic form with respect to the derivatives of a vector-valued function.

It is known (see e.g. [1], [2]) that the extremals of the integral

$$(1) \qquad \int \left[\left(\sum_{h=1}^{n} b_h(x) \frac{\partial w}{\partial x_h} \right)^2 + \sum_{h=1}^{n} \left(\frac{\partial w}{\partial x_h} \right)^2 \right] dx$$

(where $b_h(x)$ are measurable bounded functions) are continuous, in fact Hölder continuous.

This note contains a simple example showing that such a result cannot be in general extended to the case where the real function $w(x)$ is replaced by a vector-valued function $u(x) \equiv (u_1(x), \ldots, u_n(x))$ with n real components, and the integral (1) is replaced by the integral

$$(2) \qquad \int \left[\left(\sum_{h,k}^{1,n} b_{hk}(x) \frac{\partial u_h}{\partial x_k} \right)^2 + \sum_{h,k}^{1,n} \left(\frac{\partial u_h}{\partial x_k} \right)^2 \right] dx.$$

I wish to thank MARIO MIRANDA for his helpful collaboration.

1. - We denote by $\mathbf{R}^n$ the n-dimensional Euclidean space (with $n \geq 3$), by $x \equiv (x_1, \ldots, x_n)$ the generic point of this space, by $|x|$ the distance from x to the origin of $\mathbf{R}^n$, by $u(x) \equiv (u_1(x), \ldots, u_n(x))$ a vector-valued function with n real entries defined in $\mathbf{R}^n$.

We shall say that $u(x)$ is an *extremal* for the integral (2) if it is measurable, if it has square-integrable first derivatives in every compact set of $\mathbf{R}^n$, and if, for every function $g(x) \equiv (g_1(x), \ldots, g_n(x))$, continuous in $\mathbf{R}^n$ with all partial

---

[‡]Editor's note: translation into English of the paper "Un esempio di estremali discontinue per un problema variazionale di tipo ellittico", published in Boll. Un. Mat. Ital., (4) 1 (1968), 135–137.

[*]Supported in part by Grant Af-Eoar 67-38

derivatives of every order and having compact support K, there holds

$$(3) \qquad J(u,g) = \int_K \left\{ \left[\sum_{h,k}^{1,n} b_{hk} \frac{\partial u_h}{\partial x_k} \right] \left[\sum_{h,k}^{1,n} b_{hk} \frac{\partial g_h}{\partial x_k} \right] + \sum_{h,k}^{1,n} \frac{\partial u_h}{\partial x_k} \frac{\partial g_h}{\partial x_k} \right\} dx = 0.$$

Our example is a particular case of the integral (2), namely

$$(4) \qquad \int \left\{ \left[(n-2) \sum_{h=1}^{n} \frac{\partial u_h}{\partial x_h} + n \sum_{h,k}^{1,n} \frac{x_h x_k}{|x|^2} \frac{\partial u_h}{\partial x_k} \right]^2 + \sum_{h,k}^{1,n} \left(\frac{\partial u_h}{\partial x_k} \right)^2 \right\} dx.$$

The EULER equation relative to the integral (4) is given by

$$(5) \qquad \begin{aligned} &(n-2) \frac{\partial}{\partial x_h} \left[(n-2) \sum_{t=1}^{n} \frac{\partial u_t}{\partial x_t} + n \sum_{s,t}^{1,n} \frac{x_t x_s}{|x|^2} \frac{\partial u_t}{\partial x_s} \right] + \\ &+ n \sum_{k=1}^{n} \frac{\partial}{\partial x_k} \left[\frac{x_h x_k}{|x|^2} \left((n-2) \sum_{t=1}^{n} \frac{\partial u_t}{\partial x_t} + n \sum_{s,t}^{1,n} \frac{x_t x_s}{|x|^2} \frac{\partial u_t}{\partial x_s} \right) \right] + \\ &+ \triangle_2 u_h = 0 \quad \text{(for } h = 1, 2, \ldots, n). \end{aligned}$$

We have the following

LEMMA – *If $u(x) \equiv (u_1(x), \ldots, u_n(x))$ is a vector-valued function with n real components, continuous with all of its derivatives when $|x| \neq 0$, whose first derivatives are square-integrable on every compact set of $\mathbf{R}^n$, and which satisfies, for $|x| \neq 0$, the Euler equations (5), then $u(x)$ is an extremal for the integral (4).*

Proof. Let $g(x) \equiv (g_1(x), \ldots, g_n(x))$ be a vector-valued function defined in $\mathbf{R}^n$, continuous with all of its derivatives and having bounded support, let $\alpha(x)$ be a real function, infinitely differentiable and with compact support; moreover, let $\alpha(x) = 1$ in a whole neighborhood of the origin of $\mathbf{R}^n$. Define, for every positive number t,

$$(6) \qquad g_t^*(x) = g(x)[1 - \alpha(tx)].$$

From the EULER equations it follows that

$$(7) \qquad J(u, g_t^*) = 0,$$

whereas, by the assumption that $n \geq 3$, we have

$$(8) \qquad \lim_{t \to +\infty} \int_{\mathbf{R}^n} \sum_{h=1}^{n} \left| \frac{\partial g_t^*(x)}{\partial x_h} - \frac{\partial g(x)}{\partial x_h} \right|^2 dx = 0;$$

from (7), (8) the claim follows.

THEOREM – *Letting*

$$\alpha = -\frac{n}{2}\left[1 - \frac{1}{\sqrt{(2n-2)^2+1}}\right] \tag{9}$$

the function $u(x) = x|x|^\alpha = (x_1|x|^\alpha, \ldots, x_n|x|^\alpha)$ *is an extremal for the integral (4), nevertheless it is not continuous at the origin of* $\mathbf{R}^n$ *(not even bounded).*

Proof. By elementary computations one finds

$$\triangle_2(x_h|x|^\alpha) = (\alpha n + \alpha^2)x_h|x|^{\alpha-2},$$

$$\sum_{t=1}^{n}\frac{\partial}{\partial x_t}(x_t|x|^\alpha) = (n+\alpha)|x|^\alpha,$$

$$\sum_{s,t}^{1,n}\frac{x_t x_s}{|x|^2}\frac{\partial}{\partial x_s}(x_t|x|^\alpha) = (\alpha+1)|x|^\alpha,$$

$$\frac{\partial}{\partial x_h}|x|^\alpha = \alpha x_h|x|^{\alpha-2},$$

$$\sum_{k=1}^{n}\frac{\partial}{\partial x_k}(x_h x_k|x|^{\alpha-2}) = (n+\alpha-1)x_h|x|^{\alpha-2};$$

therefore the EULER equations (5) are satisfied for $|x| \neq 0$ as soon as we let $u(x) = x|x|^\alpha$ and the exponent α satisfies the condition

$$(2n-2)^2\left(\alpha+\frac{n}{2}\right)^2 + (\alpha n + \alpha^2) = 0. \tag{10}$$

Recalling the previous Lemma and the assumption that $n \geq 3$, the claim follows.

REMARK – *From the above theorem we immediately deduce that the function* $f(x) = |x|^{\alpha+2}$, *with* α *given by (9), is an extremal for the integral*

$$\int\left[\left((n-2)\triangle_2 f + n\sum_{h,k}^{1,n}\frac{x_h x_k}{|x|^2}\frac{\partial^2 f}{\partial x_h \partial x_k}\right)^2 + \sum_{h,k}^{1,n}\left(\frac{\partial^2 f}{\partial x_h \partial x_k}\right)^2\right]dx; \tag{11}$$

it is discontinuous at the origin when $n \geq 5$.

Bibliography

[1] E. DE GIORGI, *Sulla differenziabilità e l'analiticità delle estremali degli integrali multipli regolari.* Mem. Acc. Sc. Torino, 1957.

[2] J. NASH, *Continuity of the solutions of parabolic and elliptic equations.* Am. J. Math., 80, 1958.

Un esempio di estremali discontinue per un problema variazionale di tipo ellittico[‡]

Ennio De Giorgi (Pisa)[*]

Sunto. Si dà un esempio di estremali discontinue, per un integrale in cui la funzione integranda è una forma quadratica definita positiva nelle derivate prime di una funzione vettoriale.

È noto (vedi per es. [1], [2]) che le estremali dell'integrale

$$(1) \qquad \int \left[\left(\sum_{h=1}^{n} b_h(x) \frac{\partial w}{\partial x_h} \right)^2 + \sum_{h=1}^{n} \left(\frac{\partial w}{\partial x_h} \right)^2 \right] dx$$

(ove le $b_h(x)$ sono funzioni misurabili e limitate) sono funzioni continue, ed anzi hölderiane.

In questa nota vi è un semplice esempio che mostra come questo risultato non si estende in generale quando alla funzione reale $w(x)$ si sostituisca una funzione vettoriale $u(x) \equiv (u_1(x), \dots, u_n(x))$ ad n componenti reali ed all'integrale (1) l'integrale

$$(2) \qquad \int \left[\left(\sum_{h,k}^{1,n} b_{hk}(x) \frac{\partial u_h}{\partial x_k} \right)^2 + \sum_{h,k}^{1,n} \left(\frac{\partial u_h}{\partial x_k} \right)^2 \right] dx.$$

Ringrazio Mario Miranda la cui collaborazione mi è stata assai utile.

1. - Indicheremo con $\mathbf{R}^n$ lo spazio ad n dimensioni reali (con $n \geq 3$), con $x \equiv (x_1, \dots, x_n)$ il generico punto di tale spazio, con $|x|$ la distanza di x dall'origine di $\mathbf{R}^n$, con $u(x) \equiv (u_1(x), \dots, u_n(x))$ una funzione vettoriale ad n componenti reali definita in $\mathbf{R}^n$.

Diremo che $u(x)$ è *estremale* dell'integrale (2) se è misurabile, ha derivate prime di quadrato sommabile in ogni compatto di $\mathbf{R}^n$ e se per ogni funzione $g(x) \equiv (g_1(x), \dots, g_n(x))$ continua in $\mathbf{R}^n$ insieme alle sue derivate parziali di

---

[‡]Editor's note: published in Boll. Un. Mat. Ital., (4) 1 (1968), 135–137.

[*]Supported in part by Grant Af-Eoar 67-38

ogni ordine ed a supporto compatto K risulta

$$(3) \qquad \mathcal{J}(u,g) = \int_K \left\{ \left[\sum_{h,k}^{1,n} b_{hk} \frac{\partial u_h}{\partial x_k} \right] \left[\sum_{h,k}^{1,n} b_{hk} \frac{\partial g_h}{\partial x_k} \right] + \sum_{h,k}^{1,n} \frac{\partial u_h}{\partial x_k} \frac{\partial g_h}{\partial x_k} \right\} dx = 0.$$

L'esempio esposto riguarda un caso particolare dell'integrale (2), precisamente l'integrale

$$(4) \qquad \int \left\{ \left[(n-2) \sum_{h=1}^{n} \frac{\partial u_h}{\partial x_h} + n \sum_{h,k}^{1,n} \frac{x_h x_k}{|x|^2} \frac{\partial u_h}{\partial x_k} \right]^2 + \sum_{h,k}^{1,n} \left(\frac{\partial u_h}{\partial x_k} \right)^2 \right\} dx.$$

L'equazione di EULERO relativa all'integrale (4) è

$$(n-2)\frac{\partial}{\partial x_h} \left[(n-2) \sum_{t=1}^{n} \frac{\partial u_t}{\partial x_t} + n \sum_{s,t}^{1,n} \frac{x_t x_s}{|x|^2} \frac{\partial u_t}{\partial x_s} \right] +$$

$$(5) \qquad +n \sum_{k=1}^{n} \frac{\partial}{\partial x_k} \left[\frac{x_h x_k}{|x|^2} \left((n-2) \sum_{t=1}^{n} \frac{\partial u_t}{\partial x_t} + n \sum_{s,t}^{1,n} \frac{x_t x_s}{|x|^2} \frac{\partial u_t}{\partial x_s} \right) \right] +$$

$$+\triangle_2 u_h = 0 \quad (\text{per } h = 1, 2, \ldots, n).$$

Vale il

LEMMA – *Se $u(x) \equiv (u_1(x), \ldots, u_n(x))$ è una funzione vettoriale ad n componenti reali, continua con tutte le sue derivate per $|x| \neq 0$, dotata di derivate prime di quadrato sommabile su ogni compatto di $\mathbf{R}^n$, e soddisfacente, per $|x| \neq 0$ le equazioni di Eulero (5), allora $u(x)$ è estremale dell'integrale (4).*

Dim. Sia $g(x) \equiv (g_1(x), \ldots, g_n(x))$ una funzione vettoriale definita in $\mathbf{R}^n$, continua insieme a tutte le sue derivate ed a supporto compatto, $\alpha(x)$ una funzione reale pure infinitamente derivabile ed a supporto compatto; sia inoltre $\alpha(x) = 1$ in tutto un intorno dell'origine di $\mathbf{R}^n$. Poniamo allora, per ogni numero positivo t,

$$(6) \qquad g_t^*(x) = g(x)[1 - \alpha(tx)].$$

Dalle equazioni di EULERO segue

$$(7) \qquad \mathcal{J}(u, g_t^*) = 0,$$

mentre, per l'ipotesi $n \geq 3$, si ha

$$(8) \qquad \lim_{t \to +\infty} \int_{\mathbf{R}^n} \sum_{h=1}^{n} \left| \frac{\partial g_t^*(x)}{\partial x_h} - \frac{\partial g(x)}{\partial x_h} \right|^2 dx = 0;$$

dalle (7), (8) segue la tesi.

TEOREMA – *Posto*

$$\alpha = -\frac{n}{2}\left[1 - \frac{1}{\sqrt{(2n-2)^2 + 1}}\right] \tag{9}$$

la funzione $u(x) = x|x|^\alpha = (x_1|x|^\alpha, \ldots, x_n|x|^\alpha)$ *è estremale dell'integrale (4) e non è continua nell'origine di* $\mathbf{R}^n$ *(nè limitata).*

Dim. Con facili calcoli si trova

$$\triangle_2(x_h|x|^\alpha) = (\alpha n + \alpha^2)x_h|x|^{\alpha-2},$$

$$\sum_{t=1}^{n} \frac{\partial}{\partial x_t}(x_t|x|^\alpha) = (n+\alpha)|x|^\alpha,$$

$$\sum_{s,t}^{1,n} \frac{x_t x_s}{|x|^2} \frac{\partial}{\partial x_s}(x_t|x|^\alpha) = (\alpha+1)|x|^\alpha,$$

$$\frac{\partial}{\partial x_h}|x|^\alpha = \alpha x_h|x|^{\alpha-2},$$

$$\sum_{k=1}^{n} \frac{\partial}{\partial x_k}(x_h x_k|x|^{\alpha-2}) = (n+\alpha-1)x_h|x|^{\alpha-2};$$

quindi le equazioni di EULERO (5) sono soddisfatte per $|x| \neq 0$ non appena si ponga $u(x) = x|x|^\alpha$ e l'esponente α verifichi l'equazione

$$(2n-2)^2\left(\alpha + \frac{n}{2}\right)^2 + (\alpha n + \alpha^2) = 0. \tag{10}$$

Ricordando il Lemma e l'ipotesi iniziale $n \geq 3$ segue la tesi.

OSSERVAZIONE – *Dal teorema ora dimostrato si deduce subito che la funzione* $f(x) = |x|^{\alpha+2}$, *con* α *dato dalla (9), è estremale dell'integrale*

$$\int\left[\left((n-2)\triangle_2 f + n\sum_{h,k}^{1,n}\frac{x_h x_k}{|x|^2}\frac{\partial^2 f}{\partial x_h \partial x_k}\right)^2 + \sum_{h,k}^{1,n}\left(\frac{\partial^2 f}{\partial x_h \partial x_k}\right)^2\right]\,dx; \tag{11}$$

essa è discontinua nell'origine per $n \geq 5$.

Bibliografia

[1] E. DE GIORGI, *Sulla differenziabilità e l'analiticità delle estremali degli integrali multipli regolari.* Mem. Acc. Sc. Torino, 1957.

[2] J. NASH, *Continuity of the solutions of parabolic and elliptic equations.* Am. J. Math., 80, 1958.

Minimal Cones and
the Bernstein Problem[‡]

E. Bombieri, E. De Giorgi, E. Giusti

I. Introduction

The main purpose of this paper is to prove that there exist complete analytic minimal graphs of sufficiently large dimension, which are not hyperplanes.

In [8], FLEMING gave a new proof of BERNSTEIN's theorem that a complete minimal graph in $\mathbf{R}^3$ must be a plane, by pointing out that the falsity of BERNSTEIN's theorem would imply the existence of a minimal cone in $\mathbf{R}^3$, a contradiction.

The next step forward was taken by DE GIORGI [6], who improved FLEMING's argument by showing that the falsity of the extension of the BERNSTEIN theorem in $\mathbf{R}^n$ would imply the existence of a minimal cone in $\mathbf{R}^{n-1}$, thus extending the BERNSTEIN theorem to complete minimal graphs in $\mathbf{R}^4$.

In [1] ALMGREN proved the non-existence of minimal cones in $\mathbf{R}^4$, and in his recent paper SIMONS [12] extended ALMGREN's theorem up to $\mathbf{R}^7$. From these results it follows, among other things, an interior regularity theorem for the solution of the codimension one PLATEAU problem in $\mathbf{R}^n$ for $n \leq 7$, and the extension of BERNSTEIN's theorem through dimension 8.

In the same paper, SIMONS gave examples of locally stable cones of codimension one in $\mathbf{R}^{2m}$, for each $m \geq 4$, and raised the question whether these cones were global minima of the area function.

In this paper, we shall prove also that these cones are of least area in the large.

SIMONS' example is as follows.

Let $S_m(r) \subset \mathbf{R}^m$ be the sphere

$$x_1^2 + x_2^2 + \cdots + x_m^2 = r^2$$

and let $C_{2m}(r)$ be the truncated cone

$$x_1^2 + x_2^2 + \cdots + x_m^2 = x_{m+1}^2 + x_{m+2}^2 + \cdots + x_{2m}^2 < r^2$$

of codimension one in $\mathbf{R}^{2m}$ and with boundary $S_m(r) \times S_m(r) \subset S_{2m}(\sqrt{2}r)$. It can be proved that $C_{2m}(r)$ has mean curvature zero at every point except at the origin, which is singular.

[‡]Editor's note: published in Invent. Math., **7** (1969), 243–268.

Let $\mu_\epsilon : \mathbf{R}^{2m} \to \mathbf{R}^{2m}$ be a Lipschitz map which reduces to the identity on $S_{2m}(\sqrt{2}r)$ and such that

$$\|\mu_\epsilon - I\|_{\mathrm{lip}} < \epsilon$$

where I is the identity map. Then Simons proves that if $m \geq 4$ there exists $\epsilon_0(m) > 0$ such that for $0 < \epsilon < \epsilon_0(m)$:

$$\mathrm{Area}(C_{2m}(r)) < \mathrm{Area}(\mu_\epsilon(C_{2m}(r))),$$

unless $\mu_\epsilon = I$.

We shall work in the frame of oriented boundaries of least area in the sense of De Giorgi [5] and Miranda [9]; our approach can be shown to be equivalent to that of Federer and Fleming [7] by means of integral currents.

Let E be a set in $\mathbf{R}^n$ and let φ_E be its characteristic function. The essential boundary $\mathcal{F}_e E$ of E is the set

$$\mathcal{F}_e E = \left\{ y \in \mathbf{R}^n \,\Big|\, 0 < \liminf_{r \to 0} r^{-n} \int_{B(y,r)} \varphi_E \, dx \leq \limsup_{r \to 0} r^{-n} \int_{B(y,r)} \varphi_E \, dx < \omega_n \right\}$$

where $B(y,r)$ is the open ball of radius r, centered at y and ω_n is the measure of $B(0,1)$.

Let $A \subset \mathbf{R}^n$ be an open set; then E has an oriented boundary of least area with respect to A, if:

(i) $\varphi_E \in BV_{\mathrm{loc}}(A)$,

(ii) for each $g \in BV_{\mathrm{loc}}(A)$ with compact support $K \subset A$ we have

$$\int_K |D\varphi_E| \leq \int_K |D(\varphi_E + g)|$$

where, for $f \in BV_{\mathrm{loc}}(A)$, $\int_K |Df|$ is the total variation over K of the vector-valued measure Df (gradient of f in the sense of distribution theory).

If $A = \mathbf{R}^n$ we shall say that E has an oriented boundary of least area.

It is shown (De Giorgi [5], Miranda [9]) that the essential boundary of a set E of oriented boundary of least area is the union of an analytical hypersurface having mean curvature zero at every point and a singular set N of Hausdorff measure $H_{n-1}(N) = 0$.

In this paper we prove the following result:

Theorem A. – *Let $m \geq 4$. The set $E \subset \mathbf{R}^{2m}$ defined by*

$$x_1^2 + x_2^2 + \cdots + x_m^2 \geq x_{m+1}^2 + x_{m+2}^2 + \cdots + x_{2m}^2$$

has an oriented boundary of least area.

Theorem A proves that if $m \geq 4$ the cones $C_{2m}(r)$ are of least area in the large among all hypersurfaces in $\mathbf{R}^{2m}$ having $S_m(r) \times S_m(r)$ as boundary, and thus provides a counterexample to interior regularity for parametric minimal hypersurfaces of dimension 7 or more.

Theorem B. – *In $n \geq 9$ there exists complete minimal graphs in $\mathbf{R}^n$ which are not hyperplanes.*

Theorem B and SIMONS' result give a complete solution to BERNSTEIN's problem in $\mathbf{R}^n$.

Section II contains some auxiliary results needed for the proof of Theorem A and a theorem (Theorem 1) connecting the problem of minimal hypersurfaces with the problem of functions of least gradient.

In Section III, the proof of Theorem A is completed by reducing it to a question about the behaviour of solutions of an ordinary differential equation of the first order; this is solved by means of the classical methods of BENDIXON and POINCARÉ.

In Section IV, we give our proof of Theorem B. It is self-contained and depends on usual tools in the theory of elliptic partial differential equations, including classical results on the Dirichlet problem for the minimal surface equation.

Though the motivation for our choices of subsolutions and supersolutions of the minimal surface equation will be apparent from the results obtained in our proof of Theorem A, we emphasize here that the theory of minimal cones and minimal surfaces is not needed in this section. The reader interested in the more classical aspects of non-linear elliptic partial differential equations may read Section IV of this paper independently of the other two.

II. Functions of least gradient and oriented boundaries of least area

Let $A \subset \mathbf{R}^n$ be an open set and let $f \in BV_{\mathrm{loc}}(A)$. We say that f has least gradient with respect to A if for every $g \in BV_{\mathrm{loc}}(A)$ with compact support $K \subset A$:

$$\int_K |Df| \le \int_K |D(f+g)|.$$

If $A = \mathbf{R}^n$ we say that f has least gradient.

The following lemma is analogous to a result of FLEMING ([8], §5, p. 83).

LEMMA 1. – *Let $A \subset \mathbf{R}^n$ be an open set and let $f \in H^{1,1}_{\mathrm{loc}}(A)$. Suppose that*

(i) $$H_n\left(\{x \in A \mid |\nabla f| = 0\}\right) = 0,$$

where we have written $\nabla f = \mathrm{grad}\, f$; also let N be a closed set in A such that

(ii) $$H_{n-1}(N) = 0,$$

where H_{n-1} denotes the $(n-1)$-dimensional Hausdorff measure in $\mathbf{R}^n$;

(ii) $$\int_{A\setminus N} |\nabla f|^{-1} \sum_{i=1}^{n} \frac{\partial f}{\partial x_i} \frac{\partial \varphi}{\partial x_i}\, dx = 0$$

for every $\varphi \in C_0^1(A \setminus N)$.

Then f has least gradient with respect to A.

Proof. We have to show that for every $g \in BV_{\mathrm{loc}}(A)$ with compact support $K \subset A$ we have

(1) $$\int_K |Df| \le \int_K |D(f+g)|.$$

Let us first suppose that $g \in C_0^1(A)$. Then in the case in which N is empty (1) follows from (iii) and the convexity of $|Df|$.

Now suppose $N \neq \emptyset$. The set $N_K = N \cap K$ is compact and $H_{n-1}(N_K) = 0$; hence for each $\epsilon > 0$ there exists a finite covering of N_K with open balls $B_j = B(x_j, r_j)$, $x_j \in N_K$, $j = 1, 2, ..., M$, $r_1 \leq r_2 \leq \cdots \leq r_M$, such that

$$\sum_{j=1}^{M} r_j^{n-1} < \epsilon$$

Let

$$\alpha_j(x) = \begin{cases} 1 & \text{if } x \notin 2B_j \\ r_j^{-1}|x - x_j| - 1 & \text{if } x \in 2B_j \setminus B_j \\ 0 & \text{if } x \in B_j, \end{cases}$$

and let

$$\alpha(x) = \min_j \alpha_j(x).$$

We have

$$\int_K |Df| \leq \int_K |D(f + \alpha g)|$$

by our previous remark and because $\operatorname{supp}(\alpha g) \subset A \setminus N$.

Now

$$\int_K |D(f + \alpha g)| \leq \int_K |D(f + g)| + \int_K (1 - \alpha)|Dg| + \int_K |g||D\alpha|,$$

$$\int_K (1 - \alpha)|Dg| \leq (\sup_K |\operatorname{grad} g|) \int_K (1 - \alpha) \, dx,$$

$$\int_K |g||D\alpha| \leq (\sup_K |g|)$$

$$\times \left\{ \int_{2B_1} |D\alpha| + \int_{2B_2 \setminus 2B_1} |D\alpha| + \cdots + \int_{2B_M \setminus \cup_{j=1}^{M-1} 2B_j} |D\alpha| \right\}$$

$$\leq (\sup_K |g|) 2^n \omega_n \sum_{j=1}^{M} r_j^{n-1} < \operatorname{cost} \epsilon.$$

Hence

$$\lim_{\epsilon \to 0} \int_K |D(f + \alpha g)| \leq \int_K |D(f + g)|$$

and (1) is proved if $g \in C_0^1(A)$.

In order to get the full result for $g \in BV_{\mathrm{loc}}(A)$, g with compact support, some additional argument is needed.

Let $g \in BV_{\mathrm{loc}}(A)$ with compact support K and let $Dg = G_1 + G_2$ where G_1 is completely continuous and G_2 is the singular part of Dg, with support N_g of measure zero. Then we have:

$$(2) \qquad \int_K |D(f + g)| = \int_K |Df + G_1| + \int_K |G_2|$$

because $f \in H_{\mathrm{loc}}^{1,1}(A)$.

Let $g_\epsilon = g * \psi_\epsilon$, where ψ_ϵ is a mollifier; then $g_\epsilon \in C_0^1(A)$ and

$$\int_{K_\epsilon} |Df| \le \int_{K_\epsilon} |D(f + g_\epsilon)| \le \int_{K_\epsilon} |Df + G_1 * \psi_\epsilon| + \int_A |G_2 * \psi_\epsilon|,$$

where $K_\epsilon = \{x \in A | \operatorname{dist}(x, K) < \epsilon\}$.

Letting $\epsilon \to 0$ our lemma follows from (2).   q.e.d.

LEMMA 2. – *Let $A \subset \mathbf{R}^n$ be an open set and let $S \subset A$ be a closed set in A such that*

$$H_{n-1}(S \cap K) < +\infty$$

for every compact set $K \subset A$.

Let $f \in C^2(A \setminus S)$, let

$$|\nabla f| \ne 0 \qquad \text{for } x \in A \setminus S$$

and suppose that

(i) $$\nu_i(x) = |\nabla f|^{-1} \frac{\partial f}{\partial x_i}, \qquad i = 1, 2, ..., n$$

has a continuous extension to the whole of A;

(ii) $$\sum_{i=1}^n \frac{\partial \nu_i}{\partial x_i} = 0 \qquad \text{on } A \setminus S.$$

Then we have

$$\int_A \sum_{i=1}^n \nu_i \frac{\partial \varphi}{\partial x_i} \, dx = 0$$

for every $\varphi \in C_0^1(A)$.

Proof. Let $\varphi \in C_0^1(A)$, $K = \operatorname{supp}(\varphi)$, $S_K = S \cap K$. The set S_K is compact and $H_{n-1}(S_K) = L < +\infty$, whence for every $r > 0$ there exists a finite covering of S_K with open balls

$$B_j = B(x_j, r_j), \qquad j = 1, 2, ..., M, \quad x_j \in S_K, \quad r \ge r_1 \ge r_2 \ge \cdots \ge r_M,$$

such that

$$\sum_{j=1}^M r_j^{n-1} \le 2L + 1 = L_1.$$

Let

$$\beta_j(x) = \begin{cases} 1 & \text{if } x \in B_j \\ 2 - r_j^{-1}|x - x_j| & \text{if } x \in 2B_j \setminus B_j \\ 0 & \text{if } x \notin 2B_j \end{cases}$$

and let

$$\gamma_1(x) = \beta_1(x),$$

$$\gamma_j(x) = \max\{\beta_j - \max(\beta_1, \beta_2, ..., \beta_{j-1}); 0\}, \qquad 2 \leq j \leq M.$$

Then we have

$$\operatorname{supp}(\gamma_j) = 2B_j$$

$$\gamma(x) = \sum_{j=1}^{M} \gamma_j(x) = \max_j \beta_j(x),$$

and in particular

$$\gamma(x) = 1 \qquad \text{if } x \in \bigcup_j B_j,$$

$$\gamma(x) = 0 \qquad \text{if } x \notin \bigcup_j 2B_j;$$

lastly

$$|\nabla \gamma_j| \leq 2/r_j.$$

We have

$$\int_A \sum_{i=1}^{n} \nu_i \frac{\partial \varphi}{\partial x_i} \, dx = \int_A \sum_{i=1}^{n} \nu_i \frac{\partial [(1-\gamma)\varphi]}{\partial x_i} \, dx + \int_A \sum_{i=1}^{n} \nu_i \frac{\partial (\gamma\varphi)}{\partial x_i} \, dx;$$

the first integral in the right hand side of this equation is 0, because $\operatorname{supp}((1-\gamma)\varphi) \subset A \setminus S$ and because of condition (iii) of Lemma 2. Hence

$$(3) \qquad \int_A \sum_{i=1}^{n} \nu_i \frac{\partial \varphi}{\partial x_i} \, dx = \int_A \gamma \sum_{i=1}^{n} \nu_i \frac{\partial \varphi}{\partial x_i} \, dx + \sum_{i=1}^{n} \sum_{j=1}^{M} \int_A \varphi \nu_i \frac{\partial \gamma_j}{\partial x_i} \, dx.$$

We have

$$(4) \qquad \left| \int_A \gamma \sum_{i=1}^{n} \nu_i \frac{\partial \varphi}{\partial x_i} \, dx \right| \leq \max_K(|\nu||\nabla\varphi|) \int_A \gamma \, dx \leq \operatorname{cost} \sum_{j=1}^{M} r_j^n < \operatorname{cost} L_1 r.$$

Also

$$(5) \qquad \left| \int_A \varphi \nu_i \frac{\partial \gamma_j}{\partial x_i} \, dx \right| = \left| \int_{2B_j} (\varphi(x)\nu_i(x) - \varphi(x_j)\nu_i(x_j)) \frac{\partial \gamma_j}{\partial x_i} \, dx \right|.$$

The function $\varphi \nu_i$ is uniformly continuous in K and for every $\epsilon > 0$ there exists $r > 0$ such that if $|x - y| < 2r$ then

$$|\varphi(x)\nu_i(x) - \varphi(y)\nu_i(y)| < \epsilon.$$

From this remark and from (5) and the inequality $|\nabla \gamma_j| \leq 2/r_j$ we get

$$(6) \qquad \sum_{i=1}^{n} \sum_{j=1}^{M} \left| \int_A \varphi \nu_i \frac{\partial \gamma_j}{\partial x_i} \, dx \right| \leq \operatorname{cost} \epsilon \sum_{j=1}^{M} r_j^{n-1} < \operatorname{cost} L_1 \epsilon.$$

The result of Lemma 2 now follows from inequalities (3), (4), (6) and letting $r, \epsilon \to 0$.   q.e.d.

The next result gives the connection between functions of least gradient and sets with an oriented boundary of least area.

THEOREM 1. – *Let f be a function of least gradient with respect to A and let*

$$E_\lambda = \{x \in A \mid f(x) \geq \lambda\}.$$

Then the set E_λ has an oriented boundary of least area with respect to A.

Proof. Let φ_λ be the characteristic function of E_λ. By Theorem 1.6 of [9] we have $\varphi_\lambda \in BV_{\text{loc}}(A)$ for almost all λ, also

$$(7) \qquad \int_K |Df| = \int_{-\infty}^{\infty} \left(\int_K |D\varphi_\lambda| \right) d\lambda$$

for each compact set $K \subset A$.

Let t be a real constant and let

$$\begin{aligned} f_1 &= \max(f - t, 0), \\ f_2 &= \min(f, t). \end{aligned}$$

Then $f_1, f_2 \in BV_{\text{loc}}(A)$, $f = f_1 + f_2$ and by Eq. (7)

$$\int_K |Df| = \int_K |Df_1| + \int_K |Df_2|.$$

Let $g \in BV_{\text{loc}}(A)$ with compact support K; we have

$$\int_K |Df_1| + \int_K |Df_2| = \int_K |Df| \leq \int_K |D(f + g)| \leq \int_K |D(f_1 + g)| + \int_K |Df_2|$$

which shows that f_1, f_2 have least gradient with respect to A. It follows that the functions

$$\varphi_{\lambda,\epsilon} = \frac{1}{\epsilon} \min\{\epsilon; \max(f - \lambda), 0)\}$$

have least gradient with respect to A for every λ, ϵ.

Now if $H_n(\{x \in A \mid f(x) = \lambda\}) = 0$ we have for each compact set $K \subset A$

$$\lim_{\epsilon \to 0} \int_K |\varphi_{\lambda,\epsilon} - \varphi_\lambda| \, dx = 0$$

whence by Theorem 3 of MIRANDA [11] it follows that $\varphi_\lambda \in BV_{\text{loc}}(A)$ and that φ_λ has least gradient with respect to A.

If $H_n(\{x \in A \mid f(x) = \lambda\}) > 0$, there exists a sequence $\{\lambda_m\}$, such that $\lambda_m < \lambda$, $\lambda_m \to \lambda$, $H_n(\{x \in A \mid f(x) = \lambda_m\}) = 0$ and for each compact set $K \subset A$

$$\lim_{m \to +\infty} \int_K |\varphi_{\lambda_m} - \varphi_\lambda| \, dx = 0,$$

whence by our previous result and Theorem 3 of Miranda, Theorem 1 follows. q.e.d.

III. Proof of Theorem A

In order to prove our Theorem A it will be sufficient, in view of Theorem 1 of the previous section, to exhibit a function f of least gradient (with respect to $\mathbf{R}^{2m}$) and having the cone $x_1^2 + x_2^2 + \cdots + x_m^2 = x_{m+1}^2 + x_{m+2}^2 + \cdots + x_{2m}^2$ as a level hypersurface.

This section contains the construction of such a function f.

Heuristic arguments suggest looking for a function f of least gradient which has the same orthogonal symmetries of the cone. Hence let us define new variables

$$
\begin{aligned}
u &= (x_1^2 + x_2^2 + \cdots + x_m^2)^{\frac{1}{2}}, \\
v &= (x_{m+1}^2 + x_{m+2}^2 + \cdots + x_{2m}^2)^{\frac{1}{2}}
\end{aligned}
$$

and the function $F = F(u, v)$ defined by

$$
F(u, v) = f(x),
$$

and let Ω be an open set of the (u, v)-plane contained in the first quadrant and let $\tilde{\Omega}$ be the corresponding open set in $\mathbf{R}^{2m}$.

If we suppose that

(i)
$$
f \in C^2(\tilde{\Omega}),
$$

(ii)
$$
|\nabla f| \neq 0 \qquad \text{in } \tilde{\Omega}
$$

then the Euler equation for f in $\tilde{\Omega}$:

$$
\sum_{i=1}^{2m} \frac{\partial}{\partial x_i} \left(\frac{\dfrac{\partial f}{\partial x_i}}{|\nabla f|} \right) = 0
$$

is easily shown to be equivalent to

$$
\frac{\partial}{\partial u} \left[\frac{(uv)^{m-1} \frac{\partial F}{\partial u}}{|\nabla F|} \right] + \frac{\partial}{\partial v} \left[\frac{(uv)^{m-1} \frac{\partial F}{\partial v}}{|\nabla F|} \right] = 0
$$

which in turn is transformed in the new equation

$$
(8) \qquad F_v^2 F_{uu} - 2 F_u F_v F_{uv} + F_u^2 F_{vv} + p \left(\frac{F_u}{u} + \frac{F_v}{v} \right) (F_u^2 + F_v^2) = 0
$$

where

$$
F_u = \frac{\partial F}{\partial u}, \dots \qquad \text{and} \qquad p = m - 1.
$$

Our method is based on reconstructing the function F from the knowledge of its level curves.

Suppose the generic level curve $F = \text{cost}$ has a parametric representation $u = u(t)$, $v = v(t)$ with u, v twice continuously differentiable and $u'(t) \neq 0$. On eliminating F from Eq. (8) with the help of

$$dF = 0,$$

which holds true along the generic level curve of F, we get the differential equation

$$(9) \qquad u''v' - u'v'' + p[(u')^2 + (v')^2]\left(\frac{u'}{v} - \frac{v'}{u}\right) = 0.$$

It is useful to define the angular parameters

$$\varphi \;=\; \arctan\frac{v}{u},$$
$$\theta \;=\; \arctan\frac{v'}{u'},$$

which are left invariant by the homotheties

$$u(t) \;\rightarrow\; \lambda u(t),$$
$$v(t) \;\rightarrow\; \lambda v(t),$$

$\lambda > 0$.

We then get the relation

$$(10) \qquad 2p\cos(\theta + \varphi)\,d\varphi - \sin 2\varphi \sin(\theta - \varphi)\,d\theta = 0$$

and putting

$$\sigma \;=\; \theta - 3\varphi + \frac{\pi}{2},$$
$$\psi \;=\; \theta + \varphi - \frac{\pi}{2},$$

we obtain, after a suitable choice of the parameter t, $-\infty < t < +\infty$, the differential system

$$(11) \qquad \frac{d\sigma}{dt} \;=\; -\frac{3}{2}\sin\sigma - \left(2p + \frac{3}{2}\right)\sin\psi$$
$$\frac{d\psi}{dt} \;=\; \frac{1}{2}\sin\sigma - \left(2p - \frac{1}{2}\right)\sin\psi.$$

This system is easily studied in the phase space (σ, ψ) with the methods of BENDIXON and POINCARÉ (see for instance [2] for the general theory of these systems); also it is convenient to consider p as a continuous parameter.

If $p > \frac{3}{2} + \sqrt{2}$ the singular points of (11) can be classified as follows:

(A) $$(\sigma, \psi) = (h\pi, k\pi), \qquad h, k \in \mathbf{Z},$$

$h + k$ odd.

The singular point is a saddle point, with principal directions having tangent

$$
\begin{aligned}
t_1 &= 1, \\
t_2 &= -\frac{1}{4p + 3},
\end{aligned}
$$

(B) $$(\sigma, \psi) = (h\pi, k\pi), \qquad h, k \in \mathbf{Z},$$

$h + k$ even.

In this case we have a nodal point with principal direction having tangent

$$t_1 = \frac{2p - 2 - \sqrt{\Delta}}{4p + 3}$$

and exceptional direction having tangent

$$t_2 = \frac{2p - 2 + \sqrt{\Delta}}{4p + 3},$$

where $\Delta = 4p^2 - 12p + 1$. This nodal point is stable if h is even, unstable if h is odd.

We remark that this classification must be changed if $p < \frac{3}{2} + \sqrt{2}$; for example if $\frac{3}{2} - \sqrt{2} < p < \frac{3}{2} + \sqrt{2}$ then $\Delta < 0$ and the singular points (B) are focal points.

LEMMA 3. – *Let $p > \frac{3}{2} + \sqrt{2}$. There exists an integral curve γ_0, $\gamma_0(t) = (\sigma_0(t), \psi_0(t))$ of the differential system (11), with the following properties*

(i) $$
\begin{aligned}
\gamma_0(-\infty) &= (\pi, 0), \\
\gamma_0(+\infty) &= (0, 0);
\end{aligned}
$$

(ii) $$0 < \psi_0(t) < \sigma_0(t) < \pi$$

for $-\infty < t < +\infty$;

(iii) *the straight lines $\psi - \sigma = c, 0 < c < \pi$, meet γ_0 exactly once.*

Proof. Let us first verify the existence of an integral curve γ_0 satisfying (i) and (ii) of Lemma 3 provided $p > p_0$, for some sufficiently large p_0.

Let T denote the triangle

$$T = \left\{ 0 \le \psi \le \frac{\sigma}{2} \le \frac{\pi}{2} \right\}$$

with sides

$$l_1 = \{0 < \sigma < \pi, \psi = 0\},$$
$$l_2 = \left\{\sigma = \pi, 0 < \psi \le \frac{\pi}{2}\right\},$$
$$l_3 = \left\{0 < \psi = \frac{\sigma}{2} < \frac{\pi}{2}\right\}$$

and let us look at the behaviour on these sides of the vector field in (σ, ψ) determined by the system (11).

On l_1 we have

$$\frac{d\sigma}{dt} = -\frac{3}{2}\sin\sigma < 0,$$
$$\frac{d\psi}{dt} = \frac{1}{2}\sin\sigma > 0.$$

On l_2 we have

$$\frac{d\sigma}{dt} = -\left(2p + \frac{3}{2}\right)\sin\psi < 0,$$
$$\frac{d\psi}{dt} = -\left(2p - \frac{1}{2}\right)\sin\psi < 0.$$

Finally on l_3 we have

$$\frac{d\sigma}{dt} = -\frac{3}{2}\sin 2\psi - \left(2p + \frac{3}{2}\right)\sin\psi = -\sin\psi\left[3\cos\psi + \left(2p + \frac{3}{2}\right)\right] < 0,$$
$$\frac{d\psi}{dt} = \frac{1}{2}\sin 2\psi - \left(2p - \frac{1}{2}\right)\sin\psi = -\sin\psi\left[-\cos\psi + \left(2p - \frac{1}{2}\right)\right] < 0$$

and

$$\frac{d\psi}{d\sigma} = \frac{2p - \frac{1}{2} - \cos\psi}{2p + \frac{3}{2} + 3\cos\psi} \ge \frac{2p - \frac{3}{2}}{2p + \frac{9}{2}} > \frac{1}{2}$$

for

$$p > \frac{15}{4} = p_0.$$

In all cases, if $p > p_0 = \frac{15}{4}$, the tangent vector through a regular point of the boundary of T is directed towards the interior of T. It follows that if γ, $\gamma(t) = (\sigma(t), \psi(t))$ is an integral curve such that, for some point t_0, $\gamma(t_0) \in T$, then $\gamma(t) \in T$ for each $t \ge t_0$.

The point $(\pi, 0)$ is a saddle point thus there are two integral curves ending at, and two integral curves starting from, the point $(\pi, 0)$. The two integral curves ending at $(\pi, 0)$ are

$$\sigma = \psi + \pi, \qquad (0 < \psi < \pi)$$

and

$$\sigma = \psi + \pi, \qquad (-\pi < \psi < 0).$$

Hence the other two integral curves starting from $(\pi, 0)$ have tangent there

$$t_2 = -\frac{1}{4p + 3} < 0$$

whence one of them starts inside the triangle T. By our previous remark, this integral curve γ_0 remains trapped in T and so verifies (i) and (ii) of Lemma 3.

In order to eliminate the condition $p > p_0$ we proceed as follows.

The vector field determined by (11) depends continuously on p provided $p > \frac{3}{2} + \sqrt{2}$; it follows that there exists an unique integral curve γ_0 satisfying (i) of Lemma 3, which depends continuously on the parameter p. Let θ_p be the variation in the argument of the tangent vector to γ_0 at the point $\gamma_0(t)$, as t varies from $-\infty$ to $+\infty$. Then θ_p is finite and depends continuously on p and it is easily seen that

$$\lim_{p \to +\infty} \theta_p = 0.$$

If $p > p_0 = \frac{15}{4}$ the integral curve γ_0 is contained inside the triangle T and it follows that γ_0 ends at $(0,0)$ with the principal direction having tangent

$$t_1 = \frac{2p - 2 - \sqrt{\Delta}}{4p + 3};$$

by an obvious continuity argument, this remains true for $p > \frac{3}{2} + \sqrt{2}$. It follows that

$$\theta_p = \arctan\left(\frac{2p - 2 - \sqrt{\Delta}}{4p + 3}\right) - \arctan\left(-\frac{1}{4p + 3}\right) + \pi k$$

where k is an integer and arctan has its principal value. If p is sufficiently large, $k = 0$ but as θ_p depends continuously on p for $p > \frac{3}{2} + \sqrt{2}$ and as

$$0 < \arctan\left(\frac{2p - 2 - \sqrt{\Delta}}{4p + 3}\right) - \arctan\left(-\frac{1}{4p + 3}\right) < \pi$$

for $p > \frac{3}{2} + \sqrt{2}$, it follows that $k = 0$ in every case and

$$0 < \theta_p < \pi \qquad \text{for} \qquad p > \frac{3}{2} + \sqrt{2}.$$

A simple consideration of the behaviour of the vector field of the system (11) in the square Q: $-\pi < \sigma < \pi$, $-\pi < \psi < \pi$ shows that γ_0 is always contained in Q; looking at the behaviour of this vector field along the lines $\sigma = 0$, $\psi = 0$ one sees easily that γ_0 stays in the square Q': $0 < \sigma < \pi$, $0 < \psi < \pi$, for otherwise we would contradict the inequality $0 < \theta_p < \pi$.

Finally, γ_0 cannot cross the line $\sigma = \psi$; for otherwise it would cross this line at least twice because γ_0 is contained in the square Q': $0 < \sigma < \pi$, $0 < \psi < \pi$. If this were the case, by Rolle's theorem there would be a point t_0 such that

$$\frac{d\psi_0}{d\sigma_0}(t_0) = 1.$$

At such a point we have

$$\frac{1}{2}\sin\sigma_0 - \left(2p - \frac{1}{2}\right)\sin\psi_0 = -\frac{3}{2}\sin\sigma_0 - \left(2p + \frac{3}{2}\right)\sin\psi_0$$

that is

$$\sin\sigma_0 + \sin\psi_0 = 0,$$

a contradiction with the fact that $(\sigma_0, \psi_0) \in Q'$.

Hence γ_0 does not cross the line $\sigma = \psi$ if $p > \frac{3}{2} + \sqrt{2}$, which proves (ii) of Lemma 3.

The same argument proves (iii) of Lemma 3; for if γ_0 were to cross the line $\psi - \sigma = c$, $0 < c < \pi$, at least twice, there would be, again by Rolle's theorem, a point t_c such that

$$\frac{d\psi_0}{d\sigma_0}(t_c) = 1$$

and we would get the same contradiction as before.   q.e.d.

The curve γ_0 gives rise to a family of homothetic curves Γ_λ, $\lambda > 0$, in the (u, v) plane. By Lemma 3 these curves are analytic and contained in the domain $T_1 = \{0 \le v < u\}$, and there is exactly one curve of the family $\{\Gamma_\lambda\}$ passing through any given point of T_1.

Our aim is to construct a homogeneous function $F(u, v)$ of degree $2\alpha > 0$ in T_1 which has the family $\{\Gamma_\lambda\}$ as its family of level curves; then by the definition of $\{\Gamma_\lambda\}$ the function F will satisfy the Eq. (8) in $\mathring{T}_1$.

By Lemma 3, (iii) the lines $\varphi = c$, where $0 \le c < \frac{\pi}{4}$ intersect a given Γ_λ exactly once. Now let Γ_1 be the curve issuing from the point $(1, 0)$; by our previous remark we may take φ as parameter on Γ_1 and we get a parametric representation for Γ_1

$$\begin{aligned} u &= u_0(\varphi), \\ v &= v_0(\varphi) \end{aligned}$$

where

$$\tan\varphi = \frac{v_0}{u_0}.$$

If we normalize F by the requirement

$$F(u, v) = 1 \quad \text{along } \Gamma_1,$$

then if 2α is the degree of homogeneity of F we obtain if $(u, v) \in T_1$:

$$F(u, v) = (u^2 + v^2)^\alpha G\left(\arctan\frac{v}{u}\right)$$

where G is determined by $F(u_0, v_0) \equiv 1$, from which it follows

$$G(t) = [u_0^2(t) + v_0^2(t)]^{-\alpha}, \qquad 0 \le t < \frac{\pi}{4}.$$

Our results imply

$$\lim_{t \to \frac{\pi}{4}-} G(t) = 0,$$

whence the function F so defined can be extended to a continuous function in the first quadrant and analytic for $u \neq v$, by means of the relation

$$F(u, v) = -F(v, u) \qquad \text{for } (v, u) \in T_1.$$

This function F is a solution of Eq. (8) in the open set

$$\Omega = \{0 < u, \, 0 < v, \, u \neq v\};$$

also

$$
\begin{aligned}
F(u, u) &= 0 & \text{if } u > 0. \\
|\nabla F| &\neq 0 & \text{if } (u, v) \in \Omega.
\end{aligned}
$$

In order to apply Lemma 2 to the function $f(x) = F(u, v)$ we have to show that the vector $(\nabla F)/|\nabla F|$ can be extended to a continuous vector in the whole domain $\{(u, v) \mid u > 0, \, v > 0\}$.

We have

$$
\begin{aligned}
F_u &= (u^2 + v^2)^{\alpha-1}(2\alpha u G - v G'), \\
F_v &= (u^2 + v^2)^{\alpha-1}(2\alpha v G + u G'), \\
|\nabla F| &= (u^2 + v^2)^{\alpha-\frac{1}{2}}(4\alpha^2 G^2 + G'^2)^{\frac{1}{2}},
\end{aligned}
$$

whence it is enough to show that the functions

$$G(4\alpha^2 G^2 + G'^2)^{-\frac{1}{2}} \quad \text{and} \quad G'(4\alpha^2 G^2 + G'^2)^{-\frac{1}{2}}$$

have a limit if $t \to \pi/4$.

If $\varphi < \pi/4$ we get

$$
\begin{aligned}
G'(\varphi) &= -2\alpha(u_0^2 + v_0^2)^{-\alpha-1}(u_0 u_0' + v_0 v_0') \\
&= -2\alpha G(\varphi) \frac{u_0 u_0' + v_0 v_0'}{u_0^2 + v_0^2};
\end{aligned}
$$

on the other hand

$$\varphi = \arctan \frac{v_0}{u_0}$$

and it follows that

$$u_0^2 + v_0^2 = u_0 v_0' - v_0 u_0'$$

whence

$$G'(\varphi) = -2\alpha G(\varphi) \frac{1 + \dfrac{v_0}{u_0}\dfrac{v_0'}{u_0'}}{\dfrac{v_0'}{u_0'} - \dfrac{v_0}{u_0}}$$

and finally

$$(12) \qquad G'(\varphi) = 2\alpha \frac{G(\varphi)}{\tan(\varphi - \vartheta)}$$

where

$$\vartheta = \arctan \frac{v_0'}{u_0'}.$$

Let us determine the constant α in such a way that

$$(13) \qquad \lim_{\varphi \to \frac{\pi}{4}} G'(\varphi) = k \neq 0;$$

it will then follow that $(\nabla F)/|\nabla F|$ is a continuous non-zero vector in $\{(u, v)|\ u > 0, v > 0\}$.

The integral curve γ_0 considered in Lemma 3 had tangent at the origin

$$t_1 = \frac{2p - 2 - \sqrt{\Delta}}{4p + 3},$$

whence the asymptotic relation

$$\psi \sim t_1 \sigma$$

holds as $\sigma \to 0$, and we get

$$\vartheta - \varphi \sim 2 \frac{1 + t_1}{1 - t_1} \left(\frac{\pi}{4} - \varphi \right)$$

as $\varphi \to \pi/4$.

On the other hand we have

$$\begin{aligned} v_0 &= (\tan \varphi) u_0, \\ v_0' &= (\tan \vartheta) v_0 \end{aligned}$$

whence

$$(\tan \vartheta) u_0' = (\tan \varphi) u_0' + (1 + \tan^2 \varphi) u_0.$$

From this equation and the previous asymptotic relation we easily find

$$\frac{u_0'}{u_0} \sim \frac{1 - t_1}{2(1 + t_1)} \left(\frac{\pi}{4} - \varphi \right)^{-1}$$

and finally

$$u_0 \sim c \left(\frac{\pi}{4} - \varphi \right)^{-\frac{1 - t_1}{2(1 + t_1)}}$$

where c is a positive constant. It follows that

$$G(\varphi) \sim (2c^2)^{-\alpha} \left(\frac{\pi}{4} - \varphi \right)^{\frac{1 - t_1}{1 + t_1} \alpha}$$

and recalling Eq. (12) we get

$$G'(\varphi) \sim k \left(\frac{\pi}{4} - \varphi\right)^{\frac{1-t_1}{1+t_1}\alpha - 1}$$

for some constant $k \neq 0$. If

$$\alpha = \frac{1 + t_1}{1 - t_1} = \frac{2p + 1 - \sqrt{\Delta}}{4}$$

we obtain Eq. (13), as required.

It is easily seen from our previous discussion that the function

$$f(x) = F(u, v)$$

has a locally bounded gradient in $\mathbf{R}^{2m}$ and satisfies the hypotheses of Lemma 2 if one takes

$$
\begin{aligned}
A &= \{x \in \mathbf{R}^{2m} \mid u \neq 0, v \neq 0\}, \\
S &= \{x \in \mathbf{R}^{2m} \mid u = v\}.
\end{aligned}
$$

Hence Lemma 2 applies and we are able to apply Lemma 1 where now

$$
\begin{aligned}
A &= \mathbf{R}^{2m}, \\
N &= \{x \in \mathbf{R}^{2m} \mid uv = 0\};
\end{aligned}
$$

note that

$$H_{n-1}(N) = 0.$$

We conclude that $f(x)$ has least gradient in $\mathbf{R}^{2m}$. Now Simons' cone $x_1^2 + x_2^2 + \cdots + x_m^2 = x_{m+1}^2 + x_{m+2}^2 + \cdots + x_{2m}^2$ coincides with the level hypersurface $f(x) = 0$ because $F(u, v) = 0$ if and only if $u = v$, and Theorem A follows from Theorem 1.   q.e.d.

IV. Proof of Theorem B

We begin with the simple remark that it is enough to prove Theorem B in the case $n = 2m \geq 8$, because if $f(x_1, ..., x_n)$ is a solution in $\mathbf{R}^n$ of the equation

$$(14) \qquad \sum_{i=1}^{n} \frac{\partial}{\partial x_i} \left(\frac{\frac{\partial f}{\partial x_i}}{\sqrt{1 + |\nabla f|^2}} \right) = 0$$

for minimal graphs over $\mathbf{R}^n$, then it is also a solution of the same equation for each $n' > n$.

Let us look for solutions f of (14) with $n = 2m$ and which are invariant with respect to the group

$$G = SO(m) \times SO(m)$$

of automorphisms of the cone

$$C_{2m} = \{x \in \mathbf{R}^{2m} \mid x_1^2 + x_2^2 + \cdots + x_m^2 = x_{m+1}^2 + x_{m+2}^2 + \cdots + x_{2m}^2\}.$$

We may write

$$f(x) = F(u, v)$$

where

$$
\begin{aligned}
u &= (x_1^2 + x_2^2 + \cdots + x_m^2)^{\frac{1}{2}}, \\
v &= (x_{m+1}^2 + x_{m+2}^2 + \cdots + x_{2m}^2)^{\frac{1}{2}}
\end{aligned}
$$

and now Eq. (14) can be written

$$\mathcal{E}F \equiv (1 + F_v^2)F_{uu} - 2F_u F_v F_{uv} + (1 + F_u^2)F_{vv}$$

(15)
$$+ p\left(\frac{F_u}{u} + \frac{F_v}{v}\right)(1 + F_u^2 + F_v^2) = 0$$

where $p = m - 1$.

Let

$$T = \{(u, v) \mid u \geq 0,\, v \geq 0\}, \quad T_1 = \{(u, v) \mid 0 \leq v < u\}, \quad T_2 = \{(u, v) \mid 0 \leq u < v\}.$$

We shall prove in the Appendix that the function

$$F_1 = (u^2 - v^2)(u^2 + v^2)^{\alpha - 1}$$

where

$$\alpha = \frac{2p + 1 - \sqrt{\Delta}}{4} > 1, \qquad \Delta = 4p^2 - 12p + 1$$

verifies

(16)
$$\mathcal{E}F_1 > 0 \qquad \text{in } \mathring{T}_1,$$

(17)
$$\mathcal{E}F_1 < 0 \qquad \text{in } \mathring{T}_2,$$

while the function

$$F_2 = H\left\{(u^2 - v^2) + (u^2 - v^2)(u^2 + v^2)^{\alpha - 1}\left[1 + A\left|\frac{u^2 - v^2}{u^2 + v^2}\right|^{\lambda - 1}\right]\right\}$$

where λ is any real number subjected to

$$\frac{2p + 1}{2p + 2}\alpha < \lambda < \min\left(\alpha, \frac{p}{\alpha^2}\right)$$

and H is defined by

$$H(z) = \int_0^z \exp\left(B \int_{|w|}^{\infty} \frac{dt}{t^{2 - \lambda}(1 + t^{2\alpha\lambda - 2\alpha})}\right) dw$$

and $A = A(\lambda, p)$, $B = B(\lambda, p)$ are sufficiently large positive constants, verifies for $p \geq 3$

$$(18) \qquad \mathcal{E}F_2 < 0 \qquad \text{in } \mathring{T}_1,$$

$$(19) \qquad \mathcal{E}F_2 > 0 \qquad \text{in } \mathring{T}_2,$$

Moreover

$$(20) \qquad 0 < F_1 < F_2 \qquad \text{in } T_1,$$

$$(21) \qquad F_2 < F_1 < 0 \qquad \text{in } T_2,$$

$$(22) \qquad F_1 = F_2 = 0 \qquad \text{if } u = v,$$

and

$$\frac{F_u}{u}, \frac{F_v}{v} \qquad (F = F_1, F_2)$$

have a continuous extension to the whole of T.

Let

$$f_1(x) = F_1(u, v),$$
$$f_2(x) = F_2(u, v),$$
$$D_1 = \{x \in \mathbf{R}^{2m} \mid 0 \leq v < u\},$$
$$D_2 = \{x \in \mathbf{R}^{2m} \mid 0 \leq u < v\};$$

from (16) and (18) we readily get

$$(23) \qquad \int_{D_1} \sum_{i=1}^{2m} \frac{\partial f_1}{\partial x_i} \frac{\partial \varphi}{\partial x_i} (1 + |\nabla f_1|^2)^{-\frac{1}{2}} \, dx \leq 0$$

for every $\varphi \in C_0^\infty(D_1)$, $\varphi \geq 0$,

$$(24) \qquad \int_{D_1} \sum_{i=1}^{2m} \frac{\partial f_2}{\partial x_i} \frac{\partial \varphi}{\partial x_i} (1 + |\nabla f_2|^2)^{-\frac{1}{2}} \, dx \geq 0$$

for every $\varphi \in C_0^\infty(D_1)$, $\varphi \geq 0$, and analogous inequalities in D_2.

Let us consider the Dirichlet problem

$$(25) \qquad \sum_{i=1}^{2m} \frac{\partial}{\partial x_i} \left(\frac{\frac{\partial f}{\partial x_i}}{\sqrt{1 + |\nabla f|^2}} \right) = 0, \qquad f \in C^2(B_R)$$

$$f = f_1 \qquad \text{on } \partial B_R.$$

The function $f_1|_{\partial B_R}$ is of class C^2 and Theorem 1.2 of [10] gives existence and uniqueness of the solution of (14). The function f_1 is invariant by the group G whence so is the solution $f^{(R)}$ of (25) and we may write

$$f^{(R)}(x) = F^{(R)}(u, v);$$

also we have $F_j(u, v) = -F_j(v, u)$ $(j = 1, 2)$ and it follows that

$$F^{(R)}(u, v) = -F^{(R)}(v, u)$$

and

(26) $$f^{(R)}(x) = 0$$

for $x \in C_{2m} \cap B_R$.

From (22) and (26) we get

$$f_1(x) \le f^{(R)}(x) \le f_2(x)$$

for $x \in \partial(B_R \cap D_1)$. Hence we may apply the inequalities (23), (24) and the well-known maximum principle for solutions of (14) and obtain

$$f_1(x) \le f^{(R)}(x) \le f_2(x)$$

for $x \in \bar{B}_R \cap \bar{D}_1$. In the same way one finds

$$f_2(x) \le f^{(R)}(x) \le f_1(x)$$

for $x \in \bar{B}_R \cap \bar{D}_2$, and we conclude that

(27) $$|f_1(x)| \le |f^{(R)}(x)| \le |f_2(x)|$$

for $x \in \bar{B}_R$.

Now consider the sequence $f^{(k)}(x)$, $k = 1, 2, \dots$. By inequality (27) and Theorem A of [4] (see also [3]) we find that for every $k \ge h$ we have

(28) $$|\nabla f^{(k)}(x)| \le c_1 \exp\left(c_2 \frac{1}{2h} \sup_{B_{2h}} |f_2|\right) = c(h)$$

for $x \in B_h$; the essential point of this inequality is that the right hand side is independent of k. Hence there is a subsequence $\{f^{(k_\nu)}(x)\}$, $\nu = 1, 2, \dots$ of the sequence $\{f^{(k)}(x)\}$ which is uniformly convergent in the closure of the unit ball $\bar{B}_1$ to a function $f(x)$. By Remark 6 of [11] and the inequality (28) this function $f(x)$ is analytic in B_1 and verifies there the minimal surface Eq. (14).

Next, we prove that the sequence $\{f^{(k_\nu)}(x)\}$, is uniformly convergent in every compact set $K \subset \mathbf{R}^{2m}$. If this were not the case, there would exist, by the same argument as before, two subsequences uniformly convergent in a ball B_R, $R > 1$, to two distinct functions f' and f'', both analytic in B_R and solution of the minimal surface Eq. (14) there. This however is impossible, because $f' = f'' = f$ in B_1.

Arguing as before we conclude that the function $f(x)$, limit of the sequence $\{f^{(k_\nu)}(x)\}$, is analytic in $\mathbf{R}^{2m}$ and verifies there the Eq. (14) and the inequalities

$$|f_1(x)| \leq |f(x)| \leq |f_2(x)|.$$

We have

$$\limsup_{|x| \to +\infty} |f_1(x)|/|x|^{2\alpha} = 1,$$

$\alpha > 1$. Hence $f(x)$ cannot be a polynomial of the first degree, and the proof of Theorem B is complete.

Appendix

Here we prove the statements (16), ..., (22) about the two functions F_1, F_2.

It is convenient to split the differential operator $\mathcal{E}$ defined by (15) in two parts as $\mathcal{E} = \mathcal{E}_0 + \mathcal{D}$, where

$$\mathcal{E}_0 F = F_v^2 F_{uu} - 2F_u F_v F_{uv} + F_u^2 F_{vv} + p\left(\frac{F_u}{u} + \frac{F_v}{v}\right)(F_u^2 + F_v^2)$$

is the *least gradient operator* already considered in Section III, (8), and where

$$\mathcal{D}F = F_{uu} + F_{vv} + p\left(\frac{F_u}{u} + \frac{F_v}{v}\right)$$

is related to the Laplace operator.

This decomposition suggests looking for functions F_1, F_2 having the same asymptotic behaviour as the function F constructed in Section III, for large values of $|\nabla F|$.

With conditions (16), (17) and (22) to fulfill, the simplest choice is the function

$$F_1 = (u^2 - v^2)(u^2 + v^2)^{\alpha - 1}$$

where

$$\alpha = \frac{2p + 1 - \sqrt{\Delta}}{4}.$$

In order to prove that F_1 is a subsolution for the differential operator $\mathcal{E}$ in the domain $T_1 = \{(u, v)\mid 0 \leq v < u\}$ it is useful to define new independent variables

$$r = u^2 + v^2,$$
$$t = \frac{u^2 - v^2}{u^2 + v^2}, \qquad 0 < t \leq 1.$$

If we write

$$F(u, v) = G(r, t)$$

we find

$$
\text{(29)} \quad
\begin{aligned}
\tfrac{1}{8}\mathcal{E}_0 F
&= (1-t^2)\left\{2G_t^2 G_{rr} - 4G_t G_r G_{rt} + 2G_r^2 G_{tt} + \tfrac{2p+3}{r}G_r G_t^2 - 2p\tfrac{t}{r^2}G_t^3\right\} \\
&\quad + (2p+1)rG_r^3 - (2p+2)tG_r^2 G_t,
\end{aligned}
$$

$$
\text{(30)} \quad
\tfrac{1}{8}\mathcal{D}F = 2(p+1)\left(G_r - \tfrac{t}{r}G_t\right) + 2r\left[G_{rr} + \tfrac{1}{r^2}(1-t^2)G_{tt}\right]
$$

and in the special case in which

$$
G(r,t) = r^\alpha f(t)
$$

where

$$
\alpha = \frac{2p+1-\sqrt{\Delta}}{4},
$$

one finds

$$
\text{(31)} \quad
\begin{aligned}
\tfrac{1}{8}\mathcal{E}_0 F
&= r^{3\alpha-2}\{(1-t^2)[2pf f'^2 - 2ptf'^3 + 2\alpha^2 f^2 f''] \\
&\qquad\quad +(2p+1)\alpha^3 f^3 - (2p+2)\alpha^2 t f^2 f'\} \\
&= r^{3\alpha-2}M_1(t),
\end{aligned}
$$

$$
\text{(32)} \quad
\tfrac{1}{8}\mathcal{D}F = 2r^{\alpha-1}[\alpha(p+\alpha)f - (p+1)tf' + (1-t^2)f''].
$$

If $f(t) = t$ we have in $\mathring{T}_1$

$$
\begin{aligned}
\tfrac{1}{8}\mathcal{E}_0 F &= 2\alpha^2(\alpha^2-1)r^{3\alpha-2}t^3 > 0, \\
\tfrac{1}{8}\mathcal{D}F &= 2[\alpha(p+\alpha) - (p+1)]r^{\alpha-1}t > 0
\end{aligned}
$$

because $0 < t \le 1$ and $2\alpha^2 - (2p+1)\alpha + 2p = 0$.
Now take

$$
G(r,t) = rt + r^\alpha f(t);
$$

we obtain

$$
\text{(33)} \quad
\tfrac{1}{8}\mathcal{E}_0 F = r^{3\alpha-2}M_1(t) + r^{2\alpha-1}M_2(t) + r^\alpha M_3(t) - rt,
$$

where M_1 is given by (31) and

$$
\text{(34)} \quad
\begin{aligned}
M_2(t) &= (1-t^2)[(4p-2)\alpha f f' - (4p-3+4\alpha)tf'^2 + 4\alpha t f f''] \\
&\quad +(4p+1)\alpha^2 t f^2 - (4p+4)\alpha t^2 f f',
\end{aligned}
$$

$$(35) \qquad \begin{aligned} M_3(t) &= (1 - t^2)[\alpha(2p - 3 + 2\alpha)f - (2p - 2 + 4\alpha)tf' + 2t^2 f''] \\ &\quad -(2p + 2)t^3 f' + (2p - 1)\alpha t^2 f. \end{aligned}$$

We choose $f(t) = t + At^\lambda$, supposing $0 < t \le 1$, where λ is such that $\lambda > 1$. We write $M_1(t)$ as a polynomial in A and get

$$\begin{aligned} M_1(t) &= A^3 t^{3\lambda - 2}\{2\lambda(\lambda - 1)(\alpha^2 - p\lambda)(1 - t^2) + \alpha^2[(2p + 1)\alpha - (2p + 2)\lambda]t^2\} \\ &\quad + A^2 t^{2\lambda - 1}\{4\lambda(\lambda - 1)(\alpha^2 - p)(1 - t^2) \\ &\quad + \alpha^2[(6p + 3)\alpha - (2p + 2)(1 + 2\lambda)]t^2\} \\ &\quad + At^\lambda\{2(\lambda - 1)(\alpha^2\lambda - p)(1 - t^2) + \alpha^2[(6p + 3)\alpha - (2p + 2)(2 + \lambda)]t^2\} \\ &\quad + 2\alpha^2(\alpha^2 - 1)t^3. \end{aligned}$$

Now if

$$(36) \qquad \frac{2p + 1}{2p + 2}\alpha < \lambda < \frac{p}{\alpha^2}$$

it is easily seen that

$$M_1(t) \le -c_1(A^3 t^{3\lambda - 2} + A^2 t^{2\lambda - 1} + At^\lambda) + c_2(A^2 t^{2\lambda + 1} + At^{\lambda + 2} + t^3)$$

for some positive constants c_1, c_2. It follows from this inequality that if in addition $\lambda \le 3$ then

$$M_1(t) \le -c_3(A^3 t^{3\lambda - 2} + A^2 t^{2\lambda - 1} + At^\lambda)$$

if only $A \ge c_4(p, \lambda)$, for some positive c_3.

In exactly the same way one deals with $M_2(t)$ and $M_3(t)$. If λ is in the range (36) and $\lambda \le 3$, $A \ge c_4$ (possibly changing the constant c_4) we still obtain

$$\begin{aligned} M_2(t) &\le -c_5(A^2 t^{2\lambda - 1} + At^\lambda); \\ M_3(t) &\le -c_6(At^\lambda + t), \end{aligned}$$

provided $p \ge 3$. This condition comes out in handling the terms of $M_2(t)$ which are independent of A, where we need now

$$(4p - 6)\alpha - (4p - 3) \le 0,$$

and this inequality is true only for $p \ge 3$.

We conclude that with this choice of G one has in $\overset{\circ}{T}_1$:

$$(37) \qquad \frac{1}{8}\mathcal{E}_0 F \le -c_7(r^{3\alpha - 2}t^\lambda + rt)$$

for some positive c_7, provided $A \ge c_4$, $p \ge 3$ and

$$\frac{2p + 1}{2p + 2}\alpha < \lambda < \frac{p}{\alpha^2}, \qquad \lambda \ge 3.$$

A similar but much simpler estimate gives in $\mathring{T}_1$:

$$(38) \qquad \frac{1}{8}DF \leq c_8 r^{\alpha-1} t^{\lambda-2}.$$

Now let $H(z)$ be a monotone increasing function for $z > 0$ and of class C^2 there. Then for each positive F we have identically

$$\mathcal{E}_0(H(F)) = H'(F)^3 \cdot \mathcal{E}_0 F,$$

$$\mathcal{D}(H(F)) = H'(F) \cdot DF + 16r H''(F)\left(G_r^2 + \frac{1-t^2}{r^2}G_t^2\right).$$

In the special case $G = rt + r^\alpha(t + At^\lambda)$ we easily get, if $\lambda < \alpha$ and if $0 < t \leq 1$:

$$r\left(G_r^2 + \frac{1-t^2}{r^2}G_t^2\right) \geq r + r^{2\alpha-1},$$

whence if we suppose

$$(39) \qquad H'(z) \geq 1, \qquad H''(z) \leq 0$$

we obtain by (37) and (38)

$$\begin{aligned}
\mathcal{E}(H(F)) \;<\; & -c_7(r^{3\alpha-2}t^\lambda + rt)H'(F)^3 \\
& c_8 r^{\alpha-1}t^{\lambda-2}H'(F) + 16(r + r^{2\alpha-1})H''(F).
\end{aligned}$$

This gives by (39):

$$\begin{aligned}
\mathcal{E}(H(F)) \;<\; & -r^{\alpha-\lambda}t^{\lambda-2}[c_7 H'(F)^3 r^{2\alpha-\lambda}t^2 - c_8 H'(F)] \\
\;<\; & 0
\end{aligned}$$

provided $0 < t \leq 1$ and

$$r^{2\alpha-1}t^2 > c_9 = c_8/c_7.$$

Now suppose $0 < t \leq 1$ and

$$(40) \qquad r^{2\alpha-1}t^2 \leq c_9.$$

In order to get $\mathcal{E}(H(F)) < 0$ in this domain too, and hence $\mathcal{E}(H(F)) < 0$ in the whole of $\mathring{T}_1$, it is enough to fulfill the condition

$$16(r + r^{2\alpha-1})H''(F) < -c_8 r^{\alpha-1}t^{\lambda-2}H'(F),$$

which we may write

$$-\frac{H''(F)}{H'(F)} \geq c_{10}\frac{r^{\alpha-1}t^{\lambda-2}}{r + r^{2\alpha-1}}.$$

We have

$$\frac{r + r^{2\alpha-1}}{r^{\alpha-1}t^{\lambda-2}} = r^{2-\alpha}t^{2-\lambda} + r^\alpha t^{2-\lambda}.$$

If $r \geq 1$, then $F < c_{11} r^\alpha t$; it follows that

$$
\begin{aligned}
r^{2-\alpha} t^{2-\lambda} + r^\alpha t^{2-\lambda} \;&\geq\; r^\alpha t^{2-\lambda} \\
&= (r^\alpha t)^{2-\lambda+2\alpha(\lambda-1)} (t^2 r^{2\alpha-1})^{-\alpha(\lambda-1)} \\
&\geq c_{12} F^{2-\lambda+2\alpha(\lambda-1)} [\min(1, F)]^{-2\alpha(\lambda-1)} \\
&\geq c_{13} (F^{2-\lambda} + F^{2-\lambda+2\alpha(\lambda-1)})
\end{aligned}
$$

because of the inequalities (40), $r \geq 1$ and $r^\alpha t < F < c_{11} r^\alpha t$.

If $r < 1$, we have $F < c_{14} rt < c_{14}$ and using the condition $\lambda < \alpha$ we still obtain

$$
\begin{aligned}
r^{2-\alpha} t^{2-\lambda} + r^\alpha t^{2-\lambda} \;&\geq\; (rt)^{2-\lambda} \\
&\geq c_{15} F^{2-\lambda} \\
&\geq c_{16} (F^{2-\lambda} + F^{2-\lambda+2\alpha(\lambda-1)}).
\end{aligned}
$$

The conclusion is that if

$$
-\frac{H''(z)}{H'(z)} \geq c_{17} z^{\lambda-2} (1 + z^{2\alpha(\lambda-1)})^{-1}
$$

for some large c_{17} and if

$$
H'(z) \geq 1,
$$

then we have in $\overset{\circ}{T}_1$ the inequality

$$
\mathcal{E}(H(F)) < 0
$$

where $F(u, v) = rt + r^\alpha (t + At^\lambda)$. The conditions we have put on the way are

$$
\frac{2p+1}{2p+2} \alpha < \lambda < \min\left(\frac{p}{\alpha^2}, \alpha \right)
$$

(note that $1 < \lambda \leq 3$ is included in it), and $p \geq 3$, $A \geq c_4$. Note also that the interval for λ is not empty, already for $p \geq 3$.

If one takes

$$
\begin{aligned}
H'(z) \;&=\; \exp\left(c_{17} \int_z^\infty \frac{dt}{t^{2-\lambda}(1 + t^{2\alpha(\lambda-1)})} \right), \\
H(z) \;&=\; \int_0^z H'(w)\, dw
\end{aligned}
$$

then if we define F_2 by $F_2 = H(F)$ we get

$$
\mathcal{E} F_2 < 0 \quad \text{in } \overset{\circ}{T}_1,
$$

as required.

This choice of $H(z)$ is admissible, because

$$
(41) \qquad \int_0^\infty \frac{dt}{t^{2-\lambda}(1 + t^{2\alpha(\lambda-1)})} < +\infty;
$$

note that $2 - \lambda < 1$, $2 - \lambda + 2\alpha(\lambda - 1) > 1$. It is also clear that $0 < F_1 < F_2$ in $\mathring{T}_1$, and $F_1 = F_2 = 0$ on $u = v$, this because of (41).

Thus we have proved that if $p \geq 3$ the two functions F_1, F_2 defined in Section IV satisfy (16), (18), (20), (22). The proof of (17), (19), (21) is also immediate because of the antisymmetry of F_1, F_2 with respect to the line $u = v$.   q.e.d.

Bibliography

[1] F. J. ALMGREN JR., *Some interior regularity theorems for minimal surfaces and an extension of Bernstein's theorem.* Ann. of Math. **85**, 277–292 (1966).

[2] G. BIRKHOFF, AND G. C. ROTA, *Ordinary differential equations.* Ginn & Co., Boston (1962).

[3] E. BOMBIERI, *Nuovi risultati sulle ipersuperfici minimali non parametriche.* Rend. Sem. Mat. Fis. Milano **38**, 2–12 (1968).

[4] E. BOMBIERI, E. DE GIORGI, AND M. MIRANDA, *Una maggiorazione a priori relativa alle ipersuperfici minimali non parametriche.* Archive for Rat. Mech. and Analysis (1968). [‡]

[5] E. DE GIORGI, *Frontiere orientate di misura minima.* Sem. Mat. Sc. Norm. Sup. Pisa A.A. 1960/61. [‡]

[6] E. DE GIORGI, *Una estensione del teorema di Bernstein.* Annali Sc. Norm. Sup. Pisa **19**, 79–85 (1965).

[7] H. FEDERER, AND W. H. FLEMING, *Normal and integral currents.* Ann. of Math. **72**, 458–520 (1960).

[8] W. H. FLEMING, *On the oriented Plateau problem.* Rend. Circolo Mat. Palermo **9**, 69–89 (1962).

[9] M. MIRANDA, *Sul minimo dell'integrale del gradiente di una funzione.* Annali Sc. Norm. Sup. Pisa **19**, 627–665 (1965).

[10] M. MIRANDA, *Un teorema di esistenza e unicità per il problema dell'area minima in n variabili.* Annali Sc. Norm. Sup. Pisa **19**, 233–250 (1965).

[11] M. MIRANDA, *Comportamento delle successioni convergenti di frontiere minimali.* Rend. Sem. Mat. Padova **38**, 238–257 (1967).

[12] J. SIMONS, *Minimal varieties in Riemannian manifolds.* Ann. of Math. **88**, 62–105 (1968).

[‡] Editor's note: see Archive for Rat. Mech. and Analysis **32**, 255–267 (1969)
[‡] Editor's note: in this volume

An a priori estimate related to nonparametric minimal surfaces[‡]

E. Bombieri, E. De Giorgi, M. Miranda

I. Introduction

Let $\Omega \subset \mathbf{R}^n$ be an open set, let $n \geq 3$ and let $u \in \mathcal{C}^2(\Omega)$ be a solution of the minimal surfaces equation

$$\text{(1)} \qquad \sum_{i=1}^{n} D_i \left(\frac{D_i u}{(1 + |Du|^2)^{\frac{1}{2}}} \right) = 0.$$

In this paper we prove the following

THEOREM. – *Let $u \in \mathcal{C}^2(\Omega)$ be a solution of equation (1), let $x_0 \in \Omega$, let $\rho < \operatorname{dist}(x_0, \partial\Omega)$ and suppose that*

$$\text{(2)} \qquad u(x) > 0 \qquad \text{if} \quad |x - x_0| < \rho.$$

Then we have

$$\text{(3)} \qquad |Du(x_0)| \leq c_1 \exp \left(c_2 \frac{u(x_0)}{\rho} \right),$$

where c_1 and c_2 are absolute constants depending only on n.

We note that in the case $n = 2$ a formula of the same kind as (3) is already known. Through successive improvements (see R. FINN [3], [4], [5], H. JENKINS-J. SERRIN [6], J. SERRIN [12], [13]; see also the paper of J. C. C. NITSCHE [11]) the following result has been reached ($n = 2$):

$$\text{(4)} \qquad (1 + |Du(x_0)|^2)^{\frac{1}{2}} \leq \exp \left(\frac{\pi}{2} \frac{u(x_0)}{\rho} \right),$$

under the same hypotheses of our theorem. R. FINN remarked in [5] that the constant $\pi/2$ in (4) cannot be improved.

From (3) and a result of J. MOSER [10] the following results concerning the extension of Bernstein's theorem can be readily deduced.

(A) If $u \in \mathcal{C}^2(\mathbf{R}^n)$ is a solution of (1) and if $u > 0$ in $\mathbf{R}^n$, then

$$u = \text{constant};$$

[‡]Editor's note: translation into English of the paper "Una maggiorazione apriori per le ipersuperficie minimali non parametriche", published in Arch. Rational Mech. Anal., **32** (1969), 255–267.

(B) if $u \in C^2(\mathbf{R}^n)$ is a solution of (1) and if we have in $\mathbf{R}^n$

$$u(x) > -K(|x| + 1)$$

for some $K > 0$, then u is a polynomial of degree one.

The existence of an estimate from above of the type

$$(5) \qquad |Du(x_0)| \leq f\left(\frac{u(x_0)}{\rho}\right)$$

was announced by E. DE GIORGI [1] at the "Convegno di Analisi Funzionale" (Rome, Italy, March 1968). The considerations made at that time, being purely existential, did not allow an explicit knowledge of the function f; however results (A) and (B) can still be obtained from (5).

II. Notation and preliminary lemmas

We shall use the following notations.

$\mathbf{R}^n$ is the real n-dimensional space, $n \geq 3$, with coordinates $(x_1, x_2, \ldots, x_n)$.

$\mathbf{R}^{n+1}$ is the real $(n + 1)$-dimensional space with coordinates (x, y), $x \in \mathbf{R}^n$.

H_n, H_{n-1} denote respectively the Hausdorff n and $(n-1)$-dimensional measures in $\mathbf{R}^{n+1}$.

$\Omega \subset \mathbf{R}^n$ is an open set; $u \in C^2(\Omega)$ is a function satisfying equation (1).

S is the minimal hypersurface in $\mathbf{R}^{n+1}$ given by the representation

$$S = \{(x, y) \in \Omega \times \mathbf{R} \mid y = u(x)\}.$$

$\pi : \Omega \times \mathbf{R} \to \Omega$ is the projection defined by $\pi(x, y) = x$.

If $v : \Omega \to \mathbf{R}$ is a function, then $v^* = \pi^*(v)$ is the function $v^* : \Omega \times \mathbf{R} \to \mathbf{R}$ defined by

$$v^*(x, y) = v(x).$$

If $A \subset \mathbf{R}^n$ is an open set and v is a function defined on A we shall write

$$\int_A v \, dx = \int_A v^* \, dH_n.$$

For any $x_0 \in \Omega$ and any $\rho < \mathrm{dist}(x_0, \partial\Omega)$, $B_\rho = B_\rho(x_0)$ is the open ball

$$B_\rho = \{x \in \mathbf{R}^n \mid |x - x_0| < \rho\}.$$

We further set

$$C_\rho = C_\rho(x_0) = \{(x, y) \in S \mid x \in B_\rho\}$$

and moreover we let

$$S_\rho = S_\rho(x_0) = \{(x, y) \in S \mid |x - x_0|^2 + |y - u(x_0)|^2 < \rho^2\}.$$

$c_1, c_2, \ldots, c_k, \ldots$ denote constants depending only on n; in case of dependence on other parameters we shall write $c(x, y, z, t, \ldots)$.

$D_1, D_2, \ldots, D_{n+1}$ are the differential operators in $\mathbf{R}^{n+1}$

$$D_i = \frac{\partial}{\partial x_i} \quad i = 1, 2, \ldots, n, \qquad D_{n+1} = \frac{\partial}{\partial y}.$$

We define now a function $\tilde{u} = \tilde{u}(x, y)$ in $\Omega \times \mathbf{R}$ as

$$\tilde{u}(x, y) = y - u(x)$$

and for $i = 1, 2, \ldots, n+1$ we set

$$\nu_i = \frac{D_i \tilde{u}}{|D\tilde{u}|} \quad \text{with} \quad |Df| = \left[\sum_{i=1}^{n+1} (D_i f)^2 \right]^{\frac{1}{2}}.$$

Moreover, for $i = 1, 2, \ldots, n+1$, we set

$$\delta_i = D_i - \nu_i \sum_{h=1}^{n+1} \nu_h D_h.$$

Operators δ_i are themselves differential operators on the space of functions defined on $\Omega \times \mathbf{R}$. For any $\tilde{u}$ we have

$$[\delta_i, \delta_j] = \delta_i \delta_j - \delta_j \delta_i = \sum_{h=1}^{n+1} (\nu_i \delta_j \nu_h - \nu_j \delta_i \nu_h) \delta_h,$$

moreover

$$\sum_{i=1}^{n+1} \nu_i \delta_i = 0.$$

Equation (1) can now be written as

$$\sum_{i=1}^{n+1} \delta_i \nu_i = 0.$$

From the commutation formulae and from the equation above we obtain (Miranda [8])

$$\sum_{i=1}^{n+1} \delta_i \delta_i \nu_h + \left[\sum_{i,j=1}^{n+1} (\delta_i \nu_j)^2 \right] \nu_h = 0$$

for $h = 1, 2, \ldots, n+1$.

We note that the differential operators δ_i just defined satisfy the following relations

$$(6) \qquad \int_S \delta_i \alpha \, dH_n = 0 \quad \text{for} \quad i = 1, 2, \ldots, n+1$$

for any function $\alpha \in \mathcal{C}_0^1(\Omega \times \mathbf{R})$; regarding this we refer to MIRANDA [8], Lemma 2. Relation (6) can be interpreted by saying that the differential operators δ_i can be integrated by parts on S with respect to the measure dH_n.

The following result specifies, in a limit case, a theorem of MIRANDA [7], Theorem 2.

LEMMA 1. – *Let Ω be a convex set and let $g \in H_0^{1,p}(\Omega \times \mathbf{R})$ with $p \geq 1$. Then if $n > p$ we have*

$$(7) \qquad \left(\int_S |g|^{\frac{np}{n-p}} \, dH_n \right)^{\frac{n-p}{n}} \leq c(p,n) \int_S |\delta g|^p \, dH_n,$$

where

$$|\delta g|^2 = \sum_{i=1}^{n+1} (\delta_i g)^2.$$

Moreover if $p = 1$ we have

$$c(1,n) = \beta(n),$$

where $\beta(n)$ is the isoperimetric constant of FEDERER and FLEMING [2].

Proof. Without loss of generality we can suppose that $g \in \mathcal{C}_0^1(\Omega \times \mathbf{R})$ and we set by definition

$$A(t) = S \cap \left\{ (x,y) \in \Omega \times \mathbf{R} \mid |g(x,y)| > t \right\}.$$

From the isoperimetric inequality of FEDERER and FLEMING [2], Rem. 6.6 (see also MIRANDA [7], Theorem 1 and proof) we have

$$(8) \qquad H_n \left(A(t) \right)^{\frac{n-1}{n}} \leq \beta(n) H_{n-1} \left(\partial A(t) \right)$$

where $\beta(n)$ is the isoperimetric constant. From [7], Lemma 1, we obtain

$$H_{n-1}(\partial A(t)) \leq -\frac{d}{dt} \int_{A(t)} |\delta g| \, dH_n$$

at all points where the right hand side of the inequality exists. By combining this result with (8) we get

$$(9) \qquad \int_S |\delta g| \, dH_n \geq \int_0^{+\infty} H_{n-1}(\partial A(t)) \, dt \geq \frac{1}{\beta(n)} \int_0^{+\infty} H_n(A(t))^{\frac{n-1}{n}} \, dt.$$

Since $H_n(A(t))$ is non-increasing with respect to t, we have

$$(10) \qquad \int_0^{+\infty} H_n(A(t)) t^{\frac{1}{n-1}} \, dt \leq \frac{n-1}{n} \left(\int_0^{+\infty} H_n(A(t))^{\frac{n-1}{n}} \, dt \right)^{\frac{n}{n-1}}.$$

On the other hand we have

$$(11) \qquad \frac{n}{n-1} \int_0^{+\infty} H_n(A(t)) t^{\frac{1}{n-1}} \, dt = \int_S |g|^{\frac{n}{n-1}} \, dH_n,$$

whence we obtain Lemma 1 by comparison of inequalities (9), (10), (11) in the case $p = 1$. The general case can be deduced from the case $p = 1$ by using $|g|^{\frac{(n-1)p}{n-p}}$ in place of $|g|$ and by applying the Hölder inequality to the right-hand side.

For the sake of completeness we recall here the quick proof of inequality (10), due to Hardy, Littlewood and Pólya. Let $f(t)$ be non-negative and non-increasing, then

$$tf(t) \leq \int_0^t f(y)\, dy,$$

whence, if $q > 1$, we have

$$q(tf(t))^{q-1} f(t) \leq q \left(\int_0^t f(y)\, dy \right)^{q-1} f(t) = \frac{d}{dt} \left(\int_0^t f(y)\, dy \right)^q.$$

We integrate the latter inequality from 0 to $+\infty$ to obtain

$$(12) \qquad \int_0^{+\infty} f(t)^q t^{q-1}\, dt \leq \frac{1}{q} \left(\int_0^{+\infty} f(t)\, dt \right)^q.$$

In (12) we now set $q = \frac{n}{n-1}$, $f(t) = H_n(A(t))^{\frac{n-1}{n}}$ and obtain (10).

Let $x_0 \in \Omega$, $\rho < \mathrm{dist}(x_0, \partial\Omega)$, let λ, μ be two real numbers with $\lambda < \mu$ and define

$$(13) \qquad C_\rho(\lambda, \mu) = C_\rho(x_0; \lambda, \mu) = C_\rho(x_0) \cap \{(x, y) \in \Omega \times \mathbf{R} \mid \lambda < y \leq \mu\}.$$

Lemma 2. – *Let g be a non-negative function defined on $\Omega \times \mathbf{R}$ and suppose that there exist $K > 0$, $0 < \alpha \leq 1$, β such that for all $x_0 \in \Omega$ and any $\rho < \alpha\, \mathrm{dist}(x_0, \partial\Omega)$ we have*

$$(14) \qquad \int_{S_\rho(x_0)} g\, dH_n \leq K\rho^\beta.$$

Then for all $x_0 \in \Omega$ and any $\rho < \dfrac{\alpha}{2}\, \mathrm{dist}(x_0, \partial\Omega)$ we have

$$(15) \qquad \int_{C_\rho(\lambda,\mu)} g\, dH_n \leq c_3 K \left(\frac{\mu - \lambda}{\rho} + 1 \right) \rho^\beta.$$

Proof. It is enough to prove that under the hypotheses of Lemma 2 we have

$$(16) \qquad \int_{C_\rho(x_0;\lambda,\mu)} g\, dH_n \leq c_4 K \rho^\beta$$

when

$$\mu - \lambda \leq \frac{\rho}{2} \quad \text{and} \quad \rho < \frac{\alpha}{2}\mathrm{dist}(x_0, \partial\Omega).$$

We consider the set

$$\pi(C_\rho(x_0; \lambda, \mu)),$$

projection of $C_\rho(x_0; \lambda, \mu)$ on B_ρ. There exist* c_5 points x_i, $1 \leq i \leq c_5$ such that

(i)
$$x_i \in \pi(C_\rho(x_0; \lambda, \mu))$$

(ii)
$$\pi(C_\rho(x_0; \lambda, \mu)) \subset \bigcup_{i \leq c_5} B_{\rho/2}(x_i).$$

Using condition $\mu - \lambda \leq \frac{\rho}{2}$ it easily follows from (i) and (ii) that

(iii)
$$C_\rho(x_0; \lambda, \mu) \subset \bigcup_{i \leq c_5} S_\rho(x_i);$$

moreover, from $x_i \in B_\rho$, $\rho < \frac{\alpha}{2}\mathrm{dist}(x_0, \partial\Omega)$ we have

$$\rho < \alpha\,\mathrm{dist}(x_i, \partial\Omega), \qquad 1 \leq i \leq c_5.$$

Applying (14) to any $S_\rho(x_i)$ and using (iii) we obtain (16) and the lemma.

In order to state the next lemma it is convenient to introduce the following notation: x, λ, α are points in $\mathbf{R}^m$; $\alpha > 0$, $\lambda > 0$ stand for $\alpha_i \geq 0$, $\lambda_i \geq 0$ for $i = 1, 2, \ldots, m$;

$$\lambda^\alpha = \prod_{i=1}^m \lambda_i^{\alpha_i}; \qquad |\alpha| = \sum_{i=1}^m |\alpha_i|.$$

LEMMA 3. – *Let $T \subset \mathbf{R}^m$ be a convex set, let $f : T \to \mathbf{R}$ be a non-negative function such that*

(17)
$$f(x + \lambda) \leq f(x)$$

for $x \in T$, $x + \lambda \in T$, $\lambda \geq 0$. Let $x_0 \in T$ and $\lambda > 0$, let $\sigma > 1$ be a real number and suppose $x_0 + \frac{\lambda}{\sigma - 1} \in T$. We consider the sequence $\{x_k\}$ defined by

$$x_k = x_{k-1} + \lambda\sigma^{-k}, \qquad k \geq 1.$$

We suppose that for any $k \geq 1$ the inequality

(18)
$$f(x_k) \leq a \frac{\sigma^{k|\alpha|}}{\lambda^\alpha} f(x_{k-1})^\beta$$

holds true, where $\alpha \geq 0$ and $\beta > 1$.

Then, if

(19)
$$f\left(x_0 + \frac{\lambda}{\sigma - 1}\right) > 0$$

---

*Remark. The constant c_5 is not related to the nature of the set $\pi(C_\rho(x_0; \lambda, \mu))$ and the conclusion (i), (ii) is valid for any set contained in B_ρ.

we have

$$(20) \qquad f(x_0) > \sigma^{-\frac{|\alpha|\beta}{(\beta-1)^2}} \left(\frac{\lambda^\alpha}{a}\right)^{\frac{1}{\beta-1}}.$$

Proof. From the convexity of T if $x_0 \in T$, $x_0 + \frac{\lambda}{\sigma-1} \in T$ then $x_k \in T$ when $k \geq 0$. A standard proof by induction on k gives, using (18),

$$(21) \qquad f(x_{k+1})^{\beta-k-1} \leq \left(\frac{a}{\lambda^a}\right)^{\sum_1^k \beta^{-j}} \sigma^{|\alpha| \sum_1^k j\beta^{-j}} f(x_0);$$

now $f(x_{k+1}) \geq f\left(x_0 + \frac{\lambda}{\sigma-1}\right) > 0$ due to (17) and (19), and taking the limit in (21) as $k \to +\infty$ we have the required result.

III. Behaviour of the supersolutions of the operator $\sum\limits_{i=1}^{n+1} \delta_i\delta_i$

Let $u \in C^2(\Omega)$ be a solution of (1) and let $\tilde{u}$, ν_i, δ_i, S be the same as in the previous section. We set

$$v = (1 + |Du|^2)^{\frac{1}{2}}, \qquad w = \log v$$

and let v^*, w^* be the extensions of v, w to $\Omega \times \mathbf{R}$. We have $v^* = \frac{1}{\nu_{n+1}}$.

Fix $x_0 \in \Omega$, $\rho < \text{dist}(x_0, \partial\Omega)$ and t, $t_0 = \nu_{n+1}(x_0, u(x_0)) < t \leq 1$; we denote by $A_{x_0}(\rho, t) = A(\rho, t)$ the connected component of the set

$$S_\rho(x_0) \cap \{(x, y) \in B_\rho \times \mathbf{R} \mid \nu_{n+1}(x, y) < t\}$$

containing the point $(x_0, u(x_0))$. We shall also write

$$\alpha_{x_0}(\rho, t) = \alpha(\rho, t) = H_n(A(\rho, t)).$$

LEMMA 4. – *Let $x_0 \in \Omega$, $\rho < \text{dist}(x_0, \partial\Omega)$, $\nu_{n+1}(x_0, u(x_0)) = t_0 \leq t \leq 1$. Then*

$$(22) \qquad \alpha_{x_0}(\rho, t) \geq c_6 \rho^n \left(\frac{t - t_0}{t}\right)^n.$$

Proof. For simplicity we write $g = \nu_{n+1}$. We have (Miranda [8], equation (3.6)) $g \in C^1(\Omega \times \mathbf{R})$ and

$$(23) \qquad 0 < g \leq 1,$$

$$(24) \qquad \sum_{i=1}^{n+1} \delta_i\delta_i g \leq 0.$$

We denote by $g_{\rho,t}$ the function in $\Omega \times \mathbf{R}$ defined as

$$g_{\rho,t}(x,y) = \begin{cases} t - g(x,y) & \text{if } (x,u(x)) \in A(\rho,t) \\ 0 & \text{elsewhere,} \end{cases}$$

and let φ be a Lipschitz function in $\mathbf{R}^{n+1}$ with compact support in the set

$$|x - x_0|^2 + |y - u(x_0)|^2 < \rho^2.$$

Then the function $\varphi^2 g_{\rho,t}$ is non-negative, Lipschitz in $\Omega \times \mathbf{R}$ and with compact support contained in

$$|x - x_0|^2 + |y - u(x_0)|^2 < \rho^2.$$

Multiplying inequality (24) by $\varphi^2 g_{\rho,t}$, integrating over S and recalling (6) we obtain

$$\int_S \varphi^2 \sum_{i=1}^{n+1} \delta_i g_{\rho,t} \delta_i g \, dH_n \geq -2 \int_S \varphi g_{\rho,t} \sum_{i=1}^{n+1} \delta_i \varphi \delta_i g \, dH_n$$

whence, using the definition of $g_{\rho,t}$, we have

$$\int_{A(\rho,t)} \varphi^2 |\delta g_{\rho,t}|^2 \, dH_n \leq 2 \int_{A(\rho,t)} \varphi g_{\rho,t} |\delta \varphi| |\delta g_{\rho,t}| \, dH_n,$$

and using the CAUCHY inequality

$$(25) \qquad \int_{A(\rho,t)} \varphi^2 |\delta g_{\rho,t}|^2 \, dH_n \leq 4 \int_{A(\rho,t)} g_{\rho,t}^2 |\delta \varphi|^2 \, dH_n.$$

We have $|\delta(\varphi g_{\rho,t})|^2 \leq 2\varphi^2 |\delta g_{\rho,t}|^2 + 2g_{\rho,t}^2 |\delta \varphi|^2$ and, since $0 \leq g_{\rho,t} \leq t$ in $A(\rho,t)$, using (25) we end up with

$$\int_{A(\rho,t)} |\delta(\varphi g_{\rho,t})|^2 \, dH_n \leq 10t^2 \int_{A(\rho,t)} |\delta \varphi|^2 \, dH_n.$$

From this last inequality and from Lemma 1 we have

$$\left(\int_{A(\rho,t)} (\varphi g_{\rho,t})^{\frac{2n}{n-2}} \, dH_n \right)^{\frac{n-2}{n}} \leq 10c(2,n)t^2 \int_{A(\rho,t)} |\delta \varphi|^2 \, dH_n.$$

By a suitable choice of the function φ we have

$$(26) \qquad \left(\int_{A(\rho-\lambda,t)} g_{\rho,t}^{\frac{2n}{n-2}} \, dH_n \right)^{\frac{n-2}{n}} \leq c_7 \frac{t^2}{\lambda^2} \alpha(\rho,t)$$

for $0 < \lambda < \rho$. We have $g_{\rho,t} \geq \mu$ in $A(\rho-\lambda, t-\mu)$, whence, using (26), we obtain

$$(27) \qquad \alpha(\rho - \lambda, t - \mu)^{\frac{n-2}{n}} \leq c_7 \frac{t^2}{\lambda^2 \mu^2} \alpha(\rho,t)$$

for $t_0 < t \leq 1$, $0 < \lambda < \rho$, $0 < \mu < t - t_0$.

From (27) and Lemma 3 we easily derive Lemma 4.

Let $x_0 \in \Omega$, $\rho < \operatorname{dist}(x_0, \partial\Omega)$ and let $\lambda < w(x_0)$. We set

$$w_\lambda = \max(w - \lambda, 0)$$

and let

$$W_{x_0}(\rho, \lambda) = W(\rho, \lambda)$$

be the connected component containing the point x_0 of the set

$$B_\rho(x_0) \cap \{x \in \mathbf{R}^n \mid w_\lambda(x) > 0\}.$$

Recalling that $w^* = \log \frac{1}{\nu_{n+1}}$ we obtain

$$A(\rho, e^{-\lambda}) \subset S_\rho(x_0) \cap \pi^{-1}(W(\rho, \lambda)).$$

Finally, let

$$w_{\rho,\lambda} = w_{x_0,\rho,\lambda} = \begin{cases} w_\lambda & \text{if } x \in W(\rho, \lambda) \\ 0 & \text{elsewhere} \end{cases}$$

and let $w^*_{\rho,\lambda}$ be the extension of $w_{\rho,\lambda}$ to $\Omega \times \mathbf{R}$.

Lemma 5. – *Let $x_0 \in \Omega$, $\rho < \operatorname{dist}(x_0, \partial\Omega)$, let $0 \leq \lambda < \mu < w(x_0) - 1$ and let*

$$x \in W_{x_0}\left(\frac{\rho}{4}, \mu\right).$$

Then we have

$$(28) \qquad \mu - \lambda \leq 1 + c_8 \rho^{-n} \int_{S_{\rho/4}(x)} w^*_{x_0,\rho,\lambda}\, dH_n.$$

Proof. We can suppose that $\lambda < \mu - 1$. Then

$$\int_{S_{\rho/4}(x)} w^*_{\rho,\lambda}\, dH_n = \int_{S_{\rho/4}(x) \cap \pi^{-1}[W(\rho,\lambda)]} (w^* - \lambda)\, dH_n$$

$$\geq \int_{S_{\rho/4}(x) \cap A_{x_0}(\rho, e^{-\mu+1})} (\mu - \lambda - 1)\, dH_n.$$

Now

$$S_{\rho/4}(x) \cap A_{x_0}(\rho, e^{-\mu+1}) \supset A_x\left(\frac{\rho}{4}, e^{-\mu+1}\right)$$

since $x \in W\left(\frac{\rho}{4}, \mu\right)$ by hypothesis. Since $w(x) \geq \mu$ when $x \in W\left(\frac{\rho}{4}, \mu\right)$, from Lemma 4 it follows

$$\int_{S_{\rho/4}(x)} w^*_{\rho,\lambda}\, dH_n \geq (\mu - \lambda - 1)\alpha_x\left(\frac{\rho}{4}, e^{-\mu+1}\right) \geq c_9(\mu - \lambda - 1)\left(\frac{\rho}{4}\right)^n.$$

We then readily get (28).

IV. Proof of the Theorem

We shall prove the Theorem in this equivalent form:

THEOREM. – *If*

$$u(x_0) = 0, \qquad \rho < \operatorname{dist}(x_0, \partial\Omega)$$

then

$$(3') \qquad |Du(x_0)| \le c_1 \exp\left(c_2 \frac{1}{\rho} \sup_{B_\rho(x_0)} (u)\right).$$

Proof. We set

$$\tau(\rho, \lambda) = \tau_{x_0}(\rho, \lambda) = 1 + \frac{1}{\rho} \sup_{W(\rho,\lambda)} (u)$$

where $W(\rho, \lambda)$ has been defined in the previous section; since $u(x_0) = 0$, we have $\tau(\rho, \lambda) \ge 1$. From equation (1) it follows

$$(29) \qquad \int_{B_\rho(x_0)} \frac{1}{v} \sum_{h=1}^{n} D_h u D_h \psi \, dx = 0$$

for all $\psi \in H_0^{1,1}(B_\rho(x_0))$. We set

$$\psi = \varphi w_{\rho,\lambda} e^{su}$$

where $\varphi \in C_0^1(B_\rho(x_0))$ and where s is a positive constant which will be chosen later. Then from (29) we easily get

$$(30) \quad s \int_{B_\rho} \varphi w_{\rho,\lambda} e^{su} \frac{|Du|^2}{v} \, dx \le \int_{B_\rho} w_{\rho,\lambda} e^{su} |D\varphi| \, dx + \int_{B_\rho} \varphi e^{su} |Dw_{\rho,\lambda}| \, dx.$$

Noting that $\frac{1}{v} + \frac{|Du|^2}{v} = v$ and, choosing φ in a suitable way, we have, using (30),

$$(31) \quad \begin{aligned} s \int_{B_{\rho/2}} w_{\rho,\lambda} e^{su} v \, dx \ &\le\ \frac{4}{\rho} \int_{B_\rho} \frac{1}{v} w_{\rho,\lambda} e^{su} v \, dx \\ &\quad + s \int_{B_\rho} \frac{1}{v^2} w_{\rho,\lambda} e^{su} v \, dx + \int_{B_\rho} e^{su} |Dw_{\rho,\lambda}| \, dx. \end{aligned}$$

From the relations

$$D_i = \delta_i + \nu_i \sum_{h=1}^{n+1} \nu_h D_h, \quad i = 1, 2, \ldots, n+1, \quad D_{n+1} w_{\rho,\lambda}^* = 0,$$

we easily obtain

$$|Dw_{\rho,\lambda}^*| \le \sqrt{2} |\delta w_{\rho,\lambda}^*| \cdot v^*,$$

and since we have

$$\int_\Omega g \cdot v \, dx = \int_S g^* \, dH_n$$

for any $g : \Omega \to \mathbf{R}$, from (31) we get

(32)

$$
\begin{aligned}
s \int_{C_{\rho/2}} w^*_{\rho,\lambda} e^{su^*} \, dH_n \;\leq\; & \frac{4}{\rho} \int_{C_\rho} \frac{w^*_{\rho,\lambda}}{v^*} e^{su^*} \, dH_n \\
& + s \int_{C_\rho} \frac{w^*_{\rho,\lambda}}{(v^*)^2} e^{su^*} \, dH_n + \sqrt{2} \int_{C_\rho} e^{su^*} |\delta w^*_{\rho,\lambda}| \, dH_n \\
=\; & \frac{4}{\rho} I_1 + s I_2 + \sqrt{2} I_3.
\end{aligned}
$$

Upper bound for I_1. We have

$$v^* = e^{w^*},$$

whence, if $\lambda \geq 0$,

$$\frac{w^*_{\rho,\lambda}}{v^*} \leq w^* e^{-w^*} < 1.$$

It follows that

(33)
$$I_1 \leq \int_{C_\rho \cap \pi^{-1}(W(\rho,\lambda))} e^{su^*} \, dH_n.$$

For the sake of brevity we shall write $\tau = \tau(\rho,\lambda)$. We have $u^* \leq \rho(\tau - 1)$ in the set

$$C_\rho \cap \pi^{-1}(W(\rho,\lambda))$$

and, by denoting by $C_\rho(\alpha,\beta)$ the set

$$C_\rho(\alpha,\beta) = C_\rho \cap \{(x,y) \in \Omega \times \mathbf{R} \mid \alpha < y \leq \beta\},$$

we obtain from (33)

$$I_1 \leq \sum_{-\infty < k \leq \tau - 1} e^{ks\rho} \int_{C_\rho((k-1)\rho, k\rho)} dH_n.$$

From Lemma 2 and from [8], Theorem 3, we deduce

$$I_1 \leq c_{10} \rho^n \sum_{-\infty < k \leq \tau - 1} e^{ks\rho},$$

for $\rho < \frac{1}{2} \operatorname{dist}(x_0, \partial\Omega)$. Since

$$\sum_{-\infty < k \leq \tau - 1} e^{ks\rho} < \tau e^{\tau s\rho} + \frac{1}{s\rho}$$

we finally obtain

$$(34) \qquad I_1 < c_{10} \left(\tau e^{\tau s \rho} + \frac{1}{s\rho} \right) \rho^n.$$

Upper bound for I_2. Since $v^* \geq 1$, it follows

$$(35) \qquad I_2 \leq I_1.$$

Upper bound for I_3. With a similar reasoning as for the upper bound for I_1, since $|\delta w^*_{\rho,\lambda}| \leq |\delta w^*|$, we have

$$I_3 \leq \sum_{-\infty < k \leq \tau - 1} e^{ks\rho} \int_{C_\rho((k-1)\rho, k\rho)} |\delta w^*| \, dH_n.$$

For any $x \in \Omega$, $\rho < \frac{1}{2} \operatorname{dist}(x_0, \partial\Omega)$, using Theorem 3 and Lemma 3 of [8] we have

$$\int_{S_\rho(x)} |\delta w^*| \, dH_n \leq \left(\int_{S_\rho(x)} dH_n \right)^{\frac{1}{2}} \left(\int_{S_\rho(x)} |\delta w^*|^2 \, dH_n \right)^{\frac{1}{2}} \leq c_{11} \rho^{n-1}.$$

By applying again Lemma 2, we conclude that

$$(36) \qquad I_3 \leq c_{11} \left(\tau e^{\tau s \rho} + \frac{1}{s\rho} \right) \rho^{n-1}$$

for $\rho < \frac{1}{4} \operatorname{dist}(x_0, \partial\Omega)$. Take

$$(37) \qquad s = \frac{1}{\rho\tau};$$

then from (32), (34), (35), (36) we have for any $\lambda \geq 0$

$$(38) \qquad s \int_{C_{\rho/2}} w^*_{\rho,\lambda} e^{su^*} \, dH_n \leq c_{12} \tau \rho^{n-1}$$

provided $\rho < \frac{1}{4} \operatorname{dist}(x_0, \partial\Omega)$ and where s is defined according to (37). We now estimate the left-hand side of (38) from below. Suppose $w(x_0) > 1$. Then, for any μ with $0 \leq \lambda < \mu < w(x_0) - 1$ we have

$$\int_{C_{\rho/2}} w^*_{\rho,\lambda} e^{su^*} \, dH_n \geq \sum_{0 \leq k \leq \frac{1}{4}\left[\tau\left(\frac{\rho}{4},\mu\right)-1\right]} e^{-\frac{1}{2}s\rho} \int_{C_{\rho/2}((k-\frac{1}{2})\rho,(k+\frac{1}{2})\rho)} w^*_{\rho,\lambda} \, dH_n.$$

The set $C_{\rho/2}(-\frac{1}{2}\rho, \frac{1}{2}\rho)$ always contains the set $S_{\rho/4}(x_0)$ and we certainly have $x_0 \in W\left(\frac{\rho}{4}, \mu\right)$.

Now, since $u(x_0) = 0$ and since $W\left(\frac{\rho}{4}, \mu\right)$ is connected, for

$$0 < k \leq \frac{1}{4}\left[\tau\left(\frac{\rho}{4}, \mu\right) - 1\right] - 1$$

the set $C_{\rho/2}((k-\frac{1}{2})\rho, (k+\frac{1}{2})\rho)$ contains a set $S_{\rho/4}(x_k)$ with $x_k \in W\left(\frac{\rho}{4}, \mu\right)$. Let $N \geq 0$ be the largest integer which is not larger than $\frac{1}{4}\left[\tau\left(\frac{\rho}{4}, \mu\right) - 1\right] - 1$; from the previous relations we have

$$\int_{C_{\rho/2}} w^*_{\rho,\lambda} e^{su^*}\, dH_n \geq \sum_{k=0}^{N} e^{-\frac{1}{2}s\rho} \int_{S_{\rho/4}(x_k)} w^*_{\rho,\lambda}\, dH_n$$

with $x_k \in W\left(\frac{\rho}{4}, \mu\right)$, whence, from Lemma 5, we have

$$\int_{C_{\rho/2}} w^*_{\rho,\lambda} e^{su^*}\, dH_n \geq \frac{1}{c_8} e^{-\frac{1}{2}s\rho} \rho^n (\mu - \lambda - 1)(N+1).$$

It is clear that $N + 1 \geq \frac{1}{8}\tau\left(\frac{\rho}{4}, \mu\right)$ and recalling that $s\rho = \frac{1}{\tau} < 1$ we have, using (38) and the previous inequality,

$$\frac{1}{8c_8} \rho^n (\mu - \lambda - 1)\tau\left(\frac{\rho}{4}, \mu\right) \leq c_{12}\tau^2(\rho, \lambda)\rho^n$$

for any $\rho < \frac{1}{4}\operatorname{dist}(x_0, \partial\Omega)$, whence

$$(\mu - \lambda)\tau\left(\frac{\rho}{4}, \mu\right) \leq 8c_8 c_{12}\tau^2(\rho, \lambda) + \tau\left(\frac{\rho}{4}, \mu\right).$$

Since $\lambda < \mu$ we have

$$\tau\left(\frac{\rho}{4}, \mu\right) = 1 + \frac{4}{\rho} \sup_{W\left(\frac{\rho}{4}, \mu\right)} (u) \leq 4\tau(\rho, \lambda) \leq 4\tau^2(\rho, \lambda)$$

and using the last two inequalities it follows

$$(39) \qquad\qquad \tau\left(\frac{\rho}{4}, \mu\right) \leq \frac{c_{13}}{(\mu - \lambda)}\tau^2(\rho, \lambda)$$

for any $0 \leq \lambda < \mu < w(x_0) - 1$, $\rho < \frac{1}{4}\operatorname{dist}(x_0, \partial\Omega)$.

By using Lemma 3 we obtain

$$\tau(\rho, \lambda) > c_{14}[w(x_0) - 1 - \lambda]$$

and if $\lambda = 0$ we have

$$(40) \qquad\qquad w(x_0) < c_{15}\tau(\rho, 0) \quad \text{per} \quad \rho < \frac{1}{4}\operatorname{dist}(x_0, \partial\Omega).$$

Inequality (40) clearly holds also when $w(x_0) \leq 1$, and the restriction $\rho < \frac{1}{4}\operatorname{dist}(x_0, \partial\Omega)$ can be avoided by observing that

$$\tau(\rho, 0) \leq 4\tau(4\rho, 0),$$

whence

$$w(x_0) < 4c_{15}\tau(\rho, 0) \quad \text{for} \quad \rho < \operatorname{dist}(x_0, \partial\Omega).$$

By taking the exponential we get (3') and the Theorem is completely proved.

V. Final remarks

Let $\rho < \frac{1}{2}\operatorname{dist}(x_0, \partial\Omega)$, let $u(x_0) = 0$, $M = M_\rho = \sup_{B_\rho}(u)$. Suppose $M > \rho$; by reasoning like in the lower bound of the left hand side of (38) we can prove that

$$
\begin{aligned}
A(\rho) \;&=\; H_n(C_\rho \cap \{(x,y) \in \Omega \times \mathbf{R} \mid u(x) > 0\}) \\[2mm]
&=\; \int_{C_\rho(0,+\infty)} dH_n \;\geq\; \sum_{1 \leq k \leq M/\rho} \int_{S_{\rho/2}(z_k)} dH_n
\end{aligned}
$$

for suitably chosen points $z_k \in B_{\rho/2}$. From Theorem 3.8 of [9] we have

$$
\int_{S_{\rho/2}(z_k)} dH_n > c_{16}(\rho/2)^n
$$

hence se have, also for $M \leq \rho$,

$$
\frac{M}{\rho} < c_{17}\frac{A(\rho)}{\rho^n} + 1.
$$

By using (3') we obtain

$$
(41) \qquad |Du(x_0)| \leq c_{18}\exp\left(c_{19}\frac{A(\rho)}{\rho^n}\right).
$$

and it is easy to show that this inequality, which was proved for $\rho < \frac{1}{2}\operatorname{dist}(x_0, \partial\Omega)$, is also true for $\rho < \operatorname{dist}(x_0, \partial\Omega)$, possibly with a different choice of the constant c_{19}.

Inequality (41) is similar to an inequality of R. FINN [4] in the case $n = 2$.

Bibliography

[1] E. DE GIORGI, *Maggiorazioni a priori relative alle ipersuperfici minimali.* Atti del Convegno di Analisi Funzionale, Roma (1968).

[2] H. FEDERER, W. H. FLEMING, *Normal and integral currents.* Ann. of Math. **72** (1960).

[3] R. FINN, *A property of minimal surfaces.* Proc. Nat. Acad. Sci. **39** (1953).

[4] R. FINN, *On equations of minimal surfaces type.* Ann. of Math. **60** (1954).

[5] R. FINN, *New estimates for equations of minimal surface type.* Arch. Rational Mech. Anal. **14** (1963).

[6] H. JENKINS, J. SERRIN, *Variational problems of minimal surface type, I.* Arch. Rational Mech. Anal. **12** (1963).

[7] M. MIRANDA, *Disuguaglianze di Sobolev sulle ipersuperfici minimali.* Rend. Sem. Mat. Univ. Padova **38** (1967).

[8] M. MIRANDA, *Una maggiorazione integrale per le curvature delle ipersuperfici minimali.* Rend. Sem. Mat. Univ. Padova **38** (1967).

[9] M. MIRANDA, *Sul minimo dell'integrale del gradiente di una funzione.* Ann. Scuola Norm. Sup. Pisa **19** (1965).

[10] J. MOSER, *On Harnack's theorem for elliptic differential equations.* Comm. Pure Appl. Math **14** (1961).

[11] J. C. C. NITSCHE, *On new results in the theory of minimal surfaces.* Bull. Amer. Math. Soc. **71** (1965).

[12] J. SERRIN, *A priori estimates for solutions of the minimal surface equation.* Arch. Rational Mech. Anal. **14** (1963).

[13] J. SERRIN, *Addendum to "A priori estimates for solutions of the minimal surface equation".* Arch. Rational Mech. Anal. **28**, 149–154 (1968).

Istituto Matematico Università, Pisa
Scuola Normale Superiore, Pisa
Istituto Matematico Università, Pisa
and
University of Minnesota, Minneapolis

(Communicated on July 30, 1968)

A representation formula for analytic functions in $\mathbf{R}^n$[‡]

Ennio De Giorgi (Pisa), Lamberto Cattabriga (Ferrara)

Summary. An integral representation formula for analytic functions in the whole n-dimensional real Euclidean space is proved.

In this paper we prove a representation formula for analytic functions in the real Euclidean n-dimensional space that we found to be useful in the proof of the existence of an analytic solution in the real plane of every partial differential equation in two variables with constant coefficients and analytic right hand side.

We couldn't find this formula in the literature. As far as we know, similar representations of analytic functions of real variables by means of suitable potentials have been considered by F. MANTOVANI and S. SPAGNOLO [2]. Our result is contained in the theorem stated below. It is preceded by two preliminary lemmata.

We denote by $(x_1, \ldots, x_n, t_1, \ldots, t_m)$ or briefly by (x, t) a point of the real $n + m$-dimensional Euclidean space $\mathbf{R}^{n+m}$. We also set

$$|x| = \left(\sum_{j=1}^{n} x_j^2 \right)^{\frac{1}{2}}, \qquad |t| = \left(\sum_{j=1}^{m} t_j^2 \right)^{\frac{1}{2}},$$

$$|\operatorname{grad} u| = \left(\sum_{j=1}^{n} \left| \frac{\partial u}{\partial x_j} \right|^2 + \sum_{j=1}^{m} \left| \frac{\partial u}{\partial t_j} \right|^2 \right)^{\frac{1}{2}},$$

$$\Delta u = \sum_{j=1}^{n} \frac{\partial^2 u}{\partial x_j^2} + \sum_{j=1}^{m} \frac{\partial^2 u}{\partial t_j^2}.$$

LEMMA I. – *Given a positive nonincreasing function $\varphi(s)$ of the real variable s, there exists a positive nonincreasing infinitely differentiable function $\psi(s)$, such that*

$$\psi(s) + \left| \frac{d\psi}{ds} \right| + \left| \frac{d^2\psi}{ds^2} \right| \leq \varphi(s)$$

for every real s.

[‡]Editor's note: translation into English of the paper "Una formula di rappresentazione per funzioni analitiche in $\mathbf{R}^n$", published in Boll. Un. Mat. Ital., (4) 4 (1971), 1010–1014.

Proof. Let us consider a nonnegative infinitely differentiable function $\gamma(s)$ in $\mathbf{R}$, identically zero for $|s| \geq 1$ and such that

$$0 < \int_{-\infty}^{+\infty} \left(\gamma + \left| \frac{d\gamma}{ds} \right| + \left| \frac{d^2\gamma}{ds^2} \right| \right) ds < 1.$$

The function

$$\psi(s) = \int_{-\infty}^{+\infty} \gamma(s-t)\varphi(t+1)dt$$

has then the required properties.

Lemma II. – *Let w be a complex valued function which is continuous with its first and second order derivatives in $\mathbf{R}^{n+m}$, with $n \geq 1$, $m > 2$. If there exists a positive constant σ such that*

$$\int_{\mathbf{R}^m} \left(|w(x,t)| + |grad\, w(x,t)| + |\Delta w(x,t)| \right) dt \leq \sigma$$

then for every $x \in \mathbf{R}^n$

$$w(x,0) = -\frac{1}{(n+m-2)\omega_{n+m}} \int_{\mathbf{R}^{n+m}} \frac{\Delta w(\xi,t)}{\left[\sum_{j=1}^{n} (x_j - \xi_j)^2 + |t|^2 \right]^{(n+m-2)/2}} d\xi\, dt$$

where ω_{n+m} is the measure of the boundary of the unit ball in $\mathbf{R}^{n+m}$.

Proof. To begin with, let us remark that from the assumptions it follows that for every $\rho > 0$ one has that

$$\int_{|x|^2+|t|^2 \leq \rho^2} \left(|w| + |grad w| + |\Delta w| \right) dx\, dt \leq \Omega_n \rho^n \sigma,$$

where Ω_n denotes the volume of the unit ball in $\mathbf{R}^n$. Now let $\gamma : \mathbf{R} \to \mathbf{R}$ be an infinitely differentiable function such that

$$\gamma(s) = 1 \text{ for } s \leq 0, \qquad \gamma(s) = 0 \text{ for } s \geq 1.$$

For every $\rho > 0$ we consider the function

$$\beta(x,t) = w(x,t)\gamma\left((|x|^2 + |t|^2)^{\frac{1}{2}} - \rho \right).$$

Then well known results of potential theory ensure that

$$\beta(x,0) = -\frac{1}{(n+m-2)\omega_{n+m}} \int_{\mathbf{R}^{n+m}} \frac{\Delta\beta(\xi,t)}{\left[\sum_{j=1}^{n} (x_j - \xi_j)^2 + |t|^2 \right]^{(n+m-2)/2}} d\xi\, dt.$$

From this equality we obtain the thesis as $\rho \to +\infty$.

THEOREM. – *Let $f(x)$ be a real or complex valued analytic function in $\mathbf{R}^n$, $n \geq 1$; let φ be a positive nonincreasing function defined in $\mathbf{R}$ and let m be a natural number greater than two. Then there exist a function $g : \mathbf{R}^{n+1} \to \mathbf{C}$ infinitely differentiable in $\mathbf{R}^{n+1}$ and a positive nonincreasing infinitely differentiable function ψ in $\mathbf{R}$ such that*

$$(1) \qquad f(x) = \int_{\mathbf{R}^{n+1}} \frac{g(\xi, \tau)}{\left[\displaystyle\sum_{j=1}^{n} (x_j - \xi_j)^2 + \tau^2 \right]^{(n+m-2)/2}} \, d\xi \, d\tau \quad \text{for every } x \in \mathbf{R}^n,$$

$$\operatorname{supp} g \subset \left\{ (\xi, \tau) \in \mathbf{R}^{n+1}; \ \psi\left(|\xi|^2\right) \leq \tau \leq 2\psi\left(|\xi|^2\right) \right\},$$

$$\varphi\left(|\xi|^2\right) \geq \int_0^{+\infty} |g(\xi, \tau)| \, d\tau \quad \text{for every } \xi \in \mathbf{R}^n.$$

Proof. Let us consider the $n + m$-dimensional Euclidean space $\mathbf{R}^{n+m} = \mathbf{R}^n \times \mathbf{R}^m$, $x = (x_1, \ldots, x_n) \in \mathbf{R}^n$, $t = (t_1, \ldots, t_m) \in \mathbf{R}^m$. By well known results on the Cauchy problem, there exist an open set $A \subset \mathbf{R}^{n+m}$ containing the hyperplane $t_m = 0$ and a function $u(x, t)$ which is harmonic in A and satisfies the initial conditions

$$\begin{cases} f(x) = u(x_1, \ldots, x_n, t_1, \ldots, t_{m-1}, 0), \\[2mm] 0 = \dfrac{\partial u}{\partial t_m}(x_1, \ldots, x_n, t_1, \ldots, t_{m-1}, 0). \end{cases}$$

Then there exists a continuous, positive and nonincreasing function $\alpha(s)$, such that the set

$$A(s) = \left\{ (x, t) \in \mathbf{R}^{n+m}; \ |x|^2 \leq s, \ |t| \leq \alpha\left(|x|^2\right) \right\}$$

is contained in A for every $s \geq 0$. Applying Lemma I we can find a nonincreasing infinitely differentiable function ψ in $\mathbf{R}$ such that

$$(2) \qquad \begin{cases} 0 < \psi(s) \leq [\varphi(s)]^2, \quad 2\psi(s) \leq \alpha(s), \\[2mm] \psi(s) + \left|\dfrac{d\psi}{ds}\right| + \left|\dfrac{d^2\psi}{ds^2}\right| \leq [p + s + \beta(s)]^{-4} \end{cases}$$

for every $s \geq 0$, where p is an arbitrary real number greater or equal than one, and

$$\beta(s) = \max\left\{ |u| + |\operatorname{grad} u|; \ (x, t) \in A(s) \right\}.$$

Now let $\gamma(s)$ be an infinitely differentiable function in $\mathbf{R}$ such that

$$\gamma(s) = 0 \text{ for } s \geq 4, \qquad \gamma(s) = 1 \text{ for } s \leq 1,$$

$$0 \leq \gamma(s) \leq 1, \quad \left|\dfrac{d\gamma}{ds}\right| \leq 1, \quad \left|\dfrac{d^2\gamma}{ds^2}\right| \leq 1 \quad \text{for every } s \in \mathbf{R}.$$

If we consider the function

$$
w(x,t) = \begin{cases} u(x,t)\gamma\left(\dfrac{|t|^2}{\left[\psi\left(|x|^2\right)\right]^2}\right) & \text{for } (x,t) \in A \\[1em] 0 & \text{for } (x,t) \in \mathbf{R}^{n+m} \setminus A \end{cases}
$$

we have that

$$
\begin{aligned}
\Delta w &= 0 && \text{for } |t| \leq \psi\left(|x|^2\right), \\
w = \Delta w &= 0 && \text{for } |t| \geq 2\psi\left(|x|^2\right), \\
f(x) &= w(x,0) && \text{for every } x \in \mathbf{R}^n.
\end{aligned}
$$

By the inequalities (2) we can apply Lemma II to the function w and therefore we can express f by means of the Newtonian potential with density Δw. Therefore, the representation formula (1) holds with

$$
g(\xi,\tau) = -\frac{1}{(n+m-2)\omega_{n+m}} \int_{|t|=\tau} \Delta w(\xi,t)dt
$$

for $\tau > 0$ and $g(\xi,\tau) = 0$ for $\tau \leq 0$. Finally, it is easy to see that the last inequality stated in the theorem is satisfied, provided the number p appearing in (2) is large enough.

Bibliography

[1] E. De Giorgi – L. Cattabriga, *Una dimostrazione diretta dell'esistenza di soluzioni analitiche nel piano reale di equazioni a derivate parziali a coefficienti costanti*, Boll. Un. Mat. Ital., **4** (1971), 1015–1027.

[2] F. Mantovani – S. Spagnolo, *Funzionali analitici reali e funzioni armoniche*, Ann. Scuola Norm. Sup. Pisa, **18** (1964), 475–512.

Una formula di rappresentazione per funzioni analitiche in $\mathbf{R}^n$[‡]

Ennio De Giorgi (Pisa), Lamberto Cattabriga (Ferrara)

Summary. – *An integral representation formula for analytic functions in the whole n-dimensional real euclidean space is proved*

In questo lavoro proviamo una formula di rappresentazione per funzioni analitiche nello spazio euclideo reale n-dimensionale che abbiamo trovato utile nella dimostrazione dell'esistenza di una soluzione analitica nel piano reale di ogni equazione a derivate parziali in due variabili con coefficienti costanti e termine noto analitico [1].

Non siamo riusciti a trovare questa formula nella letteratura. A nostra conoscenza analoghi procedimenti di rappresentazione di funzioni analitiche di variabili reali mediante opportuni potenziali sono stati considerati da F. MANTOVANI e S. SPAGNOLO [2]. Il nostro risultato è contenuto nel teorema enunciato più sotto. Esso è preceduto da due lemmi preliminari.

Indicheremo con $(x_1, \ldots, x_n, t_1, \ldots, t_m)$ o brevemente con (x, t) un punto dello spazio euclideo reale $n + m$-dimensionale $\mathbf{R}^{n+m}$. Porremo inoltre

$$|x| = \left(\sum_{j=1}^{n} x_j^2 \right)^{\frac{1}{2}}, \qquad |t| = \left(\sum_{j=1}^{m} t_j^2 \right)^{\frac{1}{2}},$$

$$|\operatorname{grad} u| = \left(\sum_{j=1}^{n} \left| \frac{\partial u}{\partial x_j} \right|^2 + \sum_{j=1}^{m} \left| \frac{\partial u}{\partial t_j} \right|^2 \right)^{\frac{1}{2}},$$

$$\Delta u = \sum_{j=1}^{n} \frac{\partial^2 u}{\partial x_j^2} + \sum_{j=1}^{m} \frac{\partial^2 u}{\partial t_j^2}.$$

LEMMA I. – *Data una funzione positiva non crescente $\varphi(s)$ della variabile reale s, esiste una funzione positiva non crescente $\psi(s)$ indefinitamente differenziabile, tale che*

$$\psi(s) + \left| \frac{d\psi}{ds} \right| + \left| \frac{d^2\psi}{ds^2} \right| \leq \varphi(s)$$

per ogni s reale.

[‡]Editor's note: published in Boll. Un. Mat. Ital., (4) 4 (1971), 1010–1014.

Dim. Consideriamo una funzione non negativa $\gamma(s)$ indefinitamente differenziabile in $\mathbf{R}$, identicamente nulla per $|s| \geq 1$ e tale che

$$0 < \int_{-\infty}^{+\infty} \left(\gamma + \left| \frac{d\gamma}{ds} \right| + \left| \frac{d^2\gamma}{ds^2} \right| \right) ds < 1.$$

La funzione

$$\psi(s) = \int_{-\infty}^{+\infty} \gamma(s-t)\varphi(t+1)dt$$

ha allora le proprietà richieste.

Lemma II. – *Sia w una funzione a valori complessi continua con le sue derivate prime e seconde in $\mathbf{R}^{n+m}$, con $n \geq 1$, $m > 2$. Se esiste una costante positiva σ tale che*

$$\int_{\mathbf{R}^m} \left(|w(x,t)| + |\operatorname{grad} w(x,t)| + |\Delta w(x,t)| \right) dt \leq \sigma$$

allora per ogni $x \in \mathbf{R}^n$

$$w(x,0) = -\frac{1}{(n+m-2)\omega_{n+m}} \int_{\mathbf{R}^{n+m}} \frac{\Delta w(\xi,t)}{\left[\displaystyle\sum_{j=1}^{n} (x_j - \xi_j)^2 + |t|^2 \right]^{(n+m-2)/2}} \, d\xi \, dt$$

ove ω_{n+m} è la misura della frontiera della sfera unitaria in $\mathbf{R}^{n+m}$.

Dim. Osserviamo anzitutto che dalle ipotesi segue che per ogni $\rho > 0$ è

$$\int_{|x|^2+|t|^2 \leq \rho^2} \left(|w| + |\operatorname{grad} w| + |\Delta w| \right) dx \, dt \leq \Omega_n \rho^n \sigma,$$

ove Ω_n indica il volume della sfera unitaria in $\mathbf{R}^n$. Sia ora $\gamma : \mathbf{R} \to \mathbf{R}$ una funzione indefinitamente differenziabile tale che

$$\gamma(s) = 1 \text{ per } s \leq 0, \qquad \gamma(s) = 0 \text{ per } s \geq 1.$$

Per ogni $\rho > 0$ consideriamo la funzione

$$\beta(x,t) = w(x,t)\gamma\left((|x|^2 + |t|^2)^{\frac{1}{2}} - \rho \right).$$

Risultati ben noti di teoria del potenziale assicurano allora che

$$\beta(x,0) = -\frac{1}{(n+m-2)\omega_{n+m}} \int_{\mathbf{R}^{n+m}} \frac{\Delta\beta(\xi,t)}{\left[\displaystyle\sum_{j=1}^{n} (x_j - \xi_j)^2 + |t|^2 \right]^{(n+m-2)/2}} \, d\xi \, dt.$$

Da questa eguaglianza per $\rho \to +\infty$ si ottiene la tesi.

TEOREMA. – *Sia $f(x)$ una funzione a valori reali o complessi analitica in* $\mathbf{R}^n$, $n \geq 1$; *sia φ una funzione positiva non crescente definita in $\mathbf{R}$ ed m un numero naturale maggiore di due. Esistono allora una funzione $g : \mathbf{R}^{n+1} \to \mathbf{C}$ indefinitamente differenziabile in $\mathbf{R}^{n+1}$ ed una funzione positiva non crescente ψ indefinitamente differenziabile in $\mathbf{R}$ tali che*

$$(1) \qquad f(x) = \int_{\mathbf{R}^{n+1}} \frac{g(\xi,\tau)}{\left[\sum_{j=1}^{n}(x_j - \xi_j)^2 + \tau^2\right]^{(n+m-2)/2}} \, d\xi \, d\tau \quad \text{per ogni } x \in \mathbf{R}^n,$$

$$\operatorname{supp} g \subset \left\{ (\xi,\tau) \in \mathbf{R}^{n+1}; \ \psi\left(|\xi|^2\right) \leq \tau \leq 2\psi\left(|\xi|^2\right) \right\},$$

$$\varphi\left(|\xi|^2\right) \geq \int_0^{+\infty} |g(\xi,\tau)| d\tau \quad \text{per ogni } \xi \in \mathbf{R}^n.$$

Dim. Consideriamo lo spazio euclideo reale $n + m$-dimensionale $\mathbf{R}^{n+m} = \mathbf{R}^n \times \mathbf{R}^m$, $x = (x_1,\ldots,x_n) \in \mathbf{R}^n$, $t = (t_1,\ldots,t_m) \in \mathbf{R}^m$. Ben noti risultati sul problema di Cauchy assicurano che esiste un insieme aperto $A \subset \mathbf{R}^{n+m}$ contenente l'iperpiano $t_m = 0$ e una funzione $u(x,t)$ armonica in A soddisfacente le condizioni iniziali

$$\begin{cases} f(x) = u(x_1,\ldots,x_n,t_1,\ldots,t_{m-1},0), \\ 0 = \dfrac{\partial u}{\partial t_m}(x_1,\ldots,x_n,t_1,\ldots,t_{m-1},0). \end{cases}$$

Esisterà allora una funzione $\alpha(s)$ continua, positiva e non crescente, tale che l'insieme

$$A(s) = \left\{ (x,t) \in \mathbf{R}^{n+m}; \ |x|^2 \leq s, \ |t| \leq \alpha\left(|x|^2\right) \right\}$$

sia contenuto in A per ogni $s \geq 0$. Applicando il Lemma I possiamo trovare una funzione non crescente ψ indefinitamente differenziabile in $\mathbf{R}$ soddisfacente le condizioni

$$(2) \qquad \begin{cases} 0 < \psi(s) \leq [\varphi(s)]^2, \ \ 2\psi(s) \leq \alpha(s), \\ \psi(s) + \left|\dfrac{d\psi}{ds}\right| + \left|\dfrac{d^2\psi}{ds^2}\right| \leq [p + s + \beta(s)]^{-4} \end{cases}$$

per ogni $s \geq 0$, ove p è un arbitrario numero reale maggiore od uguale ad uno e

$$\beta(s) = \max\left\{ |u| + |\operatorname{grad} u|; \ (x,t) \in A(s) \right\}.$$

Sia ora $\gamma(s)$ una funzione indefinitamente differenziabile in $\mathbf{R}$ tale che

$$\gamma(s) = 0 \text{ per } s \geq 4, \qquad \gamma(s) = 1 \text{ per } s \leq 1,$$

$$0 \leq \gamma(s) \leq 1, \quad \left|\dfrac{d\gamma}{ds}\right| \leq 1, \quad \left|\dfrac{d^2\gamma}{ds^2}\right| \leq 1 \quad \text{per ogni } s \in \mathbf{R}.$$

Se noi consideriamo la funzione

$$w(x,t) = \begin{cases} u(x,t)\gamma\left(\dfrac{|t|^2}{\left[\psi\left(|x|^2\right)\right]^2}\right) & \text{per } (x,t) \in A \\[4mm] 0 & \text{per } (x,t) \in \mathbf{R}^{n+m} \setminus A \end{cases}$$

avremo

$$\Delta w = 0 \qquad \text{per } |t| \leq \psi\left(|x|^2\right),$$
$$w = \Delta w = 0 \quad \text{per } |t| \geq 2\psi\left(|x|^2\right),$$
$$f(x) = w(x,0) \quad \text{per ogni } x \in \mathbf{R}^n.$$

Le maggiorazioni (2) ci consentono ora di applicare il Lemma II alla funzione w ed esprimere quindi la funzione f mediante il potenziale newtoniano avente Δw come densità. La formula di rappresentazione (1) varrà allora con

$$g(\xi,\tau) = -\frac{1}{(n+m-2)\omega_{n+m}} \int_{|t|=\tau} \Delta w(\xi,t)dt$$

per $\tau > 0$ e $g(\xi,\tau) = 0$ per $\tau \leq 0$. Infine si vede facilmente che affinché l'ultima diseguaglianza richiesta dal teorema sia soddisfatta basta scegliere il numero p che compare nella (2) abbastanza grande.

Bibliografia

[1] E. De Giorgi – L. Cattabriga, *Una dimostrazione diretta dell'esistenza di soluzioni analitiche nel piano reale di equazioni a derivate parziali a coefficienti costanti*, Boll. Un. Mat. Ital., **4** (1971), 1015–1027.

[2] F. Mantovani – S. Spagnolo, *Funzionali analitici reali e funzioni armoniche*, Ann. Scuola Norm. Sup. Pisa, **18** (1964), 475–512.

A direct proof of the existence of analytic solutions in the real plane of partial differential equations with constant coefficients[‡]

Ennio De Giorgi (Pisa), Lamberto Cattabriga (Ferrara)

Summary. A direct proof of the following theorem is given: let $f(x,y)$ be an analytic function in $\mathbf{R}^2$ and let P be a linear partial differential operator with (real or complex) constant coefficients. Then there exists at least one function $u(x,y)$ analytic in $\mathbf{R}^2$ which satisfies the equation $Pu = f(x,y)$ for every $(x,y) \in \mathbf{R}^2$. A conjecture about possible counterexamples to this result in $\mathbf{R}^3$ is also formulated.

Aim of this paper is to provide a direct proof of the existence of analytic solutions in the two dimensional space $\mathbf{R}^2$ of a linear partial differential equation with constant coefficients and analytic right hand side. Such a proof doesn't require advanced mathematical tools, but only some elementary facts about the Fourier transform and ordinary differential equations with constant coefficients. Moreover, this proof uses a representation formula for analytic functions proved in [3]. Nevertheless, for the reader's convenience, we quoted in the references some of the papers dealing with this or similar topics in a more general context. In particular we point out that during the preparation of this paper we learned about the papers [10, 11, 12] by T. KAWAI, which seem to contain the more recent and general results on this subject.

The result of this paper is contained in the theorem stated in section 3. Section 1 contains some elementary propositions, and section 2 contains two preliminary lemmata. In section 4 we state a conjecture about possible counterexamples to our result in the three dimensional case.

1. – PROPOSITION I. – *If λ, a, γ_h, $h = 0, \ldots, n-1$, are complex numbers such that*

$$\lambda^n + \sum_{h=0}^{n-1} \gamma_h a^{n-h} \lambda^h = 0,$$

then

$$|\lambda| \le |a| \left(1 + \sum_{h=0}^{n-1} |\gamma_h|\right).$$

[‡]Editor's note: translation into English of the paper "Una dimostrazione diretta dell'esistenza di soluzioni analitiche nel piano reale di equazioni a derivate parziali a coefficienti costanti", published in Boll. Un. Mat. Ital., (4) 4 (1971), 1015–1027.

PROPOSITION II. – *Let $P(z) = (z - \mu_1) \cdots (z - \mu_n)$ and $Q(z)$ be two polynomials of degree n in the complex variable z. Let λ be a solution of the equation $Q(z) = 0$ and let a, ϵ be two positive numbers such that*

$$a \geq |\mu_h|, \ h = 1, \ldots, n, \qquad\qquad \epsilon \leq 2^{-n},$$

$$|Q(z) - P(z)| \leq \left(a|z|^{n-1} + a^n\right) \epsilon$$

for every z. Then

$$\min_h |\lambda - \mu_h| \leq 2a\epsilon^{1/n}.$$

Proof. Setting $\sigma = \min_h |\lambda - \mu_h|$, we have that

$$\sigma^n \leq |P(\lambda)| = |P(\lambda) - Q(\lambda)| \leq \left(a(a + \sigma)^{n-1} + a^n\right) \epsilon.$$

If by contradiction $\sigma > a$, then

$$\sigma^n < a 2^n \epsilon \sigma^{n-1} < a \sigma^{n-1}.$$

This proves that $0 \leq \sigma \leq a$, hence $\sigma^n \leq 2^n a^n \epsilon$.

PROPOSITION III. – *Given the ordinary differential equation with complex constant coefficients*

$$(1.1) \qquad\qquad \frac{d^n w}{dy^n} + \sum_{h=0}^{n-1} b_h \frac{d^h w}{dy^h} = q(y)$$

let $\lambda_1, \ldots, \lambda_n$ be the roots of its characteristic equation

$$\lambda^n + \sum_{h=0}^{n-1} b_h \lambda^h = 0.$$

Let r and η be real numbers, and let s, α be positive numbers such that

$$|\lambda_j| \leq \alpha, \quad |\operatorname{Re} \lambda_j - r| \geq s + 1, \quad j = 1, \ldots, n,$$

where $\operatorname{Re} \lambda_j$ denotes the real part of λ_j, and

$$|q(y)| \leq \exp\left[ry - s|y - \eta|\right]$$

for every real y. Then there exists a unique solution w of equation (1.1) such that

$$\lim_{y \to +\infty} w(y) \exp[-ry] = \lim_{y \to -\infty} w(y) \exp[-ry] = 0.$$

Such a solution satisfies the inequalities

$$\left| \frac{d^h w}{dy^h} \right| \leq \exp[ry - s|y - \eta|](\alpha + 1)^h, \quad h = 0, \ldots, n,$$

for every real y.

Proof. For $n = 1$ we consider either the solution of (1.1) given by

$$w(y) = \exp[-b_0 y] \int_{-\infty}^{y} \exp[b_0 t] q(t) dt \quad \text{if } \operatorname{Re} b_0 + r \geq s + 1$$

or the solution given by

$$w(y) = -\exp[-b_0 y] \int_{y}^{\infty} \exp[b_0 t] q(t) dt \quad \text{if } \operatorname{Re} b_0 + r \leq -s - 1.$$

In both cases it is easy to see that

$$|w(y)| \leq \exp[ry - s|y - \eta|] \quad \text{for every real } y.$$

From this inequality, using equation (1.1), if follows that

$$\left| \frac{dw}{dy} \right| \leq (|b_0| + 1) \exp[ry - s|y - \eta|] \leq (\alpha + 1) \exp[ry - s|y - \eta|]$$

for every real y.

For $n > 1$ we consider the system

$$\begin{cases} \dfrac{dw_h}{dy} = \lambda_h w_h + w_{h-1}, & h - 2, \ldots, n \\[2ex] \dfrac{dw_1}{dy} = \lambda_1 w_1 + q(y) \end{cases}$$

and we remark that if $(w_1, \ldots, w_n)$ is a solution of this system, then w_n is a solution of (1.1).

PROPOSITION IV. – *Given the differential equation (1.1), we still denote by $\lambda_1, \ldots, \lambda_n$ the roots of its characteristic equation, and we assume that s, γ are two positive numbers such that*

$$s \geq |\lambda_h| + 1, \qquad h = 1, \ldots, n,$$

and

$$|q(y)| \leq \exp[s|y|]$$

for every y with $|y| \leq \gamma$. Then the solution w of (1.1) with initial data

$$\frac{d^h w(0)}{dy^h} = 0, \qquad h = 0, \ldots, n - 1,$$

satisfies for $|y| \leq \gamma$ the inequalities

$$\left| \frac{d^h w}{dy^h} \right| \leq s^h \exp[s|y|], \qquad h = 0, \ldots, n.$$

Proof. It is easy to prove the proposition for $n = 1$. For $n > 1$, the proof can be achieved as in the preceding proposition.

Proposition V. – *Let $u(x,y)$ and $f(x,y)$ be two indefinitely differentiable functions in an open set $B \subset \mathbf{R}^2$. Assume that in B we have*

$$(1.2) \qquad \frac{\partial^n u}{\partial y^n} + \sum_{h=0}^{n-1} c_h \frac{\partial^n u}{\partial y^h \partial x^{n-h}} + \sum_{0 \le h+k \le n-1} c_{hk} \frac{\partial^{h+k} u}{\partial y^h \partial x^k} = f(x,y)$$

and

$$(1.3) \qquad \left| \frac{\partial^{s+r} f}{\partial x^s \partial y^r} \right| \le (s+r)! \, \alpha^s \beta^r \qquad \text{for } s,r = 0,1,\ldots$$

with

$$(1.3') \qquad \left| \frac{\partial^{s+r} u}{\partial x^s \partial y^r} \right| \le (s+r)! \, \alpha^s \beta^r \qquad \text{for } s = 0,1,\ldots; r = 0,\ldots,n$$

with

$$\alpha > 1, \qquad \beta > \alpha \left(1 + \sum_{h=0}^{n-1} |c_h| + \sum_{0 \le h+k \le n-1} |c_{hk}| \right);$$

then $(1.3')$ is satisfied for every natural number r.

Proof. It is enough to take the derivative of (1.2), and then to proceed by induction on r.

Proposition VI. – *Let y, η, τ be real numbers with $\eta \ge 4$, $\tau > 0$. Then the Fourier transform of the function $[x^2 + (y-\eta)^2 + \tau^2]^{-2}$, i.e. the function*

$$q(p,y,\tau) = \int_{-\infty}^{+\infty} \exp[-ixp][x^2 + (y-\eta)^2 + \tau^2]^{-2} dx$$

satisfies the inequalities

$$(1.4) \qquad 0 < q(p,y,\tau) \le \exp\left[-\frac{|p|}{4}\tau \right] \exp\left[-\frac{|p|}{2}|y-\eta| \right]$$

for $|p| > p_0(\tau) = \max\{0, 24\tau^{-1}(1 - \log\tau)\}$ for every real y;

$$(1.5) \qquad 0 < q(p,y,\tau) \le \exp\left[-\frac{|p|}{2}|y-\eta| \right]$$

for every real p and for $|y| \le \eta/4$;

$$(1.6) \qquad 0 < q(p,y,\tau) \le \frac{\pi}{2} \left(\tau^{-3} + |p|\tau^{-2} \right) \exp\left[-|p|\tau \right]$$

for every real p, y.

Proof. Inequalities (1.5) and (1.6) easily follow from

$$q(p,y,\tau) = \frac{\pi}{2} \left\{ [(y-\eta)^2 + \tau^2]^{-3/2} + |p|[(y-\eta)^2 + \tau^2]^{-1} \right\} \cdot$$

$$\cdot \exp\left[-|p|[(y-\eta)^2 + \tau^2]^{1/2} \right].$$

From this equality it follows also that

$$0 < q(p,y,\tau) \le \exp\left[-\frac{|p|}{4}\tau\right]\exp\left[-\frac{|p|}{2}|y-\eta|\right]\exp\left[-\frac{|p|}{4}\tau\right]\frac{\pi}{2}\tau^{-3}(1+|p|\tau).$$

Since $\exp[-(|p|/4)\tau](\pi/2)\tau^{-3}(1+|p|\tau) < 1$ for $|p| \ge \max\{0, 24\tau^{-1}(1-\log\tau)\}$, estimate (1.4) is proved.

2. $-$ LEMMA I. $-$ *Let us consider the differential equation with real or complex constant coefficients*

(2.1)
$$\frac{\partial^n w(p,y,\tau)}{\partial y^n} + \sum_{h=0}^{n-1} b_h p^{n-h}\frac{\partial^h w(p,y,\tau)}{\partial y^h} + \sum_{0 \le h+k \le n-1} b_{hk}p^k\frac{\partial^h w(p,y,\tau)}{\partial y^h} = q(p,y,\tau)$$

where $q(p,y,\tau)$ is the function considered in Proposition VI, with $\eta \ge 4$, $\tau > 0$. Let us denote by $\mu_1,\ldots,\mu_n$ the roots of the equation $\mu^n + \sum_{h=0}^{n-1} b_h \mu^{n-h} = 0$ and

let us set $m = 2 + \sum_{h=0}^{n-1}|b_h| + \sum_{0 \le h+k \le n-1}|b_{hk}|$. Let δ be a positive number such that

(2.2)
$$0 < \delta < 1/4$$

and

(2.3)
$$\delta \le |Re\,\mu_h|/2$$

for every h such that $Re\,\mu_h \ne 0$.

Then there exists a unique solution w of (2.1) satisfying the following conditions

$$\frac{\partial^h w(p,0,\tau)}{\partial y^h} = 0, \qquad h = 0,\ldots,n-1,$$

for $|p| < p_0(\tau) + p_1(m,\delta)$, and

$$\lim_{y\to+\infty} \exp[-\delta|p|y]w = \lim_{y\to-\infty} \exp[-\delta|p|y]w = 0$$

for $|p| \ge p_0(\tau)+p_1(m,\delta)$, where $p_0(\tau) = \max\{0, 24\tau^{-1}(1-\log\tau)\}$ and $p_1(m,\delta) = (4m/\delta)^n$. Moreover this solution satisfies the estimates

(2.4)
$$\left|\frac{\partial^h w(p,y,\tau)}{\partial y^h}\right| \le a(|p|,|y|)m^h(1+|p|)^h, \qquad h = 0,\ldots,n,$$

where for every real p

$$a(|p|,|y|) = \begin{cases} \exp[-3\delta|p|]\exp[m|y|] & \text{if } |y| \le \eta/4m \\[2em] \exp\left[-\frac{\tau}{4}|p|\right]\left[\exp\left[2m(4m+1)|y||p|^{1-1/n}\right]+ \right. \\[1em] \left. +\frac{\pi}{2}\left(\tau^{-3}+(p_0+p_1)\tau^{-2}\right)\exp\left[m(1+p_0+p_1)|y|\right]\right] & \text{if } |y| \ge \eta/4m. \end{cases}$$

Proof. Let $\lambda_1(p), \ldots, \lambda_n(p)$ be the roots of the characteristic equation of (2.1)

$$\lambda^n + \sum_{h=0}^{n-1} b_h p^{n-h} \lambda^h + \sum_{0 \leq h+k \leq n-1} b_{hk} p^k \lambda^h = 0.$$

By Proposition I we have that

$$|\mu_h| \leq m - 1, \qquad h = 1, \ldots, n,$$

hence applying Proposition II

$$\min_h \left| p^{-1} \lambda_j(p) - \mu_h \right| \leq 2(m-1)|p|^{-1/n}, \qquad j = 1, \ldots, n$$

for $|p| \geq 2^n$. By this inequality, and by (2.2) and (2.3) it follows that

$$(2.5) \qquad |\mathrm{Re}\lambda_j(p) - \delta|p|| \geq \delta|p| - 2m|p|^{1-1/n} + 1 \geq \delta|p|/2 + 1$$

for $|p| \geq p_1(m, \delta) = (4m/\delta)^n$.

By (1.4) and (2.5) we can apply Proposition III to equation (2.1) with $|p| \geq p_0 + p_1$. Therefore, for every such p there exists a unique solution w_1 of (2.1) such that

$$\lim_{y \to +\infty} \exp[-\delta|p|y]w_1 = \lim_{y \to -\infty} \exp[-\delta|p|y]w_1 = 0.$$

Now let w_2 be the solution of (2.1) satisfying the initial conditions

$$\frac{\partial^h w_2(p, 0, \tau)}{\partial y^h} = 0, \qquad h = 0, \ldots, n-1.$$

Let us define $w(p, y, \tau)$ as follows

$$w(p, y, \tau) = \begin{cases} w_1(p, y, \tau) & \text{if } |p| \geq p_0(\tau) + p_1(m, \delta) \\ w_2(p, y, \tau) & \text{if } |p| < p_0(\tau) + p_1(m, \delta) \end{cases}$$

and let us prove (2.4). First of all, we note that by Proposition I we have that

$$(2.6) \qquad |\lambda_j(p)| \leq (m-1)(1 + |p|), \qquad j = 0, \ldots, n,$$

for every real p. Now let $|y| \leq \eta/4m$; if $|p| \geq p_0 + p_1$, then by (1.4), (2.5), (2.6) and assumption $\eta \geq 4$, Proposition III applied to equation (2.1) with $r = \delta|p|$ and $s = \delta|p|/2$ gives the estimates

$$(2.7) \qquad \left| \frac{\partial^h w(p, y, \tau)}{\partial y^h} \right| \leq \exp\left[-\frac{9}{2}\delta|p| \right] m^h (1 + |p|)^h, \qquad h = 0, \ldots, n.$$

On the other hand, if $|p| < p_0 + p_1$, by (1.5) and (2.6), applying Proposition IV to equation (2.1), one has

$$(2.7') \qquad \left| \frac{\partial^h w(p, y, \tau)}{\partial y^h} \right| \leq \exp\left[-\frac{3}{4}|p| \right] \exp[m|y|]m^h (1 + |p|)^h, \qquad h = 0, \ldots, n.$$

Comparing (2.7) and (2.7'), and keeping into account (2.2), one can conclude that if $|y| \leq \eta/4m$ one has

$$(2.7'') \qquad \left|\frac{\partial^h w(p,y,\tau)}{\partial y^h}\right| \leq \exp\left[-3\delta|p|\right]\exp[m|y|]m^h(1+|p|)^h, \qquad h = 0,\dots,n$$

for every real p.

Let us assume now that $|y| \geq \eta/4m$. If $|p| \geq p_0 + p_1$, then by (1.4), (2.5), (2.6) we can apply Proposition III to equation (2.1) with $r = \delta|p|$, $s = \delta|p| - 2m|p|^{1-1/n}$, $\alpha = (m-1)(1+|p|)$ obtaining

$$(2.8) \qquad \left|\frac{\partial^h w(p,y,\tau)}{\partial y^h}\right| \leq \exp\left[-|p|\left[\frac{\tau}{4} - 2m(4m+1)|y||p|^{-1/n}\right]\right] \cdot m^h(1+|p|)^h, \qquad h = 0,\dots,n.$$

Finally, by (1.6) and (2.6), applying once again Proposition IV to equation (2.1), for $|p| < p_0 + p_1$ we obtain

$$(2.8') \qquad \left|\frac{\partial^h w(p,y,\tau)}{\partial y^h}\right| \leq \frac{\pi}{2}\left(\tau^{-3} + |p|\tau^{-2}\right)\exp\left[-|p|\tau\right] \cdot \exp[m(1+|p|)|y|]m^h(1+|p|)^h, \qquad h = 0,\dots,n.$$

From (2.7''), (2.8) and (2.8'), estimate (2.4) follows.

LEMMA II. – *Given the real numbers ξ, η, τ, with $\eta \geq 4$, $\tau > 0$, there exists a solution v of the partial differential equation with real or complex coefficients*

$$(2.9) \qquad \frac{\partial^n v}{\partial y^n} + \sum_{h=0}^{n-1} c_h \frac{\partial^n v}{\partial y^h \partial x^{n-h}} + \sum_{0 \leq h+k \leq n-1} c_{hk} \frac{\partial^{h+k} v}{\partial y^h \partial x^k} = \left[(x-\xi)^2 + (y-\eta)^2 + \tau^2\right]^{-2}$$

such that for every $(x,y) \in \mathbf{R}^2$ and every $s = 0,1,\dots; r = 0,\dots,n$

$$\left|\frac{\partial^{s+r} v}{\partial x^s \partial y^r}\right| \leq \gamma_1(s+r)!\alpha_1^s\beta_1^r \qquad \text{if } |y| \leq \eta/4m$$

$$\left|\frac{\partial^{s+r} v}{\partial x^s \partial y^r}\right| \leq \gamma_2(s+r)!\alpha_2^s\beta_2^r \qquad \text{if } |y| \geq \eta/4m,$$

with

$$\gamma_1 = 4\delta^{-1}\exp\left[m|y| + \delta\right], \qquad \alpha_1 = 2\delta^{-1}, \qquad \beta_1 = 2m\delta^{-1},$$

$$\alpha_2 = 32\tau^{-1}, \qquad \beta_2 = 4m(1 + 8\tau^{-1}),$$

$$\gamma_2 = 4\int_0^{+\infty} \exp\left[-\frac{\tau}{16}p\right]\left[\exp\left[2m(4m+1)|y|p^{1-1/n}\right] + \frac{\pi}{2}\left(\tau^{-3} + (p_0+p_1)\tau^{-2}\right)\exp\left[m(1+p_0+p_1)|y|\right]\right]dp,$$

where m and δ are positive constants which depend only on the coefficients of (2.9).

Proof. We can assume $\xi = 0$ in the right hand side of (2.9), since we can always reduce ourselves to this case by means of a translation. The Fourier transform with respect to x brings equation (2.9) in (2.1) with $b_h = c_h i^{n-h}$, $b_{hk} = c_{hk} i^k$. By Lemma I there exists a solution w of this equation satisfying (2.4). Therefore, for every real p, and every $r = 0, \ldots, n$ we have that

$$\left| \frac{\partial^r w(p, y, \tau)}{\partial y^r} \right| \leq \exp\left[m|y| + \delta \right] (m\delta^{-1})^r r! \exp\left[-2\delta|p| \right] \qquad \text{if } |y| \leq \eta/4m,$$

$$\left| \frac{\partial^r w(p, y, \tau)}{\partial y^r} \right| \leq \left[2m(1 + 8\tau^{-1}) \right]^r r! \exp\left[-\frac{\tau}{8}|p| \right] \cdot$$

$$\cdot \left[\exp\left[2m(4m + 1)|y|\, |p|^{1-1/n} \right] + \frac{\pi}{2} \left(\tau^{-3} + (p_0 + p_1)\tau^{-2} \right) \cdot \right.$$

$$\left. \cdot \exp\left[m(1 + p_0 + p_1)|y| \right] \right]$$

if $|y| \geq \eta/4m$, where m and δ are the constants of Lemma I, referred now to the equation whose solution is w. Such estimates allow us to conclude that the function

$$v(x, y, \tau) = (2\pi)^{-1} \int_{-\infty}^{+\infty} \exp[ixp] w(p, y, \tau) dp$$

has all the required properties.

3. – Theorem. – *Let $f(x, y)$ be an analytic function in $\mathbf{R}^2$, and let P be a linear differential operator with (real or complex) constant coefficients. Then there exists at least one function $u(x, y)$ which is analytic in $\mathbf{R}^2$ and satisfies the equation*

$$(3.1) \qquad\qquad\qquad Pu = f(x, y)$$

for every $(x, y) \in \mathbf{R}^2$.

Proof. Let φ be a positive nonincreasing function defined in $\mathbf{R}$. By a result proved in [3], there exist a function $g : \mathbf{R}^3 \to \mathbf{C}$, indefinitely differentiable in $\mathbf{R}^3$ and a positive nonincreasing function ψ, indefinitely differentiable in $\mathbf{R}$, such that for the given function f one has

$$(3.2) \quad f(x, y) = \int_{\mathbf{R}^3} \frac{g(\xi, \eta, \tau)}{\left[(x - \xi)^2 + (y - \eta)^2 + \tau^2 \right]^2} d\xi d\eta d\tau \quad \text{for every } (x, y) \in \mathbf{R}^2,$$

with

$$\operatorname{supp} g \subset \left\{ (\xi, \eta, \tau); \psi(\xi^2 + \eta^2) \leq \tau \leq 2\psi(\xi^2 + \eta^2) \right\},$$

$$\varphi(\xi^2 + \eta^2) \geq \int_0^{+\infty} |g(\xi, \eta, \tau)| d\tau \qquad \text{for every } (\xi, \eta) \in \mathbf{R}^2.$$

Moreover, let us assume that

$$(3.3) \qquad\qquad\qquad \int_{\mathbf{R}^2} \varphi(\xi^2 + \eta^2) d\xi d\eta < +\infty.$$

First of all, let us consider the particular case where the function $g(\xi, \eta, \tau)$ in (3.2) satisfies the further conditions

$$g(\xi, \eta, \tau) = 0 \quad \text{for } \eta \leq 4, \qquad g(\xi, \eta, \tau) = 0 \quad \text{for } |\xi| \geq \eta$$

and equation (3.1) has the form (1.2). In this case the function

$$u(x, y) = \int_{\mathbf{R}^2} d\xi\, d\eta \int_0^{+\infty} g(\xi, \eta, \tau) v(x, y, \xi, \eta, \tau)\, d\tau$$

where $v(x, y, \xi, \eta, \tau)$ is the solution of (2.9) considered in Lemma II, is a solution of (3.1) in $\mathbf{R}^2$. Moreover, from Lemma II it easily follows that

$$\left| \frac{\partial^{s+r} u}{\partial x^s \partial y^r} \right| \leq \gamma_1(y)(s+r)! \alpha_1^s \beta_1^r \int_{|\xi| \leq \eta,\ \eta \geq 4m|y|} d\xi\, d\eta \int_0^{+\infty} |g(\xi, \eta, \tau)|\, d\tau +$$

$$+ \gamma_2(y, \tau_0)(s+r)! \alpha_2^s(\tau_0) \beta_2^r(\tau_0) \int_{|\xi| \leq \eta,\ \eta < 4m|y|} d\xi\, d\eta \int_0^{+\infty} |g(\xi, \eta, \tau)|\, d\tau$$

for $s = 0, 1, \ldots;\ r = 0, \ldots, n$, where $\tau_0 = \tau_0(y) = \min\{\psi(\xi^2 + \eta^2); \eta \leq 4m|y|, |\xi| \leq \eta\}$. By Proposition V it follows that the function u is analytic in $\mathbf{R}^2$.

In the general case we cover $\mathbf{R}^2$ with a finite number of open sets A_0, A_1, $\ldots$, A_7 such that A_0 is bounded, and if (x, y), (x', y') both belong to the same A_h, $h > 0$, then

$$xx' + yy' > (x^2 + y^2)^{1/2}(x'^2 + y'^2)^{1/2}/2.$$

Let us consider then the representation of f given in (3.2), and let us assume that (3.3) is satisfied. The function g in (3.2) can be written as the sum of eight functions g_h such that

$$\operatorname{supp} g_h \subset \left\{ (\xi, \eta, \tau); (\xi, \eta) \in A_h,\ \psi(\xi^2 + \eta^2) \leq \tau \leq 2\psi(\xi^2 + \eta^2) \right\} \quad h = 0, \ldots, 7.$$

Setting

$$f_h(x, y) = \int_{\mathbf{R}^3} \frac{g_h(\xi, \eta, \tau)}{\left[(x - \xi)^2 + (y - \eta)^2 + \tau^2 \right]^2}\, d\xi\, d\eta\, d\tau, \quad h = 0, \ldots, 7,$$

we note that every equation $Pu_h = f_h$ can be reduced to the particular situation considered at the beginning of the proof by simple rotations and translations.

4. $-$ It seems that the result stated in section 3 is false when the number of independent variables is greater than two. For example, we strongly doubt about the existence of an analytic solution in $\mathbf{R}^3$ of the equations

$$\frac{\partial^2 u}{\partial x^2} + \frac{\partial^2 u}{\partial y^2} = f(x, y, t), \qquad \frac{\partial u}{\partial t} + \frac{\partial^2 u}{\partial x^2} + \frac{\partial^2 u}{\partial y^2} = f(x, y, t)$$

where

$$f(x, y, t) = \sum_{h,k=1}^{\infty} \exp\left[\varphi(h, k)(ix + ih^2 t - (y - h)^2 - 1) \right],$$

$$\varphi(h, k) = \left[([p(h)]^k)! \right]!$$

and $p(h)$ is the $(h+1)$-th prime number.

Bibliography

[1] K. G. Andersson, *Analyticity of fundamental solutions,* Ark. Mat., **8** (1970), 73–81.

[2] K. G. Andersson, *Propagation of analyticity of solutions of partial differential equations with constant coefficients,* Ark. Mat., **8** (1971), 277–302.

[3] E. De Giorgi – L. Cattabriga, *Una formula di rappresentazione per funzioni analitiche in* $\mathbf{R}^n$, Boll. Un. Mat. Ital., **4** (1971), 1010–1014.

[4] L. Ehrenpreis, *Some applications of the theory of distributions to several complex variables,* Seminar on Analytic Functions, Princeton 1957, 65–79.

[5] L. Ehrenpreis, *Solutions of some problems of division IV,* Amer. J. Math., **82** (1960), 522–588.

[6] L. Ehrenpreis, *Fourier analysis in several complex variables,* Wiley Interscience, 1970.

[7] L. Ehrenpreis, *Fourier analysis on non convex sets,* conference held at the "Convegno sulle equazioni ipoellittiche e gli spazi funzionali", Roma, 25–28 gennaio 1971.

[8] T. Kawai, *Construction of a local elementary solution for linear partial differential operators (I),* Proc. Japan Acad., **47** (1971), 19–23.

[9] T. Kawai, *Construction of a local elementary solution for linear partial differential operators (II),* Proc. Japan Acad., **47** (1971), 147–152.

[10] T. Kawai, *On the global existence of real analytic solutions of linear differential equations (I),* Proc. Japan Acad., **47** (1971), 537–540.

[11] T. Kawai, *On the global existence of real analytic solutions of linear differential equations (II),* Proc. Japan Acad. **47** (1971), 643–647.

[12] T. Kawai, *On the global existence of real analytic solutions of linear differential equations,* conference held at the "Symposium of the theory of hyperfunctions and analytic functionals", Kyoto Univ., 27–30 settembre 1971.

[13] B. Malgrange, *Existence et approximation des solutions des équations aux dérivées partielles et des équations de convolution,* Ann. Inst. Fourier (Grenoble), **6** (1956), 271–355.

[14] A. Martineau, *Sur la topologie des espaces de fonctions holomorphes,* Math. Ann., **163** (1966), 62–88.

[15] F. Trèves, *Linear partial differential equations with constant coefficients,* Gordon and Breach, 1966, Chap.9.

[16] F. Trèves – M. Zerner, *Zones d'analyticité des solutions élémentaires,* Bull. Soc. Math. France, **95** (1967), 155–191.

Una dimostrazione diretta dell'esistenza di soluzioni analitiche nel piano reale di equazioni a derivate parziali a coefficienti costanti[‡]

Ennio De Giorgi (Pisa), Lamberto Cattabriga (Ferrara)

Sunto. A direct proof of the following theorem is given: let $f(x, y)$ be a function analytic in $\mathbf{R}^2$ and let P be a linear partial differential operator with (real or complex) constant coefficients. Then there exists at least one function $u(x, y)$ analytic in $\mathbf{R}^2$ which satisfies the equation $Pu = f(x, y)$ for every $(x, y) \in \mathbf{R}^2$. A conjecture about possible counterexamples to this result in $\mathbf{R}^3$ is also formulated.

Scopo di questo lavoro è dare una dimostrazione diretta della esistenza di soluzioni analitiche nello spazio bidimensionale $\mathbf{R}^2$ di una equazione differenziale a derivate parziali lineare ed a coefficienti costanti con termine noto analitico. Tale dimostrazione non richiede metodi matematici avanzati, ma soltanto poche notizie elementari sulla trasformata di Fourier e sulle equazioni differenziali ordinarie a coefficienti costanti. Essa inoltre utilizza una formula di rappresentazione per funzioni analitiche provata in un precedente lavoro [3]. Tuttavia per comodità del lettore abbiamo indicato nella bibliografia qualcuno dei lavori che trattano queste o simili questioni in un contesto più generale. In particolare segnaliamo che durante la redazione del lavoro abbiamo avuto notizia dei lavori [10, 11, 12] di T. Kawai che ci sembrano contenere i risultati più recenti e generali finora noti su questo argomento.

Il risultato di questo lavoro è contenuto nel teorema enunciato nel n. 3. Il n. 1 contiene alcune proposizioni di carattere elementare ed il n. 2 due lemmi preliminari. Nel n. 4 è formulata una congettura su possibili controesempi del risultato ottenuto, relativi al caso tridimensionale.

1. – Proposizione I. – *Siano λ, a, γ_h, $h = 0, \ldots, n-1$, numeri complessi tali che*

$$\lambda^n + \sum_{h=0}^{n-1} \gamma_h a^{n-h} \lambda^h = 0,$$

allora

$$|\lambda| \leq |a| \left(1 + \sum_{h=0}^{n-1} |\gamma_h| \right).$$

[‡]Editor's note: published in Boll. Un. Mat. Ital., (4) 4 (1971), 1015–1027.

PROPOSIZIONE II. – *Siano $P(z) = (z - \mu_1) \cdots (z - \mu_n)$ e $Q(z)$ due polinomi di ordine n nella variabile complessa z. Sia λ una radice dell'equazione $Q(z) = 0$ ed a, ϵ due numeri positivi tali che*

$$a \geq |\mu_h|, \ \ h = 1, \ldots, n, \qquad\qquad \epsilon \leq 2^{-n},$$

$$|Q(z) - P(z)| \leq \left(a|z|^{n-1} + a^n\right) \epsilon$$

per ogni z, allora

$$\min_h |\lambda - \mu_h| \leq 2a\epsilon^{1/n}.$$

Dim. Posto $\sigma = \min\limits_h |\lambda - \mu_h|$, è

$$\sigma^n \leq |P(\lambda)| = |P(\lambda) - Q(\lambda)| \leq \left(a(a + \sigma)^{n-1} + a^n\right) \epsilon.$$

Se fosse $\sigma > a$, avremmo

$$\sigma^n < a2^n \epsilon \sigma^{n-1} < a\sigma^{n-1}.$$

Deve perció essere $0 \leq \sigma \leq a$ e ciò implica $\sigma^n \leq 2^n a^n \epsilon$.

PROPOSIZIONE III. – *Data l'equazione ordinaria a coefficienti complessi costanti*

$$(1.1) \qquad \frac{d^n w}{dy^n} + \sum_{h=0}^{n-1} b_h \frac{d^h w}{dy^h} = q(y)$$

siano $\lambda_1, \ldots, \lambda_n$ le radici della sua equazione caratteristica

$$\lambda^n + \sum_{h=0}^{n-1} b_h \lambda^h = 0.$$

Supponiamo che r ed η siano numeri reali ed s, α numeri positivi tali che

$$|\lambda_j| \leq \alpha, \quad |\mathrm{Re}\,\lambda_j - r| \geq s + 1, \quad j = 1, \ldots, n,$$

ove $\mathrm{Re}\,\lambda_j$ indica la parte reale di λ_j, e

$$|q(y)| \leq \exp\left[ry - s|y - \eta|\right]$$

per ogni y reale. Allora esiste una ed una sola soluzione w della equazione (1.1) tale che

$$\lim_{y \to +\infty} w(y) \exp[-ry] = \lim_{y \to -\infty} w(y) \exp[-ry] = 0.$$

Tale soluzione soddisfa inoltre le maggiorazioni

$$\left| \frac{d^h w}{dy^h} \right| \leq \exp[ry - s|y - \eta|](\alpha + 1)^h, \quad h = 0, \ldots, n,$$

per ogni y reale.

Dim. Per $n = 1$ consideriamo la soluzione di (1.1) data da

$$w(y) = \exp[-b_0 y] \int_{-\infty}^{y} \exp[b_0 t] q(t) dt \quad \text{se } \operatorname{Re} b_0 + r \geq s + 1$$

oppure quella data da

$$w(y) = -\exp[-b_0 y] \int_{y}^{\infty} \exp[b_0 t] q(t) dt \quad \text{se } \operatorname{Re} b_0 + r \leq -s - 1.$$

In entrambi i casi si vede facilmente che

$$|w(y)| \leq \exp[ry - s|y - \eta|] \quad \text{per ogni } y \text{ reale.}$$

Da questa maggiorazione, usando l'equazione (1.1), segue allora che

$$\left| \frac{dw}{dy} \right| \leq (|b_0| + 1) \exp[ry - s|y - \eta|] \leq (\alpha + 1) \exp[ry - s|y - \eta|]$$

per ogni y reale.

Per $n > 1$ consideriamo il sistema

$$\begin{cases} \dfrac{dw_h}{dy} = \lambda_h w_h + w_{h-1}, & h = 2, \ldots, n \\[2mm] \dfrac{dw_1}{dy} = \lambda_1 w_1 + q(y) \end{cases}$$

ed osserviamo che se $(w_1, \ldots, w_n)$ è una soluzione di questo sistema, allora w_n è soluzione di (1.1).

Proposizione IV. – *Data l'equazione differenziale (1.1), di cui indichiamo ancora con $\lambda_1, \ldots, \lambda_n$ le radici della sua equazione caratteristica, supponiamo che s, γ siano due numeri positivi tali che*

$$s \geq |\lambda_h| + 1, \qquad h = 1, \ldots, n,$$

e

$$|q(y)| \leq \exp[s|y|]$$

per ogni y con $|y| \leq \gamma$. Allora per la soluzione w della (1.1) soddisfacente le condizioni iniziali

$$\frac{d^h w(0)}{dy^h} = 0, \qquad h = 0, \ldots, n-1,$$

valgono per $|y| \leq \gamma$ le maggiorazioni

$$\left| \frac{d^h w}{dy^h} \right| \leq s^h \exp[s|y|], \qquad h = 0, \ldots, n.$$

Dim. La proposizione si prova facilmente per $n = 1$. Per $n > 1$ la dimostrazione si conduce come quella della proposizione precedente.

PROPOSIZIONE V. – *Siano $u(x,y)$ ed $f(x,y)$ due funzioni indefinitamente differenziabili in un insieme aperto $B \subset \mathbf{R}^2$. Supponiamo che in B siano soddisfatte l'equazione a derivate parziali a coefficienti costanti*

$$(1.2) \qquad \frac{\partial^n u}{\partial y^n} + \sum_{h=0}^{n-1} c_h \frac{\partial^n u}{\partial y^h \partial x^{n-h}} + \sum_{0 \le h+k \le n-1} c_{hk} \frac{\partial^{h+k} u}{\partial y^h \partial x^k} = f(x,y)$$

e le maggiorazioni

$$(1.3) \qquad \left| \frac{\partial^{s+r} f}{\partial x^s \partial y^r} \right| \le (s+r)! \alpha^s \beta^r \qquad per\ s, r = 0, 1, \dots$$

$$(1.3') \qquad \left| \frac{\partial^{s+r} u}{\partial x^s \partial y^r} \right| \le (s+r)! \alpha^s \beta^r \qquad per\ s = 0, 1, \dots; r = 0, \dots, n$$

con

$$\alpha > 1, \qquad \beta > \alpha \left(1 + \sum_{h=0}^{n-1} |c_h| + \sum_{0 \le h+k \le n-1} |c_{hk}| \right);$$

allora (1.3') vale per ogni numero naturale r.

Dim. Basta derivare la (1.2) e procedere per induzione rispetto all'indice r.

PROPOSIZIONE VI. – *Siano y, η, τ numeri reali e sia $\eta \ge 4$, $\tau > 0$. Per la trasformata di Fourier della funzione $[x^2 + (y-\eta)^2 + \tau^2]^{-2}$, cioè per la funzione*

$$q(p,y,\tau) = \int_{-\infty}^{+\infty} \exp[-ixp][x^2 + (y-\eta)^2 + \tau^2]^{-2} dx$$

valgono le maggiorazioni

$$(1.4) \qquad 0 < q(p,y,\tau) \le \exp\left[-\frac{|p|}{4}\tau \right] \exp\left[-\frac{|p|}{2}|y - \eta| \right]$$

per $|p| > p_0(\tau) = \max\{0, 24\tau^{-1}(1 - \log \tau)\}$ qualunque sia y reale;

$$(1.5) \qquad 0 < q(p,y,\tau) \le \exp\left[-\frac{|p|}{2}|y - \eta| \right]$$

per ogni p reale e per $|y| \le \eta/4$;

$$(1.6) \qquad 0 < q(p,y,\tau) \le \frac{\pi}{2}\left(\tau^{-3} + |p|\tau^{-2} \right) \exp\left[-|p|\tau \right]$$

per ogni p, y reali.

Dim. Le maggiorazioni (1.5) e (1.6) seguono facilmente dalla

$$q(p,y,\tau) = \frac{\pi}{2}\left\{ [(y-\eta)^2 + \tau^2]^{-3/2} + |p|[(y-\eta)^2 + \tau^2]^{-1} \right\} \cdot$$

$$\cdot \exp\left[-|p|[(y-\eta)^2 + \tau^2]^{1/2} \right].$$

Da quest'ultima eguaglianza segue inoltre che

$$0 < q(p,y,\tau) \le \exp\left[-\frac{|p|}{4}\tau\right] \exp\left[-\frac{|p|}{2}|y-\eta|\right] \exp\left[-\frac{|p|}{4}\tau\right] \frac{\pi}{2}\tau^{-3}(1+|p|\tau).$$

Notando che per $|p| \ge \max\{0, 24\tau^{-1}(1-\log\tau)\}$ è $\exp[-(|p|/4)\tau](\pi/2)\tau^{-3}(1+|p|\tau) < 1$, si ha la (1.4).

2.   – Lemma I. – *Consideriamo l'equazione differenziale a coefficienti costanti reali o complessi*

(2.1)

$$\frac{\partial^n w(p,y,\tau)}{\partial y^n} + \sum_{h=0}^{n-1} b_h p^{n-h} \frac{\partial^h w(p,y,\tau)}{\partial y^h} + \sum_{0 \le h+k \le n-1} b_{hk} p^k \frac{\partial^h w(p,y,\tau)}{\partial y^h} = q(p,y,\tau)$$

ove $q(p,y,\tau)$ è la funzione considerata nella Proposizione VI, con $\eta \ge 4$, $\tau > 0$. Indichiamo con $\mu_1,\ldots,\mu_n$ le radici della equazione $\mu^n + \sum_{h=0}^{n-1} b_h \mu^{n-h} = 0$ e

poniamo $m = 2 + \sum_{h=0}^{n-1} |b_h| + \sum_{0 \le h+k \le n-1} |b_{hk}|$. Sia δ un numero positivo tale che

(2.2)

$$0 < \delta < 1/4$$

e

(2.3)

$$\delta \le |Re\,\mu_h|/2$$

per ogni h tale che $Re\,\mu_h \ne 0$.

    Allora esiste una ed una sola soluzione w della (2.1) soddisfacente le condizioni

$$\frac{\partial^h w(p,0,\tau)}{\partial y^h} = 0, \qquad h = 0,\ldots,n-1,$$

per $|p| < p_0(\tau) + p_1(m,\delta)$ e

$$\lim_{y \to +\infty} \exp[-\delta|p|y]w = \lim_{y \to -\infty} \exp[-\delta|p|y]w = 0$$

per $|p| \ge p_0(\tau) + p_1(m,\delta)$, ove $p_0(\tau) = \max\{0, 24\tau^{-1}(1-\log\tau)\}$ e $p_1(m,\delta) = (4m/\delta)^n$. Per questa soluzione valgono inoltre le maggiorazioni

(2.4)

$$\left|\frac{\partial^h w(p,y,\tau)}{\partial y^h}\right| \le a(|p|,|y|)m^h(1+|p|)^h, \qquad h = 0,\ldots,n,$$

ove per ogni p reale

$$a(|p|,|y|) = \begin{cases} \exp[-3\delta|p|]\exp[m|y|] & se\ |y| \le \eta/4m \\[2ex] \exp\left[-\dfrac{\tau}{4}|p|\right]\left[\exp\left[2m(4m+1)|y||p|^{1-1/n}\right] + \right. \\[1ex] \left. +\dfrac{\pi}{2}\left(\tau^{-3}+(p_0+p_1)\tau^{-2}\right)\exp\left[m(1+p_0+p_1)|y|\right]\right] & se\ |y| \ge \eta/4m. \end{cases}$$

Dim. Siano $\lambda_1(p), \ldots, \lambda_n(p)$ le radici della equazione caratteristica di (2.1)

$$\lambda^n + \sum_{h=0}^{n-1} b_h p^{n-h} \lambda^h + \sum_{0 \leq h+k \leq n-1} b_{hk} p^k \lambda^h = 0.$$

Per la Proposizione I abbiamo

$$|\mu_h| \leq m - 1, \qquad h = 1, \ldots, n,$$

onde applicando la Proposizione II

$$\min_h |p^{-1} \lambda_j(p) - \mu_h| \leq 2(m-1)|p|^{-1/n}, \qquad j = 1, \ldots, n$$

per $|p| \geq 2^n$. Da questa disuguaglianza e da (2.2) e (2.3) segue che

$$(2.5) \qquad |\mathrm{Re}\lambda_j(p) - \delta|p|| \geq \delta|p| - 2m|p|^{1-1/n} + 1 \geq \delta|p|/2 + 1$$

per $|p| \geq p_1(m, \delta) = (4m/\delta)^n$.

In virtù di (1.4) e (2.5) possiamo applicare la Proposizione III all'equazione (2.1) con $|p| \geq p_0 + p_1$. Per ognuno di tali p esiste quindi una ed una sola soluzione w_1 di (2.1) tale che

$$\lim_{y \to +\infty} \exp[-\delta|p|y]w_1 = \lim_{y \to -\infty} \exp[\delta|p|y]w_1 = 0.$$

Sia ora w_2 la soluzione di (2.1) che soddisfa le condizioni iniziali

$$\frac{\partial^h w_2(p, 0, \tau)}{\partial y^h} = 0, \qquad h = 0, \ldots, n-1.$$

Definiamo $w(p, y, \tau)$ come segue

$$w(p, y, \tau) = \begin{cases} w_1(p, y, \tau) & \text{per } |p| \geq p_0(\tau) + p_1(m, \delta) \\ w_2(p, y, \tau) & \text{per } |p| < p_0(\tau) + p_1(m, \delta) \end{cases}$$

e procediamo a provare (2.4). Osserviamo anzitutto che per la Proposizione I è

$$(2.6) \qquad |\lambda_j(p)| \leq (m-1)(1 + |p|), \qquad j = 0, \ldots, n,$$

per ogni p reale. Sia ora $|y| \leq \eta/4m$; se $|p| \geq p_0 + p_1$, allora per le (1.4), (2.5), (2.6) e per l'ipotesi $\eta \geq 4$, la Proposizione III applicata alla equazione (2.1) con $r = \delta|p|$ e $s = \delta|p|/2$ fornisce le maggiorazioni

$$(2.7) \qquad \left| \frac{\partial^h w(p, y, \tau)}{\partial y^h} \right| \leq \exp\left[-\frac{9}{2}\delta|p| \right] m^h (1 + |p|)^h, \qquad h = 0, \ldots, n.$$

D'altra parte se $|p| < p_0 + p_1$ per le (1.5) e (2.6), applicando la Proposizione IV alla equazione (2.1), si ha che

$$(2.7') \qquad \left| \frac{\partial^h w(p, y, \tau)}{\partial y^h} \right| \leq \exp\left[-\frac{3}{4}|p| \right] \exp[m|y|]m^h (1 + |p|)^h, \qquad h = 0, \ldots, n.$$

Confrontando (2.7) e (2.7') e tenendo conto di (2.2) si conclude allora che se $|y| \leq \eta/4m$ è

$$(2.7") \qquad \left| \frac{\partial^h w(p, y, \tau)}{\partial y^h} \right| \leq \exp\left[-3\delta|p|\right] \exp[m|y|] m^h (1 + |p|)^h, \qquad h = 0, \ldots, n$$

per ogni p reale.

Supponiamo ora $|y| \geq \eta/4m$. Se $|p| \geq p_0 + p_1$, allora per le (1.4), (2.5), (2.6) possiamo applicare la Proposizione III alla equazione (2.1) con $r = \delta|p|$, $s = \delta|p| - 2m|p|^{1-1/n}$, $\alpha = (m-1)(1 + |p|)$ ottenendo così

$$\left| \frac{\partial^h w(p, y, \tau)}{\partial y^h} \right| \leq \exp\left[-|p|\left[\frac{\tau}{4} - 2m(4m+1)|y||p|^{-1/n}\right]\right] \cdot$$
$$(2.8) \qquad\qquad\qquad\qquad \cdot m^h (1 + |p|)^h, \qquad h = 0, \ldots, n.$$

Finalmente per (1.6) e (2.6) applicando di nuovo la Proposizione IV alla equazione (2.1), otteniamo per $|p| < p_0 + p_1$

$$\left| \frac{\partial^h w(p, y, \tau)}{\partial y^h} \right| \leq \frac{\pi}{2} \left(\tau^{-3} + |p|\tau^{-2}\right) \exp\left[-|p|\tau\right] \cdot$$
$$(2.8') \qquad\qquad\qquad \cdot \exp[m(1 + |p|)|y|] m^h (1 + |p|)^h, \qquad h = 0, \ldots, n.$$

Dalle (2.7"), (2.8) e (2.8') segue allora la (2.4).

Lemma II. – *Dati i numeri reali ξ, η, τ, con $\eta \geq 4$, $\tau > 0$, esiste una soluzione v della equazione a derivate parziali a coefficienti reali o complessi*
(2.9)
$$\frac{\partial^n v}{\partial y^n} + \sum_{h=0}^{n-1} c_h \frac{\partial^n v}{\partial y^h \partial x^{n-h}} + \sum_{0 \leq h+k \leq n-1} c_{hk} \frac{\partial^{h+k} v}{\partial y^h \partial x^k} = \left[(x-\xi)^2 + (y-\eta)^2 + \tau^2\right]^{-2}$$

tale che per ogni $(x, y) \in \mathbf{R}^2$ ed ogni $s = 0, 1, \ldots; \; r = 0, \ldots, n$

$$\left| \frac{\partial^{s+r} v}{\partial x^s \partial y^r} \right| \leq \gamma_1 (s+r)! \alpha_1^s \beta_1^r \qquad se \; |y| \leq \eta/4m$$

$$\left| \frac{\partial^{s+r} v}{\partial x^s \partial y^r} \right| \leq \gamma_2 (s+r)! \alpha_2^s \beta_2^r \qquad se \; |y| \geq \eta/4m,$$

con

$$\gamma_1 = 4\delta^{-1} \exp\left[m|y| + \delta\right], \qquad \alpha_1 = 2\delta^{-1}, \qquad \beta_1 = 2m\delta^{-1},$$

$$\alpha_2 = 32\tau^{-1}, \qquad \beta_2 = 4m(1 + 8\tau^{-1}),$$

$$\gamma_2 = 4 \int_0^{+\infty} \exp\left[-\frac{\tau}{16}p\right] \left[\exp\left[2m(4m+1)|y|p^{1-1/n}\right] + \right.$$
$$\left. + \frac{\pi}{2}\left(\tau^{-3} + (p_0 + p_1)\tau^{-2}\right) \exp\left[m(1 + p_0 + p_1)|y|\right]\right] dp,$$

ove m e δ sono quantità positive dipendenti soltanto dai coefficienti di (2.9).

Dim. Possiamo supporre $\xi = 0$ al secondo membro di (2.9), poiché a questo caso possiamo sempre ricondurci con una traslazione. La trasformata di Fourier rispetto ad x porta l'equazione (2.9) nella (2.1) con $b_h = c_h i^{n-h}$, $b_{hk} = c_{hk} i^k$. Per il Lemma I esiste una soluzione w di questa equazione che soddisfa le (2.4). Per ogni p reale e per ogni $r = 0, \ldots, n$ avremo pertanto

$$\left| \frac{\partial^r w(p,y,\tau)}{\partial y^r} \right| \leq \exp\left[m|y| + \delta \right] (m\delta^{-1})^r r! \exp\left[-2\delta|p| \right] \qquad \text{se } |y| \leq \eta/4m,$$

$$\left| \frac{\partial^r w(p,y,\tau)}{\partial y^r} \right| \leq \left[2m(1 + 8\tau^{-1}) \right]^r r! \exp\left[-\frac{\tau}{8}|p| \right] \cdot$$

$$\cdot \left[\exp\left[2m(4m+1)|y|\,|p|^{1-1/n} \right] + \frac{\pi}{2} \left(\tau^{-3} + (p_0 + p_1)\tau^{-2} \right) \cdot \right.$$

$$\left. \cdot \exp\left[m(1 + p_0 + p_1)|y| \right] \right]$$

se $|y| \geq \eta/4m$, ove m e δ sono le quantità così indicate nel Lemma I, riferite ora alla equazione di cui w è la soluzione che abbiamo indicato più sopra. Queste maggiorazioni ci consentono di concludere che la funzione

$$v(x,y,\tau) = (2\pi)^{-1} \int_{-\infty}^{+\infty} \exp[ixp] w(p,y,\tau)\,dp$$

ha tutte le proprietà richieste.

3. $-$ Teorema. $-$ *Sia $f(x,y)$ una funzione analitica in $\mathbf{R}^2$ e P un operatore differenziale lineare a coefficienti (reali o complessi) costanti. Esiste allora almeno una funzione $u(x,y)$ analitica in $\mathbf{R}^2$ che soddisfa l'equazione*

$$(3.1) \qquad\qquad Pu = f(x,y)$$

per ogni $(x,y) \in \mathbf{R}^2$.

Dim. Sia φ una funzione positiva e non crescente definita in $\mathbf{R}$. In virtù di un risultato provato in [3], esiste una funzione $g : \mathbf{R}^3 \to \mathbf{C}$ indefinitamente differenziabile in $\mathbf{R}^3$ ed una funzione ψ indefinitamente differenziabile in $\mathbf{R}$, positiva e non crescente tali che per la data funzione f vale la

$$(3.2) \quad f(x,y) = \int_{\mathbf{R}^3} \frac{g(\xi,\eta,\tau)}{\left[(x-\xi)^2 + (y-\eta)^2 + \tau^2 \right]^2} d\xi d\eta d\tau \quad \text{per ogni } (x,y) \in \mathbf{R}^2,$$

con

$$\operatorname{supp} g \subset \left\{ (\xi,\eta,\tau); \psi(\xi^2 + \eta^2) \leq \tau \leq 2\psi(\xi^2 + \eta^2) \right\},$$

$$\varphi(\xi^2 + \eta^2) \geq \int_0^{+\infty} |g(\xi,\eta,\tau)|\,d\tau \qquad \text{per ogni } (\xi,\eta) \in \mathbf{R}^2.$$

Supporremo inoltre che

$$(3.3) \qquad\qquad \int_{\mathbf{R}^2} \varphi(\xi^2 + \eta^2)\,d\xi d\eta < +\infty.$$

Consideriamo anzitutto il caso particolare in cui la funzione $g(\xi, \eta, \tau)$ in (3.2) soddisfi le ulteriori condizioni

$$g(\xi, \eta, \tau) = 0 \quad \text{per } \eta \leq 4, \qquad g(\xi, \eta, \tau) = 0 \quad \text{per } |\xi| \geq \eta$$

e l'equazione (3.1) abbia la forma (1.2). In questo caso la funzione

$$u(x,y) = \int_{\mathbf{R}^2} d\xi d\eta \int_0^{+\infty} g(\xi, \eta, \tau) v(x, y, \xi, \eta, \tau) d\tau$$

ove $v(x, y, \xi, \eta, \tau)$ è la soluzione di (2.9) considerata nel Lemma II, è una soluzione di (3.1) in $\mathbf{R}^2$. Dal Lemma II segue inoltre facilmente che

$$\left| \frac{\partial^{s+r} u}{\partial x^s \partial y^r} \right| \leq \gamma_1(y)(s+r)! \alpha_1^s \beta_1^r \int_{|\xi| \leq \eta, \ \eta \geq 4m|y|} d\xi d\eta \int_0^{+\infty} |g(\xi, \eta, \tau)| d\tau +$$

$$+ \gamma_2(y, \tau_0)(s+r)! \alpha_2^s(\tau_0) \beta_2^r(\tau_0) \int_{|\xi| \leq \eta, \ \eta < 4m|y|} d\xi d\eta \int_0^{+\infty} |g(\xi, \eta, \tau)| d\tau$$

per $s = 0, 1, \ldots; r = 0, \ldots, n$, ove $\tau_0 = \tau_0(y) = \min \left\{ \psi(\xi^2 + \eta^2); \eta \leq 4m|y|, |\xi| \leq \eta \right\}$. Ne segue, per la Proposizione V, che la funzione u è analitica in $\mathbf{R}^2$.

Nel caso generale ricopriamo $\mathbf{R}^2$ con un numero finito di insiemi aperti A_0, $A_1, \ldots, A_7$ tali che A_0 è limitato e se (x, y), (x', y') appartengono entrambi allo stesso A_h, $h > 0$, allora

$$xx' + yy' > (x^2 + y^2)^{1/2}(x'^2 + y'^2)^{1/2}/2.$$

Consideriamo quindi la rappresentazione di f data da (3.2) e supponiamo che sia soddisfatta (3.3). La funzione g in (3.2) può essere scritta come somma di otto funzioni g_h tali che

$$\operatorname{supp} g_h \subset \left\{ (\xi, \eta, \tau); (\xi, \eta) \in A_h, \ \psi(\xi^2 + \eta^2) \leq \tau \leq 2\psi(\xi^2 + \eta^2) \right\} \quad h = 0, \ldots, 7.$$

Posto allora

$$f_h(x, y) = \int_{\mathbf{R}^3} \frac{g_h(\xi, \eta, \tau)}{\left[(x - \xi)^2 + (y - \eta)^2 + \tau^2 \right]^2} d\xi d\eta d\tau, \quad h = 0, \ldots, 7,$$

osserviamo che ciascuna delle equazioni $Pu_h = f_h$ può essere ricondotta alla particolare situazione considerata all'inizio della dimostrazione mediante semplici rotazioni e traslazioni.

4. – Sembra che il risultato enunciato nel n. 3 non valga quando il numero di variabili indipendenti è maggiore di due. Per esempio abbiamo forti dubbi sull'esistenza di una soluzione analitica in $\mathbf{R}^3$ delle equazioni

$$\frac{\partial^2 u}{\partial x^2} + \frac{\partial^2 u}{\partial y^2} = f(x, y, t), \qquad \frac{\partial u}{\partial t} + \frac{\partial^2 u}{\partial x^2} + \frac{\partial^2 u}{\partial y^2} = f(x, y, t)$$

ove

$$f(x, y, t) = \sum_{h,k=1}^{\infty} \exp \left[\varphi(h, k)(ix + ih^2 t - (y - h)^2 - 1) \right],$$

$$\varphi(h, k) = \left[([p(h)]^k)! \right]!$$

e $p(h)$ è l'$(h+1)$-esimo numero primo.

Bibliografia

[1] K. G. ANDERSSON, *Analyticity of fundamental solutions*, Ark. Mat., **8** (1970), 73–81.

[2] K. G. ANDERSSON, *Propagation of analyticity of solutions of partial differential equations with constant coefficients*, Ark. Mat., **8** (1971), 277–302.

[3] E. DE GIORGI – L. CATTABRIGA, *Una formula di rappresentazione per funzioni analitiche in* $\mathbf{R}^n$, Boll. Un. Mat. Ital., **4** (1971), 1010–1014.

[4] L. EHRENPREIS, *Some applications of the theory of distributions to several complex variables*, Seminar on Analytic Functions, Princeton 1957, 65–79.

[5] L. EHRENPREIS, *Solutions of some problems of division IV*, Amer. J. Math., **82** (1960), 522–588.

[6] L. EHRENPREIS, *Fourier analysis in several complex variables*, Wiley Interscience, 1970.

[7] L. EHRENPREIS, *Fourier analysis on non convex sets*, Conferenza tenuta al Convegno sulle equazioni ipoellittiche e gli spazi funzionali, Roma, 25–28 gennaio 1971.

[8] T. KAWAI, *Construction of a local elementary solution for linear partial differential operators (I)*, Proc. Japan Acad., **47** (1971), 19–23.

[9] T. KAWAI, *Construction of a local elementary solution for linear partial differential operators (II)*, Proc. Japan Acad., **47** (1971), 147–152.

[10] T. KAWAI, *On the global existence of real analytic solutions of linear differential equations (I)*, Proc. Japan Acad., **47** (1971), 537–540.

[11] T. KAWAI, *On the global existence of real analytic solutions of linear differential equations (II)*, Proc. Japan Acad. **47** (1971), 643–647.

[12] T. KAWAI, *On the global existence of real analytic solutions of linear differential equations*, Conferenza tenuta al "Symposium of the theory of hyperfunctions and analytic functionals", Kyoto Univ., 27–30 settembre 1971.

[13] B. MALGRANGE, *Existence et approximation des solutions des équations aux dérivées partielles et des équations de convolution*, Ann. Inst. Fourier (Grenoble), **6** (1956), 271–355.

[14] A. MARTINEAU, *Sur la topologie des espaces de fonctions holomorphes*, Math. Ann., **163** (1966), 62–88.

[15] F. TRÈVES, *Linear partial differential equations with constant coefficients*, Gordon and Breach, 1966, Chap.9.

[16] F. Trèves – M. Zerner, *Zones d'analyticité des solutions élémentaires*, Bull. Soc. Math. France, **95** (1967), 155–191.

Convergence of the energy integrals for second order elliptic operators [‡]

E. De Giorgi – S. Spagnolo (Pisa)[*]

Summary. In this work we study the convergence of the energy integrals of solutions to elliptic equations in divergence form; in particular, we state relations between this kind of convergence and convergence in the mean of the solutions.

Introduction

In this paper we study the G-convergence of elliptic operators (see [7] and [8]) in relation with the convergence of the corresponding energy integrals.

We begin by recalling the definition of G-convergence: given two numbers λ_0 and Λ_0 with $0 < \lambda_0 < \Lambda_0$, let us consider the class $\mathcal{E}(\lambda_0, \Lambda_0)$ of all elliptic operators of the form

$$A = \sum_{ij=1}^{n} D_i(a_{ij}(x)D_j)$$

whose coefficients a_{ij} are measurable functions on $\mathbf{R}^n$ with

$$\begin{cases} a_{ij}(x) = a_{ji}(x), \\[2mm] \lambda_0|\xi|^2 \leq \sum_{ij} a_{ij}(x)\xi_i\xi_j \leq \Lambda_0|\xi|^2 \end{cases}$$

for every x and $\xi = (\xi_1, \xi_2, \ldots, \xi_n)$ in $\mathbf{R}^n$.

Given $\lambda > 0$, $f \in L^2$ and $A \in \mathcal{E}(\lambda_0, \Lambda_0)$, we denote by $(\lambda I - A)^{-1}f$ ($I=$ identity operator) the unique solution $u \in H^1$ of the equation

$$-Au + \lambda u = f \quad \text{in } \mathbf{R}^n.$$

Definition (see [8]; compare with the "abstract" definition in [1]). Let A_k ($k = 1, 2, 3, \ldots$) and A be operators in $\mathcal{E}(\lambda_0, \Lambda_0)$; we say that A_k G-converges to A if for some $\lambda > 0$ it holds

$$(1) \qquad (\lambda I - A_k)^{-1}f \to (\lambda I - A)^{-1}f \quad \text{in } L^2, \ \forall f \in L^2.$$

[‡]Editor's note: translation into English of the paper "Sulla convergenza degli integrali dell'energia per operatori ellittici del secondo ordine", published in Boll. Un. Mat. Ital., **(4) 8** (1973), 391–411.

[*]Research supported by C.N.R. within the G.N.A.F.A.

Notice that (1) holds for every $\lambda > 0$ as soon as it holds for some $\lambda > 0$.

We also remark that in the most relevant examples of G-convergence we cannot expect convergence in H^1 of the solutions, but we have convergence of the energy integrals (see Theorems 2.3 and 2.4 of §2).

We now briefly sketch the plan of this paper: in §1 we recall the main properties of G-convergence (the corresponding proofs can be found in [7] and [8]), and some results on elliptic equations we will need in the following.

In §2 we prove the following results:

THEOREM 2.3. – *Let A_k, A in $\mathcal{E}(\lambda_0, \Lambda_0)$ and u_k, u in $L^2(\Omega)$ (Ω is a bounded open set of $\mathbf{R}^n$) be such that $\{A_k\} \xrightarrow{G} A$, $\{u_k\} \to u$ in $L^2(\Omega)$ and $\{A_k u_k\} \to Au$ in $H^{-1}(\Omega)$. We then have*

$$(2) \qquad \lim_{k \to +\infty} \sum_{ij} \int_S a_{ij,k}(x) D_i u_k D_j u_k \, dx = \sum_{ij} \int_S a_{ij}(x) D_i u D_j u \, dx$$

for every measurable set S such that $\overline{S} \subseteq \Omega$ (where $a_{ij,k}$ and a_{ij} are the coefficients of A_k, A, respectively).

THEOREM 2.4 – *If $\{A_k\} \xrightarrow{G} A$ and u_k, u denote the solutions of the Dirichlet problems*

$$\begin{cases} A_k u_k = f \text{ in } \Omega, & u_k - w \in H_0^1(\Omega) \\ Au = f \text{ in } \Omega, & u - w \in H_0^1(\Omega) \end{cases}$$

with $f \in H^{-1}(\Omega)$, $w \in H^1(\Omega)$ and Ω bounded open subset of $\mathbf{R}^n$, then (2) holds with $\overline{S} \subseteq \Omega$, and also with $S \equiv \Omega$ provided Ω has Lipschitz-continuous boundary.

In §3 we introduce the quantity

$$\gamma_A(\Omega, \xi) = (\text{meas } \Omega)^{-1} \min \left\{ \sum_{ij} \int_\Omega a_{ij}(x) D_i u D_j u \, dx : \left(u - \sum_i \xi_i x_i\right) \in H_0^1(\Omega) \right\}$$

whenever $A \in \mathcal{E}(\lambda_0, \Lambda_0)$, Ω is a bounded open subset of $\mathbf{R}^n$, $\xi = (\xi_1, \ldots, \xi_n) \in \mathbf{R}^n$.

One can check that $\gamma_A(\Omega, \xi)$ is a quadratic form in ξ with eigenvalues between λ_0 e Λ_0, and that it coincides with the integral mean of the energy over Ω of the solution $\overline{u}$ of the problem

$$A\overline{u} = 0 \text{ in } \Omega, \qquad \overline{u} = \sum_i \xi_i x_i \text{ on } \partial\Omega.$$

The knowledge of $\gamma_A(\Omega, \xi)$ for every cube Ω of $\mathbf{R}^n$ suffices to determine the coefficients a_{ij} of A:

LEMMA 3.2 – *For almost every $x \in \mathbf{R}^n$ one has*

$$\lim_{\nu \to \infty} \gamma_A(\Omega_\nu, \xi) = \sum_{ij} a_{ij}(x) \xi_i \xi_j, \quad \xi \in \mathbf{R}^n$$

where $\{\Omega_\nu\}$ is an arbitrarily chosen sequence of cubes such that

$$\lim_{\nu \to \infty} \{\operatorname{diam} \Omega_\nu\} = 0 \text{ and } x \in \overline{\Omega}_\nu \text{ for every } \nu.$$

It is then natural to search for a correspondence between G-convergence and the convergence of the forms γ_A: after all, Theorem 2.4 ensures that if $\{A_k\} \xrightarrow{G} A$, we then have

$$\lim_{k \to \infty} \gamma_{A_k}(\Omega, \xi) = \gamma_A(\Omega, \xi) \text{ for every cube } \Omega \text{ and for every } \xi.$$

Conversely:

THEOREM 3.3 – *Consider the countable net of cubes in $\mathbf{R}^n$*

$$\Omega_{\nu,h} = \{x \in \mathbf{R}^n : \ 2^{-\nu}h_i < x_i < 2^{-\nu}(h_i + 1), \ \text{for each } i\}$$

where $\nu \geq 1$ is an integer and $h = (h_1, \ldots, h_n)$ is an n-tuple of integers.

Let $\{A_k\}$ be a sequence of operators in $\mathcal{E}(\lambda_0, \Lambda_0)$ such that the following limit exists:

$$\tau(\nu, h; \xi) \equiv \lim_{k \to \infty} \gamma_{A_k}(\Omega_{\nu,h}, \xi), \ \text{for each } \nu, \ h, \ \xi.$$

Then, there exists an operator A in $\mathcal{E}(\lambda_0, \Lambda_0)$ such that $\{A_k\}$ G-converges to A, and $\gamma_A(\Omega_{\nu,h}, \xi) = \tau(\nu, h; \xi)$.

We remark that, in order to obtain some more quantitative estimates, it could be useful to have an improved version of the above theorem, for instance as in the following:

CONJECTURE – Let $\Omega_0 = \{x \in \mathbf{R}^n : \ 0 < x_i < 1, \ \forall i\}$ be the unit cube, and let A and B be operators in $\mathcal{E}(\lambda_0, \Lambda_0)$. Then, for every integer $\overline{\nu} \geq 1$ we have

$$|\gamma_A(\Omega_0, \xi) - \gamma_B(\Omega_0, \xi)| \leq C \sum_{\nu=\overline{\nu}}^{\infty} \sum_{i=1}^{n} \sum_{h_i=0}^{2\nu-1} 4^{-n\nu} |\gamma_A(\Omega_{\nu,h}, \xi) - \gamma_B(\Omega_{\nu,h}, \xi)|^\alpha$$

with $C = C(\lambda_0, \Lambda_0, \overline{\nu})$, $\alpha = \alpha(\lambda_0, \Lambda_0, \overline{\nu}) > 0$.

We finish by pointing out the following, stronger statement of the local nature of G-convergence:

THEOREM 3.4 – *Let $\{A_k\} \xrightarrow{G} A$, $\{B_k\} \xrightarrow{G} B$, and suppose that for every k the coefficients of A_k coincide with those of B_k on a measurable set S. Then the coefficients of A coincide with those of B almost everywhere over S.*

In §4 we will apply the above results to the study of periodic operators, and in particular we will throw new light on the mathematical meaning of some results by Sanchez-Palencia (see [6]).

1. Notation and preliminaries.

We denote by $x = (x_1, \ldots, x_n)$ a generic point in $\mathbf{R}^n$ and by $|x| = (\sum_i x_i^2)^{1/2}$ the modulus of x.

For each scalar function $f \in L^1_{loc}(\Omega)$, with Ω open subset of $\mathbf{R}^n$, we denote by $D_i f \equiv \partial f / \partial x_i$ the derivatives in the sense of distributions, and by $Df \equiv (D_1 f, \ldots, D_n f)$ the gradient of f. We put

$$\|Df\|_{L^p(\Omega)} = \left(\sum_i \int_\Omega |D_i f|^p \, dx \right)^{1/p} , \quad \|Df\|_{L^\infty(\Omega)} = \sum_i \sup_{x \in \Omega} |D_i f(x)|.$$

We will consider the following vector spaces over $\mathbf{R}$:

$\mathcal{D}(\Omega) = \{\text{functions in } C^\infty \text{ with compact support in } \Omega\},$

$H^{1,p}(\Omega) = \text{Banach space of the functions } f \in L^p(\Omega) \text{ with } Df \in L^p(\Omega)$

$H^{1,p}_0(\Omega) = \text{closure of } \mathcal{D}(\Omega) \text{ in } H^{1,p}(\Omega)$

$H^1_{loc}(\Omega) = \{\text{functions } f \in L^1_{loc}(\Omega) \text{ such that } f \in H^{1,2}(\Omega') \text{ for every } \Omega' \subset\subset \Omega\}$

$H^{-1}(\Omega) = \text{dual Banach space of } H^{1,2}_0(\Omega) \text{ (represented as a space of distributions)}.$

We will finally write (for short) $H^1(\Omega)$ instead of $H^{1,2}(\Omega)$; L^p, H^1_{loc}, etc. instead of $L^p(\mathbf{R}^n), H^1_{loc}(\mathbf{R}^n), \ldots$; $\int f \, dx$ instead of $\int_{\mathbf{R}^n} f \, dx$ and $\sum_i \alpha_i, \sum_{ij} \alpha_{ij}$ instead of $\sum_{i=1}^n \alpha_i, \sum_{ij=1}^n \alpha_{ij}$.

Let us recall now the main properties of G-convergence for elliptic equations:

1.1. ([8] p. 585, taking into account Theorem 7 of [8], p. 594). *If $\{A_k\} \xrightarrow{G} A$ and u_k, u in $H^1_0(\Omega)$, with Ω bounded open subset of $\mathbf{R}^n$, are such that $\{A_k u_k\} \to Au$ in $H^{-1}(\Omega)$, then $\{u_k\} \to u$ in $L^2(\Omega)$.*

1.2. ([8] p. 587, 588). *If $\{A_k\}$ and A are in $\mathcal{E}(\lambda_0, \Lambda_0)$, then*

$$(a) \Rightarrow (b) \Rightarrow (c), \text{ where}:$$

(a) $\{a_{ij,k}\} \to a_{ij}$ in L^1_{loc}, $\forall i, j$
(b) $\{(\lambda I - A_k)^{-1} f\} \to (\lambda I - A)^{-1} f$ in H^1, $\forall f \in L^2$
(c) $\{A_k\} \xrightarrow{G} A$
(and where we denoted by $a_{ij,k}$, a_{ij} the coefficients of A_k, A).

1.3. *(compactness: [8], p. 582). From every sequence $\{A_k\}$ in $\mathcal{E}(\lambda_0, \Lambda_0)$, it is possible to extract a subsequence $\{A_{k_\nu}\}$ which G-converges to an operator in $\mathcal{E}(\lambda_0, \Lambda_0)$.*

1.4. *(locality: [8], p. 595). If $A_k \xrightarrow{G} A$, $B_k \xrightarrow{G} B$ and, for every k, the coefficients of A_k coincide with the corresponding coefficients of B_k in an open set Ω in $\mathbf{R}^n$, then the coefficients of A equal those of B almost everywhere in Ω.*

1.5. ([8], p. 592). *If $A_k \xrightarrow{G} A$ and $\{u_k\} \subset H^1_{loc}(\Omega)$ is such that $A_k u_k = 0$ in Ω, and $\{u_k\} \to u$ in $L^1_{loc}(\Omega)$, then $Au = 0$ in Ω.*

We finish by recalling some estimates, valid for the solutions u of the equation

$$(3) \qquad\qquad -Au + u = f + \sum_i D_i g_i$$

with $A \in \mathcal{E}(\lambda_0, \Lambda_0)$ and $f, g_1, \dots, g_n$ in L^∞.

1.6. (STAMPACCHIA: [9], p. 224). *Let $u \in H^1$ be the solution of (3) in $\mathbf{R}^n$. We have:*

$$(4) \qquad\qquad \|u\|_{L^\infty} \le c(\lambda_0, \Lambda_0)\Big(\|f\|_{L^\infty} + \sum_i \|g_i\|_{L^\infty}\Big).$$

1.7. (MEYERS: [3], p. 198 and p. 200). *Let $u \in H^1$ denote the solution of (3) in $\mathbf{R}^n$. We have*

$$(5) \qquad\qquad \|Du\|_{L^{2+\eta}(\Omega')} \le c(\lambda_0, \Lambda_0, \Omega')\Big(\|f\|_{L^\infty} + \sum_i \|g_i\|_{L^\infty}\Big)$$

for each bounded set Ω', for some $\eta = \eta(\lambda_0, \Lambda_0) > 0$.

Moreover, if Ω is a bounded open set with Lipschitz-continuous boundary and $u \in H^1_0(\Omega)$ is the solution of (3) in Ω, it holds

$$(6) \qquad \|Du\|_{L^{2+\eta}(\Omega)} \le c(\lambda_0, \Lambda_0, \Omega)\Big(\|f\|_{L^\infty(\Omega)} + \sum_i \|g_i\|_{L^\infty(\Omega)}\Big).$$

2. - Energy integrals.

LEMMA 2.1 – *Given $A \in \mathcal{E}(\lambda_0, \Lambda_0)$ and $w \in H^1$, we denote by w_ε $(\varepsilon > 0)$ the solution in H^1 of the equation*

$$(7) \qquad\qquad -\varepsilon A w_\varepsilon + w_\varepsilon = w \text{ in } \mathbf{R}^n.$$

We then have the following estimates

$$(8) \qquad\qquad \|Dw - Dw_\varepsilon\|_{L^2} \le \sqrt{\Lambda_0/\lambda_0}\|Dw\|_{L^2}$$

$$(9) \qquad\qquad \|w - w_\varepsilon\|_{L^2} \le \sqrt{\Lambda_0}\sqrt{\varepsilon}\|Dw\|_{L^2}$$

and, in case $Dw \in L^\infty$,

$$(10) \qquad\qquad \|w - w_\varepsilon\|_{L^\infty} \le c(\lambda_0, \Lambda_0)\sqrt{\varepsilon}\|Dw\|_{L^\infty}.$$

Finally, if $Aw|_\Omega \in L^2(\Omega)$ for some open set Ω in $\mathbf{R}^n$, then we have

$$(11) \qquad \|Dw - Dw_\varepsilon\|_{L^2(\Omega')} \le c(\lambda_0, \Lambda_0, \Omega', \Omega)\sqrt{\varepsilon}\big(\|Dw\|_{L^2} + \|Aw\|_{L^2(\Omega)}\big)$$

for every Ω' with $\overline{\Omega}' \subset \Omega$ and $\mathrm{dist}(\Omega', \partial\Omega) > 0$.

Proof. By putting $v_\varepsilon = \varepsilon^{-1}(w_\varepsilon - w)$, from (7) we get

$$(12) \qquad\qquad -\varepsilon A v_\varepsilon + v_\varepsilon = Aw$$

and multiplying this equation by v_ε and integrating, we easily obtain the estimates (8) and (9).

To obtain (10), it is enough to apply the dilation $x \mapsto \sqrt{\varepsilon}x$ to (12), and then use inequality (4).

Let us finally prove (11): if Ω' is such that $\overline{\Omega}' \subset \Omega$ and $\mathrm{dist}(\Omega', \partial\Omega) > 0$ we construct a function $\tau \in C^\infty$ such that $\tau \equiv 1$ in a neighborhood of $\overline{\Omega}'$, and $\tau \equiv 0$ in a neighborhood of $\mathbf{R}^n \setminus \Omega$ (if $\Omega = \mathbf{R}^n$, we simply put $\tau \equiv 1$); we then easily check that $\tau^2 v \in H_0^1(\Omega)$ whenever $v \in H^1$.

Multiplying (12) by $\tau^2 v_\varepsilon$ and integrating, we then obtain (by putting $f = Av|_\Omega$, and denoting by a_{ij} the coefficients of A):

$$\varepsilon \sum_{ij} \int_\Omega \tau^2 a_{ij}(x) D_i v_\varepsilon D_j v_\varepsilon \, dx + \int_\Omega \tau^2 v_\varepsilon^2 \, dx =$$

$$= \int_\Omega \tau^2 f v_\varepsilon \, dx - 2\varepsilon \sum_{ij} \int_\Omega a_{ij}(x) D_i v_\varepsilon \cdot \tau v \cdot D_j \tau \ \ end.$$

From this, by using (8) we infer:

$$\lambda_0 \varepsilon \int_\Omega \tau^2 |Dv_\varepsilon|^2 \, dx + \int_\Omega \tau^2 v_\varepsilon^2 \, dx \le$$

$$\le \left| \int_\Omega \tau^2 v_\varepsilon^2 \, dx \right|^{1/2} \left| \left(\int_\Omega \tau^2 f^2 \, dx \right)^{1/2} + 2\Lambda_0 \sup_{\mathbf{R}^n} |D\tau| \sqrt{\Lambda_0/\lambda_0} \left(\int |Dw|^2 \, dx \right)^{1/2} \right|$$

and thus

$$\left(\int_\Omega \tau^2 v_\varepsilon^2 \, dx \right)^{1/2} \le c(\lambda_0, \Lambda_0, \tau)(\|f\|_{L^\infty(\Omega)} + \|Dw\|_{L^2})$$

and also

$$\lambda_0 \varepsilon \int_\Omega \tau^2 |Dv_\varepsilon|^2 \, dx \le c^2(\lambda_0, \Lambda_0, \tau)(\|f\|_{L^\infty(\Omega)} + \|Dw\|_{L^2})^2.$$

Estimate (11) follows immediately from the last inequality.

LEMMA 2.2 – *Let A_k $(k = 1, 2, 3, \ldots)$ and A be in $\mathcal{E}(\lambda_0, \Lambda_0)$, and denote by u_k, u the solutions in H^1 of the equations in $\mathbf{R}^n$*

$$(13) \qquad\qquad -A_k u_k + \lambda u_k = f_k$$

$$(14) \qquad\qquad -Au + \lambda u = f$$

with $\lambda > 0$, f_k, f in H^{-1}.

Then, if $\{A_k\} \xrightarrow{G} A$ and $\{f_k\} \to f$ in H^{-1}, it holds

$$(15) \qquad \lim_{k \to \infty} \sum_{ij} \int_S a_{ij,k}(x) D_i u_k D_j u_k \, dx = \int_S a_{ij}(x) D_i u D_j u \, dx$$

for every measurable subset S of $\mathbf{R}^n$ (where $a_{ij,k}$ and a_{ij} are the coefficients of A_k and A respectively).

Proof. As the quadratic forms over H^1 defined by

$$u \mapsto \sum_{ij} \int_S a_{ij,k}(x) D_i u_k D_j u_k \, dx$$

and the linear maps $(\lambda I - A_k)^{-1} : H^{-1} \to H^1$ are all equicontinuous, it is easy to see that we may assume (with no loss of generality) that $f_k = f$, $\forall k$, (instead of $\{f_k\} \to f$ in H^{-1}). Actually, being $\mathcal{D}$ a dense subspace of H^{-1}, we may as well assume

$$f_k \equiv f \in \mathcal{D}, \ \forall k.$$

By the very definition of G-convergence we then have

$$(16) \qquad \{u_k\} \to u \in L^2,$$

whence in particular, for every $\varepsilon > 0$:

$$(17) \qquad \int_{|x| \geq N(\varepsilon)} u_k^2 \, dx \leq \varepsilon.$$

If we multiply (13) by $\tau^2 u_k$, for some C^∞ function τ such that $0 \leq \tau \leq 1$, $|D\tau| \leq 1$, $\tau(x) = 0$ for $|x| \leq N(\varepsilon)$ and $\tau(x) = 1$ for $|x| \geq N(\varepsilon) + 1$, we obtain, taking (17) into account:

$$(18) \qquad \int_{|x| \geq N(\varepsilon)+1} |Du_k|^2 \, dx \leq \varepsilon C(\lambda_0, \Lambda_0, \lambda, f).$$

From (18) we can then prove our thesis (15) in the general case, as soon as we prove it for a bounded set S. Note that proving (15) for a bounded set S, amounts to show that for every $\varphi \in \mathcal{D}$ it holds

$$(19) \qquad \lim_{k \to \infty} \sum_{ij} \int a_{ij,k}(x) D_i u_k D_j u_k \cdot \varphi \, dx = \sum_{ij} \int a_{ij}(x) D_i u D_j u \cdot \varphi \, dx.$$

Indeed, if we choose a sequence $\{\varphi_\nu\} \subset C^\infty$ of functions with equibounded supports, such that $\{\varphi_\nu\}$ converges to the characteristic function of S in L^p for every $p < \infty$, by Meyers' estimate (5) we get

$$\lim_{\nu \to \infty} \sum_{ij} \int a_{ij,k}(x) D_i u_k D_j u_k \cdot \varphi_\nu \, dx = \sum_{ij} \int_S a_{ij,k}(x) D_i u_k D_j u_k \, dx,$$

where the limit is uniform with respect to k.

Let us then prove (19).

For a fixed $\varphi \in \mathcal{D}$, let $\varphi_{k,\varepsilon}$, φ_ε denote the solutions in H^1 of the following equations in $\mathbf{R}^n$ ($\varepsilon > 0$):

$$(20) \qquad -\varepsilon A_k \varphi_{k,\varepsilon} + \varphi_{k,\varepsilon} = \varphi, \qquad -\varepsilon A \varphi_\varepsilon + \varphi_\varepsilon = \varphi.$$

By (10) in the previous lemma, we have

$$(21) \qquad \lim_{\varepsilon \to 0} \sum_{ij} \int a_{ij,k}(x) D_i u_k D_j u_k (\varphi - \varphi_{k,\varepsilon}) \, dx = 0,$$

uniformly with respect to k.

On the other hand, by applying (4) to equations (20) and (13) (with $f_k = f \in \mathcal{D}$), we get the estimates

$$(22) \qquad \|\varphi_{k,\varepsilon}\|_{L^\infty} \leq c(\lambda_0, \Lambda_0, \lambda) \|\varphi\|_{L^\infty}$$

$$(23) \qquad \|u_k\|_{L^\infty} \leq c(\lambda_0, \Lambda_0, \lambda) \|f\|_{L^\infty}.$$

From this, we infer in particular that the functions $u_k \varphi_{k,\varepsilon}$ and u_k^2 are in H^1, and thus the following decomposition makes sense:

$$\sum_{ij} \int a_{ij,k}(x) D_i u_k D_j u_k \varphi_{k,\varepsilon} \, dx =$$

$$= \sum \int a_{ij,k}(x) D_i u_k D_j(u_k \varphi_{k,\varepsilon}) \, dx - \frac{1}{2} \sum_{ij} \int a_{ij,k}(x) D_j \varphi_{k,\varepsilon} D_i u_k^2 \, dx$$

$$= \int f u_k \varphi_{k,\varepsilon} \, dx - \lambda \int u_k^2 \varphi_{k,\varepsilon} \, dx + \frac{1}{2\varepsilon} \int u_k^2 (\varphi_{k,\varepsilon} - \varphi) \, dx.$$

But now, as $\{A_k\} \xrightarrow{G} A$, we have that $\{u_k\} \to u$ in L^2 and $\{\varphi_{k,\varepsilon}\} \xrightarrow{k} \varphi_\varepsilon$ in L^2 ($\forall \varepsilon > 0$). Thus, taking into account (22) and (23), we may check that each of the terms in the above decomposition of $\sum_{ij} \int a_{ij,k}(x) D_i u_k D_j u_k \varphi_{k,\varepsilon} \, dx$ converges to the corresponding term in the similar decomposition of $\sum_{ij} \int a_{ij}(x) D_i u D_j u \varphi_\varepsilon \, dx$ as $k \to \infty$, for any $\varepsilon > 0$.

In other words, we have shown that

$$(24)$$
$$\lim_{k \to +\infty} \sum_{ij} \int a_{ij,k}(x) D_i u_k D_j u_k \varphi_{k,\varepsilon} \, dx = \sum_{ij} \int a_{ij}(x) D_i u D_j u \cdot \varphi_\varepsilon \, dx, \quad \forall \varepsilon > 0.$$

By combining (24) with (21), we obtain (19) and the lemma is proved.

From the two lemmas just proved, we obtain the following theorem:

THEOREM 2.3. – *Let $\{A_k\}$ ($k = 1, 2, 3, \ldots$) and A be in $\mathcal{E}(\lambda_0, \Lambda_0)$ with $\{A_k\} \xrightarrow{G} A$, and let $u_k, u \in H^1(\Omega)$, Ω bounded open set in $\mathbf{R}^n$, be such that*

$$(25) \qquad \{u_k\} \to u \quad in \ L^2(\Omega)$$

(26) $$\{A_k u_k\} \to Au \quad \text{in } H^{-1}(\Omega).$$

Then, denoting by $a_{ij,k}$ and a_{ij} the coefficients of A_k, A, we have

(27) $$\lim_{k \to \infty} \sum_{ij} \int_S a_{ij,k}(x) D_i u_k D_j u_k \, dx = \sum_{ij} \int_S a_{ij}(x) D_i u D_j u \, dx$$

for each measurable set S with $\overline{S} \subset \Omega$.

Proof. By arguing as at the beginning of the proof of Lemma 2.2, we see that we can replace hypothesis (26) with the following one with no loss of generality:

$$A_k u_k = Au = f, \quad \forall k; \; f \in \mathcal{D}(\Omega).$$

Now, we can approximate u_k, u by the solutions in H^1, denoted by $u_{k,\varepsilon}$ and u_ε, of the following equations in $\mathbf{R}^n$ ($\varepsilon > 0$):

$$-\varepsilon A u_{k,\varepsilon} + u_{k,\varepsilon} = \tau u, \qquad -\varepsilon A u_\varepsilon + u_\varepsilon = \tau u$$

where τ is a function in $\mathcal{D}(\Omega)$ which is equal to 1 on a neighborhood Ω' of S.

By estimate (11) in Lemma 2.1 we then get

$$\|Du_k - Du_{k,\varepsilon}\|_{L^2(S)} \leq c(\lambda_0, \Lambda_0, S, \Omega')\sqrt{\varepsilon}(\|f\|_{L^2(\Omega')} + \|D(\tau u_k)\|_{L^2})$$
$$\|Du - Du_\varepsilon\|_{L^2(S)} \leq c(\lambda_0, \Lambda_0, S, \Omega')\sqrt{\varepsilon}(\|f\|_{L^2(\Omega')} + \|D(\tau u)\|_{L^2}),$$

and thus, estimating $\|D(\tau u)\|_{L^2}$ by means of Caccioppoli inequality and using hypothesis (26), it follows that

(28) $$\begin{cases} \lim_{\varepsilon \to 0} \|Du_{k,\varepsilon} - Du_k\|_{L^2(S)} = 0, & \text{uniformly in } k, \\ \lim_{\varepsilon \to 0} \|Du_\varepsilon - Du\|_{L^2(S)} = 0. \end{cases}$$

On the other hand, Lemma 2.2 ensures that $\forall \varepsilon > 0$ it holds

(29) $$\lim_{k \to \infty} \sum_{ij} \int_S a_{ij,k}(x) D_i u_{k,\varepsilon} D_j u_{k,\varepsilon} \, dx = \sum_{ij} \int_S a_{ij}(x) D_i u_\varepsilon D_j u_\varepsilon \, dx.$$

Thus, combining (29) with (28) we obtain our thesis.

THEOREM 2.4 – *Let $\{A_k\}$ ($k = 1, 2, 3, \ldots$) and A in $\mathcal{E}(\lambda_0, \Lambda_0)$ with $\{A_k\} \xrightarrow{G} A$, and denote by u_k, u the solutions of the following Dirichlet problems in Ω:*

(30) $$A_k u_k = f, \quad u_k - w \in H_0^1(\Omega)$$

(31) $$Au = f, \quad u - w \in H_0^1(\Omega)$$

where $f \in H^{-1}(\Omega)$, $w \in H^1(\Omega)$ and Ω is a bounded open set in $\mathbf{R}^n$.

We then have (denoting by $a_{ij,k}$ and a_{ij} the coefficients of A_k, A):

(32) $$\{u_k\} \to u \quad \text{in } L^2(\Omega)$$

$$(33) \qquad \lim_{k \to \infty} \sum_{ij} \int_S a_{ij,k}(x) D_i u_k D_j u_k \, dx = \sum_{ij} \int_S a_{ij}(x) D_i u D_j u \, dx$$

for every measurable S such that $\overline{S} \subset \Omega$.

If in addition Ω has a Lipschitz boundary, (33) holds for every (measurable) set $S \subseteq \Omega$.

Proof. To show (32), we notice that the solutions u_k to (30) fulfill the following estimate

$$(34) \qquad \|u_k\|_{H^1(\Omega)} \le c(\lambda_0, \Lambda_0, \Omega)(\|f\|_{H^{-1}(\Omega)} + \|w\|_{H^1(\Omega)}),$$

so that the sequence $\{u_k - w\}$ is relatively compact both in the weak topology of $H_0^1(\Omega)$, and in the strong topology of $L^2(\Omega)$ (by Rellich's Theorem).

Thus, to prove that $\{u_k\}$ converges to u in $L^2(\Omega)$, it suffices to show that, whenever $\{u_{k_\nu}\}$ is a subsequence of $\{u_k\}$ with $\{u_{k_\nu}\} \to v$ in $L^2(\Omega)$ and $v - w \in H_0^1(\Omega)$, we necessarily have $v = u$.

To this purpose, let us consider the functions $\tilde{u}_k, \tilde{u} \in H_0^1(\Omega)$ which solve the equations $A_k \tilde{u}_k = A\tilde{u} = f$ in Ω.

As $\{A_k\} \xrightarrow{G} A$, by Proposition 1.1 we have $\{\tilde{u}_k\} \to \tilde{u}$ in $L^2(\Omega)$ and thus $\{\tilde{u}_{k_\nu} - u_{k_\nu}\} \to \tilde{u} - v$ in $L^2(\Omega)$. Moreover, being $A_{k_\nu}(\tilde{u}_{k_\nu} - u_{k_\nu}) = 0$ in Ω, $\forall \nu$, by Proposition 1.5 we have $A(\tilde{u} - v) = 0$ in Ω, that is, $Av = f$ in Ω.

This implies $v = u$ because the solution of (31) is unique.

We have thus proved (32) and, by Theorem 2.3, also (33) with S relatively compact in Ω.

Suppose now that $\partial\Omega$ is Lipschitz-continuous: under this hypothesis the space $C^1(\overline{\Omega})$ (of the functions of class C^1 in Ω, which are continuous up to $\partial\Omega$ together with their first derivatives) is dense in $H^1(\Omega)$.

Thus, by (34) we may suppose that the data f and w of our problems are chosen in such a way that $f \in \mathcal{D}(\overline{\Omega})$, $w \in C^1(\overline{\Omega})$.

In this situation, by applying Meyers' estimate (6) to the equation $A_k(u_k - w) = f - A_k w$ in Ω, we obtain

$$\|D u_k\|_{L^{2+\eta}(\Omega)} \le c(\lambda_0, \Lambda_0, \Omega)(\|f\|_{L^\infty(\Omega)} + \|Dw\|_{L^\infty(\Omega)})$$

with $\eta = \eta(\lambda_0, \Lambda_0) > 0$. As a consequence, $\forall \varepsilon > 0$ there exists $\delta(\varepsilon) > 0$ such that

$$(35) \qquad \sum_{ij} \int_T |a_{ij,k}(x) D_i u_k D_j u_k - a_{ij}(x) D_i u D_j u| \, dx \le \varepsilon$$

for every $T \subset \Omega$ such that $\operatorname{meas}(T) \le \delta(\varepsilon)$.

(35) allows then to extend (33) from the case where S is relatively compact in Ω to the general case.

3. - The quadratic forms $\gamma_A(\Omega, \xi)$.

Let $A \in \mathcal{E}(\lambda_0, \Lambda_0)$, Ω be an open cube in $\mathbf{R}^n$ and $\xi = (\xi_1, \ldots, \xi_n) \in \mathbf{R}^n$.

It is well known that the Dirichlet problem

$$
(36) \qquad \begin{cases} Au = 0 & \text{in } \Omega \\ \big(u(x) - \sum_i \xi_i x_i\big) \in H_0^1(\Omega) \end{cases}
$$

has a unique solution u, characterized by the fact that, among all functions which agree with $\sum_i \xi_i x_i$ on $\partial\Omega$, it is the one having minimal energy with respect to A.

Denoting by a_{ij} the coefficients of A, we define

$$
\gamma_A(\Omega, \xi) = (\text{meas } \Omega)^{-1} \min \Big\{ \sum_{ij} \int_\Omega a_{ij}(x) D_i u D_j u \; dx \; : \; \big(u - \sum_i \xi_i x_i\big) \in H_0^1(\Omega)\Big\}.
$$

It is easy to check that $\gamma_A(\Omega, \xi)$ is a quadratic form in the variable ξ. Moreover:

$$
\lambda_0 |\xi|^2 \leq \gamma_A(\Omega, \xi) \leq \Lambda_0 |\xi|^2, \quad \forall \xi \in \mathbf{R}^n.
$$

Indeed, on one hand, by choosing $u \equiv \sum_i \xi_i x_i$, we get

$$
\gamma_A(\Omega, \xi) \leq (\text{meas } \Omega)^{-1} \sum_{ij} \int_\Omega a_{ij}(x) \xi_i \xi_j \; dx \leq \Lambda_0 |\xi|^2.
$$

On the other hand, if we put $u = v + \sum_i \xi_i x_i$ with $v \in H_0^1(\Omega)$ (and thus $\int_\Omega D_i v \; dx = 0$), we get:

$$
\sum_{ij} \int_\Omega a_{ij}(x) D_i u D_j u \; dx \geq \lambda_0 \int_\Omega |Du|^2 \; dx
$$

$$
= \lambda_0 \left[\int_\Omega |Dv|^2 \; dx + \sum_i \xi_i \int_\Omega D_i v \; dx + \int_\Omega |\xi|^2 \; dx \right] \geq \lambda_0 \cdot \text{meas } \Omega \cdot |\xi|^2.
$$

We also readily see that, if the coefficients a_{ij} are constant, then

$$
\gamma_A(\Omega, \xi) = \sum_{ij} a_{ij} \xi_i \xi_j.
$$

It is also easy to verify that $\gamma_A(\Omega, \xi)$ depends continuously on the coefficient matrix of A (with fixed Ω and ξ), as this matrix varies in $[L^1(\Omega)]^{n^2}$.

The following Lemma shows the Hölderian nature of this continuity:

LEMMA 3.1 – *For A, B in $\mathcal{E}(\lambda_0, \Lambda_0)$, Ω open cube of $\mathbf{R}^n$, $\xi \in \mathbf{R}^n$, it holds:*

$$
(37) \qquad |\gamma_A(\Omega, \xi) - \gamma_B(\Omega, \xi)|
$$

$$
\leq c(\lambda_0, \Lambda_0) \left[(\text{meas } \Omega)^{-1} \sum_{ij} \int_\Omega |a_{ij}(x) - b_{ij}(x)| \; dx \right]^{\eta/(2+\eta)} |\xi|^2
$$

where $\eta = \eta(\lambda_0, \Lambda_0) > 0$ and a_{ij}, b_{ij} are the coefficients of A, B.

Proof. Let $\overline{u}$ be the solution of problem (36): from Meyers estimate (6) (having reduced Ω to the unit cube by means of a translation and a dilation) we get

$$(38) \qquad \int_\Omega |D\overline{u}|^{2+\eta}\, dx \le c(\lambda_0, \Lambda_0)(\text{meas } \Omega)|\xi|^{2+\eta}$$

with $\eta = \eta(\lambda_0, \Lambda_0) > 0$.

On the other hand, from the definition of $\gamma_A(\Omega, \xi)$ and by Schwarz-Hölder inequality, we obtain:

$$\gamma_B(\Omega, \xi) - \gamma_A(\Omega, \xi) \le (\text{meas } \Omega)^{-1} \sum_{ij} \int_\Omega (b_{ij}(x) - a_{ij}(x)) D_i\overline{u} D_j\overline{u}\, dx$$

$$\le \quad (\text{meas } \Omega)^{-1} \left[\sum_{ij} \int_\Omega |b_{ij} - a_{ij}|^{(2+\eta)/\eta}\, dx \right]^{\eta/(2+\eta)} \cdot \left[\int_\Omega |D\overline{u}|^{2+\eta} dx \right]^{2/(2+\eta)}.$$

From this inequality we then get the thesis (changing A with B in case $\gamma_B(\Omega, \xi) < \gamma_A(\Omega, \xi)$, and taking into account (38) and the fact that $|b_{ij}(x) - a_{ij}(x)| \le 2\Lambda_0$).

LEMMA 3.2 – *Let $A \in \mathcal{E}(\lambda_0, \Lambda_0)$, and let $x_0 \in \mathbf{R}^n$ be a Lebesgue point for each of the coefficients a_{ij} of A. Then, for every sequence $\{\Omega_\nu\}$ of open cubes in $\mathbf{R}^n$ with $x_0 \in \overline{\Omega}_\nu$, $\forall\nu$, and $\lim\limits_{\nu\to\infty}\{\text{diam }(\Omega_\nu)\} = 0$, we have:*

$$(39) \qquad \lim_{\nu\to\infty} \gamma_A(\Omega_\nu, \xi) = \sum_{ij} a_{ij}(x_0)\xi_i\xi_j, \quad \forall\xi \in \mathbf{R}^n.$$

Proof. Putting $B = \sum\limits_{ij} a_{ij}(x_0) D_i D_j$, as B is an operator with constant coefficients we have:

$$\gamma_B(\Omega, \xi) = \sum_{ij} a_{ij}(x_0)\xi_i\xi_j.$$

By (37) and denoting by I_ν the ball with center x_0 and radius equal to diam (Ω_ν), we get:

$$\left| \gamma_A(\Omega_\nu, \xi) - \sum_{ij} a_{ij}(x_0)\xi_i\xi_j \right| \le$$

$$\le c(\lambda_0, \Lambda_0) \left[(\text{meas } \Omega_\nu)^{-1} \sum_{ij} \int_{\Omega_\nu} |a_{ij}(x) - a_{ij}(x_0)|\, dx \right]^{\eta/(2+\eta)} |\xi|^2$$

$$\le c(\lambda_0, \Lambda_0) \left[(\text{meas } I_\nu)^{-1} \sum_{ij} \int_{I_\nu} |a_{ij}(x) - a_{ij}(x_0)|\, dx \right]^{\eta/(2+\eta)} |\xi|^2,$$

from which (39) follows by letting $\nu \to \infty$.

We are now in a position to prove the following theorem, which is in a sense the converse of Theorem 2.4:

THEOREM 3.3 –   *Let A_k $(k = 1, 2, 3, \ldots)$ in $\mathcal{E}(\lambda_0, \Lambda_0)$ be such that there exists the limit*

$$\lim_{k \to \infty} \gamma_{A_k}(\Omega_{\nu,h}, \xi) \equiv \tau(\nu, h; \xi) \tag{40}$$

for every $\xi \in \mathbf{R}^n$, for every ν integer ≥ 1 and for every $h = (h_1, \ldots, h_n)$ n-tuple of relative integers, where we denoted:

$$\Omega_{\nu,h} = \{x \in \mathbf{R}^n \mid 2^{-\nu} h_i < x_i < 2^{-\nu}(h_i + 1), \ \forall i\}. \tag{41}$$

Then there exists an operator $A \in \mathcal{E}(\lambda_0, \Lambda_0)$ such that

$$\{A_k\} \xrightarrow{G} A$$
$$\gamma_A(\Omega_{\nu,h}, \xi) = \tau(\nu, h; \xi), \quad \forall \nu, \forall h, \forall \xi.$$

Proof. Since $\mathcal{E}(\lambda_0, \Lambda_0)$ is a compact set with respect to G-convergence, it is enough to prove that, whenever A, B are operators in $\mathcal{E}(\lambda_0, \Lambda_0)$ with

$$\gamma_A(\Omega_{\nu,h}, \xi) = \gamma_B(\Omega_{\nu,h}, \xi)$$

for every cube $\Omega_{\nu,h}$ of the net (41), we have $A = B$.

Indeed, Theorem 2.4 ensures that, as soon as $\{A_k\} \xrightarrow{G} A$, it holds $\lim_k \gamma_{A_k}(\Omega, \xi) = \gamma_A(\Omega, \xi)$ for every cube Ω and for every $\xi \in \mathbf{R}^n$.

Let x_0 be a Lebesgue point for each of the coefficients a_{ij}, b_{ij} of the operators A and B and, for every integer $\nu \geq 1$, let Ω_ν be a cube of the net (41) with edge $2^{-\nu}$ and such that $x_0 \in \overline{\Omega}_\nu$.

We then have

$$\gamma_A(\Omega_\nu, \xi) = \gamma_B(\Omega_\nu, \xi), \quad \forall \nu, \forall \xi,$$

and, passing to the limit for $\nu \to \infty$ and recalling Lemma 3.2, we obtain

$$\sum_{ij} a_{ij}(x_0)\xi_i\xi_j = \sum_{ij} b_{ij}(x_0)\xi_i\xi_j, \ \forall \xi,$$

that is

$$a_{ij}(x_0) = b_{ij}(x_0), \quad \forall i, j.$$

The fact that $A = B$ is then a consequence of the fact that almost every point of $\mathbf{R}^n$ is a Lebesgue point for the functions a_{ij} and b_{ij}.

Notice that the coefficients of the operator A appearing in the above theorem can be recovered from $\tau(\nu, h; \xi)$ in formula (40) by means of (29).

The next Theorem is a further consequence of Lemmas 3.1 and 3.2, and is a stronger version of Proposition 1.4:

THEOREM 3.4 –   *Let A_k, B_k $(k = 1, 2, 3, \ldots)$ and A, B be operators in $\mathcal{E}(\lambda_0, \Lambda_0)$, with coefficients $a_{ij,k}$, $b_{ij,k}$, a_{ij} and b_{ij} respectively.*

Suppose that $\{A_k\} \xrightarrow{G} A$, $\{B_k\} \xrightarrow{G} B$ and

$$(42) \qquad a_{ij,k}(x) = b_{ij,k}(x), \quad \forall x \in S, \forall k, \forall i, j,$$

with S measurable subset of $\mathbf{R}^n$.

Then we have also:

$$a_{ij}(x) = b_{ij}(x) \text{ almost everywhere on } S, \ \forall i, j.$$

Proof. Let x_0 be a Lebesgue point if the functions a_{ij} and b_{ij} $(i, j = 1, \ldots, n)$ which is also a point of density 1 for the set S; let $\{\Omega_\nu\}$ be a sequence of cubes in the net (41) such that Ω_ν has edge $2^{-\nu}$ and $x_0 \in \overline{\Omega}_\nu$, $\forall \nu$.

We then get, by (37) and hypothesis (42):

$$|\gamma_{A_k}(\Omega_\nu, \xi) - \gamma_{B_k}(\Omega_\nu, \xi)|$$

$$\leq \quad c(\lambda_0, \Lambda_0) \left[(\operatorname{meas} \Omega_\nu)^{-1} \sum_{ij} \int_{\Omega_\nu \setminus S} |a_{ij,k}(x) - b_{ij,k}(x)| \, dx \right]^{\eta/(2+\eta)} |\xi|^2$$

$$\leq \quad c'(\lambda_0, \Lambda_0) \left[(\operatorname{meas} \Omega_\nu)^{-1} \cdot \operatorname{meas} (\Omega_\nu \setminus S) \right]^{\eta/(2+\eta)} |\xi|^2.$$

By letting $k \to \infty$ we then get, $\forall \nu$,

$$|\gamma_A(\Omega_\nu, \xi) - \gamma_B(\Omega_\nu, \xi)| \leq \left[(\operatorname{meas} \Omega_\nu)^{-1} \cdot \operatorname{meas} (\Omega_\nu \setminus S) \right]^{\eta/(2+\eta)} |\xi|^2,$$

and then, by letting $\nu \to \infty$ (as x_0 is a point of density 1 for S),

$$\sum_{ij} (a_{ij}(x_0) - b_{ij}(x_0)) \xi_i \xi_j = 0, \ \forall \xi \in \mathbf{R}^n.$$

As almost every point $x_0 \in S$ fulfills the conditions we required at the beginning of the proof, we obtain the thesis.

4. - Periodic solutions.

Given n linearly independent vectors $p_{(1)}, \ldots, p_{(n)}$ in $\mathbf{R}^n$ and the open parallelogram P with edges $x_0 + p_{(1)}, \ldots, x_0 + p_{(n)}$, we will call *P-periodic* those functions $\psi : \mathbf{R}^n \to \mathbf{R}^m$ for which

$$\psi(x + p_{(i)}) = \psi(x), \qquad \forall x \in \mathbf{R}^n, \forall i.$$

Let A be an operator in $\mathcal{E}(\lambda_0, \Lambda_0)$ with *P*-periodic coefficients a_{ij}, and let ξ be in $\mathbf{R}^n$. It can be proved (see [4]) that there is a unique solution $u \in H^1_{loc}$ (up to additive constants) of the problem

$$(43) \qquad \begin{cases} Au = 0 & \text{on } \mathbf{R}^n, \\ Du \text{ is } P\text{-periodic}, \\ (\operatorname{meas} P)^{-1} \int_P Du \, dx = \xi. \end{cases}$$

It is easy to check that a function $u \in H^1_{loc}$ has a P-periodic gradient if and only if the function $u(x) - \sum_i \xi_i x_i$ is in turn P-periodic, where ξ_i denotes the mean value of the derivative $D_i u$ on the "period" P.

Thus we can show, by passing to the corresponding variational problem, that the mean value on the period P of the energy, with respect to A, of every solution of (43) is given by $\mu_A(\xi)$, where

$$\mu_A(\xi) =$$

$$= (\text{meas } P)^{-1} \min \left\{ \sum_{ij} \int_P a_{ij}(x) D_i u D_j u \, dx : \left(u(x) - \sum_i \xi_i x_i \right) is \ P\text{-}periodic \right\}.$$

If the coefficients a_{ij} are constant, it can be checked that $\mu_A(\xi)$ is a quadratic form in ξ ($\xi \in \mathbf{R}^n$), with eigenvalues lying between λ_0 and Λ_0, and that $\mu_A(\xi) = \sum_{ij} a_{ij} \xi_i \xi_j$.

Moreover:

THEOREM 4.1 – *Let A_k $(k = 1, 2, 3, \ldots)$ and A in $\mathcal{E}(\lambda_0, \Lambda_0)$ be P-periodic operators, with P open parallelogram of $\mathbf{R}^n$. Then from $\{A_k\} \xrightarrow{G} A$ it follows that*

$$\lim_{k \to \infty} \mu_{A_k}(\xi) = \mu_A(\xi), \quad \forall \xi \in \mathbf{R}^n.$$

Proof. By calling $\bar{u}_k$ (respectively $\bar{u}$) the unique solution of the problem (43) relative to A_k (respectively, to A) with vanishing mean value on the period P, and denoting by $a_{ij,k}$, a_{ij} the coefficients of A_k, A, we have

$$\mu_{A_k}(\xi) \quad = \quad (\text{meas } P)^{-1} \sum_{ij} \int_P a_{ij,k}(x) D_i \bar{u}_k D_j \bar{u}_k \, dx$$

$$\mu_A(\xi) \quad = \quad (\text{meas } P)^{-1} \sum_{ij} \int_P a_{ij}(x) D_i \bar{u} D_j \bar{u} \, dx.$$

Then the thesis is a consequence of Theorem 2.3 (with $S = P$, Ω bounded open set containing $\overline{P}$), provided we verify that

$$\{\bar{u}_k\} \to \bar{u} \quad \text{in } L^2_{loc},$$

that is, as $\bar{u}_k(x) - \sum_i \xi_i x_i$ is a P-periodic function, that

(45) $$\{\bar{u}_k\} \to \bar{u} \quad \text{in } L^2(P).$$

To prove (45), notice that, as

$$(\text{meas } P)^{-1} \int_P |D\bar{u}_k|^2 \, dx \leq \frac{1}{\lambda_0} \mu_{A_k}(\xi) \leq \frac{\Lambda_0}{\lambda_0} |\xi|^2$$

and $\int_P \overline{u}_k = 0$, by Poincaré inequality we infer that the sequence $\{\overline{u}_k\}$ is bounded in $H^1(P)$.

Thus, by the usual compactness argument, it suffices to show that, if $\{\overline{u}_{k_\nu}\}$ is a subsequence of $\{\overline{u}_k\}$ such that

$$\{\overline{u}_{k_\nu}\} \overset{\nu}{\longrightarrow} w \text{ in } L^2(P) \text{ and weakly in } H^1(P),$$

then it necessarily holds $w = \overline{u}$.

We immediately see that Dw is P-periodic, and its mean value equals ξ on P, while w has vanishing mean value on P.

On the other hand, by Proposition 1.5 we have $Aw = 0$ on $\mathbf{R}^n$. But then $w = \overline{u}$, because of the uniqueness of the solution of (43) with mean value zero in P.

THEOREM 4.2 – *Let A be an operator of $\mathcal{E}(\lambda_0, \Lambda_0)$ with P-periodic coefficients a_{ij}, with P open parallelogram of $\mathbf{R}^n$, and let*

$$A_\varepsilon = \sum_{ij} D_i\left(a_{ij}(x/\varepsilon)D_j\right), \quad \varepsilon > 0$$

$$A_0 = \sum_{ij} \mu_{ij} D_i D_j$$

where (μ_{ij}) is the symmetric matrix defined by

$$\sum_{ij} \mu_{ij}\xi_i\xi_j = \mu_A(\xi), \qquad \forall \xi \in \mathbf{R}^n.$$

We then have, as $\varepsilon \to 0^+$:

$$\{A_\varepsilon\} \overset{G}{\longrightarrow} A_0.$$

Proof. As the set $\mathcal{E}(\lambda_0, \Lambda_0)$ is compact with respect to G-convergence, we only have to show that from

$$\{A_{\varepsilon_k}\} \overset{G}{\longrightarrow} A', \quad \text{with } \{\varepsilon_k\} \overset{k}{\longrightarrow} 0,$$

it follows $A' = A_0$.

Now, A' has constant coefficients, exactly as A.

Indeed, it is the G-limit of a sequence $A_k \equiv A_{\varepsilon_k}$ of periodic operators whose periods $P_k \equiv \varepsilon P$ are such that $\lim_{k \to \infty} \{\text{diam } P_k\} = 0$ *.

---

*For every $b \in \mathbf{R}^n$ we can indeed build a sequence of vectors $\{b_\nu\} \to b$ in such a way that $\tau_{b_\nu}(A_{k_\nu}) = A_{k_\nu}$ for every ν (where we denoted by $\tau_b A$ the operator obtained from A by means of a b-translation of the coefficients). On the other hand, recalling the definition of G-convergence, it is easy to check that, as $\{A_{k_\nu}\} \overset{G}{\longrightarrow} A'$ and $\{b_\nu\} \to b$, we also have $\{\tau_{b_\nu}(A_{k_\nu})\} \overset{G}{\longrightarrow} \tau_b(A')$. Hence $\tau_b(A') = A'$ for every $b \in \mathbf{R}^n$, and A' is constant.

Now, showing that the two operators with constant coefficients A' and A coincide amounts to show that $\mu_{A'}(\xi) = \mu_{A_0}(\xi)$, that is, by the definition of A_0, that

$$(46) \qquad \mu_{A'}(\xi) = \mu_A(\xi).$$

Now, by Theorem 4.1 we have

$$\mu_{A'}(\xi) = \lim_{k \to \infty} \mu_{A_{\varepsilon_k}}(\xi),$$

while, by means of a scaling argument, it is easy to verify that

$$\mu_A(\xi) = \mu_{A_\varepsilon}(\xi), \quad \forall \varepsilon > 0.$$

We thus get (46) and the theorem is proved.

FINAL REMARKS.

4.3 – Theorem 4.2 clarifies the meaning, first pointed out by SANCHEZ-PALENCIA in [6], of the quadratic form $\mu_A(\xi) = \sum_{ij} \mu_{ij}\xi_i\xi_j$ (or, better, of the operator $A_0 = \sum_{ij} \mu_{ij}D_iD_j$), as the "limiting macroscopic behaviour" of the periodic operator A in the asymptotic limit as $\varepsilon \to 0^+$.

Notice that by Theorem 3.3 we obtain, when A is a periodic operator:

$$(47) \qquad \lim_{\varepsilon \to 0} \gamma_A \left(\frac{\Omega}{\varepsilon}, \xi \right) = \mu_A(\xi), \quad \forall \ cube \ \Omega \ in \ \mathbf{R}^n;$$

and this formula connects the quadratic forms $\gamma_A(\Omega, \xi)$ and $\mu_A(\xi)$. Compare (47) with (39).

4.4 – A particular case for which $\mu_A(\xi)$ can be explicitly computed from the coefficients of A, is when A is an *isotropic* operator, that is of the form

$$(48) \qquad A = \sum_i D_i(a(x)D_i),$$

where the coefficient $a(x)$ is as follows:

$$(49) \qquad a(x) = a_1(x_1) \cdot \ldots \cdot a_n(x_n)$$

with $\sqrt[n]{\lambda_0} \le a_i(t) \le \sqrt[n]{\Lambda_0}$, $a_i(t + p_i) = a_i(t)$ $(p_i > 0)$.

In this case, by employing Theorem 4.2 and the G-convergence criterion for operators of the form $\{(48),(49)\}$ found in [2], p. 663, we see that

$$\mu_A(\xi) = \sum_i \mu_i \xi_i^2$$

where

$$\mu_i = \frac{1}{p_1 \cdots p_n} \int_0^{p_1} a_1 \, dt \ldots \int_0^{p_{i-1}} a_{i-1} \, dt \left(\int_0^{p_i} a_i^{-1} \, dt \right)^{-1} \int_0^{p_{i+1}} a_{i+1} \, dt \ldots \int_0^{p_n} a_n \, dt.$$

In the case of one variable ($n = 1$), all operators in $\mathcal{E}(\lambda_0, \Lambda_0)$ are of the form $\{(48),(49)\}$.

4.5 – As in the case of $\gamma_A(\Omega, \xi)$, we may define, for each A in $\mathcal{E}(\lambda_0, \Lambda_0)$ (not necessarily periodic) and for every cube Ω of $\mathbf{R}^n$, a quadratic form $\mu_A(\Omega, \xi)$. It is enough to put

$$\mu_A(\Omega, \xi) = \mu_{A_\Omega}(\xi)$$

where A_Ω denotes the Ω-periodic operator which coincides with A on Ω.

For $\mu_A(\Omega, \xi)$ we have similar results as those proved in §3 for $\gamma_A(\Omega, \xi)$, with completely analogous proofs: in particular, we have the equivalent result of Lemmas 3.1 and 3.2 and of Theorem 3.3.

We finally mention the following relation

$$(50) \qquad \mu_A(\Omega, \xi) \leq \gamma_A(\Omega, \xi),$$

which is due to the fact that every u with $\left(u(x) - \sum_i \xi_i x_i\right) \in H_0^1(\Omega)$ has also the property that $u(x) - \sum_i \xi_i x_i$ is Ω-periodic.

Bibliography

[1] P. Marcellini, *Su una convergenza di funzioni convesse*. Boll. Un. Mat. Ital. **8** (1973), pp. 137–158.

[2] A. Marino, S. Spagnolo, *Un tipo di approssimazione dell'operatore $\sum_{ij} D_i(a_{ij}(x)D_j)$ con operatori $\sum_j D_j(\beta(x)D_j)$*, Ann. Scuola Norm. Sup. Pisa, **23** (1969), pp. 657–673.

[3] N.G. Meyers, *An L^p estimate for the gradient of solutions of second order elliptic divergence equations*, Ann. Scuola Norm. Sup. Pisa **17** (1963), pp. 189–206.

[4] E. Sanchez-Palencia, *Solutions périodiques par rapport aux variables d'espace et applications*, C.R. Acad. Sci. Paris, Sér. A, **271** (1970), pp. 1129–1132.

[5] E. Sanchez-Palencia, *Equations aux dérivées partielles dans un type de milieux hétérogènes*, C.R. Acad. Sci. Paris, Sér. A, **272** (1971), pp. 1410–1413.

[6] E. Sanchez-Palencia, *Comportement local et macroscopique d'un type de milieux phisiques hétérogènes*, Internat. J. Engrg. Sci., **13** (1975), no. 11, pp. 923–940.

[7] S. Spagnolo *Sul limite delle soluzioni di problemi di Cauchy relativi all'equazione del calore*, Ann. Scuola Norm. Sup. Pisa, **21** (1967), pp. 657–699.

[8] S. SPAGNOLO *Sulla convergenza di soluzioni di equazioni paraboliche ed ellittiche*, Ann. Scuola Norm. Sup. Pisa, **22** (1968), pp. 571-597.

[9] G. STAMPACCHIA, *Le problème de Dirichlet pour les équations elliptiques du second ordre à coefficients discontinus*, Ann. Inst. Fourier (Grenoble), **15** (1965), pp. 189–257.

Sulla convergenza degli integrali dell'energia per operatori ellittici del secondo ordine[‡]

E. De Giorgi – S. Spagnolo (Pisa)[*]

Sunto. In this work we study the convergence of the energy-integrals of solutions to elliptic equations in divergence form; in particular we state relations between this kind of convergence and convergence in the mean of the solutions.

Introduzione

In questo lavoro viene studiata la G-convergenza di operatori ellittici (vedi [7] e [8]) in rapporto alla convergenza dei corrispondenti integrali dell'energia.

Cominciamo col richiamare la definizione di G-convergenza: fissati due numeri λ_0 e Λ_0 tali che $0 < \lambda_0 < \Lambda_0$, consideriamo la classe $\mathcal{E}(\lambda_0, \Lambda_0)$ di tutti gli operatori differenziali del tipo

$$A = \sum_{ij=1}^{n} D_i(a_{ij}(x)D_j)$$

i cui coefficienti a_{ij} sono delle funzioni misurabili su $\mathbf{R}^n$ con

$$\begin{cases} a_{ij}(x) = a_{ji}(x), \\[2mm] \lambda_0|\xi|^2 \leq \sum_{ij} a_{ij}(x)\xi_i\xi_j \leq \Lambda_0|\xi|^2 \end{cases}$$

per ogni x e $\xi = (\xi_1, \xi_2, \ldots, \xi_n)$ in $\mathbf{R}^n$.

Assegnati $\lambda > 0$, $f \in L^2$ e $A \in \mathcal{E}(\lambda_0, \Lambda_0)$, indichiamo con $(\lambda I - A)^{-1}f$ ($I=$ operatore identico) l'unica soluzione $u \in H^1$ dell'equazione

$$-Au + \lambda u = f \quad su \ \mathbf{R}^n.$$

Definizione (vedi [8]; cfr. con [1] per una definizione "astratta"). Siano A_k ($k = 1, 2, 3, \ldots$) ed A operatori di $\mathcal{E}(\lambda_0, \Lambda_0)$; si dice che A_k G-converge verso A se per qualche $\lambda > 0$ si ha

$$(1) \qquad (\lambda I - A_k)^{-1}f \to (\lambda I - A)^{-1}f \quad in \ L^2, \ \forall f \in L^2$$

[‡]Editor's note: published in Boll. Un. Mat. Ital., **(4) 8** (1973), 391–411.
[*]Lavoro eseguito con contributo del C.N.R. nell'ambito del G.N.A.F.A.

Si noti che la (1) è valida per ogni $\lambda > 0$ non appena vale per qualche $\lambda > 0$.

Notiamo anche che negli esempi più siginificativi di G-convergenza non si verifica la convergenza H^1 delle soluzioni, pur avendosi la convergenza delle energie (nella forma precisata dai teoremi 2.3 e 2.4 del §2).

Possiamo ora accennare al contenuto di questo articolo. Nel §1 vengono richiamate le principali proprietà della G-convergenza (le corrispondenti dimostrazioni si trovano in [7] e [8]), ed anche alcuni risultati sulle equazioni ellittiche che verranno poi utilizzati.

Nel §2 si provano i risultati seguenti:

TEOREMA 2.3. – *Siano A_k, A in $\mathcal{E}(\lambda_0, \Lambda_0)$ e u_k, u in $L^2(\Omega)$, con Ω aperto limitato di $\mathbf{R}^n$, tali che $\{A_k\} \xrightarrow{G} A$, $\{u_k\} \to u$ in $L^2(\Omega)$, $\{A_k u_k\} \to Au$ in $H^{-1}(\Omega)$. Si ha allora*

$$(2) \qquad \lim_{k \to +\infty} \sum_{ij} \int_S a_{ij,k}(x) D_i u_k D_j u_k \, dx = \sum_{ij} \int_S a_{ij}(x) D_i u D_j u \, dx$$

per ogni S misurabile tale che $\overline{S} \subseteq \Omega$ (dove $a_{ij,k}$ e a_{ij} sono i coefficienti di A_k, A).

TEOREMA 2.4 – *Se $\{A_k\} \xrightarrow{G} A$ e u_k, u sono le soluzioni dei problemi di Dirichlet*

$$\begin{cases} A_k u_k = f \ su \ \Omega, & u_k - w \in H_0^1(\Omega) \\ Au = f \ su \ \Omega, & u - w \in H_0^1(\Omega) \end{cases}$$

con $f \in H^{-1}(\Omega)$, $w \in H^1(\Omega)$ e Ω aperto limitato di $\mathbf{R}^n$, allora si ha la (2) con $\overline{S} \subseteq \Omega$, e anche con $S \equiv \Omega$ se Ω ha frontiera lipschitziana.

Nel §3 viene introdotta l'espressione

$$\gamma_A(\Omega, \xi) = (\text{mis } \Omega)^{-1} \min \left\{ \sum_{ij} \int_\Omega a_{ij}(x) D_i u D_j u \, dx : \ (u - \sum_i \xi_i x_i) \in H_0^1(\Omega) \right\}$$

per $A \in \mathcal{E}(\lambda_0, \Lambda_0)$, Ω aperto limitato di $\mathbf{R}^n$, $\xi = (\xi_1, \ldots, \xi_n) \in \mathbf{R}^n$.

Si verifica che $\gamma_A(\Omega, \xi)$ è una forma quadratica in ξ, con autovalori compresi tra λ_0 e Λ_0, coincidente con l'energia media su Ω della soluzione $\overline{u}$ del problema

$$A\overline{u} = 0 \ su \ \Omega, \qquad \overline{u} = \sum_i \xi_i x_i \ su \ \partial\Omega.$$

La conoscenza della $\gamma_A(\Omega, \xi)$ per ogni cubo Ω di $\mathbf{R}^n$ è sufficiente per individuare i coefficienti a_{ij} di A:

LEMMA 3.2 – *Per quasi ogni $x \in \mathbf{R}^n$ si ha*

$$\lim_{\nu \to \infty} \gamma_A(\Omega_\nu, \xi) = \sum_{ij} a_{ij}(x) \xi_i \xi_j, \quad \xi \in \mathbf{R}^n$$

dove $\{\Omega_\nu\}$ è un arbitraria successione di cubi tali che

$$\lim_{\nu \to \infty} \{\text{diam } \Omega_\nu\} = 0 \ e \ x \in \overline{\Omega}_\nu \ per \ ogni \ \nu.$$

È allora naturale pensare di tradurre la G-convergenza in termini di convergenza delle forme γ_A. Del resto il teorema 2.4 assicura che se $\{A_k\} \overset{G}{\longrightarrow} A$ si ha

$$\lim_{k \to \infty} \gamma_{A_k}(\Omega, \xi) = \gamma_A(\Omega, \xi) \quad \textit{per ogni cubo } \Omega \textit{ e per ogni } \xi.$$

Viceversa:

TEOREMA 3.3 – *Consideriamo la rete numerabile di cubi di* $\mathbf{R}^n$

$$\Omega_{\nu,h} = \{x \in \mathbf{R}^n : \ 2^{-\nu} h_i < x_i < 2^{-\nu}(h_i + 1), \ \textit{per ogni } i\}$$

con ν *intero* ≥ 1, $h = (h_1, \ldots, h_n)$ n-*upla di interi relativi.*

Sia $\{A_k\}$ *una successione di operatori in* $\mathcal{E}(\lambda_0, \Lambda_0)$ *tale che esista il limite*

$$\tau(\nu, h; \xi) \equiv \lim_{k \to \infty} \gamma_{A_k}(\Omega_{\nu,h}, \xi), \quad \textit{per ogni } \nu, \ h, \ \xi.$$

Allora esiste un operatore A *in* $\mathcal{E}(\lambda_0, \Lambda_0)$ *per cui* $\{A_k\}$ G-*converge ad* A, *e* $\gamma_A(\Omega_{\nu,h}, \xi) = \tau(\nu, h; \xi)$.

Osserviamo che al fine di ottenere delle valutazioni quantitative sarebbe utile dimostrare qualche precisazione del precedente teorema, come ad esempio la seguente:

CONGETTURA – Sia $\Omega_0 = \{x \in \mathbf{R}^n : \ 0 < x_i < 1, \ \forall i\}$ il cubo unitario, e siano A e B due operatori in $\mathcal{E}(\lambda_0, \Lambda_0)$. Allora per ogni intero $\overline{\nu} \geq 1$ si ha

$$|\gamma_A(\Omega_0, \xi) - \gamma_B(\Omega_0, \xi)| \leq C \sum_{\nu=\overline{\nu}}^{\infty} \sum_{i=1}^{n} \sum_{h_i=0}^{2\nu-1} 4^{-n\nu} |\gamma_A(\Omega_{\nu,h}, \xi) - \gamma_B(\Omega_{\nu,h}, \xi|^{\alpha}$$

con $C = C(\lambda_0, \Lambda_0, \overline{\nu})$, $\alpha = \alpha(\lambda_0, \Lambda_0, \overline{\nu}) > 0$.

Infine notiamo il seguente rafforzamento della proprietà di località della G-convergenza:

TEOREMA 3.4 – *Se* $\{A_k\} \overset{G}{\longrightarrow} A$, $\{B_k\} \overset{G}{\longrightarrow} B$ *e, per ogni* k, *i coefficienti di* A_k *coincidono con quelli di* B_k *su un insieme misurabile* S, *allora i coefficienti di* A *coincidono con quelli di* B *quasi ovunque su* S.

Nel §4 applichiamo i teoremi precedenti allo studio degli operatori periodici, mettendo fra l'altro in luce il contenuto matematico di alcuni risultati di Sanchez Palencia (vedi [6]).

1. Notazioni e richiami

Indicheremo con $x = (x_1, \ldots, x_n)$ il generico punto di $\mathbf{R}^n$ e $|x| = (\sum_i x_i^2)^{1/2}$ il modulo di x.

Per ogni funzione scalare $f \in L^1_{loc}(\Omega)$, con Ω aperto di $\mathbf{R}^n$, indicheremo $D_i f \equiv \partial f / \partial x_i$, le derivate nel senso delle distribuzioni e con $Df \equiv (D_1 f, \ldots, D_n f)$ il gradiente della f. Porremo

$$\|Df\|_{L^p(\Omega)} = \left(\sum_i \int_\Omega |D_i f|^p \, dx \right)^{1/p}, \quad \|Df\|_{L^\infty(\Omega)} = \sum_i \sup_{x \in \Omega} |D_i f(x)|.$$

Considereremo i seguenti spazi lineari (sul corpo reale):

$\mathcal{D}(\Omega) = \{$funzioni C^∞ a supporto compatto in $\Omega\}$,

$H^{1,p}(\Omega) =$spazio di Banach delle $f \in L^p(\Omega)$ con $Df \in L^p(\Omega)$

$H_0^{1,p}(\Omega) =$aderenza di $\mathcal{D}(\Omega)$ in $H^{1,p}(\Omega)$

$H_{loc}^1(\Omega) = \{$funzioni $f \in L_{loc}^1(\Omega)$ tali che $f \in H^{1,2}(\Omega')$ per ogni $\Omega' \subset\subset \Omega\}$

$H^{-1}(\Omega) =$ spazio di Banach duale di $H_0^{1,2}(\Omega)$ (rappresentato come spazio di distribuzioni).

Scriveremo infine (per ragioni di brevità) $H^1(\Omega)$ al posto di $H^{1,2}(\Omega)$; L^p, H_{loc}^1, etc. invece di $L^p(\mathbf{R}^n), H_{loc}^1(\mathbf{R}^n), \ldots$; $\int f \, dx$ al posto di $\int_{\mathbf{R}^n} f \, dx$ e $\sum_i \alpha_i, \sum_{ij} \alpha_{ij}$ al posto di $\sum_{i=1}^n \alpha_i, \sum_{ij=1}^n \alpha_{ij}$.

Richiamiamo ora le principali proprietà della G-convergenza relativa alle equazioni ellittiche:

1.1 ([8] p. 585, tenendo conto del Teor. 7 di [8], p. 594). *Se $\{A_k\} \xrightarrow{G} A$ e u_k, u in $H_0^1(\Omega)$, con Ω aperto limitato di $\mathbf{R}^n$, sono tali che $\{A_k u_k\} \to Au$ in $H^{-1}(\Omega)$, allora si ha $\{u_k\} \to u$ in $L^2(\Omega)$.*

1.2. *([8] p. 587,588). Se $\{A_k\}$ e A stanno in $\mathcal{E}(\lambda_0, \Lambda_0)$, si ha*

$$(a) \Rightarrow (b) \Rightarrow (c), \quad dove :$$

(a) $\{a_{ij,k}\} \to a_{ij}$ in L_{loc}^1, $\forall i, j$
(b) $\{(\lambda I - A_k)^{-1} f\} \to (\lambda I - A)^{-1} f$ in H^1, $\forall f \in L^2$
(c) $\{A_k\} \xrightarrow{G} A$

(e dove $a_{ij,k}$, a_{ij} sono i coefficienti di A_k, A).

1.3. *(compattezza: [8], p. 582). Da ogni successione $\{A_k\}$ in $\mathcal{E}(\lambda_0, \Lambda_0)$ si può estrarre una sottosuccessione $\{A_{k_\nu}\}$ G-convergente verso un operatore di $\mathcal{E}(\lambda_0, \Lambda_0)$.*

1.4. *(località: [8], p. 595). Se $A_k \xrightarrow{G} A$, $B_k \xrightarrow{G} B$ e, per ogni k, i coefficienti di A_k coincidono su un aperto Ω di $\mathbf{R}^n$ con i corrispondenti coefficienti di B_k, allora i coefficienti di A coincidono con quelli di B quasi-ovunque su Ω.*

1.5. *([8], p. 592). Se $A_k \xrightarrow{G} A$ e $\{u_k\} \subset H_{loc}^1(\Omega)$ è tale che $A_k u_k = 0$ su Ω e $\{u_k\} \to u$ in $L_{loc}^1(\Omega)$, allora si ha $Au = 0$ su Ω.*

Infine riportiamo alcune maggiorazioni riguardanti le soluzioni u dell'equazione

$$(3) \qquad -Au + u = f + \sum_i D_i g_i$$

con $A \in \mathcal{E}(\lambda_0, \Lambda_0)$ e $f, g_1, \dots, g_n$ in L^∞.

1.6. (STAMPACCHIA: [9], p. 224). *Sia $u \in H^1$ la soluzione di (3) su $\mathbf{R}^n$, allora si ha:*

$$(4) \qquad \|u\|_{L^\infty} \le c(\lambda_0, \Lambda_0)\Big(\|f\|_{L^\infty} + \sum_i \|g_i\|_{L^\infty}\Big)$$

1.7. (MEYERS: [3], p. 198 e p. 200). *Sia $u \in H^1$ la soluzione di (3) su $\mathbf{R}^n$, allora si ha*

$$(5) \qquad \|Du\|_{L^{2+\eta}(\Omega')} \le c(\lambda_0, \Lambda_0, \Omega')\Big(\|f\|_{L^\infty} + \sum_i \|g_i\|_{L^\infty}\Big)$$

per ogni Ω' limitato, per qualche $\eta = \eta(\lambda_0, \Lambda_0) > 0$. Se poi Ω è un aperto limitato con frontiera lipschitziana e $u \in H_0^1(\Omega)$ è la soluzione di (3) su Ω, si ha

$$(6) \qquad \|Du\|_{L^{2+\eta}(\Omega)} \le c(\lambda_0, \Lambda_0, \Omega)\Big(\|f\|_{L^\infty(\Omega)} + \sum_i \|g_i\|_{L^\infty(\Omega)}\Big).$$

2. - Gli integrali dell'energia

LEMMA 2.1 – *Dati $A \in \mathcal{E}(\lambda_0, \Lambda_0)$ e $w \in H^1$, indichiamo con w_ε ($\varepsilon > 0$) la soluzione in H^1 dell'equazione*

$$(7) \qquad -\varepsilon A w_\varepsilon + w_\varepsilon = w \ \text{su} \ \mathbf{R}^n.$$

Si hanno allora le stime seguenti

$$(8) \qquad \|Dw - Dw_\varepsilon\|_{L^2} \le \sqrt{\Lambda_0/\lambda_0}\,\|Dw\|_{L^2}$$

$$(9) \qquad \|w - w_\varepsilon\|_{L^2} \le \sqrt{\Lambda_0}\,\sqrt{\varepsilon}\,\|Dw\|_{L^2}$$

e, nel caso che sia $Dw \in L^\infty$,

$$(10) \qquad \|w - w_\varepsilon\|_{L^\infty} \le c(\lambda_0, \Lambda_0)\sqrt{\varepsilon}\,\|Dw\|_{L^\infty}.$$

Se infine è $Aw|_\Omega \in L^2(\Omega)$ per qualche aperto Ω di $\mathbf{R}^n$, si ha

$$(11) \qquad \|Dw - Dw_\varepsilon\|_{L^2(\Omega')} \le c(\lambda_0, \Lambda_0, \Omega', \Omega)\sqrt{\varepsilon}\big(\|Dw\|_{L^2} + \|Aw\|_{L^2(\Omega)}\big)$$

per ogni Ω' con $\overline{\Omega}' \subset \Omega$ e $\text{dist}(\Omega', \partial\Omega) > 0$.

Dim. Ponendo $v_\varepsilon = \varepsilon^{-1}(w_\varepsilon - w)$, la (7) si trasforma nella

$$(12) \qquad\qquad -\varepsilon A v_\varepsilon + v_\varepsilon = Aw$$

e da questa, moltiplicando per v_ε e integrando, si ricavano facilmente le stime (8) e (9).

Per ottenere la (10), basta eseguire nell'equazione (12) l'omotetia $x \mapsto \sqrt{\varepsilon} x$ e utilizzare la maggiorazione (4).

Proviamo infine la (11): se Ω' è tale che $\overline{\Omega}' \subset \Omega$ e $\mathrm{dist}(\Omega', \partial\Omega) > 0$ si costruisce una funzione C^∞, τ, tale che $\tau \equiv 1$ in un intorno di $\overline{\Omega}'$ e $\tau \equiv 0$ in un intorno di $\mathbf{R}^n \setminus \Omega$ (nel caso in cui $\Omega = \mathbf{R}^n$, basta prendere $\tau \equiv 1$); si verifica allora che $\tau^2 v$ appartiene a $H_0^1(\Omega)$ per ogni $v \in H^1$.

Moltiplicando la (12) per $\tau^2 v_\varepsilon$ e integrando, si ottiene allora (posto $f = Av|_\Omega$ e detti a_{ij} i coefficienti di A):

$$\varepsilon \sum_{ij} \int_\Omega \tau^2 a_{ij}(x) D_i v_\varepsilon D_j v_\varepsilon \, dx + \int_\Omega \tau^2 v_\varepsilon^2 \, dx =$$

$$= \int_\Omega \tau^2 f v_\varepsilon \, dx - 2\varepsilon \sum_{ij} \int_\Omega a_{ij}(x) D_i v_\varepsilon \cdot \tau v \cdot D_j \tau \, dx.$$

Da ciò segue, utilizzando la (8):

$$\lambda_0 \varepsilon \int_\Omega \tau^2 |Dv_\varepsilon|^2 \, dx + \int_\Omega \tau^2 v_\varepsilon^2 \, dx \le$$

$$\le \left| \int_\Omega \tau^2 v_\varepsilon^2 \, dx \right|^{1/2} \left| \left(\int_\Omega \tau^2 f^2 \, dx \right)^{1/2} + 2\Lambda_0 \sup_{\mathbf{R}^n} |D\tau| \sqrt{\Lambda_0/\lambda_0} \left(\int |Dw|^2 \, dx \right)^{1/2} \right|$$

da cui

$$\left(\int_\Omega \tau^2 v_\varepsilon^2 \, dx \right)^{1/2} \le c(\lambda_0, \Lambda_0, \tau)(\|f\|_{L^\infty(\Omega)} + \|Dw\|_{L^2})$$

e quindi anche

$$\lambda_0 \varepsilon \int_\Omega \tau^2 |Dv_\varepsilon|^2 \, dx \le c^2(\lambda_0, \Lambda_0, \tau)(\|f\|_{L^\infty(\Omega)} + \|Dw\|_{L^2})^2.$$

La (11) deriva subito da quest'ultima disuguaglianza.

LEMMA 2.2 – *Siano A_k $(k = 1, 2, 3, \ldots)$ ed A in $\mathcal{E}(\lambda_0, \Lambda_0)$ e siano u_k, u le soluzioni in H^1 delle equazioni in $\mathbf{R}^n$*

$$(13) \qquad\qquad -A_k u_k + \lambda u_k = f_k$$

$$(14) \qquad\qquad -Au + \lambda u = f$$

con $\lambda > 0$, f_k, f in H^{-1}.

Allora, se $\{A_k\} \xrightarrow{G} A$ e $\{f_k\} \to f$ in H^{-1}, si ha

$$(15) \qquad \lim_{k \to \infty} \sum_{ij} \int_S a_{ij,k}(x) D_i u_k D_j u_k \, dx = \int_S a_{ij}(x) D_i u D_j u \, dx$$

per ogni S misurabile $\subset \mathbf{R}^n$ (dove $a_{ij,k}$ e a_{ij} sono i coefficienti di A_k, A rispettivamente).

Dim. Dal momento che le forme quadratiche su H^1

$$u \longmapsto \sum_{ij} \int_S a_{ij,k}(x) D_i u_k D_j u_k \, dx$$

e le applicazioni lineari $(\lambda I - A_k)^{-1} : H^{-1} \to H^1$ sono equicontinue, è facile vedere che non è restrittivo supporre che sia $f_k = f$, $\forall k$, (invece di $\{f_k\} \to f$ in H^{-1}); anzi, essendo $\mathcal{D}$ denso in H^{-1}, si può supporre che sia

$$f_k \equiv f \in \mathcal{D}, \ \forall k.$$

Per la definizione stessa di G-convergenza si ha allora che

$$(16) \qquad\qquad \{u_k\} \to u \, \in \, L^2$$

da cui in particolare, per ogni $\varepsilon > 0$:

$$(17) \qquad \int_{|x| \geq N(\varepsilon)} u_k^2 \, dx \leq \varepsilon$$

Se poi si moltiplica la (13) per $\tau^2 u_k$, con τ funzione C^∞ tale che $0 \leq \tau \leq 1$, $|D\tau| \leq 1$, $\tau(x) = 0$ per $|x| \leq N(\varepsilon)$ e $\tau(x) = 1$ per $|x| \geq N(\varepsilon) + 1$, si ottiene, tenendo conto della (17):

$$(18) \qquad \int_{|x| \geq N(\varepsilon)+1} |Du_k|^2 \, dx \leq \varepsilon C(\lambda_0, \Lambda_0, \lambda, f).$$

La (18) consente di ottenere la (15), cioè la tesi, nel caso generale non appena la si sia dimostrata per S limitato. Osserviamo a questo punto che provare la (15) per ogni limitato S equivale a provare che, per ogni $\varphi \in \mathcal{D}$, si ha

$$(19) \qquad \lim_{k \to \infty} \sum_{ij} \int a_{ij,k}(x) D_i u_k D_j u_k \cdot \varphi \, dx = \sum_{ij} \int a_{ij}(x) D_i u D_j u \cdot \varphi \, dx.$$

Infatti, costruita una successione $\{\varphi_\nu\}$ di funzioni C^∞ aventi supporti equilimitati, tale che $\{\varphi_\nu\}$ converga in ogni L^p, $p < \infty$, verso la funzione caratteristica di S, si ha, utilizzando la stima (5) di Meyers,

$$\lim_{\nu \to \infty} \sum_{ij} \int a_{ij,k}(x) D_i u_k D_j u_k \cdot \varphi_\nu \, dx = \sum_{ij} \int_S a_{ij,k}(x) D_i u_k D_j u_k \, dx$$

uniformemente rispetto a k.

Proveremo dunque la (19).

Fissato φ in $\mathcal{D}$, siano $\varphi_{k,\varepsilon}$, φ_ε le soluzioni in H^1 delle equazioni su $\mathbf{R}^n$ ($\varepsilon > 0$):

$$(20) \qquad \begin{aligned} -\varepsilon A_k \varphi_{k,\varepsilon} + \varphi_{k,\varepsilon} &= \varphi \\ -\varepsilon A \varphi_\varepsilon + \varphi_\varepsilon &= \varphi. \end{aligned}$$

Utilizzando la (10) del Lemma precedente, si ha

$$(21) \qquad \lim_{\varepsilon \to 0} \sum_{ij} \int a_{ij,k}(x) D_i u_k D_j u_k (\varphi - \varphi_{k,\varepsilon})\, dx = 0,$$

uniformemente rispetto a k.

D'altra parte, applicando la (4) alle equazioni (20) e (13) (in cui si suppone $f_k = f \in \mathcal{D}$), si ottengono le stime

$$(22) \qquad \|\varphi_{k,\varepsilon}\|_{L^\infty} \le c(\lambda_0, \Lambda_0, \lambda)\|\varphi\|_{L^\infty}$$

$$(23) \qquad \|u_k\|_{L^\infty} \le c(\lambda_0, \Lambda_0, \lambda)\|f\|_{L^\infty}.$$

Da queste segue in particolare che le funzioni $u_k \varphi_{k,\varepsilon}$ e u_k^2 stanno in H^1; pertanto ha senso fare la decomposizione seguente:

$$\begin{aligned} \sum_{ij} \int & a_{ij,k}(x) D_i u_k D_j u_k \varphi_{k,\varepsilon}\, dx = \\ &= \sum \int a_{ij,k}(x) D_i u_k D_j (u_k \varphi_{k,\varepsilon})\, dx - \frac{1}{2} \sum_{ij} \int a_{ij,k}(x) D_j \varphi_{k,\varepsilon} D_i u_k^2\, dx \\ &= \int f u_k \varphi_{k,\varepsilon}\, dx - \lambda \int u_k^2 \varphi_{k,\varepsilon}\, dx + \frac{1}{2\varepsilon} \int u_k^2 (\varphi_{k,\varepsilon} - \varphi)\, dx. \end{aligned}$$

Ma, essendo $\{A_k\} \xrightarrow{G} A$, si ha $\{u_k\} \to u$ in L^2 e $\{\varphi_{k,\varepsilon}\} \xrightarrow{k} \varphi_\varepsilon$ in L^2 ($\forall \varepsilon > 0$); pertanto si può verificare, tenendo conto delle stime (22) e (23), che ogni termine del precedente sviluppo di $\sum_{ij} \int a_{ij,k}(x) D_i u_k D_j u_k \varphi_{k,\varepsilon}\, dx$ converge (per $k \to \infty$, $\forall \varepsilon > 0$) verso il termine corrispondente dell'analogo sviluppo di $\sum_{ij} \int a_{ij}(x) D_i u D_j u \varphi_\varepsilon\, dx$.

Si è, in altri termini, provato che

$$(24)$$
$$\lim_{k \to +\infty} \sum_{ij} \int a_{ij,k}(x) D_i u_k D_j u_k \varphi_{k,\varepsilon}\, dx = \sum_{ij} \int a_{ij}(x) D_i u D_j u \cdot \varphi_\varepsilon\, dx, \quad \forall \varepsilon > 0.$$

Combinando la (24) con la (21), si ottiene la (19) e quindi il lemma è dimostrato.

Dai due lemmi precedenti si ricava il seguente teorema:

Teorema 2.3. – *Siano* $\{A_k\}$ *(k = 1, 2, 3, …) ed A in* $\mathcal{E}(\lambda_0, \Lambda_0)$ *con* $\{A_k\} \overset{G}{\longrightarrow} A$ *e siano* $u_k, u \in H^1(\Omega)$, Ω *aperto limitato di* $\mathbf{R}^n$, *tali che*

$$(25) \qquad \{u_k\} \to u \quad in \ L^2(\Omega)$$

$$(26) \qquad \{A_k u_k\} \to Au \quad in \ H^{-1}(\Omega).$$

Si ha allora, detti $a_{ij,k}$ *e* a_{ij} *i coefficienti di* A_k *e* A,

$$(27) \qquad \lim_{k\to\infty} \sum_{ij} \int_S a_{ij,k}(x) D_i u_k D_j u_k \, dx = \sum_{ij} \int_S a_{ij}(x) D_i u D_j u \, dx$$

per ogni S *misurabile tale che* $\overline{S} \subset \Omega$.

Dim. Procedendo come all'inizio della dim. del Lemma 2.2, si vede che non è restrittivo sostituire l'ipotesi (26) con la seguente:

$$A_k u_k = Au = f, \quad \forall k; \ f \in \mathcal{D}(\Omega).$$

Possiamo ora approssimare u_k ed u con le soluzioni in H^1, $u_{k,\varepsilon}$ e u_ε delle equazioni su $\mathbf{R}^n$ $(\varepsilon > 0)$:

$$
\begin{aligned}
-\varepsilon A u_{k,\varepsilon} + u_{k,\varepsilon} &= \tau u \\
-\varepsilon A u_\varepsilon + u_\varepsilon &= \tau u
\end{aligned}
$$

dove τ è una funzione in $\mathcal{D}(\Omega)$ che vale 1 in un intorno Ω' di S.

Per la stima (11) del Lemma 2.1 si ha allora

$$
\begin{aligned}
\|Du_k - Du_{k,\varepsilon}\|_{L^2(S)} &\leq c(\lambda_0, \Lambda_0, S, \Omega')\sqrt{\varepsilon}(\|f\|_{L^2(\Omega')} + \|D(\tau u_k)\|_{L^2}) \\
\|Du - Du_\varepsilon\|_{L^2(S)} &\leq c(\lambda_0, \Lambda_0, S, \Omega')\sqrt{\varepsilon}(\|f\|_{L^2(\Omega')} + \|D(\tau u)\|_{L^2}),
\end{aligned}
$$

da cui, maggiorando $\|D(\tau u)\|_{L^2}$ per mezzo della disuguaglianza di Caccioppoli e utilizzando l'ipotesi (25), segue:

$$(28) \qquad
\begin{cases}
\displaystyle \lim_{\varepsilon\to 0} \|Du_{k,\varepsilon} - Du_k\|_{L^2(S)} = 0, & \text{uniform. rispetto a } k, \\
\displaystyle \lim_{\varepsilon\to 0} \|Du_\varepsilon - Du\|_{L^2(S)} = 0.
\end{cases}
$$

D'altra parte il Lemma 2.2 asserisce che, $\forall \varepsilon > 0$, si ha

$$(29) \qquad \lim_{k\to\infty} \sum_{ij} \int_S a_{ij,k}(x) D_i u_{k,\varepsilon} D_j u_{k,\varepsilon} \, dx = \sum_{ij} \int_S a_{ij}(x) D_i u_\varepsilon D_j u_\varepsilon \, dx.$$

Pertanto combinando la (29) con le (28) si ottiene la tesi.

TEOREMA 2.4 – *Siano* $\{A_k\}$ $(k = 1, 2, 3, \ldots)$ *ed* A *in* $\mathcal{E}(\lambda_0, \Lambda_0)$ *con* $\{A_k\} \xrightarrow{G} A$ *e siano* u_k, u *le soluzioni dei problemi di Dirichlet su* Ω:

$$(30) \qquad A_k u_k = f, \quad u_k - w \in H_0^1(\Omega)$$

$$(31) \qquad A u = f, \quad u - w \in H_0^1(\Omega)$$

con $f \in H^{-1}(\Omega)$, $w \in H^1(\Omega)$ *e* Ω *aperto limitato di* $\mathbf{R}^n$.

Si ha allora (detti $a_{ij,k}$ *e* a_{ij} *i coefficienti di* A_k *e* A):

$$(32) \qquad \{u_k\} \to u \quad in \ L^2(\Omega)$$

$$(33) \qquad \lim_{k \to \infty} \sum_{ij} \int_S a_{ij,k}(x) D_i u_k D_j u_k \, dx = \sum_{ij} \int_S a_{ij}(x) D_i u D_j u \, dx.$$

per ogni S *misurabile tale che* $\overline{S} \subset \Omega$.

Se poi Ω *ha frontiera lipschitziana, la (33) vale per ogni insieme (misurabile)* $S \subseteq \Omega$.

Dim. Per provare la (32), osserviamo che le soluzioni u_k di (30) verificano le maggiorazioni

$$(34) \qquad \|u_k\|_{H^1(\Omega)} \le c(\lambda_0, \Lambda_0, \Omega)(\|f\|_{H^{-1}(\Omega)} + \|w\|_{H^1(\Omega)}),$$

cosicché la successione $\{u_k - w\}$ è relativamente compatta tanto nella topologia debole di $H_0^1(\Omega)$ quanto (teor. di Rellich) nella topologia forte di $L^2(\Omega)$.

Pertanto per poter affermare che $\{u_k\}$ converge in $L^2(\Omega)$ verso u è sufficiente provare che, se $\{u_{k_\nu}\}$ è una sottosuccessione di $\{u_k\}$ tale che $\{u_{k_\nu}\} \to v$ in $L^2(\Omega)$ con $v - w \in H_0^1(\Omega)$, allora si ha necessariamente $v = u$.

Consideriamo a questo proposito le funzioni $\tilde{u}_k$, $\tilde{u} \in H_0^1(\Omega)$ che risolvono su Ω le equazioni $A_k \tilde{u}_k = A \tilde{u} = f$.

Essendo $\{A_k\} \xrightarrow{G} A$ si ha (prop. 1.1) $\{\tilde{u}_k\} \to \tilde{u}$ in $L^2(\Omega)$ e quindi $\{\tilde{u}_{k_\nu} - u_{k_\nu}\} \to \tilde{u} - v$ in $L^2(\Omega)$; essendo poi $A_{k_\nu}(\tilde{u}_{k_\nu} - u_{k_\nu}) = 0$ su Ω, $\forall \nu$, si ha (prop. 1.5) $A(\tilde{u} - v) = 0$ su Ω, cioè $Av = f$ su Ω.

Ma allora si ha $v = u$ per l'unicità della soluzione di (31).

Si è in tal modo provata la (32) e quindi, per il Teorema 2.3, anche la (33) con S relativamente compatto in Ω.

Supponiamo ora che $\partial\Omega$ sia lipschitziana: in tale ipotesi lo spazio $C^1(\overline{\Omega})$ (delle funzioni di classe C^1 su Ω che sono continue fin su $\partial\Omega$ assieme alle loro derivate prime) è denso in $H^1(\Omega)$. Pertanto non è restrittivo, data la (34), supporre che i dati f e w dei nostri problemi siano scelti in modo che sia $f \in \mathcal{D}(\overline{\Omega})$, $w \in C^1(\overline{\Omega})$.

In tal caso, applicando all'equazione $A_k(u_k - w) = f - A_k w$ su Ω, la stima (6) di Meyers, si ottiene

$$\|Du_k\|_{L^{2+\eta}(\Omega)} \le c(\lambda_0, \Lambda_0, \Omega)(\|f\|_{L^\infty(\Omega)} + \|Dw\|_{L^\infty(\Omega)})$$

con $\eta = \eta(\lambda_0, \Lambda_0) > 0$. Di conseguenza $\forall \varepsilon > 0$ esiste $\delta(\varepsilon) > 0$ per cui

$$(35) \qquad \sum_{ij} \int_T |a_{ij,k}(x)D_i u_k D_j u_k - a_{ij}(x)D_i u D_j u|\, dx \leq \varepsilon$$

per ogni $T \subset \Omega$ tale che $\mathrm{mis}(T) \leq \delta(\varepsilon)$.

La (35) permette di estendere la (33) al caso generale partendo dal caso che S sia relativamente compatto in Ω.

3. - Le forme quadratiche $\gamma_A(\Omega, \xi)$

Sia $A \in \mathcal{E}(\lambda_0, \Lambda_0)$, Ω un cubo aperto di $\mathbf{R}^n$ e $\xi = (\xi_1, \dots, \xi_n) \in \mathbf{R}^n$.
Allora è ben noto che il problema di Dirichlet

$$(36) \qquad \begin{cases} Au = 0 & su\ \Omega \\ (u(x) - \sum_i \xi_i x_i) \in H_0^1(\Omega) & \end{cases}$$

possiede un'unica soluzione u, la quale si caratterizza, fra tutte le funzioni che coincidono con $\sum_i \xi_i x_i$ su $\partial\Omega$, come quella avente la minima energia rispetto ad A.

Detti a_{ij} i coefficienti di A, porremo

$$\gamma_A(\Omega, \xi) = (\mathrm{mis}\ \Omega)^{-1} \min \Big\{ \sum_{ij} \int_\Omega a_{ij}(x)D_i u D_j u\, dx : \ \big(u - \sum_i \xi_i x_i\big) \in H_0^1(\Omega) \Big\}.$$

Si può allora verificare facilmente che $\gamma_A(\Omega, \xi)$ è una forma quadratica nella variabile ξ; inoltre si ha:

$$\lambda_0 |\xi|^2 \leq \gamma_A(\Omega, \xi) \leq \Lambda_0 |\xi|^2, \quad \forall \xi \in \mathbf{R}^n.$$

Infatti da un lato, scegliendo $u \equiv \sum_i \xi_i x_i$, si ottiene

$$\gamma_A(\Omega, \xi) \leq (\mathrm{mis}\ \Omega)^{-1} \sum_{ij} \int_\Omega a_{ij}(x)\xi_i \xi_j\, dx \leq \Lambda_0 |\xi|^2.$$

D'altro canto, posto $u = v + \sum_i \xi_i x_i$ con $v \in H_0^1(\Omega)$ (e quindi $\int_\Omega D_i v\, dx = 0$), si ha:

$$\sum_{ij} \int_\Omega a_{ij}(x)D_i u D_j u\, dx \geq \lambda_0 \int_\Omega |Du|^2\, dx$$

$$= \lambda_0 \left[\int_\Omega |Dv|^2\, dx + \sum_i \xi_i \int_\Omega D_i v\, dx + \int_\Omega |\xi|^2\, dx \right] \geq \lambda_0 \cdot \mathrm{mis}\ \Omega \cdot |\xi|^2.$$

Si vede anche immediatamente che se i coefficienti a_{ij} sono costanti si ha

$$\gamma_A(\Omega, \xi) = \sum_{ij} a_{ij}\xi_i\xi_j.$$

Si può inoltre verificare facilmente che $\gamma_A(\Omega, \xi)$ dipende con continuità dalla matrice dei coefficienti di A (per Ω e ξ fissati) quando questa varia in $[L^1(\Omega)]^{n^2}$.

Il Lemma seguente mette in evidenza il carattere hölderiano di tale continuità:

LEMMA 3.1 – *Per A, B in $\mathcal{E}(\lambda_0, \Lambda_0)$, Ω cubo aperto di $\mathbf{R}^n$, $\xi \in \mathbf{R}^n$, si ha:*

$$|\gamma_A(\Omega, \xi) - \gamma_B(\Omega, \xi)|$$

$$(37) \qquad \leq c(\lambda_0, \Lambda_0) \left[(\text{mis } \Omega)^{-1} \sum_{ij} \int_\Omega |a_{ij}(x) - b_{ij}(x)|\, dx \right]^{\eta/(2+\eta)} |\xi|^2$$

dove $\eta = \eta(\lambda_0, \Lambda_0) < 0$ e a_{ij}, b_{ij} sono i coefficienti di A e B.

Dim. Sia $\overline{u}$ la soluzione del problema (36): dalla stima (6) di Meyers si ottiene (riducendosi al caso del cubo unitario tramite una traslazione e un'omotetia)

$$(38) \qquad \int_\Omega |D\overline{u}|^{2+\eta}\, dx \leq c(\lambda_0, \Lambda_0)(\text{mis } \Omega)|\xi|^{2+\eta}$$

con $\eta = \eta(\lambda_0, \Lambda_0) > 0$.

D'altra parte si ha, in base alla definizione di $\gamma_A(\Omega, \xi)$ e tramite la disuguaglianza di Schwarz-Hölder:

$$\gamma_B(\Omega, \xi) - \gamma_A(\Omega, \xi) \leq (\text{mis } \Omega)^{-1} \sum_{ij} \int_\Omega (b_{ij}(x) - a_{ij}(x)) D_i\overline{u}D_j\overline{u}\, dx$$

$$\leq (\text{mis } \Omega)^{-1} \left[\sum_{ij} \int_\Omega |b_{ij} - a_{ij}|^{(2+\eta)/\eta}\, dx \right]^{\eta/(2+\eta)} \cdot \left[\int_\Omega |D\overline{u}|^{2+\eta}\, dx \right]^{2/(2+\eta)}.$$

Da questa disuguaglianza si ricava la tesi (scambiando A con B se $\gamma_B(\Omega, \xi)$ è minore di $\gamma_A(\Omega, \xi)$ e tenendo conto della (38) e del fatto che $|b_{ij}(x) - a_{ij}(x)| \leq 2\Lambda_0$).

LEMMA 3.2 – *Sia $A \in \mathcal{E}(\lambda_0, \Lambda_0)$ e $x_0 \in \mathbf{R}^n$ un punto di Lebesgue per ogni coefficiente A_{ij} di A. Si ha allora, per ogni successione $\{\Omega_\nu\}$ di cubi aperti di $\mathbf{R}^n$ tale che $x_0 \in \overline{\Omega}_\nu$, $\forall\nu$, e $\lim_{\nu\to\infty} \{\text{diam}\,(\Omega_\nu)\} = 0$:*

$$(39) \qquad \lim_{\nu\to\infty} \gamma_A(\Omega_\nu, \xi) = \sum_{ij} a_{ij}(x_0)\xi_i\xi_j, \quad \forall\xi \in \mathbf{R}^n.$$

Dim. Posto $B = \sum_{ij} a_{ij}(x_0) D_i D_j$ si ha (B è a coefficienti costanti):

$$\gamma_B(\Omega, \xi) = \sum_{ij} a_{ij}(x_0)\xi_i\xi_j.$$

Pertanto applicando la (37), detta I_ν la palla di centro x_0 e raggio uguale a diam (Ω_ν), si ha:

$$\left|\gamma_A(\Omega_\nu, \xi) - \sum_{ij} a_{ij}(x_0)\xi_i\xi_j\right|$$

$$\leq c(\lambda_0, \Lambda_0)\left[(\text{mis } \Omega_\nu)^{-1}\sum_{ij}\int_{\Omega_\nu}|a_{ij}(x) - a_{ij}(x_0)|\,dx\right]^{\eta/(2+\eta)}|\xi|^2$$

$$\leq c(\lambda_0, \Lambda_0)\left[(\text{mis } I_\nu)^{-1}\sum_{ij}\int_{I_\nu}|a_{ij}(x) - a_{ij}(x_0)|\,dx\right]^{\eta/(2+\eta)}|\xi|^2,$$

da cui, per $\nu \to \infty$, si ricava la (39).

Possiamo ora dimostrare il seguente teorema che rappresenta, in un certo senso, l'inverso del teorema 2.4:

TEOREMA 3.3 – *Siano A_k ($k = 1, 2, 3, \ldots$) in $\mathcal{E}(\lambda_0, \Lambda_0)$ tali che esiste*

$$(40)\qquad\qquad \lim_{k\to\infty}\gamma_{A_k}(\Omega_{\nu,h}, \xi) \equiv \tau(\nu, h; \xi)$$

per ogni $\xi \in \mathbf{R}^n$, per ogni ν intero ≥ 1, per ogni $h = (h_1, \ldots, h_n)$ n-upla di interi relativi, dove si è posto:

$$(41)\qquad\qquad \Omega_{\nu,h} = \{x \in \mathbf{R}^n\mid 2^{-\nu}h_i < x_i < 2^{-\nu}(h_i + 1),\ \forall i\}.$$

Allora esiste un operatore $A \in \mathcal{E}(\lambda_0, \Lambda_0)$ tale che

$$\{A_k\}\xrightarrow{G}A$$
$$\gamma_A(\Omega_{\nu,h}, \xi) = \tau(\nu, h; \xi), \quad \forall\nu, \forall h, \forall\xi.$$

Dim. Dal momento che $\mathcal{E}(\lambda_0, \Lambda_0)$ è un insieme compatto rispetto alla G-convergenza, è sufficiente provare che, se A, B sono due operatori in $\mathcal{E}(\lambda_0, \Lambda_0)$ tali che

$$\gamma_A(\Omega_{\nu,h}, \xi) = \gamma_B(\Omega_{\nu,h}, \xi)$$

per ogni cubo $\Omega_{\nu,h}$ della rete (41), allora si ha $A = B$.

Infatti, il Teor. 2.4 assicura che, se $\{A_k\}\xrightarrow{G}A$, $\lim_k \gamma_{A_k}(\Omega, \xi) = \gamma_A(\Omega, \xi)$ per ogni cubo Ω ed ogni $\xi \in \mathbf{R}^n$.

Sia x_0 un punto di Lebesgue per tutti i coefficienti a_{ij}, b_{ij} degli operatori A e B e sia, per ogni intero $\nu \geq 1$, Ω_ν un cubo della rete (41) avente lato $2^{-\nu}$ e tale che $x_0 \in \overline{\Omega}_\nu$.

Si ha allora

$$\gamma_A(\Omega_\nu, \xi) = \gamma_B(\Omega_\nu, \xi), \quad \forall \nu, \forall \xi,$$

da cui, passando al limite per $\nu \to \infty$ e applicando il Lemma 3.2, si ottiene

$$\sum_{ij} a_{ij}(x_0)\xi_i\xi_j = \sum_{ij} b_{ij}(x_0)\xi_i\xi_j, \quad \forall \xi$$

ovvero

$$a_{ij}(x_0) = b_{ij}(x_0), \quad \forall i, j.$$

Il fatto che sia $A = B$ è allora conseguenza del fatto che quasi ogni punto di $\mathbf{R}^n$ è di Lebesgue per le funzioni a_{ij} e b_{ij}.

Si noti che i coefficienti dell'operatore A, che compare nel precedente teorema, si possono costruire a partire dai $\tau(\nu, h; \xi)$ della formula (40), utilizzando la (29).

Il seguente Teorema è un'altra conseguenza dei Lemmi 3.1 e 3.2 e costituisce un rafforzamento della prop. 1.4:

TEOREMA 3.4 – *Siano A_k, B_k $(k = 1, 2, 3, \ldots)$ e A, B in $\mathcal{E}(\lambda_0, \Lambda_0)$, con coefficienti $a_{ij,k}$, $b_{ij,k}$, a_{ij} e b_{ij} rispettivamente.*

Supponiamo che sia $\{A_k\} \xrightarrow{G} A$, $\{B_k\} \xrightarrow{G} B$ e

$$(42) \qquad a_{ij,k}(x) = b_{ij,k}(x), \quad \forall x \in S, \forall k, \forall i, j,$$

con S insieme misurabile di $\mathbf{R}^n$.

Allora si ha anche:

$$a_{ij}(x) = b_{ij}(x) \text{ quasi ovunque su } S, \ \forall i, j.$$

Dim. Sia x_0 un punto di Lebesgue per le funzioni a_{ij} e b_{ij} $(i, j = 1, \ldots, n)$ avente densità 1 rispetto all'insieme S; sia $\{\Omega_\nu\}$ una successione di cubi appartenenti alla rete (41) tali che Ω_ν ha lato $2^{-\nu}$ e $x_0 \in \overline{\Omega}_\nu$, $\forall \nu$.

Si ha allora, utilizzando la (37) e l'ipotesi (42):

$$|\gamma_{A_k}(\Omega_\nu, \xi) - \gamma_{B_k}(\Omega_\nu, \xi)|$$

$$\leq \quad c(\lambda_0, \Lambda_0)\left[(\text{mis } \Omega_\nu)^{-1}\sum_{ij}\int_{\Omega_\nu \setminus S}|a_{ij,k}(x) - b_{ij,k}(x)|\,dx\right]^{\eta/(2+\eta)}|\xi|^2$$

$$\leq \quad c'(\lambda_0, \Lambda_0)\left[(\text{mis } \Omega_\nu)^{-1} \cdot \text{mis }(\Omega_\nu \setminus S)\right]^{\eta/(2+\eta)}|\xi|^2 \leq \cdot$$

Facendo tendere $k \to \infty$ si ottiene allora, $\forall \nu$,

$$|\gamma_A(\Omega_\nu, \xi) - \gamma_B(\Omega_\nu, \xi)| \leq \left[(\text{mis } \Omega_\nu)^{-1} \cdot \text{mis }(\Omega_\nu \setminus S)\right]^{\eta/(2+\eta)}|\xi|^2,$$

da cui, per $\nu \to \infty$ (x_0 è punto di densità 1 per S),

$$\sum_{ij}(a_{ij}(x_0) - b_{ij}(x_0))\xi_i\xi_j = 0, \quad \forall \xi \in \mathbf{R}^n.$$

Poiché quasi ogni punto $x_0 \in S$ verifica le condizioni richieste all'inizio della dimostrazione, si ha la tesi.

4. - Soluzioni periodiche

Assegnati n vettori di $\mathbf{R}^n$ linearmente indipendenti $p_{(1)}, \ldots, p_{(n)}$ ed un parallelogramma aperto P di lati $x_0 + p_{(1)}, \ldots, x_0 + p_{(n)}$, si chiameranno P-*periodiche* quelle funzioni $\psi : \mathbf{R}^n \to \mathbf{R}^m$ tali che

$$\psi(x + p_{(i)}) = \psi(x), \qquad \forall x \in \mathbf{R}^n, \forall i.$$

Sia A un operatore in $\mathcal{E}(\lambda_0, \Lambda_0)$ con coefficienti a_{ij} P-periodici e sia ξ in $\mathbf{R}^n$. Si può verificare (vedi [4]) che esiste una ed una sola (a meno di costanti additive) soluzione $u \in H^1_{loc}$ del problema

$$(43) \qquad \begin{cases} Au = 0 & su \ \mathbf{R}^n, \\ Du \ \text{è } P\text{-periodico}, \\ (\text{mis } P)^{-1} \int_P Du \, dx = \xi. \end{cases}$$

Si vede inoltre che una funzione $u \in H^1_{loc}$ ha gradiente P-periodico se e solo se la funzione $u(x) - \sum_i \xi_i x_i$ è a sua volta P-periodica, dove ξ_i è la media della derivata $D_i u$ sul "periodo" P.

Pertanto si può mostrare, passando al corrispondente problema variazionale, che l'energia media, rispetto ad A, sul periodo P di ogni soluzione di (43) è $\mu_A(\xi)$, dove

$$\mu_A(\xi) =$$

$$= (\text{mis } P)^{-1} \min \left\{ \sum_{ij} \int_P a_{ij}(x) D_i u D_j u \, dx : \left(u(x) - \sum_i \xi_i x_i\right) \text{è } P\text{-periodica} \right\}.$$

Si verifica che $\mu_A(\xi)$ è una forma quadratica in ξ ($\xi \in \mathbf{R}^n$) con autovalori compresi tra λ_0 e Λ_0 e che si ha $\mu_A(\xi) = \sum_{ij} a_{ij}\xi_i\xi_j$ se i coefficienti a_{ij} sono costanti.

Si ha inoltre:

TEOREMA 4.1 – *Siano A_k ($k = 1, 2, 3, \ldots$) ed A in $\mathcal{E}(\lambda_0, \Lambda_0)$, degli operatori P-periodici, dove P è un parallelogramma aperto di $\mathbf{R}^n$. Allora da $\{A_k\} \xrightarrow{G} A$ segue*

$$\lim_{k \to \infty} \mu_{A_k}(\xi) = \mu_A(\xi), \quad \forall \xi \in \mathbf{R}^n.$$

Dim. Detta $\overline{u}_k$ (risp. $\overline{u}$) l'unica soluzione del problema (43) relativo ad A_k (risp. ad A) avente media nulla sul periodo P, e indicati con $a_{ij,k}$, a_{ij} i coefficienti di A_k, A, si ha

$$\mu_{A_k}(\xi) \;=\; (\text{mis } P)^{-1} \sum_{ij} \int_P a_{ij,k}(x) D_i\overline{u}_k D_j\overline{u}_k \, dx$$

$$\mu_A(\xi) \;=\; (\text{mis } P)^{-1} \sum_{ij} \int_P a_{ij}(x) D_i\overline{u} D_j\overline{u} \, dx.$$

Pertanto la tesi è conseguenza del Teor. 2.3 (con $S = P$, Ω aperto limitato contenente $\overline{P}$) a patto di verificare che

$$\{\overline{u}_k\} \to \overline{u} \quad \text{in } L^2_{loc},$$

ovvero, dal momento che $\overline{u}_k(x) - \sum_i \xi_i x_i$ è P-periodica, che

$$(45) \qquad\qquad \{\overline{u}_k\} \to \overline{u} \quad \text{in } L^2(P).$$

Per provare la (45) si osservi che, essendo

$$(\text{mis } P)^{-1} \int_P |D\overline{u}_k|^2 \, dx \le \frac{1}{\lambda_0}\mu_{A_k}(\xi) \le \frac{\Lambda_0}{\lambda_0}|\xi|^2$$

e $\int_P \overline{u}_k = 0$, si ha (diseguaglianza di Poincaré) che $\{\overline{u}_k\}$ è una successione limitata in $H^1(P)$.

Pertanto seguendo il solito procedimento di compattezza sarà sufficiente dimostrare che, se $\{\overline{u}_{k_\nu}\}$ è una sottosuccessione di $\{\overline{u}_k\}$ tale che

$$\{\overline{u}_{k_\nu}\} \xrightarrow{\;nu\;} w \quad \text{in } L^2(P) \text{ e in } H^1(P) - \text{debole},$$

allora si ha necessariamente $w = \overline{u}$.

Ora si vede subito che Dw è P-periodica ed ha media uguale a ξ su P, mentre w ha media nulla su P.

D'altra parte, per la prop. 1.5, si ha $Aw = 0$ su $\mathbf{R}^n$. Ma allora dovrà essere $w = \overline{u}$, data l'unicità della soluzione del probl. (43) con media nulla su P.

TEOREMA 4.2 – *Sia A un operatore di $\mathcal{E}(\lambda_0, \Lambda_0)$ con coefficienti P-periodici a_{ij}, dove P è un parallelogramma aperto di $\mathbf{R}^n$, e sia*

$$A_\varepsilon \;=\; \sum_{ij} D_i\left(a_{ij}(x/\varepsilon)D_j\right), \quad \varepsilon > 0$$

$$A_0 \;=\; \sum_{ij} \mu_{ij} D_i D_j$$

dove (μ_{ij}) è la matrice simmetrica definita da

$$\sum_{ij} \mu_{ij}\xi_i\xi_j = \mu_A(\xi), \qquad \forall \xi \in \mathbf{R}^n.$$

Allora si ha, per $\varepsilon \to 0^+$:

$$\{A_\varepsilon\} \xrightarrow{G} A_0.$$

Dim. Poiché $\mathcal{E}(\lambda_0, \Lambda_0)$ è compatto rispetto alla G-convergenza, ci limiteremo a provare che da

$$\{A_{\varepsilon_k}\} \xrightarrow{G} A', \quad con \ \{\varepsilon_k\} \xrightarrow{k} 0,$$

segue necessariamente $A' = A_0$.

Ora A' ha, al pari di A_0, coefficienti costanti; infatti esso è il G-limite di una successione di operatori $A_k \equiv A_{\varepsilon_k}$ periodici con periodi $P_k \equiv \varepsilon P$ tali che $\lim_{k\to\infty} \{\mathrm{diam}\, P_k\} = 0$ *.

A questo punto, provare che i due operatori a coefficienti costanti A' ed A coincidono equivale a provare che $\mu_{A'}(\xi) = \mu_{A_0}(\xi)$, ovvero, per definizione di A_0, che

$$(46) \qquad \mu_{A'}(\xi) = \mu_A(\xi).$$

Ora per il Teorema 4.1 si ha

$$\mu_{A'}(\xi) = \lim_{k\to\infty} \mu_{A_{\varepsilon_k}}(\xi),$$

mentre, eseguendo un'omotetia, è facile verificare che

$$\mu_A(\xi) = \mu_{A_\varepsilon}(\xi), \quad \forall \varepsilon > 0.$$

Si ottiene pertanto la (46) e quindi il teorema è provato.

Osservazioni finali

4.3 – Il teorema 4.2 chiarisce il significato asintotico messo in luce da Sanchez-Palencia in [6], della forma quadratica $\mu_A(\xi) = \sum_{ij} \mu_{ij}\xi_i\xi_j$ (o meglio dell'operatore $A_0 = \sum_{ij} \mu_{ij}D_iD_j$) come "comportamento asintotico macroscopico limite" dell'operatore periodico A.

Si noti che, utilizzando il teorema 3.3 si ottiene, quando A è un operatore periodico:

$$(47) \qquad \lim_{\varepsilon\to 0} \gamma_A\left(\frac{\Omega}{\varepsilon}, \xi\right) = \mu_A(\xi), \quad \forall \ cubo\ \Omega\ di\ \mathbf{R}^n;$$

---

*Per ogni $b \in \mathbf{R}^n$ si può infatti costruire una successione di vettori $\{b_\nu\} \to b$ tale che $\tau_{b_\nu}(A_{k_\nu}) = A_{k_\nu}$ per ogni ν (dove si è indicato con A_b l'operatore ottenuto da A tramite una b-traslazione dei coefficienti). D'altra parte è facile verificare, utilizzando la definizione di G-convergenza, che essendo $\{A_{k_\nu}\} \xrightarrow{G} A'$ e $\{b_\nu\} \to b$, si ha anche $\{\tau_{b_\nu}(A_{k_\nu})\} \xrightarrow{G} \tau_b(A')$. In conlusione si ha $\tau_b(A') = A'$ per ogni $b \in \mathbf{R}^n$, cioè A' è costante.

formula che fornisce il legame fra le forme quadratiche $\gamma_A(\Omega, \xi)$ e la $\mu_A(\xi)$. Si confronti la (47) con la (39).

4.4 – Un caso particolare in cui la $\mu_A(\xi)$ è calcolabile esplicitamente in termini dei coefficienti di A è quello in cui A è un operatore *isotropo*, cioè del tipo

$$(48) \qquad A = \sum_i D_i(a(x)D_i),$$

ed il coefficiente $a(x)$ è della forma seguente:

$$(49) \qquad a(x) = a_1(x_1) \cdot \ldots \cdot a_n(x_n)$$

con $\sqrt[n]{\lambda_0} \leq a_i(t) \leq \sqrt[n]{\Lambda_0}$, $a_i(t + p_i) = a_i(t)$ $(p_i > 0)$.

In tal caso si vede, utilizzando il Teor. 4.2 e il criterio di G-convergenza per operatori del tipo $\{(48),(49)\}$ che si trova in [2], p. 663, che

$$\mu_A(\xi) = \sum_i \mu_i \xi_i^2$$

dove

$$\mu_i = \frac{1}{p_1 \ldots p_n} \int_0^{p_1} a_1 \, dt \ldots \int_0^{p_{i-1}} a_{i-1} \, dt$$
$$\left(\int_0^{p_i} a_i^{-1} \, dt \right)^{-1} \int_0^{p_{i+1}} a_{i+1} \, dt \ldots \int_0^{p_n} a_n \, dt.$$

Nel caso di una sola variabile $(n = 1)$ tutti gli operatori di $\mathcal{E}(\lambda_0, \Lambda_0)$ sono del tipo $\{(48),(49)\}$.

4.5 – Come nel caso delle $\gamma_A(\Omega, \xi)$, si può definire, per ogni A in $\mathcal{E}(\lambda_0, \Lambda_0)$ (anche non periodico) ed ogni cubo Ω di $\mathbf{R}^n$, una forma quadratica $\mu_A(\Omega, \xi)$. Basta porre

$$\mu_A(\Omega, \xi) = \mu_{A_\Omega}(\xi)$$

dove A_Ω indica l'operatore Ω-periodico coincidente con A su Ω.

Per le $\mu_A(\Omega, \xi)$ valgono risultati analoghi (con dimostrazioni analoghe) a quelli provati nel §3 per le $\gamma_A(\Omega, \xi)$: in particolare l'equivalente dei Lemmi 3.1 e 3.2 e del Teorema 3.3.

Si noti infine la relazione seguente

$$(50) \qquad \mu_A(\Omega, \xi) \leq \gamma_A(\Omega, \xi),$$

dovuta al fatto che ogni u tale che $(u(x) - \sum_i \xi_i x_i) \in H_0^1(\Omega)$ è anche tale che $u(x) - \sum_i \xi_i x_i$ è Ω-periodica.

Bibliografia

[1] P. Marcellini, *Su una convergenza di funzioni convesse.* Boll. Un. Mat. Ital. **8** (1973), pp. 137–158.

[2] A. Marino, S. Spagnolo, *Un tipo di approssimazione dell'operatore* $\sum_{ij} D_i(a_{ij}(x)D_j)$ *con operatori* $\sum_{j} D_j(\beta(x)D_j)$, Ann. Scuola Norm. Sup. Pisa, **23** (1969), pp. 657–673.

[3] N.G. Meyers, *An L^p estimate for the gradient of solutions of second order elliptic divergence equations,* Ann. Scuola Norm. Sup. Pisa **17** (1963), pp. 189–206.

[4] E. Sanchez-Palencia, *Solutions périodiques par rapport aux variables d'espace et applications,* C.R. Acad. Sci. Paris, Sér. A, **271** (1970), pp. 1129–1132.

[5] E. Sanchez-Palencia, *Equations aux dérivées partielles dans un type de milieux hétérogènes,* C.R. Acad. Sci. Paris, Sér. A, **272** (1971), pp. 1410–1413.

[6] E. Sanchez-Palencia, *Comportement local et macroscopique d'un type de milieux phisiques hétérogènes,* Internat. J. Engrg. Sci., **13** (1975), no. 11, pp. 923–940.

[7] S. Spagnolo *Sul limite delle soluzioni di problemi di Cauchy relativi all'equazione del calore,* Ann. Scuola Norm. Sup. Pisa, **21** (1967), pp. 657–699.

[8] S. Spagnolo *Sulla convergenza di soluzioni di equazioni paraboliche ed ellittiche,* Ann. Scuola Norm. Sup. Pisa, **22** (1968), pp. 571–597.

[9] G. Stampacchia, *Le problème de Dirichlet pour les équations elliptiques du second ordre à coefficients discontinus,* Ann. Inst. Fourier (Grenoble), **15** (1965), pp. 189–257.

On the convergence of some sequences of area-like integrals[‡][†]

by Ennio De Giorgi (Pisa)

Summary. A new kind of convergence for energy integrals has been investigated in some recent papers [1], [2],. . ., [8].

Instead of energy integrals, area-like integrals are considered here. The main result is the sequential compactness theorem included in §1.

§1.

For every positive integer n, we denote by $\mathbf{R}^n$ the n-dimensional Euclidean space, which will be considered as a vector space endowed with the usual norm $|(x_1,\ldots,x_n)| = \sqrt{x_1^2 + \ldots + x_n^2}$. We will denote by Ap_n the class of all *bounded* open subsets of $\mathbf{R}^n$, by $C^1(\mathbf{R}^n)$ the class of the functions which are continuous in $\mathbf{R}^n$ together with their first order derivatives; for every $A \in Ap_n$, $C_0^1(A)$ will denote the class of the functions in $C^1(\mathbf{R}^n)$ whose support is contained in A. For every $u \in C^1(\mathbf{R}^n)$, we denote by $Du = (D_1u,\ldots,D_nu)$ the gradient of u. Finally, all integrals in this paper are computed with respect to the usual Lebesgue measure in $\mathbf{R}^n$.

For every positive integer n and for every real number $s \geq 1$, we will denote by $\mathcal{F}_{n,s}$ the class of the real valued, measurable functions in $\mathbf{R}^{2n+1}$ which satisfy the following conditions:

a) *For every choice of $x \in \mathbf{R}^n$, $y \in \mathbf{R}$, $z \in \mathbf{R}^n$, we have*

$$(1) \qquad |z| \leq f(x,y,z) \leq s(1 + |y| + |z|).$$

b) *For every $x \in \mathbf{R}^n$, $y',y'' \in \mathbf{R}$, $z',z'' \in \mathbf{R}^n$ it holds*

$$(2) \qquad |f(x,y',z') - f(x,y'',z'')| \leq s(|y' - y''| + |z' - z''|).$$

We have then the following result (whose proof will be given in §5):

THEOREM. – *Let be given a sequence of functions*

$$(1) \qquad f_1, f_2, f_3,\ldots.$$

belonging to the class $\mathcal{F}_{n,s}$.

[‡]Editor's note: translation into English of the paper "Sulla convergenza di alcune successioni d'integrali del tipo dell'area", published in Rend. Mat., **(6) 8** (1975), 277–294.
[†]Dedicated to Mauro Picone, for his 90th birthday.

Then there exist an increasing sequence of positive integers

$$(2) \qquad 1 \leq h_1 < h_2 < h_3 < \ldots < h_k < \ldots$$

and a function $f \in \mathcal{F}_{n,s}$ which enjoy the following properties:

a) for every $A \in Ap_n$, $u \in C^1(\mathbf{R}^n)$ and for every sequence

$$(3) \qquad u_1, u_2, u_3, \ldots$$

satisfying the conditions

$$(4) \qquad u_k \in C^1(\mathbf{R}^n) \ for \ every \ k, \qquad \lim_{k \to \infty} \int_A |u - u_k| = 0,$$

it holds

$$(5) \qquad \liminf_{k \to +\infty} \int_A f_{h_k}(x, u_k(x), Du_k(x)) \, dx \geq \int_A f(x, u(x), Du(x)) \, dx;$$

b) for every $A \in Ap_n$ and every $w \in C^1(\mathbf{R}^n)$, there exists a sequence

$$(6) \qquad w_1, w_2, w_3, \ldots$$

satisfying the following conditions:

$$(7) \qquad w_k - w \in C_0^1(A) \ for \ all \ k, \qquad \lim_{k \to +\infty} \max(|w_k - w|) = 0,$$

$$(8) \qquad \lim_{k \to +\infty} \int_A f_{h_k}(x, w_k(x), Dw_k(x)) \, dx = \int_A f(x, w(x), Dw(x)) \, dx.$$

§2.

Beside the class of functions $\mathcal{F}_{n,s}$, it will be useful to introduce a strictly related class of functionals denoted by $\mathcal{G}_{n,s}$.

To this end, let us consider the class Q_n of the open intervals in $\mathbf{R}^n$ of the form $\{(x_1, \ldots, x_n) : a_h < x_h < b_h, \ h = 1, \ldots, n\}$, where a_h, b_h are rational numbers, and denote by $\mathcal{B}_n$ the class of the open pluri-intervals, obtained as a finite union of intervals belonging to Q_n. For every measurable set L in $\mathbf{R}^n$, we denote by $\mathrm{meas}(L)$ its Lebesgue measure.

For every positive integer n and every real number $s \geq 1$, we denote by $\mathcal{G}_{n,s}$ the class of all functionals g defined on $\mathcal{B}_n \times C^1(\mathbf{R}^n)$ which satisfy the following conditions.

a) For every $B \in \mathcal{B}_n$ and $u \in C^1(\mathbf{R}^n)$ we have

$$(1) \qquad \int_B |Du| \leq g(B, u) \leq s \int_B (1 + |u| + |Du|).$$

b) *For every $B \in \mathcal{B}_n$ and every pair of functions u', u'' belonging to $C^1(\mathbf{R}^n)$, it holds*

$$(2) \qquad |g(B, u') - g(B, u'')| \leq s \int_B |u' - u''| + |Du' - Du''|.$$

c) *If B, B' and B'' are in $\mathcal{B}_n$ and satisfy*

$$(3) \qquad B' \cap B'' = \emptyset, \;\; B \supset B' \cup B'', \;\; \mathrm{meas}(B \setminus (B' \cup B'')) = 0$$

then, for every $u \in C^1(\mathbf{R}^n)$, we have

$$(4) \qquad g(B, u) = g(B', u) + g(B'', u).$$

We immediately see that, if $f \in \mathcal{F}_{n,s}$ and, for $B \in \mathcal{B}_n$ and $u \in C^1(\mathbf{R}^n)$, we define

$$g(B, u) = \int_B f(x, u, Du),$$

then $g \in \mathcal{G}_{n,s}$.

In order to state a converse of this fact, we begin with the following

LEMMA I. – *Let φ', φ'' be continuous, non negative functions defined on $\mathbf{R}^n$; let γ be a set function defined on $\mathcal{B}_n$, satisfying the following conditions for every $B \in \mathcal{B}_n$:*

$$(5) \qquad \int_B \varphi'(x)\, dx \leq \gamma(B) \leq \int_B \varphi''(x)\, dx;$$

moreover, if B, B', B'' are pluri-intervals in $\mathcal{B}_n$ such that

$$(6) \qquad B' \cap B'' = \emptyset; \;\; B \supset B' \cap B'', \;\; \mathrm{meas}((B \setminus (B' \cup B'')) = 0,$$

it holds

$$(7) \qquad \gamma(B) = \gamma(B') + \gamma(B'').$$

Under these conditions, there exists a measurable function φ which satisfies, for every $x \in \mathbf{R}^n$,

$$(8) \qquad \varphi'(x) \leq \varphi(x) \leq \varphi''(x)$$

and such that

$$(9) \qquad \gamma(B) = \int_B \varphi(x)\, dx$$

for every $B \in \mathcal{B}_n$.

Proof. Let Q_n^* be the class of the intervals of the form

$$\{(x_1,\ldots,x_n): \ a_h < x_h \leq b_h, \ h = 1,\ldots,n\}$$

with rational a_h, b_h. For every $T \in Q_n^*$, we put

$$(10) \qquad \gamma^*(T) = \gamma(T \setminus \partial T)$$

where ∂T denotes the boundary of T.

From (7), we infer that the set function γ^* is additive, and as φ' is non-negative, from (5) it follows that

$$(5^*) \qquad 0 \leq \int_T \varphi'(x)\, dx \leq \gamma^*(T) \leq \int_T \varphi''(x)\, dx.$$

From well-known theorems in integration theory, we infer that γ^* is countably additive, and it can be extended to all Lebesgue measurable sets. From standard results on the differentiation of set functions absolutely continuous with respect to Lebsegue measure, it follows immediately that there exists a function φ satisfying (8) and

$$(9^*) \qquad \gamma^*(T) = \int_T \varphi(x)\, dx.$$

Equation (9) follows then from (9^*) and (10).

We can now prove a lemma connecting the classes $\mathcal{F}_{n,s}$ and $\mathcal{G}_{n,s}$.

LEMMA II. – *If $\tau \in \mathcal{G}_{n,s}$, there exists a function $f \in \mathcal{F}_{n,s}$ which satisfies, for every $B \in \mathcal{B}_n$ and every $u \in C^1(\mathbf{R}^n)$, the equation*

$$(11) \qquad \tau(B,u) = \int_B f(x, u(x), Du(x))\, dx.$$

Proof. The set of all real polynomials with rational coefficients is countable: we can thus arrange them in a sequence

$$(12) \qquad P_1, P_2, \ldots.$$

If we put $\gamma_h(B) = \tau(B, p_h)$, from conditions a), c) in the definition of $\mathcal{G}_{n,s}$ and Lemma I, there exists a measurable function φ_h satisfying

$$(13) \qquad \gamma_h(B) = \int_B \varphi_h(x)\, dx, \ \forall B \in \mathcal{B}_n$$

$$(14) \qquad |Dp_h(x)| \leq \varphi_h(x) \leq s(1 + |p_h(x)| + |Dp_h(x)|).$$

Given two integers h, k, by (13) and condition b) in the definition of $\mathcal{G}_{n,s}$, it follows

$$(15) \qquad |\varphi_h(x) - \varphi_k(x)| \leq s(|p_h(x) - p_k(x)| + |Dp_h(x) - Dp_k(x)|)$$

almost everywhere in $\mathbf{R}^n$. There is then a set M with measure 0, such that (14) and (15) hold for every choice of h, k and for every $x \in (\mathbf{R}^n \setminus M)$.

On the other hand, if we fix $x \in (\mathbf{R}^n \setminus M)$, the set described by

$$(p_h(x), D_1 p_h(x), \ldots, D_n p_h(x))$$

as h varies in $\mathbf{N}$, is a dense subset of $\mathbf{R}^{n+1}$. Then, by (14), (15) and the definition of $\mathcal{F}_{n,s}$, there exists a unique function $f \in \mathcal{F}_{n,s}$ which satisfies

$$(16) \qquad \begin{cases} f(x, p_h(x), Dp_h(x)) = \varphi_h(x), & \text{for } x \in \mathbf{R}^n \setminus M, \\ f(x, p_h(x), Dp_h(x)) = |Dp_h(x)| & \text{for } x \in M. \end{cases}$$

Equation (11) follows from (13), (16) in case $u(x)$ is a polynomial with rational coefficients. On the other hand, the sets in $\mathcal{B}_n$ are bounded, and polynomials with rational coefficients approximate uniformly the functions in $C^1(\mathbf{R}^n)$ together with their first derivatives on every bounded set. By recalling the defining properties of the classes $\mathcal{F}_{n,s}$, $\mathcal{G}_{n,s}$, we then conclude that (11) holds for every $u \in C^1(\mathbf{R}^n)$ and every open set $B \in \mathcal{B}_n$.

§3.

We now introduce some functionals approximating those belonging to $\mathcal{G}_{n,s}$. We put, for every $g \in \mathcal{G}_{n,s}$, $u \in C^1(\mathbf{R}^n)$, $B \in \mathcal{B}_n$, and for every real number $\lambda > 0$,

$$1) \qquad \theta'(g, B, u, \lambda) = \inf\left\{ g(B, u + \eta) + \lambda \int_B |\eta|; \ \eta \in C^1(\mathbf{R}^n) \right\},$$

$$2) \qquad \theta''(g, B, u, \lambda) = \inf\left\{ g(B, u + \eta) + \lambda \int_B |\eta|; \ \eta \in C_0^1(B) \right\}.$$

Taking into account the definitions of θ', θ'' and of the class $\mathcal{G}_{n,s}$, we readily verify the following propositions.

PROPOSITION I – *For $\lambda' < \lambda''$ we always have*

$$\theta'(g, B, u, \lambda') \leq \theta'(g, B, u, \lambda'') \leq \lambda'' \theta'(g, B, u, \lambda')/\lambda',$$
$$\theta''(g, B, u, \lambda') \leq \theta''(g, B, u, \lambda'') \leq \lambda'' \theta''(g, B, u, \lambda')/\lambda'.$$

PROPOSITION II – *If for every $u \in C^1(\mathbf{R}^n)$, $B \in \mathcal{B}_n$ it holds $g'(B, u) \leq g''(B, u)$, then*

$$\theta'(g', B, u, \lambda) \leq \theta'(g'', B, u, \lambda),$$
$$\theta''(g', B, u, \lambda) \leq \theta''(g'', B, u, \lambda).$$

PROPOSITION III – *The following inequalities are always true:*

$$0 \leq \theta'(g, B, u, \lambda) \leq \theta''(g, B, u, \lambda) \leq g(B, u) \leq s \int_B (1 + |u| + |Du|).$$

PROPOSITION IV – *For every $g \in \mathcal{G}_{n,s}$ and every pair of functions $u', u'' \in C^1(\mathbf{R}^n)$ we have*

$$|\theta'(g, B, u', \lambda) - \theta'(g, B, u'', \lambda)| \leq s \int_B (|u' - u''| + |Du' - Du''|),$$

$$|\theta''(g, B, u', \lambda) - \theta''(g, B, u'', \lambda)| \leq s \int_B (|u' - u''| + |Du' - Du''|).$$

PROPOSITION V – *If B, B_1, B_2 belong to $\mathcal{B}_n$ and satisfy the following conditions*

$$(3) \qquad B_1 \cap B_2 = \emptyset, \ \ B \supset B_1 \cup B_2, \ \ \mathrm{meas}(B \setminus (B_1 \cup B_2)) = 0,$$

then

$$(4) \qquad \theta'(g, B, u, \lambda) \geq \theta'(g, B_1, u, \lambda) + \theta'(g, B_2, u, \lambda),$$

$$(5) \qquad \theta''(g, B, u, \lambda) \leq \theta''(g, B_1, u, \lambda) + \theta''(g, B_2, u, \lambda).$$

From the semicontinuity of the functional $\int_B |Du|$, we deduce the following proposition:

PROPOSITION VI – *For each $B \in \mathcal{B}_n$ and each $u \in C^1(\mathbf{R}^n)$, let $g(B, u) = \int_B |Du|$. Then we have*

$$\lim_{\lambda \to +\infty} \theta'(g, B, u, \lambda) = \lim_{\lambda \to +\infty} \theta''(g, B, u, \lambda) = \int_B |Du|.$$

The proof of the following lemma is a bit more complex:

LEMMA I – *Let K be a nonempty compact subset of $\mathbf{R}^n$, let $K \subseteq B \in \mathcal{B}_n$, δ denote the distance of K from $\mathbf{R}^n \setminus B$. We then have*

$$\theta''(g, B, u, \lambda) - \theta'(g, B, u, \lambda) \leq$$
$$(6) \qquad \leq \frac{s^2}{\lambda \delta} \int_B (1 + |u| + |Du|) + s \int_{B \setminus K} (1 + |u| + |Du|),$$

for every $g \in \mathcal{G}_{n,s}$, $u \in C^1(\mathbf{R}^n)$, $\lambda > 0$.

Proof. Fix g, u, λ and a number $\epsilon > 0$. By the definition of θ', we may find $\eta \in C^1(\mathbf{R}^n)$ such that

$$
(7) \qquad \theta'(g, B, u, \lambda) + \epsilon > g(B, u + \eta) + \lambda \int_B |\eta|.
$$

Chosen $\nu > 1$, we define

$$
(8) \qquad B_0 = K, \ B_h = \{x \in \mathbf{R}^n : \ \mathrm{dist}(x, K) < \frac{h\delta}{\nu}\}, \qquad \text{for } h = 1, 2, \ldots, \nu.
$$

We clearly have

$$
(9) \qquad K = B_0 \subset B_1 \subset \ldots \subset B_\nu \subset B,
$$

and there are ν functions $\psi_1, \ldots, \psi_\nu$ which satisfy

$$
(10) \qquad
\begin{cases}
\psi_h \in C_0^1(B_h); \ 0 \le \psi_h(x) \le 1 \quad \text{for every } x \in B_h, \\[2mm]
\psi_h(x) = 1 \quad \text{for every } x \in B_{h-1}; \\[2mm]
|D\psi_h(x)| \le (\nu + 1)/\delta \quad \text{for every } x \in B_h.
\end{cases}
$$

From the definition of θ'' and (9), (10) it follows

$$
(11) \quad \theta''(g, B, u, \lambda) \le g(B, u + \eta\psi_h) + \lambda \int_B |\eta - \psi_h| \le g(B, u + \eta\psi_h) + \lambda \int_B |\eta|.
$$

We now estimate the difference

$$
(12) \qquad g(B, u + \eta\psi_h) - g(B, u + \eta).
$$

To this end, consider a function $f \in \mathcal{F}_{n,s}$ associated with g as in Lemma II, §2. Recalling (9), (10) we have

$$
(13) \qquad g(B, u + \eta\psi_h) - g(B, u + \eta) =
$$

$$
= \int_B f(x, u + \eta\psi_h, Du + D(\eta\psi_h)) - \int_B f(x, u + \eta, Du + D\eta)
$$

$$
= \int_{B \setminus B_h} (f(x, u, Du) - f(x, u + \eta, Du + D\eta)) +
$$

$$
+ \int_{B_h \setminus B_{h-1}} (f(x, u + \eta\psi_h, Du + D(\eta\psi_h))
$$

$$
- f(x, u + \eta, Du + D\eta)).
$$

Recalling (1), §1 in the definition of the class $\mathcal{F}_{n,s}$ we find

$$
(14) \qquad \int\limits_{B\setminus B_h} (f(x,u,Du) - f(x,u+\eta, Du+D\eta)) \le
$$

$$
\le \int\limits_{B\setminus B_h} f(x,u,Du) \le \int\limits_{B\setminus K} f(x,u,Du) \le s \int\limits_{B\setminus K} (1+|u|+|Du|).
$$

On the other hand by (2), §1 and (10) we have

$$
\int\limits_{B_h\setminus B_{h-1}} f(x,u+\eta\psi_h, Du+D(\eta\psi_h)) - f(x,u+\eta, Du+D\eta) \le
$$

$$
\le s \int\limits_{B_h\setminus B_{h-1}} \left(|\eta| + |D\eta| + |\eta|\frac{\nu+1}{\delta} \right)
$$

(15)

and comparing (7), (11), (13), (14), (15), (16) we get

$$
\theta''(g,B,u,\lambda) - \theta'(g,B,u,\lambda) - \epsilon <
$$

$$
< s \int\limits_{B\setminus K} (1+|u|+|Du|) + s \int\limits_{B\setminus K} \left(\frac{|\eta|}{\nu} + \frac{|D\eta|}{\nu} + |\eta|\frac{\nu+1}{\nu\delta} \right).
$$

(17)

From (17) and the arbitrariness of ν, if follows

$$
(18) \qquad \theta''(g,B,u,\lambda) - \theta'(g,B,u,\lambda) \le \epsilon + s \int\limits_{B\setminus K} \left(1+|u|+|Du|+\frac{|\eta|}{\delta} \right).
$$

On the other hand from (7) and Prop. III, it follows

$$
(19) \qquad \lambda \int\limits_{B\setminus K} |\eta| \le \lambda \int\limits_{B} |\eta| \le \epsilon + s \int\limits_{B} (1+|u|+|Du|)
$$

and from (18), (19) and the arbitrariness of ϵ, (6) follows.

§4.

From the definitions and results in §3, we get the following two lemmas.

Lemma I – *Let $\{g_1, g_2, g_3, \dots\}$ be a sequence of functionals in $\mathcal{G}_{n,s}$ such that for every set $B \in \mathcal{B}_n$, every positive rational number q and every polynomial p in n variables and with rational coefficients, the following two limits exist:*

$$
(1) \qquad \lim_{k\to\infty} \theta'(g_k, B, p, q), \qquad \lim_{k\to\infty} \theta''(g_k, B, p, q).
$$

Then the following limits exist for every positive λ and for every $u \in C^1(\mathbf{R}^n)$, $B \in \mathcal{B}_n$:

$$(2) \qquad \lim_{k \to \infty} \theta'(g_k, B, u, \lambda), \qquad \lim_{k \to \infty} \theta''(g_k, B, u, \lambda).$$

Moreover, if we define

$$(3') \qquad \lim_{k \to \infty} \theta'(g_k, B, u, \lambda) = \tau'(B, u, \lambda),$$

$$(3'') \qquad \lim_{k \to \infty} \theta''(g_k, B, u, \lambda) = \tau''(B, u, \lambda),$$

then the following two limits exist, are finite and coincide:

$$(4) \qquad \lim_{\lambda \to +\infty} \tau'(B, u, \lambda), \qquad \lim_{\lambda \to +\infty} \tau''(B, u, \lambda).$$

Finally, if we put

$$(5) \qquad \tau(B, u) = \lim_{\lambda \to +\infty} \tau'(B, u, \lambda) = \lim_{\lambda \to +\infty} \tau''(B, u, \lambda),$$

then $\tau \in \mathcal{G}_{n,s}$.

Proof. Polynomials with rational coefficients approximate functions in $C^1(\mathbf{R}^n)$ uniformly, together with their first derivatives, on bounded sets. Recalling Propositions I, IV in §3, we get the existence of the limits (2). By the same Proposition I in §3, we see that $\tau'(B, u, \lambda)$ and $\tau''(B, u, \lambda)$ are non-increasing functions of λ, so that the limits (4) exist. They are finite by Proposition III of §3. From Lemma I of §3 and Proposition III of §3 we get also

$$(6) \qquad 0 \le \lim_{\lambda \to +\infty} \tau''(B, u, \lambda) - \lim_{\lambda \to +\infty} \tau'(B, u, \lambda) \le s \int_{B \setminus K} (1 + |u| + |Du|),$$

for every $u \in C^1(\mathbf{R}^n)$, $B \in \mathcal{B}_n$ and for every non-empty compact set $K \subset B$. From (6) and the arbitrariness of K, it follows that the limits (4) are equal.

From Propositions II, III, VI of §3 and (1) of §1, we get

$$(7) \qquad \int_B |Du| \le \tau(B, u) \le s \int_B (1 + |u| + |Du|).$$

Moreover, given three sets B, B_1, B_2 in $\mathcal{B}_n$ which verify

$$(8) \qquad B_1 \cap B_2 = \emptyset, \ B \supset B_1 \cup B_2, \ \mathrm{meas}(B \setminus (B_1 \cup B_2)) = 0,$$

from Proposition V of §3 we get, for every $u \in C^1(\mathbf{R}^n)$,

$$(9) \qquad \tau(B, u) = \tau(B_1, u) + \tau(B_2, u).$$

Finally from Proposition IV in §3 we infer that

$$(10) \qquad |\tau(B, u') - \tau(B, u'')| \le s \int_B (|u' - u''| + |Du' - Du''|)$$

for every pair of functions $u', u'' \in C^1(\mathbf{R}^n)$ and every set $B \in \mathcal{B}_n$. Comparing (7), (8), (9), (10) with the definition of $\mathcal{G}_{n,s}$, we conclude that $\tau \in \mathcal{G}_{n,s}$.

We can now prove a further lemma, which is very similar to Theorem I of §1.

LEMMA II – *Let be given a sequence of functions*

$$(10') \qquad f_1, f_2, f_3, \ldots$$

belonging to the class $\mathcal{F}_{n,s}$.

It is then possible to find an increasing sequence of positive integers

$$(11) \qquad 1 \le h_1 < h_2 < h_3 < \ldots < h_k < \ldots$$

and a function $f \in \mathcal{F}_{n,s}$ with the following properties:

a) *For every $A \in Ap_n$, $u \in C^1(\mathbf{R}^n)$, and every sequence $\{u_1, u_2, u_3, \ldots\}$ satisfying*

$$(12) \qquad u_k \in C^1(\mathbf{R}^n) \ for \ every \ k, \ \lim_{k \to \infty} \int_A |u - u_k| = 0,$$

we have

$$(13) \qquad \int_A f(x, u, Du) \le \liminf_{k \to \infty} \int_A f_{h_k}(x, u_k, Du_k).$$

b) *For every $A \in Ap_n$ and every $w \in C^1(\mathbf{R}^n)$ there exists a sequence $\{\psi_1, \psi_2, \psi_3, \ldots\}$ satisfying:*

$$(14) \qquad \psi_k - w \in C_0^1(A) \ for \ every \ k, \ \lim_{k \to \infty} \int_A |\psi_k - w| = 0,$$

$$(15) \qquad \int_A f(x, w(x), Dw(x)) \, dx = \lim_{k \to \infty} \int_A f_{h_k}(x, \psi_k(x), D\psi_k(x)) \, dx.$$

Proof. Being $\mathcal{B}_n$ a countable family, like the set of all polynomial in n variables with rational coefficients, we can find a sequence of integers satisfying (11) such that, if we put

$$(16) \qquad g_k(B, u) = \int_B f_{h_k}(x, u(x), Du(x)) \, dx,$$

then the sequence $\{g_1, g_2, \ldots\}$ satisfies the hypotheses of Lemma I, §4. Let us consider the functional τ defined by (3'), (3"), (5) of Lemma I, §4, and a function $f \in \mathcal{F}_{n,s}$ associated with τ as in (11) of Lemma II, §2.

If we fix a number λ, an open set A in $\mathbf{R}^n$, a set B contained in A and belonging to $\mathcal{B}_n$, a sequence $\{u_1, u_2, \ldots\}$ of functions satisfying (12), thanks to the definition of the functional θ' in §3, (16) and property (1) of the class $\mathcal{F}_{n,s}$, we get

$$(17) \qquad \liminf_{k \to \infty} \int_A f_{h_k}(x, u_k(x), Du_k(x))\, dx \geq \liminf_{k \to \infty} \int_B f_{h_k}(x, u_k, Du_k) =$$

$$= \liminf_{k \to \infty} g_k(B, u_k) \geq \lim_{k \to \infty} \theta'(g_k, B, u, \lambda).$$

From (3'), (5) in Lemma I of §4 and (11) in Lemma II of §2, it follows that

$$(18) \qquad \int_B f(x, u, Du) = \tau(B, u) \leq \liminf_{k \to \infty} \int_A f_{h_k}(x, u_k, Du_k).$$

On the other hand, as A is a bounded open set and $u \in C^1(\mathbf{R}^n)$, for every $\epsilon > 0$ we may find B in such a way that

$$(19) \qquad B \subseteq A,\ B \in \mathcal{B}_n, \ \int_{A \setminus B} (1 + |u| + |Du|) < \epsilon.$$

From (19), by property (1) of §1 in the definition of the class $\mathcal{F}_{n,s}$, we infer

$$(20) \qquad \int_{A \setminus B} f(x, u, Du) < s\epsilon,$$

and (13) follows from (18), (20) and the arbitrariness of ϵ.

To prove b), we begin by fixing a positive number $\epsilon > 0$, an open set $A \in Ap_n$, a function $w \in C^1(\mathbf{R}^n)$, a set B_ϵ satisfying

$$(21) \qquad B_\epsilon \in \mathcal{B}_n,\ B_\epsilon \subseteq A, \ \int_{A \setminus B_\epsilon} (1 + |w| + |Dw|) < \epsilon,$$

and a positive number λ such that

$$(22) \qquad \epsilon\lambda > 1 + s \int_{B_\epsilon} (1 + |w| + |Dw|).$$

From the definition of θ'' in §3, there exists a sequence

$$\{\psi_{\epsilon,1}, \psi_{\epsilon,2}, \ldots, \psi_{\epsilon,k}, \ldots\}$$

which satisfies for every k the conditions

$$(23) \qquad (\psi_{\epsilon,k} - w) \in C_0^1(B_\epsilon) \subset C_0^1(A)$$

$$(24) \qquad g_k(B_\epsilon, \psi_{\epsilon,k}) + \lambda \int_{B_\epsilon} |w - \psi_{\epsilon,k}| \leq \frac{1}{k} + \theta''(g_k, B_\epsilon, w, \lambda).$$

From (22), (23), (24), by Prop. III in §3 and (1) in §2 we get

$$(25) \qquad \int_A |\psi_{\epsilon,k} - w| = \int_{B_\epsilon} |\psi_{\epsilon,k} - w| < \epsilon.$$

On the other hand, by Proposition I of §3, θ'' is a non decreasing function of λ and thus, from Lemma I of §4 we get:

$$(26) \qquad \lim_{k \to \infty} \theta''(g_k, B_\epsilon, w, \lambda) \leq \tau(B_\epsilon, w) = \int_{B_\epsilon} f(x, w, Dw).$$

Within $A \setminus B_\epsilon$, by (23), the functions w and $\psi_{\epsilon,k}$ coincide, and thus by (1) of §2 in the definition of $\mathcal{G}_{n,s}$ and by (21) we get

$$(27) \qquad \int_{A \setminus B_\epsilon} f_{h_k}(x, \psi_{\epsilon,k}, D\psi_{\epsilon,k}) < s\epsilon.$$

From (16), (24), (26), (27) and (1) of §2 we see that

$$(28) \quad \limsup_{k \to \infty} \int_A f_{h_k}(x, \psi_{\epsilon,k}, D\psi_{\epsilon,k}) \leq \int_{B_\epsilon} f(x, w, Dw) + s\epsilon \leq \int_A f(x, w, Dw) + s\epsilon.$$

By (23), (25), (28) and taking into account the arbitrariness of ϵ, a diagonal argument gives a sequence $\{\psi_1, \psi_2, \psi_3, \ldots, \psi_k, \ldots\}$ satisfying (14) and

$$(29) \qquad \limsup_{k \to \infty} \int_A f_{h_k}(x, \psi_k, D\psi_k) \leq \int_A f(x, w, Dw).$$

From (29), as we already proved a), (15) follows immediately.

§5.

We can now prove Theorem I of §1.

Proof. Taking into account the statement of Lemma II, §4, it is enough to build, starting from the sequence $\{\psi_1, \psi_2, \ldots\}$ which satisfies (14), (15) in Lemma II of §4, a new sequence $\{w_1, w_2, \ldots\}$ such that conditions (7), (8) in Theorem I of §1 are fulfilled. To this aim, we fix A, w, ψ_k, f, f_{h_k} as in Lemma II, §4 and we put, for every positive index k:

$$(1) \qquad \int_A |\psi_k - w| = \rho_k^2; \quad \rho_k \geq 0.$$

The sequence $\{\rho_1, \rho_2, \ldots\}$ is uniquely determined by (1) and, by (14) in Lemma II, §4, we have

$$(2) \qquad \lim_{k \to \infty} \rho_k = 0.$$

We also put, for every positive index k

$$(3) \qquad A_k = \{x \in A : \ |\psi_k - w| > \rho_k\}.$$

From (1) we deduce that, for every k,

$$(4) \qquad \operatorname{meas} A_k < \rho_k.$$

As the set A is bounded, the functions of the class $C^1(\mathbf{R}^n)$ stay bounded there together with their first order derivatives. By well known properties of the Lebesgue integral, we may find a sequence of real numbers $\{\epsilon_1, \epsilon_2, \epsilon_3, \ldots\}$ satisfying

$$(5) \qquad 0 < \epsilon_k \ \text{for every } k, \ \lim_{k \to \infty} \epsilon_k = 0$$

and such that, if we put

$$(6) \qquad E_k = \{x \in A : \ \rho_k < |\psi_k - w| < \rho_k + \epsilon_k\},$$

then, for every k,

$$(7) \qquad \int_{E_k} |D\psi_k - Dw| < \rho_k.$$

Take now a sequence $\{\beta_1, \beta_2, \ldots\}$ of functions of the real variable t, which satisfy for every positive integer k the following conditions

$$(8) \qquad \begin{cases} \beta_k \in C^1(\mathbf{R}), \ 0 < d\beta_k/dt \leq 1 \quad \text{for every } t \in \mathbf{R}, \\ \beta_k(t) = t \ \text{for } |t| \leq \rho_k, \ d\beta_k/dt = 0 \ \text{for } |t| \geq \rho_k + \epsilon_k \end{cases}$$

and put, for every k,

$$(9) \qquad w_k = w + \beta_k(\psi_k - w).$$

From (8), (9) we get, for every k,

$$(10) \qquad \int_A f_{h_k}(x, w_k, Dw_k) - f_{h_k}(x, \psi_k, D\psi_k) =$$

$$= \int_{A_k} f_{h_k}(x, w_k, Dw_k) - f_{h_k}(x, \psi_k, D\psi_k) =$$

$$\int_{A_k \setminus E_k} f_{h_k}(x, w_k, Dw_k) - f_{h_k}(x, \psi_k, D\psi_k)$$

$$+ \int_{E_k} f_{h_k}(x, w_k, Dw_k) - f_{h_k}(x, \psi_k, D\psi_k).$$

From (1), (6), (7), (8), (9) we get

$$(11) \qquad \int_{E_k} (|w_k - \psi_k| + |Dw_k - D\psi_k|) \le \rho_k^2 + \rho_k$$

and by property (2), §1 of the class $\mathcal{G}_{n,s}$ we deduce

$$(12) \qquad \int_{E_k} f_{h_k}(x, w_k, Dw_k) - f_{h_k}(x, \psi_k, D\psi_k) \le s(\rho_k^2 + \rho_k).$$

From (3), (6), (8), (9), by property (1), §1 of the class $\mathcal{F}_{n,s}$ it follows

$$(13) \qquad \int_{A_k \setminus E_k} f_{h_k}(x, w_k, Dw_k) - f_{h_k}(x, \psi_k, D\psi_k) \le$$

$$\int_{A_k \setminus E_k} f_{h_k}(x, w_k, Dw_k) \le s \int_{A_k \setminus E_k} (1 + \rho_k + \epsilon_k + |w| + |Dw|).$$

On the other hand, as A is bounded and $w \in C^1(\mathbf{R}^n)$, the functions $|w|$, $|Dw|$ are bounded in A and from (2), (4), (5), (10), (12), (13) we get

$$(14) \qquad \limsup_{k \to \infty} \int_A f_{h_k}(x, w_k, Dw_k) \le \lim_{k \to \infty} \int_A f_{h_k}(x, \psi_k, D\psi_k).$$

Condition (7) of §1 follows from (14), §4, and from (2), (5), (8), (9). From (4), §1, and (14), by property a) in Lemma II, §4, (which coincides with a) of Theorem I in §1) we finally deduce (8) of §1.

Bibliography

[1] E. De Giorgi, S. Spagnolo: *Sulla convergenza degli integrali dell'energia per operatori ellittici del secondo ordine*, Bollettino U.M.I. (4) 8 (1973); pp. 391–411.

[2] P. Marcellini: *Su una convergenza di funzioni convesse*, Bollettino U.M.I. (4) 8 (1973); pp. 137–158.

[3] A. Marino, S. Spagnolo: *Un tipo di approssimazione dell'operatore $\sum_{i,j} D_i(a_{ij}(x)D_j)$ con operatori $\sum_j D_j(\beta D_j)$*, Ann. Scuola Norm. Sup. Pisa 23 (1969); pp. 657–673.

[4] E. Sanchez-Palencia: *Solutions périodiques par rapport aux variables d'espace et applications*, C.R. Acad. Sci. Paris Sér A. 271 (1970); pp. 1129–1132.

[5] E. SANCHEZ-PALENCIA: *Equations aux dérivées partielles dans un type de milieux hétérogènes,* C.R. Acad. Sci. Paris Sér A. 272 (1971); pp. 1410–1413.

[6] E. SANCHEZ-PALENCIA: *Comportament local et macroscopique d'un tipe de milieux hétérogènes,* to appear on Internat. J. Energ. Sci.

[7] S. SPAGNOLO: *Sul limite delle soluzioni di problemi di Cauchy relativi all'equazione del calore,* Ann. Scuola Norm. Sup. Pisa, 21 (1967); pp. 657–699.

[8] S. SPAGNOLO: *Sulla convergenza delle soluzioni di equazioni paraboliche e ellittiche,* Ann. Scuola Norm. Sup. Pisa 22 (1968); pp. 571–597.

Sulla convergenza di alcune successioni d'integrali del tipo dell'area[‡][†]

E. De Giorgi (Pisa)

Sunto. A new kind of convergence for energy integrals has been investigated in some recent papers [1],[2],...,[8].

Instead of energy integrals, area-like integrals are considered here. The main result is the sequential compactness theorem included in §1.

§1.

Per ogni intero positivo n, indicheremo con $\mathbf{R}^n$ lo spazio euclideo ad n dimensioni reali, che considereremo sempre come spazio vettoriale, con la solita norma $|(x_1,\ldots,x_n)| = \sqrt{x_1^2 + \ldots + x_n^2}$. Indicheremo con Ap_n la classe degli aperti *limitati* di $\mathbf{R}^n$, con $C^1(\mathbf{R}^n)$ la classe delle funzioni continue in $\mathbf{R}^n$ assieme alle loro derivate prime; per ogni $A \in Ap_n$ indicheremo con $C_0^1(A)$ la classe delle funzioni di $C^1(\mathbf{R}^n)$ il cui supporto è contenuto in A. Per ogni $u \in C^1(\mathbf{R}^n)$ indicheremo con $Du = (D_1 u,\ldots,D_n u)$ il gradiente di u. Infine ricordiamo che tutti gl'integrali che compaiono in questo lavoro sono fatti rispetto all'usuale misura di Lebesgue in $\mathbf{R}^n$.

Per ogni intero positivo n e per ogni numero reale $s \geq 1$, chiameremo $\mathcal{F}_{n,s}$ la classe delle funzioni reali definite in $\mathbf{R}^{2n+1}$, ivi misurabili, e soddisfacenti le condizioni seguenti:

a) Per ogni scelta di $x \in \mathbf{R}^n$, $y \in \mathbf{R}$, $z \in \mathbf{R}^n$, risulta

$$(1) \qquad |z| \leq f(x,y,z) \leq s(1 + |y| + |z|).$$

b) Per ogni $x \in \mathbf{R}^n$, $y', y'' \in \mathbf{R}$, $z', z'' \in \mathbf{R}^n$ si ha

$$(2) \qquad |f(x,y',z') - f(x,y'',z'')| \leq s(|y' - y''| + |z' - z''|).$$

Vale allora il seguente teorema (la cui dimostrazione verrà esposta nel §5):

Teorema. – *Sia data una successione di funzioni*

$$(1) \qquad f_1, f_2, f_3, \ldots.$$

appartenenti alla classe $\mathcal{F}_{n,s}$.

[‡]Editor's note: published in Rend. Mat., **(6) 8** (1975), 277–294.
[†]A Mauro Picone nel suo novantesimo compleanno

Esistono allora una successione crescente di interi positivi

$$(2) \qquad 1 \le h_1 < h_2 < h_3 < \ldots < h_k < \ldots$$

ed una funzione $f \in \mathcal{F}_{n,s}$ che godono delle proprietà seguenti:

a) *Per ogni $A \in Ap_n$, $u \in C^1(\mathbf{R}^n)$ e per ogni successione*

$$(3) \qquad u_1, u_2, u_3, \ldots$$

verificante le condizioni

$$(4) \qquad u_k \in C^1(\mathbf{R}^n) \text{ per ogni } k, \qquad \lim_{k \to \infty} \int_A |u - u_k| = 0,$$

risulta

$$(5) \qquad \liminf_{k \to +\infty} \int_A f_{h_k}(x, u_k(x), Du_k(x))\, dx \ge \int_A f(x, u(x), Du(x))\, dx;$$

b) *per ogni $A \in Ap_n$ e per ogni $w \in C^1(\mathbf{R}^n)$, esiste una successione*

$$(6) \qquad w_1, w_2, w_3, \ldots$$

verificante le condizioni seguenti:

$$(7) \qquad w_k - w \in C_0^1(A) \text{ per ogni } k, \qquad \lim_{k \to +\infty} \max(|w_k - w|) = 0,$$

$$(8) \qquad \lim_{k \to +\infty} \int_A f_{h_k}(x, w_k(x), Dw_k(x))\, dx = \int_A f(x, w(x), Dw(x))\, dx.$$

§2.

Accanto alla classe di funzioni $\mathcal{F}_{n,s}$ conviene introdurre una classe di funzionali $\mathcal{G}_{n,s}$ ad essa strettamente collegata. A tale scopo cominciamo col considerare la classe Q_n degli intervalli aperti di $\mathbf{R}^n$ del tipo $\{(x_1, \ldots, x_n) : a_h < x_h < b_h, h = 1, \ldots, n\}$, ove a_h, b_h sono numeri razionali e indichiamo con $\mathcal{B}_n$ la classe dei plurintervalli aperti che si ottengono come unione di un numero finito di intervalli della classe Q_n. Per ogni insieme L misurabile di $\mathbf{R}^n$, indicheremo con $\text{mis}(L)$ la sua misura di Lebesgue.

Per ogni intero positivo n ed ogni numero reale $s \ge 1$, indicheremo con $\mathcal{G}_{n,s}$ la classe dei funzionali g definiti in $\mathcal{B}_n \times C^1(\mathbf{R}^n)$ e verificanti le condizioni seguenti.

a) *Per ogni $B \in \mathcal{B}_n$ ed ogni $u \in C^1(\mathbf{R}^n)$ risulta*

$$(1) \qquad \int_B |Du| \le g(B, u) \le s \int_B (1 + |u| + |Du|).$$

b) Per ogni $B \in \mathcal{B}_n$ ed ogni coppia u', u'' di funzioni appartenenti a $C^1(\mathbf{R}^n)$ risulta

$$(2) \qquad |g(B, u') - g(B, u'')| \leq s \int_B |u' - u''| + |Du' - Du''|.$$

c) Se B, B' e B'' appartengono a $\mathcal{B}_n$ e verificano le

$$(3) \qquad B' \cap B'' = \emptyset, \ B \supset B' \cup B'', \ \mathrm{mis}(B \setminus (B' \cup B'')) = 0$$

allora, per ogni $u \in C^1(\mathbf{R}^n)$, si ha

$$(4) \qquad g(B, u) = g(B', u) + g(B'', u).$$

È immediato verificare che, se $f \in \mathcal{F}_{n,s}$ e, per ogni $B \in \mathcal{B}_n$,

$$g(B, u) = \int_B f(x, u, Du),$$

allora $g \in \mathcal{G}_{n,s}$. Per invertire questo risultato conviene premettere il seguente

LEMMA I. – *Siano φ', φ'' due funzioni definite in $\mathbf{R}^n$ ed ivi continue e mai negative; sia γ una funzione d'insieme definita in $\mathcal{B}_n$, verificante per ogni $B \in \mathcal{B}_n$ le condizioni*

$$(5) \qquad \int_B \varphi'(x) \, dx \leq \gamma(B) \leq \int_B \varphi''(x) \, dx;$$

dati comunque tre plurintervalli B, B', B'' appartenenti a $\mathcal{B}_n$ e verificanti le

$$(6) \qquad B' \cap B'' = \emptyset; \ B \supset B' \cap B'', \ \mathrm{mis}((B \setminus (B' \cup B'')) = 0,$$

risulti

$$(7) \qquad \gamma(B) = \gamma(B') + \gamma(B'').$$

In queste condizioni esiste una funzione misurabile φ verificante, per ogni $x \in \mathbf{R}^n$, le relazioni

$$(8) \qquad \varphi'(x) \leq \varphi(x) \leq \varphi''(x)$$

e tale che, per ogni $B \in \mathcal{B}_n$, sia

$$(9) \qquad \gamma(B) = \int_B \varphi(x) \, dx.$$

Dim. Detta Q_n^* la classe degli intervalli semiaperti del tipo

$$\{(x_1, \ldots, x_n) : \ a_h < x_h \leq b_h, \ h = 1, \ldots, n\}$$

con a_h, b_h razionali, poniamo, per ogni $T \in Q_n^*$,

$$(10) \qquad \gamma^*(T) = \gamma(T \setminus \partial T)$$

ove ∂T è la frontiera di T.

Dalle (7) segue l'additività della funzione d'intervallo γ^*, ed essendo φ' non negativa, dalla (5) segue

$$(5^*) \qquad 0 \leq \int_T \varphi'(x)\, dx \leq \gamma^*(T) \leq \int_T \varphi''(x)\, dx.$$

Noti teoremi della teoria dell'integrazione ci assicurano allora la numerabile additività di γ^* e la sua prolungabilità a tutti gli insiemi misurabili secondo Lebesgue; dai teoremi sulla derivazione delle funzioni d'insieme assolutamente continue rispetto alla misura di Lebesgue, segue subito l'esistenza di una funzione misurabile φ verificante le (8) e la

$$(9^*) \qquad \gamma^*(T) = \int_T \varphi(x)\, dx.$$

Dalla (9*), (10) segue subito la (9).

Possiamo ora provare il lemma che precisa il collegamento fra le classi $\mathcal{F}_{n,s}$ e $\mathcal{G}_{n,s}$.

LEMMA II. – *Se* $\tau \in \mathcal{G}_{n,s}$, *esiste una funzione* $f \in \mathcal{F}_{n,s}$ *verificante, per ogni* $B \in \mathcal{B}_n$ *ed ogni* $u \in C^1(\mathbf{R}^n)$, *la relazione*

$$(11) \qquad \tau(B, u) = \int_B f(x, u(x), Du(x))\, dx.$$

Dim. I polinomi reali in n variabili a coefficienti razionali sono un'infinità numerabile e quindi possiamo disporli in una successione

$$(12) \qquad P_1, P_2, \ldots.$$

Posto $\gamma_h(B) = \tau(B, p_h)$ per le condizioni a), c) della definizione di $\mathcal{G}_{n,s}$ ed il lemma I esiste una funzione misurabile φ_h verificante le condizioni

$$(13) \qquad \gamma_h(B) = \int_B \varphi_h(x)\, dx, \; \textit{per ogni } B \in \mathcal{B}_n$$

$$(14) \qquad |Dp_h(x)| \leq \varphi_h(x) \leq s(1 + |p_h(x)| + |Dp_h(x)|).$$

Dati due interi h, k, per le (13) e la condizione b) della definizione di $\mathcal{G}_{n,s}$, risulta

$$(15) \qquad |\varphi_h(x) - \varphi_k(x)| \leq s(|p_h(x) - p_k(x)| + |Dp_h(x) - Dp_k(x)|)$$

quasi ovunque in $\mathbf{R}^n$; quindi esisterà un insieme M di misura nulla tale che per $x \in (\mathbf{R}^n \setminus M)$ le (14), (15) risultano verificate per ogni scelta di h, k.

D'altra parte, fissato $x \in (\mathbf{R}^n \setminus M)$, l'insieme descritto dal punto

$$(p_h(x), D_1 p_h(x), \ldots, D_n p_h(x))$$

al variare di h è denso in $\mathbf{R}^{n+1}$; quindi, per le (14), (15) e la definzione di $\mathcal{F}_{n,s}$, esiste una sola funzione $f \in \mathcal{F}_{n,s}$ verificante le condizioni

$$(16) \qquad \begin{cases} f(x, p_h(x), D p_h(x)) = \varphi_h(x), & per \ x \in \mathbf{R}^n \setminus M, \\ f(x, p_h(x), D p_h(x)) = |D p_h(x)| & per \ x \in M. \end{cases}$$

Dalle (13), (16) segue la (11) nel caso in cui $u(x)$ è un polinomio a coefficienti razionali; d'altra parte gli insiemi della classe $\mathcal{B}_n$ sono limitati e i polinomi a coefficienti razionali approssimano, uniformemente negl'insiemi limitati, le funzioni della classe $C^1(\mathbf{R}^n)$ con le loro derivate prime; ricordando le proprietà delle classi $\mathcal{F}_{n,s}$, $\mathcal{G}_{n,s}$ possiamo allora concludere che la (11) vale per ogni $u \in C^1(\mathbf{R}^n)$ ed ogni aperto $B \in \mathcal{B}_n$.

§3.

Introduciamo ora alcuni funzionali approssimanti i funzionali che appartengono a $\mathcal{G}_{n,s}$. Poniamo, per ogni $g \in \mathcal{G}_{n,s}$, $u \in C^1(\mathbf{R}^n)$, $B \in \mathcal{B}_n$, e per ogni numero reale $\lambda > 0$,

$$1) \qquad \theta'(g, B, u, \lambda) = \inf\{g(B, u + \eta) + \lambda \int_B |\eta|; \ \eta \in C^1(\mathbf{R}^n)\},$$

$$2) \qquad \theta''(g, B, u, \lambda) = \inf\{g(B, u + \eta) + \lambda \int_B |\eta|; \ \eta \in C_0^1(B)\}.$$

Tenendo conto delle definizioni di θ', θ'' e della classe $\mathcal{G}_{n,s}$, si verificano subito le seguenti proposizioni.

PROPOSIZIONE I – *Per $\lambda' < \lambda''$ si ha sempre*

$$\theta'(g, B, u, \lambda') \le \theta'(g, B, u, \lambda'') \le \lambda'' \theta'(g, B, u, \lambda')/\lambda',$$
$$\theta''(g, B, u, \lambda') \le \theta''(g, B, u, \lambda'') \le \lambda'' \theta''(g, B, u, \lambda')/\lambda'.$$

PROPOSIZIONE II – *Se per ogni $u \in C^1(\mathbf{R}^n)$ ed ogni $B \in \mathcal{B}_n$ risulta $g'(B, u) \le g''(B, u)$, allora si ha sempre*

$$\theta'(g', B, u, \lambda) \le \theta'(g'', B, u, \lambda),$$
$$\theta''(g', B, u, \lambda) \le \theta''(g'', B, u, \lambda).$$

PROPOSIZIONE III – *Valgono sempre le relazioni*

$$0 \leq \theta'(g, B, u, \lambda) \leq \theta''(g, B, u, \lambda) \leq g(B, u) \leq s \int_B (1 + |u| + |Du|).$$

PROPOSIZIONE IV – *Per ogni $g \in \mathcal{G}_{n,s}$ ed ogni coppia di funzioni $u', u'' \in C^1(\mathbf{R}^n)$ si ha sempre*

$$|\theta'(g, B, u', \lambda) - \theta'(g, B, u'', \lambda)| \leq s \int_B (|u' - u''| + |Du' - Du''|),$$

$$|\theta''(g, B, u', \lambda) - \theta''(g, B, u'', \lambda)| \leq s \int_B (|u' - u''| + |Du' - Du''|).$$

PROPOSIZIONE V – *Se B, B_1, B_2 appartengono a $\mathcal{B}_n$ e verificano le condizioni*

$$(3) \qquad B_1 \cap B_2 = \emptyset, \ B \supset B_1 \cup B_2, \ \mathrm{mis}(B \setminus (B_1 \cup B_2)) = 0,$$

allora si ha sempre

$$(4) \qquad \theta'(g, B, u, \lambda) \geq \theta'(g, B_1, u, \lambda) + \theta'(g, B_2, u, \lambda),$$

$$(5) \qquad \theta''(g, B, u, \lambda) \leq \theta''(g, B_1, u, \lambda) + \theta''(g, B_2, u, \lambda).$$

Dalla semicontinuità del funzionale $\int_B |Du|$ segue poi la proposizione seguente:

PROPOSIZIONE VI – *Per ogni $B \in \mathcal{B}_n$ ed ogni $u \in C^1(\mathbf{R}^n)$ sia $g(B, u) = \int_B |Du|$; allora anche*

$$\lim_{\lambda \to +\infty} \theta'(g, B, u, \lambda) = \lim_{\lambda \to +\infty} \theta''(g, B, u, \lambda) = \int_B |Du|.$$

Più complessa appare la dimostrazione del

LEMMA I – *Sia K un compatto non vuoto di $\mathbf{R}^n$, sia $K \subseteq B \in \mathcal{B}_n$, sia δ la distanza di K da $\mathbf{R}^n \setminus B$. Risulta allora*

$$\theta''(g, B, u, \lambda) - \theta'(g, B, u, \lambda) \leq$$
$$(6) \qquad \leq \frac{s^2}{\lambda \delta} \int_B (1 + |u| + |Du|) + s \int_{B \setminus K} (1 + |u| + |Du|),$$

per ogni $g \in \mathcal{G}_{n,s}$, $u \in C^1(\mathbf{R}^n)$, $\lambda > 0$.

Dim. Fissati g, u, λ, prendiamo un numero $\epsilon > 0$; per la definizione di θ' potremo allora trovare una $\eta \in C^1(\mathbf{R}^n)$ tale che risulti

$$(7) \qquad \theta'(g, B, u, \lambda) + \epsilon > g(B, u + \eta) + \lambda \int_B |\eta|.$$

Scelto poi un intero $\nu > 1$, poniamo per definizione

$$(8) \qquad B_0 = K, \ B_h = \{x \in \mathbf{R}^n : \ \mathrm{dist}(x, K) < \frac{h\delta}{\nu}\}, \ per \ h = 1, 2, \ldots, \nu.$$

Avremo evidentemente

$$(9) \qquad K = B_0 \subset B_1 \subset \ldots \subset B_\nu \subset B,$$

ed esistono ν funzioni $\psi_1, \ldots, \psi_\nu$ verificanti le condizioni

$$(10) \qquad \begin{cases} \psi_h \in C_0^1(B_h); \ 0 \le \psi_h(x) \le 1 \quad \text{per ogni } x \in B_h, \\[2mm] \psi_h(x) = 1 \quad \text{per ogni } x \in B_{h-1}; \\[2mm] |D\psi_h(x)| \le (\nu + 1)/\delta \quad \text{per ogni} x \in B_h. \end{cases}$$

Dalla definizione di θ'' e dalle (9), (10) segue

$$(11) \quad \theta''(g, B, u, \lambda) \le g(B, u + \eta\psi_h) + \lambda \int_B |\eta - \psi_h| \le g(B, u + \eta\psi_h) + \lambda \int_B |\eta|.$$

Vogliamo ora valutare la differenza

$$(12) \qquad g(B, u + \eta\psi_h) - g(B, u + \eta).$$

Consideriamo a tale scopo una funzione $f \in \mathcal{F}_{n,s}$ associata a g dal lemma II del §2; ricordando le (9), (10) avremo

$$(13) \qquad g(B, u + \eta\psi_h) - g(B, u + \eta) =$$
$$= \int_B f(x, u + \eta\psi_h, Du + D(\eta\psi_h)) - \int_B f(x, u + \eta, Du + D\eta) =$$
$$= \int_{B \setminus B_h} (f(x, u, Du) - f(x, u + \eta, Du + D\eta)) +$$
$$+ \int_{B_h \setminus B_{h-1}} (f(x, u + \eta\psi_h, Du + D(\eta\psi_h))$$
$$- f(x, u + \eta, Du + D\eta)).$$

Ricordando la (1) del §1 nella definizione della classe $\mathcal{F}_{n,s}$ si trova

$$(14) \qquad \int_{B\setminus B_h} (f(x,u,Du) - f(x,u+\eta,Du+D\eta)) \leq$$

$$\leq \int_{B\setminus B_h} f(x,u,Du) \leq \int_{B\setminus K} f(x,u,Du) \leq s \int_{B\setminus K} (1+|u|+|Du|);$$

d'altra parte per le (2) del §1 e le (10) si ha

$$\int_{B_h\setminus B_{h-1}} f(x,u+\eta\psi_h, Du+D(\eta\psi_h)) - f(x,u+\eta,Du+D\eta) \leq$$

$$\leq s \int_{B_h\setminus B_{h-1}} \left(|\eta|+|D\eta|+|\eta|\frac{\nu+1}{\delta}\right)$$

$$(15)$$

e confrontando le (7), (11), (13), (14), (15), (16) troviamo

$$\theta''(g,B,u,\lambda) - \theta'(g,B,u,\lambda) - \epsilon <$$

$$< s \int_{B\setminus K} (1+|u|+|Du|) + s \int_{B\setminus K} \left(\frac{|\eta|}{\nu} + \frac{|D\eta|}{\nu} + |\eta|\frac{\nu+1}{\nu\delta}\right).$$

$$(17)$$

Dalla (17), per l'arbitrarietà di ν, segue

$$(18) \qquad \theta''(g,B,u,\lambda) - \theta'(g,B,u,\lambda) \leq \epsilon + s \int_{B\setminus K} \left(1+|u|+|Du|+\frac{|\eta|}{\delta}\right);$$

d'altra parte dalle (7) e dalla Prop. III segue

$$(19) \qquad \lambda \int_{B\setminus K} |\eta| \leq \lambda \int_B |\eta| \leq \epsilon + s \int_B (1+|u|+|Du|)$$

e dalle (18), (19) per l'arbitrarietà di ϵ segue la (6).

§4.

Dalle definizioni e risultati del §3 discendono i due lemmi seguenti

LEMMA I – *Sia $\{g_1, g_2, g_3, \ldots\}$ una successione di funzionali appartenenti a $\mathcal{G}_{n,s}$ e, per ogni insieme $B \in \mathcal{B}_n$, ogni numero razionale positivo q ed ogni polinomio in n variabili a coefficienti razionali p, esistano i due limiti*

$$(1) \qquad \lim_{k\to\infty} \theta'(g_k,B,p,q), \qquad \lim_{k\to\infty} \theta''(g_k,B,p,q).$$

Allora esistono, per ogni numero positivo λ, e per ogni $u \in C^1(\mathbf{R}^n)$, $B \in \mathcal{B}_n$ i limiti

$$(2) \qquad \lim_{k\to\infty} \theta'(g_k, B, u, \lambda), \qquad \lim_{k\to\infty} \theta''(g_k, B, u, \lambda).$$

Inoltre, posto

$$(3') \qquad \lim_{k\to\infty} \theta'(g_k, B, u, \lambda) = \tau'(B, u, \lambda),$$

$$(3'') \qquad \lim_{k\to\infty} \theta''(g_k, B, u, \lambda) = \tau''(B, u, \lambda),$$

esistono e sono finiti i due limiti

$$(4) \qquad \lim_{\lambda\to+\infty} \tau'(B, u, \lambda), \qquad \lim_{\lambda\to+\infty} \tau''(B, u, \lambda),$$

ed i loro valori coincidono; se allora poniamo

$$(5) \qquad \tau(B, u) = \lim_{\lambda\to+\infty} \tau'(B, u, \lambda) = \lim_{\lambda\to+\infty} \tau''(B, u, \lambda),$$

risulta $\tau \in \mathcal{G}_{n,s}$.

Dim. I polinomi a coefficienti razionali approssimano uniformemente le funzioni della classe $C^1(\mathbf{R}^n)$, con le loro derivate prime, sugl'insiemi limitati; ricordando allora le proposizioni I, IV del §3, si trova l'esistenza dei limiti (2). Per la stessa Prop. I del §3, si vede che $\tau'(B, u, \lambda)$ e $\tau''(B, u, \lambda)$ sono funzioni non decrescenti di λ e quindi esistono i limiti (4) che risultano finiti per la Prop. III del §3. Dal Lemma I del §3 e dalla Prop. III del §3 segue poi

$$(6) \qquad 0 \leq \lim_{\lambda\to+\infty} \tau''(B, u, \lambda) - \lim_{\lambda\to+\infty} \tau'(B, u, \lambda) \leq s \int_{B\setminus K} (1 + |u| + |Du|),$$

per ogni $u \in C^1(\mathbf{R}^n)$, $B \in \mathcal{B}_n$ e per ogni compatto non vuoto $K \subset B$; dalle (6), per l'arbitrarietà di K, segue la coincidenza dei limiti (4).

Dalle proposizioni II, III, VI del §3 e dalla (1) del §1 segue

$$(7) \qquad \int_B |Du| \leq \tau(B, u) \leq s \int_B (1 + |u| + |Du|);$$

dati poi tre insiemi B, B_1, B_2 appartenenti a $\mathcal{B}_n$ e verificanti le ipotesi

$$(8) \qquad B_1 \cap B_2 = \emptyset, \ B \supset B_1 \cup B_2, \ \mathrm{mis}(B \setminus (B_1 \cup B_2)) = 0,$$

dalla Prop. V del §3 segue, per ogni $u \in C^1(\mathbf{R}^n)$,

$$(9) \qquad \tau(B, u) = \tau(B_1, u) + \tau(B_2, u).$$

Infine dalla Prop. IV del §3 segue,

$$(10) \qquad |\tau(B, u') - \tau(B, u'')| \leq s \int_B \left(|u' - u''| + |Du' - Du''| \right)$$

per ogni coppia di funzioni $u', u'' \in C^1(\mathbf{R}^n)$ ed ogni insieme $B \in \mathcal{B}_n$. Confrontando le (7), (8), (9), (10) con la definizione di $\mathcal{G}_{n,s}$ si conclude che $\tau \in \mathcal{G}_{n,s}$.

Possiamo ora provare un ulteriore lemma molto vicino al Teor. I del §1.

LEMMA II – *Sia data una successione di funzioni*

$$(10') \qquad\qquad f_1, f_2, f_3, \ldots$$

appartenenti alla classe $\mathcal{F}_{n,s}$.

Esistono allora una successione crescente di interi positivi

$$(11) \qquad\qquad 1 \leq h_1 < h_2 < h_3 < \ldots < h_k < \ldots$$

ed una funzione $f \in \mathcal{F}_{n,s}$ che godono delle proprietà seguenti:

a) Per ogni $A \in Ap_n$, $u \in C^1(\mathbf{R}^n)$, e per ogni successione $\{u_1, u_2, u_3, \ldots\}$ verificante le condizioni

$$(12) \qquad u_k \in C^1(\mathbf{R}^n) \text{ per ogni } k, \quad \lim_{k \to \infty} \int_A |u - u_k| = 0,$$

risulta

$$(13) \qquad \int_A f(x, u, Du) \leq \liminf_{k \to \infty} \int_A f_{h_k}(x, u_k, Du_k).$$

b) Per ogni $A \in Ap_n$ ed ogni $w \in C^1(\mathbf{R}^n)$ esiste una successione $\{\psi_1, \psi_2, \psi_3, \ldots\}$ verificante le condizioni seguenti:

$$(14) \qquad \psi_k - w \in C_0^1(A) \text{ per ogni } k, \quad \lim_{k \to \infty} \int_A |\psi_k - w| = 0,$$

$$(15) \qquad \int_A f(x, w(x), Dw(x))\, dx = \lim_{k \to \infty} \int_A f_{h_k}(x, \psi_k(x), D\psi_k(x))\, dx.$$

Dim. Essendo la famiglia $\mathcal{B}_n$ numerabile, ed essendo pure numerabile l'insieme dei polinomi in n variabili a coefficienti razionali, esisterà una successione d'interi verificante le (11) tale che, posto

$$(16) \qquad g_k(B, u) = \int_B f_{h_k}(x, u(x), Du(x))\, dx,$$

la successione $\{g_1, g_2, \ldots\}$ verifichi le ipotesi del Lemma I del §4. Consideriamo allora il funzionale τ definito dalle (3'),(3"), (5) del Lemma I del §4 ed una funzione $f \in \mathcal{F}_{n,s}$ associata a τ dalla (11) del Lemma II del §2.

Fissati comunque un numero λ, un aperto A di $\mathbf{R}^n$, un insieme B contenuto in A ed appartenente a $\mathcal{B}_n$, una successione di funzioni $\{u_1, u_2, \ldots\}$ verificanti le (12), avremo, per la definizione del funzionale θ' data nel §3, le (16) e la proprietà (1) della classe $\mathcal{F}_{n,s}$,

$$\liminf_{k \to \infty} \int_A f_{h_k}(x, u_k(x), Du_k(x))\, dx \geq \liminf_{k \to \infty} \int_B f_{h_k}(x, u_k, Du_k) =$$

$$= \liminf_{k \to \infty} g_k(B, u_k) \geq \lim_{k \to \infty} \theta'(g_k, B, u, \lambda);$$

(17)

ne segue, per le (3'), (5) del Lemma I del §4 e le (11) del Lemma II del §2,

$$(18) \qquad \int_B f(x, u, Du) = \tau(B, u) \leq \liminf_{k \to \infty} \int_A f_{h_k}(x, u_k, Du_k).$$

D'altra parte essendo A un aperto limitato ed $u \in C^1(\mathbf{R}^n)$, comunque si fissi $\epsilon > 0$ è possibile trovare B in modo che risulti

$$(19) \qquad B \subseteq A,\ B \in \mathcal{B}_n,\ \int_{A \setminus B} (1 + |u| + |Du|) < \epsilon;$$

dalle (19), per la proprietà (1) del §1 della classe $\mathcal{F}_{n,s}$ segue

$$(20) \qquad \int_{A \setminus B} f(x, u, Du) < s\epsilon,$$

e quindi, dalle (18), (20), per l'arbitrarietà di ϵ segue la (13).

Per provare la proprietà b) cominciamo col fissare un numero positivo $\varepsilon > 0$, un aperto $A \in Ap_n$, una funzione $w \in C^1(\mathbf{R}^n)$, un insieme B_ϵ verificante le

$$(21) \qquad B_\epsilon \in \mathcal{B}_n,\ B_\epsilon \subseteq A,\ \int_{A \setminus B_\epsilon} (1 + |w| + |Dw|) < \epsilon,$$

ed un numero positivo λ verificante la

$$(22) \qquad \epsilon \lambda > 1 + s \int_{B_\epsilon} (1 + |w| + |Dw|).$$

Dalla definizione di θ'' data nel §3 segue l'esistenza di una successione

$$\{\psi_{\epsilon,1}, \psi_{\epsilon,2}, \ldots, \psi_{\epsilon,k}, \ldots\}$$

verificante per ogni k le condizioni

$$(23) \qquad (\psi_{\epsilon,k} - w) \in C_0^1(B_\epsilon) \subset C_0^1(A)$$

$$(24) \qquad g_k(B_\epsilon, \psi_{\epsilon,k}) + \lambda \int_{B_\epsilon} |w - \psi_{\epsilon,k}| \leq \frac{1}{k} + \theta''(g_k, B_\epsilon, w, \lambda).$$

Dalle (22), (23), (24), per la Prop. III del §3 e la (1) del §2 segue

$$(25) \qquad \int_A |\psi_{\epsilon,k} - w| = \int_{B_\epsilon} |\psi_{\epsilon,k} - w| < \epsilon;$$

d'altra parte, per la Prop. I del §3, θ'' è non decrescente rispetto a λ e quindi, per il Lemma I del §4 avremo:

$$(26) \qquad \lim_{k \to \infty} \theta''(g_k, B_\epsilon, w, \lambda) \leq \tau(B_\epsilon, w) = \int_{B_\epsilon} f(x, w, Dw).$$

D'altra parte in $A \setminus B_\epsilon$, per la (23), coincidono w e $\psi_{\epsilon,k}$, e quindi, per la proprietà (1) del §2 della classe $\mathcal{G}_{n,s}$ e la (21) avremo

$$(27) \qquad \int_{A \setminus B_\epsilon} f_{h_k}(x, \psi_{\epsilon,k}, D\psi_{\epsilon,k}) < s\epsilon.$$

Dalle (16), (24), (26), (27) e dalla (1) del §2 segue

$$(28) \quad \limsup_{k \to \infty} \int_A f_{h_k}(x, \psi_{\epsilon,k}, D\psi_{\epsilon,k}) \leq \int_{B_\epsilon} f(x, w, Dw) + s\epsilon \leq \int_A f(x, w, Dw) + s\epsilon.$$

Osservando le (23), (25), (28) e tenendo conto dell'arbitrarietà di ϵ, è facile verificare che, con un procedimento diagonale, è possibile costruire una successione $\{\psi_1, \psi_2, \psi_3, \ldots, \psi_k, \ldots\}$ verificante le (14) e la

$$(29) \qquad \limsup_{k \to \infty} \int_A f_{h_k}(x, \psi_k, D\psi_k) \leq \int_A f(x, w, Dw).$$

Dalla (29), avendo già dimostrato la proprietà a), segue poi subito la (15).

§5.

Possiamo ora dimostrare il Teorema I del §1.

Dim. Tenendo presente l'enunciato del Lemma II del §4, basterà costruire, a partire dalla successione $\{\psi_1, \psi_2, \ldots\}$ verificante le (14), (15) del Lemma II del §4, una successione $\{w_1, w_2, \ldots\}$ verificante le condizioni (7), (8) del Teor. I del

§1. A tal fine, fissati A, w, ψ_k, f, f_{h_k} come nell'enunciato del Lemma II del §4, poniamo, per ogni indice intero positivo k:

$$(1) \qquad \int_A |\psi_k - w| = \rho_h^2; \quad \rho_h \geq 0;$$

le (1) individuano univocamente la successione $\{\rho_1, \rho_2, \ldots\}$ e, per la (14) del Lemma II del §4, risulta

$$(2) \qquad \lim_{k \to \infty} \rho_k = 0.$$

Poniamo inoltre, per ogni indice intero positivo k

$$(3) \qquad A_k = \{x \in A : \ |\psi_k - w| > \rho_k\};$$

dalle (1) segue, per ogni k,

$$(4) \qquad \operatorname{mis} A_k < \rho_k.$$

Essendo A limitato sono ivi limitate le funzioni della classe $C^1(\mathbf{R}^n)$ e le loro derivate prime; quindi per note proprietà dell'integrale di Lebesgue possiamo trovare una successione di numeri reali $\{\epsilon_1, \epsilon_2, \epsilon_3, \ldots\}$ verificante le

$$(5) \qquad 0 < \epsilon_k \quad \text{per ogni } k, \quad \lim_{k \to \infty} \epsilon_k = 0$$

tale che, posto

$$(6) \qquad E_k = \{x \in A : \ \rho_k < |\psi_k - w| < \rho_k + \epsilon_k\},$$

risulti, per ogni k,

$$(7) \qquad \int_{E_k} |D\psi_k - Dw| < \rho_k.$$

Prendiamo poi una successione di funzioni $\{\beta_1, \beta_2, \ldots\}$ di una variabile reale t verificanti per ogni intero positivo k le condizioni

$$(8) \qquad \begin{cases} \beta_k \in C^1(\mathbf{R}), \ 0 < d\beta_k/dt \leq 1 \quad \text{per ogni } t \in \mathbf{R}, \\[2mm] \beta_k(t) = t \text{ per } |t| \leq \rho_k, \ d\beta_k/dt = 0 \text{ per } |t| \geq \rho_k + \epsilon_k \end{cases}$$

e poniamo per ogni k,

$$(9) \qquad w_k = w + \beta_k(\psi_k - w).$$

Dalle (8), (9) segue, per ogni k,

$$(10) \qquad \int_A f_{h_k}(x, w_k, Dw_k) - f_{h_k}(x, \psi_k, D\psi_k) =$$

$$= \int_{A_k} f_{h_k}(x, w_k, Dw_k) - f_{h_k}(x, \psi_k, D\psi_k) =$$

$$\int_{A_k \backslash E_k} f_{h_k}(x, w_k, Dw_k) - f_{h_k}(x, \psi_k, D\psi_k)$$

$$+ \int_{E_k} f_{h_k}(x, w_k, Dw_k) - f_{h_k}(x, \psi_k, D\psi_k).$$

Dalle (1), (6), (7), (8), (9) segue

$$(11) \qquad \int_{E_k} (|w_k - \psi_k| + |Dw_k - D\psi_k|) \le \rho_k^2 + \rho_k$$

e quindi, per la proprietà (2) del §1 della classe $\mathcal{G}_{n,s}$ avremo

$$(12) \qquad \int_{E_k} f_{h_k}(x, w_k, Dw_k) - f_{h_k}(x, \psi_k, D\psi_k) \le s(\rho_k^2 + \rho_k).$$

Dalle (3), (6), (8), (9), per la proprietà (1) del §1 della classe $\mathcal{F}_{n,s}$ segue

$$(13) \qquad \int_{A_k \backslash E_k} f_{h_k}(x, w_k, Dw_k) - f_{h_k}(x, \psi_k, D\psi_k) \le$$

$$\int_{A_k \backslash E_k} f_{h_k}(x, w_k, Dw_k) \le s \int_{A_k \backslash E_k} (1 + \rho_k + \epsilon_k + |w| + |Dw|).$$

D'altra parte, essendo A un aperto limitato ed essendo $w \in C^1(\mathbf{R}^n)$, le funzioni $|w|$, $|Dw|$ sono limitate in A e quindi, dalle (2), (4), (5), (10), (12), (13) segue

$$(14) \qquad \limsup_{k \to \infty} \int_A f_{h_k}(x, w_k, Dw_k) \le \lim_{k \to \infty} \int_A f_{h_k}(x, \psi_k, D\psi_k).$$

Le (7) del §1 seguono dalle (14) del §4 e dalle (2), (5), (8), (9); dalle (4) del §1 e dalla (14), per la proprietà a) del Lemma II del §4 (che coincide con la a) del Teorema I del §1) segue poi la (8) del §1.

Bibliografia

[1] E. De Giorgi, S. Spagnolo: *Sulla convergenza degli integrali dell'energia per operatori ellittici del secondo ordine*, Bollettino U.M.I. (4) 8 (1973); pp. 391–411.

[2] P. Marcellini: *Su una convergenza di funzioni convesse*, Bollettino U.M.I. (4) 8 (1973); pp. 137–158.

[3] A. Marino, S. Spagnolo: *Un tipo di approssimazione dell'operatore* $\sum_{i,j} D_i(a_{ij}(x)D_j)$ *con operatori* $\sum_j D_j(\beta D_j)$, Ann. Scuola Norm. Sup. Pisa 23 (1969); pp. 657–673.

[4] E. Sanchez-Palencia: *Solutions périodiques par rapport aux variables d'espace et applications*, C.R. Acad. Sci. Paris Sér A. 271 (1970); pp. 1129–1132.

[5] E. Sanchez-Palencia: *Equations aux dérivées partielles dans un type de milieux hétérogènes*, C.R. Acad. Sci. Paris Sér A. 272 (1971); pp. 1410–1413.

[6] E. Sanchez-Palencia: *Comportament local et macroscopique d'un tipe de milieux hétérogènes*, to appear on Internat. J. Energ. Sci.

[7] S. Spagnolo: *Sul limite delle soluzioni di problemi di Cauchy relativi all'equazione del calore*, Ann. Scuola Norm. Sup. Pisa, 21 (1967); pp. 657–699.

[8] S. Spagnolo: *Sulla convergenza delle soluzioni di equazioni paraboliche e ellittiche*, Ann. Scuola Norm. Sup. Pisa 22 (1968); pp. 571–597.

Γ-convergence and G-convergence[‡]

E. De Giorgi (Pisa)[*]

Summary. In this paper I suggest a general definition of Γ-convergence and of G-convergence which I hope will be useful to extend the previous convergences so that several works, in which many different notations are used, may be framed in this theory.

Introduction

In this note I outline the overall plan of a project aimed at the unification of several researches concerning various types of convergence, suggested by many questions in the calculus of variations, measure theory and partial differential equations.

For the sake of clarity, this paper has the usual structure (with definitions, lemmas, remarks). Nevertheless, it should not be regarded not as a collection of results, but rather as a mere working hypothesis: not only the correctness of all stated results needs to be verified, but we should also explore whether the definition we introduced (which could be changed if necessary) will eventually lead to the hoped settlement of a large field in Analysis.

This note should thus be regarded as an invitation to investigate a field of research, which I feel may have a great interest.

1. - Γ-convergence

We denote by $\overline{\mathbf{R}}$ the real line, extended with the points at infinity $-\infty$ and $+\infty$, and equipped with the usual order relation and topology.

Moreover, given an arbitrary set S, we will indicate by $\mathcal{P}(S)$ the set of its parts, and we will consider topologies over S given by means of a family of open sets: such a family τ will be a family of parts of S with the usual properties: (a) $S \in \tau$, $\emptyset \in \tau$; (b) τ is stable under finite intersection and under union, both finite and of arbitrary infinite cardinality. We will not a priori assume any separation axiom.

Under these conditions, we may begin with defining the simple Γ-limits.

DEFINITION I – *Let $f : S \to \overline{\mathbf{R}}$, $\xi \in S$, and let τ be a topology on S. Then we define*

$$\Gamma(\tau^+) - \lim_{x \to \xi} f(x) = \inf_{A \in \mathcal{A}_\tau(\xi)} \sup_{x \in A} f(x),$$

[‡]Editor's note: translation into English of the paper "Γ-convergenza e G-convergenza", published in Boll. Un. Mat. Ital., **A (5) 14** (1977), 213–220.

[*]Scuola Normale Superiore, Pisa.

where $\mathcal{A}_\tau(\xi)$ denotes the family of open neighborhoods of ξ, that is

$$\mathcal{A}_\tau(\xi) = \{A: \ A \in \tau, \ \xi \in A\}.$$

In a similar way, we define

$$\Gamma(\tau^-) - \lim_{x \to \xi} f(x) = \sup_{A \in \mathcal{A}_\tau(\xi)} \inf_{x \in A} f(x).$$

REMARK I – One immediately checks that

$$\Gamma(\tau^+) - \lim_{x \to \xi}(-f(x)) = -\Gamma(\tau^-) - \lim_{x \to \xi} f(x).$$

REMARK II – We always have

$$\Gamma(\tau^-) - \lim_{x \to \xi} f(x) \le \Gamma(\tau^+) - \lim_{x \to \xi} f(x).$$

Shorthand notation I – When we have

$$\Gamma(\tau^+) - \lim_{x \to \xi} f(x) = \Gamma(\tau^-) - \lim_{x \to \xi} f(x) = \lambda,$$

we will write for short

$$\Gamma(\tau) - \lim_{x \to \xi} f(x) = \lambda.$$

REMARK III – The limits we have just introduced are nothing else than the usual maximum limit, minimum limit and limit. Nevertheless, this redefinition will be useful to understand the following extensions.

Indeed, we can define multiple Γ-limits in the following way.

DEFINITION I-BIS – *Let n sets $S_1, \ldots, S_n$ and n topologies $\tau_1, \ldots, \tau_n$ over those sets be given. Let moreover $f : S_1 \times \cdots \times S_n \to \overline{\mathbf{R}}$. Then we put*

$$\Gamma(\tau_1^\pm, \ldots, \tau_{n-1}^\pm, \tau_n^+) - \lim_{\substack{x_1 \to \xi_1 \\ \cdots \\ x_n \to \xi_n}} f(x_1, \ldots, x_n)$$

$$= \inf_{A \in \mathcal{A}_{\tau_n}(\xi_n)} \Gamma(\tau_1^\pm, \ldots, \tau_{n-1}^\pm) - \lim_{\substack{x_1 \to \xi_1 \\ \cdots \\ x_{n-1} \to \xi_{n-1}}} \sup_{x_n \in A} f(x_1, \ldots, x_n)$$

and similarly

$$\Gamma(\tau_1^\pm, \ldots, \tau_{n-1}^\pm, \tau_n^-) - \lim_{\substack{x_1 \to \xi_1 \\ \cdots \\ x_{n-1} \to \xi_{n-1}}} f(x_1, \ldots, x_n) =$$

$$\sup_{A \in \mathcal{A}_{\tau_n}(\xi_n)} \Gamma(\tau_1^\pm, \ldots, \tau_{n-1}^\pm) - \lim_{\substack{x_1 \to \xi_1 \\ \cdots \\ x_{n-1} \to \xi_{n-1}}} \inf_{x_n \in A} f(x_1, \ldots, x_n).$$

EXAMPLES. – For $n = 2$

$$\Gamma(\tau_1^+, \tau_2^+) - \lim_{\substack{x_1 \to \xi_1 \\ x_2 \to \xi_2}} f(x_1, x_2) = \inf_{A_2 \in \mathcal{A}_{\tau_2}(\xi_2)} \inf_{A_1 \in \mathcal{A}_{\tau_1}(\xi_1)} \sup_{x_1 \in A_1} \sup_{x_2 \in A_2} f(x_1, x_2),$$

$$\Gamma(\tau_1^+, \tau_2^-) - \lim_{\substack{x_1 \to \xi_1 \\ x_2 \to \xi_2}} f(x_1, x_2) = \sup_{A_2 \in \mathcal{A}_{\tau_2}(\xi_2)} \inf_{A_1 \in \mathcal{A}_{\tau_1}(\xi_1)} \sup_{x_1 \in A_1} \inf_{x_2 \in A_2} f(x_1, x_2).$$

For $n = 3$

$$\Gamma(\tau_1^+, \tau_2^-, \tau_3^+) - \lim_{\substack{x_1 \to \xi_1 \\ x_2 \to \xi_2 \\ x_3 \to \xi_3}} f(x_1, x_2, x_3) =$$

$$\inf_{A_3 \in \mathcal{A}_{\tau_3}(\xi_3)} \sup_{A_2 \in \mathcal{A}_{\tau_2}(\xi_2)} \inf_{A_1 \in \mathcal{A}_{\tau_1}(\xi_1)} \sup_{x_1 \in A_1} \inf_{x_2 \in A_2} \sup_{x_3 \in A_3} f(x_1, x_2, x_3).$$

REMARK IV. – Remark I can be easily extended to multiple Γ-limits: it suffices to change the signs of all topologies. Remark II can also be extended: if a $+$ sign is changed with a $-$ sign, and the signs attached to all other topologies are left unchanged, the Γ-limits cannot grow. Finally, Remark III can be superseded with the following.

REMARK III-BIS. – The limits $\Gamma(\tau_1^+, \ldots, \tau_n^+) - \lim$ and $\Gamma(\tau_1^-, \ldots, \tau_n^-) - \lim$ agree with the usual maximum and minimum limit, relative to the usual product topology. *On the other hand, Γ-limits with different signs seem to be a completely new concept in topology.*

Shorthand notation I-bis – Even for multiple Γ-limits, when the limit does not depend on the sign associated with one of the topologies, the sign can be omitted. For instance, if

$$\Gamma(\tau_1^+, \tau_2^-) - \lim_{\substack{x_1 \to \xi_1 \\ x_2 \to \xi_2}} f(x_1, x_2) = \Gamma(\tau_1^+, \tau_2^+) - \lim_{\substack{x_1 \to \xi_1 \\ x_2 \to \xi_2}} f(x_1, x_2) = \lambda$$

we will write

$$\Gamma(\tau_1^+, \tau_2) - \lim_{\substack{x_1 \to \xi_1 \\ x_2 \to \xi_2}} f(x_1, x_2) = \lambda.$$

Shorthand notation II – Whenever for every $(\xi_1, \ldots, \xi_n) \in S_1 \times \cdots \times S_n$ we have

$$\Gamma(\tau_1^\pm, \ldots, \tau_{n-1}^\pm, \tau_n^\pm) - \lim_{\substack{x_1 \to \xi_1 \\ \cdots \\ x_n \to \xi_n}} f(x_1, \ldots, x_n) = g(\xi_1, \ldots, \xi_n)$$

we will briefly write

$$\Gamma(\tau_1^\pm, \ldots, \tau_n^\pm) f = g.$$

2. - G-convergence.

From the definition of Γ-limit, we easily obtain that of G-limit.

Definition I – *Given a set $E \subseteq S_1 \times \ldots \times S_n$, we denote by χ_E its characteristic function, i.e.*

$$\chi_E(x_1, \ldots, x_n) = \begin{cases} 1 & if \ (x_1, \ldots, x_n) \in E \\ 0 & if \ (x_1, \ldots, x_n) \in (S_1 \times \cdots \times S_n) \setminus E \end{cases}$$

Moreover, let $L \subseteq S_1 \times \cdots \times S_n$. We will write

$$G(\tau_1^{\pm}, \ldots, \tau_n^{\pm})E = L$$

when

$$\Gamma(\tau_1^{\pm}, \ldots, \tau_n^{\pm})\chi_E = \chi_L.$$

Definition II. – *Let $n + m$ sets $Y_1, \ldots, Y_m$, $S_1, \ldots, S_n$ be given, and let $\theta_1, \ldots, \theta_m$ be topologies on $Y_1, \ldots, Y_m$ and let $\tau_1, \ldots, \tau_n$ be topologies on $S_1, \ldots, S_n$. Let $F : Y_1 \times \cdots \times Y_m \to \mathcal{P}(S_1 \times \cdots \times S_n)$. Given $(\eta_1, \ldots, \eta_m) \in Y_1 \times \cdots \times Y_m$, we will write*

$$G(\theta_1^{\pm}, \ldots, \theta_m^{\pm}; \tau_1^{\pm}, \ldots, \tau_n^{\pm}) - \lim_{\substack{y_1 \to \eta_1 \\ \cdots \\ y_m \to \eta_m}} F(y_1, \ldots, y_m) = L$$

if, for every $(\xi_1, \ldots, \xi_n) \in S_1 \times \cdots \times S_n$, it holds

$$\Gamma(\theta_1^{\pm}, \ldots, \theta_m^{\pm}; \tau_1^{\pm}, \ldots, \tau_n^{\pm}) - \lim_{\substack{y_1 \to \eta_1 \\ \cdots \\ y_m \to \eta_m \\ x_1 \to \xi_1 \\ \cdots \\ x_n \to \xi_n}} \chi_{F(y_1, \ldots, y_m)}(x_1, \ldots, x_n) = \chi_L(\xi_1, \ldots, \xi_n).$$

Remark I. – Notice that $G(\tau^+)E$ is the usual closure of E with respect to the topology τ, while $G(\tau^-)$ is the interior of E. Similarly, $G(\tau_1^+, \ldots, \tau_n^+)$ and $G(\tau_1^-, \ldots, \tau_n^-)$ are the closure and the interior of E with respect to the product topology.

Remark II. – The G-limits $G(\theta^+; \tau^+) - \lim$ and $G(\theta^-; \tau^+) - \lim$ correspond, respectively, to the maximum and minimum limit in the sense of Kuratowski.

Remark III. – If $f : Y \to S$ and for $\eta \in Y$ we have

$$G(\theta; \tau^+) - \lim_{y \to \eta} \{f(y)\} = \{\lambda\}$$

(where $\{\lambda\}$ and $\{f(y)\}$ denote the sets whose unique element is λ and $f(y)$ respectively) then it holds

$$\lim_{y \to \eta} f(y) = \lambda,$$

where the limit is understood in the usual sense of limit for a map from the topological space (Y, θ) to (S, τ).

3. - Γ-limits of functions defined on subsets

DEFINITION I – *Let S be a set, τ a topology on S, $E \subset S$, $f : E \to \overline{\mathbf{R}}$. Then we will define for every $\xi \in G(\tau^+)E$*

$$\Gamma(\tau^+) - \lim_{x \to \xi} f(x) = \inf_{A \in \mathcal{A}_\tau(\xi)} \sup_{x \in E \cap A} f(x),$$

and similarly

$$\Gamma(\tau^-) - \lim_{x \to \xi} f(x) = \sup_{A \in \mathcal{A}_\tau(\xi)} \inf_{x \in E \cap A} f(x).$$

This definition can be extended inductively, thanks to the following lemma.

LEMMA. – *Let n sets $S_1, \ldots, S_n$ be given together with topologies $\tau_1, \ldots, \tau_n$, and let $E \subseteq S_1 \times \cdots \times S_n$. For every $(x_1, \ldots, x_{n-1}) \in S_1 \times \cdots \times S_{n-1}$ we put*

$$E'_{(x_1,\ldots,x_{n-1})} = \{t \in S_n : (x_1, \ldots, x_{n-1}, t) \in E\}.$$

Moreover, for every $A \in \tau_n$ we define

$$E''_A = \{(x_1, \ldots, x_{n-1}) \in S_1 \times \cdots \times S_{n-1} : E'_{(x_1,\ldots,x_{n-1})} \cap A \neq \emptyset\}.$$

Further, let $(\xi_1, \ldots, \xi_n) \in G(\tau_1^+, \ldots, \tau_n^+)E$ and let $\xi_n \in A \in \tau_n$. Then we have $(\xi_1, \ldots, \xi_{n-1}) \subset G(\tau_1^+, \ldots, \tau_{n-1}^+)E''_A$.

Once proved, the above lemma ensures that the following definition makes sense.

DEFINITION II – *Let $S_1, \ldots, S_n$ n be sets equipped with the topologies $\tau_1, \ldots, \tau_n$. Let $E \subset S_1 \times \cdots \times S_n$ and $\xi \in G(\tau_1^+, \ldots, \tau_n^+)E$. Let $f : E \to \overline{\mathbf{R}}$. Then we define*

$$\Gamma(\tau_1^\pm, \ldots, \tau_{n-1}^\pm, \tau_n^+) - \lim_{\substack{x_1 \to \xi_1 \\ \cdots \\ x_n \to \xi_n}} f(x_1, \ldots, x_n) =$$

$$\inf_{A \in \mathcal{A}_{\tau_n}(\xi_n)} \Gamma(\tau_1^\pm, \ldots, \tau_{n-1}^\pm) - \lim_{\substack{x_1 \to \xi_1 \\ \cdots \\ x_{n-1} \to \xi_{n-1}}} \sup_{x_n \in A \cap E'(x_1,\ldots,x_{n-1})} f(x_1, \ldots, x_n),$$

and similarly

$$\Gamma(\tau_1^\pm, \ldots, \tau_{n-1}^\pm, \tau_n^-) - \lim_{\substack{x_1 \to \xi_1 \\ \cdots \\ x_n \to \xi_n}} f(x_1, \ldots, x_n) =$$

$$\sup_{A \in \mathcal{A}_{\tau_n}(\xi_n)} \Gamma(\tau_1^\pm, \ldots, \tau_{n-1}^\pm) - \lim_{\substack{x_1 \to \xi_1 \\ \cdots \\ x_{n-1} \to \xi_{n-1}}} \inf_{x_n \in A \cap E'(x_1,\ldots,x_{n-1})} f(x_1, \ldots, x_n).$$

Bibliography

[1] A. AMBROSETTI, C. SBORDONE, Γ-*convergenza e G-convergenza per problemi non lineari di tipo ellittico* Boll. Un. Mat. Ital. 5 13-A (1976), pp. 352-362.

[2] I. BABUŠKA, *Solutions of interface problems by homogenization, I, II*, Inst. for Fluid Dynam. and Appl. Math., Univ. of Maryland (1974).

[3] A. BENSOUSSAN, J.L. LIONS, G. PAPANICOLAU, *Some asymptotic results for solutions of variational inequality with highly oscillatory coefficients*, di prossima pubblicazione.

[4] L. BOCCARDO, I. CAPUZZO DOLCETTA, *G-convergenza e problema di Dirichlet unilaterale*, Boll. Un. Mat. Ital., (4) 12 (1975), pp. 115-123.

[5]] L. BOCCARDO, P. MARCELLINI, *Sulla convergenza delle soluzioni di disequazioni variazionali*, Ann. Mat. Pura e Appl. (4) 110 (1976), pp. 137-159.

[6] G. BUTTAZZO, *Su un tipo di convergenza variazionale*, Tesi di laurea in Matematica, Univ. di Pisa, 1976.

[7] L. CARBONE, *Sur la Γ-convergence des intégrales du type de l'énergie à gradient borné*, Comm. Partial Diff. Equations 2 (1977) n.6, pp. 627-651.

[8] E. DE GIORGI, *Sulla convergenza di alcune successioni di integrali del tipo dell'area*, Rend. Mat. 8 (1975), pp. 277-294.

[9] E. DE GIORGI, T. FRANZONI, *Su un tipo di convergenza variazionale*, Atti Accad. Naz. Lincei Rend. Cl. Sci. Fis. Mat. Natur., (8) 58 n.6 (1975), pp. 842-850.

[10] E. DE GIORGI, G. LETTA, *Convergenza debole per funzioni crescenti di insieme*, di prossima pubblicazione.

[11] E. DE GIORGI, S. SPAGNOLO, *Sulla convergenza degli integrali dell'energia per operatori ellittici del II ordine*, Boll. Un. Mat. Ital. (4) 8 (1973), pp. 391-411.

[12] J.L. JOLY, *Une famille de topologies et des convergences sur l'ensemble des fonctionnelles convexes*, Thèse présentée à la Faculté de Sciences de Grenoble.

[13] J.L. JOLY, *Une famille de topologies sur l'ensemble des fonctions convexes pour lequelles la polarité est bicontinue*, J. Math. Pures Appl., 52 (1973), pp. 421-441.

[14] C. KURATOWSKI, *Topologie*, vol. I, Warszawa, 1958.

[15] P. MARCELLINI, *Su una convergenza di funzioni convesse*, Boll. Un. Mat. Ital. 8 (1973), pp.137-158.

[16] P. MARCELLINI, *Un teorema di passaggio al limite per la somma di funzioni convesse*, Boll. Un. Mat. Ital. 11 (1975), pp. 107-124.

[17] A. Marino, S. Spagnolo, *Un tipo di approssimazione dell'operatore* $\sum_{ij} D_i(a_{ij}D_j)$ *con operatori* $\sum_j D_j(\beta D_j)$, Ann. Scuola. Norm. Sup. Pisa 23 (1969), pp. 657-673.

[18] L. Modica, S. Mortola, *Un esempio di* Γ-*convergenza*, Boll. Un. Mat. Ital. (5) 14 (1977), pp. 285-299.

[19] U. Mosco, *Convergence of convex sets and of solutions of variational inequalities*, Advances in Math. 3 (1969), pp. 510-585.

[20] U. Mosco, *On the continuity of the Young-Fenchel transform*, J. Math. Anal. Apl. 35 (1971), pp. 518-535.

[21] F. Murat, *Théorèmes de non exsistence pour des problèmes de contrôle dans les coefficients*, C. R. Acad. Sci. Paris Sér. A-B 274 (1972), pp. 395-398.

[22] Y. Reshetnyak, *General theorems on semicontinuity and on convergence with a functional*, Sibirsk. Mat. Ž. 8 (1967), pp. 1051-1069.

[23] E. Sanchez-Palencia, *Solutions périodiques par rapport aux variables d'espace et applications*, C.R. Acad. Sci. Paris, Sér. A-B, 271 (1970), pp. 1129-1132.

[24] E. Sanchez-Palencia, *Equations aux dérivées partielles dans un type de milieux hétérogènes*, C.R. Acad. Sci. Paris, Sér. A, 272 (1971), pp. 1410-1413.

[25] E. Sanchez-Palencia, *Comportement local et macroscopique d'un type de milieux phisiques hétérogènes*, Internat. J. Engrg. Sci., 13 (1975), no. 11, pp. 923-940.

[26] C. Sbordone, *Alcune questioni di convergenza per operatori differenziali del II ordine*, Boll. Un. Mat. Ital. 10 (1974), pp. 672-682.

[27] C. Sbordone, *Sulla G-convergenza di equazioni ellittiche e paraboliche*, Ricerche Mat. 24 (1975), pp. 76-136.

[28] C. Sbordone, *Su alcune applicazioni di un tipo di convergenza variazionale*, Ann. Scuola Norm. Sup. Pisa (4) n.2 (1975), pp. 617-638.

[29] C. Sbordone, *Su un limite di integrali polinomiali del calcolo delle variazioni*, di prossima pubblicazione.

[30] P.K. Senatorov, *The stability of a solution of Dirichlet's problem for an elliptic equation with respect to perturbations in measure of its coefficients*, Differencial'nye Uravnenija 6 (1970), pp. 1725-1726.

[31] P.K. Senatorov, *The stability of the eigenvalues and eigenfunctions of a Sturm-Liouville problem*, Differencial'nye Uravnenija 7 (1971), pp. 1667-1671.

[32] S. Spagnolo, *Sul limite delle soluzioni di problemi di Cauchy relativi all'equazione del calore*, Ann. Scuola Norm. Sup. Pisa 21 (1967), pp. 657-699.

[33] S. Spagnolo, *Una caratterizzazione degli operatori differenziabili autoaggiunti del II ordine a coefficienti misurabili e limitati*, Rend. Sem. Mat. Univ. Padova 39 (1967), pp. 56-64.

[34] S. Spagnolo, *Sulla convergenza di soluzioni di equazioni paraboliche ed ellittiche*, Ann. Scuola Norm. Sup. Pisa 22 (1968), pp. 577-597.

[35] S. Spagnolo, *Some convergence problems*, Symp. Math. 18 (1976) (Conv. Alta Mat., Roma, Marzo 1974).

[36] S. Spagnolo, *Convergence in energy for elliptic operators, Numerical solutions of Partial Differential Equations III*, SYNSPADE (1976), pp. 469-499.

[37] L. Tartar, *Convergence d'operateurs differentiels*, Analisi convessa e applicazioni, Quaderni dei gruppi di ricerca matematica del C.N.R., Ist. Mat. G. Castelnuovo, Roma.

[38] L. Tartar, *Problèmes de contrôle des coefficients dans les equations aux dérivées partielles*, Control theory, numerical methods and computer systems modelling (Intern. Symposium, IRIA LABORIA, Rocquencourt, 1974), pp. 420-426 Lecture Notes in Economy and Mathematical Systems, vol. 107, Springer, Berlin, 1975.

[39] T. Zolezzi, *Su alcuni problemi fortemente ben posti di controllo ottimo*, Ann. Mat. Pura Appl. 95 (1973), pp. 147-170.

[40] T. Zolezzi, *On convergence of minima*, Boll. Un. Mat. Ital. 8 (1973), pp. 246-257.

Γ-convergenza e G-convergenza[‡]

E. De Giorgi (Pisa)[*]

Sunto. In this paper I suggest a general definition of Γ-convergence and of G-convergence which I hope will be useful to extend the previous convergences so that several works, in which many different notations are used, may be framed in this theory.

Introduzione

In questa nota espongo le linee generali di un tentativo mirante ad unificare diverse ricerche concernenti tipi di convergenza suggeriti da diverse questioni di calcolo delle variazioni, teoria della misura, equazioni alle derivate parziali.

Anche se per ragioni di chiarezza espositiva il lavoro viene presentato nella forma usuale (definizioni, lemmi, osservazioni) esso non è un fascicolo di risultati ma una semplice ipotesi di lavoro, che deve essere verificata sia per quanto riguarda la validità dei risultati enunciati, sia per quanto riguarda la possibilità di giungere, attraverso le definizioni proposte a titolo provvisorio (ed eventualmente modificabili), alla sperata sistemazione di un ampio campo dell'Analisi.

Questa nota è perciò soltanto un invito ad approfondire un settore di ricerche che mi sembra di notevole interesse.

1. - La Γ-convergenza

Indicheremo con $\overline{\mathbf{R}}$ la retta reale completata con l'aggiunta dei due elementi impropri $-\infty$ e $+\infty$, con l'ordinamento e la topologia usuali.

Dato poi un insieme arbitrario S, indicheremo con $\mathcal{P}(S)$ l'insieme delle sue parti e considereremo topologie su S date mediante famiglie di aperti: una tale famiglia τ sarà una famiglia di parti di S che gode delle proprietà consuete: (a) $S \in \tau$, $\emptyset \in \tau$; (b) τ è stabile per intersezione finita e per unione finita o infinita di potenza arbitraria. Non imporremo a priori nessun assioma di separazione.

In queste condizioni possiamo cominciare a definire i Γ-limiti semplici.

DEFINIZIONE I – *Sia $f : S \to \overline{\mathbf{R}}$, $\xi \in S$ e τ una topologia su S. Porremo allora*

$$\Gamma(\tau^+) - \lim_{x \to \xi} f(x) = \inf_{A \in \mathcal{A}_\tau(\xi)} \sup_{x \in A} f(x)$$

ove $\mathcal{A}_\tau(\xi)$ è la famiglia degli intorni aperti di ξ, cioè

$$\mathcal{A}_\tau(\xi) = \{A : \ A \in \tau, \ \xi \in A\}.$$

[‡]Editor's note: published in Boll. Un. Mat. Ital., **A (5) 14** (1977), 213–220.
[*]Scuola Normale Superiore, Pisa.

Analogamente si pone

$$\Gamma(\tau^-) - \lim_{x \to \xi} f(x) = \sup_{A \in \mathcal{A}_\tau(\xi)} \inf_{x \in A} f(x).$$

OSSERVAZIONE I – È subito visto che

$$\Gamma(\tau^+) - \lim_{x \to \xi}(-f(x)) = -\Gamma(\tau^-) - \lim_{x \to \xi} f(x).$$

OSSERVAZIONE II – Si ha sempre

$$\Gamma(\tau^-) - \lim_{x \to \xi} f(x) \le \Gamma(\tau^+) - \lim_{x \to \xi} f(x).$$

Notazione abbreviata I – Quando risulta

$$\Gamma(\tau^+) - \lim_{x \to \xi} f(x) = \Gamma(\tau^-) - \lim_{x \to \xi} f(x) = \lambda$$

scriveremo più brevemente

$$\Gamma(\tau) - \lim_{x \to \xi} f(x) = \lambda.$$

OSSERVAZIONE III – I limiti ora introdotti non sono altro che gli usuali massimo limite, minimo limite e limite; tuttavia questa loro ridefinizione è utile per meglio comprendere le estensioni successive.

Possiamo definire infatti i Γ-limiti multipli nel modo seguente.

DEFINIZIONE I-BIS – *Siano dati n insiemi $S_1, \dots, S_n$ ed n topologie $\tau_1, \dots, \tau_n$ su tali insiemi e sia $f : S_1 \times \cdots \times S_n \to \overline{\mathbf{R}}$. Porremo allora*

$$\Gamma(\tau_1^\pm, \dots, \tau_{n-1}^\pm, \tau_n^+) - \lim_{\substack{x_1 \to \xi_1 \\ \dots \\ x_n \to \xi_n}} f(x_1, \dots, x_n)$$

$$= \inf_{A \in \mathcal{A}_{\tau_n}(\xi_n)} \Gamma(\tau_1^\pm, \dots, \tau_{n-1}^\pm) - \lim_{\substack{x_1 \to \xi_1 \\ \dots \\ x_{n-1} \to \xi_{n-1}}} \sup_{x_n \in A} f(x_1, \dots, x_n)$$

e analogamente

$$\Gamma(\tau_1^\pm, \dots, \tau_{n-1}^\pm, \tau_n^-) - \lim_{\substack{x_1 \to \xi_1 \\ \dots \\ x_{n-1} \to \xi_{n-1}}} f(x_1, \dots, x_n) =$$

$$\sup_{A \in \mathcal{A}_{\tau_n}(\xi_n)} \Gamma(\tau_1^\pm, \dots, \tau_{n-1}^\pm) - \lim_{\substack{x_1 \to \xi_1 \\ \dots \\ x_{n-1} \to \xi_{n-1}}} \inf_{x_n \in A} f(x_1, \dots, x_n).$$

ESEMPI. – Per $n = 2$

$$\Gamma(\tau_1^+, \tau_2^+) - \lim_{\substack{x_1 \to \xi_1 \\ x_2 \to \xi_2}} f(x_1, x_2) = \inf_{A_2 \in \mathcal{A}_{\tau_2}(\xi_2)} \inf_{A_1 \in \mathcal{A}_{\tau_1}(\xi_1)} \sup_{x_1 \in A_1} \sup_{x_2 \in A_2} f(x_1, x_2),$$

$$\Gamma(\tau_1^+, \tau_2^-) - \lim_{\substack{x_1 \to \xi_1 \\ x_2 \to \xi_2}} f(x_1, x_2) = \sup_{A_2 \in \mathcal{A}_{\tau_2}(\xi_2)} \inf_{A_1 \in \mathcal{A}_{\tau_1}(\xi_1)} \sup_{x_1 \in A_1} \inf_{x_2 \in A_2} f(x_1, x_2).$$

Per $n = 3$

$$\Gamma(\tau_1^+, \tau_2^-, \tau_3^+) - \lim_{\substack{x_1 \to \xi_1 \\ x_2 \to \xi_2 \\ x_3 \to \xi_3}} f(x_1, x_2, x_3) =$$

$$\inf_{A_3 \in \mathcal{A}_{\tau_3}(\xi_3)} \sup_{A_2 \in \mathcal{A}_{\tau_2}(\xi_2)} \inf_{A_1 \in \mathcal{A}_{\tau_1}(\xi_1)} \sup_{x_1 \in A_1} \inf_{x_2 \in A_2} \sup_{x_3 \in A_3} f(x_1, x_2, x_3).$$

OSSERVAZIONE IV. – Si estende facilmente ai Γ-limiti multipli l'osservazione I: basta cambiare i segni di tutte le topologie. Si estende pure l'osservazione II: se un segno $+$ è sostituito dal segno $-$, mentre i segni associati alle altre topologie restano invariati, i Γ-limiti non possono aumentare. Per quanto riguarda l'osservazione III, ad essa si può sostituire la seguente.

OSSERVAZIONE III-BIS. – I $\Gamma(\tau_1^+, \dots, \tau_n^+)$–lim e i $\Gamma(\tau_1^-, \dots, \tau_n^-)$–lim coincidono con gli usuali massimo limite e minimo limite relativi alla usuale topologia prodotto. *Invece i Γ-limiti multipli con segni diversi sembrano costituire un concetto del tutto nuovo rispetto alla topologia usuale.*

Notazione abbreviata I-bis – Anche nei Γ-limiti multipli, quando il limite è indipendente dai segni associati ad una delle topologie, tale segno può essere omesso. Per esempio se

$$\Gamma(\tau_1^+, \tau_2^-) - \lim_{\substack{x_1 \to \xi_1 \\ x_2 \to \xi_2}} f(x_1, x_2) = \Gamma(\tau_1^+, \tau_2^+) - \lim_{\substack{x_1 \to \xi_1 \\ x_2 \to \xi_2}} f(x_1, x_2) = \lambda$$

allora scriveremo

$$\Gamma(\tau_1^+, \tau_2) - \lim_{\substack{x_1 \to \xi_1 \\ x_2 \to \xi_2}} f(x_1, x_2) = \lambda.$$

Notazione abbreviata II – Quando per ogni $(\xi_1, \dots, \xi_n) \in S_1 \times \cdots \times S_n$ risulta

$$\Gamma(\tau_1^\pm, \dots, \tau_{n-1}^\pm, \tau_n^\pm) - \lim_{\substack{x_1 \to \xi_1 \\ \cdots \\ x_n \to \xi_n}} f(x_1, \dots, x_n) = g(\xi_1, \dots, \xi_n)$$

scriveremo più brevemente

$$\Gamma(\tau_1^\pm, \dots, \tau_n^\pm) f = g.$$

2. - La G-convergenza

Dalla definizione di Γ-limite si passa facilmente a quella di G-limite.

DEFINIZIONE I – *Consideriamo un insieme $E \subseteq S_1 \times \dots \times S_n$ ed indichiamo con χ_E la sua funzione caratteristica, cioè*

$$\chi_E(x_1, \dots, x_n) = \begin{cases} 1 & se \ (x_1, \dots, x_n) \in E \\ 0 & se \ (x_1, \dots, x_n) \in (S_1 \times \cdots \times S_n) \setminus E \end{cases}$$

Sia poi $L \subseteq S_1 \times \cdots \times S_n$. Scriveremo

$$G(\tau_1^\pm, \ldots, \tau_n^\pm)E = L$$

quando

$$\Gamma(\tau_1^\pm, \ldots, \tau_n^\pm)\chi_E = \chi_L.$$

Definizione II. – *Siano dati $n+m$ insiemi $Y_1, \ldots, Y_m,\ S_1, \ldots, S_n$ e siano rispettivamente $\theta_1, \ldots, \theta_m$ topologie su $Y_1, \ldots, Y_m$ e $\tau_1, \ldots, \tau_n$ topologie su $S_1, \ldots, S_n$. Sia $F : Y_1 \times \cdots \times Y_m \to \mathcal{P}(S_1 \times \cdots \times S_n)$. Fissato $(\eta_1, \ldots, \eta_m) \in Y_1 \times \ldots \times Y_m$, scriveremo*

$$G(\theta_1^\pm, \ldots, \theta_m^\pm; \tau_1^\pm, \ldots, \tau_n^\pm) - \lim_{\substack{y_1 \to \eta_1 \\ \cdots \\ y_m \to \eta_m}} F(y_1, \ldots, y_m) = L$$

se, per ogni $(\xi_1, \ldots, \xi_n) \in S_1 \times \cdots \times S_n$, risulta

$$\Gamma(\theta_1^\pm, \ldots, \theta_m^\pm; \tau_1^\pm, \ldots, \tau_n^\pm) - \lim_{\substack{y_1 \to \eta_1 \\ \cdots \\ y_m \to \eta_m \\ x_1 \to \xi_1 \\ \cdots \\ x_n \to \xi_n}} \chi_{F(y_1, \ldots, y_m)}(x_1, \ldots, x_n) = \chi_L(\xi_1, \ldots, \xi_n).$$

Osservazione I. – Notiamo che $G(\tau^+)E$ è la usuale chiusura di E rispetto alla topologia τ, mentre $G(\tau^-)$ rappresenta la parte interna di E. Analogamente $G(\tau_1^+, \ldots, \tau_n^+)$ e $G(\tau_1^-, \ldots, \tau_n^-)$ rappresentano chiusura e parte interna di E rispetto alla topologia prodotto.

Osservazione II. – Il $G(\theta^+; \tau^+) - \lim$ e il $G(\theta^-; \tau^+) - \lim$ corrispondono rispettivamente al massimo limite e al minimo limite secondo Kuratowski.

Osservazione III. – Se $f : Y \to S$ e risulta per $\eta \in Y$

$$G(\theta; \tau^+) - \lim_{y \to \eta}\{f(y)\} = \{\lambda\}$$

(ove $\{\lambda\}$ e $\{f(y)\}$ sono gli insiemi che hanno come unico elemento rispettivamente λ e $f(y)$) allora si ha

$$\lim_{y \to \eta} f(y) = \lambda$$

ove questo limite va inteso nel senso usuale dei limiti di un'applicazione dallo spazio topologico (Y, θ) allo spazio topologico (S, τ).

3. - Γ-limiti di funzioni definite su sottoinsiemi

Definizione I – *Sia S un insieme, τ una topologia su S, $E \subset S$, $f : E \to \overline{\mathbf{R}}$. Porremo allora per ogni $\xi \in G(\tau^+)E$*

$$\Gamma(\tau^+) - \lim_{x \to \xi} f(x) = \inf_{A \in \mathcal{A}_\tau(\xi)} \sup_{x \in E \cap A} f(x)$$

e analogamente

$$\Gamma(\tau^-) - \lim_{x \to \xi} f(x) = \sup_{A \in \mathcal{A}_\tau(\xi)} \inf_{x \in E \cap A} f(x).$$

Questa definizione può estendersi per induzione dopo aver dimostrato il lemma seguente.

LEMMA. – *Siano dati n insiemi $S_1, \ldots, S_n$ con le topologie $\tau_1, \ldots, \tau_n$ e sia $E \subseteq S_1 \times \cdots \times S_n$. Poniamo per ogni $(x_1, \ldots, x_{n-1}) \in S_1 \times \cdots \times S_{n-1}$*

$$E'_{(x_1, \ldots, x_{n-1})} = \{t \in S_n : (x_1, \ldots, x_{n-1}, t) \in E\}.$$

Poniamo inoltre per ogni $A \in \tau_n$

$$E''_A = \{(x_1, \ldots, x_{n-1}) \in S_1 \times \cdots \times S_{n-1} : E'_{(x_1, \ldots, x_{n-1})} \cap A \neq \emptyset\}.$$

Sia inoltre $(\xi_1, \ldots, \xi_n) \in G(\tau_1^+, \ldots, \tau_n^+)E$ e sia $\xi_n \in A \in \tau_n$. Risulta allora $(\xi_1, \ldots, \xi_{n-1}) \in G(\tau_1^+, \ldots, \tau_{n-1}^+)E''_A$.

Dopo aver provato il lemma precedente possiamo assicurare che ha senso la definizione seguente.

DEFINIZIONE II – *Siano $S_1, \ldots, S_n$ n insiemi con le topologie $\tau_1, \ldots, \tau_n$. Sia $E \subset S_1 \times \cdots \times S_n$ e $\xi \in G(\tau_1^+, \ldots, \tau_n^+)E$. Sia $f : E \to \overline{\mathbf{R}}$. Porremo allora*

$$\Gamma(\tau_1^\pm, \ldots, \tau_{n-1}^\pm, \tau_n^+) - \lim_{\substack{x_1 \to \xi_1 \\ \cdots \\ x_n \to \xi_n}} f(x_1, \ldots, x_n) =$$

$$\inf_{A \in \mathcal{A}_{\tau_n}(\xi_n)} \Gamma(\tau_1^\pm, \ldots, \tau_{n-1}^\pm) - \lim_{\substack{x_1 \to \xi_1 \\ \cdots \\ x_{n-1} \to \xi_{n-1}}} \sup_{x_n \in A \cap E'(x_1, \ldots, x_{n-1})} f(x_1, \ldots, x_n),$$

e analogamente

$$\Gamma(\tau_1^\pm, \ldots, \tau_{n-1}^\pm, \tau_n^-) - \lim_{\substack{x_1 \to \xi_1 \\ \cdots \\ x_n \to \xi_n}} f(x_1, \ldots, x_n) =$$

$$\sup_{A \in \mathcal{A}_{\tau_n}(\xi_n)} \Gamma(\tau_1^\pm, \ldots, \tau_{n-1}^\pm) - \lim_{\substack{x_1 \to \xi_1 \\ \cdots \\ x_{n-1} \to \xi_{n-1}}} \inf_{x_n \in A \cap E'(x_1, \ldots, x_{n-1})} f(x_1, \ldots, x_n).$$

Bibliografia

[1] A. AMBROSETTI, C. SBORDONE, Γ-*convergenza e G-convergenza per problemi non lineari di tipo ellittico* Boll. Un. Mat. Ital. 5 13-A (1976), pp. 352-362.

[2] I. BABUŠKA, *Solutions of interface problems by homogenization, I, II,* Inst. for Fluid Dynam. and Appl. Math., Univ. of Maryland (1974).

[3] A. BENSOUSSAN, J.L. LIONS, G. PAPANICOLAU, *Some asymptotic results for solutions of variational inequality with highly oscillatory coefficients*, di prossima pubblicazione.

[4] L. BOCCARDO, I. CAPUZZO DOLCETTA, *G-convergenza e problema di Dirichlet unilaterale*, Boll. Un. Mat. Ital., (4) 12 (1975), pp. 115-123.

[5]] L. BOCCARDO, P. MARCELLINI, *Sulla convergenza delle soluzioni di disequazioni variazionali*, Ann. Mat. Pura e Appl. (4) 110 (1976), pp. 137-159.

[6] G. BUTTAZZO, *Su un tipo di convergenza variazionale*, Tesi di laurea in Matematica, Univ. di Pisa, 1976.

[7] L. CARBONE, *Sur la Γ-convergence des intégrales du type de l'énergie à gradient borné*, Comm. Partial Diff. Equations 2 (1977) n.6, pp. 627-651.

[8] E. DE GIORGI, *Sulla convergenza di alcune successioni di integrali del tipo dell'area*, Rend. Mat. 8 (1975), pp. 277-294.

[9] E. DE GIORGI, T. FRANZONI, *Su un tipo di convergenza variazionale*, Atti Accad. Naz. Lincei Rend. Cl. Sci. Fis. Mat. Natur., (8) 58 n.6 (1975), pp. 842-850.

[10] E. DE GIORGI, G. LETTA, *Convergenza debole per funzioni crescenti di insieme*, di prossima pubblicazione.

[11] E. DE GIORGI, S. SPAGNOLO, *Sulla convergenza degli integrali dell'energia per operatori ellittici del II ordine*, Boll. Un. Mat. Ital. (4) 8 (1973), pp. 391-411.

[12] J.L. JOLY, *Une famille de topologies et des convergences sur l'ensemble des fonctionnelles convexes*, Thèse présentée à la Faculté de Sciences de Grenoble.

[13] J.L. JOLY, *Une famille de topologies sur l'ensemble des fonctions convexes pour lequelles la polarité est bicontinue*, J. Math. Pures Appl., 52 (1973), pp. 421-441.

[14] C. KURATOWSKI, *Topologie*, vol. I, Warszawa, 1958.

[15] P. MARCELLINI, *Su una convergenza di funzioni convesse*, Boll. Un. Mat. Ital. 8 (1973), pp.137-158.

[16] P. MARCELLINI, *Un teorema di passaggio al limite per la somma di funzioni convesse*, Boll. Un. Mat. Ital. 11 (1975), pp. 107-124.

[17] A. MARINO, S. SPAGNOLO, *Un tipo di approssimazione dell'operatore* $\sum_{ij} D_i(a_{ij}D_j)$ *con operatori* $\sum_{j} D_j(\beta D_j)$, Ann. Scuola. Norm. Sup. Pisa 23 (1969), pp. 657-673.

[18] L. MODICA, S. MORTOLA, *Un esempio di Γ-convergenza*, Boll. Un. Mat. Ital. (5) 14 (1977), pp. 285-299.

[19] U. MOSCO, *Convergence of convex sets and of solutions of variational inequalities*, Advances in Math. 3 (1969), pp. 510-585.

[20] U. MOSCO, *On the continuity of the Young-Fenchel transform*, J. Math. Anal. Apl. 35 (1971), pp. 518-535.

[21] F. MURAT, *Théorèmes de non exsistence pour des problèmes de contrôle dans les coefficients*, C. R. Acad. Sci. Paris Sér. A-B 274 (1972), pp. 395-398.

[22] Y. RESHETNYAK, *General theorems on semicontinuity and on convergence with a functional*, Sibirsk. Mat. Ž. 8 (1967), pp. 1051-1069.

[23] E. SANCHEZ-PALENCIA, *Solutions périodiques par rapport aux variables d'espace et applications*, C.R. Acad. Sci. Paris, Sér. A-B, 271 (1970), pp. 1129-1132.

[24] E. SANCHEZ-PALENCIA, *Equations aux dérivées partielles dans un type de milieux hétérogènes*, C.R. Acad. Sci. Paris, Sér. A, 272 (1971), pp. 1410-1413.

[25] E. SANCHEZ-PALENCIA, *Comportement local et macroscopique d'un type de milieux phisiques hétérogènes*, Internat. J. Engrg. Sci., 13 (1975), no. 11, pp. 923-940.

[26] C. SBORDONE, *Alcune questioni di convergenza per operatori differenziali del II ordine*, Boll. Un. Mat. Ital. 10 (1974), pp. 672-682.

[27] C. SBORDONE, *Sulla G-convergenza di equazioni ellittiche e paraboliche*, Ricerche Mat. 24 (1975), pp. 76-136.

[28] C. SBORDONE, *Su alcune applicazioni di un tipo di convergenza variazionale*, Ann. Scuola Norm. Sup. Pisa (4) n.2 (1975), pp. 617-638.

[29] C. SBORDONE, *Su un limite di integrali polinomiali del calcolo delle variazioni*, di prossima pubblicazione.

[30] P.K. SENATOROV, *The stability of a solution of Dirichlet's problem for an elliptic equation with respect to perturbations in measure of its coefficients*, Differencial'nye Uravnenija 6 (1970), pp. 1725-1726.

[31] P.K. SENATOROV, *The stability of the eigenvalues and eigenfunctions of a Sturm-Liouville problem*, Differencial'nye Uravnenija 7 (1971), pp. 1667-1671.

[32] S. SPAGNOLO, *Sul limite delle soluzioni di problemi di Cauchy relativi all'equazione del calore*, Ann. Scuola Norm. Sup. Pisa 21 (1967), pp. 657-699.

[33] S. Spagnolo, *Una caratterizzazione degli operatori differenziabili autoaggiunti del II ordine a coefficienti misurabili e limitati*, Rend. Sem. Mat. Univ. Padova 39 (1967), pp. 56-64.

[34] S. Spagnolo, *Sulla convergenza di soluzioni di equazioni paraboliche ed ellittiche*, Ann. Scuola Norm. Sup. Pisa 22 (1968), pp. 577-597.

[35] S. Spagnolo, *Some convergence problems*, Symp. Math. 18 (1976) (Conv. Alta Mat., Roma, Marzo 1974).

[36] S. Spagnolo, *Convergence in energy for elliptic operators*, Numerical solutions of Partial Differential Equations III, SYNSPADE (1976), pp. 469-499.

[37] L. Tartar, *Convergence d'operateurs differentiels*, Analisi convessa e applicazioni, Quaderni dei gruppi di ricerca matematica del C.N.R., Ist. Mat. G. Castelnuovo, Roma.

[38] L. Tartar, *Problèmes de contrôle des coefficients dans les equations aux dérivées partielles*, Control theory, numerical methods and computer systems modelling (Intern. Symposium, IRIA LABORIA, Rocquencourt, 1974), pp. 420-426 Lecture Notes in Economy and Mathematical Systems, vol. 107, Springer, Berlin, 1975.

[39] T. Zolezzi, *Su alcuni problemi fortemente ben posti di controllo ottimo*, Ann. Mat. Pura Appl. 95 (1973), pp. 147-170.

[40] T. Zolezzi, *On convergence of minima*, Boll. Un. Mat. Ital. 8 (1973), pp. 246-257.

On hyperbolic equations with coefficients depending on time only[‡†‡]

Ferruccio Colombini, Ennio De Giorgi, Sergio Spagnolo[*]

Introduction

The second order hyperbolic equation

$$\frac{\partial^2 u}{\partial t^2} - \sum_{i,j}^{1,n} \frac{\partial}{\partial x_i}\left(a_{ij}(x,t)\frac{\partial u}{\partial x_j}\right) = f(x,t)$$

on $\mathbf{R}_x^n \times [0,T]$ has been studied in general only in the case when the coefficients a_{ij}, besides the usual assumptions of symmetry and coercivity, are assumed to be bounded measurable with respect to x and Lipschitz continuous with respect to t.

Under this assumption, the Cauchy problem for the above equation has been studied by many authors, who obtained (see Lions [5], Lions et Magenes [6], Hurd et Sattinger [4]) several existence and uniqueness results in suitable Sobolev spaces. An assumption slightly weaker than the Lipschitz continuity in t of the coefficients was considered by De Simon et Torelli ([2]) who proved an existence and uniqueness result in the case when the coefficients $a_{ij}(x,t)$ are of bounded variation on $[0,T]$ as functions of t with values in the Banach space $L^\infty(\mathbf{R}_x^n)$.

On the other hand, one could try to solve the above Cauchy problem even when the coefficients $a_{ij}(x,t)$ are merely locally integrable functions on $\mathbf{R}_x^n \times [0,T]$. But an elementary example ([4], where the coefficients have the form $a(x,t) = \alpha(x-t)$ with $\alpha(\xi)$ equal to α_1 for $\xi < 0$, and to α_2 for $\xi > 0$, where $0 < \alpha_2 < 1 < \alpha_1$) shows that there is no hope in general to prove the existence of solutions, even in the space of distributions.

However, one might try to compensate the roughness of the coefficients with respect to t with a suitable smoothness in x, and thus to prove the existence and uniqueness of solutions even for coefficients discontinuous in t but very smooth in x.

[‡]Editor's note: translation into English of the paper "Sur les équations hyperboliques avec des coefficients qui ne dépendent que du temps", published in Ann. Sc. Norm. Sup. Pisa Cl. Sci., **(4) 6** (1979), 511–559.

[†]With the support of G.N.A.F.A. (C.N.R.).

[‡]A communication concerning the results of this paper appeared on Comptes Rendus de l'Académie des Sciences, t. 286, Série A (1978)

[*]Scuola Normale Superiore - Pisa. Communicated on July 19, 1978

The simplest case we can consider is the case of coefficients a_{ij} depending only on t: the present paper is devoted to the study of this situation.

Thus, in the following we shall consider the following problem.

PROBLEM. Let $a_{ij}(t)$ be real valued functions, integrable on $[0, T]$ and satisfying the coercivity assumption

$$\sum_{i,j}^{1,n} a_{ij}(t)\xi_i\xi_j \geq \lambda_0|\xi|^2 \qquad (\lambda_0 > 0), \qquad \forall \xi \in \mathbf{R}^n.$$

Given three functions (or functionals) $\varphi(x)$, $\psi(x)$ and $f(x, t)$, find $u(x, t)$ such that

$$(1) \qquad \frac{\partial^2 u}{\partial t^2} - \sum_{i,j}^{1,n} a_{ij}(t)\frac{\partial^2 u}{\partial x_i \partial x_j} = f(x, t) \qquad \text{on } \mathbf{R}^n \times [0, t],$$

$$(2) \qquad u(x, 0) = \varphi(x) \text{ and } \frac{\partial u}{\partial t}(x, 0) = \psi(x) \qquad \text{on } \mathbf{R}^n.$$

The data $\varphi(x)$, $\psi(x)$ and $f(x, t)$ will be chosen in X_1, X_2 and $L^1([0, T], X_3)$ respectively, where X_1, X_2 and X_3 are spaces of Gevrey ultradistributions on $\mathbf{R}^n$, or holomorphic functionals, with continuous inclusions $X_1 \hookrightarrow X_2 \hookrightarrow X_3$, such that the derivatives $\partial/\partial x_j$, $j = 1, \ldots, n$, are continuous operators from X_1 to X_2 and from X_2 to X_3.

Then we investigate the existence of a solution u in the space $L^\infty([0, T], X_1)$, and equation (1) must be interpreted in the sense of vector valued distributions defined on $]0, T[$ with values in X_3. On the other hand we see from equation (1) that the solution u actually belongs to the space $C^1([0, T], X_3)$, so that the initial conditions (2) have a well defined meaning.

In most cases under considerations here, the three spaces X_1, X_2 and X_3 will coincide.

Here we shall prove several results of existence and uniqueness for problem $\{(1),(2)\}$, even when the coefficients $a_{ij}(t)$ are very rough, namely when they are merely integrable functions on $[0, T]$. Of course, in general we will only obtain functionals as solutions, unless the data φ, ψ and f are assumed to be very smooth.

More precisely, we shall prove (th. 3) that if $\varphi(x)$, $\psi(x)$ and $f(x, t)$ are real analytic functionals on $\mathbf{R}_x^n$ (e.g., when they are integrable functions with compact support) then problem $\{(1),(2)\}$ has a unique solution $u(x, t)$, real analytic in x for each t, while (th. 4) if $\varphi(x)$, $\psi(x)$ and $f(x, t)$ are real analytic functions in x, then there exists a unique solution $u(x, t)$, real analytic in x, for all t.

Hence we can solve problem $\{(1),(2)\}$ both in the class X of real analytic functions on $\mathbf{R}^n$ and in the class X' of linear continuous functionals on X.

When the coefficients $a_{ij}(t)$ are slightly smoother a similar phenomenon occurs, moreover in this case we can choose as X a suitable class of infinitely differentiable functions on $\mathbf{R}^n$ containing the class of real analytic functions (so that the "distance" between X and X' decreases).

For instance, if a_{ij} are Hölder continuous of order α and if $1 \leq s < 1/(1-\alpha)$ we obtain (th. 4) a solution $u(x,t)$ which is a Gevrey ultradistribution of order s in x, provided the data are ultradistributions of order s in x; while, when the data are Gevrey functions of order s in x, the solution is also a Gevrey function (of the same order) in x.

Further, if the a_{ij} satisfy the condition

$$|a_{ij}(t + \tau) - a_{ij}(t)| \leq A|\tau| \left(|\log |\tau|| + 1\right)$$

for some constant A, then (th. 4) there exists a solution $u(x,t)$, distribution in x, provided the data are distributions in x; on the other hand, when the data are functions of class C^∞ in x, then the solution is also of class C^∞ in x.

Finally, by suitable counterexamples (th. 10) we show that it is not possible to improve the above results. In particular, it is impossible in general to solve the Cauchy problem $\{(1),(2)\}$ in Sobolev spaces when the coefficients are not of bounded variation.

Besides existence, we also prove a priori estimates on the solution. These estimates are useful to study a sequence of problems of the form $\{(1),(2)\}$ with coefficients $a_{ij,k}$ $(k = 1, 2, \ldots)$, non uniformly smooth, which converge to a_{ij} as $k \to \infty$, and one is interested in proving the convergence of the corresponding solutions.

Indeed, we shall prove (th. 8) that, if $\{a_{ij,k}\} \to a_{ij}$ in L^1 and the data are real analytic functionals (resp. real analytic functions) in x, then the solutions converge in the sense of analytic functionals (resp. of real analytic functions) in x.

Moreover, if the $a_{ij,k}$ are equi-Hölder continuous functions, then the corresponding solutions converge in the space of Gevrey ultradistributions (or functions) in x, provided the data are Gevrey ultradistributions (or functions) in x.

The proof of the existence of solutions is divided into the following steps:

1) We prove an existence theorem in the space of holomorphic functionals, or in the space of entire analytic functions, without any coercivity or regularity assumption on the coefficients (such a result, very close to the theorem of Cauchy-Kovalevski, is well known (see [11], or [1]); in any case, we give here a proof in the Appendix).

2) We consider the case of data $\varphi(x)$, $\psi(x)$ and $f(x,t)$ vanishing for $|x| \geq r$; we use here the Fourier-Laplace transform with respect to the variable x (which becomes the variable $\zeta \in \mathbf{C}^n$) to reduce problem $\{(1),(2)\}$ to a family, depending on the parameter ζ, of ordinary differential equations.

3) We estimate the growth in ζ $(|\zeta| \to \infty)$ of the solution of the transformed problems, by applying a result (Lemma 1) on second order ordinary differential equations; using these estimate and the Paley-Wiener theorem, we see that the solution, originally constructed in the space of holomorphic functionals, is actually much smoother.

4) Using a duality argument, we drop the technical assumption that the data φ, ψ and f have compact support in x.

In conclusion, we remark that when the data are periodic in x an alternative approach is to use an expansion in Fourier series. Then the computations are much simpler, and for this reason we shall include them here as an independent result (§3), although they are just a special case of the results exposed in the following sections.

Finally, we remark that the coercivity assumption on the quadratic form of the coefficients is not necessary, and can be replaced by the weaker assumption of non-negativity, if one restricts to the case of solutions which are real analytic functions or functionals (§7).

1. - Notation and basic results

Let Ω be an open set of $\mathbf{R}_x^n$. We shall need the following topological vector spaces on the complex field $\mathbf{C}$.

$\mathscr{H}$ entire functions on $\mathbf{R}^n$.

$\mathscr{A}(\Omega)$ analytic functions on Ω.

$\mathscr{E}_s(\Omega)$ Gevrey functions of order s on Ω $(s \geq 1)$.

$\mathscr{D}_s(\Omega)$ Gevrey functions of order s, with compact support in Ω $(s \geq 1)$.

$\mathscr{E}(\Omega)$ infinitely differentiable functions on Ω.

$\mathscr{D}(\Omega)$ infinitely differentiable functions, with compact support in Ω.

$\mathscr{H}'$ holomorphic functionals on $\mathbf{C}^n$.

$\mathscr{A}'(\Omega)$ real analytic functionals on Ω.

$\mathscr{D}'_s(\Omega)$ Gevrey ultradistributions of order s on Ω $(s \geq 1)$.

$\mathscr{E}'_s(\Omega)$ Gevrey ultradistributions of order s, with compact support in Ω.

When Ω is $\mathbf{R}^n$, we shall write simply $\mathscr{A}$, $\mathscr{E}_s$, $\mathscr{D}_s$, $\mathscr{E}$, $\mathscr{D}$, $\mathscr{A}'$, $\mathscr{D}'_s$, $\mathscr{E}'_s$ instead of $\mathscr{A}(\Omega)$, $\mathscr{E}_s(\Omega)$, $\mathscr{D}_s(\Omega)$, ...

For the topology and the main properties of these spaces, see the references Lions-Magenes ([6]), Gelfand-Shilov ([3]), Roumieu ([9], [10]) and Martineau ([7]). Here we recall only the following facts:

A function on $\mathbf{R}^n$ (complex valued) is called *entire* if it can be prolonged to a holomorphic function on the whole $\mathbf{C}^n$.

A function u, infinitely differentiable on Ω, is called *a Gevrey function of order s* (s real ≥ 1) if for all $K \subset\subset \Omega$ there exist M and A such that

$$|D^r u(x)| \leq M A^{|r|} |r|^{s|r|}, \qquad \forall x \in K, \qquad \forall r.$$

For $s = 1$ we have $\mathscr{E}_s(\Omega) = \mathscr{A}(\Omega)$ and $\mathscr{E}'_s(\Omega) = \mathscr{A}'(\Omega)$, while $\mathscr{D}_s(\Omega) = \mathscr{D}'_s(\Omega) = \{0\}$.

Let $w \in \mathscr{A}'(\Omega)$ and $K \subset\subset \Omega$. We say that the *support* of w is contained in K (supp $(w) \subseteq K$) if, $\forall \{u_k\} \subseteq \mathscr{A}(\Omega)$ such that $\{u_k\} \to 0$ in $\mathscr{A}(U)$ for some open neighborhood U of K, we have $\{\langle w, u_k \rangle\} \to 0$ $(k \to \infty)$.

Each of the topological vector spaces considered above is complete, reflexive and a *Montel* space, in the sense that any bounded sequence admits a converging subsequence.

We shall also use the Sobolev spaces $H^s(\Omega) \equiv H^{s,2}(\Omega)$ (s real) and the spaces

$$H^s_{\mathrm{loc}}(\Omega) = \{u \in \mathscr{D}'(\Omega) \ : \ \varphi u \in H^s(\Omega), \ \forall \varphi \in \mathscr{D}(\Omega)\}$$

$$H^s_c(\Omega) = \{u \in \mathscr{D}'(\Omega) \ : \ \mathrm{supp}\ (u) \text{ is compact in } \Omega\}.$$

In the study of evolution equations it is usual to consider functions (or functionals) $u(x,t)$ depending on $n+1$ variables $x = (x_1, \ldots, x_n) \in \Omega$ and $t \in [0,T]$, where T is a real number > 0.

Such a $u(x,t)$ will be considered, in general, as a function defined on $[0,T]$ and with values in a suitable space X of functions (or functionals) on Ω.

Then, let X be a locally convex complete space, and let $u : [0,T] \to X$. We say that u is *integrable* on $[0,T]$ if there exists a sequence $\{u_k\}$ of functions on $[0,T]$, with values in X, constant on each element of a finite partition of $[0,T]$ in measurable parts, and such that, if $k \to \infty$,

$$\{u_k(t)\} \to u(t) \text{ in } X, \qquad \text{a.e. on } [0,T],$$

and

$$\int_0^T \mu(u_k(t) - u(t))dt \to 0, \qquad \forall \text{ seminorm } \mu \text{ on } X.$$

If u is integrable on $[0,T]$, we set

$$\int_0^T u(t)dt = \lim_{k \to \infty} \int_0^T u_k(t)dt.$$

We denote by $L^1([0,T], X)$ the topological vector space of functions, with values in X, which are integrable on $[0,T]$, and we endow it with the seminorms

$$\left\{ u \mapsto \int_0^T \mu(u(t))dt \ : \ \mu \text{ seminorm on } X \right\}.$$

We define in a similar way the space $L^p([0,T], X)$, with p real ≥ 1 or $p = \infty$, of functions $u : [0,T] \to X$, integrable on $[0,T]$ and such that $t \to \mu(u(t))$ belongs to $L^p([0,T])$, for each seminorm μ on X.

We shall also consider the spaces

$$\begin{aligned}
H^{k,1}([0,T], X) &= \{u \in L^1([0,T], X) \ : \ u', u'', \ldots, u^{(k)} \in L^1([0,T], X)\}, \\
C([0,T], X) &= \{u : [0,T] \to X \ : \ u \text{ continuous on } [0,T]\}, \\
C^k([0,T], X) &= \{u \in C([0,T], X) \ : \ u', u'', \ldots, u^{(k)} \in C([0,T], X)\},
\end{aligned}$$

where k is an integer ≥ 1.

Finally, we remark that, if X is a subspace of $\mathscr{D}'(\Omega)$ or of $\mathscr{D}'_s(\Omega)$ for $s > 1$, each u in $L^1([0,T],X)$ can be considered also as a distribution, or an ultra-distribution, on the cylinder $\Omega \times {]}0,T{[}$. We shall write then $\partial u/\partial t$ instead of u'.

PALEY-WIENER THEOREM. – *For all $w \in \mathscr{H}'$, we define the* Fourier transform *of w, by the formula*

$$w(\zeta) = \langle w, h_\zeta \rangle \qquad (\zeta = \xi + i\eta \in \mathbf{C}^n)$$

where

$$h_\zeta(z) = \exp(-i(\zeta, z)) \qquad (z \in \mathbf{C}^n).$$

We shall write also, for $u \in L^1([0,T], \mathscr{H}')$,

$$\widehat{u}(\zeta, t) = \langle u(t), h_\zeta \rangle.$$

We know that $\widehat{w}(\zeta)$ is an entire function of exponential growth. Moreover, we have the following results.

I) Let w be in $\mathscr{H}'$.

i) w is in $\mathscr{A}'$, with support contained in the ball of $\mathbf{R}^n$, $\{|x| \le \rho\}$, if and only if, $\forall \varepsilon > 0$, $\exists C_\varepsilon$ such that

$$|\widehat{w}(\xi + i\eta)| \le C_\varepsilon \exp(\varepsilon|\xi| + (\rho + \varepsilon)|\eta|), \qquad \forall \zeta \in \mathbf{C}^n, \ |\zeta| \ge 1.$$

ii) w is in $\mathscr{E}'_s$ ($s \ge 1$) if and only if, $\forall \varepsilon > 0$, $\exists C_\varepsilon$ such that

$$|\widehat{w}(\xi)| \le C_\varepsilon \exp(\varepsilon|\xi|^{1/s}), \qquad \forall \xi \in \mathbf{R}^n, \ |\xi| \ge 1.$$

iii) w is in $\mathscr{E}'$ if and only if $\exists k$, $\exists C$ such that

$$|\widehat{w}(\xi)| \le C(1 + |\xi|^2)^{k/2}, \qquad \forall \xi \in \mathbf{R}^n, \ |\xi| \ge 1.$$

iv) w is in H^s_c (s real) if and only there exists $\gamma \in L^2(\mathbf{R}^n)$ such that

$$|\widehat{w}(\xi)| \le \gamma(\xi)(1 + |\xi|^2)^{-s/2}, \qquad \forall \xi \in \mathbf{R}^n, \ |\xi| \ge 1.$$

v) w is in $\mathscr{D}$ if and only if, $\forall k > 0$, $\exists C_k$ such that

$$|\widehat{w}(\xi)| \le C_k(1 + |\xi|^2)^{-k/2}, \qquad \forall \xi \in \mathbf{R}^n, \ |\xi| \ge 1.$$

vi) w is in $\mathscr{D}_s$ ($s > 1$) if and only if $\exists a > 0$, $\exists C$, such that

$$|\widehat{w}(\xi)| \le C_k \exp(-a|\xi|^{1/s}), \qquad \forall \xi \in \mathbf{R}^n, \ |\xi| \ge 1.$$

II) Let $\{w_\nu\}$ be a subset of $\mathscr{H}'$: $\{w_\nu\}$ is bounded in $\mathscr{A}'$ if and only if each $\widehat{w}_\nu$ satisfies the inequality of I-(i), uniformly with respect to ν.

Let $\{w_\nu\}$ be a bounded subset of $\mathscr{A}'$: $\{w_\nu\}$ is bounded in $\mathscr{E}'_s$ (resp. in $\mathscr{E}'$,...) if and only if each $\widehat{w}_\nu$ satisfies the inequality of I-(ii) (resp. (iii),...) uniformly with respect to ν.

III)

i) Let $u \in L^1([0,T], \mathscr{A}')$. Then $\exists \rho > 0$ such that, $\forall \varepsilon > 0$, $\exists C_\varepsilon$ such that

$$\int_0^T |\widehat{u}(\xi + i\eta)| dt \leq C_\varepsilon \exp(\varepsilon|\xi| + (\rho + \varepsilon)|\eta|), \qquad \forall \zeta \in \mathbf{C}^n.$$

ii) If $u \in L^1([0,T], \mathscr{A}'(B_r))$, where B_r denotes the ball $\{x \in \mathbf{R}^n : |x| < r\}$, then $\widehat{u}$ satisfies the inequality of (i) with $\rho = r$.

iii) If $u \in L^1([0,T], \mathscr{E}'_s)$ (resp. $u \in L^1([0,T], \mathscr{E}')$,...) then $\widehat{u}$ satisfies an inequality analogous to that of (i) but of the type I-(ii) (resp. of the type I-(iii),...).

iv) If $\{u_\nu\}$ is a bounded subset of $L^1([0,T], \mathscr{A}')$, then each $\{\widehat{u}_\nu\}$ satisfies the inequality of (i) uniformly with respect to ν.

[For a proof of part *III* we refer to the Appendix, part A].

We recall now a result of existence and uniqueness of solutions of problem $\{(1),(2)\}$ in the domain of holomorphic functionals or of entire functions. Notice that we make no assumption of hyperbolicity on the equation. [For a proof of this theorem we refer to the Appendix, part B].

THEOREM 1 (SEE [11], [1]). – *Let us consider problem $\{(1),(2)\}$ with coefficients $a_{ij}(t)$ in $L^1([0,T])$.*

i) If φ and ψ are in $\mathscr{H}'$ and f in $L^1([0,T], \mathscr{H}')$, then the problem has one and only one solution u in $H^{2,1}([0,T], \mathscr{H}')$.

ii) If φ and ψ are in $\mathscr{H}$ and f in $L^1([0,T], \mathscr{H})$, then the problem has one and only one solution u in $H^{2,1}([0,T], \mathscr{H})$.

2. - A result on ordinary differential equations

Using the Fourier transform, or (in the case of periodic data) the expansion in Fourier series, the study of the solutions to problem $\{(1),(2)\}$ can be reduced to the study of the solutions to a family of ordinary equations of the following form:

$$(3) \qquad v''(t) + \alpha(t)v(t) = g(t) \qquad (0 \leq t \leq T),$$

with $\alpha(t)$ and $g(t)$ integrable functions on $[0,T]$.

More precisely, we try to estimate all the solutions $v(t)$ of the equation (3) in terms of the initial values $v(0)$ and $v'(0)$, of the datum $g(t)$ and of the coefficient $\alpha(t)$.

The most direct estimate can be obtained by reducing equation (3) to a first order system and by applying Gronwall's lemma:

$$(4) \qquad |v(t)| \leq \left(|v(0)| + |v'(0)| + \int_0^T |g(s)| ds \right) \exp \left(\int_0^t (1 + |\alpha(s)|) ds \right).$$

In the case when $\alpha(t)$ is Lipschitz continuous and strictly positive, we can obtain a different estimate; indeed, it is sufficient to study the behaviour of the *energy* of the solution v, to obtain:

$$|v(t)| \leq \left(\sqrt{\alpha(0)}|v(0)| + |v'(0)| + \int_0^T |g(s)|ds \right) \frac{1}{\sqrt{\alpha(t)}} \exp \left(\frac{1}{2} \int_0^t \frac{|\alpha'(s)|}{\alpha(s)} ds \right).$$

In the following, we shall need a more general estimate than the preceding ones, which is valid without any assumption of positivity or of Lipschitz continuity of the coefficient $\alpha(t)$:

Lemma 1. – *Let us consider equation (3) with $\alpha(t)$ and $g(t)$ integrable functions on $[0, T]$ and let us assume that:*

$$\alpha(t) = \beta(t) + \gamma(t)$$

where $\beta(t)$ is a real valued function belonging to the space $H^{1,1}([0, T])$ and satisfying the inequality:

$$\beta(t) > 0, \qquad \forall t \in [0, T].$$

Let $v(t)$ be a solution of (3) in $H^{2,1}([0, T])$ and let

$$(5) \qquad\qquad E(t) = \beta(t)|v(t)|^2 + |v'(t)|^2$$

be the "β-energy" associated to v. Then we have:

$$(6) \quad \sqrt{E(t)} \leq \left(\sqrt{E(0)} + \int_0^t |g(s)|ds \right) \times$$

$$\times \exp \left(\frac{1}{2} \int_0^t \frac{|\beta'(s)|}{\beta(s)} ds + \frac{1}{2} \int_0^t \frac{|\gamma(s)|}{\sqrt{\beta(s)}} ds \right).$$

Proof. Since $v(t)$ is in $H^{2,1}([0, T])$, we see that $E(t)$ is in $H^{1,1}([0, T])$; by differentiating we obtain:

$$E'(t) = \beta'(t)|v(t)|^2 + 2\beta(t)\,\mathrm{Re}\,(\overline{v'(t)}v(t)) + 2\,\mathrm{Re}\,(\overline{v'(t)}v''(t)).$$

Replacing $v''(t)$ by $g(t) - \beta(t)v(t) - \gamma(t)v(t)$, we deduce the relation:

$$E'(t) = \beta'(t)|v(t)|^2 - 2\,\mathrm{Re}\,(\gamma(t)\overline{v'(t)}v(t)) + 2\,\mathrm{Re}\,(\overline{v'(t)}g(t)),$$

whence:

$$E' \leq \frac{|\beta'|}{\beta}\beta|v|^2 + \frac{|\gamma|}{\sqrt{\beta}}(\beta|v|^2 + |v'|^2) + 2|v'|\,|g|$$

and hence:

$$E' \leq \left(\frac{|\beta'|}{\beta} + \frac{|\gamma|}{\sqrt{\beta}} \right) + 2|g|\sqrt{E}.$$

By the Gronwall lemma, we get then (6).

REMARK. In the following we shall need to estimate the solution of a family of equations of the form (3) with coefficients $\alpha(t) \equiv \alpha(\xi, t)$ depending (at least in the one dimensional case $n = 1$) on a real parameter ξ. More precisely, we have:

$$\alpha(\xi, t) = a(t)\xi^2$$

where $a(t)$ is in general only integrable and such that $a(t) \geq \lambda_0 > 0$, and we want to estimate the growth of the solution $v(\xi, t)$ for $|\xi| \to \infty$.

Now, using (4) only we obtain:

$$|v(\xi, t)| \leq K \exp(c|\xi|^2)$$

but this estimate is too weak for our purposes.

On the other hand, since $a(t)$ is in $L^1([0, T])$, one can construct for all $\varepsilon > 0$ a Lipschitz continuous function $b_\varepsilon(t) \geq \lambda_0$ such that $\|a - b_\varepsilon\|_{L^1([0,T])} \leq \varepsilon$. We can then apply lemma 1 with $\beta(t) = b_\varepsilon \xi^2$ and we obtain the following estimate:

$$|v(\xi, t)| \leq K_\varepsilon \exp(\varepsilon|\xi|), \qquad \forall \xi > 0, \ \forall \varepsilon > 0.$$

3. - Periodic solutions in x

In this section, we shall solve problem $\{(1),(2)\}$ in the particularly simple case when the number n of space variables is equal to 1, the datum $f(x, t)$ is equal to zero and the initial data $\varphi(x)$ and $\psi(x)$ are 2π-periodic functions on the real line:

$$(7) \qquad \frac{\partial^2 u}{\partial t^2} - a(t)\frac{\partial^2 u}{\partial x^2} = 0 \qquad \text{in } \mathbf{R} \times [0, T],$$

$$(8) \qquad u(x, 0) = \varphi(x) \text{ and } \frac{\partial u}{\partial t}(x, 0) = \psi(x) \qquad \text{in } \mathbf{R}.$$

The coefficient $a(t)$ is an integrable function on $[0, T]$ such that

$$(9) \qquad a(t) \geq \lambda_0 > 0 \qquad \forall t \in [0, T].$$

Here we shall consider only the solutions $u(x, t)$ of this problem which are periodic in x, $\forall t \in [0, T]$. However, we shall see in the following sections that the problem has a unique solution.

THEOREM 2. – *Let us consider problem $\{(7),(8)\}$ with $a(t)$ integrable on $[0, T]$ and satisfying (9), and with data $\varphi(x)$ and $\psi(x)$ 2π-periodic on $\mathbf{R}$.*

i) If $\varphi(x)$ and $\psi(x)$ are analytic functions on $\mathbf{R}$, there exists one and only one solution $u(x, t)$ of class C^1 on $\mathbf{R} \times [0, T]$, 2π-periodic and analytic with respect to x for all $t \in [0, T]$.

ii) Assume that $a(t)$ is Hölder continuous of order α on $[0, T]$, where $0 < \alpha < 1$, and let $1 \leq s < 1/(1-\alpha)$. If $\varphi(x)$ and $\psi(x)$ are Gevrey functions on $\mathbf{R}$ of order s, there exists one and only one solution $u(x, t)$ of class C^1 on $\mathbf{R} \times [0, T]$, 2π-periodic and Gevrey of order s with respect to x for all $t \in [0, T]$.

iii) Assume that

$$|a(t+\tau) - a(t)| \le L\tau(|\log \tau| + 1), \qquad \forall \tau > 0.$$

If $\varphi(x)$ and $\psi(x)$ are infinitely differentiable functions on $\mathbf{R}$, there exists one and only one solution $u(x,t)$ of class C^1 on $\mathbf{R} \times [0,T]$, 2π-periodic and infinitely differentiable in x for all $t \in [0,T]$.

PROOF. By expanding the initial data φ and ψ in Fourier series one can write

$$\varphi(x) = \sum_{h=-\infty}^{+\infty} A_h \exp(ihx) \qquad \text{and} \qquad \psi(x) = \sum_{h=-\infty}^{+\infty} B_h \exp(ihx).$$

We look for solutions of $\{(7),(8)\}$ of the form

$$u(x,t) = \sum_{h=-\infty}^{+\infty} v_h(t) \exp(ihx).$$

Thus the function $v_h(t)$ must solve the problem

$$(10) \qquad \begin{cases} v_h''(t) + h^2 a(t) v_h(t) = 0 & \text{in } [0,T] \\[2mm] v_h(0) = A_h \text{ and } v_h'(0) = B_h. \end{cases}$$

It is well known that this problem admits one and only one solution of class C^1 on $[0,T]$. In particular, we have the uniqueness of 2π-periodic solutions in x for problem $\{(7),(8)\}$. Moreover, if $\varphi(x)$ and $\psi(x)$ are real valued functions ($\overline{A_h} = A_{-h}$ and $\overline{B_h} = B_{-h}$) we see that the solution is real valued.

To prove the existence of a solution to $\{(7),(8)\}$, we need to estimate the growth of $v_h(t)$ as $|h| \to \infty$. To this end, we apply Lemma 1 to the problems (10) with a suitable decomposition (depending on h) of the functions $a(t)$:

$$a(t) = b_h(t) + c_h(t), \qquad \text{where } b_h(t) > 0, \ h \ne 0.$$

Lemma 1 gives then

$$(11) \qquad \sqrt{E(t)} \le \sqrt{E(0)} \cdot \exp\left(\frac{1}{2} \int_0^T \frac{|b_h'(s)|}{|b_h(s)|} ds + \frac{|h|}{2} \int_0^T \frac{|c_h(s)|}{\sqrt{b_h(s)}} ds \right)$$

where

$$(12) \qquad E_h(t) = h^2 b_h(t) |v_h(t)|^2 + |v_h'(t)|^2.$$

Now, we can choose $b_h(t)$ as follows

$$b_h(t) = (\widetilde{a} * \rho_h)(t)$$

where $\widetilde{a}$ is an extension (to be precised in the following) of $a(t)$ on $\mathbf{R}$, while $\rho_h(t)$ is an infinitely differentiable function on $\mathbf{R}$ such that $0 \leq \rho_h(t) \leq c_0|h|$, $\rho_h(t) \equiv 0$ outside the interval $[-1/|h|, 0]$, and

$$\int_{-\infty}^{+\infty} \rho_h(t)dt = 1, \qquad \int_{-\infty}^{+\infty} |\rho_h'(t)|dt \leq c_0|h|.$$

We have then, for $|h| \geq 1/T$,

$$(13) \qquad b_h(t) \geq \lambda_0, \qquad |b_h(0)| \leq c_0|h| \int_0^T |a(s)|ds.$$

Moreover

$$b_h'(s) = \int_{-\infty}^{+\infty} (\widetilde{a}(s-\tau) - \widetilde{a}(s))\rho_h'(\tau)d\tau, \qquad c_h(s) = \int_{-\infty}^{+\infty} (\widetilde{a}(s) - \widetilde{a}(s-\tau))\rho_h(\tau)d\tau.$$

Hence

$$(14) \qquad \int_0^T (|b_h'(s)| + |h||c_h(s)|)ds \leq c_0|h| \sup_{|\tau| \leq 1/|h|} \int_{-\infty}^{+\infty} |\widetilde{a}(s-\tau) - \widetilde{a}(s)|ds.$$

Introducing (12), (13) and (14) in (11), we obtain, for $|h| \geq 1/T$,

$$|v_h(t)| + |v_h'(t)| \leq C(\lambda_0) \left(|h||A_h| \int_0^T |a|ds + |B_h| \right) \times$$

$$(15) \qquad \times \exp \left(c_0|h| \sup_{|\tau| \leq 1/|h|} \int_{-\infty}^{+\infty} |\widetilde{a}(s-\tau) - \widetilde{a}(s)|ds \right).$$

Let us now consider case (i).

Since $a(t)$ is an integrable function on $[0, T]$, if we define the extension $\widetilde{a}$ of a by choosing $\widetilde{a} \equiv \lambda_0$ outside the interval $[0, T]$, we see that

$$\sup_{|\tau| \leq 1/|h|} \int_{-\infty}^{+\infty} |\widetilde{a}(s-\tau) - \widetilde{a}(s)|ds \to 0, \qquad \text{as } |h| \to \infty.$$

On the other hand $\varphi(x)$ and $\psi(x)$ are analytic, hence (see th. 12 in the Appendix)

$$|A_h| + |B_h| \leq M \exp(-\delta|h|), \qquad \text{with } \delta > 0.$$

As a consequence (15) gives

$$|v_h(t)| + |v_h'(t)| \leq \widetilde{M} \exp(-\delta|h|/2), \qquad \text{for } |h| \text{ large enough,}$$

whence it follows that $u(x,t)$ is a C^1 function on $\mathbf{R} \times [0, T]$ and (th. 12, part b) analytic in x for all $t \in [0, T]$.

Let us consider case (ii).

Since $a(t)$ is a Hölder continuous function on $[0, T]$ of order α one can extend it to a Hölder continuous function of the same order α on $\mathbf{R}$ by choosing $\tilde{a}(t) \equiv a(0)$ if $t < 0$ and $\tilde{a}(t) \equiv a(T)$ if $t > T$. We have then

$$\sup_{|\tau| \leq 1/|h|} \int_{-\infty}^{+\infty} |\tilde{a}(s - \tau) - \tilde{a}(s)| ds \leq L|h|^{-\alpha}.$$

On the other hand, $\varphi(x)$ and $\psi(x)$ are Gevrey functions of order s and hence (th. 12)

$$|A_h| + |B_h| \leq M \exp(-\delta|h|^{1/s}), \qquad \text{with } \delta > 0.$$

As a consequence, if $1/s$ is larger than $1 - \alpha$, we have by (15)

$$|v_h(t)| + |v_h'(t)| \leq \widetilde{M} \exp\left[-\frac{\delta}{2}|h|^{1/s}\right]$$

for $|h|$ large enough, whence (th. 12, part b) the thesis.

Finally, let us consider case (iii).

With the same definition of $\tilde{a}(t)$ as above, we have

$$\sup_{|\tau| \leq 1/|h|} \int_{-\infty}^{+\infty} |\tilde{a}(s - \tau) - \tilde{a}(s)| ds \leq L|h|^{-1}(\log|h| + 1).$$

Since $\varphi(x)$ and $\psi(x)$ are infinitely differentiable, we have (th. 12)

$$|A_h| + |B_h| \leq M(p)|h|^{-p}, \qquad \forall p > 0,$$

and hence, for $|h|$ large enough,

$$|v_h(t)| + |v_h'(t)| \leq \widetilde{M}(p)|h|^{-p+L+1}, \qquad \forall p > 0,$$

whence the thesis.

We shall show later (§7) that the assumptions of th. 1 cannot be weakened. In particular it may occur that problem $\{(7),(8)\}$ has no function (or functional) solution even if the initial data φ and ψ are infinitely differentiable functions and the coefficient $a(t)$ is Hölder continuous.

4. - Solutions with compact support

We turn now to the study of problem $\{(1),(2)\}$ in the general case ($n \geq 1, \varphi, \psi$ and f not necessarily periodic). Of course, regarding the coefficients $a_{ij}(t)$, we shall always make the assumption that they are real valued functions, integrable on $[0, T]$, and such that

$$(16) \qquad \begin{cases} a_{ij}(t) = a_{ji}(t) \\ \sum a_{ij}(t)\xi_i\xi_j \geq \lambda_0|\xi|^2, \qquad \forall \xi \in \mathbf{R}^n, \end{cases}$$

with $\lambda_0 > 0$.

In this section, we begin by considering the case when the data $\varphi(x)$, $\psi(x)$ and $f(x,t)$ are functions or functionals with compact support in x on $\mathbf{R}^n$ and when the solution $u(x,t)$ of the problem has also compact support in x.

Instead of the expansion in Fourier series we can now use the Fourier transform with respect to x. Problem $\{(1),(2)\}$ is then transformed in the family, depending of the parameter $\zeta \in \mathbf{C}^n$, of Cauchy problems

$$
(17) \qquad
\begin{cases}
v''(t) + \left(\sum a_{ij}(t)\zeta_i\zeta_j\right) v(t) = \widehat{f}(\zeta,t) & \text{in } [0,T] \\[2ex]
v(0) = \widehat{\varphi}(\zeta) \text{ and } v'(0) = \widehat{\psi}(\zeta)
\end{cases}
$$

where $v(t) = \widehat{u}(\zeta,t)$.

In view of the Paley-Wiener theorem, we have to estimate the growth of the solution v to problem (17), as $|\zeta| \to \infty$.

This will be done in the following Lemma, where (by applying Lemma 1 to problem (17)) we shall obtain estimates (19) and (22). In these estimates the dependence on the matrix of the coefficients

$$
a(t) = [a_{ij}(t)]_{i,j=1,\ldots n}
$$

is of a special interest.

This dependence will be expressed through the functional

$$
(18) \qquad \omega(a,\delta) = \sup_{0\leq\tau\leq\delta} \int_0^{T-\tau} |a(t+\tau) - a(t)|\,dt, \qquad 0 < \delta < T,
$$

where $|b|$ denotes the norm of a matrix b.

This functional represents the "modulus of continuity" for $\tau \to 0^+$ of the map $\tau \to a(t+\tau)$ with values in the space $\left[L^1([0,T])\right]^{n^2}$.

We know that, when $\delta \to 0$, $\omega(a,\delta) \to 0$ uniformly for $a \in \mathscr{K}$, for any compact subset $\mathscr{K}$ of $\left[L^1([0,T])\right]^{n^2}$.

LEMMA 2. – *Let u be the solution in $H^{2,1}([0,T],\mathscr{H}')$ (see th. 1) of problem $\{(1),(2)\}$, with coefficients a_{ij} real valued and integrable on $[0,T]$ and data φ, ψ in $\mathscr{H}'$ and f in $L^1([0,T],\mathscr{H}')$.*

Let $\omega(a,\delta)$ be defined by (18) (where $a(t) \equiv [a_{ij}(t)]$) and let $c_0,c_1,\ldots,$ be constants depending on λ_0 and T.

If a_{ij} satisfy (16), we have, for $|\zeta| \geq 1$ and $t \in [0,T]$,

$$
(19) \quad |\zeta||\widehat{u}(\zeta,t)| + \left|\frac{\partial\widehat{u}}{\partial t}(\zeta,t)\right| \leq M \exp\left(c_1\omega\left(a,\frac{T}{2|\zeta|}\right)|\zeta| + \int_0^t |a(s)|^{1/2}ds\,|\eta|\right)
$$

where

$$
(20) \quad M = \exp\left(c_0 \int_0^T |a(s)|\,ds\right) \times
$$

$$
\times \left[\left(|\zeta| \int_0^{T/2|\zeta|} |a(s)|\,ds\right)^{1/2} |\zeta||\widehat{\varphi}(\zeta)| + |\widehat{\psi}(\zeta)| + \int_0^T |\widehat{f}(\zeta,s)|\,ds\right].
$$

Notice that in the particular case when a_{ij} are bounded, i.e.

$$(21) \qquad \lambda_0|\xi|^2 \leq \sum_{i,j}^{1,n} a_{ij}(t)\xi_i\xi_j \leq \Lambda_0|\xi|^2 \qquad \forall \xi \in \mathbf{R}^n,$$

(19) can be improved to

$$(22) \qquad |\zeta||\widehat{u}(\zeta,t)| + \left|\frac{\partial\widehat{u}}{\partial t}(\zeta,t)\right| \leq M_1 \exp\left(c_1\omega\left(a,\frac{T}{2|\zeta|}\right)|\zeta| + t\sqrt{\Lambda_0}|\eta|\right)$$

where

$$M_1 = \exp\left(c_2\Lambda_0\right)\left[\sqrt{|\Lambda_0|}|\zeta||\widehat{\varphi}(\zeta)| + |\widehat{\psi}(\zeta)| + \int_0^T |\widehat{f}(\zeta,s)|ds\right].$$

Proof. Let

$$m_a = \frac{1}{T}\int_0^T a(t)dt$$

the average on $[0,T]$ of the matrix $a(t) \equiv [a_{ij}(t)]_{i,j=1,\ldots,n}$ of the coefficients. Let us introduce the Lipschitz continuous functions

$$\rho_\zeta(t) = \begin{cases} 2(\delta - |2t - \delta|)\delta^{-2} & \text{si } 0 \leq t \leq \delta, \\ 0 & \text{si } t > \delta, \end{cases}$$

where

$$\delta = T(2|\zeta|)^{-1}$$

and $\zeta = \xi + i\eta$ is an element of $\mathbf{C}^n$ such that $|\zeta| \neq 0$.
We have then

$$(23) \qquad |\rho_\zeta(t)| \leq \frac{4}{T}|\zeta|,$$

$$(24) \qquad \int_0^{+\infty} \rho_\zeta(t)dt = 1, \qquad \int_0^{+\infty} |\rho_\zeta'(t)|dt = \frac{8}{T}|\zeta|, \qquad \int_0^{+\infty} \rho_\zeta'(t)dt = 0.$$

In view of Lemma 1, we decompose $a(t)$ in the following way

$$a(t) = b(\zeta,t) + c(\zeta,t)$$

where

$$b(\zeta,t) = \int_0^{+\infty} \widetilde{a}(t+\tau)\rho_\zeta(\tau)d\tau$$

and

$$\widetilde{a}(t) = \begin{cases} a(t) & \text{if } 0 \leq t \leq T, \\ m_a & \text{if } t > T. \end{cases}$$

We see then that the matrix $b(\zeta, t)$ is Lipschitz continuous with respect to t and that, if the $a_{ij}(t)$ satisfy (16),

$$(25) \qquad (b(\zeta, t)\mu, \mu) \geq \lambda_0 |\mu|^2, \qquad \forall \mu \in \mathbf{R}^n.$$

Then we may apply Lemma 1 to problem (17), where $v(t) = \widehat{u}(\zeta, t)$ and

$$\alpha(t) = (a(t)\zeta, \zeta) \equiv \sum_{i,j} a_{ij}(t)\zeta_i\zeta_j$$

$$\beta(t) = (b(\zeta, t)\xi, \xi) + (b(\zeta, t)\eta, \eta)$$

$$\gamma(t) = \alpha(t) - \beta(t).$$

The thesis of the Lemma will then be a consequence of (6) and of a suitable estimate of the function

$$\frac{1}{2}\left(\frac{|\beta'(t)|}{\beta(t)} + \frac{|\gamma(t)|}{\sqrt{\beta(t)}}\right).$$

In order to estimate this function, first of all we notice that $\gamma(t)$ can be written in the form

$$\gamma(t) = 2i(b(\zeta, t)\xi, \eta) - 2(b(\zeta, t)\eta, \eta) - (c(\zeta, t)\zeta, \zeta)$$

and that any symmetric and nonnegative matrix b satisfies the inequality

$$|(b\eta, \eta) - i(b\xi, \eta)|^2 \leq |b||\eta|^2[(b\xi, \xi) + (b\eta, \eta)].$$

We have thus

$$|\gamma(t)| \leq 2|b(\zeta, t)|^{1/2}|\eta|\sqrt{\beta(t)} + |(c(\zeta, t)\zeta, \zeta)|.$$

On the other hand we have

$$\beta'(t) = (b'(\zeta, t)\xi, \xi) + (b'(\zeta, t)\eta, \eta)$$

where $b'(\zeta, t)$ denotes the derivative of $b(\zeta, t)$ with respect to t, and we know, by (25), that $\beta(t) \geq \lambda_0 |\zeta|^2$.

In conclusion, we obtain the following inequality:

$$(26) \qquad \frac{1}{2}\left(\frac{|\beta'(t)|}{\beta(t)} + \frac{|\gamma(t)|}{\sqrt{\beta(t)}}\right) \leq \frac{1}{2\lambda_0}|b'(\zeta, t)| + |b(\zeta, t)|^{1/2}|\eta| + \frac{1}{2\sqrt{\lambda_0}}|c(\zeta, t)||\zeta|.$$

In view of (6), we need to estimate the integral in t of the second term of (26).

Now, from the definition of $b(\zeta, t)$ and of $c(\zeta, t)$ we deduce, recalling (24),

$$(27) \qquad b'(\zeta, t) = -\int_0^{+\infty} (\widetilde{a}(t) - \widetilde{a}(t + \tau))\rho'_\zeta(\tau)d\tau \qquad (0 \leq t \leq T)$$

$$(28) \qquad c'(\zeta, t) = \int_0^{+\infty} (\tilde{a}(t) - \tilde{a}(t + \tau))\rho_\zeta(\tau)d\tau \qquad (0 \leq t \leq T)$$

while by $|b| \leq |a| + |c|$ we get

$$(29) \qquad |b(\zeta, t)|^{1/2} \leq |a(t)|^{1/2} + |c(\zeta, t)||b(\zeta, t)|^{-1/2}.$$

But

$$\int_0^T |\tilde{a}(t + \tau) - \tilde{a}(t)|dt = \int_0^{T-\tau} |a(t + \tau) - a(t)|dt + \int_{T-\tau}^T |a(t) - m_a|dt$$

and hence (see the Appendix, part D) if $0 \leq \delta \leq T/2$

$$\sup_{0 \leq \tau \leq \delta} \int_0^T |\tilde{a}(t + \tau) - \tilde{a}(t)|dt \leq 2\omega(a, \delta) + 2\frac{\delta}{T} \int_0^T |a(t)|dt.$$

Using this inequality with $\delta = T(2|\zeta|)^{-1}$, the theorem of Fubini-Tonelli and (24), formulas (27) and (28) give (for $|\zeta| \geq 1$)

$$(30) \qquad \int_0^T |b'(\zeta, t)|dt \leq \frac{8}{T}|\zeta| \left(2\omega(a, T(2|\zeta|)^{-1}) + |\zeta|^{-1} \int_0^T |a(t)|dt \right),$$

$$(31) \qquad \int_0^T |c(\zeta, t)|dt \leq 2\omega(a, T(2|\zeta|)^{-1}) + |\zeta|^{-1} \int_0^T |a(t)|dt,$$

while (29) gives, since $|b| \geq \lambda_0$,

$$(32) \int_0^t |b(\zeta, s)|^{1/2}ds \leq$$

$$\leq \int_0^t |a(s)|^{1/2}ds + \frac{1}{\sqrt{\lambda_0}} \left(2\omega(a, T(2|\zeta|)^{-1}) + |\zeta|^{-1} \int_0^T |a(s)|ds \right).$$

We introduce the three estimates obtained above in (26) and we apply Lemma 1. Then (6) gives, for $|\zeta| \geq 1$,

$$\sqrt{E(t)} \leq \left(\sqrt{E(0)} + \int_0^T |\hat{f}(\zeta, s)|ds \right) \exp \left(c_0 \int_0^T |a(s)|ds \right) \times$$

$$\times \exp \left[c_1\omega(a, T(2|\zeta|)^{-1})|\zeta| + \int_0^t |a(s)|^{1/2}ds|\eta| \right]$$

where

$$E(t) = [(b(\zeta, t)\xi, \xi) + (b(\zeta, t)\eta, \eta)]|\hat{u}(\zeta, t)|^2 + \left| \frac{\partial \hat{u}}{\partial t}(\zeta, t) \right|^2.$$

To obtain (19), and thus to conclude the proof of the Lemma, it is sufficient to observe that

$$(33) \qquad \sqrt{E(t)} \geq \sqrt{\lambda_0}|\zeta||\widehat{u}(\zeta,t)| + \left|\frac{\partial \widehat{u}}{\partial t}(\zeta,t)\right|$$

and

$$\sqrt{E(0)} \leq |b(\zeta,0)|^{1/2}|\zeta||\widehat{\varphi}(\zeta)| + |\widehat{\psi}(\zeta)|.$$

Now we have

$$b(\zeta,0) = \int_0^{T/2|\zeta|} a(t)\rho_\zeta(t)dt$$

whence (recalling (23))

$$|b(\zeta,0)| \leq \frac{4}{T}|\zeta| \int_0^{T/2|\zeta|} |a(t)|dt$$

and hence

$$(34) \qquad \sqrt{E(0)} \leq \frac{2}{\sqrt{T}} \left(|\zeta| \int_0^{T/2|\zeta|} |a(t)|dt\right)^{1/2} |\zeta||\widehat{\varphi}(\zeta)| + |\widehat{\psi}(\zeta)|.$$

Keeping into account (33) and (34) we obtain then (19).

We can now prove the main result of this section.

THEOREM 3. – *Let us consider problem $\{(1),(2)\}$, with coefficients $a_{ij}(t)$ real valued and integrable on $[0,T]$, satisfying assumption (16).*

a) Assume that the data φ and ψ belong to the space $\mathscr{A}'$ of real analytic functionals and that f belongs to $L^1([0,T],\mathscr{A}')$.

Then the solution (see th. 1) u of the problem belongs to $H^{2,1}([0,T],\mathscr{A}')$.

Moreover, if φ and ψ are in $\mathscr{A}'(B(\rho))$, where $B(\rho)$ denotes the open ball of $\mathbf{R}^n$ of center 0 and radius ρ, and f is in $L^1([0,T],\mathscr{A}'(B(\rho)))$, then, $\forall \tau \in]0,T]$, u belongs to $H^{2,1}([0,\tau],\mathscr{A}'(B(\rho_\tau)))$, where:

$$(35) \qquad \rho_\tau = \rho + \int_0^\tau |a(t)|^{1/2}dt.$$

In particular, $\forall t \in [0,T]$, $u(t)$ is supported in $B(\rho_t)$, which can also be expressed by saying that u is supported in the cone Γ of $\mathbf{R}^n \times [0,T]$, where

$$\Gamma = \{(x,t) \ : \ |x| < \rho_t\}.$$

b) If the coefficients $a_{ij}(t)$ satisfy additional regularity assumptions, we have the following further results.

i) Assume there exist α $(0 < \alpha < 1)$ and $A > 0$ such that:

$$(36) \qquad \int_0^{T-\tau} |a(t+\tau) - a(t)|dt \leq A\tau^\alpha \qquad (0 \leq \tau \leq T/2)$$

and let

$$1 \le s < \frac{1}{1-\alpha}.$$

We have then:

If φ and ψ are in $\mathscr{E}'_s$ and f in $L^1([0,T],\mathscr{E}'_s)$, the solution u belongs to $H^{2,1}([0,T],\mathscr{E}'_s)$;

If φ and ψ are in $\mathscr{D}_s$ and f in $L^1([0,T],\mathscr{D}_s)$, the solution u belongs to $H^{2,1}([0,T],\mathscr{D}_s)$.

ii) Assume there exists $A > 0$ such that:

$$(37) \qquad \int_0^{T-\tau} |a(t+\tau) - a(t)|\,dt \le A\tau(|\log\tau| + 1) \qquad (0 < \tau \le T/2).$$

We have then:

If φ and ψ are in $\mathscr{E}'$ and f in $L^1([0,T],\mathscr{E}')$, the solution u belongs to $H^{2,1}([0,T],\mathscr{E}')$;

If φ and ψ are in $\mathscr{D}$ and f in $L^1([0,T],\mathscr{D})$, the solution u belongs to $H^{2,1}([0,T],\mathscr{D})$.

iii) Assume there exists $A > 0$ such that:

$$(38) \qquad \int_0^{T-\tau} |a(t+\tau) - a(t)|\,dt \le A\tau \qquad (0 \le \tau \le T/2),$$

that is to say (see [8]) that a_{ij} coincide almost everywhere with functions of bounded variation on $[0,T]$.

Then, if φ is in H_c^{s+1}, ψ in H_c^s and f in $L^1([0,T], H_c^s)$ for some real s, the solution u belongs to $L^\infty([0,T], H_c^{s+1})$ while $\partial u/\partial t$ belongs to $L^\infty([0,T], H_c^s)$ and $\partial^2 u/\partial^2 t$ is in $L^1([0,T], H_c^{s-1})$.

Proof.

a) By Lemma 2 we have, for $|\zeta| \ge 1$,

$$(39) \quad |\zeta||\widehat{u}(\zeta,t)| + \left|\frac{\partial\widehat{u}}{\partial t}(\zeta,t)\right| \le M(a,\zeta)\exp\left(\int_0^t |a(s)|^{1/2}ds|\eta|\right)\exp(\varepsilon_0(a,\zeta)|\zeta|)$$

where $M(a,\zeta)$ is defined by (20) and where we have written:

$$(40) \qquad \varepsilon_0(a,\zeta) = c_1\omega(a, T(2|\zeta|)^{-1}).$$

We recall that $c_0, c_1, c_2 \ldots$ denote several constants depending on λ_0, T and that the function $\omega(a,\delta)$ is defined by (18).

Assume now that φ, ψ are in $\mathscr{A}'$ and f in $L^1([0,T],\mathscr{A}')$. By the Paley-Wiener theorem (part I-(i) and part III-(i)) there exists a number $\rho > 0$ such that, $\forall\varepsilon > 0$, $\exists C_\varepsilon > 0$ such that, $\forall\zeta \in \mathbf{C}^n$ $(|\zeta| \ge 1)$:

$$(41) \qquad \max\left\{|\widehat{\varphi}(\zeta)|, |\widehat{\psi}(\zeta)|, \int_0^T |\widehat{f}(\zeta,t)|dt\right\} \le C_\varepsilon\exp(\varepsilon|\zeta| + \rho|\eta|).$$

Introducing this inequality in (20), (39) gives:

$$(42) \quad |\zeta||\widehat{u}(\zeta,t)| + \left|\frac{\partial\widehat{u}}{\partial t}(\zeta,t)\right| \leq$$

$$\leq C_\varepsilon \exp\left(c_0 \int_0^T |a(t)|dt\right) \exp\left(\int_0^t |a(s)|^{1/2}ds|\eta|\right) \times$$

$$\times \left[2 + |\zeta|^{\frac{3}{2}}\left(\int_0^T |a(t)|dt\right)^{\frac{1}{2}}\right] \exp[(\varepsilon_0(a,\zeta) + \varepsilon)|\zeta| + \rho|\eta|]$$

for all ζ such that $|\zeta| \geq 1$ and all $\varepsilon > 0$.

We define now:

$$(43) \quad L_\varepsilon(a) = \exp\left(c_0 \int_0^T |a(t)|dt\right) \times$$

$$\times \sup_{|\zeta|\geq 1}\left\{\left[2 + |\zeta|^{\frac{3}{2}}\left(\int_0^T |a(t)|dt\right)^{\frac{1}{2}}\right] \exp[(\varepsilon_0(a,\zeta) - \varepsilon)|\zeta|]\right\}.$$

Since the $a_{ij}(t)$ are integrable functions on $[0,T]$, we know that $\omega(a,\delta) \to 0$ if $\delta \to 0^+$ and hence, (recall the definition (40) of $\varepsilon_0(a,\zeta)$):

$$\varepsilon_0(a,\zeta) \to 0 \qquad \text{if } |\zeta| \to \infty,$$

which implies $L_\varepsilon(a) < \infty$.

Now, inequality (42) implies:

$$(44) \quad |\widehat{u}(\zeta,t)| + \left|\frac{\partial\widehat{u}}{\partial t}(\zeta,t)\right| \leq C_\varepsilon L_\varepsilon(a) \exp\left(2\varepsilon|\zeta| + \left(\int_0^t |a(s)|^{1/2}ds + \rho\right)|\eta|\right)$$

for all $\varepsilon > 0$ and $|\zeta| \geq 1$.

Inequality (44), taking into account the the Paley-Wiener theorem (part II), ensures that the holomorphic functionals $u(t)$ and $(\partial u/\partial t)(t)$ belong to $\mathscr{A}'$, for all t in $[0,T]$, and, moreover, that u and $\partial u/\partial t$ belong to $C([0,T],\mathscr{A}')$.

By equation (1) we finally obtain that $\partial^2 u/\partial t^2$ belongs to $L^1([0,T],\mathscr{A}')$.

Assume now that φ and ψ are in $\mathscr{A}'(B(\rho))$ and f is in $L^1([0,T],\mathscr{A}'(B(\rho)))$. Then, by the Paley-Wiener theorem (part I-(i) and part III-(ii)) we have that $\widehat{\varphi}$, $\widehat{\psi}$ and $\widehat{f}$ satisfy (41) and hence $\widehat{u}$ and $\partial\widehat{u}/\partial t$ satisfy (44). As a consequence (Paley-Wiener theorem, part II) u and $\partial u/\partial t$ belong to $C([0,T],\mathscr{A}(B(\rho_\tau)))$ where ρ_τ is defined by (35), $\forall\tau$ in $]0,T]$. By equation (1) we conclude that u is in $H^{2,1}([0,T],\mathscr{A}'(B(\rho_\tau)))$.

b) In view of the Paley-Wiener theorem (part I-(ii),..., I-(vi)) it is sufficient to consider (39) for $\eta = 0$, hence for $\zeta \equiv \xi$. Then (39) becomes

$$(45) \quad |\xi||\widehat{u}(\xi,t)| + \left|\frac{\partial\widehat{u}}{\partial t}(\xi,t)\right| \leq M(a,\xi)\exp(\varepsilon_0(a,\xi)|\xi|), \qquad |\xi| \geq 1,$$

where $M(a,\xi)$ and $\varepsilon_0(a,\xi)$ are defined by (20) and (40) respectively.

Let us now consider case (i).

If the matrix $a(t)$ of the coefficients satisfies assumption (36), then the function $\omega(a,\delta)$ defined by (18) satisfies

$$\omega(a,\delta) \leq A\delta^\alpha$$

and hence

$$(46) \qquad \varepsilon_0(a,\xi) \leq c_1 (T/2)^\alpha A |\xi|^{-\alpha}.$$

Now, if the data φ and ψ are in $\mathscr{E}'_s$ and f is in $L^1([0,T],\mathscr{E}'_s)$, for s real ≥ 1, we have (Paley-Wiener theorem, parts I-(ii) and III-(iii)) that $\forall \varepsilon > 0 \; \exists C'_\varepsilon > 0$ such that, $\forall \xi \in \mathbf{R}^n$,

$$(47) \qquad \max\left\{ |\widehat{\varphi}(\xi)|, |\widehat{\psi}(\xi)|, \int_0^t |\widehat{f}(\xi,t)|dt \right\} \leq C'_\varepsilon \exp(\varepsilon|\xi|^{1/s}) \qquad (|\xi| \geq 1),$$

whence, as for part (a) proved above, we obtain

$$(48)\; |\xi||\widehat{u}(\xi,t)| + \left|\frac{\partial\widehat{u}}{\partial t}(\xi,t)\right| \leq C'_\varepsilon \exp\left(c_0 \int_0^T |a(t)|dt \right) \times$$

$$\times \left[2 + |\xi|^{3/2} \left(\int_0^T |a(t)|dt \right)^{1/2} \right] \exp[\varepsilon_0(a,\xi) + \varepsilon|\xi|^{1/s}]$$

for $|\xi| \geq 1$.

Then, from the inequalities (46) and (48), we obtain

$$|\widehat{u}(\xi,t)| + \left|\frac{\partial\widehat{u}}{\partial t}(\xi,t)\right| \leq C'_\varepsilon L'_\varepsilon(a) \exp(2\varepsilon|\xi|^{1/s}) \qquad (|\xi| \geq 1)$$

where

$$L'_\varepsilon(a) = \exp\left(c_0 \int_0^T |a(t)|dt \right) \times$$

$$\times \sup_{|\xi|\geq 1} \left\{ \left[2 + |\xi|^{\frac{3}{2}} \left(\int_0^T |a(t)|dt \right)^{\frac{1}{2}} \right] \exp[c_1 (T/2)^\alpha A |\xi|^{1-\alpha} - \varepsilon|\xi|^{1/s}] \right\}.$$

Now, for $1 \leq s < 1/(1-\alpha)$ we have $L'_\varepsilon(a) < +\infty$.

This completes the proof of part (i) when the data φ and ψ are in $\mathscr{E}'_s$ and f is in $L^1([0,T],\mathscr{E}'_s)$.

The proof of the remaining cases of part (b) of the theorem are completely analogous to the case already proved.

We will only notice that, when the coefficients $a_{ij}(t)$ are bounded on $[0,T]$ (which is the case if $a(t)$ satisfies assumption (38)), it is convenient to use inequality (22) instead of (19). We finally remark that, if $a(t)$ satisfies (38) and

(16), then it satisfies (21) with $\Lambda_0 = A + T^{-1} \int_0^T |a(t)|dt$. Hence, in case (iii), (22) gives an estimate of the solution u with constants depending only on the initial data and on λ_0, A and $\int_0^T |a(t)|dt$.

REMARK 1. Assume that the data φ and ψ of problem $\{(1),(2)\}$ belong to $\mathscr{E}_s'$ and that f belongs to $C([0,T],\mathscr{E}_s')$, for some s strictly larger that 1, and assume that the coefficients $a_{ij}(t)$ satisfy assumption (36) with $\alpha > 1 - 1/s$.

Then, if φ, ψ and f vanish for $|x| > \rho$ the solution u also vanishes outside the cone $\Gamma = \{|x| < \rho_\tau\}$, where ρ_τ is defined in (35).

PROOF. Since $f : [0,T] \to \mathscr{E}_s'$ is continuous and $f(t)$ belongs to $\mathscr{E}_s'(B(\delta))$, $\forall t \in [0,T]$, $\forall \delta > \rho$, it follows that f is in $C([0,T],\mathscr{E}_s'(B(\sigma))$ and hence in $L^1([0,T],\mathscr{A}'(B(\sigma)) \; \forall \sigma > \rho$. We can then apply part (a) of Theorem 3.

REMARK 2. Assumption (36) is satisfied in particular if the coefficients $a_{ij}(t)$ are Hölder continuous on $[0,T]$ of order α or, more generally, if

$$\text{(49)} \qquad\qquad a_{ij}(t) = a_{ij}^{(1)}(t) + a_{ij}^{(2)}(t)$$

with $a_{ij}^{(1)}(t)$ Hölder continuous of order α and $a_{ij}^{(2)}(t)$ of bounded variation on $[0,T]$.

However there exist functions (e.g. $t^{\alpha-1}$ with $0 < \alpha < 1$) which satisfy (36) but are not bounded, hence they are not of the type (49).

REMARK 3. Assumption (38) is satisfied in particular when the coefficients $a_{ij}(t)$ are Lipschitz continuous on $[0,T]$.

REMARK 4. If the coefficients $a_{ij}(t)$ satisfy the following assumption (stronger than (36))

$$\text{(50)} \qquad \lim_{\tau \to 0^+} \tau^{-\alpha} \int_0^{T-\tau} |a(t+\tau) - a(t)|dt = 0 \qquad (0 < \alpha < 1),$$

then the conclusion of theorem 3 (part b, i)) is valid also for $s = 1/(1 - \alpha)$.

Notice that condition (50) for $\alpha = 0$ is satisfied by all integrable functions on $[0,T]$, and this proves again part a) of the theorem.

PROOF. It is sufficient to observe that, under assumption (50), the function $\varepsilon_0(a,\xi)$ (see (40)) appearing in inequality (45) is such that $\varepsilon_0(a,\xi)|\xi|^\alpha \to 0$ if $|\xi| \to \infty$.

As a consequence, the function $\exp(\varepsilon_0(a,\xi)|\xi| - \varepsilon|\xi|^{1/s})$ is bounded on $\mathbf{R}^n$, $\forall \varepsilon > 0$, not only for $s < 1/(1 - \alpha)$ but also for $s = 1/(1 - \alpha)$.

REMARK 5. Assume we are in the case b, ii) of theorem 3. The constant A appearing in assumption (37) is proportional to the loss of regularity of the solution u with respect to the regularity of the data φ, ψ and f.

More precisely, if φ belongs to $H_c^{s+1+\varepsilon}$, for some real s and $\varepsilon > 0$ and if ψ belongs to H_c^s and f to $L^1([0,T],H_c^s)$, then the solution u of the problem belongs to $L^\infty([0,T],H_c^{s+1-\bar{c}A})$ with $\bar{c} = \bar{c}(\lambda_0,T)$ while $\partial u/\partial t$ belongs to $L^\infty([0,T],H_c^{s-\bar{c}A})$ and $\partial^2 u/\partial t^2$ belongs to $L^1([0,T],H_c^{s-1-\bar{c}A})$.

If in addition, besides satisfying (37), the coefficients $a_{ij}(t)$ are bounded on $[0, T]$, then the preceding result holds also for $\varepsilon = 0$.

PROOF. Under assumption (37), inequality (19), with $\eta = 0$, $\zeta = \xi$ and $|\xi| = 1$, gives

$$(51) \qquad |\xi||\widehat{u}(\xi, t)| + \left|\frac{\partial \widehat{u}}{\partial t}(\xi, t)\right| \leq M(a, \xi)\overline{c}^A |\xi|^{\overline{c}A}$$

where $\overline{c} = \overline{c}(\lambda_0, T)$ and $M(a, \xi)$ is defined by (20).

In order to obtain a precise estimate of $M(a, \xi)$, it is convenient to notice (see the Appendix, part D) that

$$|\xi| \int_0^{T/2|\xi|} |a(t)| dt \leq \int_0^T |a(t)| dt + |\xi|\omega(a, T(2|\xi|)^{-1})$$

if $|\xi| \geq 1$. As a consequence we have, by (37),

$$(52) \qquad M(a, \xi) \leq C\left[(1 + A\log|\xi|)^{1/2}|\xi||\widehat{\varphi}(\xi)| + |\widehat{\psi}(\xi)| + \int_0^T |\widehat{f}(\xi, t)| dt\right]$$

where the constant C depends on λ_0, T and on $\int_0^T |a(t)| dt$.

We introduce then (52) in (51).

Finally, when the $a_{ij}(t)$ are bounded and hence satisfy (21), we observe that

$$M(a, \xi) \leq C_1\left(|\xi||\widehat{\varphi}(\xi)| + |\widehat{\psi}(\xi)| + \int_0^T |\widehat{f}(\xi, t)| dt\right)$$

with $C_1 = C_1(\lambda_0, \Lambda_0, T)$.

5. - Solutions with non compact support

In this section we propose to extend the results of §4 to the case when the data φ, ψ and f of problem $\{(1),(2)\}$ do not have compact support in $\mathbf{R}^n$. In particular we shall obtain a result of existence and uniqueness in the space of real analytic functions in x.

THEOREM 4 – (*Existence*). *Let us consider problem $\{(1),(2)\}$ with real valued and integrable coefficients $a_{ij}(t)$ on $[0, T]$, satisfying assumption (16).*

i) If the data φ and ψ are in the space $\mathscr{A}$ of real analytic functions and f is in $L^1([0, T], \mathscr{A})$, there exists a solution u of the problem belonging to $H^{2,1}([0, T], \mathscr{A})$.

ii) Assume that the coefficients $a_{ij}(t)$ satisfy (36) with $0 < \alpha < 1$. We have then, for $1 \leq s < 1/(1 - \alpha)$:

if φ and ψ are in $\mathscr{E}_s$ and f is in $L^1([0, T], \mathscr{E}_s)$, there exists a solution u in $H^{2,1}([0, T], \mathscr{E}_s)$;

if φ and ψ are in $\mathscr{D}'_s$ and f is in $L^1([0, T], \mathscr{D}'_s)$, there exists a solution u in $H^{2,1}([0, T], \mathscr{D}'_s)$.

iii) Assume that $a_{ij}(t)$ satisfy (37). We have then:

if φ and ψ are in $\mathscr{E}$ and f is in $L^1([0,T],\mathscr{E})$, there exists a solution u in $H^{2,1}([0,T],\mathscr{E})$;

if φ and ψ are in $\mathscr{D}'$ and f is in $L^1([0,T],\mathscr{D}')$, there exists a solution u in $H^{2,1}([0,T],\mathscr{D}')$.

iv) Assume that the $a_{ij}(t)$ satisfy (38) and let s be a real number. Then, if φ is in H_{loc}^{s+1}, ψ is in H_{loc}^{s} and f is in $L^1([0,T],H_{\mathrm{loc}}^{s})$, there exists a solution u in $C([0,T],H_{\mathrm{loc}}^{s+1})$; moreover $\partial u/\partial t$ belongs to $L^2([0,T],H_{\mathrm{loc}}^{s})$ and $\partial^2 u/\partial t^2$ belongs to $L^1([0,T],H_{\mathrm{loc}}^{s-1})$.

PROOF.

(i) Let φ and ψ be in $\mathscr{A}$ and f in $L^1([0,T],\mathscr{A})$.

Then there exist (see the Appendix, part E) two families $\{\varphi_\nu\}$, $\{\psi_\nu\}$ of elements of $\mathscr{H}$ and a family $\{f_\nu\}$ of elements of $L^1([0,T],\mathscr{H})$ (where ν runs on a partially ordered directed set of indices) such that:

$$(53) \qquad \begin{cases} \{\varphi_\nu\} \to \varphi & \text{in } \mathscr{A} \\[2mm] \{f_\nu\} \to f & \text{in } L^1([0,T];\mathscr{A}) \end{cases}$$

By theorem 1, part (ii), there exists a solution u_ν in $H^{2,1}([0,T];\mathscr{H})$ to the following problem:

$$(54) \qquad \frac{\partial^2 u_\nu}{\partial t^2} - \sum_{i,j}^{1,n} a_{ij}(t)\frac{\partial^2 u_\nu}{\partial x_i \partial x_j} = f_\nu \qquad \text{in } \mathbf{R}^n \times [0,T],$$

$$(55) \qquad u_\nu(0) = \varphi_\nu, \quad \text{and} \quad \frac{\partial u_\nu}{\partial t}(0) = \psi_\nu \qquad \text{in } \mathbf{R}^n.$$

Let g be an element of $L^1([0,T],\mathscr{A}')$ and let us denote by v_g the solution (see th. 3) in $H^{2,1}([0,T],\mathscr{A}')$ of the dual problem:

$$(56) \qquad \frac{\partial^2 v_g}{\partial t^2} - \sum_{i,j}^{1,n} a_{ij}(t)\frac{\partial^2 v_g}{\partial x_i \partial x_j} = g \qquad \text{in } \mathbf{R}^n \times [0,T],$$

$$(57) \qquad v_g(T) = 0, \quad \text{and} \quad \frac{\partial v_g}{\partial t}(T) = 0 \qquad \text{in } \mathbf{R}^n.$$

We can now multiply (in the duality $\langle,\rangle$ between $\mathscr{A}'$ and $\mathscr{A}$) equation (54) by v_g and (56) by u_ν, and integrate on $[0,T]$. We then obtain:

$$(58) \qquad \int_0^T \langle g(t), u_\nu(t)\rangle\, dt = \int_0^T \langle v_g(t), f_\nu(t)\rangle\, dt + \langle v_g(0), \psi_\nu\rangle - \left\langle \frac{\partial v_g}{\partial t}(0), \varphi_\nu \right\rangle.$$

Now, assume that the datum g varies in a bounded subset $\mathscr{B}$ of $L^1([0,T],\mathscr{A}')$. By inequality (19) and the Paley-Wiener theorem (part II) we then see that $v_g(0)$

and $(\partial v_g/\partial t)(0)$ vary in a bounded subset of $\mathscr{A}'$, while v_g varies in a bounded subset of $C([0,T],\mathscr{A}')$.

By (53) and (58), we then obtain that, as $\nu \to \infty$,

$$\left\{\int_0^T \langle g(t), u_\nu(t)\rangle dt\right\} \to \int_0^T \langle v_g(t), f(t)\rangle dt + \langle v_g(0), \psi\rangle - \left\langle \frac{\partial v_g}{\partial t}(0), \varphi\right\rangle$$

uniformly for g in $\mathscr{B}$.

Then there exists u in $C([0,T],\mathscr{A})$ such that, if $\nu \to \infty$,

$$\{u_\nu\} \to u \qquad \text{in } L^\infty([0,T],\mathscr{A}).$$

By (54) it also follows that, as $\nu \to \infty$,

$$\left\{\frac{\partial^2 u_\nu}{\partial t^2}\right\} \to \sum a_{ij}(t)\frac{\partial^2 u}{\partial x_i \partial x_j} + f \qquad \text{in } L^1([0,T],\mathscr{A}).$$

We conclude that the function u belongs to $H^{2,1}([0,T],\mathscr{A})$ and is a solution of problem $\{(1),(2)\}$.

(ii)-(iii) The proof is identical to the proof of case (i). It is sufficient to choose g in $L^1([0,T],\mathscr{E}'_s)$ (or in $L^1([0,T],\mathscr{D}_s)$) in case (ii) and g in $L^1([0,T],\mathscr{E}')$ (or in $L^1([0,T],\mathscr{D})$) in case (iii).

(iv) We must now choose g varying in a bounded subset $\mathscr{B}$ of $L^1([0,T],H_c^{-s-1})$ and apply inequality (22) (with $\eta = 0$) to the solution v_g of problem $\{(56),(57)\}$. Since assumption (38) implies $\omega(a, T(2|\zeta|)^{-1}) \leq AT(2|\zeta|)^{-1}$, it follows that $v_g(0)$ varies in a bounded subset of H_c^{-s} and $(\partial v_g/\partial t)(0)$ in a bounded subset of H_c^{-s-1}, while v_g varies in a bounded subset of $L^\infty([0,T],H_c^{-s})$.

Hence it will be sufficient to choose $\{\varphi_\nu\}$, $\{\psi_\nu\}$ and $\{f_\nu\}$ such that $\{\varphi_\nu\} \to \varphi$ in H_{loc}^{s+1}, $\{\psi_\nu\} \to \psi$ in H_{loc}^s and $\{f_\nu\} \to f$ in $L^1([0,T],H_{\text{loc}}^s)$.

This proves that there exists u in the space $C([0,T],H_{\text{loc}}^{s+1})$ such that $\{u_\nu\} \to u$ in $C([0,T],H_{\text{loc}}^{s+1})$.

Then by equation (54) it follows that:

$$\left\{\frac{\partial^2 u_\nu}{\partial t^2}\right\} \to \sum a_{ij}(t)\frac{\partial^2 u}{\partial x_i \partial x_j} + f \qquad \text{in } L^1([0,T],H_{\text{loc}}^{s-1}),$$

and hence that u is a solution to problem $\{(1),(2)\}$ and that $\partial^2 u/\partial t^2$ is in $L^1([0,T],H_{\text{loc}}^{s-1})$. By interpolation we finally see that $\partial u/\partial t$ belongs to $L^1([0,T], H_{\text{loc}}^s)$.

REMARK 6. Assume the hypotheses of theorem 3 or theorem 4 are satisfied.

If the data φ, ψ and $f(t)$ are real valued functionals, or functions, then the solution $u(t)$ (given by theorem 3 in the case of compact support, and by theorem 4 otherwise) is also real valued.

PROOF. First of all, we recall that a real analytic functional (resp. a Gevrey ultradistribution) χ is real valued if $\langle \chi, w\rangle$ is real for all real analytic functions w (resp. Gevrey functions with compact support) on $\mathbf{R}^n$.

When χ is compactly supported, this is equivalent to say that:

$$\widehat{\chi}(-\xi) = \overline{\widehat{\chi}(\xi)}. \qquad \forall \xi \in \mathbf{R}^n.$$

Now, if the data φ, ψ and $f(t)$ are in $\mathscr{A}'$ and u is the solution given by theorem 3, we know that $\widehat{u}(\xi,t)$ is the solution of problem (17). But if φ, ψ and $f(t)$ are real valued, we obtain that $\overline{\widehat{u}(-\xi,t)}$ is another solution of (17), whence the thesis.

If φ, ψ and $f(t)$ are ultradistributions or functions with non compact support, we may couple problem $\{(1),(2)\}$ with problem $\{(56),(57)\}$, where $g(t)$ is chosen real valued. By the preceding argument $v_g(t)$ is then real valued and, as a consequence, (see (58)) $\int_0^T \langle g(t), u(t)\rangle dt$ is in $\mathbf{R}$. Since g is arbitrary, we conclude that $u(t)$ is real valued.

THEOREM 5. – *(Continuous dependence of the solution on the data). Let us consider problem* $\{(1),(2)\}$ *with real valued and integrable coefficients* $a_{ij}(t)$ *on* $[0,T]$, *satisfying (16).*

For ν varying in a directed partially ordered set of indices, let u_ν be a solution of the problem with data φ_ν, ψ_ν, f_ν (in other words, u_ν is a solution of problem $\{(54),(55)\}$*).*

Denote by B^0 an open ball in $\mathbf{R}^n$ with radius $\rho > 0$ and B^t the open ball with the same centre as B^0 and radius $\rho(t)$, where

$$\rho(t) = \rho - \int_0^t |a(s)|^{1/2} ds.$$

i) Assume that u_ν belongs to $H^{2,1}([0,T],\mathscr{A})$, $\forall \nu$.

If $\{\varphi_\nu\}$ and $\{\psi_\nu\}$ converge to zero in $\mathscr{A}(B^0)$ and $\{f_\nu\} \to 0$ in $L^1([0,T],\mathscr{A}(B^0))$, as $\nu \to \infty$, then

$$\{u_\nu\} \to 0 \ in \ H^{2,1}([0,\tau],\mathscr{A}(B^\tau))$$

for all $\tau \in]0,T]$ such that $\rho(\tau) > 0$.

ii) Assume that $a_{ij}(t)$ satisfy (36) with $0 < \alpha < 1$ and that u_ν belongs to $H^{2,1}([0,T],\mathscr{E}_s)$ (resp. to $H^{2,1}([0,T],\mathscr{D}_s')$) with $1 < s < 1/(1-\alpha)$, $\forall \nu$.

If $\{\varphi_\nu\}$ and $\{\psi_\nu\}$ converge to zero in $\mathscr{E}_s(B^0)$ (resp. in $\mathscr{D}_s'(B^0)$) and $\{f_\nu\} \to 0$ in $L^1([0,T],\mathscr{E}_s(B^0))$ (resp. in $L^1([0,T],\mathscr{D}_s'(B^0))$), then, $\forall \tau \in]0,T]$,

$$\{u_\nu\} \to 0 \ in \ H^{2,1}([0,\tau],\mathscr{E}_s(B^\tau))$$

(resp. $\{u_\nu\} \to 0$ in $H^{2,1}([0,\tau],\mathscr{D}_s'(B^\tau))$).

iii) Assume that $a_{ij}(t)$ satisfy (37) and that u_ν belongs to $H^{2,1}([0,T],\mathscr{E})$ (resp. to $H^{2,1}([0,T],\mathscr{D}'))$, $\forall \nu$.

If $\{\varphi_\nu\}$ and $\{\psi_\nu\}$ converge to zero in $\mathscr{E}(B^0)$ (resp. in $\mathscr{D}'(B^0)$) and $\{f_\nu\} \to 0$ in $L^1([0,T],\mathscr{E}(B^0))$ (resp. in $L^1([0,T],\mathscr{D}'(B^0))$), then, $\forall \tau \in]0,T]$,

$$\{u_\nu\} \to 0 \ in \ H^{2,1}([0,\tau],\mathscr{E}(B^\tau))$$

(resp. $\{u_\nu\} \to 0$ in $H^{2,1}([0,\tau],\mathscr{D}'(B^\tau))$).

iv) Assume that $a_{ij}(t)$ satisfy (38) and u_ν belongs to $H^{2,1}([0,T], H^{s+1}_{\text{loc}})$ with s real, $\forall \nu$.

If $\{\varphi_\nu\} \to 0$ in $H^{s+1}_{\text{loc}}(B^0)$ $\{\psi_\nu\} \to 0$ in $H^s_{\text{loc}}(B^0)$ and $\{f_\nu\} \to 0$ in $L^1([0,T], H^s_{\text{loc}}(B^0))$, then, $\forall \tau \in]0,T]$,

$$\{u_\nu\} \to 0 \text{ in } L^\infty([0,\tau], H^{s+1}_{\text{loc}}(B^\tau)),$$

while $\{(\partial u_\nu / \partial t)\} \to 0$ in $L^2([0,\tau], H^s_{\text{loc}}(B^\tau))$ and $\{(\partial^2 u_\nu / \partial t^2)\} \to 0$ in $L^1([0,\tau], H^{s-1}_{\text{loc}}(B^\tau))$.

Proof.

(i) Fix τ in $]0,T]$ such that $\rho(\tau) > 0$.

Let g be in $L^1([0,\tau], \mathscr{A}(B^\tau))$ and denote by v_g the solution (see th. 3) in $H^{2,1}([0,T], \mathscr{A}')$ of the dual problem:

$$(59) \qquad \frac{\partial^2 v_g}{\partial t^2} - \sum_{i,j}^{1,n} a_{ij}(t) \frac{\partial^2 v_g}{\partial x_i \partial x_j} = g \qquad \text{in } \mathbf{R}^n \times [0,\tau],$$

$$(60) \qquad v_g(\tau) = 0, \quad \text{and} \quad \frac{\partial v_g}{\partial t}(\tau) = 0 \qquad \text{in } \mathbf{R}^n.$$

Multiplying equation (59) by u_ν and (54) by v_g and integrating on $[0,\tau]$ we obtain the identity:

$$(61) \qquad \int_0^\tau \langle g(t), u_\nu(t)\rangle dt = \int_0^\tau \langle v_g(t), f_\nu(t)\rangle dt + \langle v_g(0), \psi_\nu\rangle - \left\langle \frac{\partial v_g}{\partial t}(0), \varphi_\nu \right\rangle.$$

Since g has been chosen in $L^1([0,T], \mathscr{A}'(B^\tau))$, theorem 3, part a, gives that the solution v_g belongs to $H^{2,1}([0,T], \mathscr{A}'(B^0))$.

As a consequence, (61) implies that, for $\nu \to \infty$,

$$(61) \qquad \left\{\int_0^\tau \langle g(t), u_\nu(t)\rangle dt\right\} \to 0, \qquad \forall g \in L^1([0,\tau], \mathscr{A}'(B^\tau)).$$

Now, if g varies in a bounded subset of $L^1([0,\tau], \mathscr{A}'(B^\tau))$, we know (see (44)) that v_g varies in a bounded subset of $H^{2,1}([0,\tau], \mathscr{A}'(B^0))$. We deduce that the convergence (62) is uniform with respect to g.

We have then that $\{u_\nu\} \to 0$ in $C([0,\tau], \mathscr{A}'(B^\tau))$ and hence (recalling that u_ν satisfy the equation (1)) that $\{u_\nu\} \to 0$ in $H^{2,1}([0,\tau], \mathscr{A}'(B^\tau))$.

Parts (ii), (iii) and (iv) can be proved in a similar way.

Remark 7. Theorem 4 is still true if we replace "family depending on ν which converges to zero as $\nu \to \infty$" by "bounded family".

From theorem 5 one can derive without difficulty the following result on the "domain of dependence" of solutions of the equation (1), which in turn implies the uniqueness of solutions to problem $\{(1),(2)\}$.

Theorem 6. – *(Domain of dependence and uniqueness). Let us consider problem $\{(1),(2)\}$ with real valued and integrable coefficients $a_{ij}(t)$ on $[0,T]$ satisfying (16).*

a) Let u be a solution in $H^{2,1}([0,T],\mathscr{A}')$ and let φ,ψ be in $\mathscr{A}'$ and f in $L^1([0,T],\mathscr{A}')$ such that $\varphi(x) \equiv \psi(x) \equiv f(x,t) \equiv 0$ for $|x - x_0| < \rho$. Then $u(x,t) \equiv 0$ for $|x - x_0| < \rho - \int_0^t |a(s)|^{1/2}ds$.

b) The same result holds (without additional assumptions on the coefficients) when the data φ and ψ belong to $\mathscr{D}'_s$ for some $s > 1$ and f is in $L^1([0,T],\mathscr{D}'_s)$, provided u belongs to $H^{2,1}([0,T],\mathscr{D}'_s)$.

Proof.

(a) Let $\rho(t) = \rho - \int_0^t |a(s)|^{1/2}ds$ and let $\tau \in]0,T]$ be such that $\rho(\tau) > 0$.

The assumption that the real analytic functional $u(\tau)$ is equal to zero on the open ball $B(x_0, \rho(\tau))$ centered at x_0 with radius $\rho(\tau)$, means that

$$(63) \qquad \langle u(\tau), w_\nu \rangle \to 0 \qquad (\text{if } \nu \to \infty)$$

for any sequence $\{w_\nu\}$ of entire functions such that

$$(64) \qquad w_\nu \to 0 \qquad \text{in } \mathscr{A}(\mathbf{R}^n \setminus \overline{B(x_0, \rho(\tau) - \varepsilon)})$$

for some $\varepsilon > 0$.

Now, if $\{w_\nu\}$ is a sequence of entire functions satisfying (64), denote by v_ν the solution in $H^{2,1}([0,\tau],\mathscr{A})$ of the problem:

$$(65) \qquad \frac{\partial^2 v_\nu}{\partial t^2} - \sum_{i,j}^{1,n} a_{ij}(t)\frac{\partial^2 v_\nu}{\partial x_i \partial x_j} = 0 \qquad \text{in } \mathbf{R}^n \times [0,\tau],$$

$$(66) \qquad v_\nu(\tau) = 0, \quad \text{and} \quad \frac{\partial v_\nu}{\partial t}(\tau) = w_\nu \qquad \text{in } \mathbf{R}^n.$$

Then theorem 5, part (i), implies that $\{v_\nu\} \to 0$ in $H^{2,1}([0,\tau],\mathscr{A}(\mathbf{R}^n \setminus B(x_0, \rho - \varepsilon)))$.

On the other hand, by coupling problem $\{(65),(66)\}$ with problem $\{(1),(2)\}$, we obtain the identity

$$\langle u(\tau), w_\nu \rangle = \left\langle \varphi, \frac{\partial v_\nu}{\partial t}(0) \right\rangle - \langle \psi, v_\nu(0) \rangle - \int_0^\tau \langle f(t), v_\nu(t) \rangle dt.$$

We thus obtain (63).

(b) If u is a solution in $H^{2,1}([0,T],\mathscr{D}'_s)$, where $s > 1$, of problem $\{(1),(2)\}$ with data φ and ψ in $\mathscr{D}'_s$ and f in $L^1([0,T],\mathscr{D}'_s)$, one can replace u by $\tilde{u} = \eta(x)u$ where $\eta(x)$ is a Gevrey function of order s with compact support in $\mathbf{R}^n$, such that $\eta \equiv 1$ on the ball $B(x_0,\rho)$. Then we see that $\tilde{u}$ is a solution of problem $\{(1);(2)\}$ with data $\tilde{\varphi}$, $\tilde{\psi}$ and $\tilde{f}$ which coincide with φ, ψ and f (and hence vanish) for $|x - x_0| < \rho$ and belong respectively to $\mathscr{E}'_s$ (and hence to $\mathscr{A}'$) and to $L^1([0,T],\mathscr{E}'_s)$ (and hence to $L^1([0,T],\mathscr{A}'))$.

We are thus reduced to the case already proved.

Theorem 7. – *(Local existence). Let us consider problem $\{(1),(2)\}$ with coefficients $a_{ij}(t)$ real valued and integrable on $[0,T]$, satisfying (16).*

a) Given data φ and ψ, integrable functions on an open ball B^0 in $\mathbf{R}^n$ with center x_0 and radius ρ, and f in $L^1([0,T],\mathscr{A}(B^0))$, there exists one and only one solution $u(x,t)$ of problem $\{(1),(2)\}$ on the cone of Γ of $\mathbf{R}^n \times [0,T]$, where

$$\Gamma = \left\{ (x,t) \ : \ |x - x_0| < \rho - \int_0^t |a(s)|^{1/2}ds \right\},$$

which belongs to $H^{2,1}([0,\tau],\mathscr{A}(B^\tau))$, $\forall \tau \in]0,T]$, where B^τ is the open ball of center x_0 and radius $\rho(\tau)$ given by

$$\rho(\tau) = \rho - \int_0^\tau |a(t)|^{1/2}dt.$$

In particular the function $u(x,t)$ is of class C^1 in Γ and is analytic in x on B^t, $\forall t \in [0,T]$.

b) When the coefficients $a_{ij}(t)$ satisfy in addition assumptions (36), (37) or (38), we have similar results of existence (and uniqueness) of the solutions on the cone of Γ, corresponding to the preceding results of global existence (see th. 4, parts (ii), (iii) and (iv)).

Proof.

(a) Since $\mathscr{A}$ is dense in $\mathscr{A}(B^0)$, and hence (see the Appendix, part E) $L^1([0,T],\mathscr{A})$ is dense in $L^1([0,T],\mathscr{A}(B^0))$, given data φ, ψ and f, we can find families $\{\varphi_\nu\}$, $\{\psi_\nu\}$ in $\mathscr{A}$ and $\{f_\nu\}$ in $L^1([0,T],\mathscr{A})$, where ν runs on a directed partially ordered set, such that, if $\nu \to \infty$,

$$\{\varphi_\nu\} \to \varphi \text{ and } \{\psi_\nu\} \to \psi \text{ in } \mathscr{A}(B^0),$$

$$\{f_\nu\} \to f \text{ in } L^1([0,T],\mathscr{A}(B^0)).$$

Now let $\{u_\nu\}$ be the solution (see th. 4) in $H^{2,1}([0,T],\mathscr{A})$ of problem $\{(1),(2)\}$ with data φ_ν,ψ_ν and f_ν.

By theorem 5 applied to the families (depending on the indices (ν,μ)) $\{\varphi_\nu - \varphi_\mu\}$, $\{\psi_\nu - \psi_\mu\}$, $\{f_\nu - f_\mu\}$, we have that $\{u_\nu\}$ is a Cauchy sequence in $C([0,\tau], \mathscr{A}(B^\tau))$ for all τ in $]0,T]$ and hence $\{u_\nu\}$ converges to a solution u of problem $\{(1),(2)\}$ on Γ.

Uniqueness can be proved as in part (b) of theorem 6.

(b) We can proceed similarly as in case (a), or, alternatively, we can reduce to the theorem of global existence by multiplying the data φ, ψ and f by suitable Gevrey functions (of the right order) with compact support in B^0 and equal to 1 for $|x - x_0| \le \rho - \varepsilon$ $(\varepsilon > 0)$, and then apply theorem 6 on the domain of dependence.

6. - Convergence

In this section we consider a sequence of problems of the form $\{(1),(2)\}$, i.e.:

$$(67) \qquad \frac{\partial^2 u_k}{\partial t^2} - \sum_{i,j}^{1,n} a_{ij,k}(t)\frac{\partial^2 u_k}{\partial x_i \partial x_j} = f_k \qquad \text{in } \mathbf{R}^n \times [0,T],$$

$$(68) \qquad u_k(0) = \varphi_k, \quad \text{and} \quad \frac{\partial u_k}{\partial t}(0) = \psi_k \qquad \text{in } \mathbf{R}^n,$$

where $k = 1, 2, 3, \ldots$ and the $a_{ij,k}(t)$ are real valued and integrable on $[0,T]$ and satisfy (16) uniformly with respect to k (that is to say, with λ_0 independent of k).

If the sequence $\{a_{ij,k}(t)\}$ converges (as $k \to \infty$) to $a_{ij}(t)$ in $L^1([0,T])$, $\forall i,j = 1,\ldots,n$, and the sequences $\{\varphi_k\}$, $\{\psi_k\}$ and $\{f_k\}$ converge to φ, ψ and f respectively, in a sense to be precised, it is natural to ask whether $\{u_k\}$ also converges to the solution u of the limit problem

$$(69) \qquad \frac{\partial^2 u}{\partial t^2} - \sum_{i,j}^{1,n} a_{ij}(t)\frac{\partial^2 u}{\partial x_i \partial x_j} = f \qquad \text{in } \mathbf{R}^n \times [0,T],$$

$$(70) \qquad u_k(0) = \varphi, \quad \text{and} \quad \frac{\partial u}{\partial t}(0) = \psi \qquad \text{in } \mathbf{R}^n.$$

THEOREM 8. $-$ *Let us consider the sequence of problems $\{(67),(68)\}$ and assume that the coefficients $a_{ij,k}(t)$, real and integrable on $[0,T]$, satisfy (16) uniformly with respect to k and that, $\forall i,j = 1,\ldots,n$,*

$$(71) \qquad \{a_{ij,k}(t)\} \to a_{ij} \qquad in \ L^1([0,T]) \ for \ k \to \infty.$$

Denote by u_k the solution of problem $\{(67),(68)\}$ and by u the solution of the limit problem $\{(69),(70)\}$.

i) Assume that, for $k \to \infty$,

$$\{\varphi_k\} \to \varphi \ and \ \{\psi_k\} \to \psi \quad in \ \mathscr{A} \ (resp. \ in \ \mathscr{A}')$$

and that

$$\{f_k\} \to f \quad in \ L^1([0,T],\mathscr{A}) \ (resp. \ in \ L^1([0,T],\mathscr{A}')).$$

Then:

$$\{u_k\} \to u \quad in \ C^1([0,T],\mathscr{A}) \ (resp. \ in \ C^1([0,T],\mathscr{A}')).$$

ii) Assume that the coefficients $a_{ij,k}(t)$ satisfy (36) with constants α and A independent of k and that

$$\{\varphi_k\} \to \varphi \ and \ \{\psi_k\} \to \psi \quad in \ \mathscr{E}_s \ (resp. \ in \ \mathscr{D}'_s)$$

and

$$\{f_k\} \to f \quad in \ L^1([0,T],\mathscr{E}_s) \ (resp. \ in \ L^1([0,T],\mathscr{D}'_s))$$

where s is a real number such that $1 \le s < 1/(1-\alpha)$.

    Then:

$$\{u_k\} \to u \quad in \ C^1([0,T],\mathscr{E}_s) \ (resp. \ in \ C^1([0,T],\mathscr{D}'_s))).$$

    iii) Assume that the coefficients $a_{ij,k}(t)$ satisfy (37) with a constant A independent of k and that

$$\{\varphi_k\} \to \varphi \ and \ \{\psi_k\} \to \psi \quad in \ \mathscr{E} \ (resp. \ in \ \mathscr{D}')$$

and

$$\{f_k\} \to f \quad in \ L^1([0,T],\mathscr{E}) \ (resp. \ in \ L^1([0,T],\mathscr{D}')).$$

    Then:

$$\{u_k\} \to u \quad in \ C^1([0,T],\mathscr{E}) \ (resp. \ in \ C^1([0,T],\mathscr{D}'))).$$

    iv) Assume that the coefficients $a_{ij,k}(t)$ satisfy (38) with a constant A independent of k and that

$$\{\varphi_k\} \to \varphi \quad in \ H^{s+1}_{\mathrm{loc}}, \qquad \{\psi_k\} \to \psi \quad in \ H^s_{\mathrm{loc}}$$

and

$$\{f_k\} \to f \quad in \ L^1([0,T],H^s_{\mathrm{loc}})$$

for some real s.

    Then:

$$\{u_k\} \to u \quad in \ C([0,T],H^s_{\mathrm{loc}})$$

and

$$\left\{\frac{\partial u_k}{\partial t}\right\} \to \frac{\partial u}{\partial t} \quad in \ C([0,T],H^{s-1}_{\mathrm{loc}}).$$

Proof. Let us consider case (i). If $\{\varphi_k\} \to \varphi$ and $\{\psi_k\} \to \psi$ in $\mathscr{A}'$ and $\{f_k\} \to f$ in $L^1([0,T],\mathscr{A}')$, we see (Paley-Wiener theorem, part II) that $\widehat{\varphi}_k$, $\widehat{\psi}_k$, $\widehat{f}_k$ satisfy the inequality (41) with constants C_ε and ρ independent of k (see the proof of th. 3, to which we shall refer in the following).

On the other hand, the $a_{ij,k}$ belong to a compact subset of $L^1([0,T])$, so that the functions

$$\omega(a_k,\delta) = \sup_{0 \le \tau \le \delta} \int_0^{T-\tau} |a_k(t+\tau) - a_k(t)|\,dt \qquad (0 \le \delta \le T)$$

are equibounded on $[0,T]$ and

$$\omega(a_k,\delta) \to 0, \qquad if \ \delta \to 0^+,$$

uniformly with respect to k.

As a consequence, the function $\varepsilon_0(a_k, \zeta)$ (defined by (40)) is equibounded for $|\zeta| \geq 1$ and converges to zero for $|\zeta| \to \infty$, uniformly with respect to k. It follows that the constants $L_\varepsilon(a_k)$ defined by (43) are bounded with respect to k, for all $\varepsilon > 0$, so that $\widehat{u}$ and $\partial\widehat{u}/\partial t$ satisfy (44) with constants independent of k.

Hence (Paley-Wiener theorem, part II) the sequences $\{u_k\}$ and $\{\partial u_k/\partial t\}$ are bounded in $C([0,T], \mathscr{A}')$. By equation (67) we finally obtain (remarking that the $a_{ij,k}$ are equi-integrable on $[0,T]$, and that the f_k are equi-integrable on $[0,T]$ with values in $\mathscr{A}'$) that the $\{\partial^2 u_k/\partial t^2\}$ are equi-integrable on $[0,T]$ with values in $\mathscr{A}'$, hence the $\{\partial u_k/\partial t\}$ are equicontinuous on $[0,T]$ with values in $\mathscr{A}'$.

But $\mathscr{A}'$ is a Montel space and hence $\{\partial u_k/\partial t\}$ is relatively compact in $C([0,T], \mathscr{A}')$.

In conclusion we have proved that $\{u_k\}$ is a relatively compact sequence in $C^1([0,T], \mathscr{A}')$. Now, we see easily that if u_* is the limit of a subsequence of $\{u_k\}$ which converges in $C^1([0,T], \mathscr{A}')$ then u_* is a solution of the limit problem $\{(69),(70)\}$.

But this problem has a unique solution u, hence $u_* = u$; whence the thesis in the case of analytic functionals.

Assume now that $\{\varphi_k\} \to \varphi$ and $\{\psi_k\} \to \psi$ in $\mathscr{A}$, while $\{f_k\} \to f$ in $L^1([0,T], \mathscr{A})$.

Introduce the dual problem (see the proof of theorem 4)

$$(72) \qquad \frac{\partial^2 v_{g,k}}{\partial t^2} - \sum_{i,j}^{1,n} a_{ij,k}(t) \frac{\partial^2 v_{g,k}}{\partial x_i \partial x_j} = g \qquad \text{in } \mathbf{R}^n \times [0,T],$$

$$(73) \qquad v_{g,k}(T) = 0, \quad \text{and} \quad \frac{\partial v_{g,k}}{\partial t}(T) = 0 \qquad \text{in } \mathbf{R}^n$$

where g is a fixed element in $L^1([0,T], \mathscr{A}')$.

By coupling problem $\{(72),(73)\}$ with problem $\{(67),(68)\}$ we obtain (see (58)) that $\left\{ \int_0^T \langle g(t), u_k(t)\rangle dt \right\}$ is bounded and hence that $\{u_k\}$ is bounded in $C([0,T], \mathscr{A})$.

Moreover, by equation (67) we have that the $\{\partial^2 u_k/\partial t^2\}$ are equi-integrable on $[0,T]$ with values in $\mathscr{A}$.

The conclusion is then the same as in the case of analytic functionals.

To prove case (ii), it is convenient to consider first the case when the data φ_k, ψ_k and f_k have support in x (as functions or functionals) contained in a compact subset of $\mathbf{R}^n$ independent of k. This case can be proved in a way completely analogous to the case of analytic functionals.

Then we can drop the assumption of compact supports by the usual duality argument.

Cases (iii) and (iv) are proved in a similar way.

7. - Counterexamples

Theorem 9. – *(Counterexamples on convergence). Let us consider the sequence of problems $(k = 1, 2, 3, \dots)$*

$$(74) \qquad \frac{\partial^2 u_k}{\partial t^2} - a_k(t)\frac{\partial^2 u_k}{\partial x^2} = 0 \qquad in \ \mathbf{R} \times [0, T],$$

$$(75) \qquad u_k(x, 0) = 0, \quad and \quad \frac{\partial u_k}{\partial t}(x, 0) = \psi(x) \qquad in \ \mathbf{R},$$

where (see [1])

$$(76) \qquad a_k(t) = 1 - 4\varepsilon_k \sin(2kt) - \varepsilon_k^2(1 - \cos(2kt))^2$$

and

$$\psi(x) = \sum_{h=1}^{\infty} c_h \sin(hx).$$

(The sequences $\{\varepsilon_k\}$ and $\{c_h\}$ will be chosen in the following).

   i) Let ε and δ be two numbers > 0, $\varepsilon \leq 1/10$, and define:

$$\varepsilon_k = \varepsilon, \qquad \forall k,$$

$$c_h = \exp[-\delta h].$$

Then the coefficients $a_k(t)$ are of the form

$$a_k(t) = a(kt),$$

and they stay in a bounded subset of $L^\infty([0, T])$.

   In addition we have, for $k \to \infty$,

$$\{a_k\} \to 1 - \frac{3}{2}\varepsilon^2, \qquad weakly \ in \ L^1([0, T]).$$

   Moreover, the initial datum $\psi(x)$ is an analytic function on $\mathbf{R}$.

   However, for $t > \delta/\varepsilon$ (with the exception of t equal to an integer multiple of π), the sequence $\{u_k(t)\}$ is not bounded in $\mathscr{D}'$, neither in $\mathscr{D}'_r$ for any $r > 1$.

   ii) Let

$$\varepsilon_k = (\log k)^{-1},$$

$$c_h = \exp(-h(\log h)^{-2}).$$

We have then, for $k \to \infty$

$$(77) \qquad \{a_k\} \to 1 \qquad in \ L^\infty([0, T]).$$

   Moreover, the initial datum $\psi(x)$ is a Gevrey function of order s for all $s > 1$.

However, the sequence $\{u_k(t)\}$ is not bounded in $\mathscr{D}'_r$, for any $r > 1$ and $t > 0$ $(t \neq \nu\pi, \ \nu \ integer)$.

iii) Let $0 < \alpha < 1$ and let

$$\varepsilon_k = k^{-\alpha},$$

$$c_h = \exp(-h^{1-\alpha}(\log h)^{-1}).$$

Then we have again (77). In addition, the functions $a_k(t)$ are equi-Hölder continuous of order α.

Moreover, $\psi(x)$ is a Gevrey function of order s for all $s > 1/(1-\alpha)$.

However, the sequence $\{u_k(t)\}$ is not bounded in $\mathscr{D}'_r$, for any $r \geq 1/(1-\alpha)$ and $t > 0$ $(t \neq \nu\pi, \ \nu \ integer)$.

iv) Let

$$\varepsilon_k = k^{-1}(\log k)^3,$$

$$c_h = \exp(-(\log h)^2).$$

Then we have again (77). Moreover, the functions $a_k(t)$ are equi-Hölder continuous of order α, for all $\alpha < 1$, and $\psi(x)$ is an infinitely differentiable function.

However, the sequence $\{u_k(t)\}$ is not bounded in $\mathscr{D}'$, for any $t > 0$ $(t \neq \nu\pi, \ \nu \ integer)$.

PROOF. We expand the solution $u_k(x,t)$ of problem $\{(74),(75)\}$ in a Fourier series of the form

$$u_k(x,t) = \sum_{h=1}^{\infty} v_{k,h}(t)\sin(hx).$$

The coefficients $v_{k,h}(t)$ must then satisfy the Cauchy problem:

$$\begin{cases} v''_{k,h}(t) + h^2 a_k(t) v_{k,h}(t) = 0 & \text{in }]0, T[, \\[2ex] v_{k,h}(0) = 0 \text{ and } v'_{k,h}(0) = c_h. \end{cases}$$

Now, if $a_k(t)$ is the function defined by (76), we can solve explicitly this problem when $k = h$, and we obtain the following solution:

$$v_{k,k}(t) = \frac{1}{k}c_k \sin(kt)\cdot\exp\left[\varepsilon_k\left(kt - \frac{1}{2}\sin(2kt)\right)\right].$$

Using theorem 12 of the Appendix and after some simple verifications we conclude the proof.

THEOREM 10. – *(Counterexamples on existence). Let us consider the problem*

$$(78) \qquad \frac{\partial^2 u}{\partial t^2} - a(t)\frac{\partial^2 u}{\partial x^2} = 0 \qquad \text{in } \mathbf{R} \times [0, T],$$

$$(79) \qquad u(x,0) = \varphi(x), \quad \text{and} \quad \frac{\partial u_k}{\partial t}(x,0) = \psi(x) \qquad \text{in } \mathbf{R},$$

with $T > 1$.

Let us consider the following partition of the interval $[0, 1[$:

$$[0, 1[= \bigcup_{k=0}^{\infty} I_k,$$

where

$$I_k = [1 - 2^{-k}, 1 - 2^{-k-1}[\qquad (k = 0, 1, 2, \ldots).$$

Let

$$a(t) = \begin{cases} b(\varepsilon_k, \eta_k t) = 0, & \text{if } t \in I_k, \\ \\ 1, & \text{if } t \geq 1 \end{cases}$$

where

$$b(\varepsilon, \tau) = 1 - 4\varepsilon \sin 2\tau - \varepsilon^2 (1 - \cos 2\tau)^2$$

and

$$\eta_k = 4\pi 4^{k!}$$

(the real numbers ε_k will be chosen in the following).

i) Let:

$$\varepsilon_k = \frac{1}{8}(\log \eta_k)^{-1}.$$

Then the function $a(t)$ is continuous on $[0, T]$; however, one can find $\varphi(x)$ and $\psi(x)$, Gevrey functions of order s for all $s > 1$, such that problem $\{(78), (79)\}$ has no solution in $H^{2,1}([0, T], \mathscr{D}'_r)$, for any is $r > 1$.

ii) Let:

$$\varepsilon_k = \frac{1}{8}\eta_k^{-\alpha}, \qquad \text{where } 0 < \alpha < 1.$$

Then $a(t)$ is Hölder continuous of order α on $[0, T]$; however, one can find $\varphi(x)$ and $\psi(x)$, functions of order s for all $s > 1/(1 - \alpha)$, such that problem $\{(78), (79)\}$ has no solution in $H^{2,1}([0, T], \mathscr{D}'_r)$, for any $r > 1/(1 - \alpha)$.

iii) Let:

$$\varepsilon_k = \frac{1}{8}\eta_k^{-1}(\log \eta_k)^2.$$

Then $a(t)$ is Hölder continuous on $[0, T]$ of order α for all $\alpha < 1$; however, one can find $\varphi(x)$ and $\psi(x)$, infinitely differentiable functions, such that problem $\{(78), (79)\}$ has no solution in $H^{2,1}([0, T], \mathscr{D}')$.

Proof. We prove only case (i), the remaining cases being similar.

In order to construct the functions $\varphi(x)$ and $\psi(x)$, we construct instead a solution $v(x, t)$ of equation (78) in $\mathbf{R} \times [0, 1[$ such that

(80) $$v \in H^{2,1}([0, 1 - \delta], \mathscr{E}_s), \qquad \forall s > 1, \forall \delta > 0,$$

while

(81) $$\frac{\partial v}{\partial t}(t) \text{ is not bounded in } \mathscr{D}'_s \text{ as } t \to 1^-, \qquad \forall r > 1.$$

Once this function v has been constructed, it will be sufficient to choose

$$\varphi(x) = v(x,0) \quad \text{and} \quad \psi(x) = \frac{\partial v}{\partial t}(x,0),$$

to conclude the proof of the theorem.

Indeed, if there exists a solution $u(x,t)$ of $\{(78),(79)\}$ belonging to $H^{2,1}([0,T], \mathscr{D}'_r)$ for some $r > 1$, this solution should coincide (by theorem 6 on uniqueness) with $v(x,t)$ for $0 < t < 1$, and this would contradict (81).

The function $v(x,t)$ will be constructed in the form

$$(82) \qquad v(x,t) = \sum_{h=1}^{\infty} v_h(t) \sin(\eta_k t),$$

where $v_k(t)$ is a solution of the ordinary equation:

$$(83) \qquad v_h''(t) + \eta_h^2 a(t) v_h = 0 \quad \text{in }]0,1[.$$

To select one solution of equation (83) we impose the following initial conditions:

$$(84) \qquad v_h(t_h) = 0 \quad \text{and} \quad v_h'(t_h) = 1,$$

where t_h denotes the center of the interval I_h (hence $t_h = 1 - 3 \cdot 2^{-h-2}$).

But, through the change of variable $t \to \tau \equiv \eta_h(t-t_h)$, we can solve explicitly problem $\{(83),(84)\}$ in the interval I_h, and we obtain the solution

$$(85) \qquad v_h(t) = \frac{1}{\eta_h} w(\varepsilon_h, \eta_h(t - t_h)), \quad t \in I_h$$

where

$$w(\varepsilon, t) = (\sin \tau) \cdot \exp\left[\varepsilon\left(\tau - \frac{1}{2}\sin(2\tau)\right)\right].$$

Now, denote by t_h' and t_h'' the two endpoints of the interval I_h (hence $t_h' = 1 - 2^{-h}$ and $t_h'' = 1 - 2^{-h-1}$).

We have then (since $\sin(\eta_h t) = 0$ and $\cos(\eta_h t) = 1$ for $t = t_h'$ and $t = t_h''$)

$$(86) \qquad v_h(t_h') = 0 \quad \text{and} \quad v_h'(t_h') = \varepsilon_h \exp(-\varepsilon_h \eta_h 2^{-h-2}),$$

$$(87) \qquad v_h(t_h'') = 0 \quad \text{and} \quad v_h'(t_h'') = \varepsilon_h \exp(\varepsilon_h \eta_h 2^{-h-2}).$$

By our choice of ε_h and η_h, we check easily (using th. 12 of the Appendix) that the sequence $\{(\partial v/\partial t)(t_h'')\}$ is not bounded in $\mathscr{D}'_r$, for any $r > 1$.

We have thus proved (81).

To prove (80), we apply Lemma 1 on the interval $[0, t_h']$ (after reversing the direction of t), with $\alpha(t) = h^2 a(t)$, $\beta(t) = a(t)$ and $\gamma(t) = 0$.

Since in this interval we have $|a'(t)| \leq C\varepsilon_{h-1}\eta_{h-1}$, (6) gives

$$\eta_h|v_h(t)| + |v_h'(t)| \leq \varepsilon_h \exp(-\varepsilon_h\eta_h 2^{-h-2} + \widetilde{C}\varepsilon_{h-1}\eta_{h-1}).$$

By the choice of ε_h, η_h, we see then (thanks to th. 12) that $v(t)$ and $(\partial v/\partial t)(t)$ belong to $\mathscr{E}_s$ for all $s > 1$ and for all $t \in [0, 1[$. By equation (78) we finally obtain (80).

8. - The weakly hyperbolic case

The results obtained up to this point, concerning existence and convergence of solutions of hyperbolic problems, have been proved under the assumption that the coefficients be coercive:

$$(88) \qquad \sum a_{ij}(t)\xi_i\xi_j \geq \lambda_0|\xi|^2 \qquad \forall \xi \in \mathbf{R}^n, \ (\lambda_0 > 0).$$

Now, it is possible to show that this assumption can be replaced by the weaker assumption:

$$(89) \qquad \sum a_{ij}(t)\xi_i\xi_j \geq 0, \qquad \forall \xi \in \mathbf{R}^n,$$

at least in the case of solutions in $\mathscr{A}$ or in $\mathscr{A}'$.

Theorem 11. – *Let us consider problem $\{(1),(2)\}$ with coefficients a_{ij}, real valued and integrable, satisfying (89).*

If the data φ and ψ belong to $\mathscr{A}$ (resp. to $\mathscr{A}'$) and f belongs to $L^1([0,T],\mathscr{A})$ (resp. to $L^1([0,T],\mathscr{A}'))$, then there exists one and only one solution of the problem in $H^{2,1}([0,T],\mathscr{A})$ (resp. in $H^{2,1}([0,T],\mathscr{A}'))$.

Moreover, the result on the convergence of solutions proved in th. 8, part (i), under assumption (88), is still true under assumption (89).

Proof. The thesis can be proved as in the case of coercive coefficients, but now we must replace estimate (19) by a modified version of it.

We can prove indeed, under the weaker assumption (89) of nonnegativity, the following inequality for the solution u in $H^{2,1}([0,T],\mathscr{H}')$ of problem $\{(1),(2)\}$, with holomorphic functionals as data:

$$(90) \qquad \sqrt{\varepsilon(\zeta)}|\zeta||\widehat{u}(\zeta,t)| + \left|\frac{\partial\widehat{u}}{\partial t}(\zeta,t)\right| \leq$$

$$\leq M \exp\left(\int_0^t |a(s)|^{\frac{1}{2}}ds|\eta|\right) \exp c_1 \left[\left(\frac{\omega(\zeta)}{\varepsilon(\zeta)} + \sqrt{\varepsilon(\zeta)}\right)|\zeta| + \frac{\int_0^T |a|ds}{\varepsilon(\zeta)}\right]$$

where $\varepsilon(\zeta)$ is an arbitrary function such that $0 < \varepsilon(\zeta) \leq 1$, and where

$$M = c_2\left[\left(1 + |\zeta|\int_0^{T/2|\zeta|} |a(s)|ds\right)^{\frac{1}{2}} |\zeta||\widehat{\varphi}(\zeta)| + |\widehat{\psi}(\zeta)| + \int_0^T |\widehat{f}(\zeta,s)|ds,\right.$$

for all t in $[0,T]$ and $|\zeta| \geq 1$. We have denoted by c_1 and c_2 two constants depending on T and we have written (see (18))

$$\omega(\zeta) = \sup_{0<\tau\leq T/2|\zeta|} \int_0^{T-\tau} |a(t+\tau) - a(t)|dt.$$

To prove (90), it is sufficient to reconsider the proof of Lemma 2 by choosing now:

$$b(\zeta, t) = \int_0^{+\infty} \tilde{a}(t + \tau)\rho_\zeta(\tau)d\tau + \varepsilon(\zeta)$$

(where $\tilde{a}(t)$ and $\rho_\zeta(\tau)$ are defined as in the proof of Lemma 2), so that:

$$b(\zeta, t) \geq \varepsilon(\zeta) > 0.$$

Once (90) is proved, we choose (for instance)

$$\varepsilon(\zeta) = \min\{1, \sqrt{\omega(\zeta)} + |\zeta|^{\frac{1}{2}}\}$$

and we proceed as in the proof of th. 3, part (a).

Appendix

A) PROOF OF THE PALEY-WIENER THEOREM, PART III.

  i) Let $u \in L^1([0,T], \mathscr{A}')$. Then the set

$$\mathscr{S}_u = \{\lambda(t)u \mid \lambda \in L^\infty([0,T]), \ |\lambda(t)| \leq 1\}$$

is bounded in $L^1([0,T], \mathscr{A}')$. As a consequence, the set

$$S_u = \left\{\int_0^T v(t)dt \mid v \in \mathscr{S}_u\right\}$$

is bounded in $\mathscr{A}'$ (indeed, the linear mapping $v \mapsto \int_0^T v(t)dt$ from $L^1([0,T], \mathscr{A}')$ to $\mathscr{A}'$ is continuous).

From part II of the theorem, and from the fact that the Fourier transform of the functional $\int_0^T v(t)dt$ is equal to $\int_0^T \hat{v}(\zeta, t)dt$, we deduce that there exists a $\rho > 0$ such that, $\forall \varepsilon > 0$, $\exists C_\varepsilon$ such that

$$\left|\int_0^T \lambda(t)\hat{u}(\zeta, t)dt\right| \leq C_\varepsilon(\varepsilon|\xi| + (\rho + \varepsilon)|\eta|),$$

$\forall \lambda \in L^\infty([0,T])$, with $|\lambda(t)| \leq 1$.

  We have thus proved III-(i).

  Cases (ii), (iii), (iv) are proved in a similar way.

B) PROOF OF THEOREM 1.

  i) We can reduce problem $\{(1),(2)\}$ to a problem of the following form:

$$(91) \qquad w'(t) - \sum_{j=1}^n A_j(t)\frac{\partial w}{\partial x_j}(t) = F(t) \qquad \text{in }]0,T[, \qquad w(0) = w_0,$$

where w and F are functions on $[0,T]$ with values in $[\mathscr{H}']^{n+1}$, $A_1(t),\ldots,A_n(t)$ are scalar $(n+1) \times (n+1)$ matrices, integrable on $[0,T]$, and $w_0 \in [\mathscr{H}']^{n+1}$.

Problem (91) is equivalent to the integral equation

$$w(t) = w_0 + \int_0^t \left[\sum_{j=1}^n A_j(s) \frac{\partial w}{\partial x_j}(s) + F(s) \right] ds.$$

We define, for k integer ≥ 1,

$$w_k(t) = w_0 + \int_0^t \left[\sum_{j=1}^n A_j(s) \frac{\partial w_{k-1}}{\partial x_j}(s) + F(s) \right] ds.$$

We have then, for $k \geq 2$,

$$w_k(t) - w_{k-1}(t) = \int_0^t \sum_{j_1} A_{j_1}(t_1) \int_0^{t_1} \sum_{j_2} A_{j_2}(t_2) \ldots \int_0^{t_{k-2}} \sum_{j_{k-1}} A_{j_{k-1}}(\tau) \times$$

$$\times \int_0^\tau \left[\sum_{j_k} A_{j_k}(s) \frac{\partial^k w_0}{\partial x_{j_1} \ldots \partial x_{j_k}} + F(s) \right] ds \, d\tau \, dt_{k-2} \ldots dt_1.$$

For each convex domain D of $\mathbf{C}^n$ and $w \in (\mathscr{H}')^N$, we write:

$$\|w\|_D = \sup\{|\langle w, f \rangle|, \ f \in (\mathscr{H}')^N, \ |f(z)| \leq 1 \text{ on } D\}.$$

Moreover let

$$\beta(t) = \left| \sum_{j=1}^n A_j(t) \right|.$$

Fix now two convex domains D_0 and D of $\mathbf{C}^n$ such that $D_0 \subset\subset D$. We have then (using the Cauchy formula for holomorphic functionals)

$$\|w_k(t) - w_{k-1}(t)\|_D = \int_0^t \beta(t_1) \int_0^{t_1} \beta(t_2) \ldots \int_0^{t_{k-2}} \beta(\tau) \times$$

$$\times \int_0^\tau \left[\beta(s) \frac{\|w_0\|_{D_0} k!}{\rho^k} + \|F(s)\|_D \right] ds \, d\tau \, dt_{k-2} \ldots dt_1,$$

where

$$\rho = \operatorname{dist}(D_0, \mathbf{C}^n \setminus D).$$

But

$$\int_0^t \beta(t_1) \int_0^{t_1} \beta(t_2) \ldots \int_0^{t_{\nu-1}} \beta(s) ds \, dt_{\nu-1} \ldots dt_1 = \left[\int_0^t \beta(s) ds \right]^\nu / \nu!,$$

hence we obtain

$$\|w_k(t) - w_{k-1}(t)\|_D \leq \frac{\left[\int_0^t \beta(s) ds \right]^k}{\rho^k} \|w_0\|_{D_0} + \frac{\left[\int_0^t \beta(s) ds \right]^{k-1}}{(k-1)!} \int_0^T \|F(s)\|_D ds.$$

It follows that, for $0 \leq t \leq T$, the series $\sum_{k=1}^{\infty} \|w_k(t) - w_{k-1}(t)\|_D$ converges, provided that the domain D is large enough with respect to D_0, namely provided that

$$\operatorname{dist}(D_0, \mathbf{C}^n \setminus D) > \int_0^T \beta(s)ds.$$

We now show the uniqueness of solutions. Let w be such that

$$w(t) = \int_0^t \sum_{j=1}^n A_j(s) \frac{\partial w}{\partial x_j}(s)ds \qquad (0 \leq t \leq T)$$

and let D_0, D and ρ be as above.

We have then

$$\|w(t)\|_D \leq \left[\int_0^t \beta(s)ds \right]^k \sup_{0 \leq s \leq t} \|w(s)\|_{D_0} \frac{1}{\rho^k}.$$

But this inequality is true for all k, hence, if D and D_0 are such that $\int_0^t \beta(s)ds < \rho$, we obtain $\|w(t)\|_D = 0$.

ii) In the case of entire functions, the proof of the existence and uniqueness is similar to the case of holomorphic functionals. In this case we use the following norms on the space $\mathscr{H}^N$

$$\|f\|_D = \sup_{z \in D} |f(z)| \qquad (D = \text{domain of } \mathbf{C}^n)$$

and we choose the domains D_0, D such that $D \subset\subset D_0$.

C) THEOREM 12. – *(Paley-Wiener theorem for Fourier series).*
a) i) Let w be an analytic function on $\mathbf{R}$, 2π-periodic. Then, in the sense of the space $\mathscr{A}$, the following expansion holds:

$$(92) \qquad w(x) = \sum_{h=-\infty}^{+\infty} b_h \exp(ihx)$$

where b_h are complex numbers such that

$$(93) \qquad |b_h| \leq M \exp(-\delta|h|) \qquad (\delta > 0).$$

ii) Let w be a Gevrey function of order $s \geq 1$ on $\mathbf{R}$, 2π-periodic. Then, in the sense of $\mathscr{E}_s$, the expansion (92) holds with

$$(94) \qquad |b_h| \leq M \exp(-\delta|h|^{1/s}) \qquad (\delta > 0).$$

iii) Let w be an infinitely differentiable function on $\mathbf{R}$, 2π-periodic. Then, in the sense of $\mathscr{E}$, the expansion (92) holds with

$$(95) \qquad |b_h| \leq M(p)|h|^{-p}, \qquad \forall p > 0.$$

iv) Let w be a distribution on $\mathbf{R}$, 2π-periodic. Then, in the sense of $\mathscr{D}'$, the expansion (92) holds with

$$(96) \qquad\qquad |b_h| \le M|h|^p \qquad (p > 0).$$

v) Let w be a Gevrey ultradistribution of order $s \ge 1$ on $\mathbf{R}$, 2π-periodic. Then, in the sense of $\mathscr{D}'_s$, the expansion (92) holds with

$$(97) \qquad\qquad |b_h| \le M_\varepsilon \exp(\varepsilon|h|^{1/s}), \qquad \forall \varepsilon > 0.$$

b) Conversely, if $\{b_h\}$ is a sequence of complex numbers satisfying inequality (93) (resp. (94), resp. (95), resp. (96), resp. (97)), then the series (92) converges in the space $\mathscr{A}$ (resp. in $\mathscr{E}_s$, resp. in $\mathscr{E}$, resp. in $\mathscr{D}'$, resp. in $\mathscr{D}'_s$).

c) If $\{w_k\}$ is a bounded sequence in the space $\mathscr{A}$, then each w_k admits the expansion (92) with coefficients $b_{h,k}$ satisfying (93) uniformly with respect to k.
The reverse is also true.
Similar results are true for the other spaces considered in part (a).

D) Proposition 1. –– *Let $a(t) = [a_{ij}(t)]_{i,j=1,\dots,n}$ be in $\left(L^1([0,T])\right)^{n^2}$ and let:*

$$\omega(a,\delta) = \sup_{0 \le \tau \le \delta} \int_0^{T-\tau} |a(t+\tau) - a(t)|dt.$$

We have then, for $0 \le \tau \le T/2$ and $0 \le \bar{s} \le T - \tau$,

$$\int_{\bar{s}}^{\bar{s}+\tau} |a(t)|dt \le \frac{2\tau}{T} \int_0^T |a(t)|dt + \omega(a,\tau)$$

and

$$\int_{\bar{s}}^{\bar{s}+\tau} |a(t) - m_a|dt \le \frac{2\tau}{T} \int_0^T |a(t)|dt + \omega(a,\tau)$$

where m_a denotes the average of $a(t)$ on $[0,T]$.

Proof. Let m be in $\mathbf{C}^{n^2}$ and let

$$\theta(s) = \int_s^{s+\tau} |a(t) - m|dt, \qquad 0 \le s \le T - \tau.$$

We have then

$$\theta(\bar{s}) - \theta(s) = \int_{\bar{s}}^{s} (|a(t) - m| - |a(t+\tau) - m|)dt,$$

whence

$$\theta(\bar{s}) \le \theta(s) + \left| \int_{\bar{s}}^{s} |a(t) - a(t+\tau)|dt \right| \le \theta(s) + \omega(a,\tau).$$

Integrating with respect to s on $[0, T - \tau]$, we obtain

$$(T - t)\theta(\bar{s}) \leq \int_0^{T-\tau} \theta(s)ds + (T - \tau)\omega(a, \tau).$$

Now we have by Fubini-Tonelli theorem

$$\int_0^{T-\tau} \theta(s)ds \leq \tau \int_0^T |a(t) - m|dt.$$

From this relation, choosing $m = 0$ and $m = m_a$, we obtain the thesis.

E) PROPOSITION 2. – *Let X be one of the following topological vector spaces:*

$$\mathscr{A}, \ \mathscr{E}_s, \ \mathscr{D}'_s, \ \mathscr{E}, \ \mathscr{D}', \ H^r_{\mathrm{loc}},$$

where s is a real number > 1 and r any real number.
Then $\mathscr{H}$ is dense in X and $L^1([0,T], \mathscr{H})$ is dense in $L^1([0,T], X)$.

PROOF. The density of $\mathscr{H}$ in X is well known. Choose now u in $L^1([0,T], X)$. By definition, we know that there exists a sequence $\{u_k\}$ of step functions with values in X, converging to u in $L^1([0,T], X)$. On the other hand, since $\mathscr{H}$ is dense in X, it is easy to construct, for all k, a sequence $u_k^{(h)}$ in $L^1([0,T], \mathscr{H})$ such that $\{u_k^{(h)}\} \to u_k$ in $L^1([0,T], X)$, for $h \to \infty$. From these facts the thesis follows immediately.

Bibliography

[1] F. COLOMBINI - S. SPAGNOLO, *On the convergence of solutions of hyperbolic equations.* Comm. Partial Differential Equations, **3** (1978), pp. 77–103.

[2] L. DE SIMON - G. TORELLI, *Linear second order differential equations with discontinuous coefficients in Hilbert spaces.* Ann. Scuola Norm. Sup. Pisa, **1** (1974), pp. 131–154.

[3] J.M. GELFAND - G.E. SHILOV, *Generalized functions.* Vol. II and III, Academic Press, New York, 1967 (édition originelle: Moscou, 1958).

[4] A.E. HURD - D.H. SATTINGER, *Questions of existence and uniqueness for hyperbolic equations with discontinuous coefficients.* Trans. Amer. Math. Soc., **132** (1968), pp. 159–174.

[5] J.L.LIONS, *Equations différentielles opérationelles and problèmes aux limites.* Springer, Berlin, 1961.

[6] J.L.LIONS - E. MAGENES, *Problèmes aux limites non homogènes and applications.* Dunod, Paris, 1968.

[7] A. Martineau, *Sur les fonctionnelles analytiques et la transformation of Fourier-Borel.* J. Analyse Math., **11** (1963), pp. 1–164.

[8] W.T. Reid, *On the approximation of integrable functions by functions of bounded variation.* Ann. Scuola Norm. Sup. Pisa, **14** (1960), pp. 133–140.

[9] C. Roumieu, *Sur quelques extensions de la notion de distributions.* Ann. Sci. École Norm. Sup., **77** (1960), pp. 47–121.

[10] C. Roumieu, *Ultradistributions définies sur $\mathbf{R}^n$ et sur certaines classes de variétés différentiables.* J. Analyse Math., **10** (1962-63), pp. 153–192.

[11] F. Treves, *Basic linear partial differential equations.* Academic Press, New York, 1975.

Convergence problems for functionals and operators[‡]

Ennio de Giorgi, Scuola Normale Superiore, Pisa

Introduction

Aim of this lecture is to give a (partial) scheme of the connections between Γ- and G-convergence and several results and problems in calculus of variations, measure theory, partial differential equations, variational inequalities.

I am concerned only with general indications: I hope they will be developed in a systematic way later on. The bibliography has a simply indicative purpose, as well, without any claim of completeness. Moreover, the study of the relations between Γ- and G-convergence and other similar notions is not yet complete, so I cannot exclude that some of the problems I will point out as open questions have already been treated in another form by different theories.

I wish to thank G. Buttazzo, G. Dal Maso, L. Modica and S. Mortola, who worked together on these notes, which I hope will be useful as a first basis for a future systematic work. In view of such a design, I will be grateful to all the people that will communicate suggestions, remarks or bibliographical references to the following address: Luciano Modica, Istituto di Matematica "L. Tonelli", via Derna 1, I-56100 Pisa.

1. Definitions and abstract properties

We recall (from [18], [38], [39]) the basic definitions of Γ- and G-convergence.

DEFINITION 1.1 – *Let (X, τ) be a topological space, $\xi \in X$, $f : X \to \overline{\mathbf{R}}$. Then we define*

$$\Gamma(\tau^+) \lim_{x \to \xi} f(x) = \inf_{A \in J_\tau(\xi)} \sup_{x \in A} f(x),$$

where $J_\tau(\xi)$ is the class of open neighborhoods of ξ. Analogously we define

$$\Gamma(\tau^-) \lim_{x \to \xi} f(x) = \sup_{A \in J_\tau(\xi)} \inf_{x \in A} f(x).$$

By $\overline{\mathbf{R}} = \mathbf{R} \cup \{-\infty, +\infty\}$ and $\overline{\mathbf{N}} = \mathbf{N} \cup \{\infty\}$ we denote respectively the usual extension of real numbers and integers.

[‡]Editor's note: published in Proc. Int. Meeting on Recent Methods in Nonlinear Analysis (Rome, 1978), E. DE GIORGI, E. MAGENES, U. MOSCO EDS, Pitagora, Bologna, 1979, 131–188.

We observe that the limits just defined agree with the usual superior and inferior limit; nevertheless, their redefinition is useful to better grasp the next extensions.

We note that the function $g(\xi) = \Gamma(\tau^-) \lim_{x \to \xi} f(x)$ is the greatest lower semicontinuous function less than or equal to f; in the following, we shall use for it the symbol $g = \mathrm{sc}^-(\tau)f$.

Now we define multiple Γ-limits.

DEFINITION 1.1. BIS – *Let $(X_1, \tau_1), \ldots, (X_n, \tau_n)$ be n topological spaces, $(\xi_1, \ldots, \xi_n) \in X_1 \times \cdots \times X_n$, $f : X_1 \times \cdots \times X_n \to \overline{\mathbf{R}}$. Then we define*

$$\Gamma(\tau_1^\pm, \ldots, \tau_{n-1}^\pm, \tau_n^+) \lim_{\substack{x_1 \to \xi_1 \\ \cdots \\ x_n \to \xi_n}} f(x_1, \ldots, x_n) =$$

$$= \inf_{A \in J_{\tau_n}(\xi_n)} \Gamma(\tau_1^\pm, \ldots, \tau_{n-1}^\pm) \lim_{\substack{x_1 \to \xi_1 \\ \cdots \\ x_{n-1} \to \xi_{n-1}}} \sup_{x_n \in A} f(x_1, \ldots, x_n),$$

and analogously

$$\Gamma(\tau_1^\pm, \ldots, \tau_{n-1}^\pm, \tau_n^-) \lim_{\substack{x_1 \to \xi_1 \\ \cdots \\ x_n \to \xi_n}} f(x_1, \ldots, x_n) =$$

$$= \sup_{A \in J_{\tau_n}(\xi_n)} \Gamma(\tau_1^\pm, \ldots, \tau_{n-1}^\pm) \lim_{\substack{x_1 \to \xi_1 \\ \cdots \\ x_{n-1} \to \xi_{n-1}}} \inf_{x_n \in A} f(x_1, \ldots, x_n).$$

For example, if $n = 2$ we have

$$\Gamma(\tau_1^+, \tau_2^-) \lim_{\substack{x_1 \to \xi_1 \\ x_2 \to \xi_2}} f(x_1, x_2) = \sup_{A_2 \in J_{\tau_2}(\xi_2)} \inf_{A_1 \in J_{\tau_1}(\xi_1)} \sup_{x_1 \in A_1} \inf_{x_2 \in A_2} f(x_1, x_2)$$

$$\Gamma(\tau_1^-, \tau_2^-) \lim_{\substack{x_1 \to \xi_1 \\ x_2 \to \xi_2}} f(x_1, x_2) = \sup_{A_2 \in J_{\tau_2}(\xi_2)} \sup_{A_1 \in J_{\tau_1}(\xi_1)} \inf_{x_1 \in A_1} \inf_{x_2 \in A_2} f(x_1, x_2).$$

We note that, while $\Gamma(\tau_1^+, \ldots, \tau_n^+)$ limits and the $\Gamma(\tau_1^-, \ldots, \tau_n^-)$ limits agree with the usual superior and inferior limits with respect to the product topology on $X_1 \times \cdots \times X_n$, multiple Γ-limits with different signs seem to be a new concept in topology. To simplify the writing, we use the following short notation: when the Γ-limit is independent of the signs associated with one of the topologies, then the sign is omitted. For example, if

$$\Gamma(\tau_1^+, \tau_2^-) \lim_{\substack{x_1 \to \xi_1 \\ x_2 \to \xi_2}} f(x_1, x_2) = \Gamma(\tau_1^-, \tau_2^-) \lim_{\substack{x_1 \to \xi_1 \\ x_2 \to \xi_2}} f(x_1, x_2) = \lambda$$

we write

$$\Gamma(\tau_1, \tau_2^-) \lim_{\substack{x_1 \to \xi_1 \\ x_2 \to \xi_2}} f(x_1, x_2) = \lambda.$$

Moreover, if the first space (X_1, τ_1) is a parameter space (for example $\overline{\mathbf{N}}$ or $\overline{\mathbf{R}}$) endowed with its usual topology and if the Γ-limit is independent of the sign associated with τ_1, then we shall often omit the symbol τ_1. For example, if $\{f_h\}$

is a sequence of functions on a topological space (X, τ), with values in $\overline{\mathbf{R}}$, we write

$$\Gamma(\tau^-) \lim_{\substack{h \to +\infty \\ x \to \xi}} f_h(x)$$

instead of

$$\Gamma(\vartheta, \tau^-) \lim_{\substack{h \to +\infty \\ x \to \xi}} f_h(x),$$

where ϑ denotes the usual topology of $\overline{\mathbf{N}}$.

We note that

$$\Gamma(\tau^+) \lim_{\substack{h \to +\infty \\ x \to \xi}} f_h(x) = -\Gamma(\tau^-) \lim_{\substack{h \to +\infty \\ x \to \xi}} \left(-f_h(x)\right),$$

hence each property of $\Gamma(\tau^-)$-limits has an immediate translation into a property of $\Gamma(\tau^+)$-limits. From now on we consider only $\Gamma(\tau^-)$-limits, because, as we shall see in Section 3, these Γ-limits find some applications in convergence of minima in many problems of the calculus of variations.

If (X, τ) is a topological space satisfying the first countability axiom, we may characterize the $\Gamma(\tau^-)$ limit of a sequence $\{f_h\}$ of functions from X to $\overline{\mathbf{R}}$ by sequences in X, as it is explained in the following proposition.

THEOREM 1.2. – (see De Giorgi-Franzoni [39] or Buttazzo [18]). *If (X, τ) is a topological space satisfying the first countability axiom, and $\{f_h\}$ is a sequence of functions from X to $\overline{\mathbf{R}}$, then*

$$\Gamma(\tau^-) \lim_{\substack{h \to +\infty \\ x \to \xi}} f_h(x) = \lambda$$

if and only if

(i) *for every sequence $\{x_h\}$, that converges to ξ in (X, τ), one has*

$$\lambda \leq \liminf_{h \to +\infty} f_h(x_h);$$

(ii) *there exists a sequence $\{x_h\}$, that converges to ξ in (X, τ), such that*

$$\lambda = \lim_{h \to +\infty} f_h(x_h).$$

The first important property of $\Gamma(\tau^-)$ limits is provided by the following theorem.

THEOREM 1.3. – (see De Giorgi-Franzoni [39] or Buttazzo [18]). *Let (X, τ) be a topological space and let $\{f_h\}$ be a sequence of functions from X to $\overline{\mathbf{R}}$. If we have, for every $\xi \in X$,*

$$f(\xi) = \Gamma(\tau^-) \lim_{\substack{h \to +\infty \\ x \to \xi}} f_h(x)$$

then f is lower semicontinuous in (X, τ).

The second important result is the following compactness theorem.

THEOREM 1.4. – (see De Giorgi-Franzoni [39]). *If (X, τ) is a topological space with a countable basis for open sets, then from each sequence $\{f_h\}$ of functions from X to $\overline{\mathbf{R}}$ it is possible to select a subsequence $\{f_{h_k}\}$ so that, for every $\xi \in X$, there exists*

$$\Gamma(\tau^-) \lim_{\substack{k \to +\infty \\ x \to \xi}} f_{h_k}(x).$$

From the definition of Γ-limit, it is easy to pass to the definition of G-limit.

DEFINITION 1.5 – *Let $(X_1, \tau_1), \ldots, (X_n, \tau_n), (Y_1, \vartheta_1), \ldots, (Y_m, \vartheta_m)$ be $(n + m)$ topological spaces, and*

$$F : X_1 \times \cdots \times X_n \to \mathcal{P}(Y_1 \times \cdots \times Y_m).$$

Now, for a fixed $(\xi_1, \ldots, \xi_n) \in X_1 \times \cdots \times X_n$, we write

$$G(\tau_1^\pm, \ldots, \tau_n^\pm; \vartheta_1^\pm, \ldots, \vartheta_m^\pm) \lim_{\substack{x_1 \to \xi_1 \\ \cdots \\ x_n \to \xi_n}} F(x_1, \ldots, x_n) = L$$

if, for every $(\eta_1, \ldots, \eta_m) \in Y_1 \times \cdots \times Y_m$, one has

$$\Gamma(\tau_1^\pm, \ldots, \tau_n^\pm, \vartheta_1^\pm \ldots, \vartheta_m^\pm) \lim_{\substack{x_1 \to \xi_1 \\ \cdots \\ x_n \to \xi_n \\ y_1 \to \eta_1 \\ \cdots \\ y_m \to \eta_m}} \chi_{F(x_1, \ldots, x_n)}(y_1, \ldots, y_m) = \chi_L(\eta_1, \ldots, \eta_m),$$

where, for every $A \in \mathcal{P}(Y_1 \times \cdots \times Y_m)$, χ_A denotes the characteristic function of A (i.e. $\chi_A(y) = 1$ if $y \in A$, $\chi_A(y) = 0$ otherwise).

We note that in particular cases of the preceding definition one may recover well-known mathematical objects. For example, $G(\vartheta^+)E$ is the closure of E in the topology ϑ, $G(\vartheta^-)E$ is the largest open set contained in E and analogously $G(\vartheta_1^+, \ldots, \vartheta_m^+)E$ and $G(\vartheta_1^-, \ldots, \vartheta_m^-)E$ are respectively the closure and the interior part of E in the product topology on $Y_1 \times \cdots \times Y_m$. Finally, if $F : X \to \mathcal{P}(Y)$, the $G(\tau^+; \vartheta^+)$-limit and the $G(\tau^-; \vartheta^-)$-limit correspond to the superior and inferior limits for sets used by Kuratowski [44].

2. Γ-limits of integral functionals

When in the calculus of variations we consider Γ-limits of integrals, we must answer the following question: when a Γ-limit of integrals is an integral? Usually, the answer is not trivial. The next theorem will be very useful for the correct formulation of these problems.

THEOREM 2.1 – (see De Giorgi-Franzoni [39]). *If (X, τ) is a topological space and $\{F_h\}$ is a sequence of functions from X to $\overline{\mathbf{R}}$, then, if $F_h \geq \tilde{F}_h \geq \mathrm{sc}^-(\tau)F_h$, one has*

$$\Gamma(\tau^-) \lim_{\substack{h \to +\infty \\ v \to u}} F_h(v) = \Gamma(\tau^-) \lim_{\substack{h \to +\infty \\ v \to u}} \tilde{F}_h(v).$$

This theorem allows us to replace, in the practical calculus of the Γ-limits, a sequence $\{\tilde{F}_h\}$ with another sequence $\{F_h\}$ such that

$$\mathrm{sc}^-(\tau)F_h \leq \tilde{F}_h \leq F_h.$$

For example, if we want to study sequences of functionals as

$$\int_\Omega f_h(x, u, Du, \dots, D^\nu u)\,dx,$$

it is not necessary to fix a priori the "best space" on which one defines the functionals (Sobolev spaces $H^{k,p}(\Omega)$; $BV(\Omega)$; Orlicz spaces; or others), but one may define F_h on a large class $T(\Omega)$ (as $L^1_{\mathrm{loc}}(\Omega)$), so that F_h takes the value $+\infty$ except on a small class $S(\Omega)$ (as $C^\infty(\Omega)$) on which F_h agrees with the previous integral. Afterwards, one studies the Γ-limit of $\{F_h\}$ and finally one checks the subset of $T(\Omega)$ on which the Γ-limit takes a finite value: in many cases this subset is the "best space" usually considered in calculus of variations and functional analysis. In this section and in the next ones we shall follow this path, though in most of the original papers the problems were set in the "best space", generally a Banach space such that F_h is continuous in the strong topology and coercive in the weak topology. Our new way of acting has no disadvantages in standard problems because of Theorem 2.1, and becomes rather advantageous when the "best space" for F_h is different from the "best space" for the Γ-limit, as it happens, for example, in the cases studied by L. Modica and S. Mortola [59], [60].

In applications, the following situation is very common: one has a sequence $\{F_h(u, \Omega)\}$ of functionals defined for every bounded open subset Ω of $\mathbf{R}^n$ and for every u belonging to a topological vector space $T(\Omega)$ (as $L^1_{\mathrm{loc}}(\Omega)$, $L^p(\Omega)$, $H^{k,p}(\Omega)$ or others) as

$$(2.1) \qquad F_h(u, \Omega) = \begin{cases} \displaystyle\int_\Omega f_h(x, u, Du, \dots, D^\nu u)\,dx & \text{if } u \in S(\Omega) \\[2ex] +\infty & \text{if } u \in T(\Omega) \setminus S(\Omega), \end{cases}$$

where f_h are nonnegative functions, measurable in x and continuous in the other variables, $D^k u$ denotes the vector of the k-derivatives of u and $S(\Omega)$ is the subspace of $T(\Omega)$ (as $C^\infty(\Omega)$, $C^k(\Omega)$, $H^{k,p}(\Omega)$ or others), formed generally by more regular functions than the elements of $T(\Omega)$. A condition we suppose on $T(\Omega)$ is the continuity of the restriction map from $T(\Omega)$ in $T(\Omega')$, whenever Ω' is an open subset of Ω.

With these hypotheses, by the compactness theorem 1.4 one infers that there exists a subsequence of $\{F_h\}$ (that, for simplicity, we continue to denote by $\{F_h\}$) such that, for every bounded open subset Ω of $\mathbf{R}^n$ and for every $u \in T(\Omega)$,

$$(2.2) \qquad \sup_{\Omega' \subset\subset \Omega} \Gamma(\overline{\mathbf{N}}^-, T(\Omega')^-) \lim_{\substack{h \to +\infty \\ v \to u}} F_h(v, \Omega') =$$

$$= \sup_{\Omega' \subset\subset \Omega} \Gamma(\overline{\mathbf{N}}^+, T(\Omega')^-) \lim_{\substack{h \to +\infty \\ v \to u}} F_h(v, \Omega').$$

We write $\Omega' \subset\subset \Omega$ if the closure $\overline{\Omega'}$ of Ω' is a compact subset of Ω.

Let $F(u, \Omega)$ be the functional defined by (2.2). By the results of Dal Maso [35] and De Giorgi-Letta [40], (2.2) is equivalent to saying that, for every $u \in T(\Omega)$, the class of the open subsets A of Ω for which

$$(2.3) \qquad F(u, A) = \Gamma\big(T(A)^-\big) \lim_{\substack{h \to +\infty \\ v \to u}} F_h(v, A)$$

is rich, that is, whenever we fix a family $(A_t)_{t \in [0,1]}$ of open subsets of Ω, with $A_t \subset\subset A_s$ if $t < s$, then, for all $t \in [0, 1]$ except for a countable set of t, (2.3) holds with $A = A_t$.

It is natural, at this point, to ask whether the functional $F(u, \Omega)$ may be put in an integral form and eventually to precise such an integral or, at least, to obtain a number of qualitative properties of $F(u, \Omega)$. Several works have been devoted to this problem and may be framed in the scheme we are exposing.

Let us consider a sequence $\{F_h(u, \Omega)\}$ as defined in (2.1), such that (2.2) holds. We are interested in the following properties.

PROPERTY 1. – *For every open bounded subset Ω of $\mathbf{R}^n$ (or, at least, for the sets Ω with a smooth boundary) one has, for every $u \in T(\Omega)$*

$$F(u, \Omega) = \Gamma\big(T(\Omega)^-\big) \lim_{\substack{h \to +\infty \\ v \to u}} F_h(v, \Omega).$$

PROPERTY 2. – *For every $u \in T(\Omega)$, the application $A \mapsto F(u, A)$ defined on the open subsets A of Ω is the trace of a measure defined on the Borel subsets of Ω.*

PROPERTY 3. – *There exists a non-negative function g, measurable in the first variable and continuous in the other variables, such that*

$$F(u, \Omega) = \int_\Omega g(x, u, Du, \dots, D^\nu u) dx$$

for every $u \in S'(\Omega)$, where $S'(\Omega) \subset T(\Omega)$ is a suitable class of regular functions (it may happen that $S'(\Omega) = S(\Omega)$).

PROPERTY 4. – *If*

$$\tilde{F}(u, \Omega) = \begin{cases} F(u, \Omega) & \text{if } u \in S'(\Omega) \\ +\infty & \text{if } u \in T(\Omega) \setminus S'(\Omega) \end{cases}$$

then $F(u, \Omega) = \mathrm{sc}^-\big(T(\Omega)\big)\tilde{F}(u, \Omega)$. In other words, F may be obtained by semi-continuity from its values on $S'(\Omega)$.

For perturbation problems it is interesting to check the following Property 5 for a suitable choice of the topological space $T'(\Omega)$ contained in $T(\Omega)$.

PROPERTY 5. – *For every $u \in T'(\Omega)$, we have*

$$\Gamma\big(T(\Omega)^-\big) \lim_{\substack{h \to +\infty \\ v \to u}} F_h(v, \Omega) = \Gamma\big(T'(\Omega)^-\big) \lim_{\substack{h \to +\infty \\ v \to u}} F_h(v, \Omega).$$

An example verifying all the five properties is the one given by E. De Giorgi [37] and C. Sbordone [78]: the integrands f_h are such that

$$f_h : \Omega \times \mathbf{R} \times \mathbf{R}^n \to \mathbf{R}$$
$$|p|^\alpha \leq f_h(x, u, p) \leq s(1 + |u|^\alpha + |p|^\alpha)$$
$$|f_h^{1/\alpha}(x, u_1, p_1) - f_h^{1/\alpha}(x, u_2, p_2)| \leq s(|u_1 - u_2| + |p_1 - p_2|)$$

for some $\alpha \geq 1$ and $s > 0$ (independent of h, x, u, p). In the case $\alpha = 1$ and $S(\Omega) = S'(\Omega) = C^\infty(\Omega)$, $T(\Omega) = L^1_{\mathrm{loc}}(\Omega)$, Property 4 holds only for $u \in H^{1,1}(\Omega)$ and not, generally, for $u \in L^1_{\mathrm{loc}}(\Omega)$, Property 5 holds for $T'(\Omega) = L^p(\Omega)$ ($1 \leq p < \infty$) or $T'(\Omega) = C^0(\Omega)$.

In the case $\alpha > 1$, all properties hold, without restrictions, with the same choice of T, S, T', S'. An interesting example of the case $\alpha = 2$ is given by the "quadratic" functionals

$$\int_\Omega \sum_{i,j=1}^n a_{ij}^{(h)} D_i u D_j u \, dx,$$

where $a_{ij}^{(h)} \in L^\infty(\Omega)$ and

$$\lambda |\xi|^2 \leq \sum_{i,j=1}^n a_{ij}^{(h)} \xi_i \xi_j \leq \Lambda |\xi|^2$$

for some $0 < \lambda < \Lambda$ (independent of h) and for every $\xi \in \mathbf{R}^n$.

It should be noted that the Γ-limit of a sequence of quadratic functionals of this kind is also a quadratic functional of the same type.

Other theorems in this direction have been proved in several papers: I quote G. Buttazzo, M. Tosques [21] (functionals involving higher order derivatives), L. Carbone, C. Sbordone [27] (functionals as $\int_\Omega f(x, Du)dx$), M. Carriero, E. Pascali [28] (integrands such that $0 \leq f(x, u, p) \leq s(1 + |u| + |p|)$), G. Dal Maso [35] (abstract conditions for properties 1 and 2), P. Marcellini, C. Sbordone [52], [54] (nonuniformly elliptic quadratic functionals).

Now, let us examine some sequences of functionals which do not verify all the five properties listed above.

EXAMPLE 1. – (see G. Buttazzo, G. Dal Maso [20]). If $n \geq 4$, $a, b \in \mathbf{R}^n$, $S(a, b) = \{ta + (1 - t)b : t \in [0, 1]\}$,

$$\varphi_h(x) = \begin{cases} 1 & \text{if dist}\,(x, S(a, b)) < 1/h \\ 0 & \text{otherwise} \end{cases}$$

$$F_h(u, \Omega) = \int_\Omega \left[|x - a|^{1-n} + |x - b|^{1-n} + h^{n-1}\varphi_h(x) \right] |Du|^2 dx$$

then, by passing to the $\Gamma(L^1_{\mathrm{loc}}(\Omega)^-)$-limit, one obtains, for every $u \in C^1(\Omega)$,

(i)   if $S(a, b) \subset \Omega$ then

$$F(u, \Omega) = \int_{\Omega} \left[|x - a|^{1-n} + |x - b|^{1-n} \right] |Du|^2 dx + C \big(u(b) - u(a) \big)^2,$$

(ii)   if $S(a, b) \cap (\mathbf{R}^n \setminus \overline{\Omega}) \neq \emptyset$ then

$$F(u, \Omega) = \int_{\Omega} \left[|x - a|^{1-n} + |x - b|^{1-n} \right] |Du|^2 dx.$$

From (i) and (ii) it follows that $F(u, \Omega)$ is not a measure.

EXAMPLE 2. – (see P. Marcellini, C. Sbordone [52]). If $n = 1$ and

$$a_h(x) \;=\; \begin{cases} 1 & \text{if } |x| \geq 1/h \\ 1/h & \text{if } |x| < 1/h \end{cases}$$

$$F_h(u, \Omega) \;=\; \int_{\Omega} a_h(x) \Big(\frac{du}{dx} \Big)^2 dx,$$

$$T(\Omega) \;=\; L^1_{\mathrm{loc}}(\Omega),$$

then

$$F(u,]-1, 1[) = \begin{cases} \displaystyle\int_{-1}^{1} \Big(\frac{du}{dx} \Big)^2 dx \;\; + \frac{1}{2} \big(u(0^+) - u(0^-) \big)^2 \\ \qquad\qquad\qquad \text{if } u \in W^{1,2}(]-1, 0[\cup]0, 1[) \\[2mm] +\infty \qquad\qquad\qquad \text{otherwise,} \end{cases}$$

where $u(0^+)$ and $u(0^-)$ denote respectively the right and the left limit of u at 0. It follows that if $S'(\Omega) = C^1(\Omega)$ Property 4 holds only for functions that are continuous in 0.

EXAMPLE 3. – The functionals studied in L. Modica, S. Mortola [59]

$$F_h(u, \Omega) = \int_{\Omega} \Big[\frac{|Du|^2}{h} + h(1 - u^2)^2 \Big] dx$$

have a $\Gamma \big(L^1_{\mathrm{loc}}(\Omega)^- \big)$-limit that takes the value $+\infty$ on all regular non-constant functions, being finite only for functions $u \in BV(\Omega)$ such that $|u| = 1$. In this case also, Property 4 does not hold when $S'(\Omega) \subset C^0(\Omega)$.

EXAMPLE 4. – (see L. Carbone, C. Sbordone [27]). Let us consider in $\mathbf{R}^3$ the functionals

$$F_h(u, \Omega) = \int_{\Omega} a_h(x) |Du|^2 dx$$

where

$$a_h(x) = \begin{cases} h^2 & \text{if } x_1^2 + x_2^2 < 1/h^2 \\ 1 & \text{otherwise.} \end{cases}$$

If Ω is the half-cylinder $\{(x_1, x_2, x_3) : x_1^2 + x_2^2 < 1,\ x_2 > 0,\ |x_3| < 1\}$, $T(\Omega) = L^1(\Omega)$, $T'(\Omega) = C^0(\overline{\Omega})$, then, for every Lipschitz function u,

$$\Gamma\big(T(\Omega)^-\big) \lim_{\substack{h \to +\infty \\ v \to u}} F_h(v, \Omega) = \int_\Omega |Du|^2 dx$$

$$\Gamma\big(T'(\Omega)^-\big) \lim_{\substack{h \to +\infty \\ v \to u}} F_h(v, \Omega) = \int_\Omega |Du|^2 dx + \frac{\pi}{2} \int_{-1}^{1} \left| \frac{\partial u}{\partial x_3}(0, 0, t) \right| dt$$

and Property 5 does not hold.

The five properties remain interesting even when the sequence $\{F_h(u, \Omega)\}$ does not depend on h: it is the problem of a semicontinuous extension of a single functional of type $\int_\Omega f(x, u, Du, \dots, D^\nu u) dx$ (see J. Serrin [81]). About this subject, I want to point out the following conjectures:

CONJECTURE I. – *It should be interesting to know whether the first four properties hold when $T(\Omega) = L^1_{\text{loc}}(\Omega)$, $S'(\Omega) = S(\Omega) = C^\infty(\Omega)$, f is a nonnegative function, measurable in x, continuous in the other variables, bounded on bounded sets.*

CONJECTURE II. – *It should be interesting to know whether, adding to the previous hypotheses the global continuity of f and the strict convexity of f in the highest order derivatives of u, Property 3 holds with $g = f$.*

To show the reason of the hypothesis of global continuity we quote an example.

EXAMPLE. – (see L. Carbone, C. Sbordone [27]). Let B be an open dense subset of $\mathbf{R}$, with finite measure, and let $\chi_B(x)$ be its characteristic function; let

$$F(u, \Omega) = \int_\Omega \big(1 - \chi_B(x)\big) \left(\frac{du}{dx} \right)^2 dx.$$

Then $\mathrm{sc}^- \big(L^1_{\text{loc}}(\Omega) \big) F(u, \Omega) \equiv 0$.

Also the hypothesis of strict convexity seems to be essential, as an example of N. Aronszajn shows: for this example we refer to C. Pauc [65], pg. 54.

We note that Conjecture II has been proved by J. Serrin [81], Th. 12, when the highest order derivative is the first one (i.e., $\nu = 1$).

3. Γ-convergence and calculus of variations

We start by establishing an abstract proposition: as regards the convergence of minima and of minimum points, it becomes clear that the main field of application of Γ-convergence is the calculus of variations.

THEOREM 3.1. – *(see De Giorgi-Franzoni [39]) Let (X, τ) be a topological space, and let $\{F_h\}$ be a sequence of functions defined on X, with values in $\overline{\mathbf{R}}$. Suppose that there exist an open subset V of X and a compact subset K of X such that*

$$(3.1) \qquad \inf_{x \in K} F_h(x) = \inf_{x \in V} F_h(x) \qquad \textit{for every } h \in \mathbf{N}$$

and moreover that there exists, for every $x \in V$,

$$F(x) = \Gamma(\tau^-) \lim_{\substack{h \to +\infty \\ y \to x}} F_h(y).$$

Then

$$\min_{x \in K} F(x) = \min_{x \in V} F(x) = \lim_{h \to +\infty} \left[\inf_{x \in V} F_h(x)\right].$$

Moreover, if $\{x_h\}$ is a sequence in V, that converges in the topology τ to $x_0 \in V$, such that

$$\lim_{h \to +\infty} \left[F_h(x_h) - \inf_{x \in V} F_h(x)\right] = 0,$$

then

$$F(x_0) = \min_{x \in V} F(x).$$

When Theorem 3.1 applies, hypothesis (3.1) is generally guaranteed by the following equicoerciveness condition: for every $p > 0$, there exists a compact subset K_p of X such that

$$\bigcup_{h=1}^{+\infty} \left\{x \in X : \; F_h(x) \le p\right\} \subset K_p.$$

In most papers, X is a function space and

$$F_h(u) = \int_\Omega f_h(x, u, Du)dx,$$

hence, the information given by Theorem 3.1 may be enriched using regularity properties of f_h, studying Euler equations for F_h, choosing in the most opportune way the topology on X, etc.

Now, we pass to some interesting examples of problems satisfying the hypotheses of Theorem 3.1.

EXAMPLE 3.2. – In the paper [37], which, in a certain sense, may be considered the starting point of Γ-convergence theory, we studied some sequences $\{F_h\}$ of "area-like" functionals, more exactly

$$F_h(u) = \int_\Omega f_h(x, u, Du)dx,$$

where

$$|p| \le f_h(x, u, Du) \le s\big(1 + |u| + |p|\big)$$

$$|f_h(x, u_1, p_1) - f_h(x, u_2, p_2)| \le s\big(|u_1 - u_2| + |p_1 - p_2|\big)$$

and $s > 0$ is independent of h. Suppose that

$$\int_\Omega f(x, u, Du)dx = F(u) = \Gamma\big(L^1_{loc}(\Omega)^-\big) \lim_{\substack{h \to +\infty \\ v \to u}} F_h(v).$$

To have equicoerciveness in $L^1_{\text{loc}}(\Omega)$, it is useful to consider these other functionals

$$G_h(u) = \text{sc}^- \left(L^1_{\text{loc}}(\Omega)\right)\left[F_h(u) + \int_\Omega |u| dx\right],$$

for which (see §4)

$$\Gamma\left(L^1_{\text{loc}}(\Omega)^-\right) \lim_{\substack{h \to +\infty \\ v \to u}} G_h(v) = F(u) + \int_\Omega |u| dx = G(u).$$

Theorem 3.1 immediately yields that the minima of G_h in $L^1_{\text{loc}}(\Omega)$ converge to the minimum of G and analogously for minimum points.

EXAMPLE 3.3. – Quadratic functionals have been object of a lot of works, chiefly because they are related to the theory of G-limits of elliptic linear differential operators (see S. Spagnolo [85]). Let us consider

$$F_h(u) = \int_\Omega \left[\sum_{i,j=1}^n a_{ij}^{(h)} D_i u D_j u + u^2 + 2\varphi u\right] dx$$

where $a_{ij}^{(h)} \in L^\infty(\Omega)$, $a_{ij}^{(h)} = a_{ji}^{(h)}$, $\varphi \in L^2(\Omega)$ and the $a_{ij}^{(h)}$ satisfy the usual equiuniform ellipticity condition

$$\lambda|\xi|^2 \le \sum_{i,j=1}^n a_{ij}^{(h)} \xi_i \xi_j \le \Lambda|\xi|^2, \qquad \forall \xi \in \mathbf{R}^n, \forall h \in \mathbf{N},$$

with $0 < \lambda < \Lambda$. If $\{F_h\}_h$ $\Gamma\left(L^2(\Omega)^-\right)$-converges, then the Γ-limit (see E. De Giorgi, S. Spagnolo [41], S. Spagnolo [83]) is

$$F(u) = \int_\Omega \left[\sum_{i,j=1}^n a_{ij} D_i u D_j u + u^2 + 2\varphi u\right] dx$$

with a_{ij} satisfying the same ellipticity inequality as the $a_{ij}^{(h)}$.

In this case the hypotheses of Theorem 3.1 are verified as a consequence of the equicoerciveness in $L^2(\Omega)$, provided Ω is regular.

Then, if $\{u_h\}$ is a sequence of minima for $\{F_h\}$, this sequence converges in $L^2(\Omega)$ to a function u which gives the minimum of F. This result is equivalent to the convergence of the solutions of a Neumann problem for the equations

$$\sum_{i,j=1}^n D_i\left(a_{ij}^{(h)} D_j u_h\right) - u_h = \varphi$$

to the solution of the same problem for the equation

$$\sum_{i,j=1}^n D_i\left(a_{ij} D_j u\right) - u = \varphi.$$

Example 3.4. – This is the homogeneization problem (see P. Marcellini, C. Sbordone [53], P. Marcellini [50]). Here we briefly expose a general case of non-linear homogeneization (see L. Carbone, C. Sbordone [27]). Let $f : \mathbf{R} \times \mathbf{R}^n \to \overline{\mathbf{R}}$ be a function such that

$f(x, p)$ is periodic in x, with period 1, and convex in p,

$$m(x)|p|^\alpha \leq f(x, p) \leq M(x)(1 + |p|^\alpha) \qquad \alpha \geq 2$$

with m, M periodic, with period 1, and $m^{-1} \in L^{\alpha-1}_{\mathrm{loc}}(\mathbf{R}^n)$, $M \in L^1_{\mathrm{loc}}(\mathbf{R}^n)$; the functionals

$$F_h(u) = \int_\Omega \big(f(hx, Du) + |u| \big) dx$$

are equicoercive in $L^1_{\mathrm{loc}}(\Omega)$, there exists the Γ-limit

$$F(u) = \int_\Omega \big(f_0(Du) + |u| \big) dx$$

where f_0 is convex, is a solution of a particular minimum problem and satisfies the same inequalities as f. One deduces from Theorem 3.1 the convergence of the solutions of the "heterogeneous" minimum problems to a solution of the "homogeneous" minimum problem.

Up to here, we have examined some examples of "free" problems, which means that we seek for minima without imposing a priori boundary conditions. But even several boundary problems can be attacked by Γ-convergence: we shall devote to this subject the next section[*].

Before I conclude this section, I want to sketch rapidly the convergence problem for the "local minima".

Let A be an open subset of $\mathbf{R}^n$ and let, for every open set $\Omega \subset\subset A$, $F(u, \Omega)$ be a given functional defined on $L^1_{\mathrm{loc}}(\Omega)$. We say that u is a *local minimum* of F on A if for every $\Omega \subset\subset A$ and for every $\varphi \in L^1_{\mathrm{loc}}(\Omega)$, whose support is contained in Ω, one has

$$F(u, \Omega) \leq F(u + \varphi, \Omega).$$

The following question can be naturally raised: does a sequence $\{u_h\}$ of local minima of a Γ-convergent sequence $\{F_h\}$ converge to a local minimum of the Γ-limit F? To answer this question it is often convenient to look for a "joint" condition as the following one: let Ω_1, Ω_2 be two open bounded subsets of A, with $\overline{\Omega}_1 \subset A$, $\overline{\Omega}_2 \subset A$ and Ω be an open subset of $\Omega_1 \cup \Omega_2$, with $\overline{\Omega} \subset \Omega_1 \cup \Omega_2$; for every two sequences $\{u_h\}$ and $\{v_h\}$ that converge respectively in $L^1_{\mathrm{loc}}(\Omega_1)$ and in $L^1_{\mathrm{loc}}(\Omega_2)$ to the same function $u \in L^1_{\mathrm{loc}}(\Omega_1 \cup \Omega_2)$, there exists a third sequence $\{w_h\}$, that converges to u in $L^1_{\mathrm{loc}}(\Omega)$, so that

$$w_h(x) = \begin{cases} u_h(x) & \text{if } x \in \Omega \setminus \Omega_2 \\ v_h(x) & \text{if } x \in \Omega \setminus \Omega_1 \end{cases}$$

---

[*] In the same section we can allow "perturbations" which are more complicated than $|u|$ or $(u^2 + u\varphi)$, added to the main functional or to obstacles, etc.

$$\liminf_{h \to +\infty} F_h(w_h, \Omega) \leq \limsup_{h \to +\infty} F_h(u_h, \Omega_1) + \limsup_{h \to +\infty} F_h(v_h, \Omega_2).$$

An application of these conditions will be shown in the lecture by L. Modica, where the "joint" condition is proved for the functionals

$$F_h(u, \Omega) = \int_\Omega \left[\frac{|Du|^2}{h} + h(1 - u^2)^2 \right] dx.$$

It is possible to prove that these functionals Γ-converge to a functional related to the Caccioppoli-De Giorgi perimeter (see L. Modica, S. Mortola [59], [60]) and, by these conditions, the local minima of functionals $\{F_h\}$ converge to local minimizers of the perimeter.

4. Perturbations, obstacles and boundary problems

If (X, τ) is a topological space and $\{F_h\}$, $\{P_h\}$ are two sequences of functions from X to $\overline{\mathbf{R}}$, one has (see E. De Giorgi, T. Franzoni [39])

$$(4.1) \qquad \Gamma(\tau^-) \lim_{\substack{h \to +\infty \\ v \to u}} \left[F_h(v) + P_h(v) \right]$$

$$\geq \Gamma(\tau^-) \lim_{\substack{h \to +\infty \\ v \to u}} F_h(v) + \Gamma(\tau^-) \lim_{\substack{h \to +\infty \\ v \to u}} P_h(v).$$

In general the strict inequality holds. We also note that an analogous result holds for the product $F_h(u)P_h(u)$, provided $F_h \geq 0$, $P_h \geq 0$.

Equality holds in (4.1) if there exists

$$\Gamma(\tau) \lim_{\substack{h \to +\infty \\ v \to u}} P_h(v),$$

hence, in particular, if $P_h = P$ for every $h \in \mathbf{N}$ and P is continuous in (X, τ). Then

$$(4.2) \qquad \Gamma(\tau^-) \lim_{\substack{h \to +\infty \\ v \to u}} \left[F_h(v) + P(v) \right] = P(u) + \Gamma(\tau^-) \lim_{\substack{h \to +\infty \\ v \to u}} F_h(v).$$

From this fact one infers that a number of different topologies on X, that give origin to the same Γ-limit, permit to have (4.2) for a larger class of perturbations P.

In the calculus of variations (see §3), it is convenient to prove the existence of the Γ-limit in a weak topology, so that the F_h are equicoercive and minima convergence Theorem 3.1 holds; on the other hand, if the Γ-limit does not change when we pass to stronger topologies, a larger number of perturbations may be added to F_h so that (4.2) holds.

Let us consider, for example, a sequence

$$F_h(u) = \int_\Omega \sum_{i,j=1}^n a_{ij}^{(h)} D_i u D_j u \, dx$$

with

$$\lambda|\xi|^2 \leq \sum_{i,j=1}^{n} a_{ij}^{(h)}\xi_i\xi_j \leq \Lambda|\xi|^2$$

and suppose that, for every $u \in L^1_{\mathrm{loc}}(\Omega)$

$$F(u) = \Gamma\big(L^1_{\mathrm{loc}}(\Omega)^-\big) \lim_{\substack{h \to +\infty \\ v \to u}} F_h(v).$$

It is known that, in this case, Property 5 of §2 holds, namely, for every $u \in L^p(\Omega)$, one has

$$F(u) = \Gamma\big(L^p_{\mathrm{loc}}(\Omega)^-\big) \lim_{\substack{h \to +\infty \\ v \to u}} F_h(v).$$

Now we consider the perturbation

$$P(u) = \int_{\Omega} \big[u^4 + u^3 g(x,u)\big]\,dx$$

where $g : \mathbf{R}^{n+1} \to \mathbf{R}$ is a continuous bounded function. Then, for every $u \in L^1_{\mathrm{loc}}(\Omega)$,

$$F(u) + P(u) = \Gamma\big(L^1_{\mathrm{loc}}(\Omega)^-\big) \lim_{\substack{h \to +\infty \\ v \to u}} \big[F_h(v) + P(v)\big].$$

Indeed, it is obvious that, if $u \notin L^4(\Omega)$ both sides of the equality are $+\infty$; on the contrary, if $u \in L^4(\Omega)$, one has

$$\begin{aligned}
F(u) + P(u) \;&\leq\; \Gamma\big(L^1_{\mathrm{loc}}(\Omega)^-\big) \lim_{\substack{h \to +\infty \\ v \to u}} \big[F_h(v) + P(v)\big] \\
&\leq\; \Gamma\big(L^4_{\mathrm{loc}}(\Omega)^-\big) \lim_{\substack{h \to +\infty \\ v \to u}} \big[F_h(v) + P(v)\big] \\
&=\; P(u) + \Gamma\big(L^4_{\mathrm{loc}}(\Omega)^-\big) \lim_{\substack{h \to +\infty \\ v \to u}} F_h(v) = F(u) + P(u).
\end{aligned}$$

It follows from Theorem 3.1 that the *minima* of $F_h(u) + P(u)$ converge *to the minimum* of $F(u) + P(u)$.

We note that this result is not an immediate consequence either of differential equations methods, as g is only continuous and bounded, or of results about convex functionals, as g is not necessarily convex.

It is worthwhile to observe that (4.1) may be valid even if P is not continuous in the topology τ. But, for these cases, we haven't yet abstract, easily applicable conditions. Similar matters appear in a natural way for problems with constraints or obstacles: it is usual to attack these questions by adding to the functionals F_h a term $P(u)$, equal to 0 for the admissible u and equal to $+\infty$ for the non-admissible u. For example we recall two known results. In the first one, due to L. Boccardo, P. Marcellini [16], the functionals are quadratic and equiuniformly elliptic

$$F_h(u) = \int_{\Omega} \sum_{i,j=1}^{n} a_{ij}^{(h)} D_i u D_j u\,dx$$

and P represents an obstacle, i.e.

$$P(u) = \begin{cases} 0 & \text{if } u \geq \psi \\ +\infty & \text{otherwise,} \end{cases}$$

where ψ is a given function. Then, if

$$F(u) = \Gamma\big(L^2(\Omega)^-\big) \lim_{\substack{h \to +\infty \\ v \to u}} F_h(v),$$

it is also true that

$$F(u) + P(u) = \Gamma\big(L^2(\Omega)^-\big) \lim_{\substack{h \to +\infty \\ v \to u}} \big[F_h(v) + P(v)\big],$$

and (4.1) holds. Some generalizations of this result to more general functionals and to sequences of obstacles may be found in P. Marcellini, C. Sbordone [53] and L. Carbone, C. Sbordone [26].

In the second example (see L. Carbone [23]) $P(u)$ is a constraint on the gradient of u; indeed

$$F_h(u) = \int_0^1 \left[2 + \frac{\sin(h!2\pi x)}{|\sin(h!2\pi x)|}\right] \left(\frac{du}{dx}\right)^2 dx,$$

$$P(u) = \begin{cases} 0 & \text{if } \left|\dfrac{du}{dx}\right| \leq 1 \text{ on }]0, 1[\\ \\ +\infty & \text{otherwise.} \end{cases}$$

One obtains

$$\Gamma\big(L^2(0,1)^-\big) \lim_{\substack{h \to +\infty \\ v \to u}} F_h(v) = \frac{3}{2} \int_0^1 \left(\frac{du}{dx}\right)^2 dx,$$

while

$$\Gamma\big(L^2(0,1)^-\big) \lim_{\substack{h \to +\infty \\ v \to u}} \big[F_h(v) + P(v)\big] = \int_0^1 \psi\left(\frac{du}{dx}\right) dx,$$

where ψ is the convex function defined by

$$\psi(t) = \begin{cases} (3/2)t^2 & \text{if } |t| \leq 2/3 \\ \\ -6t^2 - 6t + 2 & \text{if } 2/3 \leq |t| \leq 1 \\ \\ +\infty & \text{if } |t| > 1. \end{cases}$$

So in this case (4.1) does not hold. For further more general results, see L. Carbone [25].

To treat penalization problems, it is useful to calculate the Γ-limit of

$$F(u) + P_h(u),$$

where F is the functional to be minimized and $\{P_h\}$ is the sequence of penalty terms, whose Γ-limit will be 0 on admissible functions and $+\infty$ on non-admissible functions. In this case, the difficulty is to prove (4.1), as well: obviously, this crucial equality is true if the P_h are continuous and the sequence $\{P_h\}$ is non-decreasing.

We note that non-trivial problems of Γ-limit with constraint appear even if the functionals F_h and P_h do not depend on the index h: this is the problem, analogous to the one considered in §2, to construct the greatest lower semicontinuous functional less than or equal to an assigned functional.

Let us consider, for example, the functional

$$F(u) = \begin{cases} \displaystyle\int_\Omega |Du|\,dx & \text{if } u \in C^1(\Omega) \\[2ex] +\infty & \text{otherwise} \end{cases}$$

and the constraint

$$P(u) = \begin{cases} 0 & \text{if } u \in C^1(\Omega) \text{ and } g_1 \le u \le g_2 \\[2ex] +\infty & \text{otherwise,} \end{cases}$$

with g_1 and g_2 assigned functions. The search for $\mathrm{sc}^-\big(L^1_{\mathrm{loc}}(\Omega)\big)[F + P]$ includes trace theorems for BV_{loc}, as well as thin obstacle problems (for which see F. Colombini, E. De Giorgi, L. C. Piccinini [30]). We note that, in this case, if

$$\overline{F} = \mathrm{sc}^-\big(L^1_{\mathrm{loc}}(\Omega)\big)F,$$

it can happen that

$$\mathrm{sc}^-\big(L^1_{\mathrm{loc}}(\Omega)\big)[F + P] = \overline{F} + P',$$

where P' may assume values different from $0, +\infty$. For example, if S is a regular hypersurface contained in Ω, $g : S \to \mathbf{R}$ is a continuous function,

$$(4.3) \qquad P(u) = \begin{cases} 0 & \text{if } u \in C^1(\Omega) \text{ and } u = g \text{ on } S \\[2ex] +\infty & \text{otherwise} \end{cases}$$

then, for every $w \in C^0(\Omega)$

$$(4.4) \qquad P'(w) = 2\int_S |w - g|\,d\sigma,$$

where σ is the usual $(n-1)$-dimensional measure on S.

In this regard, we set the following general question: if F is the functional

$$\int_\Omega f(x, u, Du, \dots, D^\nu u)\,dx$$

and P is a constraint as in (4.3), τ is a suitable topology, does there exist P' such that

$$\mathrm{sc}^-(\tau)[F + P] = \overline{F} + P',$$

where $\overline{F} = \mathrm{sc}^-(\tau)F$ and P' is a constraint, that is it takes only the values $0, +\infty$?

The answer is affirmative when P is defined as in (4.3), $F(u) = \int_\Omega |Du|^2 dx$ and $\tau = L^1_{\mathrm{loc}}(\Omega)$.

The previous question is related to a general method to set the problem of the minimum of

$$(4.5) \qquad \int_\Omega f(x, u, Du)dx$$

under the Dirichlet condition $u = 0$ on $\partial\Omega$. For example, if

$$(4.6) \qquad |p|^\alpha \le f(x, u, p) \le s\left(1 + |u| + |p|\right)^\alpha, \qquad \alpha \ge 1,$$

and f is measurable in x, continuous in u and convex in p, the classical method suggests to study the problem in the Sobolev space $H_0^{1,\alpha}(\Omega)$ (if $\alpha > 1$); but, if $\alpha = 1$, this way is unsuccessful and, moreover, without (4.6) it is not always possible to find a priori the "best space" for the problem. Then, the best way is to consider

$$F(u) = \begin{cases} \displaystyle\int_\Omega f(x, u, Du)dx & \text{if } u \in C^\infty(\Omega) \\[2ex] +\infty & \text{if } u \in L^1_{\mathrm{loc}}(\Omega) \setminus C^\infty(\Omega) \end{cases}$$

and the constraint

$$(4.7) \qquad P(u) = \begin{cases} 0 & \text{if } u \in C_0^\infty(\Omega) \\[2ex] +\infty & \text{if } u \in L^1_{\mathrm{loc}}(\Omega) \setminus C_0^\infty(\Omega) \end{cases}$$

and then to look for the minimum points of $\mathrm{sc}^-\left(L^1_{\mathrm{loc}}(\Omega)\right)[F + P]$. It will be

$$\mathrm{sc}^-\left(L^1_{\mathrm{loc}}(\Omega)\right)[F + P] = \overline{F} + P',$$

where $\overline{F} = \mathrm{sc}^-\left(L^1_{\mathrm{loc}}(\Omega)\right)F$ and P' is a functional depending on F; the class of the $u \in L^1_{\mathrm{loc}}(\Omega)$ such that $\overline{F}(u) + P'(u) < +\infty$ will be the "best space" for the Dirichlet problem. In many cases one finds a large class $\mathcal{F}$ of functionals F, for which P' is the same; for example, for all the functionals (4.5) satisfying (4.6), with $\alpha > 1$, one obtains

$$P'(u) = \begin{cases} 0 & \text{if } u \in H_0^{1,\alpha}(\Omega) \\[2ex] +\infty & \text{if } u \in L^1_{\mathrm{loc}}(\Omega) \setminus H_0^{1,\alpha}(\Omega). \end{cases}$$

After these considerations it is easy to study the convergence of minima for Dirichlet variational problems with Γ-convergence methods.

Let us suppose to have a sequence $\{F_h\}$ of functionals so that

$$(4.8.\text{h}) \qquad F_h(u) = \begin{cases} \displaystyle\int_\Omega f_h(x, u, Du)\,dx & \text{if } u \in C^\infty(\Omega) \\[2ex] +\infty & \text{otherwise} \end{cases}$$

and a functional

$$(4.8) \qquad F(u) = \begin{cases} \displaystyle\int_\Omega f(x, u, Du)\,dx & \text{if } u \in C^\infty(\Omega) \\[2ex] +\infty & \text{otherwise.} \end{cases}$$

Let

$$\overline{F} = \text{sc}^- \left(L^1_{\text{loc}}(\Omega)\right) F$$

and suppose that

$$\Gamma\left(L^1_{\text{loc}}(\Omega)^-\right) \lim_{\substack{h \to +\infty \\ v \to u}} F_h(v) = \overline{F}(u).$$

Saying that the minima of the semicontinuous extension of functionals (4.8.h), with the Dirichlet condition $u = 0$ on $\partial\Omega$, converge to the analogous minimum for (4.8), means that

$$(4.9) \qquad \min \text{sc}^- \left(L^1_{\text{loc}}(\Omega)\right)[F_h + P] \to \min \text{sc}^- \left(L^1_{\text{loc}}(\Omega)\right)[F + P],$$

where P is the constraint defined in (4.7). In order to prove (4.9), if $\{F_h + P\}$ is equicoercive in $L^1_{\text{loc}}(\Omega)$, it suffices to prove that

$$\Gamma\left(L^1_{\text{loc}}(\Omega)^-\right) \lim_{\substack{h \to +\infty \\ v \to u}} \left[F_h(v) + P(v)\right] = \text{sc}^- \left(L^1_{\text{loc}}(\Omega)\right)[F + P](u),$$

and then to apply Theorem 3.1.

Similar results have been obtained in several papers: we quote C. Sbordone [78] and P. Marcellini, C. Sbordone [54]. With the same methods we may study boundary data different from 0.

We conclude by proposing some conjectures and problems. We begin with a conjecture, that perhaps is related to Mechanics.

Conjecture I. – *Let $\{F_h\}$ be a sequence of integral functionals, quadratic in u and in the first and second derivatives, such that (see G. Buttazzo, M. Tosques [21])*

$$\text{dist}\,(\Omega', \mathbf{R}^n \setminus \Omega)^\gamma \int_{\Omega'} \sum_{|\alpha|=2} |D^\alpha u|^2 dx \leq F_h(u, \Omega) \leq s \int_\Omega \left(1 + \sum_{|\alpha|\leq 2} |D^\alpha u|^2\right) dx$$

whenever $\Omega' \subset\subset \Omega$, with $\gamma \geq 0$, $s \geq 1$, dist $(\Omega', \mathbf{R}^n \setminus \Omega) < 1$. Suppose that there exists

$$F(u, \Omega) = \Gamma\left(L^1_{\text{loc}}(\Omega)^-\right) \lim_{\substack{h \to +\infty \\ v \to u}} F_h(v, \Omega)$$

and consider the obstacle

$$P(u, \Omega) = \begin{cases} 0 & \text{if } u \geq \psi \text{ on } \Omega \\ +\infty & \text{otherwise,} \end{cases}$$

where ψ is a regular function.

Is it true that

$$F(u, \Omega) + P(u, \Omega) = \Gamma\left(L^1_{\text{loc}}(\Omega)^-\right) \lim_{\substack{h \to +\infty \\ v \to u}} \left[F_h(v, \Omega) + P(v, \Omega)\right] \ ?$$

Another kind of problem is the following one. Let F be a functional and let $\{P_h\}$ be a sequence of constraints such that

$$\overline{F}(u) + P(u) = \Gamma(\tau^-) \lim_{\substack{h \to +\infty \\ v \to u}} \left[F(v) + P_h(v)\right],$$

where $\overline{F} = \mathrm{sc}^-(\tau)F$. The question is to see whether P is a constraint, that is whether it takes only the values $0, +\infty$. In this connection there is the following conjecture.

CONJECTURE II. – *Let us consider the functional*

$$F(u, \Omega) = \begin{cases} \displaystyle\int_\Omega |Du|^2 dx & \text{if } u \in C^\infty(\Omega) \\ +\infty & \text{otherwise,} \end{cases}$$

and the obstacles

$$P_h(u, \Omega) = \begin{cases} 0 & \text{if } u \geq 0 \text{ on } A_h \cap \Omega \\ +\infty & \text{otherwise,} \end{cases}$$

where $A_h = \{x \in \mathbf{R}^n : \inf_{k \in \mathbf{Z}^n} |x - \frac{k}{h}| < p_h\}$ and $\{p_h\}$ is a sequence such that $\lim_{h \to +\infty} p_h = 0$. *The problem consists in finding under which hypotheses there exists*

$$\Gamma\left(L^1_{\text{loc}}(\Omega)^-\right) \lim_{\substack{h \to +\infty \\ v \to u}} \left[F(v, \Omega) + P_h(v, \Omega)\right] = \overline{F}(u, \Omega) + P(u, \Omega)$$

and, with respect to the speed of convergence of $\{p_h\}$, in realizing the difference between the following three cases:

(i) $P \equiv 0$

(ii) $P(u, \Omega) = c \displaystyle\int_\Omega \left(u - |u|\right)^2 dx \ \text{ with } 0 < c < +\infty$

(iii) $P(u, \Omega) = \begin{cases} 0 & \text{if } u \geq 0 \text{ it } \Omega \\ +\infty & \text{otherwise.} \end{cases}$

5. G-limits of differential equations

Γ-convergence enables us to frame several convergence problems in the calculus of variations, whereas G-convergence is suitable for convergence problems in the theory of differential equations. The most useful G-limits in this field are $G(\mathbf{N}; \sigma^+, \tau^+)$ and $G(\mathbf{N}; \sigma^-, \tau^+)$, which, as in §1, we denote respectively by $G(\sigma^+, \tau^+)$ and $G(\sigma^-, \tau^+)$.

We observe that, if (X, σ) and (Y, τ) are two topological spaces and $\{A_h\}$ is a sequence of subsets of $X \times Y$, then $G(\sigma^+, \tau^+) \lim_{h \to +\infty} A_h$ is the limit, in the sense of Kuratowski [44], of $\{A_h\}$ in the topology product of topologies σ, τ. Therefore, if (X, σ) and (Y, τ) satisfy the first countability axiom, one has that

$$A = G(\sigma^+, \tau^+) \lim_{h \to +\infty} A_h$$

if and only if

(i) $(x, y) \in A \implies \exists\, x_h \to x,\ y_h \to y$ such that $(x_h, y_h) \in A_h$ definitively;

(ii) $\exists\, x_h \to x,\ y_h \to y$ such that $(x_h, y_h) \in A_h$ for infinitely many $h \implies (x, y) \in A$.

Analogously, under the same hypotheses,

$$B = G(\sigma^-, \tau^+) \lim_{h \to +\infty} A_h$$

if and only if

(i) $(x, y) \in B \implies \forall\, x_h \to x \ \exists\, y_h \to y$ such that $(x_h, y_h) \in A_h$ definitively;

(ii) $\forall\, x_h \to x \ \exists\, y_h \to y$ such that $(x_h, y_h) \in A_h$ for infinitely many $h \implies (x, y) \in B$.

We observe that, if (X, σ) and (Y, τ) have a countable basis, compactness theorems as Theorem 1.3 hold for $G(\sigma^+, \tau^+)$-limits and $G(\sigma^-, \tau^+)$-limits.

G-limits enable us to frame convergence problems in differential equations as follows: let us consider a sequence of differential equations in an open subset Ω of $\mathbf{R}^n$

(5.1) $$f_h(x, u, Du, \dots, D^\nu u) = v,$$

where the free term v belongs to a separable metric space V and the solutions u belong to another separable metric space U; if E_h is the set of $(v, u) \in V \times U$ so that (5.1) holds, let us consider

$$(5.2) \qquad G(V^+, U^+) \lim_{h \to +\infty} E_h = A$$

$$(5.3) \qquad G(V^-, U^+) \lim_{h \to +\infty} E_h = B$$

(by passing possibly to a subsequence, such limits exist). Let us suppose that A and B are graphs of differential operators, that is there exist f and g such that

$$A = \{(v, u) \in V \times U : \ f(x, u, Du, \dots, D^\nu u) = v\}$$
$$B = \{(v, u) \in V \times U : \ g(x, u, Du, \dots, D^\nu u) = v\}.$$

(5.3) implies that u is a solution of

$$(5.4) \qquad g(x, u, Du, \dots, D^\nu u) = v$$

if and only if, *for every sequence* $\{v_h\}$ of free terms that converges to v, there exists a sequence $\{u_h\}$ converging to u of solutions of

$$(5.5) \qquad f_h(x, u_h, Du_h, \dots, D^\nu u_h) = v_h,$$

(5.2) implies that, if u is a solution of

$$(5.6) \qquad f(x, u, Du, \dots, D^\nu u) = v$$

then there exists a sequence $\{u_h\}$ converging to u, such that the sequence $\{v_h\}$ defined by

$$v_h = f_h(x, u_h, Du_h, \dots, D^\nu u_h)$$

converges to v; on the other hand if $\{u_h\}$ is a sequence of solutions of (5.5), with free terms $\{v_h\}$, and $u_h \to u$, $v_h \to v$, then (u, v) is a solution of (5.6).

In particular, an interesting case is when the G-limits (5.2) and (5.3) agree, that is there exists

$$G(V, U^+) \lim_{h \to +\infty} E_h = E.$$

In this case, if u is a solution of the equation $(v, u) \in E$, then, for every sequence $\{v_h\}$ of free terms converging to v, there exists a sequence $\{u_h\}$ converging to u and such that u_h verifies (5.5); on the other hand, if $\{u_h\}$ is a sequence of solutions of (5.5) with free term v_h, and $u_h \to u$, $v_h \to v$, then $(v, u) \in E$.

A meaningful example (see S. Spagnolo [85] and E. De Giorgi, S. Spagnolo [41], L. Tartar [88]) is obtained by considering a sequence of equiuniformly elliptic operators

$$E_h = \sum_{i,j=1}^{n} D_j \big(a_{ij}^{(h)}(x) D_i\big),$$

where

$$\lambda_0|\xi|^2 \le \sum_{i,j=1}^{n} a_{ij}^{(h)}(x)\xi_i\xi_j \le \Lambda_0|\xi|^2, \quad 0 < \lambda_0 \le \Lambda_0.$$

Then, if $V = L^2(\Omega)$ and $U = H^{1,2}(\Omega)$ with the $L^2(\Omega)$-topology, identifying operators with their graphs, we obtain that (at least for a subsequence of $\{E_h\}$) there exists

$$G(V,U^+) \lim_{h\to+\infty} E_h = E$$

with

$$E = \sum_{i,j=1}^{n} D_j\big(a_{ij}(x)D_i\big).$$

The scheme exposed so far is suitable to the study of the convergence of the whole set of solutions; further on it may be complicated, if one considers boundary problems or initial value problems (Cauchy, Dirichlet, Neumann, Darboux, mixed problems).

For example, one may consider triples (v,γ,u), where the vector function $\gamma = (\gamma_1,\ldots,\gamma_k)$ belongs to to a suitable class Z, such that (5.1) holds and moreover

$$(5.7) \qquad g_i^{(h)}(x,u,Du,\ldots,D^\nu u) = \gamma_i \quad \text{on } S \quad (i=1,\ldots,k),$$

S being generally a hypersurface in $\overline{\Omega}$. Then, for sets $E_h \subset V \times Z \times U$ formed by the (v,γ,u) satisfying (5.1) and (5.7), one may calculate some different G-limits: the $G(V^\pm, Z^\pm, U^+) - \lim_{h\to+\infty} E_h$.

Now the problems are similar to those proposed in §2, §4 for the Γ-limits of integrals, namely:

(1) See whether the G-limits of differential equations, with or without boundary or initial value conditions, are differential equations with or without conditions of the same kind.

(2) Find some results as in §4 for problems with obstacles, constraints, perturbations.

(3) Consider the behaviour of G-limits when different topologies are introduced, and the possible stability of G-limits in these different topologies.

One meets also new problems, with respect to those of Γ-convergence: the possible coincidence of $G(V^-,U^+)$ and $G(V^+,U^+)$ limits and the analogous fact for $G(V^\pm,Z^\pm,U^+)$-limits.

This is a very large range of problems: even though the known results are more numerous than in the Γ-convergence of integrals, they do not yet permit to provide a compendious scheme, similar to the one attempted in §2, 3, 4. Moreover, most papers impose conditions that appear very special from the point of view of G-convergence, as to fix the free term v or the initial conditions

γ (which is equivalent to reducing the spaces V, Z to a single point, or to endow V, Z with the discrete topology).

Another kind of problem is the relation between Γ-limits of integral functionals and G-limits of their Euler equations. In an abstract manner, these problems have been studied by A. Ambrosetti, C. Sbordone [1], where the functional satisfy a more general, yet similar, condition than convexity. This condition is easily satisfied in the case of equiuniformly elliptic quadratic integrands (see S. Spagnolo [82], ..., [85]). Therefore, it is yet an open question to find conditions more different from convexity, so that one may treat, for example, problems of G-convergence, when $h \to +\infty$, of Euler equations for functionals

$$F_h(u) = \int_\Omega \left[|Du|^2 + f(hu) \right] dx,$$

with f continuous with its first derivative and periodic (the functionals are Γ-convergent by G. Buttazzo, G. Dal Maso [19]).

Finally, one may consider the relation between G-convergence of elliptic equations in n space variables and depending on a parameter t

$$F_h(t, x, u, D_x u, \dots, D_x^\nu u) = w,$$

where D_x^k denote derivatives with respect to the space variables $(x_1, \dots, x_n)$, and G-convergence of parabolic and hyperbolic equations

$$\frac{\partial u}{\partial t} = F_h + w, \qquad \frac{\partial^2 u}{\partial t^2} = F_h + w.$$

I do not even try to list the known results, also because other lectures in this *Colloquio* concern this subject. I recall only that the study of G-limits for equations as

$$\frac{\partial^2 u}{\partial t^2} = \sum_{i,j}^n D_j \left(a_{ij}^{(h)}(t) D_i \right),$$

where the $a_{ij}^{(h)}$ converge in Hölder norm or in L^p, forces to use functional spaces as Gevrey classes or analytic functional spaces (see F. Colombini, E. De Giorgi, S. Spagnolo [31]), which are different from the spaces considered in other G-convergence problems.

I conclude with three open questions.

(1) Study the G-convergence as $h \to +\infty$ of

$$\frac{\partial^2 u}{\partial t^2} = \Delta u + h(|u| - u)$$

for every number of space variables. This problem is related to vibrations with obstacles (see [80]).

(2) Study the G-convergence as $h \to +\infty$ of

$$\frac{\partial u}{\partial t} = \Delta u + h f(hu)$$

where f is continuous, periodic, with null mean value. (This problem is related to the one considered in [19]).

(3)   Let us consider a solution $u \in C^2(\mathbf{R}^n)$ of

$$\Delta u = u^3 - u$$

such that $|u| \leq 1$, $\frac{\partial u}{\partial x_n} > 0$ in the whole $\mathbf{R}^n$. Is it true that, for every $\lambda \in \mathbf{R}$, the sets $\{u = \lambda\}$ are hyperplanes, at least if $n \leq 8$? (This problem is related to [59], [60]).

6. About other possible developments

Beside the numerous open questions indicated in the preceding sections, I will point out here, even if in an informal way, many other possible directions of research. Nevertheless, for some of these I can not make any reasonable prediction.

First, one may investigate more deeply the concept of Γ-limit: it consists in a sequel of inf and sup, hence one may study what happens by changing the order of these operations. Moreover, it should be interesting to extend the theory of sequential Γ-limits, begun by G. Moscariello [62]. Finally, there is the problem of the topological meaning of multiple Γ-limits, of the compatibility of two metrics (see G. Buttazzo [18], E. De Giorgi, T. Franzoni [39]), that is an important feature of the proofs for Property 3 of §2.

One should also exploit the possible applications of $\Gamma(\tau^-, \vartheta^+)$-limits to minimax problems, by looking for theorems of convergence, analogous to Theorem 3.1 for $\Gamma(\tau^-)$-convergence. The relation between Γ- and G-convergence and the methods of finite differences, other computational problems or problems of mechanics and of mathematical physics (see, for example, I. Babuska [3], [4], E. Sanchez Palencia [69], ..., [75]), are other important fields of research. Γ-convergence of measures and of monotone set functions has been applied mainly to integral functionals, as in §2: it cannot be excluded that applications may be found in probability and in stochastic differential equations.

Finally, G-convergence could be useful in order to study properties of partial differential systems with constant coefficients (as the "compatibility", see E. De Giorgi [36]), first obtaining these properties for non-critical open sets and operators, hence by passing to critical sets and operators through a G-limit process.

Bibliography

[1]   A. Ambrosetti, C. Sbordone, Γ-*convergenza e G-convergenza per problemi non lineari di tipo ellittico*, Boll. Un. Mat. Ital. (5) **13-A** (1976), 352–362.

[2] H. Attouch, Y. Konishi, *Convergence d'opérateurs maximaux monotones et inéquations variationnelles,* C. R. Acad. Sci. Paris Sér. A **282** (1976), 467–469.

[3] I. Babuska, *Solution of interface problems by homogenization,* SIAM J. Math. Anal. **7** (1976), 603–634 (I) and 635–645 (II).

[4] I. Babuska, *Homogenization and its applications. Mathematical and computational problems,* Proc. 3rd Symp. Num. Sol. Partial Differential Equations, College Park (1975), B. Hubbard (ed.), Acad. Press 1976, 89–116.

[5] A. Bensoussan, J.L. Lions, G. Papanicolau, *Sur quelques phénomènes asymptotiques stationnaires,* C. R. Acad. Sci. Paris, Sér. A **281** (1975), 89–95.

[6] A. Bensoussan, J.L. Lions, G. Papanicolau, *Sur quelques phénomènes asymptotiques d'évolution,* C. R. Acad. Sci. Paris, Sér. A **281** (1975), 317–322.

[7] A. Bensoussan, J.L. Lions, G. Papanicolau, *Sur des nouveaux problèmes asymptotiques,* C. R. Acad. Sci. Paris, Sér. A **282** (1976), 1277–1282.

[8] A. Bensoussan, J.L. Lions, G. Papanicolau, *Asymptotic methods in periodic structures,* North-Holland, Amsterdam, 1978.

[9] M. Biroli, *Sur la G-convergence pour des inéquations quasi-variationnelles,* C. R. Acad. Sci. Paris, Sér. A **284** (1977), 947–950.

[10] M. Biroli, *Sur la G-convergence pour des inéquations quasi-variationnelles,* Boll. Un. Mat. Ital. (5) **14-A** (1977), 540–550.

[11] M. Biroli, *An estimate on convergence for an homogeneization problem for variational inequalities,* Rend. Sem. Mat. Univ. Padova, **58** (1977), 9–22. ‡

[12] M. Biroli, *G-convergence for elliptic equations, variational inequalities and quasi variational inequalities,* Rend. Sem. Mat. Univ. Milano, **47** (1977), 269–328.

[13] L. Boccardo, I. Capuzzo Dolcetta, *G-convergenza e problema di Dirichlet unilaterale,* Boll. Un. Mat. Ital. (4) **12** (1975), 115–123.

[14] L. Boccardo, I. Capuzzo Dolcetta, *Stabilità delle soluzioni di disequazioni variazionali ellittiche e paraboliche quasi lineari,* Annali Univ. Ferrara, Sez. VII, **24** (1978), 99–111.

---

‡Editor's note: this reference and many other ones of this list have been updated

[15] L. Boccardo, I. Capuzzo Dolcetta, *Disequazioni quasi variazionali con funzioni d'ostacolo quasi limitate: esistenza e G-convergenza*, Boll. Un. Mat. Ital. (5) **15-B** (1978), 370–385.

[16] L. Boccardo, P. Marcellini, *Sulla convergenza delle soluzioni di disequazioni variazionali*, Ann. Mat. Pura Appl. (4) **110** (1976), 137–159.

[17] G. Buttazzo, *Su un tipo di convergenza variazionale*, Tesi di Laurea, Pisa, 1976.

[18] G. Buttazzo, *Su una definizione generale dei Γ-limiti*, Boll. Un. Mat. Ital. (5) **14-B** (1977), 722–744.

[19] G. Buttazzo, G. Dal Maso, *Γ-limit of a sequence of non-convex and non-equi-Lipschitz integral functionals*, Ricerche Mat., **27** (1978), 235–251.

[20] G. Buttazzo, G. Dal Maso, *Γ-limits of integral functionals*, J. Analyse Math. **37** (1980), 145–185.

[21] G. Buttazzo, M. Tosques, *Γ-convergenza per alcune classi di funzionali*, Annali Univ. Ferrara, Sez. VII, **23** (1977), 257–267.

[22] O. Calligaris, *Sulla convergenza delle soluzioni di problemi di evoluzione associati a sottodifferenziali di funzioni convesse*, Boll. Un. Mat. Ital. (5) **13-B** (1976), 778–806.

[23] L. Carbone, *Sur la Γ-convergence des intégrales du type de l'énergie à gradient borné*, J. Math. Pures Appl. (9) **56** (1977), 79–84.

[24] L. Carbone, *Γ-convergence d'intégrales sur des fonctions avec des constraintes sur le gradient*, Comm. Partial Differential Equations **2** (1977), 627–651.

[25] L. Carbone, *Sull'omogeneizzazione di un problema variazionale con vincoli sul gradiente*, Atti Accad. Naz. Lincei, Rend. Cl. Sci. Fis. Mat. Natur., (8) **63** (1978), 10–14.

[26] L. Carbone, C. Sbordone, *Un teorema di compattezza per la Γ-convergenza di funzionali non coercitivi*, Atti Accad. Naz. Lincei, Rend. Cl. Sci. Fis. Mat. Natur., (8) **62** (1977), 744–748.

[27] L. Carbone, C. Sbordone, *Some properties of Γ-limits of integral functionals*, Ann. Mat. Pura Appl. (4) **122** (1979), 1–60.

[28] M. Carriero, E. Pascali, *Γ-convergenza di integrali non negativi maggiorati da funzionali del tipo dell'area*, Annali Univ. Ferrara, Sez. VII, **24** (1978), 51–64.

[29] D. Cioranescu, J. Saint Jean Paulin, *Homogénéisation dans des ouverts à cavités*, C. R. Acad. Sci. Paris Sér. A **284** (1977), 857–860.

[30] F. COLOMBINI, E. DE GIORGI, L. C. PICCININI, *Frontiere orientate di misura minima e questioni collegate,* Quaderno della Scuola Normale Superiore di Pisa, 1972.

[31] F. COLOMBINI, E. DE GIORGI, S. SPAGNOLO, *Existence et unicité des solutions des équations hyperboliques du second ordre à coefficients ne dépendant que du temps,* C. R. Acad. Sci. Paris Sér. A **286** (1978), 1045–1048.

[32] F. COLOMBINI, S. SPAGNOLO, *Sur la convergence des solutions d'équations paraboliques avec des coefficients qui dépendent du temps,* C. R. Acad. Sci. Paris Sér. A **282** (1976), 735–737.

[33] F. COLOMBINI, S. SPAGNOLO, *Sur la convergence des solutions d'équations paraboliques,* J. Math. Pures Appl. **56** (1977), 263–306.

[34] F. COLOMBINI, S. SPAGNOLO, *On the convergence of solutions of hyperbolic equations,* Comm. Partial Differential Equations **3** (1978), 77–103.

[35] G. DAL MASO, *Alcuni teoremi sui Γ-limiti di misure,* Boll. Un. Mat. Ital. (5) **15-B** (1978), 182–192.

[36] E. DE GIORGI, *Sulle soluzioni globali di alcuni sistemi di equazioni differenziali,* Boll. Un. Mat. Ital. (4) **11** suppl. no.3 (1975), 77–79.

[37] E. DE GIORGI, *Sulla convergenza di alcune successioni di integrali del tipo dell'area,* Rend. Mat. (4) **8** (1975) 277–294.

[38] E. DE GIORGI, *Γ-convergenza e G-convergenza,* Boll. Un. Mat. Ital. (5) **14-A** (1977), 213–220.

[39] E. DE GIORGI, T. FRANZONI, *Su un tipo di convergenza variazionale,* Atti Accad. Naz. Lincei, Rend. Cl. Sci. Fis. Mat. Natur., (8) **58** (1975), 842–850.

[40] E.DE GIORGI, G.LETTA, *Une notion générale de convergence faible pour des fonctions croissantes d'ensemble,* Ann. Scuola Norm. Sup. Pisa Cl. Sci. (4) **4** (1977), 61–99.

[41] E. DE GIORGI, S. SPAGNOLO, *Sulla convergenza degli integrali dell'energia per operatori ellittici del 2° ordine,* Boll. Un. Mat. Ital. (4) **8** (1973), 391–411.

[42] G. DUVAUT, *Comportement microscopique d'une plaque perforée périodiquement,* Springer Lecture Notes in Mathematics **594** (1977), 131–145.

[43] J. L. JOLY, *Une famille de topologies sur l'ensemble des fonctions convexes pour lesquelles la polarité est bicontinue,* J. Math. Pures Appl. **52** (1973), 137–158.

[44] C. KURATOWSKI, *Topologie,* Warszawa, 1958.

[45] J. L. Lions, *Cours au Collège de France*, Paris 1976/77.

[46] J. L. Lions, *Cours au Collège de France*, Paris 1977/78.

[47] J. L. Lions, *Asymptotic behaviour of variational inequalities with highly oscillating coefficients*, in : *Applications of methods of Functional Analysis to problems in Mechanics.* Springer Lecture Notes in Mathematics **503** (1976), 30-55.

[48] P. Marcellini, *Su una convergenza di funzioni convesse*, Boll. Un. Mat. Ital. (4) **8** (1973), 137–158.

[49] P. Marcellini, *Un teorema di passaggio al limite per la somma di funzioni convesse*, Boll. Un. Mat. Ital. (4) **11** (1975), 107–124.

[50] P. Marcellini, *Periodic solutions and homogeneization of non linear variational problems*, Ann. Mat. Pura Appl. (4) **117** (1978), 139–152.

[51] P. Marcellini, *Convergence of second order linear elliptic operators*, Boll. Un. Mat. Ital. (5) **16-B** (1979). 278–290.

[52] P. Marcellini, C. Sbordone, *An approach to the asymptotic behaviour of elliptic-parabolic operators*, J. Math. Pures Appl. (9) **56** (1977), 157–182.

[53] P. Marcellini, C. Sbordone, *Sur quelques questions de G-convergence et d'homogénéisation non linéaire*, C. R. Acad. Sci. Paris Sér. A **284** (1977), 535–537.

[54] P. Marcellini, C. Sbordone, *Homogeneization of non uniformly elliptic operators*, Applicable analysis, **68** (1978), 101–114.

[55] P. Marcellini, C. Sbordone, *Dualità e perturbazione di funzionali integrali*, Ricerche Mat. **26** (1977), 383–421.

[56] A. Marino, S. Spagnolo, *Un tipo di approssimazione dell'operatore* $\sum_{i,j} D_i(a_{ij}D_j)$ *con operatori* $\sum_j D_j(\beta D_j)$, Ann. Scuola Norm. Sup. Pisa Cl. Sci. (3) **23** (1969), 657–673.

[57] M. Matzeu, *Su un tipo di continuità dell'operatore sub-differenziale*, Boll. Un. Mat. Ital. (5) **14-B** (1977), 480–490.

[58] W. H. Mc Connell, *On the approximation of elliptic operators with discontinuous coefficients*, Ann. Scuola Norm. Sup. Pisa Cl. Sci. (4) **3** (1976), 139–156.

[59] L.Modica, S. Mortola, *Un esempio di* Γ-*convergenza*, Boll. Un. Mat. Ital. (5) **14-B** (1977), 285–299.

[60] L.Modica, S. Mortola, *Il limite nella* Γ-*convergenza di una famiglia di funzionali ellittici*, Boll. Un. Mat. Ital. (5) **14-A** (1977), 526–529.

[61] G. MOSCARIELLO, *Su una convergenza di successioni di integrali del Calcolo delle Variazioni*, Atti Accad. Naz. Lincei, Rend. Cl. Sci. Fis. Mat. Natur., (8) **61** (1976), 368-375.

[62] G. MOSCARIELLO, *Γ-convergenza negli spazi sequenziali*, Rend. Accad. Sci. Fis. Mat. Napoli, (4) **43** (1976), 333–350.

[63] U. MOSCO, *Convergence of convex sets and of solutions of variational inequalities*, Advances in Math. **3** (1969), 510–585.

[64] F. MURAT, *Sur l'homogénéisation d'inéquations elliptiques du 2ème ordre, relatives au convexe* $K(\psi_1, \psi_2) = \{v \in H_0^1(\Omega) | \psi_1 \leq v \leq \psi_2 \, p.p. \, dans \, \Omega\}$, Publications Univ. Paris VI, Thèse d'état, 1976.

[65] C. PAUC, *La méthode métrique en calcul des variations*, Hermann, Paris, 1941.

[66] L. C. PICCININI, *G-convergence for ordinary differential equations with Peano phænomenon*, Rend. Sem. Mat. Univ. Padova, **58** (1977), 65–86.

[67] L. C. PICCININI, *Homogeneization for ordinary differential equations*, Rend. Circ. Mat. Palermo, (2) **27** (1978), 95–112.

[68] A. PROFETI, B. TERRENI, *Uniformità per una convergenza di operatori parabolici nel caso dell'omogeneizzazione*, Boll. Un. Mat. Ital. (5) **16-B** (1979), 826–841.

[69] J. SANCHEZ HUBERT, E. SANCHEZ PALENCIA, *Sur certaines problèmes physiques d'homogénéisation donnant lieu à des phénomènes de relaxation*, C. R. Acad. Sci. Paris Sér. A **286** (1978), 903–906.

[70] E. SANCHEZ PALENCIA, *Comportement limite d'un problème de transmission à travers une plaque mince et faiblement conductrice*, C. R. Acad. Sci. Paris Sér. A **270** (1970), 1026–1028.

[71] E. SANCHEZ PALENCIA, *Solutions périodiques par rapport aux variables d'éspace et applications*, C. R. Acad. Sci. Paris Sér. A **271** (1970), 1129–1132.

[72] E. SANCHEZ PALENCIA, *Equations aux dérivées partielles dans un type de milieux hétérogènes*, C. R. Acad. Sci. Paris Sér. A **272** (1970), 1410–1413.

[73] E. SANCHEZ PALENCIA, *Problèmes de perturbation liés aux phénomènes de conduction à travers des couches minces de grande resistivité*, J. Math. Pures Appl. **53** (1974), 251–270.

[74] E. SANCHEZ PALENCIA, *Comportement locale et macroscopique d'un type de milieux physiques hétérogènes*, Internat. J. Engineering. Sci. **12** (1974), 331–351.

[75] E. Sanchez Palencia, *Méthode d'homogénéisation pour l'etude de materiaux hétérogènes. Phénomènes de memoire.* Rend. Sem. Mat. Univ. e Politec. Torino, **36** (1977-78), 15–26.

[76] C. Sbordone, *Alcune questioni di convergenza per operatori differenziali del 2° ordine,* Boll. Un. Mat. Ital. (4) **10** (1974), 672–682.

[77] C. Sbordone, *Sulla G-convergenza di equazioni ellittiche e paraboliche,* Ricerche Mat. **24** (1975), 76–136.

[78] C. Sbordone, *Su alcune applicazioni di un tipo di convergenza variazionale,* Ann. Scuola Norm. Sup. Pisa Cl. Sci. (4) **2** (1975), 617–638.

[79] C. Sbordone, *Sur une limite d'intégrales polynômials du calcul des variations,* J. Math. Pures Appl. (9) **56** (1977), 67–77.

[80] M. Schatzman, *A class of hyperbolic unilateral problems with constraints: the vibrating string with obstacle,* Euromech 100, Meeting on: Mathematical Theory of Elasticity, Pisa, May 1978.

[81] J. Serrin, *On the definition and properties of certain variational integrals,* Trans. Amer. Math. Soc., **101** (1961), 139–167.

[82] S. Spagnolo, *Sul limite delle soluzioni di problemi di Cauchy relativi all'equazione del calore,* Ann. Scuola Norm. Sup. Pisa Cl. Sci. (3) **21** (1967), 657–699.

[83] S. Spagnolo, *Sulla convergenza di soluzioni di equazioni paraboliche ed ellittiche,* Ann. Scuola Norm. Sup. Pisa Cl. Sci. (3) **22** (1968), 577–597.

[84] S. Spagnolo, *Some convergence problems,* Symp. Math. (Conv. Ist. Sup. Alta Mat. Roma, March 1974) **18** (1976), 391–398.

[85] S. Spagnolo, *Convergence in energy for elliptic operators,* Proc. 3rd Symp. Num. Sol. Partial Differential Equations, College Park (1975), B. Hubbard (ed.), Acad. Press 1976, 469–498.

[86] S. Spagnolo, *Convergence of parabolic equations,* Boll. Un. Mat. Ital. (5) **14-B** (1977), 547–568.

[87] L. Tartar, *Convergence d'opérateurs différentiels,* in: *Analisi convessa e applicazioni,* Roma 1974, Quaderno di ricerca del CNR, 101–104.

[88] L. Tartar, *Cours Peccot au Collège de France,* Paris 1977.

On a type of variational convergence[‡][†]

Ennio De Giorgi, Tullio Franzoni

Summary. The main properties of a new type of convergence, useful in several topics in calculus of variations are presented here. Proofs and comparisons with known results will appear in forthcoming papers.

Definitions and general properties

DEFINITION 1.1 – Let X be a topological space with topology τ, let $\overline{\mathbf{R}}$ be the extended real line endowed with the usual structure of ordered set and of compact topological space, and let (f_h) be a sequence of functions from X to $\overline{\mathbf{R}}$. We define the following functions from X to $\overline{\mathbf{R}}$:

$$
\begin{aligned}
a) \quad & \Gamma^-(\tau)\min\lim_{h\to\infty} f_h(x) = \sup_{U\in\mathcal{U}_x(\tau)}\ \min\lim_{h\to\infty}\ \inf_{y\in U} f_h(y) \\
& \qquad\qquad\qquad = \sup_{U\in\mathcal{U}_x(\tau)}\ \sup_{h}\ \inf_{p\ge h}\ \inf_{y\in U} f_p(y);
\end{aligned}
$$

$$
b) \quad \Gamma^-(\tau)\max\lim_{h\to\infty} f_h(x) = \sup_{U\in\mathcal{U}_x(\tau)}\ \max\lim_{h\to\infty}\ \inf_{y\in U} f_h(y);
$$

$$
c) \quad \Gamma^+(\tau)\min\lim_{h\to\infty} f_h(x) = \inf_{U\in\mathcal{U}_x(\tau)}\ \min\lim_{h\to\infty}\ \sup_{y\in U} f_h(y);
$$

$$
d) \quad \Gamma^+(\tau)\max\lim_{h\to\infty} f_h(x) = \inf_{U\in\mathcal{U}_x(\tau)}\ \max\lim_{h\to\infty}\ \sup_{y\in U} f_h(y),
$$

where $\mathcal{U}_x(\tau)$ is the class of all (open) neighborhoods of x with respect to the topology τ.

If no confusion arises, we will omit the dependence on τ in the notation, thus we will simply write $\Gamma^-\min\lim_{h\to\infty} f_h$ and similarly for the other limits. Moreover, if a point $x \in X$ is such that $\Gamma^-(\tau)\min\lim_{h\to\infty} f_h(x) = \Gamma^-(\tau)\max\lim_{h\to\infty} f_h(x)$, this common value will be denoted by $\Gamma^-(\tau)\lim_{h\to\infty} f_h(x)$, and we will say that there exists the Γ^- limit of the sequence (f_h) at the point x, or equivalently that the sequence (f_h) is Γ^- convergent at the point x; the meaning of the symbol $\Gamma^+(\tau)\lim_{h\to\infty} f_h(x)$ will be similar. In addition, if a point x is such that there exist the two limits $\Gamma^-(\tau)\lim_{h\to\infty} f_h(x)$, $\Gamma^+(\tau)\lim_{h\to\infty} f_h(x)$ and they coincide, we

---

[‡]Editor's note: translation into English of the paper "Su un tipo di convergenza variazionale", published in the Atti Accad. Naz. Lincei Cl. Sci. Fis. Mat. Natur. **8** 58 (1975), 842–850.

[†]Communicated in the session of June 11th, 1975, by G. Stampacchia.

will denote their common value by $\Gamma(\tau) \lim_{h\to\infty} f_h(x)$, and we will say that there exists the Γ-limit of the sequence (f_h) at x, or equivalently that the sequence (f_h) Γ-converges at x.

Remark 1.2 – If we denote by δ the discrete topology on X, then

$$\min_{h\to\infty} \lim f_h(x) = \Gamma^-(\delta) \min_{h\to\infty} \lim f_h(x) = \Gamma^+(\delta) \min_{h\to\infty} \lim f_h(x),$$

$$\max_{h\to\infty} \lim f_h(x) = \Gamma^-(\delta) \max_{h\to\infty} \lim f_h(x) = \Gamma^+(\delta) \max_{h\to\infty} \lim f_h(x).$$

Remark 1.3 – If σ is a topology on X and σ is stronger than τ, then the following inequalities hold:

$$\Gamma^-(\tau) \min \lim \leq \Gamma^-(\sigma) \min \lim \leq \min \lim \leq \Gamma^+(\sigma) \min \lim \leq \Gamma^+(\tau) \min \lim;$$

$$\Gamma^-(\tau) \max \lim \leq \Gamma^-(\sigma) \max \lim \leq \max \lim \leq \Gamma^+(\sigma) \max \lim \leq \Gamma^+(\tau) \max \lim.$$

Remark 1.4 – The following equalities follow immediately from the definitions:

$$\Gamma^- \min_{h\to\infty} \lim f_h = -\Gamma^+ \max_{h\to\infty} \lim(-f_h);$$

$$\Gamma^- \max_{h\to\infty} \lim f_h = -\Gamma^+ \min_{h\to\infty} \lim(-f_h).$$

This means that any property of Γ^- limits can be translated into a corresponding property of Γ^+ limits (and conversely). Therefore, in the following, we will limit ourselves to state the properties of Γ^- limits.

Proposition 1.5 – *Let σ be a topology on X and suppose that σ is stronger than τ. Let L be a subset of X which is σ-dense in X, and assume that the functions f_h are continuous with respect to σ. Then, setting*

$$g_h(x) = \begin{cases} f_h(x) & \text{if } x \in L \\ +\infty & \text{if } x \in X \setminus L, \end{cases}$$

we have

$$\Gamma^-(\tau) \min_{h\to\infty} \lim (g_h) = \Gamma^-(\tau) \min_{h\to\infty} \lim (f_h);$$

$$\Gamma^-(\tau) \max_{h\to\infty} \lim (g_h) = \Gamma^-(\tau) \max_{h\to\infty} \lim (f_h).$$

Remark 1.6 – If (f_{h_p}) is a subsequence of (f_h) we have

$$\Gamma^- \min_{h\to\infty} \lim f_h \leq \Gamma^- \min_{p\to\infty} \lim f_{h_p} \leq \Gamma^- \max_{p\to\infty} \lim f_{h_p} \leq \Gamma^- \max_{h\to\infty} \lim f_h.$$

PROPOSITION 1.7 – *Let x be a point of X, and let $\xi < \Gamma^- \max\lim_{h\to\infty} f_h(x)$. Then there exists a subsequence (f_{h_p}) of (f_h) which is Γ^- convergent at x and is such that*

$$\Gamma^- \max_{h\to\infty} \lim f_h(x) \geq \Gamma^- \max_{p\to\infty} \lim f_{h_p}(x) > \xi.$$

PROPOSITION 1.8 – *The functions $\Gamma^- \min\lim_{h\to\infty} f_h$ and $\Gamma^- \max\lim_{h\to\infty} f_h$ are lower semicontinuous.*

REMARK 1.9 – If the sequence is constant, i.e., there exists f such that $f_h = f$ for any h, then

$$\Gamma^-(\tau) \lim_{h\to\infty} (f_h) = \text{s.c.}^-(\tau)\,(f),$$

where $\text{s.c.}^-(\tau)\,(f)$ is the largest lower semicontinuous function (with respect to the topology τ) smaller or equal than f. If in addition f is continuous (with respect to τ) then we have

$$\Gamma \lim_{h\to\infty} (f_h) = f.$$

PROPOSITION 1.10 – *Let $\varphi : \overline{\mathbf{R}} \to \overline{\mathbf{R}}$ be a continuous nondecreasing function. Then*

$$\varphi \circ \Gamma^- \min\lim_{h\to\infty} (f_h) = \Gamma^- \min\lim_{h\to\infty} (\varphi \circ f_h),$$

and similar formulas hold for the other Γ-limits.

Therefore, in general, it is not restrictive to consider equibounded functions.

PROPOSITION 1.11 – *Let $\psi : \overline{\mathbf{R}} \times \overline{\mathbf{R}} \to \overline{\mathbf{R}}$ be a function which is continuous and nondecreasing in both variables, and let (f_h), (g_h) be two sequences of functions from X to $\overline{\mathbf{R}}$. Let us set*

$$\psi_h(x) = \psi(f_h(x), g_h(x)).$$

Then the following inequalities hold:

$$\psi\left(\Gamma^- \min\lim_{h\to\infty} f_h(x), \Gamma^- \min\lim_{h\to\infty} g_h(x)\right)$$

$$\leq \Gamma^- \min\lim_{h\to\infty} \psi_h(x) \leq \psi\left(\Gamma^- \min\lim_{h\to\infty} f_h(x), \Gamma^+ \max\lim_{h\to\infty} g_h(x)\right),$$

$$\psi\left(\Gamma^- \max\lim_{h\to\infty} f_h(x)), \Gamma^- \min\lim_{h\to\infty} g_h(x)\right)$$

$$\leq \Gamma^- \max\lim_{h\to\infty} \psi_h(x) \leq \psi\left(\Gamma^- \max\lim_{h\to\infty} f_h(x), \Gamma^+ \max\lim_{h\to\infty} g_h(x)\right),$$

and also

$$\Gamma^- \min\lim_{h\to\infty} \psi_h(x) \leq \psi\left(\Gamma^- \max\lim_{h\to\infty} f_h(x), \Gamma^+ \min\lim_{h\to\infty} g_h(x)\right).$$

In particular, if the following limits exist, we have

$$\psi\left(\Gamma^-\lim_{h\to\infty}f_h(x),\Gamma^-\lim_{h\to\infty}g_h(x)\right)$$

$$\leq\Gamma^-\lim_{h\to\infty}\psi_h(x)\leq\psi\left(\Gamma^-\lim_{h\to\infty}f_h(x),\Gamma^+\lim_{h\to\infty}g_h(x)\right),$$

$$\Gamma^-\lim_{h\to\infty}\psi_h(x)=\psi\left(\Gamma^-\lim_{h\to\infty}f_h(x),\Gamma\lim_{h\to\infty}g_h(x)\right).$$

This proposition can be applied in the computation of the sum, products, maximum, minimum of Γ-limits, etc.

Applications to Calculus of Variations

PROPOSITION 2.1 – *Let A be an open subset of X. If for any $x\in A$ the $\Gamma^-\lim_{h\to\infty}f_h(x)$ exists, then*

$$\inf_{x\in A}\Gamma^-\lim_{h\to\infty}f_h(x)\geq\max\lim_{h\to\infty}\left(\inf_{x\in A}f_h(x)\right).$$

LEMMA 2.2 – *Let x be a point of X, and let (x_h) be a sequence of points of X converging to x. Then*

$$\Gamma^-\min\lim_{h\to\infty}f_h(x)\leq\min\lim_{h\to\infty}f_h(x_h),$$

$$\Gamma^-\max\lim_{h\to\infty}f_h(x)\leq\max\lim_{h\to\infty}f_h(x_h).$$

PROPOSITION 2.3 – *Let K be a sequentially compact subset of X, and assume that for any $x\in K$ the limit $\Gamma^-\lim_{h\to\infty}f_h(x)$ exists. Then*

$$\min_{x\in K}\Gamma^-\lim_{h\to\infty}f_h(x)\leq\min\lim_{h\to\infty}\left(\inf_{x\in K}f_h(x)\right).$$

COROLLARY 2.4 – *Let K be a sequentially compact subset of X, and let A be an open subset of X containing K. Let (f_h) be a sequence of functions satisfying the condition*

$$\inf_{x\in K}f_h(x)=\inf_{x\in A}f_h(x).$$

If the sequence (f_h) Γ^- converges at any point of A, setting $f(x)=\Gamma^-\lim_{h\to\infty}f_h(x)$ for any $x\in A$, we have

$$\min_{x\in A}f(x)=\min_{x\in K}f(x)=\lim_{h\to\infty}\left(\inf_{x\in K}f_h(x)\right)=\lim_{h\to\infty}\left(\inf_{x\in A}f_h(x)\right).$$

Moreover, if (u_h) is a sequence of points of X which converges to $u \in A$, and if $f_h(u_h)$ converges to $\min\limits_{x \in A} f(x)$, then $f(u) = \min\limits_{x \in A} f(x)$.

As a particular case, this happens if $f_h(u_h) = \min\limits_{x \in A} f_h(x)$.

Γ-convergence and extended distances

PROPOSITION 3.1 – *Let x be a point of X having a countable system of neighborhoods. Then*

1) $\Gamma^- \min\lim\limits_{h \to \infty} f_h(x) = \xi \iff$

    i) for any sequence (x_h) converging to x we have

$$\min\lim_{h \to \infty} f_h(x_h) \geq \xi;$$

    ii) there exists a sequence (x_h) converging to x such that

$$\min\lim_{h \to \infty} f_h(x_h) = \xi.$$

2) $\Gamma^- \max\lim\limits_{h \to \infty} f_h(x) = \xi \iff$

    i) for any sequence (x_h) converging to x we have

$$\max\lim_{h \to \infty} f_h(x_h) \geq \xi;$$

    ii) there exists a sequence (x_h) converging to x such that

$$\max\lim_{h \to \infty} f_h(x_h) = \xi.$$

3) $\Gamma^- \lim\limits_{h \to \infty} f_h(x) = \xi \iff$

    i) for any sequence (x_h) converging to x we have

$$\min\lim_{h \to \infty} f_h(x_h) \geq \xi;$$

    ii) there exists a sequence (x_h) converging to x such that

$$\lim_{h \to \infty} f_h(x_h) = \xi.$$

PROPOSITION 3.2 – *Let x be a point of X having a countable system of neighborhoods. Then for any sequence (f_h) there exists a subsequence (f_{h_p}) such that*

$$\Gamma^- \lim_{p \to \infty} (f_{h_p})(x) = \Gamma^- \min\lim_{h \to \infty} f_h(x).$$

Spaces in which any point has a countable system of neighborhoods are particularly important. Among them we will consider, in this section, those whose topology is induced by an extended distance.

DEFINITION 3.3 – We say that a function $\delta : X \times X \to \overline{\mathbf{R}}$ is an extended distance on X if the following properties hold:

1) $\delta(x, y) \geq 0$, $\delta(x, x) = 0$,

2) $\delta(x, y) = \delta(y, x)$,

3) $\delta(x, z) \leq \delta(x, y) + \delta(y, z)$,

for any $x, y, z \in X$ (hence, differently from the usual distance, the function δ can assume the value $+\infty$, and in addition $\delta(x, y) = 0$ does not necessarily imply $x = y$).

An extended distance δ induces a topology on X, in which a system of neighborhoods of x is given by the sets of the form $\{y \in X : \delta(x, y) < \varepsilon\}$, where ε is a positive number.

We will denote by $\Gamma(\delta) \min \lim\limits_{h \to \infty} f_h(x)$, and similarly for the others limits, the Γ-limits performed with respect to the topology induced by the extended distance δ.

Definition 3.4 – Let δ and σ be two extended distances on X. We say that δ and σ are compatible if there exists a constant $c \geq 0$ such that for any $x, y, z \in X$ there exists $t \in X$ such that

i) $\delta(t, y) \leq \delta(x, z) + c \min\left(\delta(x, z), \sigma(x, y)\right)$,

ii) $\sigma(t, z) \leq \sigma(x, y) + c \min\left(\delta(x, z), \sigma(x, y)\right)$.

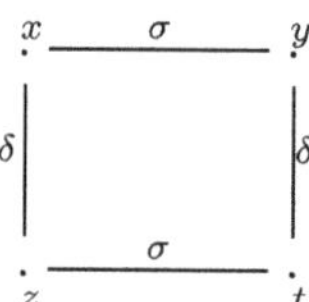

Remark 3.5 –

a) Any extended distance is compatible with itself;

b) if δ and σ are two extended distances on an Abelian group G, and are translation invariant (i.e., $\delta(x + t, y + t) = \delta(x, y)$, $\sigma(x + t, y + t) = \sigma(x, y)$ for any $x, y, t \in G$), then they are compatible;

c) if σ is an arbitrary extended distance and δ is the discrete extended distance (i.e., $\delta(x, x) = 0$, $\delta(x, y) = +\infty$ if $x \neq y$), then σ and δ are compatible;

d) we explicitly observe that compatibility is not a transitive relation.

Theorem 3.6 – *Let δ and σ be two compatible extended distances on X. Let (f_h) be a sequence of δ-equilipschitz functions, i.e., there exists a real constant $K > 0$ such that*

$$|f_h(x) - f_h(y)| \leq K\delta(x, y) \qquad \forall h, \ \forall x, y \in X.^*$$

---

*Here and in the following we assume that $|f(x) - f(y)| = 0$ if $f(x) = f(y) = +\infty$ or if $f(x) = f(y) = -\infty$.

If for any $x \in X$ the limit $\Gamma^-(\sigma) \lim_{h\to\infty} f_h(x)$ exists, then such a Γ-limit is δ-lipschitz with the same constant K, i.e.,

$$|\Gamma^-(\sigma) \lim_{h\to\infty} f_h(x) - \Gamma^-(\sigma) \lim_{h\to\infty} f_h(y)| \le K\delta(x,y) \qquad \forall x,y \in X.$$

THEOREM 3.7 – *Let δ and σ be two compatible extended distances on X. Assume that X has a countable base of open sets for the topology induced by σ. Let (f_h) be a sequence of σ-equilipschitz functions, i.e., there exists a real constant $K > 0$ such that*

$$|f_h(x) - f_h(y)| \le K\sigma(x,y) \qquad \forall h, \ \forall x,y \in X.$$

Then (f_h) admits a subsequence (f_{h_p}) which $\Gamma^-(\delta)$ converges at any point of X.

THEOREM 3.8 – *Let δ and σ be two compatible extended distances on X. Let (f_h) and (g_h) be two sequences of δ and σ-equilipschitz functions respectively, i.e., there exists a real constant $K > 0$ such that*

$$|f_h(x) - f_h(y)| \le K\delta(x,y) \qquad \forall h, \ \forall x,y \in X.$$

$$|g_h(x) - g_h(y)| \le K\sigma(x,y) \qquad \forall h, \ \forall x,y \in X.$$

Let α be another extended distance on X satisfying

$$\alpha(x,y) \le \delta(x,y) \qquad \forall x,y \in X,$$

$$\alpha(x,y) \le \sigma(x,y) \qquad \forall x,y \in X,$$

and such that there exists a constant $H > 0$ such that

$$\alpha(x,y) \ge H \inf_{z\in X} \left(\delta(x,z), \sigma(z,y)\right).$$

Finally, let $\psi : \overline{\mathbf{R}} \times \overline{\mathbf{R}} \to \overline{\mathbf{R}}$ be a function which is continuous and nondecreasing in both variables. If we set $\psi_h(x) = \psi(f_h(x), g_h(x))$, and if $x \in X$ is a point where both $\Gamma^-(\sigma) \lim_{h\to\infty} f_h(x)$ and $\Gamma^-(\delta) \lim_{h\to\infty} g_h(x)$ exist, then also $\Gamma^-(\alpha) \lim_{h\to\infty} \psi_h(x)$ exists and we have

$$\Gamma^-(\alpha) \lim_{h\to\infty} \psi_h(x) = \psi\left(\Gamma^-(\sigma) \lim_{h\to\infty} f_h(x), \Gamma^-(\delta) \lim_{h\to\infty} g_h(x)\right).$$

Final remarks

The above theory seems to be suitable to incorporate various results contained in the papers quoted in the bibliography (in particular section 3 is the abstract formulation of some procedures used in [4]); however, it allows also to consider problems of very different nature. For instance, it would be very interesting to show if the following conjecture is true or false:

Let $X = C_0^1(\mathbf{R}^n)$ be the space of continuous functions on $\mathbf{R}^n$ with compact support, having continuous first derivatives. Let

$$\delta(u, v) = \int_{\mathbf{R}^n} |u - v| \, dx.$$

Let

$$Du = \left(\frac{\partial u}{\partial x_1}, \ldots, \frac{\partial u}{\partial x_n} \right)$$

be the gradient of u. Let

$$F_h(u) = \int_{\mathbf{R}^n} \left\{ h^{-2}|Du|^2 + h^2\left[1 - \cos(hu(x))\right] \right\} \, dx,$$

and let

$$F(u) = \int_{\mathbf{R}^n} |Du| \, dx.$$

Then there exists a positive constant c such that

$$\Gamma^-(\delta) \lim_{h \to \infty} F_h(u) = cF(u) \qquad \forall u \in C_0^1(\mathbf{R}^n).$$

Bibliography

[1] I. Babuška, *Solutions of interface problems by homogenisation, I-II.* Inst. for Fluid Dynamics and Appl. Math., Univ. of Maryland, 1974.

[2] L. Boccardo and P. Marcellini, *Sulla convergenza delle soluzioni di disequazioni variazionali.* Ann. Mat. Pura Appl. (4) Vol. 110, pp. 137-159, 1976.

[3] L. Carbone, *Sur la Γ-convergence des intégrales du type de l'énergie sur des fonctions à gradient borné.* J. Math. Pures Appl. (9) Vol. 56, pp. 79-84, 1977.

[4] E. De Giorgi, *Sulla convergenza di alcune successioni di integrali del tipo dell'area.* Rend. Mat. (6) Vol. 8, pp. 277-294, 1975.

[5] E. De Giorgi and S. Spagnolo, *Sulla convergenza degli integrali dell'energia per operatori ellittici del 2° ordine.* Boll. Un. Mat. Ital. (4) Vol. 8, pp. 391-411, 1973.

[6] J. L. Joly, *Une famille de topologies sur l'ensemble des fonctions convexes pour lequelles la polarité est bicontinue.* J. Math. Pures Appl. Vol. 52, pp. 421-441, 1973.

[7] J. L. JOLY, *Une famille de topologies et des convergences sur l'ensemble des fonctionelle convexes.* Thesis presented at the Facultè de Sciences de Grenoble.

[8] C. KURATOWSKI, *Topologie. Vol. I (French) 4ème éd.* Monografie Matematyczne, Tom. 20. Państwowe Wydawnictwo Naukowe, Warsaw, 1958.

[9] J. L. LIONS, A. BENSOUSSAN AND A. PAPANICOLAU, *Some asymptotic results of variational inequalities with highly oscillatory periodic coefficients.* To appear.

[10] P. MARCELLINI, *Su una convergenza di funzioni convesse.* Boll. Un. Mat. Ital. (4) Vol. 8, pp. 137-158, 1973.

[11] P. MARCELLINI, *Un teorema di passaggio al limite per la somma di funzioni convesse.* Boll. Un. Mat. Ital. (4) Vol. 11, pp. 107-204, 1975.

[12] A. MARINO E S. SPAGNOLO, *Un tipo di approssimazione dell'operatore* $\sum_{ij} D_i(a_{ij} D_j)$ *con operatori* $\sum_j D_j(\beta D_j)$. Ann. Scuola Norm. Sup. Pisa Cl. Sc. (3) Vol. 3, pp. 657-673, 1969.

[13] U. MOSCO, *Convergence of convex sets and of solutions of variational inequalities.* Adv. in Math. Vol. 3, pp. 510-585, 1969.

[14] U. MOSCO, *On the continuity of the Young-Fenchel transformation.* J. Math. Anal. Appl. Vol. 35, pp. 518-535, 1971.

[15] E. SANCHEZ-PALENCIA, *Solutions périodiques par rapport aux variables d'espace et applications.* C. R. Acad. Sci. Paris Sér. A Vol. 271, pp. 1129-1132, 1970.

[16] E. SANCHEZ-PALENCIA, *Equations a derivées partielles dans un type de milieux hétérogènes.* C. R. Acad. Sci. Paris Sér. A Vol. 272, pp. 1410-1413, 1970.

[17] E. SANCHEZ-PALENCIA, *Comportement local et macroscopique d'un type de milieux phisiques hétérogènes.* Internat. J. Engrg. Sci. Vol. 12, pp. 331-351, 1974.

[18] C. SBORDONE, *Sulla G-convergenza di equazioni ellittiche e paraboliche.* Ricerche Mat. Vol. 24, pp. 76-136, 1975.

[19] C. SBORDONE, *Su alcune applicazioni di un tipo di convergenza variazionale.* Ann. Scuola Norm. Sup. Pisa Cl. Sc. (4) Vol. 2, pp. 617-638, 1975.

[20] S. SPAGNOLO, *Sul limite delle soluzioni di problemi di Cauchy relativi all'equazione del calore.* Ann. Scuola Norm. Sup. Pisa Cl. Sc. (3) Vol. 21, pp. 657-699, 1967.

[21] S. Spagnolo, *Sulla convergenza di soluzioni di equazioni paraboliche ed ellittiche.* Ann. Scuola Norm. Sup. Pisa Cl. Sc. (3) Vol. 22, pp. 577-597, 1968.

[22] L. Tartar, *Convergence d'operateurs differentiels.* Analisi Convessa e Applicazioni, Quaderni dei gruppi di ricerca matematica del C.N.R., Ist. Mat. G. Castelnuovo, Roma, 1974.

[23] T. Zolezzi, *Su alcuni problemi fortemente ben posti di controllo ottimo.* Ann. Mat. Pura Appl. (4) Vol. 95, pp. 147-160, 1973.

[24] T. Zolezzi, *On convergence of minima.* Boll. Un. Mat. Ital. (4) Vol. 8, pp. 246-257, 1973.

Evolution problems in metric spaces and steepest descent curves[‡][†]

Ennio De Giorgi - Antonio Marino - Mario Tosques

Summary. In this note a definition of "Maximal decreasing curve" is given, which extends the usual notion of solution of an evolution problem of the type, for example, of the heat equation.

This definition seems the right one in order to study many limit cases of evolution problems which have been settled only in the convex case.

Introduction

In the study of Γ-convergence and G-convergence, it is of great interest the connection between the Γ-convergence of a sequence of functionals and the convergence of the solutions to the corresponding evolution equations. Indeed, the first results concerning the Γ-convergence of quadratic functionals of elliptic type were obtained through the study of the corresponding evolution equations (see [1]). Subsequently, the Γ-convergence theory has undergone a remarkable development and for most of the results obtained, it has not been possible, so far, to establish an analogous connection with evolution equations (see [2], [3], [4]). Apparently this is due, at least in part, to the fact that no sufficiently general notion of "evolution problem" is available. The purpose of this work is exactly to look for a suitable framework for this problem, by suggesting a few possible extensions of that notion.

To this aim, we also investigate the main properties of the solutions and give some local existence theorems.

The proofs of the results stated here will be the object of further works, where a more complete analysis will be carried out.

Steepest descent curves in metric spaces

In the following X denotes a metric space with metric d, while $f : X \to \mathbf{R} \cup \{+\infty\}$ is a lower semicontinuous function.

First of all, we associate to each u in X an extended real number in $\mathbf{R} \cup \{+\infty\}$ which, in the case when X is $\mathbf{R}^n$ and f is a differentiable function, reduces to the norm of the gradient of f at u.

[‡] Editor's note: translation into English of the paper "Problemi di evoluzione in spazi metrici e curve di massima pendenza", published in Atti Accad. Naz. Lincei, Cl. Sci. Fis. Mat. Nat., **(8) 68** (1980), 180–187.

[†] Communicated by the Correspondent Ennio De Giorgi, in the session of March 8, 1980.

DEFINITION 1.1. Let u in X be such that $f(u) < +\infty$. We define the *descending slope of f at u* the number

$$|\nabla f|u = \begin{cases} \max\left\{0, \max\lim_{v \to u} \dfrac{f(u) - f(v)}{d(u, v)}\right\} & \text{if } u \text{ is not an isolated point of } X \\ 0 & \text{if } u \text{ is an isolated point of } X. \end{cases}$$

We consider now a curve in X, which in the case when X is $\mathbf{R}^n$ and f is a differentiable function, reduces to the solution of the equation

$$U' = -\mathrm{grad} f(U).$$

DEFINITION 1.2. Let I be an interval contained in $\mathbf{R}$, open to the right. We say that a curve $U : I \to X$ is a *curve of steepest pointwise descent for f* if
    a) U is continuous
    b) $f \circ U$ is nonincreasing and $f \circ U(t) < +\infty$ for $t \in I$, $t > \inf I$
    c) $\max\lim\limits_{\tau \to 0^+} \dfrac{d(U(t + \tau), U(t))}{\tau} \leq |\nabla f|U(t)$ if $t \in I$ with $f \circ U(t) < +\infty$
    d) $\max\lim\limits_{\tau \to 0^+} \dfrac{f \circ U(t + \tau) - f \circ U(t)}{\tau} \leq -(|\nabla f|U(t))^2$ if $t \in I$ with $f \circ U(t) < +\infty$.

Let us now consider a different type of curve, which evolves on the graph of f by following, in some sense, the direction where f decreases at the fastest rate. Indeed, we shall see that to prove the existence of such curves it is sufficient to assume conditions which are weaker than the ones needed in the case of curves with steepest descent for f.

DEFINITION 1.3. Denote by $\mathcal{G}_f : X \times \mathbf{R} \to \mathbf{R} \cup \{+\infty\}$ the following function:

$$\mathcal{G}_f(v, r) = \begin{cases} r & \text{if } r \geq f(v) \\ +\infty & \text{if } r < f(v). \end{cases}$$

On $X \times \mathbf{R}$ we consider the metric

$$\delta((v_1, r_1), (v_2, r_2)) = \{d(v_1, v_2)^2 + (r_1 - r_2)^2\}^{\frac{1}{2}}.$$

Notice that a curve of steepest pointwise descent for $\mathcal{G}_f$ in $X \times \mathbf{R}$ is always Lipschitz continuous, with a Lipschitz constant 1.

THEOREM 1.4. – *Let $U : I \to X$ be a curve of steepest pointwise descent for f, absolutely continuous and such that $f \circ U$ is finite on I.*
    Then there exist an interval J, open to the right, and a non decreasing function $\psi : J \to I$, surjective and such that, writing

$$V(s) = U \circ \psi(s), \qquad R(s) = \psi(s) - s + f \circ U(t_0)$$

the curve (V, R) is a curve of steepest pointwise descent for $\mathcal{G}_f$, and $f \circ V$ is nonincreasing. Moreover, if $f \circ V$ is continuous then $R = f \circ V$.

THEOREM 1.5. – *Let J be an interval, open to the right, and let (V, R) : $J \to X \times \mathbf{R}$ be a curve of steepest pointwise descent for $\mathcal{G}_f$, such that $f \circ V$ is nonincreasing and $R(s) = f \circ V(s)$ for at least one point s of J.*

Then there exist an interval I, open to the right, and a function $\varphi : I \to J$, strictly increasing, such that the curve $U = V \circ \varphi$ is a curve of steepest pointwise descent for f, absolutely continuous, with $f \circ U$ finite on I. Moreover, if $f \circ V = R$ then $f \circ U$ is continuous.

Notice that if (V, R) is a curve of steepest pointwise descent for $\mathcal{G}_f$ then the condition that $f \circ V$ be nonincreasing is equivalent to the property:

If at some point s_0 we have $R(s_0) > f \circ V(s_0)$, then:

$$V(s) = V(s_0), \quad R(s) = R(s_0) - (s - s_0) \quad \text{in a neighborhood of } s_0.$$

A further useful definition is the following:

DEFINITION 1.6. Let I be an interval of $\mathbf{R}$, with nonempty interior. We shall say that a curve $U : I \to X$ is a *curve of steepest descent in mean for f* if:
 a) U is continuous
 b) $f \circ U(t) < +\infty$ if $t \in I$ with $t > \inf I$
 c) $d(U(t_2), U(t_1)) \leq \int_{t_1}^{t_2} |\nabla f| U(\tau) d\tau$ if $t_1 < t_2$ in I
 d) $f \circ U(t_2) - f \circ U(t_1) \leq -\int_{t_1}^{t_2} (|\nabla f| U(\tau))^2 d\tau$ if $t_1 < t_2$ in I.

The purpose of the following Theorems (1.7) and (1.8) is to compare the definitions (1.1) and (1.6).

THEOREM 1.7. – *Let $U : I \to X$ be a curve in X. U is a curve of steepest descent in mean for f if and only if*
 a) U is continuous and in addition absolutely continuous on $\{t | t \in I, t > \inf I\}$
 b) $f \circ U$ is nonincreasing and $f \circ U(t) < +\infty$ for $t > \inf I$
 c) $\displaystyle \lim_{\tau \to 0^+} \frac{d(U(t + \tau), U(t))}{\tau} = |\nabla f| U(t)$ for almost every t in I
 d) $\displaystyle \lim_{\tau \to 0^+} \frac{f \circ U(t + \tau) - f \circ U(t)}{\tau} = -(|\nabla f| U(t))^2$ for almost every t in I.

THEOREM 1.8. – *If U is a curve of steepest pointwise descent which is absolutely continuous, then it is a curve of steepest descent in mean.*

If U is a curve of steepest descent in mean for f and if $|\nabla f| U(t)$ is lower semicontinuous then U is a curve of steepest pointwise descent for f.

A few general properties

Theorem 2.1. – *(Composition with a monotone function).*
Let $U : I \to X$ be a curve of steepest descent in the mean for f. Let $\gamma : J \to \mathbf{R}$ be a nondecreasing function on the interval J open to the left and containing $f \circ U(I)$. Assume that
(2.2)

$$\begin{cases} \dfrac{1}{(D^-\gamma)(f \circ U(t))} & \text{is integrable on every interval } [a,b] \text{ contained in } I \\[2ex] (D^-\gamma)(f \circ U(t)) & \text{is different from } +\infty \text{ for at least a point } t \text{ in } I. \end{cases}$$

Then there exists an increasing function $\varphi : I_1 \to I$ such that $U_1 = U \circ \varphi$ is a curve of steepest descent in the mean for $\gamma \circ f$. Moreover, U_1 and U have the same image. (Here $(D^-\gamma)(x)$ denotes the upper left derivative of γ at the point x of J).

Notice that the assumptions (2.2) are fulfilled as soon as one of the following conditions holds:

a) γ is Lipschitz continuous and $0 < c_1 \leq \gamma' \leq c_2$ for two constants c_1, c_2;

b) $\dfrac{1}{\gamma'}$ is integrable on compact subsets of J and $|\nabla f|U(t) \geq c > 0$.

Theorem 2.3. – *(Criterion for continuity and monotonicity of $f \circ V$).*
Assume that f satisfies the following assumption:

(2.4)

$$\begin{cases} \text{for any } u_0 \text{ in } X \text{ there exist a neighborhood } U \text{ of } u_0 \text{ and a map} \\ R(\cdot, u_0) : U \to \mathbf{R} \text{ such that} \\[2ex] f(u_0) \geq f(u) - (|\nabla f|u)d(u, u_0) - R(u, u_0) \\[2ex] \text{if } u \in U \text{ with } f(u) < +\infty, \text{ and } \lim_{u \to u_0} R(u, u_0) = 0. \end{cases}$$

Then, if (V, R) is of steepest pointwise descent for $\mathcal{G}_f$ and $R(s_0) = f \circ V(s_0)$, we have $R(s) = f \circ V(s)$ for $s \in J \cap [s_0, +\infty[$.

We remark that all continuous functions on X satisfy (2.4).
Moreover, when X is a Hilbert space and f is e.g. convex, then (2.4) is again fulfilled.

Theorem 2.5. – *(Criterion for finite slope along a trajectory).*
Suppose that f satisfies the following assumptions:
for any u_0 in X with $f(u_0) < +\infty$, there exist $\delta > 0$ and a nondecreasing continuous function $\omega : [0, 2\delta[\to \mathbf{R}$, with the properties

(2.6)

$$\begin{cases} f(v) \geq f(u) - (|\nabla f|u)d(u, v) - \omega(d(u, v))d(u, v) \\ \quad \text{if } u, v, \in B(u_0, \delta) \quad \text{and} \quad f(u) < +\infty, \\ \dfrac{\omega(\sigma)}{\sigma} \text{ is integrable on } [0, \delta[. \end{cases}$$

Then, if $(V, R) : [\alpha, \beta[\to X \times \mathbf{R}$ is of steepest descent for $\mathcal{G}_f$ and $R(\alpha) = f \circ V(\alpha)$, we have:

a) $R = f \circ V$

b) $|\nabla f| V(s)$ is bounded on any interval $[a, b] \subset]\alpha, \beta[$

c) $\lim\limits_{s \to s_0^+} |\nabla f| V(s) = |\nabla f| V(s_0)$ for any s_0 in $[\alpha, \beta[$

d) $\max\lim\limits_{\sigma \to 0^-} \dfrac{f \circ V(s_0 + \sigma) - f \circ V(s_0)}{\sigma} \leq (f \circ V)'_+(s_0)$ if $s_0 \in]\alpha, \beta[$.

We remark that assumptions (2.6) are certainly fulfilled if X is e.g. an Hilbert space and f is the sum of a convex function and a function of class $C^{1,\alpha}(X)$ $(\alpha > 0)$.

Existence theorems

We now state some sufficient conditions which ensure the existence of a curve of steepest descent in mean for $\mathcal{G}_f$ or f.

THEOREM 3.1. – *(Existence).*
Assume that:
a) the restriction of f to $\{u | f(u) < +\infty\}$ is continuous;
b) f is "locally coercive", i.e.,
for any $\rho > 0$, for any real c and for any u in X with $f(u) < +\infty$, the set $B(u, \rho) \cap \{v | f(b) \leq c, v \in X\}$ is compact;
c) for any u in X with $f(u) < +\infty$, the following condition holds

$$\min_{\substack{v \to u \\ f(v) \leq f(u)}} \lim |\nabla f| v \geq |\nabla f| u.$$

Then for all u_0 in X with $f(u_0) < +\infty$ there exists $U : [0, T[\to X$ of steepest descent in mean for f such that $U(0) = u_0$.

The following existence result can be applied to a large class of functions, not necessarily continuous, containing e.g. the sums of a convex function and a function of class C^1.

THEOREM 3.2. – *(Existence).*
Assume f has the following properties
a) f is locally coercive
b) for any u_0 in X with $f(u_0) < +\infty$ there exist a neighborhood U of u_0 and a continuous $\omega : U \times U \to \mathbf{R}$ such that $f(v) \geq f(u) - (|\nabla f| u + \omega(u, v))d(u, v)$ if $u, v \in U$ and $f(u) < +\infty$, $\omega(u, u) = 0$.

Then for any u_0 in X with $f(u_0) < +\infty$ there exists a curve $U : [0, T[\to X$ of steepest pointwise descent for f, absolutely continuous, such that $U(0) = u_0$ and $f \circ U$ is continuous.

Evolution equations in Banach spaces

We now give some results concerning curves of steepest descent in mean in the framework of Banach spaces.

Notice that our notion of curve of steepest descent contains as special cases the curves considered by several authors as solutions to some classes of evolution equations (see e.g. [5]) and our definition of subdifferential and gradient is an extension of theirs.

Let X be a real Banach space, uniformly convex, with differentiable norm, and such that the norm of X' is strictly convex.

Recall that we always consider lower semicontinuous functions $f : X \to \mathbf{R} \cup \{+\infty\}$.

DEFINITION 4.1. Let u be in X with $f(u) < +\infty$. We say that f is subdifferentiable at u if there exists α in X' such that

$$\min_{v \to u} \lim \frac{f(v) - f(u) - \langle \alpha, v - u \rangle}{\|v - u\|} \geq 0.$$

We define the subdifferential of f at u as the set of all such α, and we denote it by $\partial^- f(u)$.

If $\partial_0^- f(u)$ is the element of $\partial^- f(u)$ with smallest norm, we define the gradient of f at u as the vector $\operatorname{grad} f(u)$ with the property

$$\langle \partial_0^- f, \operatorname{grad} f(u) \rangle = \|\partial_0^- f(u)\|^2, \qquad \|\operatorname{grad} f(u)\| = \|\partial_0^- f(u)\|;$$

when $\partial^- f(u) = \emptyset$, we write $\|\operatorname{grad} f(u)\| = +\infty$.

DEFINITION 4.2. We say that f belongs to the class $\mathcal{K}(X)$ if for any u_0 in X with $f(u_0) < +\infty$, we can find a neighborhood U of u_0 and a continuous function $\omega : U \times U \to \mathbf{R}$ such that

$$f(v) \geq f(u) + \langle \alpha, v - u \rangle - \omega(u, v)\|v - u\|$$

for all $u, v \in U$ with $f(u) < +\infty$ and $\omega(u, u) = 0$ and all $\alpha \in \partial^- f(u)$.

REMARK 4.3. Let f be weakly lower semicontinuous. Consider the following properties:

$\alpha)$ $f \in \mathcal{K}(X)$

$\beta)$ $\|\operatorname{grad} f(u)\|$ is lower semicontinuous on $\{u | f(u) < +\infty\}$

$\gamma)$ $|\nabla f|u = \|\operatorname{grad} f(u)\|$ for all u with $f(u) < +\infty$.

Then we have $\alpha) \Rightarrow \beta) \Rightarrow \gamma)$.

REMARK 4.4. If $U : I \to X$ is of steepest pointwise descent, or in average, for f, and if condition $\gamma)$ holds, then we have:

$$(4.5) \qquad U'_+(t) = -\operatorname{grad} f(U(t)), \qquad (f \circ U(t))'_+ = -\|\operatorname{grad} f(U(t))\|^2$$

for almost any t in I.

THEOREM 4.6. – *(Existence)*.
Assume that:
a) f is locally coercive
b) $f \in \mathcal{K}(X)$.
Then, for any u_0 in X with $f(u_0) < +\infty$ there exists $U : [0, T[\to X \;\; (T > 0)$ of steepest pointwise descent for f, absolutely continuous and with $U(0) = u_0$, such that for almost any t there exist $U'_+(t)$, $\operatorname{grad} f(U(t))$, and for these values of t conditions (4.5) hold.
Moreover, when X is of finite dimensional, conditions (4.5) hold on the whole of $[0, T[$.

THEOREM 4.7. – *(Composition with a monotone function)*.
Assume f satisfies a) and b) in (4.6). Given $u_0 \in X$ with $f(u_0) < +\infty$, let J be an interval, open to the left, containing $f(u_0)$ and let $\gamma : J \to \mathbf{R}$ be a nondecreasing function such that $\frac{1}{\gamma'}$ is integrable on J. Moreover, assume 0 does not belong to $\partial^- f(u_0)$.
Then there exists an absolutely continuous $U : [0, T[\to X \;\; (T > 0)$, such that $U'_+(t)$ and $\operatorname{grad}(\gamma \circ f)(U(t))$ exist for almost any t, and in addition

$$U'_+(t) = -\operatorname{grad}(\gamma \circ f)(U(t)) \quad \text{for almost any } t \text{ in } [0, T[$$
$$U(0) = u_0$$
$$(\gamma \circ f \circ U)'_+(t)) = -\|\operatorname{grad}(\gamma \circ f)(U(t))\|^2 \quad \text{for almost any } t \text{ in } [0, T[.$$

In particular, U is of steepest descent in mean for $\gamma \circ f$.

Bibliography

[1] S. SPAGNOLO, *Sulla convergenza di soluzioni di equazioni paraboliche ed ellittiche*. Ann. Scu. Norm. Sup. Pisa **22** (1968), pp. 577–597.

[2] L. MODICA - S. MORTOLA, *Il limite nella Γ-convergenza di una famiglia di funzionali ellittici*. Boll. Un. Mat. Ital. **14**-A (1977), pp. 526–529.

[3] G. BUTTAZZO - G. DAL MASO, *Γ-limit of a sequence of non-convex and non-equi-Lipschitz integral functionals*. Ricerche di Matematica **27** (1978), pp. 235–251.

[4] A. AMBROSETTI - C. SBORDONE, *Γ-convergenza e G-convergenza per problemi non lineari di tipo ellittico*. Boll. Un. Mat. Ital. **15**-A (1976), pp. 352–362.

[5] H. BREZIS, *Operateurs Maximaux Monotones*. Notes de Matematica **50**, North-Holland.

Generalized limits in Calculus of Variations[‡]

Ennio De Giorgi

The present work, together with the following paper of G. Dal Maso and L. Modica[*], may be enclosed in the framework of the applications of G-convergence and Γ-convergence theory to the Calculus of Variations. In this report I wish to present this research field and I will begin by stating and briefly illustrating some recent results in the Calculus of Variations which I think are understandable and meaningful also to readers not acquainted with G-convergence and Γ-convergence theory, even if this theory plays an important role in the proofs of these results.

The first result is the following.

THEOREM 1. – *Let $a(t)$ be a 2π-periodic function defined on $\mathbf{R}$ such that*

$$0 < \lambda \le a(t) \le \Lambda < +\infty$$

for every $t \in \mathbf{R}$ and for some $\lambda, \Lambda \in \mathbf{R}$. Denote by α the harmonic mean of $a(t)$, that is

$$\frac{1}{\alpha} = \frac{1}{2\pi} \int_0^{2\pi} \frac{1}{a(t)} dt.$$

Then

$$\lim_{h \to +\infty} \min_u \int_a^b \left[a(h\,t)\left(\frac{du}{dt}\right)^2 + g(t, u(t)) \right] dt = \min_u \int_a^b \left[\alpha\left(\frac{du}{dt}\right)^2 + g(t, u(t)) \right] dt$$

whenever $a, b \in \mathbf{R}$ with $a < b$ and g is a continuous function on $\mathbf{R}^2$ such that

$$g(t, u) \ge p\,u^2 - q \quad \forall (t, u) \in \mathbf{R}^2$$

for some real constants $p > 0$, $q > 0$.

This theorem is concerned with the study of the limit of minima of a given sequence of integral functionals. It is also interesting to consider the opposite problem, that is to approximate a minimum of some given functional by the minima of some sequences of integrals of different type. For example, to approximate anisotropic functionals by sequences of isotropic functionals as in the following theorem.

[‡]Editor's note: published in Topics in Functional Analysis, Quaderno Sc. Norm. Sup. Pisa, Editrice Tecnico Scientifica, Pisa, 1981, 117–148.

[*]G. DAL MASO AND L. MODICA, *A general theory of variational functionals*, in *Topics in Functional Analysis, 1980–81*, Quaderni, Scuola Norm. Sup. Pisa, Pisa, 1981, pp. 149–221.

THEOREM 2. – *Let a_{ij} $(i, j = 1, \ldots, n)$ be bounded and measurable functions on $\mathbf{R}^n$ such that*

$$0 < \lambda |\xi|^2 \leq \sum_{i,j=1}^{n} a_{ij}(x)\xi_i\xi_j \leq \Lambda |\xi|^2 < +\infty \quad \forall x \in \mathbf{R}^n, \ \forall \xi \in \mathbf{R}^n \setminus \{0\}$$

for some $\lambda, \Lambda \in \mathbf{R}$. Then there exists a sequence φ_h of positive bounded real functions on $\mathbf{R}^n$ such that

$$\min_{u} \int_{\Omega} \left[\sum_{i,j=1}^{n} \left(a_{ij} \frac{\partial u}{\partial x_i}(x) \frac{\partial u}{\partial x_j}(x) \right) + g(x, u(x)) \right] dx =$$

$$= \lim_{h \to +\infty} \min_{u} \int_{\Omega} \left[\varphi_h(x) |Du(x)|^2 + g(x, u(x)) \right] dx$$

whenever Ω is a bounded open subset of $\mathbf{R}^n$ and g is a continuous function on $\mathbf{R}^{n+1}$ such that

$$g(x, u) \geq p\, u^2 - q \quad \forall (x, u) \in \mathbf{R}^n \times \mathbf{R}$$

for some real constants $p > 0$, $q > 0$.

These two theorems show a remarkable phenomenon: if we want to save the minima in a limit of integral functionals, it is often necessary to take as limit functional something rather different from the usual limit. For example in the former, if we fix the function u, it is easy to obtain

$$\lim_{h \to +\infty} \int_a^b \left[a(h\, t) \left(\frac{du}{dt} \right)^2 \right] dt = \beta \int_a^b \left(\frac{du}{dt} \right)^2 dt,$$

where

$$\beta = \frac{1}{2\pi} \int_0^{2\pi} a(t)dt,$$

so the functional

$$F_\infty(u) = \alpha \int_a^b \left(\frac{du}{dt} \right)^2 dt$$

which is in Theorem 1 the "limit functional" of the sequence of functionals

$$F_h(u) = \alpha \int_a^b a(h\, t) \left(\frac{du}{dt} \right)^2 dt$$

is different from the usual limit because $\beta > \alpha$ (if $a(t)$ is not constant). Analogously, in the latter, if the eigenvalues of the matrix $(a_{ij}(x))$ are not all equal, the sequence of isotropic functionals

$$F_h(u) = \int_{\Omega} \varphi_h(x) |Du(x)|^2 dx$$

"converges" in Theorem 2 to the anisotropic functional

$$F_\infty(u) = \int_\Omega \sum_{i,j=1}^n a_{ij} \frac{\partial u}{\partial x_i} \frac{\partial u}{\partial x_j} dx.$$

In both cases, according to the definition we shall give later, the functional F_∞ is the Γ-limit as $h \to +\infty$ of the functionals F_h, so it becomes clear that this possibility of changing the form of the limit with respect to the usual limit is an important feature of Γ-convergence and G-convergence theory that makes it substantially more general than other theories for approximation problems in the Calculus of Variations.

The proofs of Theorems 1 and 2 may be easily obtained by combining the results of A. Marino and S. Spagnolo [190], [135], [270], which are sufficient when $g(x, u) = f(x)u + p u^2$ and f is continuous and bounded in Ω, with the results proved in a work [130], [131] of T. Franzoni and myself, where we consider functions g that are only continuous. This possibility of considering non-convex and non-differentiable perturbations is typical of the applications of Γ-convergence theory in the Calculus of Variations.

The third result, in which the nature of the functionals taken into considerations is modified by taking the limit in a even more drastic way than above, is the following.

THEOREM 3. – *Let Ω be an open bounded subset of $\mathbf{R}^n$ with smooth boundary and let $g : \Omega \times \mathbf{R} \to \mathbf{R}$ be a continuous function such that*

$$g(x, u) \geq p u^2 - q \quad \forall (x, u) \in \Omega \times \mathbf{R}$$

for some real constants $p > 0$, $q > 0$.
Then, for every $a > 0$

$$\lim_{h \to +\infty} \min_u \int_\Omega \left[\frac{|Du(x)|^2}{h} + h(a^2 - u^2(x))^2 + g(x, u(x)) \right] dx =$$

$$= \min_{E \subseteq \Omega} \left[\int_E g(x, a)dx + \int_{\Omega \setminus E} g(x, -a)dx + \frac{8}{3}a^3 P(E, \Omega) \right],$$

where $P(E, \Omega)$ denotes the perimeter of E in Ω, which agrees with the elementary $(n-1)$ dimensional measure of $\partial E \cap \Omega$ if E has smooth boundary, while in general it is given by

$$P(E, \Omega) = \sup \left\{ \int_E \operatorname{div} \psi \, dx : \psi \in \left[C_0^1(\Omega) \right]^n , \ |\psi| \leq 1 \right\}.$$

We recall that $P(E, \Omega) \leq H_{n-1}(\partial E \cap \Omega)$, where H_{n-1} denotes the $(n-1)$ dimensional Hausdorff measure. More precisely, if $P(E, \Omega) < +\infty$ then $P(E, \Omega) = H_{n-1}(\partial^* E \cap \Omega)$ for every E, where

$$\partial^* E = \left\{ x \in \partial E : \lim_{\rho \to 0^+} \frac{|E \cap B_\rho|}{|B_\rho|} = \frac{1}{2} \right\},$$

$B_\rho = \{y \in \mathbf{R}^n : |y - x| < \rho\}$ and $|\cdot|$ denotes the Lebesgue measure in $\mathbf{R}^n$ (see M. Miranda [200], [201], [202]). We recall also that for the set E_0 which realizes the minimum of

$$\int_E g(x, a)dx + \int_{\Omega \setminus E} g(x, -a)dx + \frac{8}{3}a^3 P(E, \Omega)$$

we have (see U. Massari [194], E. Giusti [152], M. Giaquinta [150])

$$H_{n-1}(\partial E_0 \setminus \partial^* E_0) = 0$$

so the same E_0 realizes the minimum of

$$\int_E g(x, a)dx + \int_{\Omega \setminus E} g(x, -a)dx + \frac{8}{3}a^3 H_{n-1}(\partial E \cap \Omega).$$

The proof of Theorem 3 may be obtained by combining the results of L. Modica and S. Mortola [207], [208] and of the work [130], [131] with the theory of the minimal oriented boundaries (see, for example, the books [128], [153]).

By passing to the Euler-Lagrange equations, Theorem 3 suggests a possible connection between the non-linear partial differential equation $\Delta u = f(u)$ and the sets whose boundaries have a prescribed mean curvature. I think this aspect should be deeply analyzed because, perhaps, it could open new ways in the study of both problems. An interesting, difficult and still open conjecture about this subject is the following.

Let u be a function of class C^2, solution in the whole of $\mathbf{R}^n$ of the equation

$$\Delta u = u^3 - u$$

such that $\partial u / \partial x_n > 0$ on $\mathbf{R}^n$. Then all level sets $\{x \in \mathbf{R} : u(x) = \text{constant}\}$ of u are hyperplanes.

The first steps in the direction of the proof of this conjecture are in papers of L. Modica and S. Mortola [205], [209].

The next result may be viewed as a mathematical interpretation of a Faraday's grid. Let us image in the space $\mathbf{R}^3$ a system C_h of infinitely many parallel wires of diameter $2\rho_h$ arranged in the plane xy orthogonally to the y-axis and separated by a distance $2\delta_h > 2\rho_h$:

$$C_h = \bigcup_{n \in \mathbf{Z}} \left\{(x, y, z) \in \mathbf{R}^3 : x^2 + (y - 2n\delta_h)^2 \leq \rho_h^2\right\}.$$

Now, suppose that all these wires have the same electric potential V and that u_h be the solution of the problem

$$\begin{cases} \Delta u_h = 0 & \text{in } \mathbf{R}^3 \setminus C_h \\ u_h = V & \text{on } \partial C_h \\ \lim_{|P| \to +\infty} u_h(P) = 0. \end{cases}$$

Then the following result holds.

THEOREM 4. – *Suppose that the sequences (ρ_h) and (δ_h) converge to zero and there exists*

$$K = \lim_{h \to +\infty} (-\pi \delta_h \log \rho_h)^{-1}.$$

Then the sequence (u_h) converges in $L^2_{loc}(\mathbf{R}^3)$ to the solution u_∞ of the problem

$$\begin{cases} \Delta u_\infty = 0 \quad in \ \mathbf{R}^3 \\ \lim_{x \to 0^+} \left[\dfrac{\partial u_\infty}{\partial x}(x,y,z) - \dfrac{\partial u_\infty}{\partial x}(-x,y,z) \right] = K(V - u_\infty(0,y,z)) \ \forall (y,z) \in \mathbf{R}^2 \\ \lim_{|P| \to +\infty} u_\infty(P) = 0, \end{cases}$$

where the second condition has the meaning $V = u_\infty(0,y,z)$ if $K = +\infty$.

The proof of Theorem 4 may be obtained by the methods of the theory of the limits of functionals with constraints or obstacles studied by D. Cioranescu, F. Murat [97], L. Carbone, F. Colombini [85], G. Dal Maso, P. Longo [129], [112], [114], H. Attouch, C. Picard [18]. The opposite problem, for example the approximation of a penalization term by obstacle-type constraints, is shown in the following result.

THEOREM 5. – *Let χ_1 and χ_2 be two functions of class C^3 defined on $\mathbf{R}^{n+1}$ such that, for every $(x,t) \in \mathbf{R}^n \times \mathbf{R}$*

$$\chi_1(x,t) \geq 0, \ \frac{\partial \chi_1}{\partial t}(x,t) \geq 0, \ \frac{\partial^2 \chi_1}{\partial t^2}(x,t) \geq 0, \ \frac{\partial^3 \chi_1}{\partial t^3}(x,t) \geq 0;$$

$$\chi_2(x,t) \geq 0, \ \frac{\partial \chi_2}{\partial t}(x,t) \leq 0, \ \frac{\partial^2 \chi_2}{\partial t^2}(x,t) \geq 0, \ \frac{\partial^3 \chi_2}{\partial t^3}(x,t) \leq 0.$$

Denote $\chi = \chi_1 + \chi_2$. Then there exist two sequences (φ_h), (ψ_h) of functions defined on $\mathbf{R}^n$ such that

$$\min_u \int_\Omega \left[|Du(x)|^2 + \chi(x,u(x)) + g(x,u(x)) \right] dx =$$

$$= \lim_{h \to +\infty} \min_u \left\{ \int_\Omega \left[|Du(x)|^2 + g(x,u(x)) \right] dx : \varphi_h \leq u \leq \psi_h \right\}$$

for every Ω and g as in Theorem 3.

Many results proved by Γ-convergence methods have an interesting physical interpretation, as we said with regard to Theorems 2 and 4, for example the study of the global properties of a body composed of small particles of different substance. In particular, Theorem 1 is a first elementary example of homogenization, that is the mathematical theory of the heterogeneous materials with highly periodic structure. We refer the reader to the books of A. Bensoussan, J.L. Lions, G. Papanicolaou [38], and of J.L. Lions [137 bis]. A number of other problems from physics and engineering suitable of being tackled by methods of Γ-convergence is available in the book of E. Sanchez-Palencia [252].

If we want to give at least some ideas about the methods of the proofs of Theorems 1–5, as well as of many similar results, we have to give a short information about G-convergence and Γ-convergence. We shall follow essentially the formulation proposed in the seminars held during the last months of 1981 in Bressanone [125], Catania [126], S. Margherita Ligure [127], which takes into account the work [130], [131] and the subsequent considerations on the application of these convergence theories to several fields of mathematics such as differential equations, control theory, probability, generalized subgradients, min-max problems, steepest descent curves (see the bibliography at the end of the paper). This formulation is based upon the notion of operator of type $G(\overline{\mathbf{R}})$.

Before we introduce this notion, we want to remark that in the following we shall consider functions f with values in $\overline{\mathbf{R}} = \mathbf{R} \cup \{+\infty, +\infty\}$ whose domain $\mathrm{dom}f$ is a subset (possibly void) of a fixed universe, but that will not be further specified.

An operator g which transforms functions with values in $\mathbf{R}$ into functions with values in $\mathbf{R}$ is said to be an operator of type $G(\overline{\mathbf{R}})$ if the following three conditions hold:

a) $\mathrm{dom}f_1 \subseteq \mathrm{dom}f_2 \Rightarrow \mathrm{dom}\, g(f_1) \subseteq \mathrm{dom}\, g(f_2)$;

b) $\mathrm{dom}f_1 = \mathrm{dom}f_2, f_1 \leq f_2 \Rightarrow g(f_1) \leq g(f_2)$;

c) if $\vartheta : \overline{\mathbf{R}} \to \overline{\mathbf{R}}$ is a strictly increasing continuous function, then $g(\vartheta \circ f) = \vartheta \circ g(f)$ for every function f.

The notion of operator of type $G(\overline{\mathbf{R}})$ is stable for composition; moreover, condition (c) applied with $\vartheta(t) = \arctan t$ allows in some cases to limit oneself to considering only bounded functions.

Typical operators of type $G(\overline{\mathbf{R}})$ are the lower limit and the upper limit in a fixed topological space X. Note that the domain of $\liminf(f)$ and $\limsup(f)$ is the set of the cluster points in X of $\mathrm{dom}f \cap X$. These two operators may be obtained as products of simpler basic operators that we are going to define and that play an important role in the definition of the Γ-limits and also of other generalized limits.

Let X be a topological space and denote by $I_X(x)$ the family of all neighborhoods of $x \in X$ in X. By $G(+, X+)$, $G(+, X-)$, $G(-, X+)$, $G(-, X-)$ we denote the four operators of type $G(\mathbf{R})$ which transform a function f respectively into the functions $G(+, X+)f$, $G(+, X-)f$, $G(-, X+)f$, $G(-, X-)f$ defined by the following relations:

$$\begin{cases} \mathrm{dom}\, G(+, X+)f = \{A \mid A \subseteq X, A \cap \mathrm{dom}f \neq \emptyset\} \\ [G(+, X+)f]\,(A) = \sup_{x \in A \cap \mathrm{dom}f} f(x) \end{cases}$$

$$\begin{cases} \mathrm{dom}\, G(+, X-)f = \mathrm{dom}\, G(+, X+)f \\ [G(+, X-)f]\,(A) = \inf_{x \in A \cap \mathrm{dom}f} f(x) \end{cases}$$

$$\begin{cases} \mathrm{dom}\, G(-, X+)f = \{x \mid x \in X, I_X(x) \subseteq \mathrm{dom}f\} \\ [G(-, X+)f]\,(A) = \inf_{A \in I_X(x)} f(A) \end{cases}$$

$$\begin{cases} \mathrm{dom}\, G(-, X-)f = \mathrm{dom}\, G(-, X+)f \\ [G(-, X-)f]\,(x) = \sup_{A \in I_X(x)} f(A). \end{cases}$$

Note that, substantially, the operators $G(+, X\pm)$ transform functions of the point $x \in X$ into functions of the set $A \subseteq X$, while the operators $G(-, X\pm)$ act in the opposite way. Note also that the operators $G(+, X\pm)$ do not depend on the topological structure of X.

As we said, the lower limit and the upper limit in a topological space X are products of operators $G(\pm, X\pm)$: indeed,

$$\liminf_{x \to x_0} f(x) = [G(-, X-)G(+, X-)f](x_0),$$

$$\limsup_{x \to x_0} f(x) = [G(-, X+)G(+, X+)f](x_0).$$

These two limits are the simplest examples of Γ-limits. In order to define the less obvious and more interesting cases of G-convergence and Γ-convergence, it is necessary to consider, beside each operator of type $G(\overline{\mathbf{R}})$, also the corresponding "partial operators" of the same type.

When one has a function f of n variables, that is $\operatorname{dom} f$ is contained in a product $D_1 \times \cdots \times D_n$ of n sets, and an operator g of type $G(\overline{\mathbf{R}})$, we define for every $1 \le h \le n$ the partial operator g_h which consists in applying g to the function of a single variable obtained by fixing all the variables but the h-th variable. This process is analogous to the introduction of the partial derivative starting from the derivative of the functions of one real variable. Precisely, to say that

$$(g_h f)(x_1, \ldots, x_{h-1}, \lambda, x_{h+1}, \ldots, x_n) = t$$

is equivalent to saying that, if we fix $x_1, \ldots, x_{h-1}, x_{h+1}, \ldots, x_n$ and we define the function φ of a single variable by

$$\varphi(\xi) = f(x_1, \ldots, x_{h-1}, \xi, x_{h+1}, \ldots, x_n),$$

we have

$$[g(\varphi)](\lambda) = t.$$

It is easy to check that g_h is an operator of type $G(\overline{\mathbf{R}})$.

Now, we define the Γ-limits. They are the following products of partial $G_i(\pm, X_i\pm)$:

$$\Gamma(X_1\alpha_1, \ldots, X_k\alpha_k) =$$
$$= G_k(-, X_k\alpha_k) \cdots G_k(-, X_1\alpha_1) \, G_k(+, X_1\alpha_1) \cdots G_k(+, X_k\alpha_k),$$

where k is a positive integer, $X_1, \ldots, X_k$ are topological spaces and $\alpha_1, \ldots, \alpha_k$ are symbols chosen anyway between $+$ and $-$.

If $k = 1$, one finds again the lower limit and the upper limit. A particular case with $k = 2$ gives the definition of $\Gamma(X^-)$ convergence of a sequence of functions. Let $X_1 = \overline{\mathbf{N}} = \mathbf{N} \cup \{+\infty\}$ be equipped with the usual topology, let $X_2 = X$ be a topological space and let $f^*(h, x) = f_h(x)$ be a function defined on $\mathbf{N} \times X$ with values in $\overline{\mathbf{R}}$. Then we have

$$\left[\Gamma(\overline{\mathbf{N}}+, X-)f^*\right](+\infty, x_0) =$$
$$= \left[G_2(-, X-)\, G_1(-, \overline{\mathbf{N}}+)\, G_1(+, \overline{\mathbf{N}}+)\, G_2(+, X-)f^*\right](+\infty, x_0) =$$
$$= \sup_{U \in I_X(x_0)} \inf_{V \in I_{\overline{\mathbf{N}}}(+\infty)} \sup_{h \in V} \inf_{x \in U} f_h(x) =$$
$$= \sup_{U \in I_X(x_0)} \limsup_{h \to +\infty} \inf_{x \in U} f_h(x).$$

Following [120] and [121] we indicate the last expression by

$$\Gamma^-(X) \limsup_{h \to +\infty} f_h(x_0)$$

or also by

$$\Gamma(\overline{\mathbf{N}}^+, X^-) \lim_{\substack{h \to +\infty \\ x \to x_0}} f_h(x).$$

Similarly

$$\left[\Gamma(\overline{\mathbf{N}}-, X-)f^*\right](+\infty, x_0) = \Gamma^-(X) \liminf_{h \to +\infty} f_h(x_0) =$$
$$= \Gamma(\overline{\mathbf{N}}^-, X^-) \lim_{\substack{h \to +\infty \\ x \to x_0}} f_h(x).$$

Finally, if f_∞ is a function such that

$$f_\infty(x) = \left[\Gamma(\overline{\mathbf{N}}+, X-)f^*\right](+\infty, x) = \left[\Gamma(\overline{\mathbf{N}}-, X-)f^*\right](+\infty, x) \qquad \forall x \in X,$$

we say that the sequence (f_h) $\Gamma(X^-)$ converges to f_∞ and we write

$$f_\infty = \Gamma(X^-) \lim_{h \to +\infty} f_h.$$

Several topological concepts may be viewed as cases of Γ-convergence of characteristic functions of sets. If X is a topological space and we let for $E \subseteq X$

$$J_E(x) = \begin{cases} 0 & \text{if } x \in E \\ +\infty & \text{if } x \in X \setminus E \end{cases}$$

we obtain,

$$\Gamma(X-)J_E = J_{\overline{E}}, \quad \Gamma(X+)J_E = J_{E^0},$$

where $\overline{E}$ and E^0 denote respectively the closure and the interior part of E in X. If (x_h) is a sequence converging in X to x_∞ then

$$J_{\{x_\infty\}} = \Gamma(X^-) \lim_{h \to +\infty} J_{\{x_h\}},$$

where, of course, $\{x\}$ is the set with the unique element x. More generally, if (E_h) is a sequence of subsets of X and if there exists

$$\Gamma(X^-) \lim_{h \to +\infty} J_{\{E_h\}} = f$$

then $f = J_{\{E_\infty\}}$ and E_∞ is one of the limits of (E_h) introduced by K. Kuratowski [169].

Perhaps it would be interesting to develop a systematic study of the topological interpretations of the multiple Γ-limits of characteristic functions. For instance, we note that the Γ-convergence of a sequence of functions with values in $\overline{\mathbf{R}}$ agrees with the Γ-convergence of the characteristic functions of the epigraphs. In this direction, it could be useful to consider a more general definition of operator of type G that does not concern only functions with values in $\overline{\mathbf{R}}$ but also functions with values in an arbitrary complete lattice and that has been suggested by some recent research by R. Peirone [227].

By a complete lattice $S = (L, \alpha)$ we mean a set L and a relation $\alpha \subseteq L \times L$ such that, if we denote

$$x \leq_\alpha y \Leftrightarrow (x, y) \in \alpha$$

$$l_1 = \min_\alpha A \Leftrightarrow \emptyset \neq A \subseteq L, \ l_1 \in A, \ : \ l_1 \leq_\alpha x;$$

$$l_2 = \max_\alpha A \Leftrightarrow \emptyset \neq A \subseteq L, \ l_2 \in A, \ : \ x \leq_\alpha l_2;$$

the following four conditions are fulfilled:

(a) $\forall x \in L : \quad x \leq_\alpha x$

(b) $\forall x, y \in L : \quad x \leq_\alpha y, \ y \leq_\alpha x \Rightarrow x = y;$

(c) $\forall x, y, z \in L : x \leq_\alpha y, \ y \leq_\alpha z \Rightarrow x \leq_\alpha z;$

(shortly, α partially orders L);

(d) $\forall A \subseteq L, \ A \neq \emptyset \ \exists \ l_1, l_2 \in L :$

$$l_1 = \max_\alpha \{x \in L \mid \forall a \in A, \ x \leq_\alpha a\}$$

$$l_2 = \min_\alpha \{x \in L \mid \forall a \in A, \ a \leq_\alpha x\}$$

(shortly, every non-void subset of L has greatest lower bound and a least upper bound). Naturally, we shall denote the elements l_1 and l_2 in condition (d) by

$$l_1 = \inf_\alpha A = S\text{-}\inf A; \quad l_2 = \sup_\alpha A = S\text{-}\sup A.$$

Notice that by (d) there exist

$$m_1 = \inf_\alpha L = \min_\alpha L; \quad m_2 = \sup_\alpha L = \max_\alpha L.$$

If (L_1, α_1) and (L_2, α_2) are two complete lattices, we say that a map $\varphi : L_1 \to L_2$ is a complete homomorphism if for any subset A of L_1 we have

$$\inf_{\alpha_2} [\varphi(A)] = \varphi(\inf_{\alpha_1} A); \quad \sup_{\alpha_2} [\varphi(A)] = \varphi(\sup_{\alpha_1} A).$$

In particular a complete homomorphism φ is order preserving in the sense that

$$x_1 \leq_{\alpha_1} x_2 \Rightarrow \varphi(x_1) \leq_{\alpha_2} \varphi(x_2).$$

Now, we define the (universal) operators of type G. An operator g which transforms pairs (S, f) of a complete lattice $S = (L, \alpha)$ and of a function f with values in L into functions $g(S, f)$ with values in L is of type G if the following conditions are verified whenever $S_1 = (L_1, \alpha_1)$, $S_2 = (L_2, \alpha_2)$ are complete lattices and f_1, f_2 are functions with values respectively in L_1 and L_2:

(a) $\operatorname{dom} f_1 \subseteq \operatorname{dom} f_2 \Rightarrow \operatorname{dom} g(S_1, f_1) \subseteq \operatorname{dom} g(S_2, f_2)$

(b) $S_1 = S_2$, $\operatorname{dom} f_1 = \operatorname{dom} f_2$, $f_1 \leq_{\alpha_1} f_2 \Rightarrow g(S_1, f_1) \leq_{\alpha_1} g(S_2, f_2)$

(c) for every complete homomorphism $\varphi : L_1 \to L_2$,

$$\varphi \circ g(S_1, f_1) = g(S_2, \varphi \circ f_1).$$

Among the universal operators of type G, interesting examples are given by the following operators. Let us fix a function τ with values sets or families of sets and define

(1) $$G(\exists, \tau, +)(S, f) = f_1$$

if and only if

$$\begin{cases} \operatorname{dom} f_1 = \{x \mid \tau(x) \cap \operatorname{dom} f \neq \emptyset\} = \{x \mid \tau(x) \neq \emptyset, \ \exists y \in \tau(x) \cap \operatorname{dom} f\} \\ f_1(x) = S\text{-}\sup \{f(t) \mid t \in \tau(x)\}, \end{cases}$$

(2) $$G(\exists, \tau, -)(S, f) = f_2$$

if and only if

$$\begin{cases} \operatorname{dom} f_2 = \operatorname{dom} f_1 \\ f_2(x) = S\text{-}\inf \{f(t) \mid t \in \tau(x)\}, \end{cases}$$

(3) $$G(\forall, \tau, +)(S, f) = \gamma_1$$

if and only if

$$\begin{cases} \operatorname{dom} f_3 = \{x \mid \tau(x) \neq \emptyset, \ \tau(x) \subseteq \operatorname{dom} f\} = \{x \mid \tau(x) \neq \emptyset, \ y \in \operatorname{dom} f \ \forall y \in \tau(x)\} \\ f_3(x) = S\text{-}\sup \{f(t) \mid t \in \tau(x)\}, \end{cases}$$

(4) $$G(\forall, \tau, -)(S, f) = f_4$$

if and only if

$$\begin{cases} \operatorname{dom} f_4 = \operatorname{dom} f_3 \\ f_4(x) = S\text{-}\inf \{f(t) \mid t \in \tau(x)\}. \end{cases}$$

Notice that the operator $G(\forall, \tau, \pm)$ are simply the restrictions of the corresponding operators $G(\exists, \tau, \pm)$ on smaller domains.

Finally, observe that, if X is a topological space,

$$\begin{aligned} \operatorname{dom} \tau &= \mathcal{P}(X), & \tau(A) &= A \\ \operatorname{dom} \vartheta &= X, & \vartheta(x) &= I_X(x) \end{aligned}$$

and $S = (\overline{R}, \leq)$, then

$$G(\exists, \tau, +)(S, \cdot) = G(+, X+)(\cdot)$$
$$G(\exists, \tau, -)(S, \cdot) = G(+, X-)(\cdot)$$
$$G(\forall, \vartheta, +)(S, \cdot) = G(-, X-)(\cdot)$$
$$G(\forall, \vartheta, -)(S, \cdot) = G(-, X+)(\cdot)$$

hence we have obtained by the operators $G(\exists, \tau, \pm)$ and $G(\forall, \tau, \pm)$ a generalization of operators $G(\pm, X\pm)$: I think this generalization could help to investigate many aspects of the theory of generalized limits.

In the applications of Γ-convergence theory to the Calculus of Variations and in particular in the proofs of Theorems 1-5, the following five propositions have a particular weight.

PROPOSITION 1. – *Let X be a topological space and let f_h ($h = 1, 2, \ldots, \infty$) be functions defined in X with values in $\overline{\mathbf{R}}$. Suppose that*

(i)
$$f_\infty = \Gamma(X^-) \lim_{h \to +\infty} f_h$$

Then f_∞ is lower semicontinuous in X,

(a)
$$x_h \to x_\infty \ \text{in} \ X \Rightarrow \liminf_{h \to +\infty} f_h(x_h) \geq f_\infty(x_\infty)$$

and, for every open subset A of X,

$$\inf_{x \in A} f_\infty(x) \geq \limsup_{h \to +\infty} \inf_{x \in A} f_h(x)$$

while, for every compact subset K of X,

$$\min_{x \in K} f_\infty(x) \leq \liminf_{h \to +\infty} \inf_{x \in K} f_h(x).$$

Two straightforward consequences of the previous proposition are the following results, widely employed in the applications of Γ-convergence theory to the Calculus of Variations.

PROPOSITION 2. – *Let X, f_h be as in Proposition 1 and suppose (i) holds. Moreover suppose that the functions f_h are "equicoercive" in the sense that, for every $\lambda \in \mathbf{R}$ there exists a compact subset K_λ of X such that*

$$\{x \in X \ : \ f_h(x) \leq \lambda\} \subseteq K_\lambda \quad \forall h \in \mathbf{N}.$$

Then the function f_∞ has a minimum point in X and

(ii)
$$\min_{x \in X} f_\infty(x) = \lim_{h \to +\infty} \inf_{x \in X} f_h(x).$$

PROPOSITION 3. – *Let X, f_h be as in Proposition 1 and suppose (i) and (ii) hold. If for every $h \in \mathbf{N}$ there exists $x_h \in X$ so that*

$$f_h(x_h) = \min_{x \in X} f_h(x), \quad \lim_{h \to +\infty} x_h = x_\infty$$

then x_∞ is a minimum point of f_∞ in X.

Another useful result, complementary to (a) of Proposition 1, is the following.

PROPOSITION 4. – *Let X, f_h be as in Proposition 1. Suppose* (i) *holds and let x_∞ be a point in X with a countable base of neighborhoods. Then*

(b) $$\exists\, x_h \to x_\infty \text{ in } X \; : \; \lim_{h \to +\infty} f_h(x_h) = f_\infty(x_\infty).$$

It is worthwhile noting that properties (a) in Proposition 1 and (b) in Proposition 4 determine univocally the value $f_\infty(x_\infty)$, hence (a) and (b) are an alternative definition of Γ-limit in metric spaces (see G. Moscariello [213]).

Note also that, in the absence of equicoerciveness (see Proposition 2), Γ-convergence does not ensure automatically the convergence of minima (ii). However, some useful information may be still obtained: for instance, a case of convergence of minima for non-uniformly elliptic forms has been recently studied by S. Mortola and A. Profeti [211].

In Theorems 1, 2, 3, 5 a perturbation

$$\int g(x,u)dx$$

independent of h was added to the "principal" functionals F_h: this possibility is ensured by the following proposition.

PROPOSITION 5. – *Let X, f_h be as in Proposition 1 and suppose* (i) *holds. Then*

$$\Gamma(X^-) \lim_{h \to +\infty} (f_h + g) = f_\infty + g$$

whenever $g : X \to \mathbf{R}$ is a continuous function.

The last proposition has an obvious meaning because it shows that Γ-convergence is unaffected by adding a fixed continuous perturbation. However, as in the case of equicoerciveness, Γ-convergence gives some information also for the addition of non-continuous perturbations as for instance obstacle type perturbation: examples may be found in the papers of L. Boccardo and P. Marcellini [61] and of L. Carbone and S. Salerno [86]. Boccardo and Marcellini studied the Γ-convergence of the functionals

$$\int_\Omega f_h(x)|Du(x)|^2 dx$$

under the constraint $u \le \varphi$ with φ continuous function, Carbone and Salerno studied the Γ-convergence of the same functionals under the constraints $|Du| \le \psi$.

It should be remarked that in the applications of Γ-convergence to the Calculus of Variations, the greatest difficulty is not to prove the existence of the Γ-limit, because there are many general compactness results, but to prove that the Γ-limit has an integral representation. More precisely, if we consider a sequence $(F_h(u,\Omega))$ of functionals depending on the function u and on the open set Ω (let us think of the integral functionals

$$\int_\Omega f(x, u(x), Du(x))dx \ ),$$

a first crucial step is to ascertain whether the Γ-limit $F_\infty(u,\Omega)$ is a measure if considered as a function of Ω. Afterwards, the second step is to ascertain whether $F_\infty(u,\Omega)$ is an integral functional: this happens in the case treated in Theorems 1, 2, 5 but not in Theorem 3.

For the first step it is useful to have some results about the functionals depending on a set, considered in a very particular and actual case in the paper [119], later on developed by many authors (e.g. G. Letta [132], G. Buttazzo, G. Dal Maso [70], L. Carbone, C. Sbordone [88]) and now systematically treated in the following paper of G. Dal Maso and L. Modica.

Some refined results of Γ-convergence theory are useful for the second step, for instance those about the compatible metrics (see [130], [131]) as the following proposition.

PROPOSITION 6. – *Let G be a commutative group equipped with two distance functions d_1 and d_2 both translation invariant. If $f_h : G \to \mathbf{R}$ ($h \in \mathbf{N}$) is a sequence of functions equiuniformly Lipschitz continuous with respect to d_1 and there exists*

$$f_\infty = \Gamma((G, d_2)^-) \lim_{h \to +\infty} f_h$$

then f_∞ is Lipschitz continuous with respect to d_1.

Note that f_∞ may be only lower semicontinuous in (G, d_2).

A particular case of the problem quoted above are the relaxation problems, that is, given a functional

$$\int_\Omega f(x, u(x), Du(x))dx$$

and a functional space S, to ask whether there exists a function g such that

$$\int_\Omega g(x, u(x), Du(x))dx = \liminf_{\substack{\nu \to u \\ S}} \int_\Omega f(x, \nu(x), D\nu(x))dx.$$

In Theorems 1–5 we have avoided for the sake of simplicity to assign boundary conditions on u in the minimum problems. There are many papers concerned with Γ-convergence for boundary value problems, in particular for fixed or variable Dirichlet data. We want to give an example of Γ-convergence with variable Dirichlet data (see [134]).

Let us consider the area functional

$$(1) \qquad\qquad F_h = \int_B \sqrt{1 + |Du|^2}dx$$

with $B = \{x \in \mathbf{R}^2 : |x| < 1\}$ under the condition

$$(2) \qquad\qquad u(\cos\vartheta, \sin\vartheta) = f(h\,\vartheta)$$

with f continuous and 2π-periodic function. More precisely, F_h is defined by (1) if $u \in C^1(B) \cap C^0(\overline{B})$ and fulfills (2), while $F_h(u) = +\infty$ for all other functions of $L^1(B)$. Then there exists

$$F_\infty = \Gamma(L^1(B)^-) \lim_{h \to +\infty} F_h$$

and, letting

$$E(u) = \{(x, y) \in B \times \mathbf{R} : y \geq u(x)\}$$

we have

$$F_\infty(u) = P(E(u), B \times \mathbf{R}) + \lim_{h \to +\infty} \int_0^{2\pi} |f(h\,\vartheta) - \overline{u}(\cos\vartheta, \sin\vartheta)|\, d\vartheta,$$

where $P(E, \Omega)$ is the functional introduced in Theorem 3 and

$$\overline{u}(x) = \limsup_{\rho \to 0^+} \frac{1}{|B_\rho(x) \cap B|} \int_{B_\rho(x) \cap B} u(y)\, dy.$$

Recalling the remarks after Theorem 3, it is obvious that, if $u \in C^1(B) \cap C^0(\overline{B})$ then $\overline{u} = u$ and

$$P(E(u), B \times \mathbf{R}) = \int_B \sqrt{1 + |Du|^2}\, dx.$$

The functions giving the minimum of many variational problems under several different boundary conditions fall within the class of the so-called locally minimizing functions. The study of locally minimizing functions and of the relations between Γ-convergence and convergence of these functions is the second goal of the following paper of G. Dal Maso and L. Modica.

Roughly speaking, the locally minimizing functions of an integral functional F on an open set Ω are the functions which minimize F on every compact subset K of Ω with respect to the perturbations with support in K. An example is given by harmonic functions in an open set Ω. All these functions, even those u for which

$$\int_\Omega |Du|^2 dx = +\infty,$$

are locally minimizing in Ω the functional

$$F(u, A) = \int_A |Du|^2 dx.$$

Finally, we want to observe that up to this time the applications of Γ-convergence theory to the Calculus of Variations are concerned mainly with minimizing functions. The research could be extended to the study of other critical points of functionals. In this field there are still few papers: we mention the works of A. Ambrosetti, and C. Sbordone [3], A. Marino and M. Tosques [133], [191], [192]. We conclude by pointing out that a list of papers and books that consider problems treated explicitly by Γ-convergence and G-convergence theory, or also very similar problems studied by other methods, is furnished at the end of this paper. However, I think that G-convergence and Γ-convergence theory is rather flexible so that, probably, only a few of its possible applications have been explored so far.

We add to the references of this paper many other papers and books related, even not closely, with Γ-convergence and G-convergence theory. We apologize for the gaps and the mistakes in the bibliography and we thank in advance those readers who will point them out.

By [ ]* we indicate those works in which Γ-convergence or G-convergence are explicitly mentioned.

For the reader's convenience and without any claim of completeness, we begin by grouping the items of the bibliography by subject.

Subject Index of the Bibliography

General articles, surveys, conferences, books: 13, 32, 38, 64, 121, 123, 124, 125, 126, 143, 169, 173 bis, 210, 252, 272, 286, 290.

Abstract Γ-convergence theory: 13, 67, 92, 94, 120, 125, 130, 131, 136, 138, 213, 227.

Other convergences linkable with Γ-convergence: 10, 11, 15, 20, 29, 30, 95, 132, 137, 140, 214, 268.

Convex analysis, generalized differentiation, tangency: 10, 11, 15, 79, 139, 143, 158, 159, 174, 175, 176, 196, 214, 240, 258, 268.

Limits of measures and integrals: 19, 55, 69, 70, 71, 76, 87, 89, 110, 111, 115, 116, 117, 119, 135, 148, 183, 207, 208, 211, 212, 232, 256, 257.

Semicontinuity, direct methods in the Calculus of Variations: 2, 80, 96, 147, 149, 151, 154, 170, 178, 185, 239, 265, 266, 287.

Convergence for partial differential equations and variational inequalities: 1, 3, 7, 24, 25, 31, 34, 35, 36, 37, 40, 41, 42, 43, 45, 46, 56, 57, 60, 61, 62, 77, 78, 105, 106, 107, 108, 135, 155, 156, 157, 163, 164, 165, 166, 172, 179, 182, 186, 187, 190, 193, 197, 216, 217, 218, 221, 233, 234, 236, 238, 243, 244, 245, 246, 247, 248, 250, 253, 254, 255, 263, 264, 267, 269, 270, 271, 272, 273, 274, 275, 278, 279, 284, 285, 286.

Convergence for ordinary differential equations: 226, 228, 229, 230, 231, 276, 277.

Homogenization, asymptotic developments: 8, 9, 21, 22, 23, 24, 25, 26, 27, 44, 47, 48, 49, 50, 51, 52, 59, 63, 65, 66, 98, 99, 100, 104, 118, 141, 142, 145, 146, 160, 161, 162, 173, 177, 181, 184, 188, 199, 220, 229, 231, 235, 237, 241, 242, 249, 251, 280, 282.

Control theory: 33, 72, 161, 171, 215, 289.

Mini-max problems: 20, 92, 93, 258.

Problems with constraints: 12, 17, 18, 53, 54, 58, 81, 82, 83, 84, 85, 86, 97, 112, 113, 114, 122, 129, 219.

Bounce problems: 4, 5, 68, 73, 74, 75, 90, 91, 101, 102, 103, 109, 259, 260, 261.

Minimal surfaces: 6, 128, 134, 150, 152, 153, 194, 195, 200, 201, 202, 203, 204, 205, 209.

Steepest-descent curves, evolution equations: 14, 39, 133, 189, 191, 192, 262.

Probability, stochastic equations: 28, 167, 168, 206, 222, 223, 224, 225, 283, 286.

Bibliography [‡]

[1]* E. ACERBI, *Convergence of second-order elliptic operators in complete form.* Boll. Un. Mat. Ital., (5) **18 - B** (1981), 539 - 556.

[2] E. ACERBI, N. FUSCO, *Semicontinuity problems in the calculus of variations.* Archive for Rat. Mech. and Anal. **86** (1984), 125–145.

[3]* A. AMBROSETTI, C. SBORDONE, Γ-*convergenza e* G-*convergenza per problemi non lineari di tipo ellittico.* Boll. Un. Mat. Ital., (5) **13-A** (1976), 352–262.

[4] L. AMERIO, *Continuous solutions of the problem of a string vibrating against an obstacle.* Rend. Sem. Mat. Univ. Padova, **59** (1978), 67–96.

[5] L. AMERIO, G. PROUSE, *Study of the motion of a string vibrating against an obstacle.* Rendiconti di Matematica (6) **8** (1975), 563–585.

[6] G. ANZELLOTTI, M. GIAQUINTA, *Funzioni BV e tracce.* Rend. Sem. Mat. Padova **60** (1978), 1–22.

[7] A. AROSIO, *Asymptotic behaviour as $t \to +\infty$ of solutions of linear parabolic equations.* Comm. Partial Diff. Eq. **4** (1979), 769–794.

[8] M. ARTOLA, G. DUVAUT, *Homogénéisation d'une plaque renforcé.* C.R. Acad. Sci. Paris, Sér. A, **284** (1977), 707–710.

[9] M. ARTOLA, G. DUVAUT, *Homogénéisation d'une classe de problèmes non linéaires.* C.R. Acad. Sci. Paris, Sér. A, **288** (1979), 775–778.

[10] H. ATTOUCH, *Convergence de fonctions convexes, des sous-differentiales et semigroupes associés.* C.R. Acad. Sci. Paris, Sér. A, **284** (1977), 539–542.

[11] H. ATTOUCH, *Convergence de fonctionnelles convexes.* Proceed. Journ. d'Anal. nonlinéaire, Besançon, 1977, Lecture Notes in Math., Springer, **665** (1978), 1–40.

[‡]Editor's note: some references have been updated

[12]* H. ATTOUCH, *Convergence des solutions d'inéquations variationelles avec obstacles.* Proceed. Int. Meeting on "Recent Methods in Nonlinear analysis", Rome, May 8-12, 1978, ed. by E. De Giorgi, E. Magenes, U. Mosco, Pitagora ed., Bologna (1979) 101–114; C.R. Acad. Sci. Paris, Sér. A, **227** (1978),1001–1003.

[13]* H. ATTOUCH, *Sur la Γ-convergence.* Séminaire Brezis-Lions, Collège de France (1979).

[14]* H. ATTOUCH, *Famille d'opérateurs maximaux monotones et mésurabilité.* Ann. Mat. Pura Appi. (4) **120** (1979), 35–111.

[15]* H. ATTOUCH, Y. KONISHI, *Convergence d'opérateurs maximaux monotones et inéquations variationnelles.* C.R. Acad. Sci. Paris, Sér. A, **282** (1976), 467–469.

[16]* H. ATTOUCH, C. PICARD, *Problèmes variationnels et théorie du potentiel non linéaire.* Ann. Fac. Sci. Toulouse **1** (1979), 89–136.

[17]* H. ATTOUCH, C. PICARD, *Convergence d'inéquations variationnelles avec obstacles et espaces fonctionnels attachés aux éléments précisés.* Applicable Anal. **12** (1981), 287–306.

[18]* H. ATTOUCH, C. PICARD, *Asymptotic Analysis of variational problems with constraints of obstacle type.* Publications Mathématiques d'Orsay **82**, 7. Université de Paris-Sud, Département de Mathématique, Orsay, 1982. 95 pp.

[19]* H. ATTOUCH, C. SBORDONE, *Asymptotic limits for perturbed functionals of the Calculus of Variations.* Ricerche Mat. 29 (1980), 85–124.

[20]* H. ATTOUCH, R.J. WETS, *A convergence theory for saddle functions.* Trans. Amer. Math. Soc. **280** (1983), 1–41.

[21]* I. BABUSKA, *Solution of interface problems by homogenization.* SIAM J. Math. Anal. **7** (1976), 603–645; **8** (1977), 923–937.

[22]* I. BABUSKA, *Homogenization and its application. Mathematical and computational problems.* Proceed. 111 Symp. Numer. Solut. Partial Diff. Eq., College Park, 1975, ed. by B. Hubbard, Acadernic Press, New York, (1976), 89–116.

[23]* I. BABUSKA , *Homogenization approach in Engineering.* Lecture Notes in Econ. and Math. systems, Springer, **134** (1976), 137–153.

[24]* L.A. BAGIROV, M.A. SHUBIN, *On the stabilization of solutions of Cauchy problems for parabolic equations with coefficients that are almost periodic in the space-variables.* Differential'nye Uravneniya **11** (1975), 2205–2209; transl. in Differential Equations **11** (1975), 1637–1640.

[25] N.S. BAHVALOV, *Averaging of partial differential equations with rapidly oscillating coefficients.* Soviet Math. Doklady **16** (1975), 351–355; transl. from Doklady Akad. Nauk SSSR **221** (1975), 516–519.

[26] N.S. BAHVALOV, *Averaging of non linear partial differential equations with rapidly oscillating coefficients.* Soviet Math. Doklady **16** (1975), 1469–1473; transl. from Doklady Akad. Nauk SSSR **225** (1975), 249–252.

[27] N.S. BAHVALOV, *Homogenization and perturbations problems.* Proceed. Meeting on "Computing Methods in Applied Sciences and Engineering", Versailles, Dec. 1979, ed. by R. Glowinski, J.L. Lions, North-Holland (1980), 645–658.

[28] N.S. BAHVALOV, A.A. ZLOTNIK, *Coefficient stability of differential equations with random coefficients.* Soviet Math. Dokl. **19** (1978), 1171–1175; transl. from Doklady Akad. Nauk SSSR **242** (1978), 745–748.

[29] G. BEER, *A natural topology for upper semicontinuous functions and a Baire category dual for convergence in measure.* Pacific J. Math. **96** (1981), 251–263.

[30]* G. BEER, *Upper semicontinuous functions and the Stone approximation theorem.* J. Approx. Theory **34** (1982), 1–11.

[31] J.J. BELOV, *On some systems of linear differential equations with a small parameter.* Soviet. Math. Dokl. **17** (1976), 1675–1679; transl. from Doklady Akad. Nauk. SSSR **231** (1976), 1037–1040.

[32] J.J. BENEDETTO, *Real variable and integration.* Mathematische Leitfäden, Teubner, Stuttgart (1976).

[33] M.L. BENNATI, *Perturbazioni variazionali di problemi di controllo.* Boll. Un. Mat. Ital. (5) **16-B** (1979), 910–922.

[34]* A. BENSOUSSAN, J.L. LIONS, *Inéquations quasi variationnelles dépendant d'un paramètre.* Ann. Sc. Norm. Sup. Pisa Cl. Sci. (4), **4** (1977), 231–255.

[35]* A. BENSOUSSAN, J.L. LIONS, G. PAPANICOLAOU, *Sur quelques phénomènes asymptotiques stationnaires.* C.R. Acad. Sci. Paris, Sér. A, **281** (1975), 89–95.

[36]* A. BENSOUSSAN, J.L. LIONS, G. PAPANICOLAOU, *Sur quelques phénomènes asymptotiques d'évolutions.* C.R. Acad. Sci. Paris, Sér. A, **281** (1975), 317–322.

[37]* A. BENSOUSSAN, J.L. LIONS, G. PAPANICOLAOU, *Sur des nouveaux problèmes asymptotiques.* C.R. Acad. Sci. Paris, Sér. A, **282** (1976), 1277–1282.

[38] A. BENSOUSSAN, J.L. LIONS, G. PAPANICOLAOU, *Asymptotic methods in periodic structures.* North-Holland, Amsterdam (1978).

[39] A.L. Berdichevskii, V.L. Berdichevskii, *The flow of an ideal fluid around a periodic system of obstacles.* Izv. Akad. Nauk **6** (1978), 3–18.

[40] V.L. Berdichevskii, *On the averaging of periodic structures.* Prikl. Mat. Mekh. 41 (1977), 993–1006; transl. in J. Appl. Math. Mech. **41** (1977), 1010–1023.

[41] V.L. Berdichevskii, *Spatial averaging of periodic structures.* Soviet Phys. Dokl. 20 (1975), 334–335; transl. from Doklady Akad. Nauk SSSR **222** (1975), 565–567.

[42]* M. Biroli, *Sur la G-convergence pour des inéquations quasi-variationnelles.* C.R. Acad. Sci. Paris, Sér. A, **284** (1977), 947–950.

[43]* M. Biroli, *Sur la G-convergence pour des inéquations quasi-variationnelles.* Boll. Un. Mat. Ital. (5) **14-A** (1977), 540–550.

[44]* M. Biroli, *An estimate on convergence for an homogenization problem for variational inequalities.* Rend. Sem. Mat. Univ. Padova **58** (1977), 9–22.

[45]* M. Biroli, *G-convergence for elliptic equations, variational inequalities and quasi-variational inequalities.* Rend. Sem. Mat. Univ. Milano **47** (1977), 269–328.

[46]* M. Biroli, *An estimate on convergence of approximations by iterations.* Ann. Mat. Pura Appl. (4) **112** (1979), 83–101.

[47]* M. Biroli, *Homogenization for variational and quasi-variational inequalities.* Proceed. Meeting on "Free boundary problems", Pavia, Sett. Ott. 1979, Istituto Nazionale di Alta Matematica, Roma, vol. II (1980), 45–61.

[48]* M. Biroli, *Homogenization for variational and quasi-variational inequalities.* Rend. Sem. Mat. Fis. Milano **52** (1982), 375–391 (1985).

[49]* M. Biroli, *Homogenization for parabolic variational and quasi-variational inequalities.* Proceed. Meeting on "Differential Equations", Rousse, June 1981.

[50]* M. Biroli, *Homogenization for variational and quasi-variational inequalities.* Proceed. Meeting on "Equadiff 5", Bratislava, August 1981.

[51]* M. Biroli, S Marchi, *An estimate in homogenization for elliptic equations with L^∞ coefficients.* Quaderno I.A.C. n. 92 (1978).

[52]* M. Biroli, S Marchi, T. Norando, *Homogenization estimates for quasi-variational inequalities.* Boll. Un. Mat. Ital. **18-A** (1981), 267–275.

[53]* M. Biroli, U. Mosco, *Stability and homogenization for nonlinear variational inequalities with irregular obstacles and quadratic growth.* Nonlinear Anal. **7** (1983), 41–60.

[54] L. BOCCARDO, *Alcuni problemi al contorno con vincoli unilaterali dipendenti da un parametro.* Boll. Un. Mat. Ital. (4) **8** (1973), 97–109.

[55]* L. BOCCARDO, *Perturbazioni di funzionali del calcolo delle variazioni.* Ricerche Mat. **29** (1980), 213–242.

[56]* L. BOCCARDO, I. CAPUZZO DOLCETTA, *G-convergenza e problema di Dirichlet unilaterale.* Boll. Un. Mat. Ital. (4) **12** (1975), 115–123.

[57]* L. BOCCARDO, I. CAPUZZO DOLCETTA, *Stabilità delle soluzioni di disequazioni variazionali ellittiche e paraboliche quasi lineari.* Ann. Univ. Ferrara, Sez. VII, **24** (1978), 99–111.

[58]* L. BOCCARDO, I. CAPUZZO DOLCETTA, *Disequazioni quasi variazionali con funzioni d'ostacolo quasi limitate: esistenza e G-convergenza.* Boll. Un. Mat. Ital. (5) **15-B** (1978), 370–385.

[59]* L. BOCCARDO, I. CAPUZZO DOLCETTA, M. MATZEU, *Nouveaux résultats d'existence, perturbation et homogénéisation de quelques problèmes non linéaires.* Proceed. Int. meeting on "Recent Methods in Nonlinear Analysis", Rome, May 8-12, 1978, ed. by E. De Giorgi, E. Magenes, U. Mosco, Pitagora ed., Bologna (1979), 115–130.

[60]* L. BOCCARDO, F. DONATI, *Existence and stability results for solutions of some strongly nonlinear constrained problems.* Nonlinear Analysis, Th. Meth. Appl. **5** (1981), 975–988.

[61] L. BOCCARDO, P. MARCELLINI, *Sulla convergenza delle soluzioni di disequazioni variazionali.* Ann. Mat. Pura Appl. (4) **110** (1976), 137–159.

[62] L. BOCCARDO, F. MURAT, *Nouveaux résultats de convergence dans des probleèmes unilatéraux.* Collège de France, Seminar on Nonlinear partial differential equations, 1979-80, ed. by H. Brezis, J.L. Lions, Research Notes in Math., Pitman, London (1981).

[63]* L. BOCCARDO, F. MURAT, *Homogénéisation de problèmes quasi-linéaires.* Pubbl. I.R.M.A., Lille, 3 (7), (1981).

[64] N.N. BOGOLIUBOV, Y.A. MITROPOLSKI, *Asymptotic methods in the theory of non-linear oscillations.* Gordon and Breach, New York (1962).

[65] J.F. BOURGAT, *Numerical experiments of the homogenization method for operators with periodic coefficients.* Proceed. Int. Colloq. on "Computing Methods in Applied Sciences and Engineering", Versailles, Dec. 1977, Lecture Notes Math. **704** Springer (1979), 330–356.

[66] R. BRIZZI, J.P. CHALOT, *Homogénéisation de frontière.* Thèse, Université de Nice (1978).

[67]* G. BUTTAZZO, *Su una definizione generale dei Γ-limiti.* Boll. Un. Mat. Ital. (5) **14-B** (1977), 722–744.

[68]* G. Buttazzo, *Il problema del rimbalzo su varietà riemanniane*. Atti del Convegno su "Studio di problemi-limite in Analisi Funzionale", Bressanone, Settembre 1981, ed. da P. Patuzzo, S. Steffè, C.L.E.U.P., Padova (1982).

[69] G. Buttazzo, G. Dal Maso, *Γ-limit of a sequence of nonconvex and non-equilipschitz integral functionals*. Ricerche Mat. **27** (1978), 235–251.

[70]* G. Buttazzo, G. Dal Maso, *Γ-limits of integral functionals*. J. Analyse Math. **37** (1980), 145–185.

[71]* G. Buttazzo, G. Dal Maso, *Integral representation on $W^{1,\alpha}(\Omega)$ and $BV(\Omega)$ of limits of variational integrals*. Atti Accad. Naz. Lincei, Rend. Cl. Sci. Mat. Fis. Natur. (8) **66** (1979), 338–344.

[72]* G. Buttazzo, G. Dal Maso, *Γ-convergence and optimal control problems*. J. Optimization Theory Appl. **38** (1982), 385–407.

[73]* G. Buttazzo, D. Percivale, *Sull'approssimazione del problema del rimbalzo unidimensionale*. Ricerche Mat. **30** (1981), 217–231.

[74]* G. Buttazzo, D. Percivale, *The bounce problem on n-dimensional Riemannian manifolds*. Atti Accad. Naz. Lincei Rend. Cl. Sci. Fis. Mat. Natur. **70** (1981), 246–250.

[75] G. Buttazzo, D. Percivale, *On the approximation of the elastic bounce problem on Riemannian manifolds*. J. Differential Equations **47** (1983), 227–245.

[76]* G. Buttazzo, M. Tosques, *Gamma-convergenza per alcune classi di funzionali*. Ann. Univ. Ferrara, Sez. VII, **23** (1977), 257–267.

[77] G. Buttazzo , *The asymptotics of solutions of some model problems in chemical kinetics, taking diffusion into account*. Soviet. Math. Dokl. **19** (1978), 1079–1083 transl. from Doklady Akad. Nauk SSSR **242** (1978).

[78] L.A. Caffarelli, A. Friedman, *Reinforcement problem in elasto-plasticity*. Rocky Mountain J. Math. **10** (1980), 155–184.

[79] O. Caligaris, *Sulla convergenza delle soluzioni di problemi di evoluzione associati a sottodifferenziali di funzioni convesse*. Boll. Un. Mat. Ital. (5) **13-B** (1976), 778–806.

[80] O. Caligaris, F. Ferro, P. Oliva, *Sull'esistenza dei minimo per problemi di Calcolo delle Variazioni*. Boll. Un. Mat. Ital. (5) **14-B** (1977), 340–369.

[81]* L. Carbone, *Sur la Gamma-convergence des integrals du type de l'énergie à gradient borné*. J. Math. Pures Appl. (9) **56** (1977), 79–84.

[82]* L. CARBONE, Γ^--*convergence d'integrales sur des functions avec des contraintes sur le gradient*. Comm. Partial Diff. Equation **2** (1977) 627–651.

[83]* L. CARBONE, *Sull'omogeneizzazione di un problema variazionale con vincoli sul gradiente*. Atti Accad. Naz. Lincei, Rend. Cl. Sci. Mat. Fis. Natur. (8) **63** (1978), 10 - 14.

[84]* L. CARBONE, *Sur un problème d'homogénéisation avec des contraintes sur le gradient*. J. Math. Pures Appl. **58** (1979), 275–297.

[85]* L. CARBONE, F. COLOMBINI, *On convergence of functionals with unilateral constraints*. J. Math. Pures Appl. **59** (1980), 465–500.

[86]* L. CARBONE, S. SALERNO, *On a problem of homogenization with quickly oscillating constraints on the gradient*. J. Math. Anal. Appl. **90** (1982), no. 1, 219–250.

[87] L. CARBONE, C. SBORDONE, *Un teorema di compattezza per la Γ-convergenza di funzionali non coercitivi*. Atti Accad. Naz. Lincei, Rend. Cl. Sci. Mat. Fis. Natur. (8) **62** (1977), 744 - 748.

[88]* L. CARBONE, C. SBORDONE, *Some properties of Γ-limits of integral functionals*. Ann. Mat. Pura Appl. (4) **122** (1979), 1–60.

[89]* M. CARRIERO, E. PASCALI, Γ-*convergenza di integrali non negativi maggiorati da funzionali del tipo dell'area*. Ann. Univ. Ferrara **24** (1978), 51–64.

[90]* M. CARRIERO, E. PASCALI, *Il problema del rimbalzo unidimensionale e sue approssimazioni con penalizzazioni non convesse*. Rendiconti di Matematica (4) **13** (1980), 541–554.

[91] M. CARRIERO, E. PASCALI, *Uniqueness of the one-dimensional bounce problem as a generic property in $L^1([0,T]; \mathbf{R})$*. Boll. Un. Mat. Ital. (6) **1-A** (1982), 87–91.

[92]* E. CAVAZZUTI, Γ-*convergenze multiple, convergenza dei punti di sella e di max-min*. Boll. Un. Mat. Ital. (6) **1-B** (1982), 251–274.

[93]* E. CAVAZZUTI, *Multiple Γ-convergence and application io convergence of saddle points and max-min points*. Proceed. Meeting on "Recent Methods in Non-linear Analysis and Applications", S.A.F.A. IV, Napoli March 21-28, 1980, ed. by A. Canfora, S. Rionero, C. Sbordone, G. Trombetti, Liguori ed., Napoli (1981), 333–340.

[94]* E. CAVAZZUTI, *Alcune caratterizzazioni della Γ-convergenza multipla*. Ann. Mat. Pura Appl. (4) **132** (1982), 69–112 (1983).

[95]* G. CICOGNA, *Un'analisi "categoriale" delle nozioni di convergenza*. Boll. Un. Mat. Ital. (5) **17-A** (1980), 531–536.

[96] D. CIORANESCU, *Calcul des variations sur des sour-espaces variables.* C.R. Acad. Sci. Paris, Sér. A, **291** (1980), 19–22; 87–90.

[97]* D. CIORANESCU, F. MURAT, *Un terme étrange venu d'ailleurs.* Proceedings of the Seminar "Nonlinear partial differential equations and their applications", Collège de France, ed. by H. Brezis and J.L. Lions, Research Notes in Mathematics, Pitman, London (1982), vol. II, III.

[98] D. CIORANESCU, J. SAINT JEAN PAULIN, *Homogénéisation dans des ouverts à cavités.* C.R. Acad. Sci. Paris, Sér. A, **284** (1977), 857–860.

[99] D. CIORANESCU, J. SAINT JEAN PAULIN, *Homogénéisation de problèmes d'évolution dans des ouverts à cavités.* C.R. Acad. Sci. Paris, Sér. A, **286** (1978), 889–902.

[100] D. CIORANESCU, J. SAINT JEAN PAULIN, *Homogenization in open sets with holes.* J. Math. Pures Appl. **71** (1979), 590–607.

[101] C. CITRINI, *Controesempi all'unicità del moto di una corda in presenza di una parete.* Atti Accad. Naz. Lincei, Rend. Cl. Sci. Fis. Mat. Natur. (8) **67** (1979), 179–185.

[102] C. CITRINI, *Discontinuous solutions of a nonlinear hyperbolic equation with unilateral constraint.* Manuscripta Math. **29** (1979), 323–352.

[103] C. CITRINI, B. D'ACUNTO, *Sull'urto tra due corde.* Ricerche Mat. **28** (1979), 375–398.

[104] M. CODEGONE, *Problèms d'homogénéisation en théorie de la diffraction.* C.R. Acad. Sci. Paris, S/'er. A, **288** (1979), 387–389.

[105]* M. CODEGONE, *Scattering of elastic waves through a heterogenous medium.* Math. Methods in the Appl. Sciences **2** (1980), 271–287.

[106]* F. COLOMBINI, S. SPAGNOLO, *Sur la convergence des solutions d'équations paraboliques avec des coefficients qui dépendent du temps.* C.R. Acad. Sci. Paris, Sér. A, **282** (1976), 735–737.

[107]* F. COLOMBINI, S. SPAGNOLO, *Sur la convergence des solutions d'équations paraboliques.* J. Math.. Pures Appl. **56** (1977), 263–306.

[108]* F. COLOMBINI, S. SPAGNOLO, *On the convergence of solutions of hyperbolic equations.* Comm. Partial Diff. Equations **3** (1978), 77–103.

[109] B. D'ACUNTO, *Sull'urto elastico di una corda in un caso non lineare.* Ricerche Mat. **27** (1978), 301–317.

[110]* G. DAL MASO, *Alcuni teoremi sui Γ-limiti di misure.* Boll. Un. Mat. Ital. (5) **15-B** (1978), 182–192.

[111]* G. DAL MASO, *Integral representation on* $BV(\Omega)$ *of* Γ-*limits of variational integrals.* Manuscripta Math. **30** (1980), 387–416.

[112]* G. DAL MASO, *Limiti di soluzioni di problemi variazionali con ostacoli bilateri.* Atti Accad. Naz. Lincei, Rend. Cl. Sci. Mat. Fis. Natur. (8) **69** (1980), 333–337 (1982).

[113]* G. DAL MASO, *Limiti di problemi di minimo con ostacoli.* Atti dei Convegno su "Studio dei problemi-limite in Analisi Funzionale", Bressanone, Settembre 1981, ed. da P. Patuzzo, S. Steffè, C.L.E.U.P., Padova (1982).

[114]* G. DAL MASO, P. LONGO, Γ-*limits of obstacles.* Ann. Mat. Pura Appl. (4) **128** (1980), 1–50.

[115]* G. DAL MASO, L. MODICA, *Convergenza dei minimi locali.* Pubblicazione interna. Scuola Normale Superiore, Pisa (1979).

[116]* G. DAL MASO, L. MODICA, *Un criterio generale per la convergenza dei minimi locali.* Rendiconti di Matematica (7) **1** (1981), 81–94.

[117]* G. DAL MASO, L. MODICA, *Sulla convergenza dei minimi locali.* Boll. Un. Mat. Ital. (6) **1-A**, (1982), 55–61.

[118] A. DAMLAMIAN, *Homogénéisation du problème de Stefan.* C.R. Acad. Sci. Paris, Sér. A, **289** (1979),9–11.

[119]* E. DE GIORGI, *Sulla convergenza di alcune successioni di integrali del tipo dell'area.* Rendiconti di Matematica (4) **8** (1975), 277–294.

[120]* E. DE GIORGI, Γ-*convergenza e G-convergenza.* Boll. Un. Mat. Ital. (5) **14-A** (1977), 213 - 220.

[121]* E. DE GIORGI, *Convergence problems for functionals and operators.* Proceed. Int. Meeting on "Recent Methods in Nonlinear Analysis", Rome, May 8 - 12, 1978, ed. by E. De Giorgi, E. Magenes, U. Mosco, Pitagora ed., Bologna (1979), 131–188.

[122]* E. DE GIORGI, Γ-*limiti di ostacoli.* Atti S.A.F.A. IV, Napoli, March 21–28, 1980, ed. by A. Canfora, S. Rionero, C. Sbordone, G. Trombetti, Liguori, Napoli (1981), 51–84.

[123]* E. DE GIORGI, *New problems in* Γ-*convergence and G-convergence.* Proceed. Meeting on "Free Boundary problems", Pavia Sept. Oct. 1979, Istituto Nazionale di Alta Matematica, Roma, vol. II (1980), 183–194.

[124]* E. DE GIORGI, *Quelques problèms de* Γ-*convergence.* Proceed. Meeting on "Computing Methods in Applied Sciences and Engineering", Versailles, Dec. 1979, ed. by R. Glowinski, J.L. Lions, North-Holland (1980), 637–643.

[125]* E. De Giorgi, *Operatori elementari di limite ed applicazioni al Calcolo delle Variazioni*. Atti del Convegno su "Studio di problemi-limite in Analisi Funzionale", Bressanone, Settembre 1981, ed. da P. Patuzzo e S. Steffè, C.L.E.U.P., Padova (1982).

[126]* E. De Giorgi, G. Buttazzo, *Limiti generalizzati e loro applicazioni alle equazioni differenziali*. Matematiche (Catania) **36** (1981), 53–64.

[127]* E. De Giorgi, G. Dal Maso, Γ-*convergence and calculus of variations*, in *Mathematical Theories of Optimization (S. Margherita Ligure, 1981)*, T. Zolezzi ed., 121-143, Lecture Notes in Math. **979**, Springer, Berlin, 1983.

[128] E. De Giorgi, F. Colombini, L.C. Piccinini, *Frontiere orientate di misura minima e questioni collegate*. Quaderni della Scuola Normale Superiore di Pisa (1972).

[129]* E. De Giorgi, G. Dal Maso, P. Longo, Γ-*limiti di ostacoli*. Atti Accad. Naz. Lincei, Rend. Cl. Sci. Mat. Fis. Natur. (8) **68** (1980), 481–487.

[130]* E. De Giorgi, T. Franzoni, *Su un tipo di convergenza variazionale*. Atti Accad. Naz. Lincei, Rend. Cl. Sci. Mat. Fis. Natur. (8) **58** (1975), 842–850.

[131]* E. De Giorgi, T. Franzoni, *Su un tipo di convergenza variazionale*. Rend. Sem. Mat. Brescia **3** (1979), 63–101.

[132]* E. De Giorgi, G. Letta, *Une notion générale de convergence faible pour des fonctions croissantes d'ensemble*. Ann. Sc. Norm. Sup. Pisa Cl. Sci. (4) **4** (1977) 61–99.

[133]* E. De Giorgi, A. Marino, M. Tosques, *Problemi di evoluzione in spazi metrici e curve di massima pendenza*. Atti Accad. Naz. Lincei, Rend. Cl. Sci. Fis. Mat. Natur. (8) **68** (1980), 180–187.

[134]* E. De Giorgi, L. Modica, Γ-*convergenza e superfici minime*. Pubblicazione interna Scuola Normale Superiore, Pisa (1979).

[135]* E. De Giorgi, S. Spagnolo, *Sulla convergenza degli integrali dell'energia per operatori ellittici del II ordine*. Boll. Un. Mat. Ital. (4) **8** (1973), 391–411.

[136]* I. Del Prete, M. Lignola, *On the variational properties of $\Gamma^-(d)$-convergence*. Ricerche Mat. **31** (1982), 179–190.

[137] Z. Denkowski, *The convergence of generalized sequences of sets and functions in locally convex spaces. I, II*. Zeszyty Nauk. Uniw. Jagielloń. Prace Mat. No. **22** (1981), 37–58, 59–72.

[138] M. Dolcher, *Topologie a strutture di convergenza*. Ann. Sc. Norm. Sup. Pisa, Sci. Fis. Mat. (3) **14** (1960), 63–92.

[139]* S. DOLECKI, *Tangency and differentiation: some applications of convergence theory.* Ann. Mat. Pura Appl. (4) **130** (1982), 223–255.

[140]* S. DOLECKI, G. SALINETTI, R.J. WETS, *Convergence of functions: equi-semicontinuity.* Trans. Amer. Math. Soc. **276** (1983), 409–430.

[141] G. DUVAUT, *Comportement microscopique d'une plaque perforée périodiquement.* I. Lecture Notes in Mathematics, Springer, **594** (1977), 131–145.

[142] G. DUVAUT, A.M. METELLUS, *Homogénéisation d'une plaque mince.* C.R. Acad. Sci. Paris, Sér. A, **283** (1976), 947–950.

[143] I. EKELAND, R. TEMAM, *Convex analysis and variational problems.* North-Holland, Amsterdam (1978).

[144] F. FERRO, *Integral characterization of functionals defined on spaces of BV functions.* Rend. Sem. Mat. Univ. Padova **61** (1979), 177–201.

[145] F. FLEURY, G. PASA, D. POLYSENSHI, *Homogénéisation de corps composites.* C.R. Acad. Sci. Paris, Sér. B, **289** (1979), 241–244.

[146] C. FOIAS, L. TARTAR, *Opérateurs de Toeplits généralisés et homogénéisation.* C.R. Acad. Sci. Paris, Sér. A, **291** (1980), 15–18.

[147]* N. FUSCO, *Dualità e semicontinuità per integrali del tipo dell'area.* Rend. Accad. Sci. Fis. Mat. Napoli (4) **46** (1979) 81–90.

[148]* N. FUSCO, *Γ-convergenza unidimensionale.* Boll. Un. Mat. Ital. (5) **16-B** (1979), 74–86.

[149] N. FUSCO, *Quasi-convessità e semicontinuità per integrali multipli di ordine superiore.* Ricerche Mat. **29** (1980), 307–324.

[150] M. GIAQUINTA, *Regolarità delle superfici $BV(\Omega)$ con curvatura media assegnata.* Boll. Un. Mat. Ital. (4) **8** (1973), 567–578.

[151] M. GIAQUINTA, G. MODICA, J. SOUCEK, *Functionals with linear growth in the Calculus of Variations.* Comment. Math. Univ. Carolinae **20** (1979), 143–172.

[152] E. GIUSTI, *On the equations of surface of prescribed mean curvature.* Invent. Math. **46** (1978), 111–137.

[153] E. GIUSTI, *Minimal surfaces and functions of bounded variation.* Notes on Pure Mathematics, vol. 10, Australian National Univ. Canberra (1977).

[154] C. GOFFMAN, J. SERRIN, *Sublinear functions of measures and variational integrals.* Duke Math. J. **31** (1964), 159–178.

[155] A.K. GUSHCHIN, V.P. MIKHAILOV, *On the stabilization of solutions of Cauchy problems for a parabolic equation.* Differential'nye Uravnenyia **7** (1971) 297–311; transl. in Differential Equations **7** (1971), 232–242.

[156] Z. Hashin, S. Shtrikman, *A variational approach to the theory of the elastic behaviour of multiphase materials.* J. Mech. Phys. Solids **11** (1963), 127–140.

[157] H.P. Hery, E. Sanchez-Palencia, *Phénomènes de transmission à travers des couches minces de conductivité élevée.* J. Math. Anal. Appl. **47** (1974), 284–309.

[158] J.L. Joly, *Une famille de topologies sur l'ensemble des fonctions convexes pour lesquelles la polarité est bicontinue.* J. Math. Pures Appl. **52** (1973) 137–158.

[159] J.L. Joly, F. De Thelin, *Convergence of convex integrals in L^p space.* J. Math. Anal. Appl. **54** (1976), 230–244.

[160]* S. Kesavan, *Homogénéisation et valeurs propres.* C.R. Acad. Sci. Paris, Sér. A, **285** (1977), 229–232.

[161] S. Kesavan, *Homogénéisation d'un problème de controle optimal.* C.R. Acad. Sci. Paris, Sér. A, **285** (1977), 441–444.

[162]* S. Kesavan, *Homogenization of elliptic eigenvalues problem.* Appl. Math. Optim. **5** (1979), 163–167; 197–216.

[163]* K'en T'en Ngoan, *On the convergence of solutions of boundary value problems for sequences of elliptic equations.* Uspekhi Mat. Nauk. **32** (1977), 183–184.

[164]* Kha T'en Ngoan, *On the convergence of solutions of boundary value problems for sequences of elliptic systems.* Moskow Univ. Math. Bull. **32** (1977), 66–74; transl. from Vestnik Moskov Univ. Mat. Mekh. 1977, 83–92.

[165]* E. Ya. Khruslov, *The asymptotic behavior of solutions of the second boundary-value problem under the collapsing of the boundary of the domain.* Math. USSR Sb. **35** (1979), 266–282; transl. from Mat. Sb. **106** (1978), 604–621.

[166]* S.M. Kozlov, *The averaging of differential operators with almost periodic rapidly oscillating coefficients.* Math. USSR Sb. **35** (1978), 481–498; transl. from Mat. Sb. **107** (1978), 199–217.

[167]* S.M. Kozlov, *Averaging random structures.* Soviet Math. Dokl. **19** (1978), 950–954, transl. from Dokl. Akad. Nauk SSSR **241** (1978), 1016–1019.

[168]* S.M. Kozlov, *Averaging of random operators.* Math. USSR Sb. **37** (1980), 167–180; transl. from Mat. Sb. **109** (1979), 188–192.

[169] K. Kuratowski, *Topology.* Academic Press, New York (1966).

[170] A. LICHNEWSKY, *Sur un problème de Calcul des Variations intervenant dans l'étude de la plasticité.* C.R. Acad. Sci. Paris, Sér. A, **287** (1978), 1069–1072.

[171] J.L. LIONS, *Perturbations singulières dans les Problèmes aux limites et en Controle optimal.* Lecture Notes in Mathematics, Springer, **323** (1973).

[172] J.L.LIONS, *Asymptotic behaviour of variational inequalities with highly oscillating coefficients.* Lectures Notes in Mathematics, Springer, **503** (1976), 30–55.

[173] J.L. LIONS, *Homogénéisation non locale.* Proceed. Int. Meeting on "Recent Methods in Nonlinear Analysis", Rome, May 8-12, 1978, ed. By E. De Giorgi, E. Magenes, U. Mosco, Pitagora ed., Bologna (1979), 189–204.

[173bis] J.L. LIONS, *Some Methods in the Mathematical Analysis of Systems and Their Control.* Science Press, Beijing, China; Gordon and Breach Inc. New York (1981).

[174] P.L. LIONS, *Two remarks on the convergence of convex functions and monotone operators.* Nonlinear Analysis **2** (1978), 553–562.

[175] P. MARCELLINI, *Su una convergenza di funzioni convesse.* Boll. Un. Mat. Ital. (4) **8** (1973), 137–158.

[176] P. MARCELLINI, *Un teorema di passaggio al limite per la somma di funzioni convesse.* Boll. Un. Mat. Ital. (4) **11** (1975), 107–124.

[177]* P. MARCELLINI, *Periodic solutions and homogenization of nonlinear variational problems.* Ann. Mat. Pura Appl. (4) **117** (1978), 139–152.

[178]* P. MARCELLINI, *Some problems of semicontinuity.* Proceed. Int. Meeting on "Recent Methods in Nonlinear Analysis", Rome, May 8-12, 1978, ed. by E. De Giorgi, E. Magenes, U. Mosco, Pitagora ed., Bologna (1979), 205–222.

[179]*P. MARCELLINI, *Convergence of Second Order Linear Elliptic Operators.* Boll. Un. Mat. Ital. (5) **16-B**, (1979), 278–290.

[180]* P. MARCELLINI, C. SBORDONE, *Relaxation of nonconvex Variational problems.* Atti Accad. Naz. Lincei, Rend. Cl. Sci. Fis. Mat. Natur. (8) **63** (1977), 341–344.

[181]* P. MARCELLINI, C. SBORDONE, *Sur quelques questions de G-convergence et d'homogénéisation non linéaire.* C.R. Acad. Sci. Paris Sér. A **284** (1977), 535–537

[182]* P. MARCELLINI,C. SBORDONE, *An approach to the asymptotic behavior of elliptic-parabolic operators.* J. Math. Pures Appl. (9) **56** (1977), 157–182.

[183]* P. Marcellini, C. Sbordone, *Dualità e perturbazioni di funzionali integrali*. Ricerche Mat. **26** (1977), 383–421.

[184]* P. Marcellini, C. Sbordone, *Homogenization of non uniformly elliptic operators*. Applicable Analysis **68** (1978), 101–114.

[185]* P. Marcellini, C. Sbordone, *Semicontinuity problems in the Calculus of Variations*. Nonlinear Anal. **4** (1980), 241–257.

[186] V.A. Marchenko, E.J. Khruslov, *Problèms aux limites dans des domains à frontière finement granulée (Russian)*. Naukova Dumka, Kiev (1974).

[187] V.A. Marchenko, E.J. Khruslov, *New results in the theory of boundary value problems for regions with close-grained boundaries*. Uspekhi Mat. Nauk **33** (1978), 127.

[188]* S. Marchi, *Stime di omogeneizzazione per equazioni ellittiche*. Proceed. Meeting on "Recent Methods in Nonlinear Analysis and Applications", S.A.F.A. IV, Napoli, March 21–28, 1980, ed. by A. Canfora, S. Rionero, C. Sbordone, G. Trombetti, Liguori ed., Napoli (1981), 357–361.

[189] A. Marino, D. Scolozzi, *Geodetiche con ostacolo*. Boll. Un. Mat. Ital. **2-B** (1983), 1–31.

[190] A. Marino, S. Spagnolo, *Un tipo di approssimazione dell'operatore* $\sum_{ij} D_j(a_{ij}D_j)$ *con operatori* $\sum_j D_j(\beta D_j)$. Ann. Sc. Norm. Sup. Pisa, Cl. Sci. (3) **23** (1969), 657–673.

[191]* A. Marino, M. Tosques, *Curves of maximal slope for a certain class of nonregular functions*. Boll. Un. Mat. Ital. **1-B** (1982), 143–170.

[192]* A. Marino, M. Tosques, *Existence and properties of the curves of maximal slope*. To appear.

[193] V.G. Markov, O.A. Oleinik, *On propagation of heat in one-dimensional disperse media*. Prikl. Mat. Mekh. 30 (1975), 1073–1081; transl. in J. Appl. Math. Mech. **39** (1975), 1028–1037.

[194] U. Massari, *Frontiere di curvatura media assegnata in* L^p. Rend. Sem. Mat. Univ. Padova **53** (1975), 37–52.

[195] U. Massari, L. Pepe, *Successioni convergenti di ipersuperfici di curvatura media assegnata*. Rend. Sem. Mat. Univ. Padova **53** (1975), 37–52.

[196]* M. Matzeu, *Su un tipo di continuità dell'operatore subdifferenziale*. Boll. Un. Mat. Ital. (5) **14-B** (1977), 480–490.

[197]* W.H. Mc Connell, *On the approximation of elliptic operators with discontinuous coefficients*. Ann. Sc. Norm. Sup. Pisa, Cl. Sci. (4) **3** (1976), 139–156.

[198] N.G. MEYERS, J. SERRIN, $H = W$. Proc. Nat. Acad. Sci. U.S.A. **51** (1964), 1055–1056.

[199] F. MIGNOT, J.P. PUEL, P.M. SUQUET, *Homogenization and bifurcation.* In ternational J. Engineering Sci. **18** (1980), 409–414.

[200] M. MIRANDA, *Distribuzioni aventi derivate misure. Insiemi di perimetro finito.* Ann. Sc. Norm. Sup. Pisa, Cl. Sci. Fis. Mat. (3) **18** (1964), 27–56.

[201] M. MIRANDA, *Un teorema di esistenza ed unicità per il problema dell'area in n variabili.* Ann. Sc. Norm. Sup. Pisa, Cl. Sci. Fis. Mat. (3) **19** (1965), 233–249.

[202] M. MIRANDA, *Sul minimo dell'integrale del gradiente di una funzione.* Ann. Sc. Norm. Sup. Pisa, Cl. Sci. Fis. Mat. (3) **19** (1965), 627–665.

[203] M. MIRANDA, *Comportamento delle successioni convergenti di frontiere minimali.* Rend. Sem. Mat. Univ. Padova **38** (1967), 238–257.

[204] M. MIRANDA, *Dirichlet problem with L^1 data for the non-homogeneous minimal surfaces equation.* Indiana Univ. Math. J. **24** (1974), 227–241.

[205]* L. MODICA, *Γ-convergence to minimal surfaces problem and global solutions to $\Delta u = 2(u^3 - u)$.* Proceed. Int. Meeting on "Recent Methods in Nonlinear Analysis", Rome, May, 8-12, 1978, ed. by E. De Giorgi, E. Magenes, U. Mosco, Pitagora ed., Bologna (1979), 223–244.

[206]* L. MODICA, *Omogeneizzazione con coefficienti casuali.* Atti dei Convegno su "Studio di problemi-limite in Analisi Funzionali", Bressanone, Settembre 1981, ed. da P. Patuzzo, S. Steffè, C.L.E.U.P., Padova (1982).

[207]* L. MODICA, S. MORTOLA, *Un esempio di Γ-convergenza.* Boll. Un. Mat. Ital. (5) **14-B** (1977), 285–299.

[208] - L. MODICA, S. MORTOLA, *Il limite nella Γ-convergenza di una famiglia di funzionari ellittici.* Boll. Un. Mat. Ital. (5) **14-A** (1977), 526–529.

[209] L. MODICA, S. MORTOLA, *Some Entire Solutions in the Plane of Nonlinear Poisson Equations.* Boll. Un. Mat. Ital. (5) **17-B** (1980), 614–622.

[210] C.B. MORREY *Multiple Integrals in the Calculus of Variations.* Springer Verlag, Berlin (1966).

[211]* S. MORTOLA, A. PROFETI, *On the convergence of minimum points for non-equi-coercive functionals.* Comm. Partial Differential Equations **7** (1982), 645–673.

[212]* G. MOSCARIELLO, *Su una convergenza di successione di integrali del Calcolo delle Variazioni.* Atti Accad. Naz. Lincei, Rend. Cl. Sci. Fis. Mat. Natur. (8) **61** (1976), 368–375.

[213]* G. Moscariello, Γ-*convergenza negli spazi sequenziali.* Rend. Accad. Sci. Fis. Mat. Napoli (4) **43** (1976), 333–350.

[214] U. Mosco, *Convergence of convex sets and of solutions of variational inequalities.* Advances in Math. **3** (1969), 510–585.

[215] F. Murat, *Contre-exemple pour divers problèmes où le contrôle intervient dans les coéfficients.* Ann. Mat. Pura Appl. (4) **112** (1977), 49–68.

[216]* F. Murat, *Compacité par compensation.* Ann. Sc. Norm. Sup. Pisa, Cl. Sci. (4) **5** (1978), 489–507.

[217]* F. Murat, *Compacité par compensation II.* Proceed. Int. Meeting on "Recent Methods in Nonlinear Analysis", Rome, May 8-12, 1978, ed. by E. De Giorgi, E. Magenes, U. Mosco, Pitagora ed., Bologna (1979), 245–256.

[218] F. Murat, *Compacité par compensation: condition nécéssaire et suffisante de continuité faible sous une hypothèse de rang constant.* Ann. Sc. Norm. Sup. Pisa, Cl. Sci. (4) **8** (1981), 69–102.

[219]* F. Murat, *Sur l'homogénéisation d'inéquations elliptiques du 2ème ordre, relative au convexe* $K(\psi_1, \psi_2) = \{\nu \in H_0^1(\Omega) \mid \psi_1 \leq \nu \leq \psi_2 \ p.p. \ dans \ \Omega\}$. Publication 76013. Laboratoire d'analyse numerique. Universitè Pierre et Marie Curie, Paris, 1976.

[220]* T. Norando, *Stime di omogeneizzazione per disequazioni quasi-variazionali.* Proceed. Meeting on "Recent Methods in Nonlinear Analysis and Applications", S.A.F.A. IV, Napoli March 21-28, 1980, ed. by A. Canfora, S. Rionero, C. Sbordone, G. Trombetti, Liguori ed., Napoli (1981), 367–370.

[221]* O.A. Oleinik, *On the convergence of solutions of elliptic and parabolic equations under weak convergence of coefficients.* Uspekhi Mat. Nauk **30** (1975), 257–258.

[222] G.C. Papanicolaou, *Introduction to the asymptotic analysis of stochastic equations.* Mod. Model. Contin. Phenom., Lect. Appl. Math. **16** (1977), 109–147.

[223] G.C. Papanicolaou, *Heat conduction in a one-dimensional random medium.* Commun. Pure Appl. Math. **31** (1978), 583–592.

[224] G.C. Papanicolaou, *A limit theorem for turbulent diffusion.* Commun. Math. Phys. **65** (1979), 97–128.

[225] G.C. Papanicolaou and S.R.S. Varadhan, *Boundary value problems with rapidly oscillating random coefficients.* Coll. Math. Soc. Janos Bolyai (1979).

[226]* P. PATUZZO GRECO, *Sulla G-convergenza delle equazioni differenziali ordinarie nel caso di domini illimitati.* Boll. Un. Mat. Ital. (5) **16-B** (1979), 466–479.

[226]* R. PEIRONE, *Γ-convergenza per funzioni a valori vettoriali.* Tesi di laurea, Genova, 1981.

[228]* L.C. PICCININI, *G-convergence for ordinary differential equations with Peano phenomenon.* Rend. Sem. Mat. Univ. Padova **58** (1977), 65–86.

[229]* L.C. PICCININI, *Homogenization for ordinary differential equations.* Rend. Cir. Mat. Palermo (2) **27** (1978), 95–112.

[230]* L.C. PICCININI, *G-convergence for ordinary differential equations.* Proceed. Int. Meeting on "Recent Methods in Nonlinear Analysis", Rome, May 8-12, 1978, ed. by E. De Giorgi, E. Magenes, U. Mosco, Pitagora ed., Bologna (1979), 257–280.

[231]* L.C. PICCININI, *Ergodic properties in the theory of homogenization.* Atti dei Convegno su "Studio di problemi-limite in Analisi Funzionale", Bressanone, Set. 1981, ed. da P. Patuzzo e S. Steffé, C.L.E.U.P., Padova (1982).

[232] G. PIERI, *Variational perturbations of the linear-quadratic problemi.* J. Optimization Theory Appl. **22** (1977), 63–77.

[233]* G. PORRU, *Su un teorema di compattezza rispetto alla G-convergenza per operatori ellittici.* Rend. Sem. Fac. Sci. Univ. Cagliari **48** (1978), 67–80.

[234] V.I.- PRJAZINSKII, V.G. SUSKO, *Small parameter aymptotics of some solutions of the Cauchy problem for a quasi-linear parabolic equation.* Soviet Math. Dokl. **20** (1979), 698–700; transl. from Doklady Akad. Nauk **247** (1979).

[235]* A. PROFETI, B. TERRENI, *Uniformità per una convergenza di operatori parabolici nel caso dell'omogeneizzazione.* Boll. Un. Mat. Ital. (5) **16-B** (1979), 826–841.

[236]* A. PROFETI, B. TERRENI, *Su alcuni metodi per lo studio della convergenza di equazioni paraboliche.* Boll. Un. Mat. Ital. (5) **17-B** (1980), 675–691.

[237]* P. PUEL, *Homogénéisation d'un problème de bifurcation.* Proceed. Int. Meeting on "Recent Methods in Nonlinear Analysis", Rome, May 8-12, 1978, ed. by E. De Giorgi, E. Magenes, U. Mosco, Pitagora ed., Bologna (1979), 280–310.

[238] V.D. REPNIKOV, S.D. EIDEL'MAN, *A new proof of the stabilization theorem for solutions of the Cauchy problem for the heat-conduction equation.* Mat. Sb. **73** (1967), 155–159; transl. in Math. USSR-Sb. **2** (1967), 135–139.

[239] Y. Reshetniak, *General theorems on semicontinuity and on convergence with a functional.* Siberian Math. J. **8** (1968), 801–816; transl. from Sibirsk Mat. Z. **8** (1967), 1051–1069.

[240] R.T. Rockafellar, *Generalized directional derivatives and subgradients of nonconvex functions.* Can. J. Math. **32** (1980), 257–280.

[241] J. Saint Jean Paulin, *Homogénéisation et perturbations.* C.R. Acad. Sci. Paris, Sér. A, **291** (1980), 23–26.

[242] J. Sanchez-Hubert, *Etude de certaines équations intégrodifférentielles issues de la théorie de l'homogénéisation.* Boll. Un. Mat. Ital. (5) **16-B** (1979), 857–875.

[243] J. Sanchez-Hubert, E. Sanchez-Palencia, *Sur certains problèms physiques d'homogénéisation donnant lieu à des phénomènes de relaxation.* C.R. Acad. Sci. Paris, Sér. A, **286** (1978), 903–906.

[244] E. Sanchez-Palencia, *Comportement limite d'un problème de trasmission à travers une plaque mince e faiblement conductrice.* C.R. Acad. Sci. Paris, Sér. A, **270** (1970), 1026–1028.

[245] E. Sanchez-Palencia, *Solutions périodiques par rapport aux variables d'espace et applications.* C.R. Acad. Sci. Paris, Sér. A, **271** (1970), 1129–1132.

[246] E. Sanchez-Palencia, *Equations aux derivées partielles dans un type de milieux hétérogènes.* C.R. Acad. Sci. Paris, Sér. A, **272** (1970), 1410–1413.

[247] E. Sanchez-Palencia, *Problèmes de perturbations liées aux phénomènes de conduction à travers des couches minces de grande resistivité.* J. Math. Pures Appl. **53** (1974), 251–270.

[248] E. Sanchez-Palencia, *Comportement locale et macroscopique d'un type de milieux physiques hétérogènes.* International J. Engineering Sci. **12** (1974), 331–351.

[249] E. Sanchez-Palencia, *Perturbations spectrales.* Lectures Notes in Mathematics, Springer, **294**, (1977), 437–455.

[250]* E. Sanchez-Palencia, *Méthode d'homogénéisation pour l'etude de matériaux hétérogènes. Phénomènes de mémoire.* Rend. Sem. Mat. Univ. e Politecn. Torino **36** (1977-78), 15–26.

[251] E. Sanchez-Palencia, *Phénomènes de relaxation dans les écoulements des fluides visqueux dans des milieux poreux.* C.R. Acad. Sci. Paris, Sér. A, **286** (1978), 185–188.

[252]* E. Sanchez-Palencia, *Nonhomogeneous media and vibration theory.* Lecture Notes in Physics, Springer, **127** (1980).

[253] E. SANCHEZ-PALENCIA, M. LODO HIDALGO, *Perturbation of spectral properties for a class of stiff problems.* Proceed. Meeting on "Computing Methods in Applied Sciences and Engineering", Versailles, Dec. 1979, ed. by R. Glowinski, J.L. Lions, North-Holland (1980), 683–697.

[254]* C. SBORDONE, *Alcune questioni di convergenza per operatori differenziali del 20 ordine.* Boll. Un. Mat. Ital. (4) **10** (1974), 672–682.

[255]* C. SBORDONE, *Sulla G-convergenza di equazioni ellittiche e paraboliche.* Ricerche Mat. **24** (1975), 76–136.

[256]* C. SBORDONE, *Su alcune applicazioni di un tipo di convergenza variazionale.* Ann. Sc. Norm. Sup. Pisa, Cl. Sci. (4) **2** (1975), 617–638.

[257]* C. SBORDONE, *Sur un limite d'integrales polynomials du calcul de variations.* J. Math. Pures Appl. (9) **56** (1977), 67–77.

[258] C. SBORDONE *Closure and duality of saddle functions.* Manuscript.

[259] M. SCHATZMAN, *A class of hyperbolic unilateral problems with constraints: the vibrating string with obstacle.* Proceed. Meeting on "Mathematical theory of elasticity", Euromec **100**, Pise, May 1978.

[260] M. SCHATZMAN, *A Hyperbolic problem of second order with unilateral constraints.* J. Math. Pures Appl. **73** (1980), 138–191.

[261] M. SCHATZMAN, *Un problème hypérbolique du deuxième ordre avec contrainte unilatérale: le corde vibrante avec obstacle ponctuel.* J. Differential Equations **36** (1980), 295–334.

[262] D. SCOLOZZI, *Geodetiche su sottovarietà con bordo di R^n.* Boll. Un. Mat. Ital. **4-A** (1985), 451–457.

[263] P.K. SENATOROV, *The stability of a solution of Dirichlet's problems for an elliptic equation with respect to perturbation in measure of its coefficients.* Differential'nye Uravnenija **6** (1970), 1725–1726; transl. in Differential Equations **6** (1970), 1312–1313.

[264] P.K. SENATOROV, *The stability of the eigenvalues and eigenfunctions of a Sturm-Liouville problem* Differential'nye Uravnenija 7 (1971), 1667–1671; transl. in Differential Equations 7 (1971), 1266–1269.

[265]* J. SERRIN, *A new definition of the integral for non-parametric problems in the Calculus of Variations.* Acta Math. **102** (1959), 23–32.

[266] J. SERRIN, *On the definition and properties of certain variational integrals.* Trans. Amer. Math. Soc. **101** (1961), 139–167.

[267]* L. SIMON, *On G-convergence of elliptic operators.* Indiana Math. Univ. J. **28** (1979), 587–594.

[268]* Y. Sonntag, *Convergence au sens de U. Mosco*, Thèse, Université de Provence, Marseille (1980).

[269]* S. Spagnolo, *Sul limite delle soluzioni di problemi di Cauchy relativi all'equazione del calore*. Ann. Sc. Norm. Sup. Pisa, Cl. Sci. (3) **21** (1967), 657–699.

[270]* S. Spagnolo, *Sulla convergenza delle soluzioni di equazioni paraboliche ed ellittiche*. Ann. Sc. Norm. Sup. Pisa, Cl. Sci. Fis. Mat. (3) **22** (1968), 575–597.

[271]* S. Spagnolo, *Some convergence problems*. Symp. Math. **18** (1976), 391–398.

[272]* S. Spagnolo, *Convergence in energy for elliptic operators*. Proceed. III Symp. Numer. Solut. Partial Diff. Eq. College Park, 1975, ed. by B. Hubbard Academic Press (1976), 469–498.

[273]* S. Spagnolo, *Convergence of parabolic equations*. Boll. Un. Mat. Ital. (5) **14-B** (1977), 547–568.

[274]* S. Spagnolo, *Convergence de solutions d'équations d'évolution*. Proceed. Int. Meeting on "Recent Methods in Nonlinear Analysis", Rome, May 8-12, 1978, ed. by E. De Giorgi, E. Magenes, U. Mosco, Pitagora ed., Bologna (1979), 311–328.

[275]* S. Spagnolo, *Some type of perturbation for evolution equations*. Proceed. Meeting on "Computing Methods in Applied Sciences and Engineering", ed. by R. Clowinski, J.L. Lions, North-Holland, Amsterdam (1980), 675–682.

[276]* S. Steffè, *Un teorema di omogeneizzazione per equazioni differenziali ordinarie a coefficienti quasi-periodici*. Boll. Un. Mat. Ital. (5) 17-B (1980), 757–764.

[277]* S. Steffè, Atti del Convegno su "Studio di problemi-limite in Analisi Funzionale", Bressanone, Sett. 1981, ed. da P. Patuzzo e S. Steffè, C.L.E.U.P., Padova (1982).

[278]* L. Tartar, *Convergence d'operateurs differentiels*. Quaderno di ricerca del C.N.R. "Analisi convessa e applicazioni", Roma, (1974), 101–104.

[279]* L. Tartar, Cours Péccot au Collège de France, Paris (1977).

[280]* L. Tartar, *Homogénéisation en hydrodinamique*. Lecture Notes in Mathematics, Springer, 594 (1977), 474–481.

[281] M.M. Vainberg, *Variational methods for the study of non-linear operators*. Holden-Day, San Francisco (1964).

[282] M. VANNINATHAN, *Homogénéisation des valeurs propres dans les milieux perforés.* C.R. Acad. Sci. Paris, Sér. A, **287** (1978), 403–406.

[283]* V.V. YURINSKIJ, *Averaging elliptic equations with random coefficients.* Sib. Math. J. 20 (1980), 611–623; transl. from Sibirskii Mat. Zhurnal **21** (1980), 209–223.

[284] V.V. ZHIKOV, *On the stabilization of solutions of parabolic equations.* Mat. Sb. 104 (1977), 597–616; transl. in Math. USSR-Sb. **33** (1977), 519–537.

[285]* V.V. ZHIKOV, *A point stabilization criterion for second-order parabolic equations with almost-periodic coefficients.* Mat. Sb. 110 (1979); transl. in Math. USSR-Sb **38** (1981), 279–292.

[286]* V.V. ZHIKOV, S.M. KOZLOV, O.A. OLEINIK, KHA T'EN NGOAN, *Averaging and G-convergence of differential operators.* Uspeki Math. Nauk 34 (1979), 65–133; transl. in Russian Math. Surveys **34** (1979), 69–147.

[287] T. ZOLEZZI, *On convergence of minima.* Boll. Un. Mat. Ital. **8** (1973), 246–257.

[288] T. ZOLEZZI, *Characterization of some variational perturbations of the abstract linear-quadratic problems* SIAM J. Control Optim. **16** (1978), 106–121.

[289]* T. ZOLEZZI, *Variational perturbations of linear optimal control problems.* Proceed Int. Meeting on "Recent Methods in Nonlinear Analysis", Rome, May 8-12, 1978, ed. by E. De Giorgi, E. Magenes, U. Mosco, Pitagora ed., Bologna (1979), 329–338.

[290]* T. ZOLEZZI, *Some approximation and convergence problems in optimization.* Serdica **9** (1983), 400–406 (1984).

A framing theory for the Foundations of Mathematics[‡]

Ennio De Giorgi, Marco Forti[*]

Summary. We propose a "natural" axiomatic theory of the Foundations of Mathematics (Theory Q) where, in addition to the membership relation (between elements and classes), pairs, sets, natural numbers, n–tuples and operations are also introduced as primitive by means of suitable ground classes. Moreover, the theory Q allows an easy introduction of other mathematical and logical entities.

The theory Q is finitely axiomatized in §2, using a first–order language with a binary relation $\in$ (membership) and five constants Kur, Ins, $gsuc$, $Grpl$, $Grop$ (ground classes), and it is shown to be equiconsistent with Gödel–Bernays class theory; in fact, in §3, both these theories are mutually interpreted inside each other.

§1. Introduction

The goal of this Note is to propose a theory of the foundations of Mathematics which, while following generally the classical, set–theoretic presentation of Von Neumann–Bernays–Gödel, and while remaining finitely axiomatizable like the latter, differs from it in two essential points: first, it admits the existence of many elements (which we call, following Zermelo, *urelements*) which are neither sets nor classes, but are not even simple atoms without properties, as they can be other fundamental objects of mathematics (pairs, natural numbers, n–tuples, operations); second, by giving up the axiom of foundations, it does not exclude that many "big classes" (the universe, the membership relation, etc.) and many "big operations" (composition, application, etc.) be also elements, and hence that they can belong to sets and be components of n–tuples.

The first feature seems to be desirable because we intend to propose a "natural" theory, where definitions and axioms are as close as possible to ancient tradition, common intuition and the current language of mathematical practice, while avoiding artificial codings which replace the main mathematical objects (such as natural numbers, pairs and operations) with models of them built by skillful set–theoretic constructions; moreover, we feel it important to distinguish the notion of *operation* from the one of graph, which allows us later to fit directly in our setting also many non–extensional theories (recursive functions, λ-calculus, etc.).

[‡]Editor's note: translation into English of the paper "Una teoria quadro per i Fondamenti della Matematica", published in Atti Accad. Naz Lincei, Cl. Sci. Fis. Mat. Nat., **(8) 79** (1985), 55–67.

[*]Note presented by E. De Giorgi to the Accademia dei Lincei on November 22, 1985.

We note also that the notion of *pair* introduced as a primitive concept, can be easily generalized to the one of n–tuple, so as to obtain without ambiguity n–tuples for every choice of n, including the cases $n = 1$ and $n = 0$, as well as to define better the notions of power of a class and of Cartesian product, while keeping distinct, for instance, $A^1 \times A^n$ from $A^n \times A^1$ and from A^{n+1}.

Moreover, we leave open the possibility to "activate" other urelements whenever one wants to introduce as primitive other fundamental concepts of mathematics and logic (such as categories, structures, variables, properties or qualities, etc.).

The second choice gives us, first, a greater freedom when dealing with the postulate of *foundation* and the principles of *free construction*; for instance, one can keep foundation for sets and admit at the same time operations acting on themselves, and/or "self–pairs" of the kind $x = (x, x)$, or instead, admit only well–founded pairs, but accept "self–singletons" $x = \{x\}$ (see [4], [1]).

Secondly, if one gives up with the axioms of foundation, one can admit the existence of many "big" *classes* which are *elements*, but not *sets*, and which are considerably interesting when studying problems of *self–reference*; this possibility will be used in [3], where we will introduce in the frame theory outlined here, several "self–reference axioms" postulating the existence, as "urelementary operations", of many common mathematical operations (composition, inversion and transposition of functions, union, intersection and difference of classes, etc.) and which imply the presence of many elements which are distinguished classes (the universe, the membership relation, etc.).

Again we note that the frame theory presented here is *finitely axiomatized* and hence, as it contains no axiom scheme, in order to be read and understood, it requires a minimum of prerequisites and linguistic concepts; finally, its *consistency*, differently from the one of the extensions proposed in [3], follows easily from the one of ZF; in fact, we will see in the third section that it is possible to build a model for it within the theory $ABCD$ of Gödel–Bernays (see [5]).

It is impossible to cite here all the bibliography more or less related to this note; so we will just cite the texts which are necessary for direct reference. Instead, we feel it important to say that the problem of a frame theory for the foundations of mathematics, suitable also for a natural formulation of the problems of self–reference, universality and free construction, has been widely discussed starting from the beginning of the '80's in the course held by De Giorgi at the Scuola Normale Superiore of Pisa (see [2]). So we thank all the participants and in particular we mention M. Boffa, M. Clavelli, F. Honsell, G. Longo and M. Sciuto, who brought and continue to bring contributions to the development of the theory.

§2. The axioms of the theory Q

We will expose our theory Q in a semiformal way, putting some intuitive consideration before each group of axioms; we have finitely many axioms, written in the usual first order language, with a symbol for binary relation $\in$ (to be intended as the *membership* relation between an *element* and a *class*), and five constants *Kur, Ins, gsuc, Grpl, Grop* (to be interpreted by the *fundamental classes* which introduce and qualify *pairs, sets, natural numbers, n–tuples* and *operations*).

We will give step by step the definitions which we use, also to fix the notation, and in doing this we will keep as much as possible the most common conventions.

Generally we will use the **boldface** font for the *defined predicates* (the ones of being an element, a class, etc.), whereas we will reserve the *italic* font for the *objects* we consider (classes, elements, etc.); instead, we will exploit the usual mathematical liberty of using various fonts (uppercase, lowercase, Greek etc.) to denote classes, elements, operations, etc.

2.1. *Elements, classes and sets.*

The axioms of this section express the idea that the objects studied by the theory Q are *elements* (**El**) or *classes* (**Cl**); moreover, some classes, called *elementary classes*, are also elements (**Clel**), whereas other classes, called *proper*, are not (**Clpr**); and some elements, called *urelements* following Zermelo, are not classes (**Ur**).

Among the elementary classes, *sets* are particularly easy to handle; they form the *fundamental class Ins*.

Classes enjoy the property of *extensionality* (axiom Q2) and are stable under binary *union* and *complement* (axiom Q3).

DEFINITION 1.[‡]

$$\mathbf{El}\ x \Leftrightarrow \exists y(x \in y) \qquad\qquad (x \text{ is an } \textit{element})$$

$$\mathbf{Cl}\ x \Leftrightarrow \exists y(y \in x) \vee x \in Ins \qquad (x \text{ is a } \textit{class})$$

$$\mathbf{Clel}\,x \Leftrightarrow \mathbf{El}\ x \wedge \mathbf{Cl}\ x \qquad\qquad (x \text{ is an } \textit{elementary class})$$

$$\mathbf{Ur}\,x \Leftrightarrow \mathbf{El}\ x \wedge \neg\mathbf{Cl}\ x \qquad\qquad (x \text{ is an } \textit{urelement})$$

$$\mathbf{Trans}\,x \Leftrightarrow \mathbf{Cl}\ x \wedge \forall y,z(z \in y \in x \to z \in x) \quad (x \text{ is a } \textit{transitive} \text{ class})$$

$$x \subseteq y \Leftrightarrow \mathbf{Cl}\,x \wedge \mathbf{Cl}\,y \wedge \forall z(z \in x \Rightarrow z \in y) \quad (x \text{ is } \textit{included} \text{ in } y)$$

$$\mathbf{Cl}\,y \wedge x = -y \Leftrightarrow \mathbf{Cl}\,x \wedge \forall z \qquad (x \text{ is the } \textit{complement} \text{ of } y)$$
$$(z \in x \Leftrightarrow \mathbf{El}\,z \wedge z \notin y)$$

[‡]Editor's note: in the definition of the complement we added the condition "**El** z" which seems to be necessary.

$$x = y \cap z \Leftrightarrow \mathbf{Cl}\, x, y, z \wedge \forall t) \qquad\qquad (x \text{ is the } intersection \text{ of } y \text{ and } z)$$
$$(t \in x \Leftrightarrow t \in y \wedge t \in z$$

$$x = y \cup z \Leftrightarrow \mathbf{Cl}\, x, y, z \wedge \forall t \qquad\qquad (x \text{ is the } union \text{ of } y \text{ and } z)$$
$$(t \in x \Leftrightarrow t \in y \vee t \in z)$$

$$\mathbf{Cl}\, y, z \wedge x = y \setminus z \Leftrightarrow \mathbf{Cl}\, x \wedge \forall t \qquad\qquad (x \text{ is the } difference \text{ of } z \text{ from } y)$$
$$(t \in x \Leftrightarrow t \in y \wedge t \notin z)$$

$$x = \{y\} \Leftrightarrow \forall t(t \in x \Leftrightarrow t = y) \qquad\qquad (x \text{ is the } singleton \text{ of } y)$$

$$x = \{y, z\} \Leftrightarrow \forall t(t \in x \Leftrightarrow t = y \vee t = z) \quad (x \text{ is the } doubleton \text{ of } y \text{ and } z).$$

AXIOM Q1. – *Only classes and elements exist.*

$(Q1)$ $\qquad\qquad \forall x.\mathbf{Cl}\, x \vee \mathbf{El}\, x.$

AXIOM Q2. – *Classes having the same elements are equal.*

$(Q2)$ $\qquad\qquad X \subseteq Y \wedge Y \subseteq X \Rightarrow X = Y.$ *(Axiom of* extensionality*).*

From the Axiom of extensionality it follows that union, intersection, double-ton, etc., *if they exist*, are *uniquely determined*; so, for instance, the doubleton $\{x, x\}$ equals the singleton $\{x\}$. To guarantee *existence* we need the following axioms.

AXIOM Q3. – *Every class has a complement, and any two classes have an intersection.*

$(Q3.1)$ $\qquad\qquad \mathbf{Cl}\, X \Rightarrow \exists Y (Y = -X).$

$(Q3.2)$ $\qquad\qquad \mathbf{Cl}\, X \wedge \mathbf{Cl}\, Y \Rightarrow \exists Z.(Z = X \cap Y).$

From Axiom Q3 it follows that there is the *empty class* $\emptyset$ having no elements, and the *universal class* V to which all elements belong, and that two classes always have the *union* and the *difference*.

Note that from the definition above it follows that all *sets* are *elementary classes*, that is $x \in Ins \Rightarrow \mathbf{Clel}\, x.$

2.2. *Pairs and graphs.*

We will start from the idea that the notion of pair is primitive, and we will denote by (x, y) the *pair* having x as *first* and y as *second element*. So, in the theory Q, pairs will be urelements not to be confused with *doubletons*, which are sets, and in particular with *Kuratowski doubletons* (sets of the kind $\{\{x\}, \{x, y\}\}$).

The link between pairs and Kuratowski doubletons is ensured by the *fundamental class Kur* whose elements are all the doubletons $\{(x, y), \{\{x\}, \{x, y\}\}\}$.

The *classes of pairs* deserve particular interest as they can be considered as *graphs of relations between elements* (**Grel**); among them we can isolate the *functional graphs* (**Gfun**) and the *injective graphs* (**Gfinj**).

Axiom Q5 guarantees the *existence of the graphs* of the most important relations and operations, whereas Axiom Q6 completes the properties of *stability of classes.*

AXIOM Q4. – *Any two elements have a unique pair.*

$(Q4.1)$ $\qquad \mathbf{El}\, x \wedge \mathbf{El}\, y \Rightarrow \exists u(\mathbf{Ur}\, u \wedge \{u, \{\{x\}, \{x, y\}\}\} \in Kur).$

$(Q4.2)$ $\qquad z \in Kur \Rightarrow \exists u, x, y(\mathbf{Ur}\, u \wedge \{u, \{\{x\}, \{x, y\}\}\} = z).$

$(Q4.3)$ $\qquad x \in Kur \wedge y \in Kur \wedge x \neq y \Rightarrow x \cap y = \emptyset$

Existence and uniqueness of pairs allow us to introduce the following notions:

DEFINITION 2.

$u = (x, y) \Leftrightarrow \{u, \{\{x\}, \{x, y\}\}\} \in Kur$
(*u* is the *pair* whose *first element* is *x* and whose *second element* is *y*)

$z = x \times y \Leftrightarrow \mathbf{Cl}\, x, y, z \wedge \forall t(t \in z \Leftrightarrow \exists u \in x \exists v \in y(t = (u, v))$
(*z* is the *Cartesian product* of *x* and *y*)

$z = \hat{x}(y) \Leftrightarrow \mathbf{Cl}\, x, y, z \wedge \forall t(t \in z \Leftrightarrow \exists u \in y((u, t) \in x)$
(*z* is the *image* of *y through* the class *x*)

$\mathbf{Grel}\, x \Leftrightarrow \mathbf{Cl}\, x \wedge \forall u \in x \exists y, z(u = (y, z))$
(*x* is a *relational graph* or simply *graph*)

$\mathbf{Gfun}\, x \Leftrightarrow \mathbf{Grel}\, x \wedge \forall u, v, w((u, v) \in x \wedge (u, w) \in x \Rightarrow v = w)$
(*x* is a *functional graph*)

$\mathbf{Gfinj}\, x \Leftrightarrow \mathbf{Gfun}\, x \wedge \forall u, v, w((u, v) \in x \wedge (w, v) \in x \Rightarrow u = w)$

(*x* is an *injective functional graph*).

AXIOM Q5. – *The following relational graphs exist:*

$(Q5.1)$ $\qquad E = \{(x, y) \,|\, \mathbf{El}\, x, y \wedge x \in y\}$
(the graph of the *membership* between an *element* and an *elementary class*)

$(Q5.2)$ $\qquad Gp_1^2 = \{((x, y), x) \,|\, \mathbf{El}\, x, y\}$
$\qquad\qquad\qquad\qquad\qquad$ (the graph of the *first projection of pairs*)

$(Q5.3)$ $\qquad Ginv = \{((x, y), (y, x)) \,|\, \mathbf{El}\, x, y\}$
$\qquad\qquad\qquad\qquad\qquad$ (the graph of the *inversion of pairs*)

$(Q5.4)$ $\qquad Gsc = \{(((x, y), z), (x, (y, z))) \,|\, \mathbf{El}\, x, y, z\}$
$(Q5.5)$ $\qquad Gsc' = \{(((x, y), z), ((x, z), y)) \,|\, \mathbf{El}\, x, y, z\}$
$\qquad\qquad\qquad\qquad\qquad$ (the graphs of the *permutations of composed pairs*).

AXIOM Q6. – *The Cartesian product of two classes and the image of the one through the other always exist.*

$$(Q6.1) \qquad \mathbf{Cl}\, X \wedge \mathbf{Cl}\, Y \Rightarrow \exists Z(Z = X \times Y).$$

$$(Q6.2) \qquad \mathbf{Cl}\, X \wedge \mathbf{Cl}\, Y \Rightarrow \exists Z(Z = \hat{X}(Y)).$$

We note that, from Axiom Q4, it follows that the *doubleton* of any two elements always exists and is an elementary class; for this reason we omitted the usual Axiom on doubletons in section 2.1. That all doubletons are sets will follow later from the replacement Axiom (see section 2.3).

It is easy to see that, except for using primitive pairs instead of Kuratowski doubletons, the axioms Q1–6 are essentially the ones of the groups AB of the theory of Von Neumann–Bernays–Gödel (see [5]): of course, extensionality is given in such a way to allow the existence of urelements.

So, it would be possible to state and prove, from axioms Q1–6, a *predicative comprehension principle* for classes ("theorem of Bernays"); instead we prefer just to list (and fix notations about) some classes which we will need in the sequel: their existence, anyway, can be easily deduced from the axioms.

The following classes exist:

$$V \times V = V^2 \qquad \text{(the class of } pairs \text{)}$$

$$Gid = \{(x, x) \mid x \in V\} \qquad \text{(the graph of the } identity \text{ or } equality \text{)}$$

$$Gp_2^2 = \{((x, y), y) \mid x \in V\} \qquad \text{(the graph of the } second\ projection\ of\ pairs \text{)}$$

$$Clel = \{x \mid \mathbf{Clel}\, x\} \qquad \text{(the class of } elementary\ classes \text{)}$$

$$Grel = \{x \in V \mid \mathbf{Grel}\, x\} \qquad \text{(the class of } elementary\ graphs \text{)}$$

$$Gfun = \{x \in V \mid \mathbf{Gfun}\, x\} \qquad \text{(the class of } elementary\ functional\ graphs \text{)}$$

$$Gfinj = \{x \in V \mid \mathbf{Gfinj}\, x\} \qquad \text{(the class of } elementary\ injective\ graphs. \text{)}$$

If X is any class, the following classes exist:

$$\bigcup X = \{y \mid \exists x \in X(y \in x)\} \qquad \text{(the } union \text{ of the classes belonging to } X)$$

$$\bigcap X = \{y \mid \forall x \in X(y \in x)\} \qquad \text{(the } intersection \text{ of the classes belonging to } X)$$

$$\wp(X) = \{y \in Clel \mid y \subseteq X\} \qquad \text{(the class of } parts, \text{ or } power \text{ of } X)$$

$$TC(X) = \{y \mid \forall t(\mathbf{trans}\, t \wedge t \supseteq X \Rightarrow y \in t)\} \qquad \text{(the } transitive\ closure \text{ of } X).$$

If G is a relational graph, the following classes exist:

$$Img(G) = \hat{G}(V) \qquad \text{(the } image \text{ of } G)$$

$$Dom\, G = \widehat{Gp_1^2}(G) \qquad \text{(the } domain \text{ of } G)$$

$$G^{-1} = \widehat{Ginv}(G) \qquad \text{(the } inverse\ graph \text{ of } G)$$

and for every element x

$$\tilde{G}(x) = \{y \mid (x, y) \in G\} = \hat{G}(\{x\}) \quad \text{(the \textit{fiber} of } G \text{ in } x\text{)}.$$

If F and G are graphs, the following graphs exist as well:

$$F \circ G = \{(x, y) \mid \exists z((x, z) \in G \wedge (z, y) \in F)\} \quad \text{(the \textit{composition} of } F \text{ and } G\text{)}$$

$$\langle F, G \rangle = \{(x, (y, z)) \mid (x, y) \in F \wedge (x, z) \in G\} \quad \text{(the \textit{fibered product} di } F \text{ e } G\text{)}.$$

Note that $\langle F, G \rangle$ is determined by the property

$$\widetilde{\langle F, G \rangle}(x) = \tilde{F}(x) \times \tilde{G}(x) \text{ for every } x.$$

Following common usage, we will usually write $y = F(x)$ instead of $(x, y) \in F$ when F is a functional graph.

2.3. *Sets.*

As mentioned in section 2.1, *sets* are elementary classes particularly easy to handle. The idea that sets can be handled "easily and safely" is expressed by the postulates of this section, which are essentially identical to the ones of the group C of [5].

However, we recall that from our definitions it follows only the inclusion $Ins \subseteq Clel$, hence we did *not* assume that *all* elementary classes are *sets*, but the latter ones are introduced by means of the *fundamental class Ins*. As a matter of fact, almost all axioms introduced in [3] imply the existence of distinguished classes which are elements, but they cannot consistently belong to Ins.

Axiom Q7. – *The class of sets is stable under replacement, union and power.*

(Q7.1) $\quad x \in Ins \wedge \mathbf{Gfun}F \Rightarrow \hat{F}(x) \in Ins$ (Axiom of *replacement*)

(Q7.2) $\quad x \in Ins \cap \wp(Ins) \Rightarrow \bigcup x \in Ins$ (Axiom of *union*)

(Q7.3) $\quad x \in Ins \Rightarrow \wp(x) \in Ins$ (Axiom of *powerset*).

Axiom Q8. – *There is a set having the empty set as element and closed under singleton addition.*

(Q8) $\quad \exists x \in Ins(\emptyset \in x \wedge \forall y(y \in x \Rightarrow y \cup \{y\} \in x))$ (Axiom of *infinity*).

As mentioned in the section 2.2, we omitted the Axiom of *doubleton* which follows from Q4.1, Q7.1 and Q.8. Note also that we did not insert neither the Axiom of *choice* nor the Axiom of *foundation*.

As a matter of fact, the Axiom of choice for *sets* becomes *provable* in the theory Q provided, for instance, one assumes the existence of an operation associating to each urelementary operation a *right inverse* (see below, sect. 2.6 and [3]).

On the other hand, quite a lot of axioms of the note [3] are *incompatible* with foundation, taking into account that all singletons and doubletons are sets, even if they have as elements "big" classes which are not sets; actually we cannot even prove that the *transitive closure* of a set is a *set* as well. So it is useful to consider the following classes:

DEFINITION 3.

$Itrans = \{x \in Ins \mid \mathbf{Trans}\, x\}$
(the *transitive* sets)

$Ibtrans = \{x \in Itrans \mid \forall y \in \wp(x) \setminus \wp(Ur) \exists z \in y (z \cap (y \setminus Ur)) = \emptyset\}$
(the *well transitive* sets)

$Ifond = \{x \in Ins \mid TC(x) \in Ins\}$
(the *founded* sets)

$Ibfond = \{x \in Ins \mid TC(x) \in Ibtrans\}$
(the *well founded* sets).

2.4. *Natural numbers.*

As anticipated, we do not want to identify natural numbers with *finite Von Neumann ordinals*. Hence, *natural numbers* will be a set $\mathbf{N}$ of urelements, whose properties are expressed by the *Peano axioms*; the link with set theory is ensured by the *fundamental class gsuc* (the graph of the *successor* operation): *gsuc* is a set on which we assume the axioms Q9–10.

AXIOM Q9. – *gsuc is an injective functional graph and a set of pairs of urelements (which are not pairs).*

$(Q9) \qquad gsuc \in Gfinj \cap Ins \cap \wp((Ur \setminus V^2)^2).$

The set $\mathbf{N}$ of natural numbers is the *domain* of *gsuc*: $\mathbf{N} = Dom\, gsuc$.

The *existence* of a unique natural number (0, the *zero*) such that the image of *gsuc* is $\mathbf{N} \setminus \{0\}$, as well as the *principle of induction*, are ensured by:

AXIOM Q10. – *Existence of zero and principle of induction.*

$(Q10) \qquad \exists z \in \mathbf{N}(\widehat{gsuc}(\mathbf{N}) = \mathbf{N} \setminus \{z\} \wedge \forall X(z \in Z \wedge \widehat{gsuc}(X) \subseteq X \Rightarrow \mathbf{N} \subseteq X)).$

The *natural ordering* and the *arithmetic operations* are intended to be defined on natural numbers in the usual way, starting from the successor operation.

Moreover, it is easy to define the *canonical bijection* between $\mathbf{N}$ and the Von Neumann ordinal $\omega = \bigcap \{X \in Ins \mid \emptyset \in X \wedge \forall t(t \in X \Rightarrow (t \cup \{t\}) \in X)\}$.

Finally note that, by using axioms Q9–10 on natural numbers and of the replacement Q7.1, it is easy to prove the axiom of infinity Q8, which has been stated anyway in order to keep sections 2.1–3 self–contained.

2.5. *Uruples.*

After introducing natural numbers, the notion of pair can be extended to the one of n–tuple $(x_1, \ldots, x_n)$ for every $n \in \mathbf{N}$.

The *fundamental class Grpl* provides a link between the *functional graph* $\{(1, x_1), \ldots, (n, x_n)\}$ and the corresponding n–*tuple* $(x_1, \ldots, x_n)$. Axiom Q11 states also the *existence* and *uniqueness* of the n–tuples of any elements for any

$n \in \mathbf{N}$; in particular, there will be a unique 0–tuple whose graph is $\emptyset$ (denoted by φ) and a 1–tuple with graph $\{(1, x)\}$ for every element x (denoted by $[x]$ to avoid ambiguities).

n–tuples are urelements, and for this reason the class of all n–tuples for all $n \in \mathbf{N}$ will be called the class of *uruples* (*Urpl*).

Finally with axiom Q12 we will require that 2–tuples coincide with the *pairs* already introduced in section 2.2.

Axiom Q11. – *Grpl is an injective functional graph, which provides a bijection between the class of functional graphs whose domain is an initial segment of* $\mathbf{N}$, *and a class of urelements (which are not natural numbers).*

$$(Q11) \qquad \mathbf{Gfinj}\, Grpl \wedge Dom\, Grpl \subseteq Ur \setminus \mathbf{N} \wedge$$

$$\wedge\, Img\, Grpl = \{g \in Gfun \,|\, \exists n \in \mathbf{N}(Dom\, g = \{1, \ldots, n\})\}.$$

Axiom Q12. – *The pair* (x, y) *coincides with the 2–tuple with graph* $\{(1, x), (2, y)\}$.

$$(Q12) \qquad \forall x, y \in V((x, y), \{(1, x), (2, y)\}) \in Grpl.$$

Definition 4.

$$Urpl = Dom\, Grpl \qquad\qquad\qquad\qquad \text{(the class of the } uruples\text{)}$$

$$V^n = \{u \in Urpl \,|\, Dom\, Grpl(u) = \{1, \ldots, n\}\} \quad \text{(the class of the } n\text{–tuples)}.$$

If $u \in V^n$, we will say that n is the *length* of u ($n = lng\, u$), that $Grpl(u)$ is the *graph* of u ($graf\, u$), that $Img\, Grpl(u)$ is the *support*, or the set of the *components* of u ($supp\, u$) and that $(Grpl(u))(i)$, for $1 \leq i \leq n$, is the i–*th component* of u (u_i).

Except for the cases $n = 0$ and $n = 1$, where we will use the notation φ and $[u_1]$, we will write $(u_1, \ldots, u_n)$ to denote the n–*tuple* with i–*th component* u_i, $1 \leq i \leq n$.

If $u \in V^n$ and $v \in V^m$, the *concatenation* of u and v is the $n + m$–tuple $w = u \cdot v$ whose components are $w_i = u_i$ for $1 \leq i \leq n$ and $w_i = v_{i-n}$ for $n + 1 \leq i \leq n + m$.

If $u \in V^n$, $v \in V^k$ and *sost* $v \subseteq \{1, \ldots, n\}$, the *composition* of u with v is the k–tuple $w = u \circ v$ whose components are $w_i = u_{v_i}$ for $1 \leq i \leq k$.

$A^n = \{u \in V^n \mid supp\ u \subseteq A\}$
(the n–th power of the class A)

$\underbrace{A \times B} = \{\underbrace{u \cdot v} \mid (u, v) \in A \times B\}$
(for $A, B \subseteq Urpl$: the *concatenated product*)

$\langle G_1, \ldots, G_n \rangle = \{(x, (y_1, \ldots, y_n)) \mid (x, y_i) \in G_i, 1 \leq i \leq n\}$
(**Grel** $G_1, \ldots, G_n$: the n–ary fibered product of graphs)

$Gp_h = \{(u, u_h) \in Urpl \times V \mid lng\ u \geq h\}$
(the graph of the h–th projection)

$Gp_h^n = Gp_h \cap (V^n \times V)$
(the graph of the h–th projection of n–tuples).

Note that

$$A^0 = \{\varphi\},\ A^2 = A \times A,\ A \times B = \underbrace{A^1 \times B^1},\ \underbrace{A^n \times B^0} = \underbrace{B^0 \times A^n} = A^n,\ \text{but}$$

$A^1 \neq A$ and in general $\underbrace{A^n \times A^m} = A^{n+m} \neq A^n \times A^m$.

2.6. *Urelementary operations.*

The concept which has a privileged role in the further extensions of the frame theory is that of *urelementary operation*, which we will consider as primitive and clearly distinct from the one of *functional graph*.

So we will suppose that, among urelements, there are many operations which form a class (denoted by $Urop$); if $f \in Urop$, we will write $f\ x = y$ (or also $f(x) = y$) to mean that *at x the operation f is defined* and gives y as a *result* (or *value*).

The link with class theory is ensured by the *fundamental class $Grop$*, which allows one to generate the *functional graph associated* to each urelementary operation; however, we will have to formulate the relative Axiom Q13 in a different way than the corresponding Axiom Q11 for uruples, as on one hand we do not want to postulate *extensionality* of operations, and on the other hand we do not want to exclude the possibility (which we will use in [3]) that *urelementary operations* exist, whose graph is a *proper class*.

The particular handiness of sets appears also in the *Comprehension Axiom* Q14, which ensures the existence of urelementary operations associated to any functional graph which is a *set*.

Note that the need of admitting operations whose graph is *empty* prevents us to simply choose the fundamental class in a way that the graph of any urelementary operation is the fiber of $Grop$ corresponding to it. *A posteriori* it will turn out that $Grop$ is a graph whose *domain* is $Urop$ and whose *fiber* in f is the graph of f if this is *not* empty, and $\{\emptyset\}$ otherwise.

AXIOM Q13. – *Grop is a graph whose domain is included in* $Ur \backslash (\mathbf{N} \cup Urpl)$, *and whose fibers different from* $\{\emptyset\}$ *are functional graphs.*

$$(Q13) \qquad \mathbf{Grel}\, Grop \wedge Dom\; Grop \subseteq (\mathbf{N} \cup Urpl) \wedge$$
$$\wedge\, \forall f(\tilde{Grop}(f) = \{\emptyset\} \vee \mathbf{Gfun}\, \tilde{Grop}(f)).$$

AXIOM Q14. – *Among the fibers of Grop there are $\{\emptyset\}$ and all functional graphs which are sets.*

$$(Q14) \qquad \forall g \in Ins \cap Gfun\; \exists f \in Dom\; Grop(g = \tilde{Grop}(f) \cap V^2).$$

DEFINITION 5.

$Urop = Dom\; Grop$ (the class of the *urelementary operations*).

For $f \in Urop$ we will write $y = f\,x$ to mean $(f,(x,y)) \in Grop$ and we will let:

$$graph\; f = \tilde{Grop}(f) \cap V^2 = \{(x,y)\,|\,y = f\,x\} \qquad \text{(the } graph \text{ of } f)$$

$$dom\; f = Dom\; graf\; f = \{x\,|\,\exists y(y = f\,x)\} \qquad \text{(the } domain \text{ of } f)$$

$$img\; f = Img\; graf\; f = \{y\,|\,\exists x(y = f\,x)\} \qquad \text{(the } image \text{ of } f)$$

$$\hat{f}(C) = \widehat{graf}\; f(C) \qquad \text{(For every class } C\text{: the } image \text{ of } C \text{ through } f).$$

We will say that $f, g \in Urop$ are *equivalent* (in symbols $f \simeq g$) if they have the same graph.

The lack of an axiom of *extensionality* for *operations* prevents us from defining operations by simply assigning a graph to them, even when Axiom Q14, or some stronger comprehension principle (see [3]) guarantees its existence; in particular, at this point there is no possibility of defining composition, transposition, inversion of urelementary operations.

However, this can be obtained in two ways: we can either postulate directly the possibility of performing certain manipulations on urelementary operations, or assume some selection principle: but we will not add any such axiom to the theory Q, whereas we will propose many of them in [3].

2.7. *Atomic urelements.*

The last axiom assumed here is intended to ensure a sufficiently large *reserve* of urelements for any future need, for instance, as mentioned in §1, to introduce in a natural way other important primitive concepts of mathematics and logic, like cardinals, real numbers, categories, properties or qualities, structures, languages, etc. So we will call *atomic urelements* those urelements which have not been characterized as uruples, operations and natural numbers, and we will postulate:

AXIOM Q15. – *There is an embedding of the universe into the atomic urelements.*

$$(Q15) \qquad \exists J(\mathbf{Gfinj}J \wedge Dom\; J = V \wedge Img\; J \subseteq Ur \setminus (\mathbf{N} \cup Urpl \cup Urop)).$$

So we will call *Theory Q* the theory determined by the axioms Q1–15: this is the *framing theory* which will form, from now on, the premise and the formal environment where we will insert the self–referent axioms of [3] and the future developments of our program of organization of the foundations of mathematics.

§3. The consistency of the frame theory Q

It is not difficult to say that, from the viewpoint of *consistency*, the theory Q is exactly *equivalent* to Zermelo–Fraenkel's set theory ZF. More precisely, many classes definable in the theory Q provide *inner models* of ZF: for instance, it is enough to consider the class

$$U = \{x \in Ibfond \,|\, TC(x) \cap Ur = \emptyset\},$$

with the membership relation restricted to $U \times U$, to get a natural model of ZF within Q.

Now we are going to define an inner model of Q within Gödel–Bernays theory $ABCD$, which is known to be a conservative extension of ZF (cfr. [5]).

Given a model $\mathcal{M} = (V, C, \in)$ di $ABCD$ we define a model of Q

$$\mathcal{M}'=(V, C', \in', Ins, Kur, gsuc, Grpl, Grop)$$

in the following way:

$$C' = (C \setminus V) \cup \{]x, \underline{0}[\,|\, x \in V\};$$

$$x \in' y \Leftrightarrow (x \in y \in C \setminus V) \vee (y =]z, \underline{0}[\wedge x \in z);$$

$$Ins = \{[x, \underline{0}[\,|\, x \in V\};$$

$$Kur = \{]\{]\{]\underline{1}, x[,]\underline{2}, y[\}, \underline{2}[,]\{]\}x\}, \underline{0}[,]\{x, y\}, \underline{0}[\}\underline{0}[\}\underline{0}[\,|\, x, y \in V\};$$

$$gsuc =]\{]\{]\underline{1}, \underline{n}[,]\underline{2}, \underline{n+1}[\}, \underline{2}[\,|\, \underline{n} \in \omega\}, \underline{0}[;$$

$Grop = \{]\{]\underline{1},]g, \underline{1}[[,]\underline{2},]\{]\underline{1}, x[,]\underline{2}, y[\}, \underline{2}[[\}, \underline{2}[\,|\,]x, y[\in g\},$ g functional graph of $\mathcal{M}$;

$$Grpl = \{]\{]\underline{1},]\{]\underline{1}, x_1[, \ldots,]n, x_n[\}, \underline{2}[[,$$

$$]\underline{2},]\{]\{]\underline{1}, \underline{1}[,]\underline{2}, x_1[\}, \underline{2}[, \ldots,]\{]\underline{1}, \underline{n}[,]\underline{2}, x_n[\},$$

$$\underline{2}[\}, \underline{0}[\}, \underline{2}[\,|\, x_1, \ldots, x_n \in V\}.$$

N. B.: in the definitions above, which are posed *within* the theory $ABCD$, we denoted by $]x, y[$ the Kuratowski doubleton, and by $\underline{0}, \underline{1}, \ldots, \underline{n}, \ldots$ the finite Von Neumann ordinals.

With these definitions we get:

(i) $\mathcal{M}'$ has the same *elements* and the same *proper classes* of $\mathcal{M}$, in particular the same *universal class* V;

(ii) $]x, \underline{0}[$ is in $\mathcal{M}'$ the *set* having the elements of x in $\mathcal{M}$;

(iii) the *natural number* n of $\mathcal{M}'$ is the ordinal $\underline{n}$ of $\mathcal{M}$ (which is not a set in $\mathcal{M}'$), so $\mathbf{N} =]\omega, \underline{0}[$;

(iv) $]g, \underline{1}[$ is the *urelementary operation* of $\mathcal{M}'$ which acts like the function of graph g in $\mathcal{M}$;

(v) $]\{]\underline{1}, x_1[, \ldots,]\underline{n}, x_n[\}, \underline{2}[$ is the *n–tuple* $(x_1, \ldots, x_n)$ in $\mathcal{M}'$ (in particular $]\emptyset, \underline{2}[= \varphi)$;

(vi) in $\mathcal{M}'$, $Ins = Clel$ and $Urop$ is in a *bijective* correspondence with $Gfun$.

We leave to the reader the proof, easy and direct, that axioms Q1–15 hold in $\mathcal{M}'$: it does not essentially differ from the case of "permutation models" widely treated in [6].

Finally, given a model of the theory Q, we get immediately a *natural model* of $ABCD$ having as a *universe* the class U defined above, and as *classes* the subclasses of U; so we have shown the *equivalence* of the theories Q and $ABCD$ not only in the sense of *consistency*, but even in the stronger one of *mutual interpretability*.

Bibliography

[1] M. Clavelli, *Nuove presentazioni dei fondamenti della matematica*. Tesi di laurea, Pisa, 1984.

[2] E. De Giorgi – M. Forti, *Premessa a nuove teorie assiomatiche dei fondamenti della matematica*. Dip. Matematica, Pisa, Quad. n. 54, 1984.

[3] E. De Giorgi – M. Forti – V.M. Tortorelli[‡], *Sul problema dell'autoriferimento*. Atti Accad. Naz Lincei, Cl. Sci. Fis. Mat. Nat. (8) 80 (1986), 363–372.

[4] M. Forti – F. Honsell, *Set theory with free construction principles*. Ann. Scuola Norm. Sup. Pisa, Cl. Sci. (4) 10 (1983), 493–522.

[5] K. Goedel, *The consistency of the Axiom of Choice and of the Generalized Continuum Hypothesis*. Ann. Math. Stud., Princeton, 3, 1940.

[6] L. Rieger, *A contribution to Gödel's axiomatic set theory, I.* Czech. Math. Journ. 7 (1957), 323–357.

[‡]Editor's note: this reference has been updated

A self–reference oriented theory
for the foundations of mathematics[‡†]

M. Clavelli, E. De Giorgi, M. Forti, V.M. Tortorelli [*]

Introduction

We present in this paper a theory for the foundations of Mathematics, called Ample Theory (shortly A-Theory), which is, in our opinion, suitable for a simple and natural formulation of many interesting problems of self-reference.

The Ample theory is not intended as the foundation of a "new mathematics", nor even as a "new foundation" of traditional mathematics. We simply remark that the existing classical set theories prevent contradictions and ambiguities through strong hierarchical structures, thus reducing the self-referential power. Self-reference is in turn largely present in ordinary language, which however is often ambiguous and inconsistent.

Our theory is an attempt to put together, in a consistent framework, the main mathematical structures and some basic terms of the every day language used in the description of these structures. Thus we hope we can save both the solid hierarchy of classical set-theories and the wide self-referential power of natural language (note that most Italian grammars are written in Italian!).

Actually the A-Theory has built in as inner objects many fundamental properties of objects (to be a relation, an operation, a collection, etc.), many relations between objects (equality, membership, cardinal ordering, etc.), and many operations on objects (projections, evaluation, cardinality, etc.).

The self-referential feature of the A-Theory, which is itself formulated according to conservative criteria, could be strengthened in several ways. However, if we consider a comprehensive list of axioms giving interesting and pleasant strengthenings of the A-Theory (e.g. the individually consistent "strong axioms" of Chapter VI), it is easily seen that the whole list is inconsistent with the A-Theory, and we are left with the problem of isolating large relatively consistent sublists.

Therefore the axioms of Chapter VI are not to be intended to constitute a part of the A-Theory, which is completely outlined in the first five chapters, but as a collection of "interesting" consistency problems which any future strengthening of the Theory has to deal with.

[‡]Editor's note: published in Analyse Mathématique et Applications, Gauthier Villars, Paris, 1988, 67–115.

[†]Dedicated to J.L. Lions on his 60th birthday

[*]Research partially supported by 40% and 60% M.P.I. grants

A deep analysis of this kind of strengthenings offers a wide spectrum of complex questions (some of which were indicated in [8]), covering most foundational problems of classical theories such as Quine's NF (see [17]), Curry's illative combinatory logic (see [6]), Martin-Loef's types theory (see [16]).

This analysis involves difficult problems both of a technical and of an expository nature. Therefore we hope that we can bring these problems to the attention of working researchers from any branch of mathematics; this is actually the main goal of our paper.

The exposition of the A-Theory given in Chapters I-V is semiformal, but we point out that it can be finitely axiomatized as a first order theory, in a language with just one ternary predicate and one constant: we sketch such a formalization in the Appendix to Chapter I.

In Chapter I we deal with some concepts of the common language used in the description of the mathematical structures (pairs, qualities, relations and operations); the axioms of this chapter are inspired by the idea of providing a highly self-referential theory.

In Chapter II the fundamental structures of mathematics (collections, systems and cardinal numbers) are introduced in the frame created by the ones considered in the first chapter. These mathematical structures are developed in the following chapters III and IV, where we tried to remain as close as possible to the language and conventions of traditional mathematics.

Although the axiom system developed in Chapters III and IV is weak enough to allow any kind of different extensions and strenghtenings of the theory, we nevertheless decided to provide in Chapter V a great number of universes, which are intended, à la Grothendieck (see [14]), as controlled fields within which the usual mathematical theories can be developed without any trouble.

They provide, within the A-Theory, natural inner models of the frame theory of [7], as well as of most classical theories for the foundations of mathematics, such as Zermelo-Fraenkel set-theory, Gödel-Bernays, Kelley-Morse or Fraenkel-Mostowski class theories and λ-calculus (see [10]). We also outline in Chapter V a sort of propositional calculus, which will be useful for future developments of logic structures as inner objects of the A-Theory.

Notice that the general objects of Chapter I are non-extensional, whereas the corresponding mathematical objects of Chapter II are extensional: we emphasize this distinction, since non-extensional objects seem particularly suitable to describe themselves and other objects, whereas extensional ones are more tractable and easy to describe as results of various operations. Moreover an analysis of the known antinomies shows that it is a good criterion to divide the objects into two kinds, those having high descriptive power and those which can be freely manipulated. We decided to have the largest expressivity of non-extensional objects (in particular relations), and the largest manageability of extensional objects (collections and systems).

Following this criterion many relations, qualities and operations have to be "hand made", while collections, systems and universes can also be "manufactured".

In addition we are inspired by the following heuristic principles:

(i) *follow confidently the naïve intuition (and the conventions adopted by the simplest manuals of pure and applied mathematics) in dealing with finite or not "too infinite" objects;*

(ii) *be extremely cautious when dealing with objects of the "largest infinite size".*

It is fairly easy to specify the notion of "maximal infinity". We will do this within a rather weak cardinal axiomatization, which is consistent with all classical versions of the axiom of choice as well as with strong negations of this axiom.

More difficult is the problem of finding the "best rules" to be adopted when dealing with maximal infinite objects, avoiding both excessive confidence leading to contradictory theories and excessive fear excluding the existence of innocuous large objects and useful self-referential properties. In waiting for this problem to be better understood we have been very cautious, and this explains the great number of special axioms concerning singular relations, qualities and operations stated in the first chapters. Strong general axioms, like those of Chapter VI, besides making more tractable non-extensional objects, and more expressive extensional ones, can greatly reduce the number of particular axioms and objects one has to introduce. Some limits of this simplifying and unifying programme are marked by the antinomies pointed out by some examples in Chapter VI.

Another limit is given by various kinds of fundamental objects which are present in the A-Theory; in fact, we tried to maintain at the same conceptual level several notions which in traditional theories are chosen as the unique primitive concept: collections in [10], operations in [6] and [21], relations in [22]. The very name of the Ample Theory arises from the ample spaces it allows for strengthenings, which arise either from emphasizing one or another of these traditional basic concepts, or introducing other basic structures (e.g. categories, sheaves, languages, etc.). We consider the great number of axioms and fundamental objects a worthwhile price to pay for granting such an ampleness.

It is impossible to quote all bibliographical references which more or less directly influenced our approach to the foundational problems: they include classical works by H. Curry [6], G. Frege [11], [12], J. Von Neumann [21], B. Russell [22] as well as more recent technical papers in the areas of set theory with a universal set (inter alia [4], [13], [17]), λ-calculus and constructive mathematics ([2], [9], [16], [18]). Thus we restrict ourselves to quoting in the sequel those papers which are needed for direct reference.

The main issues of this paper have been discussed at De Giorgi's Seminar on Logic and Foundations, held at the Scuola Normale Superiore, Pisa, since 1980. We are deeply indebted to all partecipants for useful discussions, and in particular to M. Boffa, F. Honsell, G. Longo and M. Sciuto.

I. Qualities, Relations, Operations and Pairs

I.0. Basic conventions

We give in this chapter a semi-formal introduction to the A-Theory, which clarifies the motivations of the axioms we will introduce in the following sections. A complete formalization in first order predicate calculus will be done in the appendix to this chapter.

We call *objects* all the inner entities of the Ample Theory. We begin by considering the following kinds of objects:

qualities (or properties), (binary) *relations, operations* (or transformations), (ordered) *pairs*.

These concepts play a fundamental role both in common language and in most scientific theories. We do not give their definitions, since we consider them as primitive. In fact, our assumptions allow these objects to behave as close as possible to their common usage and meaning.

Therefore we assume that there are objects which are qualities, relations, operations and pairs, and we stipulate the following conventions:

(1) When q is a *quality*, we write qx with the intended meaning that *q is a quality of x*, or that *x has the quality q*; we write also $\not q z$ when *not qx*.

(2) When r is a *relation*, we write xry with the intended meaning that *x is in the relation r with y*, or *x is related by the relation r to y*; we write also $x \not r y$ when *not $x \, r \, y$*.

(3) When f is an *operation*, we write $y = fx$ whith the intended meaning that *f transforms x in y*; we also say that fx is the *result* of the operation f *applied* to the object x.

(4) When z is a *pair*, we write $z = (x, y)$ with the intended meaning that *z is the pair having x as first and y as second component*.

I.1. First self-referential axioms

It is a typical feature of the A-Theory that many basic properties of objects are objects themselves, as well as many basic relations between objects. Hence in order to deal with the above concepts inside the A-Theory, we now introduce eleven "fundamental" objects. They are the five qualities of being an *object*, a *quality*, a *relation*, an *operation*, a *pair*

$$qobj, \quad qqual, \quad qrel, \quad qop, \quad qpair,$$

and the six fundamental relations concerning *objects, qualities, relations, operations* and *pairs*

$$robj, \quad rqual, \quad rrel, \quad rop, \quad rfst \text{ and } rsnd,$$

whose meaning will be clarified by the axioms below.

Notice that we employ symbols beginning with lower case q and r, to denote qualities and relations, respectively. We shall maintain this notational convention throughout the paper.

The behaviour of each of these objects in the Ample Theory is as close as possible to their natural meaning. E.g. when we write $qqual\ q$, $qrel\ r$ we intend the assertions "q is a quality" and "r is a relation", respectively. When we write $x\ rqual\ q$, we intend the assertion "q is a quality of x" i.e. $q\,x$, and when we write $(x,y)\ rrel\ r$, we intend "x is related by the relation r to y" i.e. $x\,r\,y$. Similar descriptions can be given for the remaining objects: they are summarized in the following axioms.

Axiom.1.A   *qobj, qqual, qrel, qop* and *qpair are qualities.*
robj, rqual, rrel, rop, rfst and *rsnd are relations.*

Axiom.1.B   *Both qobj x and x robj y hold for all objects x and y.*

Axiom.1.C   *The object q is a quality if and only if qqual q.*
x rqual q if and only if q is a quality and q x.

Axiom.1.D   *The object r is a relation if and only if qrel r.*
(x,y) rrel r if and only if r is a relation and x r y.

Axiom.1.E   *The object f is an operation if and only if qop f.*
(x,y) rop f if and only if f is an operation and y=fx.

Axiom.1.F   *The object z is a pair if and only if qpair z.*
z= (x,y) if and only if x rfst z and y rsnd z.

We also give an axiom expressing that the operations are *single valued*, and three more axioms which ensure that all ordered pairs of objects are objects, and that *qpair, rfst* and *rsnd* describe the usual behaviour of ordered pairs.

Axiom.1.G   *If (x,y) rop f and (x,z) rop f then y=z.*

Axiom.1.H   *For any two objects x and y there exists a unique object z such that r rfst z and y rsnd z.*

Axiom.1.I   *If the object z is a pair there are unique x and y such that x rfst z and y rsnd z.*

Axiom.1.J   *If x rfst z or x rsnd z for some x, then qpair z.*

We take these objects and the axioms I.A-J as a basis for all further developments of the A-Theory. In order to formalize this part of the Ample Theory in first order predicate calculus, it suffices to introduce a ternary predicate $\mathcal{P}$ such that $\mathcal{P}(x, r, y)$ corresponds to the use of $x\,r\,y$ fixed by convention (2), together with eleven constants, corresponding to the fundamental objects listed above (cf. Appendix).

We now consider the six fundamental pairs
$$\mathcal{S}_0 = (qobj, robj), \quad \mathcal{S}_{01} = (qpair, rfst), \quad \mathcal{S}_{02} = (qpair, rsnd),$$
$$\mathcal{S}_1 = (qqual, rqual), \quad \mathcal{S}_2 = (qrel, rrel), \quad \mathcal{S}_3 = (qop, rop),$$
whose first components are qualities and second components are relations. Each of them has the property that *any object, to which the relation occurring in the second component associates something, enjoys the corresponding quality occurring in the first component.*

This fact leads us to introduce another basic object, *qqrs* (the quality of being a *quality-relation-structure*, shortly *qr-structure*), ruled by the axiom:

Axiom.1.K   *qqrs is a quality, qqrs $\mathcal{S}$ if and only if $\mathcal{S} = (q, r)$,*
where qqual q, qrel r, and if $y\,r\,x$ for some y, then $q\,x$.

I.2. Other fundamental objects

We can consider, beside the basic qualities and relations introduced in Section 1, many other relations qualities and operations, relevant for further developments of the A-Theory. First we consider the relations of *equality* and *non-equality*

rid and *rnid,*

and the operations *identity, diagonalization, first* and *second projections* of pairs

id, diag, p_1, p_2.

These objects behave in the natural way. Two objects are in the relation *rid* or in the relation *rnid* if and only if they are *equal* or they are *different*, respectively. The operations *id* and *diag* transform any object x in the object *itself* and in the pair (x, x), respectively; the projections p_1, p_2 associate to any pair its *first*, respectively *second* component. More formally

Axiom.2.A   *x rid y if and only if $x=y$; and x rnid y if and only if $x \neq y$.*

Axiom.2.B   *id $x=x$ and diag $x=(x,x)$ for any object x.*

Axiom.2.C   *$p_1 z$ is the first and $p_2 z$ is the second component of any pair*
$z = (x, y)$, i.e. $p_1(x, y) = x, p_2(x, y) = y$.

It would be useful to have of a lot of basic natural qualities, relations and operations; we introduce here only few of them, and we also avoid the introduction of operations generating infinitely many "large" objects, such as the compositions of operations or relations. We deal with this topic in Chapter VI, in the context of different strengthenings of the Ample Theory. Thus we begin by introducing now only few objects, namely

qinjop	(the quality of being an *injective operation*),
rdom, rval	(the relations of *domain* and *range* of operations),
invrel, invop	(*inversion* of relations and of injective operations),
eval	(*evaluation* of an operation at an object),
conj	(*conjunction* of qualities).

The meaning of these objects being the natural one, we simply list the axioms ruling their actions.

Axiom.2.D. *qinjop is a quality.*
qinjop f if and only if f is an operation and fx=fy implies x=y.

As usual, we say that the operation f is *injective* whenever *qinjop f*.

Axiom.2.E. *rdom and rval are relations.*
x rdom f if and only if f is an operation and y=fx for some y;
y rval f if an only if f is an operation and y=fx for some x.

We say that x *lies* in the *domain* of an operation f, or that f is *defined at the argument* x (written $f \downarrow x$) if and only if x *rdom* f. We also say that y lies in the *range* of f, or y is a *value* of f (written $f \uparrow y$) if and only if y *rval* f.

Axiom.2.F. *invrel is an operation.*
invrel $\downarrow$ r and invrel $\uparrow$ r' if an only if r and r' are relations.
x(invrel r) y if and only if y r x.
Moreover invrel(invrel r)=r for any relation r.

Axiom.2.G. *invop is an operation.*
invop $\downarrow$ f and invop $\uparrow$ g if and only if f and g are injective operations.
y=(invop f) x if an only if x=fy.
Moreover invop(invop f)=f for any injective operation f.

We call *invrel r* the *inverse relation* of r, and *invop f* the *inverse operation* of f, and we usually denote them by r^{-1} and f^{-1}.

Axiom.2.H. *eval is an operation.*
eval $\downarrow$ z if and only if z=(f,x), qop f and f $\downarrow$ x; in this case
eval(f,x)=f x.

Axiom.2.I. *conj is an operation.*
conj $\downarrow$ z if and only if z is a pair of qualities.
conj(q_1, q_2)x holds if and only if both $q_1 x$ and $q_2 x$ hold.

We write $q_1 \wedge q_2$ for $conj(q_1, q_2)$, and call it the *conjunction* of q_1 and q_2.

Axiom.2.J. *(idempotency, commutativity and associativity)*
If q_1, q_2 and q_3 are qualities, then $q_1 \wedge q_1 = q_1, q_1 \wedge q_2 = q_2 \wedge q_1$,
and $q_1 \wedge (q_2 \wedge q_3) = (q_1 \wedge q_2) \wedge q_3$.

I.3. Some qualities of ordering relations

The ordering relations (to be *less than, previous to, better than,* etc.) play a distinguished role among the relations considered both in common language and in scientific theories. We introduce here some properties concerning these types of relations, namely:

$qrefl$ *(reflexivity),*
$qtrans$ *(transitivity),*
$qsym$ *(symmetry),*
$qantis$ *(anti-symmetry),*
$qcon$ *(connectedness).*

The axioms below give the meaning of these qualities. We also tacitly assume the axiom stating that all the above objects are qualities of relations.

Axiom.3.A *A relation r is reflexive, i.e. $qrefl\ r$, if and only if $x\ r\ y$ implies*
both $x\ r\ x$ and $y\ r\ y$.

Axiom.3.B *A relation r is transitive, i.e. $qtrans\ r$, if and only if $x\ r\ y$ and*
$y\ r\ z$ imply $x\ r\ z$.

Axiom.3.C *A relation r is symmetric, i.e. $qsym\ r$, if and only if $x\ r\ y$ implies*
$y\ r\ x$.

Axiom.3.D *A relation r is anti-symmetric, i.e. $qantis\ r$, if and only if $x=y$*
whenever $x\ r\ y$ and $y\ r\ x$.

Axiom.3.E *A relation r is connected, i.e. $qcon\ r$, if and only if r is reflexive*
and $x\ r\ x$, $y\ r\ y$ imply $x\ r\ y$ or $y\ r\ x$.

We can now obtain by conjunction the derived qualities of being a *preordering*, a *partial ordering*, a *linear* (or total) *ordering*, and an *equivalence relation*.

Definition

$$qpreo = qrefl \wedge qtrans \quad \text{(preordering)}$$
$$qpo = qpreo \wedge qantis \quad \text{(partial ordering)}$$
$$qlo = qpo \wedge qcon \quad \text{(linear ordering)}$$
$$qeq = qpreo \wedge qsym \quad \text{(equivalence relation).}$$

I. Appendix: A formalization in first order predicate calculus

As remarked above, we can formalize the A-Theory in first order predicate calculus by means of a ternary predicate $\mathbf{P}$, whose intended meaning is that $\mathbf{P}(x, r, y)$ holds if and only if the object r is a relation and $x \, r \, y$, and by means of a suitable number of constants, corresponding to the basic qualities and relations.

For sake of readability, we denote by
$$\amalg_0, \amalg_1, \amalg_2, \amalg_3, \amalg_{00}$$
the constants corresponding to *qobj, qqual, qrel, qop, qpair* and by
$$\mathbf{r}_0, \mathbf{r}_1, \mathbf{r}_2, \mathbf{r}_3, \mathbf{r}_{01}, \mathbf{r}_{02}$$
the constants corresponding to *robj, rqual, rrel, rop, rfst, rsnd.*

Then we state the following axioms:

FA0 $\qquad \mathcal{P}(x, \mathbf{r}_0, y) \wedge \mathcal{P}(x, \mathbf{r}_1, \amalg_0)$

FA1 $\qquad \mathcal{P}(x, \mathbf{r}_1, y) \to \mathcal{P}(x, \mathbf{r}_1, \amalg_1)$

FA2 $\quad \mathcal{P}(x, y, z) \to \mathcal{P}(y, \mathbf{r}_1, \amalg_2) \wedge \exists w \mathcal{P}(w, \mathbf{r}_2, y)$
$\mathcal{P}(w, \mathbf{r}_2, y) \leftrightarrow \exists x \exists z (\mathcal{P}(x, \mathbf{r}_{01}, w) \wedge \mathcal{P}(z, \mathbf{r}_{02}, w) \wedge \mathcal{P}(x, y, z))$

FA3 $\quad \mathcal{P}(x, \mathbf{r}_3, y) \to \mathcal{P}(x, \mathbf{r}_1, \amalg_{00}) \wedge \mathcal{P}(y, \mathbf{r}_1, \amalg_3)$
$\mathcal{P}(x, \mathbf{r}_3, y) \wedge \mathcal{P}(z, \mathbf{r}_3, y) \wedge \mathcal{P}(w, \mathbf{r}_{01}, x) \wedge \mathcal{P}(w, \mathbf{r}_{01}, z) \wedge$
$\wedge \mathcal{P}(v, \mathbf{r}_{02}, x) \to \mathcal{P}(v, \mathbf{r}_{02}, z)$

FA4 $\quad \exists! z (\mathcal{P}(x, \mathbf{r}_{01}, z) \wedge \mathcal{P}(y, \mathbf{r}_{02}, z))$
$\mathcal{P}(x, \mathbf{r}_1, \amalg_{00}) \to \exists! y \exists! z (\mathcal{P}(y, \mathbf{r}_{01}, x) \wedge \mathcal{P}(z, \mathbf{r}_{02}, x))$
$\mathcal{P}(x, \mathbf{r}_{01}, y) \vee \mathcal{P}(x, \mathbf{r}_{02}, y) \to \mathcal{P}(y, \mathbf{r}_1, \amalg_{00})$

The axioms stated above yield in particular the theorem:

Theorem

$$\mathcal{P}(\amalg_0, \mathbf{r}_1, \amalg_1) \wedge \mathcal{P}(\amalg_1, \mathbf{r}_1, \amalg_1) \wedge \mathcal{P}(\amalg_2, \mathbf{r}_1, \amalg_1) \wedge \mathcal{P}(\amalg_3, \mathbf{r}_1, \amalg_1) \wedge \mathcal{P}(\amalg_{00}, \mathbf{r}_1, \amalg_1) \wedge$$
$$\mathcal{P}(\mathbf{r}_0, \mathbf{r}_1, \amalg_2) \wedge \mathcal{P}(\mathbf{r}_1, \mathbf{r}_1, \amalg_2) \wedge \mathcal{P}(\mathbf{r}_2, \mathbf{r}_1, \amalg_2) \wedge \mathcal{P}(\mathbf{r}_3, \mathbf{r}_1, \amalg_2) \wedge$$
$$\mathcal{P}(\mathbf{r}_{01}, \mathbf{r}_1, \amalg_2) \wedge \mathcal{P}(\mathbf{r}_{02}, \mathbf{r}_1, \amalg_2) \,.$$

We can now give the formal definitions corresponding to the conventions (1)-(4) stipulated at the beginning of Chapter I.

Definition

(1) *q is a quality* if and only if $\mathcal{P}(q, \mathbf{r}_1, \amalg_1)$; in this case
 $q\,x$ if and only if $\mathcal{P}(x, \mathbf{r}_1, q)$

(2) *r is a relation* if and only if $\mathcal{P}(r, \mathbf{r}_1, \amalg_2)$; in this case
 $x\,r\,y$ if and only if $\mathcal{P}(x, r, y)$

(3) *f is an operation* if and only if $\mathcal{P}(f, \mathbf{r}_1, \amalg_3)$; in this case
 $y = fx$ if and only if $\exists z(\mathcal{P}(x, \mathbf{r}_{01}, z) \wedge \mathcal{P}(y, \mathbf{r}_{02}, z) \wedge \mathcal{P}(z, \mathbf{r}_3, f))$

(4) *z is a pair* if and only if $\mathcal{P}(z, \mathbf{r}_1, \amalg_{00})$; in this case
 $z = (x, y)$ if and only if $\mathcal{P}(x, \mathbf{r}_{01}, z) \wedge \mathcal{P}(y, \mathbf{r}_{02}, z)$

The above theorem can now be written: $\amalg_1\amalg_0$, $\amalg_1\amalg_1$, $\amalg_1\amalg_2$, $\amalg_1\amalg_3$, $\amalg_1\amalg_{00}$ and $\amalg_2\mathbf{r}_0$, $\amalg_2\mathbf{r}_1$, $\amalg_2\mathbf{r}_2$, $\amalg_2\mathbf{r}_3$, $\amalg_2\mathbf{r}_{01}$, $\amalg_2\mathbf{r}_{02}$.

After this definition, the axioms FA0-4 can be easily recognized as naturally formalizing the axioms 1.A-J of Section 1. The formalization of the remaining axioms of Chapter I can be easily obtained by introducing a new constant for each basic object considered and by giving a formal counterpart of the corresponding axiom. E.g., the axiom 1.K is formalized by the axiom FA5 below, using the above definition and a new constant $\mathbf{s}$, corresponding to the quality $qqrs$ of being a qr-structure.

FA5 $\qquad x\mathbf{r}_1\mathbf{s} \leftrightarrow \exists y \exists z(x = (y, z) \wedge \amalg_1 y \wedge \amalg_2 z \wedge \forall u \forall v(uzv \rightarrow yv))$.

Similarly, axiom 2.A can be formalized using two new constants $\mathbf{e}, \mathbf{d}$, corresponding to the relations rid (equality) and $rnid$ (difference):

FA6 $\qquad (\mathcal{P}(x, \mathbf{e}, y), \leftrightarrow x = y) \wedge (\mathcal{P}(u, \mathbf{d}, v) \leftrightarrow u \neq v)$.

The same translation can be carried out for all the axioms of the first and of the following chapters.

Note that we can avoid the introduction of more and more constants, by means of only one constant $\mathbf{rg}$, representing the *relation generator of natural numbers and of fundamental constants*. To this aim, we can state the axioms

G1 $\quad \exists! x(\exists y \mathcal{P}(x, \mathbf{rg}, y) \wedge \neg \exists z \mathcal{P}(z, \mathbf{rg}, x))$

G2 $\quad \exists y \mathcal{P}(x, \mathbf{rg}, y) \rightarrow \exists! z(\exists w \mathcal{P}(z, \mathbf{rg}, w) \wedge \mathcal{P}(x, \mathbf{rg}, z))$

G3 $\quad \exists y \mathcal{P}(x, \mathbf{rg}, y) \rightarrow \exists! v(\neg \exists w \mathcal{P}(v, \mathbf{rg}, w) \wedge \mathcal{P}(x, \mathbf{rg}, v))$

G4 $\quad \mathcal{P}(x, \mathbf{rg}, y) \wedge \mathcal{P}(z, \mathbf{rg}, y) \rightarrow x = z$.

Then we can obtain an arbitrary number of defined terms by

Definition

0 is the unique x verifying G1.

If $\exists y \mathcal{P}(x, \mathbf{rg}, y)$ holds, then $x + 1$ is the unique z verifying G2.

If $\exists y \mathcal{P}(x, \mathbf{rg}, y)$ holds, then c_x is the unique v verifying G3.

We could of course use $c_0, \ldots, c_{12}$, etc. instead of the constant $\mathrm{II}_i, \mathbf{r}_i$ and $\mathbf{s}, \mathbf{d}, \mathbf{e}$, etc. in the axioms FA0-6, etc., thus achieving a formalization of the A-Theory in the first order language having as non logical symbols only the ternary predicate $\mathcal{P}$ and the constant $\mathbf{rg}$.

II. Collections, Systems and Cardinals

II.0. Introduction

The fundamental structures introduced in Chapter I give a general frame, suitable for most scientific theories.

Having introduced qualities, relations, operations and pairs as very fundamental concepts, every new primitive notion can be defined by means of suitable qr-structures. We consider in this chapter some fundamental concepts of mathematics: *numbers, sets, classes, sequences, maps,* etc.

Hence we introduce three qr-structures:

$$\mathcal{S}_4 = (qcoll, rcoll), \quad \mathcal{S}_5 = (qsys, rsys), \quad \mathcal{S}_6 = (qcard, rcard)$$

the *structure of collections*, the *structure of systems* and the *structure of cardinal numbers*.

The objects enjoying the quality *qcoll* are *collections* in the usual acceptation, the relation *rcoll* giving the usual membership. Thus x *rcoll* C means that x is a member of the collection C and it will be written $x \in C$, following the usual notation. Sets and classes will be introduced in the sequel as particular collections.

Similarly the objects enjoying the property *qsys* are *indexed systems* in the natural acceptation, and among them we shall consider *n-tuples, sequences* and *maps*.

The relation *rsys* is such that (i, x) *rsys* S means that i is an *index of* the system S and that *in S x is indexed by* (associated to) i, shortly written $i\, S\, x$.

Of particular interest are the *univalent systems*, i.e. those systems where at most one object is associated to an index. Thus we introduce the quality of being an *univalent system*, denoted by *qunsys*, and we write in general $x = S_i$ for $i\, S\, x$, when S is univalent.

Finally the *cardinal numbers* (or cardinals) are axiomatically introduced as objects enjoying the quality *qcard, rcard* being the natural ordering relation between cardinals. We assume that *rcard* is a partial ordering, and we write $\alpha \leq \beta$ (*α is less than or equal to β*) for α *rcard* β and $\alpha < \beta$ (*α is a strictly less than β*) for $\alpha \leq \beta$, and $\alpha \ngeq \beta$.

II.1. Basic axioms on collections and systems

We state below the axioms on collections and systems.

Axiom.4.A $\mathcal{S}_4 = (qcoll, rcoll)$ *and* $\mathcal{S}_5 = (qsys, rsys)$ *are qr-structures.*

Axiom.4.B *qunsys is a quality of systems; qunsys T if and only if T is a system and for any i there is at most one x such that $i\ T\ x$.*

Collections and systems in their natural acceptation are *extensional* objects corresponding to qualities and relations: univalent systems can as well be viewed as extensional counterparts of operations. Thus we state the extensionality of the structures $\mathcal{S}_4$ and $\mathcal{S}_5$, by the following

Axiom.4.C *Two collections C, D are equal if and only if they have the same members. (i.e. $(x \in C \leftrightarrow x \in D) \to C = D$)*

Axiom.4.D *Two systems S, T are equal if and only if they have the same indices and each index is associated to the same objects in both of them. (i.e. $(i\ S\ x \leftrightarrow i\ T\ x) \to S = T)$.*

We introduce also the relation of (extensional) *inclusion between collections, rincl,* with an axiom giving its meaning:

Axiom.4.E *C rincl D if and only if C and D are collections and every member of C is also a member of D.*

We shall usually write $C \subseteq D$ for C *rincl* D (C is *included* or is a *subcollection* of D), and $C \supseteq D$ for C *rincl*^{-1}D (C *includes* D).

II.2. First elementary operations on collections and systems

The axioms of extensionality allow to define *singletons, doubletons, unitary systems,* the *inverse* of a system and the *projection of a univalent system at an index.*

Definition

C is the *singleton* of an object x (written $C = \{x\}$) if and only if x is the only member of the collection C.

C is the *doubleton* of two objects x and y (written $C = \{x, y\}$) if and only if x and y are the only members of the collection C.

S is a *unitary system* if and only if S is univalent and it has only one index. We denote by $\left(\begin{smallmatrix} i \\ x \end{smallmatrix}\right)$ the unitary system S such that $x = S_i$.

The system T is the inverse of the system S (written $T = S^{-1}$) when $x\ T\ y$ if and only if $y\ S\ x$.

The object x is the *projection* of the univalent system S at the index i when $x = S_i$.

We assume that there are objects of the A-Theory which are operations producing singletons, doubletons, unitary systems, etc., and we denote them by

sing, doub (*singleton* and *doubleton*),
gus (the *generator* of unitary systems),
invsys (the *inversion* of systems),
psys (the *projection* of an univalent system at one of its indices).

We state the corresponding axioms.

Axiom.5.A *doub associates to any pair (x,y) the doubleton $\{x,y\}$,
i.e. doub $(x,y) = \{x,y\}$.
sing associates to any object x its singleton $\{x\}$, i.e. sing $x = \{x\}$.*

Axiom.5.B *gus associates to any pair (x,y) the unitary system $\begin{pmatrix} x \\ y \end{pmatrix}$.*

Axiom.5.C *invsys associates to any system S its inverse, invsys $S = S^{-1}$.*

Axiom.5.D *psys(i,S)=y if and only if qunsys S and $y = S_i$.*

We also assume

Axiom.5.E *There is a collection C without any member, and a system S without any index. (i.e. for any two objects x and y $x \notin C$ and $x \not\$ y$).*

We call *empty collection*, denoted by $\emptyset$, the collection without any member, and *empty system*, denoted by $\oslash$, the system without indices: their uniqueness follows from the axioms of extensionality 4.C-D. Note that one can generate infinitely many collections and systems starting from $\emptyset$ and $\oslash$ and applying the operations *sing* and *gus*.

Other basic operations generating collections and systems are introduced in Chapter IV, according to a conservative "principle of limitation of size" which will be specified in Chapter III Section 3, when a precise notion of cardinality is available.

II.3. Cardinal numbers

We introduce in the A-Theory the cardinal numbers as a primitive structure $\mathcal{S}_6 = (qcard, rcard)$ characterized by suitable axioms.

The use of cardinals as a natural *measure* of the *size* of objects, will be coded in Chapter III. Thus our approach to cardinals is similar to the algebraic theory of Tarski (cf. [19]), rather than to the usual set theoretical approach based on *equipotency*.

The algebraic properties of cardinal numbers are not fully described in terms of the natural ordering *rcard*, hence we introduce as objects of the A-Theory the operation of *cardinal addition, sum,* the relation of *strict cardinal ordering, rscard,* and the particular cardinals $0, 1, \aleph_0$ and $ת$ (the cardinals *zero, one, aleph zero,* and *tav*).

$0, 1$, and $\aleph_0$ have the natural properties of the corresponding cardinals of the standard set theories. Moreover, we introduce beside the least infinite cardinal $\aleph_0$, also a largest cardinal, denoted by $\beth$, the last letter of the hebrew alphabet. The cardinal plays a distinguished role in our theory, in connection with the principle of limitation of size (see Chapter III Section 3; for a comparison with other theories having a largest cardinal, see Chapter VI Section 9).

We adopt in this paragraph the convention that Greek lower case letters denote generic cardinals, and we write $\alpha + \beta$ for $sum(\alpha, \beta)$, and $\alpha \leq \beta, \alpha < \beta$ for $\alpha \, rcard \, \beta$, $\alpha \, rscard \, \beta$.

The fundamental properties of cardinal numbers are expressed by the following axioms.

Axiom.6.A   *$\mathcal{S}_6$ is a qr-structure and rcard is a partial ordering.*
rscard is a relation;
$\alpha \, rscard \, \beta$ if and only if $\alpha \, rcard \, \beta$ and $\beta \, rcard \, \alpha$.

Axiom.6.B   *sum is an operation, defined at all pairs of cardinals, whose values are cardinals.*

Axiom.6.C   *(commutativity and associativity)*
$\alpha + \beta = \beta + \alpha$ and $(\alpha + \beta) + \gamma = \alpha + (\beta + \gamma)$ for all cardinals α, β and γ.

Axiom.6.D   *$\alpha \leq \beta$ if and only if there is a cardinal γ such that $\alpha + \gamma = \beta$.*

Axiom.6.E   *1 is the least cardinal different from 0.*
$\alpha + 0 = \alpha$ for any cardinal α.

Axiom.6.F   *If α and β are cardinals different from $\beth$, then $\alpha + \beta$ is strictly less than $\beth$.*

Note that one gets immediately that $0 \leq \alpha \leq \beth$ and $\alpha + \beth = \beth$ for any cardinal α.

II.4. Natural numbers

Before stating the axioms on natural numbers we give the following definition:

Definition

A cardinal number α is *finite* if and only if $\alpha < \aleph_0$.
A cardinal number α is *infinite* if and only if $\alpha \geq \aleph_0$.

We notice that the axioms do not exclude the existence of cardinals which are neither finite nor infinite; thus, in particular, cardinals associated to Dedekind-finite sets are not excluded a priori.

We identify the *natural numbers* with the finite cardinals, and we introduce them by assuming the existence of the *collection of the natural numbers*, $\mathbb{N}$, satisfying the following axioms.

Axiom.7.A *There is a collection $\mathbb{N}$ such that $n \in \mathbb{N}$ if and only if n is a finite cardinal number.*

Axiom.7.B *If $n \in \mathbb{N}$ then $n \neq n + 1 \in \mathbb{N}$.*

Axiom.7.C *Let r be a relation, n a natural number and x an object, such that $n\ r\ x$.*
Then there is a least $h \in \mathbb{N}$ verifying $h\ r\ x$.

Axiom.7.D *Let r be a relation, n and m natural numbers and x an object, such that $n\ r\ x$ and $n \leq m$.*
Then there is a largest $k \leq m$ such that $k\ r\ x$.

It follows from axioms 6 and 7 that any non-finite cardinal is strictly larger than any natural number.

II.5. n-tuples and sequences

Natural numbers being available, we can define *n-tuples* and *sequences* as univalent systems whose indices are initial segments of the positive natural numbers.

We assume that, among the objects of the A-Theory, there are the qualities $qseq$ and $qfseq$, and the relation *rtuple* verifying the following axioms:

Axiom.8.A *rtuple is a relation, qseq and qfseq are qualities.*

Axiom.8.B *n rtuple S if and only if $n \in \mathbb{N}$, qunsys S, and the indices of S are exactly the natural numbers i such that $1 \leq i \leq n$.*

Axiom.8.C *qseq S if and only if qunsys S and the indices of S are either all the natural numbers, or all natural numbers different from 0.*

Axiom.8.D *qfseq S if and only if there is an n such that n rtuple S.*

We say that the object S is a *sequence* (respectively a *finite sequence*) if and only if $qseq\ S$ (respectively $qfseq\ S$). We say that S is a *n-tuple*, or a finite sequence of *length n*, if and only if n *rtuple* S. In particular, the empty system $\oslash$ is the unique 0-tuple.

We also introduce the operation *conc (concatenation of finite sequences)*.

Axiom.8.E *conc is an operation defined at all pairs of finite sequences.*
If S is a m-tuple and T is a n-tuple, then conc(S,T) is a (m+n)-tuple such that:
$$conc(S,T)_i = S_i \quad for\ \ 1 \leq i \leq m$$
$$conc(S,T)_{m+j} = T_j \quad for\ \ 1 \leq j \leq n$$

We denote the concatenation of two finite sequences S and T by $S \circ T$.

II.6. Structural consistency

We never dealt hitherto with the question as to whether the same object can participate (or necessarily participates) to more than one fundamental structure.

Most foundational theories are reductionist, in the sense that only one or two kinds of objects are really different, the other basic objects being identified with special cases of the previous ones. E.g. the usual set-theoretical foundations identify pairs with Kuratowski doubletons, relations, operations and systems with their graphs and natural numbers with their initial segments. On the contrary, the Frame Theory Q of [7] assumes as an axiom that the various types of entities considered are pairwise disjoint.

We shall briefly discuss this topic in Chapter VI Section 2; however in order to prevent undesired clashings, it seems convenient to assume the following axiom of *structural consistency*:

Axiom.9.A   *Let $\mathcal{S}_i = (q_i, r_i)$ $(1 \leq i \leq 5)$ be the fundamental structures of qua-lities, relations, operations, collections and systems.*
     If $(q_i \wedge q_j)x$, then $y \, r_i \, x$ if and only if $y \, r_j \, x$ $(1 \leq i,j \leq 5)$.

For obvious reasons, we exclude from the axiom of structural consistency the trivial structure $\mathcal{S}_0$, and both the structures of pairs $\mathcal{S}_{01}$ and $\mathcal{S}_{02}$; we also exclude the structure $\mathcal{S}_6$ of cardinals, since we will not prevent, say, an identification à la Frege-Russell with collections of equipollent collections.

However, it seems natural and mathematically profitable to identify ordered pairs with 2-tuples, and we therefore take *in this case* the reductionist attitude, by postulating

Axiom.9.B   *The pair $z=(x, y)$ is the univalent system, with indices 1 and 2, verifying $z_1 = x$ and $z_2 = y$.*

The axiom 9.B allows an unambiguous use of the notation $(z_1, \ldots, z_n)$ for any n-tuple z.

III. Extension and Cardinality

III.1. Relative extension

Cardinals have been introduced in Chapter II as an algebraic structure, without any reference to their main role, which is to give a measure of the size of structured objects (i.e. of the amount of objects belonging to a collection, or enjoying a quality, or laying in the domain of an operation, etc.).

In order to enforce this use of cardinal numbers in a uniform way with respect to the various structures, it seems convenient to consider the *relational pairs*, i.e. pairs whose first component is a relation: the relational pair (r, x) can be viewed as coding the *extension of x with respect to the relation r*.

We introduce the quality of being a *relational pair, qrepa*, and the operation *collection-extension, cext*, which associates to a relational pair (r, x) the collection giving the *relative extension* of x with respect to r, *provided that such a collection is among the objects* of the A-Theory.

Axiom.10.A   *qrepa is a quality. qrepa z if and only if z= (x,y) and qrel x.*

Axiom.10.B   *cext is an operation. cext z=C if an only if z=(r,x) and qrel r, qcoll C, and $t \in C$ exactly when t r x.*

Following the standard notation, whenever $cext \downarrow (r,x)$ we denote the collection $cext(r,x)$ by $\{t|t\ r\ x\}$, thus $C = \{t|t \in C\}$ for any collection C.

Notice that when $C = cext(r,x)$ and $C' = cext(r',x')$, one has $C' \supseteq C$ if and only if $t\ r\ x$ implies $t\ r'x'$.

Since this relation between relational pairs is useful also when the collection-extension does not exist, we introduce the preordering of *generalized inclusion, rginc,* and the *relation of extensional equivalence, rexteq.*

Before stating the axioms characterizing $rginc$ and $rexteq$ we fix our notation.

If r and r' are relations we write $(r,x)\preceq(r',x')$ for $(r,x)rginc(r',x')$, and we say that (r,x) is *extensionally included* in (r',x'); we also write $(r,x) \approx (r',x')$ for both $(r,x)\preceq(r',x')$ and $(r',x')\preceq(r,x)$ and we say that (r,x) and (r',x') are *extensionally equivalent.*

Axiom.10.C   *rginc is a relation between relational pairs.*
        (r, x) rginc (r, x') if and only if t r x implies t r'x'.

Axiom.10.D   *rexteq is a relation.*
        z rexteq z' if and only if $z'\preceq z$ and $z\preceq z'$.

It follows from the axiom 10.B-C that $rginc$ is a preordering and $resteq$ an equivalence.

It is convenient to introduce particular operations giving the extensions corresponding to the canonical structures, namely:

ext   (the *extension* of a quality),
graph   (the *graph* of an operation),
dom   (the *domain* of a operation),
img   (the *range*, or image of an operation),
grsys   (the *graph* of a system),
grrel   (the *graph* of a relation).

We assume the natural axioms on these operations:

Axiom.10.E   *ext, graph, dom, img, grsys and grrel are operations.*

Axiom.10.F   *ext q=C if and only if q is a quality and C=cext(rqual,q);*
        graph f=C if and only if f is an operation and C=cext(rop,f);
        dom f=C if and only if f is an operation and C=cext(rdom,f);
        img f=C if and only if f is an operation and C=cext(rval,f);
        grsys S=C if and only if S is a system and C=cext(rsys,S);
        grrel r=C if and only if S is a relation and C=cext(rrel,r).

By the axiom 4.D, two systems S and T are equal if $(rsys, S) \approx (rsys, T)$; hence it seems convenient to introduce, beside *cext*, the operation *syext* which associates to suitable relational pairs the corresponding systems, *provided they exist*:

Axiom.10.G   *syext is an operation;*
$\qquad$ *syext $z=S$ if and only if S is a system and $z \approx (rsys, S)$.*

We shall discuss further the domains of the operations *cext* and *syext* in Section 3, after introducing the notion of general cardinality.

III.2. Relative cardinality

We have introduced cardinal numbers in order to obtain a measure of the size of objects depending on the *relative extensions*. To this aim we assume the existence of the operation *gcard (general cardinality)*, and we state the following axiom:

Axiom.11.A   *gcard is an operation.*
$\qquad$ *gcard is defined at z if and only if z is a relational pair, and the values of gcard are cardinal numbers.*

We call $gcard(r, x)$ the general cardinality of (r, x), or the *relative cardinality of x with respect to r*, and we denote it by $(\overline{\overline{r,x}})$; when the relation r is unambiguously fixed by the context, we shall often write simply $\overline{\overline{x}}$ instead of $(\overline{\overline{r,x}})$, and call it the cardinality of x.

The following axioms ensure that *gcard* actually measures the relative extensions.

Axiom.11.B $gcard(r, x) \leq gcard(r', y)$ *whenever* $(r, x) \preceq (r', y)$.

Axiom.11.C $gcard(rdom, f) = gcard(rop, f)$, *for any operation f.*

Axiom.11.D $gcard\ (rrel, r) = gcard(rrel, r^{-1})$, *for any relation r.*
$\qquad gcard(rsys, S) = gcard(rsys, S^{-1})$, *for any system S.*
$\qquad gcard(rop, f) = gcard(rop, f^{-1})$, *for any injective operation f.*

When C is a collection we call cardinality of C the cardinal $gcard(rcoll, x)$, and we usually denote it by $|C|$. It is easily seen that $gcard(r, x) = |cext(r, x)|$ whenever $cext(r, x)$ exists.

For this reason and also for its historical importance, it seems better to single out the "natural" cardinality of collections. Thus we introduce the operation *card*, and we postulate

Axiom.11.E   *card is an operation.*
$\qquad$ *card $C = \alpha$ if and only if C is a collection and $\alpha = gcard\ (rcoll, C)$.*

We classically identify "equicardinality" with "equipotency", but only for collections with cardinalities strictly less than $\daleth$:

Axiom.11.F *If $|A| = |B| < \daleth$, then there is an injective operation f such that $A=dom\ f$ and $B=img\ f$.*

Notice that in this case we have $|A| = |B| = gcard(rop, f)$. We assume axiom 11.F only below $\daleth$, according to our conservative criterion: in fact, we will not postulate the existence of similar connections between objects of maximal cardinality.

After introducing in Chapter IV the operations of union, intersection and Cartesian product on collections, we shall relate them with cardinal operations. Here we simply state an axiom giving objects of cardinality $0, 1, 2, \aleph_0$ and $\daleth$.

Axiom.11.G $|\emptyset| = 0, |\{x\}| = 1, |\{x, y\}| = 2$ *if* $x \neq y, |\mathbb{N}| = \aleph_0$ *and* $\overline{\overline{qobj}} = \daleth$.

It follows from the axioms 10.B-G that $gcard(r, x) = 0$ if and only if there is no object t verifying $t\ r\ x$, and $gcard(r, x) = 1$ if and only if there is exactly one t such that $t\ r\ x$.

We conclude with an "inaccessibility" property of $\daleth$, inspired by the replacement axiom, in addition to axiom 6.F of Chapter Il Section 3.

Axiom.11.H *If $gcard(rop,f) < \daleth$, then $gcard(rval,f) < \daleth$.*

III.3. The principle of limitation of size

Since we have now at our disposal the general cardinality of any relational pair, we can make explicit the main role of the cardinal $\daleth$ in our theory, by formulating a "Principle of Limitation of Size" as a general performability criterion for operations.

Principle of Limitation of Size:

A fundamental operation can be performed at an argument, provided the cardinality of the result is a priori bounded by $\daleth$.

In order to simplify the formalization of the axioms expressing this principle we introduce the qualities *qsmall* and *qlarge*, of being a *small object* and a *large object* respectively. These qualities are ruled by the following axiom:

Axiom.12.A   *qsmall and qlarge are qualities.*
   Let $S_i = (q_i, r_i)$ $(1 \leq i \leq 5)$ be the fundamental structures of qualities, relations, operations, collections and systems. If $q_i x$ then: qsmall x if and only if $gcard(r_i, x) < \daleth$; qlarge x if and only if $gcard(r_i, x) = \daleth$.

We say that an object is *small* if it enjoys *qsmall*, that it is *large* if it enjoys *qlarge*: in particular a collection C is small if $|C| < \daleth$, large otherwise.

As first application of our principle of limitation of size we state the:

Axiom.12.B   *If qrepa z and gcard $z < \daleth$, then cext $\downarrow z$.*
   If moreover $z \preceq (rqual, qpair)$, then syext $\downarrow z$.

It follows from axioms 12.A-B that the operations *ext, graph, dom, img, grel, grsys* are always defined at small arguments.

IV. Operations on Collections and Systems

IV.1. Operations on collections

Introducing the A-Theory we pointed out that non-extensional objects are very expressive, while extensional ones are both easily tractable and freely "manufacturable". Following this criterion, we introduce here many operations (non-extensional objects) which perform on collections and systems the classical set-theoretical and function-theoretical constructions.

In the formulation of the axioms describing the behaviour of these operations we follow our prudential principle of limitation of size, hence we give axioms stating that *any such operation can be performed on an object, provided the result is a priori small.*

The possibility of applying these operations when the results can be large, is neither excluded nor assumed in the A-Theory: various necessary and sufficient conditions are considered in Chapter VI.

In order to simplify the enunciation of the axioms of this chapter and of the following ones, we call *collection of collections (of systems, of pairs, of ...)*, a collection whose *members are all collections (systems, pairs, ...)*. Similarly we call *system of collections (of systems, of operations, of ...)*, a univalent system whose *projections are all collections (systems, operations, ...)*.

The operations union, intersection and Cartesian product are usually performed on univalent systems of collections; to this aim we introduce the operations *union (un), intersection (int)* and *Cartesian product (cart)* described by the following axioms.

Axiom.13.A   *(union, intersection, and Cartesian product of a system of collections) un, int, and cart are operations.*
un $S=C$ if and only if S is a univalent system of collections and C is the collection whose members are exactly the objects belonging to S_i for some index i.
int $S=C$ if and only if S is a univalent system of collections and C is the collection whose members are exactly the objects belonging to S_i for each index i.
cart $S=C$ if and only if S is a univalent system of collections and C is the collection whose members are exactly the univalent systems T having the same indices of S and such that $T_i \in S_i$ for any index i.

Whenever the left hand side is defined we put $un\,S = \cup_i S_i$, $int\,S = \cap_i S_i$, $cart\,S = \mathcal{P}i_i S_i$; if moreover $S = (C, D)$ is a pair, we put $un\,S = C \cup D$, $int\,S = C \cap D$ and $cart\,S = C \times D$; if S is a n-tuple of collections all equal to C, then we put $cart\,S = C^n$.

Applying the principle of limitation of size we state the following axiom:

Axiom.13.B   *If S is a small univalent system of small collections, then both un and cart are defined at S and $\cup_i S_i$, $\mathcal{P}i_i S_i$ are small.*
If S is a univalent system of collection and $|S_i| < \daleth$ for some index i, then int is defined at S.
Moreover $|C \cup D| + |C \cap D| = |C| + |D|$ for any two collections C and D.

Union and intersection are often performed also on collections of collections, therefore we introduce also the operations *unc* and *intc*.

Axiom.13.C   *(union and intersection of a collection of collection)*
unc and intc are operations.
unc $C=D$ if and only if C is a collection of collections, and D is a collection whose members are exactly the objects belonging to some member of C.
intc $C=D$ if and only if C is a collection of collections, and D is a collection whose members are exactly the objects belonging to each member of C.

Axiom.13.D   *If C is a small collection of small collection, then unc $\downarrow C$.*
If C is a collection of collections, and C has a small member, then intc $\downarrow C$.

Whenever the left hand side is defined, we put $unc\,C = \cup C$, $intc\,C = \cap C$; notice that if A and B are collections, then $unc\{A, B\} = un(A, B) = A \cup B$, and $intc\{A, B\} = int(A, B) = A \cap B$.

Note that if C satisfies the conditions of the axiom 13.D, then $\cup C$ is small.

Next we introduce the operation of difference between two collections, the operation of power of two collections, and the power-set operation $\wp$ giving the collection of the parts of a collection.

Axiom.13.E   *diff, pow and $\wp$ are operations. (i.e. qop diff, qop pow, qop $\wp$).*

Axiom.13.F   *(difference)*
diff $S=E$ if and only if S is a pair of collections, and E is a collection whose members are exactly the objects belonging to S_1 and not belonging to S_2.

Axiom.13.G   *(power)*
pow $S=E$ if and only if S is a pair of collection, and E is a collection whose members are exactly the univalent systems, whose indices are all member of S_1, and whose projections belong to S_2.

Axiom.13.H   *(power set operation)*
$\wp C = D$ if and only if C and D are collections and the members of D are exactly the subcollections of C.

Whenever the left hand side is defined, we put $diff(C, D) = C \backslash D$, and $pow(C, D) = C^D$.

Axiom.13.I   *If both C and D are small, then pow is defined at (C,D). If C is small, then $diff{\downarrow}\,(C, D)$, $\wp \downarrow C$ and $\wp C$ is small.*

Axioms 12.B and 13.B yield that C^D is small if both C and D are small.

The clause $|\wp C| < \beth$ if $|C| < \beth$ could be omitted from axiom 13.I, but this would require axioms giving the "natural" bijection between $\wp C$ and $\{0, 1\}^C$, which we will introduce in the sequel.

Several other clauses of the axioms of this chapter, asserting that an operation is defined at small arguments and gives small results, could be omitted on the ground of axioms 15.C of Section 3, 16.C-D of Chapter V. We prefer, however, to give here an axiomatization independent of the existence of universes.

We conclude this section with some remarks on the results of the above operations when the empty collection $\emptyset$ and the empty system $\oslash$ are involved.

First of all $intc\,\emptyset$ and $int\,\oslash$ exist if and only if there is the *total collection* Obj of all objects. Moreover one has $unc\,\emptyset = un\,\oslash = \emptyset, \emptyset^A = \emptyset$ if A is a non empty collection, and $cart\,\oslash = A^\emptyset = \{\oslash\}$, for any collection A.

IV.2. Operations on systems

One of the most important operations on systems is *compsys*, the composition of pairs of systems, satisfying

Axiom.14.A *compsys is an operation.*
compsys $S=T$ if and only if T is a system, S is a pair of systems, and $i\,T\,x$ exactly when $i\,S_2\,z$ and $z\,S_1\,x$ for some z.

Whenever $compsys(U,V)$ exists, we denote it by $U \circ V$, and we call it the composition of the system U with the system V.

Notice that if both U and V are univalent, so are $U \circ V$ and $(U \circ V)_i = U_{(V_i)}$.

A sufficient condition for the existence of the composition of two systems is given by:

Axiom.14.B *The composition of two systems U and V exists, and is small, if V is small and U is either univalent or small.*

Two kinds of products are naturally involved with systems: therefore we introduce the operations *fibsys* and *tensys* called fibred and tensor (or binary) product, satisfying

Axiom.14.C.1 *fibsys and tensys are operations.*
C.2 *fibsys transforms univalent systems of systems into systems. fibsys $S=T$ if and only if iTx exactly when x is a univalent system with the same indices as S and $i\,S_j x_j$ for any index j of S.*
C.3 *tensys transforms univalent systems of systems into systems. tensys $S=T$ if and only if iTx exactly when i and x are univalent systems with the same indices as S and $i_j S_j x_j$ for any index j of S.*

We denote the fibred product of a system S by $\langle S_i \rangle_i$, and its tensor product by $\otimes_i S_i$. When S is an n-tuple we also write $\langle S_1, \ldots, S_n \rangle$ and $S_1 \otimes \ldots \otimes S_n$ for *fibsys S* and *tensys S*.

If all systems S_i are univalent, then so are both $\langle S_i \rangle_i$ and $\otimes_i S_i$. In particular one has

$$\langle S_1, \ldots, S_n \rangle_j = ((S_1)_j, \ldots, (S_n)_j)$$
$$(S_1 \otimes \ldots \otimes S_n)_{(j_1, \ldots, j_n)} = ((S_1)_{j_1}, \ldots (S_n)_{j_n})$$

(note that in the first expression j must be a common index of all systems S_i, for $1 \le i \le n$, while in the second one j_i runs over all indices of S_i).

Axiom.14.D *Let S be a small univalent system of small systems: then tensys is defined at S, and tensys S is small.*
Moreover, for fibsys being defined at S it is sufficient that at least one projection of S is small.

The last group of operations on systems we introduce is

aggsys (*aggregation* of systems),
fagsys (*fibred aggregation* of systems),
sepsys (*separation* of systems),
transys (*transposition* of systems).

The axioms ruling them are

Axiom.14.E.1	*aggsys, fagsys, sepsys, and transys are operations.*

E.2	*aggsys transforms univalent systems of systems into systems indexed by pairs.*
aggsys $S=T$ if and only if $(i,j)\ T\ z$ exactly when $j\ S_i\ z$.

E.3	*fagsys transforms univalent systems of systems into systems.*
fagsys $S=T$ if and only if $i\ T\ z$ exactly when $i\ S_i\ z$.

E.4	*sepsys transforms systems indexed by pairs into univalent systems of non empty systems.*
sepsys $S=T$ if and only if $j\ T_i\ z$ exactly when $(i,j)\ S\ z$.

E.5	*transys transforms univalent systems of systems into univalent systems of non empty systems.*
transys $S=T$ if and only if $j\ T_i\ z$ exactly when $i\ S_j\ z$.

Note that the separation and the transposition of a system are defined only up to empty systems, thus we have inserted the clause of having no empty projection in the axiom 14.E.4-5.

If U is a univalent system of univalent systems, then *aggsys* $U=A$, *fagsys* $U=F$, *sepsys* $U=S$, and *transys* $U=T$ are all univalent systems provided they exist; moreover, for suitable indices i, j

$$A_{(i,j)} = (U_i)_j, \quad F_i = (U_i)_i, \quad (S_i)_j = U_{(i,j)}, \quad (T_i)_j = (U_j)_i \ .$$

Finally, we remark that if *fibsys* $\oslash$ exists, it is the univalent system having all objects as indices, and all projections equal to $\oslash$; on the other hand *tensys* $\oslash$ is the unitary system $\left(\begin{smallmatrix} \oslash \\ \oslash \end{smallmatrix} \right) = gus(\oslash, \oslash)$, and *aggsys* $\oslash = fagsys\ \oslash = sepsys\ \oslash = transys\ \oslash = \oslash$.

We conclude with the sufficient conditions derived from the principle of limitation of size:

Axiom.14.F	*Let S be a small univalent system of small systems: then aggsys, fagsys and transys are defined at S, and their results are small. Let S be a small pair-indexed system: then sepsys is defined at S and its result is a small system.*

IV.3. Restrictions

We introduce here two operations acting on pairs of a non-extensional and an extensional object, namely the *restriction of an operation to a collection, oprest,* and the *bilateral restriction of a relation to a pair of collections, birest.*

Axiom.15.A	*oprest is an operation, and if oprest $\downarrow z$, then $z=(f,C)$ where f is an operation and C a collection.*
If oprest $(f,C)=g$, then g is an operation such that $y= gx$ if and only if $x \in C$ and $y = fx$.

Axiom.15.B *birest is an operation, and if birest $\downarrow$ z, then $z=(r,(C,D))$ where r*
is a relation and C, D are collections.
If $birest(r,(C,D))=r'$, then r' is a relation such that $x\,r'\,y$ if and
only if $x\,r\,y$, $x \in C$ and $y \in D$.

Whenever $oprest(f,C)$ is defined, we call it the *restriction of f to the collection C*, written $f_{|C}$; similarly we call *bilateral restriction of r to C and D*, written ${}_{C|}r_{|D}$, the relation *birest$(r,(C,D))$*.

We give the usual sufficient conditions:

Axiom.15.C *If f is an operation, C is a collection and either f or*
C are small, then oprest $\downarrow$ (f,C).
If r is a relation, C and D are collections and either r or
$C \times D$ are small, then birest $\downarrow$ $(r,(C,D))$.

Note that, when the restrictions are defined, one has $\overline{\overline{f}}_{|C} \leq \overline{\overline{f}}$, $\overline{\overline{f}}_{|C} \leq |C|$, and ${}_{C|}\overline{\overline{r}}_{|D} \leq \overline{\overline{r}}$, ${}_{C|}\overline{\overline{r}}_{|D} \leq |C \times D|$; hence the conditions of axiom 15.C are sufficient to guarantee that the restrictions are small.

Since operations and relations are non-extensional objects, it seems convenient to give some coherence constraints.

Axiom.15.D $(f_{|C})_{|D} = f_{|C \cap D}$; ${}_{D|}({}_{C|}r_{|C'})_{|D'} = {}_{C \cap D|}r_{|C' \cap D'}$; $({}_{C|}r_{|C'})^{-1} =$
${}_{C'|}r^{-1}{}_{|C}$.
If $(rdom,f) \preceq (rcoll,C)$, then $f_{|C} = f$.
If $x\,r\,y$ implies $x \in C$ and $y \in D$, then ${}_{C|}r_{|D} = r$.

We now introduce an operation giving the images: we call it *hat*, since following [7] we denote by $\hat{f}(C)$ the image of a collection C under an operation f.

Axiom.15.E *qop hat, and hat $x=y$ if and only if $x=(f,C)$, f is an operation,*
C and y are collections, and $t \in y$ exactly when there is $s \in C$
such that $t=fs$.

Whenever $hat(f,C)$ exists, we call it the *image of C under f*, written $\hat{f}(C)$; note that $hat(f,C) = img\,f_{|C}$ whenever the latter exists.

We can finally introduce domains, codomains and fields of relations by means of the operations *domrel*, *codrel* and *fieldrel*, and index-collections and images of systems, by means of *indsys* and *imgsys*.

Axiom.15.F *domrel $r=C$ if and only if C is a collection whose members are*
exactly the objects x such that $x\,r\,y$ for some y.
codrel $r=C$ if and only if $C=$domrel r^{-1}.
fieldrel $r=C$ if and only if $C=$domrel $r \cup$ codrel r.
indsys $S=C$ if and only if C is a collection whose members are
exactly the indices of S.
imgsys $S=C$ if and only if $C=$indsys S^{-1}.

It follows from the previous axioms that all small relations have small domains, codomains and fields, and that all small systems have small index-collections and images, which are given by the following equalities:

$$domrel\ r = \hat{p}_1\,(grrel\ r),$$
$$codrel\ r\ = \hat{p}_2\,(grrel\ r),$$
$$indsys\ S = \hat{p}_1\,(grsys\ S),$$
$$imgsys\ S = \hat{p}_1\,(grsys\ S).$$

V. Universes and Propositions

V.1. Universes

Traditional mathematical activity can be carried out over structures of bounded cardinalities, using the operations introduced in the previous chapters.

We introduce *universes* à la Grothendieck as suitable fields where such bounds are available. Thus we consider the qualities *quniv* of being a *universe* and *qinac* of being an *inaccessible cardinal*, satisfying the following axioms:

Axiom.16.A   *qinac is a quality of cardinals.*

         qinac α if and only if whenever a univalent system has general cardinality strictly less than α and its projections are collections having cardinalities strictly less than α, then both the Cartesian product and the union of the system have cardinalities strictly less than α.

Thus in particular $\daleth$ is inaccessible.

Note that the condition of inacessibility for α implies that if C, D are collections and $|C| < \alpha, |D| < \alpha$, then also $|C \times D| < \alpha, |C^D| < \alpha$, and $|\wp C| < \alpha$.

Axiom.16.B   *quniv is a quality of collections.*

         quniv x implies qinac $|x|$ and $\aleph_0 < |x| < \daleth$.

We introduce some particular subcollections of a universe V, which "relativize" to V the basic concepts of the A-Theory, and three fundamental concepts of the contemporary mathematics, those of *set, map,* and *correspondence*.

Definition

V-qual is the collection of the qualities belonging to V whose extensions are included in V. Its members are the *V-qualities*.

V-rel is the collection of the relations belonging to V whose domains and images are included in V. Its members are the *V-relations*.

V-op is the collection of the operations belonging to V whose domains and images are included in V. Its members are the *V-operations*.

V-coll is the collection of the collections belonging to V and included in V. Its members are the *V-collections*.

V-sys is the collection of the systems belonging to V whose domains and images are included in V. Its members are the *V-systems*.

V-unsys is the subcollection of *V-sys* whose members are univalent systems.

V-card is the collection of all the cardinals belonging to V, which are the general cardinalities of relational pairs, whose first component is a *V-relation*. Its members are the *V-cardinals*.

V-set is the subcollection of *V-coll* whose members have cardinalities strictly less than $|V|$. Its members are the *V-sets*.

V-corr is the subcollection of *V-sys* whose members have cardinality strictly less than $|V|$. Its members are the *V-correspondences*.

V-map is the intersection of *V-unsys* and *V-corr*. Its members are the *V-maps*.

When f is a *V-map* and x is an index of f, we follow the common usage and denote f_x also by fx and $f(x)$.

Axiom.16.C *Every subcollection of V having cardinality strictly less than $|V|$ is a V-set, and also the extension of some V-quality.*
Every system S having domain and image included in V, and cardinality strictly less than $|V|$, is a V-correspondence; more-over (rsys,S)≈(rrel,r) for some V-relation r.
If f is a V-map then (rsys,f)≈(rop,g) for some V-operation g.
If E is a V-set then its cardinality is a V-cardinal.

We conclude this section with an axiom which guarantees that every small object can be freely handled inside some universe:

Axiom.16.D   *Every small collection is a V-set for some universe V.*

Note that a universe is a natural model of the Frame Theory Q of [7], provided it verifies the supplementary condition that *V-op, V-coll, V-map,* and $\mathbb{N}$ are pairwise disjoint. In particular any universe is a natural model of the theory ZF_0A (Zermelo-Fraenkel set theory with urelements and without foundation): therefore the hereditarily well-founded sets give a model of ZF.

V.2. Propositions

In this section we introduce inside the A-Theory the structure of *propositions*, as a basis of a sort of classical propositional calculus. This frame is suitable to various strengthenings, and allows embeddings of many logical and linguistic structures as inner objects of the A-Theory. Here we restrict ourselves to a weak axiomatization: wider developments can be subject of separate papers.

The qr-structure of propositions is $\mathcal{S}_7 = (qprop, rprop)$, and $rprop$ relates a "truth value" to each proposition. Therefore we introduce beside $\mathcal{S}_7$ the qualities $qtrue$ and $qfalse$, of b*eing a true proposition* and of *being a false proposition*. These objects are ruled by the following axioms:

Axiom.16.AA $\quad$ *$\mathcal{S}_7$ is a qr-structure.*
$\qquad\qquad$ *If qprop p then there is exactly one v such that v rprop p, and v is equal either to 1 or to 0.*

Axiom.16.BB $\quad$ *Both qtrue and qfalse are qualities.*
$\qquad\qquad$ *qtrue p if and only if 1 rprop p.*
$\qquad\qquad$ *qfalse p if and only if 0 rprop p.*

We say that an object p is a proposition whenever $qprop$ p, and we call *truth value* of p the unique v such that v $rprop$ p.

We say that a proposition p *is true* if $qtrue$ p, and that p *is false* if $qfalse$ p.

Next we introduce the operations corresponding to the usual propositional connectives, namely

$\quad$ *et* $\quad$ (*conjunction* of a system of propositions),
$\quad$ *vel* $\quad$ (*disjunction* of a system of propositions),
$\quad$ *non* $\quad$ (*negation* of a proposition).

Their truth values are ruled by the natural axioms:

Axiom.16.CC $\quad$ *non is an operation defined at each proposition, having proposi-*
$\qquad\qquad$ *tions as values.*
$\qquad\qquad$ *non p=s implies that s is true exactly when p is false.*

Axiom.16.DD $\quad$ *Both et and vel are operations whose values are propositions, and*
$\qquad\qquad$ *they are defined only at univalent systems of propositions.*
$\qquad\qquad$ *If vel P=p then p is true exactly when some projection of P is true.*
$\qquad\qquad$ *If et P=p then p is true exactly when all projections of P are true.*

Axiom.16.EE $\quad$ *Both vel and et are defined at any small univalent system of*
$\qquad\qquad$ *propositions.*

We put $vel\ P = \vee_i P_i$, and $et\ P = \wedge_i P_i$, whenever the left hand side is defined, and we write $\neg p$ for *non p*.

When P is a pair of propositions we also write $P_1 \vee P_2$ and $P_1 \wedge P_2$, for *vel* P and *et* P, and $P_1 \to P_2$ for $(\neg P_1) \vee P_2$, $P_1 \leftrightarrow P_2$ for $(P_1 \to P_2) \wedge (P_2 \to P_1)$.

Finally we introduce the relation $rbid$, which connects any triple (x, r, y), where r is a relation, to the proposition *bidding* that x *is in the relation r with y*.

Axiom.16.FF   *rbid is a relation.*
If z rbid p then p is a proposition and z is a triple whose second projection is a relation.
For any relation r and any pair (x,y) there is a unique proposition p such that (x,r,y) rbid p.

The unique proposition p such that $(x, r, y) rbid\, p$ is said to *bid x r y*, and it is denoted by "x r y". Moreover whenever q is a quality we write also "q x" for "x rqual q".

Notice that we have not assumed that "x r y"="$x'r'y'$" implies $(x, r, y) = (x', r', y')$; we only postulate the natural condition on their truth-values, namely:

Axiom.16.GG   "x r y" *is true if and only if x r y.*

VI. Strengthenings of the A-Theory

In the previous chapters we have given a basic axiom system, which can be strengthened in various directions. We present in this chapter several groups of axioms which indicate different interesting ways of extending the A-Theory.

The extension of the A-Theory obtained by assuming simultaneously all the axioms of this chapter is clearly contradictory: classical antinomies can already be derived by adding small groups of these axioms to the A-Theory.

Two kinds of problems naturally arise: that of isolating groups of contradictory axioms, either generating in new ways classical antinomies or leading to new antinomies, and that of selecting large groups of axioms which, added to the A-Theory, lead to systems whose consistency can be reduced to that of some classical foundational theory (e.g. Zermelo-Fraenkel set theory, possibly strengthened by means of some large cardinal assumption).

There is a large number of such problems, reflecting several aspects of self-reference. These problems can be of different type and degree of difficulty: some of them are the subject of a specific research in progress. In this chapter we simply suggest them (at least implicitly) by giving a list of strengthening axioms followed by a few comments.

It is interesting to notice that while some of these axioms are surely relatively consistent and even unavoidable (at least relativized to some universe) in order to develop modern mathematics, other axioms leave a very problematic consistency status, and are introduced here in order to suggest investigations on extremal capabilities of self-reference.

VI.1. Restrictive axioms

As pointed out before, we have been very cautious in introducing large non-extensional objects, and we have assumed no axiom implying the existence or large collections and systems.

A final word in this restrictive direction would be to assume:

RA.1 *All collections and systems are small, and all but a finite number of qualities, relations and operations are small.*

This axiom excludes operations generating infinitely many large objects, such as the composition of any pair of operations. One could even fix a precise finite bound for the number of qualities, relations and operations having maximal cardinality, say 1,000.

A less restrictive axiom, which allows for operations generating infinitely many large objects but maintaining a good control over them, is:

RA.2 *qlarge is a small quality. (i.e. qsmall qlarge).*

VI.2. Axioms of identification and separation of kinds

A totally different kind of restrictive axioms deals with the connections between the various types of fundamental objects of the A-Theory. As pointed out at the end of Chapter II, we assumed only the axiom of "structural consistency", so as to rule out the existence of "ambiguous" objects which are, say, simultaneously collections and qualities, but behave differently when related to *rcoll* and *rqual*.

The set-theoretical reductionism being well known, we propose an alternative possibility of identifying different kinds of objects, beginning with the following *identification axiom:*

IA.1 *Any collection is a quality; any operation and any system is a relation.*

A radical reductionism is reached by adding to IA.1 the following axiom IA.2, where the *relations* are identified with *qualities of pairs* (i.e. with their graphs), the *cardinals* with *qualities of qualities* (e.g. à la Frege-Russell with the quality of all equicardinal qualities) and the *propositions* with their *truth-qualities.*

IA.2 *Any relation is a quality of pairs, any cardinal number a quality of qualities, and there are only two propositions, identified with the qualities qtrue and qfalse (and qtrue qtrue, qfalse qfalse).*

Thus, all structured objects are qualities. Of course, one could also choose a different kind of basic objects, e.g. collections or relations.

The opposite attitude is also possible, and we can postulate instead of IA.1-2 the following *separation axiom:*

SA.1 *Let $q_i, 1 \leq i \leq 7$, be the fundamental qualities qqual, qrel, qop, qcoll, qsys, qcard, qprop: then $q_i \wedge q_j$ is an empty quality whenever $i \neq j$.*

Note that the identification axioms IA.1-2 have the somehow unpleasant feature of excluding the possibility of consistently codifying paradoxical extensions, playing different keyboards. E.g. we do not exclude, in the A-Theory, the

"R*ussell quality of collections*" *qrussc* verifying

    qrussc C if and only if qcoll C and $C \notin C$,

which cannot have a collection-extension, and the "Russell collection of qualities"

    Russq= $\{q| qqual\ q \wedge q\ rq\!\!\not|al\ q\}$

which cannot be the extension of any quality.

However, the task of coding by means of relational pairs all "describable extensions" is obviously hopeless for no "*Russell relational pair*" *(russ,b)* verifying

    (r,x) russ b $\leftrightarrow$ (r, x) $\not\in$ x for every relational pair (r,x)

can exist. This generalizes in our formalism the antinomy (3) of Principia Mathematica (see [22], Introduction, p.60).

VI.3. Axioms of comprehension

In order to formulate some strong axioms of comprehension in our frame, we introduce first the following qualities of qr-structures:

    qextst   (to be an *extensional qr-structure*)
    qabscom   (to be an *absolutely comprehensive qr-structure*)

whose meaning is given by the following axioms:

CA.1   *qextst and qabscom are qualities of qr-structures.*

CA.2   *qextst (q,r) if and only if $((r, u) \approx (r, v) \wedge qu \wedge qv) \rightarrow u = v$.*

CA.3   *qabscom (q,r) if and only if $qrepa\ z \rightarrow \exists x(qx \wedge (r, x) \approx z)$.*

We say that a qr-structure $\mathcal{S}$ is *extensional* if *qextst* $\mathcal{S}$, and that $\mathcal{S}$ is a*bsolutely comprehensive* if *qabscom* $\mathcal{S}$.

The simplest form of comprehension axioms are:

CA.4   *There exists an absolutely comprehensive qr-structure.*

CA.5   *There exists an extensional absolutely comprehensive qr-structure.*

More informative are the following axioms, which postulate that some fundamental structures are comprehensive:

CA.6   *(qqual, rqual) is absolutely comprehensive.*

CA.7   *(qcoll, rcoll) is absolutely comprehensive.*

Instead of considering all relational pairs, one can take into account only those of the kinds $(rrel, r)$ or (rop, f). Thus we introduce the qualities:

$qrepabin$ (to be a *binary* relational pair),
$qrepaop$ (to be an *operative* relational pair),
$qbincom$ (to be a *binarily comprehensive* qr-structure),
$qopcom$ (to be an *operatively comprehensive* qr-structure),

whose meaning is given by the following axioms:

CA. 8   *qrepabin and qrepaop are qualities of relationals pairs,*
         qbincom and qopcom are qualities of qr–structures.

CA.9   *qrepabin z if and only if qrepa $z \wedge z \preceq (rqual, qpair)$.*

CA.10   *qrepaop z if and only if qrepabin $z \wedge (z = (r, x) \wedge (u, v)rx \wedge (u, w)r$
        $x \to v = w)$.*

CA.11   *qbincom (q,r) if and only if qrepabin $z \to \exists x(q\, x \wedge (r, x) \approx z)$.*

CA.12   *qopcom (q,r) if and only if qrepaop $z \to \exists x(q\, x \wedge (r, x) \approx z)$.*

The corresponding comprehension axioms are:

CA.13.1   *(qrel,rrel) is binarily comprehensive.*
  13.2   *(qop,rop) is operatively comprehensive.*

CA.14.1   *(qsys,rsys) is binarily comprehensive.*
  14.2   *(qunsys,rsys) is operatively comprehensive.*

One can introduce the operation general extension, *gext*, giving the *relative extension* of a relational pair with respect to an *extensional* qr-structure:

CA.15   *qop gext $\wedge (rdom, gext) \approx (rqual, qextst) \wedge (rval, gext) \approx (rqual, qop)$.*
        gext $(q,\ r)z = y \leftrightarrow (qrepa\ z \wedge z \approx (r, y))$.

The operation *gext $\mathcal{S}$* will be denoted by $\mathcal{S}$-*ext* and called the *relative extension operation* with respect to the qr-structure $\mathcal{S}$.

Note that $cext = (qcoll, rcoll)$-*ext and* $syext = (qsys, rsys)$-*ext,* thus the axiom of comprehension CA.7 says that any relational pair has a collection-extension, while CA.14.1-2 say that any binary (respectively operative) relational pair has a system-extension.

VI.4. Selectors and choice operations

It is impossible to determine completely non-extensional objects, like relations, qualities and operations, by simply describing their behaviour. This being often useful, we introduce the operations *selq, selr and selop* (the *selectors* of qualities, relations and operations) and we assume the following axioms of *selection*:

AS.1 $qop\ selq \wedge (rdom, selq) \approx (rqual, qqual).$
$qqual\ x \to (rqual, x) \approx (rqual, selq\ x).$
$(qqual\ x \wedge qqual\ y \wedge (rqual, x) \approx (rqual, y)) \to selq\ x = selq\ y.$

AS.2 $qop\ selr \wedge (rdom, selr) \approx (rqual, qrel).$
$qrel\ x \to (rrel, x) \approx (rrel, selr\ x).$
$(qrel\ x \wedge qrel\ y \wedge (rrel, x) \approx (rrel, y)) \to selr\ x = selr\ y.$

AS.3 $qop\ selop \wedge (rdom, selop) \approx (rqual, qop).$
$qop\ x \to (rop, x) \approx (rop, selop\ x).$
$(qop\ x \wedge qop\ y \wedge (rop, x) \approx (rop, y)) \to selop\ x = selop\ y.$

We can also introduce a *general selector, gsel,* and a *homogeneons selector, hsel,* which associate to any (possibly non-extensional) structure $\mathcal{S}$ a corresponding *specific $\mathcal{S}$-selector, $\mathcal{S}$-gsel=gsel $\mathcal{S}$,* and a *homogeneous $\mathcal{S}$-selector, $\mathcal{S}$-hsel=hsel $\mathcal{S}$.* The general selector *gsel* is ruled by:

AS.4 $qop\ gsel \wedge (rdom, gsel) \approx (rqual, qqrs) \wedge (rval, gsel) \preceq (rqual, qop).$
$gsel\ (q,r){\downarrow}\ z \leftrightarrow \exists x(qx \wedge (r, x) \approx z).$
$(gsel(q,r)z = y \wedge z' \approx z) \to gsel(q,r)z' = y.$

The *homogeneous (q,r)-selector, (q,r)-hsel=hsel(q,r),* is obtained by restricting the action of the specific (q, r)-selector (q, r)-*gsel* to relational pairs (r, x), according to the following axiom:

AS.5 $qop\ hsel \wedge (rdom, hsel) \approx (rqual, qqrs) \wedge (rval, gsel) \preceq (rqual, qop).$
$hsel\ (q,r){\downarrow}\ x \leftrightarrow q\ x.$
$hsel(q,r)x = gsel(q,r)(r, x).$

Notice that the specific $\mathcal{S}$-selectors corresponding to the fundamental structures of qualities, relations and operations when applied to pairs $(rcoll, C)$, give a sort of *inverses* of the operations *ext, grrel,* and *graph* of Chapter III Section 1. In fact, they select a quality of given extension and a relation or an operation of given graph. Moreover, the previously introduced selectors *selq, selr* and *selop* can be viewed as the homogeneous $\mathcal{S}$-selectors corresponding to the fundamental structures of qualities, relations and operations. Note also that if $\mathcal{S}$ is an extensional structure, then $\mathcal{S}$-*gsel* and $\mathcal{S}$-*gext* are equivalent operations.

We can also introduce a *choice operation, choice,* which selects an element out of any non-empty collection:

AC.1 $qop\ choice \wedge (choice \downarrow x \leftrightarrow qcoll\ x \wedge x \neq \emptyset) \wedge (choice\ x = y \to y \in x).$

Since we have not assumed in the A-Theory that the structure of collections is absolutely comprehensive, one can strengthen AC.1 by assuming a *generalized choice operator, gchoice,* defined at all non empty relational pairs:

AC.2 *qop gchoice$\wedge$*
$(gchoice \downarrow z \leftrightarrow (qrepa\, z \wedge z \not\approx (rcoll, \emptyset))) \wedge (gchoice(r,x))rx$.

Instead of AC.1-2 many weaker forms of the axiom of choice can be postulated. We point out only the classical *principle of depending choice*, which seems necessary in modern analysis and topology:

DC *If r is a relation, C a non-empty collection, and for any $x \in C$ there is $y \in C$ such that $x\, r\, y$, then there exists a sequence s of elements of C such that $s_n\, r\, s_{n+1}$, for any $n \in \mathbb{N}$.*

VI.5. Strong axioms on operations

We have given many operations on systems in Chapter IV Section 2, and we have remarked there that they transform univalent systems into univalent systems. We can introduce the corresponding operations on operations, namely *comp (composition), tens (tensor product), trans (transposition), agg (aggregation), fagg (fibred aggregation)* and *sep (separation)*.

We make explicit only the axioms on composition; the axioms ruling the remaining operations can be obtained simply by translating the corresponding operations on systems (see Chapter IV Section 2).

Note that the axioms on operations, like AO.1 below, contain clauses, like 1.3, which seem to be opportune and cannot be derived in the absence of extensionality for operations; for the same reason 1.2 is formulated as a one-sided implication.

AO.1 *(composition)*

   1.1 *comp is an operation transforming any pair of operations into an operation.*

   1.2 *comp $(f,g) = h \rightarrow (((h \downarrow x \leftrightarrow (g \downarrow x \wedge f \downarrow gx)) \wedge hx = f(gx)))$.*

   1.3 *comp is associative and comp (id,f)=comp (f,id)=f.*

Similar axioms on *sep, fib, trans, agg, fagg* and *tens* allow the construction of a great variety of operations. E.g. applying the separation *sep* to the projection *psys*, we obtain *(sep psys)$i = gp_i$*, the *general projection* of univalent systems, verifying $gp_i S = S_i$ whenever i is an index of the univalent system S. We can postulate directly *gp* if *sep* is not available; note that p_1 and p_2 are equivalent to the restrictions of gp_1 and gp_2 to all pairs.

We also introduce directly three interesting operations associating an operation to any object, namely Dirac's δ, Curry's K, and the *pairing operator, opair*:

AO.2 *δ is an operation defined at any object x.*
   δx is an operation defined at y if and only if y is an operation defined at x, and $(\delta x)y = yx$.

AO.3   *K is an operation defined at any object x.*
Kx is an operation, and $(Kx)y=x$ for any object y.

AO.4   *opair is an operation defined at any object x.*
opair x is an operation and (opair $x)y=(x, y)$ for any object y.

VI.6. Strong axioms on qualities and relations

A few interesting manipulations on qualities and relations are missing a corresponding object in the A-Theory. In fact, there is no operation giving the negation of a quality or of a relation, or the domain of a relation or the conjunction of two relations. One would like to overcome this lack in the self-referential power of the A-Theory by introducing the relation *rdomrel* and the operation *notq*, *notr* and *conjr* by the axioms:

AQ.1   *(negation of qualities)*
$qop\ notq \wedge (notq \downarrow q \leftrightarrow qqual\ q) \wedge qqual(notq\ q) \wedge ((notq\ q)x \leftrightarrow \bar{q}x)$.
$notq(notq\ q)=q$.

AR.1   *(negation of relations)*
$qop\ notr \wedge (notr \downarrow r \leftrightarrow qrel\ r) \wedge qrel(notr\ r) \wedge (x(notr\ r)y \leftrightarrow x\bar{r}y)$.
$notr(notr\ r) = r \wedge notr(r^{-1}) = (notr\ r)^{-1}$.

AR.2   *(conjunction of relations)*
$qop\ conjr.\ conjr \downarrow z \leftrightarrow (z = (r',r) \wedge qrel\ r' \wedge qrel\ r)$.
$conjr(r',r) = r'' \rightarrow (x\,r\,y \leftrightarrow x\,r'\,y \wedge x\,r''\,y)$.
$conjr(r',r'') = conjr\ (r'',r') \wedge conjr\ (r',conjr(r'',r'''))$
$= conjr(conjr(r',r''),r''')$.
$conjr(r'^{-1},r''^{-1}) = (conjr(r',r''))^{-1} \wedge conj(r,r) = r$.

AR.3   *(domain of relations)*
$qrel\ rdomrel \wedge (x\ rdomrel\ r \leftrightarrow qrel\ r \wedge (r^{-1},x) \not\approx (rcoll,\emptyset))$.

We denote $conjr(r',r'')$ by $r' \wedge r''$, and call it the *conjunction* of r' and r''. We denote also *not q* and *not r* by $\bar{q}$ and $\bar{r}$, and call then the *negation of q* and the *negation of r*, respectively.

Unfortunately, it is easily seen that the A-Theory extended by the axioms AR.1-3 yields the following contradiction:

Antinomy I: $r = rid \wedge rdo\bar{m}rel \rightarrow (rrr \leftrightarrow r\bar{r}r)$.

Another interesting operation on relations is the *composition, comprel,* which can be introduced by postulating:

AR.4 *(composition of relations)*
$$qop\ comprel \wedge (comprel \downarrow z \leftrightarrow (z = (r',r) \wedge qrel\ r' \wedge qrel\ r).$$
$$comprel(r',r'') = r \rightarrow qrel\ r \wedge (x\ r\ y \leftrightarrow \exists t(x\ r''\ t \wedge t\ r'\ y)).$$
$$comprel(r',comprel(r'',r''')) = comprel(comprel(r',r''),r''').$$
$$(comprel(r',r''))^{-1} = comprel(r''^{-1},r'^{-1})$$
$$comprel(r,rid) = comprel(rid,r) = r.$$

We denote the composition $comprel(r',r'')$ by $r' \circ r''$.

Again, one can obtain a contradiction from the A-Theory together with the axioms AR.1, 2, 4, without using *rdomrel*, namely.

Antinomy II: $r = (r\!\!\!/el \wedge rfst^{-1}) \circ rsnd \rightarrow (rrr \leftrightarrow r\!\!\!/r)$.

(The relation r holds between any two non-empty relations x, y if and only if $x\ r\ y \leftrightarrow y\!\!\!/x$).

We leave open the question as to which weakenings of the above axioms give suitable extensions of the A-Theory. E.g. one can admit AR.1, and postulate only that for every partial ordering r there is the corresponding *strict-order*, $r \wedge (\!\!/^{-1})$.

An interesting relation which can be directly postulated is a *linear ordering of all objects*, *rlobj*, verifying:

AR.5 $\qquad qlo\ rlobj,\ and\ (rrel,rid) \preceq (rrel,rlobj).$

Assuming AR.5 one can replace the structure of objects $\mathcal{S}_0$ by the more informative one $\mathcal{S}'_0 = (qobj, rlobj)$.

VI.7. Strong axioms on collections and systems

The axioms CA.7 and CA.14.1 in Section 3 can also be viewed as strong assumptions providing the existence of large collections and systems. In fact CA yields that *cext* is defined at every relational pair and CA.14.1 that *syext* is defined at every binary relational pair, thus giving at once collections corresponding to the extensions of all fundamental qualities, and systems corresponding to the system-extensions of all fundamental relations. E.g.

Obj=ext qobj	*Obj²=ext qpair*	*Sobj=syext(rqual,qpair)*
Qual=ext qqual	*Rel=ext qrel*	*Op=ext qop*
Coll=ext qcoll	*Sys=ext qsys*	*Unsys=ext qunsys*
Card=ext qcard	*Univ=ext quniv*	*Prop=ext qprop.*

Instead of CA.7 and CA.14.1, we can assume individually only part of the above collections and systems. Doing this, we avoid the unfair consequences of the codability of the A-Theory in terms of collections alone (see Sections 1-2 of this chapter).

Other natural axioms are obtained by assuming that all operations of Chapter IV (*unions, intersections, composition, transposition,* etc.) are defined at any collection or system of collections, and system of systems, eliminating the smallness condition throughout the axioms of Chapter IV. This gives an immediate contradiction if it is assumed together with CA.7-CA.14.1, but can be relatively consistent with some weakening of these comprehension axioms.

We single out just one interesting weakening of CA.7-CA.14.1, which makes the collections into a model of Quine's New Foundations:

NF.1 *domsys and grsys are defined at any system, and fibsys at any pair of systems.*
NF.2 *diff, cart and int are defined at any pair of collections.*
NF.3 *syext is defined at any collection of pairs.*
NF.4 *grrel is defined at rid and rincl.*
NF.5 *There is an operation τ defined at any system and verifying:*

$$\tau S = T \leftrightarrow (x\,S\,y \leftrightarrow \{x\}T\{y\}) \ .$$

The axioms NF.1-5 imply that there is the collection *Obj* of all objects, and that *Obj* is closed under all Hailperin's operations (see [15]): since operations like τ or *sing* cannot consistently have a graph, we obtain only a model of NFU (a very special one, indeed, since all small collections are Cantorian).

In order to get a model NF with full extensionality, and in fact of a very strong extension of both ZF and NF recently proposed by A. Oberschelp, we only need a universal hereditary collection; to this aim we can assume:

NF.6 *There is a collection $C = \wp C$.*

VI.8. Restrictions and domains

The restriction-operations are particularly useful when dealing with universes. In addition to axioms postulating the performability of the operations *oprest* and *birest* of Chapter IV Section 3 also at large arguments, we can introduce other kinds of restrictions, namely:

oprestq	(the *restriction of a quality* to a collection),
birestop	(the *bilateral restriction of an operation*),
oplrestrel, oprrestrel	(the *left* and *right restrictions of relations*),
oplrestsys, oprrestsys	(the *left* and *right restrictions of systems*).

R.1   *oprestq, birestop, oplrestrel, oprrestrel, oplrestsys, oprrestsys*
 are operations.

R.2   *oprestq $\downarrow z \leftrightarrow (z = (q, C) \wedge qqual\,q \wedge qcoll\,C)$.*
 oprest$(q, C) = q' \rightarrow (qqual\,q' \wedge (q'x \leftrightarrow qx \wedge x \in C))$.

R.3 $birestop \downarrow z \leftrightarrow (z = (f, (C, D)) \wedge qop\, f \wedge qcoll\, C \wedge qcoll\, D)$.
   $birestop(f, (C, D)) = g \rightarrow (qop\, g \wedge (y = gx \leftrightarrow y = f\,x \wedge y \in C \wedge x \in D))$.

R.4 $oplrestrel \downarrow z \leftrightarrow oprrestrel \downarrow z \leftrightarrow (z = (r, C) \wedge qrel\, r \wedge qcoll\, C)$.
   $oplrestrel(r, C) = r' \rightarrow (qrel\, r' \wedge (x\, r'\, y \leftrightarrow x\, r\, y \wedge x \in C))$.
   $oprrestrel(r, C) = (oplrestrel(r^{-1}, C))^{-1}$.

R.5 $oplrestsys \downarrow z \leftrightarrow oprrestsys \downarrow z \leftrightarrow (z = (S, C) \wedge qsys\, S \wedge qcoll\, C)$.
   $oplrestsys(S, C) = S' \rightarrow (qsys\, S' \wedge (xS'y \leftrightarrow x\, S\, y \wedge x \in C))$.
   $oprrestsys(S, C) = (oplrestsys(S^{-1}, C))^{-1}$.

We will denote by $_{C|}x$, $x_{|C}$, $_{C|}x_{|D}$ the left, right, and bilateral restrictions of an object x.

It is convenient to add an axiom of coherence to the axioms R.1-4, which gives the restrictions of non-extensional objects. We give only an instance of such an axiom of coherence for the restriction of qualities:

R.6 $(q_{|C})_{|D} = (q_{|D})_{|C} = q_{|C \cap D} \cdot (q \wedge q')_{|C} = q_{|C} \wedge q'_{|C}$.
   $(rqual, q) \approx (rcoll, C) \rightarrow q_{|C} = q$.

One can also introduce an operation *grest* of *general restriction*, which applied to a qr-structure $\mathcal{S} = (q, r)$ gives the $\mathcal{S}$-*restriction*, $\mathcal{S}$-*rest*, which in turn is defined on pairs (x, z) (where x enjoys the quality q and z is a relational pair), and gives the restriction of x to z.

R.7 $qop\, grest \wedge (rdom, grest) \approx (rqual, qqrs) \wedge (rval, grest) \approx (rqual, qop)$
   $grets(q, r) = f \rightarrow (f \downarrow y \leftrightarrow y = (x, z) \wedge q\, x \wedge qrepa\, z)$.
   $(grest(q, r) = f \wedge f(x, (r', x')) = u) \rightarrow (q\, u \wedge (t\, r\, u \leftrightarrow t\, r\, x \wedge t\, r'\, x'))$.

For sake of brevity we omit further axioms on *grest*, but we state an axiom that directly yields the widest action of the operation *hat*.

HA.1 *hat is defined at all pairs (f, C), where f is an operation and C a*
   collection.

Various axioms of this kind could be also introduced on the operations giving domains and codomains of relations, operations and systems (see Chapter II Section 1 and Chapter IV Section 3). These axioms are actually either particular instances of the general comprehension axiom CA.15, or consequences of CA.15 together with the axiom AR.3. However, we remark that an easy contradiction can be obtained by simply assuming that any operation has a domain, any pair operations has a composition, and any two collections have the intersection (see [8] and Section 2 of this chapter).

Antinomy III: $dom(id_{|\{\emptyset\}} \circ int \circ \langle id, sing \rangle) = \{x | qcoll\, x \wedge x \notin x\} = Russ$.

VI.9. Cardinals

We already remarked that only few algebraic and order-theoretic properties of cardinal numbers can be derived from our basic axioms; however assuming the following decomposition property suffices to make cardinals a *"Weak Cardinal Algebra"* (WCA of [20]), thus providing many interesting arithmetic properties:

C.1   *If $\alpha + \beta = \gamma + \delta$, then $\alpha = \pi_1 + \pi_2$, $\beta = \pi_3 + \pi_4$, $\gamma = \pi_1 + \pi_3$ and $\delta = \pi_2 + \pi_4$, for suitable cardinals $\pi_1, \pi_2, \pi_3, \pi_4$. If moreover $\alpha = \gamma$, then also π_1 can be taken equal to α.*

If we want that our cardinal numbers have all properties of a *"Cardinal Algebra"* (CA of [21]), it suffices to introduce an operation *sumseq* defined at sequences of cardinal numbers, postulating:

C.2   *For any sequences s of cardinals, sumseq s is defined at s, and sumseq satisfies the axioms* III, IV, VI *and* VII *of* [21, p.3-4].

It is well known that these properties hold for cardinals in the standard set-theories. An axiom making our setting of cardinals very close to the usual ones is:

C.3   *Any cardinal number is the cardinality of some relation pair.*

Axiom C.3 says in particular that all cardinals below $\daleth$ are cardinalities of collections, and, together with the axioms 6.D, 11.F, 13.D, implies that their cardinal ordering is the usual "injective" ordering. On the contrary, we did not give any "functional" meaning to the property of having cardinality $\daleth$. We can complete this process of standardization by postulating:

C.4   *If $(\overline{\overline{r,x}}) = \daleth$ then there is an injective operation f such that $(rval, f) \approx (r, x)$ and $(rdom, f) \approx (robj, x)$*

The axiom C.4 admits also a weaker surjective version postulating:

C.4*   *If $(\overline{\overline{r,x}}) = \daleth$ then there is an operation f such that $(rdom, f) \approx (r, x)$ and $(rval, f) \approx (robj, x)$*

An axiom yielding that the cardinalities of collections are ordered by injections, but allows also "non-standard" cardinals (which can be useful in particular universes) is the following strengthening of the axiom 6.D of Chapter II Section 3:

C.3*   *If $|A| \le |B|$, then there is a collection C such that $|A| + |C| = |B|$.*

VI.10. λ-calculus

Here we introduce three operations *gap*, S and ε which endow the objects with a natural structure of λ-*model*: namely, *gap* is an extension of *eval*, to be interpreted as a *general application*, S is the *Curry's combinator*, and ε a *selector* with respect to this notion of application.

$\Lambda.1 \qquad qop\, gap \wedge (rdom, gap) \approx (rqual, qpair) \wedge (rop, eval) \preceq (rop, gap) \ .$

$\Lambda.2 \qquad qop\, S \wedge qop\, Sx \wedge qop(Sx)y \wedge ((Sx)y)z = gap(gap(x, z), gap(y, z)) \ .$

$\Lambda.3 \qquad qop\, \varepsilon \wedge qop\, \varepsilon x \wedge (\varepsilon x)y = gap(x, y) \wedge (\varepsilon x = \varepsilon y \rightarrow (rop, \varepsilon x) \approx (rop, \varepsilon y)) \ .$

Any model of the A-Theory verifying the axioms Λ1-3 and A0.3 is a λ-*model* (see [1]), and is "natural" in the sense that a "true" operation applied to any of its "true" arguments gives the "correct" result.

$\Lambda.2$ and $\Lambda.3$ can be omitted, if one has transposition, compositions, separations, etc., but the axioms Λ.1-3 are most interesting when these operations are not available, since they allow to define suitable λ-*terms* giving their counterparts with respect to the generalized application *gap*, which is in fact the *true evaluation* of an operation at any of its arguments.

We can give to the general application *gap* a complete descriptive power by coding the fundamental relation of the A-Theory by means of two objects c_1, and c_2 satisfying the following axiom:

$\Lambda.4 \qquad\qquad\qquad c_1\, x\, r\, y = c_2\, x\, r\, y \rightarrow qrel\, r \wedge x\, r\, y \ .$

(We simply write xy for $gap(x, y)$ and we associate to the left, following the common usage of λ-calculus).

Since the action of *gap* is almost completely undetermined on pairs whose first components are not operations, it seems natural to postulate that on structured objects it is consistent with the traditional coding of the same concepts as λ-terms. E.g. we can postulate that $id = SKK$, and that pairs and integers operate according to their Church's equivalents:

$\Lambda.5 \qquad (x, y) = S(SI(K\, x))(K\, y) \wedge 0 = K\, id \wedge n + 1 = S(trans\, comp)n \ .$

VI.11. Free construction and general recursion axioms

Axioms derived from a general "free construction principle" (also called "axioms of universality") have been introduced and studied by various authors. For a general treatment in the frame of the theory Q of [7], we refer to [5]. We give here only few examples of axioms of universality which are expressed in terms of qr-structures: note that the role of the isomorphisms of [5] and [3] is played here by suitable injective operations which, in some sense, respect the corresponding structures.

LC.1   *(universality for qualties)*
If $(q,\ r)$ is a qr-structure then there exists an injective operation f such that $qz \leftrightarrow qqual\ fx$, $x\ r\ y \leftrightarrow fx\ rqual\ fy$, and $(rqual,\ fy) \approx (rval,\ f)$.

The axioms LC.2 and LC.3 deal with particular qr-structures, namely the qr-structures enjoying the quality of being *binary qr-structures (qbinst)* and the quality of being *operative qr-structure (qopst)*, respectively. These two qualities are ruled by the axioms QR.1-3:

QR.1   *qbinst and qopst are qualities of qr-structures.*

QR.2   *$qbinst(q, r) \leftrightarrow \forall x\, qrepabin(r, x)$.*

QR.3   *$qopst(q, r) \leftrightarrow \forall x\, qrepaop(r, x)$.*

LC.2   *(universality for relations)*
If (q,r) is a binary qr-structure then there exists an injective operation f such that $qx \leftrightarrow qrel\ fx$, $(x, y)\ r\ z \rightarrow (fx, fy)rrel\ fz$, and $u\ fz\ v \rightarrow \exists x \exists y(u = fx \wedge v = fy)$.

LC.3   *(universality for operations)*
If (q,r) is an operative qr-structure then there exists an injective operation f such that $qx \leftrightarrow qop\ fx$, $(x, y)rz \leftrightarrow (fx, fy)\ rop\ fz$, and $v = (fz)\,u \leftrightarrow \exists x \exists y(u = fx \wedge v = fy)$.

Using the same translation, Boffa's axiom of universality U (see [3]) becomes:

LC.4   *(universality for collections)*
If (q,r) is an operative qr-structure then there exists an injective operation f such that $qx \leftrightarrow qcoll\ fx$, $x\,r\,y \leftrightarrow fx \in fy$ and $(rcoll, fy) \approx (rval, f)$.

A complementary principle for constructing operations is given in the following axiom, expressing in our formalism the general recursion principle (see [2]).

GR   *For any operation g there is an operation f such that $g(f,x)=y$ if and only if $fx=y$.*

We note that the axiom GR *excludes* many large operations defined at all pairs of operations. E.g. assuming GR, there is no operation g verifying:

$$g(x, x) = \begin{cases} x & \text{if } x \text{ is an empty operation, or } xx \neq x \\ \text{some object different from } x & \text{otherwise,} \end{cases}$$

since any operation f verifying GR would verify also $ff = f$ if and only if $ff \neq f$.

VI.12. Strong axioms inside universes

In dealing with various consistency problems it is often useful to consider the relativization of the considered axioms to some universe. E.g. we can relativize the A-Theory to a universe V by postulating that the restrictions $q_{|V}$, $_V|r_{|V}$, $_V|f_{|V}$ of all fundamental qualities, relations and operations introduced in Chapters I-V, *are elements of* V. After doing this, it is possible to relativize all the axioms of Sections 1-11. E.g. the axiom of comprehension CA.7 becomes:

CU.V $\hspace{5cm}$ $\hat{c}ext(V\text{-}rel \times V) = V\text{-}coll$.

Recall that $\hat{c}ext(V\text{-}rel \times V) = \{x | \exists r \exists y (r \in V\text{-}rel \wedge y \in V \wedge \forall t (t \in x \leftrightarrow t\, r\, y))\}$. The axiom of general selection AS.4 can be phrased similarly:

SU.V $\hspace{5cm}$ $\hat{g}sel(V\text{-}qual \times V\text{-}rel) \subseteq V\text{-}op$.

Many antinomies can be expressed by assertions of the form $gex(q,r)z \notin V$, e.g.:

(∗) $\hspace{4cm}$ $gext(qop, rop)(rcoll, G) \notin V$, where

$$G = diag(V\text{-}op) \backslash \hat{p}_1((graph\ diag^{-1}{}_{|V} \cap grrel\ _V|rop_{|V}) .$$

In the usual notation $G = \{(f, f) | f \in V\text{-}op \wedge (f, f) \notin graph\ f\}$, and (∗) expresses the fact that there is no V-operation g such that $gf = f$ if and only if $f \in V\text{-}op$ and either $f \downarrow f$ or $ff \neq f$.

Perhaps the general form of all interesting axioms of this kind on universes is

$$A \cdot (f, g, U) \quad f_1 U \in U, \ldots, f_n U \in U, \ \hat{g}_1(U) \subseteq U, \ldots, \hat{g}_n(U) \subseteq U$$

where $f_1, \ldots, f_n$, $g_1, \ldots, g_n$, are suitable operations, e.g. built up by composing and transposing the restrictions of basic operations to some universe V such that U is a V-set: the use of the larger universe V is needed to allow any kind of manipulations of operations.

In particular we can postulate the existence of a universe U whose U-collections are models of Quine's NF, by choosing the operations corresponding to the axioms NF.1-6 of Scetion 7.

Similarly, the general form of the relativization of antinomies to universes may be expressed by:

(∗∗) $\hspace{5cm}$ $fU \notin U$.

A different kind of axioms can be obtained by strengthening the inaccessibility property of a universe, e.g.

SI   *There is a universe V such that any V-set is also a U-set for some universe U which is a V-set.*

Finally, interesting properties of free construction can be postulated inside universes (see [5]). If this is done inside a universe satisfying SI, one obtains universes verifying the relativization of any consistent sublist of axioms given in this section.

Glossary and list of symbols

Qualities

qabscom	to be an absolutely comprehensive qr-structure	VI.3
qantis	anti-symmetry	I.3
qbincom	to be a binarily comprehensive relational pair	VI.3
qbinst	to be a binary qr-structure	VI.11
qcard	to be a cardinal	II.0-3
qcoll	to be a collection	II.0-1
qcon	connectedness	I.3
qeq	to be an equivalence relation	I.3
qextst	to be an extensional qr-structure	VI.3
qfalse	to be a false proposition	V.2
qfseq	to be a finite sequence	II.5
qinac	to be an inaccessible cardinal	V.1
qinjop	to be an injective operation	I.2
qlarge	to be large	III.3
qlo	to be a linear ordering	I.3
qobj	to be an object	I.1
qop	to be an operation	I.1
qopcom	to be an operatively comprehensive qr-structure	VI.3
qopst	to be an operative qr-structure	VI.11
qpair	to be a pair	I.1
qpo	to be a partial ordering	I.3
qpreo	to be a preordering	I.3
qprop	to be a proposition	V.2
qqrs	to be a qr-structure	I.1
qqual	to be a quality	I.1
qrefl	reflexivity	I.3
qrel	to be a relation	I.1
qrepa	to be a relational pair	III.1
qrepabin	to be a binary relational pair	VI.3
qrepaop	to be an operative relational pair	VI.3
qrussc	the Russell quality of collections	VI.2
qseq	to be a sequence	II.5
qsmall	to be small	III.3
qsym	symmetry	I.3
qsys	to be a system	II.0-1
qtrans	transitivity	I.3
qtrue	to be a true proposition	V.2
quniv	to be a universe	V.1
qunsys	to be a univalent system	II.0-1

Relations

rbid	bidding relation of propositions	V.2
rcard	to be less than or equal to (between cardinals)	II.0-3
rcoll	to be a member of	II.0-1
rdom	to be in the domain of (for operations)	I.2
rdomrel	to be in the domain of (for relations)	VI.6
resteq	extensional equivalence	III.1
rfirst	to be the first component of	I.1
rginc	general inclusion	III.1
rid	equality	I.2
rincl	inclusion between collections	II.1
rnid	non-equality	I.2
robj	the fundamental relation of objects	I.1
rop	the fundamental relation of operations	I.1
rprop	the fundamental relation of proposition	V.2
rqual	the fundamental relation of qualities	I.1
rrel	the fundamental relation of relations	I.1
rscard	strict cardinal ordering	II.3
rsnd	to be the second component of	I.1
rsys	the fundamental relation of systems	II.0-1
rtuple	the relation associating n to n-uples	II.5
rval	to be a value of	I.2

Operations

agg	aggregation of operations	VI.5
aggsys	aggregation of systems	IV.2
birest	bilateral restriction of a relation	IV.3
birestop	bilateral restriction of an operation	VI.8
card	cardinality of collections	III.2
cart	Cartesian product of systems of collections	IV.1
cext	collection-extension	III.1
choice	choice operation	VI.4
codrel	codomain of relations	IV.3
comp	composition of operations	VI.5
comprel	composition of relations	VI.6
compsys	composition of systems	IV.2
conc	concatenation of finite sequences	II.5
conj	conjunction of qualities	I.2
conjr	conjunction of relations	VI.6
δ	Dirac's δ	VI.5
diag	diagonalization	I.2
diff	difference between two collections	IV.1

dom	domain of operations	III.1
$domrel$	domain of relations	IV.3
$doub$	doubleton	II.2
ε	selector w.r. to gap	VI.10
et	conjunction of a system of propositions	V.2
$eval$	evaluation of an operation at an object	I.2
ext	extension of qualities	III.1
$fagg$	fibred aggregation of operations	VI.5
$fagsys$	fibred aggregation of systems	IV.2
$fibsys$	fibred prodouct of systems	IV.2
$fieldrel$	field of relations	IV.3
gap	general application	VI.10
$gcard$	general cardinality	III.2
$gchoice$	generalized choice operator	VI.4
$gext$	general extension	VI.3
gp_i	general projection at the index i	VI.5
$graph$	graph of operations	III.1
$grest$	general restriction	VI.8
$grrel$	graph of relations	III.1
$grsys$	graph of systems	III.1
$gsel$	general selector	VI.4
gus	generator of unitary systems	II.2
hat	generator of images	IV.3
$hsel$	homogeneous selector	VI.4
K	Curry's combinator K	VI.5
id	identity	I.2
img	range of an operation	III.1
$imgsys$	image of systems	IV.3
$indsys$	indices-collection of systems	IV.3
int	intersection of systems of collections	IV.1
$intc$	intersection of collections of collections	IV.1
$invop$	inversion of operations	I.2
$invrel$	inversion of relations	I.2
$invsys$	inversion of systems	II.2
non	negation of propositions	V.2
$notq$	negation of qualities	VI.6
$notr$	negation of relations	VI.6
$opair$	pairing operator	VI.
$oplrestrel$	left restriction of relations	VI.8
$oplrestsys$	left restriction of systems	VI.8
$oprrestrel$	right restriction of relations	VI.8
$oprrestsys$	right restriction of systems	VI.8
$oprest$	restriction of an operation	IV.3
$oprestq$	restriction of a quality	VI.8

$\mathcal{P}$	power-set operator	IV.1
p_1	first projection of a pair	I.2
p_2	second projection of a pair	I.2
pow	power of two collections	IV.1
psys	projection of univalent systems	II.2
S	Curry's combinator S	VI.10
selop	selector of operations	VI.4
selq	selector of qualities	VI.4
selr	selector of relations	VI.4
sep	separation of operations	VI.5
sepsys	separation of systems	IV.3
$\mathcal{S}$-*ext*	relative extension of S	VI.3
$\mathcal{S}$-*gsel*	general S-selector	VI.4
$\mathcal{S}$-*hsel*	homogeneous S-selector	VI.4
$\mathcal{S}$-*rest*	S-restriction	VI.8
sing	singleton	II.2
sum	cardinal addition	II.3
syext	system-extension	III.1
τ	type-rising operator	VI.7
tens	tensor product of operations	VI.5
tensys	tensor product of systems	IV.2
trans	transposition of operations	VI.5
transys	transposition of systems	IV.2
un	union of systems of collections	IV.1
unc	union of collections of collections	IV.1
vel	disjunction of a system of propositions	V.2

Collections

Card	cardinals	VI.7
Coll	collections	VI.7
$\mathbb{N}$	natural numbers	II.4
Prop	propositions	VI.7
Obj	total collection	IV.1, VI.7
Obj^2	pairs	VI.7
Op	operations	VI.7
Qual	qualities	VI.7
Rel	relations	VI.7
Russq	Russell collection of qualities	VI.2
Sys	systems	VI.7
Univ	universes	VI.7
Unsys	univalent systems	VI.7
V-card	*V*-cardinals	V.1
V-corr	*V*-correspondences	V.1

V-*coll*	V-collections	V.1
V-*map*	V-maps	V.1
V-*op*	V-operations	V.1
V-*qual*	V-qualities	V.1
V-*rel*	V-relations	V.1
V-*set*	V-sets	V.1
V-*sys*	V-systems	V.1
V-*unsys*	univalent V-systems	V.1

Structures

$\mathcal{S}_0$	the fundamental structure of objects	I.1
$\mathcal{S}_{01}$	the first fundamental structure of pairs	I.1
$\mathcal{S}_{02}$	the second fundamental structure of pairs	I.1
$\mathcal{S}_1$	the fundamental structure of qualities	I.1
$\mathcal{S}_2$	the fundamental structure of relations	I.1
$\mathcal{S}_3$	the fundamental structure of operations	I.1
$\mathcal{S}_4$	the fundamental structure of collections	II.0-1
$\mathcal{S}_5$	the fundamental structure of systems	II.0-1
$\mathcal{S}_6$	the fundamental structure of cardinals	II.0-3
$\mathcal{S}_7$	the fundamental structure of propositions	V.2

Other symbols

$\aleph_0$	aleph zero	II.3
$\alpha \le \beta$	α is less than or equal to β (for cardinals)	II.0-3
$\alpha < \beta$	α is less than β (for cardinal)	II.0-3
$\alpha + \beta$	cardinal addition	II.3
$C \backslash D$	the difference of two collections C and D	IV.1
$C \times D$	the Cartesian product of two collections	IV.1
C^n	the n-th power of collections	IV.1
C^D	power of two collections	IV.1
$\hat{f}(C)$	the image of C under f	IV.3
$f \downarrow x$	f is defined at the argument x	I.2
$f \uparrow y$	y is a value of f	I.2
f^{-1}	the inverse operation of f	I.2
$f_{|C}$	the restriction of the operation f	IV.3
$i\,S\,x$	x is indexed by i in the system S	II.0
1	one I App.,	II.3
$p_1 \wedge p_2$	the conjunction of two propositions	V.2
$\wedge_i p_i$	the conjunction of a system of propositions	V.2
$p_1 \vee p_2$	the disjunction of two propositions	V.2
$\vee_i p_i$	the disjunction of a system of propositions	V.2

$\neg p$	the negation of a proposition	V.2
$q_1 \wedge q_2$	the conjunction of two qualities	I.2
r^{-1}	the inverse relation of r	I.2
$_C|^r|_D$	the bilateral restriction of a relation r	IV.3
$_C|^x|_D$	the bilateral restriction of an object x	VI.8
$_C|x$	the left restriction of an object x	VI.8
$x|_D$	the right restriction of an object x	VI.8
$\oslash$	the empty system	II.2
$\emptyset$	the empty collection	II.2
$Sobj$	the total system	VI.7
S^{-1}	the inverse of a system S	II.2
S_i	the projection at i of the univalent system S	II.0
$S \overset{\circ}{\smile} T$	the concatenation of two finite sequences	II.5
$S \circ T$	the composition of two systems	IV.2
$\cup_i S_i$	the union of a system of collections	IV.1
$\cup C$	the union of a collection of collections	IV.1
$A \cup B$	the union of two collections	IV.1
$\cap_i S_i$	the intersection of a system of collections	IV.1
$\cap C$	the intersection of a collection of collections	IV.1
$A \cap B$	the intersection of two collections	IV.1
$\daleth$	tav (the largest cardinal number)	II.3
0	zero (the least cardinal number)	I App., II.3
$\otimes_i S_i$	the tensor product of a system of systems	IV.2
$S \otimes T$	the tensor product of two systems	IV.2
$\langle S_i \rangle_i$	the fibred product of a system of systems	IV.2
$\langle S, T \rangle$	the fibred product of two systems	IV.2
$\binom{x}{y}$	the unitary system $gus(x,y)$	II.2
$x \approx y$	x is extensionally equivalent to y	III.1
$x \preceq y$	x is extensionally included in y	III.1
$(\overline{\overline{r,x}})$	relative cardinality of x with respect to r	III.2
$\overline{\overline{x}}$	relative cardinality of x	III.2
$|C|$	the cardinality of a collection	III.2

Bibliography

[1] H.P. Barendregt, *The Lambda Calculus*, Amsterdam 1981.

[2] M. Beeson, *Towards a Computation System Based on Set Theory*, Preprint CSLI -Stanford, (1987).

[3] M. Boffa, *Sur la théorie des ensembles sans axiome de fondement*, Bull. Soc. Math. Belg., 21 (1969), 16-56.

[4] A. Church, *Set theory with a Universal Set*, in *Proceedings of the Tarski Symposium* (L. Henkin et al. eds.), Proc. of Symp. P. math. XXV, Providence 1974, 297-308.

[5] M. Clavelli, *Principii di libera costruzione*, Atti Acc. Naz. Lincei Rend. Cl. Sci. Fis. Mat. Nat., (8) 81 (1986), to appear.

[6] H.B. Curry & R. Feys, *Combinatory Logic*, Amsterdam 1985.

[7] E. De Giorgi & M. Forti, *Una teoria quadro per i fondamenti della Matematica*, Atti Acc. Naz. Lincei Rend. Cl. Sci. Fis. Mat. Nat., (8) 79 (1985), 55-67.

[8] E. De Giorgi, M. Forti & V.M. Tortorelli, *Sul problema dell'autoriferimento*, Atti Acc. Naz. Lincei Rend. Cl. Sci. Fis. Mat. Nat., (8) 80 (1986), 363-372.

[9] S. Fefermann, *Constructive Theories of Functions and Classes*, in *Logic Colloquium 1978* (M. Boffa et al. eds.), Amsterdam 1979, 159-224.

[10] A.A. Fraenkel, Y. Bar-Hillel & A. Lévy, *Foundations of Set Theory*, Amsterdam 1973.

[11] G. Frege, *Grundgesetze der Arithmetik, begriffsschriftlich abgeleitet, Vol.1* Jena 1893, Vol.2 Pohle, Jena 1903, (reprinted Olms, Hildesheim 1962).

[12] G. Frege, *Begriffsschrift, eine der arithmetischen nachgebildete Formelsprache des reinen Denkens*, Halle 1879, (reprinted in *Begriffsschrift und andere Aufsätze* (I.Angelelli ed.), Olms, Hildesheim 1964).

[13] J.M. Glubrecht, A. Oberschelp & G. Todt, *Klassenlogik*, Zürich 1983.

[14] A. Grothendieck & J.L. Verdier, *Prefaisceaux, Exposé I in Théorie des Topos, S.G.A. 4* (A. Grothendieck et al. eds.), Lec. Notes Math. 269, Berlin 1972, 1-94.

[15] T. Hailperin, *A set of axioms for Logic*, J. Symb. Logic, 9 (1944), 1-19.

[16] P. MARTIN-LÖF, *Constructive Mathematics and Computer Programming,* in *Logic, Methodology and Philosophy of Science VI* (L.J. Cohen et al. eds.), Amsterdam 1982, 153-175.

[17] W.V.O. QUINE, *New Foundations for Mathematical Logic,* Amer. Math. Montly, 44 (1973), 70-80.

[18] D. SCOTT, *Combinators and Classes,* in λ-*calculus and Computer Science Theory* (C.Böhm ed.), Lec. Notes Comp. Sc. 37, Berlin 1975, 1-26.

[19] A. TARSKI, *Cardinal Algebras,* New York 1949.

[20] J. TRUSS, *Convex Sets of Cardinals,* Proc. London Math. Soc., (3) 27 (1973), 577-599.

[21] J. VON NEUMANN, *Eine Axiomatisierung der Mengenlehre,* J. f. Math., 154 (1925), 219-240.

[22] A.N. WHITEHEAD & B. RUSSELL, *Principia Mathematica,* Cambridge 1925.

M. Clavelli, E. De Giorgi, V.M. Tortorelli: Scuola Normale Superiore
Pisa, Italy

M. Forti: Dipartimento di Matematica, Università di Cagliari
Cagliari, Italy

Introduction to free-discontinuity problems
Papers in honor of Luigi Radicati[‡]

E. De Giorgi

Introduction

In offering a paper to a scholar who, among his many merits, has that of a clear understanding of the connections between Mathematics and Physics, I would have liked to present a work that could equally be of interest to mathematicians and physicists. In other words, I would have liked to write a paper that, starting from considering some physical phenomena that are easy to observe, would continue through an ample and free mathematical elaboration, in the end reaching a conclusion equally interesting both from the mathematical and the physical viewpoint. Up to now, I have not been able to reach such a conclusion. Hence, I will limit myself to illustrating some initial considerations on a type of problems that I believe to be of interest both to the mathematician and to the physicist; that is, those problems of the Calculus of Variations for which it is reasonable to think that "piecewise-smooth" solutions exist. For the time being, I can only say that such problems constitute, to my knowledge, a region of Mathematics, both pure and applied, for the most part unexplored and I think that the exploration of this region requires the collaboration of researchers from different areas (for example, Measure Theory, Differential Geometry, Calculus of Variations, Differential Equations, Functional Analysis, Topology, etc.). I hope that the brief considerations presented in this work could encourage many mathematicians and physicists to work in a field where probably no one may consider himself a complete expert and hence where everyone may bring interesting ideas.

This paper is divided into a number of conversations, and I would like to stress the fact that these are real conversations, whose summary has been written by different interlocutors. Hence, the reader must not expect neither uniformity of notation nor absence of repetitions. Even the references, due to the various interlocutors, are more or less ample in the various conversations. Even less the reader should read these conversations as a development of an organic program addressed to a precise category of mathematicians and physicists possessing certain prerequisites other than the simple desire to talk with other researchers.

In these conversations I present some ideas that must still be analyzed thoroughly, programs that I am not able to complete, problems that I cannot solve

---

[‡]Editor's note: translation into English of the paper "Introduzione ai problemi di discontinuità libera", published in Symmetry in Nature, a volume in honour of Luigi A. Radicati di Brozolo, vol. I, Scuola Norm. Sup., Pisa (1989), 265–285.

but I strongly wish would be taken into consideration by other mathematicians or physicists more capable than I am to evaluate the interests and the difficulties, to indicate possible previous work, to recognize analogous problems already treated in the mathematical or physical literature.

I think that presently this is the type of mathematical work more suited to me. Nevertheless, I am comforted by the fact that my progressive detachment from 'written mathematics' of the more traditional type does not decrease, but maybe increases my interest to converse with mathematician and physicist friends.

To those friends that are more difficult for me to meet, I recommend to write to Prof. Antonio Leaci - Dipartimento di Matematica - Università di Lecce - 73100 - Lecce, communicating to him their remarks, comments, and, above all, possible references, indicating which, among my many conjectures, look easier to prove or disprove.

Conversation 1

Some classes of piecewise-smooth manifolds

I have found considerable difficulties in trying to translate into precise definitions the vague idea of a piecewise-smooth manifold. In this conversation I present some definitions of more or less regular piecewise-smooth manifolds that seem to me suited to the description of possible solutions to many problems that I am going to present in the following conversations. Such definitions have no claim of originality. On the contrary, I would be very grateful to those who would point out to me definitions already present in the literature and more or less close to these ones.

We denote with $\mathcal{A}(\mathbf{R}^n)$ the family of all open sets of $\mathbf{R}^n$ and with $\mathcal{K}(\mathbf{R}^n)$ the family of all compact sets of $\mathbf{R}^n$. For all $A \in \mathcal{A}(\mathbf{R}^n)$ and for all $p \geq 1$ we denote with $L^p(A)$ the space of the measurable real functions with summable p–th power in A. Moreover, we denote with $C^\alpha(A)$ the space of the real functions continuous in A together with their derivatives up to order α if $\alpha \in \mathbf{N}$, to any order if $\alpha = \infty$, the space of the analytic real functions in A if $\alpha = \omega$. Given a real function v defined in a set $E \subset \mathbf{R}^n$, we set, for $\alpha \in \mathbf{N}$, $\alpha = \infty$, or $\alpha = \omega$,

$$\operatorname{dom} C^\alpha(v) = \cup\{A \in \mathcal{A}(\mathbf{R}^n); v \in C^\alpha(A)\}.$$

For all $h \in \mathbf{N}$, $x \in \mathbf{R}^h$, $\mathbf{R}^h o > 0$, we denote with $B^h_{\mathbf{R}} o(x)$ the sphere $\{y \in \mathbf{R}^h; |y - x| < \mathbf{R}^h o\}$, with $B^h_{\mathbf{R}} o$ the sphere $B^h_{\mathbf{R}} o(0)$, and with ω_h the measure of B_1. We denote with χ_E the characteristic function of the set E and with $\operatorname{supp} f$ the support of the function f.

In the sequel we consider Ω in $\mathcal{A}(\mathbf{R}^n)$ as fixed.

DEFINITION 1.1 – For $h \in \mathbf{N}, h > 0$, we define $V_h C^\alpha(\Omega)$ the class of all sets $E \subset \mathbf{R}^n$ such that $E \cap \Omega = \overline{E} \cap \Omega$ and for all $x \in E \cap \Omega$ there exist $A \in \mathcal{A}(\mathbf{R}^n)$, $B \in \mathcal{A}(\mathbf{R}^h)$, $\varphi \in (C^\alpha(B))^n$, $\psi \in (C^\alpha(A))^h$ such that

$$x \in A, \quad \psi(\varphi(y)) = y \qquad \text{for all } y \in B, \quad E \cap A = \varphi(B).$$

Moreover, we set $E \in V_0 C^\alpha(\Omega)$ for all α if and only if E is locally finite in Ω.

Note that $\emptyset \in V_h C^\alpha(\Omega)$ for all $h \in \mathbf{N}$, and for all α.

DEFINITION 1.2 – For $h \in \mathbf{N}, h > 0$, we define $V_h BC^\alpha(\Omega)$ the class of all sets $E \subset \mathbf{R}^n$ such that $E \cap \Omega = \overline{E} \cap \Omega$ and for all $x \in E \cap \Omega$ there exist $A \in \mathcal{A}(\mathbf{R}^n)$, $B \in \mathcal{A}(\mathbf{R}^h)$, $\varphi \in (C^\alpha(B))^n$, $\psi \in (C^\alpha(A))^h$ and $z \in \mathbf{R}^h$ such that

$$x \in A, \quad \psi(\varphi(y)) = y \qquad \text{for all } y \in B,$$

$$E \cap A = \{\varphi(y); \ y \in B, \ \langle y, z \rangle \geq 0\}.$$

DEFINITION 1.3 – We define $F_h C^\alpha(\Omega)$ the class of all functions w such that for all $x \in \Omega$ there exist $A \in \mathcal{A}(\mathbf{R}^n)$, $\nu \in \mathbf{N}$, $E_1, \ldots, E_\nu \in V_h C^\alpha(A)$ for which $x \in A$ and

$$w(y) = \sum_{i=1}^{\nu} \chi_{E_i}(y) \qquad \text{for all } y \in A.$$

DEFINITION 1.4 – We define $F_h BC^\alpha(\Omega)$ the class of all functions w such that for all $x \in \Omega$ there exist $A \in \mathcal{A}(\mathbf{R}^n)$, $\nu \in \mathbf{N}$, $E_1, \ldots, E_\nu \in V_h BC^\alpha(A)$ for which $x \in A$ and

$$w(y) = \sum_{i=1}^{\nu} \chi_{E_i}(y) \qquad \text{for all } y \in A.$$

DEFINITION 1.5 – We define $V_h PC^\alpha(\Omega)$ the class of all sets $E \subset \mathbf{R}^n$ such that there exist $E_0 \subset E_1 \subset \ldots \subset E_h = E$ such that

$$E_0 \in V_0 C^\alpha(\Omega), \quad E_i \in V_i C^\alpha(\Omega \setminus E_{i-1}), \quad i = 1, \ldots, h;$$

there exist $f_1, \ldots, f_h$ such that

$$\operatorname{supp} f_i = \overline{E_i \setminus E_{i-1}},$$

$$f_1 \in F_1 BC^\alpha(\Omega), \quad f_i \in F_i BC^\alpha(\Omega \setminus E_{i-2}), \quad i = 2, \ldots, h.$$

Particular examples of sets $E \in V_h PC^\omega(\Omega)$ are convex h-dimensional poly-hedra.

Conversation 2

Partitioning problems. Soap bubble clusters

For the reader's convenience we briefly recall the definition of Hausdorff measure and the concept of rectifiable set, referring for example to [FH1] for an ample treatment of these subjects and for proofs.

DEFINITION 2.1 – Let (X, τ) be a topological space and let $\mathcal{P}(X)$ be the set of the parts of X. By a regular Borel measure we mean a map $\mu : \mathcal{P}(X) \to [0, +\infty]$ such that

1) μ is countably subadditive; that is, for all sequences $(E_i)_i$ of subsets of X we have

$$\mu\left(\bigcup_{i=1}^{\infty} E_i\right) \leq \sum_{i=1}^{\infty} \mu(E_i) \; ;$$

2)   Borel sets of X are Carathéodory measurable; that is, for all Borel sets $B \subset X$ we have

$$\mu(E) = \mu(E \cap B) + \mu(E \setminus B) \quad \text{for all } E \subset X \; ;$$

3)   for all $E \subset X$ we have

$$\mu(E) = \inf \{\mu(B); \quad E \subset B, \; B \text{ Borel set} \} \; .$$

DEFINITION 2.2 – Let (M, σ) be a metric space, and let $h > 0$ be a real number. We define the h-dimensional Hausdorff measure with respect to the distance σ by setting for all $E \subset M$

$$\mathcal{H}_\sigma^h(E) = \frac{2^{1-h} \pi^{h/2}}{h\Gamma(h/2)} \; \lim_{\epsilon \to 0^+} \; \inf \left\{ \sum_{i=1}^{\infty} (\mathrm{diam} E_i)^h \; ; \; E \subset \bigcup_{i=1}^{\infty} E_i \; , \; \mathrm{diam} E_i < \epsilon \right\},$$

where $\Gamma(s) = \int_0^\infty e^{-t} \, t^{s-1} \, dt$ for $s > 0$, and $\mathrm{diam} E = \sup\{\sigma(x, y); x, y \in E\}$.
Moreover, we define $\mathcal{H}_\sigma^0$ as the point-counting measure; that is,

$$\mathcal{H}_\sigma^0(E) = \begin{cases} \text{number of elements of E} & if E is finite \\ +\infty & otherwise. \end{cases}$$

Hausdorff measures with respect to the Euclidean distance in $\mathbf{R}^n$ will be denoted, as usual, simply by $\mathcal{H}^h$.
The following well-known results hold.

1.   For $h \geq 0$ the measure $\mathcal{H}_\sigma^h$ is Borel regular.

2.   If (M, σ) is the Euclidean space $\mathbf{R}^n$, equipped with the Euclidean distance, then $\mathcal{H}^n$ coincides with the n-dimensional Lebesgue measure. Moreover, for all h, the measure $\mathcal{H}^h$ is invariant under translations and for all $r > 0$ we have $\mathcal{H}^h(rE) = r^h \mathcal{H}^h(E)$ for all $E \subset \mathbf{R}^n$.

3.   If $0 < \mathcal{H}_\sigma^h(E) < +\infty$ then $\mathcal{H}_\sigma^k(E) = 0$ for all $k > h$ and $\mathcal{H}_\sigma^k(E) = +\infty$ for all $k < h$.

We note, in addition, that the measure $\mathcal{H}^h$ coincides with every reasonable definition of h-dimensional measure on the regular h-dimensional submanifolds of $\mathbf{R}^n$.

With the previous result 3 in mind, it is reasonable to give the following definition.

DEFINITION 2.3 – The Hausdorff dimension of the subset E of the metric space (M, σ) is the real number

$$\dim_{\mathcal{H}}(E) = \inf \left\{ h \geq 0; \ \mathcal{H}^h_\sigma(E) = 0 \right\} .$$

We finally recall the definition of rectifiable set given in [FH1].

DEFINITION 2.4 – Let (M, σ) be a metric space, and let $E \subset M$, $h \in \mathbf{N}$. We say that E is h-rectifiable if a Lipschitz function exists that maps a bounded set of $\mathbf{R}^h$ on E. We say that E is countably h-rectifiable if E is a countable union of h-rectifiable sets. We say that E is countably $(\mathcal{H}^h_\sigma, h)$-rectifiable if a countably h-rectifiable set F exists such that $\mathcal{H}^h_\sigma(E \setminus F) = 0$. We say that E is $(\mathcal{H}^h_\sigma, h)$-rectifiable if it is countably $(\mathcal{H}^h_\sigma, h)$-rectifiable and $\mathcal{H}^h_\sigma(E) < +\infty$.

The following characterization of the countably $(\mathcal{H}^h, h)$-rectifiable subsets of $\mathbf{R}^n$ holds.

THEOREM 2.5 – *A subset $E \subset \mathbf{R}^n$ is countably $(\mathcal{H}^h, h)$-rectifiable if and only if a sequence $(S_i)_i$ of h-dimensional manifolds of class C^1 exists such that*

$$\mathcal{H}^h \left(E \setminus \bigcup_{i=1}^{\infty} S_i \right) = 0 \quad .$$

We finally recall that the following *area formula* holds.

THEOREM 2.6 – *Let $h, n \in \mathbf{N}$ with $h \leq n$, let $A \subset \mathbf{R}^h$ be a open set and let $f \in (C^1(A))^n$. Denoted by $J(f)$ the Jacobian matrix of f and by $J(f)^*$ its transposed matrix, we have*

$$\int_A \sqrt{\det(J(f)^* J(f))} \, dy = \int_{\mathbf{R}^n} \mathcal{H}^0(f^{-1}(x)) \, d\mathcal{H}^h(x) .$$

A partitioning problem that we may regard as freely inspired by considerations on soap bubble clusters has been recently studied by F. J. Almgren in [AF]. As a particular case of the results in that paper we have the following theorem.

THEOREM 2.7 – *Let $\nu \in \mathbf{N}$ and let $a_1, \ldots, a_\nu$ be positive real numbers; then there exists*

$$\min \left\{ \mathcal{H}^{n-1} \left(\bigcup_{i=1}^{\nu} \partial A_i \right) ; A_i \in \mathcal{A}(\mathbf{R}^n), \ \mathcal{H}^n(A_i) = a_i \right\} .$$

Using the definitions introduced in Conversations 1 we can formulate the following conjecture.

CONJECTURE 2.8 – For all $n \geq 2$, if the open sets $A_1, \ldots, A_\nu$ give the minimum in Theorem 2.7, then $(\bigcup_{i=1}^{\nu} \partial A_i) \in V_{n-1} PC^\omega(\mathbf{R}^n)$.

To our knowledge, until now Conjecture 2.8 (easy to prove in the case $n = 2$) has been proved for $n = 3$ by J.E. Taylor in [TJ], while we think it is still open for $n > 3$ and considerable difficulties are expected for $n \geq 8$ (see [FH2]), a case that seems very interesting even though far from the physical phenomenon of the soap bubble cluster.

In this respect, I note that often an interesting mathematical problem may arise through the most daring elaborations even from a very superficial observation of certain phenomena, and even from the development of a theory that may result completely inconsistent from the physical standpoint. On the other hand the study of some mathematical problems, vaguely inspired by a certain phenomenon, may be an incentive to the research of mathematical models more suited to that phenomenon.

A problem of the same type of that considered by J.E. Taylor, but maybe a little closer to the physical problem of the soap bubble cluster, is the following one:

find the

$$\min\left\{\mathcal{H}^{n-1}\left(\bigcup_{i=1}^{\nu}\partial A_i\right) + p_0\sum_{i=1}^{\nu}\mathcal{H}^n(A_i) - \sum_{i=1}^{\nu}p_i\log\mathcal{H}^n(A_i)\right\},$$

where $p_0,\dots,p_\nu$ *are given positive numbers and the open sets* A_i *are pairwise disjoint.*

Beside the analytical problems considered above and, more or less freely, inspired by the physical problems of soap bubble clusters, it would be interesting to look for the best mathematical models for the equilibrium states and for the dynamics of soap bubble clusters in presence of different forces, constraints, etc. Clearly, such models should be looked for in the mathematical-physical literature (or invented, in case they have not been yet invented).

Beside the problems inspired by soap bubbles there are many other "partitioning problems" that come up in approximation theory. For example, if we consider the approximation of a function g with piecewise-constant functions the following theorem proved in [CT] holds.

THEOREM 2.9 – *Let* $n \geq 2$, $\Omega \in \mathcal{A}(\mathbf{R}^n)$, $p \geq 1$ *and let* $g \in L^p(\Omega) \cap L^\infty(\Omega)$. *Then there exists a pair* (K, u) *solution of the problem*

$$\min_{K,u}\left\{\mathcal{H}^{n-1}(K\cap\Omega) + \int_{\Omega\setminus K}|u - g|^p dx\right\}$$

when K *varies among all closed sets of* $\mathbf{R}^n$, $u \in C^1(\Omega\setminus K)$ *and* $\nabla u \equiv 0$ *in* $\Omega\setminus K$.

Passing to the approximation with piecewise-affine functions or piecewise-polynomial functions we have the following conjecture.

CONJECTURE 2.10 – Let $h \geq 0$, $k \geq 2$, $n \geq 2$ be integers, $\Omega \in \mathcal{A}(\mathbf{R}^n)$, $p \geq 1$ and let $g \in L^p(\Omega) \cap L^\infty(\Omega)$. Then there exists a pair (K, u) solution of the problem

$$\min_{K,u}\left\{\mathcal{H}^{n-1}(K\cap\Omega) + \int_{\Omega\setminus K}\left(\sum_{i=1}^{h+k-1}|\nabla^i u|^2 + |u - g|^p\right) dx\right\}$$

when K varies among all closed sets of $\mathbf{R}^n$, $u \in C^h(\Omega) \cap C^{h+k}(\Omega \setminus K)$ and $\nabla^{h+k} u \equiv 0$ in $\Omega \setminus K$.

Bibliography

[AF] F.J. Almgren, *Existence and regularity almost everywhere of solutions to elliptic variational problems with constraints*, Mem. Am. Math. Soc. **165** (1976).

[AT] F. Almgren, J.E. Taylor, *Geometry of soap films*, Sci. Am. **235** (1976), 82–89.

[CT] G. Congedo, I. Tamanini, *On the existence of solutions to a problem in multidimensional segmentation*. Ann. Inst. H. Poincaré Anal. Non Linéaire **8** (1991), 175–195.

[FH1] H. Federer, *Geometric Measure Theory*, Springer, Berlin, 1969.

[FH2] H. Federer, *The singular set of area minimizing rectifiable currents with codimension one and of area minimizing flat chains modulo two with arbitrary codimension*, Bull. Am. Math. Soc. **76** (1970), 767–771.

[MF] F. Morgan, *Soap bubbles and soap films*, in Joseph Malkevitch and Donald McCarthy, ed., *Mathematical Vistas: New and Recent Publications in Mathematics from the New York Academy of Sciences*, Vol. 607, 1990, 98–106.

[TJ] J.E. Taylor, *The structure of singularities in soap-bubble-like and soap-film-like minimal surfaces*, Ann. of Math. **103** (1976), 489–524.

Conversation 3

**A problem related to Image Segmentation
and other free discontinuity problems**

Recently, various researches on Image Segmentation problems in the framework of theoretical Computer Science (see [MS]) have led to the consideration of problems admitting solutions that are regular except for an *a priori* unknown discontinuity set. Wishing to provide a very easy example of such a situation we consider the following existence theorem, proved in [DGCL].

Theorem 3.1 – *Let $n \geq 2$, $p \geq 1$, $\Omega \in \mathcal{A}(\mathbf{R}^n)$, $g \in L^p(\Omega) \cap L^\infty(\Omega)$; then there exists*

$$(3.1) \qquad \min_{K,u} \left\{ \int_{\Omega \setminus K} |\nabla u|^2 dx + \int_{\Omega \setminus K} |u - g|^p dx + \mathcal{H}^{n-1}(K \cap \Omega) \right\},$$

where K is a closed set of $\mathbf{R}^n$ and $u \in C^1(\Omega \setminus K)$.

It can be proved, in addition, that if the pair (K, u) gives the minimum in (3.1), then there exist a closed set $K' \subset K$ and a function $u' \in C^1(\Omega \setminus K')$ such that $\mathcal{H}^{n-1}[(K \setminus K') \cap \Omega] = 0$, $u' = u$ in $\Omega \setminus K$, the pair (K', u') is still minimizing and for all $x \in K' \cap \Omega$ we have

$$(3.2) \qquad \liminf_{\rho \to 0} \rho^{1-n} \mathcal{H}^{n-1}(K' \cap B_\rho(x)) > 0.$$

The conjecture that the set K' possesses, beside the property expressed by (3.2), a much higher degree of regularity is indeed well grounded (see [DG]).

CONJECTURE 3.2 – Let (K, u) be a minimizing pair as in Theorem 3.1; then there exist n closed sets $K_0 \subset K_1 \subset \ldots \subset K_{n-1} \subset K$ such that $\mathcal{H}^{n-1}(K \setminus K_{n-1}) = 0$, $K_i \in V_i C^1(\Omega \setminus K_{i-1})$ for $i = 1, \ldots, n-1$, and $K_0 \in V_0 C^1(\Omega)$.

A problem that seems related to the reconstruction of blurred images will be considered in Conversation 6 (see Conjecture 6.4).

The terminology "free-discontinuity problems" has been introduced in [DG] to denote problems similar to that considered in Theorem 3.1. A free-discontinuity problem in an open set $\Omega \subset \mathbf{R}^n$ is a variational problem whose solution is a pair (K, u), where K is a closed set and $u \in C^\alpha(\Omega \setminus K)$.

Other existence theorems for free-discontinuity problems have been recently proved in [CL1] and [CL2].

THEOREM 3.3 – Let $n \geq 2$, let Ω be a bounded domain in $\mathbf{R}^n$, with $\mathcal{H}^{n-1}(\partial\Omega) < +\infty$; let M be a closed set such that $\mathcal{H}^{n-1}(M) = 0$ and $\partial\Omega \in V_{n-1} C^1(\mathbf{R}^n \setminus M)$; let $w \in C^1(\partial\Omega \setminus M) \cap L^\infty(\partial\Omega \setminus M)$. Then there exists

$$(3.3) \qquad \min_{K, u} \left\{ \int_{\Omega \setminus K} |\nabla u|^2 dx + \mathcal{H}^{n-1}(K \cap \overline{\Omega}) \right\},$$

where K is a closed set of $\mathbf{R}^n$ and $u \in C^1(\Omega \setminus K) \cap C^0(\overline{\Omega} \setminus (M \cup K))$ with $u = w$ on $\partial\Omega \setminus (M \cup K)$.

In the following theorem we consider a problem where the function u takes the values in the sphere $S^k = \{z \in \mathbf{R}^{k+1}; |z| = 1\}$ (for analogous problems see [BCL]).

THEOREM 3.4 – Let $n \geq 2$, $k \in \mathbf{N}$, $p \geq 1$, $\Omega \in \mathcal{A}(\mathbf{R}^n)$, $g \in [L^p(\Omega) \cap L^\infty(\Omega)]^{k+1}$; then there exists

$$\min_{K, u} \left\{ \int_{\Omega \setminus K} |\nabla u|^2 dx + \int_{\Omega \setminus K} |u - g|^p dx + \mathcal{H}^{n-1}(K \cap \Omega) \right\},$$

where K is a closed set of $\mathbf{R}^n$ and $u \in [C^1(\Omega \setminus K)]^{k+1}$ with $|u| = 1$.

It may be interesting both from the viewpoint of numerical computations and from a mathematical-physical viewpoint to approximate free-discontinuity problems with standard problems of the Calculus of Variations and of the theory of partial differential equations.

For example we may consider the following conjecture.

CONJECTURE 3.5 – There exist two sequences (φ_h), (ψ_h), with $\varphi_h \in C^1(\mathbf{R})$, $\psi_h \in C^1(\mathbf{R})$ such that for all $n \geq 2$, $k \in \mathbf{N}$, $\Omega \in \mathcal{A}(\mathbf{R}^n)$ with $\partial\Omega \in V_{n-1}C^1(\mathbf{R}^n)$, $g \in [L^2(\Omega) \cap L^\infty(\Omega)]^{k+1}$ we have

$$\lim_{h \to +\infty} \min\left\{ \int_\Omega \left(\varphi_h(|u|) + \psi_h(|u|)|\nabla u| + |u - g|^2 \right) dx; \ u \in [C^1(\Omega)]^{k+1} \right\}$$

$$= \min_{K,u}\left\{ \int_{\Omega \setminus K} |\nabla u|^2 dx + \int_{\Omega \setminus K} |u - g|^2 dx + \mathcal{H}^{n-1}(K \cap \Omega); \right.$$

$$\left. K \ \text{closed set} \ , \ u \in [C^1(\Omega \setminus K)]^{k+1}, \ |u| = 1 \right\}.$$

Bibliography

[BCL] H. BREZIS, J.M. CORON, E.H.LIEB, *Harmonic maps with defects*, Comm. Math. Phys. **107** (1986), 679–705.

[CL1] M. CARRIERO, A. LEACI, *Existence theorem for a Dirichlet problem with free discontinuity set*, Nonlinear Anal. **15** (1990), 661–677.

[CL2] M. CARRIERO, A. LEACI, S^k-*valued maps minimizing the L^p-norm of the gradient with free discontinuities.* Ann. Scuola Norm. Sup. Pisa Cl. Sci. (4) **18** (1991), 321–352.

[DG] E. DE GIORGI, *Free discontinuity problems in Calculus of Variations, Frontiers in pure and applied Mathematics, a collection of papers dedicated to J.L.Lions on the occasion of his 60th birthday* (R. Dautray ed.), North-Holland, Amsterdam, 1991.

[DGCL] E. DE GIORGI, M. CARRIERO, A. LEACI, *Existence theorem for a minimum problem with free discontinuity set.* Arch. Rational Mech. Anal. **108** (1989), 195–218.

[MS] D. MUMFORD, J. SHAH, *Optimal approximations by piecewise smooth functions and associated variational problems.* Comm. Pure Appl. Math. **17** (1989), 577–685.

Conversation 4

Geometric-distributional derivatives

If we examine the proofs of the theorems stated in conversations 2 and 3, for which we refer to the cited papers, we note that those proofs are essentially based on the theory of the perimeters of Caccioppoli sets, on Geometric Measure Theory, on the theory of Varifolds and on the theory of functions with bounded variation (with particular regard to the theory of special functions with bounded variation introduced in [DGA]). It seems fairly natural then to look for the most useful ideas for the unification and development of those theories. In that direction a very recent attempt, whose validity is, maybe, too early to judge, is given by the introduction of the concept of geometric-distributional derivative that we now describe.

DEFINITION 4.1 – Let μ be a real-valued (not necessarily positive) measure defined on the bounded Borel sets of $\mathbf{R}^n$. If $|\mu|$ is its total variation, we then have $|\mu|(K) < +\infty$ for all compact sets K. Let $E \subset \mathbf{R}^n$, $f : E \to \mathbf{R}^k$ and let $x \in \mathbf{R}^n$ be such that

$$\lim_{\rho \to 0} \frac{|\mu|(B_\rho(x) \cap E)}{|\mu|(B_\rho(x))} = 1;$$

we say that $a \in \mathbf{R}^k$ is the μ-approximate limit of f in x, and we set $a = \mu\text{-ap} \lim_{y \to x} f(y)$ if:

$$\lim_{\rho \to 0} \frac{1}{|\mu|(B_\rho(x))} \int_{B_\rho(x) \cap E}^{*} (|f(y) - a| \wedge 1) \, d|\mu|(y) = 0$$

where $\int^{*}$ denotes the upper integral.

REMARK 4.2 – This definition of approximate limit is slightly more general than the definition of Lebesgue point and is inspired by the definition of (μ, V)-ap lim given by H.Federer (see [FH1], 2.9.12).

DEFINITION 4.3 – Let μ and f be as in Definition 4.1, and let $\tilde{f}(x) = \mu\text{-ap} \lim_{y \to x} f(y)$; moreover, for $w \in \mathcal{L}(\mathbf{R}^n, \mathbf{R}^k)$ (that is, $w : \mathbf{R}^n \to \mathbf{R}^k$, w linear) let:

$$\psi_x(w, y) = \begin{cases} \dfrac{|f(y) - \tilde{f}(x) - w(y - x)|}{|y - x|} & \text{if } y \neq x, \\[2mm] 0 & \text{if } y = x. \end{cases}$$

We set $w \in {}_\mu\mathcal{D}f(x)$ if

$$\mu\text{-ap} \lim_{y \to x} \psi_x(w, y) = 0.$$

Note that ${}_\mu\mathcal{D}f(x)$ is a convex and closed subset of $\mathcal{L}(\mathbf{R}^n, \mathbf{R}^k)$, that we think equipped with the Hilbert norm

$$|w|^2 = \sum_{i=1}^{n} \sum_{j=1}^{k} \langle w(e_i), e_j' \rangle^2$$

where (e_i) (resp., (e_j')) is an orthonormal basis of $\mathbf{R}^n$ (resp., $\mathbf{R}^k$). This allows the following definition.

DEFINITION 4.4 – If $_\mu\mathcal{D}f(x) \neq \emptyset$ then we denote by $_\mu\nabla f(x)$ the element of minimal norm in $_\mu\mathcal{D}f(x)$; that is, we set $w = {_\mu\nabla} f(x)$ if $w \in {_\mu\mathcal{D}}f(x)$ and $|w| \leq |v|$ for all $v \in {_\mu\mathcal{D}}f(x)$.

DEFINITION 4.5 – Let μ be as in Definition 4.1. For all x in the support of μ we denote by $N\mu(x)$ and by $T\mu(x)$, respectively, the normal space and the tangent space to μ at x, defined as:

$$N\mu(x) = \{z \in \mathbf{R}^n : \langle _\mu\nabla f(x), z \rangle = 0 \ \text{ for all } \ f \in C^1(\mathbf{R}^n)\};$$

$$T\mu(x) = \text{ the orthogonal complement of } N\mu(x).$$

REMARK 4.6 – For all x in the support of μ and for all $\varphi : \mathbf{R}^n \to \mathbf{R}^n$ we denote by $\mathcal{P}_{T\mu}\varphi(x)$ the projection of $\varphi(x)$ on $T\mu(x)$. Then for all $f \in C^1(\mathbf{R}^n)$ we set

$$_\mu\nabla f(x) = \mathcal{P}_{T\mu}\nabla f(x).$$

REMARK 4.7 – If $E \in V_h C^1(\mathbf{R}^n)$, $\mu(B) = \mathcal{H}^h(B \cap E)$, $f \in C^1(\mathbf{R}^n)$, then $T\mu(x)$ is the usual tangent space to E and $_\mu\nabla f(x)$ is the usual tangential gradient of f along E.

DEFINITION 4.8 – Let μ be as in Definition 4.1, and let γ be a vector measure with n components satisfying the same conditions of μ. We say that γ is the geometric-distributional derivative of μ and we set $\gamma = GDD\mu$ if for all $f \in C_0^1(\mathbf{R}^n)$ we have

$$(4.1) \qquad \int_{\mathbf{R}^n} {_\mu\nabla_i} f d\mu + \int_{\mathbf{R}^n} f \, d\gamma_i = 0 \qquad \text{for all } i = 1, ..., n.$$

Formula (4.1) should be compared with the divergence theorem (see for example [SL]).

For further developments in the study of the properties of the geometric-distributional derivative it would be important to prove or disprove the following conjecture.

CONJECTURE 4.9 – If μ is a measure satisfying the conditions of Definition 4.1 and $GDD\mu$ exists then there exist $n + 1$ functions $f_0, f_1, ..., f_n$ such that for all bounded Borel sets $B \subset \mathbf{R}^n$ we have:

$$\mu(B) = \sum_{i=0}^{n} \int_B f_i(x) \, d\mathcal{H}^i(x).$$

REMARK 4.10 – If $\mu(B) = \int_B u(x)dx$, where u is a function with bounded variation in $\mathbf{R}^n$, then (4.1) reduces to

$$\int_{\mathbf{R}^n} u(x)\nabla f(x)dx + \int_{\mathbf{R}^n} f \, Du = 0$$

in the sense of distributions (see [MM] and also [GE]).

REMARK 4.11 – If $a, b \in \mathbf{R}^n$, $E = \{at + (1-t)b, 0 \le t \le 1\}$, and $\mu(B) = \mathcal{H}^1(B \cap E)$, then (4.1) reduces to

$$\int_{\mathbf{R}^n} {}_\mu \nabla f d\mu = \frac{b-a}{|b-a|} \left(f(b) - f(a) \right).$$

REMARK 4.12 – If $E \in V_1 C^2(\mathbf{R}^n) \cap \mathcal{K}(\mathbf{R}^n)$ and $\mu(B) = \mathcal{H}^1(B \cap E)$, then, denoting by $c(x)$ the curvature vector at the point $x \in E$, formula (4.1) reduces to

$$\int_{\mathbf{R}^n} {}_\mu \nabla f d\mu + \int_E f(x)c(x)d\mathcal{H}^1(x) = 0.$$

REMARK 4.13 – The considerations above may be clearly localized by considering measures defined on the Borel subsets of an arbitrary open set of $\mathbf{R}^n$.

REMARK 4.14 – The concept of geometric-distributional derivative can be extended without difficulties to vector measures and hence we may pass to the geometric-distributional derivatives of order higher than the first. These derivatives will be tensor measures and will be denoted by GDD^i.

After the introduction of the geometric-distributional derivatives of higher order we can consider the following conjecture.

CONJECTURE 4.15 – Let P be a polyhedron in $\mathbf{R}^n$, and let $\mu(B) = \mathcal{H}^n(B \cap P)$ for all Borel sets $B \subset \mathbf{R}^n$. Then for all i there exist the geometric-distributional derivatives $GDD^i \mu$ and in addition $GDD^{n+1} \mu = 0$.

REMARK 4.16 – It would be interesting to study the behaviour of the geometric-distributional derivative with respect to changes of coordinates.

CONJECTURE 4.17 – Let μ be such that $GDD\mu$ exists, and let $\varphi : \mathbf{R}^n \to \mathbf{R}$ be a Lipschitz function; then for μ-a.e. $x \in \mathbf{R}^n$ ${}_\mu \nabla \varphi(x)$ exists and we have

$$\int_{\mathbf{R}^n} {}_\mu \nabla \varphi d\mu + \int_{\mathbf{R}^n} \varphi GDD\mu = 0.$$

CONJECTURE 4.18 – Let μ be such that $GDD\mu$ exists, and let $\varphi : \mathbf{R}^n \to \mathbf{R}$ be a Lipschitz function; then there exists $GDD(\varphi\mu)$ and we have:

$$GDD(\varphi\mu) = \varphi \, GDD\mu + ({}_\mu \nabla \varphi)\mu.$$

Bibliography

[DGA] E. DE GIORGI, L. AMBROSIO, *New functionals in calculus of variations*, Atti Accad. Naz. Lincei Rend. Cl. Sci. Fis. Mat. Natur. (8) **82** (1988), 199–210 (1989).

[FH1] H. FEDERER, *Geometric Measure Theory*, Springer, Berlin, 1969.

[GE] E.Giusti, *Minimal Surfaces and Functions of Bounded Variation*, Birkhäuser, 1984.

[MM] M.Miranda, *Distribuzioni aventi derivatives misure. Insiemi di perimetro localmente finito*, Ann. Sc. Norm. Sup. Pisa **18** (1964), 27-56.

[SL] L.Simon, *Lectures on Geometric Measure Theory*, Proc. Centre Math. Anal. **3** Australian Natl. Univ., 1983.

Conversation 5

Tangential derivatives
and derivatives of the square of the distance from a manifold

In the theory of geometric-distributional derivatives an important role is played by the operator $_\mu\nabla$. We have noted that if μ is given by the formula $\mu(B) = \mathcal{H}^h(E \cap B)$ with $E \in V_h C^1(\mathbf{R}^n)$, then $_\mu\nabla f(x)$ is the orthogonal projection of $\nabla f(x)$ onto the tangent space to E at x and this observation may be extended (by remark 4.13) to the case of $E \in V_h C^1(\Omega)$.

In the sequel we equally use the notation $_\mu\nabla f(x)$, $_E\nabla f(x)$ or $_w\nabla f(x)$ in the case when $\mu(B) = \mathcal{H}^h(B \cap E)$ and we have $\mathcal{H}^h(B_\rho(x) \cap E) > 0$ for all $\rho > 0$ or $\mu(B) = \int_B w d\mathcal{H}^h$ and we have $\int_{B_\rho(x)} w d\mathcal{H}^h > 0$ for all $\rho > 0$ and $\int_{B_\rho(x)} w d\mathcal{H}^h < +\infty$ for some $\rho > 0$.

For the study of higher-order derivatives it may be of some use the comparison with the tangential derivatives with respect to a regular manifold introduced by using the squared distance function from the manifold.

Let E be a closed set in $\mathbf{R}^n$; we set

$$DS\, E(x) = \frac{1}{2}[\mathrm{dist}(x, E)]^2.$$

For notational simplicity, for all functions w and for all measures μ, we set

$$DS\, w = DS\, \mathrm{supp} w, \qquad DS\, \mu = DS\, \mathrm{supp} \mu.$$

Remark 5.1 – Setting

$$Amb V_h C^\alpha(E) = \cup\{A \in \mathcal{A}(\mathbf{R}^n); E \in V_h C^\alpha(A)\},$$

if $\Omega = Amb V_h C^\alpha(E)$ then $E \in V_h C^\alpha(\Omega)$.

Remark 5.2 – If $E = \cup_{h=0}^n E_h$ with $E_h \in V_h C^\omega(\mathbf{R}^n)$ and $E_i \cap E_j = \emptyset$ for $i \neq j$ then there exists an open set Ω containing E such that $DS\, E \in C^\omega(\Omega)$.

Conjecture 5.3 – If $DS\, E \in C^\omega(\Omega)$ (resp., C^∞) then $E = \cup_{h=0}^n E_h$ with $E_h \in V_h C^\omega(\Omega)$ (resp., C^∞) and $E_i \cap E_j = \emptyset$ for $i \neq j$.

Remark 5.4 – It is quite easy to check that the following formulas hold.

If $DS\ E$ is of class C^1 in a neighborhood U of the set E then

$$|\nabla DS\ E|^2 = 2\ DS\ E \qquad \text{in } U.$$

Let $f \in (C^\omega(\mathbf{R}^h))^{n-h}$, with $1 \leq h < n$, be such that $|f(0)| = |\nabla f(0)| = 0$; let

$$E = \{(x_1, \ldots, x_n) \in \mathbf{R}^n; x_{i+h} = f_i(x_1, \ldots, x_h), 1 \leq i \leq n - h\}.$$

Then

$$DS\ E(0) = 0, \quad \nabla DS\ E(0) = 0,$$
$$\partial_{x_i}\partial_{x_k} DS\ E(0) = 0 \qquad \text{for } i \neq k,$$
$$\partial_{x_i}\partial_{x_i} DS\ E(0) = 0 \qquad \text{for } i \leq h,$$
$$\partial_{x_i}\partial_{x_i} DS\ E(0) = 1 \qquad \text{for } h < i \leq n,$$
$$\partial_{x_{i+h}}\partial_{x_j}\partial_{x_k} DS\ E(0) = -\partial_{x_j}\partial_{x_k} f_i(0) \qquad \text{for } 1 \leq i \leq n - h, \quad 1 \leq j, k \leq h.$$

REMARK 5.5 – Let $E \in V_h C^\infty(\mathbf{R}^n)$, let $A \in \mathcal{A}(\mathbf{R}^n)$ be such that $DS\ E \in C^2(A)$ and let f be a function of class $C^1(A)$; for all $x \in E \cap A$ we consider $_E\nabla f(x)$; then we have

$$_E\nabla_i f(x) = \partial_{x_i} f(x) - \sum_{j=1}^n \partial^2_{x_i x_j} DS\ E(x)\ \partial_{x_j} f(x).$$

Moreover, setting $p_i(x) = x_i$ for $i = 1, \ldots, n$ and $Id_n = (p_1, \ldots, p_n)$, we have

$$(5.1) \qquad\qquad \partial^2_{x_i x_j} DS\ E + {}_E\nabla_i p_j = \delta_{ij}.$$

It follows that for all k, n there exist polynomials $\varphi_{k,n}$ such that

$$_E\nabla^k(Id_n) = \varphi_{k,n}(\nabla^2(DS\ E), \ldots, \nabla^{k+1}(DS\ E)).$$

For $k = 1$ (5.1) allows to express $\nabla^2 DS\ E$ in terms of $_E\nabla Id_n$; the possibility of extending this result is stated in the following conjecture.

CONJECTURE 5.6 – For all k, n there exist polynomials $\psi_{k,n}$ such that

$$\nabla^{k+1}(DS\ E) = \psi_{k,n}(\ _E\nabla(Id_n), \ldots, \ _E\nabla^k(Id_n)).$$

It would be interesting to study the algebraic properties of $\varphi_{k,n}$ (and possibly of $\psi_{k,n}$). More in general it would be interesting to try to express all geometric-differential properties of a manifold E embedded in $\mathbf{R}^n$ through the derivatives $_E\nabla^k(Id_n)$ (or $\nabla^k(DS\ E)$).

From the viewpoint of the Calculus of Variations the formula of the first variation of the area is particularly interesting. In order to show how it can be written using the differential operators introduced above, we give the following definition.

DEFINITION 5.7 – Given $w \in F_h C^\omega(\mathbf{R}^n)$ and $\varphi \in C^2(\mathbf{R}^n)$, we define the tangential divergence and the tangential Laplacian of φ along w, respectively, by

$$_w \mathrm{div}\, \varphi = \sum_{i=1}^{n} {}_w\nabla_i \varphi_i \quad \text{and} \quad {}_w\triangle\varphi = \sum_{i=1}^{n} {}_w\nabla_i({}_w\nabla_i\varphi).$$

Let $w \in F_h C^\omega(\Omega)$; for all Ω' with closure compactly contained in Ω and for all $\psi \in (C_0^1(\Omega'))^n$ the following formula for the first variation of the area holds

$$\frac{\mathrm{d}}{\mathrm{d}t}\left[\int_{\Omega'} w(Id_n + t\psi)\, d\mathcal{H}^h\right]_{t=0} =$$

$$= -\int_{\Omega'} ({}_w\mathrm{div}\,\psi)\, w\, d\mathcal{H}^h = \int_{\Omega'} \langle {}_w\triangle Id_n, \psi\rangle\, w d\mathcal{H}^h.$$

The formulas giving the derivatives of $DS\,w$ for hypersurfaces in implicit form and for the sections of a manifold seem also interesting. In order to write them in a shorter form we set for all $w \in F_h C^\omega(\mathbf{R}^n)$

$$\nu(w)(x) = \nabla^2 DSw(x).$$

Moreover, if $a, b \in \mathbf{R}^n$, we set $(a \otimes b)_{ij} = a_i b_j$, $i, j = 1, \ldots, n$.

REMARK 5.8 – Let $f \in C^\omega(\mathbf{R}^n)$ and, for all $t \in \mathbf{R}$, let $E_t = \{x \in \mathbf{R}^n; f(x) = t\}$. For almost every $t \in \mathbf{R}$ we have

$$E_t \in V_{n-1} C^\omega(\mathbf{R}^n)\ ,$$

and

$$\nu(\chi_{E_t}) = \frac{\nabla f \otimes \nabla f}{|\nabla f|^2}\ .$$

If, in addition, $w \in F_h C^\omega(\mathbf{R}^n)$ with $h > 0$, then for almost every $t \in \mathbf{R}$ we have

$$w\chi_{E_t} \in F_{h-1} C^\omega(\mathbf{R}^n)\ ,$$

and for $\mathcal{H}^{h-1}$-almost every $x \in \mathrm{supp}\, w \cap E_t$ we have

$$\nu(w\chi_{E_t})(x) = \nu(w)(x) + \frac{{}_w\nabla f(x) \otimes {}_w\nabla f(x)}{|{}_w\nabla f(x)|^2}\ .$$

Bibliography

[CB] B.Y. CHEN, *Geometry of Submanifolds*, Marcel Dekker, 1973.

[CE] J. CHEEGER, D.G. EBIN, *Comparison Theorems in Riemannian Manifolds*, North Holland, 1973.

[DR] G. De Rham, *Variétés différentiables*, Hermann, 1960.

[SM] M. Spivak, *A Comprehensive Introduction to Differential Geometry, I-V*, Publish or Perish, 1970-1975.

Conversation 6
Variational problems involving tangential derivatives

The concepts introduced in the previous conversations allow to formulate interesting variational problems in which the unknown is a manifold. A problem that exhibits a strong coerciveness is considered in the following conjecture.

CONJECTURE 6.1 – Let $f \in C^\omega(\mathbf{R}^n)$ be such that $\lim_{|x| \to +\infty} f(x) = +\infty$, $i \in \mathbf{N}$, $i \geq 2$. Then there exists

$$\min\left\{\int_{\mathbf{R}^n} |_w\nabla^i(Id_n)|^2 w \, d\mathcal{H}^h + \left(\int_{\mathbf{R}^n} w \, d\mathcal{H}^h\right)^2 + \int_{\mathbf{R}^n} fw \, d\mathcal{H}^h\right\}$$

in the class of functions $w \in F_h C^\omega(\mathbf{R}^n)$.

We may ask whether an existence result of the same type holds under weaker coerciveness conditions, for example as in the following conjecture.

CONJECTURE 6.2 – Let $f \in C^\omega(\mathbf{R}^n)$ be such that $\lim_{|x| \to +\infty} f(x) = +\infty$, $\varepsilon > 0$ and $p > 0$. Then there exists

$$\min\left\{\int_{\mathbf{R}^n} \varepsilon \, |_w\triangle Id_n|^2 \, w \, d\mathcal{H}^h + \int_{\mathbf{R}^n} fw \, d\mathcal{H}^h\right\}$$

in the class of the functions $w \in F_h C^\omega(\mathbf{R}^n)$ such that $\int_{\mathbf{R}^n} w \, d\mathcal{H}^h \leq p$.

We may also consider variational problems with free discontinuity.

CONJECTURE 6.3 – Let $f \in C^\omega(\mathbf{R}^n)$ be such that $\lim_{|x| \to +\infty} f(x) = +\infty$, $\varepsilon > 0$, $p > 0$ and $\lambda > 0$. Then there exists

$$\min\left\{\int_{\mathbf{R}^n} \varepsilon \, |_w\triangle Id_n|^2 \, w \, d\mathcal{H}^h + \int_{\mathbf{R}^n} fw \, d\mathcal{H}^h + \lambda \, \mathcal{H}^{h-1}(K)\right\}$$

in the class of pairs (K, w), where $K \in \mathcal{K}(\mathbf{R}^n)$ and $w \in F_h C^\omega(\mathbf{R}^n \setminus K)$ satisfies the condition $\int_{\mathbf{R}^n} w \, d\mathcal{H}^h \leq p$.

Problems similar to those considered in the previous conjectures may be stated for $f \in C^\alpha$ with $\alpha \in \mathbf{N}$ or $\alpha = \infty$.

A problem that seems related to the reconstruction of blurred images is considered in the following conjecture.

CONJECTURE 6.4 – For all $\varphi \in L^1(\mathbf{R}^n, \mathcal{H}^h)$ we set

$$T_{h,\lambda}\varphi(x) = \int_{\mathbf{R}^n} \varphi(\xi)e^{-\lambda|x-\xi|^2}\, d\mathcal{H}^h(\xi);$$

then for all $g \in L^2(\mathbf{R}^n)$, $i \geq 2$, $\alpha, \beta > 0$ there exists

$$\min_{K,w,f,\lambda} \left\{ \int_{\mathbf{R}^n} \left(1 + |_w \nabla^i Id_n|^2\right) w\, d\mathcal{H}^h + \int_{\mathbf{R}^n} f^2 w\, d\mathcal{H}^h + \right.$$

$$\left. + \alpha \mathcal{H}^{h-1}(K) + \beta \int_{\mathbf{R}^n} |g - T_{h,\lambda}(fw)|^2\, dx \right\}$$

where $K \subset \mathbf{R}^n$ is a closed set, $w \in F_h C^i(\mathbf{R}^n \setminus K)$, f is a function such that $fw \in L^1(\mathbf{R}^n, \mathcal{H}^h)$ and $\lambda > 0$.

Bibliography

[AA] W.K. ALLARD, F.J. ALMGREN (EDS.), *Geometric Measure Theory and the Calculus of Variations*, Proc. Symp. Pure Math. **44** Am. Math. Soc., 1986.

[HJ] J.E. HUTCHINSON, *Second fundamental form for varifolds and the existence of surfaces minimizing curvature*, Indiana Univ. J. **35** (1986), 45-71.

Conversation 7

**Free-discontinuity problems related to the
notions of boundary and geodesic**

We give the definition of h-dimensional boundary of a measure μ.

DEFINITION 7.1 – Let μ be a measure as in Definition 4.1; we define the h-dimensional boundary of μ as the set $\partial_h \mu$ of points x satisfying the following property:

$$\text{there exists } \lim_{\rho \to 0^+} \cos\left(\frac{2\pi}{\omega_{h+1}\rho^{h+1}} \int_{B_\rho(x)} d\mu\right) = -1.$$

REMARK 7.2 – If $E \in V_{h+1}BC^1(\mathbf{R}^n)$ and if $\mu(B) = \mathcal{H}^{h+1}(E \cap B)$ then $\partial_h \mu \in V_h C^1(\mathbf{R}^n)$ and it coincides with the elementarily defined boundary of E.

We may adopt the notation $\partial_h w$ and $\partial_h E$ to denote the h-dimensional boundaries of the measures μ respectively defined by the formulas

$$\mu(B) = \int_B w d\mathcal{H}^{h+1}$$

and

$$\mu(B) = \mathcal{H}^{h+1}(E \cap B).$$

REMARK 7.3 – If $w = \sum_{i=1}^{\nu} w_i$, with $w_i \in F_{h+1}BC^1(\Omega)$, then the set $\partial_h w$ coincides with the set of the points x that belong to $\partial_h w_i$ for an odd number of values of the index i. With this consideration in mind, it would be very interesting to compare the notion of boundary defined in this way with the theory of currents modulo 2 (see [ZW]).

We may now state the following free-discontinuity problem related to the Plateau problem:

Find reasonable hypotheses on the set $S \subset \mathbf{R}^n$ so that for all closed sets K satisfying the condition $\mathcal{H}^h(K) < +\infty$, for all open sets $\Omega \subset \mathbf{R}^n$ and for all numbers $\lambda > 0$ there exists the

$$(7.1) \qquad \min_{w,E} \left\{ \int_\Omega w(x) d\mathcal{H}^{h+1}(x) + \lambda \mathcal{H}^h(E \cap \Omega) \right\}.$$

when E varies among all closed sets, $w \in F_{h+1}C^1(\Omega \setminus E)$, $\operatorname{supp} w \subset S$, $\partial_h w \cap (\Omega \setminus E) = K \cap (\Omega \setminus E)$.

CONJECTURE 7.4 – The minimum in (7.1) exists when $S \in V_r C^1(\mathbf{R}^n)$ with $h < r \le n$ or when S is the boundary of a polyhedron or is obtained as a finite union and intersection of boundaries of polyhedra.

The study of problem (7.1) suggests a rethinking of the concept of geodesic of which we now indicate the main lines.

DEFINITION 7.5 – Let $\Omega \in \mathcal{A}(\mathbf{R}^n)$, and $S \subset \mathbf{R}^n$. We say that a function w is $(\Omega, S, \mathcal{H}^{h+1})$-minimal if it satisfies the following conditions:

1.   there exists $K \subset \mathbf{R}^n$ closed with $\mathcal{H}^h(K \cap \Omega) = 0$ and $w \in F_{h+1}BC^1(\Omega \setminus K)$;

2. $\mathcal{H}^h(\partial_h w \cap \Omega) = 0$;

3. $\operatorname{supp} w \subset S$;

4.   if g and K' satisfy the conditions analogous to 1, 2, 3, and in addition there is a compact set $\Omega' \subset \Omega$ such that $w = g$ in $\Omega \setminus \Omega'$ then

$$\int_\Omega w d\mathcal{H}^{h+1} \le \int_\Omega g d\mathcal{H}^{h+1}.$$

CONJECTURE 7.6 – If w is $(\Omega, S, \mathcal{H}^{h+1})$-minimal then for $\mathcal{H}^{h+1}$-a.e. $x \in \Omega$ we have either $w(x) = 0$ or $w(x) = 1$.

DEFINITION 7.7 – We say that w is $(\mathcal{H}^{h+1}, S)$-geodesic if $\operatorname{supp} w \subset S$ and moreover for all $x \in \operatorname{supp} w$ there exist $\Omega, w_1, \ldots, w_\nu$ such that $w = \sum_{1=1}^{\nu} w_i$, $x \in \Omega$, and the w_i are $(\Omega, S, \mathcal{H}^{h+1})$-minimal.

It would be very interesting to find conditions on S such that closed geodesics exist in the various dimensions. The case in which S is a regular Riemannian manifold is considered in [PJ].

Bibliography

[PJ] J.T. Pitts, *Existence and Regularity of Minimal Surfaces on Riemannian Manifolds*, Princeton University Press, 1981.

[ZW] W.P. Ziemer, *Integral currents mod 2*, Trans. AMS **105** (1962), 496-524.

Conversation 8
Generalized Euler equations and evolution equations

The formula of the first variation of the area might be generalized by substituting the functional of the area with other functionals; to this end, we introduce the Euler operator Eu through the following definition.

DEFINITION 8.1 – Let $\Omega \in \mathcal{A}(\mathbf{R}^n)$ and let a functional $F(f, \Omega)$ be given; for all $g \in [C_0^\infty(\Omega)]^n$ we set

$$Eu(F, f, g, \Omega) = \left[\frac{d}{dt} F(f \circ (Id_n + tg), \Omega) \right]_{t=0}.$$

REMARK 8.2 – If $F(f, \Omega) = \int_\Omega f d\mathcal{H}^h$, with $f = \chi_E$ for $E \in V_h C^\omega(\Omega)$, then

$$Eu(F, f, g, \Omega) = \int_\Omega \langle {}_f \triangle Id_n, g \rangle f d\mathcal{H}^h$$

for all $g \in [C_0^\infty(\Omega)]^n$.

The generalization of the Euler formulas suggests analogous generalizations for the evolution problems corresponding to the area functional. To this end, it is convenient to associate to a functional defined on manifolds embedded in $\mathbf{R}^n$ the related evolution manifold embedded in $\mathbf{R}^{n+1}$ via the following Definition 8.3.

DEFINITION 8.3 – Let $\Omega \in \mathcal{A}(\mathbf{R}^{n+1})$. Let $f : \Omega \to \mathbf{R}$ be the characteristic function of a set $E \in V_{h+1} C^\omega(\Omega)$. We say that f is an evolution manifold for the functional F if, setting $f_t = f \cdot \chi_{\{x_{n+1}=t\}}$, for all $x \in \text{supp} f \cap \Omega$ there exists a unique $\psi(x) \in \mathbf{R}^{n+1}$ such that

$$|\psi(x)| = \lim_{\varepsilon \to 0} \frac{1}{\varepsilon} \text{dist}(x, \text{supp} f_{x_{n+1}+\varepsilon}),$$

$$\lim_{\varepsilon \to 0} \frac{1}{\varepsilon} \text{dist}[x + \varepsilon \psi(x), \text{supp} f_{x_{n+1}+\varepsilon}] = 0$$

and such that for all Ω' with closure compactly contained in Ω, for all $g \in [C_0^\infty(\Omega)]^{n+1}$ and for all $t \in \mathbf{R}$ we have

$$Eu(F, f_t, g, \Omega') = \int_{\Omega'} \sum_{i=1}^n \psi_i \, g_i \, f_t \, d\mathcal{H}^h.$$

It would be interesting to compare this definition with other ones already existing in the literature (for example in the cited works below).

Bibliography

[BK] K.A. BRAKKE, *The Motion of a Surface by its Mean Curvature*, Princeton University Press, 1978.

[GH] M. GAGE, R.S. HAMILTON, *The heat equation shrinking convex plane curves*, J. Diff. Geom., **23** (1986), 69-96.

[GM] M. GRAYSON, *Shortening embedded curves*, Ann. of Math., **129** (1989), 71-113.

[HG] G. HUISKEN, *Flow by mean curvature of convex surfaces into spheres*, J. Diff. Geom., **20** (1984), 237-266.

Conclusions

The problems presented in the previous conversations are only some examples of problems for which the existence of piecewise-regular solutions can reasonably be expected. We have limited ourselves to considering problems of the Calculus of Variations, even though there are, probably, differential equations not derived from variational problems for which the research of piecewise-regular solutions would be interesting. In the study of such equations it might be nonetheless useful to use variational methods that substitute the research of solutions of an equation of the type $f(x, u, \nabla u, \ldots, \nabla^i u) = 0$ by the research of the minima of integrals of the type $\int_\Omega (f(x, u, \nabla u, \ldots, \nabla^i u))^2 \, dx$.

Existence theorem for a minimum problem with free discontinuity set[‡]

E. De Giorgi, M. Carriero & A. Leaci[*]

Summary. We study the variational problem

$$\min\left\{\int_\Omega k|\nabla u|^2 dx + \mu \int_\Omega k|u - g|^q dx + \lambda H_{n-1}(K \cap \Omega) : \right.$$

$$\left. K \subset \mathbf{R}^n \text{ closed set, } u \in C^1(\Omega \setminus K)\right\},$$

where Ω is an open set in $\mathbf{R}^n$, $n \geq 2$, $g \in L^q(\Omega) \cap L^\infty(\Omega)$, $1 \leq q < +\infty$, $0 < \lambda, \mu < +\infty$ and H_{n-1} is the $(n-1)$-dimensional Hausdorff measure.

1. Introduction

In this paper the following theorem is proved.

THEOREM 1.1. – *Let $n \in \mathbf{N}$, $n \geq 2$, let $\Omega \subseteq \mathbf{R}^n$ be an open set, $1 \leq q < +\infty$, $0 < \lambda < +\infty$, $0 < \mu < +\infty$, $g \in L^q(\Omega) \cap L^\infty(\Omega)$; then there exists at least one pair (K, u) minimizing the functional G defined for every closed set $K \subset \mathbf{R}^n$ and for every $u \in C^1(\Omega \setminus K)$ by*

$$G(K, u) = \int_{\Omega \setminus K} k|\nabla u|^2 dx + \mu \int_{\Omega \setminus K} k|u - g|^q dx + \lambda H_{n-1}(K \cap \Omega),$$

where H_{n-1} is the $(n-1)$-dimensional Hausdorff measure.

This theorem was announced under slightly more restrictive hypotheses, during the Meeting in honour of J.L. Lions held in Paris from 6 to 10 June 1988 (see [10]). It provides the beginning of a positive answer to a two-dimensional problem of image segmentation in Computer Vision Theory posed by D. MUMFORD and J. SHAH in [21].

The proof includes investigation of the properties of some functional defined on certain classes of functions with bounded variation (the classes SBV(Ω) recently introduced in [11] for possible schematizations for various problems in mathematical physics (see [4], [6], [8], [14], [20], [22])).

The result here stated, besides sufficing to prove the above cited theorem, are a first step in the study of the regularity of functions which minimize the functionals defined in SBV(Ω) classes (see [11], [13], [7], [3]).

The plan of the exposition is the following.

[‡]Editor's note: published in Arch. Rational Mech. Anal., **108** (1989), 195–218.
[*]*Communicated by* E. GIUSTI

The second section recalls the definition and some properties of the class SBV(Ω) for use in this paper.

In a third section a Poincaré-Wirtinger type inequality for functions of class SBV in a ball (see Theorem 3.1) and some consequences are proved.

The fourth section is devoted to the study of properties of quasi-minima for some functionals defined on SBV(Ω): in particular the behavior of a sequence of quasi-minima is studied (see Theorem 4.8) and a theorem concerning the singular set S_u is proved (see Theorem 4.12).

In the last section the functional G considered in Theorem 1.1 is compared with another functional $\overline{G}$ defined on SBV$_{\mathrm{loc}}(\Omega)$. Finally, Theorem 1.1 is established by proving first the existence of at least one minimizer of $\overline{G}$ (see Lemma 5.1), then that such a minimizer, by Theorem 4.12, gives also the minimum of G (see Lemma 5.2 and Conclusion).

2. Preliminary results for functions in SBV(Ω)

Given an open set $\Omega \subseteq \mathbf{R}^n$, we define, following [11], the class of special functions of bounded variation SBV(Ω) and we point out some of their properties.

For a given set $E \subseteq \mathbf{R}^n$ we denote by $H_k(E)$ ($0 \le k \le n$) its k-dimensional Hausdorff measure, by $|E|$ its Lebesgue outer measure, by $\overline{E}$ its topological closure and by ∂E its topological boundary.

We denote by $B_\varrho(x)$ the ball $\{y \in \mathbf{R}^n; |y - x| < \varrho\}$, and we set $B_\varrho = B_\varrho(0)$, $\omega_n = |B_1|$.

Let $u : \Omega \to \mathbf{R}$ be a Borel function; for $x \in \Omega$, $z \in \widetilde{\mathbf{R}} = \mathbf{R} \cup \{\infty\}$ we set (following [11]) $z = \operatorname*{aplim}_{y \to x} u(y)$, (the approximate limit of u at x, denoted also by $\widetilde{u}(x)$) if

$$g(z) = \lim_{\varrho \to 0} |B_\varrho|^{-1} \int_{B_\varrho} g(u(x + y))\, dy$$

for every $g \in C^0(\widetilde{\mathbf{R}})$; if $z \in \mathbf{R}$ this definition is equivalent to 2.9.12 in [15].

The set

$$S_u = \left\{ x \in \Omega;\ \operatorname*{aplim}_{y \to x} u(y)\ \text{does not exist} \right\}$$

is a Borel set of negligible Lebesgue measure. Let $x \in \Omega \setminus S_u$ be such that $\widetilde{u}(x) \in \mathbf{R}$; we say that u is approximately differentiable at x if there exists a vector $\nabla u(x) \in \mathbf{R}^n$ (the approximate gradient of u at x) such that

$$\operatorname*{aplim}_{y \to x} \frac{|u(y) - \widetilde{u}(x) - \nabla u(x) \cdot (y - x)|}{|y - x|} = 0.$$

For every $u \in L^1_{\mathrm{loc}}(\Omega)$ we define (see [19])

$$\int_{\Omega \setminus K} |Du| = \sup \left\{ \int_{\Omega \setminus K} u \operatorname{div} \varphi \, dx;\ \varphi \in C^1_0(\Omega; \mathbf{R}^n), |\varphi| \le 1 \right\}.$$

By $BV(\Omega)$ we denote the Banach space of all functions u of $L^1(\Omega)$ with $\int_{\Omega \setminus K} |Du| < +\infty$; moreover, by $BV_{\mathrm{loc}}(\Omega)$ we denote the space of all functions which belong to $BV(\Omega')$ for every open set $\Omega' \subset\subset \Omega$ (*i.e.* $\overline{\Omega}'$ is compact and $\overline{\Omega}' \subset \Omega$). It is well known that $u \in BV(\Omega)$ if and only if $u \in L^1(\Omega)$ and its distributional derivative Du is a bounded vector measure.

If $E \subseteq \mathbf{R}^n$ is a Borel set, we define the perimeter of E in Ω as $P(E, \Omega) = \int_{\Omega \setminus K} |D1_E|$ where 1_E is the characteristic function of E.

For the main properties of functions with bounded variation we refer *e.g.* to [12], [15], [17], [18]. Here we recall only that for every $u \in BV(\Omega)$ the following properties hold:

S_u is countably $(n-1)$-rectifiable (see [9], or [15], 4.5.9(16));

$H_{n-1}(\{x \in \Omega \,; \tilde{u}(x) = \infty\}) = 0$ (see [15], 4.5.9(3));

∇u exists a.e. on Ω and coincides with a Radon-Nikodym derivative of Du with respect to the Lebesgue measure (see [15], 4.5.9(26));

the coarea formula holds (see *e.g.* [15], Theorem 1.23)

$$(2.1) \qquad \int_{\Omega \setminus K} |Du| = \int_{-\infty}^{+\infty} P(\{u < t\}, \Omega)\, dt;$$

moreover, for H_{n-1} almost all $x \in S_u$ there exist $\nu(x) \in \partial B_1$, $u_+(x) \in \mathbf{R}$, $u_-(x) \in \mathbf{R}$ with $u_+(x) > u_-(x)$ such that

$$\lim_{\varrho \to 0} \varrho^{-n} \int_{\{y \in B_\varrho \,; y \cdot \nu(x) > 0\}} |u(x+y) - u_+(x)|\, dy = 0,$$

$$\lim_{\varrho \to 0} \varrho^{-n} \int_{\{y \in B_\varrho \,; y \cdot \nu(x) < 0\}} |u(x+y) - u_-(x)|\, dy = 0,$$

and

$$(2.2) \qquad \int_{\Omega \setminus K} |Du| \geq \int_{\Omega \setminus K} |\nabla u|\, dx + \int_{S_u \cap \Omega} (u_+ - u_-)\, dH_{n-1}$$

(see [15], 4.5.9(17), (22), (15)).

Following [11], we define a class of special functions of bounded variation which are characterized by a property stronger than (2.2).

DEFINITION 2.1 – We define $SBV(\Omega)$ as the class of all functions $u \in BV(\Omega)$ such that

$$(2.3) \qquad \int_{\Omega \setminus K} |Du| = \int_{\Omega \setminus K} |\nabla u|\, dx + \int_{S_u \cap \Omega} (u_+ - u_-)\, dH_{n-1}.$$

By $SBV_{\mathrm{loc}}(\Omega)$ we denote the class of all functions which belong to $SBV(\Omega')$ for every open set $\Omega' \subset\subset \Omega$.

We remark that the well-known Cantor-Vitali function has bounded variation but does not satisfy (2.3).

REMARK 2.2 – Let $u \in \mathrm{BV}(\Omega)$ and set $u_a = (u \wedge a) \vee (-a)$ for $0 < a < +\infty$. The following properties hold:

$$|\nabla u_a| \leq |\nabla u| \quad \text{a.e. on } \Omega;$$

$$H_{n-1}\left((S_{u_a} \setminus S_u) \cap \Omega\right) = 0;$$

$$\int_{\Omega \setminus K} |Du_a| \leq \int_{\Omega \setminus K} |Du|;$$

$$\int_{\Omega \setminus K} |\nabla u|\, dx = \lim_{a \to +\infty} \int_{\Omega \setminus K} |\nabla u_a|\, dx;$$

$$H_{n-1}(S_u \cap \Omega) = \lim_{a \to +\infty} H_{n-1}(S_{u_a} \cap \Omega);$$

$$\int_{\Omega \setminus K} |Du| = \lim_{a \to +\infty} \int_{\Omega \setminus K} |Du_a|.$$

Moreover, for $u \in \mathrm{BV}(\Omega)$,

$u \in \mathrm{SBV}(\Omega)$ if and only if $u_a \in \mathrm{SBV}(\Omega)$ for every $0 < a < +\infty$;

and more generally

$u \in \mathrm{SBV}(\Omega)$ if and only if $\varphi(u) \in \mathrm{SBV}(\Omega)$ for every $\varphi : \mathbf{R} \to \mathbf{R}$ uniformly Lipschitz continuous with $\varphi(0) = 0$.

We point out some properties of the functions in $\mathrm{SBV}(\Omega)$; for further results we refer to [11], [2], [3].

LEMMA 2.3 – *Let $\Omega \subseteq \mathbf{R}^n$ be open and $u \in L^\infty)\Omega) \cap L^1(\Omega)$. Let $K \subset \mathbf{R}^n$ be closed and assume*

$$u \in C^1(\Omega \setminus K),$$

$$\int_{\Omega \setminus K} k|\nabla u|\, dx < +\infty,$$

$$H_{n-1}(K \cap \Omega) < +\infty.$$

Then

$$u \in \mathrm{SBV}(\Omega) \quad \text{and} \quad S_u \cap \Omega \subseteq K.$$

Proof. Since K is a closed set with $H_{n-1}(K \cap \Omega) < +\infty$, for every $h \in \mathbf{N}$ there is a locally finite covering of K with balls $B^h_{\varrho_i}(x_i)$ such that $\varrho_i < 1/h$ and

$$\sum_{i=1}^{+\infty} H_{n-1}\left(\partial B^h_{\varrho_i}(x_i)\right) \leq (n\Omega_n/\Omega_{n-1})\left(H_{n-1}(K \cap \Omega) + 1\right). \text{ Set}$$

$$u_h(x) = \begin{cases} u(x) & \text{for } x \in \Omega \setminus \bigcup_i B^h_{\varrho_i}(x_i) \\ 0 & \text{elsewhere in } \Omega; \end{cases}$$

then $u_h \in \mathrm{BV}(\Omega)$, $u_h \to u$ in $L^1(\Omega)$ and

$$\int_{\Omega \setminus K} |Du_h| \leq \int_{\Omega \setminus K} k|\nabla u|\, dx + 2\left(n\omega_n/\omega_{n-1}\right)\|u\|_{L^\infty(\Omega)}\left(H_{n-1}(K \cap \Omega) + 1\right);$$

thus, by the lower semicontinuity of the total variation, u belongs to $BV(\Omega)$ and clearly $S_u \cap \Omega \subseteq K$. Since (see [2], Proposition 3.1)

$$\int_{(K\cap\Omega)\setminus S_u} |Du| = 0,$$

we obtain

$$\begin{aligned}
\int_{\Omega\setminus K} |Du| &= \int_{\Omega\setminus K} k|Du| + \int_{K\cap\Omega} |Du| \\
&= \int_{\Omega\setminus K} k|\nabla u| + \int_{S_u\cap K\cap\Omega} (u_+ - u_-)\, dH_{n-1} \\
&\leq \int_{\Omega\setminus K} |\nabla u|\, dx + \int_{S_u\cap\Omega} (u_+ - u_-)\, dH_{n-1}.
\end{aligned}$$

Therefore, taking (2.2) into account, we conclude that u is in $SBV(\Omega)$. $\qquad\square$

We denote by $W^{1,p}(\Omega)$ $(p \geq 1)$ the Sobolev space of functions $u \in L^p(\Omega)$ such that $Du \in L^p(\Omega; \mathbf{R}^n)$. Then we remark that, for $u \in SBV(\Omega)$,

(2.4)

$$u \in W^{1,p}(\Omega) \text{ if and only if } H_{n-1}(S_u\cap\Omega) = 0 \text{ and } \int_{\Omega\setminus K} (|\nabla u|^p + |u|^p)\, dx < +\infty$$

(see *e.g.* [15], 4.5.9(30)).

In this paper we use the following semicontinuity theorem in $SBV(\Omega)$, which is an obvious consequence of a result by L. Ambrosio (see Theorems 2.1 and 3.4 of [3]) and Remark 2.2.

THEOREM 2.4 – *Let $p > 1$. Let $u_h \in SBV(\Omega)$ be such that*

$$u_h \to u \text{ in } L^1_{\text{loc}}(\Omega),$$

$$\sup_{h\in\mathbf{N}} \left\{ \int_{\Omega\setminus K} |\nabla u_h|^p\, dx + H_{n-1}(S_{u_h} \cap \Omega) \right\} < +\infty,$$

$$\sup_{h\in\mathbf{N}} \left\{ \int_{\Omega\setminus K} |Du_h| + \int_{\Omega\setminus K} |u_h|\, dx \right\} < +\infty.$$

Then
   (i) $u \in SBV(\Omega)$,
   (ii) $H_{n-1}(S_u \cap \Omega) \leq \liminf_h H_{n-1}(S_{u_h} \cap \Omega)$,

   (iii) $\int_{\Omega\setminus K} |\nabla u|^p\, dx \leq \liminf_h \int_{\Omega\setminus K} |\nabla u_h|^p\, dx.$

We remark that the conclusions just asserted do not follow if $p = 1$, because in this case it is possible to approximate every $u \in BV(\Omega)$ by a sequence of smooth functions (which are also functions of class $SBV(\Omega)$).

Later on in this work we shall often consider, for a given function $u \in \mathrm{SBV}_{\mathrm{loc}}(\Omega)$ the $(n-1)$-dimensional density at $x \in \Omega$

$$\lim_{\varrho \to 0} \varrho^{1-n} \left[\int_{B_\varrho(x)} |\nabla u|^p \, dx + H_{n-1}\left(S_u \cap B_\varrho(x)\right) \right].$$

Its main property is stated in Theorem 3.6. First we prove, in Lemma 2.6, a simpler property; to this end we use a result from measure theory, here stated as Lemma 2.5, which is established *e.g.* in [12], Cap. III, Teorema 3.3, or in [15], 2.10.19(3).

LEMMA 2.5 – *Let $n \in \mathbf{N}$ and $0 \le k \le n$. If ν is a positive Borel measure on $\mathbf{R}^n$ and if $B \subseteq \mathbf{R}^n$ is a bounded Borel set with the following properties*

$$\nu(B) = \inf\left\{\nu(A); \ A \ \text{open}, \ B \subseteq A\right\}$$

and

$$\limsup_{\varrho \to 0}(\omega_k \varrho^k)^{-1} \nu\left(B_\varrho(x)\right) \ge t > 0 \ \text{whenever} \ x \in B,$$

then

$$\nu(B) \ge t H_k(B).$$

LEMMA 2.6 – *Let $n \in \mathbf{N}$, $n \ge 2$, $p \ge 1$ and let $\Omega \subseteq \mathbf{R}^n$ be open. Let $u \in \mathrm{SBV}(\Omega)$ such that*

$$\int_K |\nabla u|^p \, dy + H_{n-1}\left(S_u \cap K\right) < +\infty$$

for every compact set $K \subset \Omega$. Then

$$\lim_{\varrho \to 0} \varrho^{1-n} \left[\int_{B_\varrho(x)} |\nabla u|^p \, dy + H_{n-1}\left(S_u \cap B_\varrho(x)\right) \right] = 0$$

for H_{n-1} almost all $x \in \Omega \setminus S_u$.

Proof. Let $K \subset \Omega$ be a compact set. For every $t > 0$, we define the following Borel set

$$K_t = \left\{ x \in K \setminus S_u; \ \limsup_{\varrho \to 0} \varrho^{1-n} \left[\int_{B_\varrho(x)} |\nabla u|^p \, dy + H_{n-1}(S_u \cap B_\varrho(x)) \right] \ge \omega_{n-1} t \right\}.$$

From Lemma 2.5 we have that

$$t H_{n-1}(K_t) \le \int_{K_t} |\nabla u|^p \, dy + H_{n-1}\left(S_u \cap K_t\right) = \int_{K_t} |\nabla u|^p \, dy < +\infty.$$

Hence $|K_t| = 0$, and so we infer that $H_{n-1}(K_t) = 0$. Since t is arbitrary the assertion follows. $\qquad\square$

3. A Poincaré-Wirtinger type inequality in $\mathrm{SBV}(\Omega)$ and some consequences

Let B be a ball in $\mathbf{R}^n$, $n \geq 2$. We prove a Poincaré-Wirtinger type inequality for functions in the space $\mathrm{SBV}(B)$; to this end we begin by defining, for every measurable function $u : B \to \mathbf{R}$,

$$u_*(s, B) = \inf \left\{ t \in \mathbf{R}; \ |\{u < t\} \cap B| \geq s \right\} \ \text{for } 0 \leq s \leq |B|,$$

$$\mathrm{med}(u, B) = u_*(\tfrac{1}{2}|B|, B) \ \text{(the least median of } u \text{ in } B);$$

moreover for every $u \in \mathrm{SBV}(B)$ such that $(2\gamma_n H_{n-1}(S_u \cap B))^{n/(n-1)} < \tfrac{1}{2}|B|$ we set

$$\tau'(u, B) \ = \ u_* \left((2\gamma_n H_{n-1}(S_u \cap B))^{n/(n-1)}, B \right),$$

$$\tau''(u, B) \ = \ u_* \left(|B| - (2\gamma_n H_{n-1}(S_u \cap B))^{n/(n-1)}, B \right),$$

where γ_n is the isoperimetric constant relative to the balls of $\mathbf{R}^n$, $i.e.$ for every Borel set E

$$(3.1) \qquad \min \left\{ |E \cap B|^{(n-1)/n}, \ |B \setminus E|^{(n-1)/n} \right\} \leq \gamma_n P(E, B).$$

With these definitions we can state the following theorem.

THEOREM 3.1 – *Let B be a ball in $\mathbf{R}^n$, $n \geq 2$, $1 \leq p < n$ and $p^* = np/(n-p)$. Let $u \in \mathrm{SBV}(B)$, $H_{n-1}(S_u \cap B) < \dfrac{1}{2\gamma_n}(\tfrac{1}{2}|B|)^{(n-1)/n}$, and*

$$\overline{u} = (u \wedge \tau''(u, B)) \vee \tau'(u, B).$$

Then

$$\left(\int_B |\overline{u} - \mathrm{med}(u, B)|^{p^*} \, dx \right)^{1/p^*} \leq \frac{2\gamma_n p \,(n - 1)}{n - p} \left(\int_B |\nabla u|^p \, dx \right)^{1/p}.$$

Proof. By assumption $\tau'(u, B) \leq \mathrm{med}(u, B) \leq \tau''(u, B)$. We may assume that $\mathrm{med}(u, B) = 0$ and moreover that $H_{n-1}(S_u \cap B) > 0$ (because, if $H_{n-1}(S_u \cap B) = 0$, then $\overline{u} = u \in W^{1,1}(B)$ and the assertion is well known). Since $\overline{u} \in \mathrm{SBV}(B)$ we have, by (2.3),

$$\int_B |D\overline{u}| = \int_B |\nabla \overline{u}| \, dx + \int_{S_{\overline{u}} \cap B} (\overline{u}_+ - \overline{u}_-) \, dH_{n-1},$$

and also, since $H_{n-1}((S_{\overline{u}} \setminus S_u) \cap B) = 0$,

$$(3.2) \qquad \int_B |D\overline{u}| \leq \int_B |\nabla \overline{u}| \, dx + (\tau''(u, B) - \tau'(u, B)) \, H_{n-1}(S_u \cap B).$$

By using the coarea formula (2.1) and the isoperimetric inequality (3.1), we have

$$\int_B |D\overline{u}| = \int_{\tau'(u,B)}^{\tau''(u,B)} P(\{\overline{u} < t\}, B)\, dt$$

$$\geq \frac{1}{\gamma_n} \int_{\tau'(u,B)}^{0} |\{\overline{u} < t\} \cap B|^{(n-1)/n}\, dt + \frac{1}{\gamma_n} \int_{0}^{\tau''(u,B)} |B \setminus \{\overline{u} < t\}|^{(n-1)/n}\, dt.$$

Since for $\tau'(u,B) < t < 0$,

$$\frac{1}{\gamma_n} |\{\overline{u} < t\} \cap B|^{(n-1)/n} \geq 2H_{n-1}(S_u \cap B)$$

and for $0 < t < \tau''(u,B)$,

$$\frac{1}{\gamma_n} |B \setminus \{\overline{u} < t\}|^{(n-1)/n} \geq 2H_{n-1}(S_u \cap B),$$

we have

$$\int_B |D\overline{u}| \geq 2\left(\tau''(u,B) - \tau'(u,B)\right) H_{n-1}(S_u \cap B);$$

hence, by comparison with (3.2),

$$\left(\tau''(u,B) - \tau'(u,B)\right) H_{n-1}(S_u \cap B) \leq \int_B |\nabla \overline{u}|\, dx;$$

therefore, by (3.2)

$$(3.3) \qquad \int_B |D\overline{u}| \leq 2 \int_B |\nabla \overline{u}|\, dx.$$

Finally, since

$$\left(\int_B |\overline{u}|^{n/(n-1)}\, dx\right)^{(n-1)/n} \leq \gamma_n \int_B |D\overline{u}|$$

(see *e.g.* [15] page 504), we have

$$\left(\int_B |\overline{u}|^{n/(n-1)}\, dx\right)^{(n-1)/n} \leq \gamma_n \int_B |\nabla \overline{u}|\, dx.$$

Thus the proof is complete for $p = 1$. For $1 < p < n$ we may apply in a standard way the previous inequality to the function $v = |\overline{u}|^{p^*(n-1)/n}$. $\qquad \square$

REMARK 3.2 – With the previous notation and under the same assumptions as in Theorem 3.1, we have, by the definition of $\overline{u}$ and by (3.3),

(i) $|\{\overline{u} \neq u\} \cap B| \leq 2\left(2\gamma_n H_{n-1}(S_u \cap B)\right)^{n/(n-1)}$,

(ii) $\int_B |D[\overline{u} - \operatorname{med}(u,B)]| \leq 2 \int_B |\nabla u|\, dx \leq 2|B|^{(p-1)/p} \left(\int_B |\nabla u|^p\, dx\right)^{1/p}.$

REMARK 3.3 – If $p \geq n$, for every $q \geq 1$, by the preceding theorem and by the Hölder inequality, we have

$$\|\overline{u} - \operatorname{med}(u, B)\|_{L^q(B)} \leq \frac{2\gamma_n q(n-1)}{n} |B|^{\frac{1}{n} + \frac{1}{q} - \frac{1}{p}} \|\nabla u)\|_{L^p(B)}.$$

REMARK 3.4 – Theorem 3.1 may be extended to bounded open sets, more general than balls, for which an isoperimetric inequality similar to (3.1) holds (*e.g.* bounded open sets having the cone property).

Now, by using Theorem 3.1 we prove a compactness result in $\operatorname{SBV}(B)$.

THEOREM 3.5 – *Let $B \subset \mathbf{R}^n$ be a ball, $u_h \in \operatorname{SBV}(B)$, $p > 1$, and let*

$$\sup_{k \in \mathbf{N}} \int_B |\nabla u_h|^p \, dx < +\infty,$$

$$\lim_h H_{n-1}(S_{u_h} \cap B) = 0.$$

Then
(i) the functions $\overline{u}_h - \operatorname{med}(u_h, B)$, defined as in Theorem 3.1, are uniformly bounded in $\operatorname{BV}(B)$;
(ii) there exist a subsequence (u_{h_i}) and a function $u_\infty \in W^{1,p}(B)$ such that

$$\lim_i [u_{h_i} - \operatorname{med}(u_{h_i}, B)] = u_\infty \quad \text{a.e. on } B.$$

Proof.
(i) In the proof of the assertion we may assume $1 < p < n$. For h large enough, by Theorem 3.1, we have

$$(3.4) \qquad \int_B |\overline{u}_h - \operatorname{med}(u_h, B)|^{p^*} \, dx \leq \text{const.};$$

moreover, by Remark 3.2(ii),

$$\int_B |D[\overline{u}_h - \operatorname{med}(u_h, B)]| \leq \text{const.}$$

Therefore the functions $\overline{u}_h - \operatorname{med}(u_h, B)$ are uniformly bounded in $\operatorname{BV}(B)$.
(ii) From (3.4) in the case $1 < p < n$ or from Remark 3.3 in the case $p \geq n$, by virtue of the compactness theorem in $\operatorname{BV}(B)$ (see *e.g.* [17], Theorem 1.19) there exist a subsequence (u_{h_i}) and $u_\infty \in \operatorname{BV}(B)$ such that

$$(3.5) \qquad \lim_i [\overline{u}_{h_i} - \operatorname{med}(u_{h_i}, B)] = u_\infty$$

in $L^r(B)$ for every $1 \leq r < np/(n-p)$ if $1 < p < n$, and in $L^r(B)$ for every $r \geq 1$ if $p \geq n$. Now, by Theorem 2.4, u_∞ belongs to $\operatorname{SBV}(B)$; moreover

$$H_{n-1}(S_{u_\infty} \cap B) \leq \liminf_i H_{n-1}(S_{\overline{u}_{h_i}} \cap B) = 0,$$

and

$$(3.6) \qquad \int_B |\nabla u_\infty|^p \, dx \leq \liminf_i \int_B |\nabla u_{h_i}|^p \, dx < +\infty,$$

thus, by (2.4), u_∞ is in $W^{1,p}(B)$. Finally, since from Remark 3.2(i)

$$|\{\bar{u}_{h_i} \neq u_{h_i}\} \cap B| \leq 2 \left(2\gamma_n H_{n-1}(S_{u_{h_i}} \cap B)\right)^{n/(n-1)},$$

from (3.5) it follows that, possibly by restriction again to a subsequence,

$$\lim_i \left[u_{h_i} - \mathrm{med}(u_{h_i}, B)\right] = u_\infty \ \text{ a.e. on } B. \qquad \square$$

Using again Theorem 3.1, we conclude this section by proving a condition ensuring that, given $u \in \mathrm{SBV}(\Omega)$ and fixed $x \in \Omega$, the point x does not belong to the singular set S_u (*i.e.* u is approximately continuous at x). Some similar conditions have been studied for a given $u \in W^{1,p}(\Omega)$ in [16] and for $u \in \mathrm{BV}(\Omega)$ in [15], 4.5.9(20).

THEOREM 3.6 – *Let $n \in \mathbf{N}$, $n \geq 2$, $p > 1$, $\Omega \subseteq \mathbf{R}^n$ be an open set, $x \in \Omega$ and $u \in \mathrm{SBV}_{\mathrm{loc}}(\Omega)$. If*

$$\lim_{\varrho \to 0} \varrho^{1-n} \left[\int_{B_\varrho(x)} |\nabla u|^p \, dy + H_{n-1}\left(S_u \cap B_\varrho(x)\right)\right] = 0,$$

then

$$x \notin S_u \quad \text{and} \quad \mathrm{aplim}_{y \to x} u(y) \in \mathbf{R}.$$

Proof. We may assume $p < n$ and $x = 0$. By assumption there exists $\varrho_0 > 0$ such that

$$(3.7) \qquad \int_{B_\varrho(x)} |\nabla u|^p \, dy + H_{n-1}\left(S_u \cap B_\varrho(x)\right) < \frac{1}{2\gamma_n}\left(\frac{1}{4}|B_\varrho|\right)^{(n-1)/n}$$

for every $\varrho < \varrho_0$; hence we may use Theorem 3.1 in B_ϱ. According to Theorem 3.1, we may set

$$\bar{u}_\varrho = (u \wedge \tau''(u, B_\varrho)) \vee \tau'(u, B_\varrho)$$

and

$$\bar{u}_\varrho = \mathrm{med}(u, B_\varrho).$$

We prove the existence of $\lim_{\varrho \to 0} u_\varrho = u_0 \in \mathbf{R}$.

Let α be a real number with $\frac{3}{4} < \alpha^n < 1$. Fix $r > 0$ with $r < \varrho_0$, and fix s with $\alpha r \leq s \leq r$. Then

$$(3.8) \qquad \tau'(u, B_s) \leq u_r \leq \tau''(u, B_s)$$

and

$$(3.9) \qquad \tau'(u, B_r) \leq u_s \leq \tau''(u, B_r)$$

For brevity's sake, we prove only (3.8). If there exists t with $u_r < t < \tau'(u, B_s)$, then by (3.7)

$$
\begin{aligned}
|\{u < t\} \cap B_r| \ &\leq\ |B_r \setminus B_s| + |\{u < t\} \cap B_s| \\
&\leq\ \omega_n(1 - \alpha^n)r^n + (2\gamma_n H_{n-1}(S_u \cap B_s))^{n/(n-1)} < \frac{1}{2}\omega_n r^n,
\end{aligned}
$$

and this contradicts the definition of u_r. Thus the first inequality in (3.8) is proved.

On the other hand, if t is such that $\tau''(u, B_s) < t < u_r$, then

$$
\begin{aligned}
|\{u < t\} \cap B_r| \ &\geq\ |\{u < t\} \cap B_s| \geq \omega_n s^n - (2\gamma_n H_{n-1}(S_u \cap B_s))^{n/(n-1)} \\
&\geq\ \frac{3}{4}|B_s| \geq \frac{3}{4}\omega_n \alpha^n r^n > \frac{1}{2}\omega_n r^n,
\end{aligned}
$$

a contradiction.

Now we define

$$\hat{u} = (u \wedge (\tau''(u, B_r) \wedge \tau''(u, B_s))) \vee (\tau'(u, B_r) \vee \tau'(u, B_s)).$$

By (3.8), (3.9), Theorem 3.1 and by (3.7) we have

$$
\begin{aligned}
|u_r - u_s| \ &= \left(\frac{1}{\omega_n s^n} \int_{B_s} |u_r - u_s|^{p^*} dy\right)^{1/p^*} \\
&\leq (\omega_n s^n)^{-\frac{1}{p^*}} \left[\left(\int_{B_s} |\hat{u} - u_r|^{p^*} dy\right)^{1/p^*} + \left(\int_{B_s} |\hat{u} - u_s|^{p^*} dy\right)^{1/p^*}\right] \\
&\leq (\omega_n s^n)^{-\frac{1}{p^*}} \left[\left(\int_{B_s} |\overline{u}_r - u_r|^{p^*} dy\right)^{1/p^*} + \left(\int_{B_s} |\overline{u}_s - u_s|^{p^*} dy\right)^{1/p^*}\right] \\
&\leq (\omega_n \alpha^n r^n)^{-\frac{1}{p^*}} \frac{2\gamma_n p\,(n-1)}{n-p} \left[\left(\int_{B_r} |\nabla u|^p dy\right)^{1/p} + \left(\int_{B_s} |\nabla u|^p dy\right)^{1/p}\right] \\
&\leq \text{const.}\, r^{(p-1)/p}.
\end{aligned}
$$

Hence for $h \in \mathbf{N}$ and $\alpha^{k+1} r \leq s \leq \alpha^k r$ for some $k \geq h$, we obtain

$$|u_s - u_{\alpha^h r}| \leq |u_s - u_{\alpha^k r}| + \sum_{j=h}^{k-1} |u_{\alpha^{j+1} r} - u_{\alpha^j r}| \leq \text{const} \sum_{j=h}^{k} (\alpha^j r)^{(p-1)/p}.$$

Since $\alpha < 1$, there exists $u_0 = \lim_{\varrho \to 0} u_\varrho \in \mathbf{R}$.

Finally we show that $u_0 = \operatorname*{aplim}_{y \to 0} u(y)$; in particular we show (see a characterization *e.g.* in [3], Proposition 1.1 (i)) that for every $\varepsilon > 0$

$$\text{(3.10)} \qquad \lim_{\varrho \to 0} \varrho^{-n} \left| \{|u - u_0| \geq \varepsilon\} \cap B_\varrho \right| = 0.$$

Let $\varepsilon > 0$ and let $r' > 0$ be such that

$$|u_\varrho - u_0| < \frac{\varepsilon}{2}$$

for every $\varrho < r'$. By assumption, for every $\sigma > 0$ there is an $r_\sigma > 0$ such that

$$\int_{B_\varrho} |\nabla u|^p \, dy + H_{n-1} \left(S_u \cap B_\varrho \right) < \sigma \varrho^{n-1}$$

for every $\varrho < r_\sigma$. Hence for every $\varrho < r' \wedge r_\sigma$ we have

$$\varrho^{-n} \left| \{|u - u_0| \geq \varepsilon\} \cap B_\varrho \right|$$
$$\leq \varrho^{-n} \Big[|\{u > \tau''(u, B_\varrho)\} \cap B_\varrho|$$
$$\qquad + |\{u < \tau'(u, B_\varrho)\} \cap B_\varrho| + |\{|\overline{u}_\varrho - u_0| \geq \varepsilon\} \cap B_\varrho| \Big]$$
$$\leq 2(2\gamma_n \sigma)^{n/(n-1)} + \frac{2}{\varepsilon} \varrho^{-n} \int_{B_\varrho} |\overline{u}_\varrho - u_\varrho| \, dy$$
$$\leq 2(2\gamma_n \sigma)^{n/(n-1)} + \frac{2}{\varepsilon} \varrho^{-n} (\omega_n \varrho^n)^{1-1/p^*} \left(\int_{B_\varrho} |\overline{u}_\varrho - u_\varrho|^{p^*} \, dy \right)^{1/p^*}$$
$$\leq 2(2\gamma_n \sigma)^{n/(n-1)} + \text{const.} \varrho^{(p-1)/p}.$$

Then
$$\limsup_{\varrho \to 0} \varrho^{-n} \left| \{|u - u_0| \geq \varepsilon\} \cap B_\varrho \right| \leq 2(2\gamma_n \sigma)^{n/(n-1)},$$

and by the arbitrariness of σ we may infer (3.10). $\qquad \square$

4. A limit theorem and some estimates for quasi-minima in $\mathrm{SBV}_{\mathrm{loc}}(\Omega)$

In this section we introduce some functionals defined on $\mathrm{SBV}_{\mathrm{loc}}(\Omega)$ that will allow us to prove the estimate $H_{n-1} \left((\overline{S}_u \cap \Omega) \setminus S_u \right) = 0$ for functions $u \in \mathrm{SBV}_{\mathrm{loc}}(\Omega)$ that satisfy a quasi-minimum condition. The main step in the proof of this result is Lemma 4.9, which characterizes the decay of

$$\varrho^{1-n} \left[\int_{B_\varrho(x)} |\nabla u|^2 \, dy + H_{n-1} \left(S_u \cap B_\varrho(x) \right) \right]$$

when ϱ decreases.

Definition 4.1 – Let $\Omega \subseteq \mathbf{R}^n$ be open, $u \in \mathrm{SBV}_{\mathrm{loc}}(\Omega)$, $0 < c < +\infty$. Let $K \subset \Omega$ be closed. We set

$$F(u, c, K) = \int_K |\nabla u|^2 \, dx + cH_{n-1}(S_u \cap K),$$

$$\Phi(u, c, K) = \inf \{F(v, c, K);\ v \in \mathrm{SBV}_{\mathrm{loc}}(\Omega),\ v = u \text{ in } \Omega \setminus K\};$$

moreover, if $\Phi(u, c, K) < +\infty$, we set

$$\Psi(u, c, K) = F(u, c, K) - \Phi(u, c, K).$$

Remark 4.2 – It is easy to verify that if $u \in \mathrm{SBV}_{\mathrm{loc}}(\Omega)$, $0 < c < +\infty$, and $a, b \in \mathbf{R}$, $a < b$, setting $v = (u \wedge b) \vee a$, we have

$$F(v, c, K) \le F(u, c, K), \quad \Phi(v, c, K) \le \Phi(u, c, K).$$

Remark 4.3 – If $u \in \mathrm{SBV}_{\mathrm{loc}}(\Omega) \cap L^\infty(\Omega)$ then, from the previous remark, the compactness theorem in $\mathrm{BV}(\Omega)$ and the semicontinuity Theorem 2.4, it follows that, for every closed set $K \subset \Omega$,

$$\Phi(u, c, K) = \min \{F(v, c, K);\ v \in \mathrm{SBV}_{\mathrm{loc}}(\Omega),\ v = u \text{ in } \Omega \setminus K\}.$$

The next two lemmas concerning the functionals introduced in Definition 4.1 are easy to prove.

Lemma 4.4 – *Let $u \in \mathrm{SBV}(B_r)$. For every $0 < c < \infty$ the functions*

$$\varrho \mapsto F(u, c, \overline{B}_\varrho)$$

and

$$\varrho \mapsto \Psi(u, c, \overline{B}_\varrho)$$

are non-decreasing in $(0, r)$.

Lemma 4.5 – *Let $u \in \mathrm{SBV}(B_r(x_0))$, $\varrho < r$. Set $u_\varrho = \varrho^{-1/2} u(x_0 + \varrho x)$ for every $x \in B_{r/\varrho}$, then*

$$u_0 \in \mathrm{SBV}(B_{r/\varrho}),$$

$$F(u_\varrho, c, \overline{B}_1) = \varrho^{1-n} F(u, c, \overline{B}_\varrho(x_0))$$

and

$$\Phi(u_\varrho, c, \overline{B}_1) = \varrho^{1-n} \Phi(u, c, \overline{B}_\varrho(x_0)).$$

By using the notion of approximate limit for a Borel function we prove the following lemma.

Lemma 4.6 – *Let $u, v \in \mathrm{SBV}(B_r)$, $0 < c < \infty$ and $0 < \varrho < r$. Suppose*

$$H_{n-1}(S_u \cap \partial B_\varrho) = H_{n-1}(S_v \cap \partial B_\varrho) = 0.$$

Then

$$|\Phi(u, c, \overline{B}_\varrho) - \Phi(v, c, \overline{B}_\varrho)| \leq cH_{n-1}(\{\tilde{u} \neq \tilde{v}\} \cap \partial B_\varrho).$$

Proof. Fixed $\varepsilon > 0$, let $w \in \mathrm{SBV}(B_r)$ such that $w = u$ in $B_r \setminus \overline{B}_\varrho$ and

$$F(w, c, \overline{B}_\varrho) \leq \Phi(u, c, \overline{B}_\varrho) + \varepsilon.$$

Define

$$w'(x) = \begin{cases} w(x) & \text{for } x \in \overline{B}_\varrho \\ v(x) & \text{for } x \in B_r \setminus \overline{B}_\varrho. \end{cases}$$

Hence

$$\begin{aligned} \Phi(v, c, \overline{B}_\varrho) &\leq F(w', c, \overline{B}_\varrho) \leq F(w, c, \overline{B}_\varrho) + cH_{n-1}(\{\tilde{u} \neq \tilde{v}\} \cap \partial B_\varrho) \\ &\leq \Phi(u, c, \overline{B}_\varrho) + \varepsilon + cH_{n-1}(\{\tilde{u} \neq \tilde{v}\} \cap \partial B_\varrho). \end{aligned}$$

Since ε is arbitrary, it follows that

$$\Phi(v, c, \overline{B}_\varrho) \leq \Phi(u, c, \overline{B}_\varrho) + cH_{n-1}(\{\tilde{u} \neq \tilde{v}\} \cap \partial B_\varrho);$$

as the assumption is symmetric in u and v, the lemma is proved. $\qquad\square$

LEMMA 4.7 – *Let $u, v \in \mathrm{SBV}(B_r)$, $0 < \varrho' < \varrho < r$. For every $0 < c < \infty$ and for every $0 < \lambda < 1$ it is true that*

$$\begin{aligned} \Phi(u, c, \overline{B}_\varrho) \leq \frac{1}{1 - \lambda} &\left[\Phi(v, c, \overline{B}_{\varrho'}) + F(u, c, \overline{B}_\varrho) - F(u, c, \overline{B}_{\varrho'}) \right. \\ &\left. + F(v, c, \overline{B}_\varrho) - F(v, c, \overline{B}_{\varrho'}) \right] + \frac{1}{\lambda(\varrho - \varrho')^2} \int_{B_\varrho \setminus B_{\varrho'}} |u - v|^2 \, dy. \end{aligned}$$

Proof. Let φ be the cutoff function between $B_{\varrho'}$ and B_ϱ defined by

$$\varphi(y) = \left(0 \vee \frac{|y| - \varrho'}{\varrho - \varrho'} \right) \wedge 1 \quad \text{for } y \in B_r.$$

For $\varepsilon > 0$ fixed, let $w \in \mathrm{SBV}(B_r)$ such that $w(y) = v(y)$ for $y \in B_\varrho \setminus \overline{B}_{\varrho'}$ and

$$F(w, c, \overline{B}_{\varrho'}) \leq \Phi(v, c, \overline{B}_{\varrho'}) + \varepsilon.$$

Setting $w' = \varphi u + (1 - \varphi) w$, we have $w'(y) = u(y)$ for $y \in B_r \setminus \overline{B}_\varrho$ and moreover

$$\begin{aligned} F(w', c, \overline{B}_\varrho) &= \int_{B_\varrho} |\nabla w'|^2 dy + cH_{n-1}(S_{w'} \cap \overline{B}_\varrho) \\ &\leq \frac{1}{1 - \lambda} \left[\int_{B_\varrho} |\nabla w|^2 dy + \int_{B_\varrho \setminus B_{\varrho'}} |\nabla u|^2 dy \right] + \frac{1}{\lambda(\varrho - \varrho')^2} \int_{B_\varrho \setminus B_{\varrho'}} |u - w|^2 dy \\ &\quad + cH_{n-1}(S_w \cap \overline{B}_\varrho) + cH_{n-1}(S_u \cap (\overline{B}_\varrho \setminus \overline{B}_{\varrho'})) \end{aligned}$$

$$\leq \frac{1}{1-\lambda}\left[\int_{B_{\varrho'}}|\nabla w|^2 dy + cH_{n-1}(S_w \cap \overline{B}_{\varrho'}) + \int_{B_\varrho \setminus B_{\varrho'}}|\nabla u|^2 dy\right.$$

$$+cH_{n-1}(S_u \cap (\overline{B}_\varrho \setminus \overline{B}_{\varrho'})) + \int_{B_\varrho \setminus B_{\varrho'}}|\nabla w|^2 dy$$

$$\left.+cH_{n-1}(S_w \cap (\overline{B}_\varrho \setminus \overline{B}_{\varrho'}))\right]$$

$$+\frac{1}{\lambda(\varrho-\varrho')^2}\int_{B_\varrho \setminus B_{\varrho'}}|u-v|^2 dy$$

$$\leq \frac{1}{1-\lambda}\left[\Phi(v,c,\overline{B}_{\varrho'}) + \varepsilon + F(u,c,\overline{B}_\varrho) - F(u,c,\overline{B}_{\varrho'})\right.$$

$$\left.+F(w,c,\overline{B}_\varrho) - F(w,c,\overline{B}_{\varrho'})\right] + \frac{1}{\lambda(\varrho-\varrho')^2}\int_{B_\varrho \setminus B_{\varrho'}}|u-v|^2 dy.$$

By the arbitrariness of ε it follows that

$$\Phi(u,c,\overline{B}_\varrho) \leq F(w',c,\overline{B}_\varrho)$$

$$\leq \frac{1}{1-\lambda}\left[\Phi(v,c,\overline{B}_{\varrho'}) + F(u,c,\overline{B}_\varrho) - F(u,c,\overline{B}_{\varrho'})\right.$$

$$\left.+F(v,c,\overline{B}_\varrho) - F(v,c,\overline{B}_{\varrho'})\right]$$

$$+\frac{1}{\lambda(\varrho-\varrho')^2}\int_{B_\varrho \setminus B_{\varrho'}}|u-v|^2 dy. \qquad \square$$

We are now in a position to prove the following limit theorem.

THEOREM 4.8 – *Let $n \geq 2$, $\Omega \subseteq \mathbf{R}^n$ be an open set; assume that $\overline{B}_r(x) \subset \Omega$, $u_h \in \mathrm{SBV}(\Omega)$, $c_h \in \mathbf{R}$ for every $h \in \mathbf{N}$, $u_\infty \in W^{1,2}_{\mathrm{loc}}(B_r(x))$ are such that*
(1) $\lim\limits_h c_h = +\infty$,
(2) $\lim\limits_h F(u_h, c_h, \overline{B}_\varrho(x)) = \lim\limits_h \Phi(u_h, c_h, \overline{B}_\varrho(x)) = \alpha(\varrho) < +\infty$ *for almost all* $\varrho < r$,
(3) $\lim\limits_h u_h = u_\infty$ *a.e. on* $B_r(x)$.
Then the function u_∞ is harmonic and

$$\alpha(\varrho) = \int_{B_\varrho(x)}|\nabla u_\infty|^2\, dy \quad \text{for almost all } \varrho < r.$$

Proof. Again, we may assume $x = 0$. Since $u_\infty \in W^{1,2}_{\mathrm{loc}}(B_r(x))$ we have, for every $c > 0$ and $\varrho < r$, $F(u_\infty, c, \overline{B}_\varrho) = \int_{B_\varrho}|\nabla u_\infty|^2\, dy$. From assumptions (1) and (2) it follows that

$$\lim_h H_{n-1}(S_{u_h} \cap B_\varrho) = 0$$

and

$$\sup_{h \in \mathbf{N}} \int_{B_\varrho}|\nabla u_h|^2\, dy < +\infty$$

for every $\varrho < r$, so that by (3.6)

$$F(u_\infty, c, \overline{B}_\varrho) \leq \liminf_h F(u_h, c, \overline{B}_\varrho) \leq \lim_h F(u_h, c_h, \overline{B}_\varrho) = \alpha(\varrho)$$

for almost all $\varrho < r$. Thus $\displaystyle\int_{B_\varrho} |\nabla u_\infty|^2 \, dy \leq \alpha(\varrho)$ for almost all $\varrho < r$. The assertion will be proved by the following

Main step: for almost all $\varrho < r$ and for every $v \in W^{1,2}_{\mathrm{loc}}(B_r)$ such that $v = u_\infty$ in $B_r \setminus \overline{B}_\varrho$ it follows that

$$\int_{B_\varrho} |\nabla v|^2 \, dy \geq \alpha(\varrho).$$

We begin by remarking that, by assumption (3) and by (3.5), $\overline{u}_h \to u_\infty$ in $L^2(B_r)$, where the functions $\overline{u}_h$ are as in Theorem 3.1. Moreover, first we prove that there is a subsequence of $(\overline{u}_h)$ such that

$$\lim_i \Phi(\overline{u}_{h_i}, c_{h_i}, \overline{B}_\varrho) = \alpha(\varrho) \quad \text{for almost all } \varrho < r.$$

By assumption (2) it suffices to prove that

$$(4.1) \qquad \lim_i \left[\Phi(u_{h_i}, c_{h_i}, \overline{B}_\varrho) - \Phi(\overline{u}_{h_i}, c_{h_i}, \overline{B}_\varrho) \right] = 0 \quad \text{for almost all } \varrho < r.$$

Setting $\eta_h = H_{n-1}(S_{u_h} \cap B_r)$, by assumption (2) we have

$$(4.2) \qquad\qquad\qquad c_h \eta_h \leq \text{const.},$$

and moreover, for almost all $\varrho < r$,

$$(4.3) \qquad\qquad H_{n-1}(S_{u_h} \cap \partial B_\varrho) = H_{n-1}(S_{\overline{u}_h} \cap \partial B_\varrho) = 0$$

for every $h \in \mathbf{N}$. Now, by Remark 3.2(i), for h large enough we have

$$c_h \int_0^r H_{n-1}\left(\{(\overline{u}_h) \neq u_h\} \cap \partial B_\varrho\right) d\varrho \quad = \quad c_h \left|\{(\widetilde{u}_h) \neq \widetilde{u}_h\} \cap B_r\right|$$
$$\leq \quad c_h 2(2\gamma_n \eta_h)^{n/(n-1)}.$$

By (4.2) there is a subsequence such that

$$(4.4) \qquad\qquad \lim_i c_{h_i} H_{n-1}\left(\{(\widetilde{u}_{h_i}) \neq \widetilde{u}_{h_i}\} \cap \partial B_\varrho\right) = 0$$

for almost all $\varrho < r$. By (4.3), (4.4) and by Lemma 4.6 it follows that

$$\lim_i \left[\Phi(u_{h_i}, c_{h_i}, \overline{B}_\varrho) - \Phi(\overline{u}_{h_i}, c_{h_i}, \overline{B}_\varrho) \right] = 0$$

for almost all $\varrho < r$. Hence (4.1) is true and therefore

$$(4.5) \qquad\qquad \lim_i \Phi(u_{h_i}, c_{h_i}, \overline{B}_\varrho) = \alpha(\varrho) \quad \text{for almost all } \varrho < r.$$

Since $\Phi(\overline{u}_h, c_h, \overline{B}_\varrho) \leq F(\overline{u}_h, c_h, \overline{B}_\varrho) \leq F(u_h, c_h, \overline{B}_\varrho)$ for every $h \in \mathbf{N}$, we have also

$$(4.6) \qquad \lim_i F(\overline{u}_{h_i}, c_{h_i}, \overline{B}_\varrho) = \alpha(\varrho) \quad \text{for almost all } \varrho < r.$$

We are now in a position to prove the above-described *main step*. For brevity's sake, we denote $(\overline{u}_{h_i})$ still by $(\overline{u}_h)$. Since the function $\varrho \to \alpha(\varrho)$ is non-decreasing, it is also a continuous function for almost all $\varrho < r$. Fixing a $\varrho' < r$ for which (4.5), (4.6) are satisfied and $\alpha(\cdot)$ is continuous in ϱ', we suppose that the *main step* is not true. Then there exist $\varepsilon > 0$ and $v \in W^{1,2}_{\text{loc}}(B_r)$ such that $v = u_\infty$ in $B_r \setminus \overline{B}_{\varrho'}$ and

$$\int_{B_{\varrho'}} |\nabla v|^2 \, dy < \alpha(\varrho') - \varepsilon.$$

Let $\varrho > 0$, $\varrho' < \varrho < r$, be such that again (4.5), (4.6) are satisfied; moreover $\alpha(\varrho) - \alpha(\varrho') < \varepsilon/4$ and $\int_{B_\varrho \setminus B_{\varrho'}} |\nabla v|^2 \, dy < \varepsilon/4$. For every $0 < \lambda < 1$, by Lemma 4.7, we obtain

$$\begin{aligned}
\Phi(\overline{u}_h, c_h, \overline{B}_\varrho) &\leq \frac{1}{1-\lambda} \left[\Phi(v, c_h, \overline{B}_{\varrho'}) + F(\overline{u}_h, c_h, \overline{B}_\varrho) - F(\overline{u}_h, c_h, \overline{B}_{\varrho'}) \right. \\
&\qquad \left. + F(v, c_h, \overline{B}_\varrho) - F(v, c_h, \overline{B}_{\varrho'}) \right] \\
&\qquad + \frac{1}{\lambda(\varrho - \varrho')^2} \int_{B_\varrho \setminus B_{\varrho'}} |\overline{u}_h - v|^2 \, dy \\
&\leq \frac{1}{1-\lambda} \left[\int_{B_{\varrho'}} |\nabla v|^2 \, dy + F(\overline{u}_h, c_h, \overline{B}_\varrho) - F(\overline{u}_h, c_h, \overline{B}_{\varrho'}) + \varepsilon/4 \right] \\
&\qquad + \frac{1}{\lambda(\varrho - \varrho')^2} \int_{B_\varrho \setminus B_{\varrho'}} |\overline{u}_h - v|^2 \, dy.
\end{aligned}$$

Since $\overline{u}_h \to v$ in $L^2(B_\varrho \setminus B_{\varrho'})$, letting $h \to +\infty$ we obtain

$$\alpha(\varrho) \leq \frac{1}{1-\lambda} \left[\alpha(\varrho') - \varepsilon + \varepsilon/4 + \varepsilon/4 \right];$$

hence, because of the arbitrariness of λ,

$$\alpha(\varrho) \leq \alpha(\varrho') - \varepsilon/2,$$

and this is a contradiction because the function $\varrho \to \alpha(\varrho)$ is non-decreasing. $\square$

From Theorem 4.8 we infer the following decay estimate.

LEMMA 4.9 – *For every $n \in \mathbf{N}$, $n \geq 2$, and every $0 < c < +\infty$, $0 < \alpha < 1$, and $0 < \beta < 1$, there exist $\varepsilon = \varepsilon(n, c, \alpha, \beta)$ and $\vartheta = \vartheta(n, c, \alpha, \beta)$ such that if $\Omega \subseteq \mathbf{R}^n$ is open, $\varrho > 0$, $\overline{B}_\varrho(x) \subset \Omega$ and $u \in \text{SBV}(\Omega)$ with*

$$F(u, c, \overline{B}_\varrho(x)) \leq \varepsilon \varrho^{n-1},$$

$$\Psi(u, c, \overline{B}_\varrho(x)) \leq \vartheta F(u, c, \overline{B}_\varrho(x)),$$

then

$$F(u, c, \overline{B}_{\alpha\varrho}(x)) \leq \alpha^{n-\beta} F(u, c, \overline{B}_\varrho(x)).$$

Proof. Suppose the lemma is not true. Then there exist $n \geq 2$, $c > 0$, $0 < \alpha < 1$, and $0 < \beta < 1$, two sequences (ε_h), (ϑ_h) such that $\lim_h \varepsilon_h = \lim_h \vartheta_h = 0$, a sequence (u_h) in $\mathrm{SBV}(\Omega)$, a sequence (x_h) in Ω and a sequence (ϱ_h) in $\mathbf{R}$ with $\overline{B}_{\varrho_h}(x_h) \subset \Omega$, such that

$$F(u_h, c, \overline{B}_{\varrho_h}(x)) \leq \varepsilon_h \varrho_h^{n-1},$$

$$\Psi(u_h, c, \overline{B}_{\varrho_h}(x_h)) \leq \vartheta_h F(u_h, c, \overline{B}_{\varrho_h}(x_h)),$$

and

$$(4.7) \qquad F(u_h, c, \overline{B}_{\alpha\varrho_h}(x_h)) > \alpha^{n-\beta} F(u_h, c, \overline{B}_{\varrho_h}(x_h)).$$

For each h, translating x_h into the origin and blowing up, *i.e.*, setting

$$v_h(x) = (\varrho_h \varepsilon_h)^{-1/2} u_h(x_h + \varrho_h x) \quad x \in B_1,$$

we have a sequence of functions $v_h \subset \mathrm{SBV}(B_1)$ such that (by using Lemma 4.5)

$$F(v_h, c/\varepsilon_h, \overline{B}_1) = 1,$$

$$(4.8) \qquad \Psi(v_h, c/\varepsilon_h, \overline{B}_1) \leq \vartheta_h.$$

Since $\lim_h c/\varepsilon_h = +\infty$, so that $\lim_h H_{n-1}(S_{v_h} \cap B_1) = 0$, by Theorem 3.5 there are a subsequence of (v_h), still denoted by (v_h), and a function $v_\infty \in W^{1,2}(B_1)$ such that

$$\lim_h \left[v_h - \mathrm{med}(v_h, B_1) \right] = v_\infty \quad \text{a.e. on } B_1.$$

Recalling that the functions $\varrho \to F(v_h, c/\varepsilon_h, \overline{B}_\varrho)$ $(0 < \varrho \leq 1)$ are non-decreasing, possibly by restriction to a new subsequence, we see that

$$\lim_h F(v_h, c/\varepsilon_h, \overline{B}_\varrho) \leq 1 \quad \text{for almost all } \varrho < 1,$$

and moreover, since also the functions $\varrho \to \Psi(v_h, c/\varepsilon_h, \overline{B}_\varrho)$ $(0 < \varrho \leq 1)$ are non-decreasing, from (4.8) it follows that

$$\lim_h \Psi(v_h, c/\varepsilon_h, \overline{B}_\varrho) = 0 \quad \text{for all } 0 < \varrho \leq 1,$$

Therefore by Theorem 4.8 we argue that the function v_∞ is harmonic and that

$$\limsup_h F(v_h, c/\varepsilon_h, \overline{B}_\alpha) \leq \int_{B_\alpha} |\nabla v_\infty|^2 dy.$$

Then

$$\limsup_h F(v_h, c/\varepsilon_h, \overline{B}_\alpha) \le \alpha^n \int_{B_1} |\nabla v_\infty|^2 dy \le \alpha^n.$$

On the other hand from (4.7) we have

$$\begin{aligned} F(v_h, c/\varepsilon_h, \overline{B}_\alpha) &= \varepsilon_h^{-1} \varrho_h^{1-n} F(u_h, c, \overline{B}_{\alpha\varrho_h}(x_h)) \\ &> \alpha^{n-\beta} \varepsilon_h^{-1} \varrho_h^{1-n} F(u_h, c, \overline{B}_{\alpha\varrho_h}(x_h)) = \alpha^{n-\beta}. \end{aligned}$$

Therefore we obtain a contradiction. $\qquad\square$

By the preceding Lemma 4.9 we infer the following two lemmas.

LEMMA 4.10 – *Let* $n \in \mathbf{N}$, $n \ge 2$, $0 < c < +\infty$, $0 < \alpha < 1$, $0 < \beta < 1$; *let* ε, ϑ *be as in Lemma 4.9. Let* $\Omega \subseteq \mathbf{R}^n$ *be an open set,* $\varrho > 0$, $\overline{B}_\varrho(x) \subset \Omega$, $u \in \mathrm{SBV}_{\mathrm{loc}}(\Omega)$ *and* σ, σ' *such that* $0 \le \sigma \le \varepsilon$, $0 \le \sigma' \le \varepsilon$. *If*

$$F(u, c, \overline{B}_\varrho(x)) \le \sigma \varrho^{n-1},$$

$$\Psi(u, c, \overline{B}_t(x)) \le \sigma' \vartheta(\alpha t)^{n-1} \text{ for every } t \le \varrho,$$

then

$$F(u, c, \overline{B}_{\alpha^h \varrho}(x)) \le \max\left\{ \sigma\alpha^{(1-\beta)h}, \sigma' \right\} (\alpha^h \varrho)^{n-1} \text{ for every } h \in \mathbf{N}.$$

Proof. We may suppose $x = 0$. We give the proof by induction with respect to h. For $h = 0$ the assertion is obvious. We now assume

$$F(u, c, \overline{B}_{\alpha^h \varrho}) \le \max\left\{ \sigma\alpha^{(1-\beta)h}, \sigma' \right\} (\alpha^h \varrho)^{n-1}$$

and we examine the following two cases:

$$F(u, c, \overline{B}_{\alpha^h \varrho}) < \sigma'(\alpha^{h+1} \varrho)^{n-1}$$

and

$$F(u, c, \overline{B}_{\alpha^h \varrho}) \ge \sigma'(\alpha^{h+1} \varrho)^{n-1}.$$

In the first one, it follows obviously that

$$F(u, c, \overline{B}_{\alpha^{h+1} \varrho}) < \sigma'(\alpha^{h+1} \varrho)^{n-1}.$$

In the second one, by the assumption on Ψ we obtain

$$\Psi(u, c, \overline{B}_{\alpha^h \varrho}) < \sigma' \vartheta(\alpha^{h+1} \varrho)^{n-1} \le \vartheta F(u, c, \overline{B}_{\alpha^h \varrho})$$

and by induction,

$$F(u, c, \overline{B}_{\alpha^h \varrho}) \le \varepsilon(\alpha^h \varrho)^{n-1}.$$

By Lemma 4.9 it follows that

$$\begin{aligned} F(u, c, \overline{B}_{\alpha^{h+1} \varrho}) &\le \alpha^{n-\beta} F(u, c, \overline{B}_{\alpha^h \varrho}) \le \alpha^{n-\beta} \max\left\{ \sigma\alpha^{(1-\beta)h}, \sigma' \right\} (\alpha^h \varrho)^{n-1} \\ &\le \max\left\{ \sigma\alpha^{(1-\beta)(h+1)}, \sigma' \right\} (\alpha^{h+1} \varrho)^{n-1}, \end{aligned}$$

thus the lemma is proved. $\qquad\square$

LEMMA 4.11 – *Let $n \in \mathbf{N}$, $n \geq 2$, $0 < c < +\infty$, $0 < \alpha < 1$, $0 < \beta < 1$; let ε and ϑ be as in Lemma 4.9. Let $\Omega \subseteq \mathbf{R}^n$ be open, $\varrho > 0$, $\overline{B}_\varrho(x) \subset \Omega$, $u \in \mathrm{SBV}(\Omega)$. Assume*

$$F(u, c, \overline{B}_\varrho(x)) \leq \varepsilon \varrho^{n-1},$$

$$\Psi(u, c, \overline{B}_t(x)) \leq \varepsilon \vartheta (\alpha t)^{n-1} \ \text{for every } t \leq \varrho;$$

moreover we assume that

$$(4.9) \qquad \lim_{t \to 0} t^{1-n} \Psi(u, c, \overline{B}_t(x)) = 0.$$

Then

$$\lim_{t \to 0} t^{1-n} F(u, c, \overline{B}_t(x)) = 0.$$

Proof. We again assume that $x = 0$ and first of all we shall prove that

$$(4.10) \qquad \lim_{h}(\alpha^h \varrho)^{1-n} F(u, c, \overline{B}_{\alpha^h \varrho}) = 0.$$

By Lemma 4.10 we have for every $h \in \mathbf{N}$

$$F(u, c, \overline{B}_{\alpha^h \varrho}) \leq \varepsilon (\alpha^h \varrho)^{n-1}.$$

Moreover, by assumption (4.9), for every $\sigma' > 0$ there exists $\overline{h} \in \mathbf{N}$ such that

$$(\alpha^h \varrho)^{1-n} \Psi(u, c, \overline{B}_{\alpha^h \varrho}) \leq \sigma' \ \text{ for every } h \leq \overline{h};$$

thus for every $h \geq \overline{h}$ such that $\varepsilon \alpha^{(1-\beta)h} \leq \sigma'$ we have, by virtue of Lemma 4.10,

$$(\alpha^h \varrho)^{1-n} F(u, c, \overline{B}_{\alpha^h \varrho}) \leq \sigma'.$$

Because of the arbitrariness of σ', (4.10) is proved. Now let $t < \varrho$ and $\alpha^h \varrho \leq t < \alpha^{h-1} \varrho$; then we have

$$\begin{aligned} t^{1-n} F(u, c, \overline{B}_t) &\leq (\alpha^h \varrho)^{1-n} F(u, c, \overline{B}_{\alpha^{h-1} \varrho}) \\ &= \alpha^{1-n} (\alpha^{h-1} \varrho)^{1-n} F(u, c, \overline{B}_{\alpha^{h-1} \varrho}), \end{aligned}$$

and so the assertion follows. $\qquad \square$

Finally we prove a regularity theorem for functions of the space $\mathrm{SBV}_{\mathrm{loc}}(\Omega)$ which satisfy the quasi-minimum condition (4.11).

THEOREM 4.12 – *Let $n \in \mathbf{N}$, $n \geq 2$, $0 < c < +\infty$, and let $\Omega \subseteq \mathbf{R}^n$ be open, $u \in \mathrm{SBV}_{\mathrm{loc}}(\Omega)$; assume that, for every compact set $K \subset \Omega$, $F(u, c, K) < +\infty$ and moreover that*

$$(4.11) \qquad \lim_{\varrho \to 0} \varrho^{1-n} \sup_{x \in K} \Psi(u, c, \overline{B}_\varrho(x)) = 0.$$

Set $\Omega_0 = \left\{ x \in \Omega; \ \lim_{\varrho \to 0} \varrho^{1-n} F(u, c, \overline{B}_\varrho(x)) = 0 \right\}$, then

(i) Ω_0 *is open,*
(ii) $\overline{S}_u \cap \Omega = \Omega \setminus \Omega_0$,
(iii) $H_{n-1}((\overline{S}_u \cap \Omega) \setminus S_u) = 0$.

Proof. Let $x \in \Omega_0$ and let α, β, ε and ϑ be as in Lemma 4.9. By the definition of Ω_0 and by (4.11), there exists $0 < r < \frac{1}{2}\mathrm{dist}(x, \partial\Omega)$ such that

$$(4.12) \qquad\qquad F(u, c, \overline{B}_\varrho(x)) \leq \varepsilon 2^{1-n} \varrho^{n-1},$$

and

$$(4.13) \qquad\qquad \Psi(u, c, \overline{B}_\varrho(y)) \leq \varepsilon\vartheta(\alpha\varrho)^{n-1},$$

for every $\varrho \leq r$ and for every $y \in B_r(x)$. From (4.12), we have

$$(4.14) \qquad F(u, c, \overline{B}_{r/2}(y)) \leq F(u, c, \overline{B}_r(x)) \leq \varepsilon(r/2)^{n-1}$$

for every $y \in B_{r/2}(x)$. Taking (4.14), (4.13) and the assumption (4.11) into account, from Lemma 4.11 we conclude that $B_{r/2}(x) \subset \Omega_0$. Thus Ω_0 is an open set. By Theorem 3.6 we have $S_u \subseteq \Omega \setminus \Omega_0$, and so $\overline{S}_u \cap \Omega \subseteq \Omega \setminus \Omega_0$.

We now prove (ii). Let $B_s(x) \subset \Omega \setminus \overline{S}_u$; we shall prove that $x \in \Omega_0$. Indeed $u \in W^{1,2}(B_s(x))$ and moreover, for every $t > 0$, from (4.11) it follows that

$$\lim_{\varrho \to 0} \varrho^{1-n} \Psi(tu, c, \overline{B}_\varrho(x)) = 0.$$

Since for every $\varrho < s$ we have $F(tu, c, \overline{B}_\varrho(x)) = t^2 F(u, c, \overline{B}_\varrho(x))$, for t small enough it is still possible to use Lemma 4.11 and we obtain

$$\lim_{\varrho \to 0} \varrho^{1-n} F(u, c, \overline{B}_\varrho(x)) = 0.$$

Finally (iii) follows from Lemma 2.6. $\qquad\qquad\qquad\qquad\qquad\qquad \square$

5. Proof of Theorem 1.1

We are now in a position to prove, under the assumptions of Theorem 1.1, the existence of at least one minimizer of the functional

$$G(K, u) = \int_{\Omega \setminus K} k|\nabla u|^2 dx + \mu \int_{\Omega \setminus K} k|u - g|^q dx + \lambda H_{n-1}(K \cap \Omega).$$

To this aim we introduce a new functional defined for every $u \in \mathrm{SBV}_{\mathrm{loc}}(\Omega)$ by

$$\overline{G}(u) = \int_{\Omega \setminus K} |\nabla u|^2 dx + \mu \int_{\Omega \setminus K} |u - g|^q dx + \lambda H_{n-1}(S_u \cap \Omega).$$

First we prove that in $\mathrm{SBV}_{\mathrm{loc}}(\Omega)$ at least one minimizer of the functional $\overline{G}$ exists.

LEMMA 5.1 – *Let $n \in \mathbf{N}$, $n \geq 2$, let $\Omega \subseteq \mathbf{R}^n$ be open, $1 \leq q < +\infty$, $0 < \lambda < +\infty$, $0 < \mu < +\infty$ and $g \in L^q(\Omega) \cap L^\infty(\Omega)$. Then*

$$\min\left\{\int_{\Omega\backslash K}|\nabla u|^2 dx + \mu\int_{\Omega\backslash K}|u-g|^q dx + \lambda H_{n-1}(S_u \cap \Omega); \ u \in \mathrm{SBV}_{\mathrm{loc}}(\Omega)\right\}$$

exists, and it is smaller than, or equal to,

$$\inf\left\{\int_{\Omega\backslash K}k|\nabla u|^2 dx + \mu\int_{\Omega\backslash K}k|u-g|^q dx + \lambda H_{n-1}(K \cap \Omega);\right.$$
$$\left. K \subset \mathbf{R}^n \, closed \ set, \ u \in C^1(\Omega \setminus K)\right\}.$$

Proof. Let $(u_h) \subseteq \mathrm{SBV}_{\mathrm{loc}}(\Omega)$ be a minimizing sequence for $\overline{G}$. Taking Remark 2.2 into account, we may assume $\|u_h\|_{L^\infty(\Omega)} \leq \|g\|_{L^\infty(\Omega)}$ for every $h \in \mathbf{N}$. Since

$$\lim_h\left[\int_{\Omega\backslash K}|\nabla u_h|^2 dx + \mu\int_{\Omega\backslash K}|u_h-g|^q dx + \lambda H_{n-1}(S_{u_h} \cap \Omega)\right] \leq \mu\|g\|^q_{L^q(\Omega)},$$

then, by using (2.3), for every open set $\Omega' \subset\subset \Omega$, we conclude that

$$\sup_{h\in\mathbf{N}}\int_{\Omega'}|Du_h| < +\infty;$$

hence by the compactness theorem in $\mathrm{BV}_{\mathrm{loc}}(\Omega)$ (see *e.g.* [17], Theorem 1.19) there are a subsequence, still denoted by (u_h), and a function $w \in \mathrm{BV}_{\mathrm{loc}}(\Omega)$ such that $u_h \to w$ in $L^1_{\mathrm{loc}}(\Omega)$. Since $\|w\|_{L^\infty(\Omega)} \leq \|g\|_{L^\infty(\Omega)}$ by Theorem 2.4 $w \in \mathrm{SBV}_{\mathrm{loc}}(\Omega)$, and
$$\overline{G}(w) \leq \lim_h \overline{G}(u_h),$$

thus w is a minimizer for $\overline{G}$. Now we notice that, since $g \in L^\infty(\Omega)$,

$$\inf\left\{G(K,u); \ K \text{ closed}, \ u \in C^1(\Omega\backslash K)\right\}$$
$$= \ \inf\left\{G(K,u); \ K \text{ closed}, \ u \in L^\infty(\Omega) \cap C^1(\Omega\backslash K)\right\}.$$

Indeed if $\varphi \in C^1(\mathbf{R}) \cap L^\infty(\mathbf{R})$ with $0 \leq \varphi' \leq 1$ and $\varphi(t) = t$ for $|t| \leq \|g\|_{L^\infty(\Omega)}$, then for every K closed set and $u \in C^1(\Omega \setminus K)$,

$$G(K,\varphi(u)) \leq G(K,u);$$

more precisely if $G(K,u) < +\infty$ and $u \notin L^\infty(\Omega)$, then

$$(5.1) \qquad\qquad\qquad G(K,\varphi(u)) < G(K,u);$$

hence if (K,u) minimizes G, then u is necessarily bounded. By Lemma 2.3 we infer

$$\min\left\{\overline{G}(u); \ u \in \mathrm{SBV}_{\mathrm{loc}}(\Omega)\right\} \leq \inf\left\{G(K,u); \ K \text{ closed}, \ u \in C^1(\Omega\backslash K)\right\}. \qquad \square$$

Now we state some regularity properties for a minimizer of the functional $\overline{G}$.

LEMMA 5.2 – *Let $n \in \mathbf{N}$, $n \geq 2$, let $\Omega \subseteq \mathbf{R}^n$ be open, $1 \leq q < +\infty$, $0 < \lambda < +\infty$, $0 < \mu < +\infty$ and $g \in L^q(\Omega) \cap L^\infty(\Omega)$. Let w be a minimizer of the functional $\overline{G}(u)$ among the functions $u \in \mathrm{SBV}_{\mathrm{loc}}(\Omega)$; moreover, for every $x \in \Omega \setminus S_w$, we set $\widetilde{w} = \mathrm{aplim}_{y \to x} w(y)$. Then*

$$\widetilde{w} \in C^1(\Omega \setminus \overline{S}_w) \quad \text{and} \quad H_{n-1}((\overline{S}_w \cap \Omega) \setminus S_w) = 0.$$

Proof. As in the proof of the preceding lemma, we have

$$\|w\|_{L^\infty(\Omega)} \leq \|g\|_{L^\infty(\Omega)} = M.$$

Let $\overline{B}_\varrho(x) \subset \Omega \setminus \overline{S}_w$; then $w \in W^{1,2}(B_\varrho(x))$ and it is a minimizer of the functional

$$\int_{B_\varrho(x)} |\nabla u|^2 dy + \mu \int_{B_\varrho(x)} |u - g|^q dy$$

among the functions u in $w + W_0^{1,2}(B_\varrho(x))$; thus, by well-known regularity theorems, $\widetilde{w} \in C^1(B_\varrho(x))$ (see *e.g.* [1] or [5] for the case $q > 1$ and use an approximation argument for $q = 1$).

We prove that w satisfies the hypotheses of Theorem 4.12. Let $K \subset \Omega$ be compact. It is easy to verify that $F(w, \lambda, K) < +\infty$. Now let $\varrho > 0$ with $\overline{B}_\varrho(x) \subset \Omega$ for every $x \in K$. Then, for every $v \in \mathrm{SBV}_{\mathrm{loc}}(\Omega)$ such that $v = w$ in $\Omega \setminus \overline{B}_\varrho(x)$ it follows (we may assume $H_{n-1}(S_w \cap \partial B_\varrho(x)) = 0$ and $\|w\|_{L^\infty(\Omega)} \leq M$) that

$$\begin{aligned}
F(w, \lambda, \overline{B}_\varrho(x)) \ &\leq \ F(w, \lambda, \overline{B}_\varrho(x)) + \mu \int_{B_\varrho(x)} |w - g|^q dy \\
&\leq \ F(v, \lambda, \overline{B}_\varrho(x)) + \mu \int_{B_\varrho(x)} |v - g|^q dy \\
&\leq \ F(v, \lambda, \overline{B}_\varrho(x)) + \mu(2M)^q \omega_n \varrho^n.
\end{aligned}$$

Thus because of the arbitrariness of v we have

$$\Psi(w, \lambda, \overline{B}_\varrho(x)) \leq \mu(2M)^q \omega_n \varrho^n,$$

and so also (4.11) is proved.

Finally, by Theorem 4.12, we have $H_{n-1}((\overline{S}_w \cap \Omega) \setminus S_w) = 0$. $\qquad\square$

CONCLUSION – By Lemmas 5.2 and 5.1, we infer that

if w is a minimizer for $\overline{G}$, then $(\overline{S}_w, \widetilde{w})$ gives the minimum of G and

$$G(\overline{S}_w, \widetilde{w}) = \overline{G}(w);$$

moreover, from (5.1) and by Lemma 2.3 and 5.1, we also deduce that

if the pair (K, u) minimizes G, then $u \in \mathrm{SBV}_{\mathrm{loc}}(\Omega)$, $H_{n-1}((K \setminus \overline{S}_u) \cap \Omega) = 0$ and u is a minimizer for $\overline{G}$.

Thus Theorem 1.1 is proved.

Acknowledgement. This research was supported in part by a National Research Project of the M.P.I.

Bibliography

[1] S. AGMON, A. DOUGLIS, L. NIRENBERG, *Estimates near the boundary for solutions of elliptic partial differential equations satisfying general boundary value conditions I.* Comm. Pure Appl. Math. 12 (1959), 623–727.

[2] L. AMBROSIO, *A compactness theorem for a special class of functions of bounded variation.* Boll. Un. Mat. Ital. 3-B (1989), 857–881.

[3] L. AMBROSIO, *Existence theory for a new class of variational problems.* Arch. Rational Mech. Anal. 111 (1990), 291–322.

[4] J. BLAT, J.M. MOREL, *Elliptic problems in image segmentation and their relation to fracture theory.* Recent advances in nonlinear elliptic and parabolic problems (Nancy, 1988), 216–228, Pitman Res. Notes Math. Ser. 208, Pitman, Boston.

[5] H. BREZIS, *Analyse Fonctionnelle, Théorie et Applications.* Masson, Paris, 1983.

[6] H. BREZIS, J.M. CORON, E.H. LIEB, *Harmonic maps with defects.* Comm. Math. Phys. 107 (1986), 679–705.

[7] M. CARRIERO, A. LEACI, D. PALLARA, E. PASCALI, *Euler conditions for a minimum problem with free discontinuity surfaces.* Preprint Dipartimento di Matematica, Lecce, 1988.

[8] S. CHANDRASEKHAR, *Liquid Crystals.* Cambridge University Press, Cambridge, 1977.

[9] E. DE GIORGI, *Nuovi teoremi relativi alle misure $(r-1)$-dimensionali in uno spazio ad r dimensioni.* Ricerche Mat. 4 (1955), 95–113.

[10] E. DE GIORGI, *Free discontinuity problems in Calculus of Variations,* in *Frontiers in pure and applied Mathematics, a collection of papers dedicated to J.L.Lions on the occasion of his 60th birthday* (R. Dautray ed.), North-Holland, Amsterdam, 1991.

[11] E. De Giorgi, L. Ambrosio, *Un nuovo tipo di funzionale del calcolo delle variazioni.* Atti Accad. Naz. Lincei Rend. Cl. Sci. Fis. Mat. Natur. 82 (1988), 199–210.

[12] E. De Giorgi, F. Colombini, L.C. Piccinini, *Frontiere orientate di misura minima e questioni collegate.* Quaderni S.N.S., Editrice Tecnico Scientifica, Pisa, 1972.

[13] E. De Giorgi, G. Congedo, I. Tamanini, *Problemi di regolarità per un nuovo tipo di funzionale del calcolo delle variazioni.* Atti Accad. Naz. Lincei Rend. Cl. Sci. Fis. Mat. Natur. (8) 82 (1988), 673–678

[14] J.L. Ericksen, *Equilibrium Theory of Liquid Crystals.* Adv. Liq. Cryst., Vol. 2, G.H. Brown, Ed., Academic Press, New York, 1976, 233–299.

[15] H. Federer, *Geometric Measure Theory.* Springer-Verlag, Berlin, 1969.

[16] H. Federer, W.P. Ziemer, *The Lebesgue set of a function whose distributions derivatives are p-th power summable.* Indiana Univ. Math. J. 22 (1972), 139–158.

[17] E. Giusti, *Minimal Surfaces and Functions of Bounded Variation.* Birkhäuser, Boston, 1984.

[18] U. Massari, M. Miranda, *Minimal Surfaces of Codimension One.* North-Holland, Amsterdam, 1984.

[19] M. Miranda, *Distribuzioni aventi derivate misure, insiemi di perimetro localmente finito.* Ann. Scuola Norm. Sup. Pisa 18 (1964), 27–56.

[20] J.M. Morel, S.Solimini, *Segmentation of images by variational methods (on a model of Mumford and Shah).* Revista Matem. de la Univ. Complutense de Madrid 1 (1988),169–182.

[21] D. Mumford, J. Shah, *Optimal approximations by piecewise smooth functions and associated variational problems.* Comm. Pure Appl. Math. 17 (1989), 577–685.

[22] E. Virga, *Drops of nematic liquid crystals.* Arch. Rational Mech. Anal. 107, (1989), 371–390.

Scuola Normale Superiore
Pisa

Dipartimento di Matematica
Università di Lecce

(Received January 17, 1989)

New functionals in Calculus of Variations[‡]

E. De Giorgi and L. Ambrosio

1. Definition of the functional

Recent studies on energy functionals corresponding to mixtures of different fluids, some of which may be liquid crystals, lead to investigate functionals of the type (see [4, 5, 7, 8, 15, 22, 23, 24])

$$(1) \quad F(u) = \int_\Omega f(x, u, \nabla u)\, dx + \int_{S_u} \varphi(x, tr^+(x, u, \nu), tr^-(x, u, \nu), \nu)\, dH_{n-1},$$

where $\Omega \subset \mathbf{R}^n$ is an open set, H_{n-1} is the $(n-1)$-dimensional Hausdorff measure, $u : \Omega \to \mathbf{R}^k$ is a Lebesgue-measurable function[*],

$$f : \Omega \times \mathbf{R}^k \times \mathcal{L}(\mathbf{R}^n; \mathbf{R}^k) \to\,]-\infty, +\infty], \qquad \varphi : \Omega \times \widetilde{\mathbf{R}}^k \times \widetilde{\mathbf{R}}^k \times \mathbf{S}^{n-1} \to\,]-\infty, +\infty]$$

are Borel functions, and φ satisfies the equality

$$\varphi(x, a, b, \nu) = \varphi(x, b, a, -\nu)$$

for every $x \in \Omega$, $\nu \in \mathbf{S}^{n-1}$, $a, b \in \widetilde{\mathbf{R}}^k$. Moreover, we assume that the following inequalities are satisfied

$$f(x, u, p) \geq \gamma(x), \qquad \varphi(x, a, b, \nu) \geq \beta(x)$$

with

$$\int_\Omega |\gamma|\, dx < +\infty, \qquad \int_\Omega |\beta|\, d\mathcal{H}_{n-1} < +\infty.$$

Heuristically, the first integral in (1) represents the sum of the internal energies of the fluids and possibly a potential associated to external forces. The second integral represents the interface energy of the fluids in the regions of mutual contact, and contact with a possible container (see the example in Section 4).

[‡]Editor's note: published in Nonsmooth Optimization and Related Topics, Proc. of the Fourth Course of the International School of Mathematics held in Erice, June 20-July 1, 1988, F. H. Clarke, V. F. Dem'yanov, F. Giannessi eds, Ettore Majorana International Science Series: Physical Science, 43, Plenum Press, 1989, 49–59.

[*]By $\widetilde{\mathbf{R}}^k = \mathbf{R}^k \cup \{\infty\}$ we denote the Alexandroff compactification of $\mathbf{R}^k$, and by B and $\mathbf{S}^{n-1}$ the n-dimensional unit ball and the $(n-1)$-dimensional unit sphere in $\mathbf{R}^n$, respectively. We also denote by $\mathcal{L}(\mathbf{R}^n; \mathbf{R}^k)$ the set of linear maps $L : \mathbf{R}^n \to \mathbf{R}^k$, endowed with the Hilbert norm

$$|L| = \sqrt{\sum_{i=1}^n |L(e_i)|^2},$$

$\{e_i\}_{1 \leq i \leq n}$ being an orthonormal basis of $\mathbf{R}^n$.

In order to give a precise mathematical definition of all the symbols which occur in (1), we essentially follow the notion of asymptotic limit developed in [16].

Definition 1. – If $x \in \Omega$, $z \in \widetilde{\mathbf{R}}^k$, we say that z is the approximate limit of the function u at x, and we write $z = \mathrm{ap} \lim_{y \to x} u(y)$ if

$$g(z) = \lim_{\rho \to 0^+} \fint_{B_\rho} g(u(x + \xi))\, d\xi$$

for all $g \in C^0(\widetilde{\mathbf{R}}^k)$, the space of continuous (hence bounded) real functions in $\widetilde{\mathbf{R}}^k$. In the case $z \in \mathbf{R}^k$, our definition is equivalent to the definitions in [16, 25, 26]. We also set

$$S_u = \{x \in \Omega : u \text{ has no approximate limit at } x\}.$$

Definition 2. – If $x \in \Omega$, $z \in \widetilde{\mathbf{R}}^k$, $\nu \in \mathbf{S}^{n-1}$, we say that z is the outer trace of the function u in x along the direction ν, and we write $z = tr^+(x, u, \nu)$, if

$$g(z) = \lim_{\rho \to 0^+} \fint_{B_\rho \cap \{\xi : \langle \nu, \xi \rangle > 0\}} g(u(x + \xi))\, d\xi$$

for every $g \in C^0(\widetilde{\mathbf{R}}^k)$ (see [25, 26] for similar definitions). We also define the internal trace $tr^-(x, u, \nu)$ in the following way:

$$tr^-(x, u, \nu) = tr^+(x, u, -\nu).$$

Remark 1. – The set S_u belongs to the Borel σ-algebra, is negligible, and $\mathrm{ap} \lim_{y \to x} u(y)$ is equal to $u(x)$ almost everywhere ([16, Theorem 2.9.13]). It can be easily seen that if $\mathrm{ap} \lim_{y \to x} u(y)$ exists, then

$$tr^+(x, u, \nu) = \mathrm{ap} \lim_{y \to x} u(y) = tr^-(x, u, \nu)$$

for every $\nu \in \mathbf{S}^{n-1}$. Conversely, if for some $x \in \Omega$, $\nu \in \mathbf{S}^{n-1}$ there exist $tr^+(x, u, \nu)$ and $tr^-(x, u, \nu)$ and are equal, then u has approximate limit in x. Moreover, if $x \in S_u$ and $\nu, \nu' \in \mathbf{S}^{n-1}$ are such that there exist

$$tr^+(x, u, \nu),\ tr^-(x, u, \nu),\ tr^+(x, u, \nu'),\ tr^-(x, u, \nu')$$

then necessarily $\nu = \pm\nu'$.

Definition 3. – Let $x \in \Omega$ and let $L \in \mathcal{L}(\mathbf{R}^n; \mathbf{R}^k)$; we say that u is approximately differentiable at x, L is the approximate differential of u at x, and we write $L = \nabla u(x)$, if $x \in \Omega \setminus S_u$, $z = \mathrm{ap} \lim_{y \to x} u(y) \in \mathbf{R}^k$ and

$$\mathrm{ap} \lim_{y \to x} \frac{|u(y) - z - L(y - x)|}{|y - x|} = 0.$$

2. The classes $GBV(\Omega, \mathbf{R}^k; E)$, $GSBV(\Omega, \mathbf{R}^k; E)$

We now define some classes of functions which seem to be well suited as domain of the functionals in (1). The wider class that we are going to define contains the well-known space BV of functions with bounded variation, and we denote this class by GBV (short form of generalized functions with bounded variation).

We shall also identify a class of GBV functions in which many functionals of the type (1) have minimum; we denote by $GSBV$ this class (short form of generalized special functions with bounded variation). We set

$$\mathcal{M}f(A; \mathbf{R}^k) = \{u : A \to \mathbf{R}^k : u \text{ is Lebesgue measurable}\}$$

for any open set $A \subset \mathbf{R}^n$. The space $\mathcal{M}f(A; \mathbf{R}^k)$ is endowed with the topology whose closed sets are stable with respect to the convergence almost everywhere. For any continuous function $g : \Omega \times \mathbf{R}^k \to [0, +\infty[$, for every open set $A \subset \Omega$ and every function $u \in C^1(A; \mathbf{R}^k)$ we set

$$F_g(u, A) = \int_A g(x, u)|\nabla u|\, dx.$$

The relaxed functional $\overline{F}_g(u, A)$ is defined by

$$\overline{F}_g(u, A) = \inf\{\liminf_{h \to +\infty} F_g(u_h, A) : u_h \to u \text{ almost everywhere in A}\}$$

for every open set A and every function $u \in \mathcal{M}f(A; \mathbf{R}^k)$.

DEFINITION 4. – Let $u \in \mathcal{M}f(\Omega; \mathbf{R}^k)$ and let $E \subset \Omega \times \mathbf{R}^k$ be an open set. We say that u is a function with generalized bounded variation in E, and we write $u \in GBV(\Omega, \mathbf{R}^k; E)$, if

$$\overline{F}_g(u, \Omega) < +\infty$$

for every non-negative continuous function g with compact support in E.

In order to compare $BV(\Omega)$ with the space $BV(\Omega)$, studied for instance in [16, 17, 18], the following remark is useful.

REMARK 2. – If $k = 1$, $u \in L^1(\Omega)$ and g is the function identically equal to 1, then

$$\overline{F}_g(u, \Omega) < +\infty$$

if and only if the function u belongs to $BV(\Omega)$ (see [17, 18]). Moreover, it can be shown that if $u \in GBV(\Omega, \mathbf{R}^k; E)$ then the function $\psi(x, u(x))$ belongs to $BV(\Omega)$ for every function $\psi \in C_o^1(E)$.

For every function $u \in \mathcal{M}f(\Omega; \mathbf{R}^k)$ we define $GBV\mathrm{amb}(u)$ as the union of the open sets $E \subset \Omega \times \mathbf{R}^k$ such that $u \in GBV(\Omega, \mathbf{R}^k; E)$. Moreover, we set

$$GBV\mathrm{dom}(u) = \{x \in \Omega \setminus S_u : (x, \operatorname*{ap\,lim}_{y \to x} u(y)) \in GBV\mathrm{amb}(u)\};$$

$$GBV(\Omega; \mathbf{R}^k) = \{u \in \mathcal{M}f(\Omega; \mathbf{R}^k) : GBV\mathrm{amb}(u) = \Omega \times \mathbf{R}^k\}.$$

In [3] some properties of GBV functions are studied. We now state some properties of approximate differentiability and existence of traces.

PROPOSITION 1. – *Let* $u \in \mathcal{M}f(\Omega, \mathbf{R}^k)$. *Then the approximate differential of* u *exists at almost every point of* $GBV\mathrm{dom}(u)$ *and for every non-negative function* $g \in C^0(\Omega \times \mathbf{R}^k)$ *with compact support in some open set* A *such that* $u \in GBV(\Omega, \mathbf{R}^k; A)$ *we have*
(2)
$$\int_{GBV\mathrm{dom}(u)\cap A} g(x,u)|\nabla u|\, dx = \inf\{\overline{F}_g(u, A \setminus K) : K \text{ compact, } \mathrm{meas}(K) = 0\}$$

for every open set $A \subset \Omega$.

In the case $E \supset \Omega \times (\mathbf{R}^k \setminus F)$ with F finite, in [3] the following proposition is proved.

PROPOSITION 2. – *Let* $u \in GBV(\Omega, \mathbf{R}^k; E)$. *Then* u *is approximately differentiable almost everywhere in* Ω *and for* H_{n-1} *almost every* $x \in S_u$ *there exist* $\nu \in \mathbf{S}^{n-1}$, $u', u'' \in \widetilde{\mathbf{R}}^k$ *such that*

$$u' = tr^+(x,u,\nu),\ u'' = tr^-(x,u,\nu).$$

Moreover, the set S_u *admits the following representation:*

$$S_u = \bigcup_{i=1}^{\infty} K_i \cup N,$$

where $H_{n-1}(N)=0$ *and the sets* K_i *are compact subsets of* C^1 *hypersurfaces. Finally, for any* C^1 *hypersurface* $\Gamma \subset \Omega$, *at* H_{n-1} *almost every point* x *of* Γ *there exist*

$$tr^+(x,u,\nu),\ tr^-(x,u,\nu)$$

along the direction ν *normal to* Γ.

Even if Proposition 2 implies that the functionals in (1) are well defined in $GBV(\Omega; \mathbf{R}^k)$, variational and heuristic considerations suggest to restrict the functionals to a special class of GBV functions characterized by a stronger property than (2). To this end, we give the following definition.

DEFINITION 5. – Let E be an open set in $\Omega \times \mathbf{R}^k$. We say that u is a special GBV function in E, and we write $u \in GSBV(\Omega, \mathbf{R}^k; E)$, if $u \in GBV(\Omega, \mathbf{R}^k; E)$ and the following equality holds

$$\int_{GBV\mathrm{dom}(u)} g(x,u)|\nabla u|dx$$
$$= \inf\left\{\overline{F}_g(u, \Omega \setminus K) : K \subset \Omega \text{ compact, } H_{n-1}(K) < +\infty\right\}$$

for every non-negative continuous function g with compact support in E. We also denote the class $GSBV(\Omega, \mathbf{R}^k; \Omega \times \mathbf{R}^k)$ by $GSBV(\Omega; \mathbf{R}^k)$.

Now we can illustrate with an example our initial statement that many functionals (1) have minimizers in the class $GSBV$. Let $f_1 : \Omega \times \mathbf{R}^k \to [0, +\infty[$ be a Borel function, continuous with respect to the last k variables, and satisfying the condition

$$\lim_{|y|\to+\infty} f_1(x,y) = +\infty$$

for every $x \in \Omega$. Let $\beta > 1$ and let $f_2 : \Omega \times \mathbf{R}^k \to [0,+\infty[$ be a continuous function, strictly positive outside $\Omega \times F$, F being a finite subset of $\mathbf{R}^k$. Then the functional

$$\int_\Omega [f_1(x,u) + f_2(x,u)|\nabla u|^\beta]\,dx + \int_{S_u} dH_{n-1}$$

has at least one minimizer in $GSBV(\Omega, \mathbf{R}^k; \Omega \times (\mathbf{R}^k - F))$. It is to be noted that suitable choices of f_1, f_2 guarantee that the minimizers are not locally summable; hence, they do not belong to any of the functional spaces commonly used in the Calculus of Variations and in the theory of Partial Differential Equations.

3. Semicontinuity problems

The same heuristic motivations which suggest the study of functionals (1) in the domain $GSBV$ also suggest to look for conditions on f, φ which ensure the lower semicontinuity with respect to the topology of $\mathcal{M}f(\Omega; \mathbf{R}^k)$. The search for lower semicontinuity conditions for the functionals (1) seems to be very difficult, because it contains at the same time the lower semicontinuity problems studied in non-linear elasticity [1, 21] and the lower semicontinuity problems of Geometric Measure Theory (see for instance [16], Chapter 5). We refer the reader interested in this topic to [2,3]. Here we only give purely heuristic suggestions; we think that the research of lower semicontinuity conditions could be carried on in the following three steps:

1) Study of the lower semicontinuity of the restriction of the functional F to a Sobolev space $W^{1,p}(\Omega; \mathbf{R}^k)$. This problem has been extensively studied, and a wide bibliography on the subject is collected in [12].

2) Study of the lower semicontinuity of the restriction of the functional F to the class of $GBV(\Omega; \mathbf{R}^k)$ functions whose range is a finite set. These functions necessarily belong to $GSBV(\Omega; \mathbf{R}^k)$, and it can be shown that if φ is continuous then the following conditions [3] are necessary for lower semicontinuity:

(I) for every $x \in \Omega$, $a, b, c \in \mathbf{R}^k$, $\nu \in \mathbf{S}^{n-1}$ we have

$$\varphi(x,a,b,\nu) \leq \varphi(x,a,c,\nu) + \varphi(x,c,b,\nu);$$

(II) for every $x \in \Omega$, $a, b \in \mathbf{R}^k$, $\nu \in \mathbf{S}^{n-1}$ the function

$$\varphi(p) = \begin{cases} |p|\varphi\left(x,a,b,\dfrac{p}{|p|}\right) & \text{if } p \neq 0 \\ 0 & \text{if } p = 0 \end{cases}$$

is convex in $\mathbf{R}^n$.

3) Study of the connections between f and φ. The following condition seems to be particularly interesting:

(III) for every $x \in \Omega$, $s \in \mathbf{R}^k$, $\theta \in \mathbf{R}^k$, $\nu \in \mathbf{S}^{n-1}$ we have

$$\lim_{t \to +\infty} \frac{f(x, s, t\nu \otimes \theta)}{t} = \lim_{t \to 0^+} \frac{\varphi(x, s + t\theta, s, \nu)}{t},$$

where $\nu \otimes \theta \in \mathcal{L}(\mathbf{R}^n; \mathbf{R}^k)$ is defined by

$$\nu \otimes \theta(\xi) = \langle \nu, \xi \rangle \theta$$

for every $\xi \in \mathbf{R}^n$.

Besides the semicontinuity problems, there is also the search of a characterization of the relaxed functional

$$\mathcal{F}(u) = \inf \left\{ \liminf_{h \to +\infty} F(u_h) : (u_h) \subset GSBV(\Omega; \mathbf{R}^k),\ u_h \to u \text{ almost everywhere} \right\}$$

defined for all functions $u \in \mathcal{M}f(\Omega; \mathbf{R}^k)$. Under suitable continuity assumptions, a reasonable condition which ensures that $\mathcal{F}$ be infinite in $\mathcal{M}f(\Omega; \mathbf{R}^k) \setminus GSBV(\Omega; \mathbf{R}^k)$ seems to be the following [2,3]:

$$\text{(IV)} \qquad \begin{cases} \lim_{|w| \to +\infty} \dfrac{f(x, u, w)}{|w|} = +\infty \\ \lim_{t \to 0} \dfrac{\varphi(x, s + t, s, \nu)}{|t|} = +\infty. \end{cases}$$

In the case where (IV) is not satisfied and the functional $\mathcal{F}$ is finite for some function $u \in \mathcal{M}f(\Omega; \mathbf{R}^k) \setminus GSBV(\Omega; \mathbf{R}^k)$, integral representation problems of the type considered in [9] arise. Some suggestions for this problem are given in Section 5. Finally, in connection with functionals (1) or, possibly, their relaxed functionals, problems of Γ-convergence and regularity of minimizers can be studied.

The regularity of minimizers seem to be a very difficult problem, mixing the regularity of minimal hypersurfaces with the boundary regularity of solutions of elliptic equations.

4. An example from the static theory of liquid crystals

After the purely mathematical remarks in the previous sections, we want to show by an example how the functionals (1) may occur in the study of mixtures of liquids and liquid crystals. Our example does not pretend to be a complete modelization of the physical problem, but just a suggestion of the direction where this modelization can be found. Let us consider (see figure) a container R containing an isotropic liquid L and a liquid crystal C, both incompressible. A particular configuration of the system can be described by a function u defined on the portion Ω of space occupied by R with values in $\mathbf{R}^5$, with the following constraints:

(1) $u(x) = (0,0,0,0,0)$ almost everywhere in $\Omega \setminus \Omega'$, where Ω' is the interior of R, $\Omega' \subset\subset \Omega$;

(2) $u(x) = (1,0,0,0,0)$ if in x there is the isotropic liquid L;

(3) $u(x) = (0,1,w_1,w_2,w_3)$ if in x there is the liquid crystal C, and the vector $(w_1,w_2,w_3) \in \mathbf{S}^2$ is the optic axis of the crystal in x.

It is reasonable to assume that the energy of the system is described by a functional of the type (see [4, 5, 7, 8])

$$\mathcal{E}(u) = \int_\Omega [g(x,u) + c|\nabla u|^2]\, dx + \int_{S_u} \varphi(tr^+(x,u,\nu), tr^-(x,u,\nu), \nu)\, dH_{n-1}.$$

The equilibrium configurations of the system can thus be found among the minimum or at least the stationary points of the functional $\mathcal{E}$ in the space $GSBV(\Omega; \mathbf{R}^5)$, with the constraints

$$u(x) \quad \in \quad \{(0,0,0,0,0)\} \cup \{(1,0,0,0,0)\} \sqcup$$
$$\cup\{z \in \mathbf{R}^5 : z = (0,1,w_1,w_2,w_3) \text{ with } (w_1,w_2,w_3) \in \mathbf{S}^2\}$$

$$u(x) = 0 \text{ in } \Omega \setminus \Omega'$$

$$\int_\Omega u^{(i)}\, dx = V_i \qquad \text{for } i = 1,2\,,$$

where V_1, V_2 are the prescribed volumes of L and C, respectively, $u^{(i)}$ is the i-th component of u, and

$$V_1 + V_2 = \operatorname{meas}(\Omega').$$

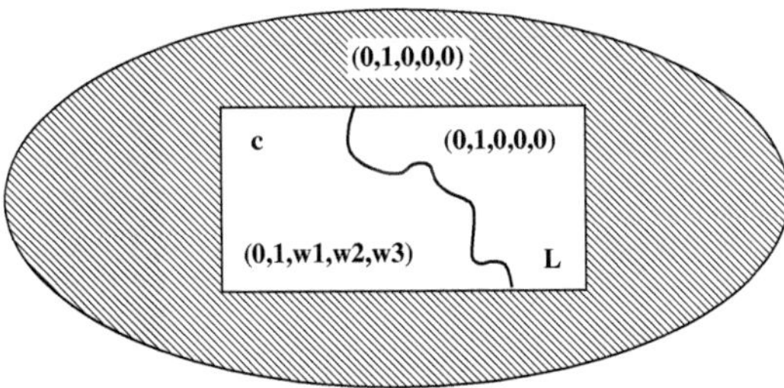

Figure 1

686   E. DE GIORGI, L. AMBROSIO

5. Representation of functionals in $GBV(\Omega, \mathbf{R}^k; E) \setminus GSBV(\Omega, \mathbf{R}^k; E)$

In this section we consider functionals defined in $GBV(\Omega, \mathbf{R}^k; E)$ that admit in $GSBV(\Omega, \mathbf{R}^k; E)$ a representation as in (1), and can be represented as follows

$$F(u) \;=\; \int_\Omega f(x, u, \nabla u)\, dx + \int_{S_u} \varphi(x, tr^+(x, u, \nu), tr^-(x, u, \nu), \nu)\, dH_{n-1}$$

$$(3) \qquad + \int_\Omega \psi(x, \tilde{u}, CDu),$$

where $\psi : \Omega \times \mathbf{R}^k \times \mathcal{L}(\mathbf{R}^n; \mathbf{R}^k) \to [0, +\infty]$ is a Borel function positively homogeneous of degree 1 in the last variable, the function $\tilde{u}$ is defined by

$$\tilde{u}(x) = \mathrm{ap}\lim_{y \to x} u(y)$$

for every $x \in \Omega \setminus S_u$ and the meaning of $\int_\Omega \psi(x, u, CDu)$ will be given in the next definition. Before giving this definition we need to state the following theorem.

THEOREM 1. – *For every* $u \in \mathcal{M}f(\Omega, \mathbf{R}^k)$ *there exist a non-negative Borel measure* μ *and a Borel function* $w : \Omega \to \mathcal{L}(\mathbf{R}^n; \mathbf{R}^k)$ *such that*

(i) $$H_{n-1}(B) < +\infty \quad \Rightarrow \quad \mu(B) = 0;$$

(ii) $$H_1(C) = 0 \quad \Rightarrow \quad \mu(\{x \in \Omega : \tilde{u}(x) \in C\}) = 0;$$

(iii) $$\mu(\Omega \setminus GBV\mathrm{dom}(u)) = 0;$$

(iv) $$|w(x)| = 1 \quad \mu \ \textit{almost everywhere in } \Omega;$$

(v) *for every non-negative function* $g \in C^0(\Omega \times \mathbf{R}^k)$ *with compact support in some open set* A *such that* $u \in GBV(\Omega, \mathbf{R}^k; A)$ *and every function* $\Theta : \mathcal{L}(\mathbf{R}^n; \mathbf{R}^k) \to [0, +\infty[$, *convex and positively homogeneous of degree 1, we have the representation formula*

$$\inf\{\overline{F}_{g,\Theta}(u, A \setminus K) : H_{n-1}(K) < +\infty\} =$$

$$= \int_{GBV\mathrm{dom}(u) \cap A} g(x, u)\Theta(\nabla u)dx + \int_{GBV\mathrm{dom}(u) \cap A} g(x, \tilde{u})\Theta(w)d\mu\,,$$

where

$$\overline{F}_{g,\Theta}(u, A)$$

$$= \inf\left\{\liminf_{h \to +\infty} \int_A g(x, u_h)\Theta(\nabla u_h)\, dx : u_h \in C^1(A; \mathbf{R}^k),\ u_h \to u \text{ a.e. in } A\right\}.$$

A still open problem is to estimate the rank of the function w; we do not know if necessarily the rank of the matrix $w(x)$ is 1 in μ almost every point x. The set function

$$\int_B w\, d\mu,$$

defined for all Borel sets B such that $\mu(B) < +\infty$, will be called the "Cantor part" of the derivative of u. This name is justified by the well-known Cantor-Vitali function; this function has a derivative in the sense of distributions which is concentrated on Cantor's middle third set, and this derivative is equal to $\int_B w \, d\mu$. We can now give the following definition.

DEFINITION 6. – Let $\psi(x, u, p)$ be a non-negative Borel function, positively homogeneous of degree 1 in p. Using a notation similar to that in [17, 18], we set

$$\int_B \psi(x, \tilde{u}, CDu) = \int_B \psi(x, \tilde{u}, w) \, d\mu$$

for every Borel set B.

Semicontinuity problems for functionals (3) can also be considered. We remark that, under rather general assumptions on the integrands f and ψ, a reasonable lower semicontinuity condition which might be added to the conditions in Section 3, is the following

$$\psi(x, u, p) = \lim_{t \to +\infty} \frac{f(x, u, tp)}{t}.$$

Acknowledgment

We wish to thank E. Virga for the many conversations which suggested to us the study of the problems illustrated in this paper.

Bibliography

[1] E. ACERBI, N. FUSCO, *Semicontinuity problems in the valculus of variations*. Archive for Rat. Mech. and Anal. **86** (1984), 125–145.

[2] L. AMBROSIO, *A compactness theorem for a special class of functions of bounded variation*. Boll. Un. Mat. Ital. 3-B (1989), 857–881.

[3] L. AMBROSIO, *Existence theory for a new class of variational problems*. Arch. Rational Mech. Anal. **111** (1990), 291–322.

[4] H. BREZIS, J.M. CORON, E.H. LIEB, *Estimations d'energie pour des applications de* $\mathbf{R}^3$ *a valeurs dans* $\mathbf{S}^2$. C.R. Acad. Sc. Paris **303** (1986), 207–210.

[5] H. BREZIS, J.M. CORON, E.H. LIEB, *Harmonic maps with defects*. Comm. Math. Phys. **107** (1986), 679–705.

[6] A.P. CALDERON, A. ZYGMUND, *On the differentiability of functions which are of bounded variation in Tonelli's sense*. Revista Union Mat. Arg. **20** (1960), 102–121.

[7] S. CHANDRASEKHAR, *Liquid Crystals*. Brown, Dienes and Labes Editors, Gordon and Breach, New York, 331–340, 1966.

[8] S. Chandrasekhar, *Liquid Crystals*. Cambridge University Press, Cambridge, 1977.

[9] G. Dal Maso, *Integral representation on $BV(\Omega)$ of Γ-limits of integral functionals*. Manuscripta Math. **30** (1980), 387–413.

[10] E. De Giorgi, *Su una teoria generale della misura $(r-1)$-dimensionale in uno spazio a r dimensioni*. Ann. Mat. Pura Appl. **36** (1954), 191–213.

[11] E. De Giorgi, *Nuovi teoremi relativi alle misure $(r-1)$-dimensionale in uno spazio a r dimensioni*. Ricerche Mat. **4** (1955), 95–113.

[12] E. De Giorgi, *Generalized limits in Calculus of Variations*. in *Topics in Functional Analysis 1980-81*, Quaderno Scuola Normale Superiore, Pisa, 1981, pp. 117–148.

[13] E. De Giorgi, *Some semicontinuity and relaxation problems*. in *Ennio De Giorgi Colloquium*, Research Notes in Mathematics 125, Pitman Publishing Inc., Boston, 1985.

[14] E. De Giorgi, G.Letta, *Une notion générale de convergence faible pour des fonctions croissantes d'ensemble*. Ann. Sc. Norm. Sup. Pisa Cl. Sci. (4) **4** (1977), 61–99.

[15] J.L. Ericksen, *Advances in liquid crystals*. Vol. 2, Glenn and Brown editors, Academic Press, New York, 1976, pp. 233–298.

[16] H. Federer, *Geometric Measure Theory*. Springer Verlag, Berlin, 1969.

[17] E. Giusti, *Minimal Surfaces and Functions of Bounded Variation*. Birkäuser, Boston, 1984.

[18] M. Miranda, *Distribuzioni aventi derivate misure. Insiemi di perimetro localmente finito*. Ann. Sc. Norm. Sup. Pisa Cl. Sci., (3) **18** (1964), 27–56.

[19] M. Miranda, *Superfici cartesiane generalizzate ed insiemi di perimetro localmente finito sui prodotti cartesiani*. Ann. Sc. Norm. Sup. Pisa Cl. Sci. (3) **18** (1964), 515–542.

[20] L. Modica, S. Mortola, *Un esempio di Γ- convergenza*. Boll. Un. Mat. Ital. (5) **14-B** (1977), 285–299.

[21] C.B. Morrey, *Multiple Integrals in the Calculus of Variations*. Springer Verlag, Berlin, 1966.

[22] D. Mumford, J. Shah, *Boundary detection by minimizing functionals*. Proc. of the Conference on Computer Vision and Pattern Recognition, San Francisco 1985.

[23] D. Mumford, J. Shah, *Optimal approximations by piecewise smooth functions and associated variational problems*. Comm. Pure Appl. Math. **17** (1989), 577–685.

[24] E. VIRGA, *Forme di equilibrio di piccole gocce di cristallo liquido.* Preprint n. 562 dell'Istituto di Analisi Numerica, Pavia, 1987.

[25] A.I. VOL'PERT, *Spaces BV and Quasi-Linear Equations.* Math. USSR Sb. **17**, 1972.

[26] A.I. VOL'PERT, S.I. HUDJAEV, *Analysis in Classes of Discontinuous Functions and Equations of Mathematical Physics.* Martinus Nijhoff Publisher, Dordrecht, 1985.

[27] H. WHITNEY, *Analytic extensions of differentiable functions defined in closed sets.* Trans. Amer. Mat. Soc. **36** (1934), 63–89.

Conjectures on limits of some quasilinear parabolic equations and flow by mean curvature[‡]

Ennio De Giorgi

Dedicated to Louis Nirenberg

In this paper we present some conjectures which connect the theory of flow by mean curvature to the asymptotic behaviour of solutions of certain quasi linear parabolic equations. A part of these conjectures has been explained in [6], and it is reproduced here for completeness' sake.

From a theorem of L. Modica and S. Mortola (see [17]) and from regularity results on sets with boundary of given mean curvature (see *e.g.* [13]) it follows that for every $g \in L^2(\mathbf{R}^n)$ such that $-\pi < g(x) < 3\pi$ the following equality holds:

$$\min_{E \subset \mathbf{R}^n} \left\{ 8\sqrt{2}\mathcal{H}^{n-1}(\partial E) + \int_{\mathbf{R}^n} |2\pi\chi_E(x) - g(x)|^2 dx \right\} =$$

$$= \lim_{p \to +\infty} \left[\min_{u \in H^{1,2}(\mathbf{R}^n)} \left\{ \int_{\mathbf{R}^n} \left(|u - g|^2 + \frac{1}{p} |\nabla u|^2 + p(1 - \cos u) \right) dx \right\} \right],$$

where χ_E is the characteristic function of the set E and $\mathcal{H}^{n-1}$ is the $(n-1)$-dimensional Hausdorff measure. This result suggests the idea of approximating *minimal surface type problems* by *minimal quadratic energy type problems* and *surfaces evolution* by *heat type equations* (see *e.g.* [3]). The following conjectures are intended to be an encouragement towards the realization of this idea.

DEFINITION 1 – *Mean curvature motion equation*

If u is a function of $n + 1$ variables $x_1, \ldots, x_n, t$ and satisfies the following equation

$$|\nabla_x u|^2 \left(\frac{\partial u}{\partial t} - \Delta_x u \right) + \sum_{h,k=1}^{n} \frac{\partial u}{\partial x_h} \frac{\partial u}{\partial x_k} \frac{\partial^2 u}{\partial x_h \partial x_k} = 0,$$

where $\nabla_x u = \left(\frac{\partial u}{\partial x_1}, \ldots, \frac{\partial u}{\partial x_n} \right)$, $\Delta_x u = \sum_{h=1}^{n} \frac{\partial^2 u}{\partial x_h^2}$, then we say that u satisfies the mean curvature motion equation (MCM-Eq.).

[‡]Editor's note: published in Partial differential equations and related subjects, Proceedings of a Conference dedicated to L. Nirenberg, (Trento, 1992), M. Miranda ed., Pitman Res. Notes in Math. 269, Longman and Wiley, 1992, 85–95.

DEFINITION 2 – *The MCM^* operator*
Let $p \in \mathbf{R}$, $f : \mathbf{R}^n \to \mathbf{R}$, $w : \mathbf{R}^n \times [0, +\infty[\to \mathbf{R}$; we write $w = MCM^*(f, p)$ iff, by definition

$$f \in L^1_{loc}(\mathbf{R}^n) \quad \text{and} \quad \int_{\mathbf{R}^n} |f(x)| e^{-\lambda |x|^2} dx < +\infty \text{ for every } \lambda > 0,$$

$$w \in C^\infty(\mathbf{R}^n \times]0, +\infty[) \text{ and } \int_0^a dt \int_{\mathbf{R}^n} |w(x,t)| e^{-\lambda |x|^2} dx$$

$$< +\infty \; \forall \, \lambda > 0, 0 < a < +\infty$$

$$\lim_{t \to 0} \int_{\mathbf{R}^n} |w(x,t) - w(x,0)| e^{-\lambda |x|^2} dx = 0 \quad \text{for every } \lambda > 0,$$

$$\begin{cases} \dfrac{\partial w}{\partial t} = \Delta_x w - p^2 \sin w & \forall (x,t) \in \mathbf{R}^n \times]0, +\infty[\\[2mm] w(x,0) = f(x) & \forall x \in \mathbf{R}^n. \end{cases}$$

REMARK 1 – The operator MCM^* is obviously invariant under translations with respect to x, and increasing with respect to f. Moreover, for every f and for every p it holds:

$$MCM^*(-f, p) = -MCM^*(f, p), \qquad MCM^*(f + 2\pi, p) = 2\pi + MCM^*(f, p).$$

DEFINITION 3 – *The MCM operator*
Let $A \subset \mathbf{R}^n$ be open, $B \subset \mathbf{R}^n \times [0, +\infty[$. We set $B = MCM(A)$ iff, by definition

$$B = \bigcup_{\substack{\varphi \in C_0^\infty(\mathbf{R}^n) \\ \mathrm{spt}\varphi \subset A}} \left\{ (x,t) \in \mathbf{R}^n \times [0, +\infty[: \liminf_{p \to +\infty} MCM^*(\varphi, p)(x,t) > 0 \right\}.$$

DEFINITION 4 – Let $E \subset \mathbf{R}^n$; we set

$$MCM(E) = \bigcap_A \{ MCM(A) : E \subset A \subset \mathbf{R}^n, \; A \text{ open} \}.$$

REMARK 2 – From definitions 3, 4 it follows for every open set $A \subset \mathbf{R}^n$:

$$MCM(A) = \bigcup_K \{ MCM(K) : K \subset A, \; K \text{ compact} \}.$$

DEFINITION 5 – Let $f : \mathbf{R}^n \to [-\infty, +\infty]$, $u : \mathbf{R}^n \times [0, +\infty[\to [-\infty, +\infty]$; we set $u = MCMf$ iff, by definition

$$MCM(\{x \in \mathbf{R}^n : f(x) > \lambda\}) = \{(x,t) \in \mathbf{R}^n \times [0, +\infty[: u(x,t) > \lambda\} \quad \forall \lambda \in \mathbf{R}.$$

REMARK 3 – By remark 2, if $(A_h)_h$ is an increasing sequence of open sets, then $\lim_{h \to +\infty} MCM(A_h) = MCM \left(\lim_{h \to +\infty} A_h \right)$ and then for every f continuous (or lower semicontinuous) there exists $MCMf$.

CONJECTURE 1 – If f is uniformly continuous, then $MCM(-f) = -MCMf$.

REMARK 4 – Probably the thesis of conjecture 1 is false if f is only continuous. A counterexample could be given by the following conjecture.

CONJECTURE 2 – If $f(x_1, x_2) = x_1^2 \exp(x_2)$, then for some $(x,t) \in \mathbf{R}^n \times]0, +\infty[$ it holds $MCMf(x,t) < -MCM(-f)(x,t)$.

REMARK 5 – From the definition it immediately follows that MCM is invariant under translations with respect to x, and increasing with respect to f. Moreover, for every f and for every constant a it holds:

$$MCM(f + a) = a + MCM(f).$$

More generally, if $g : [-\infty, +\infty] \to [-\infty, +\infty]$ and g is strictly increasing and continuous, then

$$MCM(g \circ f) = g \circ MCMf.$$

REMARK 6 – If f is continuous and φ is continuous and positive, then by remark 2 it follows

$$MCMf = \lim_{\varepsilon \to 0^+} MCM(f - \varepsilon\varphi).$$

REMARK 7 – If we are able to compute $MCMf$ for every Lipschitz continuous function f, then it is easy to deduce $MCM(A)$ for every open set A (it suffices to take as f the distance from $\mathbf{R}^n \setminus A$), hence for every set E and for every function f.

CONJECTURE 3 – If $f \in C^\infty(\mathbf{R}^n)$, $f(x) = |x|^2$, then $MCMf(x,t) = |x|^2 + 2(n-1)t$.

CONJECTURE 4 – If $f \in C^0(\mathbf{R}^{m+1})$, $a \in \mathbf{R}$, $a \neq 0$, $g \in C^0(\mathbf{R}^m)$, and $f(x_1, \ldots, x_m, x_{m+1}) = ax_{m+1} + g(x_1, \ldots, x_m)$, then $MCMf$ is continuous and there exists $\psi \in C^\infty(\mathbf{R}^m \times]0, +\infty[)$ such that for every $(x,t) \in \mathbf{R}^m \times [0, +\infty)$

$$MCMf(x_1, \ldots, x_{m+1}, t) = ax_{m+1} + \psi(x_1, \ldots, x_m, t),$$

$$(a^2 + |\nabla_x \psi|^2) \left(\frac{\partial \psi}{\partial t} - \Delta_x \psi \right) + \sum_{h,k=1}^m \frac{\partial \psi}{\partial x_h} \frac{\partial \psi}{\partial x_k} \frac{\partial^2 \psi}{\partial x_h \partial x_k} = 0.$$

REMARK 8 – Set $L_{k,t} = \{x \in \mathbf{R}^n : MCMf(x,t) = k\}$, the previous conjectures say that, as t varies, $L_{k,t}$ moves with velocity proportional to its mean curvature; this justifies the name $MCM - operator$.

For a comparison with some known results on motion of graphs by mean curvature see [8], [18].

CONJECTURE 5 – If g is uniformly continuous and $f_a(x_1, \ldots, x_m, x_{m+1}) = ax_{m+1} + g(x_1, \ldots, x_m)$, then $MCMf_a$ is continuous and

$$\lim_{a \to 0} MCMf_a(x_1, \ldots, x_{m+1}, t) = MCMg(x_1, \ldots, x_m, t)$$

uniformly on compact subsets of $\mathbf{R}^{m+1} \times [0, +\infty[$.

It would be interesting to compare the previous conjecture with the results of [10]. If conjectures 4, 5 were proved, they would show a connection between the limits of the solutions of the equation considered in definition 2 and the motion by mean curvature: such connection is further precised in the following conjecture.

CONJECTURE 6 – If $f : \mathbf{R}^n \to \mathbf{R}$ is uniformly continuous, then for every x verifying the condition

$$\frac{f(x) + \pi}{2\pi} \notin \mathbf{Z}$$

(where $\mathbf{Z}$ is the set of the integers) and for every $t > 0$ it holds

$$\lim_{p \to +\infty} MCM^*(f, p)\left(x, \frac{t}{p}\right) = 2\pi \left[\frac{f(x) + \pi}{2\pi}\right],$$

where the square brackets denote the integral part, *i.e.* $[y]$ stands for the greatest integer less than or equal to y.

Moreover for every $(x, t) \in \mathbf{R}^n \times]0, +\infty[$ verifying the condition

$$\frac{MCMf(x, t) + \pi}{2\pi} \notin \mathbf{Z}$$

it holds

$$\lim_{p \to +\infty} MCM^*(f, p)(x, t) = 2\pi \left[\frac{MCMf(x, t) + \pi}{2\pi}\right]$$

uniformly on compact subsets of the open set

$$A = \left\{ (x, t) \in \mathbf{R}^n \times]0, +\infty[: \frac{MCMf(x, t) + \pi}{2\pi} \notin \mathbf{Z} \right\}.$$

For results which present some similarities with conjectures 3 and 6, see [3], [7].

REMARK 9 – Let us point out that if f is uniformly continuous and conjecture 6 is true, then by remark 1 we have

$$MCMf = -MCM(-f) \quad \text{and} \quad 2\pi MCMf = \lim_{p \to +\infty} \int_{-\pi}^{\pi} MCM^*(f + k, p)dk,$$

so that it is possible to consider conjecture 1 as a part of conjecture 6.

A further consequence of conjecture 6 is the following result:

$$MCMf = \lim_{k \to +\infty} \liminf_{p \to +\infty} \frac{MCM^*(kf, p)}{k} = \lim_{k \to +\infty} \limsup_{p \to +\infty} \frac{MCM^*(kf, p)}{k}.$$

These limits and the limit considered in conjecture 5 give some more "operative" expressions of $MCMf$ in many situations where definition 5 appears to be too theoretic.

In particular, it would be interesting to evaluate from a numeric point of view the velocity of convergence of these limits.

REMARK 10 – From conjecture 6 it follows that in definition 3 it is possible to replace $\liminf_{p \to +\infty} MCM^*(\varphi, p)$ by $\limsup_{p \to +\infty} |MCM^*(\varphi, p)|$.

REMARK 11 – From conjecture 6 and the consequent relation $MCM(-f) = -MCMf$ it follows that if f is uniformly continuous, set for $\lambda \in \mathbf{R}$

$$A_\lambda^+ = \{x \in \mathbf{R}^n : f(x) > \lambda\}, \quad \text{and} \quad A_\lambda^- = \{x \in \mathbf{R}^n : f(x) < \lambda\},$$

for almost any $\lambda \in \mathbf{R}$ it holds

$$\mathcal{H}^{n+1}\left(\mathbf{R}^n \times]0, +\infty[\setminus \left(MCM(A_\lambda^+) \cup MCM(A_\lambda^-)\right)\right) = 0.$$

The idea that the $n-1$-dimensional measure of a surface moving along its mean curvature decreases can be formulated in the present context as a monotonicity theorem for the total variation of the gradient of $MCMf$. We recall first the following definition (see [14]).

DEFINITION 6 – Let $g \in L^1_{loc}(\mathbf{R}^n)$; we set

$$\int_{\mathbf{R}^n} |Dg| = \sup_\varphi \left\{ \int g \operatorname{div}\varphi : \varphi \in [C_0^1(\mathbf{R}^n)]^n, \ \sup_{x \in \mathbf{R}^n} |\varphi(x)| \leq 1 \right\}.$$

We say that g is of bounded variation in $\mathbf{R}^n$ if $\int_{\mathbf{R}^n} |Dg| < +\infty$.

CONJECTURE 7 – If f is continuous and also of bounded variation in $\mathbf{R}^n$, then for every $t \geq 0$ the function $MCMf(\cdot, t)$ is of bounded variation in $\mathbf{R}^n$ and the function $t \mapsto \int_{\mathbf{R}^n} |D\,MCMf(\cdot, t)|$ is continuous and not increasing.

The following conjecture concerns the regularity of $MCMf$, under regularity hypotheses on the initial data. Let us first fix some notation.

NOTATION – For $g : \mathbf{R}^n \to \mathbf{R}$ we set

$$AmbC^\infty(g) = \bigcup_A \{A \subset \mathbf{R}^n, \ A \text{ open}, \ g \in C^\infty(A)\}.$$

CONJECTURE 8 – If $f \in Lip(\mathbf{R}^n) \cap C^\infty(\mathbf{R}^n)$ then

$$\mathcal{H}^{n+1}\left((\mathbf{R}^n \times [0, +\infty[) \setminus AmbC^\infty(MCMf)\right) = 0,$$

and moreover, set $u = MCMf$, in $AmbC^\infty(u)$ it holds the equation

$$|\nabla_x u|^2 \left(\frac{\partial u}{\partial t} - \Delta_x u\right) + \sum_{h,k=1}^{n} \frac{\partial u}{\partial x_h} \frac{\partial u}{\partial x_k} \frac{\partial^2 u}{\partial x_h \partial x_k} = 0.$$

CONJECTURE 9 – Let $f \in Lip(\mathbf{R}^n)$ be such that $\lim_{|x|\to+\infty}|f(x)| = +\infty$; set $E_k = \{(x,t) \in \mathbf{R}^n \times [0,+\infty[: MCMf(x,t) = k\}$, for a.e. k it results $\mathcal{H}^n(E_k) < +\infty$.

CONJECTURE 10 – Suppose that $f \in C^0(\mathbf{R}^2)$ and that there exists a bounded open set $A \subset \mathbf{R}^2$ such that $f \in C^2(A)$, and $|\nabla_x f| \neq 0$ everywhere in A. Suppose also that for some λ it holds $F_\lambda = \{x \in \mathbf{R}^2 : f(x) = \lambda\} \subset A$; then $\{(x,t) \in \mathbf{R}^2 \times [0,+\infty[: MCMf(x,t) = \lambda\} = MCM(F_\lambda)$ and $\mathcal{H}^2(MCM(F_\lambda)) < +\infty$.

REMARK 12 – The previous conjecture is tied to some results on the curves flow by mean curvature (cf. *e.g.* [11], [12]).

A statement analogous to conjecture 10 seems to be false for $n > 2$. We set the following conjecture.

CONJECTURE 11 – Let $m \geq 2$, and let $f : \mathbf{R}^{m+1} \to \mathbf{R}$ be given by:

$$f(x) = \sqrt{\left(\sqrt{x_1^2 + \ldots + x_m^2} - 1\right)^2 + a^2 x_{m+1}^2};$$

then, set $E_\lambda = \{(x,t) \in \mathbf{R}^n \times [0,+\infty[: MCMf(x,t) = \lambda\}$, there exist a, λ such that $a > 0$, $0 < \lambda < 1$ and $\mathcal{H}^{m+1}(E_\lambda) > 0$.

REMARK 13 – Notice that for $0 < \lambda < 1$ and $a \neq 0$ the level surfaces of the function f in conjecture 11 are tori.

In order to compare our point of view with the results of L.Bronsard and R.V.Kohn (see [2], [3]), it would be interesting to consider, beside conjecture 6, even the following conjecture.

CONJECTURE 12 – Let $f \in C^\infty(\mathbf{R}^n)$ be uniformly continuous, and for every $h \in \mathbf{N}$, $u_h \in C^\infty(\mathbf{R}^n \times [0,+\infty[)$ be such that

$$\begin{cases} \dfrac{\partial u_h}{\partial t} = \Delta_x u_h - h^2(u_h^3 - u_h) & \forall(x,t) \in \mathbf{R}^n \times]0,+\infty[\\[2ex] u_h(x,0) = f(x) & \forall x \in \mathbf{R}^n. \end{cases}$$

Then for every $(x,t) \in \mathbf{R}^n \times]0,+\infty[$ such that $MCMf(x,t) > 0$ it holds

$$\lim_{h\to+\infty} u_h(x,t) = 1$$

and for every $(x,t) \in \mathbf{R}^n \times]0,+\infty[$ such that $MCMf(x,t) < 0$ it holds

$$\lim_{h\to+\infty} u_h(x,t) = -1.$$

Conjecture 12 concerns regular initial data; it can be partially extended to less regular initial data (*e.g.* characteristic functions of sets) as in the following conjecture.

CONJECTURE 13 – Let be $u_h \in C^\infty(\mathbf{R}^n \times]0, +\infty[)$, with

$$\frac{\partial u_h}{\partial t} = \Delta_x u_h - h^2(u_h^3 - u_h);$$

given $\varepsilon_h > 0$ such that $\lim_{h \to +\infty} \varepsilon_h = 0$, and set

$$A^+ = \left\{ x \in \mathbf{R}^n : \liminf_{\substack{h \to +\infty \\ \xi \to x}} u_h(\xi, \varepsilon_h) > 0 \right\},$$

it holds

$$\lim_{h \to +\infty} u_h(x, t) = 1 \quad \text{for any } t > 0 \text{ such that } (x, t) \in MCM(A^+).$$

REMARK 14 – Conjecture 13, even more than conjectures 6, 12, allows a comparison with the results of L.Bronsard and R.V.Kohn.

It is also interesting to extend the results on the limits of solutions of equations of the type

$$\frac{\partial u}{\partial t} = \Delta_x u + hg'(u)$$

to the case of systems of the type

$$\frac{\partial u}{\partial t} = \Delta_x u + h\nabla g(u);$$

in this direction, I point out the following conjecture, which is connected to the theory of evolution of harmonic maps, and in particular with the results of [4] (see also [9], [19]).

CONJECTURE 14 – Let $f : \mathbf{R}^n \to \mathbf{R}^n$ be a function of the variables $(x_1, \ldots, x_n)$, and $u :]0, +\infty[\times\mathbf{R}^n \times \mathbf{R}^n \times [0, +\infty[\to \mathbf{R}^n$ be a function of the variables $p \in]0, +\infty[$, $x \in \mathbf{R}^n$, $y \in \mathbf{R}^n$, $t \in [0, +\infty[$; suppose that $u \in [C^0(]0, +\infty[\times\mathbf{R}^n \times \mathbf{R}^n \times [0, +\infty[)]^n$ and $u \in [C^2(]0, +\infty[\times\mathbf{R}^n \times \mathbf{R}^n \times]0, +\infty[)]^n$, and that u verifies the following differential system:

$$\begin{cases} \dfrac{\partial u}{\partial t} = \Delta_x u + p(1 - |u|^2)u \\[2em] u(p, x, y, 0) = f(x) + y \qquad \forall p > 0,\ x \in \mathbf{R}^n,\ y \in \mathbf{R}^n \end{cases}$$

then, for a.e. $x, y, t \in \mathbf{R}^n \times \mathbf{R}^n \times]0, +\infty[$ the following limit exists

$$u_\infty(x, y, t) = \lim_{p \to +\infty} u(p, x, y, t) \quad \text{and moreover } |u_\infty(x, y, t)| = 1 \text{ for a.e. } x, y, t.$$

REMARK 15 – It is worthy noting that in this conjecture we consider $\lim_{p \to +\infty} u(p, x, y, t)$ and not limits of subsequences, $\lim_{h \to +\infty} u(p_h, x, y, t)$; we remark also that the role of the variable y is important in connection to the existence of $\lim_{p\to+\infty} u(p,x,y,t)$ and that probably f, y and $E \subset \mathbf{R}^n \times [0,+\infty[$ can be chosen in such a way that $\mathcal{H}^{n+1}(E) > 0$ and $\lim_{p\to+\infty} u(p,x,y,t)$ does not exist for $(x,t) \in E$.

FINAL REMARK – All the conjectures may be true or false. Many of the conjectures above seem to be similar to various results existing in literature, even though they cannot be deduced by such results until the relationships between our definitions 3, 4, 5 and the definitions used by other Authors in dealing with limit of Ginzburg–Landau dynamics and flow by mean curvature will be clarified.

Let us also notice that all the definitions and conjectures may be rearranged in various ways, *e.g.* interchanging the role of some of the objects involved, or starting by proving some corollaries or particular cases, and then passing to the proof of more general results. Proofs of some consequences or particular cases, together with various results somehow similar might be found in the mathematical literature. We have pointed out some examples of such a possibility (see conjectures 3, 5, 6, 13 and remark 7): it would be interesting to find other examples.

Please send any comment or information to one of the following addresses:

Prof. Stefano Mortola
Scuola Normale Superiore
56100 Pisa

or

Prof. Antonio Leaci
Dipartimento di Matematica
Università di Lecce
73100 Lecce

Acknowledgments The first idea about an approximation of minimal surfaces (or flow by mean curvature) problems by quasilinear equations like $\Delta u - f(u) = 0$ (or $\partial u/\partial t = \Delta_x u + f(u)$) is born in me from some conversations with L.Modica and S.Mortola.

Recently this idea has been supported by some numerical experiments performed by G.Bellettini, M.Ciampa, M.Paolini and C.Verdi.

Bibliography

[1] K.A.BRAKKE, *The motion of a surface by its mean curvature*, Princeton U.P., Princeton, N.J., 1978.

[2] L.BRONSARD-R.V.KOHN, *On the slowness of phase boundary motion in one space dimension*, Comm. Pure Appl. Math. **43** (1990), 983–997.

[3] L.BRONSARD-R.V.KOHN, *Motion by mean curvature as the singular limit of Ginzburg-Landau Dynamics*, J. Differential Equations **90** (1991), 211–237.

[4] Y.Chen, *The weak solutions to the evolution Problems of harmonic maps*, Math. Z. **201** (1989), 69–74.

[5] Y.G.Chen-Y.Giga-S.Goto, *Uniqueness and existence of viscosity solutions of generalized mean curvature flow equation*, J. Differential Geom. **33** (1991), 749–786.

[6] E.De Giorgi, *Some conjectures on flow by mean curvature*, in: Methods of real analysis and partial differential equations, M.L.Benevento-T.Bruno-C.Sbordone eds. Liguori, Napoli, 1990.

[7] P.De Mottoni-M.Schatzman, *Evolution géometrique d'interfaces*, C.R. A.S. **309** (1989), 453–458.

[8] K.Ecker-G.Huisken, *Mean curvature evolution of entire graphs*, Ann. of Math. **130** (1989), 453–471.

[9] J.Eells-J.H.Sampson, *Harmonic mappings of Riemannian manifolds*, Amer. J. Math. **86** (1964), 109–160.

[10] L.C.Evans-J.Spruck, *Motion of level sets by mean curvature I*, J. Differential Geom. **33** (1991), 635–681.

[11] M.Gage-R.S.Hamilton, *The heat equation shrinking convex plane curves*, J. Diff. Geom. **23** (1986), 69–96.

[12] M.Grayson, *Shortening embedded curves*, Ann. of Math. **129** (1989), 71–113.

[13] U.Massari, *Esistenza e regolarità delle ipersuperfici di curvatura media assegnata in* $\mathbf{R}^n$, Arch. Rational Mech. and Anal., **55** (1974), 357–382.

[14] M.Miranda, *Distribuzioni aventi derivate misure. Insiemi di perimetro localmente finito*, Ann. Sc. Norm. Sup. Pisa, s. III, **18** 1964, 27–56.

[15] L.Modica, *Γ-convergence to minimal surfaces problem and global solutions of* $\Delta u = 2(u^3 - u)$, in: Recent methods in non-linear analysis, E.De Giorgi-E.Magenes-U.Mosco eds., Pitagora, Bologna, 1979.

[16] L.Modica, *The gradient theory of phase transition and the minimal interface criterion*, Arch. Rational Mech. Anal., **98** (1987), 123–142.

[17] L.Modica-S.Mortola, *Un esempio di* Γ^-*-convergenza*, Boll. U.M.I. **14-B** (1977), 285–299.

[18] H.M.Soner, *Motion of a set by the curvature of its boundary*, J. Differential Equations **101** (1993), 313–372.

[19] M.Struwe, *The evolution of harmonic mappings with free boundaries*, Manuscripta Math. **70** (1991), 373–384.

New problems on minimizing movements[‡†]

Ennio De Giorgi

In this paper I intend to deepen the idea of *minimizing movement* which has been presented in the conference [10]. Such idea seems to be suitable to unify many problems in calculus of variations, differential equations, geometric measure theory: among others, steepest descent methods, heat equation, mean curvature flow, monotone operators, various evolution problems, etc. This paper is completely self-contained and may be read independently of the papers quoted in the bibliography; nevertheless, we remark that the main definitions of this paper may be considered as slight generalizations of the definitions given in [10] and that the paper [10] has been inspired mainly by the paper [1].

Minimizing movements are tied in various ways to penalization methods, Γ-convergence, singular perturbation, geometric measure theory, etc., hence the bibliographic indications will be unavoidably partial and far from being complete. In many cases the reader can surely find many other interesting references, as well as many interesting examples, problems, conjectures suggested by his own experience, which could be more interesting and expressive than those presented in this paper. One could think of finding general hypotheses on F and S such that the set of *minimizing movements* $MM(F, S)$ or the set of *generalized minimizing movements* $GMM(F, S)$ are nonvoid, or finite or such that their elements can be characterized by some differential equation, and/or some other meaningful condition.

I believe that the idea of minimizing movement is the natural meeting point of many problems of analysis, geometry, mathematical physics and numerical analysis, and its development will require the contribution of many researchers with different backgrounds.

I wish to thank L.Ambrosio, A.Leaci, S.Mortola and D.Pallara for their co-operation in preparing this paper.

Notation. In this paper we shall indicate by $\mathbf{Z}$ the set of signed integers, by $\overline{\mathbf{R}} = \mathbf{R} \cup \{-\infty, +\infty\}$ the extended real line, by $]a, b[$ (for $-\infty \leq a < b \leq +\infty$) the open interval $\{x \in \mathbf{R} : a < x < b\}$ and by $[x] = \max\{z \in \mathbf{Z} : z \leq x\}$ the integral part of x. Furthermore, S will denote a topological space and, for any pair of metric spaces M and M', $Lip(M, M')$ (shortened to $Lip(M)$ if $M' = \mathbf{R}$) denotes the set of M'-valued Lipschitz continuous functions on M. If $u \in Lip(M, M')$ we denote by $lip(u, M, M')$ the Lipschitz constant of u. If $M' = \mathbf{R}$, then we

---

[‡] Editor's note: published in Boundary value problems for partial differential equations and applications, volume in honour of E. Magenes for his 70th birthday, C. BAIOCCHI, J. L. LIONS EDS, Masson, 1993, 81–98.

[†] Dedicated to Enrico Magenes for his 70th birthday

shorten $lip(u, M)$ for $lip(u, M, \mathbf{R})$; when M is unambiguously determined by the context, we also write $lip(u)$ for $lip(u, M)$.

1. Minimizing movements definitions and examples in $\mathbf{R}^n$

Let us define the *minimizing movements $MM(F, S)$*.

DEFINITION 1.1 – Let be $F :]1, +\infty[\times \mathbf{Z} \times S \times S \to \overline{\mathbf{R}}$ and $u : \mathbf{R} \to S$; we say that u is a *minimizing movement associated to F, S*, and we write $u \in MM(F, S)$, if there exists $w :]1, +\infty[\times \mathbf{Z} \to S$ such that for any $t \in \mathbf{R}$

$$\lim_{\lambda \to +\infty} w(\lambda, [\lambda t]) = u(t)$$

and for any $\lambda \in]1, +\infty[$, $k \in \mathbf{Z}$

$$F\big(\lambda, k, w(\lambda, k+1), w(\lambda, k)\big) = \min_{s \in S} F\big(\lambda, k, s, w(\lambda, k)\big).$$

One can consider the following examples.

EXAMPLE 1.1 – Let be $S = \mathbf{R}^n$, $f \in Lip(\mathbf{R}^n) \cap C^2(\mathbf{R}^n)$ and let $\xi \in \mathbf{R}^n$ be given. Set

$$F(\lambda, k, x, y) = \begin{cases} |x - \xi|^2 & \text{if } k \leq 0, \\[2mm] f(x) + \lambda|x - y|^2 & \text{if } k > 0; \end{cases}$$

then, $u \in MM(F, S)$ if and only if $u \in Lip(\mathbf{R}, \mathbf{R}^n)$ and u solves

$$(1.1) \qquad \begin{cases} u(t) = \xi & \forall t \leq 0 \\[2mm] 2\dfrac{du}{dt} = -\nabla f(u) & \text{in }]0, +\infty[. \end{cases}$$

EXAMPLE 1.2 – Let be $S = \mathbf{R}^n$, $f \in Lip(\mathbf{R}^n) \cap C^2(\mathbf{R}^n)$ and let $\xi \in \mathbf{R}^n$, $\beta \in \mathbf{R}$, $1 < \beta \leq 2$, be given. Set

$$(1.2) \qquad F(\lambda, k, x, y) = \begin{cases} |x - \xi|^\beta & \text{if } k \leq 0, \\[2mm] \lambda f(x) + \big(\lambda|x - y|\big)^\beta & \text{if } k > 0; \end{cases}$$

then, $u \in MM(F, S)$ if and only if $u \in Lip(\mathbf{R}, \mathbf{R}^n)$ and u solves

$$\begin{cases} u(t) = \xi & \forall t \leq 0 \\[2mm] \beta\dfrac{du}{dt} = -\left|\dfrac{du}{dt}\right|^{2-\beta} \nabla f(u) & \text{in }]0, +\infty[. \end{cases}$$

EXAMPLE 1.3 – Let be $S = \mathbf{R}^n$, $f \in Lip(\mathbf{R}^n) \cap C^2(\mathbf{R}^n)$ with $|\nabla f(x)| \neq 0$ for any x, and let $\xi \in \mathbf{R}^n$ be given. Set

$$F(\lambda, k, x, y) = |x - \xi|^2 \qquad \text{if } k \leq 0,$$

$$F(\lambda, k, x, y) = \begin{cases} f(x) & \text{if } k > 0, \ \lambda|x - y| \leq 1, \\ \\ +\infty & \text{if } k > 0, \ \lambda|x - y| > 1; \end{cases}$$

then, $u \in MM(F, S)$ if and only if $u \in Lip(\mathbf{R}, \mathbf{R}^n)$ and u solves

$$\begin{cases} u(t) = \xi & \forall t \leq 0 \\ \\ \dfrac{du}{dt} = -\dfrac{\nabla f(u)}{|\nabla f(u)|} & \text{in }]0, +\infty[. \end{cases}$$

REMARK 1.1 – If we replace in example 1.1 the hypothesis $f \in Lip(\mathbf{R}^n) \cap C^2(\mathbf{R}^n)$ by the weaker condition $f \in Lip(\mathbf{R}^n) \cap C^1(\mathbf{R}^n)$, then neither the existence nor the uniqueness are guaranteed in $MM(F, \mathbf{R}^n)$. We can only say that any $u \in MM(F, \mathbf{R}^n)$ is a Lipschitz function satisfying the Cauchy problem (1.1), but the opposite implication does not hold. Indeed, if $n = 1$, $\xi = 0$ and $f(x) = -|\sin x|^{3/2}$, then $u \equiv 0$ is a solution of (1.1) which does not belong to $MM(F, \mathbf{R})$.

REMARK 1.2 – An example showing that $MM(F, \mathbf{R}^n)$ in example 1.1 may be empty if we only assume that $f \in Lip(\mathbf{R}^n) \cap C^1(\mathbf{R}^n)$ is the following: let $n = 1$, $\xi = 0$ and F as in example 1.1, with

$$(1.3) \qquad f(x) = \begin{cases} -|\sin x|^{3/2} + \sin^4 x \, \sin \dfrac{1}{x} & \text{if } x \neq 0 \\ \\ 0 & \text{if } x = 0. \end{cases}$$

Then, $MM(F, S) = \emptyset$.

REMARK 1.3 – Finally, if we only assume that f is a Lipschitz continuous function everywhere differentiable in $\mathbf{R}^n$ whose derivatives are discontinuous in some point and if we define F as in example 1.1, then it can happen that functions $u \in MM(F, \mathbf{R}^n)$ do not solve the Cauchy problem (1.1). Indeed, let $n = 1$, $\xi = 0$ and let F be as in example 1.1, with

$$f(x) = \begin{cases} \dfrac{x}{4} + \sin^2 x \, \sin \dfrac{1}{x} & \text{if } x \neq 0 \\ \\ 0 & \text{if } x = 0. \end{cases}$$

Then $u \equiv 0$ belongs to $MM(F, \mathbf{R})$ and does not satisfy (1.1).

Remark 1.2 and many other cases where $MM(F, S) = \emptyset$ motivate the definition of *generalized minimizing movements* $GMM(F, S)$.

DEFINITION 1.2 – Let be $F :]1, +\infty[\times \mathbf{Z} \times S^2 \to \overline{\mathbf{R}}$ and $u : \mathbf{R} \to S$; we say that u is a *generalized minimizing movement associated to F, S*, and we write

$u \in GMM(F, S)$, if there exist a sequence $\{\lambda_i\}_{i\in\mathbf{N}}$ such that $\lim \lambda_i = +\infty$ and a sequence $\{w_i\}_{i\in\mathbf{N}}$ of functions $w_i : \mathbf{Z} \to S$ such that for any $t \in \mathbf{R}$

$$\lim_{i\to+\infty} w_i([\lambda_i t]) = u(t)$$

and for any $i \in \mathbf{N}$, $k \in \mathbf{Z}$

$$F\big(\lambda_i, k, w_i(k+1), w_i(k)\big) = \min_{s\in S} F\big(\lambda_i, k, s, w_i(k)\big).$$

Remark 1.4 – It seems to me that generalized minimizing movements GMM could give some good formalization of the heuristic idea of curve of maximal slope, and that it would be interesting to compare it with other definitions already proposed (see *e.g.* [3], [12], [13]). All these definitions agree in the case $S = \mathbf{R}^n$ and $f \in Lip(\mathbf{R}^n) \cap C^2(\mathbf{R}^n)$.

Remark 1.5 – Notice that in the case considered in remark 1.2 there are two elements in $GMM(F, S)$, namely the two solutions of the Cauchy problem $u' = -\nabla f(u)$, $u(0) = 0$ that are different from 0 for any $t > 0$.

Remark 1.6 – It might seem restrictive to consider only functions $u \in GMM(F, S)$ defined on the whole real line. However it is easy include in our definitions the case of functions $u :]a, b[\to S$, where $a, b \in \mathbf{R}$, with $a < b$. For instance, if $S = \mathbf{R}^n$ and $f \in Lip(\mathbf{R}^n) \cap C^2(\mathbf{R}^n)$, then all the solutions defined in $]a, b[$ of the equation

$$2\frac{du}{dt} = -\nabla f(u)$$

are restrictions to $]a, b[$ of functions $u \in MM(F, \mathbf{R}^n)$, where

$$F(\lambda, k, x, y) = \begin{cases} 0 & \text{if } k < [\lambda a]; \\ f(x) + \lambda|x - y|^2 & \text{if } [\lambda a] \leq k \leq [\lambda b]; \\ 0 & \text{if } k > [\lambda b]. \end{cases}$$

It is also interesting to consider the following

Example 1.4 – Let be $S = \mathbf{R}^n$, $f \in Lip(\mathbf{R}^n)$ and let $\xi \in \mathbf{R}^n$, $\beta \in]1, +\infty[$ be given. Set

$$F(\lambda, k, x, y) = \begin{cases} |x - \xi|^\beta & \text{if } k \leq 0, \\ f(x) + \lambda^{\beta-1}|x - y|^\beta & \text{if } k > 0; \end{cases}$$

then $GMM(F, S) \neq \emptyset$ and if $u \in GMM(F, S)$ then $u \in Lip(\mathbf{R}, \mathbf{R}^n)$. If moreover $f \in Lip(\mathbf{R}^n) \cap C^1(\mathbf{R}^n)$ then each $u \in GMM(F, S)$ solves

$$(1.4) \qquad \begin{cases} u(t) = \xi & \forall t \leq 0 \\ \\ \beta\dfrac{du}{dt} = -\left|\dfrac{du}{dt}\right|^{2-\beta}\nabla f(u) & \text{in }]0, +\infty[\end{cases}$$

for $\beta \leq 2$, and

$$(1.5) \qquad \begin{cases} u(t) = \xi & \forall t \leq 0 \\[2mm] \beta \left| \dfrac{du}{dt} \right|^{\beta-2} \dfrac{du}{dt} = -\nabla f(u) & \text{in }]0, +\infty[\end{cases}$$

for $\beta \geq 2$.

REMARK 1.7 – In example 1.4 we remarked that if $f \in Lip(\mathbf{R}^n) \cap C^1(\mathbf{R}^n)$, then each $u \in GMM(F, S)$ solves equation (1.4)-(1.5). On the other hand, by remark 1.1, there could exist solutions of (1.4)-(1.5) which don't belong neither to $MM(F, S)$ nor to $GMM(F, S)$. If we want to gather all the solutions of (1.4), (1.5) as minimizing movements, we may consider conjectures of the following type.

CONJECTURE 1.1 – Let $f \in Lip(\mathbf{R}^n) \cap C^1(\mathbf{R}^n)$, $\xi \in \mathbf{R}^n$. Then, u is a solution of (1.1) if and only if there exists $\varphi : \mathbf{R}^n \times]1, +\infty[\to \mathbf{R}$ such that
(a) $\varphi(\cdot, \lambda) \in Lip(\mathbf{R}^n)$ for any $\lambda > 0$ and

$$\lim_{\lambda \to +\infty} lip(f - \varphi(\cdot, \lambda)) = 0;$$

(b) $u \in GMM(\Phi, \mathbf{R}^n)$, where $\Phi(\lambda, k, x, y)$ is defined by

$$\Phi(\lambda, k, x, y) = \begin{cases} |x - \xi|^2 & \text{if } k \leq 0, \\[2mm] \varphi(x, \lambda) + \lambda |x - y|^2 & \text{if } k > 0. \end{cases}$$

REMARK 1.8 – It would also be interesting to consider non-autonomous equations, taking a function f depending on $n+1$ variables, and the functionals

$$F(\lambda, k, x, y) = \begin{cases} |x - \xi|^2 & \text{if } k \leq 0, \\[2mm] \lambda \displaystyle\int_k^{k+1} f\left(x, \dfrac{\tau}{\lambda}\right) d\tau + \left(\lambda |x - y|\right)^{\beta} & \text{if } k > 0. \end{cases}$$

REMARK 1.9 – Instead of regularity conditions such as Lipschitz continuity or differentiability, in the previous examples lower semicontinuity conditions and convexity or quasiconvexity hypotheses could be introduced, or bounds of the following type:

$$\inf_{x \in \mathbf{R}^n} \frac{f(x)}{1 + |x|^{\beta}} > -\infty.$$

The same type of condition will likely play an important rôle in passing from $\mathbf{R}^n$ to infinite dimensional spaces. In this paper we shall not explore the general, abstract problems which arise in such spaces, limiting ourselves to point out in the next chapters some meaningful example of minimizing movement arising in the theory of partial differential equations and in geometric measure theory.

2. Minimizing movements and PDE

Examples 1.1, 1.2, 1.3, 1.4 show that the definitions of $MM(F,S)$ and $GMM(F,S)$ cover in the case $S = \mathbf{R}^n$ some problems of *steepest descent*. Many interesting and difficult problems can be formulated if S is a space of functions and F is an integral functional, giving problems of *gradient flow* type. For the sake of simplicity we consider in this chapter only spaces of functions defined on $\mathbf{R}^n$ and unless otherwise stated the integrals are intended on the whole of $\mathbf{R}^n$. We begin with an example related to the heat equation.

EXAMPLE 2.1 – Let be $S = H^{1,2}(\mathbf{R}^n)$ and $\varphi \in H^{1,2}(\mathbf{R}^n)$ be given; set

$$
F(\lambda, k, f, g) = \begin{cases} \displaystyle\int |f - \varphi|^2 dx & \text{if } k \le 0, \\[2em] \displaystyle\int |\nabla f|^2 + \lambda |f - g|^2 dx & \text{if } k > 0; \end{cases}
$$

then, $u \in MM(F,S)$ if and only if $u : \mathbf{R} \to S$ is continuous, $u(t) = \varphi$ for any $t \le 0$, and, setting $v(x,t) = u(t)(x)$, v solves

$$
\frac{\partial v}{\partial t} = \Delta_x v \qquad \text{in } \mathbf{R}^n \times]0, +\infty[.
$$

Beside the heat equation, other evolution equations might be considered (see *e.g.* [12], [22]), and it is likely true that in many cases the existence of functions u in $MM(F,S)$ or in $GMM(F,S)$ can be obtained in more general hypotheses than those already considered in the literature.

CONJECTURE 2.1 – Let $S = H^{1,2}(\mathbf{R}^n)$ and $\varphi \in H^{1,2}(\mathbf{R}^n)$, $1 < \beta \le 2$ be given; set

$$
F(\lambda, k, f, g) = \begin{cases} \displaystyle\int |f - \varphi|^2 dx & \text{if } k \le 0, \\[2em] \displaystyle\lambda \int |\nabla f|^2 + \left(\lambda |f - g|\right)^\beta dx & \text{if } k > 0; \end{cases}
$$

then, $u \in MM(F,S)$ if and only if $u : \mathbf{R} \to S$ is continuous, $u(t) = \varphi$ for any $t \le 0$ and, setting $v(x,t) = u(t)(x)$, v solves

$$
\frac{\beta}{2} \frac{\partial v}{\partial t} = \left| \frac{\partial v}{\partial t} \right|^{2-\beta} \Delta_x v \qquad \text{in } \mathbf{R}^n \times]0, +\infty[.
$$

In a different direction, example 2.1 may be generalized as follows.

CONJECTURE 2.2 – Let $S = H^{1,2}(\mathbf{R}^n)$, $\varphi \in H^{1,2}(\mathbf{R}^n)$ and $a_1, \ldots, a_n \in L^\infty(\mathbf{R}^n)$ be given; set

$$F(\lambda, k, f, g) = \begin{cases} \displaystyle\int |f - \varphi|^2 dx & \text{if } k \le 0, \\[4mm] \displaystyle\int |\nabla f|^2 - 2\sum_{i=1}^n a_i \frac{\partial g}{\partial x_i} f + \lambda |f - g|^2 dx & \text{if } k > 0; \end{cases}$$

then, $MM(F, S) \ne \emptyset$, and $u \in MM(F, S)$ if and only if $u : \mathbf{R} \to S$ is continuous, $u(t) = \varphi$ for any $t \le 0$ and, setting $v(x, t) = u(t)(x)$, v solves

$$\frac{\partial v}{\partial t} = \Delta_x v + \sum_{i=1}^n a_i \frac{\partial v}{\partial x_i} \qquad \text{in } \mathbf{R}^n \times]0, +\infty[.$$

The following conjecture concerns a "moving obstacle" problem.

CONJECTURE 2.3 – Let $S = H^{1,2}(\mathbf{R}^n)$ and $\varphi \in H^{1,2}(\mathbf{R}^n)$ be given; set

$$F(\lambda, k, f, g) = \int |f - \varphi|^2 dx \qquad \text{if } k \le 0,$$

and, for $k > 0$,

$$F(\lambda, k, f, g) = \begin{cases} \displaystyle\int |\nabla f|^2 + \lambda |f - g|^2 dx & \text{if } f \ge g, \\[4mm] +\infty & \text{otherwise;} \end{cases}$$

then, $u \in MM(F, S)$ if and only if $u : \mathbf{R} \to S$ is continuous, $u(t) = \varphi$ for any $t \le 0$, and, set $v(x, t) = u(t)(x)$, v solves

$$2\frac{\partial v}{\partial t} = \Delta_x v + |\Delta_x v| \qquad \text{in } \mathbf{R}^n \times]0, +\infty[.$$

REMARK 2.1 – In discussing the previous conjectures it will likely be convenient to study first the existence and the uniqueness of an element $u \in MM(F, S)$ and in a second time its regularity, in order to clarify the sense which can be given to the differential equations stated therein. The following example with a "fixed obstacle" seems to be closer to known techniques (see *e.g.* [3], [12]).

EXAMPLE 2.2 – Let be $S = H^{1,2}(\mathbf{R}^n)$, $\psi \in C_0^\infty(\mathbf{R}^n)$ and $\varphi \in H^{1,2}(\mathbf{R}^n)$ be given, with $\varphi \ge \psi$; set

$$F(\lambda, k, f, g) = \int |f - \varphi|^2 dx \qquad \text{if } k \le 0,$$

and, for $k > 0$,

$$F(\lambda, k, f, g) = \begin{cases} \displaystyle\int |\nabla f|^2 + \lambda |f - g|^2 dx & \text{if } f \geq \psi, \\[2em] +\infty & \text{otherwise;} \end{cases}$$

then $MM(F, S) \neq \emptyset$.

The introduction of $MM(F, S)$ and $GMM(F, S)$ leads also to interesting problems related to hyperbolic equations, as in the following

PROBLEM 2.1 – For $S = H^{1,2}(\mathbf{R}^n)$, and

$$F(\lambda, k, f, g) = \begin{cases} \displaystyle\int |f - \varphi|^2 dx & \text{if } k \leq 0, \\[2em] \displaystyle\int \frac{1}{\log \lambda} |\nabla f|^2 - 2 \sum_{i=1}^{n} a_i \frac{\partial g}{\partial x_i} f + \lambda |f - g|^2 dx & \text{if } k > 0 \end{cases}$$

find conditions on a_i and φ in order to obtain $MM(F, S) \neq \emptyset$ (or at least $GMM(F, S) \neq \emptyset$) and $u \in MM(F, S)$ (or $u \in GMM(F, S)$) if and only if $u(t) = \varphi$ for any $t \leq 0$ and, setting $v(x, t) = u(t)(x)$,

$$\frac{\partial v}{\partial t} = \sum_{i=1}^{n} a_i(x) \frac{\partial v}{\partial x_i}.$$

REMARK 2.2 – Problem 2.1 is inspired to the idea that by a suitable choice of F it is possible to combine time discretization with the vanishing viscosity method (see *e.g.* [23]). The possibility of such combinations is one of the best features of the minimizing movements. A similar approximation idea will be considered in conjecture 2.4.

REMARK 2.3 – In the previous problem, as well as in conjecture 2.2, it would also be interesting to consider time-dependent coefficients $a_i \in L^\infty(\mathbf{R}^{n+1})$ and the functionals

$$F(\lambda, k, f, g) = \begin{cases} \displaystyle\int |f - \varphi|^2 dx & \text{if } k \leq 0 \\[2em] \displaystyle\int_k^{k+1} d\tau \int |\nabla f|^2 - 2 \sum_{i=1}^{n} a_i\left(x, \frac{\tau}{\lambda}\right) \frac{\partial g}{\partial x_i} f + \lambda |f - g|^2 dx & \text{if } k > 0, \end{cases}$$

$$F(\lambda, k, f, g) = \begin{cases} \displaystyle\int |f - \varphi|^2 dx & \text{if } k \leq 0 \\[2em] \displaystyle\int_k^{k+1} d\tau \int \frac{1}{\log \lambda} |\nabla f|^2 - 2 \sum_{i=1}^{n} a_i\left(x, \frac{\tau}{\lambda}\right) \frac{\partial g}{\partial x_i} f + \\[1em] \quad + \lambda |f - g|^2 dx & \text{if } k > 0. \end{cases}$$

REMARK 2.4 – After studying the equation

$$\frac{\partial v}{\partial t} = \sum_{i=1}^{n} a_i(x) \frac{\partial v}{\partial x_i},$$

systems of first order equations could be studied as well, even for time-dependent coefficients.

In the context of minimizing movements it is possible to consider various examples of boundary value problems of very general type, so that, beside example 2.1 concerning the heat equation, we may consider the following problem.

PROBLEM 2.2 – For $S = H^{1,2}(\mathbf{R}^n)$, $E \subset \mathbf{R}^{n+1}$, setting $E_t = \{x \in \mathbf{R}^n : (x,t) \in E\}$ and

$$F(\lambda, k, f, g) = \int_k^{k+1} d\tau$$
$$\left(\int_{\mathbf{R}^n} \lambda |f - \varphi|^2 + |\nabla f|^2 - 2f\varphi(x,\tau)dx + (\log \lambda) \int_{\mathbf{R}^n \setminus E_{\tau/\lambda}} |f|^2 dx \right)$$

find conditions on φ and E in order to obtain $MM(F,S) \neq \emptyset$, or at least $GMM(F,S) \neq \emptyset$.

REMARK 2.5 – Setting $v(x,t) = u(t)(x)$ for $u \in MM(F,S)$, the term $\log \lambda \int_{\mathbf{R}^n \setminus E_{k/\lambda}} |f|^2 dx$ in the previous problem leads, in the regular cases (for instance $\varphi \in C_0^\infty(\mathbf{R}^{n+1})$ and $E = \{(x,t) : t > |x|^2\}$), to the equation $\partial v/\partial t = \Delta_x v + \varphi$ in the open set E with the boundary condition $v(x,t) = 0$ for any $(x,t) \in \partial E$.

REMARK 2.6 – An interesting connection between Γ-convergence, time discretization, limits of parabolic equations and mean curvature flow is the following conjecture, related to the Γ-convergence result in [24], and the problems considered in the papers [2], [4], [5], [8], [9], [14], [15], [16], [21]. The function F of conjecture 2.4 arises from a time discretization of the equation

$$\frac{\partial u}{\partial t} = \Delta u + \mu(u - u^3)$$

and a limit as $\mu \to +\infty$. The utility of such a connection has been suggested in [17].

CONJECTURE 2.4 – Let $S = L^2(\mathbf{R}^{n+1})$ and $\varphi \in Lip(\mathbf{R}^n)$ be given; set

$$F(\lambda, k, f, g) = \int_{\mathbf{R}} dy \int_{\mathbf{R}^n} |f(x,y) - \varphi(x,y) \exp\{-|x|^2 - y^2\}|^2 dx \qquad \text{if } k \leq 0,$$

and, for $k > 0$,

$$F(\lambda, k, f, g) = \int_{\mathbf{R}} dy \int_{\mathbf{R}^n} |\nabla_x f|^2 + \log \lambda (|f|^2 - 1)^2 + \lambda |f - g|^2 dx$$

if $f(\cdot, y) \in H^{1,2}(\mathbf{R}^n)$ for any $y \in \mathbf{R}$, $F(\lambda, k, f, g) = +\infty$ otherwise.
Then, $MM(F, S) \neq \emptyset$.

REMARK 2.7 – In the previous conjecture, it would also be interesting to replace $\log \lambda$ by λ^β, for some positive β and $(1 - |f|^2)^2$ by $(1 - \cos f)$ or by any other function satisfying the hypotheses of the Modica-Mortola theorem (see [24]). In the study of flow of periodic surfaces by mean curvature it would also be interesting the replacement of $L^2(\mathbf{R}^{n+1})$ by a space of periodic functions (see [9]).

3. Minimizing movements of surfaces and of rectifiable currents

Since we are going to consider possible applications of minimizing movements in geometric measure theory, where the notion of Hausdorff measure play a central rôle, we recall its definition, for the reader's convenience.

DEFINITION 3.1 – Let $E \subset \mathbf{R}^n$. For any $\alpha > 0$ we define

$$\mathcal{H}^\alpha(E) = \frac{[\Gamma(1/2)]^\alpha}{2^\alpha \Gamma(1 + \alpha/2)} \sup_{\delta > 0} \inf \left\{ \sum_{i=1}^\infty \operatorname{diam}^\alpha(E_i) : E \subset \bigcup_{i \in I} E_i, \ \operatorname{diam}(E_i) < \delta \right\},$$

where $\Gamma(r) = \int_0^\infty e^{-t} t^{r-1}\, dt$ is the Euler function. We also define $\mathcal{H}^0(E)$ to be the cardinality of E if E is a finite set, and $+\infty$ otherwise. The first problem related to flow by mean curvature is the following.

CONJECTURE 3.1 – Let S be the space of all open convex bounded subsets of $\mathbf{R}^n$, endowed with the metric $d(C_1, C_2) = \mathcal{H}^n(C_1 \triangle C_2)$; given $\alpha \in\,]0, +\infty[$ and $C \in S$, set

$$F(\lambda, k, C_1, C) = \begin{cases} \mathcal{H}^n(C_1 \triangle C) & \text{if } k \leq 0, \\[2em] \mathcal{H}^{n-1}(\partial C_1) + \lambda^\alpha \displaystyle\int_{C_1 \triangle C} \left[\operatorname{dist}(x, \partial C)\right]^\alpha d\mathcal{H}^n(x) & \text{if } k > 0; \end{cases}$$

then, there exists a unique $u \in MM(F, S)$ and, in the case $\alpha = 1$, $\partial u(t)$ moves along its mean curvature.

The preceding conjecture is connected with some results of [1], [18] and [19]. The following conjecture deals with mean curvature flow as well.

CONJECTURE 3.2 – Let be

$$S = \left\{ E \subset \mathbf{R}^n : \mathcal{H}^n(E) < +\infty, \ x \in E \Leftrightarrow \lim_{\rho \to 0+} \rho^{-n} \mathcal{H}^n(B_\rho(x) \setminus E) = 0 \right\},$$

where $B_\rho(x)$ denotes the open ball centered at x with radius ρ. We endow S with the distance $d(E_1, E_2) = \mathcal{H}^n(E_1 \triangle E_2)$; given $\alpha \in\,]0, +\infty[$, $L \in S$ such that

$\mathcal{H}^{n-1}(\partial L) < +\infty$ and $g \in L^1(\mathbf{R}^n) \cap L^n(\mathbf{R}^n)$, set

$$F(\lambda, k, E_1, E_2) = \begin{cases} \mathcal{H}^n(E_1 \triangle L) & \text{if } k \leq 0, \\[2ex] \mathcal{H}^{n-1}(\partial E_1) + \displaystyle\int_{E_1} g(x)dx + \lambda^\alpha \int_{E_1 \triangle E_2} \left[\text{dist}(x, \partial E_2)\right]^\alpha dx \\[2ex] & \text{if } k > 0; \end{cases}$$

then, $GMM(F, S) \neq \emptyset$.

REMARK 3.1 – Conjecture 3.2 is interesting even under stronger regularity hypotheses on L and g; remark also that the presence of g makes nontrivial the case $n = 1$ as well. If $g \equiv 0$, $\alpha = 1$ and $P(L) < +\infty$ (where $P(L)$ denotes the perimeter of L, see [6], [7], [11]), conjecture 3.2 is tied to some results of [1]. Even for $n = 1$, it is not clear whether conjecture 3.2 holds without the assumption $P(L) < +\infty$. If one does not want to use explicitly the notion of perimeter, one can consider the hypothesis (equivalent if $n = 1$, slightly more restrictive if $n > 1$), $\mathcal{H}^{n-1}(\partial L) < +\infty$. Very close to the one dimensional case is the case where g and L present spherical symmetry. The case of spherical symmetry is also interesting in connection with conjecture 2.4 (see [4], [19], [25]).

Problems relative to mean curvature flow on subsets $K \subset \mathbf{R}^n$ are taken into account in the following problem (see also [2], [20], [21]).

PROBLEM 3.1 – Given $s \geq 1$ and a compact set $K \subset \mathbf{R}^n$ such that $\mathcal{H}^s(K) < +\infty$, let

$$S = \left\{ E \subset K : x \in E \Leftrightarrow \lim_{\rho \to 0^+} \rho^{-s} \mathcal{H}^s(K \cap B_\rho(x) \setminus E) = 0 \right\},$$

endowed with the distance $d(E_1, E_2) = \mathcal{H}^s(E_1 \triangle E_2)$; given a set $L \in S$ we define

$$F(\lambda, k, E_1, E_2) = \mathcal{H}^s(E_1 \triangle L)$$

if $k \leq 0$ and

$$F(\lambda, k, E_1, E_2) = \mathcal{H}^{s-1}\big(\partial E_1 \cap \partial(K \setminus E_1)\big) + \int_{E_1} g \, d\mathcal{H}^s + \lambda \int_{E_1 \triangle E_2} \text{dist}(x, \partial E_2) d\mathcal{H}^s$$

if $k > 0$. Our problem is the following: find conditions on K, L, g such that $GMM(F, S) \neq \emptyset$.

In order to pass from the study of subsets of $\mathbf{R}^n$ to very general problems concerning currents in a metric space, we shall introduce some concepts which may be considered as a wide generalization of some definitions in [16], and are a development of some ideas in [26].

DEFINITION 3.2 – Let M be a metric space, let $k \geq 1$ be an integer and let $\mathbf{B}(M)$, $\mathcal{B}^\infty(M)$ be the class of Borel subsets of M and the space of bounded real valued functions $f : M \to \mathbf{R}$ respectively. We say that $G : \mathcal{B}^\infty(M) \times \left(Lip(M)\right)^k \to \mathbf{R}$ belongs to $GMT_k(M)$ (the set of *geometric measure theory functionals of dimension* k *relative to the space* M) if the following conditions are satisfied:

(a) there exist a finite Borel measure μ in M and $\alpha : \left(Lip(M)\right)^k \to \mathcal{B}^\infty(M)$ such that

$$G(f_0; f_1, \ldots, f_k) \;=\; \int_M f_0 \alpha(f_1, \ldots, f_k)\, d\mu \quad \forall f_0 \in \mathcal{B}^\infty(M),$$
$$\forall f_1, \ldots, f_k \in Lip(M);$$

(b) for any choice of $f_1, \ldots, f_k \in Lip(M)$ the following inequality holds

$$|\alpha(f_1, \ldots, f_k)|(x) \le \prod_{i=1}^{k} lip(f_i), \qquad \forall x \in M.$$

It is easy to introduce the concept of *boundary* of $H \in GMT_{k+1}(M)$ as follows.

DEFINITION 3.3 – Let $H \in GMT_{k+1}(M)$; we say that G is the boundary of H, and we write $G = \partial H$, if $G \in GMT_k(M)$ and

$$H(1; f_0, \ldots, f_k) = G(f_0; f_1, \ldots, f_k) \quad \forall f_0 \in Lip(M) \cap \mathcal{B}^\infty(M),$$
$$\forall f_1, \ldots, f_k \in Lip(M).$$

REMARK 3.2 – It is also possible to give a notion of *mass* of a functional $G \in GMT_k(M)$, taking into account that from definition 3.2 and Radon-Nikodym theorem it easily follows that for any $G \in GMT_k(M)$ there exists a pair (μ, α) verifying (a), (b) which minimizes $\mu(M)$. Denoting by $(\bar\mu, \bar\alpha)$ such a minimizing pair, for any $B \in \mathbf{B}(M)$ and for any other pair (μ, α) verifying (a), (b) in definition 3.2, $\mu(B) \ge \bar\mu(B)$ holds, hence $\bar\mu(B)$ is determined by M and G and is said *mass* of G and denoted $\|G\|_M$, or briefly $\|G\|$ if there is no ambiguity.

REMARK 3.3 – Notice that, given two metric spaces M, M' having the same elements and equivalent but different distance functions, the sets $GMT_k(M)$ and $GMT_k(M')$ coincide, but the masses $\|G\|_M$ and $\|G\|_{M'}$ may be different.

We now give the definition of *push forward* of a functional $G \in GMT_k(M)$ (cf. [16, 4.1.7, 4.1.14]).

DEFINITION 3.4 – Let M, M' be metric spaces, $\varphi : M \to M'$ be a Lipschitz continuous function, $G \in GMT_k(M)$, $H \in GMT_k(M')$. We write $H = \varphi_\#(G)$ if

$$H(f_0; f_1, \ldots, f_k) = G(f_0 \circ \varphi; f_1 \circ \varphi, \ldots, f_k \circ \varphi) \quad \forall f_0 \in \mathcal{B}^\infty(M)$$
$$\forall f_1, \ldots, f_k \in Lip(M').$$

Very interesting particular cases of $G \in GMT_k(M)$ are the *integral rectifiable currents* (see [16, 4.1.24] for the case $M = \mathbf{R}^n$), defined as follows.

DEFINITION 3.5 – We say that $G \in GMT_k(M)$ belongs to $IRC_k(M)$, the set of integral rectifiable k-currents in M, if there exist a Borel set $B \subset \mathbf{R}^k$ and a Lipschitz function $\varphi : B \to M$ such that $\mathcal{H}^k(B) < +\infty$ and $G = \varphi_\#([[B]])$, where $[[B]] \in GMT_k(\mathbf{R}^k)$ is defined by

$$[[B]](f_0; f_1, \ldots, f_k) = \int_B f_0(x) J(f)(x)\, dx \qquad \forall f_0 \in \mathcal{B}^\infty(\mathbf{R}^k),$$

$$f_1, \ldots, f_k \in Lip(\mathbf{R}^k)$$

and $J(f)$ is the Jacobian of the map $f = (f_1, \ldots, f_k)$.

We are now in a position to formulate in a general form a minimizing movement problem concerning integral rectifiable currents which may be considered as a wide generalization of classical problems of mean curvature flow.

PROBLEM 3.1 – Let $S = IRC_k(M)$ endowed of the following topology: we say that $C \subset S$ is closed if, for any sequence $\{G_h\}_{h \in \mathbf{N}} \subset C$ and any $G \in S$, the condition

$$\lim_{h \to +\infty} G_h(f_0; f_1, \ldots, f_k) = G(f_0; f_1, \ldots, f_k)$$

for any $f_0 \in Lip(M) \cap \mathcal{B}^\infty(M)$, $f_1, \ldots, f_k \in Lip(M)$ implies that $G \in C$.

Given any $\xi \in IRC_k(M)$ we define $F(\lambda, k, G, H) = \|G - \xi\|(M)$ if $k \leq 0$ and

$$\|G\|(M) + \lambda \inf_\Theta \left\{ \int_M \mathrm{dist}\,(x, \mathrm{supp}(H))\, d\|\Theta\| : \Theta \in IRC_{k+1}(M), \partial\Theta = G - H \right\}$$

if $k > 0$. Our problem is the following: find conditions on ξ, M, such that $MM(F, S) \neq \emptyset$, or $GMM(F, S) \neq \emptyset$ and, in the affirmative case, study the properties of the elements of $MM(F, S)$ or $GMM(F, S)$.

REMARK 3.4 – Finally, we remark that the theory presented in this paper contains the stationary case, in the sense that a point $x \in S$ is $MM(F, S)$-stationary, or $GMM(F, S)$-stationary, if the constant function $u \equiv x$ belongs to $MM(F, S)$ or to $GMM(F, S)$. Obviously, it would be interesting to compare, in the examples presented in this paper as well as in other examples to be formulated, the notions of MM-stationary point and GMM-stationary point with other notions of stationary point given elsewhere.

Bibliography

[1] F.ALMGREN-J.E.TAYLOR-L.WANG, *Curvature driven flows: a variational approach*, SIAM J. Control Optim. **31** (1993), 387–438.

[2] K.A.BRAKKE, *The motion of a surface by its mean curvature*, Princeton U.P., Princeton, N.J., 1978.

[3] H.BREZIS, *Operateurs maximaux monotones et semigroupes de contractions dans les espaces de Hilbert*, North-Holland, 1973.

[4] L.BRONSARD-R.V.KOHN, *Motion by mean curvature as the singular limit of Ginzburg-Landau Dynamics*, J. Diff. Eqs., **90** (1991), 211–237.

[5] Y.G.CHEN-Y.GIGA-S.GOTO, *Uniqueness and existence of viscosity solutions of generalized mean curvature flow equation,* J. Diff. Geom. **33** (1991), 749–786.

[6] E.DE GIORGI, *Su una teoria generale della misura $(r-1)$-dimensionale in uno spazio ad r dimensioni,* Ann. Mat. Pura Appl. **36** (1954), 191–212.

[7] E.DE GIORGI, *Nuovi teoremi relativi alle misure $(r-1)$-dimensionali in uno spazio ad r dimensioni,* Ric. di Mat. **4** (1955), 95–113.

[8] E.DE GIORGI, *Some conjectures on flow by mean curvature,* in: Methods of real analysis and partial differential equations, M.L.Benevento-T.Bruno-C.Sbordone eds. Liguori, Napoli, 1990.

[9] E.DE GIORGI, *Conjectures on limits of some quasilinear parabolic equations and flow by mean curvature,* in: Partial differential equations and related subjects, proc. of a Conference held in Trento, 1990.

[10] E.DE GIORGI, *Movimenti minimizzanti,* in: Aspetti e problemi della Matematica oggi, proc. of a Conference held in Lecce, 1992.

[11] E.DE GIORGI-F.C.COLOMBINI-L.C.PICCININI, *Frontiere orientate di misura minima e questioni collegate,* Quaderni Sc. Norm. Sup., Pisa, 1972.

[12] E.DE GIORGI-M.DEGIOVANNI-A.MARINO-M.TOSQUES, *Evolution equations for a class of nonlinear operators,* Atti Accad. Naz Lincei, Cl. Sci. Fis. Mat. Natur., s. VIII **75** (1983), 1–8.

[13] E.DE GIORGI-A.MARINO-M.TOSQUES, *Problemi di evoluzione in spazi metrici e curve di massima pendenza,* Atti Accad. Naz Lincei, Cl. Sci. Fis. Mat. Natur., s. VIII **68** (1980), 180–187.

[14] K.ECKER-G.HUISKEN, *Mean curvature evolution of entire graphs,* Ann. of Math. **130** (1989), 453–471.

[15] L.C.EVANS-J.SPRUCK, *Motion of level sets by mean curvature I,* J. Diff. Geom. **33** (1991), 635–681, *II,* Trans. Am. Math. Soc. **330** (1992), 321–332, *III,* J. Geom. Anal. **2** (1992), 121–150, *IV,* J. Geom. Anal. **5** (1995), no. 1, 77–114.

[16] H.FEDERER, *Geometric Measure Theory,* Springer, 1969.

[17] I.FONSECA, informal talk during the Conference "Surface tension and movement by mean curvature", Trento, July 1992.

[18] M.GAGE-R.S.HAMILTON, *The heat equation shrinking convex plane curves,* J. Diff. Geom. **23** (1986), 69–96.

[19] G.HUISKEN, *Flow by mean curvature of convex surfaces into spheres,* J. Diff. Geom. **20** (1984), 237–266.

[20] T.ILMANEN, *Generalized flow of sets by mean curvature on a manifold,* Indiana Univ. Math. J. **41** (1992), 671–705.

[21] T.ILMANEN, *Convergence of the Allen-Cahn Equation to Brakke's motion by mean curvature,* J. Differential Geom. **38** (1993), 417–461.

[22] J.L. LIONS-E.MAGENES, *Non homogeneous boundary value problems and applications,* Springer, 1972.

[23] P.L. LIONS, *Generalized solutions of Hamilton-Jacobi equations,* Pitman, Boston, 1982.

[24] L.MODICA-S.MORTOLA, *Un esempio di Γ^--convergenza,* Boll. U.M.I. **14-B** (1977), 285–299.

[25] H.M.SONER-P.E.SOUGANIDIS, *Singularities and uniqueness of cylindrically symmetric surfaces moving by mean curvature,* Comm. Partial Differential Equations **18** (1993), 859–894.

[26] S.VENTURINI, private communication.

The sapiential value of Mathematics
Lectio magistralis[‡]

E. De Giorgi

**Lecture held by Ennio De Giorgi for the
"Laurea honoris causa" in Philosophy conferred on him
by the "Università degli Studi di Lecce"**

I wish to thank all my friends from Lecce who, awarding me the degree in a branch of human knowledge seemingly detached from mathematics, have wanted to show not only their esteem and friendship for me but also their desire for an open dialogue and fruitful collaboration among scholars of different disciplines. I have just said "seemingly detached", in as much as, in the end, I believe that sciences, arts, civilised traditions of each people and civic traditions, all are, so to speak, branches of the unique tree of wisdom. The various forms of engagement in each of these disciplines are forms of "Philosophy", in an etymological sense: they are expressions of love for wisdom, and this applies to mathematics, too. Its "value in wisdom" is what I would like to illustrate.

This is not an easy task, of course: many people say "I am completely unable to understand mathematics". This is a reproach to all of us mathematicians in so far as we have been unable to show how in reality mathematics is much closer to the great vital questions and to the human universal desire for knowledge and wisdom than it is generally understood to be.

Mathematics has ancient links with all the other branches of wisdom. Ever since the age of Pythagoras the relationships existing between mathematics and music have been known. Moreover, the whole of ancient Greek history reflects how much the ancient Greeks were devoted to discovering the right and perfect mathematical ratio in masterpieces of sculpture, architecture and all the arts. On the other hand, by a comparison between ancient Greek with Hebrew and Middle Eastern culture, their heritage being a part of our one, we notice in the Holy Bible, where the establishment of such a monumental harvest of arts, technique and faith in that period, as King Solomon's Temple was recounted, a special care in considering the right and even proportions for each among the several parts leading to the unifying plan by Solomon, proposed as a model of culture, science and Hebrew and Middle Eastern wisdom.

[‡]Editor's note: translation into English of "Il valore sapienziale della matematica - Lectio Magistralis" published in Ennio De Giorgi Doctor Honoris Causa in Filosofia, Adriatica Editrice Salentina, Lecce, 1994, 19–28.

Even in the history of Italian art we find many examples of the attention paid to the link between art and mathematics, such as Piero della Francesca, Paolo Uccello, Leonardo da Vinci, etc.

If mathematics is closely linked to the world of art and work, even closer is the link with experimental sciences. You cannot conceive of physics without mathematics and this is not a new idea. Already Archimedes, who was probably the greatest scientist of antiquity, was at the same time a great mathematician and a great physicist and gave a rigorously mathematical form to the laws of physics discovered by him. For example the hydrostatic principle, whereby an object immersed in water receives a thrust proportional to the volume of the part immersed. Thus according to an old story Archimedes concretely applied this principle to show the fraud that had been committed by making a crown golden on the outside whereas the inside was made of a poorer material. The dishonest craftsman did not realize that the specific weight of gold is much greater than the specific weight of other metals. Archimedes could show that the inside of the crown was not completely made of gold without breaking the crown or ruining a work that seemed of merit, simply putting it in water and measuring the thrust it received. I do not know if this is history or legend, however it shows us an interesting aspect of mathematics, *i.e.* the ability to obtain from simple, easy experiments the solution to problems that seem to need complex, difficult experiments. The idea of not having to break the crown and to be able with a much less traumatic experiment and a mathematical calculation to establish its internal composition is after all a very simple example of how mathematics is linked with the rest of our experience.

Mathematics does not substitute observation but manages to strengthen it allowing us to draw very interesting conclusions from simpler and less expensive experiments than those that at first sight would seem necessary.

This idea of the experiment that, interpreted using intelligent mathematical analysis, gives us important information was one of the leading ideas of science at the time of Archimedes, who is considered by many as the founder of mathematical physics. Galileo Galilei was proud to consider himself a distant pupil of Archimedes. Pascal, to whom we owe among other things some fundamental discoveries in the field of hydrostatics, placed Archimedes as a model of scientific glory.

Many modern examples of the possibility of reducing costs and risks in experiments come to us from the medical research of less intrusive diagnostic techniques. It is necessary without damaging the human organism to obtain the necessary information for its cure and health and for this it is necessary to exploit with intelligence apparently weak evidence that however, elaborated with mathematical method, becomes the source of precious information. So in Medical Faculties all over the world today they are beginning to study mathematics a little more and are thinking that mathematics, which for a long time has been an essential part of the culture of the engineer, physicist and chemist, must become in our time an essential component of the culture of the doctor.

Moreover in human sciences the study of the relationships between logic, linguistics and mathematics is one of the directions that modern thought has taken.

The idea of mathematics not being a closed discipline reserved to few specialists but an open one that, although keeping its own autonomy, feeds on dialogue with the other forms of knowledge is one of the ideas that we mathematicians are not always able to illustrate sufficiently. Also because our very information, already partial in mathematics (nobody manages to follow all the branches of mathematics), is even more limited as far as possible applications and interactions with other disciplines are concerned, and is often limited to what we are sometimes told by some friendly colleague.

This poses many problems for us in spreading in some way to others our love for mathematics as an essential component of human wisdom, and making it understood that mathematics is something more than the simple ability to calculate and manipulate numbers. Of course, the theory of numbers is one of the greatest components of mathematics and starting from Pythagoras, the Egyptians, the Babilonians, the Chinese and the Indians, the study of numbers was the beginning of mathematics, which besides quantitative problems also studies problems of a qualitative nature.

Among the qualitative ideas that have emerged in our time I would emphasize the idea of non-linearity. To understand non-linearity it must be remembered that a linear mathematical problem usually corresponds to a situation (in physics, chemistry, economics, etc.) where there is a fixed ratio between cause and effect, where the sum of different causes produces a sum of respective effects. This in general happens in the majority of the phenomena only for comparatively short times and for relatively weak interactions. It is usually said that linearity gives an adequate description only on a small scale of the majority of the phenomena. However when you pass to a global large scale, for example if you consider the evolution over a long period of a complex physical system you see that the linear theory is not enough, because besides the direct consequences of whatever action we can exercise on the system there is a series of indirect consequences that add up and intertwine in various ways. For this reason even a small initial action can in the end produce very great consequences and an increase in the intensity of the action does not always correspond in the long term to a proportional increase in the result. Then sometimes there are quick changes, cracks, which with evocative language, even if a little "alarming", have been called "catastrophes" (in fact we speak of the "catastrophe theory", in more concise terms one could speak of the study of a certain kind of discontinuity). One thing that is being studied in mathematics today is in fact the problem of various kinds of discontinuity which can occur even starting from a very regular initial situation which at a certain point can have a quick change. This possibility has still to a large degree to be explored both from the qualitative and quantitative points of view.

The exploration of non linear phenomena will require a long patient dialogue between mathematics and experimental science scholars. In fact the mathematician rarely manages in a straightforward way to find the solution to the problem posed by the physicist, engineer, or doctor, etc. He must more or less content himself with understanding what the type of problem is that has been proposed to him and to guess what type of mathematical model will probably suit best and transform this intuition into precise propositions and then say to the physicist or engineer "it is likely that phenomenon you proposed to me could be schematised with this equation or this mathematical structure or with mathematical structures similar to it from the qualitative point of view". Mathematics can serve all these scholars if it is not taken to be a little machine where one pushes the problem and immediately gets the solution, but a means of better understanding the problem that is been studied, a means that the physicist or engineer must adapt to his needs and from which, if this is done with intelligence, he can draw really innovative ideas. This is the problem of the so called propaedeutic studies in mathematics for all applied sciences; it is not a question of giving once and for all a number of notions with which physicists, engineers, economists, etc. can solve all the problems, but of giving some general ideas like the conservation of energy, the evolution in time of a system, frequency, resonance and to show how they can serve in the study of many concrete problems. For example a bridge stands up well to the stress that does not have the same characteristic frequencies as the bridge itself, whereas it is much more sensitive to the stress that has these same frequencies. This phenomenon of resonance can be easily studied on a violin string: it is more complicated to study it on a very complex system like that of a bridge, ship or building, but it is already useful for the engineer to know that there is a qualitative analogy between what happens to a single violin string and what can happen to enormously complex structures.

Mathematics serves above all to have a qualitative idea of what might happen, to widen the imaginative ability of the experimental scholars or the planner capable of understanding that a certain mathematical model can be the right model to interpret certain phenomena. A prejudice that must be eliminated maintains that mathematics is interested only in quantitative aspects and not qualitative aspects of reality and that mathematics is the enemy of the imagination and freedom. In fact mathematics, if well understood, widens the imaginative capacity of a person. For example, we would not have had all the development in modern physics if the mathematical imagination had not come to the idea of infinite dimensional space. Equally, the concept of surfaces and manifolds with different curvatures makes the idea of a curved space-time possible.

This ability of mathematics to widen man's imagination is an aspect that should be more emphasized by mathematicians themselves. But there is another aspect that according to me is equally important. In the relationship between mathematics and other branches of knowledge there is always great freedom on both sides; mathematics proposes many possible models for a phenomenon and never imposes on the physicist or engineer or economist or doctor to choose a specific model. It leaves them the freedom to decide which, according to their experience and aims, can be the most suitable mathematical model for understanding a certain phenomenon. Mathematics reserves for itself the same freedom it offers everyone else. The mathematicians can draw inspiration from the most diverse sources, from physics, engineering, art, economics, law and philosophy. There is no form of knowledge that the mathematician cannot draw inspiration from. Even quite simple phenomena that have been known for some time can be a source of inspiration to original work for the mathematician. For example, the problem of many bodies that attract each other according to Newton's laws is a problem about which there is still a lot to be said.

Whatever the initial source of inspiration the mathematician must be completely free to conceive his theories and let them develop according to their internal logic and not let himself be discouraged by the fact that some developments do not have an immediate use. They impose themselves for "reasons of harmony" (it is difficult to find another word) and they are "mathematically natural". The most "harmonious" theoretical developments often with the passage of time show themselves to be the most useful for practical applications. Let us think of imaginary numbers (remember that the imaginary unit times itself gives -1): they seem to be purely ideal objects useful in certain algebraic calculations but initially they had no real meaning. Today we know that all electrical phenomena where there are oscillations must be studied using imaginary numbers that give the best representation of the characteristics of a variable circuit.

In most other cases applications of a mathematical theory can be very far from the original sources of inspiration. Let us think for example of studies in logic that for along time seemed to be linked only to the most theoretical philosophy, far from practical applications. Then between logic and computer science the contacts became ever closer. What at first seemed a purely ideal interest in understanding the nature of mathematical reasoning and then in general every kind of human reasoning is today also linked to the art of finding good programs to insert in a computer and good methods for protecting secrecy etc. .

Not being able to give an adequate idea of the breadth of the relationships between mathematics and other branches of knowledge I will try however to give some further indications. Personally the mystery of the harmony between mathematics and physical reality has always struck me very forcibly by the fact that things behave according to an internal logic very close to mathematical logic. Everything we see makes us think of a very close relationship between the harmony of physical laws and the internal harmony of mathematical thinking.

Another point that strikes me is the fact that we manage to study the finite only thinking of it immersed in an infinite framework. For example we would be not able to have a satisfactory theory of numbers with less than fifty digits after the decimal point, which however are more than enough for all practical uses, without a general theory that has its natural environment in the infinite set of all number. One of the paradoxes of mathematics is this: to study the most concrete things you have to pass through reflection of concepts that instead seem completely to go beyond our sensitive experience. This is something that makes you think.

Everything we manage to see in the finite appears incomprehensible and discordant if we do not think of it as part of a wider framework of infinite size. The fact that this infinite framework is for the most part unknown must not lead us to deny its existence. The same system of all integers is for the most part unknown to mathematicians although it is the central pillar of all mathematics and indirectly of all sciences that use mathematics in various ways. Furthermore, it was proven by Kurt Gödel and was one of the most important discoveries of this century, that a finite axiom system never manages to characterise the structure of integers. There is some very important information on this very mysterious system of integers and in the future we will be able to improve our knowledge; but whatever the amount of our progress, the majority of the properties of this system will remain mysterious to us.

It is something that according to taste can appear disappointing or comforting. For the scientist the knowledge that it will never be possible to exhaust a certain field of inquiry can be a disappointment but also a hope, which means that there will always be something to discover. Knowing that in mathematics there is still a lot to be done and that it is not already a closed science but a science that must continually be open to new problems does not cancel the major discoveries of the ancients but makes them be seen in a new light. If we think of the ancient theorem of Pythagoras we see that what seems characteristic of a very particular structure was then revealed as the model for very many structures. All in all there is not a conflict between mathematical tradition and mathematical innovation but the one enriches the other. On the one hand tradition is deeply appreciated insofar as new consequences and insights are discovered from it and on the other hand innovation becomes richer when it recognises its own roots. This is not true only for mathematics but for many other branches of knowledge. It is necessary to look not only for reconciliation but a deep synthesis between respect and love for traditions and desire to examine them closely always by more innovative thinking and one always more open to hopes for the future. And perhaps mathematics also tells us something about what is called (in a reductive way) "tolerance", whereas I would rather call it "understanding" of people, groups, sciences, cultures and different peoples. True understanding comes with a sincere love of the truth, respect for the dignity of man, the conviction that love of wisdom can be proposed but not imposed. To this love everybody can freely add by thinking of his own experience and of all his own life that it is unique and unrepeatable. This kind of comprehension is not relativism, scepticism or indifference to the truth but is rather a deep free love of wisdom and the desire that all men freely search for wisdom through their own personal conviction. In this search meeting someone humble, patient and sincere can be of great help who always tries to increase his own intellectual honesty and his own capacity to explain what he thinks in a comprehensible way. I think that it is also true for the teaching of exact sciences. Perhaps what still makes mathematics not very appealing to many people is the fact that many people feel it more as an imposition rather than a proposition.

The task that all we scholars of mathematics, physics and natural sciences and scholars of human sciences are called upon to perform is simply the rediscovery of the "value in wisdom" of our disciplines, a necessary premise for a harmonious development of all the branches of wisdom.

Il valore sapienziale della matematica
Lectio magistralis[‡]

Ennio De Giorgi

**Relazione di Ennio De Giorgi in occasione del conferimento
della Laurea honoris causa in Filosofia presso
l'Università degli Studi di Lecce**

Desidero ringraziare gli amici di Lecce che, conferendomi la laurea in un ramo del sapere umano apparentemente lontano dalla matematica, hanno voluto dimostrare non solo la loro stima ed amicizia nei miei confronti, ma anche il loro desiderio di dialogo e di collaborazione tra studiosi di discipline diverse. Ho detto apparentemente lontano perché credo che, in ultima analisi, le scienze, le arti, le tradizioni di civiltà di ogni popolo, di ogni città sono, per cosí dire, rami dell'unico albero della sapienza. Tutte le forme di impegno in queste discipline sono forme, in senso etimologico, di "filosofia", cioè manifestazioni di amore per la sapienza e questo vale anche per la matematica, di cui vorrei illustrare il "valore sapienziale".

Certamente il mio non è un compito facile: vi è tanta gente che dice "io sono totalmente incapace di capire la matematica", e questo è un rimprovero per noi matematici che non siamo stati in grado di mostrare quanto la matematica è in realtà piú vicina ai problemi della vita, al desiderio di sapere comune a tutti gli uomini, di quanto generalmente non si crede.

Il collegamento della matematica con tutti gli altri rami del sapere è antichissimo, già ai tempi di Pitagora erano noti i rapporti che ci sono tra matematica e musica e tutta la storia greca ci ricorda quale importanza attribuissero i greci nella scultura, nell'architettura e in tutte le arti alla esatta proporzione matematica delle varie parti di un'opera d'arte. Del resto confrontando due culture di cui siamo un po' eredi, cultura greca e cultura ebraica e medio–orientale, troviamo nella Bibbia, quando si parla della costruzione del tempio di Salomone, monumento massimo dell'arte, della tecnica e della fede dell'epoca, molta attenzione alle proporzioni esatte ed equilibrate con cui le varie parti venivano progettate da Salomone, presentato come il modello della cultura, della scienza, della sapienza ebraica e medio–orientale.

Anche nella storia dell'arte italiana troviamo molti esempi di attenzione al collegamento tra arte e matematica, come Piero della Francesca, Paolo Uccello, Leonardo da Vinci, ecc. .

[‡]Editor's note: published in Ennio De Giorgi Doctor Honoris Causa in Filosofia, Adriatica Editrice Salentina, Lecce, 1994, 19–28.

Se la matematica è strettamente legata al mondo dell'arte e del lavoro, ancor piú stretto è il legame con le scienze sperimentali. Non si concepisce una fisica senza matematica e questa non è una idea recente: già Archimede, che è stato probabilmente il piú grande scienziato dell'antichità, era simultaneamente grande matematico e grande fisico e diede una forma rigorosamente matematica alle leggi fisiche da lui scoperte, come il principio idrostatico per cui un oggetto immerso nell'acqua riceve una spinta proporzionale al volume della parte immersa. Anzi, secondo un antico racconto, Archimede applicò concretamente questo principio per mostrare la frode che era stata commessa facendo una corona dorata di fuori, mentre all'interno era composta di un materiale piú vile. L'artigiano disonesto non pensò al fatto che il peso specifico dell'oro è molto superiore al peso specifico di altri metalli: Archimede poté dimostrare che l'interno della corona non era tutto di oro senza rompere la corona, senza rovinare un lavoro che sembrava di pregio, semplicemente mettendola nell'acqua e misurando la spinta che riceveva. Non so se questa sia storia o leggenda, comunque ci mostra un aspetto interessante della matematica: cioè la capacità di ricavare da esperimenti semplici e facili la soluzione di problemi che sembrano richiedere esperimenti complessi e difficili. L'idea di non dover rompere la corona e di poter, con un esperimento molto meno traumatico e con un calcolo matematico, stabilire la sua composizione interna è in fondo un esempio semplicissimo di come la matematica entra in relazione con il resto delle nostre esperienze.

La matematica non sostituisce l'osservazione, però riesce a potenziarla permettendoci di trarre delle conclusioni molto interessanti da esperimenti piú facili e meno costosi di quelli che a prima vista sembrerebbero necessari.

Questa idea dell'esperimento che, interpretato mediante una intelligente analisi matematica, ci dà informazioni importanti, è stata una delle idee guida della scienza dai tempi di Archimede, che è ritenuto da molti il fondatore della fisica–matematica. Galileo Galilei si gloriava di considerarsi lontano allievo di Archimede, Pascal, a cui dobbiamo tra l'altro alcune fondamentali scoperte nel campo dell'idrostatica, poneva Archimede come modello della gloria scientifica.

Molti esempi moderni della possibilità di ridurre costi e rischi degli esperimenti ci vengono dalla ricerca medica delle tecniche di diagnosi meno invasive. Occorre, senza ledere l'organismo umano, ottenere le informazioni necessarie per la sua cura e per la sua salute e per questo bisogna sfruttare con intelligenza indizi apparentemente deboli che però, elaborati con metodi matematici, diventano fonte di informazioni preziose. Perciò nelle Facoltà di Medicina di tutto il mondo oggi si incomincia a studiare un po' piú la matematica, si pensa che la matematica, già da lungo tempo componente essenziale della cultura dell'ingegnere, del fisico, del chimico, debba diventare nei nostri tempi una componente essenziale della cultura del medico.

Del resto anche nelle scienze umane lo studio delle relazioni tra logica, linguistica e matematica è una delle direzioni di lavoro in cui si è sviluppato il pensiero moderno.

L'idea della matematica che non è conoscenza chiusa e riservata a pochi specialisti, ma conoscenza che, pur conservando la propria autonomia, si alimenta nel dialogo con le altre forme del sapere è una delle idee che noi matematici non sempre siamo in grado di illustrare adeguatamente, anche perché le nostre stesse informazioni, che già sulla matematica sono parziali (nessuno ormai riesce a seguire tutti i rami della matematica), per quanto riguarda le possibili applicazioni e interazioni con altre discipline sono ancora piú scarse, spesso si limitano a quello che ogni tanto ci viene raccontato da qualche collega amico.

Questo pone molti problemi a noi che dobbiamo in qualche modo trasmettere agli altri l'amore per la matematica, come componente essenziale della saggezza umana e far capire che la matematica è qualcosa di piú della semplice abilità di calcolo, della pura manipolazione di numeri. Certamente la teoria dei numeri è una delle grandi componenti della matematica e, a partire da Pitagora, dagli egiziani, dai babilonesi, dai cinesi, dagli indiani, lo studio dei numeri è stato l'inizio della matematica, ma questa, accanto ai problemi di tipo quantitativo studia anche problemi di tipo qualitativo.

Tra le idee qualitative che sono emerse nei nostri tempi sottolinerei l'idea di non linearità. Per capire la non linearità occorre ricordare che un problema matematico linare corrisponde in genere a una situazione (fisica, chimica, economica, ecc.) in cui c'è proporzionalità tra cause ed effetti, in cui la somma di diverse cause provoca una somma dei rispettivi effetti. Questo, in genere, succede nella maggior parte dei fenomeni soltanto per tempi abbastanza brevi e per azioni relativamente deboli. Si suol dire che la linearità dà una descrizione adeguata soltanto in piccolo della maggior parte dei fenomeni. Quando invece si passa allo studio in grande, per esempio si considera l'evoluzione in tempi lunghi di un sistema fisico complesso, vediamo che lo schema linare non basta piú, perché accanto alle conseguenze dirette di una qualsiasi azione che noi possiamo esercitare sul sistema, ci sono una serie di conseguenze indirette che vanno a sommarsi e intrecciarsi in vario modo, per cui anche delle azioni iniziali molto piccole possono dare alla fine delle conseguenze molto grandi, non sempre ad un aumento della intensità dell'azione corrisponde nel lungo termine un aumento proporzionale del risultato. Talvolta poi si hanno dei bruschi cambiamenti, dei fenomeni di frattura, che con linguaggio suggestivo, anche se un po' "allarmante", sono stati chiamati "catastrofi" (si parla infatti di "teoria delle catastrofi", forse con termini più "freddi" si potrebbe parlare di studio di certe discontinuità). Una delle cose che oggi si sta studiando in matematica è proprio il problema delle discontinuità, che possono sorgere anche partendo da una situazione iniziale molto regolare che a un certo punto può avere un cambiamento brusco e questa possibilità è ancora in gran parte da esplorare sia dal punto di vista qualitativo che dal punto di vista quantitativo.

L'esplorazione dei fenomeni non lineari richiederà un dialogo lungo e paziente tra matematici e studiosi di scienze sperimentali: infatti raramente il matematico riesce direttamente a trovare la soluzione del problema che gli viene posto dal fisico, dall'ingegnere, dal medico, ecc. Per lo piú deve accontentarsi di capire qual è il tipo di problema che gli viene proposto, intuire quale tipo di modello matematico probabilmente si adatterà meglio, trasformare questa intuizione in proposizioni precise e poi dire al fisico o all'ingegnere "è probabile che il fenomeno che tu mi proponi possa essere schematizzato con questa equazione o con questa struttura matematica o con delle strutture matematiche simili a questa dal punto di vista qualitativo". La matematica può servire a tutti questi studiosi, se non viene presa come una macchinetta in cui uno pone il problema e riceve immediatamente la soluzione ma come un mezzo per capire meglio il problema che si studia, un mezzo che il fisico o l'ingegnere deve adattare alle sue esigenze e da cui, se questo adattamento viene fatto con intelligenza, può trarre idee veramente innovative. Questo è il problema dei cosiddetti studi propedeutici di matematica per tutte le scienze applicate; non si tratta di dare una volta per tutte un bagaglio di nozioni con cui fisici, ingegneri, economisti, ecc., possono risolvere tutti i loro problemi, si tratta di dare alcune idee generali, come quelle di conservazione dell'energia, di evoluzione nel tempo di un sistema, di frequenza, di risonanza, e mostrare come esse possano servire nello studio di molti problemi concreti. Per esempio un ponte regge bene a delle sollecitazioni che non abbiano le stesse frequenze caratteristiche del ponte stesso, mentre è molto piú sensibile a delle sollecitazioni che abbiano le stesse frequenze. Questo fenomeno di risonanza può essere facilmente studiato su una singola corda di violino; è piú complicato studiarlo su un sistema molto complesso come può essere un ponte, una nave, un palazzo, ma è già utile all'ingegnere sapere che c'è una analogia qualitativa tra quello che succede a una singola corda di violino e quello che può succedere a strutture enormemente complesse.

La matematica serve soprattutto per avere un'idea qualitativa di quello che può accadere, per allargare la capacità di immaginazione dello studioso sperimentale, del progettista, capace di capire che un certo modello matematico può essere il modello giusto per comprendere certi fenomeni. Un pregiudizio che va eliminato ritiene la matematica interessata solo agli aspetti quantitativi e non agli aspetti qualitativi delle cose, pensa che la matematica sia nemica della fantasia e della libertà. In realtà la matematica, se ben compresa, allarga le capacità di immaginazione di una persona. Per esempio non avremmo avuto tutto lo sviluppo della fisica moderna se l'immaginazione matematica non fosse arrivata all'idea di spazio a infinite dimensioni; parimenti l'idea di superficie e varietà con diverse curvature rende possibile l'immaginazione di uno spazio–tempo curvo.

Questa capacità della matematica di allargare l'immaginazione dell'uomo è un aspetto che dovrebbe essere messo dagli stessi matematici un poco piú in evidenza, ma c'è un altro aspetto che secondo me è egualmente importante. Nel rapporto tra matematica e gli altri rami del sapere c'è sempre una grande libertà da entrambe le parti; la matematica propone molti possibili modelli di un fenomeno, non impone mai né al fisico, né all'ingegnere, né all'economista, né al medico di scegliere un determinato modello, gli lascia la libertà di decidere lui quale, secondo la sua esperienza, i suoi fini, può essere il modello matematico piú adatto per comprendere un certo fenomeno. La stessa libertà che la matematica offre a tutti i suoi interlocutori, riserva a sé stessa. Il matematico può trarre ispirazione dalle fonti piú diverse, dalla fisica, dall'ingegneria, dall'arte, dall'economia, dal diritto, dalla filosofia. Non c'è forma di sapere da cui il matematico non può trarre ispirazione. Anche fenomeni abbastanza semplici ormai noti da parecchio tempo possono essere fonte di ispirazione di lavori originali per il matematico. Per esempio, il problema degli N corpi che si attraggono secondo le leggi di Newton è un problema su cui ancora c'è molto da dire.

Qualunque sia la fonte iniziale di ispirazione il matematico deve sentirsi molto libero nel concepire le sue teorie e lasciarle sviluppare secondo la loro logica interna, non lasciarsi scoraggiare dal fatto che alcuni sviluppi non hanno una immediata utilizzazione, ma si impongono per "ragioni di armonia", (è difficile trovare un'altra parola), sono "matematicamente naturali". Spesso a distanza di tempo più o meno grande gli sviluppi più "armoniosi" sul piano teorico si rivelano anche i più fecondi sul piano delle applicazioni pratiche. Pensiamo ai numeri immaginari (ricordiamo che l'unità immaginaria moltiplicata per sé stessa dà -1): sembravano oggetti puramente ideali utili in certi calcoli algebrici ma non avevano inizialmente nessuna interpretazione reale; oggi noi sappiamo che tutti i fenomeni elettrici in cui si hanno delle oscillazioni vanno studiati mediante l'uso dei numeri immaginari, che dànno la migliore rappresentazione delle caratteristiche di un circuito oscillante.

In molti altri casi le applicazioni di una teoria matematica possono essere assai lontane dalle originali fonti di ispirazione. Pensiamo ad esempio agli studi di logica, che per lungo tempo sembravano legati solo alla filosofia più teorica, lontani dalle applicazioni pratiche. Poi fra logica e informatica i contatti sono diventati sempre più stretti, quello che prima sembrava un interesse puramente ideale, capire la natura del ragionamento matematico e più in generale di ogni ragionamento umano, oggi è legato anche all'arte di trovare buoni programmi da immettere in un calcolatore, buoni metodi per proteggere la segretezza, ecc.

Non potendo dare un'idea adeguata dell'ampiezza delle relazioni tra matematica ed altri rami del sapere, tenterò di dare egualmente qualche ulteriore indicazione. Personalmente, mi ha sempre molto colpito il mistero dell'armonia tra matematica e realtà fisica, il fatto che le cose si comportino con una logica interna molto vicina alla logica matematica: tutto ciò che noi vediamo ci fa pensare a uno strettissimo rapporto fra armonia delle leggi fisiche e armonia interna del pensiero matematico.

Un altro punto che mi colpisce è il fatto che noi riusciamo a studiare il finito solo pensandolo immerso in una cornice infinita. Per esempio, noi non potremmo avere una teoria soddisfacente dei numeri con meno di cinquanta cifre decimali, che pure sono largamente sufficienti per tutti gli usi pratici, senza una teoria generale che ha il suo ambiente naturale nell'insieme infinito di tutti i numeri. Uno dei paradossi della matematica è questo: per studiare le cose più concrete bisogna passare attraverso la riflessione su concetti che invece sembrano superare completamente la nostra esperienza sensibile. Questo è un dato che fa pensare: tutto ciò che noi riusciamo a vedere nel finito ci appare incomprensibile e disarmonico, se non lo pensiamo come parte di un quadro più ampio di grandezza infinita. Il fatto che questo quadro infinito sia in gran parte sconosciuto non ci deve portare a negarne l'esistenza: lo stesso insieme dei numeri interi è in gran parte sconosciuto ai matematici, eppure è il pilastro centrale di tutta la matematica e indirettamente di tutte le scienze che usano in vario modo la matematica.

È stato fra l'altro dimostrato da Kurt Gödel, ed è stata una delle piú importanti scoperte di questo secolo, che un sistema finito di assiomi non riesce mai a caratterizzare univocamente la struttura dei numeri interi. Si hanno alcune informazioni molto importanti su questo insieme abbastanza misterioso dei numeri interi, potremo in futuro migliorare le nostre conoscenze, ma qualunque sarà il grado del nostro progresso, la maggior parte delle proprietà di questo insieme resteranno per noi misteriose.

È qualcosa che secondo i gusti può apparire deludente e può apparire confortante; per lo scienziato, il sapere che non sarà mai possibile esaurire un certo campo di indagine può essere una delusione ma può essere anche una speranza: vuol dire che ci sarà sempre qualcosa da scoprire. Sapere che la matematica è ancora in gran parte da fare, non è una scienza già chiusa, ma una scienza che continuamente deve aprirsi verso problematiche nuove non cancella le maggiori scoperte dell'antichità ma le fa vedere attraverso una luce sempre nuova. Se noi pensiamo all'antico teorema di Pitagora, vediamo che quello che sembrava caratteristico di una struttura molto particolare si è poi rivelato modello di moltissime strutture. In fondo non c'è un conflitto fra tradizione e innovazione matematiche, ma l'una arricchisce l'altra. Da un lato la tradizione viene apprezzata in profondità solo nella misura in cui se ne scoprono conseguenze nuove, aspetti nuovi, dall'altro l'innovazione diviene piú ricca nella misura in cui riconosce le proprie radici. Questo non vale soltanto per la matematica, ma per molti altri rami del sapere: occorre cercare non solo una conciliazione, ma una profonda sintesi tra rispetto e amore delle tradizioni e desiderio di approfondirle attraverso un pensiero sempre piú innovativo e sempre piú aperto alle speranze del futuro. E forse anche la matematica ci dice qualcosa su ciò che viene chiamato (un po' riduttivamente) "tolleranza", mentre io chiamerei piuttosto "comprensione" verso persone, gruppi, scienze, culture e popoli diversi. La vera comprensione nasce con l'amore sincero per la verità, il rispetto per la dignità dell'uomo, la convinzione che l'amore della sapienza può essere proposto ma non imposto. A questo amore ogni persona deve giungere liberamente attraverso la meditazione sulla propria esperienza e su tutta la propria vita che è unica e irripetibile; da questo punto di vista la comprensione non è relativismo, scetticismo, indifferenza rispetto alla verità, ma è piuttosto amore della sapienza profondo e libero, desiderio che tutti gli uomini ricerchino liberamente la sapienza per propria personale convinzione. In questa ricerca può essere di grande aiuto l'incontro con chi è umile, paziente e sincero, con chi cerca sempre di accrescere la propria onestà intellettuale, la propria capacità di esporre in modo comprensibile ciò che pensa. Penso che questo vale anche per l'insegnamento delle scienze esatte: forse ciò che rende ancora per molti la matematica poco attraente è il fatto che molti la sentono piú come un'imposizione che come una proposta.

In fondo il compito a cui tutti noi, studiosi di scienze matematiche, fisiche e naturali e studiosi di scienze umane, siamo chiamati è la riscoperta del valore sapienziale delle nostre discipline, premessa necessaria per un armonico sviluppo di tutti i rami del sapere.

Fundamental principles of Mathematics[‡†]

Ennio De Giorgi

Lecture by Ennio De Giorgi at the session
Fundamental Principles of Mathematics and
Artificial Intelligence
October 28, 1994

I present in this lecture some ideas concerning a "basic theory" of the foundations of mathematics on which I have been working in the last few years, together with a small group of colleagues and friends (cfr. bibliography). The first idea of the "basic theory" is the idea that mathematics can consider *qualitatively* and not only *quantitatively* different objects. Aiming to follow the tradition, and wanting to avoid that the first axioms on qualities resemble puns, we often use the word "property" as a perfect synonym of the word "quality". Moreover, when q denotes a quality and x an object of any kind, we write

$$q\,x$$

to say that x enjoys the property q.

The first quality that we consider is the property of being a quality, which we denote by the symbol $Qqual$. Therefore we write $Qqual\,z$ to say that z is a quality. So the first axiom of our theory can be written in the form

$$Qqual\,Qqual,$$

and it is equivalent to the statement: $Qqual$ enjoys the property $Qqual$.

After $Qqual$ we introduce $Qrel$, *i.e.* the property of being a relation, Qop, *i.e.* the quality of being an operation, $Qcoll$, *i.e.* the quality of being a collection, $Qnum$, *i.e.* the quality of being a number, $Qprop$, *i.e.* the quality of being a proposition, $Qpred$, *i.e.* the quality of being a predicate, $Qver$, *i.e.* the quality of being a true proposition, $Qfals$, *i.e.* the quality of being a false proposition. For each of these species we then choose some objects that we consider fundamental. For example, among the collections we have chosen V, the collection of all collectible objects, $Coll$, the collection of all collections, Ins, the collection of all sets, $\mathbf{N}$, the collection of the natural numbers (*i.e.* of the numbers $0, 1, 2, \ldots$), $\emptyset$, the empty collection. Besides the fundamental

---

[‡]Editor's note: published in Proceedings Pontificiae Academiae Scientiarum, Scripta varia, n. 92, 269-277.
[†]Plenary Session of the Pontifical Academy of Sciences, October 25-29, 1994

objects of the various species we have to choose a certain number of fundamental axioms. From these axioms theorems will follow that, together with the axioms, should constitute the building of the theory, that we hope to be coherent and harmonious.

The freedom in which axioms are chosen in mathematics may disconcert who correctly trusts in "mathematical certainties", and considers them as cornerstones, even in a century that has seen many "scientific revolutions" and also the success in epistemology of the idea that scientific theories are "falsifiable", but not "demonstrable". I think that mathematical certainties are really indestructible, but I recall that they are mainly "hypothetical-deductive" certainties, that the statements of the theorems are of the type "if the hypothesis A holds, then the thesis B holds", "if the axiom X is accepted then the consequence Y has to be accepted".

I believe that mathematicians should make wide use of this freedom in choosing axioms but they should not abuse it. They have to avoid the introduction of not very expressive axioms, which only serve to form useless and often not very entertaining games. They should look for the "wisest" axioms with great humility, carefully listen to questions, observations, criticism of other mathematicians, and also of scholars of disciplines more or less far from mathematics; they should not consider this criticism an attack, against which their own theories have to be defended, but a precious help in the difficult path to greater wisdom.

By using the word wisdom (or knowledge) I intend to give to this term all the wealth of meanings that it has in ancient sapiential books, in the writings of the greatest philosophers, in the best cultural traditions of different nations. So I think that the "wisest axioms" are those that are suitable to increase beauty, expressive capability, communicative ability, cultural value, and usefulness of mathematics. I believe that an independent spirit and a great attitude to dialogue are helpful in doing good mathematical research and particularly in choosing good fundamental axioms. Close to both these virtues are many other "sapiential virtues", such as respect for dignity, freedom, and originality of every person, understanding and friendship among those persons that, in any time and in any country, worked and are working in the fields of science, art, and technology, with the attitude that the ancients called "philosophy", that is "love for wisdom".

I quoted those who *worked* with love for wisdom because I believe that mathematicians should love tradition, they should remember with admiration and gratitude the great mathematicians, scientists and philosophers of the past, and appreciate the inexhaustible fertility of their best ideas. Mathematicians should combine respect for tradition with willingness for innovation, and remember Shakespeare's warning: "there are more things in heaven and earth than are dreamt of in your philosophy". To these words I would like to add that there are more "things" in the minds and hearts of human beings than they themselves imagine. "Wise" mathematicians look with admiration at beauty, harmony and order of the Creation, that Galileo considered a book written in mathematical characters. They also consider with admiration how the human reason succeeds, at least partly, in deciphering this marvellous book, and how it reflects on its own very nature, discovering, through the study of logic, some fundamental rules of human reasoning.

On the other hand all those who study mathematics with seriousness and sincere passion have many occasions for recognising their own possible errors, the limited interest of many results obtained by themselves, the great importance of the many problems they are not able to solve, and sometimes are not even able to state in a clear way. Summing up, they have many opportunities of practising the virtues of humility and intellectual honesty, which seem to me important components of what one of the most ancient sapiential books of the Bible, the Book of Proverbs, calls "*timor Domini principium sapientiae*".

In order to apply Shakespeare's warning to the foundations of mathematics, mathematicians have to be open to innovations, always willing to enlarge the horizon of their researchs, so as to encompass new "things" that were not embodied by previous theories.

Having to quote an innovation which seems to me a widening of the "mathematical horizon" obtained by the "basic theory", I would point out the treatment of collections, which seems to me innovative, if compared to the usual presentations of the concept of set. In this connection I recall that in the "basic theory" sets are particular collections, that the empty collection $\emptyset$ and the collection of natural numbers $\mathbf{N}$ are sets, whereas the universal collection V, the collection of all collections $Coll$ and the collection of all sets Ins are not sets. Roughly speaking, one could say that "too big" collections are not sets, since they are not only infinite, but "infinitely bigger than every infinite set". On the contrary, there are no qualitative differences between objects belonging to the collection V and objects belonging to sets. All objects considered by our theory belong to the collection V, but for each object x there is a "singular set" (or "singleton"), to which only the object x belongs; this set is usually denoted by the symbol $\{x\}$. Apparently, singletons are the "smallest" of all sets, except for the empty set, which is the least set.

I called innovative the treatment of collections, but I have to add, for the sake of intellectual honesty, that my knowledge of the mathematical literature, and my capability in bibliographical researchs are notably inferior to those of many colleagues of mine, and I succeed in working in mathematics mainly because many colleagues and friends provide me always with interesting bibliographic informations. Therefore, while I feel relatively sure in evaluating the interest of a mathematical idea, I feel less sure in declaring its absolute novelty.

Willing to unite "wise" innovation and tradition, we reserved a place of honour in the "basic theory" to arithmetic, which has always been the "queen of mathematics", since the times of Pythagoras, Diophantus, up to Fermat and to the present days. For this reason, immediately after the set of natural numbers $\mathbf{N}$, we introduce the simplest arithmetical operation, *i.e.* the operation $Nsucc$ that associates to each natural number the successive natural number.

Together with the elementary operation $Nsucc$ we introduce the quality $Qops$, i.e. the quality of being a *simple operation*, and a plain notation. Namely, if f

is a simple operation performable on the object x, then we denote by

$$fx$$

the result obtained by performing the operation f on the object x; *e.g.*

$$Nsucc\,0 = 1, \quad Nsucc\,1 = 2, \quad Nsucc\,2 = 3, \quad Nsucc\,3 = 4, \ \ldots, \ \text{etc.} \ \ldots$$

After the operation *Nsucc* one introduces the other arithmetic operations, for which I refer to the note [4], so as not to dull this exposition by many formulae.

After introducing arithmetic, the "basic theory" can be developed following two paths: in the first path precedence is given to relations, collections and sets; in the second one, precedence is given to qualities, operations, propositions, and predicates. Essentially the first path tends to a wide extension of the usual theory of sets, whereas the second path tends to an even wider expansion of propositional and predicate calculus. In the sequel of this exposition I will hint mainly to this "second path", both because it is yet the least explored one and because along this path many problems are met that in my opinion are interesting for mathematics, logic, computer science, philosophy, epistemology, etc. In substance the problem is that of evaluating the possible answers to the question: "How can one elaborate mathematical theories that encompass, at least in principle, all traditional mathematics and logic, and also have a high degree of 'self-description'?". I use the term "self-description" of a theory to mean that the theory itself is among its own objects. Examples of self-description can be found by thinking of a dictionary, which among other words contains the word "dictionary", or of a grammar, written according to grammatical and syntactical rules.

In order to describe at least the very first steps along what we called the second path, it is necessary to introduce some very simple notations concerning the results obtained by successively performing several simple operations. Therefore we start by considering simple operations f and g, and we suppose that there are objects x, y, z satisfying $fx = y$, $gy = z$. In this case we use the notation

$$z = g(fx)$$

to mean that z has been obtained by performing first the operation f on the object x and then by performing the operation g on the result thus obtained. In this notation the case $f = g$ is not excluded, for example we have

$$Nsucc(Nsucc\,0) = Nsucc\,1 = 2, \quad Nsucc(Nsucc\,2) = Nsucc\,3 = 4, \ \ldots$$

$$Nsucc(Nsucc\,7) = Nsucc\,8 = 9, \ldots$$

It may happen that by performing the operation α on the object x one gets another operation β. We take into account this possibility by stipulating that, if α and β are operations and x, y, z are objects such that $\beta = \alpha x$, $z = \beta y$, then we use the notation

$$z = (\alpha x)y.$$

Similarly, if α, β, γ are operations and x, y, z, t are objects such that $\beta = \alpha x$, $\gamma = \beta y$, $t = \gamma z$, then we use the notation

$$t = \big((\alpha x)y\big)z.$$

One could deal similarly with the case of a greater number of operations, always maintaining the convention that operations placed inside a couple of brackets have to be performed before operations lying outside these brackets. Having so fixed this elementary notation, we can introduce two simple operations that are relevant in the theory of predicates: the operation *Predord*, which gives the *order of predicates* and the operation *Gpred*, the operation *generator of predicates*, which associates to the various objects introduced up to now predicates describing their behaviour. We impose two axioms to the operation *Predord*:

1. In order that p be a proposition, *i.e.* p enjoy the property *Qprop*, it is necessary and sufficient that p be a predicate, *i.e.* p enjoy the property *Qpred*, and that the order of p be zero, *i.e.* *Predord* $p = 0$. In other words, we identify propositions and predicates of order zero.

2. If n is a natural number, p is a predicate, and the order of p is the natural number successor of n, *i.e.*

$$Predord\, p = Nsucc\, n,$$

    then p is a simple operation and, for every object x, px is a predicate of order n.

Essentially these two axioms say that from predicates of higher orders one can get in countless ways predicates of lower orders, and eventually one gets propositions.

The second operation *Gpred* associates to each object x a predicate whose order essentially characterizes the "structural complexity" of x, and hence this order will be named also by the term "arity", suggested by the words binary, ternary, quaternary etc. Aiming to express this fact by means of an axiom, one can state the following axiom:

3. There exists a simple operation Ar such that, for every object x,

$$Ar\, x = Predord\,(Gpred\, x).$$

As far as the arity of the various objects introduced up to now is concerned, we notice that:

if $Qnum\ x$ then $Ar\, x = 0$, *i.e.* each number has arity 0,

if $Qprop\ p$ then $Ar\, p = 0$,

if $Qqual\ q$ then $Ar\, q = 1$,

if $Qcoll\ C$ then $Ar\, C = 1$.

$Qops\ f$ if and only if $Qop\ f$, $Ar\, f = 2$.

Besides simple operations, one can consider more complex operations, namely binary, ternary etc. operations. Concerning these operations, we call *Qopb* the property of being a binary operation, *Qopt* the property of being a ternary operation, etc., and we have:

$$Qopb\ g \text{ if and only if } Qop\ g,\ Ar\ g = 3,$$
$$Qopt\ h \text{ if and only if } Qop\ h,\ Ar\ h = 4, \text{ etc.}$$

A similar treatment can be applied to relations, by distinguishing binary relations, enjoying the quality *Qrelb*, ternary relations, enjoying the quality *Qrelt*, quaternary relations, enjoying the quality *Qrelq*, etc. For these objects we have:

$$Qrelb\ \alpha \text{ if and only if } Qrel\ \alpha,\ Ar\ \alpha = 2,$$
$$Qrelt\ \beta \text{ if and only if } Qrel\ \beta,\ Ar\ \beta = 3,$$
$$Qrelq\ \gamma \text{ if and only if } Qrel\ \gamma,\ Ar\ \gamma = 4, \text{ etc.}$$

These axioms show that the operation *Gpred* is essential for developing the "second path" that can be followed in building the basic theory for the foundations of mathematics. Its role is better clarified by the following supplementary axioms:

4. Let q be a quality. In order that x enjoy the property q it is necessary and sufficient that $(Gpred\ q)x$ enjoy the quality *Qver*.

5. Let f be a simple operation. In order that $fx = y$ it is necessary and sufficient that $((Gpred\ f)x)y$ enjoy the quality *Qver*.

In order to easily remind axioms 4 and 5, it is convenient to introduce, after the of bracketed notation, also the quoted notation. We write:

$$\text{``}qx\text{'' in place of } (Gpred\ q)x,\quad \text{``}fx = y\text{'' in place of } ((Gpred\ f)x)y.$$

Similar axioms hold for collections, for more complex operations (binary, ternary etc.), and for relations, so a suitable quoted notation can be adopted also for these objects.

After introducing in such a way predicates and propositions, the main operations of predicate calculus can be introduced, *e.g.* negation, disjunction, interpretation of formulae, etc. In fact all these operations turn out to be embedded in a much wider framework than the usual one. In such a framework, the principle of non-contradiction still holds, *i.e.* a proposition can not enjoy simultaneously both properties *Qver* and *Qfals*, but, having been lavish with operations that "produce propositions", we may not exclude that there are propositions that do not enjoy either of these two properties. Therefore we have to add to the quality *Qprop* the quality *QVF*, *i.e.* the quality of being a proposition that enjoys either *Qver* or *Qfals*.

This attitude is substantially equivalent to recognizing that we give a wide range of meanings to the word "proposition", that we intend to include at once "sensible, clear, unambiguous" propositions, on which a sure judgment of truth or falsehood can be expressed, together with other more ambiguous, obscure, less sensible propositions, on which it is difficult or impossible to give such a judgment. An example of such impossibilities is given by "self-negating propositions" *i.e.* by propositions p verifying the condition:

$$p = \text{``}Q\,fals\,p\text{''}.$$

Self-negating propositions have been *essentially* known since ancient times, through Liar's antinomy, which I consider the "mother of all antinomies". While it is easy to admit that there are propositions enjoying the property QVF and other propositions that do not enjoy this property, it seems difficult (perhaps impossible) to draw a good "border line" between the former and the latter ones. Perhaps this is the very reason why the study of the quality QVF seems to me to be very interesting and probably connected to many problems in logic, computer science, psychology, philosophy and linguistics, etc., and even to speculations on the the concepts of human intelligence and of "artificial intelligence".

Bibliography

[1] E. De Giorgi, M. Forti, G. Lenzi, V. M. Tortorelli: *Calcolo dei predicati e concetti metateorici in una teoria base dei fondamenti della matematica,* Atti Accademia Nazionale Lincei, Classe Scienze Fisiche, Matematiche, Naturali, Rendiconti Matematica e Applicazioni (9) 6-2 (1995), 79-92.

[2] M.Clavelli, E. De Giorgi, M. Forti, V. M. Tortorelli: *A self-reference oriented theory for the foundations of mathematics,* in Analyse Mathématique et applications- Contributions en l' honneur de Jacques-Louis Lions, Gauthier-Villars, Parigi 1988, 67-115.

[3] E. De Giorgi, M. Forti: *Una teoria quadro per i fondamenti della matematica,* Atti Accademia Nazionale Lincei, Rendiconti Classe Scienze Fisiche, Matematiche, Naturali, (8) 79 (1985), 55-67.

[4] E. De Giorgi, M.Forti, G. Lenzi: *Una Proposta di teorie base dei fondamenti della matematica,* Atti Accademia Nazionale Lincei, Rendiconti Classe Scienze Fisiche, Matematiche, Naturali, (9) 5 (1994), 11-22.

[5] M.Forti, F.Honsell: *Models of self-descriptive set theories:* Partial Differential Equations and the Calculus of Variations- Essays in honour of Ennio De Giorgi, F.Colombini et al. eds. Birkhäuser, 1989, 473-518.

[6] M.Forti, F.Honsell: *Models of self-descriptive set theories:* Partial Differential Equations and the Calculus of Variations- Essays in honour of Ennio De Giorgi, curatori F.Colombini et al. Birkhäuser, 1989, 473-518.

[7] G.LENZI: *Estensioni contraddittorie della teoria ampia,* Atti Accademia Nazionale Lincei, Rendiconti Classe Scienze Fisiche, Matematiche, Naturali, (8) 83 (1989), 13-28.

[8] G.LENZI, V.M.TORTORELLI: *Introducing Predicates into a basic Theory for the Foundations of Matematics,* Preprint Scuola Normale Superiore di Pisa, luglio 1989.

Technical complements to the lecture of E. De Giorgi[‡]

Ennio De Giorgi[†]

In order to allow a deeper technical appreciation of the lecture given at the Pontificial Academy, which has mainly a cultural approach, containing few formulae and few details of a logical-mathematical character, we have to recall, though not systematically, some notions and notations from the note [4] in the bibliography. Therefore we begin by recalling that among all the operations, *i.e.* among all the objects enjoying the quality Qop, besides simple operations there are binary operations, ternary operations, etc. . Binary operations enjoy the quality $Qopb$, ternary operations enjoy the property $Qopt$. Given a binary operation φ, we use the notation

$$\varphi\, xy = z$$

to say that the operation Φ, performed on the objects x, y gives as result the object z. *E.g.* if we consider $Nadd$, the addition between natural numbers, then we have:

$$Nadd\, xy = x + y.$$

If we consider the operation $Npot$, of raising a natural number to power, we have:

$$Npot\, xy = x^y.$$

Similarly, if we have a ternary operation ψ, *i.e.* $Qopt\ \psi$, and four objects x, y, z, t, we write:

$$\psi\, xyz = t$$

to state that the object t is the result obtained by performing the operation ψ on the objects x, y, z.

Besides simple, binary, ternary etc. operations there are also binary, ternary, quaternary, etc. relations. They are characterised by the qualities $Qrelb$, $Qrelt$, $Qrelq$, etc., respectively. When r is a binary relation and x, y are objects, we write:

$$r\, xy$$

to say that x and y are in the relation r.

Given a ternary relation ρ and three objects x, y, z, we write:

$$\rho\, xyz$$

[‡]Editor's note: translation into English of the paper "Complementi tecnici alla relazione di Ennio De Giorgi", Pontificial Academy of Sciences, October 28, 1994.

[†]Plenary Session of the Pontifical Academy of Sciences, October 28, 1994

to state that x, y, z are in the relation ρ.

Finally, given a quaternary relation τ and four objects x, y, z, t, we write:

$$\tau\, xyzt$$

to say that x, y, z, t are in the relation τ.

Next we introduce the function *arity*, which we denote by the symbol Ar, to mean the "complexity" degree of the various objects. Precisely, as already noticed in the lecture, we have:

a) If $Qqual\, q$ then $Ar\, q = 1$;

b) If $Qcoll\, C$ then $Ar\, C = 1$;

c) If $Qops\, f$ then $Ar\, f = 2$;

d) If $Qrelb\, r$ then $Ar\, r = 2$;

e) If $Qopb\, \varphi$ then $Ar\, \varphi = 3$;

f) If $Qrelt\, \rho$ then $Ar\, \rho = 3$;

g) If $Qopt\, \psi$ then $Ar\, \psi - 4$;

h) If $Qrelq\, \tau$ then $Ar\, \tau = 4$.

After the function Ar, we can introduce the simple operation $Rfond$ which is defined on positive natural numbers. Let the natural number h greater then zero be fixed: then $Rfond\, h$ is a relation of arity $h + 1$, which we call *fundamental relation* of the objects of arity h, becouse it describes the "action" of all such objects.

For instance, if q is a quality and x an arbitrary object, then
$(Rfond\, 1)\, qx$ if and only if x enjoys the quality q (in formulae, $q\, x$).
If C is a collection, then
$(Rfond\, 1)\, Cx$ if and only if x belongs to C (in formulae, $x \in C$).
If f is a simple operation, then
$(Rfond\, 2)\, fxy$ if and only if $f\, x = y$.
if r is a binary relation, then
$(Rfond\, 2)\, r\, xy$ if and only if the objects x, y are in the relation r (in our notations we write $r\, xy$).
If f is a binary operation, then
$(Rfond\, 3)\, f\, xyz$ if and only if $f\, xy = z$.

$\vdots$

Similarly one deals with objects having higher and higher arities. In order to shorten formulae, we use the following notation:

$x \uparrow\downarrow y$ instead of $(Rfond\, 1)\, xy$.
$x \uparrow\downarrow yz$ instead of $(Rfond\, 2)\, xyz$.

$x \uparrow\downarrow yzt$ instead of $(Rfond\,3)\,xyzt$, etc.

Anyway,

if $x \uparrow\downarrow y$, then $Ar\,x = 1$;
if $x \uparrow\downarrow yz$, then $Ar\,x = 2$; etc.

We extend the usual set-theoretic notation and, given objects a, b of arity 1, we write $a \subseteq b$ to say that every object x satisfying the condition $a \uparrow\downarrow x$ satisfies also $b \uparrow\downarrow x$. Similarly, given objects c, d of arity 2, we write $c \subseteq d$ to say that any two objects x, y satisfying the condition $c \uparrow\downarrow xy$ satisfy also $d \uparrow\downarrow xy$.

We notice that there are other objects having arity to 2, besides binary relations and simple operations. Among them *finite systems* have particular relevance. They form a collection denoted by the symbol Sif. This collection enjoys the property of being *extensional*, *i.e.* the following axiom holds:

Axiom of extensionality for finite systems
If $a \in Sif$, $b \in Sif$, $a \subseteq b$, $b \subseteq a$, then $a = b$.

When $S \in Sif$ and $S \uparrow\downarrow xy$ we say that x is an index of S and y is associated to the index x in the system S.

We introduce now the notion of *univocal finite system*. If only one object is associated to each index of the finite system S, we say that S is a *univocal finite system*: univocal finite systems form a collection denoted by the symbol $Siuf$. Given a system $S \in Siuf$, we will often use instead of $S \uparrow\downarrow hx$ the notation:

$$S_h = x.$$

This notation is inspired by the current notation for ordered pairs, ordered triples, ... ordered n-tuples, which, are special finite systems, as we will see soon.

The simplest among the finite systems is the empty finite system. It is denoted by the symbol $\emptyset_2$, and characterised by the property of having no object as index. In other words, $\emptyset_2$ is an empty object of arity to 2, and obviously $\emptyset_2 \subseteq S$ holds for each $S \in Sif$.

After the empty system, the "smallest" finite systems are the singular systems, which have only one index and only one corresponding object. The following axiom guarantees the existence of these systems:

Axiom of singular systems
Given (not necessarily distinct) objects a, b, there exists $S \in Siuf$ such that for all x, y

$$S \uparrow\downarrow xy \text{ if and only if } x = a,\ y = b.$$

The singular system identified by this axiom is denoted by the symbol $\binom{a}{b}$; so, for any finite system S, instead of writing $S \uparrow\downarrow ab$ we can write $\binom{a}{b} \subseteq S$.

Singular systems of the kind $\binom{1}{x}$ are called 1-tuples, and they are also denoted by the symbol $[x]$. The collection of all 1-tuples is denoted by the symbol V^1. We denote by the symbol V^0 the set having just one element, the empty system $\emptyset_2$ (also called 0-tuple).

A relevant operation which can be performed on finite systems is the binary union of finite systems, denoted by the symbol $Unbs$. It is a binary operation characterised in the following way: $Unbs$ can be performed on all finite systems, always gives a finite system as result, and if

$$S = Unbs\, S_1 S_2$$

then for all objects x, y,

$$S \mathrel{\uparrow\downarrow} xy$$

holds if and only if at least one of the relations $S_1 \mathrel{\uparrow\downarrow} xy$, $S_2 \mathrel{\uparrow\downarrow} xy$ holds.

In analogy with the usual set-theoretic notation, we write $S_1 \cup S_2$ instead of $Unbs\, S_1 S_2$. We also use the following notation

$$\begin{pmatrix} a & b \\ x & y \end{pmatrix} \text{ instead of } \begin{pmatrix} a \\ x \end{pmatrix} \cup \begin{pmatrix} b \\ y \end{pmatrix},$$

$$\begin{pmatrix} a & b & c \\ x & y & z \end{pmatrix} \text{ instead of } \begin{pmatrix} a & b \\ x & y \end{pmatrix} \cup \begin{pmatrix} c \\ z \end{pmatrix} \ldots, \text{ etc.}$$

The systems of the kind $\begin{pmatrix} 1 & 2 \\ x & y \end{pmatrix}$ are the ordered pairs, briefly denoted by the symbol (x, y). The collection of all ordered pairs is denoted by the symbol V^2. Similarly, one constructs the ordered triples (x, y, z), whose collection is denoted by the symbol V^3, and all other n-tuples. For each natural number n the collection of all *n-tuples* is denoted by the symbol V^n.

Having so recalled the main objects of the Basic Theory presented in the note [4] of the bibliography, we have to introduce a new quality that allows for distinguishing between theory and meta-theory. This quality is denoted by the symbol $Qmetaqual$, or shortly by QMQ, and the objects enjoying this property are called *metaqualities* (or *metaproperties*). Essentially, metaproperties play in the meta-theory the same role that qualities play in the theory exposed up to now. Thereby if μ is a metaproperty and x is any object we write

$$\mu\, x$$

to say that *x enjoys the metaquality μ.*
We remark that metaqualities are well distinct from qualities, and, first of all, if μ is a metaquality then its arity $Ar\,\mu$ is zero. It follows that, even if x enjoys the metaquality μ,

$$(Rfond\,1)\mu x$$

does not hold.
Hence, whilst $Rfond\,1$ describes the action of qualities, it does not describe the action of metaqualities, which, although enjoying the quality QMQ and belonging to the collection V like every other object, nevertheless occupy a separate place. In a rather anthropomorphic description, we could say that metaqualities tell us of the other objects in V and deeply analyse their *actions*, as we shall

see soon, but the other objects in V, although able to denote metaqualities and also pairs, triples, quadruples, sets of metaqualities, do not manage to describe the *action* of these "higher level" objects.

This separation between qualities and metaqualities can also be highlighted by considering the operation $Gpred$, generator of predicates, already introduced in the lecture. To this aim we state the following axiom:

If $Ar\,x = 0$, then $Gpred\,x = ((Gpred\,Rid)\,x)\,x$,

where Rid is the binary relation of identity (*i.e. $Rid\,xy$* means that x exactly coincides with y, *i.e.* x is precisely *the same thing* as y).

This fact shows that, whilst $Gpred\,z$ gives a lot of information on the action of z whenever $Ar\,z$ is a positive natural number, it gives instead no information on the action of z when z is a metaquality. In the latter case $Gpred\,z$ simply says to us that z is identical to itself.

A relevant role among metaqualities is played by the triple of metaqualities $MT = (Mver, Mfals, MVF)$, which "classifies" many propositions, by distinguishing true propositions and false propositions. The triple MT enjoys the following properties.

1. There is no object x satisfying simultaneously both conditions $Mver\,x$ and $Mfals\,x$.

   This axiom essentially corresponds to the principle of non-contradiction.

2. An object x enjoys the metaproperty MVF if and only if x enjoys one of the metaqualities $Mver$, $Mfals$.

3. If an object x enjoys the metaquality MVF, then x enjoys also the quality $Qprop$.

We do not not impose to all propositions to enjoy the metaproperty MVF, since we do not want to exclude *a priori* the existence of "ambiguous, obscure, less sensible" propositions, or the possibility that there are true or false propositions that evade the "judgment capability" of the triple MT. Informally speaking, we could say that the triple MT has an enormous judgment capability: It is able to evaluate all propositions of traditional mathematics and many others besides. However we do not expect that MT judges every possible proposition true or false; in other terms $Mver\,\omega$ ensures that ω is true in the "natural" sense of the word, but we do not want to exclude that there are true (or false) propositions that do not enjoy the metaproperty MVF. Also in this aspect we try and follow Shakespeare's warning.

However, we can give some axioms that substancially express the great strength of the triple MT, by reconsidering the operation $Gpred$, introduced in the lecture addressed at the Pontifical Academy.

Axiom 1: If x is an object such that $Ar\,x = 1$, then, for all y, $(Gpred\,x)y$ enjoys the metaquality MVF; moreover the following conditions are equivalent:

$Mver\,((Gpred\,x)\,y)$, and $\quad x \uparrow\downarrow y$.

Axiom 2: If $Ar\,x = 2$, then, for all y, z, the proposition $\quad ((Gpred\,x)y)z$ enjoys the metaquality MVF; moreover the following conditions are equivalent:

$Mver\,(((Gpred\,x)\,y)\,z)$, and $\quad x \uparrow\downarrow yz$.

Axiom 3: If $Ar\,x = 3$, then, for all y, z, t, the proposition $(((Gpred\,x)y)z)t$ enjoys the metaproperty MVF, moreover the following two conditions are equivalent:

$Mver\,((((Gpred\,x)\,y)\,z)\,t)$, $\quad x \uparrow\downarrow yzt$.

Similarly one proceeds for objects with arities 4, 5, ..., etc.

In order to avoid an utterly heavy notation, it is convenient to introduce the quoted notation, and write for instance,

"$x \uparrow\downarrow y$" instead of $(Gpred\,x)\,y$.
"$x \uparrow\downarrow yz$" instead of $((Gpred\,x)\,y)\,z$.
"$x \uparrow\downarrow yzt$" instead of $(((Gpred\,x)\,y)\,z)\,t$.
Similarly one proceeds for objects of arity greater than 3.

Moreover, if q is a quality, we also write "qx" for "$q \uparrow\downarrow x$"; if C is a collection we write "$x \in C$" for "$C \uparrow\downarrow x$"; if f is a simple operation we write "$f\,x = y$ for "$f \uparrow\downarrow xy$"; if r is a binary relation we write "$r\,xy$" for "$r \uparrow\downarrow xy$"; "$x = y$" for "$Rid \uparrow\downarrow xy$"; etc.
Notice that the notation "$(f\,x) = y$" denotes the proposition "$Rid \uparrow\downarrow (f\,x)y$" which is different from "$f\,x = y$"$=$ "$f \uparrow\downarrow xy$".

More generally, the link between the operation $Gpred$ and the operation $Rfond$ is granted by the following axiom:
For each natural number h greater than 0 and every object x such that $Ar\,x$ is equal to h:
$Gpred\,x \;=\; (Gpred\,(Rfond\,h))\,x.$

Finally, we notice that the following theorem is a straightforward consequence of the axioms exposed in the lecture addressed to the Pontifical Academy together with the axioms stated above.

Theorem. If α is a self-negating proposition, *i.e.* if α verifies the condition

$$\alpha \;=\; \text{``}Qfals\,\alpha\text{''}$$

then α does not enjoy the quality QVF, but it enjoys the metaproperty $Mfals$.

Intuitively this theorem could be understood by saying that the Liar of the classical antinomy cannot be declared as such by anybody located at its same level, whereas he is recognised as a liar by whom puts himself at a higher level.

The comparison between the triple $MT = (Mver,\ Mfals,\ MVF)$ and the triple $QT = (Qver,\ Qfals,\ QVF)$ can be summarized by saying that if a proposition ω enjoys the property $Qver$, then it also enjoys the metaproperty $Mver$, if it enjoys the property $Qfals$, then it also enjoys the metaproperty $Mfals$, hence if it enjoys the property QVF then it also enjoys the metaproperty MVF. In this way a sort of subordination of the triple QT to the triple MT is established.

We have seen that in passing from higher order predicates to propositions one needs an increasing number of operations; however, the use of systems allows for a reduction of this number. To this aim, we introduce the simple operation *Sival*, the *evaluation of predicates by means of systems*, on which we impose the following axioms:

a) If p is a predicate whose order is the natural number h, then $Sival\,p$ is a simple operation acting on the elements of V^h.

b) In particular one has:

$(Sival\,p)\emptyset_2 = p$, when $Predord\,p = 0$

$(Sival\,p)[x] = p\,x$, when $Predord\,p = 1$

$(Sival\,p)(x,y) = (p\,x)y$, when $Predord\,p = 2$

$(Sival\,p)(x,y,z) = ((p\,x)y)z = (Sival(p\,x))(y,z)$, when $Predord\,p = 3$.

c) When h is a natural number > 3 and $Qpred\,p$, $Predord\,p = h$, then

$(Sival\,p)(x_1,\ldots,x_h) = ((Sival\,(p\,x_1))(x_2,\ldots,x_h)$.

The introduction of the operation *Sival* allows the introduction of a quite simple notation to describe the actions of objects having high arities. Namely, if $S \in V^h$ and $Ar\,z = h$, we write:

"$z \downarrow\downarrow\uparrow\uparrow S$" instead of $(Sival\,(Gpred\,z))\,S$,

and we write:

$z \downarrow\downarrow\uparrow\uparrow S$ instead of $Mver$ "$z \downarrow\downarrow\uparrow\uparrow S$".

In this case, following the indications given in the lecture addressed to the Pontifical Academy, we can say that if the proposition "$z \downarrow\downarrow\uparrow\uparrow S$" enjoys the metaquality $Mver$, then it also enjoys the quality $Qver$, whereas it could enjoy the metaquality $Mfals$ without enjoying the quality $Qfals$.

We can now introduce the first operations acting on propositions, namely the binary operations *vel*, *et* together with the simple operation *non*. Following the usual notation we then write:

$a \vee b$ instead of *vel ab*; $a\&b$ instead of *et ab*; $\neg\,a$ instead of *non a*.

The behaviour of the triple MT with respect to these operations is characterised by the following axiom:

If at least one of the propositions a, b enjoys the metaquality $Mver$, then $Mver\,a \vee b$.

If both propositions a, b enjoy the metaquality $Mfals$, then $Mfals\, a \vee b$.
If at least one of the propositions a, b enjoys the metaquality $Mfals$, then $Mfals\, a\&b$.
If both propositions a, b enjoy the metaquality $Mver$, then $Mver\, a\&b$.
If $Mver\, a$, then $Mfals\,(\neg a)$; if $Mfals\, a$, then $Mver\,(\neg a)$.

We introduce now the first axioms on the binary operations $\exists$, $\forall$.
Given a collection C and a predicate p of order 1, the binary operations $\exists$, $\forall$ act on C, p, their results $\exists Cp$ and $\forall Cp$ are propositions, and their behaviour with respect to the triple MT is characterised by the following axioms:

a) If px enjoys $Mver$ for at least an element $x \in C$, then $\exists Cp$ enjoys $Mver$.

b) If px enjoys $Mfals$ for every element $x \in C$, then $\exists Cp$ enjoys $Mfals$.

c) If px enjoys $Mfals$ for at least an element $x \in C$, then $\forall Cp$ enjoys $Mfals$.

d) If px enjoys $Mver$ for every element $x \in C$, then $\forall Cp$ enjoys $Mver$.

We notice that, on the grounds of these axioms, if the collection C is the empty set $\emptyset$, then we have $Mfals\,(\exists \emptyset p)$ and $Mver\,(\forall \emptyset p)$ for each first order predicate p.

After introducing the logical operations on propositions and predicates (which in the lecture to the Pontifical Academy have been introduced as operations whose results are propositions) we can turn to the calculus on symbols of predicates and to the interpretations of formulae.
So we consider four collections: the collection *Spred* of symbols of predicates, the collection *Svar* of symbols of variables, the collection *Scost* of symbols of constants, and the collection *Form* of formulae. Besides the collections *Svar* and *Scost* we consider also the collection *Lett*, the collection of *letters*, which includes all elements of *Svar* and *Scost*.
We let the operation *Predord* be performable also on symbols of predicates, and we introduce the binary operation *Sat*, *i.e.* the saturation of predicates, satisfying the following axiom:

If l is a n-tuple of letters, $p \in Spred$, and $Predord\, p = m$, then the binary operation *Sat* can be performed on p, l and one has $Sat\, pl \in Form$.

The binary operations $\vee$ and $\&$, and the simple operation $\neg$ can also be performed on formulae. Moreover, the operations $\exists$, $\forall$ can be performed on x, f whenever x is a symbol of variable and f is a formula.

The link between logical operations performed on formulae and logical operations performed on propositions and predicates is provided by the simple operation *Inter*, whose behaviour is characterized by the following axioms:

a) Let C be a collection, s a symbol of predicates, p a predicate, $l = (l_1, \ldots, l_n)$ a n-tuple of letters, $(x_1, \ldots, x_n)$ a n-tuple of arbitrary objects. If $n =$

Predord s $=$ *Predord p*, f $=$ *Sat sl*, and the system S $=$ $\binom{l_1,\dots,l_n}{x_1,\dots,x_n}$ is univocal, then

$$Inter \text{ acts on the triple } \left(C, \binom{s}{p}, S\right)$$

and the result is a simple operation acting on f and satisfying the condition:

$$\left(Inter \left(C, \binom{s}{p}, S\right)\right) f = (Sival\, p)(x_1, \dots, x_n).$$

b) Let C be a collection, let S, L be univocal systems, let f be a formula, and let ω be a proposition. Assume that S', L' are univocal systems verifying the conditions $S' \supseteq S$, $L' \supseteq L$, and such that all indices of S' are symbols of predicates, all indices of L' are letters, and to each index of S' is associated in S' a predicate of the same order. Then, if
$$(Inter\,(C,\,S,\,L))\,f = \omega.$$
also
$$(Inter\,(C,\,S',\,L'))\,f = \omega.$$

Remark: Axiom (b) essentially tells us that whenever the information given by the systems S, L are sufficient to give the interpretation of the formula f, further information have no influence at all on this interpretation.

Given two formulae f, g, we can consider the formulae $f \vee g$, $f \& g$, $\neg f$.

The interpretation of these formulae is characterised by the following axiom:

c) If α, β are propositions verifying the conditions
$$(Inter\,(C,\,S,\,L))\,f = \alpha, \quad (Inter\,(C,\,S,\,L))\,g = \beta,$$
then
$$(Inter\,(C,\,S,\,L))\,f \vee g = \alpha \vee \beta$$
$$(Inter\,(C,\,S,\,L))\,f \& g = \alpha \& \beta$$
$$(Inter\,(C,\,S,\,L))\,\neg f = \neg \alpha.$$

Together with interpretations giving results that are propositions, when performed on a formula f, we can consider interpretations giving results that are first order predicates, when performed on pairs (x, f), where x is a symbol of variables and f is a formula. Namely:

If x is a symbol of variables, and for every object t there exists a proposition ω_t such that
$$\left(Inter\,\left(C,\,S,\,L \cup \binom{x}{t}\right)\right) f =, \omega_t,$$
then there exists a first order predicate p such that

$$(Inter\,(C,\ S,\ L))(x, f) \ = \ p,$$

and for every object t

$$pt \ = \ \omega_t.$$

The existence of such interpretations of pairs (x, f), which give first order predicates as results, allows us to interpret formulae of the type as $\exists\,xf$, $\forall\,xf$ in the following way:

If $p = (Inter\,(C,\ S,\ L))\,(x, f)$

 then

$$\exists\,Cp \ = \ (Inter\,(C,\ S,\ L))\,(x, f),$$
$$\forall\,Cp \ = \ (Inter\,(C,\ S,\ L))\,(x, f).$$

Finally, we remark that great relevance is given to interpretations connected with sets and collections in the traditional study of predicate calculus. In order to recover such results it is convenient to assume the following axiom:

Given an arbitrary collection C and a natural number n, there exists a predicate p of order n such that, for all $x \in V^n$

$$(Sival\,p)\,x \ = \ \text{``}x \in C\text{''}.$$

Among the various open problems I would like to point out in particular the "technical-cultural" problem of finding sufficiently strong axioms for the triple QT that make its behaviour as close as possible to that of the triple MT, of course avoiding contradictions.

Another group of problems could be that of finding whole families of triples, variously connected to one other, and resembling the triples MT, QT.

Finally, besides triples of qualities, one could consider *Boolean operators*, *i.e.* simple operations f that act on some propositions, give the numbers 0, 1 as results, and verify the following compatibility conditions with respect to the triple MT:

$$\text{If } f\,\omega \ = \ 1, \text{ then } Mver\,\omega; \text{ if } f\,\omega \ = \ 0, \text{ then } Mfals\,\omega.$$

Bibliography

[1] E. DE GIORGI-M. FORTI-G. LENZI-V.M.TORTORELLI:[‡] La qualità QVF e la teoria base dei fondamenti della matematica, Preprint Scuola Normale Superiore di Pisa, 1994, work in progress.

[‡]Editor's note: published with the title: "Calcolo dei predicati e concetti metateorici in una teoria base dei Fondamenti della Matematica", in Atti Accad. Naz Lincei, Cl. Sci. Fis. Mat. Nat., **(9) 5**, (1995), 79–92.

[2] M.Clavelli-E. De Giorgi-M. Forti-V. M. Tortorelli: *A self-reference oriented theory for the foundations of mathematics*, Contributions en l'honneur de J.-L. Lions, Paris 1986.

[3] E. De Giorgi-M. Forti: *Una teoria quadro per i fondamenti della matematica*, Atti Accademia Nazionale Lincei, Rendiconti Classe Scienze Fisiche, Matematiche, Naturali, serie 8, volume 79 (1985), 55–67.

[4] E. De Giorgi-M.Forti-G. Lenzi: *Una Proposta di Teorie Base dei Fondamenti della Matematica*, Atti Accademia Nazionale Lincei, Rendiconti Classe Scienze Fisiche, Matematiche, Naturali, serie 9, volume 5 (1994), 11–22.

[5] M.Forti-F.Honsell: *Models of Self-Descriptive set Theories*, in: Partial Differential Equations, Essays in honour of Ennio De Giorgi, editors F.Colombini et al. Birkhäuser, Boston 1989, 473–518.

[6] G.Lenzi: *Estensioni contraddittorie della teoria ampia*, Atti Accademia Nazionale Lincei, Rendiconti Classe Scienze Fisiche, Matematiche, Naturali, serie 8, volume 83 (1989), 13–28.

[7] G.Lenzi-V.M.Tortorelli: *Introducing Predicates into a Basic Theory for the Foundations of Matematics*, Preprint Scuola Normale Superiore di Pisa, luglio 1989.

Introducing basic theories for the Foundations of Mathematics[‡]

Ennio De Giorgi, Marco Forti, Giacomo Lenzi[*]

Summary. Some basic theories of the Foundations of Mathematics are proposed, which take as primitive concepts the notions of natural number, collection, quality, operation and relation; the operations and relations we consider can be more or less complex: the natural number indicating the degree of complexity is called *arity*. A high degree of self-reference is reached in the theories we consider.

KEYWORDS: Foundations; Operations; Relation; System; Basic–theory.

0. INTRODUCTION

Recently, various theories of the Foundations of Mathematics and a presentation of the various mathematical concepts close to the common intuition and to the mathematical tradition have been proposed with the aim of reaching a high degree of self–reference (see [8,10–12, 14, 18, 19]). In particular, in these theories the mathematical concepts were not reduced to a unique primitive concept, but the aim was rather of introducing several ones (number, quality, relation, operation, collection) related by axioms as close as possible to the common usage and to the mathematical tradition, and such as to guarantee the highest possible degree of self–reference of the theory itself. A critical analysis of these theories has suggested the opportunity to separate an initial, very small core (which we will call "basic theory") and then to "engraft" on it all the known mathematical theories. In the basic theories proposed in this *Note* we postulate for every integer n the existence of objects having the arity n (intuitively, the arity of an object is the index of its complexity). For the reader looking for philosophical interpretations of the basic theory this is equivalent to postulating the existence of very complex "realities".

The basic theories have the advantage of being quite simple and their consistency (that is, freedom from contradiction) cannot be posed in doubt, at least by those who believe in the consistency of the most elementary part of arithmetic. This reliability is a valid starting point for discussion on the consistency of the more advanced mathematical theories which can be engrafted on the basic theories. In fact, once the axioms of a basic theory are assumed, it should be easy to "engraft" on this theory, in a natural way, the most complex mathematical theories (for instance the various set theories, modal logics, standard and non–standard Analysis, etc.).

[‡]Editor's note: translation into English of the paper "Una proposta di teorie base dei Fondamenti della Matematica", published in Atti Accad. Naz Lincei, Cl. Sci. Fis. Mat. Nat., **(9) 5** (1994), 11–22.

[*]Note presented by E. De Giorgi to the Accademia dei Lincei on June 18, 1993.

Maybe it will be possible to engraft on the same basic theories also non–strictly mathematical theories; so, in order to make collaboration to this research easier for scholars of various scientific and human disciplines, we have tried to present the basic theory in an informal language, as close as possible to common language; in another paper (see [15]) it is shown that the axioms introduced here in an informal way can be formalized in the language of the first order Predicate Calculus. Finally, there is hope, maybe remote but not manifestly groundless, that some interesting, original, scientific or philosophical theories may "bud" on the basic theories exposed here. In this vein it must be noticed that the basic theories are "qualitatively open–ended" theories which do not exclude the existence of objects "qualitatively" different from the fundamental objects introduced by the theory, that is qualities, collections, natural numbers, relations and operations. We can say that the general idea which inspires this work is the essentially anti–reductionist philosophy expressed by Shakespeare with the words of Hamlet to Horace "there are more things in heaven and earth than are dreamt of in your philosophy" (Hamlet, Act I, Sc. V). Finally, we remark that, even if this *Note* takes into account many previous logical–mathematical investigations, it is written in a completely self–contained way, so that it can be read and thoroughly understood also by those who ignore all classical and modern foundational theories cited in the Bibliography [1–7, 9, 13, 16, 17, 20–23] including the foundational theories exposed in [8, 10–12, 14, 18, 19].

1. Qualities, collections, relations and operations

In the informal exposition of our theory we call *objects* all the entities which the theory talks about, we assume as *primitive* the notions of *quality, collection, relation* (binary, ternary, etc.) and *operation* (simple, binary, etc.) and we adopt the following notation:

if q is a *quality* and x is an *object*, we write $q\,x$ for "*x enjoys the quality q*";

if C is a *collection* and x is an *object* we write $x \in C$ for "*x belongs to the collection C*", or "*x is an element of the collection C*"; we write $x \notin C$ to say that x does not belong to the collection C;

if r is a *binary relation* and x and y are two objects, we write $r\,x\,y$ for "*x is in the relation r with y*; for instance, the phrase "Rome is more populated than Frascati" expresses a binary relation r between x, that is Rome, and y, that is Frascati;

if ρ is a *ternary relation* and x, y and z are *objects*, we write $\rho\,x\,y\,z$ for "*x is in the relation ρ with y and z*"; an example is the phrase "Francesco likes cheese more than ham" where x is Francesco, y is cheese and z is ham;

if τ is *quaternary relation* and x, y, ξ and η are *objects*, we write $\tau\,x\,y\,\xi\,\eta$ for "*x is in the relation τ with y, ξ and η*; an example is the phrase "the distance between Roma and Milano is smaller than the one between Paris and Moscow", with $x =$ Roma, $y =$ Milano, $\xi =$ Paris and $\eta =$ Moscow.

Likewise, for a generic n–ary relation α we will write $\alpha\, x_1 \ldots x_n$ to say that the objects are in the relation α.

We turn now to operations, of which we have important examples also in elementary arithmetic, such as the binary operations of addition of two numbers, subtraction and multiplication, or the simple operation which associates to a number n it successor $n + 1$. We note that, in general, for operations with more than one argument, the result depends on the order of the arguments; consider, for instance, the binary operation of power of natural numbers, Pow, such that $Pow\, m\, n = m^n$; we have $Pow\, 2\, 3 = 2^3 = 8$ which is different from $Pow\, 3\, 2 = 3^2 = 9$.

If f is a *simple operation* and x and y are *objects*, we write $y = f\, x$ for *"the operation f transforms x in y"*;

if φ is a *binary operation* and x, y and z are *objects*, we write $z = \varphi\, x\, y$ for *"the operation φ takes first the object x, then the object y and gives z as a result"*.

Likewise, we proceed for the h–ary operations with $h > 2$ by writing, for every h–ary operation ψ, $w = \psi\, x_1 \ldots x_h$ if w is the result of ψ calculated on the arguments $x_1 \ldots x_h$.

2. First fundamental objects

Among the objects of different kind which we consider, we mention now some "fundamental objects" and we give some axioms which specify their role in the basic theory. First we introduce some qualities corresponding to the kinds of objects considered in §1; this correspondence is stated by the axioms A1–A7.

First we introduce the quality of being a collection, $Qcoll$:

Axiom A1. – *$Qcoll$ is a quality; for every object C we have $Qcoll\, C$ if and only if C is a collection.*

We proceed likewise for the other kinds of objects; the quality of being a binary relation is $Qrelb$:

Axiom A2. – *$Qrelb$ is a quality; for every object r we have $Qrelb\, r$ if and only if r is a binary relation.*

The quality of being a ternary relation is $Qrelt$:

Axiom A3. – *$Qrelt$ is a quality; for every object ρ we have $Qrelt\, \rho$ if and only if ρ is a ternary relation.*

The quality of being a quaternary relation is $Qrelq$:

Axiom A4. – *$Qrelq$ is a quality; for every object τ we have $Qrelq\, \tau$ if and only if τ is a quaternary relation.*

The quality of being a simple operation is $Qops$:

Axiom A5. – *$Qops$ is a quality; for every object f we have $Qops\, f$ if and only if f is a simple operation.*

The quality of being a binary operation is $Qopb$:

AXIOM A6. – *Qopb is a quality; for every object φ we have Qopb φ if and only if φ is a binary operation.*

Finally, we have the quality of being a quality, called *Qqual*, characterized by the following axiom:

AXIOM A7. – *Qqual is a quality; for every object q we have Qqual q if and only if q is a quality.*

We remark that axiom A7 implies *Qqual Qqual*, that is the quality *Qqual* is enjoyed by itself; this is the first example of *self–reference* in our theory.

The next step is the introduction of some *fundamental relations* which rule the behavior of the objects of lowest arity. First we introduce the *fundamental relations of unary objects, $Rfun$*:

AXIOM A8. – *Rfun is a binary relation. For every quality q and for every object x, we have Rfun q x if and only if q x.*
For every collection C and for every object x we have Rfun C x if and only if $x \in C$.

Like for unary objects we introduce the *fundamental relation of binary objects, $Rfbin$*:

AXIOM A9. – *Rfbin is a binary relation; for every binary relation r and for every x, y we have Rfbin r x y if and only if r x y.*
For every simple operation f and every x, y we have Rfbin f x y if and only if $y = f$ x.

We note that a simple operation is actually a binary object in that it relates two objects, the argument and the result. Likewise we consider binary operations as ternary objects, etc.

The third fundamental relation is the one of ternary objects, *Rfter*:

AXIOM A10. – *Rfter is a quaternary relation; for every ternary relation ρ and for every x, y, z we have Rfter ρ x y z if and only if ρ x y z.*
For every binary operation φ and for every x, y, z we have Rfter φ x y z if and only if $z = \varphi$ x y.

After introducing *Rfbin* and *Rfter*, the functionality of simple and binary operations can be expressed by the following axiom:

AXIOM A11. – *If f is a simple operation then Rfbin f x y together with Rfbin x y' implies $y = y'$. If φ is a binary operationthen Rfter φ x y z together with Rfter φ x y z' implies $z = z'$.*

Even more complex relations than the previous ones will be introduced later, after natural numbers; in this first chapter we introduce only some more, very simple objects, beginning with the identity operation, *Id*:

AXIOM A12. – *Id is a simple operation; for every object x we have Id $x = x$.*

Finally, we introduce the universal collection, V, the empty collection, $\emptyset$, and the collection of all collections, *Coll*:

AXIOM A13. – *V, $\emptyset$ and Coll are collections. For every object x we have $x \in V$, $x \notin \emptyset$, whereas $x \in Coll$ if and only if x is a collection.*

From the previous axioms we get the self–referential relations $Coll \in Coll$, $V \in V$ and the "mutually referential" relations $V \in Coll$, $Coll \in V$.

Before stating the axiom of extensionality of collections, we introduce the usual notation for inclusion: if C, C' are collections, we write $C \subseteq C'$ if every object belonging to C belongs to C' as well. Now we give the axiom of extensionality:

AXIOM A14. – *if C and C' are collections and we have at the same time $C \subseteq C'$, $C' \subseteq C$, then $C = C'$.*

3. ELEMENTARY ARITHMETIC

In order to insert in the basic theory the most elementary part of Arithmetic, we introduce first the collection of all natural numbers, which, following the common use, will be denoted by $\mathbf{N}$; we begin with stating the axiom:

AXIOM B1. – $\mathbf{N}$ *is a collection.*

Now we establish some useful notations to write concisely axioms concerning operations. If f is a simple operation and x is an object, we write $f \uparrow x$ for *"there exists y such that $y = f\ x$"*; if φ is a binary operation and x and y are objects, we write $\varphi \uparrow x\ y$ for *"there is z such that $z = \varphi\ x\ y$"*, finally if φ is a binary operation and z is an object, we write $\varphi \downarrow z$ for *"there exist x and y such that $z = \varphi\ x\ y$"*.

Next we describe some arithmetic operations (addition, subtraction, multiplication) and we begin with the *addition of two natural numbers*, which will be called *Nadd*. We give the axioms:

AXIOM B2. – *Nadd is a binary operation.*

AXIOM B3. – *Nadd $\uparrow$ $x\ y$ if and only if $x \in \mathbf{N}$ and $y \in \mathbf{N}$. If Nadd $\downarrow$ z then $z \in \mathbf{N}$.*

When dealing with the operation *Nadd* we adopt the usual *notation*: $x + y = Nadd\ x\ y$.

The following axioms express some elementary properties of addition:

AXIOM B4. – *(associativity)* $x + (y + z) = (x + y) + z$.

AXIOM B5. – *(commutativity)* $x + y = y + x$.

AXIOM B6. – *(property of the number zero) There is $z \in \mathbf{N}$ such that for every $x \in \mathbf{N}$ we have $x = z = x$.*

The z of the previous axiom turns out to be unique by axiom B5 and will be denoted by the usual symbol 0.

The second fundamental operation on natural numbers is the *multiplication*, which will be denoted by *Nmult*. This operation satisfies the following axioms:

AXIOM B7. – *Nmult is a binary operation.*

AXIOM B8. – *Nmult $\uparrow$ $x\ y$ if and only if $x \in \mathbf{N}$ and $y \in \mathbf{N}$. If Nmult $\downarrow$ z then $x \in \mathbf{N}$.*

When dealing with the operation *Nmult* we adopt the usual *notation*: $xy = Nmult\ x\ y$. The following axioms express some elementary properties of multiplication:

Axiom B9. – *(associativity)* $(xy)z = x(yz)$.

Axiom B10. – *(commutativity)* $xy = yx$.

Axiom B11. – *(distributivity with respect to addition)* $x(y+z) = (xy)+(xz)$.

Axiom B12. – *(property of the number one)* *There is* $u \in \mathbf{N}$ *such that for every* $x \in \mathbf{N}$ *we have* $ux = x$.

The number u of the previous axiom turns out to be unique by axiom B10, and it will be denoted by the usual symbol 1. As usual we let $2 = 1 + 1$, $3 = 2 + 1$, etc.

Besides the operations of addition and multiplication, the integers have a natural ordering, in which 0 precedes 1, 1 precedes 2, etc; this ordering is referred to in the phrase *"x is less than or equal to y"*. Hence in our theory we consider the relation *Nord* (the *natural ordering* of natural numbers), which satisfies the following axioms:

Axiom B13. – *Nord is a binary relation. If Nord x y then* $x \in \mathbf{N}$ *and* $y \in \mathbf{N}$.

Following the common usage, we write $x \leq y$ for *Nord x y*. We write also $x < y$ for $x \leq y$ and $x \neq y$.

Two first properties of *Nord* are the following:

Axiom B14. – *(linearity)* *For every* x *and* y *belonging to* $\mathbf{N}$ *we have* $x \leq y$ *or* $y \leq x$.

Axiom B15. – *(antisymmetry)* *If* $x \leq y$ *and* $y \leq x$ *then* $x = y$.

Now we give an axiom which relates *Nord* with *Nadd*:

Axiom B16. – *If* $x, y \in \mathbf{N}$ *then* $x < y$ *if and only if there is* $p \in \mathbf{N}$ *such that* $p \neq 0$ *and* $y = x + p$.

From the previous axioms it follows:

1) *Nord* is transitive, that is for every $x, y, z \in \mathbf{N}$ if $x \leq y$ and $y \leq z$ then $x \leq z$,

2) for every $x, y \in \mathbf{N}$ we have $x \leq y$ if and only if there is $h \in \mathbf{N}$ such that $y = x + h$,

3) 0 is the *least* element of $\mathbf{N}$ with respect to *Nord*, that is $0 \leq x$ for every $x \in \mathbf{N}$.

Finally we postulate that one is the immediate successor of zero:

Axiom B17. – $0 < 1$ *and for no* $z \in \mathbf{N}$ *we have* $0 < z < 1$.

Besides the addition of natural numbers we can consider the subtraction in the context of natural numbers, *Nsub*, characterized by the following axioms:

Axiom B18. – *Nsub is a binary operation. Nsub* $\uparrow$ *x y if and only if* $x \in \mathbf{N}$, $y \in \mathbf{N}$ *and* $y \leq x$. *If Nsub* $\downarrow$ *z then* $z \in \mathbf{N}$.

Axiom B19. – $z = Nsub\ x\ y$ *if and only if* $x = y + z$.

We will adopt the usual notation *Nsub* $x\ y = x - y$. We note that the theory of natural numbers exposed so far is very weak (we do not have, for instance, the power and the induction axioms) but it is strong enough to exclude finite models. One might consider even weaker theories, having finite models, but they are not a comfortable basis for engrafting upon the rest of Arithmetic.

4. Arities and mutual references among fundamental relations

The introduction in §3 of the first elements of Arithmetic allows us to unify and extend the concepts introduced in §2 and to formalize the notion of "complexity" (or "arity") informally used in §§1,2. First we formalize the general notions of relation and operation (of any arity) by means of the two qualities $Qrel$ and Qop:

AXIOM C1. – *$Qrel$ and Qop are qualities. $QRel\,x$ if and only if x is a relation. $Qop\,x$ if and only if x is an operation.*

Let us turn now to the axiom on arity by introducing the operation Ar (and in particular we determine the arity of the objects considered in §§1,2,3):

AXIOM C2. – *Ar is a simple operation which enjoys the following properties:*

a) *For $x \in \mathbf{N}$ we have $Ar\,x = 0$.*

b) *For every quality q and every collection C, $Ar\,q = Ar\,C = 1$.*

c) *Let r be a relation. Then $Ar\,r = h$ if and only if r is an h-ary relation. In particular $Ar\,r = 2(3,4)$ if and only if r is binary (ternary, quaternary respectively).*

d) *Let f be an operation. Then $Ar\,f = h + 1$ if and only if r is an h-ary operation. In particular $Ar\,f = 2(3)$ if and only if f is simple (binary, respectively).*

We remark that the axioms introduced here exclude that an object may have two arities simultaneously, and may be, for instance, both a collection and an operation, or both a binary relation and a ternary relation. Instead, so far we do not exclude that an object may be both a quality and a collection, or both a binary relation and a simple operation, etc. We do not even exclude that there are objects whose arity is not a natural number. Finally we do not exclude that there are objects of arity zero other than natural numbers, that there are objects of arity one other than qualities and collections, etc. This freedom may be useful for "engrafting" on the basic theory the various branches of Mathematics, and possibly of other, non–mathematical theories. For instance, this freedom might happen to be useful for those who want to engraft infinitary logics on the basic theory.

Now we would like to realize theories which, despite being quite simple, be more rich on the complexity ground than the usual set theories, which essentially admit only objects with arity one, or than theories like Lambda–Calculus or combinatory or relational algebras, which essentially deal with objects of arity two; to this aim we stipulate that for every natural number h there is at least one object of arity h, and that objects with higher arity describe the action of objects with lower arity. So we introduce the operation which generates fundamental relations, $Rfond$:

Axiom C3. – *$Rfond$ is a simple operation which enjoys the following properties:*

   a) For every natural number $h > 0$, $Rfond\,h$ exists, and is a relation of arity $h = 1$.

   b) $(Rfond\,h)\,\alpha\,x_1\ldots x_h$ implies $Ar\,\alpha = h$.

   c) $Rfond\,1 = Rfun$, $Rfond\,2 = Rfbin$ and $Rfond\,3 = Rfter$.

   d) For every relation ρ of arity h and for every $x_1\ldots x_h$ we have: $\rho\,x_1\ldots x_h$ if and only if $(Rfond\,h)\,\rho\,x_1\ldots x_h$.

   d) For every operation f of arity $h + 1$ and for every $x_1\ldots x_h, y$ we have: $f\,x_1\ldots x_h = y$ if and only if $(Rfond\,h)\,f\,x_1\ldots x_h\,y$.

Remark. Axiom C3 states the existence of $Rfond\,h$ when h is a natural number greater than 0, but it does not exclude that $Rfond$ be defined on other arguments as well.

We will often use the notation $\alpha \downarrow\uparrow x_1\ldots x_h$ to be read "α *acts on* $x_1\ldots x_h$". From the axiom it follows that the action defined in this way is the usual one for qualities, relations, operations and collections: for instance, for every quality q and every object x we have: $q \downarrow\uparrow x$ if and only if $q\,x$.

We extend also the notation for inclusion to objects having the same arity h by letting $\alpha \subseteq \beta$ if, for any $z_1,\ldots,z_h$, $\alpha \downarrow\uparrow z_1\ldots z_h$ implies $\beta \downarrow\uparrow z - 1\ldots z_h$.

Moreover we extend to h–ary objects the notation of the upward and downward arrows, by letting $\alpha \uparrow x_1\ldots x_{h-1}$ if there is x_h such that $\alpha \downarrow\uparrow x_1\ldots x_h$, and $\alpha \downarrow x_h$ if there are $x_1,\ldots,x_{h-1}$ such that $\alpha \downarrow\uparrow x_1\ldots x_h$. In particular, this extends to all objects of arity 2 the notations $\alpha \uparrow x$ and $\alpha \downarrow y$ introduced in §3 for simple operations. By means of this notation we can express the functionality of operations by the axiom:

Axiom C4. – *If f is an operation with arity $h + 1$, then $f \downarrow\uparrow x_1\ldots x_h y$ and $f \downarrow\uparrow x_1\ldots x_h z$ imply $y = z$.*

5. Finite systems

The notion of finite system which we are going to introduce in this section subsumes the notions of ordered pair, ordered triple, ... ordered n-tuple, the notion of substitution on a finite number of elements, and several other concepts of Mathematics or real life where some objects are put in relation with some indexes. For instance, we can think of the Italian tax identifier, the rating plate of a car, or the label of a bottle of wine (this is an example of non–functional indexing, if labels are not numbered). All finite systems have arity 2, and they form a collection which will be denoted by the symbol Sif; among them we have the univalent systems, that is those systems where the index determines the indexed object in a unique way; these systems form the collection $Siuf$. So we can give the first axioms on finite systems.

AXIOM D1. – *Sif and Siuf are collections and Siuf $\subseteq$ Sif; for every $x \in Sif$ we have $Ar\, x = 2$.*

AXIOM D2. – *(Extensionality of Sif) If $x, y \in Sif$, $x \subseteq y \subseteq x$ then $x = y$.*

AXIOM D3. – *(Characterizing Siuf) $x \in Siuf$ if and only if $x \in Sif$ and, for every y, z, z', $x \downarrow\uparrow y$ and $x \downarrow\uparrow y z'$ imply $z = z'$.*

The previous axioms do not yet ensure the existence of any finite system; to ensure the latter it is useful to introduce the following axiom about existence of "singular" systems:

AXIOM D4. – *For every x, y there is $S \in Siuf$ such that $S \uparrow z$ if and only if $z = x$, and $s \downarrow t$ if and only if $y = t$.*

REMARK. By the extensionality axiom, the system considered in axiom D3 is determined by x and y, and will be denoted by the symbol $\begin{pmatrix} x \\ y \end{pmatrix}$. In order to introduce more complex systems besides the singular ones, and in particular pairs, triples, etc., it is useful to introduce the operation of union of two binary objects. The latter will be denoted by Unb and is characterized by the following axioms:

AXIOM D5. – *Unb is a binary operation. $Unb \uparrow x y$ if and only if $Ar\, x = Ar\, y = 2$; if $z = Unb\, x y$, then $Ar\, z = 2$ and for every u, v we have: $z \downarrow\uparrow u v$ if and only if at least one of $x \downarrow\uparrow u v$ and $y \downarrow\uparrow u v$ is true. When $x, y \in Sif$, then $Unb\, x y \in Sif$ as well.*

Also for binary union we can adopt the usual notations by writing $x \cup y$ for $Unb\, x y$; in particular, by the extensionality of Sif, if x, y are finite systems, then $x \subseteq y$ if and only if $x \cup y = y$.

Besides binary union, other operations on finite systems are the simple operation of inversion of systems, $Invs$, and the binary operation of composition of systems, $Comps$, characterized by the following axioms:

AXIOM D6. – *Invs is a simple operation. The three conditions $Invs \uparrow x$, $invs \downarrow x$, $x \in Sif$ are equivalent. For every $x, y \in Sif$ and for every u, v we have: $(Invs\, x) \downarrow\uparrow u v$ if and only if $x \downarrow\uparrow v u$.*

By the extensionality of Sif, the axiom determines uniquely, for every finite system x, its inverse system $Invs\, x$, which will be denoted by the symbol x^{-1}. Turning to the operation $Comps$ we have the axiom:

AXIOM D7. – *Comps is a binary operation. $Comps \uparrow x y$ if and only if $x, y \in Sif$; $Comps \downarrow z$ if and only if $z \in Sif$. finally $(Comps\, x y) \downarrow\uparrow u w$ if and only if there is t such that $x \downarrow\uparrow t w$ and $y \downarrow\uparrow u t$.*

Once again, the extensionality of Sif ensures that the composition of two finite systems x, y is uniquely determined by axiom D7; this composition will be denoted by the usual notation $x \circ y$.

Following the usual notation for n–tuples, if x is a univalent system we will write $x_h = w$ for $x \downarrow\uparrow h w$. We will also say that h is an *index* of the system x when

$x \uparrow h$, and that y is a *value* of the system x when $x \downarrow y$. We will then use the notation taken from the theory of groups of substitutions:

$$\begin{pmatrix} x & y \\ a & b \end{pmatrix} = \begin{pmatrix} x \\ a \end{pmatrix} \cup \begin{pmatrix} y \\ b \end{pmatrix},$$

$$\begin{pmatrix} x & y & z \\ a & b & c \end{pmatrix} = \begin{pmatrix} x \\ a \end{pmatrix} \cup \begin{pmatrix} y \\ b \end{pmatrix} \cup \begin{pmatrix} z \\ c \end{pmatrix}, \; etc.$$

REMARK. By composing $\begin{pmatrix} 1 \\ 1 \end{pmatrix}$ and $\begin{pmatrix} 2 \\ 2 \end{pmatrix}$ we get the empty system (unique by the extensionality axiom), which will be denoted by the symbol $\emptyset_2$. The symbol is motivated by the analogy with the empty collection $\emptyset$: in fact, $\emptyset_2$ is an empty binary object (not necessarily the unique one, for instance we could have several relations or operation with "empty graph").

As particular cases of univalent systems we have 1–tuples, that is systems of the kind $\begin{pmatrix} 1 \\ x \end{pmatrix}$, which will be denoted also by $[x]$, ordered pairs $(x, y) = \begin{pmatrix} 1 & 2 \\ x & y \end{pmatrix}$, triples $(x, y, z) = \begin{pmatrix} 1 & 2 & 3 \\ x & y & z \end{pmatrix}$, etc.

Now we introduce the collections V^n of all the n–tuples, defined by the following axiom:

AXIOM D8. – *For every n, V^n is a collection. V^0 contains only $\emptyset_2$; for every $n \in \mathbf{N}, n > 0$, we have $x \in V^n$ if and only if $x \in Siuf$ and the indexes of x are exactly the natural numbers from 1 to n.*

It is useful also to introduce an axiom about the existence of n–tuples, like the following:

AXIOM D9. – *For every n there exists the tuple $(1, \ldots, n)$.*

REMARK. Since we did not introduce any axiom of induction in the basic theories, we can prove *separately* the particular cases of axiom D9 for $n = 0, 1, 2, 3, \ldots$, but we cannot prove the axiom itself in its generality, so the latter axiom seems to be independent from the other axioms of the basic theories. We conclude this section with an axiom expressing "finiteness" of systems, by associating to them a natural number serving as "cardinality" of the system:

AXIOM D10. – *For every system s there are a natural number n and two n–tuples x, y such that $s = y \circ x^{-1}$. Moreover there is a least such number.*

The least number above will be called *cardinality* of the system and will be denoted by the symbol $Card\, s$. We note that the first part of axiom D10 is not enough for proving, from the axioms of §3, the existence of a *least* n, as we do not have the axiom of induction. Again by the lack of the induction axiom it is useful to introduce the following axiom:

AXIOM D11. – *Let $s = y \circ x^{-1}$, $x, y \in V^n$; then $Card\, s = n$ if and only if for $1 \le i < j \le n$ we have always $\begin{pmatrix} x_i \\ y_i \end{pmatrix} \neq \begin{pmatrix} x_j \\ y_j \end{pmatrix}$.*

The meaning of axiom D11 is the following: if n is the cardinality of $s = y \circ x^{-1}$, then s is the union of n distinct singular systems $\begin{pmatrix} x_1 \\ y_1 \end{pmatrix}, \ldots, \begin{pmatrix} x_n \\ y_n \end{pmatrix} y$. By the way, from the axioms D9–11 it follows that the integers from 1 to k cannot be in a bijective correspondence with the ones from 1 to h for $h \neq k$.

6. The universal relations

Now we introduce a new, highly self–referential object with high describing power, which "absorbs" all the relations $R\,fond\,h$, that is the operation generating universal relations, $Runiv$, characterized by the following axiom:

AXIOM E. – *$Runiv$ is a simple operation having the following properties:*

a) *If $h \in \mathbf{N}$, $h > 0$ and $z = Runiv\,h$, then z is a relation of arity $h + 2$.*

b) *If $(Runiv\,h) \downarrow\uparrow x_1 \ldots x_h\,y\,t$, then $x_i = i$ for $1 \leq i \leq h$.*

c) *If $(Runiv\,h) \downarrow\uparrow 1 \ldots h\,y\,t$, $Ar\,y = k \in \mathbf{N}$, $k \geq 1$, then $t \in V^k$.*

d) *If $Ar\,y = k \in \mathbf{N}$, $k \geq 1$, then $(Runiv\,h) \downarrow\uparrow 1 \ldots h\,y\,(x_1, \ldots, x_k)$ if and only if $y \downarrow\uparrow x_1 \ldots x_k$.*

The inspiring idea of the basic theories is of granting the existence of the relation $Runiv\,k$ for k sufficiently large, so that the latter, which has a very strong and "dangerous" self–referential power, have a complexity large enough to protect the basic theory, as well as the "engraftings" of the most known mathematical theories which can be done on it, from the risks of antinomies of the kinds studied in [8, 18, 19]. So the fundamental axiom of a basic theory will depend on a fixed natural number $\nu > 0$ and will ensure the existence of $Runiv\,k$ for $k \geq \nu$. So to any fixed ν, we associate the axiom:

AXIOM F_ν. – *For every $k \in \mathbf{N}$, if $k \geq \nu$, then $Runiv\,k$ exists.*

The theory having the axioms $ABCDEF_\nu$ will be called TB_ν (ν–ary basic theory). As will be shown in [15], F_ν is the "keystone" of the theory and it is essential for ensuring the finite axiomatizability and a high self–descriptive power of the theory itself. We note that from the conditions $k \geq \nu > 0$ it follows that $Runiv\,k$ has arity greater than two, hence it cannot be confused with the objects of arity 0, 1 and 2 of traditional Mathematics.

Bibliography

[1] A. ASPERTI – G. LONGO, *Categories, types and structures – An introduction to category theory for the working computer scientist*, Cambridge, Mass. 1991.

[2] H. P. BARENDREGT, *The Lambda–Calculus*, Amsterdam 1981.

[3] Y. Bar Hillel – A. A. Fraenkel – A. Levy, *Foundations of set theory*, Amsterdam 1973.

[4] M. Beeson, *Towards a computation system based on set theory*, preprint, CSLI, Stanford 1987.

[5] J. Benabou, *Fibered categories and the foundations of naïve category theory*, Journal of Symbolic Logic, 50, 1985, pp. 10–37.

[6] M. Boffa, *Sur la théorie des ensembles sans axiome de fondement*, Bull. Soc. Math. Belg., 21, 1969, 16–56.

[7] A. Church, *Set theory with a universal set.* In: L. Henkin et al. (eds.), *Proceedings of the Tarski Symposium.* Proc. of Symp. P. Math. XXV, Rhode Island 1974, 297–308.

[8] M. Clavelli – E. De Giorgi – M. Forti – V. M. Tortorelli, *A selfreference oriented theory for the Foundations of Mathematics.* In: *in "Analyse Mathématique et applications. Contributions en l'honneur de Jacques–Louis Lions.* Paris 1988, 67–115.

[9] H. B. Curry – R. Feys, *Combinatory logic*, Amsterdam 1958.

[10] E. De Giorgi, *Fondamenti della Matematica e teoria base: gli esempi delle teorie 7×2 e 7×5* (sunto di una conversazione). Lecce 1989, manuscript.

[11] E. De Giorgi – M. Forti, *Una teoria quadro per i fondamenti della Matematica.* Atti Acc. Lincei, Rend. Fis., s. 8, vol. 79, 1985, 55–67.

[12] E. De Giorgi – M. Forti, *"5×7": A Basic Theory for the Foundations of Mathematics.* Preprint di Matematica n. 74, Scuola Normale Superiore, Pisa 1990.

[13] S. Feferman, *Constructive theories of functions and classes.* In: M. Boffa et al. (eds.), *Logic Colloquium 1978.* Amsterdam 1979.

[14] M. Forti – F. Honsell, *Models of self–descriptive set theories.* In: F. Colombini et al. (eds.), *Partial Differential Equations and the Calculus of Variations – Essays in Honor of Ennio De Giorgi.* Boston 1989, 473–518.

[15] M. Forti – G. Lenzi, *Assiomi e modelli di teorie base dei Fondamenti della Matematica.* (Unpublshed manuscript)

[16] G. Frege, *Grundgesetze der Aritmetik, begriffsschriftlich abgeleitet.* Vol. I, Jena 1893; vol. II, Pohle, Jena 1903 (reprinted Olms, Hildesheim 1962).

[17] J. M. Glubrecht – A. Oberschelp – G. Todt, *Klassenlogik.* Zürich 1983.

[18] G. LENZI, *Estensioni contraddittorie della teoria Ampia.* Atti Acc. Lincei, Rend. fis., s. 8, vol. 83, 1989, 13–28.

[19] G. LENZI – V. M. TORTORELLI, *Introducing predicates into a basic theory for the foundations of Mathematics.* Preprint di Matematica n. 51, Scuola Normale Superiore, Pisa, 1989.

[20] W. V. O. QUINE, *New foundations for mathematical logic.* Amer. Math. Monthly, 44, 1973, 70–80.

[21] B. RUSSELL – A. N. WHITEHEAD, *Principia Mathematica.* Cambridge 1925.

[22] D. SCOTT, *Combinators and classes.* In: C. BÖHM (ed.), λ-*Calculus and Computer Science Theory.* LNCS, 37, Berlin 1975.

[23] J. VON NEUMANN, *Eine Axiomatisierung der Mengenlehre.* J. f. Math., 154, 1925, 219–240.

E. De Giorgi, G. Lenzi:
Scuola Normale Superiore
Piazza dei Cavalieri 7 – 56126 PISA

M. Forti:
Istituto di Matematiche Applicate "U. Dini"
Università degli studi di Pisa
Via Bonanno, 25 B – 56100 PISA

Una proposta di teoria base dei Fondamenti della Matematica[‡]

Ennio De Giorgi, Marco Forti, Giacomo Lenzi[*]

Sunto. Vengono proposte alcune teorie base dei Fondamenti della Matematica che assumono come concetti primitivi i concetti di numero naturale, collezione, qualità, operazione e relazione; le operazioni e le relazioni considerate possono essere più o meno complesse: il numero naturale che indica il grado di complessità è detto *arietà*. Nelle teorie considerate è raggiunto un alto grado di autoreferenza.

KEYWORDS: Foundations; Operations; Relation; System; Basic–theory.

0. INTRODUZIONE

Recentemente sono state proposte diverse teorie dei Fondamenti della Matematica, nelle quali si cercava di raggiungere un alto grado di autoreferenza e una presentazione dei diversi concetti matematici vicina alla comune intuizione e alla tradizione matematica (cfr. [8],[10]–[12],[14],[18], [19]). In particolare in queste teorie non si riducevano tutti i concetti matematici ad un unico concetto primitivo, ma si cercava piuttosto di introdurne diversi (numero, qualità, relazione, operazione, collezione) collegati da assiomi il più possibile vicini all'uso comune e alla tradizione matematica e tali da poter garantire il più alto grado possibile di autoriferimento della teoria stessa. Un'analisi critica di queste teorie ha messo in evidenza l'opportunità di separare un nucleo iniziale molto ridotto (che chiameremo "teoria base") su cui sia possibile "innestare" tutte le teorie matematiche note. Nelle teorie base che proponiamo in questa nota ammettiamo per ogni intero n l'esistenza di oggetti aventi arietà n (intuitivamente l'arietà di un oggetto è l'indice della sua complessità). Per chi ricerchi possibili interpretazioni filosofiche della teoria base questo equivale ad ammettere l'esistenza di "realtà" molto complesse.

Le teorie base hanno il vantaggio di essere delle teorie abbastanza semplici sulla cui coerenza (cioè non contraddittorietà) non possono esistere dubbi, almeno per chi crede nella coerenza della parte più elementare dell'aritmetica. Questa sicurezza è un valido punto di partenza per la discussione sulla coerenza delle teorie matematiche più avanzate che possono essere innestate sulle teorie base. Infatti, una volta assunti gli assiomi di una teoria base, dovrebbe essere facile "innestare" in modo naturale su tale teoria le più complesse teorie matematiche

---

[‡]Editor's note: published in Atti Accad. Naz Lincei, Cl. Sci. Fis. Mat. Nat., **(9) 5** (1994), 11–22.

[*]Nota presentata da E. De Giorgi all'Accademia dei Lincei il 18 Giugno 1993.

(per esempio le varie teorie degli insiemi, le logiche modali, l'Analisi standard e quella non–standard, ecc.)

Forse sulle stesse teorie base sarà possibile anche l'innesto di teorie non strettamente matematiche, e per facilitare la collaborazione a questa ricerca di studiosi di varie discipline scientifiche ed umanistiche, abbiamo cercato di presentare la teoria base in un linguaggio informale più vicino possibile al linguaggio comune; in altra sede (vedi [15]) si mostra che gli assiomi introdotti informalmente possono essere formalizzati nel linguaggio del Calcolo dei Predicati del primo ordine. Vi è infine la speranza, forse remota ma non manifestamente infondata, che sulle teorie base ora esposte possano in futuro "germogliare" teorie scientifiche e filosofiche interessanti ed originali. Per questo occorre notare che le teorie base sono teorie "qualitativamente aperte" che non escludono l'esistenza di oggetti "qualitativamente" diversi dagli oggetti fondamentali introdotti dalla teoria, cioè qualità, collezioni, numeri naturali, relazioni e operazioni. Possiamo dire che l'idea generale che ispira questo lavoro è la filosofia sostanzialmente antiriduzionista espressa da Shakespeare con le parole di Amleto a Orazio "vi sono più cose tra cielo e terra di quante ne sogni la tua filosofia" (Amleto, atto I, scena V). Osserviamo infine che questa nota, pur tenendo conto di molte ricerche logico–matematiche precedenti, è scritta in forma del tutto autosufficiente in modo da poter essere letta e compresa perfettamente anche da chi ignora tutte le teorie fondazionali classiche e moderne citato in Bibliografia ([1]–[7], [9], [13], [16], [17], [20]–[23]) comprese le teorie dei fondamenti esposte in [8], [10]–[12], [14], [18], [19].

1. Qualità, collezioni, relazioni e operazioni

Nell'esposizione informale della nostra teoria chiamiamo *oggetti* tutte le entità di cui parla la teoria, assumiamo come *primitive* le nozioni di *qualità*, *collezione*, *relazione* (binaria, ternaria, ecc.) e *operazione* (semplice, binaria, ecc.) e adottiamo le seguenti notazioni:

se q è una *qualità* e x è un *oggetto*, scriviamo $q\,x$ per "*x gode della qualità q*";

se C è una *collezione* e x è un *oggetto* scriviamo $x \in C$ per "*x appartiene alla collezione C*", oppure "*x è un elemento della collezione C*"; scriviamo $x \notin C$ per dire che x non appartiene alla collezione C;

se r è una *relazione binaria* e x e y sono due *oggetti*, scriviamo $r\,x\,y$ per "*x è nella relazione r con y*"; per esempio, la frase "Roma è più popolata di Frascati" esprime una relazione binaria r tra x, cioè Roma, e y, cioè Frascati;

se ρ è una *relazione ternaria* e x, y e z sono *oggetti*, scriviamo $\rho\,x\,y\,z$ per "*x è nella relazione ρ con y e z*"; un esempio è la frase "A Francesco piace più il formaggio del prosciutto" ove x è Francesco, y è il formaggio e z è il prosciutto;

se τ è una *relazione quaternaria* e x, y, ξ e η sono *oggetti*, scriviamo $\tau\,x\,y\,\xi\,\eta$ per "*x è nella relazione τ con y, ξ e η*"; un esempio è la frase "Roma è più vicina a Milano di quanto Parigi sia vicina a Mosca", con x=Roma, y=Milano, ξ=Parigi e η=Mosca.

Analogamente per una generica relazione n-aria α scriveremo $\alpha\, x_1 \ldots x_n$ per dire che gli oggetti $x_1, \ldots, x_n$ sono nella relazione α.

Passiamo alle operazioni, di cui si hanno importanti esempi anche nell'aritmetica elementare, come le operazioni binarie di addizione di due numeri, sottrazione e moltiplicazione, o l'operazio- ne semplice che associa ad un numero n il suo successore $n + 1$. Notiamo che in generale, per operazioni con più di un argomento, il risultato dipende dall'ordine degli argomenti; per es. si consideri l'operazione binaria di potenza di numeri naturali, Pot, tale che $Pot\, m\, n = m^n$; si ha $Pot\, 2, 3 = 2^3 = 8$ che è diverso da $Pot\, 3, 2 = 3^2 = 9$.

Se f è un'*operazione semplice* e x e y sono *oggetti*, scriviamo $y = f x$ per *"l'operazione f trasforma x in y"*;

se φ è un'*operazione binaria* e x, y e z sono *oggetti*, scriviamo $z = \varphi\, x y$ per *"l'operazione φ prende prima l'oggetto x, poi l'oggetto y e dà come risultato z"*.

Analogamente si procede per le operazioni h–arie con $h > 2$ scrivendo, per ogni operazione h–aria ψ, $w = \psi x_1 \ldots x_h$ se w è il risultato di ψ calcolata sugli argomenti $x_1, \ldots, x_h$.

2. Primi oggetti fondamentali

Tra gli oggetti di diversa specie considerati indichiamo ora alcuni "oggetti fondamentali" e diamo alcuni assiomi che ne precisano il ruolo nella teoria base. Introduciamo anzitutto alcune qualità corrispondenti ai tipi di oggetti considerati nel §1; questa corrispondenza è affermata dagli assiomi A1–A7.
Introduciamo anzitutto la qualità di essere una collezione, $Qcoll$:

ASSIOMA A1 – *$Qcoll$ è una qualità; per ogni oggetto C si ha $Qcoll\, C$ se e solo se C è una collezione.*

Analogamente procediamo per gli altri tipi di oggetti; la qualità di essere una relazione binaria è $Qrelb$:

ASSIOMA A2 – *$Qrelb$ è una qualità; per ogni oggetto r si ha $Qrelb\, r$ se e solo se r è una relazione binaria.*

La qualità di essere una relazione ternaria è $Qrelt$:

ASSIOMA A3 – *$Qrelt$ è una qualità; per ogni oggetto ρ si ha $Qrelt\, \rho$ se e solo se ρ è una relazione ternaria.*

La qualità di essere una relazione quaternaria è $Qrelq$:

ASSIOMA A4 – *$Qrelq$ è una qualità; per ogni oggetto τ si ha $Qrelq\, \tau$ se e solo se τ è una relazione quaternaria.*

La qualità di essere un'operazione semplice è $Qops$:

ASSIOMA A5 – *$Qops$ è una qualità; per ogni oggetto f si ha $Qops\, f$ se e solo se f è un'operazione semplice.*

La qualità di essere un'operazione binaria è $Qopb$:

ASSIOMA A6 – *$Qopb$ è una qualità; per ogni oggetto φ si ha $Qopb\, \varphi$ se e solo se φ è un'operazione binaria.*

Infine abbiamo la qualità di essere una qualità, chiamata *Qqual*, caratterizzata dall'assioma seguente:

ASSIOMA A7 – *Qqual è una qualità; per ogni oggetto q si ha $Qqual\, q$ se e solo se q è una qualità.*

Osserviamo che l'assioma A7 implica $Qqual\, Qqual$, cioè la qualità *Qqual* è goduta da se stessa; questo è il primo esempio di *autoriferimento* nella nostra teoria.

Il prossimo passo è l'introduzione di alcune *relazioni fondamentali* che descrivono il comportamento degli oggetti di arietà più bassa. Anzitutto introduciamo la *relazione fondamentale degli oggetti unari*, *Rfun*:

ASSIOMA A8 – *Rfun è una relazione binaria. Per ogni qualità q e per ogni oggetto x si ha $Rfun\, q\, x$ se e solo se $q\, x$.*
Per ogni collezione C e per ogni oggetto x si ha $Rfun\, C\, x$ se e solo se $x \in C$.

Come abbiamo fatto per gli oggetti unari introduciamo la *relazione fondamentale degli oggetti binari*, *Rfbin*:

ASSIOMA A9 – *Rfbin è una relazione ternaria. Per ogni relazione binaria r e per ogni x, y si ha $Rfbin\, r\, x\, y$ se e solo se $r\, x\, y$.*
Per ogni operazione semplice f e per ogni x, y si ha: $Rfbin\, f\, x\, y$ se e solo se $y = f\, x$.

Si noti che un'operazione semplice è in realtà un oggetto binario in quanto collega due oggetti, argomento e risultato. Analogamente consideriamo le operazioni binarie come oggetti ternari, ecc.

La terza relazione fondamentale è quella degli oggetti ternari, *Rfter*:

ASSIOMA A10 – *Rfter è una relazione quaternaria. Per ogni relazione ternaria ρ e per ogni x, y, z si ha $Rfter\, \rho\, x\, y\, z$ se e solo se $\rho\, x\, y\, z$.*
Per ogni operazione binaria φ e per ogni x, y, z si ha
$Rfter\, \varphi\, x\, y\, z$ se e solo se $z = \varphi\, x\, y$.

Dopo l'introduzione di *Rfbin* e *Rfter* l'univocità delle operazioni semplici e binarie può essere espressa dall'assioma seguente:

ASSIOMA A11 – *Se f è un'operazione semplice allora $Rfbin\, f\, x\, y$ insieme a $Rfbin\, f\, x\, y'$ implica $y = y'$.*
Se φ è un'operazione binaria allora $Rfter\, \varphi\, x\, y\, z$ insieme a $Rfter\, \varphi\, x\, y\, z'$ implica $z = z'$.

Relazioni ancora più complesse delle precedenti verranno introdotte in seguito, dopo i numeri naturali; in questo primo capitolo introduciamo solo altri oggetti molto semplici, cominciando dall'operazione identità, *Id*:

ASSIOMA A12 – *Id è un'operazione semplice. Per ogni oggetto x si ha $Id\, x = x$.*

Infine introduciamo la collezione universale, V, la collezione vuota, $\emptyset$, e la collezione delle collezioni, *Coll*:

ASSIOMA A13 – *V, $\emptyset$ e Coll sono collezioni. Per ogni oggetto x si ha $x \in V$, $x \notin \emptyset$, mentre $x \in Coll$ se e solo se x è una collezione.*

Dall'assioma precedente seguono le relazioni autoreferenziali $Coll \in Coll$, $V \in V$ e le relazioni di "mutua referenza" $V \in Coll$, $Coll \in V$.

Prima di enunciare l'assioma di estensionalità delle collezioni, introduciamo la consueta notazione dell'inclusione: se C, C' sono collezioni, scriviamo $C \subseteq C'$ se ogni oggetto appartenente a C appartiene anche a C'. Dopo di ciò diamo l'assioma di estensionalità:

Assioma A14 – *Se C e C' sono collezioni, e si ha simultaneamente $C \subseteq C'$, $C' \subseteq C$ allora $C = C'$.*

3. Aritmetica elementare

Per inserire nella teoria base la parte più elementare dell'Aritmetica introduciamo anzitutto la collezione dei numeri naturali, che seguendo l'uso comune indicheremo con $\mathbf{N}$, e perciò cominciamo con l'enunciare l'assioma:

Assioma B1 – $\mathbf{N}$ *è una collezione.*

Stabiliamo poi alcune notazioni utili per scrivere concisamente assiomi che riguardano operazioni. Se f è un'operazione semplice e x è un oggetto, scriviamo $f \uparrow x$ per "*esiste un y tale che $y = fx$*"; se φ è un'operazione binaria e x e y sono oggetti, scriviamo $\varphi \uparrow x\,y$ per "*esiste uno z tale che $z = \varphi\,x\,y$*". Se f è un'operazione semplice e y è un oggetto, scriviamo $f \downarrow y$ per "*esiste un x tale che $y = fx$*"; infine se φ è un'operazione binaria e z è un oggetto, scriviamo $\varphi \downarrow z$ per "*esistono un x e un y tali che $z = \varphi\,x\,y$*".

Descriviamo poi alcune operazioni aritmetiche (addizione, moltiplicazione, sottrazione) cominciando dall'*addizione di due numeri naturali*, che chiameremo *Nadd*, per la quale diamo gli assiomi:

Assioma B2 – *Nadd è un'operazione binaria.*

Assioma B3 – *Nadd $\uparrow x\,y$ se e solo se $x \in \mathbf{N}$ e $y \in \mathbf{N}$. Se Nadd $\downarrow z$ allora $z \in \mathbf{N}$.*

Nella trattazione dell'operazione *Nadd* adottiamo l'usuale *notazione*: $x + y = Nadd\ x\ y$.

I seguenti assiomi esprimono alcune proprietà elementari dell'addizione:

Assioma B4 – (*associatività*) $(x + y) + z = x + (y + z)$.

Assioma B5 – (*commutatività*) $x + y = y + x$.

Assioma B6 – (*proprietà del numero zero*) *Esiste $z \in \mathbf{N}$ tale che per ogni $x \in \mathbf{N}$ si ha $x + z = x$.*

Lo z dell'assioma precedente risulta unico per l'assioma B5 e sarà denotato con l'usuale simbolo 0.

La seconda operazione fondamentale sui numeri naturali è la *moltiplicazione*, che indichiamo con *Nmult* e soddisfa gli assiomi seguenti:

Assioma B7 – *Nmult è un'operazione binaria.*

Assioma B8 – *Nmult $\uparrow x\,y$ se e solo se $x \in \mathbf{N}$ e $y \in \mathbf{N}$. Se Nmult $\downarrow z$ allora $z \in \mathbf{N}$.*

Nella trattazione dell'operazione *Nmult* adottiamo l'usuale *notazione*: $xy = Nmult\ x\ y$.

I seguenti assiomi esprimono alcune proprietà elementari della moltiplicazione:

Assioma B9 – (*associatività*) $(xy)z = x(yz)$.

Assioma B10 – (*commutatività*) $xy = yx$.

Assioma B11 – (*distributività rispetto alla somma*) $x(y + z) = (xy) + (xz)$.

Assioma B12 – (*proprietà del numero uno*) *Esiste* $u \in \mathbf{N}$ *tale che per ogni* $x \in \mathbf{N}$ *si ha* $ux = x$.

Il numero u dell'assioma precedente risulta unico per l'assioma B10 e sarà denotato con l'usuale simbolo 1. Come al solito poniamo $2 = 1 + 1$, $3 = 2 + 1$ ecc.

Oltre alle operazioni di addizione e moltiplicazione, i numeri interi possiedono un ordinamento naturale, secondo il quale 0 precede 1, 1 precede 2, ecc.; a tale ordinamento ci si riferisce nella frase " x è *minore o uguale* a y". Perciò nella nostra teoria consideriamo la relazione $Nord$ (l'*ordinamento naturale* dei numeri naturali) che soddisfa gli assiomi seguenti:

Assioma B13 – $Nord$ *è una relazione binaria. Se* $Nord\,x\,y$ *allora* $x \in \mathbf{N}$ *e* $y \in \mathbf{N}$.

Secondo l'uso comune scriviamo $x \leq y$ per $Nord\,x\,y$. Scriviamo inoltre $x < y$ per $x \leq y$ ed $x \neq y$.

Due prime proprietà di $Nord$ sono le seguenti:

Assioma B14 – (*connessione*) *Per ogni* x *e* y *appartenenti a* $\mathbf{N}$ *si ha* $x \leq y$ *oppure* $y \leq x$.

Assioma B15 – (*antisimmetria*) *Se* $x \leq y$ *e* $y \leq x$ *allora* $x = y$.

Diamo ora un assioma che collega $Nord$ con $Nadd$:

Assioma B16 – *Se* $x, y \in \mathbf{N}$ *allora* $x < y$ *se e solo se esiste un* $p \in \mathbf{N}$ *tale che* $p \neq 0$ *e* $y = x + p$.

Dagli assiomi precedenti segue:

1) $Nord$ è transitiva, cioè per ogni $x, y, z \in \mathbf{N}$ se $x \leq y$ e $y \leq z$ allora $x \leq z$;

2) per ogni $x, y \in \mathbf{N}$ si ha $x \leq y$ se e solo se esiste $h \in \mathbf{N}$ tale che $y = x + h$;

3) 0 è il *minimo* elemento di $\mathbf{N}$ rispetto a $Nord$, cioè $0 \leq x$ per ogni $x \in \mathbf{N}$.

Infine postuliamo che l'uno è il successore immediato dello zero:

Assioma B17 – $0 < 1$ *e per nessuno* $z \in \mathbf{N}$ *si ha* $0 < z < 1$.

Accanto alla somma di numeri naturali possiamo considerare la sottrazione fatta nell'ambito dei naturali, $Nsub$, caratterizzata dai seguenti assiomi:

Assioma B18 – $Nsub$ *è un'operazione binaria.*
$Nsub \uparrow x\,y$ *se e solo se* $x \in \mathbf{N}$, $y \in \mathbf{N}$ *e* $y \leq x$. *Se* $Nsub \downarrow z$, *allora* $z \in \mathbf{N}$.

Assioma B19 – $z = Nsub\,x\,y$ *se e solo se* $x = y + z$.

Adotteremo poi la solita notazione $Nsub\,x\,y = x - y$.

Osserviamo che la teoria dei numeri naturali ora esposta è molto debole (mancano per es. la divisione, la potenza, l'assioma d'induzione) ma è abbastanza forte per non possedere modelli finiti. Si potrebbero considerare teorie ancora più deboli, che posseggono modelli finiti, ma esse non sono comode come base su cui innestare il resto dell'Aritmetica.

4. Arietà e mutue referenze tra relazioni fondamentali

L'introduzione nel §3 dei primi elementi dell'Aritmetica consente di unificare ed estendere i concetti introdotti nel §2 e di formalizzare la nozione di "complessità" (o "arietà") usata informalmente nei §§1,2. Anzitutto formalizziamo le nozioni generali di relazione e operazione (di arietà qualunque) mediante le due qualità $Qrel$ e Qop:

Assioma C1 – *$Qrel$ e Qop sono qualità.*

$Qrel\,x$ se e solo se x è una relazione. $Qop\,x$ se e solo se x è un'operazione.

Passiamo ora agli assiomi sull'arietà introducendo l'operazione Ar (e in particolare determiniamo l'arietà degli oggetti considerati nei §§1,2,3):

Assioma C2 – *Ar è un'operazione semplice che gode delle proprietà seguenti:*

a) *Per $x \in \mathbf{N}$ si ha $Ar\,x = 0$.*

b) *Per ogni qualità q e ogni collezione C, $Ar\,q = Ar\,C = 1$.*

c) *Sia r una relazione. Allora $Ar\,r = h$ se e solo se r è una relazione h–aria. In particolare $Ar\,r = 2\,(3,4)$ se e solo se r è binaria (ternaria, quaternaria rispettivamente).*

d) *Sia f un'operazione. Allora $Ar\,f = h + 1$ se e solo se f è un'operazione h–aria. In particolare $Ar\,f = 2\,(3)$ se e solo se f è semplice (binaria rispettivamente).*

Osserviamo che gli assiomi che abbiamo introdotto escludono che un oggetto possa avere simultaneamente due arietà, essere p. es. sia una collezione che un'operazione, o una relazione sia binaria che ternaria. Invece non è finora escluso che un oggetto sia simultaneamente una qualità e una collezione, o una relazione binaria e un'operazione semplice, ecc. Non è neppure escluso che esistano oggetti la cui arietà non è un numero naturale. Infine non è escluso che oltre ai numeri naturali ci siano altri oggetti di arietà 0, oltre alle qualità e alle collezioni altri oggetti di arietà 1, ecc. Questa libertà può essere utile per l'"innesto" sulle teorie base dei diversi rami della Matematica, ed eventualmente di altre teorie non matematiche. Per esempio questa libertà potrebbe risultare utile a chi volesse innestare sulla teoria base le logiche infinitarie.

Volendo poi realizzare teorie abbastanza semplici ma ugualmente più ricche sul piano della complessità delle usuali teorie insiemistiche, che in sostanza ammettono solo oggetti di arietà 1, e di teorie tipo Lambda–Calcolo o algebre combinatorie e relazionali, che in sostanza trattano oggetti di arietà 2, stabiliamo che per ogni numero naturale h esiste almeno un oggetto di arietà h e che gli oggetti di arietà più alta descrivono l'azione degli oggetti di arietà più bassa. Introduciamo perciò l'operazione generatrice di relazioni fondamentali, $Rfond$:

ASSIOMA C3 – *$Rfond$ è un'operazione semplice che gode delle proprietà seguenti:*

a) *Per ogni numero naturale $h > 0$ esiste $Rfond\,h$ ed è una relazione di arietà $h + 1$.*

b) *$(Rfond\,h)\,\alpha\,x_1 \ldots x_h$ implica $Ar\,\alpha = h$.*

c) *$Rfond\,1 = Rfun$, $Rfond\,2 = Rfbin$ e $Rfond\,3 = Rfter$.*

d) *Per ogni relazione ρ di arietà h e per ogni $x_1, \ldots, x_h$ si ha: $\rho x_1 \ldots x_h$ se e solo se $(Rfond\,h)\,\rho\,x_1 \ldots x_h$.*

e) *Per ogni operazione f di arietà $h + 1$ e per ogni $x_1, \ldots, x_h, y$ si ha $(Rfond\,(h + 1))\,f\,x_1 \ldots x_h\,y$ se e solo se $y = f\,x_1 \ldots x_h$.*

OSSERVAZIONE: l'assioma C3 afferma l'esistenza di $Rfond\,h$ quando h è un numero naturale maggiore di 0 ma non esclude che $Rfond$ sia definita anche su altri argomenti.

Spesso useremo la notazione: $\alpha \downarrow\uparrow x_1 \ldots x_h$, da leggersi "$\alpha$ agisce su $x_1 \ldots x_h$" invece di $(Rfond\,h)\,\alpha x_1 \ldots x_h$. Dagli assiomi segue che l'azione così definita è quella usuale per qualità, relazioni, operazioni e collezioni: ad es. per ogni qualità q e per ogni x si ha $q \downarrow\uparrow x$ se e solo se $q\,x$.

Estendiamo anche la notazione dell'inclusione tra oggetti aventi la stessa arietà h ponendo $\alpha \subseteq \beta$ se, per ogni $z_1, \ldots, z_h$, $\alpha \downarrow\uparrow z_1 \ldots z_h$ implica $\beta \downarrow\uparrow z_1 \ldots z_h$.

Estendiamo inoltre ad oggetti h–ari la notazione della freccia in alto e quella della freccia in basso, ponendo $\alpha \uparrow x_1 \ldots x_{h-1}$ se esiste un x_h tale che $\alpha \downarrow\uparrow x_1 \ldots x_h$, e $\alpha \downarrow x_h$ se esistono $x_1, \ldots, x_{h-1}$ tali che $\alpha \downarrow\uparrow x_1 \ldots x_h$. In particolare questo estende a tutti gli oggetti di arietà 2 le notazioni $\alpha \uparrow x$ e $\alpha \downarrow y$ introdotte nel §3 per le operazioni semplici. Partendo da tali notazioni possiamo esprimere l'univocità delle operazioni mediante l'assioma:

ASSIOMA C4 – *Se f è un'operazione di arietà $h + 1$, $f \downarrow\uparrow x_1 \ldots x_h\,y$, $f \downarrow\uparrow x_1 \ldots x_h\,z$ allora $y = z$.*

5. SISTEMI FINITI

La nozione di sistema finito che introduciamo in questo paragrafo comprende le nozioni di coppia ordinata, terna ordinata, quaterna ordinata,.... n-upla ordinata, la nozione di sostituzione su un numero finito di elementi e vari altri concetti della Matematica e della vita comune ove vengono messi in relazione alcuni oggetti con alcuni indicatori. Per esempio si pensi al codice fiscale, alle targhe automobilistiche o alle etichette delle bottiglie di vino (quest'ultimo è un esempio di indicizzazione non univoca, se le etichette non sono numerate). I sistemi finiti hanno tutti arietà 2, e formano una collezione che indicheremo col simbolo Sif; tra essi vi sono i sistemi univoci, cioè quelli in cui l'indicatore (o indice) determina univocamente l'oggetto indicato, che formano la collezione $Siuf$. Possiamo quindi dare i primi assiomi sui sistemi finiti.

ASSIOMA D1 – *Sif e $Siuf$ sono collezioni e $Siuf \subseteq Sif$; per ogni $x \in Sif$ si ha $Ar\, x = 2$.*

ASSIOMA D2 – *(Estensionalità di Sif) Se $x, y \in Sif$, $x \subseteq y$, $y \subseteq x$ allora $x = y$.*

ASSIOMA D3 – *(Caratterizzazione di $Siuf$) $x \in Siuf$ se e solo se $x \in Sif$ e inoltre, per ogni y, z, z', le due condizioni $x \downarrow\uparrow yz$, $x \downarrow\uparrow yz'$ implicano $z = z'$.*

Gli assiomi precedenti non garantiscono ancora l'esistenza di qualche sistema finito; per garantirla conviene introdurre il seguente assioma di esistenza di sistemi "singolari":

ASSIOMA D4 – *Per ogni x, y esiste $S \in Siuf$ tale che $S \uparrow z$ se e solo se $z = x$, e $S \downarrow t$ se e solo se $t = y$.*

OSSERVAZIONE. Per l'assioma di estensionalità il sistema considerato nell'assioma D4 è determinato da x e y e sarà indicato col simbolo $\begin{pmatrix} x \\ y \end{pmatrix}$.

Per introdurre accanto ai sistemi singolari altri sistemi più complessi e in particolare le coppie, terne, ecc. conviene introdurre l'operazione di unione di due oggetti binari. Essa sarà indicata con Unb e caratterizzata dal seguente assioma:

ASSIOMA D5 – *Unb è un'operazione binaria.*
$Unb \uparrow x y$ se e solo se $Ar\, x = Ar\, y = 2$; se $z = Unb\, x y$ allora $Ar\, z = 2$ e per ogni u, v si ha: $z \downarrow\uparrow uv$ se e solo se è verificata almeno una delle condizioni $x \downarrow\uparrow uv$ e $y \downarrow\uparrow uv$.
Quando $x, y \in Sif$, anche $Unb\, x y \in Sif$.

Anche per l'unione binaria possiamo usare le notazioni usuali e scrivere $x \cup y$ per $Unb\, x\, y$; in particolare, per l'estensionalità di Sif, se x, y sono sistemi finiti, si ha $x \subseteq y$ se e solo se $x \cup y = y$.

Oltre all'unione binaria agiscono sui sistemi finiti l'operazione semplice di inversione di sistemi, $Invs$, e l'operazione binaria di composizione di sistemi, $Comps$, caratterizzate dagli assiomi seguenti:

ASSIOMA D6 – *$Invs$ è un'operazione semplice. Sono equivalenti le tre condizioni: $Invs \uparrow x$, $Invs \downarrow x$, $x \in Sif$.*
Per ogni $x \in Sif$ e per ogni u, v si ha $(Invs\, x) \downarrow\uparrow uv$ se e solo se $x \downarrow\uparrow vu$.

Per l'estensionalità di Sif l'assioma determina univocamente per ogni sistema finito x il sistema inverso $Invs\, x$, che sarà indicato col simbolo x^{-1}.
Passando all'operazione $Comps$ abbiamo l'assioma:

ASSIOMA D7 – *$Comps$ è un'operazione binaria.*
$Comps \uparrow x y$ se e solo se $x, y \in Sif$; $Comps \downarrow z$ se e solo se $z \in Sif$.
$(Comps\, x\, y) \downarrow\uparrow uw$ se e solo se esiste un t tale che $x \downarrow\uparrow tw$ e $y \downarrow\uparrow ut$.

Anche in questo caso l'estensionalità di Sif ci assicura che la composizione di due sistemi finiti x, y è univocamente determinata dall'assioma D7; essa sarà indicata con l'usuale notazione $x \circ y$.

Seguendo le notazioni usuali nel caso delle n–uple, se x è un sistema univoco scriveremo $x_h = w$ per $x \downarrow\uparrow hw$. Diremo anche che h è un *indice* del sistema x quando $x \uparrow h$ e diremo che y è un *valore* del sistema x quando $x \downarrow y$. Useremo poi le notazioni mutuate dalla teoria dei gruppi di sostituzioni:

$$\begin{pmatrix} x & y \\ a & b \end{pmatrix} = \begin{pmatrix} x \\ a \end{pmatrix} \cup \begin{pmatrix} y \\ b \end{pmatrix},$$

$$\begin{pmatrix} x & y & z \\ a & b & c \end{pmatrix} = \begin{pmatrix} x \\ a \end{pmatrix} \cup \begin{pmatrix} y \\ b \end{pmatrix} \cup \begin{pmatrix} z \\ c \end{pmatrix}, etc.$$

OSSERVAZIONE. Componendo $\begin{pmatrix} 1 \\ 1 \end{pmatrix}$ e $\begin{pmatrix} 2 \\ 2 \end{pmatrix}$ si ottiene il sistema vuoto (unico per l'assioma di estensionalità dei sistemi) che verrà indicato col simbolo $\emptyset_2$. Il simbolo è giustificato dall'analogia con la collezione vuota $\emptyset$: $\emptyset_2$ è un oggetto binario vuoto (ma non è detto che sia l'unico, potrebbero esserci ad esempio varie relazioni oppure operazioni vuote).

Casi particolari di sistemi univoci sono le 1-uple, cioè i sistemi del tipo $\begin{pmatrix} 1 \\ x \end{pmatrix}$, che indicheremo anche con $[x]$, le coppie ordinate $(x, y) = \begin{pmatrix} 1 & 2 \\ x & y \end{pmatrix}$, le terne $(x, y, z) = \begin{pmatrix} 1 & 2 & 3 \\ x & y & z \end{pmatrix}$, etc.

Introduciamo ora le collezioni V^n delle n–uple, definite dall'assioma seguente:

ASSIOMA D8 – *Per ogni n, V^n è una collezione.*
Alla collezione V^0 appartiene soltanto $\emptyset_2$; per ogni $n \in \mathbf{N}, n > 0$ si ha $x \in V^n$ se e solo se $x \in Siuf$ e gli indici di x sono tutti e soli i numeri naturali da 1 a n.

Conviene anche introdurre un assioma di esistenza di n–uple come il seguente:

ASSIOMA D9 – *Per ogni $n \in \mathbf{N}$ esiste l'n–upla $(1, \ldots, n)$.*

OSSERVAZIONE: Non avendo introdotto nelle teorie base alcun tipo di assioma d'induzione potremmo dimostrare *separatamente* i vari casi particolari dell'assioma D9 per $n = 1, 2, 3, \ldots$ ma non possiamo dimostrare in generale l'assioma stesso, che nelle teorie base appare quindi indipendente dagli altri assiomi.

Concludiamo questo paragrafo con un assioma che esprime la "finitezza" dei sistemi associando ad ogni sistema un numero naturale che funge da "cardinalità" del sistema:

ASSIOMA D10 – *Per ogni sistema s esistono un numero naturale n e due n–uple x, y tali che $s = y \circ x^{-1}$. Inoltre tra tali n ce n'è uno minimo.*

Tale numero minimo sarà detto *cardinalità* del sistema s e indicato col simbolo $Card\, s$. Si noti che la prima parte dell'assioma D10 non è sufficiente per dimostrare, a partire dagli assiomi del §3, l'esistenza di un *minimo n*, in quanto non abbiamo l'assioma di induzione. Sempre per la mancanza dell'assioma di induzione è opportuno introdurre il seguente assioma:

Assioma D11 – *Sia $s = y \circ x^{-1}$, $x, y \in V^n$; allora Card $s = n$ se e solo se per $1 \leq i < j \leq n$ si ha sempre $\begin{pmatrix} x_i \\ y_i \end{pmatrix} \neq \begin{pmatrix} x_j \\ y_j \end{pmatrix}$.*

Il significato dell'assioma D11 è il seguente: se n è la cardinalità di $s = y \circ x^{-1}$ allora s è unione di n sistemi singolari tra loro distinti $\begin{pmatrix} x_1 \\ y_1 \end{pmatrix}, \ldots, \begin{pmatrix} x_n \\ y_n \end{pmatrix}$. Dagli assiomi D9–11 segue tra l'altro che gli interi da 1 a k non possono essere in corrispondenza biunivoca con quelli da 1 a h per $h \neq k$.

6. Le relazioni universali

Introduciamo ora un nuovo oggetto altamente autoreferente e di alto potere descrittivo che "assorbe" tutte le relazioni $R fond h$, cioè l'operazione generatrice di relazioni universali $Runiv$, caratterizzata dall'assioma seguente:

Assioma E – *$Runiv$ è un'operazione semplice avente le seguenti proprietà:*

a) *Se $h \in \mathbf{N}$, $h > 0$ e $z = Runiv\, h$ allora z è una relazione di arietà $h+2$.*

b) *Se $(Runiv\, h) \downarrow\uparrow x_1 \ldots x_h\, y\, t$ allora $x_i = i$ per $1 \leq i \leq h$.*

c) *Se $(Runiv\, h) \downarrow\uparrow 1 \ldots h\, y\, t$, $Ar\, y = k \in \mathbf{N}$, $k \geq 1$, allora $t \in V^k$.*

d) *Se $Ar\, y = k \in \mathbf{N}$, $k \geq 1$, $(Runiv\, h) \downarrow\uparrow 1 \ldots h\, y(x_1, \ldots, x_k)$ se e solo se $y \downarrow\uparrow x_1 \ldots x_k$.*

L'idea ispiratrice delle teorie base è quella di garantire l'esistenza della relazione $Runiv\, k$ per k abbastanza grande in modo che la $Runiv k$, che è dotata di un grandissimo e "pericoloso" contenuto autoreferenziale, abbia una complessità sufficiente per proteggere la teoria base, e gli "innesti" delle più note teorie matematiche che su di essa si possono compiere, dai rischi di antinomie dei tipi studiati in [8], [18], [19]. Perciò l'assioma fondamentale di una teoria base dipenderà da un prefissato numero naturale $\nu > 0$ e garantirà l'esistenza di $Runiv\, k$ per $k \geq \nu$. Fissato quindi ν, ad esso associamo l'assioma:

Assioma F_ν – *Per ogni $k \in \mathbf{N}$, se $k \geq \nu$ allora esiste $Runiv\, k$.*

La teoria avente gli assiomi ABCDEF$_\nu$ sarà chiamata TB$_\nu$ (teoria base ν-aria). Come si mostrerà in [15], F_ν è la "chiave di volta" della teoria ed è essenziale per garantire la finita assiomatizzabilità e un alto grado di autodescrizione della teoria stessa. Notiamo che dalle condizioni $k \geq \nu > 0$ segue che $Runiv\, k$ ha arietà maggiore di 2 e quindi non rischia di essere confuso con gli oggetti di arietà 0, 1 e 2 propri della Matematica tradizionale.

Research partially supported by 40% MURST funds "Logica Matematica e Applicazioni".

Bibliografia

[1] A. Asperti – G. Longo, *Categories, types and structures – An introduction to category theory for the working computer scientist*, Cambridge, Mass. 1991.

[2] H. P. BARENDREGT, *The Lambda–Calculus*, Amsterdam 1981.

[3] Y. BAR HILLEL – A. A. FRAENKEL – A. LEVY, *Foundations of set theory*, Amsterdam 1973.

[4] M. BEESON, *Towards a computation system based on set theory*, preprint, CSLI, Stanford 1987.

[5] J. BENABOU, *Fibered categories and the foundations of naïve category theory*, Journal of Symbolic Logic, 50, 1985, pp. 10–37.

[6] M. BOFFA, *Sur la théorie des ensembles sans axiome de fondement*, Bull. Soc. Math. Belg., 21, 1969, 16–56.

[7] A. CHURCH, *Set theory with a universal set*. In: L. HENKIN *et al.* (eds.), *Proceedings of the Tarski Symposium*. Proc. of Symp. P. Math. XXV, Rhode Island 1974, 297–308.

[8] M. CLAVELLI – E. DE GIORGI – M. FORTI – V. M. TORTORELLI, *A selfreference oriented theory for the Foundations of Mathematics*. In: *in "Analyse Mathématique et applications. Contributions en l'honneur de Jacques–Louis Lions. Paris 1988, 67–115.*

[9] H. B. CURRY – R. FEYS, *Combinatory logic*, Amsterdam 1958.

[10] E. DE GIORGI, *Fondamenti della Matematica e teoria base: gli esempi delle teorie 7×2 e 7×5* (sunto di una conversazione). Lecce 1989, manoscritto.

[11] E. DE GIORGI – M. FORTI, *Una teoria quadro per i fondamenti della Matematica*. Atti Acc. Lincei, Rend. Fis., s. 8, vol. 79, 1985, 55–67.

[12] E. DE GIORGI – M. FORTI, *"5×7": A Basic Theory for the Foundations of Mathematics*. Preprint di Matematica n. 74, Scuola Normale Superiore, Pisa 1990.

[13] S. FEFERMAN, *Constructive theories of functions and classes*. In: M. BOFFA *et al.* (eds.), *Logic Colloquium 1978*. Amsterdam 1979.

[14] M. FORTI – F. HONSELL, *Models of self–descriptive set theories*. In: F. COLOMBINI *et al.* (eds.), *Partial Differential Equations and the Calculus of Variations – Essays in Honor of Ennio De Giorgi*. Boston 1989, 473–518.

[15] M. FORTI – G. LENZI, *Assiomi e modelli di teorie base dei Fondamenti della Matematica*. In preparazione.

[16] G. FREGE, *Grundgesetze der Aritmetik, begriffsschriftlich abgeleitet*. Vol. I, Jena 1893; vol. II, Pohle, Jena 1903 (reprinted Olms, Hildesheim 1962).

[17] J. M. Glubrecht – A. Oberschelp – G. Todt, *Klassenlogik*. Zurich 1983.

[18] G. Lenzi, *Estensioni contraddittorie della teoria Ampia*. Atti Acc. Lincei, Rend. fis., s. 8, vol. 83, 1989, 13–28.

[19] G. Lenzi – V. M. Tortorelli, *Introducing predicates into a basic theory for the foundations of Mathematics*. Preprint di Matematica n. 51, Scuola Normale Superiore, Pisa, 1989.

[20] W. V. O. Quine, *New foundations for mathematical logic*. Amer. Math. Monthly, 44, 1973, 70–80.

[21] B. Russell – A. N. Whitehead, *Principia Mathematica*. Cambridge 1925.

[22] D. Scott, *Combinators and classes*. In: C. Böhm (ed.), λ-*Calculus and Computer Science Theory*. LNCS, 37, Berlin 1975.

[23] J. Von Neumann, *Eine Axiomatisierung der Mengenlehre*. J. f. Math., 154, 1925, 219–240.

E. De Giorgi, G. Lenzi:
Scuola Normale Superiore
Piazza dei Cavalieri 7 – 56126 Pisa

M. Forti:
Istituto di Matematiche Applicate "U. Dini"
Università degli studi di Pisa
Via Bonanno, 25 B – 56100 Pisa

An introduction of variables into the frame of the basic theories for the Foundations of Mathematics[‡]

Ennio De Giorgi, Marco Forti, Giacomo Lenzi*

Summary. We deal with the notion of *variable* inside of the axiomatic frame of the Basic Theories for the Foundations of Mathematics [9]. Variables are introduced into this frame as "unary" objects, taking values of different kinds, which can be connected by *correlations* (or correspondences), and allow local functional representations. In choosing the axioms on variables we take into account the main uses of the term "variable" in Mathematical Analysis, Mathematical Physics, Algebra, Geometry, Logic and in several exact and human sciences (Physics, Biology, Computer Science, Economics, Sociology, etc.).
KEYWORDS: Variable; Correlation; Operation; Basic theory.

0. INTRODUCTION

This *Note* is a development of the Basic theories for the Foundations of Mathematics exposed in [9]; more precisely, we introduce in this frame the concept of *variable*. This concept is widely used in many branches of pure and applied Mathematics, of experimental Sciences and even in human Sciences. The axioms exposed in this work have been suggested by various sources of inspiration, so they should be a good starting point for comparing the various ideas of variable present in many scientific and human disciplines. As it was for the basic theories, we cannot provide an exhaustive bibliography of our sources of inspiration; so we limit ourselves to signaling some works which we took particularly into account while writing this *Note* (see the Bibliography).

A direct precedent of this *Note* is the paper [5]: in this paper, the notion of variable is "engrafted" on the Ample Theory exposed in [6], while here the reference, as we said, are the basic theories of [9].

To give an intuitive idea of the experiences which inspire the axioms on variables, we could think of variables as available information on some reality which we cannot know completely. For instance, we could think that the reality is the overall physical and chemical state of people undergoing clinical tests; then variables could be radiographs, ecographs, encephalograms, cardiograms, etc. It is sure that the "values" of these variables depend on the state of the patient in a very complex, partially unknown way; so in order to have an idea of this state

---

[‡]Editor's note: translation into English of the paper "Introduzione delle variabili nel quadro delle teorie base dei Fondamenti della Matematica", published in Atti Accad. Naz Lincei, Cl. Sci. Fis. Mat. Nat., **(9) 5** (1994), 117–128.

*Note presented by E. De Giorgi to the Accademia dei Lincei on December 11, 1993.

it is useful first to look for links between the available information; these links will be schematized here in the concept of *correlation* or *correspondence*.

1. Recalling the basic theories

The basic theories of [9] are the most recent development of a research project on the Foundations of Mathematics, whose main features are the non–reductionism (the theories consider various kinds of objects rather than just numbers and sets), and a considerable degree of self–reference (the global behavior of the objects of the theory is described by some "big" objects of the theory itself). Basic theories are "open–ended theories" which can be extended in various directions; they are intended to be, as suggested by their name, a "basis" on which one can "engraft" the various branches of Mathematics and possibly other forms of human knowledge.

Among the fundamental objects introduced in [9] we have some *qualities* (or properties), *collections, relations, operations, natural numbers* and *finite systems*. Qualities and collections are unary objects; finite systems are binary; relations can be binary, ternary, quaternary, etc., and operations can be simple, binary, ternary, etc. A quality or property can be *enjoyed* by an object x; in this case we write $q\,x$. A collection can *have* an object x *as an element*, and in this case we write $x \in C$. A binary relation can *hold* between two objects x, y, and in this case we write $r\,x\,y$ (or sometimes $x\,r\,y$)[‡]; likewise, for ternary, quaternary, etc. relations r we use the notation $r\,x\,y\,z$, $r\,x\,y\,z\,t$, etc. A simple operation can *transform* an object x into an object y, and in this case we write $y = f\,x$; likewise, for binary, ternary, etc. operations f we use the notation $z = f\,x\,y$, $t = f\,x\,y\,z$, etc.

The first qualities, in the spirit of self–reference, correspond to the various kind of objects introduced; so we will have the quality $Qqual$ of being a quality, the quality $Qcoll$ of being a collection, etc. Here we have already a self–reference because the quality $Qqual$ enjoys itself, that is we have $Qqual\,Qqual$. Other objects of general purpose are the identity operation Id, such that $Id\,x = x$ for any x, the universal collection V, whose elements are all the objects, the empty collection $\emptyset$ without elements, and the collection of all collections $Coll$. Note that we have a chain of mutual membership: $V \in Coll \in Coll \in V \in V$. Collections are *extensional*, that is two collections with the same elements are equal.

After introducing collections, relations and operations we can express the most elementary part of Arithmetic by introducing the *collection* **N** of all natural numbers, the arithmetic *operations* of addition, subtraction and multiplication, and the usual ordering *relation* between natural numbers. The part of *Arithmetic* exposed in [9] is very reduced; for instance powers are not introduced, and no form of the induction principle is postulated.

Introducing the first elements of Arithmetic allows us to formalize the notion of "complexity" (or "arity") previously used in an informal way, by means of an operation *arity*, Ar. For $x \in$ **N** we have $Ar\,x = 0$; for every quality q and

---

[‡]Editor's note: when we use this notation the left hand side is never a relation, so no ambiguity arises.

every collection C, $Ar\,q = Ar\,C = 1$. Moreover, for every relation r we have $Ar\,r = h$ if and only if r is an h–ary operation, and for every operation f we have $Ar\,f = h + 1$ if and only if f is an h–ary operation. More generally, we will say that an object x is *unary* (respectively *binary*, *ternary*, etc.) if we have $Ar\,x = 1$ (2, 3, etc.) The axioms allow us to introduce, during further engraftings, other binary objects than binary relations, simple operations and finite systems which will be talked about later, etc.; this freedom is useful for further engraftings, including the present *Note*, where variables will be introduced as unary objects. In [9] we do not introduce objects whose arity is not a natural number, but this possibility is not excluded by the axioms, and it might make it easier to perform future engraftings, for instance that of infinitary logics.

In the theories exposed in [9] a good degree of self–description is obtained by introducing the two operations generating fundamental relations $Rfond$ and $Runiv$: beginning with $Rfond$, we recall that for every natural number $h > 0$, $Rfond\,h$ exists and is a *relation* of arity $h + 1$ which rules the objects of arity h. In general, if α is an object of arity h, we will use the notation

$$\alpha \downarrow\uparrow x_1 \ldots x_n,$$

to be read as "α *acts on* $x_1 \ldots x_h$", for $(Rfond\,h)\,\alpha\,x_1 \ldots x_h$. The *action* defined in this way subsumes and extends the one previously considered for qualities, relations, operations and collections; for instance:

for every quality q and for every object x we have $q \downarrow\uparrow x$ if and only if $q\,x$;

for every collection C and for every x we have $C \downarrow\uparrow x$ if and only if $x \in C$;

for every binary relation r and for every x, y we have $r \downarrow\uparrow x\,y$ if and only if $r\,x\,y$;

for every simple operation f and for every x, y we have $f \downarrow\uparrow x\,y$ if and only if $y = f\,x$;

etc.

It is useful to introduce also the following notation, where α and β are objects of the same arity $h > 0$:

$\alpha \subseteq \beta$ if, for every $z_1, \ldots, z_h$, $\alpha \downarrow\uparrow z_1 \ldots z_h$ implies $\beta \downarrow\uparrow z_1 \ldots z_h$;

$\alpha \simeq \beta$ if $\alpha \subseteq \beta \subseteq \alpha$;

$\alpha \uparrow x_1 \ldots x_{h-1}$ if there is x_h such that $\alpha \downarrow\uparrow x_1 \ldots x_{h-1}x_h$;

$\alpha \downarrow x_h$ if there are $x_1, \ldots, x_{h-1}$ such that $\alpha \downarrow\uparrow x_1 \ldots x_{h-1}x_h$.

By means of the notation above we express *functionality* and *extensionality* by means of the qualities $Qfun$ and $Qest$. Namely we will say that an $h + 1$–ary object f is *functional* (that is, it enjoys the quality $Qfun$) if $f \downarrow\uparrow x_1 \ldots x_h\,y$ and $f \downarrow\uparrow x_1 \ldots x_h\,z$ imply $y = z$. For instance, all operations are functional objects.

Moreover, we will say that a collection C is *extensional* (that is, it enjoys the quality $Qest$ if all elements of C have the same arity, and $x, y \in C$ together with $x \simeq y$ implies $x = y$.

The last fundamental kind of objects introduced in [9] is the one of *finite systems*, which contains ordered pairs, triples, quadruples, ..., n–tuples, substitutions on a finite number of elements, and various other concepts of Mathematics and real life, where some objects are put in relation with some indexes. All finite systems have arity 2, and they form an extensional collection which we will denote by the symbol Sif. We will also say that h is an *index* of the system x when $x \uparrow h$, and y is a *value* of a system x when $x \downarrow y$. Among finite systems we have the *univalent* ones, that is finite systems where the index determines uniquely the indexed object (hence, they are functional in the sense above): they form the collection $Siuf$.

In [9] we postulate that, given any two objects x, y, there is always the univalent system $\begin{pmatrix} x \\ y \end{pmatrix}$, which has only the index x and only the value y. Moreover, systems admit the operations of binary union, denoted by the usual symbol $\cup$, of inversion (symbol $^{-1}$), and of composition (symbol $\circ$).

We note that by composing, say, $\begin{pmatrix} 1 \\ 1 \end{pmatrix}$ and $\begin{pmatrix} 2 \\ 2 \end{pmatrix}$, we get the empty system (unique by the extensionality axiom on the collection Sif) which will be denoted by the symbol $\emptyset_2$, to emphasize that it is an empty object of arity 2 (so it is different from the collection $\emptyset$, which has arity 1).

Since union is available, we can construct more complex systems, for which we will use the notations taken from the theory of groups of substitutions:

$$\begin{pmatrix} x & y \\ a & b \end{pmatrix} = \begin{pmatrix} x \\ a \end{pmatrix} \cup \begin{pmatrix} y \\ b \end{pmatrix},$$

$$\begin{pmatrix} x & y & z \\ a & b & c \end{pmatrix} = \begin{pmatrix} x \\ a \end{pmatrix} \cup \begin{pmatrix} y \\ b \end{pmatrix} \cup \begin{pmatrix} z \\ c \end{pmatrix}, \ etc.$$

As particular cases of univalent systems we have:

the 1–tuples, that is the systems of the kind $\begin{pmatrix} 1 \\ x \end{pmatrix}$, which we will denote also by $[x]$,

the ordered pairs $(x, y) = \begin{pmatrix} 1 & 2 \\ x & y \end{pmatrix}$,

the triples $(x, y, z) = \begin{pmatrix} 1 & 2 & 3 \\ x & y & z \end{pmatrix}$, etc.

As we recalled above, in [9] we do not introduce the principle of induction, which has to be considered in further developments of the theory, but we postulate explicitly that for every $n \in \mathbf{N}$ there is the tuple $(1, \dots, n)$ and the collection V^n of all n–tuples.

In general, if x is a functional binary object, for instance a univalent finite system, rather than $x \downarrow\uparrow h\,w$ we will write sometimes $x_h = w$, following the usual notation for n–tuples, or $x\,h = w$, inspiring ourselves to the notation for operations.

Finally, since finite systems are available, we introduce a new, highly self–referential object with high describing power, which "absorbs" all relations $R\,fond\,h$, that is the operation generating universal relations, $Runiv$, such that

$$(Runiv\,k) \downarrow\uparrow 1\,2\ldots k\,\alpha(t_1,\ldots,t_h) \text{ if and only if } \alpha \downarrow\uparrow t_1\ldots t_h.$$

The fundamental axiom F_ν of a basic theory depends on a fixed natural number $\nu > 0$ and states the existence of $Runiv\,k$ for $k \geq \nu$. We will call TB_ν (ν–ary basic theory) the theory containing the axiom F_ν besides the other axioms exposed in the *Note* [9] (which we outlined here).

Since $Ar\,(Runiv\,k) = k + 2$, the axiom F_ν introduces only relations of arity not less than $\nu + 1 \geq 3$, hence the universal relations are not necessarily concerned by the various manipulations which one might want to perform on unary and binary objects when strengthening the "open–ended theories" TB_ν. If one wants a greater freedom of manipulating objects of arity $3, 4,,$ etc., it is useful to consider theories TB_ν with a sufficiently high index ν: in this way, we can simultaneously ensure a strong self–reference and a wide freedom of "engrafting" and manipulating objects with low arities without too much danger of antinomies (see [9,11]).

It is sure that, among the objects introduced in [9], the operation $Runiv$ seems to be the less "natural", in particular it seems artificial to have the numbers $1,\ldots,k$ in the formula expressing the action of $Runiv\,k$; but anyway, this operation is useful for granting freedom and consistency.

2. Transposition of binary objects

First we fix some shorthand terminology for finite systems. When a finite system S has only *values* of a certain kind, we will simply say that S is a finite system *of* objects of that kind: for instance, a "finite system of operations" is a finite system whose values are operations.

Moreover, since in this and the next section we will consider only finite univalent systems, in these sections we will use simply the word "system" instead of "finite univalent system"; for instance we will write "S is a system" rather than "$S \in Siuf$".

In order to prepare the engrafting of variables, we begin by introducing a new operation, the *transposition of binary objects*, denoted by $Trab$, which associates to each system of functional binary objects a functional binary object whose values are systems; for instance, if f, g are two operations defined on x, we have $(Trab\,(f,g))\,x = (f\,x, g\,x))$.

The axioms which characterize $Trab$ are the following:

Axiom A1. – *Trab is a simple operation.*

Axiom A2. – *Let S be a system whose values are functional binary objects: Then $Tab\,S$ exists and is a functional binary object whose values are nonempty* *systems. and we have the relation* $((Trab\, S)x)_i \cong (S_i)x$, *where the left hand side is defined if and only if the second one is.*

Axiom A3. – *If S is a system of systems, then $Trab\, S$ is a system of systems as well.*

In the sequel we will use the abridged notation $S^t = Trab\, S$.

We note that, by extensionality of the collection Sif, the previous axioms determine uniquely the transposition of systems of systems. For instance, let us consider the empty system $\emptyset_2$: it results $(\emptyset_2)^t = \emptyset_2$; more generally, since the values of the transposed system must always be nonempty systems, we have as well:

$$[\emptyset_2]^t = (\emptyset_2, \emptyset_2)^t = (\emptyset_2, \emptyset_2, \emptyset_2)^t = \ldots = \emptyset_2.$$

Other examples can be obtained by considering $m, n \in \mathbf{N}, m, n > 0, x = (x_1, \ldots, x_m)$, with $x_h = (a_{h1}, \ldots, a_{mk})$. Since the m–tuples of n–tuples (with n fixed) can be identified with the matrices $m \times n$, this result reminds the usual transposition of matrices, which explains the name "transposition" for the operation $Trab$. In general the transposition of a nonempty system S of n–tuples (with n constant) is an n–tuple of nonempty systems with the same indexes as S.

More complex is the situation concerning m–tuples of n–tuples with $n > 0$ non–constant; in this case the transposition, in general, will not produce an analogous object, but only a k–tuple of systems indexed by natural numbers, where k is the greatest value of n. For instance:

$$((a,b), (c,d,e))^t = ((a,c), (b,d), \begin{pmatrix} 2 \\ e \end{pmatrix}),$$

$$(\emptyset_2, (a,b), (c,d,e))^t = (\begin{pmatrix} 2 & 3 \\ a & c \end{pmatrix}, \begin{pmatrix} 2 & 3 \\ b & d \end{pmatrix}, \begin{pmatrix} 3 \\ e \end{pmatrix}).$$

Turning to the transposition of systems which are not n–tuples, we have for instance:

$$\begin{pmatrix} 2 & 4 & 6 \\ (a,b) & [c] & (d,e,f) \end{pmatrix}^t = (\begin{pmatrix} 2 & 4 & 6 \\ a & c & d \end{pmatrix}, \begin{pmatrix} 2 & 6 \\ b & e \end{pmatrix}, \begin{pmatrix} 6 \\ f \end{pmatrix}).$$

More generally, if S is any system of systems, the transposition can be iterated on S: intuitively, two transpositions should cancel each other; actually it results $S^{tt} = S$ if and only if $\emptyset_2$ is not a value of S, whereas in any case we have $S^{ttt} = S^t$.

3. Introducing variables

Now we introduce the main object of this *Note*, that is the collection Var of all *variables*, and we postulate that they are unary objects by means of the axiom:

Axiom B1. – *Var is a collection. If $v \in Var$, then $Ar\, v = 1$.*

If v is a variable and $v \downarrow\uparrow x$, we will say that x is a *value* of v.

The fundamental operation on variables is the *transposition of n–tuples of variables, $Trav$*; the first axiom concerning $Trav$ states:

AXIOM B2. – *$Trav$ is a simple operation. If S is an n–tuple of variables, then $Trav\,S$ exists and is a variable, whose values are nonempty systems.*

We use the notation $S^* = Trav\,S$. Moreover, since among variable–valued n–tuples we have the empty system, we postulate:

AXIOM B3. – *The variable $\emptyset_2^*$ has no value, that is $\emptyset_2^* \simeq \emptyset$.*

We are left with establishing the link between the transposition of variables and the one of functional binary objects. We want to ensure that "locally", that is within any system S of variables, the transposition be an image of the transposition of suitable functional binary objects, by means of a "homomorphism" τ, which "preserves values" and maps two equivalent functional objects to the same variable; so we give the axiom:

AXIOM B4. – (*Local functional representation*). *Let S be a system of variables. Then there is a system τ such that:*

1) *the indexes of τ are functional binary objects, and its values are variables;*

2) *the values of S are also values of τ;*

3) *the variable τf takes the same values as the functional object f;*

4) *if f, g are indexes of τ and $f \simeq g$, then $\tau f = \tau g$;*

5) *if $(v_1, \ldots, v_n)$ is an n–tuple of variables, with $n > 0$, and the $n+1$ variables $v_1 = \tau\, f_1, \ldots, v_n = \tau\, f_n, (v_1, \ldots, v_n)^*$ are values of S, then $(f_1, \ldots, f_n)^t$ is an index of τ, and we have: $(v_1, \ldots, v_n)^* = \tau\,(f_1, \ldots, f_n)^t$.*

Axiom B4 is the fundamental axiom of the theory of variables exposed in this *Note*: to have an intuitive motivation of it we can think of the problem of the state of Economy. The quantities which can be easily measured will be, in our setting, the *variables* of the economical system (for instance: production of wheat, the change between currencies, number of workers employed in various activities, production of cars, level of salaries, interest rate, etc.); they are the result of a great deal of heterogeneous factors (choices of each consumer or producer, political factors, etc.). The economist cannot know all these factors, but he can take a system of variables and see how they relate to each other (that is, consider *correlations* between variables of this system); studying correlations between the variables of a certain system $(x_1, \ldots, x_n)$ corresponds, in our axiomatic theory, to studying the values of $(x_1, \ldots, x_n)^*$. Moreover, the economist can propose theoretical models, essentially by introducing some functions $f_1, \ldots, f_n$ which, if the model is well chosen, take values close to the ones of $x_1, \ldots, x_n$, whereas the values of $(f_1, \ldots, f_n)^t$ are close to the ones of $(x_1, \ldots, x_n)^*$: the axiom B4 is inspired by the limit case of a perfect coincidence between the theoretical model and the experimental reality.

From axioms B1–4 we derive the following useful

CRITERION FOR EQUALITY BETWEEN VARIABLES. – *If x, y are two variables, then $x = y$ if and only if $(x, x)^* \simeq (x, y)^*$.*

We note that, by the criterion above, $\emptyset_2^*$ is the only variable without values.

The transposition of variables allows us to define the qualities of *coherence, Qcoer*, of *independence, Qind*, and the binary relations between variables expressing *dominance, Rdom* (where $y\, Rdom\, x$ means that the variable y is dominated by the variable x), *subordination, Rsub* (where $y\, Rsub\, x$ means that y is subordinated to x), and *dependency Rdip.*

These qualities and relations play a very important role in the study of variables, and they are characterized by the following axioms:

AXIOM B5. – *Qcoer is a quality. If $(x_1, \ldots, x_n)$ is an n–tuple of variables, then: Qcoer $(x_1, \ldots, x_n)$ if and only if all values of $(x_1, \ldots, x_n)^*$ are n–tuples.*

AXIOM B6. – *Qind is a quality. If $(x_1, \ldots, x_n)$ is an n–tuple of variables, then $Qind (x_1, \ldots, x_n)$ is equivalent to the folllowing condition:*

$$(x_1, \ldots, x_n)^* \downarrow\uparrow (y_1, \ldots, y_n)\, iff\, x_i \downarrow\uparrow y_i\, for\, every\, i\, with\, 1 \leq i \leq n.$$

AXIOM B7. – *Rdom, Rsub, Rdip are binary relations. If x, y are two variables, then:*

1) *$y\, Rdom\, x$ if and only if $(x, y)^* \downarrow\uparrow a$ implies $a \in V^1 \cup V^2$;*

2) *$y\, Rsub\, x$ if and only if the previous condition holds, and moreover, $(x, y)^* \downarrow\uparrow [a], (x, y)^* \downarrow\uparrow (b, c)$ imply $a \neq b$;*

3) *$y\, Rdip\, x$ if and only if both the previous conditions hold and moreover, $(x, y)^* \downarrow\uparrow (a, b), (x, y)^* \downarrow\uparrow (a, c)$ imply $b = c$.*

We note that, if (x, y) is a coherent pair of variables, then the conditions 1) and 2) of the previous axiom are necessarily satisfied.

REMARK. Whereas the notions of dependency and independence are inspired by common usage of these terms in various sciences, the notions of coherence, dominance and subordination are suggested by axiom B4. Namely, if $v_1 = \tau f_1, \ldots, v_n = \tau f_n$, then $Qcoer (v_1, \ldots, v_n)$ is equivalent to the condition: *for every object a and for every pair of integers (h, k) from 1 to n, we have $f_h \uparrow a$ if and only if $f_k \uparrow a$.*

Moreover, let $x = \tau f, y = \tau g$: the condition $y\, Rdom\, x$ is equivalent to the condition:

1*) *for every object a, $g \uparrow a \implies f \uparrow a$.*

If moreover we consider the condition

2*) *for every a, b, $g \uparrow a$ and $fa = fb$ imply $g \uparrow b$,*

one can see that $y\, Rsub\, x$ holds if and only if both 1*) and 2*) hold.

From axioms B4–7 we derive the following proposition, which generalizes the above criterion for equality:

PROPOSITION 3.1. – *Let x, y, z be variables. If $y\, Rdip\, x$, $z\, Rdip\, x$ and $(x, y)^* \simeq (x, z)^*$, then $y = z$.*

An important manipulation of variables concerns the *restriction* of a variable x, which takes only the values taken by x when a variable y takes a fixed value k. So we introduce the operation *Restr* and we give the following axiom:

AXIOM B8. – *Restr is a binary operation. For every pair (x, y) of variables and for every object k there is a variable $z = Restr\, x\, (y, k)$ such that:*

1) $z\, Rdip\, (x, y)^*$;

2) $((x, y)^*, z)^* \downarrow\uparrow (a, b)$ *if and only if* $a = (b, k)$ *and* $(x, y)^* \downarrow\uparrow (b, k)$.

By Proposition 3.1, the variable z of the axiom above is uniquely determined by the conditions 1) and 2); it will be called the *restriction* of x under the condition $y = k$ and will be denoted by $x|_{y=k}$. We note that, if y does *not* take the value k, then $x|_{y=k} = \emptyset_2^*$. Moreover, we note that, if x is dominated by y, then x is uniquely determined by its restrictions $x|_{y=k}$.

As an intuitive motivation for introducing restriction, we can think of the three variables of the kinetic theory of gases, that is temperature, volume and pressure: often one considers the correlations between two of these three variables restricted to the case when the third takes a fixed value.

An operation on variables related to restriction is *aggregation*, denoted here by *Aggr*, which satisfies the following axiom:

AXIOM B9. – *Aggr is a simple operation. If x is a variable whose values are variables, then $Aggr\, x$ exists, and is a variable dominated by x such that, for every variable k, we have: $(Aggr\, x)|_{x=k} = k|_{x=k}$.*

We conclude this section with an axiom which produces many variables taking a finite number of values.

AXIOM B10. – *Let S be a nonempty system. For every $n \in \mathbf{N}$ there exists a coherent, independent n–tuple of variables having the same values as S.*

4. GRAPHS, EXTENSIONS AND CORRELATIONS

Now we want to relate the notion of variable introduced in §3 with other aspects of the basic theory, in a way inspired to the common usage, although informal, of the word "variable" in Analysis, Algebra, Geometry, etc.; so, after axiom B4 relating variables with binary objects, we give some definitions and axioms which relate binary objects and variables with collections.

So we begin by introducing the usual concepts of *graph*, *domain* and *codomain* of a binary object by means of the operations *Graf*, *Dom*, *Cod*:

AXIOM C1. – *Graf, Dom and Cod are simple operations whose values are collections.*

If $Ar\, \alpha = 2$ and $Graf\, \alpha$ exists, then $Graf\, \alpha \subseteq V^2$, and for every (x_1, x_2) we have $(x_1, x_2) \in Graf\, \alpha$ if and only if $\alpha \downarrow\uparrow x_1 x_2$.

If $Ar\,\alpha = 2$ and $Dom\,\alpha$ exists, then for every x we have $x \in Dom\,\alpha$ if and only if $\alpha \uparrow x$.

If $Ar\,\alpha = 2$ and $Cod\,\alpha$ exists, then for every x we have $x \in Dom\,\alpha$ if and only if $\alpha \downarrow x$.

We note that the operations $Graf$, Dom, Cod are not required to be defined on *all* binary objects: this would be a very strong hypothesis, contradicting the introduction (convenient for many reasons) of many operations on collections, and in particular contradicting some consequences of axioms C6–9.

The unary analogue of graph is the *extension*, which can be introduced by means of the operation Ext:

Axiom C2. – *Ext is a simple operation whose values are collections. If $Ar\,\alpha = 1$ and $Ext\,\alpha$ exists, then for every x we have $x \in Ext\,\alpha$ if and only if $\alpha \downarrow\uparrow x$.*

After graphs, we introduce the extensional collection $Corr$ of *correlations*, or *correspondences* (whose elements are uniquely determined by their graph), characterized by the axiom:

Axiom C3. – *$Corr$ is an extensional collection of binary objects. All finite systems belong to $Corr$. Every element of $Corr$ has a graph, a domain and a codomain.*

Because of the importance of *functional correlations*, we introduce also their collection $Corfun$ described by the axiom:

Axiom C4. – *$Corfun$ is a collection whose elements are exactly those elements of $Corr$ which enjoy the quality $Qfun$.*

Now we can introduce an operation relating functional objects with variables, the *composition of functional correlations and variables*, Cov, which satisfies the following axiom:

Axiom C5. – *Cov is a binary operation. If f is a functional correlation and v is a variable, then $Cov\,f\,v$ exists, and it is a variable denoted by the symbol $f_{\#}v$. Moreover:*

1) *$f_{\#}v\,Rdip\,v$;*

2) *$(v, f_{\#}v)^* \downarrow\uparrow (a, b)$ if and only if $v \downarrow\uparrow a$ and $f\,a = b$.*

We note that the axiom above determines uniquely $f_{\#}v$ by Prop. 3.1.

By means of graph and extension we can define, when it exists, the *abstraction* of a variable w in the following way:

$$Astr\,w = r \text{ if and only if } r \text{ is a correlation and } Graf\,r = Ext\,w.$$

Moreover, if y, z are variables such that $Astr\,(y, z)^*$ exists and is a functional correlation, the *substitution of the variable x by the variable z inside the variable y* can be defined by: $Sost(z, y/x) = (Astr\,((x, z)^*)_{\#}y$.

We conclude this *Note* by mentioning some *strong axioms* (which imply the existence of many "big collections", "big variables" and "big correlations"), whose consequences would deserve, in our opinion, a deeper investigation (by the way, they have been discussed in the seminar [8]). We begin with the following axiom, which provides "independent and coherent" copies of any fixed variable:

AXIOM C6. – *Given an n–tuple of variables $(x_1, \ldots, x_n)$, there is an n–tuple of variables $(z_1, \ldots, z_n)$ such that:*

1) $Qcoer\,(z_1, \ldots, z_n)$;

2) $Qind\,(z_1, \ldots, z_n)$;

3) $z_i \simeq x_i$ *for* $1 \leq i \leq n$.

Finally, we give the axioms on the existence of the extensions of variables, and on the inversion of the operations Ext and $Graf$:

AXIOM C7. – *For every variable v there is the collection $Ext\,v$.*

AXIOM C8. – *For every collection C there is a variable v such that $Ext\,v = C$.*

AXIOM C9. – *For every collection C there is a correlation r such that, given any two objects x, y, we have $r \downarrow\uparrow x\,y$ if and only if $(x, y) \in C$.*

We note that, in axiom C9, $Graf\,r$ is the collection of all pairs belonging to C. From the first, incomplete investigations on the consequences of the "strong" axioms of this section, an incompatibility seems to emerge between axioms B9 and C8 on the one hand, and all the other axioms on the other hand; so, it seems convenient to develop a theory considering a collection $Gvar$ of "big variables", which satisfies all axioms except B9, and a collection $Lvar$ of "local" or "small" variables, which satisfies all axioms except C8. The connections between local variables and big variables should be analogous to the ones existing between sets and proper classes in theories like Gödel–Bernays.

Finally, we remark that the notion of variable introduced here seems to reflect quite faithfully the common usage of the term "variable" in Analysis, Geometry, Algebra, Mathematical Physics, Logic, Economy, etc. (see for instance [1,7]). If instead one would like to capture the probabilistic notion of "random variable", it would be necessary to enrich the notion of variable presented here with other concepts, reflecting the idea of "probability that a variable takes certain values" (see [2–4,10]).

Acknowledgement. Research partially supported by a 40% MURST grant.

Bibliography

[1] V. ARNOLD, *Méthodes Mathématiques de la Mécanique Classique.* Mir, Moscow, 1976.

[2] P. BALDI, *Calcolo delle Probabilità e Statistica.* Mc Graw–Hill, Milano, 1992.

[3] P. Billingsley, *Probability and Measure*. Wiley & Sons, New York, 1986.

[4] L. Breiman, *Probability*. Addison–Wesley, Reading, Mass., 1968.

[5] M. Clavelli, *Variabili e teoria A*. Rend. Sem. Fac. Sci. Univ. Cagliari, (2) 59 (1985), 125–130.

[6] M. Clavelli – E. De Giorgi – M. Forti – V. M. Tortorelli, *A self–reference oriented theory for the Foundations of Mathematics*. In: *Analyse Mathématique et applications. Contributions en l'honneur de Jacques–Louis Lions*. Gauthiers–Villars, Paris 1988, 67–115.

[7] R. Courant – D. Hilbert, *Methods of Mathematical Physics*. Wiley & Sons, New York, 1989.

[8] G. Lenzi, *Seminario sui Fondamenti della Matematica*, directed by Prof. Ennio De Giorgi. Scuola Normale Superiore, Pisa, a.a. 1993–94 (typewritten notes).

[9] E. De Giorgi – M. Forti – G. Lenzi, *Una proposta di teorie base dei Fondamenti della Matematica*, Rend. Mat. Acc. Lincei, s. 9, v. 5, 1994, 11–22.

[10] W. Feller, *An Introduction to Probability Theory and its applications*. Wiley& Sons, New York, 1968.

[11] G. Lenzi, *Estensioni contraddittorie della teoria Ampia*. Atti Acc. Lincei Rend. Fis., s. 8, vol. 83, 1989, 13–28.

E. De Giorgi, G. Lenzi:
Scuola Normale Superiore di Pisa
Piazza dei Cavalieri 7 – 56126 Pisa

M. Forti:
Istituto di Matematiche Applicate "U. Dini"
Università degli studi di Pisa
Via Bonanno, 25 B – 56100 Pisa

Introduzione delle variabili nel quadro delle teorie base dei Fondamenti della Matematica

Ennio De Giorgi, Marco Forti, Giacomo Lenzi[‡*]

Sunto.   Introduciamo la nozione di *variabile* nel quadro assiomatico delle Teorie Base dei Fondamenti della Matematica [9].   In tale quadro le variabili sono inserite come oggetti "unari", assumono valori di varie specie, possono essere connesse da *correlazioni* (o corrispondenze) e ammettono *rappresentazioni funzionali locali*. Gli assiomi sulle variabili sono scelti tenendo presenti gli usi più frequenti del termine "variabile" in Analisi Matematica, Fisica Matematica, Algebra, Geometria, Logica e in molte scienze esatte ed umane (Fisica, Biologia, Informatica, Economia, Sociologia, ecc.).
KEYWORDS: Variable; Correlation; Operation; Basic theory.

0. INTRODUZIONE

Questa *Nota* rappresenta uno sviluppo delle teorie base dei Fondamenti della Matematica esposte in [9]; precisamente introduciamo in tale quadro il concetto di *variabile*. Questo concetto è largamente usato in molti rami della Matematica pura e applicata, delle diverse Scienze sperimentali e delle stesse Scienze umane. Gli assiomi che esponiamo in questo lavoro sono in varia misura ispirati a fonti diverse e dovrebbero costituire quindi un buon punto di partenza per un confronto fra le diverse idee di variabile presenti in molte discipline scientifiche ed anche umanistiche.   Come già per le teorie base è impossibile fornire un'indicazione bibliografica esauriente delle fonti a cui ci siamo ispirati; ci limitiamo a indicare alcuni lavori che abbiamo tenuto particolarmente presenti nello scrivere questa *Nota* (vedi Bibliografia).

Un precedente diretto di questa *Nota* è costituito dall'articolo [5]: in esso la nozione di variabile veniva "innestata" sulla Teoria Ampia esposta in [6], mentre qui il riferimento, come abbiamo detto, sono le teorie base di [9].

Volendo dare un'idea intuitiva delle esperienze che ispirano gli assiomi sulle variabili, potremmo pensare alle variabili come informazioni disponibili riguardanti una realtà che non possiamo conoscere in modo completo. Per esempio potremmo pensare che la realtà sia lo stato complessivo fisico e psichico delle persone che si sottopongono ad esami clinici; le variabili potrebbero allora essere radiografie, ecografie, encefalogrammi, cardiogrammi, ecc. Certamente i "valori" di queste variabili dipendono in modo estremamente complesso e in parte sconosciuto dallo stato del paziente; per farsi un'idea di questo stato conviene in primo luogo

---

[‡]Editor's note: published in Atti Accad. Naz Lincei, Cl. Sci. Fis. Mat. Nat., **(9) 5** (1994), 117–128.

[*]Nota presentata da E. De Giorgi all'Accademia dei Lincei l'11 Dicembre 1993.

cercare collegamenti tra le varie informazioni disponibili; questi collegamenti verranno qui schematizzati nel concetto di *correlazione* o *corrispondenza*.

1. Richiamo delle teorie base

Le teorie base di [9] sono lo sviluppo più recente di un filone di ricerca sui Fondamenti della Matematica le cui tendenze principali sono il non riduzionismo (le teorie considerano varie specie di oggetti e non solo numeri e insiemi) e un notevole grado di autoriferimento (il comportamento globale degli oggetti della teoria è descritto da alcuni "grandi" oggetti della teoria stessa). Le teorie base sono "teorie aperte" che possono essere ampliate in molte direzioni; esse intendono essere, appunto, una "base" su cui "innestare" i vari rami della Matematica ed eventualmente di altre forme del sapere umano.

Tra gli oggetti fondamentali introdotti in [9] vi sono alcune *qualità* (o proprietà), *collezioni, relazioni, operazioni*, i *numeri naturali* ed i *sistemi finiti*. Le qualità e le collezioni sono oggetti unari; i sistemi finiti sono oggetti binari; le relazioni possono essere binarie, ternarie, quaternarie, ecc. e le operazioni possono essere semplici, binarie, ternarie, ecc. Una qualità o proprietà q può essere *goduta* da un oggetto x; in questo caso si scrive $q\,x$. A una collezione C può *appartenere* un oggetto x, e in questo caso si scrive $x \in C$. Una relazione binaria r può *sussistere* tra due oggetti x e y, e in questo caso si scrive $r\,x\,y$ (oppure talvolta $x\,r\,y$)[‡]; analogamente per le relazioni r ternarie, quaternarie, ecc. si usano le notazioni $rxyz$, $rxyzt$, ecc. Un'operazione semplice f può *trasformare* un oggetto x in un oggetto y e in questo caso si scrive $y = f\,x$; analogamente per le operazioni f binarie, ternarie, ecc. si usano le notazioni $z = f\,x\,y$, $t = f\,x\,y\,z$, ecc.

Le prime qualità, nello spirito dell'autoriferimento, corrispondono alle diverse specie di oggetti introdotti; avremo perciò la qualità $Qqual$ di essere una qualità, la qualità $Qcoll$ di essere una collezione, ecc. Qui si ha già un autoriferimento perché la qualità $Qqual$ gode di se stessa, cioè si ha $Qqual\,Qqual$. Altri oggetti di uso generale sono l'operazione identità, Id, tale che per ogni x si ha $x = Id\,x$, la collezione universale, V, cui appartengono tutti gli oggetti, la collezione vuota, $\emptyset$, cui non appartiene nessun oggetto, e la collezione delle collezioni, $Coll$. Si noti che si ha una catena di reciproche appartenenze: $V \in Coll \in Coll \in V \in V$. Le collezioni sono *estensionali*, cioè due collezioni cui appartengono gli stessi oggetti sono uguali.

Avendo introdotto le collezioni, le relazioni e le operazioni si può esporre la parte più elementare dell'Aritmetica, introducendo la *collezione* **N** dei numeri naturali, le *operazioni* aritmetiche di addizione, sottrazione e moltiplicazione, e l'usuale *relazione* d'ordine tra numeri naturali. La parte dell'Aritmetica esposta in [9] è molto ristretta; ad esempio non sono introdotte le potenze e non è postulata alcuna forma del principio d'induzione.

L'introduzione dei primi elementi dell'Aritmetica consente di formalizzare la nozione di "complessità" (o "arietà") usata informalmente in precedenza, mediante un'operazione *arietà*, Ar. Per $x \in$ **N** si ha $Ar\,x = 0$; per ogni qualità

[‡]Editor's note: quando si usa quest'ultima notazione il membro di sinistra non è mai esso stesso una relazione, quindi non insorge alcuna ambiguità.

q e ogni collezione C, $Ar\ q\ =\ Ar\ C\ =\ 1$. Inoltre per ogni relazione r si ha $Ar\ r\ =\ h$ se e solo se r è una relazione h–aria, e per ogni operazione f si ha $Ar\ f\ =\ h+1$ se e solo se f è un'operazione h–aria. In generale diremo che un oggetto x è *unario* (risp. *binario, ternario*, ecc.) se si ha $Ar\ x\ =\ 1$ (2, 3, ecc.). Gli assiomi permettono di introdurre, in occasione di successivi innesti, altri oggetti unari oltre alle collezioni e alle qualità, altri oggetti binari oltre alle relazioni binarie, alle operazioni semplici e ai sistemi finiti di cui parleremo tra poco, ecc.; questa libertà è utile in vista dei successivi innesti sulle teorie base, compresa la presente nota, dove le variabili saranno introdotte come oggetti unari. In [9] non sono introdotti oggetti la cui arietà non è un numero naturale, ma tale possibilità non è esclusa dagli assiomi e potrebbe forse agevolare futuri innesti, come ad esempio quello delle logiche infinitarie.

Nelle teorie esposte in [9] un buon grado di autodescrizione è ottenuto introducendo le due operazioni generatrici di relazioni fondamentali, $Rfond$ e $Runiv$: cominciando da $Rfond$, ricordiamo che per ogni numero naturale $h>0$ esiste $Rfond\ h$ ed è una *relazione* di arietà $h+1$ che governa gli oggetti di arietà h. In generale, se α è un oggetto di arietà h, useremo la notazione

$$\alpha \downarrow\uparrow x_1 \ldots x_h$$

da leggersi "α *agisce* su $x_1 \ldots x_h$" invece di $(Rfond\ h)\alpha x_1 \ldots x_h$. L'*azione* così definita riassume ed estende quella precedentemente considerata per qualità, relazioni, operazioni e collezioni; ad esempio:

per ogni qualità q e per ogni oggetto x si ha $q \downarrow\uparrow x$ se e solo se $q\,x$;

per ogni collezione C e per ogni x si ha $C \downarrow\uparrow x$ se e solo se $x \in C$;

per ogni relazione binaria r e per ogni x, y si ha $r \downarrow\uparrow x\,y$ se e solo se $r\,x\,y$;

per ogni operazione semplice f e per ogni x, y si ha $f \downarrow\uparrow x\,y$ se e solo se $y = f\,x$;

ecc.

Risulta comodo introdurre anche le notazioni seguenti, ove α e β sono oggetti della stessa arietà $h>0$:

$\alpha \subseteq \beta$ se, per ogni $z_1, \ldots, z_h$, $\alpha \downarrow\uparrow z_1 \ldots z_h$ implica $\beta \downarrow\uparrow z_1 \ldots z_h$;

$\alpha \simeq \beta$ se $\alpha \subseteq \beta \subseteq \alpha$;

$\alpha \uparrow x_1 \ldots x_{h-1}$ se esiste un x_h tale che $\alpha \downarrow\uparrow x_1 \ldots x_{h-1}x_h$;

$\alpha \downarrow x_h$ se esistono $x_1, \ldots, x_{h-1}$ tali che $\alpha \downarrow\uparrow x_1 \ldots x_{h-1}x_h$.

Partendo da tali notazioni esprimiamo l'*univocità*, o *funzionalità*, e l'*estensionalità* per mezzo delle qualità $Qfun$ e $Qest$. Precisamente diremo che un oggetto $h+1$-ario f è *funzionale* (gode della qualità $Qfun$) se le condizioni $f \downarrow\uparrow x_1 \ldots x_h\,y$, $f \downarrow\uparrow x_1 \ldots x_h\,z$ implicano $y = z$. Ad esempio tutte le operazioni sono oggetti funzionali. Inoltre diremo che una collezione C è *estensionale* (gode della qualità $Qest$) se tutti gli elementi di C hanno la stessa arietà e le condizioni $x, y \in C$, $x \simeq y$ implicano $x = y$. Ad esempio la collezione $Coll$ è estensionale.

L'ultima specie fondamentale di oggetti introdotta in [9] è quella dei *sistemi finiti*, che comprende le coppie ordinate, terne ordinate, quaterne ordinate,..., n-uple ordinate, le sostituzioni su un numero finito di elementi e vari altri concetti della Matematica e della vita comune ove vengono messi in relazione alcuni oggetti con alcuni indicatori. I sistemi finiti hanno tutti arietà 2, e formano una collezione estensionale che indicheremo col simbolo Sif. Diremo anche che h è un *indice* del sistema x quando $x \uparrow h$ e diremo che y è un *valore* del sistema x quando $x \downarrow y$. Tra i sistemi finiti vi sono quelli *univoci*, cioè i sistemi finiti in cui l'indice determina univocamente l'oggetto indicato (e che quindi sono funzionali nel senso precedente): essi formano la collezione $Siuf$. In [9] si postula che, dati comunque due oggetti x, y, esiste sempre il sistema univoco $\begin{pmatrix} x \\ y \end{pmatrix}$, che ha come unico indice x e come unico valore y. Inoltre i sistemi ammettono le operazioni di unione binaria, indicata con il solito simbolo $\cup$, inversione (simbolo $^{-1}$) e composizione (simbolo $\circ$).

Si noti che, ad es., componendo $\begin{pmatrix} 1 \\ 1 \end{pmatrix}$ e $\begin{pmatrix} 2 \\ 2 \end{pmatrix}$ si ottiene il sistema vuoto (unico per l'assioma di estensionalità della collezione Sif) che verrà indicato col simbolo $\emptyset_2$ per sottolineare che si tratta di un oggetto vuoto di arietà 2 (e quindi diverso dalla collezione vuota $\emptyset$ che ha arietà 1).

Disponendo dell'unione si possono costruire sistemi più complessi, per i quali useremo le notazioni mutuate dalla teoria dei gruppi di sostituzioni:

$$\begin{pmatrix} x & y \\ a & b \end{pmatrix} = \begin{pmatrix} x \\ a \end{pmatrix} \cup \begin{pmatrix} y \\ b \end{pmatrix},$$

$$\begin{pmatrix} x & y & z \\ a & b & c \end{pmatrix} = \begin{pmatrix} x \\ a \end{pmatrix} \cup \begin{pmatrix} y \\ b \end{pmatrix} \cup \begin{pmatrix} z \\ c \end{pmatrix}, \; ecc.$$

Casi particolari di sistemi univoci sono:

le 1–uple, cioè i sistemi del tipo $\begin{pmatrix} 1 \\ x \end{pmatrix}$, che indicheremo anche con $[x]$;

le coppie ordinate $(x, y) = \begin{pmatrix} 1 & 2 \\ x & y \end{pmatrix}$;

le terne $(x, y, z) = \begin{pmatrix} 1 & 2 & 3 \\ x & y & z \end{pmatrix}$, ecc.

Come abbiamo già ricordato, in [9] non viene introdotto il principio d'induzione, che dovrà essere trattato in successivi sviluppi della teoria, ma si postula esplicitamente che per ogni $n \in \mathbf{N}$ esistono l'n-upla $(1, \ldots, n)$ e la collezione V^n di tutte le n-uple.

In generale se x è un oggetto binario funzionale, per esempio un sistema finito univoco, in luogo di $x \downarrow\uparrow hw$ scriveremo talvolta $x_h = w$, seguendo le notazioni usuali nel caso delle n-uple, oppure $x\,h = w$, ispirandoci alle notazioni delle operazioni.

Infine, disponendo dei sistemi finiti, si introduce un nuovo oggetto altamente autoreferente e di alto potere descrittivo che "assorbe" tutte le relazioni $Rfond\,h$, cioè l'operazione generatrice di relazioni universali $Runiv$, tale che

$$(Runiv\ k) \downarrow\uparrow 1\ 2 \ldots k\,\alpha\,(t_1, \ldots, t_h) \text{ se e solo se } \alpha \downarrow\uparrow t_1 \ldots t_h.$$

L'assioma fondamentale F_ν di una teoria base dipende da un prefissato numero naturale $\nu > 0$ e afferma l'esistenza di $Runiv\ k$ per $k \geq \nu$. Chiameremo TB_ν (Teoria Base ν-aria) la teoria comprendente l'assioma F_ν oltre agli altri assiomi esposti nella *Nota* [9] (che qui abbiamo richiamato nelle sue grandi linee).

Poiché $Ar(Runiv\ k) = k + 2$, l'assioma F_ν introduce solo relazioni di arietà non minore di $\nu + 2 \geq 3$ e quindi le relazioni universali non sono necessariamente sottoposte alle varie manipolazioni a cui si può decidere di sottoporre tutti gli oggetti unari e binari in occasione di ulteriori rafforzamenti delle "teorie aperte" TB_ν. Volendo avere una maggiore libertà di manipolazione di oggetti di arietà 3,4 ecc. conviene considerare teorie TB_ν con un indice ν abbastanza alto: in questo modo si riesce nello stesso tempo a garantire un forte autoriferimento e un'ampia "libertà di innesto" e di manipolazione degli oggetti di arietà bassa senza correre troppi rischi di antinomie (vedi [9] e [11]).

Certamente, tra i vari oggetti introdotti in [9], l'operazione $Runiv$ sembra la meno "naturale", in particolare sembra "artificiosa" la presenza dei numeri $1 \ldots k$ nella formula che descrive l'azione di $Runiv\ k$; essa è comunque utile per garantire libertà e coerenza.

2. TRASPOSIZIONE DI OGGETTI BINARI

Stabiliamo anzitutto alcune locuzioni abbreviate riguardanti i sistemi finiti. Quando un sistema finito S ha per *valori* solo oggetti di una certa specie, diremo semplicemente che S è un sistema finito *di* oggetti di quella specie: ad esempio un "sistema finito di operazioni" è un sistema finito i cui valori sono operazioni. Poiché inoltre in questo paragrafo e nel successivo si considereranno solo sistemi univoci finiti, in questi due paragrafi useremo semplicemente la parola "sistema" invece di "sistema univoco finito"; ad esempio scriveremo che "S è un sistema" invece di "$S \in Siuf$".

Per preparare l'innesto delle variabili cominciamo con l'introdurre una nuova operazione, la *trasposizione di oggetti binari*, denotata con $Trab$, che associa a un sistema di oggetti binari funzionali un oggetto binario funzionale a valori sistemi; ad esempio se f, g sono due operazioni definite in x si ha $(Trab(f,g))x = (fx, gx)$.

Gli assiomi che caratterizzano $Trab$ sono i seguenti:

ASSIOMA A1 – *$Trab$ è un'operazione semplice.*

ASSIOMA A2 – *Sia S un sistema a valori oggetti binari funzionali; allora $Trab\ S$ esiste ed è un oggetto binario funzionale a valori sistemi non vuoti; in tal caso vale la relazione $((Trab\ S)x)_i \equiv (S_i)x$, ove il primo membro è definito se e solo se è definito il secondo.*

ASSIOMA A3 – *Se S è un sistema di sistemi allora lo è anche $Trab\ S$.*

Nel seguito useremo la notazione abbreviata $S^t = Trab\ S$.

Notiamo che, per l'estensionalità della collezione Sif, gli assiomi precedenti individuano univocamente la trasposizione dei sistemi di sistemi. Ad esempio consideriamo il sistema vuoto, $\emptyset_2$: risulta $(\emptyset_2)^t = \emptyset_2$; più in generale, poiché i valori del sistema trasposto debbono essere sempre sistemi non vuoti, si ha anche

$$[\emptyset_2]^t = (\emptyset_2, \emptyset_2)^t = (\emptyset_2, \emptyset_2, \emptyset_2)^t = \ldots = \emptyset_2.$$

Altri esempi si hanno considerando $m, n \in \mathbf{N}, m, n > 0$, $x = (x_1, \ldots, x_m)$ con $x_h = (a_{h1}, \ldots, a_{hn})$. Allora $x^t = (y_1, \ldots, y_n)$ con $y_k = (a_{1k}, \ldots, a_{mk})$. Dato che le m-uple di n-uple (con n fissato) possono essere identificate con le matrici $m \times n$, questo risultato ricorda l'usuale trasposizione di matrici, il che spiega il nome di "trasposizione" per l'operazione $Trab$. In generale la trasposizione di un sistema non vuoto S di n-uple (con n costante) è una n-upla di sistemi non vuoti aventi gli stessi indici di S.

Situazioni più complesse riguardano m-uple di n-uple con $n > 0$ variabile: in questo caso la trasposizione, in generale, non produrrà un oggetto analogo ma solo una k-upla di sistemi indicizzati da numeri naturali, ove k è il massimo degli n. Ad esempio:

$$((a, b), (c, d, e))^t = \left((a, c), (b, d), \begin{pmatrix} 2 \\ e \end{pmatrix} \right),$$

$$(\emptyset_2, (a, b), (c, d, e))^t = \left(\begin{pmatrix} 2 & 3 \\ a & c \end{pmatrix}, \begin{pmatrix} 2 & 3 \\ b & d \end{pmatrix}, \begin{pmatrix} 3 \\ e \end{pmatrix} \right).$$

Passando alla trasposizione di sistemi che non sono n–uple abbiamo per esempio:

$$\begin{pmatrix} 2 & 4 & 6 \\ (a,b) & [c] & (d,e,f) \end{pmatrix}^t = \left(\begin{pmatrix} 2 & 4 & 6 \\ a & c & d \end{pmatrix}, \begin{pmatrix} 2 & 6 \\ b & e \end{pmatrix}, \begin{pmatrix} 6 \\ f \end{pmatrix} \right).$$

Più in generale se S è un qualunque sistema di sistemi, la trasposizione può essere iterata su S: intuitivamente due trasposizioni dovrebbero cancellarsi; in realtà risulta $S^{tt} = S$ se e solo se $\emptyset_2$ non è un valore di S, mentre in ogni caso si ha $S^{ttt} = S^t$.

3. Introduzione delle variabili

Introduciamo ora l'oggetto principale di questa nota, e cioè la collezione Var delle *variabili*, e postuliamo che esse sono oggetti unari mediante l'assioma:

Assioma B1 – *Var è una collezione. Se $v \in Var$, allora $Ar\, v = 1$.*

Se v è una variabile e $v \downarrow\uparrow x$ diremo che x è un *valore* di v.

L'operazione fondamentale sulle variabili è la *trasposizione di n-uple di variabili*, $Trav$; il primo postulato riguardante $Trav$ afferma:

Assioma B2 – *$Trav$ è un'operazione semplice. Se S è una n-upla di variabili, allora $Trav\, S$ esiste ed è una variabile a valori sistemi non vuoti.*

Usiamo la notazione $S^* = Trav\, S$. Inoltre, poiché tra le n-uple a valori variabili c'è il sistema vuoto, postuliamo:

ASSIOMA B3 – *La variabile $\emptyset_2^*$ non ha valori, cioè $\emptyset_2^* \simeq \emptyset$.*

Resta da stabilire il legame tra la trasposizione di variabili e quella di oggetti binari funzionali. Vogliamo fare in modo che "localmente", cioè all'interno di ogni sistema di variabili S, la trasposizione di variabili sia un'immagine della trasposizione di opportuni oggetti binari funzionali mediante un "omomorfismo" τ, che "conserva i valori" e porta due oggetti funzionali equivalenti nella stessa variabile; diamo perciò l'assioma:

ASSIOMA B4 – *(Rappresentazione funzionale locale)*
Sia S un sistema di variabili. Allora esiste un sistema τ tale che:

1) *gli indici di τ sono oggetti binari funzionali e i suoi valori sono variabili;*

2) *i valori di S sono anche valori di τ;*

3) *la variabile τf prende gli stessi valori dell'oggetto funzionale f;*

4) *se f,g sono indici di τ e inoltre $f \simeq g$, allora $\tau f = \tau g$;*

5) *se $(v_1, \ldots, v_n)$ è una n-upla di variabili, con $n > 0$, e le $n + 1$ variabili $v_1 = \tau f_1, \ldots, v_n = \tau f_n$, $(v_1, \ldots, v_n)^*$ sono valori di S, allora $(f_1, \ldots, f_n)^t$ è un indice di τ e si ha: $(v_1, \ldots, v_n)^* = \tau(f_1, \ldots, f_n)^t$.*

L'assioma B4 è l'assioma fondamentale della teoria delle variabili esposta in questa nota: per averne una motivazione intuitiva possiamo pensare al problema dello stato dell'Economia. Le quantità che possono essere facilmente misurate saranno, nella nostra impostazione, le *variabili* del sistema economico (per esempio: produzione di grano, cambio delle monete, numero dei lavoratori impiegati in diverse attività, produzione di automobili, livello dei salari, tassi d'interesse, ecc.); esse sono il risultato di moltissimi fattori eterogenei (scelte dei singoli consumatori, dei produttori, fattori politici, climatici, ecc.). L'economista non può conoscere tutti questi fattori, ma può prendere un sistema di variabili e vedere come sono legate l'una all'altra (cioè considerare le *correlazioni* tra le variabili di questo sistema): allo studio delle correlazioni esistenti tra le variabili di un certo sistema $(x_1, \ldots, x_n)$ corrisponde, nella nostra teoria assiomatica, lo studio dei valori di $(x_1, \ldots, x_n)^*$. Inoltre l'economista può proporre dei modelli teorici introducendo in sostanza certe funzioni $f_1, \ldots, f_n$ che, se il modello è ben scelto, prendono valori vicini a quelli delle $x_1, \ldots, x_n$ mentre i valori di $(f_1, \ldots, f_n)^t$ sono vicini a quelli di $(x_1, \ldots, x_n)^*$: l'assioma B4 è ispirato all'ipotesi limite di una perfetta aderenza fra modello teorico e realtà sperimentale.

Dagli assiomi B1–4 deriva il seguente utile

CRITERIO DI EGUAGLIANZA TRA VARIABILI: – *Se x, y sono due variabili allora $x = y$ se e solo se $(x, x)^* \simeq (x, y)^*$.*

Notiamo che, in base a tale criterio, $\emptyset_2^*$ è l'unica variabile priva di valori.

La trasposizione di variabili consente di definire le qualità di *coerenza Qcoer*, di *indipendenza Qind* e le relazioni binarie tra variabili di *dominanza Rdom* (ove $y\, Rdom\, x$ significa che la variabile y è dominata dalla variabile x), di *subordinazione Rsub* ($y\, Rsub\, x$ significa che y è subordinata a x) e di *dipendenza Rdip* ($y\, Rdip\, x$ significa che y dipende da x).

Queste qualità e relazioni hanno un ruolo molto importante nello studio delle variabili e sono caratterizzate dai seguenti assiomi:

Assioma B5 – *Qcoer è una qualità. Se $(x_1, \ldots, x_n)$ è una n–upla di variabili, allora si ha: $Qcoer(x_1, \ldots, x_n)$ se e solo se tutti i valori di $(x_1, \ldots, x_n)^*$ sono n-uple.*

Assioma B6 – *Qind è una qualità. Se $(x_1, \ldots, x_n)$ è una n-upla di variabili, allora*
$Qind(x_1, \ldots, x_n)$ equivale alla condizione seguente:
$(x_1, \ldots, x_n)^* \downarrow\uparrow (y_1, \ldots, y_n)$ se e solo se si ha $x_i \downarrow\uparrow y_i$ per ogni i con $1 \leq i \leq n$.

Assioma B7 – *Rdom, Rsub, Rdip sono relazioni binarie.*
Se x, y sono due variabili, allora si ha y Rdom x se e solo se:

1) $(x,y)^* \downarrow\uparrow a$ implica $a \in V^1$ oppure $a \in V^2$;
si ha y Rsub x se e solo se oltre alla 1) si ha:

2) $(x,y)^* \downarrow\uparrow [a]$, $(x,y)^* \downarrow\uparrow (b,c)$ implicano $a \neq b$;
infine si ha y Rdip x se e solo se oltre alla 1) e alla 2) si ha:

3) $(x,y)^* \downarrow\uparrow (a,b)$, $(x,y)^* \downarrow\uparrow (a,c)$ implicano $b = c$.

Notiamo che se (x,y) è una coppia coerente di variabili allora sono necessariamente verificate le condizioni 1),2) dell'assioma precedente.

Osservazione: mentre le nozioni di dipendenza e indipendenza sono ispirate all'uso comune di questi termini in varie scienze, le nozioni di coerenza, dominanza e subordinazione sono suggerite dall'assioma B4. Precisamente, se $v_1 = \tau f_1, \ldots, v_n = \tau f_n$, allora $Qcoer(v_1, \ldots, v_n)$ equivale alla condizione:

P – *er ogni oggetto a e per ogni coppia di interi h, k compresi tra 1 ed n si ha $f_h \uparrow a$ se e solo se $f_k \uparrow a$.*

Sia inoltre $x = \tau f, y = \tau g$: la condizione y Rdom x equivale alla condizione:

1*) *per ogni oggetto a, se $g \uparrow a$ allora $f \uparrow a$.*

Se poi oltre alla 1*) consideriamo la condizione

2*) *per ogni a, b se $g \uparrow a$ e inoltre $fa = fb$ allora $g \uparrow b$,*

si vede che si ha y Rsub x se e solo se valgono simultaneamente la 1*) e la 2*).

Dagli assiomi B4–7 discende la seguente proposizione, che generalizza il precedente criterio di eguaglianza:

Proposizione 3.1 – *Siano x, y, z variabili. Se y Rdip x, z Rdip x e inoltre $(x,y)^* \simeq (x,z)^*$, allora $y = z$.*

Un'importante manipolazione di variabili riguarda la *restrizione* di una variabile x che prende solo i valori assunti da x quando una variabile y assume un certo valore k. Perciò introduciamo l'operazione *Restr* e diamo il seguente assioma:

Assioma B8 – *Restr è un'operazione binaria. Per ogni coppia (x, y) di variabili e per ogni oggetto k esiste una variabile $z = Restr\ x\ (y, k)$ tale che:*

1) z Rdip $(x,y)^*$;

2) $((x,y)^*, z)^* \downarrow\uparrow (a,b)$ se e solo se si ha $a = (b, k)$, $(x,y)^* \downarrow\uparrow (b, k)$.

Per la proposizione 3.1 la variabile z dell'assioma precedente è univocamente determinata dalle condizioni 1),2); essa sarà chiamata la *restrizione* di x sotto la condizione $y = k$ e sarà denotata da $x|_{y=k}$. Notiamo che, se y *non* prende il valore k, allora $x|_{y=k} = \emptyset_2^*$. Inoltre osserviamo che, se x è dominata da y allora x è univocamente determinata dalle sue restrizioni $x|_{y=k}$.

Come motivazione intuitiva dell'introduzione della restrizione, possiamo pensare alle tre variabili della teoria cinetica dei gas, cioè temperatura, volume e pressione: spesso si considerano le correlazioni tra due di queste tre variabili ristrette al caso in cui la terza prende un valore fisso.

Un'operazione sulle variabili legata alla restrizione è l'*aggregazione*, qui indicata con $Aggr$, che soddisfa il seguente assioma:

ASSIOMA B9 – *Aggr è un'operazione semplice.*
Se x è una variabile a valori variabili allora esiste $Aggr\ x$ ed è una variabile dominata da x tale che, per ogni variabile k, si ha $(Aggr\ x)|_{x=k} = k|_{x=k}$.

Concludiamo questo paragrafo con un assioma che produce molte variabili che prendono un numero finito di valori:

ASSIOMA B10 – *Sia S un sistema non vuoto. Per ogni $n \in \mathbf{N}$ esiste una n-upla coerente e indipendente di variabili aventi tutte gli stessi valori di S.*

4. Grafici, estensioni e correlazioni

Per collegare la nozione di variabile introdotta nel §3 agli altri aspetti della teoria base in modo ispirato all'uso corrente, anche se informale, della parola "variabile" in Analisi, Algebra, Geometria, ecc., dopo l'assioma B4 che collega variabili ed oggetti binari, diamo alcune definizioni ed alcuni assiomi che collegano gli oggetti binari e le variabili con le collezioni.

Cominciamo perciò con l'introduzione degli usuali concetti di *grafico, dominio e codominio* di un oggetto binario mediante le operazioni $Graf, Dom, Cod$:

ASSIOMA C1 – *Graf, Dom e Cod sono operazioni semplici i cui valori sono collezioni.*
Se $Ar\,\alpha = 2$ ed esiste $Graf\,\alpha$, allora $Graf\,\alpha \subseteq V^2$ e per ogni (x_1, x_2) si ha $(x_1, x_2) \in Graf\,\alpha$ se e solo se $\alpha \downarrow\uparrow x_1\,x_2$.
Se $Ar\,\alpha = 2$ ed esiste $Dom\,\alpha$, allora per ogni x si ha $x \in Dom\,\alpha$ se e solo se $\alpha \uparrow x$.
Se $Ar\,\alpha = 2$ ed esiste $Cod\,\alpha$, allora per ogni x si ha $x \in Cod\,\alpha$ se e solo se $\alpha \downarrow x$.

Notiamo che non si richiede che le operazioni $Graf, Dom, Cod$ siano definite su *tutti* gli oggetti binari: questa sarebbe un'ipotesi molto forte, contraddittoria con l'introduzione (opportuna per varie ragioni) di molte operazioni sulle collezioni e in particolare con alcune conseguenze degli assiomi C6–9.

L'analogo unario del grafico è l'*estensione*, che introduciamo mediante l'operazione Ext:

ASSIOMA C2 – *Ext è un'operazione semplice i cui valori sono collezioni. Se $Ar\,\alpha = 1$ ed esiste $Ext\,\alpha$, allora per ogni x si ha $x \in Ext\,\alpha$ se e solo se $\alpha \downarrow\uparrow x$.*

Dopo i grafici introduciamo la collezione estensionale *Corr* delle *correlazioni* o *corrispondenze* (i cui elementi sono univocamente determinati dal loro grafico) caratterizzata dall'assioma:

Assioma C3 – *Corr è una collezione estensionale di oggetti binari. I sistemi finiti appartengono a Corr. Ogni elemento di Corr ha grafico, dominio e codominio.*

Data l'importanza delle *correlazioni funzionali* introduciamo pure la loro collezione *Corfun* individuata dall'assioma:

Assioma C4 – *Corfun è una collezione i cui elementi sono tutti e soli gli elementi di Corr che godono della qualità Qfun.*

Possiamo ora introdurre un'operazione che collega correlazioni funzionali e variabili, la *composizione di correlazioni funzionali e variabili, Cov*, che soddisfa il seguente assioma:

Assioma C5 – *Cov è un'operazione binaria. Se f è una correlazione funzionale e v è una variabile allora esiste Cov f v ed è una variabile denotata col simbolo $f_{\#}v$. Inoltre:*

1) $f_{\#}v$ *Rdip* v;

2) $(v, f_{\#}v)^* \downarrow\uparrow (a, b)$ *se e solo se* $v \downarrow\uparrow a$, $fa = b$.

Notiamo che l'assioma precedente determina univocamente $f_{\#}v$ per la prop. 3.1. Mediante il grafico e l'estensione è possibile definire, ove esiste, l'*astrazione* di una variabile w nel modo seguente:

$$Astr \ w = r \text{ se e solo se } r \text{ è una correlazione e } Graf \ r = Ext \ w.$$

Se poi y, z sono variabili tali che $Astr(y, z)^*$ esiste ed è una correlazione funzionale, allora la *sostituzione della variabile x con la variabile y nella variabile z* può essere definita da: $Sost(z, y/x) = (Astr((x, z)^*))_{\#}y$.

Concludiamo questa *Nota* indicando alcuni *assiomi forti* (che implicano l'esistenza di molte "grandi collezioni", "grandi variabili" e "grandi correlazioni") le cui conseguenze meriterebbero a nostro avviso un approfondito esame (e saranno discusse tra l'altro nel seminario [8]). Cominciamo dal seguente assioma, che provvede "copie indipendenti e coerenti" di ogni variabile prefissata:

Assioma C6 – *Data una n-upla di variabili* $(x_1, \dots, x_n)$ *esiste una n-upla di variabili* $(z_1, \dots, z_n)$ *tale che:*

1) $Qcoer(z_1, \dots, z_n)$;

2) $Qind(z_1, \dots, z_n)$;

3) $z_i \simeq x_i$ *per* $1 \leq i \leq n$.

Seguono infine gli assiomi sull'esistenza dell'estensione delle variabili e sull'inversione delle operazioni *Ext* e *Graf*:

Assioma C7 – *Per ogni variabile v esiste la collezione Ext v.*

Assioma C8 – *Per ogni collezione C esiste una variabile v con Ext $v = C$.*

ASSIOMA C9 – *Per ogni collezione C esiste una correlazione r tale che, fissati comunque due oggetti x, y, si ha $r \downarrow\uparrow xy$ se e solo se $(x, y) \in C$.*

Notiamo che, nell'assioma C9, $Graf\ r$ è la collezione di tutte le coppie appartenenti a C.

Dalle prime provvisorie analisi sulle conseguenze degli assiomi "forti" del presente paragrafo sembrano emergere l'incompatibilità degli assiomi B9 e C8 con il blocco di tutti gli altri assiomi e l'opportunità di sviluppare una teoria che consideri una collezione $Gvar$ delle "grandi variabili", che soddisfa tutti gli assiomi tranne B9, e una collezione $Lvar$ di variabili "piccole" o "locali" che li soddisfa tutti tranne C8. Le connessioni fra variabili locali e grandi variabili dovrebbero essere analoghe a quelle esistenti fra insiemi e classi proprie nelle teorie del tipo Gödel–Bernays.

Osserviamo infine che la nozione di variabile ora introdotta sembra rispecchiare in modo abbastanza completo l'uso corrente del termine "variabile" in Analisi, Geometria, Algebra, Fisica Matematica, Logica, Economia, ecc. (vedi ad es. [1],[7]). Se invece si volesse recuperare la nozione probabilistica di "variabile aleatoria", sarebbe necessario arricchire la nozione di variabile qui presentata con altri concetti che rispecchino l'idea di "probabilità che una variabile assuma certi valori" (vedi [2],[3],[4],[10]).

Acknowledgement. Ricerca parzialmente finanziata dai fondi 40% M.U.R.S.T.

Bibliografia

[1] V. ARNOLD, *Méthodes Mathématiques de la Mécanique Classique.* Mir, Moscow, 1976.

[2] P. BALDI, *Calcolo delle Probabilità e Statistica.* Mc Graw–Hill, Milano, 1992.

[3] P. BILLINGSLEY, *Probability and Measure.* Wiley & Sons, New York, 1986.

[4] L. BREIMAN, *Probability.* Addison–Wesley, Reading, Mass., 1968.

[5] M. CLAVELLI, *Variabili e teoria A.* Rend. Sem. Fac. Sci. Univ. Cagliari, (2) 59 (1985), 125–130.

[6] M. CLAVELLI – E. DE GIORGI – M. FORTI – V. M. TORTORELLI, *A self–reference oriented theory for the Foundations of Mathematics.* In: *Analyse Mathématique et applications. Contributions en l'honneur de Jacques–Louis Lions.* Gauthiers–Villars, Paris 1988, 67–115.

[7] R. COURANT – D. HILBERT, *Methods of Mathematical Physics.* Wiley & Sons, New York, 1989.

[8] G. LENZI, *Seminario sui Fondamenti della Matematica,* diretto dal Prof. Ennio De Giorgi. Scuola Normale Superiore, Pisa, a.a. 1993–94 (note dattiloscritte).

[9] E. De Giorgi – M. Forti – G. Lenzi, *Una proposta di teorie base dei Fondamenti della Matematica*, Rend. Mat. Acc. Lincei, s. 9, v. 5, 1994, 11–22.

[10] W. Feller, *An Introduction to Probability Theory and its applications*. Wiley& Sons, New York, 1968.

[11] G. Lenzi, *Estensioni contraddittorie della teoria Ampia*. Atti Acc. Lincei Rend. Fis., s. 8, vol. 83, 1989, 13–28.

E. De Giorgi, G. Lenzi:
Scuola Normale Superiore di Pisa
Piazza dei Cavalieri 7 – 56126 Pisa

Marco Forti:
Istituto di Matematiche Applicate "U. Dini"
Università degli studi di Pisa
Via Bonanno, 25 B – 56100 Pisa

General Plateau problem and geodesic functionals[‡]

Ennio De Giorgi

In this lecture I will introduce the essential definitions necessary to formulate the *general Plateau problem*, consisting in the Plateau problem for surfaces in finite or infinite dimensional metric spaces. I will also give the notion of geodesic functional.

The new formulation includes as special cases many problems already considered in the mathematical literature (see *e.g.* [Alm], [Fed] and [FF]) and requires a very few prerequisites. This will be clear from this lecture, which is self-contained and perfectly understandable by any mathematician knowing the basic elements of measure theory.

I think that even partial answers to the question set in the *general Plateau problem* could have a great importance in the future development of the project, that I recently begun to develop, of a unitary theory of currents, differential forms and manifolds in metric spaces.

About this still vague project I would like to know the opinions of many friends having various mathematical interests. For this reason my lecture is above all an invitation to collaboration addressed both to mathematicians having a large experience in the field of geometric measure theory, and to mathematicians having experience in other fields, thus perhaps capable of bringing new and interesting ideas to this subject.

1. Preliminary notation

Given a topological space S, we denote by $\mathcal{B}(S)$ the family of Borel subsets of S and by $\mathcal{B}^\infty S$ the family of real valued Borel functions defined in S and bounded therein.

We denote by $\overline{M}(S)$ the set of *extended* positive measures defined in $\mathcal{B}(S)$, namely the set of functions μ defined on the Borel subsets of S, whose values can be either real nonnegative numbers or $+\infty$ and satisfying the following properties:

a) $\mu(\emptyset) = 0$,

b) they are countably additive, namely: for any sequence $\{B_h\}_{h \in \mathbf{N}}$ of Borel subsets of S with the property that $B_h \cap B_k = \emptyset$ whenever $h,\, k \in \mathbf{N}$ with

[‡]Editor's note: translation into English of the paper "Problema di Plateau generale e funzionali geodetici", published in Nonlinear Analysis – Calculus of variations (Perugia 1993), Atti Sem Mat. Fis. Univ. Modena, **43** (1995), 285–292.

$h < k$, the following equality holds:

$$\mu\left(\bigcup_{h=0}^{\infty} B_h\right) = \sum_{h=0}^{\infty}\mu(B_h).$$

In the class $\overline{M}(S)$ clearly a minimal element exists, which is the identically zero function, and a maximal element, which assumes the value $+\infty$ on all nonempty Borel sets.

We can consider the infimum and the supremum of a set $\mathcal{F} \subset \overline{M}(S)$ with respect to the usual ordering of measures. Such extremal elements will be denoted by the symbols $\overline{M}(S)\inf\mathcal{F}$, $\overline{M}(S)\sup\mathcal{F}$ and are defined by the formulae

$$\overline{M}(S)\inf\mathcal{F} \;=\; \max\{\mu \in \overline{M}(S)\big|\text{ for any } \nu \in \mathcal{F}, \mu \le \nu\},$$
$$\overline{M}(S)\sup\mathcal{F} \;=\; \min\{\mu \in \overline{M}(S)\big|\text{ for any } \nu \in \mathcal{F}, \mu \ge \nu\}.$$

When $\mathcal{F}$ is empty we clearly have

$$\overline{M}(S)\inf\emptyset = \max\overline{M}(S), \qquad \overline{M}(S)\sup\emptyset = \min\overline{M}(S).$$

The family $M(S)$ of bounded positive measures is particularly interesting among families contained in $\overline{M}(S)$. This is the family of functions $\mu \in \overline{M}(S)$ satisfying the condition $\mu(S) < +\infty$.

We shall now specialize our considerations from topological to metric spaces. Let us introduce some notation concerning Lipschitz functions; namely, given two metric spaces S and S', we shall denote by $\mathrm{lip}\,(S, S')$ the family of Lipschitz functions from S to S'; for any function $f \in \mathrm{lip}\,(S, S')$ we shall denote by $\mathrm{lip}\,(f, S, S')$ its Lipschitz constant. Whenever $S' = \mathbf{R}$ we shall use the abbreviated notation $\mathrm{lip}\,S$ in place of $\mathrm{lip}\,(S, \mathbf{R})$ and $\mathrm{lip}\,(f, S)$ in place of $\mathrm{lip}\,(f, S, \mathbf{R})$.

2. Definition of some classes of functionals and some operations on them

In our formulation of the general Plateau problem, which we shall next describe, we shall look for solutions among real functionals defined on $(k + 1)$-tuples of Lipschitz real functions, the first of which is bounded. We shall denote by $FL_k(S)$ the class of such functionals, which is is uniquely identified by the following

DEFINITION 1. – *Let S be a metric space and let k ba natural number. We shall denote by $FL_k(S)$ the class of real functionals F defined on $(k + 1)$-tuples $(f_0, f_1, ..., f_k)$ belonging to $(\mathrm{lip}\,S \cap \mathcal{B}^{\infty}S) \times (\mathrm{lip}\,S)^k$.*

We shall give now two elementary examples of functionals belonging to $FL_k(S)$; they are *prototypes* of functionals associated to oriented curves, oriented surfaces, currents in the sense of [FF], non-oriented surfaces, varifolds in the sense of [Alm].

EXAMPLE 1. – Let $k > 0$ be a natural number, let $A \subset \mathbf{R}^k$ an open set and let $g \in L^1(A)$. We may consider the two functionals $F, G \in FL_k(A)$ defined as follows:

$$F(f_0, f_1, ..., f_k) = \int_A g f_0 J(f_1, f_2, ..., f_k),$$

$$G(f_0, f_1, ..., f_k) = \int_A |g f_0 J(f_1, f_2, ..., f_k)|,$$

where $J(f_1, f_2, ..., f_k)$ denotes the Jacobian determinant, which is defined at almost all points of A. It is bounded because of the Lipschitz continuity of the functions f_1, f_2, ..., f_k. These functions are defined on the open set A, endowed with the usual metric induced by $\mathbf{R}^k$.

EXAMPLE 2. – Let $\{g_h\}$ be a sequence of real numbers satisfying the condition $\sum_{h=0}^{\infty} |g_h| < +\infty$. We can consider the two functionals F, G, belonging to $FL_0(\mathbf{N})$, defined as

$$F(f) = \sum_{h=0}^{\infty} f_h g_h, \qquad G(f) = \sum_{h=0}^{\infty} |f_h g_h|.$$

In this example we think of $\mathbf{N}$ as a metric space where the distance between two numbers is the absolute value of their difference.

We want to generalize the prototypes of examples 1) and 2), constructiong general functionals associated to curves and surfaces representable in parametric form. To this aim, we first need to give the notion of transposition of the functionals FL, which is introduced in the following

DEFINITION 2. – *Given two metric spaces S and S' and a function $\varphi \in \mathrm{lip}\,(S, S')$ we shall denote by $\varphi_\sharp$ the operator defined as follows: if $F \in FL_k(S)$, the functional $F' = \varphi_\sharp F$ belongs to $FL_k(S')$ and is defined by the formula:*

$$F'(f_0, f_1, ..., f_k) = F(f_0 \circ \varphi, f_1 \circ \varphi, ..., f_k \circ \varphi).$$

The well-posedness of this definition is ensured by the fact that the composition of two Lipschitz function is still a Lipschitz function and the composition of a bounded function and a Lipschitz function is bounded.

Starting from Examples 1), 2) and from Definition 2 we can define the class of rectifiable currents of dimension k on a metric space S, which is denoted by $CR_k(S)$, by means of the following

DEFINITION 3. – *Let S be a metric space, let k be a positive natural number and let C a functional belonging to $FL_k(S)$. We say that C is a rectifiable current and we shall write $C \in CR_k(S)$ if there exist an open set A in $\mathbf{R}^k$, a function $\varphi \in \mathrm{lip}\,(A, S)$, a function g and a functional F satisfying the conditions of Example 1 such that $C = \varphi_\sharp F$.*
On the other hand, given a functional $C \in FL_0(S)$, we say that $C \in CR_0(S)$ if there exist $\varphi \in \mathrm{lip}\,(\mathbf{N}, S)$, g and F satisfying the conditions of Example 2 such that $C = \varphi_\sharp F$.

Finally, if φ and g can be chosen in such a way that g assumes only integral values, we shall say that C is a rectifiable integral current *and we shall write $C \in CRI_k(S)$.*

In a similar way we can define functionals associated to curves, surfaces and non-oriented varieties by means of the following

DEFINITION 4. – *Let S be a metric space, let k a positive natural number and let W a functional belonging to $FL_k(S)$. We say that W is a* rectifiable variety *and we shall write $W \in VR_k(S)$ if there exist an open set A in $\mathbf{R}^k$, a function $\varphi \in \mathrm{lip}\,(A, S)$, a function g and a functional G satisfying the conditions of Example 1 such that $W = \varphi_\sharp G$.*
On the other hand, given a functional $W \in FL_0(S)$, we say that $W \in VR_0(S)$ if there exist $\varphi \in \mathrm{lip}\,(\mathbf{N}, S)$, g and G satisfying the conditions of Example 2 such that $W = \varphi_\sharp G$.
Finally, if φ and g can be chosen in such a way that g assumes only integral values, we shall say that W is a rectifiable integral variety *and we shall write $W \in VRI_k(S)$.*

We can give a notion of *mass* for functionals in the classes FL, extending the usual notions of length of a curve, area of a surface, etc. To this end we first need introduce the notion of external product between a functional and a $(k+1)$-tuple of functions by means of the following definition:

DEFINITION 5. – *Let h, k be two natural numbers, let S be a metric space and let $F \in FL_{k+h}(S)$, $f = (f_0, f_1, ..., f_k) \in (\mathrm{lip}\, S \cap \mathcal{B}^\infty S) \times (\mathrm{lip}\, S)^k$; then the* right external product $F \wedge f$ *and the* left external product $f \wedge F$ *are, by definition, the functionals belonging to $FL_h(S)$ given by the formulae*

$$(F \wedge f)(g_0, g_1, ..., g_h) \;=\; F(f_0 g_0, g_1, g_2, ... g_h, f_1, ..., f_k),$$
$$(f \wedge F)(g_0, g_1, ..., g_h) \;=\; F(f_0 g_0, f_1, f_2, ... f_k, g_1, ..., g_h).$$

The previous discussion concerning supremum and infimum within the lattice of extended positive measures and Definition 5 allows us to give the following definition of mass of a functional belonging to $FL_k(S)$.

DEFINITION 6. – *Let S be a metric space and take $F \in FL_0(S)$. We shall call* mass of F *and denote it by the symbol $\|F\|_S$, the following measure:*

$$\|F\|_S = \overline{M}(S) \inf \left\{ \mu \in M(S); \text{for all } f, g \in \mathrm{lip}\, S \cap \mathcal{B}^\infty S, \right.$$
$$\left. |F(f) - F(g)| \le \int_S |f - g|\, d\mu \right\}$$

If k is a natural number larger than 0 and $F \in FL_k(S)$ we set by definition

$$\|F\|_S = \overline{M}(S) \sup \left\{ \|F \wedge (1_S, f_1, ..., f_k)\|_S \left(\epsilon + \prod_{i=1}^{k} \mathrm{lip}\,(f_i, S) \right)^{-1}; \right.$$
$$\left. \epsilon > 0, \; f_1, ..., f_k \in \mathrm{lip}\, S \right\},$$

where the symbol 1_S denotes the function assuming identically the value 1 on the whole space S.

REMARK 1. – If $F \in FL_0(S)$, it follows from Definition 6 that $\|F\|_S < +\infty$, or $\|F\|_S = \max \overline{M}(S)$. If k is a natural number, S and S' are two metric spaces with the same elements, δ, δ' denote the distances in S and S' respectively, $\delta' \equiv \lambda\delta$ for a real positive constant λ, then it follows $FL_k(S) = FL_k(S')$ and, for any functional $F \in FL_k(S)$, we have $\|F\|_{S'} = \lambda^k\|F\|_S$.

REMARK 2. – For a given functional F in $CR_k(S) \cup VR_k(S)$ it is easy to show that $\|F\|_S(S) < +\infty$.

We conclude with the following definition of boundary of a functional $F \in FL_{k+1}(S)$.

DEFINITION 7. – *Let S be a metric space, let k be a natural number, and let $F \in FL_{k+1}(S)$. We shall call boundary of F and denote it by the symbol ∂F, the functional $F' \in FL_k(S)$ defined by means of the following formula*

$$F'(f_0, f_1, ..., f_k) = F(1_S, f_0, f_1, ..., f_k),$$

where 1_S is the function defined in S with value identically equal to 1.

REMARK 3. – The geometrical meaning of this definition of boundary is rather clear when F and ∂F belong to $CRI_{k+1}(S)$ and $CRI_k(S)$ respectively. It seems interesting to investigate the possible meanings of boundary for other functionals F belonging to $FL_{k+1}(S)$.

3. Open problems

Definitions 3, 6, 7 finally allow us the statement of the general Plateau problem:

PROBLEM 1. – Establish what are the correct hypotheses to be imposed on the natural number k and the metric space S in such a way that the following Theorem holds:
For all $C \in CRI_{k+1}(S)$ there exists the minimum of $\|F\|_S(S)$ subject to the requirements: $F \in CRI_{k+1}(S)$, $\partial F = \partial C$.

REMARK 4. – In order to prove the Theorem in the special case where S is a finite-dimensional Euclidean space it is probably sufficient to carefully compare the previous definitions with the definitions of [FF] and [Fed].
Nevertheless, it would be interesting to establish if there exist more direct proofs based on easily expressable lemmas and starting from Definitions 1, ..., 7, without explicitly referring to the theory of currents as presented in [FF] and in [Fed].
Much more uncertain (and thus much more interesting) is the case when S is an infinite-dimensional space, for example a Hilbert or a Banach space, or a subspace of such spaces with various regularity requirements.
For such choices of S, it would be interesting to investigate the validity of the Theorem even in the special case $k = 1$ or, for $k > 1$, if the mass $\|C\|_S$ has a compact support.
It would also be interesting to investigate the general Plateau problem when the space S is a Lipschitz variety having a finite or infinite dimension.

REMARK 5. – In addition to the class $CRI_{k+1}(S)$ it seems interesting to set the general Plateau problem in different reasonable classes contained in $FL_{k+1}(S)$. On the contrary it does not seem interesting to set the general Plateau problem described above in the class $VRI_{k+1}(S)$. Indeed for $W \in VRI_{k+1}(S)$ the condition $\partial S = \partial W$ appears to be too restrictive.

On the other hand it appears again interesting to investigate the geodesic functionals both in the class $CRI_{k+1}(S)$ and in the class $VRI_{k+1}(S)$. Hence we give the definition of geodesic functional.

DEFINITION 8. – *Let S be a metric space, δ denoting the distance in this space. Let k be a natural number and take $F \in FL_{k+1}(S)$. We say that F is a geodesic functional, and we write $F \in FG_{k+1}(S)$, if $0 < \|F\|_S(S) < +\infty$ and the following equality holds:*

$$\lim_{\rho \to 0^+} \frac{1}{\rho} \sup_\varphi \Big\{ \|F\|_S(S) - \|\varphi_\sharp F\|_S(S);$$
$$\varphi \in \mathrm{lip}\,(S, S'), \delta(\varphi(x), x) \le \rho \; \forall x \in S \Big\} = 0.$$

We are now in a position to write down the *existence problem for currents and generalized geodesic varieties.*

PROBLEM 2. – Give reasonable conditions on the space S ensuring the existence of functionals belonging to $FG_{k+1}(S) \cap CRI_{k+1}(S)$ and the existence of functionals belonging to $FG_{k+1}(S) \cap VRI_{k+1}(S)$.

REMARK 6. – Some very interesting study cases for Problem 2 could arise when S is the boundary of a sufficiently regular open subset of a Hilbert of Banach space, or of the complement of such an open subset.

I wish to express my thanks to prof. Stefano Mortola for his precious collaboration in the preparation of this conference.

Bibliography

[Alm] F. Almgren, *Theory of Varifolds.* Princeton (1965).

[DG1] E. De Giorgi, *Un progetto di teoria unitaria delle correnti, forme differenziali, varietà non orientate, ambientate in spazi metrici.* In preparation.

[FF] H. Federer, W. H. Fleming, *Normal and integral currents.* Ann. Math., **72** (1960).

[Fed] H. Federer, *Geometric Measure Theory.* Springer (1969).

Conjectures about some evolution problems[‡]

Ennio De Giorgi

Summary. I have greatly appreciated the invitation to write a paper in honor of John F. Nash Jr., a scientist who has been a major source of new ideas for mathematicians. His work has been a clear example of how the most original mathematical ideas are often close to the fundamental problems of other disciplines. I believe that this example is very important for any student who is still motivated by the drive that the ancients called *philosophy*, that is, love for Wisdom.

The character of my paper is exploratory. Some conjectures concerning "evolution problems" are presented. They are related to the "steepest descent", to the approximations of Newton's gravitation law, to hyperbolic nonlinear equations and to "descent movements" of manifolds.

The article identifies several questions, of which I do not know the solution, and points out a number of analogies between problems that are apparently far from each other. I believe that the study of these conjectures might provide an opportunity for scientists who are expert in different fields within pure and applied mathematics to get together and ponder the connections that exist among various mathematical concepts, such as linear vs. nonlinear behavior, stability vs. instability, or convergence of different approximation methods, and certain ideas well developed in physics, such as deterministic vs. nondeterministic behavior, predictability vs. nonpredictability, order and chaos, etc.

Introduction. In this paper I expose some conjectures about "evolution problems" related to the idea of "maximal slope descent", to Newton's universal gravitation law, to hyperbolic equations and motions of manifolds embedded in a Euclidean space.

In my opinion the study of these conjectures can help to better understand what happens when one approximates "more difficult and unstable problems" by means of "more easy ans stable" ones. In particular it should be possible to see when an approximation procedure does not converge for all "admissible data", but one has convergence and a certain stability of the limit only for a "dense set" of data or for "almost every" datum. Of course the notions of "stability", of "dense set", and of "almost every", depend on the introduction of reasonable topologies, distances or some significant measure in the family of admissible data.

Due to the heterogeneous character of these conjectures, which maybe have been treated in some paper that I ignore, I can provide only fragmentary and incomplete references, recommended by various colleagues and friends, and I didn't have enough time to examine all of them carefully. They must be considered

---

[‡]Editor's note: translation into English of the paper "Congetture riguardanti alcuni problemi di evoluzione", published in Duke Math. J., **81-(2)** (1995), 255–268.

only has a starting point for a bibliographic research needed to understand the present "state of the art".

I hope that the study of these conjectures might provide an opportunity for scientists that are expert in different fields within pure and applied mathematics to get together and ponder on the connections that exist among various mathematical concepts, such as linear vs. nonlinear behavior, stability vs. instability, or convergence of different approximation methods, and certain ideas well developed in physics, such as deterministic vs. nondeterministic behavior, predictability vs. nonpredictability, order and chaos, etc.

1. Descent problems

Let us begin by considering some problems related to the "descent method". Let us consider functionals such as

$$(1) \qquad F(w, A) = \int_A g(w, \nabla w, \dots, \nabla^k w)$$

where A is a bounded open set in $\mathbf{R}^n$, w is a $C^\infty(\mathbf{R}^n)$ function, g is a C^∞ function in the space of polynomials of n real variables and degree less or equal than k.

Under such assumptions we say that a function $\psi \in C^\infty(\mathbf{R}^n)$ is the Euler function associated to F and w, and we write

$$\psi = \mathcal{E}(F, w)$$

if for every open bounded set $A \subset \mathbf{R}^n$ and every test function $\tau \in C_0^\infty(A)$ the following condition is satisfied

$$(2) \qquad \lim_{\lambda \to 0} \frac{1}{\lambda} \left(F(w + \lambda\tau, A) - F(w, A) \right) = \int_A \tau \, \psi.$$

In some cases where the function g is not convex (or does not satisfy other conditions similar to convexity, see [4]) the study of the differential equation

$$(3) \qquad \frac{\partial u}{\partial t} = -\mathcal{E}(F, u(\cdot, t))$$

with initial conditions

$$(4) \qquad u(x, 0) = \varphi(x)$$

in the set $\mathbf{R}^n \times [0, +\infty)$ can be very difficult, and it seems reasonable to follow methods such as artificial viscosity (cf. e.g. [5]) and associate to F functionals F_ε defined, for $\varepsilon \neq 0$, by the formula

$$F_\varepsilon(w, A) = F(w, A) + \varepsilon^2 \int_A |\nabla^{k+1} w|^2$$

where as usual

$$|\nabla^s w|^2 = \sum_{j_1=1}^{n} \cdots \sum_{j_s=1}^{n} \left(\frac{\partial^s u}{\partial x_{j_1} \ldots \partial x_{j_s}}\right)^2$$

and then try to approximate the differential equation (3) with the equation

$$(5) \qquad \frac{\partial u}{\partial t} = -\mathcal{E}(F_\varepsilon, u(\cdot, t)).$$

A simple but non trivial example of this situation arises by considering the case where $k = 1$, $n = 1$ (cf. [9], [10]), and

$$g\left(w, \frac{\partial w}{\partial x}\right) = \frac{1}{2} \log\left(1 + \left|\frac{\partial w}{\partial x}\right|^2\right).$$

In this case equation (3) becomes

$$\frac{\partial u}{\partial t} = \frac{\partial}{\partial x}\left(\frac{\partial u}{\partial x}\left(1 + \left(\frac{\partial u}{\partial x}\right)^2\right)^{-1}\right) = \frac{\partial^2 u}{\partial x^2}\left(1 - \left|\frac{\partial u}{\partial x}\right|^2\right)\left(1 + \left|\frac{\partial u}{\partial x}\right|^2\right)^{-2}.$$

The study of this equation can be very difficult when the initial datum satisfies condition $\left|\frac{\partial \varphi}{\partial x}\right| > 1$ somewhere in $\mathbf{R}$; one can then consider the functional F_ε which leads to the "stabilized" equation

$$(6) \qquad \frac{\partial u}{\partial t} = \frac{\partial}{\partial x}\left(\frac{\partial u}{\partial x}\left(1 + \left(\frac{\partial u}{\partial x}\right)^2\right)^{-1}\right) - \varepsilon^2\frac{\partial^4 u}{\partial x^4}$$

and state a first conjecture.

CONJECTURE 1 – *If $\varphi \in C_0^\infty(\mathbf{R})$, and, for every $\varepsilon \neq 0$, u_ε is a solution of equation (6) with initial conditions (4) and is bounded in $\mathbf{R} \times [0, +\infty[$, then for almost every $(x, t) \in \mathbf{R} \times [0, +\infty)$ there exists the limit of $u_\varepsilon(x, t)$ as ε tends to zero.*

REMARK 1 – If Conjecture 1 were true, one could strengthen it by requiring the convergence *for every* (x, t) and also stronger convergences yielding a greater regularity of the limit. One could also require some uniformity of the convergence as the initial datum φ varies and then a continuous dependence of the limit on the datum itself. If on the contrary one could find some counterexample to Conjecture 1, this could be weakened in various ways, for example by requiring the convergence only for data in some dense subset of a metric or topological significant space. For example, one could consider spaces of analytic functions, trigonometric polynomials, functions related to Gauss' error curve, and some reasonable measure μ which can be introduced in these spaces, and such that for μ almost every datum one has convergence of u_ε.

REMARK 2 – Conjecture 1 can be weakened also by not requiring convergence for almost every point in $\mathbf{R} \times [0, +\infty)$, but requiring the existence of a positive number T such that u_ε converges almost everywhere in $\mathbf{R} \times [0, T)$. A first step toward the proof of Conjecture 1 or the search of possible counterexamples could be the study of sequences of positive numbers (ε_i) such that $\lim_{i \to +\infty} \varepsilon_i = 0$, and such that one has a more or less strong convergence of the sequence (u_{ε_i}).

REMARK 3 – One can imagine several variants of Conjecture 1, for example one can pass from functions defined in $\mathbf{R}$ to functions defined in an interval of $\mathbf{R}$, introducing reasonable boundary conditions. One can also consider, instead of compactly supported initial data and bounded solutions, data and solutions which are periodic in the space variable x, or replace the boundedness with other conditions at infinity. More generally, one can pass to the space $\mathbf{R}^n$ and set

$$F(w, A) = \frac{1}{2} \int_A \log(1 + |\nabla w|^2) \,,$$

or set in the case $n = 1$

$$F(w, A) = \int_A \left(1 - \left|\frac{\partial w}{\partial x}\right|^2\right)^2 \,,$$

and then consider the analogous functional in $\mathbf{R}^n$,

$$F(w, A) = \int_A \left(1 - |\nabla w|^2\right)^2 \,.$$

One could obtain interesting results by considering other cases, for example setting

$$F_\varepsilon(w, A) = \int_A \varepsilon^2 w^2 + \left(1 - |\nabla w|^2\right)^2 + \varepsilon^2 |\nabla^2 w|^2.$$

Finally, one can consider cases where the function g which appears in (1) depends not only on w and its derivatives, but also on the space variables $(x_1, \ldots, x_n)$ and possibly on a parameter t, which we find as time variable in (3), (5).

2. Approximate gravitational problem

We now formulate a problem related to a reasonable approximation of Newton's gravitation law (cf. [3], [6]); this problem will be called "approximate gravitational problem" and can be stated as follows.

PROBLEM 2.1 – Let μ be a positive measure defined on Borel subsets of $\mathbf{R}^6$, and let $\mu(\mathbf{R}^6) < +\infty$. We call solution of the approximate gravitational problem associated to μ a set of three functions (s, v, u) satisfying the qualitative conditions (A) and the integro-differential conditions (B) stated below.
(A)

$$s = s(\varepsilon, \xi, \eta, t) \in \left[C^1((\mathbf{R} \setminus \{0\}) \times \mathbf{R}^7)\right]^3,$$

$$v = v(\varepsilon, \xi, \eta, t) \in \left[C^1((\mathbf{R} \setminus \{0\}) \times \mathbf{R}^7)\right]^3,$$

$$u = u(\varepsilon, x, t) \in C^0((\mathbf{R} \setminus \{0\}) \times \mathbf{R}^4).$$

The scalar function u is differentiable in the variables x_1, x_2, x_3, and its gradient in such variables, $\nabla_x u$, belongs to $C^0(\mathbf{R} \setminus \{0\} \times \mathbf{R}^4)^3$.

(B) The vector valued functions s, v satisfy the differential equations

$$(1) \qquad \frac{\partial s}{\partial t} = v, \quad \frac{\partial v}{\partial t} = \alpha(\varepsilon, s, t), \quad \alpha(\varepsilon, x, t) = \nabla_x u(\varepsilon, x, t)$$

with initial conditions

$$(2) \qquad s(\varepsilon, \xi, \eta, 0) = \xi, \qquad v(\varepsilon, \xi, \eta, 0) = \eta$$

while u fulfills the integral condition

$$(3) \qquad u(\varepsilon, x, t) = \int_{\mathbf{R}^6} (|x - s(\varepsilon, \xi, \eta, t)|^2 + \varepsilon^2)^{-1/2} d\mu(\xi, \eta).$$

REMARK 1 – Intuitively, $s(\varepsilon, \xi, \eta, t)$ represents the position at time t of a particle that at time $t = 0$ lies in the point ξ with velocity η; $v(\varepsilon, \xi, \eta, t)$ represents the velocity at time t of the same particle, $\alpha(\varepsilon, x, t)$ represents the acceleration of a particle which at time t is at point x and finally u should provide, for very small ε, a reasonable approximation of the Newtonian potential. The measure μ represents a mass distribution (possibly with a given velocity) at time $t = 0$. In order to better understand the meaning of formulas (1), (2), (3), one can consider the case where μ is a sum of n measures concentrated in points, i.e. for every Borel subset $B \subset \mathbf{R}^6$ one has that $\mu(B) = \sum_{h=1}^{n} m_h \delta(\xi_h, \eta_h)(B)$, where δ denotes the usual Dirac's measure, i.e.

$$\begin{cases} \delta(\xi_h, \eta_h)(B) = 1 & \text{for } (\xi_h, \eta_h) \in B \\[2mm] \delta(\xi_h, \eta_h)(B) = 0 & \text{for } (\xi_h, \eta_h) \notin B. \end{cases}$$

In this case the potential u appearing in (3) becomes

$$u(\varepsilon, x, t) = \sum_{h=1}^{n} m_h (|x - s(\varepsilon, \xi_h, \eta_h, t)|^2 + \varepsilon^2)^{-1/2}$$

hence system $(1), (2), (3)$ seems to be a reasonable approximation of the classical n body problem.

REMARK 2 – The case where μ is not concentrated in a finite number of points is suggested by the observation of the universe where there is interaction between stars, planets, satellites, gas clouds, and other forms of "diffuse mass". In this case (3) can be written in the form

$$(3^*) \qquad u(\varepsilon, x, t) = \int_{\mathbf{R}^6} (|x - \xi|^2 + \varepsilon^2)^{-1/2} d\mu_t^*(\xi, \eta),$$

where, for every $t \in \mathbf{R}$, the measure μ_t^* is given by

$$\mu_t^*(B) = \mu\left(\left\{(\xi, \eta) \in \mathbf{R}^6 : \big(s(\varepsilon, \xi, \eta, t), v(\varepsilon, \xi, \eta, t)\big) \in B\right\}\right);$$

intuitively, for every $t \in \mathbf{R}$, μ_t^* describes the distribution and velocity at time t of the mass in the universe; the evolution of μ_t^*, as t varies, describes what would be the evolution of the universe if it would be governed by an "approximate gravitational force" computed from an "elementary potential" such as

$$(4) \qquad\qquad\qquad\qquad (|x - \xi|^2 + \varepsilon^2)^{-1/2},$$

which for small ε seems to be a reasonable approximation of the classical Newtonian gravitational potential proportional to

$$|x - \xi|^{-1}.$$

As far as I know, in the literature there is no exhaustive answer about the existence of limits as $\varepsilon \to 0$ of the functions considered in the approximate gravitational problem, and in particular the following conjecture has not been proved or disproved.

Conjecture 1 – *If (s, v, u) is a solution of the approximate gravitational problem associated to the measure μ, then, for almost every point $(x, t) \in \mathbf{R}^4$, the following limit exists and is finite*

$$\lim_{\varepsilon \to 0} u(\varepsilon, x, t).$$

Moreover, for almost every $(\xi, \eta, t) \in \mathbf{R}^7$, the following two limits exist

$$\lim_{\varepsilon \to 0} s(\varepsilon, \xi, \eta, t),$$

$$\lim_{\varepsilon \to 0} v(\varepsilon, \xi, \eta, t).$$

Remark 3 – It is likely that Conjecture 1, even if true for large classes of measures μ, is not true for every choice of the measure μ satisfying the conditions of Problem 1. Up to now, I didn't find any counterexample in the literature; I think that the construction of a counterexample should be easier in the case of measures μ satisfying

$$\mu = \sum_{h=1}^{+\infty} m_h \delta(\xi_h, \eta_h), \qquad \sum_{h=1}^{+\infty} m_h < +\infty, \qquad \sum_{h=1}^{+\infty} |\eta_h| = +\infty.$$

More difficult and more interesting should be the case

$$\mu = \sum_{h=1}^{+\infty} m_h \delta(\xi_h, \eta_h), \qquad \sum_{h=1}^{+\infty} m_h < +\infty, \qquad \sup_h \left(|\xi_h| + |\eta_h|\right) < +\infty.$$

Intuitively this corresponds to the case where at the initial time all the particles lie in a bounded region of space, and the set of their initial velocities is also bounded.

We can consider also a weaker conjecture.

CONJECTURE 2 – *If we denote by $\mathcal{M}(\mathbf{R}^6)$ the space of Borel measures μ which are positive in $\mathbf{R}^6$ and satisfy $\mu(\mathbf{R}^6) < +\infty$, endowed with the distance δ defined by*

$$\delta(\mu, \nu) = \sup_B \left(\mu(B) - \nu(B)\right) + \sup_B \left(\nu(B) - \mu(B)\right),$$

then the set of measures μ for which Conjecture 1 is true in dense in $\mathcal{M}(\mathbf{R}^6)$.

It would be also desirable to estimate the stability degree, as μ varies, of the limits of solutions of gravitational problems satisfying the thesis of Conjecture 1. Maybe such a stability degree is not very large and therefore it seems reasonable to introduce a weak enough notion of stability, called α–stability, by means of the following definition.

DEFINITION 1 – Given a measure μ, a set of three functions (s, v, u) which are solutions of the approximate gravitational problems associated to μ, and a real number $\alpha \geq 0$, we say that (s, v, u) is the α–stable solution of the gravitational problem associated to μ if the following limits exist for almost every $(x, t) \in \mathbf{R}^4$ and almost every $(\xi, \eta, t) \in \mathbf{R}^7$

$$\lim_{\varepsilon \to 0} u(\varepsilon, x, t),$$

$$\lim_{\varepsilon \to 0} s(\varepsilon, \xi, \eta, t),$$

$$\lim_{\varepsilon \to 0} v(\varepsilon, \xi, \eta, t),$$

and moreover, for every sequence of positive numbers (ε_i) such that $\lim_{i \to +\infty} \varepsilon_i = 0$, and for every sequence of positive measures (μ_i) such that

$$\lim_{i \to +\infty} \varepsilon_i^{-\alpha} \delta(\mu_i, \mu) = 0,$$

denoted by (s_i, v_i, u_i) the solution of the approximate gravitational problem associated to μ_i, for almost every $(x, t) \in \mathbf{R}^4$ one has that

$$\lim_{i \to +\infty} u_i(\varepsilon_i, x, t) = \lim_{\varepsilon \to 0} u(\varepsilon, x, t),$$

while for almost every $(\xi, \eta, t) \in \mathbf{R}^7$ one has that

$$\lim_{i \to +\infty} s_i(\varepsilon_i, \xi, \eta, t) = \lim_{\varepsilon \to 0} s(\varepsilon, \xi, \eta, t),$$

$$\lim_{i \to +\infty} v_i(\varepsilon_i, \xi, \eta, t) = \lim_{\varepsilon \to 0} v(\varepsilon, \xi, \eta, t).$$

We can now give a significant reinforcement of Conjecture 2 by stating the following.

CONJECTURE 3 – *There exist positive numbers α such that the set of measures μ such that the corresponding solutions of the approximate gravitational problem are α-stable is dense in $\mathcal{M}(\mathbf{R}^6)$.*

REMARK 4 – Should Conjecture 3 be true, it would be interesting to find the infimum of the numbers α considered therein; it is very likely that such an infimum is strictly greater than zero; since α-stability is weaker as α increases, such infimum would give us an indication on the degree of global stability of the approximate gravitational problem.

3. Variants of the approximate gravitational problem

One can imagine several variants of the gravitational problem illustrated in section 2. One can think that the force acting between two particles is the sum of an attractive and a repulsive force, and replace (3) of section 1 with

$$u(\varepsilon, x, t) = \int_{\mathbf{R}^6} (|x - s|^2 + \varepsilon^2)^{-1/2} - c^2(|x - s|^2 + \varepsilon^2)^{-1} d\mu(\xi, \eta)$$

where c is a positive constant.

One can consider, after the limits as $\varepsilon \to 0$, the limits of such limits as $c \to 0$; this corresponding to the idea that repulsive forces act only on very small scales. Finally, one can imagine more complex expressions corresponding to forces depending not only on the position, but also on the velocity of particles, e.g. taking into account integrals such as

$$\alpha(\varepsilon, x, y, t) = \int_{\mathbf{R}^6} \varphi(\varepsilon, x, y, s(\varepsilon, \xi, \eta, t), v(\varepsilon, \xi, \eta, t)) d\mu(\xi, \eta)$$

and then differential equations such as

$$\frac{\partial v}{\partial t} = \alpha(\varepsilon, s, v, t).$$

Of course, further complications could be introduced, inspired by more complex physical theories that are beyond the scope of this paper, since its aim is only to draw the attention on some "classes" of mathematical problems, of which several variants rich of physical significance could be imagined later on.

Coming back to the approximate gravitational problem and in particular to Remark 2 of section 2, we can consider the case where the mass μ^* is regularly distributed, i.e. there exists a function $\varphi = \varphi(\varepsilon, x, y, t) \in C^\infty((\mathbf{R} \setminus \{0\}) \times \mathbf{R}^7)$ such that

$$(1) \qquad \mu_t^*(B) = \int_B \varphi(\varepsilon, x, y, t) dx dy;$$

in this case the function $u(\varepsilon, x, t)$ is given by

$$u(\varepsilon, x, t) = \int_{\mathbf{R}^6} (|x - \xi|^2 + \varepsilon^2)^{-1/2} \varphi(\varepsilon, \xi, \eta, t) \, d\xi \, d\eta,$$

and moreover φ satisfies the first order linear differential equation

$$\frac{\partial \varphi}{\partial t} = -\sum_{h=1}^{3}\left(\frac{\partial \varphi}{\partial x_h}y_h + \frac{\partial \varphi}{\partial y_h}\frac{\partial u}{\partial x_h}\right),$$

whose characteristic lines are determined by equations (1), (2) of Problem 2.1. One can then state the problem of the existence of the limit as $\varepsilon \to 0$ of the function $\varphi(\varepsilon, x, y, t)$ and study the possible relations with the limits considered in Conjectures 1, 2, 3 of section 2. In such a study some interesting phenomena about more or less generalized solutions of first order partial differential equations are likely to appear.

4. Limits of variational problems which can be related to hyperbolic equations

In the search of possible approximations of difficult and unstable problems by means of more easy and stable ones, one can also consider the idea of obtaining solutions of evolution problems as the limit of solutions of minimum problems. An interesting example where this idea has been successfully applied can be found in [8]. A possible variant of the result by Ilmanen is given by the following conjecture.

CONJECTURE 1 – *Let $\varphi, \psi \in C_0^\infty(\mathbf{R}^n)$, let $k > 1$ be an integer; for every positive real number λ, let $w_\lambda = w_\lambda(x_1, \ldots, x_n, t)$ be a minimizer of the functional*

$$(1) \qquad F_\lambda(u) = \int_{\mathbf{R}^n \times [0,+\infty[} e^{-\lambda t}\left[\left|\frac{\partial^2 u}{\partial t^2}\right|^2 + \lambda^2|\nabla_x u|^2 + \lambda^2 u^{2k}\right]dx\,dt$$

in the class of all u satisfying the initial conditions

$$u(x,0) = \varphi(x), \qquad \frac{\partial u}{\partial t}(x,0) = \psi(x).$$

Then there exists $\lim\limits_{\lambda \to +\infty} w_\lambda(x,t) = w_0(x,t)$, satisfying the equation

$$(2) \qquad \frac{\partial^2 w_0}{\partial t^2} = \Delta_x w_0 - k w_0^{2k-1}.$$

REMARK 1 – Should this conjecture be false, one could think to suitable variants, e.g. by requiring that it is true only for a dense set of initial data, by replacing C^∞ with analytic data with fast decay, by taking into account periodic functions and in particular trigonometric polynomials, or also think to a functional F_λ different from (1) whose minimizers converge to solutions of (2). If on the contrary the conjecture should be true, one could think to several generalizations, starting from a functional F of the calculus of variations, and associating to it the functional

$$F_\lambda(u) = \int_{\mathbf{R}^n \times [0,+\infty[} e^{-\lambda t}\left|\frac{\partial^2 u}{\partial t^2}\right|^2 dx\,dt + \lambda^2 \int_0^{+\infty} e^{-\lambda t}F\big(u(\cdot,t)\big)\,dt.$$

5. Movements of manifolds by Eulerian functions

In order to introduce movements of manifolds of any dimension and codimension embedded in an Euclidean space $\mathbf{R}^n$, we begin to establish some notation. For every open set $A \subset \mathbf{R}^n$ we still denote by $C^\infty(A)$ the space of real functions which are continuous in A with their derivatives of any order, while in the case where the set E is not open we write $f \in C^\infty(E)$ if and only if there exist an open set $A \supset E$ and a function $f^* \in C^\infty(A)$ such that $f(x) = f^*(x)$ for every $x \in E$. Given two open sets A, Ω, we write $A \subset\subset \Omega$ if and only if the closure of A is a compact set contained in Ω. Finally, for every $E \subset \mathbf{R}^n$ we can define the function η_E in the following way:

$$(1) \qquad \eta_E(x) = \frac{1}{2}[\mathrm{dist}(x, E)]^2 = \inf\left\{\frac{|x - \xi|^2}{2}; \; \xi \in E\right\}.$$

We can now give the following notion of h-dimensional manifold of class C^∞ in an open set Ω.

DEFINITION 1 – Let Ω be an open set in $\mathbf{R}^n$, and let h be an integer, $0 \le h \le n$. We set that a set E is a h-dimensional manifold of class C^∞ in Ω and we write $E \in V_h C^\infty(\Omega)$, if the following three conditions are satisfied:

$$(a) \qquad \Omega \cap E = \{x \in \Omega : \eta_E(x) = 0\}$$

$$(b) \qquad \text{there exists an open set } A \supset \Omega \cap E \text{ such that } \eta_E \in C^\infty(A)$$

(c) for every $x \in E \cap \Omega$ the Hessian matrix $\nabla^2 \eta_E(x)$ has characteristic $n - h$.

If g is a C^∞ function defined in the set of real polynomials of degree less or equal than k of n real variables, and $E \in V_h C^\infty(\Omega)$, then for every open set $A \subset\subset \Omega$ we can consider the functional

$$(2) \qquad F(E, A) = \int_{A \cap E} g\big(\eta_E(x), \nabla \eta_E(x), \ldots, \nabla^k \eta_E(x)\big) \, d\mathcal{H}^h(x),$$

where $\mathcal{H}^h$ is the h-dimensional Hausdorff measure; since at the points of E both η_E and $\nabla \eta_E$ vanish, this functional depends in a non trivial way on g for $k \ge 2$. We now give the definition of Eulerian function associated to E and F.

DEFINITION 2 – Let E be a subset of $\mathbf{R}^n$, let Ω be an open set in $\mathbf{R}^n$, and let $E = E \cap \Omega \in V_h C^\infty(\Omega)$. Let $\psi \in (C^\infty(E))^n$, and let F be defined by (2). Then we say that the vector valued function ψ is the *Eulerian function* of F on E, and we write

$$\psi \in \mathcal{E}(F, E)$$

if for every open set $A \subset\subset \Omega$ and every $\tau \in [C_0^\infty(A)]^n$ one has that

$$\lim_{\lambda \to 0} \frac{1}{\lambda}\big(F(E_\lambda, A) - F(E, A)\big) = \int_{E \cap A} \sum_{j=1}^{n} \tau_j(x)\psi_j(x) \, d\mathcal{H}^h(x),$$

where $E_\lambda = \{x + \lambda\tau(x); \; x \in A \cap E\}$.

We can now give a notion of descent movement by the Eulerian function.

DEFINITION 3 – Let $E \subset \mathbf{R}^n$, let F be defined by (2), and let $0 < T \le +\infty$, $u = u(x_1, \ldots, x_n, t) \in [C^\infty(E \times [0, T[)]^n$. We say that u is a *descent movement* of the functional F with initial datum E, and we write $u \in \mathcal{MD}(F, E, T)$, if it satisfies the following conditions:

(A)
$$u(x, 0) = x , \qquad \text{for every } x \in E.$$

(B)   There exists a function A defined in $E \times [0, T[$ with values in the open sets of $\mathbf{R}^n$ such that for every $(x, t) \in E \times [0, T[$, one has that

$$x \in A(x, t)$$

and moreover, setting $L(x, t) = \{(x, t); \ x \in E \cap A(x, t)\}$, for every $(x, t) \in E \times [0, T[$ one has that

$$\frac{\partial u}{\partial t}(x, t) = \mathcal{E}\big(F, L(x, t)\big)\big(u(x, t)\big).$$

REMARK 1 – In the case where g reduces to a positive constant, hence $F(E, A)$ is proportional to the measure $\mathcal{H}^h(E \cap A)$, we find the classical definition of movement by mean curvature (cf. e.g. [1]); in general, descent movements do not necessarily exist for all functionals F defined by (2) and for all initial data E, however one can think to approximate the same functionals as in section 1. More precisely, if F is defined by (2), we can set, for every $\varepsilon > 0$,

$$(3) \qquad F_\varepsilon(E, A) = F(E, A) + \varepsilon^2 \int_{E \cap A} |\nabla^{k+1} \eta_E|^2 \, d\mathcal{H}^h$$

and state the following

CONJECTURE 1 – *If E is a compact set belonging to $V_h C^\infty(\mathbf{R}^n)$, and F is defined by (2), then, for every $\varepsilon \ne 0$ there exist $T_\varepsilon > 0$ and $u_\varepsilon \in [C^\infty(E \times [0, T_\varepsilon[)]^n$ such that $u_\varepsilon \in \mathcal{MD}(F_\varepsilon, E, T_\varepsilon)$.*

In the case where g reduces to a positive constant, one can imagine that T does not depend on ε as in the following

CONJECTURE 2 – *If to the assumptions of Conjecture 1 we add the assumption that $k > h$ and g has a constant positive value, then one can assume that $T_\varepsilon \equiv +\infty$ hence $u_\varepsilon \in \big[C^\infty(E \times [0, +\infty[)\big]^n$.*

REMARK 2 – Conjecture 2 provides a canonical way to approximate motions by mean curvature. Regarding the limits as $\varepsilon \to 0$ of such approximated movements, one can state problems more or less similar to those indicated in the preceding sections. To begin with, one could discuss the following conjecture which is essentially equivalent to some kind of uniqueness result for motion by mean curvature defined through a "viscosity method".

CONJECTURE 3 – *Under the assumptions of Conjecture 2, for $\mathcal{H}^{h+1}$ almost every point in $E \times [0, +\infty)$ there exists the limit*

$$\lim_{\varepsilon \to 0} u_\varepsilon(x, t).$$

REMARK 3 – Should Conjecture 3 be false, one could weaken it in different ways, for example by requiring that it is true only in the case where E belongs not only to $V_h C^\infty(\mathbf{R}^n)$, but also to $V_h C^\omega(\mathbf{R}^n)$, i.e. the function η_E of Definition 1 is analytic in the open set A. One can also ask that the thesis of Conjecture 3 is satisfied as E varies in a dense subset of the space of compact sets belonging to $V_h C^\omega(\mathbf{R}^n)$, endowed with a suitable topology.

If on the contrary Conjecture 3 should be true, one could examine some cases where the function g in (2) is not a constant. From a physical point of view, for suitable choices of the functional F, also equations such as

$$\frac{\partial^2 u}{\partial t^2} = \mathcal{E}\big(F, L(x, t)\big)\big(u(x, t)\big).$$

could be interesting.

Acknowledgments. This note is the result of several talks with various colleagues and friends, in particular I wish to thank Luigi Ambrosio, Giovanni Bellettini, Giuseppe Bertin, Ivar Ekeland, Antonio Leaci, Carlo Mantegazza, Antonio Marino, Stefano Mortola, Diego Pallara, Sergio Spagnolo and Piero Villaggio.

Bibliography

[1] L. AMBROSIO - H.M. SONER, *Level set approach to mean curvature flow in arbitrary codimension*, J. Differential Geom. **43** (1996), 693–737.

[2] H. BREZIS, *Opérateurs Maximaux Monotones et Sémi-groupes de Contractions dans les Espaces de Hilbert*, North Holland, Amsterdam, 1973.

[3] J.M. DAWSON, *Particle Simulation of Plasmas*, Rev. Modern Phys. **55** (1983), 403.

[4] M. DEGIOVANNI, A. MARINO, M. TOSQUES, *Evolution Equations with Lack of Convexity*, Nonlinear Anal. **9** (1985), 1401–1443.

[5] R.J. DI PERNA, P.L. LIONS, *Ordinary differential equations, transport theory and Sobolev spaces*, Invent. Math. **98** (1989), 511–547.

[6] M. GRILLAKIS, *Regularity for the wave equations with a critical nonlinearity*, Comm. Pure Appl. Math. **14** (1992), 749–774.

[7] R.W.HOCKNEY, J.W. EASTWOOD, *Computer Simulation Using Particles*, McGraw-Hill, New York, 1981.

[8] T.ILMANEN, *Elliptic regularization and partial regularity for motion by mean curvature*, Mem. Amer. Math. Soc. **108** (1994), # 520.

[9] R.H. MILLER, *Numerical experiments in collisionless systems*, Astrophys. Space Sci. **14** (1971), 73.

[10] P.PERONA-J.MALIK, *Scale-space and edge detection using anisotropic diffusion*, in Proceedings of the IEEE Computer Society Workshop on Computer Vision, Miami, Florida 1987, Computer Society Press of the IEEE, Washington, D.C., 16–22.

[11] P.PERONA-J.MALIK, *Scale-space and edge detection using anisotropic diffusion*, IEEE Trans. Pattern Anal. Mach. Intell. **12** (1990), 629–639.

[12] J. SHATAH, M. STRUWE, *Regularity results for nonlinear wave equations*, Ann. of Math. (2) **138** (1993), 503–518.

Congetture riguardanti alcuni problemi di evoluzione[‡]

Ennio De Giorgi

Sunto. I have greatly appreciated the invitation to write a paper in honor of John F. Nash Jr., a scientist who has been a major source of new ideas for mathematicians. His work has been a clear example of how the most original mathematical ideas are often close to the fundamental problems of other disciplines. I believe that this example is very important for any student who is still motivated by the drive that the ancients called *philosophy*, that is, love for Wisdom.

The character of my paper is exploratory. Some conjectures concerning "evolution problems" are presented. They are related to the "steepest descent", to the approximations of Newton's gravitation law, to hyperbolic nonlinear equations and to "descent movements" of manifolds.

The article identifies several questions, of which I do not know the solution, and points out a number of analogies between problems that are apparently far from each other. I believe that the study of these conjectures might provide an opportunity for scientists who are expert in different fields within pure and applied mathematics to get together and ponder the connections that exist among various mathematical concepts, such as linear vs. nonlinear behavior, stability vs. instability, or convergence of different approximation methods, and certain ideas well developed in physics, such as deterministic vs. nondeterministic behavior, predictability vs. nonpredictability, order and chaos, etc.

Introduzione. In questo lavoro esporrò alcune congetture relative a "problemi di evoluzione" legate all'idea di "discesa secondo la massima pendenza", alla legge di gravitazione universale di Newton, alle equazioni iperboliche e ai movimenti di varietà immerse in uno spazio euclideo.

È mia opinione che lo studio di queste congetture possa servire a una migliore comprensione dei fenomeni che si possono presentare quando si approssimano problemi "piú difficili e piú instabili" mediante problemi "meno difficili e piú stabili". In particolare si dovrebbe vedere quando un processo di approssimazione non converge in corrispondenza a tutti i "dati ammissibili" ma si ha convergenza e una certa stabilità del limite solo in corrispondenza a un "insieme denso" di dati oppure a "quasi tutti" i dati. Naturalmente le nozioni di "stabilità", di "insieme denso" e di "quasi tutti" sono legate all'introduzione di ragionevoli topologie, metriche o di qualche misura significativa nella famiglia dei dati ammissibili.

Dato il carattere eterogeneo di queste congetture, che forse sono state in parte trattate in lavori a me ignoti, posso dare solo indicazioni bibliografiche molto frammentarie e incomplete che mi sono state segnalate da vari colleghi e amici e che io stesso non ho ancora avuto il tempo di approfondire. Esse devono essere

---

[‡]Editor's note: published in Duke Math. J., **81-(2)** (1995), 255–268.

considerate solo come una prima traccia utile per iniziare la sistematica ricerca bibliografica necessaria per fare realmente il "punto della situazione" al momento attuale.

Spero che una tale ricerca bibliografica e la stessa riflessione sulle congetture esposte possa essere occasione per ritrovare e collegare fra loro vari lavori finora ritenuti molto lontani e forse mettere in evidenza eventuali collegamenti che si possono stabilire tra idee matematiche riguardanti linearità e non linearità, stabilità e instabilità, convergenza di differenti metodi di approssimazione e concetti della fisica come determinismo e indeterminismo, prevedibilità e imprevedibilità, ordine e caos.

1. Problemi di discesa

Cominciamo col considerare alcuni problemi collegati al "metodo della discesa". Consideriamo funzionali del tipo

$$(1) \qquad F(w, A) = \int_A g(w, \nabla w, \ldots, \nabla^k w)$$

ove A è un aperto limitato di $\mathbf{R}^n$, w è una funzione di classe $C^\infty(\mathbf{R}^n)$, g è una funzione di classe C^∞ nello spazio dei polinomi di n variabili reali e grado minore o uguale a k.

In queste ipotesi diremo che una funzione $\psi \in C^\infty(\mathbf{R}^n)$ è la funzione di Eulero associata a F e a w e scriveremo

$$\psi = \mathcal{E}(F, w)$$

se per ogni aperto limitato $A \subset \mathbf{R}^n$ e per ogni funzione test $\tau \in C_0^\infty(A)$ vale la condizione

$$(2) \qquad \lim_{\lambda \to 0} \frac{1}{\lambda} \left(F(w + \lambda\tau, A) - F(w, A) \right) = \int_A \tau\,\psi.$$

In alcuni casi in cui la funzione g non è convessa (o non soddisfa altre condizioni vicine alla convessitá, vedi [4]) lo studio dell'equazione differenziale

$$(3) \qquad \frac{\partial u}{\partial t} = -\mathcal{E}(F, u(\cdot, t))$$

con le condizioni iniziali

$$(4) \qquad u(x, 0) = \varphi(x)$$

nell'insieme $\mathbf{R}^n \times [0, +\infty)$ può presentare notevoli difficoltà e sembra opportuno seguire metodi tipo viscosità artificiale (cfr. per es. [5]) e associare al funzionale F funzionali F_ε definiti, per $\varepsilon \neq 0$, dalla formula

$$F_\varepsilon(w, A) = F(w, A) + \varepsilon^2 \int_A |\nabla^{k+1} w|^2$$

ove al solito

$$|\nabla^s w|^2 = \sum_{j_1=1}^{n} \cdots \sum_{j_s=1}^{n} \left(\frac{\partial^s u}{\partial x_{j_1} \ldots \partial x_{j_s}} \right)^2$$

e quindi tentare di approssimare l'equazione differenziale (3) con l'equazione

$$(5) \qquad \frac{\partial u}{\partial t} = -\mathcal{E}(F_\varepsilon, u(\cdot, t)).$$

Un esempio semplice ma non banale di questa situazione si ha considerando il caso $k = 1$, $n = 1$ (cfr. [9,10]),

$$g\left(w, \frac{\partial w}{\partial x} \right) = \frac{1}{2} \log \left(1 + \left| \frac{\partial w}{\partial x} \right|^2 \right).$$

In tal caso l'equazione (3) diventa

$$\frac{\partial u}{\partial t} = \frac{\partial}{\partial x} \left(\frac{\partial u}{\partial x} \left(1 + \left(\frac{\partial u}{\partial x} \right)^2 \right)^{-1} \right) = \frac{\partial^2 u}{\partial x^2} \left(1 - \left| \frac{\partial u}{\partial x} \right|^2 \right) \left(1 + \left| \frac{\partial u}{\partial x} \right|^2 \right)^{-2}.$$

Lo studio di questa equazione può presentare notevoli difficoltà quando il dato iniziale verifica in una parte di $\mathbf{R}$ la condizione $\left| \frac{\partial \varphi}{\partial x} \right| > 1$; si può allora considerare il funzionale F_ε che conduce all'equazione "stabilizzata"

$$(6) \qquad \frac{\partial u}{\partial t} = \frac{\partial}{\partial x} \left(\frac{\partial u}{\partial x} \left(1 + \left(\frac{\partial u}{\partial x} \right)^2 \right)^{-1} \right) - \varepsilon^2 \frac{\partial^4 u}{\partial x^4}$$

ed enunciare una prima congettura.

Congettura 1 – *Se $\varphi \in C_0^\infty(\mathbf{R})$, e, per ogni $\varepsilon \neq 0$, u_ε è una soluzione dell'equazione (6) con la condizione iniziale (4) ed è limitata in $\mathbf{R} \times [0, +\infty[$, allora per quasi ogni $(x, t) \in \mathbf{R} \times [0, +\infty)$ esiste il limite di $u_\varepsilon(x, t)$ per ε che tende a zero.*

Osservazione 1 – Se la Congettura 1 risultasse vera si potrebbe rafforzarla imponendo la convergenza *per ogni* (x, t) ed eventualmente anche convergenze piú forti che comportano una maggiore regolarità del limite. Si potrebbe anche richiedere una certa uniformità di convergenza al variare del dato iniziale φ e quindi una dipendenza continua del limite dal dato stesso. Se invece si trovasse qualche controesempio alla Congettura 1 essa potrebbe essere indebolita in vario modo, accontentandosi per esempio della convergenza soltanto per i dati appartenenti ad insiemi densi di qualche spazio metrico o topologico significativo. Per esempio, si potrebbero considerare spazi di funzioni analitiche, polinomi trigonometrici, funzioni legate alla curva degli errori di Gauss, e si potrebbe anche prendere in considerazione qualche misura ragionevole μ che può essere introdotta in tali spazi e sia tale che per μ quasi tutti i dati vi sia convergenza di u_ε.

OSSERVAZIONE 2 – Altri indebolimenti della Congettura 1 si potrebbero avere non richiedendo la convergenza in quasi tutti i punti di $\mathbf{R} \times [0, +\infty)$ ma richiedendo l'esistenza di un numero positivo T tale che u_ε converga in quasi tutto $\mathbf{R} \times [0, T)$. Un primo passo verso la conferma della Congettura 1 o la ricerca di eventuali controesempi potrebbe essere lo studio di successioni di numeri positivi (ε_i) tali che $\lim_{i \to +\infty} \varepsilon_i = 0$ e si abbia una convergenza piú o meno forte della successione (u_{ε_i}).

OSSERVAZIONE 3 – Si possono immaginare molte varianti della Congettura 1, per esempio si può passare dalle funzioni definite in $\mathbf{R}$ al caso di funzioni definite in un intervallo di $\mathbf{R}$ introducendo ragionevoli condizioni al contorno. Si possono pure considerare, invece che dati iniziali a supporto compatto e soluzioni limitate, dati e soluzioni periodiche nella variabile spaziale x, oppure imporre, in luogo della limitatezza, altre condizioni all'infinito. Piú in generale si può passare allo spazio $\mathbf{R}^n$ e porre

$$F(w, A) = \frac{1}{2} \int_A \log(1 + |\nabla w|^2) \,,$$

oppure nel caso $n = 1$ porre

$$F(w, A) = \int_A \left(1 - \left| \frac{\partial w}{\partial x} \right|^2 \right)^2 ,$$

e considerare successivamente il funzionale analogo in $\mathbf{R}^n$,

$$F(w, A) = \int_A \left(1 - |\nabla w|^2 \right)^2 .$$

Risultati interessanti si potrebbero avere considerando altri casi, per esempio ponendo

$$F_\varepsilon(w, A) = \int_A \varepsilon^2 w^2 + \left(1 - |\nabla w|^2 \right)^2 + \varepsilon^2 |\nabla^2 w|^2 .$$

Infine si possono considerare casi in cui la funzione g che compare nella (1) dipende, oltre che da w e dalle sue derivate, dalle variabili spaziali $(x_1, \ldots, x_n)$ ed eventualmente da un parametro t che ritroveremmo come variabile temporale nelle (3), (5).

2. Problema gravitazionale approssimato

Passiamo ora alla formulazione di un problema legato a una ragionevole approssimazione della legge di gravitazione universale di Newton (cfr [3,6]); tale problema sarà chiamato "Problema gravitazionale approssimato" e può essere formulato nel modo seguente.

PROBLEMA 2.1 – Sia μ una misura positiva definita sugli insiemi di Borel di $\mathbf{R}^6$ e sia $\mu(\mathbf{R}^6) < +\infty$. Chiameremo soluzione del Problema gravitazionale approssimato associato a μ una terna di funzioni (s, v, u) che soddisfa le condizioni qualitative (A) e le condizioni integro-differenziali (B) che ora enunciamo.

(A)

$$s = s(\varepsilon, \xi, \eta, t) \in \left[C^1((\mathbf{R} \setminus \{0\}) \times \mathbf{R}^7) \right]^3,$$

$$v = v(\varepsilon, \xi, \eta, t) \in \left[C^1((\mathbf{R} \setminus \{0\}) \times \mathbf{R}^7) \right]^3,$$

$$u = u(\varepsilon, x, t) \in C^0((\mathbf{R} \setminus \{0\}) \times \mathbf{R}^4).$$

La funzione scalare u è derivabile rispetto alle variabili x_1, x_2, x_3 e il suo gradiente, relativo a tali variabili, $\nabla_x u$ appartiene a $C^0(\mathbf{R} \setminus \{0\} \times \mathbf{R}^4)^3$.

(B)
Le funzioni vettoriali s, v verificano le equazioni differenziali

$$(1) \qquad \frac{\partial s}{\partial t} = v, \quad \frac{\partial v}{\partial t} = \alpha(\varepsilon, s, t), \quad \alpha(\varepsilon, x, t) = \nabla_x u(\varepsilon, x, t)$$

con le condizioni iniziali

$$(2) \qquad s(\varepsilon, \xi, \eta, 0) = \xi, \qquad v(\varepsilon, \xi, \eta, 0) = \eta$$

mentre u soddisfa la condizione integrale

$$(3) \qquad u(\varepsilon, x, t) = \int_{\mathbf{R}^6} (|x - s(\varepsilon, \xi, \eta, t)|^2 + \varepsilon^2)^{-1/2} d\mu(\xi, \eta).$$

Osservazione 1 – Intuitivamente $s(\varepsilon, \xi, \eta, t)$ rappresenta la posizione al tempo t di una particella che all'istante $t = 0$ si trovava nel punto ξ ed aveva velocità η; $v(\varepsilon, \xi, \eta, t)$ rappresenta la velocità al tempo t della stessa particella, $\alpha(\varepsilon, x, t)$ rappresenta l'accelerazione di una particella che nell'istante t si trovi nel punto x, infine u dovrebbe fornire, per ε molto piccolo, una ragionevole approssimazione del potenziale newtoniano. La misura μ rappresenta una distribuzione di masse (eventualmente animate da una certa velocità) all'istante $t = 0$.
Per comprendere meglio il significato delle formule (1), (2), (3), conviene considerare il caso in cui μ è somma di n misure concentrate in punti, cioè per ogni insieme di Borel $B \subset \mathbf{R}^6$ risulta $\mu(B) = \sum_{h=1}^{n} m_h \delta(\xi_h, \eta_h)(B)$ ove δ indica la usuale delta di Dirac, cioè

$$\delta(\xi_h, \eta_h)(B) = \begin{cases} 1 & \text{per } (\xi_h, \eta_h) \in B \\ \\ \delta(\xi_h, \eta_h)(B) = 0 & \text{per } (\xi_h, \eta_h) \notin B. \end{cases}$$

In tal caso il potenziale u che compare nella la formula (3) diventa

$$u(\varepsilon, x, t) = \sum_{h=1}^{n} m_h (|x - s(\varepsilon, \xi_h, \eta_h, t)|^2 + \varepsilon^2)^{-1/2}$$

e quindi il sistema costituito da $(1), (2), (3)$ appare una ragionevole approssimazione per il classico problema degli n corpi.

OSSERVAZIONE 2 – Il caso in cui μ non è concentrata in un numero finito di punti è suggerito dall'osservazione dell'universo in cui interagiscono stelle, pianeti, satelliti, nubi di gas, polveri interstellari e altre forme di "massa diffusa". In questo caso la (3) può essere anche scritta nella forma

$$(3^*) \qquad u(\varepsilon, x, t) = \int_{\mathbf{R}^6} (|x - \xi|^2 + \varepsilon^2)^{-1/2} d\mu_t^*(\xi, \eta),$$

ove, per ogni $t \in \mathbf{R}$, la misura μ_t^* è data dalla formula

$$\mu_t^*(B) = \mu\left(\{(\xi, \eta) \in \mathbf{R}^6 : \ (\sigma(\varepsilon, \xi, \eta, t), v(\varepsilon, \xi, \eta, t)) \in B\}\right);$$

intuitivamente, per ogni $t \in \mathbf{R}$, μ_t^* descrive la distribuzione e le velocità all'istante t della massa presente nell'universo; l'evoluzione di μ_t^* al variare di t sostanzialmente descrive quella che sarebbe l'evoluzione dell'universo se esso fosse governato da una "forza gravitazionale approssimata" calcolabile a partire da un "potenziale elementare" del tipo

$$(4) \qquad (|x - \xi|^2 + \varepsilon^2)^{-1/2},$$

che per ε piccolo sembra una ragionevole approssimazione del potenziale newtoniano classico proporzionale a

$$|x - \xi|^{-1}.$$

Non mi risulta che vi sia nella letteratura una risposta esauriente alle domande circa l'esistenza di ben determinati limiti per $\varepsilon \to 0$ delle funzioni considerate nel Problema gravitazionale approssimato e in particolare vi sia la verifica o la confutazione della seguente.

CONGETTURA 1 – *Se (s, v, u) è soluzione del Problema gravitazionale approssimato associato alla misura μ allora, per quasi ogni punto $(x, t) \in \mathbf{R}^4$, esiste determinato e finito il limite*

$$\lim_{\varepsilon \to 0} u(\varepsilon, x, t),$$

inoltre, per quasi ogni $(\xi, \eta, t) \in \mathbf{R}^7$, esistono i due limiti

$$\lim_{\varepsilon \to 0} s(\varepsilon, \xi, \eta, t),$$

$$\lim_{\varepsilon \to 0} v(\varepsilon, \xi, \eta, t).$$

OSSERVAZIONE 3 – È probabile che la Congettura 1, pur valendo per classi assai larghe di misure μ, non valga per ogni scelta della misura μ soddisfacente le condizioni del Problema 1. Finora non ho trovato nella letteratura alcun controesempio; penso che la sua eventuale costruzione dovrebbe risultare piú facile nel caso di misure μ verificanti le condizioni

$$\mu = \sum_{h=1}^{+\infty} m_h \delta(\xi_h, \eta_h), \qquad \sum_{h=1}^{+\infty} m_h < +\infty, \qquad \sum_{h=1}^{+\infty} |\eta_h| = +\infty.$$

Piú difficile ma piú interessante dovrebbe essere il caso

$$\mu = \sum_{h=1}^{+\infty} m_h \delta(\xi_h, \eta_h), \qquad \sum_{h=1}^{+\infty} m_h < +\infty, \qquad \sup_h \left(|\xi_h| + |\eta_h|\right) < +\infty.$$

Intuitivamente ció corrisponde al caso in cui nell'istante iniziale tutte le particelle sono racchiuse in una zona limitata dello spazio ed è limitato l'insieme delle loro velocità iniziali.

Possiamo considerare anche una congettura piú debole.

CONGETTURA 2 – *Detto* $\mathcal{M}(\mathbf{R}^6)$ *lo spazio delle misure di Borel* μ *che sono positive in* $\mathbf{R}^6$ *e verificano la condizione* $\mu(\mathbf{R}^6) < +\infty$, *dotato della distanza* δ *definita dalla relazione*

$$\delta(\mu, \nu) = \sup_B \left(\mu(B) - \nu(B)\right) + \sup_B \left(\nu(B) - \mu(B)\right),$$

è denso in $\mathcal{M}(\mathbf{R}^6)$ *l'insieme delle misure* μ *per cui vale la tesi della Congettura 1.*

Sarebbe pure desiderabile valutare il grado di stabilità al variare di μ dei limiti di soluzioni dei problemi gravitazionali che soddisfano la tesi della Congettura 1. Probabilmente tale grado di stabilità non è sempre molto elevato e quindi sembra opportuno introdurre una nozione abbastanza debole di stabilità, detta α–stabilità, mediante la seguente definizione.

DEFINIZIONE 1 – Data una misura μ, una terna di funzioni (s, v, u) soluzioni del Problema gravitazionale approssimato associato a μ ed un numero reale $\alpha \geq 0$, diremo che la (s, v, u) è soluzione α–stabile del Problema gravitazionale associato a μ se esistono per quasi ogni $(x, t) \in \mathbf{R}^4$ e quasi ogni $(\xi, \eta, t) \in \mathbf{R}^7$ i limiti

$$\lim_{\varepsilon \to 0} u(\varepsilon, x, t),$$

$$\lim_{\varepsilon \to 0} s(\varepsilon, \xi, \eta, t),$$

$$\lim_{\varepsilon \to 0} v(\varepsilon, \xi, \eta, t),$$

ed inoltre, scelta comunque una successione di numeri positivi (ε_i), verificante la condizione $\lim_{i \to +\infty} \varepsilon_i = 0$, e una successione di misure positive (μ_i) verificanti la condizione

$$\lim_{i \to +\infty} \varepsilon_i^{-\alpha} \delta(\mu_i, \mu) = 0,$$

e detta (s_i, v_i, u_i) la soluzione del Problema gravitazionale approssimato associata a μ_i, risulta per quasi tutti gli $(x, t) \in \mathbf{R}^4$

$$\lim_{i \to +\infty} u_i(\varepsilon_i, x, t) = \lim_{\varepsilon \to 0} u(\varepsilon, x, t),$$

mentre per quasi ogni $(\xi, \eta, t) \in \mathbf{R}^7$ risulta

$$\lim_{i \to +\infty} s_i(\varepsilon_i, \xi, \eta, t) = \lim_{\varepsilon \to 0} s(\varepsilon, \xi, \eta, t),$$

$$\lim_{i \to +\infty} v_i(\varepsilon_i, \xi, \eta, t) = \lim_{\varepsilon \to 0} v(\varepsilon, \xi, \eta, t).$$

Possiamo ora dare un rafforzamento significativo della Congettura 2 enunciando la seguente

CONGETTURA 3 – *Esistono numeri positivi α per i quali è denso nello spazio $\mathcal{M}(\mathbf{R}^6)$ l'insieme delle μ tali che le soluzioni del Problema gravitazionale approssimato associato a μ sono soluzioni α – stabili.*

OSSERVAZIONE 4 – Se fosse verificata la Congettura 3 sarebbe interessante trovare l'estremo inferiore dei numeri α ivi considerati; è assai probabile che tale estremo sia strettamente maggiore di zero; tenendo conto del fatto che l'α – stabilità è tanto piú debole quanto piú α è grande, questo estremo inferiore ci darebbe un'indicazione sul grado globale di stabilità del Problema gravitazionale approssimato.

3. Varianti del Problema gravitazionale approssimato

Si possono immaginare molte varianti del problema gravitazionale esposto nel §2. Si potrebbe pensare che la forza che agisce tra 2 particelle sia somma di una forza attrattiva e una forza repulsiva e porre per esempio in luogo della (3) del §1

$$u(\varepsilon, x, t) = \int_{\mathbf{R}^6} (|x - s|^2 + c^2)^{-1/2} - c^2(|x - s|^2 + \varepsilon^2)^{-1} d\mu(\xi, \eta)$$

con c costante positiva.

Si potrebbero anche considerare, dopo i limiti per $\varepsilon \to 0$, i limiti di tali limiti per $c \to 0$; ciò corrisponderebbe all'idea che le forze repulsive agiscono solo a distanze piccolissime. Infine si potrebbero immaginare espressioni piú complesse corrispondenti a forze che dipendono non solo dalla posizione ma anche dalla velocità delle particelle, per esempio prendendo in considerazione integrali del tipo:

$$\alpha(\varepsilon, x, y, t) = \int_{\mathbf{R}^6} \varphi(\varepsilon, x, y, s(\varepsilon, \xi, \eta, t), v(\varepsilon, \xi, \eta, t)) d\mu(\xi, \eta)$$

e quindi equazioni differenziali del tipo

$$\frac{\partial v}{\partial t} = \alpha(\varepsilon, s, v, t).$$

Naturalmente ulteriori complicazioni potrebbero essere introdotte ispirandosi a teorie fisiche piú complesse che non interessano per il momento dato che lo scopo di questa nota è solo di richiamare l'attenzione su alcuni "tipi" di problemi matematici di cui successivamente si potrebbero immaginare molte varianti ricche di significato fisico.

Ritornando al Problema gravitazionale approssimato e in particolare all'Osservazione 2 del §2, potremmo considerare il caso in cui la massa μ^* è distribuita con regolarità, cioè esista una funzione $\varphi = \varphi(\varepsilon, x, y, t) \in C^\infty(\mathbf{R} \setminus \{0\} \times \mathbf{R}^7)$ tale che

$$(1) \qquad \mu_t^*(B) = \int_B \varphi(\varepsilon, x, y, t) dx dy;$$

in tal caso la funzione $u(\varepsilon, x, t)$ sarà data dalla formula

$$u(\varepsilon, x, t) = \int_{\mathbf{R}^6} \left(|x - \xi|^2 + \varepsilon^2\right)^{-1/2} \varphi(\varepsilon, \xi, \eta, t)\, d\xi\, d\eta,$$

e inoltre φ soddisferà l'equazione differenziale lineare del primo ordine

$$\frac{\partial \varphi}{\partial t} = -\sum_{h=1}^{3} \left(\frac{\partial \varphi}{\partial x_h} y_h + \frac{\partial \varphi}{\partial y_h}\frac{\partial u}{\partial x_h}\right)$$

di cui le equazioni (1), (2) del Problema 2.1 individuano le linee caratteristiche. Si può porre allora il problema dell'esistenza del limite per $\varepsilon \to 0$ della funzione $\varphi(\varepsilon, x, y, t)$ e studiare i possibili collegamenti con i limiti considerati nelle Congetture 1, 2, 3 del §2. Probabilmente in tale studio potrebbero emergere fenomeni interessanti riguardanti le soluzioni piú o meno generalizzate delle equazioni differenziali alle derivate parziali del primo ordine.

4. Limiti di problemi variazionali collegabili ad equazioni iperboliche

Nella ricerca di eventuali approssimazioni di problemi difficili ed instabili con problemi più facili e più stabili si può far rientrare l'idea di ottenere soluzioni di problemi di evoluzione come limite delle soluzioni di problemi di minimo. Un esempio interessante in cui quest'idea è stata applicata con successo si trova nel lavoro [8]. Una possibile variante del risultato di Ilmanen è data dalla seguente congettura.

CONGETTURA 1 – *Siano $\varphi, \psi \in C_0^\infty(\mathbf{R}^n)$, $k > 1$ intero; per ogni numero reale positivo λ, $w_\lambda = w_\lambda(x_1, \ldots, x_n, t)$ fornisca il minimo del funzionale*

$$(1) \qquad F_\lambda(u) = \int_{\mathbf{R}^n \times [0, +\infty[} e^{-\lambda t} \left[\left|\frac{\partial^2 u}{\partial t^2}\right|^2 + \lambda^2 |\nabla_x u|^2 + \lambda^2 u^{2k}\right] dx\, dt$$

nella classe delle u soddisfacenti le condizioni iniziali

$$u(x, 0) = \varphi(x), \qquad\qquad \frac{\partial u}{\partial t}(x, 0) = \psi(x).$$

Allora esiste il $\lim_{\lambda \to +\infty} w_\lambda(x, t) = w_0(x, t)$, *soddisfacente l'equazione*

$$(2) \qquad \frac{\partial^2 w_0}{\partial t^2} = \Delta_x w_0 - k w_0^{2k-1}.$$

OSSERVAZIONE 1 – Se questa congettura risultasse falsa si potrebbe pensare a opportune varianti, per esempio richiedere che essa sia vera per un insieme denso di dati iniziali, sostituire a C^∞ dati analitici a decrescenza rapida, prendere in considerazione funzioni periodiche e in particolare polinomi trigonometrici, o addirittura pensare a un funzionale F_λ di forma diversa da quella indicata nella (1) le cui minimizzanti convergono verso soluzioni della (2). Se viceversa la congettura fosse verificata, si potrebbe pensare a un gran numero di generalizzazioni, partendo da un funzionale F del calcolo delle variazioni e associando ad esso il funzionale

$$F_\lambda(u) = \int_{\mathbf{R}^n \times [0,+\infty[} e^{-\lambda t} \left| \frac{\partial^2 u}{\partial t^2} \right|^2 dxdt + \lambda^2 \int_0^{+\infty} e^{-\lambda t} F\big(u(\cdot,t)\big)\, dt.$$

5. Movimenti di varietà secondo funzioni euleriane

Per introdurre movimenti di varietà immerse in uno spazio euclideo $\mathbf{R}^n$ di ogni dimensione e codimensione cominciamo con lo stabilire alcune notazioni. Per ogni aperto $A \subset \mathbf{R}^n$ continueremo a indicare con $C^\infty(A)$ lo spazio delle funzioni reali continue in A insieme con tutte le loro derivate di ogni ordine, mentre nel caso in cui l'insieme E non è un aperto, scriveremo $f \in C^\infty(E)$ se e soltanto se esistono un aperto $A \supset E$ ed una funzione $f^* \in C^\infty(A)$ tali che, per ogni $x \in E$, $f(x) = f^*(x)$. Dati due aperti A, Ω, scriveremo $A \subset\subset \Omega$ se e solo se la chiusura di A è un compatto contenuto in Ω. Infine, per ogni $E \subset \mathbf{R}^n$ possiamo definire la funzione η_E nel modo seguente:

$$(1) \qquad \eta_E(x) = \frac{1}{2}[\mathrm{dist}(x,E)]^2 = \inf\left\{ \frac{|x-\xi|^2}{2};\ \xi \in E \right\}.$$

Possiamo ora dare la seguente definizione di varietà h-dimensionale di classe C^∞ in un aperto Ω.

DEFINIZIONE 1 – Sia Ω un aperto di $\mathbf{R}^n$, sia h intero, $0 \le h \le n$. Diremo che un insieme E è una varietà h-dimensionale di classe C^∞ in Ω e scriveremo $E \in V_h C^\infty(\Omega)$, se sono soddisfatte le tre condizioni seguenti:

$$(a) \qquad \Omega \cap E = \{x \in \Omega : \eta_E(x) = 0\}$$

$$(b) \qquad \text{esiste un aperto } A \supset \Omega \cap E \text{ tale che } \eta_E \in C^\infty(A)$$

(c) per ogni $x \in E \cap \Omega$ la matrice hessiana $\nabla^2 \eta_E(x)$ ha caratteristica $n - h$.

Se g è una funzione di classe C^∞ definita nell'insieme dei polinomi reali di grado minore o uguale di k in n variabili reali ed $E \in V_h C^\infty(\Omega)$, allora per ogni aperto $A \subset\subset \Omega$ possiamo considerare il funzionale

$$(2) \qquad F(E,A) = \int_{A \cap E} g\big(\eta_E(x), \nabla\eta_E(x), \ldots, \nabla^k \eta_E(x)\big)\, d\mathcal{H}^h(x),$$

ove $\mathcal{H}^h$ è la misura h-dimensionale di Hausdorff; poiché nei punti di E si annullano η_E e $\nabla\eta_E$, questo funzionale dipende in modo non banale da g per $k \ge 2$.

Diamo ora la definizione di funzione euleriana associata ad E ed F.

DEFINIZIONE 2 – Sia E un sottoinsieme di $\mathbf{R}^n$, Ω un aperto di $\mathbf{R}^n$, sia $E = E \cap \Omega \in V_h C^\infty(\Omega)$. Sia $\psi \in (C^\infty(E))^n$ e sia F definito dalla (2). Diremo allora che la funzione vettoriale ψ è la *funzione euleriana* di F su E, e scriveremo

$$\psi \in \mathcal{E}(F, E)$$

se per ogni aperto $A \subset\subset \Omega$ e per ogni $\tau \in [C_0^\infty(A)]^n$ risulta

$$\lim_{\lambda \to 0} \frac{1}{\lambda}\big(F(E_\lambda, A) - F(E, A)\big) = \int_{E \cap A} \sum_{j=1}^{n} \tau_j(x)\psi_j(x)\, d\mathcal{H}^h(x),$$

ove $E_\lambda = \{x + \lambda\tau(x);\ x \in A \cap E\}$.

Possiamo dare ora una nozione di movimento di discesa secondo la funzione euleriana.

DEFINIZIONE 3 – Sia $E \subset \mathbf{R}^n$, F definito dalla (2), $0 < T \le +\infty$, $u = u(x_1, \dots, x_n, t) \in C^\infty(E \times [0, T[)$. Diremo che u è un *movimento di discesa* del funzionale F di origine E, e scriveremo $u \in \mathcal{MD}(F, E, T)$ se soddisfa le condizioni seguenti:

(A)
$$u(x, 0) = x, \qquad \text{per ogni } x \in E.$$

(B)   Esiste una funzione A definita in $E \times [0, T[$ e a valori negli aperti di $\mathbf{R}^n$ tale che per ogni $(x, t) \in E \times [0, T[$, sia

$$x \in A(x, t)$$

e inoltre, posto $L(x, t) = \{(x, t);\ x \in E \cap A(x, t)\}$, per ogni $(x, t) \in E \times [0, T[$ sia

$$\frac{\partial u}{\partial t}(x, t) = \mathcal{E}\big(F, L(x, t)\big)(u(x, t)).$$

OSSERVAZIONE 1 – Nel caso in cui g si riduca a una costante positiva, e quindi $F(E, A)$ è proporzionale alla misura $\mathcal{H}^h(E \cap A)$, ritroviamo in sostanza la definizione classica di movimento secondo la curvatura media (cfr. per es. [1]); in generale non è detto che esistano per tutti i funzionali F definiti dalla (2) e per tutti i dati iniziali E dei movimenti di discesa, ma si può pensare di approssimare gli stessi funzionali con un procedimento simile a quello seguito nel §1. Precisamente, se F è definito dalla (2), possiamo porre, per ogni $\varepsilon > 0$,

$$(3) \qquad F_\varepsilon(E, A) = F(E, A) + \varepsilon^2 \int_{E \cap A} |\nabla^{k+1} \eta_E|^2\, d\mathcal{H}^h$$

e formulare la seguente

CONGETTURA 1 – *Se E è un compatto appartenente a $V_h C^\infty(\mathbf{R}^n)$, ed F è definito dalla (2), allora, per ogni $\varepsilon \ne 0$ esistono $T_\varepsilon > 0$ ed $u_\varepsilon \in [C^\infty(E \times [0, T_\varepsilon[)]^n$ tali che $u_\varepsilon \in \mathcal{MD}(F_\varepsilon, E, T_\varepsilon)$.*

Nel caso in cui g si riduca a una costante positiva, si può ipotizzare l'indipendenza di T da ε come nella seguente

CONGETTURA 2 – *Se alle ipotesi della Congettura 1 aggiungiamo l'ipotesi $k > h$ e l'ipotesi che g abbia un valore costante positivo allora si può assumere $T_\varepsilon \equiv +\infty$ e quindi $u_\varepsilon \in \left[C^\infty(E \times [0, +\infty[) \right]^n$.*

OSSERVAZIONE 2 – In sostanza, la Congettura 2 ci fornirebbe un metodo canonico per approssimare i movimenti secondo la curvatura media. Per quanto riguarda i limiti per $\varepsilon \to 0$ di questi movimenti approssimati si porranno dei problemi piú o meno simili a quelli indicati nei paragrafi precedenti. Per cominciare si potrebbe discutere la seguente congettura che in sostanza equivale ad una specie di risultato di unicità per il movimento secondo la curvatura media definito con un "metodo di viscosità".

CONGETTURA 3 – *Nelle ipotesi della Congettura 2, esiste per $\mathcal{H}^{h+1}$ quasi tutti i punti di $E \times [0, +\infty)$ il limite*

$$\lim_{\varepsilon \to 0} u_\varepsilon(x, t)$$

OSSERVAZIONE 3 – Se la Congettura 3 risultasse falsa si potrebbe indebolirla in vario modo, chiedendo per esempio che essa valga nel caso in cui E oltre ad appartenere a $V_h C^\infty(\mathbf{R}^n)$ appartenga a $V_h C^\omega(\mathbf{R}^n)$, cioé che la funzione η_E che compare nella Definizione 1 sia analitica nell'aperto A. Si potrebbe anche chiedere che la tesi della Congettura 3 sia verificata al variare di E in un sottoinsieme denso dello spazio formato dai compatti che appartengono a $V_h C^\omega(\mathbf{R}^n)$ dotato di qualche opportuna topologia.

Se invece la Congettura 3 fosse verificata, si potrebbero esaminare alcuni casi in cui la funzione g che compare nella (2) non si riduca ad una costante. Da un punto di vista fisico, potrebbero anche avere un certo interesse per opportune scelte del funzionale F equazioni del tipo

$$\frac{\partial^2 u}{\partial t^2} = \mathcal{E}\big(F, L(x, t)\big)\big(u(x, t)\big).$$

Ringraziamenti. Questa nota é il risultato di molte conversazioni avute con vari colleghi e amici, in particolare desidero ringraziare Luigi Ambrosio, Giovanni Bellettini, Giuseppe Bertin, Ivar Ekeland, Antonio Leaci, Carlo Mantegazza, Antonio Marino, Stefano Mortola, Diego Pallara, Sergio Spagnolo e Piero Villaggio.

Bibliografia

[1] L. AMBROSIO - H.M. SONER, *Level set approach to mean curvature flow in arbitrary codimension*, J. Differential Geom. **43** (1996), 693–737.

[2] H. BREZIS, *Opérateurs Maximaux Monotones et Sémi-groupes de Contractions dans les Espaces de Hilbert*, North Holland, Amsterdam, 1973.

[3] J.M. DAWSON, *Particle Simulation of Plasmas*, Rev. Modern Phys. **55** (1983), 403.

[4] M. DEGIOVANNI, A. MARINO, M. TOSQUES, *Evolution Equations with Lack of Convexity*, Nonlinear Anal. **9** (1985), 1401–1443.

[5] R.J. DI PERNA, P.L. LIONS, *Ordinary differential equations, transport theory and Sobolev spaces*, Invent. Math. **98** (1989), 511–547.

[6] M. GRILLAKIS, *Regularity for the wave equations with a critical nonlinearity*, Comm. Pure Appl. Math. **14** (1992), 749–774.

[7] R.W.HOCKNEY, J.W. EASTWOOD, *Computer Simulation Using Particles*, McGraw-Hill, New York, 1981.

[8] T.ILMANEN, *Elliptic regularization and partial regularity for motion by mean curvature*, Mem. Amer. Math. Soc. **108** (1994), # 520.

[9] R.H. MILLER, *Numerical experiments in collisionless systems*, Astrophys. Space Sci. **14** (1971), 73.

[10] P.PERONA-J.MALIK, *Scale-space and edge detection using anisotropic diffusion*, in Proceedings of the IEEE Computer Society Workshop on Computer Vision, Miami, Florida 1987, Computer Society Press of the IEEE, Washington, D.C., 16–22.

[11] P.PERONA-J.MALIK, *Scale-space and edge detection using anisotropic diffusion*, IEEE Trans. Pattern Anal. Mach. Intell. **12** (1990), 629–639.

[12] J. SHATAH, M. STRUWE, *Regularity results for nonlinear wave equations*, Ann. of Math. (2) **138** (1993), 503–518.

Towards the axiomatic systems of the third millennium in Mathematics, Logic, and Computer Science[‡]

Ennio De Giorgi, Marco Forti and Giacomo Lenzi

Introduction

In his talk, Nelson has brilliantly presented his ideas of a mathematician who gave up what he calls "Pythagorean religion", and he shows new ways that, in his opinion, mathematicians should follow. We too are exploring, since several years, the possible future ways of mathematics, but with a different perspective (see [6-10,12-15,24,25]). Nelson, in systematically developing the formalistic perspective, gets to identify mathematics with the study of its formulae, and refuses any semantical perspective that associates numbers, sets, spaces, functions, etc. to these formulae. We acknowledge the irreplaceable rôle of formulae in elaborating and communicating mathematical ideas, as well as in comparing mathematics with other branches of knowledge. We also know that the manipulations of formulae, and in particular the deduction rules, are themselves very interesting objects of mathematics, but we do not intend to reduce the whole of mathematics to manipulation of formulae. Our attitude with respect to formulae is in some sense closer to that of an ancient navigator or explorer w.r.t. maps, who takes them as indispensable tools for navigating and for communicating the discoveries of his navigations, rather than to the attitude of the modern map collector, who takes them as valuable objects.

Nelson recalled the difficulties faced by Hilbert and Brouwer in their attempts of finding a satisfactory "philosophy", where the objects commonly considered by mathematicians can be located in a "natural" way (see [3,22,23]). We do not ignore these difficulties, but we don't forget the words of Shakespeare "there are more things in heaven and earth than are dreamt of in your philosophy": we think that it does not suit to declare an object unexisting on the sole ground that we have not found a satisfactory philosophy that accomodates it. In some sense, Nelson asks mathematicians to "dream less", e.g. not to assign a "too realistic" meaning to the beautiful theorems on infinite dimensional spaces of Hilbert and of Banach: on the contrary, we always recall the words of Shakespeare and hence we think that the scientist who aims to understand the real things existing in

[‡]Editor's note: translation into English of the paper "Verso i sistemi assiomatici del 2000 in Matematica, Logica e Informatica", published in Nuova civiltà delle macchine, **57-60** (1997), 248–259.

heaven and earth needs to "dream more", rather than less. We think that the greatest achievements of science are the outcome of some "wonderful dreams": complex numbers, infinitesimal calculus, Newton's mechanics, general relativity, quantum mechanics, etc.; we also think that a relevant part of the actual work of experimental physicists consists in looking for visible marks of some objects "dreamt of" by theoretical physicists.

We appreciate an elegant explicit solution to a nice mathematical problem, a simple and fast method of computing; we know that they can provide a great improvement on the mere proof of an existence theorem: but we do not believe that abandoning what Hilbert named "Cantor's paradise", peopled by more or less strange sets and by infinites of any order, may help in finding explicit solutions and efficient algorithms. Rather, we would like to imagine "Cantor's paradise" even more rich and colourful than Cantor conceived it, filled up with objects of any different kind: sets, collections, operations, formulae, languages with their semantical interpretations, variables, categories, standard and non-standard numbers, algorithms, propositions, predicates, classical logic and other logics, etc. Admittedly, this freedom in dreaming has a corresponding duty: to translate dreams into axioms, conjectures, theorems formulated with the maximal clarity and with the utmost exactitude (so as to make them, after all, understandable and subject to critical analysis also by scholars adopting a non-realistic, purely formalistic perspective). Not only the dialogue on these themes can go on in friendship and comprehension between "formalists" and "dreamers", but it can become a valuable source of very interesting ideas for both of them. Even better, we underline that the axioms we state below are suitable to be critically discussed, and possibly modified, enriched, improved by scholars close to any current of the philosophy of sciences. Therefore it is important that a general talk concerning mathematics, logic, computer science be formulated in much clearer way than a specialistic talk. In fact the former talk has to be understandable and subject to criticism by a great variety of interlocutors, whereas it suffices that the latter talk be understood by a restricted group of specialists. In particular, a general talk about the main basic concepts of mathematics, logic and computer science is not reserved only to a restricted number of "specialists of foundations", but it has to be accessible to all those scholars of sciences and humanities who are informed with what the ancients called philosophy, i.e. love for Wisdom.

For these reasons it seems unapproprate to classify our work under the category "foundations of mathematics": we do not intend to refund mathematics on safer grounds, we are rather looking for some new tracks through the forest of mathematics, logic and computer science, giving up none of the ingenious intuitions of the scholars that marked out the first roads in this forest, which is, in our opinion, yet mainly unexplored. To be sure, we have no desire to begin the compiling of the "Bourbaki of the third millennium", a task that should inevitably become uncontrollable in dimension and length (cfr. [28]), for it should encompass mathematics, logic, and computer science (disciplines that nowadays have to be considered together, at least for their basic ideas). We simply believe that we have singled out a first simple, clear, "natural" axiomatic basis, upon which it should be possible to engraft the various branches of these disciplines, e.g. standard and nonstandard analysis, classical logic and other logics, set theories, probability calculus, categories, algorithms, languages, syntax, semantics, etc. Upon this basis it should be possible to engraft also some fundamental ideas of other disciplines of science and humanities, giving wide breath to the reflexion upon the relations among different fields of knowledge. In fact, on the one hand one has to reflect upon the most relevant applications of mathematics and upon the ideas suggested to mathematicians by comparing with various scientific, technical, human, artistic disciplines. But, on the other hand, it seems convenient to think also about the deepest reasons of the greatest results achieved by joint work of mathematicians and other scholars.

We now pass to expound, in a hopefully simple and clear way, a first axiomatic basis upon which one can engraft in a "natural" way the fundamental notions of many disciplines of science and humanities. This basis originates from various reflexions upon the main notions of mathematics, logic and computer science (see, e.g. [1,4,22,26,27,29,30]) and from many conversations with scholars of different disciplines (mathematics, physics, logic, computer science, biology, history, philosophy, economics, theology, etc.). Starting from these reflexions, it seemed appropriate to overcome the so called "set theoretic reductionism", i.e. the trend to reduce all of mathematics to set theory. On the contrary, we try and include the mathematical theories, and, if possible, also other scientific theories, within a wider framework, where a critical comparison of the fundamental ideas of different disciplines be possible. Surely, the set theories proposed by Cantor, Zermelo, Gödel, Bernays, Von Neumann and other great mathematicians of this century (see [2]) stay among the highest expressions of the human mind (comparable, e.g., with Newtonian mechanics, general relativity, quantum mechanics, Dante's 'Commedia', Michelangelo's 'Moïses', Shakespeare's tragedies, etc.). However, in order to get a better understanding of the main ideas of mathematics, logic and computer science, it seems appropriate to put them in a more general framework, ruled by the two ideas of *quality* and *relation*. In fact, these disciplines, as well as physical, chemical, biological, economical, linguistical disciplines, etc., all consider *qualitatively different objects* and study *relations among these objects*. It seems therefore appropriate to propose a short, simple system of few general *qualities* and *relations*, as a general premise to the exposition of these disciplines and to the comparison of their main ideas. These qualities and relations should constitute solid and flexible grounds upon which qualities and relations specific to each science be inserted.

Fundamental qualities and relations

We deal in this chapter with the first basic ideas concerning qualities and relations, intended as primitive notions. Notice that with this acception of primitive notion we do not intend to answer either the psychological question as to which ideas present themselves first to the mind of a child, or the historical question

as to which ideas have been first considered by Mankind; we simply mean that
these notions cannot be reduced to other previously introduced concepts (by
means of suitable definitions).

Thus we introduce as primitive notions the idea of *"quality"*, and the idea of
"enjoying a given quality". We stipulate that given an object x of any kind and
a quality q, when we write

$$q\,x$$

we intend to say that *"x enjoys the quality q"*. In this chapter we introduce
seven fundamental qualities: $Qqal$, $Qrel$, $Qrelb$, $Qrelt$, $Qrelq$, $Qrun$, $Qrbiun$,
with the following meanings:

$Qqual\,x$ means that x is a *quality*;

$Qrel\,x$ means that x is a *relation*;

$Qrelb\,x$ means that x is a *binary relation*;

$Qrelt\,x$ means that x is a *ternary relation*;

$Qrelq\,x$ means that x is a *quaternary relation*;

$Qrun\,x$ means that x is a *univocal relation*;

$Qqual\,x$ means that x is a *biunique relation*.

These seven qualities enjoy $Qqual$, hence we can write:

Axiom 1.1 – *$Qqual\,Qqual$, $Qqual\,Qrel$, $Qqual\,Qrelb$, $Qqual\,Qrelt$, $Qqual\,Qrelq$,
$Qqual\,Qrun$, $Qqual\,Qrbiun$.*

The three qualities $Qrelb$, $Qrelt$, $Qrelq$ are particular cases of the more general
quality $Qrel$, i.e.:

Axiom 1.2 – *An element x enjoying any of the qualities $Qrelb$, $Qrelt$, $Qrelq$
enjoys the quality $Qrel$ as well.*

Notice that we are not excluding that there exist relations more complex than
binary, ternary or quaternary relations: we do not introduce them in this chapter
simply because we shall make no use of them in this paper.

After the primitive idea of enjoying a given quality, the second most important
primitive idea of this section is that of *"being in a given relation"*. Namely, given
two objects x, y of any kind, and a binary relation r, we write

$$r\,x, y$$

or sometimes

$$r\,x; y$$

to mean that *"x and y are in the relation r"*. At times, instead of saying that x
and y are in the relation r, we also say that x is in the relation r with y.

Similarly, if x, y, z are objects of any kind and ρ is a ternary relation, we write

$$\rho\,x, y, z$$

or

$$\rho\,x; y; z$$

to mean that *"x, y, z are in the relation ρ"*.

Finally, if τ is a quaternary relation and x, y, z, t are objects of any kind, we write

$$\tau\, x, y, z, t$$

or

$$\tau\, x; y; z; t$$

to mean that "x, y, z, t are in the relation τ".

In this section we introduce four fundamental relations: *Rqual, Rrelb, Rrelt, Rid*. The relation *Rqual* is a binary relation that connects *qualities* with *elements enjoying them*. Namely:

Axiom 1.3 – *Rqual is a binary relation. Given objects x, y, if*

$$Rqual\, x, y$$

holds, then x is a quality (i.e. x enjoys Qqual). Moreover, if q is a quality and x is any object, then Rqual q, x holds if and only if $q\, x$ (i.e. x enjoys the quality q).

The relation *Rrelb* is a ternary relation, and it connects *binary relations* with *objects which are in these relations*. In other words:

Axiom 1.4 – *Rrelb is a ternary relation. Given objects x, y, z, if*

$$Rrelb\, x, y, z$$

holds, then x is a binary relation (i.e. x enjoys Qrelb). Moreover, if r is a binary relation and x, y are objects of any kind, then Rrelb r, x, y holds if and only if $r\, x, y$ (i.e. x, y are in the relation r).

The relation *Rrelt* is a quaternary relation that connects *ternary relations* with *objects which are in these relations*. In other words:

Axiom 1.5 – *Rrelt is a quaternary relation. Given objects x, y, z, t, if*

$$Rrelt\, x, y, z, t$$

holds, then x is a ternary relation (i.e. x enjoys Qrelt). Moreover, if ρ is a ternary relation and x, y, z are objects of any kind, then Rrelt ρ, x, y, z holds if and only if $\rho\, x, y, z$ (i.e. x, y, z are in the relation ρ).

Finally, *Rid* is a binary relation representing *identity*. In other words:

Axiom 1.6 – *Rid is a biunique relation. In order to have*

$$Rid\, x, y$$

it is necessary and sufficient that x and y be the same object. In other words, every object is in the relation Rid with itself and only with itself.

The relation *Rid* expresses the identity between objects of *any kind*. Hence we shall write usually $x = y$ instead of *Rid* x, y.

The relation *Rid* is the simplest instance of a ralation enjoing both qualities *Qrun* and *Qrbiun*. Concerning the quality *Qrun*, we state the following axiom, which expresses the idea of *univocity* applied to *binary, ternary* and *quaternary relations*:

Axiom 1.7 – *Qrun is a quality.*
1) if Qrun x, then Qrel x;
2) if Qrelb r; Qrun r; r x,y; r x,z , then y = z;
3) if Qrelt ρ; Qrun ρ; ρ x,y,z; ρ x,y,t , then z = t;
4) if Qrelq τ; Qrun τ τ x,y,z,t; τ x,y,z,u , then t = u.

Finally *Qrbiun* is the quality of being a *binary biunique relation*, and biuniqueness is expressed by the axiom:

Axiom 1.8 – *Qrbiun is a quality.*
1) if Qrbiun x, then Qrelb x, Qrun x;
2) if Qrbiun x; r y,x; r z,x , then y = z.

Operations, collections, sets and natural numbers

In this section we give some simple examples of "engrafting" mathematical notions into the "trunk" of the fundamental axioms concerning qualities and relations. Of course, each of these engraftings should be considered only a first "budd", from which whole "branches" of mathematics can develop in various ways. In this section we introduce the notions of operation, collection, set and natural number. We give only a few basic descriptive axioms, so as to leave open the possibility of developing the respective theories in different directions.

We introduce first the notion of *operation* by means of three qualities: *Qop*, the quality of being an operation; *Qops*, the quality of being a *simple operation*; *Qopb*, the quality of being a *binary operation*, and by means of two relations *Rops* and *Ropb*, which describe the way simple and binary operations operate. These objects satisfy the following axioms:

Axiom 2.1 – *Qops, Qopb, and Qop are qualities. Rops is a ternary univocal relation, and Ropb is a quaternary univocal relation.*
1) If either Qops x or Qopb x, then Qop x. (In other words, Qops and Qopb are specializations of the generic quality Qop.)
2) for all x, y, z, if Rops x, y, z, then Qops x;
3) if Ropb x, y, z, t, then Qopb x.

Whenever f is a simple operation, we often write $y = fx$ instead of *Rops* f, x, y, and we say that f *trasforms* x into y, or that f *maps* x onto y, or else that f *operates* on x giving the *result* y. Similarly, whenever φ is a binary operation, we often write $z = \varphi\, x, y$ instead of *Ropb* φ, x, y, z. Notice that *univocity* of the relations *Rops* and *Ropb* yields directly "functionality" of simple and binary operations, i.e. the fact that, if f is a simple operation operating on an object x, the result $y = fx$ is uniquely determined. Similarly, if φ is a binary operation operating on objects x, y, the result $z = \varphi\, x, y$ is uniquely determined.

After operations, we engraft another kind of objects, often considered in mathematics, namely *collections* and *sets*, which are particular collections, widely used in modern mathematics. To this aim, we introduce the quality *Qcoll*, i.e. the quality of being a *collection*, the relation *Rcoll*, the relation of *membership to collections*, the quality *Qins* of being a *set* and the relation *Rins* of *membership to sets*. These objects satisfy the following axioms:

Axiom 2.2 – *Qcoll and Qins are qualities. Rcoll and Rins are binary relations.*
1) Qins x implies Qcoll x;
2) if Rcoll x, y, then x is a collection, i.e. Qcoll x;
3) Rins x, y if and only if Qins x, Rcoll x, y.

Following the common usage, we write $y \in x$ instead of *Rcoll* x, y, and we say that y *belongs* to x, or that y is an *element* of x. The clause 3 says essentially that *Rins* is the restriction of *Rcoll* to sets.

We introduce also the relation *Rincl*, the relation of *inclusion*, satisfying the axiom:

Axiom 2.3 – *Rincl is a binary relation.*
1) If A, B are collections, then Rincl A, B if and only if every element of A is also element of B;
2) if Qins E, Qcoll X, and Rincl X, E, then Qins X.

Following the common usage, when A, B are collections and *Rincl* A, B holds, we write $A \subseteq B$ and we say that A is *included* in B or that A is a *part* of B. The clause 2) says that any collection included in a set is itself a set.

We can now state the fundamental axiom of the theory of collections, namely the *axiom of extensionality*:

Axiom 2.4 – *If A, B are collections, $A \subseteq B, B \subseteq A$, then $A = B$.*

In order to provide first instances of collections and sets we introduce now the *universal collection* V, the binary operation *Compl* (*relative complement*) and the simple operation *Csing* (generator of *singular collections* or *singletons*).

Axiom 2.5 – *V is a collection, Csing is a simple operation, Compl is a binary operation.*
1) For any object x it holds $x \in V$.
2) For any object x there exists Csing x and it is a set whose unique element is x.
3) If A, B are collections, then Compl A, B exists, and it is a collection whose elements are all and only those elements of A that are not elements of B.

We denote by $\{x\}$ the set *Csing* x, and we call it the *singleton* of x. We denote by $A \setminus B$ the collection *Compl* A, B, and we call it the *relative complement* of B w.r.t. A, or the *difference* between A and B.

Notice that it follows from the axioms that there exists an *empty collection* $\emptyset$, which is a set, and that there exist the *intersection* $A \cap B = A \setminus (A \setminus B)$ and the *union* $A \cup B = V \setminus ((V \setminus A) \cap (V \setminus B))$ of two collections A, B. In order to build up sets starting from singletons, we also postulate:

Axiom 2.6 – *If A, B are sets, then also $A \cup B$ is a set.*

So, given objects $x, y, z, t, \ldots$ and starting from their singletons, we get the *doubleton* $\{x, y\} = \{x\} \cup \{y\}$, the *tripleton* $\{x, y, z\} = \{x, y\} \cup \{z\}$, the *quartet* $\{x, y, z, t\} = \{x, y, z\} \cup \{t\}$, etc.

As previously remarked, we give here only the first "descriptive" axioms on operations and collections, so as to leave completely free space to developments of these notions by means of axioms specifying other ways of "costructing" collections, sets and operations (see, e.g. [15,18,19]).

Notice that we have carefully avoided the introduction of an axiom of extensionality for operations: in fact we intend to embody into the notion of operation the ideas of "construction", "manufacturing", "computing procedure", even giving to this idea an extremely broad intension, which in many cases, like those considered in the axiom 2.5, is very far from "effectiveness". So we will not exclude the possibility that two operations remain *different*, notwithstanding the fact that they operate on the *same objects* and they give always the *same results* (see [19]).

We conclude this section by introducing the natural numbers, i.e. the numbers $0, 1, 2, 3, 4, \ldots$, by means of the quality $Qnnat$, of being a *natural number* and of the biunique relation $Rnsuc$, which connects each natural number with its *immediate successor*. These objects are ruled by the following axioms:

Axiom 2.7 – *Qnnat is a quality and Rnsuc is a biunique relation.*
1) Rnsuc x, y implies Qnnat x, Qnnat y.
2) There exists a unique z such that Qnnat z and for no x Rnsuc x, z.
3) If Qnnat x, then there exists y such that Rnsuc x, y.

Given a natural number x, the unique natural number y such that $Rnsuc\ x, y$ is called the *successor* of x. The axiom 2.7 suffices for characterizing the natural numbers 0 (the unique z of clause 2), 1 (the successor of 0), 2 (the successor of 1), etc.

The arithmetical theory provided by the axiom 2.7 is very weak. However it provides the possibility of defining *infinitely many* natural numbers by means of the relation $Rnsuc$. In case only a few natural numbers are needed, e.g. only $0, 1, 2, 3, 4, 5, 6, 7$, one can replace the clause 3 in axiom 2.7 by some particular instances, for instance one can assume the axioms:

2.7.3.1) There exists the natural number 1 such that Rnsuc $0; 1$;
2.7.3.2) There exists the natural number 2 such that Rnsuc $1; 2$;
2.7.3.3) There exists the natural number 3 such that Rnsuc $2; 3$;
2.7.3.4) There exists the natural number 4 such that Rnsuc $3; 4$;
2.7.3.5) There exists the natural number 5 such that Rnsuc $4; 5$;
2.7.3.6) There exists the natural number 6 such that Rnsuc $5; 6$;
2.7.3.7) There exists the natural number 7 such that Rnsuc $6; 7$.

The development of various strong theories of (standard and nonstandard) arithmetic might depend on the introduction of operations (beginning with the four operations of the primary scool), relations (natural ordering, divisibility, etc.), collections (the collection $\mathbf{N}$ of all natural numbers, the collection $\mathbf{P}$ of all primes, etc.), various forms of the induction principle, etc.

Correlations, functions and systems

In this section we consider another kind of fundamental objects of mathematics, namely the correlations, together with the special cases of functional correlations, systems and functions.

In the usual treatments, systems and functions are often identified with their graphs, obtained by means of the so called "Kuratowski pairs", i.e. sets of the type $\{\{x\}, \{x, y\}\}$. We prefer instead to consider correlations as a kind of objects quite separate from the kind of collections, so as to leave open way to various theories concerning the relations between collections and correlations: see [18] for a first example of such theories. In this section we restrict ourselves to ascertain some similarities between the theory of collections and that of correlations. In particular we exploit the possibility of extending to correlations some axioms concerning the relation *Rincl* and the operation *Compl* already considered in the previous section.

We also remark that the notion of operation is quite separate from those of functional correlation and of function, notwithstanding the fact that all of them are characterized by univocity axioms: operations do not satisfy any axiom of extensionality, insofar they bring about the intuitions of manufacturing, computing, constructing, which are essentially non-extensional. The idea of correlation, instead, is inspired to a mere "inspection of input-output tables". The attitude of an engineer who has to organize the work of a factory corresponds, in some sense, to the concept of operation; the concept of correlation is closer to the attitude of a warehouse-keeper, who has simply to register ingoing materials and outgoing products.

A first step in engrafting correlations is done by introducing the quality *Qcorr* of being a *correlation* and the relation *Rcorr* that describes the action of correlations. About these objects we state the axiom:

Axiom 3.1 – *Qcorr is a quality. Rcorr is a ternary relation. If Rcorr x, y, z, then Qcorr x.*

Given a correlation C, instead of writing *Rcorr* C, x, y we can say that x is an *index* of C, that y is a *value* of C, and that the index x and the value y are *correlated* by C.

Having engrafted correlations, we can pass to functional correlations, systems, and functions by introducing the qualities *Qcorfun*, of being a *functional correlation*, *Qsys* of being a *system*, and *Qfun* of being a *function*. These qualities satisfy the axiom:

Axiom 3.2 – *Qcorfun, Qsys and Qfun are qualities.*
1) If Qcorfun x or Qsys x, then Qcorr x.
2) Qfun x if and only if x enjoys simultaneously Qsys and Qcorfun.

We introduce also the relations *Rcorfun, Rsys, Rfun*, the restrictions of *Rcorr* to functional correlations, systems, and functions respectively. This fact is expressed by the following axiom:

Axiom 3.3 – *Rcorfun, Rsys, Rfun are ternary relations.*
1) Rcorfun F, x, y if and only if Rcorr F, x, y and Qcorfun F.

2) Rsys S, x, y if and only if Rcorr S, x, y and Qsys S.

3) Rfun f, x, y if and only if Rcorr f, x, y and Qfun f.

We can now give the axiom of *univocity* for functional correlations and functions:

Axiom 3.4 – *Rcorfun and Rfun are univocal ternary relations.*
Let F be a correlation: then Qcorfun F holds (i.e. F is a functional correlation) if and only if whenever x, y, z satisfy simultaneously the conditions Rcorr F, x, y and Rcorr F, x, z, then $y = z$.

The inclusion relation *Rincl*, already introduced in the preceding section, may concern, besides collections, also other objects like correlations, for which we give an axiom analogous to the axiom 2.3:

Axiom 3.5 – *Let F, G be correlations: then Rincl F, G if and only if, for all x, y, Rcorr F, x, y implies Rcorr G, x, y (in other words, Rincl F, G holds if and only if indices and values correlated by F are also correlated by G). Moreover, if s is a system, t is a correlation and Rincl t, s holds, then also t is a system.*

Also in the case of correlations we use a notation similar to that used for collections, i.e. $F \subseteq G$ or $G \supseteq F$ instead of *Rincl F, G*, and we say that the correlation G includes the correlation F, or that F is a part of the correlation G. We give the *axiom of extensionality* for correlations in the following way:

Axiom 3.6 – *If F and G are correlations and both $F \subseteq G, G \subseteq F$ hold, then $F = G$.*

We introduce now the relation *Rdom*, connecting a correlation with its *indices* (also called the "elements of its domain") and the relation *Rcod*, connecting a correlation with its *values* (also called the "elements of its codomain").

Axiom 3.7 – *Rdom and Rcod are binary relations. If F is a correlation, then one has:*
1) Rdom F, x if and only if there exists y such that Rcorr F, x, y;
2) Rcod F, y if and only if there exists x such that Rcorr F, x, y.

In this paper we do not postulate that for any correlation F there is a collection whose elements are all and only the indices of F, nor we postulate that there is one with all and only the values of F. On the other hand, we shall see in a moment that the relations *Rdom* and *Rcod* concern also other kinds of objects, in particular operations and relations.

However we want to ensure first the existence of a few correlations of particular relevance. To this aim we introduce the universal correlation V_2 and the operation *Fsing*, generator of *singular functions*, and we extend to correlations the operation *Compl* previously introduced. We set the axioms:

Axiom 3.8 – *1) Fsing is a binary operation. For any given objects a, b there exists Fsing $a, b = f$, and f is a function such that Rfun f, x, y holds if and only if $x = a$ and $y = b$.*
2) V_2 is a correlation such that Rcorr V_2, x, y holds for all x, y.

Axiom 3.9 – *If F, G are correlations, then Compl $F, G = H$ exists, and H is a correlation such that, for all x, y, Rcorr H, x, y if and only if Rcorr F, x, y, but not Rcorr G, x, y.*

Inspired by the algebraic notation of substitutions, we denote $\binom{a}{b}$ the singular function $F sing\, a, b$. Moreover we extend to correlations the notation of collections: the *difference* $F \setminus G = Compl\, F, G$; the *intersection* $F \cap G = F \setminus (F \setminus G)$; the *union* $F \cup G = V_2 \setminus ((V_2 \setminus F) \cap (V_2 \setminus G))$.

In this notation, the conditions $Rcorr\, F, x, y$ and $F \supseteq \binom{x}{y}$ are equivalent for any correlation F.

We can now introduce the relation $Rgraf$, connecting a correlation to the *singular functions* included in it (also called at times "elements of its graph"). More precisely, we state the axiom:

Axiom 3.10 – *$Rgraf$ is a binary relation.*
1) If F is a correlation and $Rgraf\, F, z$ holds, then z is a singular function;
2) for all x, y one has $Rgraf\, F, \binom{x}{y}$ if and only if $F \supseteq \binom{x}{y}$.

We remark that the relations $Rdom, Rcod, Rgraf$ involve, besides correlations, also relations, on wich we give the following axiom:

Axiom 3.11 – *1) If r is a binary relation and x is any object, one has $Rdom\, r, x$ if and only if there exists y such that r, x, y; one has $Rcod\, r, y$ if and only if there exists x such that r, x, y; one has $Rgraf\, r, w$ if and only if there exist x, y such that $w = \binom{x}{y}$ and $r\, x, y$.*

2) If ρ is a ternary relation, one has $Rdom\, \rho, w$ if and only if there exist x, y such that $w = \binom{x}{y}$ and there exists z such that $\rho\, x, y, z$; one has $Rcod\, \rho, z$ if and only if there exist x, y such that $\rho\, x, y, z$; one has $Rgraf\, \rho, z$ if and only if there exist a, b, c such that $z = \left(\binom{\binom{a}{b}}{c}\right)$ and $\rho\, a, b, c$.

3) If τ is a quaternary relation, then one has $Rdom\, \tau, w$ if and only if there exist a, b, c such that $w = \left(\binom{\binom{a}{b}}{c}\right)$ and there exists d such that $\tau\, a, b, c, d$; one has $Rcod\, \tau, d$ if and only if there exist a, b, c such that $\tau\, a, b, c, d$; one has $Rgraf\, \tau, z$ if and only if there exist a, b, c, d such that $z = \left(\binom{\binom{\binom{a}{b}}{c}}{d}\right)$ and $\tau\, a, b, c, d$.

A similar axiom could be given also for simple and binary operations. Notice that the relation $Rgraf$ connects quaternary relations and singular functions, which in turn have been introduced by appealing to ternary relations and binary operations; we close in this way a sort of cicle within the first three sections of this note, and an interesting example of *selfreference*, or better *mutual reference* comes off. Much harder problems concerning possible and impossible selfreferences arise from reflexions upon the next section, where we deal with propositions and predicates.

We conclude this section by pointing out two more interesting operations acting on correlations, the operation of inversion and the operation of composition; we give also some instances of interesting systems that can be constructed by union of correlations, starting from singular functions. We introduce the operations Inv (inversion) and $Comp$ (composition), fulfilling the axioms:

Axiom 3.12 – *Inv is a simple operation.*
1) If F is a correlation, then $Inv\, F = G$ exists, and G is a correlation such that $Rcorr\, G, x, y$ if and only if $Rcorr\, F, y, x$.
2) If s is a system, then $Inv\, s$ is a system.

Axiom 3.13 – *Comp is a binary operation.*
1) If F,G are correlations, then $Comp\ F,G = H$ exists, and H is a correlation; moreover one has $Rcorr\ H,x,y$ if and only if there exists z such that $Rcorr\ G,x,z$ and $Rcorr\ F,z,y$.
2) If s,t are systems, then $Comp\ s,t$ is a system.

In order to shortly indicate inverse and composite correlations we adopt the notation $F^{-1} = Inv\ F$, $F \circ G = Comp\ F,G$.

The axioms we introduced up to now allow for manufacturing "by hand" those "finite" systems that one actually want tu use. For instance, given two singular operations $\binom{a}{b}, \binom{c}{d}$, one gets by union the system $\begin{pmatrix} a & c \\ b & d \end{pmatrix}$; similarly, given a third singular function $\binom{e}{f}$ we get the system $\begin{pmatrix} a & c & e \\ b & d & f \end{pmatrix}$. In particular, for any integer n defined according to axiom 2.7, or according to the axioms 2.7.3.1 ,..., 2.7.3.7, one can consider the n-tuples as functions defined at the numbers $1, 2, \ldots, n$. In particular for avery object a one has the "1-tuple" $\binom{1}{a}$, given objects a, b one has the 2-tuple, or ordered pair $\begin{pmatrix} 1 & 2 \\ a & b \end{pmatrix}$, which will be denoted by the symbol (a, b), given objects a, b, c one has the triple $\begin{pmatrix} 1 & 2 & 3 \\ a & b & c \end{pmatrix}$, denoted by (a, b, c), and given objects a, b, c, d one has the 4-tuple $\begin{pmatrix} 1 & 2 & 3 & 4 \\ a & b & c & d \end{pmatrix}$, denoted by (a, b, c, d), etc.

Propositions and predicates

In this section we introduce some very general ideas about propositions and predicates. The axioms we are introducing are intentionally very weak, and they deal essentially with "classical" propositions and predicates, so as to leave widest freedom to subsequent development of other logics, different from classical logic, as well as to an investigation of the difficult problems arising in passing from the study of "restricted" models, whose domain is a set (as usual in model theory), to the study of "universal" models of different kinds of logics, whose domain is the universal collection V.

Therefore we begin by introducing the "most general" notion of proposition by means of the quality $Qgprop$: so we write $Qgprop\ x$ to intend that x is a *proposition*, without any further specification. Among all propositions there are in particular the "classical" propositions, which obey the rules of classical propositional calculus. These propositions are characterized by the quality $Qclp$: so $Qclp\ x$ means that x is a *classical proposition*. We can state the axiom:

Axiom 4.1 – *$Qgprop, Qclp$ are qualities. If $Qclp\ p$, then $Qgprop\ p$.*

The traditional logical operations Et (*conjunction*), Vel (*disjunction* or *alternative*), Non (*negation*) act on propositions:

Axiom 4.2 – *Et and Vel are binary operations. Non is a simple operation. If p, q are classical propositions, then Et p, q, Vel p, q, Non p exist and are classical propositions.*

Following the common usage, we denote by $p\&q$, $p \vee q$, $\neg p$ the propositions *Et* p, q, *Vel* p, q, and *Non* p.

We consider also the quality *Qtver*, of being "absolutely and totally" true, and so we write *Qtver* x to affirm that x is *true*, and we write simultaneously *Qgprop* p, *Qtver* p to affirm that p is a *true proposition*. The following axiom connects the classical propositions to the quality *Qtver*, and expresses, *inter alia*, the classical principles of *non contradiction* and of the *excluded middle*:

Axiom 4.3 – *Qtver is a quality. If p, q are classical propositions, then:*
1) p enjoys Qtver if and only if $\neg\, p$ does not enjoy Qtver.
2) $p\&q$ enjoys Qtver if and only if both p and q enjoy Qtver.
3) $p \vee q$ enjoys Qtver if and only if at least one of p and q enjoys Qtver.

When $\neg p$ enjoys the quality *Qtver*, we say also that p is *false*.
We introduce also the relation *Rseq* of *logical consequence* and the relation *Rleq* of *logical equivalence* between propositions. We give for these relations only two very general axioms, restricting ourselves, as customary, to the case of classical propositions:

Axiom 4.4 – *Let p, q, r be classical propositions. If Rseq p, q and Qtver p, then Qtver q. Moreover one has:*
1) Rleq p, q if and only if Rseq p, q, Rseq q, p.
2) Rseq p, p.
3) If Rseq p, q and Rseq q, r, then Rseq p, r.

Axiom 4.5 – *Let p, q be classical propositions. Then one has:*
1) Rseq $p\&q, p$ and Rseq $p\&q, q\&p$.
2) Rseq $p, p \vee q$ and Rseq $q \vee p, p \vee q$.
3) Rleq $p, \neg\neg p$.

Having thus introduced propositions, we can resume the classical notion of predicate "*à la* Frege" (see [20]), i.e. as an operation taking on propositions as values. Of course, many different introductions are possible (see, e.g., [13,15,25]). Here we introduce the quality *Qopreds* of being a *simple predicative operation* and the quality *Qclops* of being a *simple classical predicative operation*. The following axiom holds:

Axiom 4.6 – *Qopreds and Qclops are qualities.*
1) If Qclops x, then Qopreds x.
2) If Qopreds x, then Qops x.
3) If Qopreds x and Rops x, y, z, then Qgprop z.
4) If Qclops x and Rops x, y, z, then Qclp z.

The connections between the objects studied in the previous three sections and the propositions describing them are established by the operation *Gops*, the generator of *simple predicative operations*. This operation fulfils the following axioms:

Axiom 4.7 – *Gops is a simple operation. If τ is a quaternary relation, then:*
1) Gops τ is a simple operation;
2) for every object x, $(Gops\ \tau)x$ is a simple operation;
3) for all x, y, $((Gops\ \tau)x)y$ is a simple operation;
4) for all x, y, z, $((Gops\ \tau)x)y)z$ is a simple predicative operation;
5) for all x, y, z, t, $(((Gops\ \tau)x)y)z)t$ is a proposition;
6) for all x, y, z, t, the proposition $(((Gops\ \tau)x)y)z)t$ enjoys Qtver if and only if
$\tau\ x, y, z, t$ holds (i.e. if and only if x, y, z, t are in the relation τ).

The clause 6 of the axiom above justifies the shortened notation

$$\text{``}\tau\ x, y, z, t\text{''} = ((((Gops\ \tau)x)y)z)t.$$

After defining the behaviour of the operation *Gops* at quaternary relations it is
an easy task to pass to other kinds of objects introduced in the previous sections,
by operating in the following way:

Axiom 4.8 – *1) If ρ is a ternary relation, then Gops $\rho = (Gops\ Rrelt)\rho$.*
2) If r is a binary relation, then Gops $r = (Gops\ Rrelb)r$.
3) If q is a quality, then Gops $q = (Gops\ Rqual)q$.
4) If f is a simple operation, then Gops $f = (Gops\ Rops)f$.
5) If φ is a binary operation, then Gops $\varphi = (Gops\ Ropb)\varphi$.
6) If C is a collection, then Gops $C = (Gops\ Rcoll)C$.
7) If F is a correlation, then Gops $F = (Gops\ Rcorr)F$.

The "elementary" propositions generated by predicates obtained by means of the
operation *Gops* are statements concerning those qualities or relations to which
Gops has been applied. Hence we extend the quoted notation in the natural
way:
if q is a quality, then "qx" stands for $(Gops\ q)x$;
if r is a binary relation,, then "$r\ x, y$" stands for $((Gops\ r)x)y$;
if ρ is a ternary relation, then "$\rho\ x, y, z$" stands for $(((Gops\ \rho)x)y)z)$;
if C is a collection, then "$x \in C$" stands for $(Gops\ C)x$;
finally "$x = y$" stands for $((Gops\ Rid)x)y$.

We can now introduce the *existential quantifier* and the *universal quantifier* by
means of the operations *Exist* and *Univ*, ruled by the axiom:

Axiom 4.9 – *Univ and Exist are simple operations.*
1) If p enjoys Qclops, then both Univ p and Exist p exist and enjoy Qclp.
2) The proposition Univ p is true if and only if, whenever the proposition px
exists, it is true.
3) The proposition Exist p is true if and only if, for at least an object x, the
proposition px exists and it is true.

We shall use the more common notation $\forall p$, $\forall x.px$, $\forall y.py$, etc. instead of *Univ p*,
and similarly $\exists\ p, \exists x.px, \exists y.py$, etc. instead of *Exist p*: of course, these expressions do not refer to any specified objects x, y, etc.
Having introduced operations and predicates, and in particular the "elementary"
predicates defined through *Gops*, two problems arise, which might be analyzed

in several directions. The first problem regards the study of propositions and predicates that can be built up by means of various manipulations of operations, starting from *Gops* and from the logical operations *Et, Vel, Non, Exist, Univ.* Solutions to a similar problem have been given in [13,15,25], but obviously there are many different, equally interesting possible solutions. More difficult is the second problem, which leads us into the neighborhood of all antinomies and paradoxes: namely that of discerning which, among the propositions and predicates built up by means of *Gops*, could be *classical* propositions and predicates. Some negative results, essentially inspired by the Liar's Paradox, or by Tarski's theorem, have been proved in similar contexts, see [13,15]. The search for the strongest positive axioms that do not lead to contradiction is still open.

Concluding remarks

Besides the objects introduced in the last three sections above, one could of course consider, within the framework of the most general axioms on qualities and relations, other objects like *variables* (see [5,11]), *categories*, and possibly resume into consideration the *metaqualities*, introduced in [13] as intermediate objects between the level of qualities and relations, "premathematical" objects, and the level of "mathematical" objects like operations, numbers, correlations and collections. Each theory can be developed in several directions, and so we hope that many scholars with different cultural education and various interests take part in these developments, retaining the greatest freedom and independence, but also sharing great willingness of a sincere comparison of ideas. The section concerning propositions and predicates is probably the one that will face more problems and greater difficulties, but it might also give rise to the newest, most interesting ideas. Indeed it is in this field that one reaches the deepest common roots of mathematics, logic and informatics, and one gets close to ancient and modern paradoxes and antinomies.

The situation of the scholar who wants to study in depth these topics resembles that of a mountaineer who advances, on very narrow ridges, surrounded by deep ravines, proceeding towards high, beautiful peaks. In fact, one treads on the dangerous ground of "selfreference" and of "mutual reference" (cfr. [14,24]): these topics implicate treacherous problems, but the ultimate significance of the different forms of human knowledge stands out therefrom. In order to get a clear understanding of these questions, a frank and open discussion among scholars of diverse disciplines is necessary: a restricted debate among specialists is unappropriate. Notice that various forms of selfreference (or of mutual reference) are met not only in mathematics, logic and computer science, but also in many other branches of human knowledge. One may quote dictionaries that explain a word through other words, and that contain also the word "dictionary", grammars written according to the grammar rules, laws that regulate the lawmaking activity, economic doctrines that, if successful, describe economic phenomena that depend on the actions of agents deeply influenced by these doctrines. Other examples are the history of historiography, the painter who paints himself while painting, the theatre on the stage, the novelist or the poet who talks about the writing of a novel or about the creation of a poem, the biologist investigating the relations between eyes and brain on the grounds of observations made by her own eyes, etc.

Countless other cases of selfreference and mutual reference could be mentioned, and perhaps the respective similarities and differences could be the subject matter of a frank interdisciplinary comparison of ideas. In this spirit of frank comparison of ideas have to be intended also the statements of this paper, which are "proposed" (not "imposed") as "axioms", i.e. statements not deduced from a system of previous statements, but chosen as a possible starting point for further developments of various theories. Neither are these axioms "deduced" from the history of mathematics, from the philosophy of science, from logic, etc. On the contrary, they can be better understood by assuming an attitude as naïve as possible, and by keeping to the most common meanings that every day language assigns to the words quality and relation. Only after such a first reading of this paper it is appropriate a critical second reading, where everybody can obviously bring about their own experiences and knowledge of mathematics, logic, computer science, history, philosophy, physics, economics, etc.: these esperiences could be very important in setting forth deeper and better justified critical valuations, and hopefully in suggesting supplementary or alternative "axioms", which better represent and clarify the involved intuitive notions.

We do not expect, nor even want, that our proposal be unconditionally accepted: we believe that a vaste work of critical reflexion and friendly discussions between scholars of diverse education and attitudes is still needed, in order to attain a good axiomatic system (or some good axiomatic systems), suitable for substantially improving upon the present situation. On the other hand, we think that it is worthwhile to pursue this goal, because neither formalism, nor set theoretic reductionism, along with any other form of reductionism, seem to offer a suitable perspective for a real understanding of many problems that the culture of our time has to face.

Any investigation of the fundamental axioms of mathematics, logic and computer science, as well as of various experimental, human and philosophical sciences needs, among other things, to overcome a too restricted vision of the different specialities, and requires a broader idea of mathematical and scientific rigor. Mathematical rigor is not only carefulness of the proofs, but also engagement in exposing, in the most clear and understandable way, the problems one would want to solve, the theorems one would want to prove, the conjectures one would want to verify or refute. We think that scientific rigor ultimately consists in clearly and frankly exposing their own certainties and doubts, which problems one believes to have solved and which one would like to solve or see solved, while avoiding those confuse, obscure, uselessly complicated talks that end up in annoying even the most favourably disposed listener.

Summing up, we may conclude that any consideration on method, rigor, and meaning of science leads us in the end to the ancient intuitions of the sapiential value of humility, of "conviviality" (which keeps together sharing of knowledge, friendship, search of mutual understanding), and of trust in Wisdom, which meets all those who love and seek it.

Bibliography

[1] H. P. BARENDREGT – *The Lambda-Calculus*, Amsterdam 1981.

[2] Y. BAR HLLEL, A. A. FRÆNKEL, A. LEVY – *Foundations of set theory*, Amsterdam 1973.

[3] L.E.J. BROUWER – *Wiskunde, waarheid, werkelijkheid*, Groningen 1919.

[4] A. CHURCH – Set theory with a universal set, in *Proceedings of the Tarski Symposium* (L. Henkin *et al.*, eds.), Proc. Symposia Pure Math. XXV, A.M.S., Providence R.I. 1974, 297–308.

[5] M. CLAVELLI – Variabili e teoria A, *Rend. Sem. Fac. Sci. Univ. Cagliari* (2) **59** (1985), 125–130.

[6] M. CLAVELLI, E. DE GIORGI, M. FORTI, V.M. TORTORELLI – A self-reference oriented theory for the Foundations of Mathematics, in *Analyse Mathématique et applications – Contributions en l'honneur de Jacques-Louis Lions*, Paris 1988, 67–115.

[7] E. DE GIORGI – Contributed paper to the session *"Fundamental Principles of Mathematics"*, Plenary Session of the Pontifical Academy of Sciences (25-29 October 1994), 1994.

[8] E. DE GIORGI – *Riflessioni sui Fondamenti della Matematica*, Conference for the centennial of the society "Mathesis", Roma, 23 october 1995, typewritten manuscript.

[9] E. DE GIORGI, M. FORTI – Una teoria quadro per i fondamenti della Matematica, *Atti Acc. Naz. Lincei, Rend. Cl. Sci. Fis. Mat. Nat.* (8) **79** (1985), 55–67.

[10] E. DE GIORGI, M. FORTI – "5× 7": *A Basic Theory for the Foundations of Mathematics*, Preprint di Matematica **74**, Scuola Normale Superiore, Pisa 1990.

[11] E. DE GIORGI, M. FORTI, G. LENZI – Introduzione delle variabili nel quadro delle teorie base dei Fondamenti della matematica, *Rend. Mat. Acc. Lincei* (9) **5** (1994), 117–128.

[12] E. DE GIORGI, M. FORTI, G. LENZI – Una proposta di teorie base dei Fondamenti della Matematica, *Rend. Mat. Acc. Lincei* (9) **5** (1994), 11–22.

[13] E. De Giorgi, M. Forti, G. Lenzi, V.M. Tortorelli – Calcolo dei predicati e concetti metateorici in una teoria base dei Fondamenti della Matematica, *Rend. Mat. Acc. Lincei* (9) **6** (1995), 79–92.

[14] E. De Giorgi, M. Forti, V.M. Tortorelli – Sul problema dell'autoriferimento, *Atti Acc. Naz. Lincei, Rend. Cl. Sci. Fis. Mat. Nat.* (8) **80** (1986), 363–372.

[15] E. De Giorgi, G. Lenzi – La teoria '95. Una proposta di teoria aperta e non riduzionistica dei fondamenti della Matematica, *Atti Accad. Naz. Sci. dei XL, Mem. Mat. Appl.* **20** (1996), 7–34.

[16] E. De Giorgi, A. Preti – Anche la scienza ha bisogno di sognare, *Quotidiano di Lecce*, 6 Gennaio 1996.

[17] M. Forti, F. Honsell – Models of self-descriptive set theories, in *Partial Differential Equations and the Calculus of Variations – Essays in Honor of Ennio De Giorgi* (F. Colombini *et al.*, eds.), Boston 1989, 473–518.

[18] M. Forti, F. Honsell – *Sets and classes within the basic theories for the Foundations of Mathematics*, Quad. Ist. Mat. Appl. "U.Dini", Univ. Pisa **11**, Pisa 1994.

[19] M. Forti, F. Honsell, M. Lenisa – An axiomatization of partial n-place operations, *Math. Struct. Computer Sci.* **7** (1997), 283–302.

[20] G. Frege – *Grundgesetze der Arithmetik, begriffsschriftlich abgeleitet*, Bd. 1 Jena 1893, Bd. 2 Pohle, Jena 1903 (reprinted Olms, Hildesheim 1962).

[21] M. Grassi, G. Lenzi, V. M. Tortorelli – A formalization of a basic theory for the foundations of Mathematics, *Rend. Acc. Naz. Sci. dei XL, Mem. Mat. Appl.* **19** (1995), 129–157.

[22] A. Heyting – *Mathematische Grundlagenforschung. Intuitionismus. Beweistheorie.* Berlin 1934.

[23] D. Hilbert – Über das Unendliche, *Math. Annalen* **88** (1923), 151–165.

[24] G. Lenzi – Estensioni contraddittorie della teoria Ampia, *Atti Acc. Naz. Lincei, Rend. Cl. Sc. Fis. Mat. Nat.* (8) **83** (1989), 13–28.

[25] G. Lenzi, V. M. Tortorelli – *Introducing predicates into a basic theory for the foundations of Mathematics*, Scuola Normale Superiore, preprint di Matematica **51**, Pisa 1989.

[26] W.V.O. Quine – New foundations for mathematical logic, *Amer. Math. Monthly* **44** (1973), 70–80.

[27] B. Russell, A. N. Whitehead – *Principia Mathematica*, Cambridge 1925.

[28] M. SCHMIDT – *Hommes de science*, Hermann, Paris 1990.

[29] D. SCOTT – Combinators and classes, in λ-*Calculus and Computer Science Theory* (C. Böhm, ed.), Lecture Notes in Computer Science **37**, Berlin 1975.

[30] J. VON NEUMANN – Eine Axiomatisierung der Mengenlehre, *J. f. reine und angew. Math.* **154** (1925), 219–240.

Truth and judgments within a new axiomatic framework[‡]

Ennio De Giorgi, Marco Forti and Giacomo Lenzi

We have been glad to accept the proposal of writing an article for the journal "Con-Tratto", because we believe that the comparison of ideas among scholars of different disciplines, and in particular between mathematicians, logicians, and computer scientists on the one hand, and philosophers and theologicians on the other hand, could be of inspiration both for the former and the latter ones. In particular we think that reflecting on the main ideas of mathematics, logic and computer science, on the very meaning of these disciplines, could be an occasion of real cultural enrichment for everybody.

The first contribution that we can give to this comparison of ideas consists in expounding, in the neatest and less ambiguous form we know, namely as an axiomatic system, what we called our "dreams" in [11] and [15]. Notice that when we state axioms we do not intend to impose them to our interlocutors, but rather to submit them to their reflexions and criticism. Actually we do not pretend to elaborate alone what will come out to be the axiomatic theories of the third millennium, rather we think that a wide comparison of ideas on this topic, open to scholars of every branch of knowledge, is the best way leading to a good grounding of mathematics, logic, and computer science, as well as of all other disciplines of science and humanities.

In mentioning dreams, we remind the words of Shakespeare's Hamlet "there are more things in heaven and earth than are dreamt of in your philosopy". These words have been elected by us as the fundamental principle of our reflexions on mathematics, logic, and computer science, and they could inspire as well all scholars of scientific and human disciplines.

We do not ignore the difficulties faced by great mathematicians, like Hilbert and Brouwer, in their attempts of finding a satisfactory "philosophy", where the objects commomly considered by mathematicians can be accomodated (see [3,22,23]), but we think that it does not suit to declare an object unexisting on the sole ground that we have not found a satisfying philosophy that accomodates it. Therefore we do not accept the "formalist" or "neopositivist" solicitation to "dream less", e.g. not to assign a "too realistic" meaning to the beautiful theorems on infinite dimensional spaces of Hilbert and of Banach: on the contrary, we always recall the words of Shakespeare, and so we think that the scientist who aims to understand the real things existing in heaven and earth needs to "dream more", rather than less (see [15]). We think that the greatest

---

[‡]Editor's note: translation into English of the paper "Verità e giudizi in una nuova prospettiva assiomatica", published in Con-Tratto, Istituto Filosofico di Studi Tomistici, F. Barone, G. Basti, A. Testi (eds.), volume "Il fare della Scienza: i fondamenti e le palafitte", 1996, 233–252.

achievements of science are the outcome of some "wonderful dreams": complex numbers, infinitesimal calculus, Newtonian mechanics, general relativity, quantum mechanics, etc.; we also think that a relevant part of the actual work of experimental physicists consists in looking for visible marks of some objects "dreamt of" by theoretical physicists.

We appreciate a neat explicit solution of a nice mathematical problem, a simple and fast computing procedure, we know that they can provide a great improvement on the mere prove of an existence theorem, but we do not believe that abandoning what Hilbert named "Cantor's paradise", peopled by more or less strange sets and by infinites of any order, may help in finding explicit solutions and efficient algorithms. Rather, we would like to imagine "Cantor's paradise" even more rich and colourful than Cantor conceived it, filled up with objects of most variate kinds: sets, collections, operations, formulae, languages with their semantical interpretations, variables, categories, standard and nonstandard numbers, algorithms, informations, propositions, predicates, classical logic and other logics, etc. Admittedly, this freedom in dreaming has a corresponding duty, namely that of translating dreams into axioms, conjectures, theorems formulated with the maximal clarity and with the utmost exactitude (so as to make them, after all, understandable and subject to critical analysis also by scholars not adopting a realistic perspective, but rather a purely formalistic one). Not only the dialogue on these themes among "formalists", "realists", "idealists", "dreamers" can go on in full friendship and comprehension, but it can become a valuable source of very interesting ideas for all of them. All the axioms we shall state below are suitable to be critically discussed, and possibly modified, enriched, improved by scholars close to any current of the philosophy of science. Therefore it is important that a general argument concerning mathematics, logic, computer science be formulated in much clearer a way than a specialistic one. In fact the former argument has to be understandable and subject to criticism by a great variety of interlocutors, whereas it suffices that the latter argument be understood by a restricted group of specialists. In particular, a general talk about the main basic concepts of mathematics, logic and computer science is not reserved only to a restricted number of "specialists of foundations", but it has to be accessible to all those scholars of sciences and humanities that are informed with that sentiment the ancients called philosophy, i.e. love for Wisdom.

So we shall present, in a way that seemed to us the most accessible to scholars of diverse education, some ideas, matured in these years, about the overcoming the so called "set theoretic reductionism", i.e. the trend to reduce all of mathematics to set theory, and about the widening of the horizon of mathematics, which should describe a world much more rich and "colourful" than the world of sets, even rich and interesting in itself. The most difficult problems that we have met up to now, but perhaps also the most interesting from the philosophical standpoint, probably arise when considering the delicate connections between more "general" objects, like qualities and relations, and more "specific" objects of mathematics and logic, like operations, collections, correlations, and propositions. The central knot in these problems comes out in considering the quality

Qtver, enjoyed by *all* true propositions. We found many "technical" difficulties in searching for good axioms concerning *Qtver*; we suspect that these difficulties are only the "face appearing to a mathematician" of deeper philosophical difficulties, connected to ancient and modern antinomies, to the problems of "selfreference" (see [13,23]), to questions about the nature of "mathematical, logical, and informatical entities", to the meaning of the words "exist", "existence", which perhaps are used by a mathematician in a different way from the philosopher, the theologist, the psychologist, etc. (Think of the distance separating an existence theorem from an existential problem!). In particular, interesting philosophical meanings could be assigned to the fact that all modern antinomies seem to be "daughters" of the classical antinomy of the Liar.

In this paper we try and give a first answer to these difficulties by introducing some "intermediate" qualities between the more traditional mathematical objects and the qualities and relations that are "most difficult to handle", such as the quality *Qtver* enjoied by all true propositions. Such "intermediate" qualities will be called "metaqualities", taking into account that they are involved in questions of the so called "metamathematical" type. However, before arriving to metaqualities, we have to take up from [11] some general ideas concerning qualities and relations. These ideas provide a first axiomatic base, upon which we think possible to "engraft" in a natural way the fundamental notions of many scientific and human disciplines, in particular the topics of this paper. The choice of this basis has been suggested by various reflections on the fundamental notions of mathematics, logic, and computer science (see, e.g., [1,3,4,23,28,29,31]) and by many conversations with scholars of different disciplines (mathematics, physics, logic, computer science, biology, history, philosophy, economics, theology, etc.), which revealed the convenience of overcoming all kinds of "reductionism", trying instead to accomodate the mathematical theories, and possibly also other scientific theories, within a wider framework, suitable for a critical comparison of the fundamental ideas of the different disciplines. For instance, although the theories proposed by Cantor, Russell and Whitehead, Zermelo, Church, Curry, Gödel and Bernays, Von Neumann, and other great mathematicians of this century (see [2], [4], [28]) stay among the highest expressions of the human mind in the modern era (comparable, e.g., with Newtonian mechanics, with general relativity, with Michelangelo's Moïses, with Bach's, Mozart's, Rossini's music, with Shakespeare's theatre, with the Universal Declaration of Human Rights of December 10th, 1948, etc.), nevertheless, in order to better understand the main ideas of mathematics, logic, and computer science, it seems appropriate to put them in a more general framework, ruled by the two ideas of *quality* and *relation*. In fact these disciplines, as well as physical, chemical, biological, economic, linguistic disciplines, etc., consider *qualitatively different objects*, and study *relations among these objects*. Therefore we proposed in [11] a symple system of few fundamental *qualities* and *relations* that should constitute the solid and flexible basis upon which more specific qualities and relations of various sciences could be inserted: these qualities and relations are intended as a general premise to the axiomatic treatment of mathematics, logic, and computer science and to the comparison with other scientific and human disciplines. This premise is taken up in section 1 of this paper, which we tried to write in a plain style, so as to be accessible to scholars of variate education, but, at the same time, precise and neat enough to avoid equivocation and ambiguity.

1. Fundamental qualities and relations

We deal in this section with the first fundamental ideas concerning qualities and relations, intended as primitive notions. Notice that with this meaning of primitive notion we do not intend to answer the psycological problem as to which are the ideas that come first to the mind of the child, nor to the historical problem as to which have been the ideas that mankind has considered first; we simply intend that these notions are not reduced (through suitable definitions) to other previously introduced notions.

So we introduce as primitive notions the idea of "*quality*" and the idea of "*enjoying a given quality*". We stipulate that, given an object x of any kind, and a quality q, when writing

$$qx$$

we intend that "x enjoys the quality q".

In this section we introduce seven fundamental qualities, whose meanings are:

Qqual x means that x is a *quality*;
Qrel x means that x is a *relazione*;
Qrelb x means that x is a *binary relation*;
Qrelt x means that x is a *ternary relation*;
Qrelq x means that x is a *quaternary relation*;
Qrun x means that x is a *univocal relation*;
Qrbiun x means that x is a *biunique relation*.

We introduce also four fundamental relations:

Rqual, the binary relation connecting *qualities* with *objects enjoying them*;
Rrelb, the ternary relation connecting *binary relations* with *objects that are in these relations*;
Rrelt, the quaternary relation connecting *ternary relations* with *objects that are in these relations*;
Rid, the binary relation of *identity*.

The seven qualities enjoy *Qqual* and so we can write:

Axiom 1.1 – *Qqual Qqual, Qqual Qrel, Qqual Qrelb, Qqual Qrelt, Qqual Qrelq, Qqual Qrun, Qqual Qrbiun.*

The three qualities *Qrelb*, *Qrelt*, *Qrelq* are specializations of the more general quality *Qrel*, namely:

Axiom 1.2 – *An object x enjoying any of the qualities Qrelb, Qrelt, Qrelq enjoys also the quality Qrel.*

Notice that we do not exclude at all that the existence of more complex relations than binary, ternary, and quaternary relations: we simply do not introduce them here, because we make no use of them in this paper.

After the primitive idea of enjoying a given quality, the second most important primitive idea of this section is that of *"being in a given relation"*. More precisely, given objects x, y of any kind and a binary relation r, we write

$$r\ x, y$$

or at times

$$r\ x; y$$

to intend that "x and y are in the relation r". Instead of saying that x and y are in the relation r, we shall also say that "x is in the relation r with y". We remark that the use of a "semicolon" may be useful when the arguments are integers: for instance, if $x = 127$ and $y = 151$, writing $r127; 151$ is clearer than writing $r127, 151$, the latter hinting to the decimal notation of numbers.

Similarly, if x, y, z are objects of any kind and ρ is a ternary relation, we write

$$\rho\ x, y, z$$

or

$$\rho\ x; y; z$$

to intend that "x, y, z are in the relation ρ".

Finally, if τ is a quaternary relation and x, y, z, t are objects of any kind, we write

$$\tau\ x, y, z, t$$

or

$$\tau\ x; y; z; t$$

to intend that "x, y, z, t are in the relation τ".

We pass to the four fundamental relations *Rqual, Rrelb, Rrelt, Rid*, and we state the following axioms:

Axiom 1.3 – *Rqual is a binary relation. Given objects x, y, if*

$$Rqual\ x, y$$

then x is a quality (i.e. x enjoys Qqual). Moreover, if q is a quality and x is any object, then one has Rqual q, x if and only if qx (i.e. x enjoys the quality q).

Axiom 1.4 – *Rrelb is a ternary relation. Given objects x, y, z, if one has*

$$Rrelb\ x, y, z$$

then x is a binary relation (i.e. x enjoys Qrelb). Moreover, if r is a binary relation, and x, y are objects of any kind, then one has Rrelb r, x, y if and only if r x, y (i.e. x, y are in the relation r).

Axiom 1.5 – *Rrelt is a quaternary relation. Given objects x, y, z, t, if one has*

$$Rrelt\ x, y, z, t$$

then x is a ternary relation (i.e. x enjoys Qrelt). Moreover, if ρ is a ternary relation and x, y, z are objects of any kind, then one has Rrelt ρ, x, y, z if and only if ρ x, y, z (i.e. x, y, z are in the relation ρ).

Axiom 1.6 – *Rid is a biunique relation. In order to have*

$$Rid\ x, y$$

it is necessary and sufficient that x and y are exactly the same object. In other words, any object is in the relation Rid with itself and only with itself.

The relation *Rid* expresses the identity of an object of *any kind* with itself, and so we often write $x = y$ instead of *Rid* x, y.

The relation *Rid* is the simplest instance of a ralation enjoing both qualities *Qrun* and *Qrbiun*. Concerning the quality *Qrun*, we state the following axiom, which expresses the idea of *univocity* applied to *binary*, *ternary* and *quaternary* *relations*:

Axiom 1.7 – *Qrun is a quality.*
1) if Qrun x, then Qrel x;
2) if Qrelb r; Qrun r; r x, y; r x, z , then $y = z$;
3) if Qrelt ρ; Qrun ρ; ρ x, y, z; ρ x, y, t , then $z = t$;
4) if Qrelq τ; Qrun τ τ x, y, z, t; τ x, y, z, u , then $t = u$.

Finally *Qrbiun* is the quality of being a *biunique binary relation*, and biuniqueness is expressed by the axiom:

Axiom 1.8 – *Qrbiun is a quality.*
1) if Qrbiun x, then Qrelb x and Qrun x;
2) if Qrbiun r, r y, x, and r z, x, then $y = z$.

2. Operations and natural numbers

In this section we give some very simple examples of "engrafting" mathematical notions into the "trunk" of the fundamental axioms concerning qualities and relations. More precisely, we introduce five qualities:

Qop, the quality of being an *operation*;
Qops, the quality of being a *simple operation*;
Qopb, the quality of being a *binary operation*;
Qtops, the quality of being a *total simple operation*;
Qnnat, the quality of being a *natural number*;

four relations:

Rops, the ternary relation describing the *action of simple operations*;
Ropb, the quaternary relation describing the *action of binary operations*;
Rnsuc, the binary relation connecting a *natural number* to its *immediate successor*;
Rtsc, the ternary relation associating to pairs of *non-zero distinct natural numbers* a *swap operation*, acting "*à la Gödel*" on *iterated total simple operations*;

and two operations:

Id, the *identity* operation;
K, the operation *generator of constant operations*.
Of course, each of these engraftings should be considered only a first "budd", from which whole "branches" of mathematics can develop in various ways. In this section we give only a few basic descriptive axioms, so as to leave open the possibility of developing the respective theories in different directions.
We begin by dealing with the notion of operation.

Axiom 2.1 – *Qops, Qopb, Qop and Qtops are qualities. Rops is a univocal ternary relation and Ropb is a univocal quaternary relation.*
1) if Qops x or Qopb x, then Qop x;
2) if Rops x, y, z, then Qops x;
3) if Ropb x, y, z, t, then Qopb x;
4) Qtops x holds if and only if Qops x and for all y there exists z such that Rops x, y, z.

So *Qops* and *Qopb* are specializations of the general quality *Qop*, while *Qtops* is a specialization of *Qops*.
When f is a simple operation, instead of writing *Rops* f, x, y we often write $y = fx$, and we say that f *acts* on x giving the *risult y* , or that f *trasforms* x into y, or that f *maps* x onto y, or also that f *applied* to x gives the *value y*. Similarly, when φ is a binary operation, instead of writing *Ropb* φ, x, y, z we often write $z = \varphi\, x, y$. We remark that *univocity* of the relations *Rops* and *Ropb* yields directly "functionality" of simple and binary operations, i.e. the fact that, if f is a simple operation acting on an object x, the result $y = fx$ is uniquely determined, and similarly, if φ is a binary operation acting on the objects x, y, the result $z = \varphi\, x, y$ is uniquely determined.
At the moment we introduce only two, very simple fundamental operations, namely *Id*, the *identity* operation, and *K*, the *generator of constant operations*, by means of the the axiom:

Axiom 2.2 – *Id and K are total simple operations.*
1) Id $x = x$ for any object x.
2) For any object x, the result Kx is a total simple operation such that, for any object y, one has $(Kx)y = x$.

We introduce now the quality *Qnnat* of being a *natural number* (like 0, 1, 2, 3, 4,...) and the relation *Rnsuc* connecting each natural number to its *immediate successor*. These objects satisfy the following axiom:

Axiom 2.3 – *Qnnat is a quality, Rnsuc is a biunique relation.*
1) Rnsuc x,y implies Qnnat x, Qnnat y.
2) There exists a unique z such that Qnnat z and for no x Rnsuc x,z.
3) If Qnnat x, then there exists y such that Rnsuc x,y.

Given a natural number x, the unique natural number y such that *Rnsuc x,y* is called the *successor* of x. The axiom 2.3 suffices for characterizing, through the relation *Rnsuc*, the natural numbers 0 (the unique z of clause 2), 1 (the successor of 0), 2 (the successor of 1), etc.
The arithmetical theory provided by the axiom 2.3 is very weak, however it provides the possibility of defining *infinitely many* natural numbers, through the relation *Rnsuc*. In case only a few natural numbers are needed, e.g. only $0, 1, 2, 3, 4, 5, 6, 7$, one can replace the clause 3 in axiom 2.3 by some particular instances; e.g. one can assume the axioms:

2.7.3.1) There exists the natural number 1 such that Rnsuc $0; 1$;
2.7.3.2) There exists the natural number 2 such that Rnsuc $1; 2$;
2.7.3.3) There exists the natural number 3 such that Rnsuc $2; 3$;
2.7.3.4) There exists the natural number 4 such that Rnsuc $3; 4$;
2.7.3.5) There exists the natural number 5 such that Rnsuc $4; 5$;
2.7.3.6) There exists the natural number 6 such that Rnsuc $5; 6$;
2.7.3.7) There exists the natural number 7 such that Rnsuc $6; 7$.

The development of various stronger theories of (standard and nonstandard) arithmetic might depend on the introduction of operations (beginning with the four operations of the primary school), relations (natural ordering, divisibility, etc.), collections (the collection **N** of all natural numbers, the collection **P** of all primes, etc.), of various forms of the induction principle, etc.
In this paper we do not intend to introduce the arithmetical operations (addition, multiplication, etc.) by means of corresponding axioms. We shall only employ, as a *mere notation*, the traditional expression $y = x+1$ instead of *Rnsuc x,y*. From a psycological, as well as from a conceptual standpoint, passing from a number to its successor corresponds to the elementary practice of *counting*, which is definitely simpler than the practice of *adding*.
The comparison of the notions of set and of natural number is of great importance, both from a logic-mathematical and a philosophical standpoint. The classical way of carrying out this comparison is by assuming the existence of a set **N**, to which belong all and only the "true" natural numbers (i.e. those objects that enjoy *Qnnat*), and of biunique correspondences between **N** and sutable sets of sets, for instance the "finite Von Neumann ordinals"; or even by identifying **N** with one such set. We could name "neopythagorean" those expositions of arithmetic and set theory that satisfy this kind of axioms. From a logical standpoint, one could investigate "non-pythagorean" set theories, possibly connected to ancient and modern reflexions about the notions of actual and potential infinity.

We shall come back to this theme in section 6, here we conclude by using the natural numbers in introducing some "combinatorial" operations (Gödel swaps), which play a rôle in some "operational" treatments of predicate calculus, with which we shall deal in sections 3, 4, 5. Following the general line that takes qualities and relations as "more primitive" notions than operations, we introduce the swap operations by means of the univocal ternary relation $Rtsc$.

Axiom 2.4 – *Rtsc is a univocal ternary relation.*
1) If $Rtsc\ x, y, z$, then x, y are different non-zero natural numbers, and z is a simple operation.
2) If x, y are different non-zero natural numbers, then there exists z such that $Rtsc\ x, y, z$.
3) If one has $Rtsc\ h, k, f$ and $y = fx$, then y, x are total operations taking on total operations as values.

The unique z such that $Rtsc\ m, n, z$ is denoted by T_{mn} or $T_{m;n}$. The swap operated by the operations T_{mn} is described by the next axiom. It is conceived in such a way that it can yield the equations:

$$((T_{1;2}f)x)y = (fy)x;$$

$$(((T_{1;3}f)x)y)z = ((fz)y)x;$$

$$((((T_{1;4}f)x)y)z)t = (((ft)y)z)x;$$

and other analogous equations of the following kind:

$$(\ldots((T_{1n}f)x_1)x_2\ldots)x_n = (\ldots(fx_n)x_2\ldots)x_1.$$

We also want the equations:

$$(((T_{2;3}f)x)y)z = ((fx)z)y;$$

$$((((T_{2;4}f)x)y)z)t = (((fx)t)z)y;$$

$$((((T_{3;4}f)x)y)z)t = (((fx)y)t)z;$$

$$(((((T_{3;5}f)x)y)z)t)s = ((((fx)y)s)t)z.$$

and other analogous equations of this kind:

$$(\ldots(\ldots((T_{mn}f)x_1)\ldots)x_n)\ldots)x_m = (\ldots(\ldots(fx_1)\ldots x_m)\ldots)x_n.$$

All these equations are consequences of the following axiom:

Axiom 2.5 –
1) $T_{mn} = T_{nm}$;
2) if $T_{mn}x$ exists, then there exists also $T_{mn}(T_{mn}x)$ and it is equal to x;
3) if m, n are different natural numbers greater than 1, then $T_{mn}x$ exists if and only if both $T_{1n}x$ and $T_{1m}x$ exist;

4) if both $T_{hk}x, T_{mn}x$ exist, then there exists also $T_{hk}(T_{mn}x)$.
5) $T_{1;2}f$ exists if and only if f a total operation whose values are total operations, and in this case one has $((T_{1;2}f)x)y = (fy)x$ for all x, y;
6) if f enjoys Qtops and m, n are different non-zero natural numbers, then $T_{m+1;n+1}f$ exists if and only if $T_{mn}f$ esists and, for all x, $T_{mn}(fx)$ exists; in this case one has $(T_{m+1;n+1}f)x = T_{mn}(fx)$ for all x;
7) if $T_{mh}f, T_{nh}f, T_{mn}f$ exist, then one has $T_{mh}(T_{mn}(T_{mh}f)) = T_{hn}f$.

3. Propositions and predicative operations

In this section we introduce some general ideas about *propositions* and *predicates*. More precisely, we introduce five qualities:

Qgprop, the quality of being a *proposition of general* type;
Qclp, the quality of being a *classical proposition*;
Qtgopr, the quality of being a *total predicative operation* of *general* type;
Qclopr, the quality of being a *classical predicative operation*;
Qtver, the quality (of high level) enjoyed by all *true propositions*;

a binary relation:

Rsom, the relation interconnecting *structurally homogeneous* objects;

and six operations:

Et, the binary operation of logical *conjunction*;
Vel, the binary operation of logical *disjunction*;
Non, the simple operation of *negation*;
Exist, the simple operation of *existential quantification*;
Univ, the simple operation of *universal quantification*;
Gopr, the simple operation *generator of elementary predicative operations.*

The axioms we introduce are intentionally very weak; in particular those axioms which deal with true and false propositions mostly concern "classical" propositions and predicative operations, so as to leave greatest freedom for possible engraftings of logics different from classical logic.
So we begin by introducing the "most general" notion of proposition through the quality *Qgprop*: hence we write *Qgprop* x to intend that x is a *proposition*, without further specification. Among the propositions there are in particular "classical" propositions, subject to the rules of classical propositional calculus and characterized by the quality *Qclp*: hence *Qclp* x means that x is a *classical proposition*. We can state the axiom:

Axiom 3.1 – *Qgprop, Qclp are qualities. If Qclp x, then Qgprop x.*

The tradizional logical operations *Et* (*conjunction*), *Vel* (*disjunction* or *alternative*), *Non* (*negation*) operate on all propositions:

Axiom 3.2 – *Et and Vel are binary operations. Non is a simple operation. If p, q are propositions, then Et p, q, Vel p, q, Non p exist and are propositions.*

If p, q are classical propositions, then also Et p, q, Vel p, q are classical. Finally, Non p is a classical proposition if and only if p is a classical proposition.

Following the common usage, we denotere by $p\&q$, $p \vee q$, $\neg p$ the propositions *Et p, q, Vel p, q, Non p.*

The quality *Qtver* denotes truth in an "absolute and total sense", therefore we write *Qtver* x to intend that x is *true*, and we write both *Qgprop* p and *Qtver* p to mean that p is a *true proposition*. When p is a proposition and $\neg p$ enjoys the quality *Qtver* we also say that p is *false*.

The following axiom connects classical propositions with the quality *Qtver*, expressing, *inter alia*, the classical principles of *non contradiction* and *excluded middle*:

Axiom 3.3 – *Qtver is a quality. If p, q are classical propositions then:*
1) p enjoys Qtver if and only if $\neg\, p$ does not enjoy Qtver.
2) $p\&q$ enjoys Qtver if and only if both p and q enjoy Qtver.
3) $p \vee q$ enjoys Qtver if and only if at least one of p and q enjoys Qtver.

Having thus introduced propositions, we can resume the classical notion of predicate "*à la* Frege" (see [19]), i.e. as an operation taking on *propositions as values*. Of course, many different introductions are possible (see, e.g., [5,12,14,24,25,26]). In this paper we consider total simple operations that produce propositions either immediately, or after some iterations. To this aim, we consider *total predicative operations* of *general type* (characterized by the quality *Qtgopr*), *classical predicative operations* (characterized by the quality *Qclopr*), and the relation of *structural homogeneity, Rsom.* The qualities *Qgprop, Qclp, Qtgopr, Qclopr,* the relation *Rsom* and the swap operations T_{hk} ruled by the axiom 2.5 are connected to each other by the following axiom:

Axiom 3.4 – *Qtgopr, Qclopr are qualities, Rsom is a binary relation.*
1) For all x, y, z, if Rsom x, y and Rsom y, z, then also Rsom x, x, Rsom y, x, Rsom x, z.
2) If α is a proposition and β is any object, then Rsom α, β holds if and only if β is a proposition.
3) If Qtgopr f and g is any object, then Rsom f, g holds if and only if one has Qtgopr g and Rsom fx, gx for all x.
4) Qtgopr f holds if and only if the following condition are fulfilled:
4.1) f enjoys Qtops;
4.2) for all x, fx enjoys Qtgopr or Qgprop;
4.3) for all x, y one has Rsom fx, fy.
5) If f enjoys Qtgopr and $T_{hk}f$ exists, then Rsom $f, T_{hk}f$ holds.
6) If f enjoys Qclopr, then f enjoys Qtgopr; moreover, for all x, fx enjoys either Qclp or Qclopr.
7) If f enjoys Qclopr and $T_{hk}f$ exists, then also $T_{hk}f$ enjoys Qclopr.

To total predicative operations apply the *existential quantifier Exist* and the *universal quantifier Univ,* which satisfy the following axiom:

Axiom 3.5 – *Univ and Exist are simple operations.*
1) If f is a total predicative operation whose values are propositions, then both Univ f and Exist f exist and are propositions;

2) if f enjoys Qtgopr and, for all x, fx enjoys Qtgopr, then both Univ f and Exist f exist, enjoy Qtgopr and one has:
2.1) $(Univ\, f)x = Univ(fx)$ for all x;
2.2) $(Exist\, f)x = Exist(fx)$ for all x.
3) If f enjoys Qclopr, then Univ f and Exist f enjoy either Qclopr or Qclp.
4) If f enjoys Qclopr and all values of f are propositions, then one has:
4.1) Univ f is true if and only if the proposition fx is true for every x;
4.2) Exist f is true if and only if the proposition fx is true for some object x.

We often adhere to the more usual notation $\forall p$, $\exists p$ instead of $Univ\, p$, $Exist\, p$; when the values of p are propositions, we use also the notation $\forall x.px$, $\forall y.py$, etc., instead of $Univ\, p$, and similarly $\exists x.px$, $\exists y.py$, etc., instead of $Exist\, p$: of course, this notation does not refer to any specific object x, y, etc.

Also the operations of conjunction, disjunction and negation act on total predicative operations, according to the following commutation rules:

Axiom 3.6 – *Let f, g be total predicative operations and assume $Rsom\, f, g$. Then*
1) Non f exists, is a total predicative operation, and one has $(Non\, f)x = Non(fx)$ for all x.
2) Et f, g exists, is a total predicative operation, and one has $(Et\, f, g)x = Et\, fx, gx$ for all x.
3) Vel f, g exists, is a total predicative operation, and one has $(Vel\, f, g)x = Vel\, fx, gx$ for all x.
4) Non f is classical if and only if f is classical;
5) if both f, g are classical, then also Et f, g, Vel f, g are classical.

Also in the case of total predicative operations we adhere to the usual notazion $f\&g$, $f \vee g$, $\neg f$ instead of $Et\, f, g$, $Vel\, f, g$, $Non\, f$, respectively.

The links of qualities, relations, operations with the propositions describing their actions are provided by the operation *Gopr*, generator of *elementary predicative operations*. Not all operations generated by *Gopr* are classical: in fact we shall see in section 5 that $Gopr\, \rho$, for some relation ρ, does not enjoy *Qclopr*. The operation *Gopr* satisfies the following axioms:

Axiom 3.7 – *Gopr is a simple operation. If τ is a quaternary relation, then:*
1) Gopr τ is a total predicative operation;
2) for any object x, $(Gopr\, \tau)x$ is a total predicative operation;
3) for any x, y, $((Gopr\, \tau)x)y$ is a total predicative operation;
4) for any x, y, z, $((Gopr\, \tau)x)y)z$ is a total predicative operation;
5) for any x, y, z, t, $(((Gopr\, \tau)x)y)z)t$ is a proposition;
6) for any x, y, z, t, the proposition $(((Gopr\, \tau)x)y)z)t$ enjoys Qtver if and only if one has $\tau\, x, y, z, t$ (i.e. if and only if x, y, z, t are in the relation τ).

Clause 6 of the axiom above explains the shortened notation

$$\text{``}\tau\, x, y, z, t\text{''} = ((((Gopr\, \tau)x)y)z)t.$$

Comparing the clause 6 of the axiom 3.6 with the preceding axioms, one observes an interesting phenomenon of *mutual reference*: on the one hand, quaternary relations indirectly describe, through *Rrelt, Rrelb, Rqual, Rops*, also qualities and simple operations; on the other hand, the simple operation *Gopr* and the quality *Qtver* describe, through the axiom 3.7, the quaternary relations. This circularity justifies the choice that we have done in section 1, of stopping at quaternary relations; these relations have also the supplementary value of describing, through *Ropb*, the binary operations.

After describing the behaviour of the operation *Gopr* at quaternary relations, it is an easy task to deal with the other kinds of objects introduced in the preceding sections. We proceed as follows:

Axiom 3.8 –
1) If ρ is a ternary relation, then $Gopr\,\rho = (Gopr\,Rrelt)\rho$.
2) If r is a binary relation, then $Gopr\,r = (Gopr\,Rrelb)r$.
3) If q is a quality, then $Gopr\,q = (Gopr\,Rqual)q$.
4) If f is a simple operation, then $Gopr\,f = (Gopr\,Rops)f$.
5) If φ is a binary operation, then $Gopr\,\varphi = (Gopr\,Ropb)\varphi$.

The "elementary" propositions generated by predicates obtained through the operation *Gopr* are statements that concern the quality or the relation to which *Gopr* has been applied. Hence we extend the quoted notation in the natural way:

if q is a quality, then "qx" stands for $(Gopr\,q)x$;

if r is a binary relation, then "$r\,x,y$" stands for $((Gopr\,r)x)y$;

if ρ is a ternary relation, then "$\rho\,x,y,z$" stands for $(((Gopr\,\rho)x)y)z)$;

finally "$x = y$" stands for $((Gopr\,Rid)x)y$,

and "$x \neq y$" stands for $\neg(((Gopr\,Rid)x)y)$.

Both the relation *Rid* and the operation *Id* talk about the identity of objects; hence we state the axiom:

Axiom 3.9 – *For all objects x, y one has*

$$((Gopr\,Id)x)y = ((Gopr\,Rid)x)y = ((Gopr\,Rid)y)x.$$

Now that we have introduced propositions and predicative operations, together with the logical operations *Et, Vel, Non, Exist, Univ*, and the "elementary predicates" generated by *Gopr*, the problem arises of searching for "axioms of classicality" that do not lead to contradiction.

Negative resultats, essentially inspired by Liar's Paradox (or by Tarski's theorem), have been proved in similar contexts, see [12,14], and will be discussed here in section 5. The search for the strongest positive axioms that do not lead to contradiction is still open.

4. Truth and judging subjects

We introduce in this section four qualities:

Qjuds, the quality of being a *judging subject*;

Qcljq, the quality of being a *classical judging quality*;

Qmq, the quality of being a *metaquality*;

Qmr, the quality of being a *metarelation*;

and a binary relation:

$Rjud$, the relation connecting judging subjects with objects pertaining to their judgments.

The quality $Qtver$, expressing "absolute truth", is located at a so elevated level that it is practically impossible to subject it to the usual logico-mathematical manipulations. It seems therefore appropriate to introduce "intermediate notions" that can provide the wanted flexibility and functional capacity. So we introduce the quality $Qjuds$ of being a *judging subject*, and the corrisponding relation $Rjud$ of *bearing to judgment* by means of the axiom:

Axiom 4.1 – *$Qjuds$ is a quality, and $Rjud$ is a binary relation.*
If $Rjud\,x,y$, then $Qjuds\,x$.

Of course, the most varied kinds of judgments are possible: e.g. one may judge an object true, false, good, bad, necessary, possible, contingent, guilty, innocent, etc. In this paper we are mainly interested in judgments about truth of classical propositions, and to this aim we have introduced the quality $Qcljq$ of being a *classical judging quality*. Therefore we state the axiom:

Axiom 4.2 – *$Qcljq$ is a quality. If $Qcljq\,x$, then $Qjuds\,x$ and $Qqual\,x$.*
1) If α is a proposition and μ enjoys $Qcljq$, then:
1.1) if α enjoys μ, then α is classical and true;
1.2) $Rjud\,\mu,\alpha$ holds if and only if either α or $\neg\alpha$ enjoys μ.
2) If α,β are propositions, μ enjoys $Qcljq$, and one has $Rjud\,\mu,\alpha$, $Rjud\,\mu,\beta$, then one has also $Rjud\,\mu,\alpha\&\beta$, $Rjud\,\mu,\alpha\vee\beta$.
3) If f is a total predicative operation and μ enjoys $Qcljq$, then one has $Rjud\,\mu,f$ if and only if for all x one has $Rjud\,\mu,fx$.
4) If f is a total predicative operation, μ enjoys $Qcljq$, and one has $Rjud\,\mu,f$, then one has also $Rjud\,\mu,\exists f$, $Rjud\,\mu,\forall f$.
5) if f is a total predicative operation, $T_{hk}f$ exists, μ enjoys $Qcljq$ and one has $Rjud\,\mu,f$, then one has also $Rjud\,\mu,T_{hk}f$.

If μ enjoys $Qcljq$ and α is a proposition such that $Rjud\,\mu,\alpha$, then we say that *μ judges α*, or that *α is judged by μ*; when f is a predicative operation and $Rjud\,\mu,f$ we say that *f bears the judgment of μ*, or that *μ considers f* (in its judgments).

We now state that classical propositions and predicative operations are characterized by the fact of bearing the judgment of some classical judging quality.

Axiom 4.3 – *1) Let α be a proposition: then α enjoys $Qclp$ if and only if there exists μ, enjoying $Qcljq$, such that $Rjud\,\mu,\alpha$.*
2) A total predicative operation f enjoys $Qclopr$ if and only if there exists μ, enjoying $Qcljq$, such that $Rjud\,\mu,f$.

The connections between judging qualities, metaqualities and metarelations are established by the following axiom:

Axiom 4.4 – *Qmq and Qmr are qualities.*
1) Qmq x holds if and only if Qqual x and Gopr x enjoys Qclopr;
2) Qmr x holds if and only if Qrel x and Gopr x enjoys Qclopr.

In order to simultaneously judge several objects that can be "separately" judged, it is convenient to assume that, given two classical judging qualities there is a third quality which is "more competent":

Axiom 4.5 – *If μ_1, μ_2 enjoy Qcljq, then there exists ν enjoying Qcljq and fulfilling both conditions:*
$Rjud\,\mu_1, x$ implies $Rjud\,\nu, x$;
$Rjud\,\mu_2, x$ implies $Rjud\,\nu, x$.

Finally we want to ensure that many interesting propositions and predicative operations are classical. To this aim we state:

Axiom 4.6 – *All qualities enjoying Qcljq enjoy also Qmq; moreover the following qualities enjoy Qmq:*

$$Qqual, Qrel, Qrelb, Qrelt, Qrelq, Qop, Qops, Qopb, Qtops, Qgprop, Qtgopr;$$

the following relations enjoy Qmr:

$$Rid, Rops, Ropb, Rsom.$$

5. A theorem inspired by Liar's Paradox

By means of the elementary predicative operations generated by *Gopr*, and performing the manipulations allowed by the axioms of sections 3-4, one can actually express all "first order" propositions and predicates that involve objects considered in this paper. It is worth noticing that these capabilities already suffice to prove that not only *Gopr Qtver* cannot be a classical predicative operation, but also that a whole "hierarchy" of classical predicative qualities is required, a hierarchy that does not admit a dominating metaquality.
We begin by stating the following lemma:

Lemma 1. – *There exists a classical predicative operation F such that, for all x, y, $(Fx)y$ is a true proposition if and only if x is a total predicative operation, y is a proposition, and $y = \neg(xx)$.*

Proof. Put $P = Gopr\,Rops$, $Q_0 = Gopr\,Qgprop$, $Q_1 = Gopr\,Qtgopr$, and put:
$F_0 = K(K(KQ_0))$, hence $(((F_0 x)y)z)t = $"$Qgprop\,t$", i.e. "$t$ is a proposition";
$F_1 = T_{1;4}(K(K(KQ_1)))$, hence $(((F_1 x)y)z)t = $"$Qtgopr\,x$", i.e. "$x$ is a total predicative operation";
$F_2 = T_{1;2}(T_{2;3}(T_{3;4}(KP)))$, hence $(((F_2 x)y)z)t = $"$Rops\,x, y, z$", i.e. "the operation x applied to y gives value z";
$F_3 = T_{2;3}(T_{1;4}(K(K(P\,Id))))$, hence $(((F_3 x)y)z)t = $"$Rops\,Id, y, x$", i.e. "$x = y$";
$F_4 = K(K(P\,Non))$, hence $(((F_4 x)y)z)t = $"$Rops\,Non, z, t$", i.e. "$t = \neg z$".

Now put:

$F_5 = F_0 \& (F_1 \& (F_2 \& (F_3 \& F_4)))$;

$F_6 = T_{2;4} F_5$, hence $(((F_6 x)y)z)t$ says "x is a total predicative operation, y is a proposition, $t = x$, $z = xt$, $y = \neg z$";

$F_7 = \exists F_6$, hence $((F_7 x)y)z = \exists t.\,(((F_6 x)y)z)t$, i.e. "$x$ is a total predicative operation, y is a proposition, $z = xx$, $y = \neg z$";

$F = \exists F_7$, hence $(Fx)y = \exists z.\,((F_7 x)y)z$, i.e. "$x$ is a total predicative operation, y is a proposition, $y = \neg(xx)$".

Clearly F is classical, because $Qgprop$, $Qtgopr$ are metaqualities, and $Rops$ is a metarelation. QED

In order to see how Lemma 1 and the following theorems are inspired by Liar's Paradox, we remark that, when the proposition xx is intuitively interpreted as "what x is affirming about itself", truth of both y and $(Fx)y$ yields falsity of "what x is affirming about itself".

We shall come back to this interpretation after Theorem 2; by now we show how to get from Lemma 1 a theorem analogous to Tarski's theorem (see [29]):

Theorem 1. – *Let F be as in Lemma 1. Then the predicative operation*

$$F \,\&\, K(Gopr\,Qtver)$$

cannot be classical. It follows that the quality $Qtver$ is not a metaquality.

Proof. Assume by contradiction that the operation $F \& K(Gopr\,Qtver)$ be classical; then also the operation

$$G = \exists(F \& K(Gopr\,Qtver))$$

is classical, and one has, for all x, $Gx = \exists y.((Fx)y \,\&\, \text{"}Qtver\,y\text{"})$, i.e. "$x$ is a total predicative operation and $\neg(xx)$ is a true proposition".

Taking $x = G$, one gets in particular that GG is a true proposition (it enjoys $Qtver$) if and only if G is a total predicative operation and $\neg(GG)$ is a true proposition (enjoying $Qtver$). But this is impossible if GG is a classical proposition, for in that case one and only one between GG and $\neg(GG)$ enjoys $Qtver$. QED

As one can see by looking at axiom 3.7, in order to produce the operation $Gopr\,Qtver$ one can make use of the elementary predicative operations $Gopr\,Rrelq, Gopr\,Rrelt, Gopr\,Rrelb, Gopr\,Rqual$; so Theorem 1 yields immediately

Corollary. – *The relations $Rrelq, Rrelt, Rrelb, Rqual$ are not metarelations (i.e. they do not enjoy Qmr).*

We can apply Lemma 1 also to prove that there exists no "dominating" classical judging quality, since no classical judging quality can judge *all* classical propositions.

Theorem 2. – *Let F be as in Lemma 1 and let μ be a classical judging quality considering F, i.e. such that $Rjud\,\mu, F$. Put*

$$H = \exists(F \& K(Gopr\,\mu)).$$

Then the classical proposition HH is false, but it is not judged by μ.
It follows that μ does not consider the predicate $Gopr\,\mu$, i.e. $Rjud\,\mu, Gopr\,\mu$ does not hold.

Proof. One has, for all x, $Hx = \exists y.((Fx)y\ \&\ \text{“}\mu y\text{”})$, i.e. "$x$ is a total predicative operation and $\neg(xx)$ is a proposition enjoying μ".
Then, putting $x = H$ we get:
i) if $\mu(HH)$ holds, then HH is a true classical proposition, hence $\neg(HH)$ enjoys μ, contradiction.
ii) if $\mu(\neg(HH))$ holds, then HH is a false classical proposition; so, being H a total predicative operation, the proposition "$\mu(\neg(HH))$" must be false, contradiction.
So it is impossible for μ to judge HH, i.e. $Rjud\,\mu, (HH)$ does not hold. Therefore μ does not consider the predicate H and, given that it considers F by hypotesis, it cannot consider $Gopr\,\mu$. On the other hand, i) implies that HH cannot enjoy $Qtver$, and so, being a classical proposition, it is false. QED

Coming back to the intuitive analogy with Liar's Paradox, we could call the operation H of Teorema 2 a "μ-liar", and interpret the proposition HH as "what H says about itself". Then we could say that H says simultaneously that it is lying and that it is judged by μ as a liar; the first statement is true, the second one is false, and so their conjunction is false, by axiom 3.3. Intuitively we could say that the "μ-liar" lies, but μ is unable to belie it!

6. Final remarks

Sections 2, 3, 4 of this paper constitute the engraftings of some mathematical and logical objects into the trunk of the fundamental qualities and relations introduced in [11] and resumed in section 1. The theorems of section 5 represent the "testing" of these engraftings by facing the crucial antinomy of the Liar.
Many more engraftings are possible, and all ideas presented up to now can be developed with great freedom, in several different directions. For instance, among the judging subjects one could consider, besides judging qualities, also *judging operations* or *relations*; one has only to introduce boolean operations or univocal boolean relations, associating, through suitable axioms, the number 1 to true propositions and the number 0 to false propositions. One can also consider, besides classical logic, with two truth values, also multivalued logics, or, more generally, logics of various kinds, with their syntactical operations and their semantical interpretations.
Turning to topics more specific of mathematics, the following can be engrafted (as it has been done in [11]):
sets through the quality $Qins$ and the binary relation $Rins$;
collections through the quality $Qcoll$ and the binary relation $Rcoll$;
functions through the quality $Qfun$ and the univocal ternary relation $Rfun$;
correlations through the quality $Qcorr$ and the ternary relation $Rcorr$.
If we assume that $Qins, Qcoll, Qfun, Qcorr$ are metaqualities, and that $Rins, Rcoll, Rfun, Rcorr$ are metarelations, we obtain a good context for engrafting

various theories of sets, starting with Zermelo-Frænkel set theory and Gödel-Bernays-Von Neumann class theory.

From set theories one can pass to the study of the connections between sets and natural numbers: both "neopythagorean" and "non-pythagorean" approaches are feasible. The first important decision to be taken by whom is choosing a "neopythagorean" approach is that of assuming that $Qnnat$ is a metaquality and that $Rnsuc$ is a metarelation. This decision is indeed necessary in order to take the "true" natural numbers as a manageable basis of mathematics, logic, and computer science.

From arithmetic one can pass to analysis, constructing rational and real numbers in the usual ways. More interesting could be, perhaps, to pass directly from the general theory of sets and functions to the engrafting of *general real numbers*, comprehending standard and nonstandard reals, as objects of a *qualitatively new type*. To this aim one could introduce the quality $Qgreal$ of being a general real number and the binary operations $Gradd$ (*addition* of two general real numbers), and $Grmult$ (*multiplication* of two general real numbers). One could introduce thereafter various models of the analysis (standard and nonstandard): to each model $\mathcal{M}$ would be associated a set $\mathbf{R}^{\mathcal{M}}$, the *set of real numbers of* $\mathcal{M}$, whose elements should always enjoy $Qgreal$.

Besides standard and nonstandard analysis, classical and non-classical logics, many more branches of mathematics, logic, and computer science could be engrafted, introducing variables, categories, algorithms, probability, information, etc.. The boldest persons might try and engraft branches of other disciplines, introducing, for instance, phyisical, biological, economic laws, theories and experiments confirming or belying them, ethical or juridical principles, natural and artificial languages, etc..

Programming the exploration of the variate, rich, colourful world that should enlarge "Cantor's paradise" is a difficult task. One can only say that the success of the exploration will probably depend on the number and variety of the "explorers", on their ability of appreciating the best traditions and the greatest cultural achievements of the past, in conjunction with the faculty of imagining which innovations might be most fruitful and valuable. It seems utmost necessary to combine imagination and scientific rigor, intending rigor in the broader sense outlined in [2000], where we assert:

> Any investigation of the fundamental axioms of mathematics, logic and computer science, as well as of various experimental, human and philosophical sciences needs, among other things, to overcome a too restricted vision of the different specialities, and requires a broader idea of mathematical and scientific rigor. Mathematical rigor is not only carefulness of the proofs, but also engagement in exposing, in the most clear and understandable way, the problems one would want to solve, the theorems one would want to prove, the conjectures one would want to verify or refute. We think that scientific rigor ultimately consists in clearly and frankly exposing their own certainties and doubts, which problems one believes to have solved and which

one would like to solve or see solved, while avoiding those confuse, obscure, uselessly complicated talks that end up in annoying even the most favourably disposed listener.

Summing up, we may conclude that any consideration on method, rigor, and meaning of science leads us in the end to the ancient intuitions of the sapiential value of humility, of "conviviality" (which keeps together sharing of knowledge, friendship, search of mutual understanding), and of trust in Wisdom, which meets all those who love and seek it.

Bibliography

[1] J. BARWISE, J. ETCHEMENDY – *The Liar*, Oxford 1987.

[2] Y. BAR HLLEL, A. A. FRÆNKEL, A. LEVY – *Foundations of set theory*, Amsterdam 1973.

[3] L.E.J. BROUWER – *Wiskunde, waarheid, werkelijkheid*, Groningen 1919.

[4] A. CHURCH – A Set of Postulates for the Foundation of Logic, *Ann. Math.* **33** (1932), 346–366. Second paper, *ibid* **34** (1933), 837–864.

[5] M. CLAVELLI, E. DE GIORGI, M. FORTI, V.M. TORTORELLI – A self-reference oriented theory for the Foundations of Mathematics, in *Analyse Mathématique et applications – Contributions en l'honneur de Jacques-Louis Lions*, Paris 1988, 67–115.

[6] E. DE GIORGI – Contributed paper to the session *"Fundamental Principles of Mathematics"*, Plenary Session of the Pontifical Academy of Sciences (25-29 October 1994), 1994.

[7] E. DE GIORGI – *Riflessioni sui Fondamenti della Matematica*, Conference for the centennial of the society "Mathesis", Roma, 23 october 1995, typewritten manuscript.

[8] E. DE GIORGI, M. FORTI – Una teoria quadro per i fondamenti della Matematica, *Atti Acc. Naz. Lincei, Rend. Cl. Sci. Fis. Mat. Nat.* (8) **79** (1985), 55–67.

[9] E. DE GIORGI, M. FORTI – "5×7": *A Basic Theory for the Foundations of Mathematics*, Preprint di Matematica **74**, Scuola Normale Superiore, Pisa 1990.

[10] E. DE GIORGI, M. FORTI, G. LENZI – Una proposta di teorie base dei Fondamenti della Matematica, *Rend. Mat. Acc. Lincei* (9) **5** (1994), 11–22.

[11] E. DE GIORGI, M. FORTI, G. LENZI – Verso i sistemi assiomatici del 2000 in Matematica, Logica e Informatica, *Nuova civiltà delle macchine* **57–60** (1997), 248–259.

[12] E. DE GIORGI, M. FORTI, G. LENZI, V.M. TORTORELLI – Calcolo dei predicati e concetti metateorici in una teoria base dei Fondamenti della Matematica, *Rend. Mat. Acc. Lincei* (9) **6** (1995), 79–92.

[13] E. DE GIORGI, M. FORTI, V.M. TORTORELLI – Sul problema dell'auto-riferimento, *Atti Acc. Naz. Lincei, Rend. Cl. Sci. Fis. Mat. Nat.* (8) **80** (1986), 363–372.

[14] E. DE GIORGI, G. LENZI – La teoria '95. Una proposta di teoria aperta e non riduzionistica dei fondamenti della Matematica, *Atti Accad. Naz. Sci. dei XL, Mem. Mat. Appl.* **20** (1996), 7–34.

[15] E. DE GIORGI, A. PRETI – Anche la scienza ha bisogno di sognare, *Quotidiano di Lecce*, 6 Gennaio 1996.

[16] M. FORTI, F. HONSELL – Models of self-descriptive set theories, in *Partial Differential Equations and the Calculus of Variations – Essays in Honor of Ennio De Giorgi* (F. Colombini *et al.*, eds.), Boston 1989, 473–518.

[17] M. FORTI, F. HONSELL – *Sets and classes within the basic theories for the Foundations of Mathematics*, Quad. Ist. Mat. Appl. "U.Dini", Univ. Pisa **11**, Pisa 1994.

[18] M. FORTI, F. HONSELL, M. LENISA – An axiomatization of partial n-place operations, *Math. Struct. Computer Sci.* **7** (1997), 283–302.

[19] G. FREGE – *Grundgesetze der Arithmetik, begriffsschriftlich abgeleitet*, Bd. 1 Jena 1893, Bd. 2 Pohle, Jena 1903 (reprinted Olms, Hildesheim 1962).

[20] M. GRASSI, G. LENZI, V. M. TORTORELLI – A formalization of a basic theory for the foundations of Mathematics, *Rend. Acc. Naz. Sci. dei XL, Mem. Mat. Appl.* **19** (1995), 129–157.

[21] A. HEYTING – *Mathematische Grundlagenforschung. Intuitionismus. Beweistheorie*. Berlin 1934.

[22] D. HILBERT – Über das Unendliche, *Math. Annalen* **88** (1923), 151–165.

[23] G. LENZI – Estensioni contraddittorie della teoria Ampia, *Atti Acc. Naz. Lincei, Rend. Cl. Sc. Fis. Mat. Nat.* (8) **83** (1989), 13–28.

[24] G. LENZI (ED.) – *Seminario sui Fondamenti della Matematica di E. De Giorgi - Scuola Normale Superiore di Pisa - aa.aa. 1993/94 e segg.*. Pisa 1994, 1995, 1996 (typewritten notes).

[25] G. Lenzi, V. M. Tortorelli – *Introducing predicates into a basic theory for the foundations of Mathematics*, Scuola Normale Superiore, preprint di Matematica **51**, Pisa 1989.

[26] G. Lenzi, V. M. Tortorelli – *A basic theory with predicates*, preprint del Dipartimento of Matematica dell'Università di Pisa **1.141.913**, Pisa 1996.

[27] A. Robinson – *Non-standard Analysis*. Amsterdam 1974.

[28] B. Russell, A. N. Whitehead – *Principia Mathematica*, Cambridge 1925.

[29] D. Scott – Combinators and classes, in λ-*Calculus and Computer Science Theory* (C. Böhm, ed.), Lecture Notes in Computer Science **37**, Berlin 1975.

[30] A. Tarski – *Logic, Semantics, Metamathematics*, Oxford 1956.

[31] J. Von Neumann – Eine Axiomatisierung der Mengenlehre, *J. f. reine und angew. Math.* **154** (1925), 219–240.

Overcoming set–theoretic reductionism in search of wider and deeper mutual understanding between Mathematicians and scholars of different scientific and human disciplines[‡]

Note by ENNIO DE GIORGI[*]

In April 1996 Ennio De Giorgi transmitted a Note to the Rendiconti dell'Accademia dei Lincei, by sending it to the President and to the Vice–president of the Academy. Actually, as De Giorgi wrote in the letter enclosed, the Note itself was "of a quite unusual kind, different from the Notes oriented to the specialists, and different as well from interdisciplinary lectures, and it cannot be reduced in any way to a lecture". In this Note –as De Giorgi wrote– I expose as modestly and clearly as possible, very few ideas which in my opinion seem to be quite new, interesting, and can be understood, discussed and critically examinated by everyone who considers them with a little bit of care. (. . .) I would like that (. . .), when the Note will be discussed, (. . .), there be more interventions of the other Members than mine, and that I mostly answer to remarks, suggestions and criticisms." The Note is published now, after that De Giorgi left us, and is intended to be a further token of the respect, admiration and regret of the Academy.

Summary. We propose in this paper an open–ended, non–reductionist axiomatic framework, grounded on the primitive notions of *quality* and *relation*. In our opinion, this framework should be suitable for engrafting the main concepts of Mathematics, Logic and Computer Science. We give here only some examples dealing with the very first notions of Mathematics. We hope that a free development of this framework will foster a fruitful debate and a critical analysis of the main fundamental ideas of the different scientific and human disciplines, not restricted to "specialists of Foundations" only, but rather extended to all interested scholars.
KEY WORDS: Foundations; Non–reductionism; Quality; Relation.

Various reflections on the fundamental concepts of mathematics, Logic and Computer Science (see [1–24]) and many discussions with scholars of different disciplines (Mathematics, Physics, Logic, Computer Science, Biology, History, Philosophy, Economy, Theology, etc.) have convinced me about the opportunity of overcoming the so–called "set–theoretic reductionism", that is the trend towards reducing all Mathematics to set theories, and rather trying to insert these and

---

[‡]Editor's note: translation into English of the paper "Dal superamento del riduzionismo insiemistico alla ricerca di una più ampia e profonda comprensione tra matematici e studiosi di altre discipline scientifiche e umanistiche", published in Atti Accad. Naz. Lincei Cl. Sci. Fis. Mat. Natur. Rend. Lincei (9) Mat. Appl., **9** (1998), 71–80.

[*]Presented by E.Vesentini to the Accademia dei Lincei on February 13, 1998.

all other scientific theories in a broader framework, where one could perform a critical comparison of the fundamental ideas of the various scientific and human disciplines.

It is sure that the set theories proposed by Cantor, Zermelo, Gödel, Bernays, Von Neumann and other great mathematicians of this century, lie among the highest expressions of human spirit (they can be compared to Newtonian mechanics, general relativity, quantum mechanics, the Comedy of Dante, the Moïses of Michelangelo, the music of Bach, Mozart, the tragedies of Shakespeare, etc.), but to understand better their values, in my opinion it is useful to insert them into a broader framework dominated by the two ideas of *quality* and *relation*.

In such a frame, it should also be easier to set up a critical comparison of the fundamental ideas of Mathematics, Logic, Computer Science, and other scientific and human disciplines.

All these disciplines consider *qualitatively different objects* and study *relations between these objects*. Hence it seems reasonable to propose, as a general premise to the exposition of these disciplines and to the comparison of their fundamental ideas, a short, simple system of a few *qualities* and *relations* of general kind, which should form the solid, flexible basis on which one can insert qualities and relations more specific of the various sciences.

In fact, perhaps we should look for new solid, flexible bases for Mathematics and other sciences, new ways, less rigid, of organizing exposition, not only to solve new problems, but even to find quick rigorous ways towards classical theorems of Mathematics, for instance, Stokes' formula. This need is indirectly confirmed by a remark on this formula in the interview [26] of Henri Cartan, one of the greatest mathematicians who participated to the collective work of Bourbaki.

This system will be called "general premise to scientific theories" and will be exposed in the next chapter.

For someone this will look an unrealizable dream; in my opinion, in scientific research it is good to follow one's dreams, and at the same time communicate them with clarity, intellectual honesty, while being ready to accept criticisms of one's interlocutors. I defended this idea in the interview [25]. I will try to give a realization of it, even if modest, in this *Note*, and I hope that the latter turns out to be really understandable for scholars of every scientific and human disciplines. Even more, in my opinion, it can be understood by everyone who reads it as "naïvely" as possible, without thinking it necessary to have prerequisites of any kind, mathematical, logical, informatical, philosophical, etc. The bibliography which concludes this *Note* is intended mostly to satisfy the curiosity of those who, after reading it, ask how the ideas exposed in it were born.

CHAPTER 1

We begin with assuming as primitive concepts, that is not reducible to other previously introduced concepts (by means of suitable definitions), the idea of a *"quality"* and the idea of *"enjoying a given quality"*. We note that when talking about primitive concepts, we do not mean to answer the psychological problem of which ideas come up first in the mind of a child, nor the historical problem of which ideas were considered first by Mankind.

We stipulate also that given an object x of any kind and a quality q, when we write

$$q\,x$$

we intend to say that "x enjoys the quality q". In this premise we consider the seven fundamental qualities: $Qqal$, $Qrel$, $Qrelb$, $Qrelt$, $Qrelt$, $Qrelq$, $Qrun$, $Qrbiun$ with the following meanings:

$Qqual\,x$ means that x is a quality;

$Qrel\,x$ means that x is a relation;

$Qrelb\,x$ means that x is a binary relation;

$Qrelt\,x$ means that x is a ternary relation;

$Qrelq\,x$ means that x is a quaternary relation;

$Qrun\,x$ means that x is a univocal relation;

$Qrbiun\,x$ means that x is a bi–unique relation.

The seven qualities all enjoy $Qqual$ and then we can write:

AXIOM 1. – $Qqual\,Qqual$, $Qqual\,Qrel$, $Qqual\,Qrelb$, $Qqual\,Qrelt$, $Qqual\,Qrelq$, $Qqual\,Qrun$, $Qqual\,Qrbiun$.

All these and the following statements of this chapter are "proposed" (not "imposed") as "axioms", that is as statements not deduced by a system of previous statements, but chosen as a possible starting point for further developments of various theories. They are not even "deduced" by history of Mathematics, philosophy of Science, Logic, etc., and even more, they can be better understood by those who keep an attitude as "naïve" as possible, and think of the most common meaning that everyday language gives to the words "quality" and "relation". Only upon a naïve understanding of the premise a critical reading is useful, where everyone can bring their experience and knowledge in Mathematics, Logic, Computer Science, History, Philosophy, Physics, Economy, etc.; even more, these experiences can be very important in expressing deeper, better motivated critical assessments.

The qualities $Qrelb$, $Qrelt$, $Qrelq$ are particular cases of the more general quality $Qrel$, that is:

AXIOM 2 – *Every element x which enjoys one of the qualities $Qrelb$, $Qrelt$, $Qrelq$ enjoys the quality $Qrel$ as well.*

Instead, we do not exclude the existence of relations more complex than binary, ternary or quaternary ones, although we will not use them in this premise.

After the primitive idea of enjoying a given quality, the second most important primitive idea of this premise is that of *"being in a given relation"*. Namely, given two objects x, y, of any kind, and a binary relation r, we will write

$$r\,x, y$$

or sometimes

$$r\,x; y$$

to mean that "x and y are in the relation r". Sometimes, instead of saying that x and y are in the relation r, we will say also that x is in the relation r with y. Likewise, if x, y, z are objects of any kind and ρ is a ternary relation, we will write

$$\rho\,x, y, z$$

or

$$\rho\,x; y; z$$

to say that "x, y, z are in the relation ρ".

Finally, if τ is a quaternary relation and x, y, z, t are objects of any kind, we will write

$$\tau\,x, y, z, t$$

or

$$\tau\,x; y; z; t$$

to say that "x, y, z, t are in the relation τ".

The four fundamental relations considered in this premise are: *Rqual*, *Rrelb*, *Rrelt*, and *Rid*. The relation *Rqual* is binary, and relates qualities with objects enjoying them. Namely:

Axiom 3. – *Given two objects x, y, to have*

$$Rqual\,x, y$$

it is necessary that x be a quality (that is, x must enjoy Qqual). Moreover, if q is a quality and x is any object, the condition Rqual q, x is necessary and sufficient to have that x enjoys q.

The relation *Rrelb* is ternary, and relates binary relations with the objects related by them. In other words:

Axiom 4. – *To have*

$$Rrelb\,x, y, z$$

it is necessary that x be a binary relation. Moreover, if r is a binary relation and x, y are objects of any kind, the following two statements are equivalent:

$$r\,x, y;$$

$$Rrelb\,r, x, y.$$

Finally, *Rrelt* is a quaternary relation, and it associates ternary relations with the objects related by them. In other words:

Axiom 5. – *To have*

$$Rrelt\,x, y, z, t$$

it is necessary that x be a ternary relation. Moreover, if ρ is a ternary relation and x, y, z are objects of any kind, the following two statements are equivalent:

$$\rho\, x, y, z;$$

$$Rrelt\, \rho, x, y, z.$$

Finally, *Rid* is a binary relation representing *identity*. In other words:

AXIOM 6. – *To have*

$$Rid\, x, y$$

it is necessary and sufficient that x and y be the same object. In other words, every object is in the relation Rid with itself and only with itself.

The relation *Rid* is the simplest example of a relation enjoying both qualities *Qrun* and *Qrbiun*. For the quality *Qrun* we postulate the following axiom, expressing the idea of being univocal:

AXIOM 7 –

$$
\begin{array}{llllll}
1) & \textit{If } Qrun\ x & & & & \textit{then } Qrel\ x; \\
2) & \textit{if } Qrelb\ r; & Qrun\ r; & r\ x, y; & r\ x, z & \textit{then } y = z; \\
3) & \textit{if } Qrelt\ \rho; & Qrun\ \rho; & \rho\ x, y, z; & \rho\ x, y, t & \textit{then } z = t; \\
4) & \textit{if } Qrelq\ \tau; & Qrun\ \tau; & \tau\ x, y, z, t; & \tau\ x, y, z, u & \textit{then } t = u.
\end{array}
$$

Finally, the axioms concerning *Qrbiun* are:

AXIOM 8 –

$$
\begin{array}{lll}
1) & \textit{If } Qrbiun\ x & \textit{then } Qrelb\ x \textit{ and } Qrun\ x; \\
2) & \textit{if } Qrbiun\ x; \quad r\ y, x; \quad r\ z, x & \textit{then } y = z.
\end{array}
$$

Biunique relations are very important both within Mathematics and in applications of Mathematics, where relations between "concrete" objects and their mathematical models are important. The existence of a good biunique relation may mean a good adequacy of the mathematical model which has been chosen.

CHAPTER 2

It is easy to engraft the various branches of Mathematics on the trunk of the "general premise to scientific theories" exposed in chapter 1. This engrafting may be performed by following various strategies, for instance by introducing simultaneously several kinds of objects related with each other, or by introducing separately single kinds of objects, and then performing later the description of the relations between them. Strategies of the first kind have been followed in [1–24]; in this chapter, we give some examples of the second strategy, which seems to be more convenient for a clear and simple communication among scholars of different disciplines, and also for better exploiting the various mentalities and intuitions of mathematicians themselves.

In order to introduce in the simplest way natural numers $0, 1, 2, 3, 4, \ldots$, it is enough to introduce a quality $Qnat$, that is the quality of being a natural number, a biunique relation $Rnsuc$, that is the relation between a natural number and its successor, and the binary relation $Rnord$, which describes the usual ordering relation between natural numbers. They are related by the follolwing axioms:

Axiom N1. –

1) *If $Rnord\, x, y$ then $Qnat\, x$ and $Qnat\, y$;*

2) *given three objects x, y, z, if $Rnord\, x, y$ and $Rnord\, y, z$ then $Rnord\, y, z$;*

3) *given two natural numbers x, y, at least one of the relations $Rnord\, x, y$ and $Rnord\, y, x$ is true; they are both true if and only if $x = y$;*

4) *if $Rnsuc\, x, y$, then $Rnord\, x, y$ and $x \neq y$;*

5) *if $Qnat\, x$, then there exists y such that $Rnsuc\, x, y$;*

6) *if $Rnord\, x, y$, $x \neq y$, $Rnsuc\, x, z$ then $Rnord\, z, y$;*

7) *there is a unique z such that $Qnat\, z$ and for no x we have $Rnsuc\, x, z$.*

Remark: in order to recover the usual arithmetical notions, it is enough to admit that the statement $Rnord\, x, y$ is equivalent to the statement: x, y are natural numbers and x is less than y or equal to y, or: x, y are natural numbers and y is greater than x, or equal to x. In this case one usually writes $x \leq y$ or $y \geq x$. Likewise, instead of $Rnsuc\, x, y$ we could say that y is the immediate successor of x, or that x is the immediate predecessor of y. The only natural number with no predecessor will be denoted, as usual, by the symbol 0, its successor by the symbol 1, the immediate successor of 1 by the symbol 2, etc.

Remark: the notions introduced here form only the very first part of Arithmetic. Later on, after introducing in general simple and binary operations, one could study the four operations of elementary Arithmetic, and even later, after introducing sets, one could state the principle of induction, with which we enter more advanced Arithmetic.

As a first step of this development we engraft the concept of operation upon the trunk of the general premise. Namely, we introduce three qualities Qop, the quality of being an operation, $Qops$, the quality of being a simple operation, and $Qopb$, the quality of being a binary operation, and the relations $Rops$ and $Ropb$, which describe the way of operating of simple and binary operations. They satisfy the following axioms:

Axiom O1. –

1) *If $Qops\, x$ or $Qops\, x$ then $Qop\, x$. In other words, $Qops$ and $Qopb$ are particular cases of the general quality Qop.*

2) *$Rops$ is a univocal ternary relation.*

3) *for every choice of x, y, z, if $Rops\, x, y, z$ then $Qops\, x$.*

4) *$Ropb$ is a univocal quaternary relation.*

5) *if $Ropb\, x, y, z, t$ then $Qopb\, x$.*

Remark: when f is a simple operation, instead of writing $Rops\,f, x, y$ we will often write $y = f\,x$. When φ is a binary operation, instead of writing $Ropb\,\varphi, x, y, z$ we will often write $z = \varphi\,x, y$. After natural numbers and operations, we can introduce collections, while warning that the concept of collection is a wide generalization of the usual concept of set. To this aim we will introduce the quality $Qcoll$, that is the quality of being a collection, the relation $Rcoll$, the relation of membership to collections, and the relation $Rcin$, the relation of inclusion between collections. They satisfy the following axioms:

Axiom C1. –

1) *$Rcoll$ is a binary relation;*

2) *if $Rcoll\,x, y$ then x is a collection, that is $Qcoll\,x$.*

Following common usage, instead of writing $Rcoll\,x, y$ we will write $y \in x$, or we will say that y belongs to x or that y is an element of x.

Axiom C2. –

1) *$Rcin$ is a binary relation;*

2) *if $Rcin\,x, y$ then x, y are collections;*

3) *if A, B are collections, then $Rcin\,A, B$ holds if and only if every element of A is element of B as well.*

Following common usage we will write $A \subseteq B$ and we will say that A is contained in B, or that A is a part of B.
Now we can state the fundamental axiom of the theory of collections, the axiom of extensionality.

Axiom C3. – *If A, B are collections, $A \subseteq B$ and $B \subseteq A$, then A coincides with B.*

After introducing collections we can introduce sets by means of the quality $Qins$ and the membership relation $Rins$. They are related with $Qcoll$ and $Rcoll$ by the following axiom:

Axiom I1. –

1) *If $Qins\,x$, then $Qcoll\,x$. In other words, sets are particular collections.*

2) *If $Rins\,x, y$ then $Qins\,x$.*

3) *If $Qins\,E$ then $Rcoll\,E, x$ if and only if $Rins\,E, x$. In other words, $Rins$ is the restriction of the relation $Rcoll$ to the cases when the first object under consideration is a set.*

Remark: after introducing collections and sets, one could introduce functions, systems, correlations by means of the pairs $Qfun$, $Rfun$, $Qsys$, $Rsys$, $Qcorr$, $Rcorr$.

For a wider description of these objects we refer to the article [16], which also points out the differences existing between the notion of set and the more general notion of collection.

For a further investigation on the relations between the concepts introduced here, it is useful to consider also universal collections, or universes. To this aim we introduce the quality $Quniv$, the quality of being a universal collection, and we pose the following axiom:

AXIOM CU1. – *If $Quniv\,V$ then $Qcoll\,V$, and for every collection C we have $C \in V$ if and only if $C \subseteq V$.*

Remark: it is immediate to verify that, if V is a universal collection, then $V \in V$. Moreover one could deduce from the usual axioms of set theory, which we did not introduce here, that a universal collection cannot be a set: in fact, from these axioms one derives (Cantor's theorem) that no set can have all its subsets as elements.

All theories cited in the bibliography are suitable for being engrafted, up to appropriate modifications, on the general premise to scientific theories exposed in chapter 1. For instance, we could engraft the variables considered in [4, 10] by means of the quality $Qvar$ and the relation $Rvar$. The elements enjoying $Qvar$ will be called variables; when $Rvar\,w, x$, we will say that x is a value taken by the variable w. Finally, a wide space is left for engrafting other mathematical or logical concepts like those of proposition, predicate, language, syntax, semantics, category, nonstandard Analysis, non–classical logics, etc. On this subject I will be glad to listen suggestions and opinions of colleagues and friends.

CHAPTER 3

The axioms proposed in chapters 1, 2 should be easy to understand and criticize by scholars of different disciplines sharing that feeling which in ancient times was called philosophy, that is love for wisdom, even if they have a different degree of knowledge in the fields of Mathematics, Logic, Computer Science, History and Philosophy of Science, Epistemology, etc.

If several people will find chapters 1 and 2 too difficult to be read, I will have to infer that there is something wrong in my exposition, because I believe that the fundamental axioms of Mathematics, as of any other science, must be written so as to be clear and comprehensible by everyone who has a sincere will of understanding them. Of course I do not expect, and I do not want, an unconditional acceptance of my proposals, because I think we still need a thorough work of critical reflection and friendly discussion between scholars of different culture and opinions, in order to get to a good axiom system (or some good axiom systems) which mark a real significant progress with respect to the present situation.

On the other hand, I think that this goal is realizable, also because I think that set–theoretic reductionism, like any other reductionism, does not give an appropriate perspective for a real understanding of the many problems which the culture of our time must undertake, and which remind me the words of

Shakespeare "there are more things in heaven and earth than are dreamt of in your philosophy".

I think also that the search for good axiom systems for Mathematics and other sciences must be undertaken with that spirit of humility, hope, conviviality which, after all, characterizes sapiential mentality in any time.

More cultured people could speak of sapiential mentality better than me. I will just recall that even in one of the most ancient sapiential books of the Bible, the book of Proverbs, there is an invitation to humility and devotion to God, to being beware of those who think to be wise and reject any criticism, whereas the real wise man shows sincere gratitude to those who give him a fair, impartial criticism. In the same book, besides the invitation to humility, there is an invitation to hope: we are told about the Wisdom which was with the Lord during Creation, which is glad to reveal itself to the ones who are looking for it, and which invites everyone to his banquet. In various form, this convivial idea of communicating knowledge is found in the subsequent centuries up to the Universal Declaration of Human Rights of December 10th, 1948, which keeps the education to friendship and comprehensionas among the fundamental task of school . I believe that the secret of success in any search for the fundamental axioms of the various sciences lies just in humility, hope, conviviality, friendship and comprehension. This search requires, among other things, overcoming a too closed vision of the various specialities, and of what is usually called mathematical rigor, or scientific rigor. Mathematical rigor is not only carefulness in proofs, but even engagement in exposing in the most clear and understandable way the problems which one would want to solve, the theorems one would like to prove, and the conjectures one would want to prove or disprove. At the opposite side of this broad idea of mathematical rigor, there is the attitude of a student who, asked to "explain hypothesis and thesis of this theorem", answers "if you want I can prove it", or the one of the researcher who says "I do not talk about problems which I have not yet solved, I talk only about results that I have obtained, and not about the ones I am still investigating".

On the contrary, I think that the real rigor is the one of those who talk clearly and freely about their opinions and doubts, about the problems they solved as well as about the ones which they would like to solve or to see solved by someone, while avoiding those confuse, obscure, uselessly complicated speeches, which at the end disturb even the most well–meaning listener.

Finally, the right scientific rigor means also an impartial attitude towards tradition, admiration for the great conquers of the past, and freedom of introducing reasonable innovations; it means also avoiding the old habit of attributing one's ideas to some illustrious predecessor, together with the opposite habit of presenting as one's own original ideas, some ideas hardly different from other ones already present in the scientific literature.

Bibliography

[1] H. P. Barendregt, *The Lambda–Calculus*. Amsterdam 1981.

[2] Y. Bar Hillel – A. A. Fraenkel – A. Levy , *Foundations of set theory*. Amsterdam 1973.

[3] A. Church, *Set theory with a universal set*. In: L. Henkin *et al.* (eds.), *Proceedings of the Tarski Symposium*. Proc. of Symp. P. Math. XXV, Rhode Island 1974, 297–308.

[4] M. Clavelli, *Variabili e teoria A*. Rend. Sem. Fac. Sci. Univ. Cagliari, (2) 59, 1985, 125–130.

[5] M. Clavelli – E. De Giorgi – M. Forti – V. M. Tortorelli, *A self-reference oriented theory for the Foundations of Mathematics*. In: *Analyse Mathématique et applications. Contributions en l'honneur de Jacques-Louis Lions*. Paris 1988, 67–115.

[6] E. De Giorgi, Contribution to the session *Fundamental Principles of Mathematics*. Plenary Session of the Pontifical Academy of Sciences (25–29 October 1994), 1994.

[7] E. De Giorgi, *Riflessioni sui Fondamenti della Matematica*. Conference for the centennial of the society Mathesis, Rome, 23 ottobre 1995, manuscript.

[8] E. De Giorgi – M. Forti, *Una teoria-quadro per i fondamenti della Matematica*. Atti Acc. Lincei Rend. Fis.. s. 8, v. 79, 1985, 55–67.

[9] E. De Giorgi – M. Forti, *"5× 7": A Basic Theory for the Foundations of Mathematics*. Preprint di Matematica n. 74, Scuola Normale Superiore, Pisa 1990.

[10] E. De Giorgi – M. Forti – G. Lenzi, *Introduzione delle variabili nel quadro delle teorie base dei Fondamenti della Matematica*. Rend. Mat. Acc. Lincei, s.9, v. 5, 1994, 117–128.

[11] E. De Giorgi – M. Forti – G. Lenzi, *Una proposta di teorie base dei Fondamenti della Matematica*. Rend. Mat. Acc. Lincei, s. 9, v. 5, 1994, 11–22.

[12] E. De Giorgi – M. Forti – G. Lenzi – V. M. Tortorelli, *Calcolo dei predicati e concetti metateorici in una teoria base dei Fondamenti della Matematica*. Rend. Mat. Acc. Lincei, s. 9, v. 6, 1995, 79–92.

[13] E. De Giorgi – M. Forti – V. M. Tortorelli, *Sul problema dell'autoriferimento*. Atti Acc. Lincei Rend. Fis., s. 8, v. 80, 1986, 363–372.

[14] E. De Giorgi – G. Lenzi, *La teoria '95. Una proposta di teoria aperta e non riduzionistica dei fondamenti della Matematica.* Atti dell'Accademia dei XL, Mem. Mat. Appl., 20, 1996, 7–34.

[15] M. Forti – F. Honsell, *Models of self–descriptive set theories.* In: F. Colombini *et al.* (eds.), *Partial Differential Equations and the Calculus of Variations. Essays in Honor of Ennio De Giorgi.* Boston 1989, 473–518.

[16] M. Forti – F. Honsell, *Sets and classes within the basic theories for the Foundations of Mathematics.* Quaderni Ist. Mat. Appl. "U. Dini", 11, 1994.

[17] M. Forti – F. Honsell – M. Lenisa, *An axiomatization of partial n–place operations.* Math. Struct. Computer Sci., 7, 1997, 283–302.

[18] M. Grassi – G. Lenzi – V. M. Tortorelli, *A formalization of a basic theory for the foundations of Mathematics.* Preprint del Dipartimento di Matematica dell'Università di Pisa, n. 1.98.804, 1994.

[19] G. Lenzi, *Estensioni contraddittorie della teoria Ampia.* Atti Acc. Lincei Rend. fis., s. 8, v. 83, 1989, 13–28.

[20] G. Lenzi – V. M. Tortorelli, *Introducing predicates into a basic theory for the foundations of Mathematics.* Preprint di Matematica, n. 51, Scuola Normale Superiore, Pisa, 1989.

[21] W. V. O. Quine, *New foundations for mathematical logic.* Amer. Math. Monthly, 44, 1973, 70–80.

[22] B. Russell – A. N. Whitehead, *Principia Mathematica.* Cambridge 1925.

[23] D. Scott, *Combinators and classes.* In: C. Böhm (ed.), λ–*Calculus and Computer Science Theory.* Lec. Notes Comp. Sc., 37, Berlin 1975.

[24] J. Von Neumann, *Eine Axiomatisierung der Mengenlehre.* J. f. Reine und Angew. Math., 154, 1925, 219–240.

[25] E. De Giorgi – A. Preti, *Anche la scienza ha bisogno di sognare.* Quotidiano di Lecce, 6 Gennaio 1996.

[26] M. Schmidt, *Hommes de science.* Hermann, 1990.

Dal superamento del riduzionismo insiemistico alla ricerca di una più ampia e profonda comprensione tra matematici e altri studiosi di altre discipline scientifiche ed umanistiche[‡]

Nota di ENNIO DE GIORGI*

Nell'Aprile del 1996 Ennio De Giorgi trasmise una Nota ai Rendiconti, inviandola al Presidente ed al Vicepresidente dell'Accademia. La procedura era inconsueta. Del resto, come scriveva De Giorgi nella lettera di trasmissione, la Nota stessa era "di un tipo un po' inconsueto, diversa dalle Note di carattere specialistico e anche diversa dalle conferenze di interesse interdisciplinare, e non riducibile in alcun modo ad una conferenza". Nella Nota – scriveva De Giorgi – espongo nella forma più sobria e più chiara possibile pochissime idee che mi sembrano abbastanza nuove, abbastanza interessanti e abbastanza comprensibili, discutibili e criticabili da parte di ogni persona che le consideri con una certa attenzione. (...) Mi piacerebbe (...) che, in occasione della discussione della Nota, (...) gli interventi degli altri Soci prevalessero rispetto ai miei discorsi, che dovrebbero essere soprattutto risposta a osservazioni, suggerimenti e critiche". La Nota viene pubblicata oggi, dopo che Ennio De Giorgi ci ha lasciato, e vuole essere un'ulteriore testimonianza della stima, dell'ammirazione e del rimpianto dell'Accademia.

Sunto. Proponiamo in questa *Nota* un quadro assiomatico aperto e non riduzionista, che si basa sulle idee primitive di *qualità* e *relazione*, in cui speriamo sia possibile innestare i concetti fondamentali della Matematica, della Logica e dell'Informatica (di cui diamo solo alcuni primissimi esempi). Auspichiamo che sviluppando liberamente tale quadro sia possibile giungere ad un fruttuoso confronto critico delle idee fondamentali delle diverse discipline scientifiche ed umanistiche, non ristretto agli "specialisti dei Fondamenti", ma aperto a tutti gli studiosi interessati.

KEY WORDS: Foundations; Non–reductionism; Quality; Relation.

Varie riflessioni sui concetti fondamentali della Matematica, della Logica e dell'Informatica (vedi [1-24]) e molte conversazioni con studiosi di varie discipline (Matematica, Fisica, Logica, Informatica, Biologia, Storia, Filosofia, Economia, Teologia, ecc.) mi hanno convinto dell'opportunità di superare il cosiddetto "riduzionismo insiemistico" cioè la tendenza a ridurre tutta la Matematica alle teorie degli insiemi e cercare invece di inserire tali teorie e tutte le altre teorie

---

[‡]Editor's note: published in Atti Accad. Naz. Lincei Cl. Sci. Fis. Mat. Natur. Rend. Lincei (9) Mat. Appl., **9** (1998), 71–80.

*Presentata da E. Vesentini all'Accademia del Lincei il 13 Febbraio 1998.

scientifiche in un quadro più ampio in cui sia possibile un confronto critico delle idee fondamentali delle diverse discipline scientifiche e umanistiche.

Certamente le teorie degli insiemi proposte da Cantor, Zermelo, Gödel, Bernays, Von Neumann e altri grandi matematici di questo secolo restano tra le espressioni più elevate dello spirito umano (paragonabili per esempio alla meccanica di Newton, alla relatività generale, alla meccanica quantistica, alla Divina Commedia di Dante, al Mosè di Michelangelo, alle musiche di Bach, Mozart, alle tragedie di Shakespeare ecc.), tuttavia per comprenderne meglio il valore mi sembra utile inserirle in un quadro più generale dominato dalle due idee di *qualità* e *relazione*. In un tale quadro dovrebbe anche essere più facile impostare un confronto critico tra le idee fondamentali della Matematica, della Logica, dell'Informatica, delle altre discipline scientifiche ed umanistiche.

Tutte queste discipline infatti considerano *oggetti qualitativamente diversi* e studiano *relazioni tra tali oggetti*. Sembra quindi ragionevole proporre come premessa generale all'esposizione di tali discipline e al confronto delle loro idee fondamentali un breve e semplice sistema di poche *qualità* e *relazioni* di carattere generale che dovrebbero costituire la base solida e flessibile su cui inserire qualità e relazioni più specifiche delle varie scienze.

Infatti occorre probabilmente cercare nuove basi solide e insieme flessibili per la Matematica e le altre scienze, nuovi modi meno rigidi di organizzarne l'esposizione non solo per risolvere nuovi problemi, ma anche per trovare delle vie abbastanza rapide e rigorose per arrivare agli stessi teoremi classici della Matematica, come per esempio alla formula di Stokes. Questa necessità è indirettamente confermata anche da un'osservazione su tale formula contenuta nell'intervista [26] di Henri Cartan, uno dei maggiori matematici che ha partecipato all'opera collettiva del Bourbaki.

Chiameremo tale sistema "premessa generale alle teorie scientifiche" e lo esponiamo nel seguente capitolo.

A qualcuno tutto questo potrà sembrare un sogno irrealizzabile; a me sembra che nella ricerca scientifica qualche volta sia bene seguire i propri sogni e nello stesso tempo cercare di comunicarli con chiarezza, onestà intellettuale, disponibilità ad accettare le critiche dei propri interlocutori. Ho sostenuto quest'idea nell'intervista [25]. Cerco di darne una sia pur limitata attuazione in questa *Nota*, che spero risulti veramente comprensibile da parte di studiosi di ogni disciplina scientifica e umanistica. Mi sembra anzi che essa potrà essere compresa da chi la legge con lo spirito più "ingenuo" possibile senza pensare che siano necessari prerequisiti di qualsiasi natura matematici, logici, informatici, filosofici, ecc. La bibliografia che conclude questa *Nota* serve soprattutto a soddisfare le curiosità di chi, dopo averla letta, si chiede come siano sorte le idee in essa esposte.

Capitolo 1

Cominciamo assumendo come concetti primitivi, cioè non riconducibili (mediante opportune definizioni) ad altri concetti precedentemente introdotti, l'idea di *"qualità"* e l'idea di *"godere di una data qualità"*. Notiamo che con questa accezione di concetto primitivo non intendiamo rispondere né al problema psicologico di quali siano le idee che per prime si presentano alla mente del bambino, né al problema storico di quali siano state le idee che per prime l'umanità ha considerato.

Conveniamo pure che dati un oggetto x di qualsiasi natura ed una qualità q, quando scriveremo

$$q\ x$$

intenderemo dire che "x gode della qualità q". In questa premessa consideriamo le sette qualità fondamentali: $Qqual$, $Qrel$, $Qrelb$, $Qrelt$, $Qrelq$, $Qrun$, $Qrbiun$ aventi i seguenti significati:

$Qqual\ x$ vuol dire che x è una qualità;

$Qrel\ x$ vuol dire che x è una relazione;

$Qrelb\ x$ vuol dire che x è una relazione binaria;

$Qrelt\ x$ vuol dire che x è una relazione ternaria;

$Qrelq\ x$ vuol dire che x è una relazione quaternaria;

$Qrun\ x$ vuol dire che x è una relazione univoca;

$Qrbiun\ x$ vuol dire che x è una relazione biunivoca.

Le sette qualità godono di $Qqual$ e quindi possiamo scrivere:

ASSIOMA 1 – $Qqual\ Qqual$, $Qqual\ Qrel$, $Qqual\ Qrelb$, $Qqual\ Qrelt$, $Qqual\ Qrelq$, $Qqual\ Qrun$, $Qqual\ Qrbiun$.

Tutte queste e le altre affermazioni di questo capitolo vengono "proposte", (non "imposte") come "assiomi", cioè come affermazioni non dedotte da un sistema di affermazioni precedenti ma scelte come possibile punto di partenza per gli ulteriori sviluppi di varie teorie. Esse non sono nemmeno "dedotte " dalla storia della Matematica, dalla filosofia della Scienza, dalla Logica, ecc. ed anzi possono essere meglio comprese da chi si pone nell'atteggiamento più "ingenuo" possibile e si riferisce ai significati più comuni che il linguaggio di ogni giorno dà alle parole qualità e relazione. Solo dopo una comprensione ingenua della premessa è conveniente la sua rilettura critica in cui ciascuno può naturalmente portare le proprie esperienze e conoscenze di Matematica, Logica, Informatica, Storia, Filosofia, Fisica, Economia, ecc. anzi tali esperienze potranno essere molto importanti per formulare valutazioni critiche più profonde e meglio motivate.

Le tre qualità $Qrelb$, $Qrelt$, $Qrelq$ sono particolarizzazioni della qualità più generale $Qrel$, cioè:

ASSIOMA 2 – *Un elemento x che goda di una delle tre qualità $Qrelb$, $Qrelt$, $Qrelq$ gode anche della qualità $Qrel$.*

Non escludiamo invece che vi possano essere relazioni più complesse delle relazioni binarie, ternarie o quaternarie anche se non dovremo farne uso in questa premessa.

Dopo l'idea primitiva di godere di una certa qualità, la seconda più importante idea primitiva di questa premessa è quella di *"essere in una certa relazione."*

Precisamente, dati due oggetti x, y di qualsiasi natura e una relazione binaria r, scriveremo

$$r \ x, y$$

oppure talvolta

$$r \ x; y$$

per dire che "x ed y sono nella relazione r". Talvolta invece di dire che x ed y sono nella relazione r si dice anche che x è nella relazione r con y.

Analogamente se x, y, z sono oggetti di qualsiasi natura e ρ è una relazione ternaria, scriveremo

$$\rho \ x, y, z$$

oppure

$$\rho \ x; y; z$$

per dire che "x, y, z sono nella relazione ρ".

Infine se τ è una relazione quaternaria ed x, y, z, t sono oggetti di natura qualsiasi, scriveremo

$$\tau \ x, y, z, t$$

oppure

$$\tau \ x; y; z; t$$

per dire che x, y, z, t sono nella relazione τ.

Le quattro relazioni fondamentali considerate in questa premessa sono: *Rqual, Rrelb, Rrelt, Rid.* La relazione *Rqual* è una relazione binaria e collega le qualità con gli elementi che ne godono. Precisamente:

ASSIOMA 3 – *Dati due oggetti x, y, perché si abbia*

$$Rqual \ x, y$$

è necessario che x sia una qualità (cioè goda di Qqual). Inoltre se q è una qualità e x è un qualsiasi oggetto, la condizione Rqual q, x è necessaria e sufficiente perché x goda di q.

La relazione *Rrelb* è una relazione ternaria e collega le relazioni binarie con gli oggetti che sono in tali relazioni. In altri termini:

ASSIOMA 4 – *Perché si abbia*

$$Rrelb \ x, y, z$$

è necessario che x sia una relazione binaria. Inoltre se r è una relazione binaria, ed x, y sono oggetti di qualsiasi natura, sono equivalenti le due affermazioni:
$r \ x, y$;
Rrelb r, x, y.

Infine *Rrelt* è una relazione quaternaria e collega relazioni ternarie con oggetti che sono in tali relazioni. In altri termini:

Assioma 5 – *Perché si abbia*

$$Rrelt\ x,y,z,t$$

è necessario che x sia una relazione ternaria. Inoltre se ρ è una relazione ternaria e x,y,z sono oggetti di qualsiasi natura, sono equivalenti le due affermazioni:
Rrelt ρ,x,y,z;

$\rho\ x,y,z$.

Infine *Rid* è una relazione binaria e rappresenta l'*identità*. In altri termini:

Assioma 6 – *Affinché sia*

$$Rid\ x,y$$

occorre e basta che x ed y siano esattamente lo stesso oggetto. In altri termini ogni oggetto è nella relazione Rid con se stesso e soltanto con se stesso.

La relazione *Rid* è l'esempio più semplice di relazione che gode delle due qualità *Qrun* e *Qrbiun*. Per quanto riguarda la qualità *Qrun* poniamo il seguente assioma che esprime l'idea di univocità:

Assioma 7 –

$$
\begin{array}{llllll}
1) & se\ Qrun\ x & & & & allora\ Qrel\ x; \\
2) & se\ Qrelb\ r; & Qrun\ r; & r\ x,y; & r\ x,z & allora\ y=z; \\
3) & se\ Qrelt\ \rho; & Qrun\ \rho; & \rho\ x,y,z; & \rho\ x,y,t & allora\ z=t; \\
4) & se\ Qrelq\ \tau; & Qrun\ \tau; & \tau\ x,y,z,t; & \tau\ x,y,z,u & allora\ t=u.
\end{array}
$$

Infine gli assiomi che riguardano *Qrbiun* sono:

Assioma 8 –

$$
\begin{array}{llll}
1) & se\ Qrbiun\ x & & allora\ Qrelb\ x\ e\ Qrun\ x; \\
2) & se\ Qrbiun\ x; & r\ y,x; & r\ z,x\quad allora\ y=z.
\end{array}
$$

Le relazioni biunivoche hanno un ruolo molto importante sia all'interno della matematica sia nelle applicazioni della Matematica in cui sono importanti le relazioni tra oggetti "concreti" e loro modelli matematici. L'esistenza di una buona relazione biunivoca può significare una buona adeguatezza del modello matematico che è stato prescelto.

Capitolo 2

Sul tronco delle "premessa generale alle teorie scientifiche" esposta nel capitolo 1 è facile innestare i diversi rami della Matematica. Tale innesto può essere realizzato seguendo varie strategie, per esempio introducendo simultaneamente varie specie di oggetti tra loro collegate oppure introducendo separatamente singole specie di oggetti e procedendo in seguito alla descrizione delle relazioni che le collegano. Strategie del primo tipo sono state seguite in [1–24]; noi in questo capitolo diamo qualche semplice esempio della seconda strategia che ci sembra più adatta per una chiara e semplice comunicazione tra studiosi di diverse discipline ed anche per meglio valorizzare le diverse mentalità e le diverse intuizioni degli stessi matematici.

Per introdurre nel modo più semplice i numeri naturali, cioè i numeri $0, 1, 2, 3, 4, \ldots$, è sufficiente introdurre una qualità $Qnnat$, cioè la qualità di essere un numero naturale, una relazione biunivoca $Rnsuc$, cioè la relazione che collega ciascun numero naturale al suo immediato successore e la relazione binaria $Rnord$ che descrive l'ordinamento naturale dei numeri naturali. Esse sono collegati dai seguenti assiomi:

ASSIOMA N1 –

1) *Se vale Rnord x, y allora vale Qnnat x , Qnnat y.*

2) *Dati tre oggetti x, y, z, se Rnord x, y; Rnord y, z allora Rnord x, z.*

3) *Dati due numeri naturali x, y vale una delle due relazioni Rnord x, y , Rnord y, x. Esse si verificano simultaneamente se e solo se $x = y$.*

4) *Se Rnsuc x, y allora Rnord x, y e $x \neq y$.*

5) *Se Qnnat x, allora esiste un y tale che Rnsuc x, y.*

6) *Se Rnord x, y, $x \neq y$, Rnsuc x, z allora Rnord z, y.*

7) *Esiste un unico z tale che Qnat z e per nessun x si ha Rnsuc x, z.*

Osservazione: per ritrovare le usuali nozioni aritmetiche basta ammettere che l'affermazione $Rnord$ x, y equivale all'affermazione: x, y sono numeri naturali e x è minore di y o uguale a y, oppure x, y sono numeri naturali e y è maggiore di x o uguale a x. In tal caso si usa scrivere $x \leq y$ oppure $y \geq x$. Analogamente in luogo di $Rnsuc$ x, y potremo dire che y è il successore immediato di x oppure che x è il predecessore immediato di y. L'unico numero naturale che non ha predecessore verrà denotato come di consueto col simbolo 0, il suo successore col simbolo 1, l'immediato successore di 1 col simbolo 2, ecc.

Osservazione: le nozioni ora introdotte servono solo come primissimo inizio dell'Aritmetica. Più tardi, dopo avere introdotto in generale le operazioni semplici e binarie si potrebbero studiare le quattro operazioni dell'Aritmetica elementare e successivamente dopo l'introduzione degli insiemi formulare il principio di induzione con cui si entra nell'Aritmetica più avanzata.

Come primo passo di tale sviluppo innestiamo sul tronco della premessa generale il concetto di operazione. Precisamente introduciamo le tre qualità Qop, qualità di essere un'operazione, $Qops$, qualità di essere un'operazione semplice, $Qopb$, qualità di essere un'operazione binaria, e le relazioni $Rops$ e $Ropb$ che descrivono il modo di operare delle operazioni semplici e binarie. Esse soddisfano i seguenti assiomi:

ASSIOMA O1 –

1) *Se Qops x oppure Qopb x allora Qop x. In altri termini Qops e Qopb sono particolarizzazioni della qualità generale Qop.*

2) *Rops è una relazione ternaria univoca;*

3) *per ogni scelta di x, y, z, se Rops x, y, z allora Qops x;*

4) *Ropb è una relazione quaternaria univoca;*

5) *se Ropb x, y, z, t allora Qopb x.*

Osservazione: quando f è un'operazione semplice invece di scrivere *Rops* f, x, y scriveremo spesso $y = f\, x$. Quando φ è un'operazione binaria, invece di scrivere *Ropb* φ, x, y, z scriveremo spesso $z = \varphi\, x, y$. Dopo i numeri naturali e le operazioni possiamo introdurre le collezioni avvertendo che il concetto di collezione è un'ampia generalizzazione dell'usuale concetto di insieme. A tale scopo introdurremo la qualità *Qcoll*, cioè la qualità di essere una collezione, la relazione *Rcoll*, relazione di appartenenza a collezioni, e la relazione *Rcin*, relazione di inclusione tra collezioni. Esse soddisfano gli assiomi seguenti:

Assioma C1 –

1) *Rcoll è una relazione binaria;*

2) *se Rcoll x, y allora x è una collezione, cioè Qcoll x.*

Seguendo l'uso, invece di scrivere *Rcoll* x, y scriveremo $y \in x$ oppure diremo che y appartiene a x oppure che y è un elemento di x.

Assioma C2 –

1) *Rcin è una relazione binaria;*

2) *se Rcin x, y allora x, y sono collezioni;*

3) *se A, B sono collezioni allora Rcin A, B se e solo se ogni elemento di A è anche elemento di B.*

Secondo l'uso scriveremo $A \subseteq B$ e diremo che A è contenuto in B oppure che A è parte di B.

Possiamo ora enunciare l'assioma fondamentale della teoria delle collezioni, l'assioma di estensionalità

Assioma C3 – *Se A, B sono collezioni, $A \subseteq B, B \subseteq A$ allora A coincide con B.*

Dopo aver introdotto le collezioni si possono introdurre gli insiemi mediante la qualità *Qins* e la relazione di appartenenza *Rins*. Esse sono collegate a *Qcoll* e *Rcoll* dal seguente assioma:

Assioma I1 –

1) *Se Qins x allora Qcoll x. In altri termini gli insiemi sono particolari collezioni;*

2) *se Rins x, y allora Qins x;*

3) *se Qins E allora Rcoll E, x se e solo se Rins E, x. In altri termini Rins è la restrizione della relazione Rcoll ai casi in cui il primo oggetto preso in considerazione è un insieme.*

Osservazione: dopo aver introdotto collezioni e insiemi si potrebbero introdurre funzioni, sistemi, correlazioni mediante le coppie $Qfun, Rfun, Qsys, Rsys, Qcorr, Rcorr$.

Per un'ampia descrizione di tali oggetti si può vedere l'articolo [16] che mette anche in luce le differenze che esistono tra la nozione d'insieme e quella più generale di collezione. Per un maggiore approfondimento dei rapporti esistenti fra i concetti ora introdotti è utile anche la considerazione delle collezioni universali o universi. A tale scopo si introduce la qualità $Quniv$, la qualità di essere una collezione universale, e si pone l'assioma seguente:

ASSIOMA CU1 – *Se Quniv V allora Qcoll V e per ogni collezione C si ha* $C \in V$ *se e solo se* $C \subseteq V$.

Osservazione: è immediato verificare che se V è una collezione universale allora $V \in V$. Si potrebbe dedurre pure dagli usuali assiomi della teoria degli insiemi, che qui non abbiamo ancora introdotto, che una collezione universale non può essere un insieme: infatti da tali assiomi si deduce (teorema di Cantor) che a nessun insieme possono appartenere tutti i suoi sottinsiemi.

Tutte le teorie citate nella bibliografia si prestano ad essere innestate con opportuni adattamenti sulle premesse generali delle teorie scientifiche esposte nel capitolo 1. Ad esempio potremmo innestare le variabili considerate in [4], [10] mediante la qualità $Qvar$ e la relazione $Rvar$. Gli elementi che godono di $Qvar$ saranno detti variabili; quando si ha $Rvar \, w, x$ diremo che x è un valore assunto dalla variabile w. Infine resta ancora disponibile un ampio spazio per l'innesto di altri concetti matematici o logici come quelli di proposizione, predicato, linguaggio, sintassi, semantica, categoria, Analisi non standard, logiche non classiche, ecc. sui quali sarò molto lieto di ascoltare i suggerimenti e le opinioni di colleghi ed amici.

CAPITOLO 3

La proposta di assiomi contenuta nei capitoli 1,2 dovrebbe essere comprensibile e criticabile da parte di studiosi di diverse discipline che hanno in comune quel sentimento che gli antichi chiamarono filosofia, cioè amore della sapienza, pur avendo un livello d'informazione molto diseguale nel campo della Matematica, della Logica, dell'Informatica, della Storia e della Filosofia della Scienza, dell'Epistemologia, ecc.

Se molti troveranno la lettura dei due capitoli 1,2 troppo difficile dovrò concludere che nella mia esposizione vi è qualcosa di sbagliato, perché credo che gli assiomi fondamentali della Matematica come di ogni altra scienza debbono essere scritti in modo chiaro e comprensibile da parte di ogni persona che abbia una sincera volontà di comprenderli.

Naturalmente non mi attendo e non desidero un'approvazione incondizionata delle mie proposte, perché credo che ancora sia necessario un ampio lavoro di

riflessione critica e di discussione amichevole tra studiosi di varia formazione e
di vario orientamento, per arrivare a un buon sistema di assiomi (o ad alcuni
buoni sistemi di assiomi) che possano rappresentare un reale significativo progresso rispetto all'attuale situazione. D'altra parte penso che questo obiettivo
sia raggiungibile, anche perché credo che il riduzionismo insiemistico al pari di
ogni altra forma di riduzionismo non offra una prospettiva adeguata per una vera
comprensione dei molti problemi che la cultura del nostro tempo deve affrontare
e che mi ricordano le parole di Shakespeare "vi sono più cose in cielo e in terra
di quante ne sogni la tua filosofia".

Penso pure che la ricerca di buoni sistemi di assiomi per la Matematica e le altre
scienze deve essere affrontata con quello spirito di umiltà, speranza, convivialità
che in fondo caratterizza la mentalità sapienziale in ogni tempo.

Persone più colte di me potrebbero parlare meglio di me della mentalità sapienziale. Io mi limiterò a ricordare che già in uno dei più antichi libri sapienziali
della Bibbia, il libro dei Proverbi, vi è l'invito all'umiltà e al timor di Dio, alla
diffidenza verso chi si crede saggio e respinge ogni critica, mentre il vero saggio
mostra sincera gratitudine verso chi gli rivolge una critica obiettiva e serena.
Nello stesso libro accanto all'invito all'umiltà, vi è l'invito alla speranza: si parla
della Sapienza che era con il Signore nella creazione, che si fa trovare volentieri
da coloro che la cercano, che invita tutti gli uomini al suo banchetto. In varie
forme questa idea conviviale della comunicazione del sapere si ritrova nei secoli
successivi fino alla Dichiarazione Universale dei Diritti Umani del 10–12–1948
che pone tra i compiti fondamentali della scuola l'educazione all'amicizia ed alla
comprensione. Io credo che proprio nell'umiltà, nella speranza, nella convivialità, nell'amicizia, nella comprensione sia il segreto del successo di ogni ricerca
sugli assiomi fondamentali delle diverse scienze. Esse implicano fra l'altro il
superamento di una visione troppo chiusa delle diverse specializzazioni e di ciò
che viene chiamato comunemente rigore matematico o rigore scientifico. Il rigore matematico non è solo accuratezza nelle dimostrazioni ma anche impegno
a esporre nel modo più chiaro e comprensibile i problemi che si vorrebbero risolvere, i teoremi che si vorrebbero dimostrare, le congetture che si vorrebbero
verificare o confutare. All'opposto di questa idea ampia del rigore matematico
si trova l'atteggiamento dello studente che alla domanda "mi spieghi l'ipotesi e
la tesi di questo teorema" risponde "se vuole glielo dimostro" oppure quella del
ricercatore che dice "io non parlo dei problemi che non ho ancora risolto, parlo
solo dei risultati che ho conseguito e non di quelli che sto ancora ricercando".

Io penso invece che il vero rigore sia quello di chi parla chiaramente e liberamente delle proprie certezze e dei propri dubbi, dei problemi che ha risolto e di
quelli che vorrebbe risolvere o vedere risolti da qualcuno, evitando solo quei discorsi confusi, oscuri, inutilmente complicati che finirebbero con l'annoiare anche
l'ascoltatore meglio disposto.

Infine un giusto rigore scientifico significa anche un atteggiamento sereno nei
confronti della tradizione, ammirazione per le grandi conquiste del passato e
libertà di introdurre ragionevoli innovazioni; significa evitare l'abitudine antica
di attribuire le proprie idee a qualche predecessore illustre e insieme l'opposta

abitudine di presentare come proprie idee originali idee poco diverse da altre già presenti nella letteratura scientifica.

Bibliografia

[1] H. P. BARENDREGT, *The Lambda–Calculus*. Amsterdam 1981.

[2] Y. BAR HILLEL – A. A. FRAENKEL – A. LEVY , *Foundations of set theory*. Amsterdam 1973.

[3] A. CHURCH, *Set theory with a universal set*. In: L. HENKIN *et al.* (eds.), *Proceedings of the Tarski Symposium*. Proc. of Symp. P. Math. XXV, Rhode Island 1974, 297–308.

[4] M. CLAVELLI, *Variabili e teoria A*. Rend. Sem. Fac. Sci. Univ. Cagliari, (2) 59, 1985, 125–130.

[5] M. CLAVELLI – E. DE GIORGI – M. FORTI – V. M. TORTORELLI, *A self-reference oriented theory for the Foundations of Mathematics*. In: *Analyse Mathématique et applications. Contributions en l'honneur de Jacques-Louis Lions*. Paris 1988, 67–115.

[6] E. DE GIORGI, Contributo alla sessione *Fundamental Principles of Mathematics*. Plenary Session of the Pontifical Academy of Sciences (25–29 October 1994), 1994.

[7] E. DE GIORGI, *Riflessioni sui Fondamenti della Matematica*. Conferenza per ilm centenario della società Mathesis, Roma 23 ottobre 1995, dattiloscritto.

[8] E. DE GIORGI – M. FORTI, *Una teoria-quadro per i fondamenti della Matematica*. Atti Acc. Lincei Rend. Fis.. s. 8, v. 79, 1985, 55–67.

[9] E. DE GIORGI – M. FORTI, *"5× 7": A Basic Theory for the Foundations of Mathematics*. Preprint di Matematica n. 74, Scuola Normale Superiore, Pisa 1990.

[10] E. DE GIORGI – M. FORTI – G. LENZI, *Introduzione delle variabili nel quadro delle teorie base dei Fondamenti della Matematica*. Rend. Mat. Acc. Lincei, s.9, v. 5, 1994, 117–128.

[11] E. DE GIORGI – M. FORTI – G. LENZI, *Una proposta di teorie base dei Fondamenti della Matematica*. Rend. Mat. Acc. Lincei, s. 9, v. 5, 1994, 11–22.

[12] E. DE GIORGI – M. FORTI – G. LENZI – V. M. TORTORELLI, *Calcolo dei predicati e concetti metateorici in una teoria base dei Fondamenti della Matematica*. Rend. Mat. Acc. Lincei, s. 9, v. 6, 1995, 79–92.

[13] E. De Giorgi - M. Forti - V. M. Tortorelli, *Sul problema dell'autoriferimento*. Atti Acc. Lincei Rend. Fis., s. 8, v. 80, 1986, 363–372.

[14] E. De Giorgi - G. Lenzi, *La teoria '95. Una proposta di teoria aperta e non riduzionistica dei fondamenti della Matematica*. Atti dell'Accademia dei XL, Mem. Mat. Appl., 20, 1996, 7–34.

[15] M. Forti - F. Honsell, *Models of self-descriptive set theories*. In: F. Colombini *et al.* (eds.), *Partial Differential Equations and the Calculus of Variations. Essays in Honor of Ennio De Giorgi*. Boston 1989, 473–518.

[16] M. Forti - F. Honsell, *Sets and classes within the basic theories for the Foundations of Mathematics*. Quaderni Ist. Mat. Appl. "U. Dini", 11, 1994.

[17] M. Forti - F. Honsell - M. Lenisa, *An axiomatization of partial n–place operations*. Math. Struct. Computer Sci., 7, 1997, 283–302.

[18] M. Grassi - G. Lenzi - V. M. Tortorelli, *A formalization of a basic theory for the foundations of Mathematics*. Preprint del Dipartimento di Matematica dell'Università di Pisa, n. 1.98.804, 1994.

[19] G. Lenzi, *Estensioni contraddittorie della teoria Ampia*. Atti Acc. Lincei Rend. fis., s. 8, v. 83, 1989, 13–28.

[20] G. Lenzi - V. M. Tortorelli, *Introducing predicates into a basic theory for the foundations of Mathematics*. Preprint di Matematica, n. 51, Scuola Normale Superiore, Pisa, 1989.

[21] W. V. O. Quine, *New foundations for mathematical logic*. Amer. Math. Monthly, 44, 1973, 70–80.

[22] B. Russell - A. N. Whitehead, *Principia Mathematica*. Cambridge 1925.

[23] D. Scott, *Combinators and classes*. In: C. Böhm (ed.), *λ–Calculus and Computer Science Theory*. Lec. Notes Comp. Sc., 37, Berlin 1975.

[24] J. Von Neumann, *Eine Axiomatisierung der Mengenlehre*. J. f. Reine und Angew. Math., 154, 1925, 219–240.

[25] E. De Giorgi - A. Preti, *Anche la scienza ha bisogno di sognare*. Quotidiano di Lecce, 6 Gennaio 1996.

[26] M. Schmidt, *Hommes de science*. Hermann, 1990.

The Editors and Springer-Verlag thank the original publishers of Ennio de Giorgi's articles for permission to publish them here.

The articles marked (E) were originally published in English. All other articles were first published in Italian and have been translated into English for this volume.

Accademia delle Scienze di Torino: 7
Accademia Nazionale dei Lincei: 1, 8, 13, 24, 25, 27, 37, 38, 43
Accademia Pontificia: 35(E), 36
Adriatica Editrice Salentina: 34
Atti Seminario Matematico e Fisico dell'Università di Modena: 39
Istituto Filosofico di Studi Tomistici: 42
Taylor & Francis: 32(E)
Dipartimento di Matematica e Applicazioni dell'Università degli studi di Napoli Federico II: 5
Duke University Press: 40
Dunod Editeur: 28(E), 33(E)
Edizioni ETS: 9, 10, 26(E)
Fondazione Annali di Matematica Pura ed Applicata: 2, 3
Mir, Moscow: 12
Nuova civiltà delle macchine: 41
Pitagora Editrice: 23
Scuola Normale Superiore: 11, 22, 29
Springer Science and Business Media: 31(E)
Unione Matematica Italiana: 6, 14, 17, 18, 19, 21
Università degli Studi di Roma "La Sapienza", Dipartimento di Matematica: 4, 20
Frontispiece: Courtesy of Foto Frassi, Pisa